Topography and Sea-floor Bathymetry

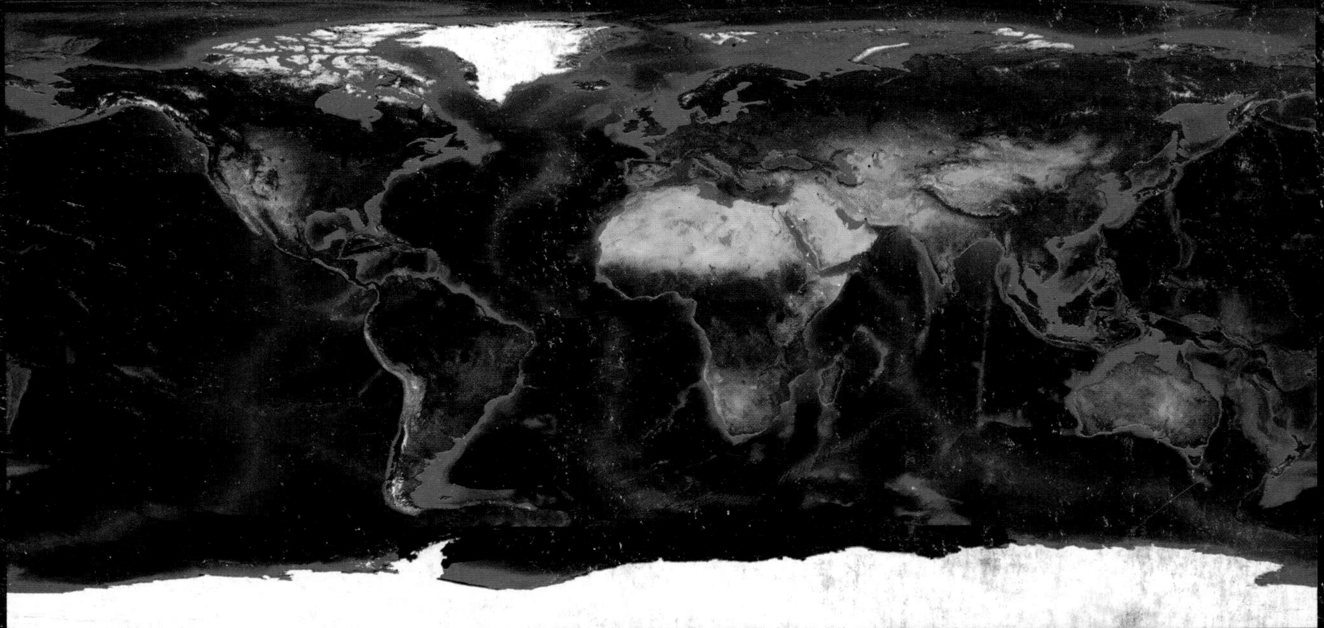

NASA/GSFC, GEBCO

Earth at Night, City Lights

Defense Meteorological Satellite Program (DMSP) NASA/GSFC

Geosystems

Combination of satellite *Terra* MODIS sensor image and *GOES* satellite image produces a true-color view of North and South America from 35,000 km (22,000 mi) in space. [Courtesy of MODIS Science Team, Goddard Space Flight Center, NASA and NOAA.]

Edges of a cumulonimbus cloud scatter sunlight in dramatic crepuscular rays. [Bobbé Christopherson.]

AN INTRODUCTION TO PHYSICAL GEOGRAPHY

Geosystems

EIGHTH EDITION

ROBERT W. CHRISTOPHERSON

Prentice Hall

Boston Columbus Indianapolis New York San Francisco Upper Saddle River
Amsterdam Cape Town Dubai London Madrid Milan Munich Paris Montréal Toronto
Delhi Mexico City São Paulo Sydney Hong Kong Seoul Singapore Taipei Tokyo

Geography Editor: Christian Botting
Development Editor: Ginger Birkeland
VP/Executive Director, Development: Carol Trueheart
Project Manager, Editorial: Anton Yakovlev
Marketing Manager: Maureen McLaughlin
Assistant Editor: Kristen Sanchez
Editorial Assistant: Christina Ferraro
Marketing Assistant: Nicola Houston
Managing Editor, Geosciences and Chemistry: Gina M. Cheselka
Project Manager, Science: Shari Toron
Full-Service Project Management/Composition: Suganya Karrupasamy/Element LLC
Vice-President, Higher Education, Element LLC: Cindy Miller
Production Assistant to the Author: Bobbé Christopherson
Senior Manufacturing and Operations Manager: Nick Sklitsis
Operations Specialist: Maura Zaldivar
Art Director: Mark Ong
Interior and Cover Designer: Randall Goodall
Senior Technical Art Specialist: Connie Long
Illustrations: Precision Graphics, JC Morgan
Photo Research Manager: Elaine Soares
Photo Researcher: Dian Lofton
Media Producer: Kate Brayton
Senior Media Producer: Angela Bernhardt
Associate Media Producer: Kristin Mayo
Production Coordinator, Media: Shannon Kong
Media Supervisor: Liz Winer
Copy Editor: Sherry Goldbecker
Proofreader: Jeff Georgeson

Cover Credits — Front: *Kvinberget Peak, south of Minebutka, Duvefjorden, Spitsbergen, Arctic Ocean. Dipping strata of red-brown quartzites, weathered to produce talus slopes*, Bobbé Christopherson. Back: Earth photo courtesy of NASA. All other photos by Bobbé Christopherson.
Dedication Page Quote: Barbara Kingsolver, *Small Wonder* (New York: Harper Collins Publishers, 2002), p. 39.

Credits and acknowledgments borrowed from other sources and reproduced, with permission, in this textbook appear on the appropriate page within the text.

Library of Congress Cataloging-in-Publication Data

Christopherson, Robert W.
Geosystems : an introduction to physical geography / Robert W. Christopherson. – 8th ed.
p. cm.
Includes index.
ISBN-13: 978-0-321-70622-5
ISBN-10: 0-321-70622-6
1. Physical geography. I. Title.
GB54.5.C48 2012
910′.02–dc22

2010020275

Printed in the United States
10 9 8 7 6 5 4 3 2 1

ISBN-10: 0-321-70622-6
ISBN-13: 978-0-321-70622-5

Prentice Hall
is an imprint of

www.pearsonhighered.com

DEDICATION

To all the students and teachers of Earth, our home planet, and a sustainable future.

To the children and grandchildren, for it is their future and planet.

The land still provides our genesis,
however we might like to forget
that our food comes from dank,
muddy Earth, that the oxygen in our lungs was
recently inside a leaf, and that every
newspaper or book we may pick up is made
from the hearts of trees that died for the
sake of our imagined lives. What you hold in
your hands right now, beneath these words,
is consecrated air and time and sunlight.

—Barbara Kingsolver

About Our Sustainability Initiatives

This book is carefully crafted to minimize environmental impact. The materials used to manufacture this book originated from sources committed to responsible forestry practices. The paper is Forest Stewardship Council™ (FSC®) certified. The printing, binding, cover, and paper come from facilities that minimize waste, energy consumption, and the use of harmful chemicals.

Pearson closes the loop by recycling every out-of-date text returned to our warehouse. We pulp the books, and the pulp is used to produce items such as paper coffee cups and shopping bags. In addition, Pearson aims to become the first climate neutral educational publishing company.

The future holds great promise for reducing our impact on Earth's environment, and Pearson is proud to be leading the way. We strive to publish the best books with the most up-to-date and accurate content, and to do so in ways that minimize our impact on Earth.

BRIEF CONTENTS

CONTENTS

PART II

The Water, Weather, and Climate Systems 160

Welcome to the Eighth Edition of *Geosystems, An Introduction to Physical Geography*. This edition builds on the widespread success of the first seven editions, as well as the companion texts, *Elemental Geosystems*, now in its sixth edition, and *Geosystems, Canadian Edition*, Second Edition. Students and teachers appreciate the systems organization, scientific accuracy, integration of figures and text, clarity of the summary and review sections, and overall relevancy. *Geosystems* continues to evolve with the geospatial sciences, presenting the latest science and geophysical events, in student-friendly language that tells a story.

The goal of physical geography is to explain the spatial dimension of Earth's dynamic systems—its energy, air, water, weather, climate, tectonics, landforms, rocks, soils, plants, ecosystems, and biomes. Understanding human-Earth relations is part of the challenge of physical geography—to create a holistic (or complete) view of the planet and its inhabitants. Welcome to physical geography!

New to the Eighth Edition

Nearly every page of *Geosystems*' Eighth Edition contains updated material, new features, and new content in text and figures—far too many to list here. See the visual Walkthrough of new features at the end of this Preface. Following is a sampling of the Eighth Edition's features:

- Many figures are new or recast to improve student learning. There are more than **600 new photos and images** bringing real-world scenes into the classroom. Our photo and remote-sensing program exceeds 750 items, integrated throughout the text.
- The 4 Part Openers and 21 Chapter Openers are redesigned with **24 new photos**, the only exception being Chapter 12, with its new integrated map of the ocean floor. Each of these four parts, described on the back cover, features a **new systems schematic** that portrays the organization of the content; check these out on pages 36, 160, 294, and 524.
- ***Key Learning Concepts*** appear at the outset of each chapter, many rewritten for clarity. Each chapter concludes with a ***Key Learning Concepts Review***, which summarizes the chapter using the opening objectives. New to this edition are integrated representative figures, photos, illustrations, and images highlighting each concept in the review.
- New to each chapter is an opening case study feature, ***Geosystems Now***. These original, unique essays help focus interest on the chapter content. *Geosystems Now* topics for each chapter include:

1. Where Is Four Corners, Exactly?
2. Chasing the Subsolar Point
3. Humans Help Define the Atmosphere
4. Albedo Impacts, a Limit on Future Arctic Shipping
5. Temperature Change Affects St. Kilda's Sheep
6. Ocean Currents Bring Invasive Species
7. Earth's Lakes Provide an Important Warming Signal
8. The Front Lines of Intense Weather
9. Water Budgets, Climate Change, and the Southwest
10. A Look at Puerto Rico's Climate at a Larger Scale
11. Earth's Migrating Magnetic Poles
12. The San Jacinto Fault Connection
13. Human-Caused Mass Movement at the Kingston Steam Plant, Tennessee
14. Removing Dams and Restoring Salmon on the Elwha River, Washington
15. Increasing Desertification and Political Action—A Global Environmental Issue
16. What Once Was Bayou Lafourche
17. Ice Shelves and Tidewater Glaciers Give Way to Warming
18. High Latitude Soils Emit Greenhouse Gases
19. Species' Distributions Shift with Climate Change
20. Invasive Species Arrive at Tristan da Cunha
21. Seeing Earth and Ourselves from Space

Be sure to read these as you begin each chapter. For instance, "Ocean Currents Bring Invasive Species," in Chapter 6, tracks an oil-drilling platform left adrift in a storm in the South Atlantic Ocean. Months later the platform ran aground in Tristan da Cunha. The *Geosystems Now* in Chapter 20 picks up the story with "Invasive Species Arrive at Tristan da Cunha." The rig was contaminated with non-native species, introducing them to fragile Tristanian marine ecosystems. The *Geosystems* approach involves such connections and linkages across the chapters and among Earth systems.

- New within each chapter are ***Geo Reports***, placed along the bottom of pages. *Geo Reports* offer facts, events related to the discussion in the chapter, student action items, and new sources of information, among other features. There are 78 *Geo Reports* throughout the book, with topics such as: water in the Solar System, why we always see the same side of the Moon, measuring Earth's rotation, polar region warming, Iceland volcanic ash and aircraft flights, iceberg hazards, tornado and tropical storm records, how water is measured, the Intergovernmental Panel on Climate Change and the Nobel Prize, Earth's weight, how large earthquakes affect the entire Earth system, what a bayou is, a saltier Mediterranean Sea, killer waves, high-latitude ice losses, soil losses, ongoing extinction rates, and sources of food and medicine in the rain forest, among many others.

- New to this edition and placed throughout the chapters are 53 carefully crafted ***Critical Thinking*** items to take you to the next level of learning, offering opportunities to check conceptual understanding and placing you in charge of further inquiry. Topics include finding your location, tracking seasonal change, analyzing your ozone column, assembling your physical geography profile, measuring wind through observation, following clouds, assessing hazard perception, finding your climate, reducing climate forcing, touring the ocean floor, looking at landslide potential, assessing the post-Katrina Gulf Coast, relating to rising sea levels, living at the polar station, observing ecosystem disturbances, and tracking shifting climates, among others. These bring relevancy to learning physical geography.
- Placed at the end of the text for each chapter is a new ***Geosystems Connection*** feature. In a brief paragraph, we review what we covered in the chapter and what is about to unfold in the next chapter, "bridging" from one chapter to the next. The final *Geosystems Connection* in Chapter 21 bridges to what comes next beyond this course.
- Climate change science is well established and affects systems in every chapter of *Geosystems*. Part of this revision effort further updates our **extensive climate change coverage** throughout the chapters. We present a new section on "Reasons for Concern" to organize the climate discussion. The record year of warmth for land-surface temperatures was 2005, with 2007 and 2009 close behind. Through August 2010, new monthly temperature records for land and ocean were set. For the Southern Hemisphere, 2009 was the warmest in the modern record. The decade of 2001–2010 was the warmest decade in the entire record. As an integrative spatial science, physical geography is well equipped to identify and analyze related impacts to Earth's systems. *Geosystems* presents all the aspects of climate change and has done so since its first edition in 1992.
- *Geosystems* continues to embed **Internet URLs** within the text. More than 200 appear in this edition; many are new links, and all are revised and checked. These allow you to pursue topics of interest in greater depth or to obtain the latest information about weather and climate, tectonic events, floods, and the myriad other subjects covered in the book.
- New to this edition is ***MasteringGeography***™. Used by over 1 million science students each year, the *Mastering* platform is the most effective and widely used online homework, tutorial, and assessment system for the sciences. *Geosystems* is supported by *MasteringGeography*™ assignable activities that include geoscience animations, *Encounter Geosystems* Google Earth™ multimedia, Thinking Spatially and Data Analysis; there are figure labeling tasks followed by data analysis tasks, and MapMaster™ interactive maps, as well as a robust student Study Area with many digital resources, including a Pearson eText version of *Geosystems*. Available at www.masteringgeography.com.

The *Geosystems* Learning/Teaching Package

The Eighth Edition provides a complete physical geography program for students and teachers.

For You, the Student:

- ***MasteringGeography*™ with Pearson eText.** Now available with *Geosystems*' Eighth Edition, ***MasteringGeography***™ offers:
  - Assignable activities that include geoscience animations, *Encounter Geosystems* Google Earth™ multimedia, Thinking Spatially and Data Analysis, geography videos from Television for the Environment's global *Earth Report* series, and MapMaster™ interactive maps.
  - Additional end-of-chapter questions, *Test Bank* questions, and reading quizzes.
  - A student Study Area with geoscience animations, satellite loops, author notebooks, photo galleries, geography videos, MapMaster™ interactive maps, *In the News* RSS feeds, web links, career links, physical geography case studies, flashcard glossary, quizzes, and more.
  - Pearson eText for *Geosystems*' Eighth Edition, which gives you access to the text whenever and wherever you can access the Internet and includes powerful interactive and customization functions. Available at www.masteringgeography.com.
- The *MasteringGeography*™ Student Access Code Card (0-321-73038-0) offers access to the *MasteringGeography*™ site for *Geosystems*' Eighth Edition.

- ***Applied Physical Geography—Geosystems in the Laboratory***, Eighth Edition (0-321-73214-6) by Charlie Thomsen and Robert Christopherson of American River College. This spiral-bound geography lab manual contains 20 lab exercises that are divided into logical sections. Each exercise comes with a list of key terms and learning concepts. We integrate links to Google Earth™ KMZ files available at www.mygeoscienceplace.com into the exercises, allowing you to actually experience and manipulate topographic maps in digital elevation mode relief as you work through problems. This revised edition comes with updated topographic maps to view with stereolenses, up-to-date Google Earth™ activities, and a revised chapter on geographic information systems that contains activities that require you to use ArcGISExplorer. (A complete answer key for the Eighth Edition is available for teachers to download from www.pearsonhighered. com/irc.)
- ***Encounter Geosystems* Workbook and Premium Web Site** (0-321-63699-6). *Encounter Geosystems* by Charlie Thomsen, is a print workbook and website that provides rich, interactive explorations of physical geography concepts through Google Earth™. All chapter explorations are available in print format as well as online quizzes, accommodating different classroom needs. All worksheets are accompanied with corresponding Google Earth™ media files, available for download from www.mygeoscienceplace.com.

- ***Goode's World Atlas 22nd Edition* (0-321-65200-2).** *Goode's World Atlas* has been the world's premiere educational atlas since 1923—and for good reason. It features over 250 pages of maps, from definitive physical and political maps to important thematic maps that illustrate the spatial aspects of many important topics. The 22nd edition includes 160 pages of new, digitally produced reference maps as well as new thematic maps on global climate change, sea-level rise, emissions, polar ice fluctuations, deforestation, extreme weather events, infectious diseases, water resources, and energy production.
- ***Dire Predictions* (0-13-604435-2).** Periodic reports from the Intergovernmental Panel on Climate Change evaluate the risk of climate change brought on by humans. But the sheer volume of scientific data remains inscrutable to the general public, particularly to those who may still question the validity of climate change. In just over 200 pages, this practical text presents and expands upon the essential findings in a visually stunning and undeniably powerful way to the lay reader. Scientific findings that provide validity to the implications of climate change are presented in clear-cut graphic elements, striking images, and understandable analogies.

For You, the Teacher:

Geosystems is designed to give you flexibility in presenting your course. The text is comprehensive in that it is true to each scientific discipline from which it draws subject matter. This diversity is a strength of physical geography, yet makes it difficult to cover the entire book in a semester. *Geosystems* is organized to help you customize your presentation. You should feel free to use the text based on your specialty or emphasis, rearranging parts and chapters as desired. The following materials are available to assist you. Have a great class!

- ***Instructor Resource Manual*** (online download only) by Charlie Thomsen. The *Instructor Resource Manual*, intended as a resource for both new and experienced teachers, includes lecture outlines and key terms, additional source materials, teaching tips, and a complete annotation of chapter review questions. Available for download from www.pearsonhighered.com/irc.
- **TestGen® Test Bank (online download only)** by Charlie Thomsen and Robert Christopherson. TestGen® is a computerized test generator that lets you view and edit *Test Bank* questions, transfer questions to tests, and print tests in a variety of customized formats. This *Test Bank* includes approximately 3,000 multiple-choice, true/false, and short answer/essay questions. New to this edition, all of the questions are correlated against the National Geography Standards and Bloom's Taxonomy to help you better map the assessments against both broad and specific teaching and learning objectives. The *Test Bank* is also available in Microsoft Word®, and is importable into Blackboard and WebCT. Available for download from www.pearsonhighered.com/irc.
- **Blackboard Test Bank (online download only).** The Blackboard Test Bank provides questions for import into the Blackboard Learning System. Available for download from www.pearsonhighered.com/irc.
- ***Instructor Resource Center on DVD*** (0-321-73030-5). *The Instructor Resource Center on DVD* provides everything you need, where you want it. It helps make you more effective by saving you time and effort. All digital resources can be found in one well-organized, easy to-access place, and include
  - All textbook images as JPGs, PDFs, and PowerPoint™ slides;
  - Preauthored lecture PowerPoint presentations, which outline the concepts of each chapter with embedded art and can be customized by you to fit your lecture requirements;
  - Classroom Response System "Clicker" questions in PowerPoint format, which are correlated against the National Geography Standards and Bloom's Taxonomy;
  - TestGen® software, providing questions and answers, for both MACs and PCs;
  - Electronic files of the *Instructor Resource Manual* and *Test Bank*;
  - Instructor Resource content available completely online via the Instructor Resources section of www.mygeoscienceplace.com and www.pearsonhighered.com/irc.
- **Television for the Environment *Earth Report* Geography Videos on DVD (0-32-166298-9).** This three-DVD set is designed to help students visualize how human decisions and behavior have affected the environment and how individuals are taking steps toward recovery. With topics ranging from the poor land management promoting the devastation of river systems in Central America to the struggles for electricity in China and Africa, these 13 videos from Television for the Environment's global *Earth Report* series recognize the efforts of individuals around the world to unite and protect the planet.
- ***Aspiring Academics: A Resource Book for Graduate Students and Early Career Faculty* (0-13-604891-9).** Drawing on several years of research, this set of essays is designed to help graduate students and early career faculty begin their careers in geography and related social and environmental sciences. This teaching aid stresses the interdependence of teaching, research, and service in faculty work—and the importance of achieving a healthy balance in professional and personal life—and does not view these as a collection of unrelated tasks. Each chapter provides accessible, forward-looking advice on topics that often cause the most stress in the first years of a college or university appointment.
- ***Teaching College Geography: A Practical Guide for Graduate Students and Early Career Faculty* (0-13-605447-1).** This resource provides a starting point for becoming an effective geography teacher from the very first day of class. Divided in two parts, the first set of chapters addresses "nuts-and-bolts" teaching issues in the context of the new technologies, student demographics, and institutional expectations that are

the hallmarks of higher education in the 21st century. The second part explores other important issues: being an effective teacher in the field, supporting critical thinking with geographic information system and mapping technologies, engaging learners in large geography classes, and promoting awareness of international perspectives and geographic issues.

- **AAG Community Portal for Aspiring Academics and Teaching College Geography.** This web site is intended to support community-based professional development in geography and related disciplines. Here you will find activities providing extended treatment of the topics covered in both books. The activities can be used in workshops, graduate seminars, brown bags, and mentoring programs offered on campus or within an academic department. You can also use the discussion boards and contributions tool to share advice and materials with others. Available at www.pearsonhighered.com/aag/.

Author Acknowledgments

I thank my family for believing in this work, especially considering the next generation: Chavon, Bryce, Payton, Brock, Trevor, Blake, Chase, Téyenna, and Cade. When I look into our grandchildren's faces, I see why we work toward a sustainable future—thus, the dedication for this edition.

I give special gratitude to all the students during my 30 years of teaching at American River College, for it is in the classroom crucible that the *Geosystems* books are forged. Special continued thanks to Charlie Thomsen for his creative work and collaboration on the *Encounter Geosystems* book, the *Applied Physical Geography* lab manual, and ancillaries—an honor to be his colleague. My thanks go to the many authors and scientists who publish research that enriches my work. Thanks to all the students and teachers across the globe who shared their thoughts with me through e-mails—an appreciated dialogue.

Thanks to all the colleagues who served as reviewers on one or more editions of each book or who offered helpful suggestions in conversations at our national and regional geography meetings; and thanks to Lisa DeChano-Cook and Stephen Cunha for special reviews for this edition. I am grateful for the generosity of ideas and sacrifice of time. Here is a master list of our reviewers on all the *Geosystems* textbooks.

Philip P. Allen, *Frostburg State University*
Ted J. Alsop, *Utah State University*
Ward Barrett, *University of Minnesota*
Steve Bass, *Mesa Community College*
Stefan Becker, *University of Wisconsin–Oshkosh*
Daniel Bedford, *Weber State University*
David Berner, *Normandale Community College*
Franco Biondi, *University of Nevada–Reno*
Peter D. Blanken, *University of Colorado–Boulder*
Patricia Boudinot, *George Mason University*
Anthony Brazel, *Arizona State University*
David R. Butler, *Southwest Texas State University*
Mary-Louise Byrne, *Wilfred Laurier University*
Ian A. Campbell, *University of Alberta–Edmonton*
Randall S. Cerveny, *Arizona State University*
Fred Chambers, *University of Colorado–Boulder*
Muncel Chang, *Butte College* Emeritus
Jordan Clayton, *Georgia State University*
Andrew Comrie, *University of Arizona*
C. Mark Cowell, *Indiana State University*
Richard A. Crooker, *Kutztown University*
Stephen Cunha, *Humboldt State University*
Armando M. da Silva, *Towson State University*
Dirk H. de Boer, *University of Saskatchewan*
Dennis Dahms, *University of Northern Iowa*
Shawna Dark, *California State University–Northridge*
Lisa DeChano-Cook, *Western Michigan University*
Mario P. Delisio, *Boise State University*
Joseph R. Desloges, *University of Toronto*
Lee R. Dexter, *Northern Arizona University*
Don W. Duckson, Jr., *Frostburg State University*
Christopher H. Exline, *University of Nevada–Reno*
Michael M. Folsom, *Eastern Washington University*
Mark Francek, *Central Michigan University*
Glen Fredlund, *University of Wisconsin–Milwaukee*
William Garcia, *University of North Carolina–Charlotte*
Doug Goodin, *Kansas State University*
David E. Greenland, *University of North Carolina–Chapel Hill*
Duane Griffin, *Bucknell University*
Barry N. Haack, *George Mason University*
Roy Haggerty, *Oregon State University*
John W. Hall, *Louisiana State University–Shreveport*
Vern Harnapp, *University of Akron*
John Harrington, *Kansas State University*
Blake Harrison, *Southern Connecticutt University*
Jason "Jake" Haugland, *University of Colorado, Boulder*
Gail Hobbs, *Pierce College*
Thomas W. Holder, *University of Georgia*
David H. Holt, *University of Southern Mississippi*
David A. Howarth, *University of Louisville*
Patricia G. Humbertson, *Youngstown State University*
David W. Icenogle, *Auburn University*
Philip L. Jackson, *Oregon State University*
J. Peter Johnson, Jr., *Carleton University*
Gabrielle Katz, *Appalachian State University*
Guy King, *California State University–Chico*
Ronald G. Knapp, *SUNY–The College at New Paltz*
Peter W. Knightes, *Central Texas College*
Thomas Krabacher, *California State University–Sacramento*
Hsiang-te Kung, *University of Memphis*
Richard Kurzhals, *Grand Rapids Junior College*
Steve Ladochy, *California State University–Los Angeles*
Charles W. Lafon, *Texas A & M University*
Paul R. Larson, *Southern Utah University*
Robert D. Larson, *Southwest Texas State University*
Elena Lioubimtseva, *Grand Valley State University*
Joyce Lundberg, *Carleton University*
W. Andrew Marcus, *Montana State University*

Brian Mark, *Ohio State University*
Nadine Martin, *University of Arizona*
Elliot G. McIntire, *California State University–Northridge*
Norman Meek, *California State University–San Bernardino*
Leigh W. Mintz, *California State University–Hayward* Emeritus
Sherry Morea-Oaks, Boulder, CO
Debra Morimoto, *Merced College*
Patrick Moss, *University of Wisconsin–Madison*
Lawrence C. Nkemdirim, *University of Calgary*
Andrew Oliphant, *San Francisco State University*
John E. Oliver, *Indiana State University*
Bradley M. Opdyke, *Michigan State University*
Richard L. Orndorff, *University of Nevada–Las Vegas*
Patrick Pease, *East Carolina University*
James Penn, *Southeastern Louisiana University*
Greg Pope, *Montclair State University*
Robin J. Rapai, *University of North Dakota*
Philip D. Renner, *American River College* Emeritus
William C. Rense, *Shippensburg University*
Leslie Rigg, *Northern Illinois University*
Dar Roberts, *University of California–Santa Barbara*
Wolf Roder, *University of Cincinnati*
Robert Rohli, *Louisiana State University*
Bill Russell, *L.A. Pierce College*
Dorothy Sack, *Ohio University*
Randall Schaetzl, *Michigan State University*
Glenn R. Sebastian, *University of South Alabama*
Daniel A. Selwa, *U.S.C. Coastal Carolina College*
Debra Sharkey, *Cosumnes River College*
Peter Siska, *Austin Peay State University*
Lee Slater, *Rutgers University*
Thomas W. Small, *Frostburg State University*
Daniel J. Smith, *University of Victoria*
Richard W. Smith, *Hartford Community College*
Stephen J. Stadler, *Oklahoma State University*
Michael Talbot, *Pima Community College*
Paul E. Todhunter, *University of North Dakota*
Susanna T.Y. Tong, *University of Cincinnati*
Liem Tran, *Florida Atlantic University*
Suzanne Traub-Metlay, *Front Range Community College*
Alice V. Turkington, *University of Kentucky*
Jon Van de Grift, *Metropolitan State College of Denver*
David Weide, *University of Nevada–Las Vegas*
Thomas B. Williams, *Western Illinois University*
Brenton M. Yarnal, *Pennsylvania State University*
Catherine H. Yansa, *Michigan State University*
Stephen R. Yool, *University of Arizona*
Don Yow, *Eastern Kentucky University*
Susie Zeigler-Svatek, *University of Minnesota*

After all these years, the strength of a publishing team remains ever essential. Thanks to President Paul Corey for his leadership since 1990. Thanks to my new editor Christian Botting for his guidance, publishing partnership, and welcomed great ideas and conversation, as we build our relationship; and to the geosciences staff Anton Yakovlev, Kristen Sanchez, Christina Ferraro, and Crissy Dudonis for their careful attention. Thanks to Managing Editor Gina Cheselka, Project Manager Shari Toron, and Vice-President of Development Carol Trueheart; organizational talent such as theirs is so valued. I shall always remember a satellite phone call to Gina Cheselka in New Jersey during our expedition to Antarctica, fur seals all about barking and staging mock attacks, and King Penguins staring at me, as I discussed details for this edition with Gina—a feeling of support from across the globe.

My appreciation to Pearson Art Project Manager Connie Long and to Jay McElroy and to designers Mark Ong and Randall Goodall for such skill in a complex art program and book design. And thanks to all the staff for allowing us to participate in the entire publishing process—I call them my "*Geosystems Team*." To Maureen McLaughlin, Marketing Manager, and the many sales representatives who spend months in the field communicating the *Geosystems* approach, as always, thanks and safe travels to them—may the wind be at their backs!

My thanks for production coordination to Vice-President of Higher Education Cindy Miller of Element LLC for such friendship and sustaining care through five books. Much appreciation to my Production Editor Suganya Karuppasamy for her ability to respond to my feedback as she oversees manuscript, copy editing, complex compositing, and page proofs; she is dedicated to the work and so talented. To copy editor Sherry Goldbecker, proofreader Jeff Georgeson, and indexer Robert Swanson I give an author's thanks for quality work and helping me craft the book. Much gratitude for the invaluable work of my developmental editor Ginger Birkeland, she is truly a valuable collaborator and colleague.

As you read this book and look at the cover, you will learn from more than 415 content-specific, beautiful photographs made by my wife, photographer, and expedition partner, Bobbé Christopherson. Her contribution to the success of *Geosystems* is obvious and begins with the spectacular cover photo and title-page photo and continues through four part-opening and all chapter-opening photos. She made the cover photo while in a Zodiac raft on expedition at 80° 22′ N latitude. In addition, Bobbé processed all the photos and satellite imagery in this text. Please visit the photo galleries at the *MasteringGeography*™ Study Area and learn more from her camera work. Bobbé is my colleague, wife, and best friend.

Physical geography teaches us a holistic view of the intricate supporting web that is Earth's environment and our place in it. Dramatic global change is under way in human–Earth relations as we alter physical, chemical, and biological systems. My attention to climate change science and applied topics is in response to the impacts we are experiencing and the future we are shaping. All things considered, this is a critical time for you to be enrolled in a physical geography course! The best to you in your studies—and ***carpe diem!***

Robert W. Christopherson
P. O. Box 128
Lincoln, California 95648-0128
E-mail: bobobbe@aol.com

Unique Earth systems organization

***Geosystems: An Introduction to Physical Geography,* Eighth Edition** is organized around the natural flow of energy, materials, and information, presenting subjects in the same sequence in which they occur in nature—an organic approach that is unique in this discipline. Within a one-of-a-kind Earth systems organization, ***Geosystems,* Eighth Edition** combines student-friendly writing, current applications, outstanding art and photos, compound figures that teach, and a strong media and assessment program.

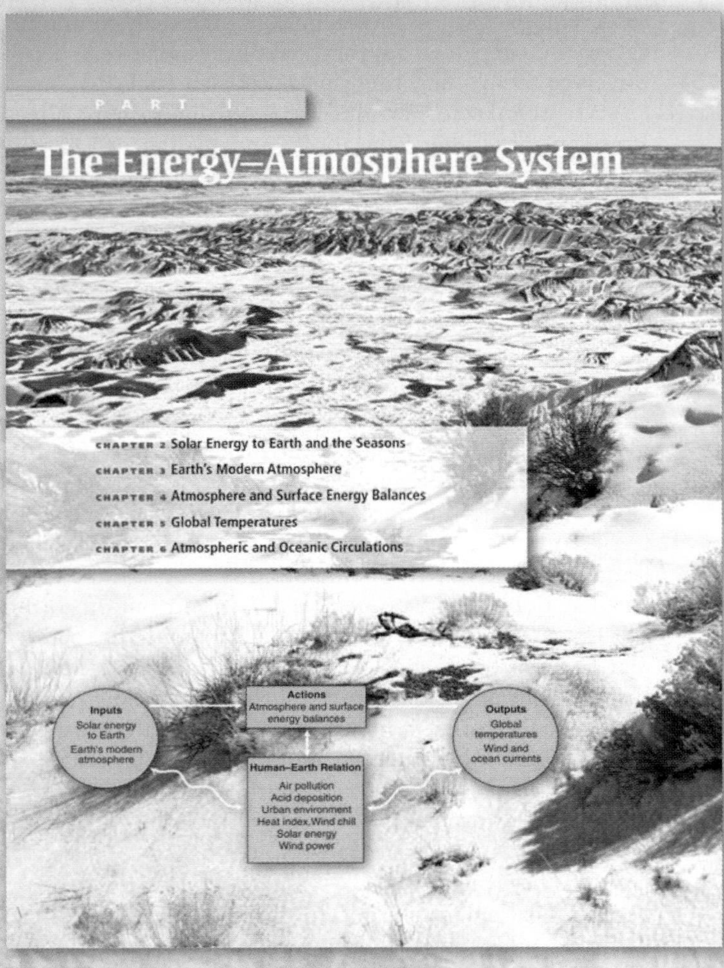

Part Opener Design

New part opener design, comprising illustrations and photos, emphasizes the chapter content connections to each system:

- Energy-Atmosphere System (Chapters 2–6)
- The Water, Weather, and Climate Systems (Chapters 7–10)
- The Earth-Atmosphere Interface (Chapters 11–17)
- Soils, Ecosystems, and Biomes (Chapters 18–21)

Global temperature patterns are a significant output of the energy–atmosphere system. We examined the complex interactions of several factors that produce these patterns and studied maps of their distributions. In the next chapter, we shift to another output in this energy–atmosphere system, that of global circulation of winds and ocean currents. We look at the forces that interact to produce these movements of air and water. Such events as the Gulf of Mexico oil disaster that began in April 2010 make this knowledge important as we see how ocean circulation spreads the toxic debris for thousands of miles.

NEW! Geosystems Connection

Geosystems Connection at the end of each chapter precedes the Key Learning Concepts Review and reminds students of what they have learned and where they are headed. This new feature helps reinforce the systems organization by bridging and connecting between the chapters.

Powerful pedagogical tools for students and teachers

Christopherson helps students achieve a deeper understanding of concepts through a variety of learning aids.

KEY LEARNING CONCEPTS

After reading the chapter, you should be able to:

- ***Define*** geography and physical geography in particular.
- ***Describe*** systems analysis, open and closed systems, and feedback information, and ***relate*** these concepts to Earth systems.
- ***Explain*** Earth's reference grid: latitude and longitude and latitudinal geographic zones and time.
- ***Define*** cartography and mapping basics: map scale and map projections.
- ***Describe*** remote sensing, and ***explain*** the geographic information system (GIS); both as tools used in geographic analysis.

Key Learning Concepts

Every chapter provides strong pedagogical tools and a learning path structured around the **Key Learning Concepts** that appear at the beginning of every chapter. These form the organization of the **Key Learning Concepts Review** section at the end of every chapter.

CRITICAL THINKING 3.1

Where is your tropopause?

On your next flight as the plane reaches its highest ask flight attendants to determine the temperature plane. Some planes have video screens that display air temperature. By definition, the tropopause is wl temperature –57°C (–70°F) occurs. Depending on t the year, altitude and temperature make an interes son. Is the tropopause at a higher altitude in summer or winter?

CRITICAL THINKING 9.1

Your local water budget

Select your campus, yard, or perhaps a house plant and apply the water-balance concepts. Where does the water supply originate—its source? Estimate the ultimate water supply and demand for the area you selected. For a general idea of PRECIP and POTET, find your locale on Figures 9.5 and 9.6. Consider the seasonal timing of this supply and demand, and estimate water needs and how they vary as components of the water-budget change.

NEW! Critical Thinking

New and revised **Critical Thinking** questions are integrated into each chapter, giving students the opportunity to stop and practice conceptual understanding and critical thinking as they read through the chapter. These questions challenge students to take the next step in applying their learning.

KEY LEARNING CONCEPTS REVIEW

- ***Identify*** the pathways of solar energy through the troposphere to Earth's surface: transmission, scattering, diffuse radiation, refraction, albedo (reflectivity), conduction, convection, and advection.

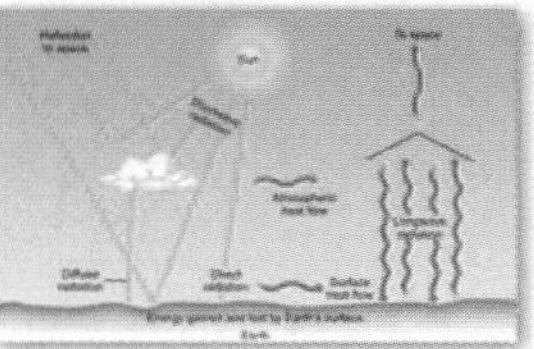

Radiant energy from the Sun that cascades through complex circuits to the surface powers Earth's biosphere. Our budget of atmospheric energy comprises shortwave radiation *inputs* (ultraviolet light, visible light, and near-infrared wavelengths) and longwave radiation *outputs* (thermal infrared).

KEY LEARNING CONCEPTS REVIEW

- ***Define*** the concept of air pressure, and ***describe*** instruments used to measure air pressure.

The weight (created by motion, size, and number of molecules) of the atmosphere is **air pressure,** which exerts an average force of approximately 1 kg/cm² (14.7 lb/in.²), from Chapter 3. A **mercury barometer** measures air pressure at the surface (mercury in a tube—closed at one end and open at the other, with the open end placed in a vessel of mercury—that changes level in response to pressure changes), as does an **aneroid barometer** (a closed cell, partially evacuated of air, that detects changes in pressure).

air pressure (p. 130)
mercury barometer (p. 132)
aneroid barometer (p. 132)

1. How does air exert pressure? Describe the basic instrument used to measure air pressure. Compare the operation of two different types of instruments discussed.
2. What is normal sea-level pressure in millimeters? Millibars? Inches? Kilopascals?

- ***Define*** wind, and ***explain*** how wind is measured, how wind direction is determined, and how winds are named.

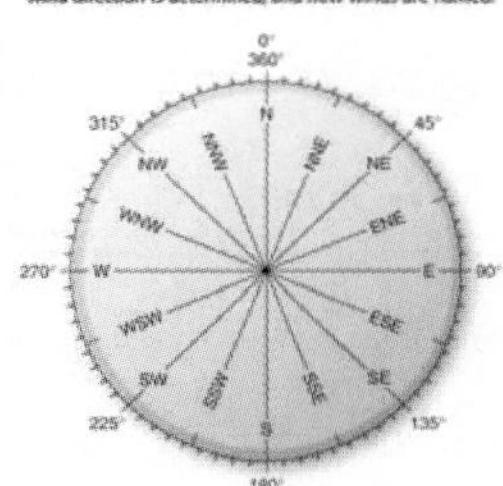

Wind is the horizontal movement of air across Earth's surface; turbulence adds wind updrafts and downdrafts, and thus a vertical component to the definition. Its speed is measured with an **anemometer** (a device with cups that are pushed by the wind) and its direction with a **wind vane** (a flat blade or surface that is directed by the wind). A descriptive scale useful in visually estimating wind speed is the traditional Beaufort wind scale.

wind (p. 132)
anemometer (p. 133)
wind vane (p. 133)

3. What is a possible explanation for the beautiful sunrises and sunsets during the summer of 1992 in North America? Relate your answer to global circulation.

Key Learning Concepts Review

The **Key Learning Concepts** are revisited at the end of each chapter, and include narrative definitions, key terms list and page numbers, review questions that require critical thinking and analysis, and now also repeat important related figures from the chapter for additional reinforcement and to accommodate different learning styles.

We turn for weather forecasts to the National Weather Service (NWS) in the United States (http://www.nws.noaa.gov/) or the Canadian Meteorological Centre, a branch of the Meteorological Service of Canada (MSC) (http://www.msc-smc.ec.gc.ca/cmc/index_e.html), to see curren satellite images and to hear weather analysis. Internationally the World Meteorological Organization coordinates weath er information (see http://www.wmo.ch/). Many sources o weather information and related topics are found on ou *Mastering Geography* Web site.

Clouds are the most variable factor influencing Earth's radiation budget, making them the subject of simulations in computer models of atmospheric behavior. The International Satellite Cloud Climatology Project (http://isccp.giss.nasa.gov/), part of the World Climate Research Programme, is presently in the midst of such research. Clouds and the Earth Radiant Energy System (CERES, http://science.larc.nasa.gov/ceres/) sensors aboard the *TRMM*, *Terra*, and *Aqua* satellites are assessing cloud effects on longwave, shortwave, and net radiation patterns as never before possible.

Embedded URLs

Christopherson has **embedded URLs** into the text for both the student and instructor to link directly online for further study and inquiry. In the Pearson eText version, these URLs are active links that direct students to the Web right from the eText page.

Compelling real-world applications and currency

Recognizing the function and significance of Geosystems in today's world boosts student comprehension and engagement. ***Geosystems*, Eighth Edition** relates the material in the text to **current real events and phenomena** and presents the most thorough and integrated treatment of **climate change science**, giving students compelling reasons for learning physical geography.

GEOSYSTEMS NOW

On the Front Lines of Intense Weather

How do you deal with the televised images of intense weather, tornadoes, or floods? Imagine what people face on the front lines of these events. Improved warnings have reduced loss of life, yet higher storm intensities and frequencies have caused huge increases in property damage.

Think of Greensburg, Kansas, southeast of Dodge City, with its 1600 residents. For tourists, the town features the world's largest hand-dug water well. Nearby stood the old water tower with the town's name inscribed before the tornado hit—here is small town, Midwest, USA.

On the evening of May 4, 2007, a frontal system moved through. As tornado warning sirens and NOAA weather radio signaled the alert, people went to their basements and storm shelters. Just before 10 P.M., an EF-5–intensity tornado roared in from the southwest. This was the first designated "5" since the Fujita Scale revision in February 2007, described in this chapter.

The monster was 2.7 km (1.7 mi) wide, and it attacked Greensburg. It shredded all that was familiar into a chaotic mess in less than 10 minutes. After the storm, people came out into the now-silent evening, lights were out, some called for help, yet nothing familiar was left. Flashlights came on, as did headlights when an owner could find keys and a surviving car. Sunrise brought shock; all the tree-lined streets were barren, trees pruned and broken by the winds. Eleven people died that night.

Heroism was all about, the National Guard mobilized, a federal disaster area was declared, the media arrived, and in a few days politicians began their tours. Residents appeared on TV, interviewed for their anecdotes. Disaster assistance processing was set up, forms were filled out, and weeks of delays began. In the void of factual information, rumors spread quickly.

In a couple of weeks, the outsiders left; streets and lots were cleared of debris, piled into a hastily prepared dump. All that was familiar was in the pile north of town. The months rolled by with no media coverage, just the daily task of dealing with shattered reality. A year later, when we asked people about that night, they readily told their stories. How do people on disaster front lines physically and mentally deal with such changes in their lives?

FIGURE GN 8.1 Most of Greensburg, Kansas, was erased by an EF-5 tornado in 2007. A new water tower stands a year later where the old one fell. [Bobbé Christopherson.]

We drive along the Bolivar Peninsula on the Gulf Coast of Texas one year after the September 13, 2008, 2 A.M. landfall of Hurricane Ike. We pass 35 km (21 mi) of leveled coastal neighborhoods. Ike's winds were hurricane force for more than 177 km (110 mi) along the coast. Debris protrudes from the barrier-island sand, power is still out in spots, and few trees are left standing. Near the western end of the island, toward the Galveston ferry dock, large piles of former homes and businesses stand above the sand. Several piles are burning, filling the air with foul smoke—and this is a year since the hurricane.

Bolivar, like Hatteras and many other barrier islands, uses septic sewage systems, which become disrupted by storm surge. In 2009, residents in Bolivar were wrestling with whether to attach to a sewer line for the first time. One developer is constructing small houses on 8.5-m (28-ft) stilts, perched in ominous preparation for the next hurricane.

We can easily expand this narrative to thoughts of New Orleans years after Katrina, Rita, Gustav, and Cindy, and now the BP oil-spill disaster; or the floods in Missouri, North Dakota, Oklahoma City, parts of Texas, Milwaukee, Chicago, and elsewhere; or the Wadena tornado in the chapter-opening photo. Again the question: How do people on the front lines cope with such disaster? Contemplate the human dimension as you work through the weather chapter.

FIGURE GN 8.2 Hurricane Ike swept away all homes and the pavement in this neighborhood. Many kilometers of coastline were destroyed on Bolivar Peninsula, Texas. [Bobbé Christopherson.]

191

NEW! *Geosystems Now* Case Studies

***Geosystems Now* chapter-opening case studies** draw students into the material by discussing an interesting, current application of chapter concepts. These unique case studies are based on firsthand experiences out in the field and on expedition by the author and his wife, who is a nature photographer, as they bear witness to dramatic, real-world applications of physical geography and Earth systems science. Examples include chasing the subsolar point in the Atlantic Ocean (Ch. 2); visiting Wadena, Minnesota, four days after an EF-04 tornado tore through town (Ch. 8); and documenting the decline in Lake Mead behind Hoover Dam (Ch. 9).

FOCUS STUDY 8.1

Atlantic Hurricanes and the Future

Forecasters at the National Hurricane Center (NHC) in Miami were busy throughout the 1995–2009 hurricane seasons. This was the most active 15-year period in the history of the NHC, with 207 named tropical storms, including 111 hurricanes (48 of which were intense, meaning category 3 or higher). This was a record level of activity and individual storm intensity despite the reduced number of tropical cyclones during the 1997 El Niño season.

Statistically, the damage caused by tropical cyclones is increasing substantially as more and more development occurs along susceptible coastlines, whereas loss of life is decreasing in most parts of the world owing to better forecasting of these storms. The journal *Science* offered this "Perspective":

> The risk of human losses is likely to remain low . . . because of a well-established warning and rescue system and ongoing improvements in hurricane prediction. A main concern is the risks of high damage costs (up to $100 billion in a single event) because of ongoing population increases in coastal areas and increasing investment in buildings and extensive infrastructure in general.*

*L. Bengtsson, "Hurricane threats," *Science* 293 (July 20, 2001): 441.

(MG) Notebook
Record-breaking
2005 Atlantic hurricane season

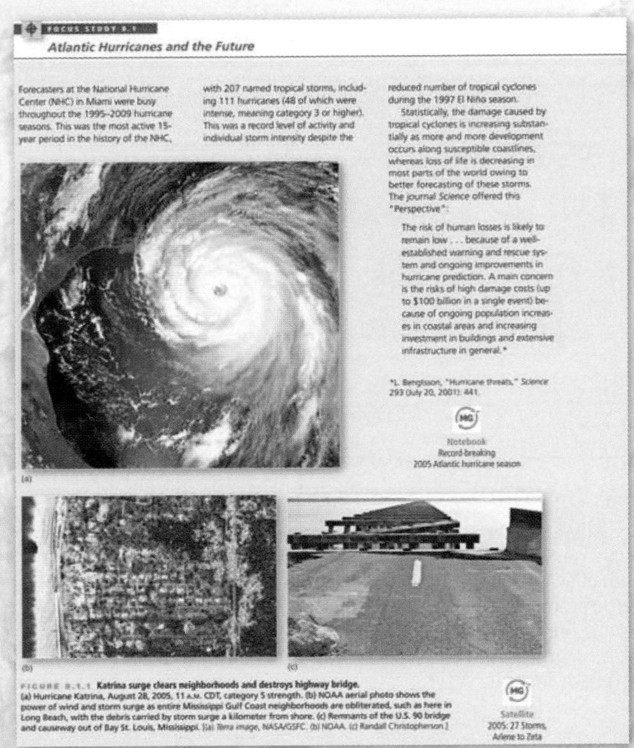

FIGURE 8.1.1 Katrina surge clears neighborhoods and destroys highway bridge. (a) Hurricane Katrina, August 28, 2005, 11 A.M. CDT, category 5 strength. (b) NOAA aerial photo shows the power of wind and storm surge as entire Mississippi Gulf Coast neighborhoods are obliterated, such as here in Long Beach, with the debris carried by storm surge a kilometer from shore. (c) Remnants of the U.S. 90 bridge and causeway out of Bay St. Louis, Mississippi. [(a) *Terra* image, NASA/GSFC. (b) NOAA. (c) Randall Christopherson.]

(MG) Satellite
2005: 27 Storms,
Arlene to Zeta

GEO REPORT 1.4 ***GPS origins***

Originally devised in the 1970s by the U.S. Department of Defense for military purposes, the present GPS is commercially available worldwide. In 2000, the Pentagon shut down its Pentagon Selective Availability security control, making commercial resolution the same as military applications. Additional frequencies were added in 2003 and 2006, which increased accuracy significantly to less than 10 m (33 ft). For a GPS overview, see http://www.colorado.edu/geography/gcraft/notes/gps/gps_f.html.

GEO REPORT 2.3 ***Why do we always see the same side of the Moon?***

Note in Figure 2.13 that the Moon both revolves around Earth and rotates on its axis in a counterclockwise direction when viewed from above Earth's North Pole. It does both these motions in the same amount of time. The Moon's speed in orbit varies slightly during the month, whereas its rotation speed is constant, so we see about 59% of the lunar surface during the month, or exactly 50% at any one moment—always the same side.

NEW! *Geo Reports*

Four to six *Geo Reports* are found in each chapter at the bottom of pages, so as not to interrupt the flow of the chapter narrative. This feature presents new and legacy factoids, examples, applications, and student action items that relate to the material in the main text narrative.

Focus Study essays

Focus Study essays present additional, detailed discussions of critical physical geography topics, provoking students' desire to learn physical geography. Students are drawn into discussions such as the Great Lakes; the several important ice cores drilled in the polar regions that push the climate record back 800,000 years; a rational look at shoreline planning and hazard assessment and perception; the increasing intensity of tropical cyclones as air and water temperatures climb to new records; and many other important topics.

Photography and art that teach

A **superlative photo program,** consisting of photographs taken by the author and his wife during expeditions, provides authoritative examples and applications of physical geography and Earth systems science. More than 350 photographs were made by Bobbé Christopherson, composed in the field specifically for use in ***Geosystems,* Eighth Edition**.

A **market-leading art program** includes compound figures that teach the most difficult-to-understand concepts, with major revisions to numerous figures in this edition that have been updated for clarity, consistent style, and to improve overall student comprehension.

Engaging students in the study of physical geography

Immerse your students in the study of physical geography with MasteringGeography. Used by over one million students, the Mastering platform is the most effective and widely used online tutorial, homework, and assessment system in the sciences.

Assignable Content:

- Geoscience Animation Activities
- *Encounter Geosystems* Google Earth Activities
- Thinking Spatially Activities
- Test Bank Questions
- End of Chapter Questions
- Reading Quiz Questions
- *Earth Report* Video Activities
- MapMaster™ Interactive Maps

For Student Self Study:

- Geoscience Animations
- Satellite Loops
- Author Notebooks
- Photo Galleries
- Career Links
- Destinations
- Quick Links
- Physical Geography Case Studies
- "In the News" RSS Feeds
- Pearson eText
- Visual Glossary
- Self Study Quizzes
- *Earth Report* Videos
- MapMaster Interactive Maps

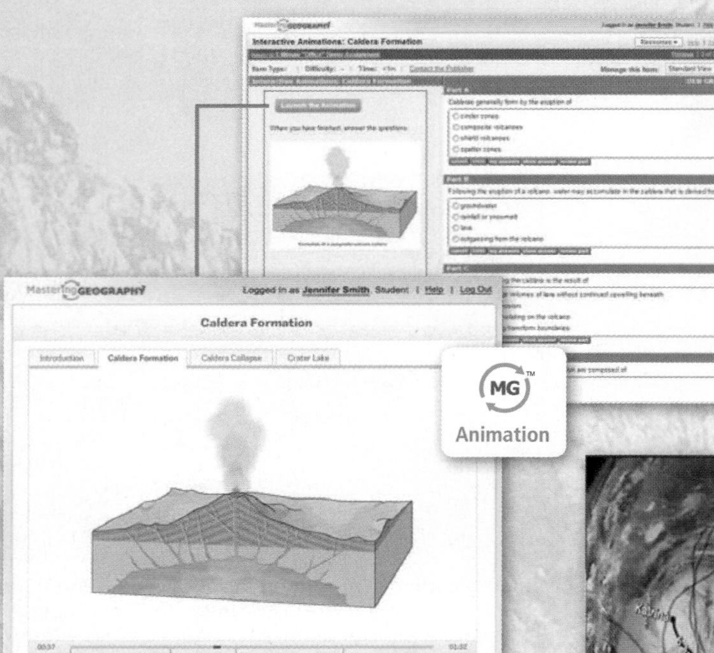

Geoscience Animations

Geoscience Animations illuminate the most difficult-to-visualize topics from across the physical geosciences, such as solar system formation, hydrologic cycle, plate tectonics, glacial advance and retreat, global warming, etc. Animations include audio narration, a text transcript, and assignable multiple-choice quizzes with hints and specific wrong-answer feedback to help guide students towards mastery of these core physical process concepts. Icons integrated throughout the text indicate to students when they can login to the Study Area of MasteringGeography to view the animations.

Satellite Loops

These satellite visualizations illustrate important physical geography concepts. Icons integrated throughout the text indicate to students when they can login to the Study Area of MasteringGeography to view the satellite loops.

Encounter Geosystems

Encounter Geosystems activities have students use the dynamic features of Google Earth™ to visualize and explore Earth's physical landscape and answer multiple-choice and short-answer questions related to core physical geography concepts. All Explorations include corresponding Google Earth KMZ media files, and questions include hints and specific wrong-answer feedback to help coach students towards mastery of the concepts.

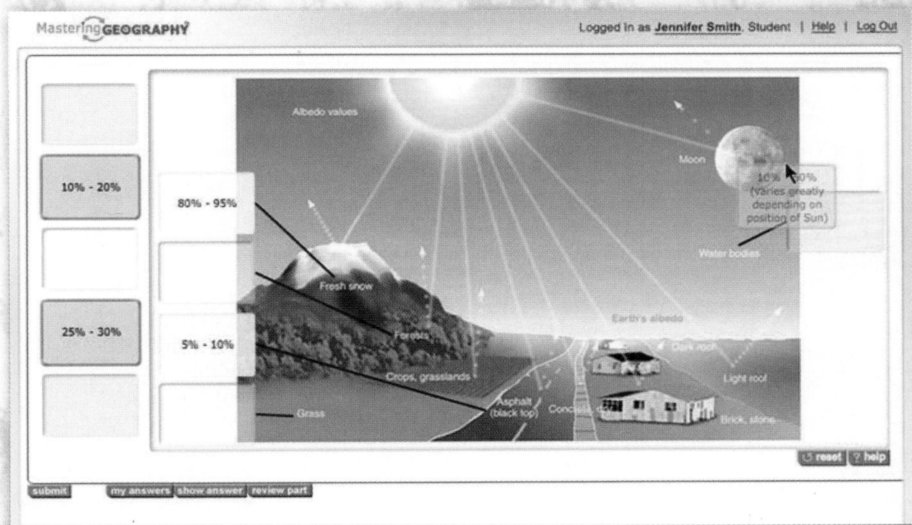

Thinking Spatially

Thinking Spatially activities help students develop spatial reasoning and critical-thinking skills by identifying and labeling features from maps, illustrations, photographs, graphs, and charts. Students then examine related data sets, answering multiple-choice questions and increasingly higher order conceptual short-answer questions, which include hints and specific wrong-answer feedback.

Earth Report Videos

These videos are designed to help students visualize how human decisions and behavior have affected the environment, and how individuals are taking steps toward recovery. With topics ranging from the poor land management promoting the devastation of river systems in Central America to the struggles for electricity in China and Africa, these 13 videos from Television for the Environment's global *Earth Report* series recognize the efforts of individuals around the world to unite and protect the planet.

Quickly monitor and display student results

With customizable, easy-to-assign, and automatically graded assessments, instructors can maximize class time and motivate students to learn outside of class and arrive prepared for lecture.

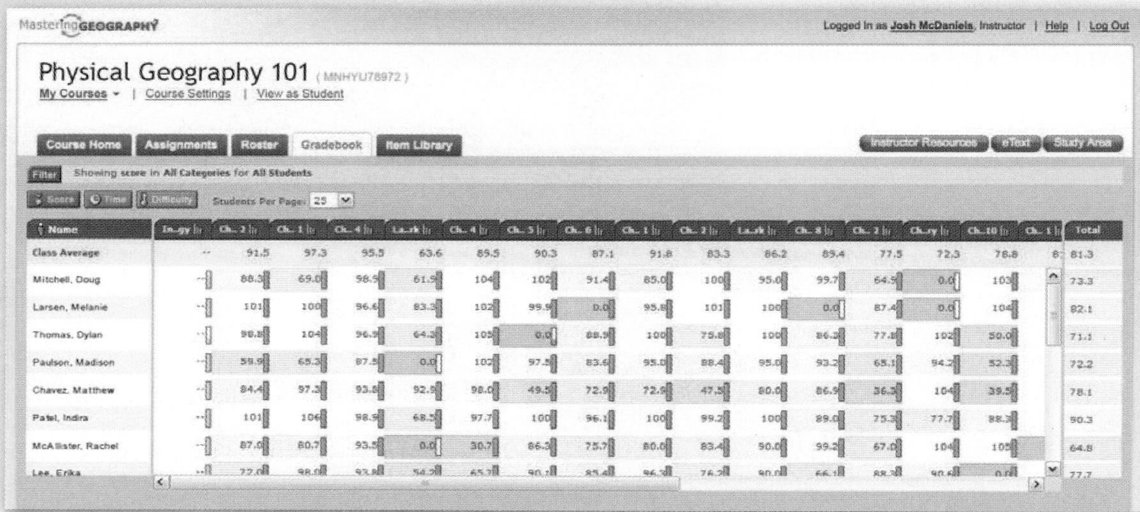

Gradebook

Every assignment is **automatically graded. Shades of red** highlight vulnerable students and challenging assignments.

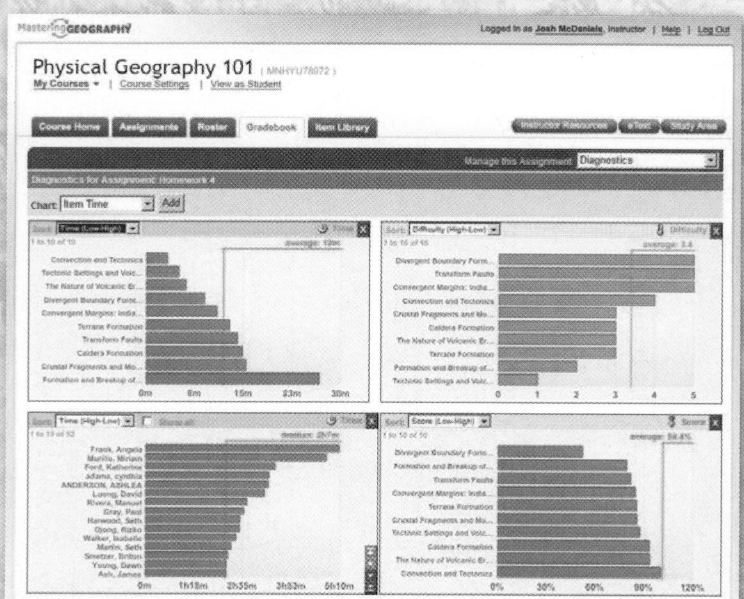

Gradebook Diagnostics

This screen gives instructors weekly diagnostics. With a single click, charts summarize the most difficult problems, vulnerable students, grade distribution, and improvements in student scores throughout the semester and year.

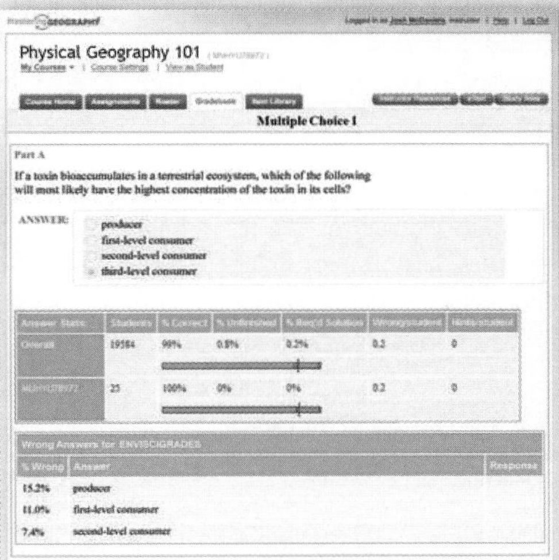

Student Performance Data

At a glance, instructors can use the color-coded gradebook to identify students who are having difficulty. Instructors can also identify the most difficult activity (and the most difficult parts of an activity) in each assignment, or critique the detailed work of anyone who needs more help. They can even compare results on any activity and any step with a previous class, or with the national average.

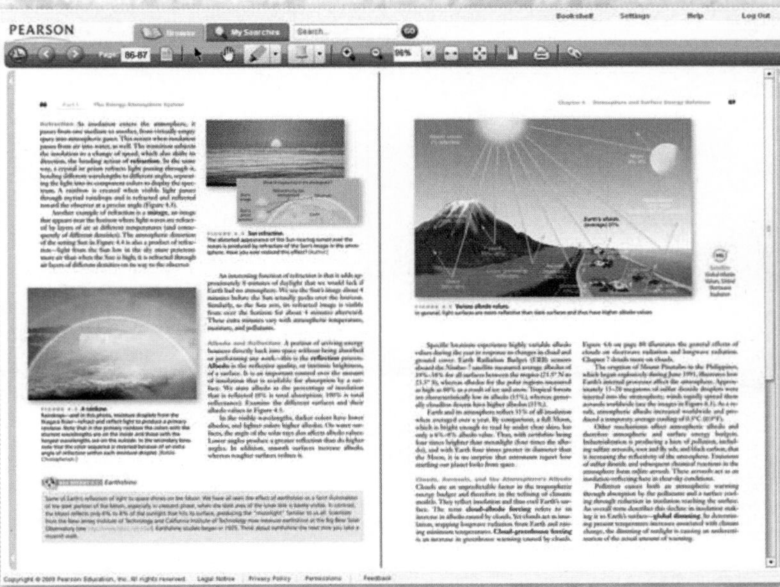

Pearson eText

Pearson eText gives students access to the text whenever and wherever they can access the Internet. The eText pages look exactly like the printed text, and include powerful interactive and customization functions. Users can create notes, highlight text in different colors, create bookmarks, zoom, click hyperlinked words and phrases to view definitions, and view as a single page or as two pages. Pearson eText also links students to associated media files, enabling them to view animations as they read the text, and offers a full-text search and the ability to save and export notes. The Pearson eText for ***Geosystems*, Eighth Edition** also includes embedded URLs in the chapter text with active links to the Internet.

Continuously Improving Content

MasteringGeography™ offers a dynamic pool of assignable content that improves with student usage. Detailed analysis of student performance statistics—including time spent, answers submitted, solutions requested, and hints used—ensures the highest quality content.

1. **We conduct a thorough analysis** of each question by reviewing student performance data that has been generated by real students.
2. **We make enhancements** to improve the clarity and accuracy of content, answer choices, and instructions for each problem.
3. **We repeat the process.**

This ongoing process helps students learn as students help the system improve.

CHAPTER 1

Essentials of Geography

Using map, GPS unit, and radar instruments, a course is set through the Antarctic Sound to the Weddell Sea, Antarctica, at 63° 31′ S latitude. A dramatic tabular iceberg dominates the horizon. The former Larsen Ice Shelf area began about 1.5° to the south. [Bobbé Christopherson.]

KEY LEARNING CONCEPTS

After reading the chapter, you should be able to:

- ***Define*** geography and physical geography in particular.
- ***Describe*** systems analysis, open and closed systems, and feedback information, and ***relate*** these concepts to Earth systems.
- ***Explain*** Earth's reference grid: latitude and longitude and latitudinal geographic zones and time.
- ***Define*** cartography and mapping basics: map scale and map projections.
- ***Describe*** remote sensing, and ***explain*** the geographic information system (GIS); both as tools used in geographic analysis.

GEOSYSTEMS NOW

Where is Four Corners, Exactly?

Amidst the high desert landscape of the American Southwest is the unique Four Corners landmark, where a monument marks the intersection of Colorado, New Mexico, Arizona, and Utah. This location is the only point in the United States where four states touch. It is a must-see destination for many visitors, a place where feet and hands can each be in a different state and families and friends can straddle the disk marking the precise location where state boundaries meet. The original surveyor, Chandler Robbins, marked the spot with a 2.1-m (7-ft) sandstone monument in 1875. The present monument and visitor plaza date to 1992 and are owned and operated by the Navajo Nation Parks and Recreation Department.

In the spring of 2009, news reports claimed that the location of Four Corners was in error, that in fact the monument was located 4 km (2.5 mi) west of the precise location. The reports claimed that new surveys by the National Geodetic Survey (NGS) using the satellite Global Positioning System (GPS) and state-of-the-art mapping technology refuted the original 1875 surveys. For such an important landmark, this was shocking news. Did the monument now need to be moved, or was the news in error?

For the average visitor, such precise measurement details are unimportant. For the scientist, engineer, or natural resource professional, such an error has wide-ranging implications. Four Corners is an established benchmark in the western United States upon which other surveys and locations are based.

The question of accuracy relates to which meridian was used for the original survey. Congress ordered Chandler Robbins in 1875 to use the Greenwich meridian, London, which later became the official 0° prime meridian, from which we measure longitude west or east on Earth. However, in the late 1800s, many western states were set with surveys using the Washington meridian, which is about 77° 3′ west longitude from Greenwich. If Robbins' survey had used the Washington meridian, the marker would be off, as described by the media reports.

A few weeks later the news stories were corrected as the details of the original survey came to light. While the monument does appear to be offset from its intended location, the actual offset is only 548 m (1800 ft) to the east rather than west. However, according to NGS, the monument is in the right spot. It turns out that the location of a physical monument is usually the ultimate authority in delineating a boundary.

Thus, once the Four Corners monument was established and accepted by all parties involved—in this case, the four territories and the U.S. Congress—it became the *exact* spot where the four states meet. The location, set for more than 125 years, is used in the modern National Spatial Reference System and is in the right place—36° 59′ 56.3150 north latitude by 109° 02′ 42.6210 west longitude (http://www.glorecords.blm.gov/).

At the heart of geographic science are such topics as place, location, latitude and longitude, the Greenwich prime meridian, and mapping. In this chapter, you work with these concepts, the "Essentials of Geography."

(a)

(b)

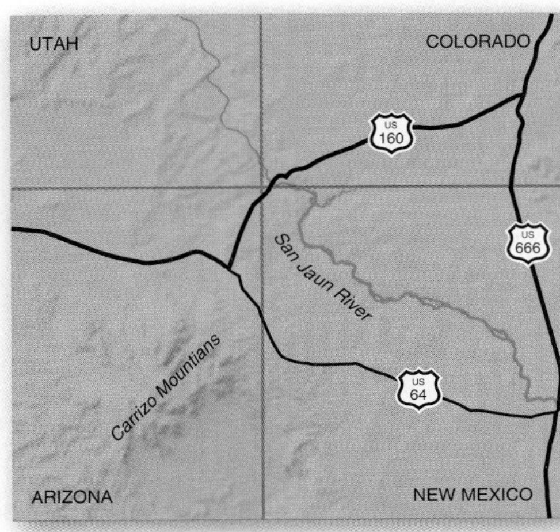

FIGURE GN 1.1 (a) The Four Corners tourist plaza just off U.S. Highway 160. (b) Survey benchmark 1992 replacement. [USGS.]

Welcome to the Eighth Edition of *Geosystems: An Introduction to Physical Geography*! Much has happened since the first edition of *Geosystems* (© 1992). Imagine the challenge society as a whole and we in physical geography face to understand how Earth's systems operate and then to forecast changes in these systems to determine what future environments we may expect. In every chapter, *Geosystems* continues to present, as it has for 20 years, updated science and information to help you understand our dynamic Earth systems.

With events such as the 2010 eruption of the Eyjafjallajökull volcano in Iceland, causing disruption of airline schedules as ash spread through prevailing winds, or the 2010 *Deepwater Horizon* oil spill disaster in the Gulf of Mexico that continues to unfold in its many spatial pollution impacts, *Geosystems* is there with content and context. This tragic oil spill alone is more than 34 times the size of the 1989 *Exxon/Valdez* spill in Alaska, discussed in Chapter 21 in many editions.

Physical geography deals with powerful Earth systems that influence our lives and includes the many ways humans impact those systems. In this second decade of the 21st century, a century that will see many changes to the natural environment, we find ourselves in an exciting time to study physical geography, learning about the building blocks that form the landscapes, seascapes, atmosphere, and ecosystems upon which we depend. Climate change science will be an overriding topic for the global society this century.

In the first edition, *Geosystems* presented the *First Assessment Report* by the Intergovernmental Panel on Climate Change (IPCC, http://www.ipcc.ch/). The IPCC completed its *Fourth Assessment Report* in 2007. The Climate Congress held an International Climate Change Conference in Copenhagen in March 2009 as an update to the IPCC with more than 80 countries and 2500 scientists (http://climatecongress.ku.dk/). The Climate Congress concluded:

> The *worst-case* IPCC (2007) projections, or even worse, are being realized. . . . Emissions are soaring, projections of sea-level rise are higher than expected, and climate impacts around the world are appearing with increasing frequency.

The Conference of the Parties held its 15th meeting (COP–15) under the U.N. Framework Convention on Climate Change (FCCC, http://unfccc.int/2860.php) in Copenhagen in December 2009 and launched the *Copenhagen Accord* in an effort to reduce emissions of greenhouse gases that enhance warming. During the span of eight *Geosystems* editions, the level of carbon dioxide, a principal greenhouse gas, rose 16% in Earth's atmosphere, and CO_2 emissions have accelerated to increasing at 3.0% per year since 2000. These changes affect systems in every chapter of *Geosystems*.

This past decade experienced the highest temperatures over land and water in the instrumental record (Figure 1.1). The year 2009 tied 2007 as the second highest year for global temperatures, just behind 2005. In the Southern Hemisphere, 2009 had the highest temperatures on record. A good summary web site is http://climate.nasa.gov/.

Financial data from 15 of the world's largest financial institutions estimated that annual weather-related damage losses (drought, floods, hail, tornadoes, derechos, tropical systems, monsoon intensity, storm surges, blizzard and ice storms, and wildfires) could exceed $1 trillion by A.D. 2040 (adjusted to current dollars)—up from $210 billion in 2005 (Hurricane Katrina alone produced $125 billion in losses). Global climate change is driving this increasing total—and at a pace that is accelerating.

The January 2010 magnitude 7.0 earthquake, with thousands of aftershocks, struck Haiti and literally brought one of the poorest countries on Earth to collapse, killing more than 200,000 people and leaving millions homeless. In Haiti, the problem is one of poor construction, weak infrastructure, and a lack of planning in response to scientific knowledge—although in Haiti poverty blocks hazard perception and preparation. Estimates set the damage in Haiti at more than double the country's entire gross domestic product. The scene repeated itself with a magni-

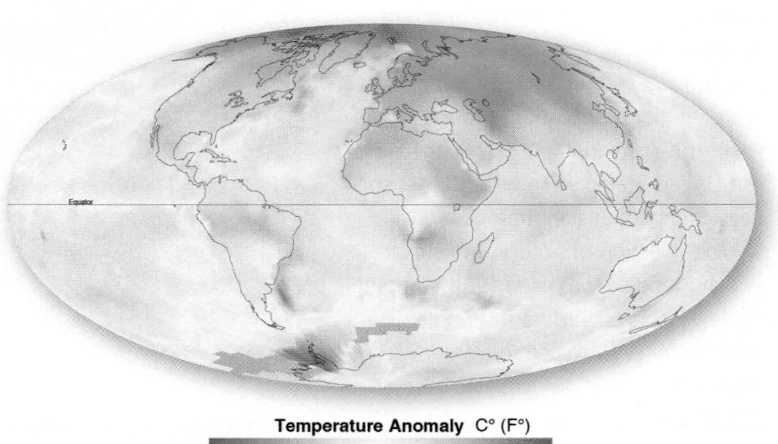

FIGURE 1.1 Record temperatures for the decade. Temperature anomalies for the years 2000–2009 compared to the 1951–1980 standard baseline. The decade experienced the highest average air temperatures since record keeping began in 1880, along with the highest temperatures in the global oceans and many lakes. [Map image courtesy of GISS NASA.]

tude 8.8 earthquake that struck the next month in Maule, Chile, southwest of Santiago. Physical geography explains the operations of Earth systems from a *spatial* perspective and gives you the tools to understand what is happening in Haiti and Chile and across the globe.

Why does the environment vary from equator to midlatitudes, between deserts and polar regions? How does solar energy influence the distribution of trees, soils, climates, and lifestyles? Why are we experiencing record levels of wildfires? Why are earthquakes and volcanoes active in certain regions and what is the risk? What produces the patterns of wind, weather, and ocean currents? Why are global sea levels on the rise? Why did the countries of the world join in the International Polar Year (IPY) study of the polar regions between 2007 and 2009? How do natural systems affect human populations, and, in turn, what impact are humans having on natural systems? Why are record levels of plants and animals facing extinction? In this book, we explore those questions, and more, through geography's unique perspective. Welcome to an exploration of physical geography!

In this chapter: Our study of geosystems—Earth systems—begins with a look at the science of physical geography and the geographic tools we use. Physical geography is key to studying entire Earth systems because of its integrative spatial approach.

Physical geographers analyze systems to study the environment. Therefore, we discuss systems and the feedback mechanisms that influence system operations. We then consider location, a key theme of geographic inquiry—the latitude, longitude, and time coordinates that inscribe Earth's surface and the new technologies in use to measure them. The study of longitude and a universal time system provides us with interesting insights into geography. Next, we examine maps as critical tools that geographers use to portray physical and cultural information. This chapter concludes with an overview of the technology that is adding exciting new dimensions to geography: remote sensing from space and computer-based geographic information systems (GIS).

The Science of Geography

Now is the era of **Earth systems science**. This science contributes to our emerging view of Earth as a complete entity—an interacting set of physical, chemical, and biological systems that produce a whole Earth. Physical geography is at the heart of Earth systems science as we answer the *spatial* questions concerning Earth's physical systems and their interaction with living things.

Geography (from *geo*, "Earth," and *graphein*, "to write") is the science that studies the relationships among natural systems, geographic areas, society, and cultural activities and the interdependence of all of these *over space*. The term **spatial** refers to the nature and character of physical space, its measurement, and the distribution of things within it.

For example, think of your own route to the classroom or library today or the way you get to work and how you use your knowledge of street patterns, traffic trouble spots, one-way streets, parking spaces, or bike rack locations to minimize walking distance. All these are spatial considerations that influence human actions.

To help in this definition, we separate geographic science into five spatial themes: **location**, **region**, **human–Earth relationships**, **movement**, and **place**, each illustrated and defined in Figure 1.2 on page 4. *Geosystems* draws on each theme.

Geographic Analysis

Within these five geographic themes, a method rather than a specific body of knowledge governs geography. The method is **spatial analysis**. Using this method, geography synthesizes (brings together) topics from many fields, integrating information to form a whole-Earth concept. Geographers view phenomena as occurring in spaces, areas, and locations. The language of geography reflects this spatial view: space, territory, zone, pattern, distribution, place, location, region, sphere, province, and distance. Geographers analyze the differences and similarities among places.

Process, a set of actions or mechanisms that operate in some special order, is central to geographic analysis. As examples in *Geosystems*, numerous processes are involved in Earth's vast water–atmosphere–weather system, or in continental crust movements and earthquake occurrences, or in ecosystem functions, or in a river's drainage basin dynamics. Geographers use spatial analysis to examine how Earth's processes interact over space or area.

Therefore, **physical geography** is the spatial analysis of all the physical elements and process systems that make up the environment: energy, air, water, weather, climate, landforms, soils, animals, plants, microorganisms, and Earth itself. As scientists, physical geographers employ the **scientific method**. Focus Study 1.1 on pages 6–7 explains this essential process of science.

The Geographic Continuum

Geography is eclectic, integrating a wide range of subject matter from diverse fields; virtually any subject can be examined geographically. Figure 1.3 on page 5 shows a continuous distribution—a continuum—along which the content of geography is arranged. Physical and life sciences are at one end, and human and cultural sciences are at the other. As the figure shows, various specialties within geography draw from these subject areas.

The continuum in Figure 1.3 reflects a basic duality, or split, within geography—*physical geography* versus *human/cultural geography*. Some societies parallel this duality. Humans at times think of themselves as exempt from physical Earth processes—like actors not paying attention to their stage, props, and lighting. We all depend on Earth's systems to provide oxygen, water, nutrients, energy, and materials to support life. The growing complexity of the

Location
Location identifies a specific address or absolute and relative position on Earth. These New Jersey road signs direct drivers on route 37 to locations.

Place
No two places on Earth are exactly alike. The Bay of Fundy, separating New Brunswick and Nova Scotia, has the highest average tides in the world—up to 17 m (55.7 ft) in height.

Region
A region is an area defined by uniform characteristics. South-central Texas is a distinct region of grasslands, large cattle ranches, irrigated feed crops, and feedlots.

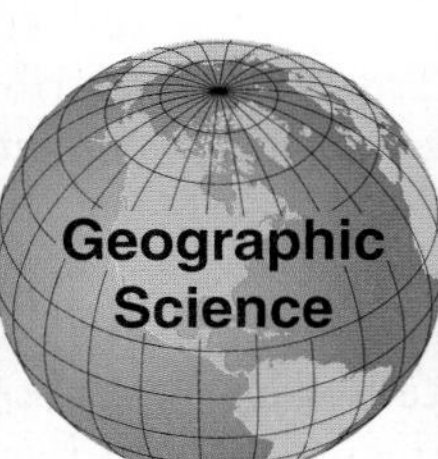

Movement
Communication, circulation, migration, and diffusion across Earth's surface represent movement in our interdependent world. Animals migrate with seasonal change; Snow Geese feed and rest along their journey.

Human–Earth Relationships
Human–environment connections include resource exploitation, hazard perception, and environmental pollution and modification. A *Terra* satellite image of the *Deepwater Horizon* oil spill in the Gulf of Mexico, highlighted by a sunglint on the gulf's surface. This human-caused spill is the largest offshore drilling disaster in history.

FIGURE 1.2 Five themes of geographic science.
Drawing from your own experience, can you think of several examples of each theme? [Photos by Bobbé Christopherson; image NASA/GSFC.]

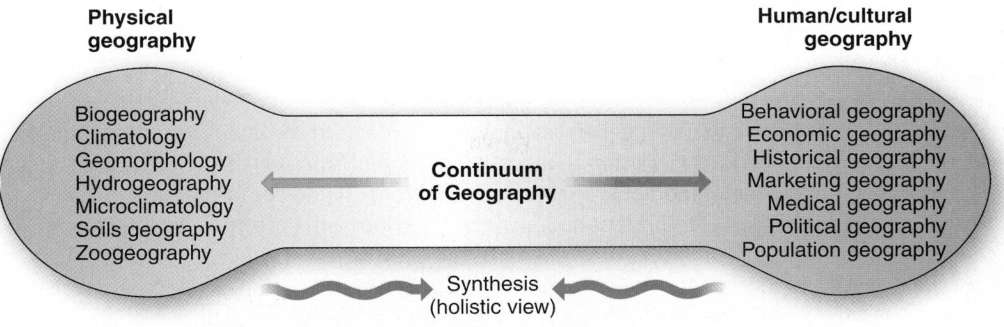

FIGURE 1.3 The content of geography. Geography derives subject matter from many different sciences. The focus of this book is physical geography, but we also integrate some human and cultural components. Movement toward the middle of the continuum suggests a synthesis of Earth topics and human topics.

human–Earth relationship requires that we shift our study of geographic processes toward the center of the continuum in Figure 1.3 to attain a more balanced perspective—such is the thrust of *Geosystems*. Many geography departments are offering courses that serve this growing synthesis approach.

Earth Systems Concepts

The word *system* is in our lives daily: "Check the car's cooling system"; "Is a broadband system available for my Internet browser?"; "How does the grading system work?"; "There is a weather system approaching." *Systems analysis* techniques began with studies of energy and temperature (thermodynamics) in the 19th century and were further developed in engineering during World War II. Systems methodology is an important analytical tool. In this book's 4 parts and 21 chapters, you find the content is organized along logical flow paths consistent with systems thinking.

Systems Theory

Simply stated, a **system** is any ordered, interrelated set of things and their attributes, linked by flows of energy and matter, as distinct from the surrounding environment outside the system. The elements within a system may be arranged in a series or interwoven with one another. A system comprises any number of subsystems. Within Earth's systems, both matter and energy are stored and retrieved, and energy is transformed from one type to another. (Remember: *Matter* is mass that assumes a physical shape and occupies space; *energy* is a capacity to change the motion of, or to do work on, matter.)

Open Systems Systems in nature are generally not self-contained: Inputs of energy and matter flow into the system, and outputs of energy and matter flow from the system. Such a system is an **open system**. Within a system, the parts function in an interrelated manner, acting together in a way that gives each system its operational character (Figure 1.4). Earth is an open system in terms of energy because solar energy enters freely and heat energy leaves, going back into space. Most natural systems are open in terms of energy.

Most Earth systems are dynamic (energetic, in motion) because of the tremendous infusion of radiant energy

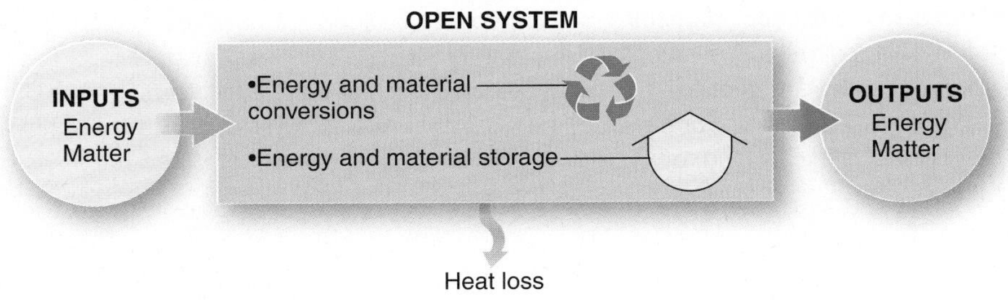

FIGURE 1.4 An open system. In an open system, inputs of energy and matter undergo conversions and are stored as the system operates. Outputs include energy, matter, and heat energy (waste) that flow from the system. See how the various inputs and outputs are related to the operation of a car—an open system. Expand your viewpoint to the entire system of auto production, from raw materials, to assembly, to sales, to car accidents, to junkyards. Can you identify other open systems that you encounter in your daily life?

The Scientific Method

The term *scientific method* may have an aura of complexity, but it should not. The scientific method is simply the application of common sense in an organized and objective manner. A scientist observes, makes a general statement to summarize the observations, formulates a hypothesis, conducts experiments to test the hypothesis, and develops a theory and governing scientific laws. Sir Isaac Newton (1642–1727) developed this method of discovering the patterns of nature, although the term *scientific method* was applied later.

Complexity dominates nature, making several outcomes possible as a system operates. Science serves an important function in understanding such uncertainty. Yet, the more knowledge we have, the more the uncertainty and awareness of other possible scenarios (outcomes and events) increase. This, in turn, demands more precise and aggressive science.

Follow the scientific method illustration in Figure 1.1.1, beginning at the top. The scientific method begins with our perception of the real world and a determination of what we know, what we want to know, and the many unanswered questions that exist.

Scientists who study the physical environment turn to nature for clues that they can observe and measure.

Step 1: *Observe and measure.* They figure out what data are needed and begin to collect those data. These observations are analyzed to identify coherent patterns that may exist. This search for patterns requires inductive reasoning, or the process of drawing generalizations from specific facts. This step is important in modern geographic science, in which the goal is to understand a whole functioning Earth rather than isolated, small compartments of information. Such understanding allows the scientist to construct models of general operations of Earth systems.

Step 2: *Form and test a hypothesis.* If patterns are discovered, the researcher may formulate a *hypothesis*—a tentative explanation for the phenomena observed. Further observations are related to the general principles established by the hypothesis. More data gathering may support or disprove the hypothesis, or predictions made according to it may prove accurate or inaccurate. These findings provide negative feedback to adjust data collection and model building and to refine the hypothesis statement. Verification of the hypothesis after exhaustive testing may lead to its elevation to the status of a *theory.*

Step 3: *Formulate a theory.* The word *theory* can be confusing as used by the media and general public. A theory is constructed on the basis of several extensively tested hypotheses. Theories represent truly broad general principles—unifying concepts that tie together the laws that govern nature. Examples include the theory of relativity, theory of evolution, atomic theory, Big Bang theory, stratospheric ozone depletion theory, and plate tectonics theory. A theory is a powerful device with which to understand both the order and the chaos (disorder) in nature. Using a theory allows predictions to be made about things not yet known, the effects of which can be tested and verified or disproved through tangible evidence. The value of a theory is the continued observation, testing, understanding, and pursuit of knowledge that the theory stimulates. A general theory reinforces our perception of the real world, acting as positive feedback.

Pure science does not make value judgments. Instead, pure science provides people and their institutions with objective information on which to base their own value judgments. Social and political judgments about the applications of science are increasingly critical as Earth's natural systems respond to the impact of modern civilization. Jane Lubchenco, National Oceanic and Atmospheric Administration Administrator, in her 1997 American Association for the Advancement of Science Presidential Address stated:

> Science alone does not hold the power to achieve the goal of greater sustainability, but scientific knowledge and wisdom are needed to help inform decisions that will enable society to move toward that end.

The growing awareness that human activity is producing global change places increasing pressure on scientists to participate in decision making. Numerous editorials in scientific journals have called for such applied science involvement.

from the Sun. This energy comes through the outermost edge of Earth's atmosphere to power terrestrial systems, being transformed along the way into various forms of energy, such as kinetic energy (of motion), potential energy (of position), or chemical or mechanical energy—setting the fluid atmosphere and ocean in motion. Eventually, Earth radiates this energy back to the cold vacuum of space as heat energy.

Closed Systems A system that is shut off from the surrounding environment so that it is self-contained is a **closed system.** Although such closed systems are rarely found in nature, Earth is essentially a closed system in terms of physical matter and resources—air, water, and material resources. The only exceptions are the slow escape of lightweight gases (such as hydrogen) from the atmosphere into space and the input of frequent, but tiny

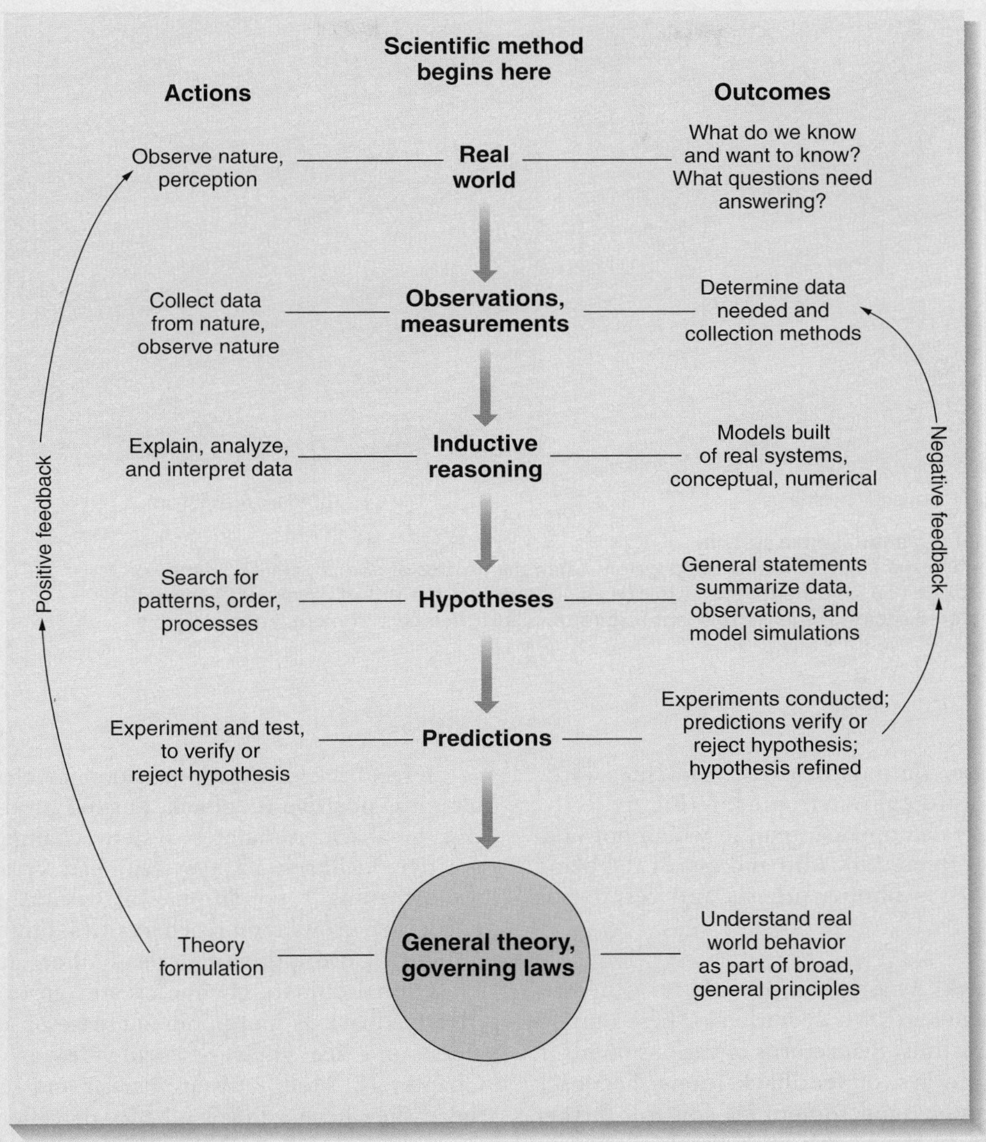

FIGURE 1.1.1 Scientific method flow chart.
The scientific method process: from perceptions, to observations, reasoning, hypothesis, and predictions, and possibly to general theory and natural laws.

meteors and cosmic dust. The fact that Earth is a closed material system makes recycling efforts inevitable if we want a sustainable global economy.

System Example Figure 1.5 on page 8 illustrates a simple open-flow system, using plant photosynthesis and respiration as an example. In photosynthesis (Figure 1.5a), plants use sunlight as an energy input and material inputs of water, nutrients, and carbon dioxide. The photosynthetic process converts these inputs to stored chemical energy in the form of plant sugars (carbohydrates). The process also releases an output from the plant system: the oxygen we breathe.

Reversing the process, plants derive energy for their operations from respiration. In respiration, the plant consumes inputs of chemical energy (carbohydrates) and

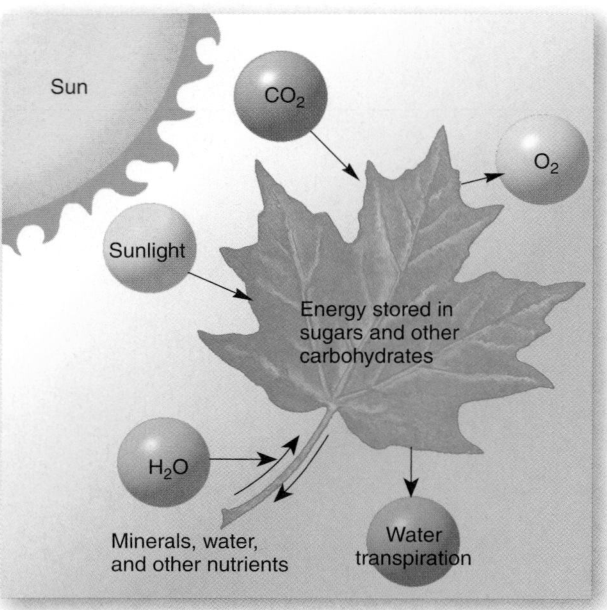

(a) Plant photosynthesis

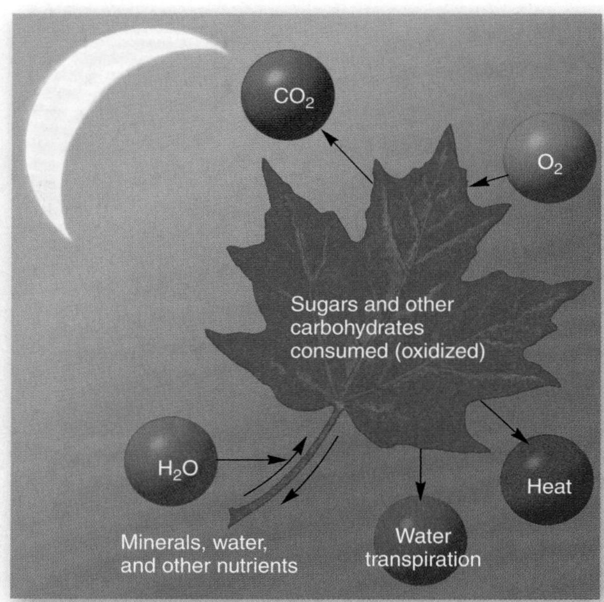

(b) Plant respiration

FIGURE 1.5 A leaf is a natural open system.
A plant leaf provides an example of a natural open system. (a) In the process of photosynthesis, plants consume light, carbon dioxide (CO_2), nutrients, and water (H_2O) and produce outputs of oxygen (O_2) and carbohydrates (sugars) as stored chemical energy. (b) Plant respiration, illustrated here at night, approximately reverses this process.

oxygen and releases outputs of carbon dioxide, water, and heat energy into the environment (Figure 1.5b). Thus, a plant acts as an open system, in which both energy and materials freely flow into and out of the plant. (Chapter 19 discusses photosynthesis and respiration processes.)

System Feedback As a system operates, it generates outputs that influence its own operations. These outputs function as "information" that returns to various points in the system via pathways, or **feedback loops**. Feedback information can guide, and sometimes control, further system operations. In the plant's photosynthetic system (see Figure 1.5), any increase or decrease in daylength (sunlight availability), carbon dioxide, or water produces feedback that causes specific responses in the plant. For example, decreasing the water input slows the growth process; increasing daylength increases the growth process, within limits.

If the feedback information discourages change in the system, it is **negative feedback**. Further production of such feedback opposes system changes. Such negative feedback informs and causes self-regulation in a natural system, stabilizing the system. Think of maintaining your healthy body weight—the human body is an open system. When you stand on the scales, the news reported about your weight helps you regulate system activity. If your weight is too high, you decrease your inputs or increase your level of exercise until your weight stabilizes.

If feedback information encourages change in the system, it is **positive feedback**. Further production of positive feedback stimulates system changes. Unchecked positive feedback in a system can create a runaway ("snowballing") condition. In natural systems, such unchecked growth can reach a critical limit, leading to instability in, disruption of, or death of organisms.

Global climate change creates an example of positive feedback as meltponds increase on ice sheets, glaciers, and ice shelves worldwide, as illustrated in Chapter 17. Meltponds are darker and reflect less sunlight; they have a lower albedo, or reflectivity. Therefore, they absorb more solar energy, which, in turn, melts more ice, which forms more meltponds, and so forth. Thus, a positive feedback loop is in operation, further enhancing the effects of higher temperatures and warming trends.

Based on satellite images and aerial surveys, scientists determined a spatial increase in meltpond distribution of 400% across the Arctic region since 2000. Overall ice-mass losses from Greenland more than doubled between 1996 and 2008. The meltponds are symptomatic of climatic warming. Likewise, as sea ice melts, the darker open water absorbs more sunlight, which produces increased air and water temperatures.

Think of the numerous devastating wildfires that occurred globally in the last decade, including record wildfires in terms of area burned in the western United States, Australia, and elsewhere, as examples of positive feedback.

As the fires burned, they dried wet shrubs and green wood around the fire, thus providing more fuel for combustion. The greater the fire, the greater the availability of fuel becomes, and thus more fire is possible—a positive feedback for the fire.

System Equilibrium Most systems maintain structure and character over time. An energy and material system that remains balanced over time, in which conditions are constant or recur, is in a *steady-state condition*. When the rates of inputs and outputs in the system are equal and the amounts of energy and matter in storage within the system are constant (or more realistically, when they fluctuate around a stable average), the system is in **steady-state equilibrium**.

However, a steady-state system may demonstrate a changing trend over time, a condition described as **dynamic equilibrium**. These changing trends of either increasing or decreasing system operations may appear gradual. Figure 1.6 illustrates these two equilibrium conditions, steady-state and dynamic.

Note that systems try to maintain their functional operations; they tend to resist abrupt change. However, a system may reach a **threshold**, or *tipping point*, where it can no longer maintain its character, so it lurches to a new operational level. This abrupt change places the system in a *metastable equilibrium*. An example of such a condition is a hillside or coastal bluff that adjusts after a sudden landslide. A new equilibrium is eventually achieved among slope, materials, and energy over time.

This threshold concept raises concern in the scientific community, especially if some natural systems reach their tipping-point limits. The relatively sudden collapse of ice shelves surrounding a portion of Antarctica and the sudden crack-up of ice shelves on the north coast of Ellesmere Island, Canada, serve as examples of systems at a threshold that are changing to a new status—that of disintegration (more on this in Chapter 17).

Thresholds can also be reached in plant and animal communities. The bleaching (death) of living coral reefs worldwide accelerated dramatically after 1997, when warming conditions in oceans combined with pollution to lead coral systems to a threshold. Today, about 50% of the corals on Earth are ailing or nearing collapse (more on this in Chapter 16). Harlequin frogs of tropical Central and South America are another example of species reaching a tipping point, with increased extinctions since 1986 (Figure 1.7 on page 10).

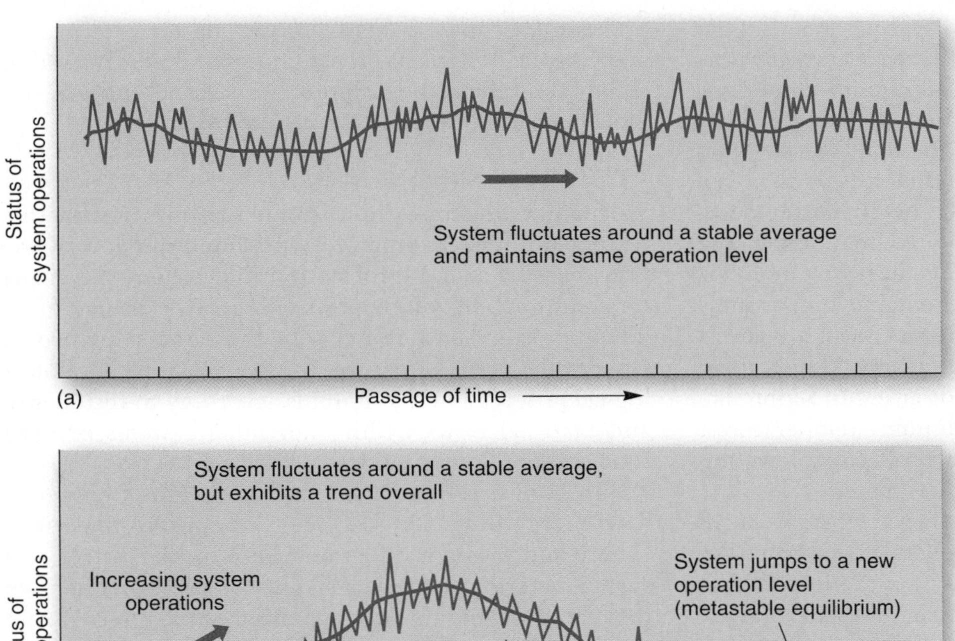

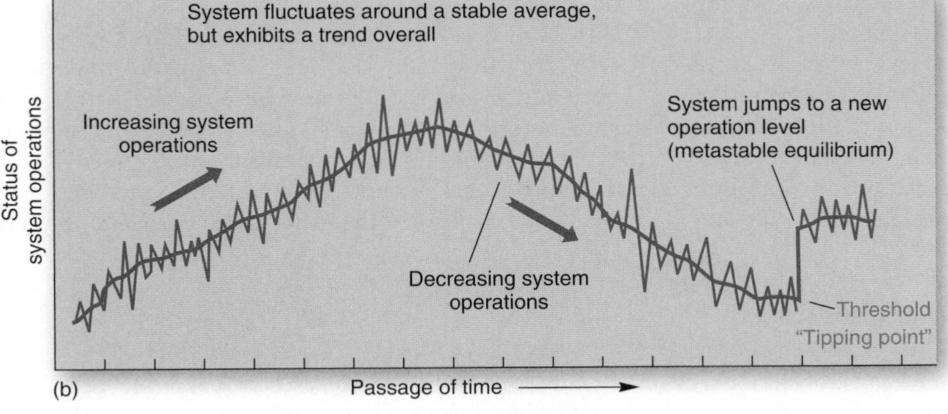

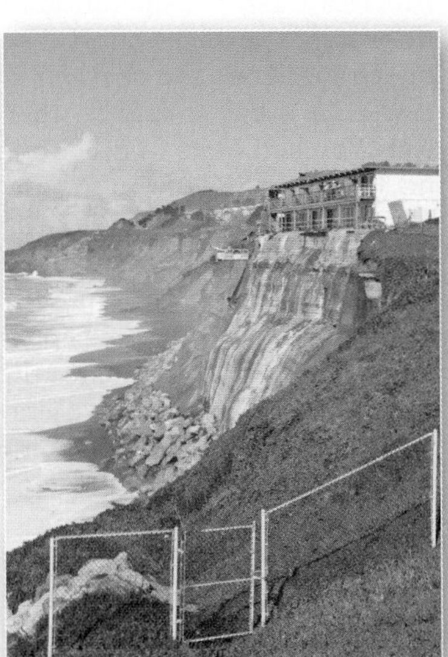

Coastal systems pass a threshold.

FIGURE 1.6 System equilibria: steady-state and dynamic.
(a) A steady-state equilibrium over time; system operations fluctuate around a stable average. Other systems are in a condition of dynamic equilibrium, with an increasing or decreasing operational trend as in (b). Landslides along the Pacific Coast south of San Francisco moved the cliff face some 30.5 m (100 ft) inland during January 2010. A threshold was reached and the bluffs collapsed. [Bobbé Christopherson.]

FIGURE 1.7 Harlequin frog extinction alarm. About two-thirds of harlequin frog species went extinct over the last two decades as a result of climate change. This is "rana dorado," or golden frog, that only survives in captivity. [©Michael & Patricia Fogden/CORBIS, All rights reserved.]

Systems Analysis and Species Extinction The key to unraveling what was happening to harlequin frog and golden toad species in the Monteverde Cloud Forest Reserve, Costa Rica, and elsewhere in Central and South America was a consideration of climate change and how this alters transmission of diseases and pathogens. One infectious agent, the *chytrid fungus* (*Bactrachochytrium dendrobactidis*), has spread as thermal-optimum conditions were reached in its mid-elevation range (1000 m to 2400 m; 3280 ft to 7874 ft) in the mountain-cloud forest.

Human-forced climate change is increasing the temperature of the ocean and atmosphere. Higher temperatures cause higher evaporation rates that affect the condensation level (see Chapter 7). As the warm, moist air moves onshore and reaches the mountains, the air mass lifts and cools, and condensation occurs at higher elevations than before (see Chapter 8). Increased clouds affect the daily temperature range: Clouds at night, acting like insulation, raise nighttime minimum temperatures, whereas clouds during the day act as reflectors, lowering daytime maximum temperatures (see Chapter 4).

The chytrid fungus does best when temperatures are between 17°C and 25°C. The new cloud cover keeps the daytime maximum below 25°C (77°F) at the forest floor. In these more favorable conditions, the disease pathogen flourishes. Harlequin frogs have moist, porous skin that the fungus penetrates, killing the frog. Between 1986 and 2006, approximately 67% of the 110 known species of harlequin frogs went extinct—extinction is forever.

From this example, can you see the power of systems analysis? Global warming intensifies the hydrologic cycle, raises air and water temperatures, and increases evaporation, all of which change the nature of cloud formations and alter daily temperature patterns.

Mount Pinatubo—Global System Impact A dramatic example of interactions between volcanic eruptions and Earth systems illustrates the strength of spatial analysis in the organization of this textbook. Mount Pinatubo in the Philippines erupted violently in 1991, injecting 15–20 million tons of ash and sulfuric acid mist into the upper atmosphere (Figure 1.8). This was the second greatest eruption during the 20th century; Mount Katmai in Alaska (1912) was the only one greater. The eruption materials from Mount Pinatubo affected Earth systems in several ways, which are noted on the map.

As you progress through this book, you see the story of Mount Pinatubo and its implications woven through eight chapters: Chapter 1 (systems theory), Chapter 4 (effects on energy budgets in the atmosphere), Chapter 6 (satellite images of the spread of debris by atmospheric winds), Chapter 10 (temporary effect on global atmospheric temperatures), Chapters 11 and 12 (volcanic processes), Chapter 17 (past climatic effects of volcanoes), and Chapter 19 (effects on net photosynthesis). Instead of simply describing the eruption, we see the linkages and global impacts of such a volcanic explosion.

Systems in *Geosystems*

To further illustrate the systems organization in this textbook, notice its organization around the flow of energy, materials, and information throughout the chapters and portions of chapters (Figure 1.9 on page 12). This book presents subjects in the same sequence in which they occur in nature. In this way, topics follow a logical progression that unfolds according to the energy and material flows within individual systems or with time and the flow of events.

In Figure 1.9b, you see the Sun begins Chapter 2. The energy flows across space to the top of the atmosphere and through the atmosphere to the surface and surface energy budgets (Chapters 3 and 4). Then we look at the outputs of temperature (Chapter 5) and winds and ocean currents (Chapter 6). Note the same logical systems flow in the other three parts and in each

GEO REPORT 1.1 *Amphibians at thresholds*

Amphibian species are a threatened group of animals, with approximately one-third of recognized species now at risk of extinction. According to the Amphibian Specialist Group of the International Union for Conservation of Nature (IUCN, http://www.iucn.org/), two new initiatives are aimed at stopping the amphibian decline: increased habitat protection for species that are found in only a single location and stepped-up efforts at testing antifungal drugs to halt the killer frog disease and ease the current amphibian extinction crisis (more at http://www.amphibians.org/).

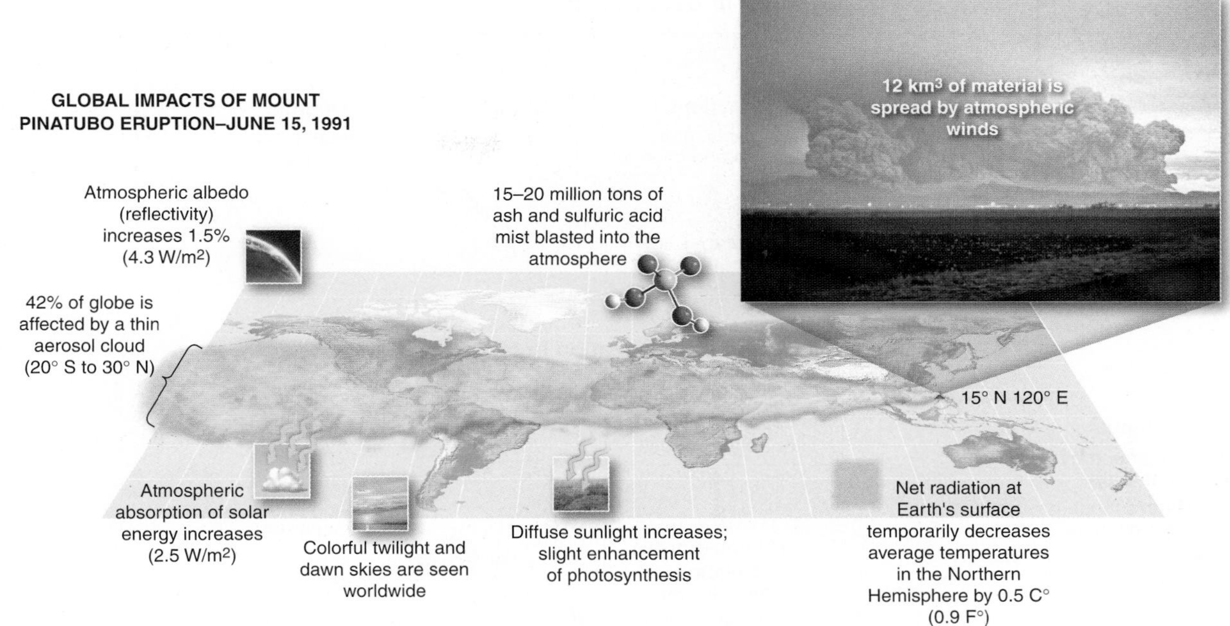

FIGURE 1.8 The eruption of Mount Pinatubo.
The 1991 Mount Pinatubo eruption affected the Earth–atmosphere system on a global scale. As you read *Geosystems,* you will find references to this eruption in many chapters. A summary of the impacts is in Chapter 12. [Inset photo by Dave Harlow, USGS.]

chapter of this text—Part II and Chapter 16 also are shown as examples.

Models of Systems A **model** is a simplified, idealized representation of part of the real world. Models are designed with varying degrees of generalization. In your life, you may have built or drawn a model of something that was a simplification of the real thing—such as a model airplane or model house.

The simplicity of a model makes a system easier to understand and to simulate in experiments (for example, the leaf model in Figure 1.5). A good example is a model of the *hydrologic system,* which models Earth's entire water system, its related energy flows, and the atmosphere, surface, and subsurface environments through which water moves (see Figure 9.1 in Chapter 9). We discuss many *system models* in this text.

Adjusting the variables in a model produces differing conditions and allows predictions of possible system operations. However, predictions are only as good as the assumptions and accuracy built into the model. A model is best viewed for what it is—a simplification to help us understand complex processes. Each of Earth's four "spheres" represents such a model of Earth's systems.

Earth's Four "Spheres" Earth's surface is a vast area of 500 million km^2 (193 million mi^2) where four immense open systems interact. Figure 1.10 on page 13 shows a simple model of three **abiotic**, or nonliving, systems overlapping to form the realm of the **biotic**, or living system. The abiotic spheres are the *atmosphere*, *hydrosphere*, and *lithosphere*. The biotic sphere is the *biosphere*.

As noted in the figure, these four spheres form the part structure in which chapters are grouped in this book. The various arrows in the figure show that content in each part and among chapters interrelates as we build our discussion of Earth's abiotic and biotic systems.

- **Atmosphere (Part I, Chapters 2–6)** The **atmosphere** is a thin, gaseous veil surrounding Earth, held to the planet by the force of gravity. Formed by gases arising from within Earth's crust and interior and by the exhalations of all life over time, the lower atmosphere is unique in the Solar System. It is a combination of nitrogen, oxygen, argon, carbon dioxide, water vapor, and trace gases.
- **Hydrosphere (Part II, Chapters 7–10)** Earth's waters exist in the atmosphere, on the surface, and in the crust near the surface. Collectively, these waters form the **hydrosphere**. That portion of the hydrosphere that is frozen is the **cryosphere**—ice sheets, ice caps and fields, glaciers, ice shelves, sea ice, and subsurface ground ice. Water of the hydrosphere exists in all three states: liquid, solid (the frozen cryosphere), and gaseous (water vapor). Water occurs in two general chemical conditions, fresh and saline (salty).
- **Lithosphere (Part III, Chapters 11–17)** Earth's crust and a portion of the upper mantle directly below the crust form the **lithosphere**. The crust is quite brittle compared with the layers deep beneath the surface, which move slowly in response to an uneven distribution of heat energy and pressure. In a broad sense, the term *lithosphere* sometimes refers

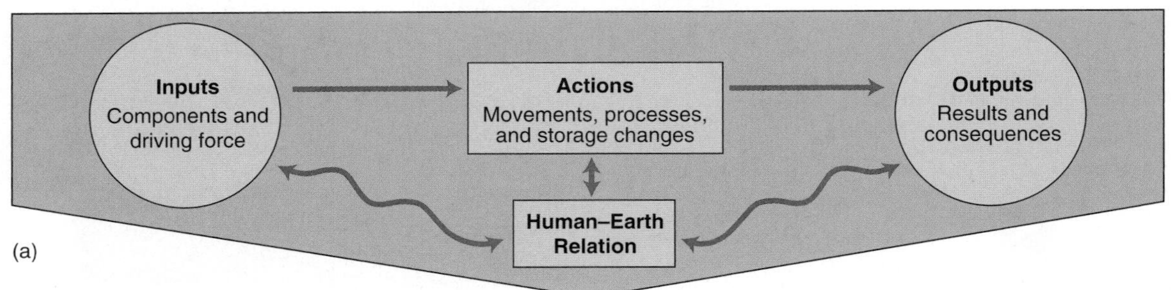

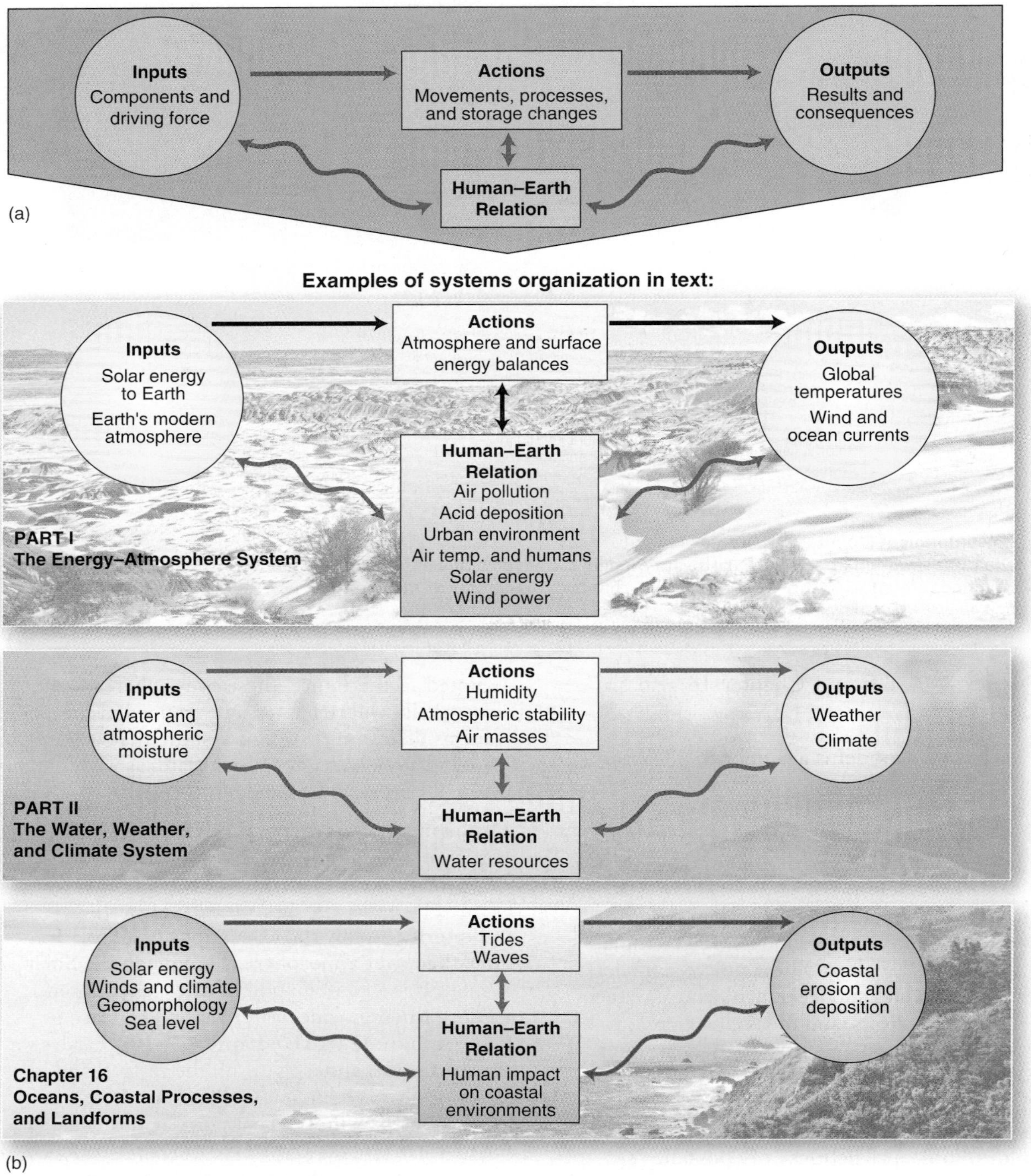

FIGURE 1.9 **The systems in *Geosystems.***
(a) The sequence of systems flow. (b) Parts I and II provide an example of this systems structure. As a chapter example, turn to Chapter 16 on coastal processes and see how the chapter is organized using the systems flow.

GEO REPORT 1.2 *Water in the Solar System*

Among the planets in the Solar System, only Earth possesses a hydrosphere and surface water in such quantity, some 1.36 billion km^3. There are discoveries of subsurface water on the Moon in its polar areas, Mars, Jupiter's moon Europa, and Saturn's moons Enceladus and Titan. In the Martian polar region, remote spacecraft are studying ground ice and patterned ground phenomena caused by freezing and thawing water, as discussed for Earth in Chapter 17. In the Universe, deep-space telescopes reveal traces of water at nebula and distant planetary objects.

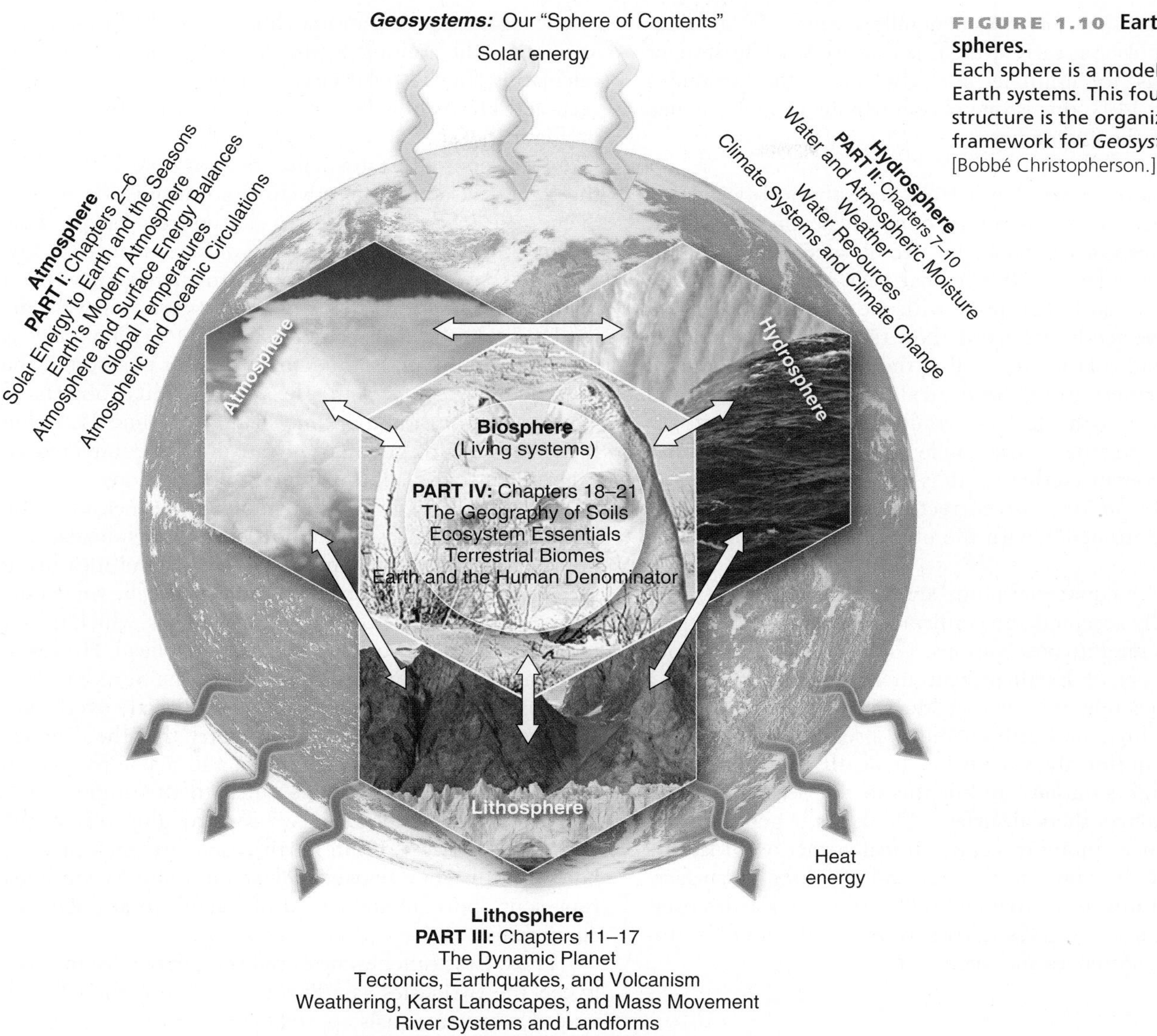

FIGURE 1.10 Earth's four spheres.
Each sphere is a model of vast Earth systems. This four-part structure is the organizational framework for *Geosystems*. [Bobbé Christopherson.]

to the entire solid planet. The soil layer is the *edaphosphere* and generally covers Earth's land surfaces. In this text, soils represent the bridge between the lithosphere (Part III) and the biosphere (Part IV).

- **Biosphere (Part IV, Chapters 18–21)** The intricate, interconnected web that links all organisms with their physical environment is the **biosphere**, or **ecosphere**. The biosphere is the area in which physical and chemical factors form the context of life. The biosphere exists in the overlap among the abiotic, or nonliving, spheres, extending from the seafloor, the upper layers of the crustal rock, to about 8 km (5 mi) into the atmosphere. Life is sustainable within these natural limits. The biosphere evolves, reorganizes itself at times, faces extinctions, and manages to flourish.

A Spherical Planet

We all have heard that some people believed Earth was flat. Yet Earth's *sphericity*, or roundness, is not as modern an idea as many think. For instance, more than two millennia ago, the Greek mathematician and philosopher Pythagoras (ca. 580–500 B.C.) determined through observation that Earth is spherical. We do not know what observations led Pythagoras to this conclusion. Can you guess at what he saw to deduce Earth's roundness?

He might have noticed ships sailing beyond the horizon and apparently sinking below the water's surface, only to arrive back at port with dry decks. Perhaps he noticed Earth's curved shadow cast on the lunar surface during an eclipse of the Moon. He might have deduced that the Sun and Moon are not just the flat disks they appear to be in the sky, but are spherical, and that Earth must be a sphere as well.

Earth's sphericity was generally accepted by the educated populace as early as the first century A.D. Christopher Columbus, for example, knew he was sailing around a sphere in 1492; this is one reason why he thought he had arrived in the East Indies.

Earth as a Geoid Until 1687, the spherical-perfection model was a basic assumption of **geodesy**, the science that determines Earth's shape and size by surveys and mathematical calculations. But in that year, Sir Isaac Newton postulated that Earth, along with the other planets, could not be perfectly spherical. Newton reasoned that the more rapid rotational speed at the equator—the equator being farthest from the central axis of the planet and therefore moving faster—would produce an equatorial bulge as centrifugal force pulled Earth's surface outward. He was convinced that Earth is slightly misshapen into an *oblate spheroid*, or more correctly, an *oblate ellipsoid* (*oblate* means "flattened"), with the oblateness occurring at the poles.

Earth's equatorial bulge and its polar oblateness are universally accepted and confirmed with tremendous precision by satellite observations. The "geoidal epoch" is our modern era of Earth measurement because Earth is a **geoid**, meaning literally that "the shape of Earth is Earth-shaped." Imagine Earth's geoid as a sea-level surface that extends uniformly worldwide beneath the continents. Both heights on land and depths in the oceans measure from this hypothetical surface. Think of the geoid surface as a balance among the gravitational attraction of Earth's mass, the distribution of water and ice along its surface, and the outward centrifugal pull caused by Earth's rotation. Figure 1.11 gives Earth's average polar and equatorial circumferences and diameters.

Measuring Earth in 247 B.C.

Eratosthenes (ca. 276–195 B.C.) served as the librarian of Alexandria in Egypt during the third century B.C. He was a Greek geographer and astronomer in a position of scientific leadership. Alexandria's library was the finest in the ancient world. Among Eratosthenes' achievements was calculating Earth's polar circumference to a high level of accuracy. Here's how he did it. As you read, follow along on Figure 1.12.

Travelers told Eratosthenes that on June 21 they saw the Sun's rays shine directly to the bottom of a well at Syene, the location of present-day Aswan, Egypt. This meant that the Sun was directly overhead on that day. North of Syene in Alexandria, Eratosthenes knew from his own observations that the Sun's rays never were directly overhead, even at noon on June 21. This day is when the noon Sun is at its northernmost position in the sky. He knew objects in Alexandria, unlike objects in Syene, always cast a noontime shadow on June 21. Using his knowledge of geometry, Eratosthenes conducted an experiment to explain the different observations.

In Alexandria at noon on June 21, he measured the angle of a shadow cast by an obelisk (a perpendicular column). Knowing the height of the obelisk and measuring the length of the Sun's shadow from its base, he solved the triangle for the angle of the Sun's noon rays, which he determined to be 7.2° off from directly overhead. However, at Syene on the same day, the angle of the Sun's rays was 0° from a perpendicular, arriving from directly overhead.

Eratosthenes knew from geometry that the distance on the ground between Alexandria and Syene formed an arc of Earth's circumference equal to the angle of the Sun's rays at Alexandria. Since 7.2° is roughly 1/50 of the 360° (360° ÷ 7.2° = 50) in Earth's total circumference, he knew the distance between Alexandria and Syene must represent approximately 1/50 of Earth's total polar circumference.

Next, Eratosthenes measured the surface distance between the two cities as 5000 stadia. A *stadium*, a Greek unit of measure, equals approximately 185 m (607 ft). He then multiplied 5000 stadia by 50 to determine that Earth's polar circumference is about 250,000 stadia. Eratosthenes' calculations convert to roughly 46,250 km

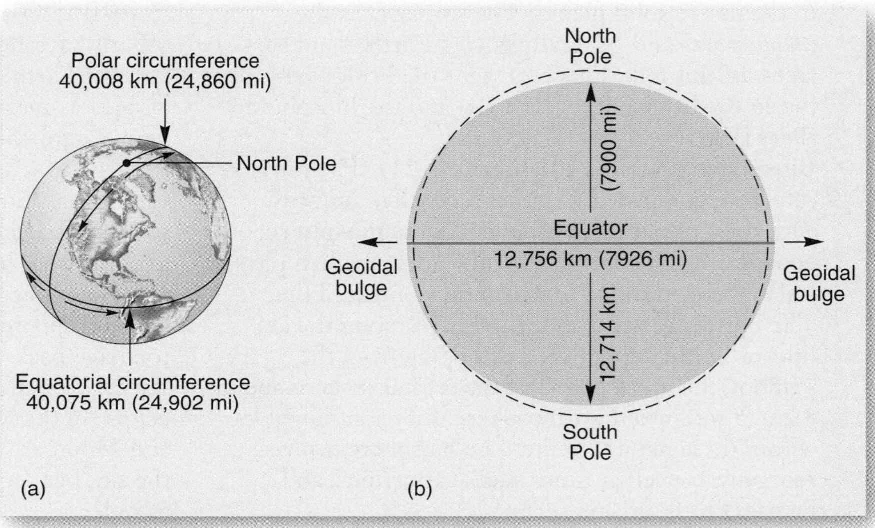

FIGURE 1.11 Earth's dimensions. Earth's equatorial and polar circumference (a) and diameter (b). The dashed line is a perfect circle for reference to Earth's geoid.

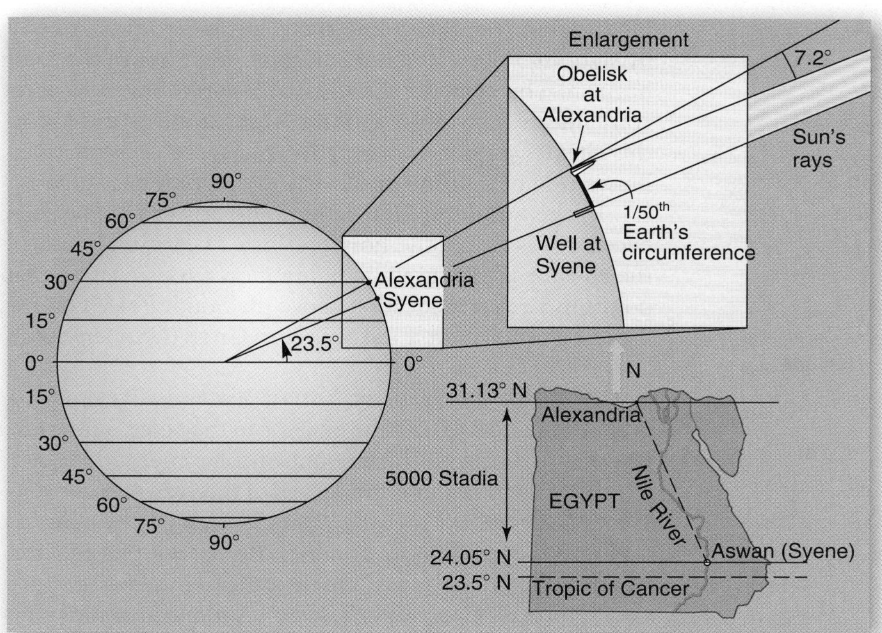

FIGURE 1.12 Eratosthenes' calculation.
Eratosthenes' work teaches the value of observing carefully and integrating all observations. Calculating Earth's circumference required application of his knowledge of Earth–Sun relationships, geometry, and geography.

(28,738 mi), which is remarkably close to the correct value of 40,008 km (24,860 mi) for Earth's polar circumference. Not bad for 247 B.C.!

Location and Time on Earth

Geographic science requires an internationally accepted, coordinated grid system to determine location on Earth. From the chapter's opening photograph, you can see the importance of determining the location of the research ship and tracking the locations of tabular icebergs. The terms *latitude* and *longitude* were in use on maps as early as the first century A.D., with the concepts themselves dating back to Eratosthenes and others.

The geographer, astronomer, and mathematician Ptolemy (ca. A.D. 90–168) contributed greatly to the development of modern maps, and many of his terms are still used today. Ptolemy divided the circle into 360 degrees (360°), with each degree having 60 minutes (60′) and each minute having 60 seconds (600) in a manner adapted from the ancient Babylonians. He located places using these degrees, minutes, and seconds. However, the precise length of a degree of latitude and a degree of longitude remained unresolved for the next 17 centuries.

Latitude

Latitude is an angular distance north or south of the equator, measured from the center of Earth (Figure 1.13a on page 16). On a map or globe, the lines designating these angles of latitude run east and west, parallel to the equator (Figure 1.13b). Because Earth's equator divides the distance between the North Pole and the South Pole exactly in half, it is assigned the value of 0° latitude. Thus, latitude increases from the equator northward to the North Pole, at 90° N latitude, and southward to the South Pole, at 90° S latitude.

A line connecting all points along the same latitudinal angle is a **parallel**. In the figure, an angle of 49° N latitude is measured, and, by connecting all points at this latitude, we have the 49th parallel. Thus, *latitude* is the name of the angle (49° N latitude), *parallel* names the line (49th parallel), and both indicate distance north of the equator.

Latitude is readily determined by observing *fixed celestial objects* such as the Sun or the stars, a method dating to ancient times. During daylight hours, the angle of the Sun above the horizon indicates the observer's latitude, after adjustment is made for the season and for the time of day. Because Polaris (the North Star) is almost directly overhead at the North Pole, persons anywhere in the Northern

GEO REPORT 1.3 ***The value of a stadium***

From the Greek *stade,* the value of a Greek stadium varied in time and place. For this calculation related to Eratosthenes' measurement, I used 185 m (607 ft) as a stadium length. This represents an average value in the Greek era. The word for the stadium measure was expanded to be the word to describe the arena in which foot races took place—thus our modern stadiums. Think of Eratosthenes' work when you see a stadium on campus.

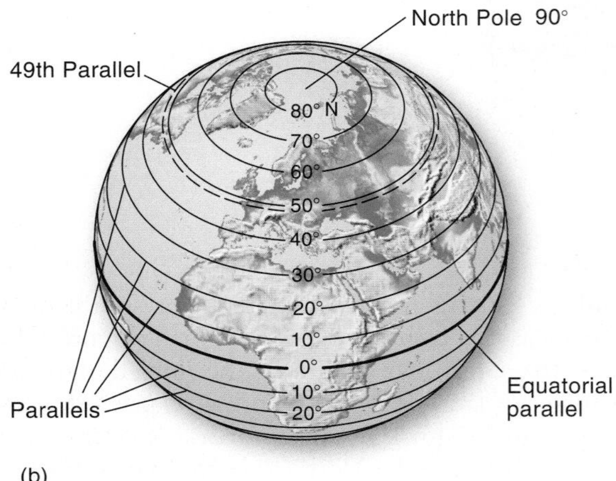

FIGURE 1.13 Parallels of latitude.
(a) Latitude is measured in degrees north or south of the equator (0°). Earth's poles are at 90°. Note the measurement of 49° latitude. (b) These angles of latitude determine parallels along Earth's surface. Do you know your present latitude?

Hemisphere can determine their latitude at night simply by sighting Polaris and measuring its angle above the local horizon. The angle of elevation of Polaris above the horizon indicates the latitude of the observation point. Go to this chapter on our *Mastering Geography* web site to see an illustration of sighting on Polaris to determine latitude.

In the Southern Hemisphere, Polaris cannot be seen because it is below the horizon. Instead, latitude south of the equator is measured by sighting on a constellation that points to a celestial location above the South Pole. This indicator constellation is the Southern Cross (*Crux Australis*).

Latitudinal Geographic Zones Natural environments differ dramatically from the equator to the poles. These differences result from the amount of solar energy received, which varies by latitude and season of the year. As a convenience, geographers identify *latitudinal geographic zones* as regions. "Lower latitudes" are those nearer the equator, whereas "higher latitudes" are those nearer the poles.

Figure 1.14 portrays these zones, their locations, and their names: *equatorial* and *tropical*, *subtropical*, *midlatitude*, *subarctic* or *subantarctic*, and *arctic* or *antarctic*. These generalized latitudinal zones are useful for reference and comparison, but they do not have rigid boundaries; rather, think of them as transitioning one to another.

CRITICAL THINKING 1.1

Latitudinal zones and temperature

Refer to the graph in Figure 5.3 that plots annual temperature data for five cities from near the equator to beyond the Arctic Circle. Note the geographic location for each of the five cities on the latitudinal geographic zone map in Figure 1.14. In which zone is each city located? To approximate locations use an atlas, web site, or encyclopedia. Roughly characterize changing temperature patterns through the seasons as you move away from the equator. Describe what you discover.

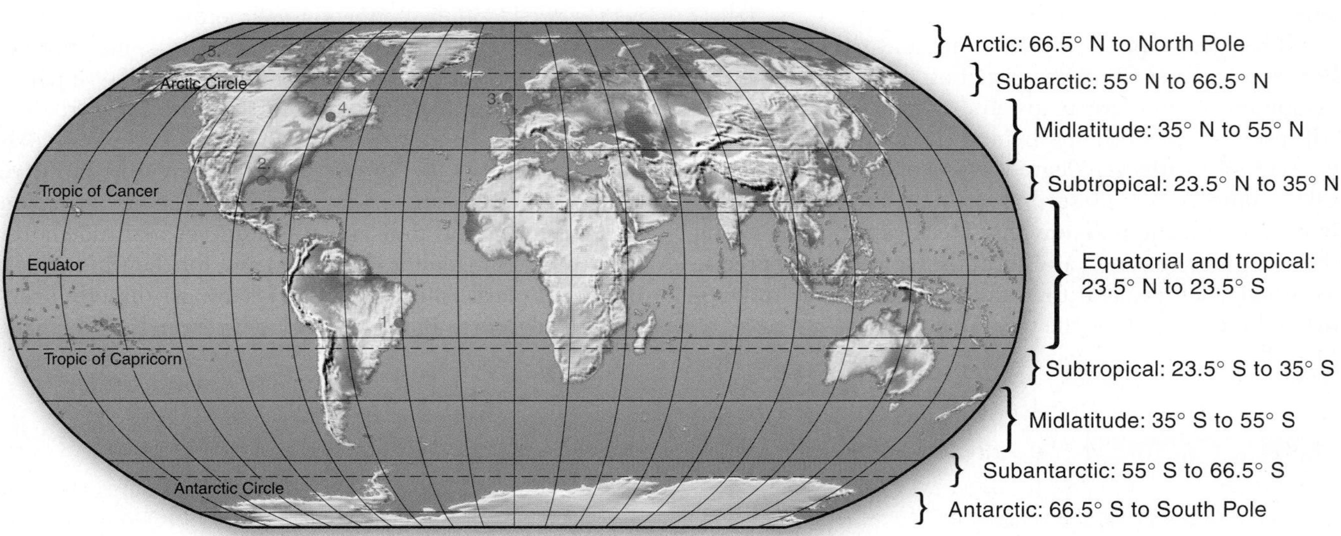

FIGURE 1.14 Latitudinal geographic zones.
Geographic zones are generalizations that characterize various regions by latitude. Think of these as transitioning into one another over broad areas. Noted cities: 1. Salvador, Brazil; 2. New Orleans, 3. Edinburgh, Scotland; 4. Montreal, Quebec; and, 5. Barrow, Alaska.

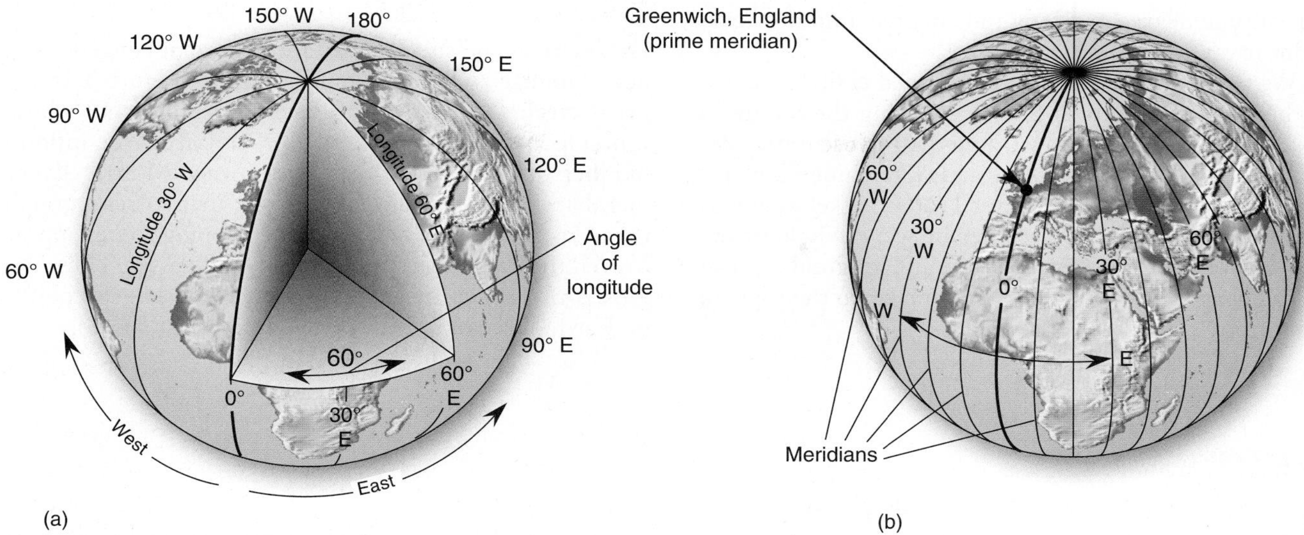

FIGURE 1.15 Meridians of longitude.
(a) Longitude is measured in degrees east or west of a 0° starting line, the prime meridian. Note the measurement of 60° E longitude. (b) Angles of longitude measured from the prime meridian determine other meridians. North America is west of Greenwich; therefore, it is in the Western Hemisphere. Do you know your present longitude?

The *Tropic of Cancer* (about 23.5° N parallel) and the *Tropic of Capricorn* (about 23.5° S parallel) are the most extreme northern and southern parallels that experience perpendicular (directly overhead) rays of the Sun at local noon. When the Sun arrives overhead at these tropics, it marks the first day of summer in each hemisphere. (Further discussion of the tropics is in Chapter 2.) The *Arctic Circle* (about 66.5° N parallel) and the *Antarctic Circle* (about 66.5° S parallel) are the parallels farthest from the poles that still experience 24 uninterrupted hours of night (including dawn and twilight) during local winter or continuous day during local summer.

Longitude

Longitude is an angular distance east or west of a point on Earth's surface, measured from the center of Earth (Figure 1.15a). On a map or globe, the lines designating these angles of longitude run north and south (Figure 1.15b). A line connecting all points along the same longitude is a **meridian**. In the figure, a longitudinal angle of 60° E is measured. These meridians run at right angles (90°) to all parallels, including the equator.

Thus, *longitude* is the name of the angle, *meridian* names the line, and both indicate distance east or west of an arbitrary **prime meridian**—a meridian designated as 0° (Figure 1.15b). Earth's prime meridian passes through the old Royal Observatory at Greenwich, England, as set by an 1884 treaty—the *Greenwich prime meridian*.

Determination of Latitude and Longitude Table 1.1 compares the length of latitude and longitude degrees. Because meridians of longitude converge toward the poles, the actual distance on the ground spanned by a degree of longitude is greatest at the equator (where meridians separate to their widest distance apart) and diminishes to zero at the poles (where meridians converge). In comparison, note the consistent distance represented by a degree of latitude from equator to poles. For

TABLE 1.1 Physical Distances Represented by Degrees of Latitude and Longitude

Latitudinal Location	Latitude Degree Length		Longitude Degree Length	
	km	mi	km	mi
90° (poles)	111.70	69.41	0	0
60°	111.42	69.23	55.80	34.67
50°	111.23	69.12	71.70	44.55
40°	111.04	69.00	85.40	53.07
30°	110.86	68.89	96.49	59.96
0°(equator)	110.58	68.71	111.32	69.17

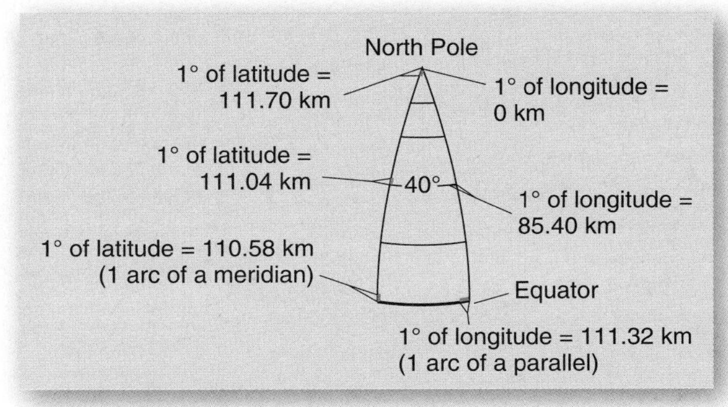

your latitude degree and longitude degree, find the nearest linear value for each.

We noted that latitude is determined easily by sighting the Sun or the North Star or by using the Southern Cross as a pointer. In contrast, a method of accurately determining longitude, especially at sea, remained a major difficulty in navigation until after 1760. The key to measuring the longitude of a place lies in accurately knowing time. The relationship between time and longitude and an exciting chapter in human discovery make up the topic of Focus Study 1.2.

Great Circles and Small Circles

Great circles and small circles are important concepts that help summarize latitude and longitude (Figure 1.16). A **great circle** is any circle of Earth's circumference whose center coincides with the center of Earth. An infinite number of great circles can be drawn on Earth. Every meridian is one-half of a great circle that passes through the poles. On flat maps, airline and shipping routes appear to arch their way across oceans and landmasses. These are *great circle routes*, the shortest distance between two points on Earth (discussion with Figure 1.26 is just ahead).

FOCUS STUDY 1.2

The Timely Search for Longitude

Unlike latitude, longitude cannot be determined readily from fixed celestial bodies. Determining longitude is particularly critical at sea, where no landmarks are visible. The problem is Earth's rotation, which constantly changes the apparent position of the Sun and stars.

In the early 1600s, Galileo explained that longitude could be measured by using two clocks. Any point on Earth takes 24 hours to travel around the full 360° of one rotation (1 day). If you divide 360° by 24 hours, you find that any point on Earth travels through 15° of longitude every hour. Thus, if time could be accurately measured at sea, a comparison of two clocks could give a value for longitude.

One clock would indicate the time back at homeport (Figure 1.2.1). The other clock would be reset at local noon each day, as determined by the highest Sun position in the sky (solar zenith). The time difference between the ship and homeport would indicate the longitudinal difference traveled: 1 hour for each 15° of longitude. The principle was sound; all that was needed was accurate clocks. Unfortunately, the pendulum clock invented by Christian Huygens in 1656, and accurate on land, did not work on the rolling deck of a ship at sea!

In 1707, the British lost four ships and 2000 men in a sea tragedy that was blamed specifically on the longitude problem. In response, Parliament passed an act in 1714—"Publik Reward . . . to Discover the Longitude at Sea"—and authorized a prize worth more than $2 million in today's dollars to the first successful inventor of an accurate seafaring clock. The Board of Longitude was established to judge any devices submitted.

John Harrison, a self-taught country clockmaker, began work on the problem in 1728 and finally produced his brilliant marine chronometer, known as *Number 4,* in 1760. The clock was tested on a voyage to Jamaica in 1761. When taken ashore and compared to land-based longitude, Harrison's ingenious *Number 4* was only 5 seconds slow, an error that translates to only 1.259′ or 2.3 km (1.4 mi)—or well within Parliament's standard. After many delays, Harrison finally received most of the prize money in his last years of life.

With his marine clocks, John Harrison tested the waters of space–time. He succeeded, against all odds, in using the fourth—temporal—dimension to link points on the three-dimensional globe. He wrested the world's whereabouts from the stars, and locked the secret in a pocket watch.*

From that time, it was possible to determine longitude accurately on land and sea, as long as everyone agreed upon a meridian to use as a reference for time comparisons, which became the Royal Observatory in Greenwich, England. In this modern era of atomic clocks and GPS satellites in mathematically precise orbits, we have far greater accuracy available for the determination of longitude on Earth.

*From Dava Sobel, *Longitude: The Story of a Lone Genius Who Solved the Greatest Scientific Problem of His Time* (New York: Walker and Co., 1995), p. 175.

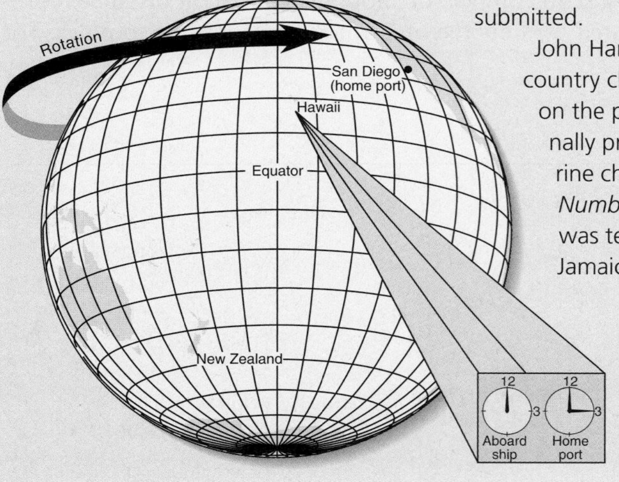

FIGURE 1.2.1 Clock time determines longitude. If the shipboard clock reads local noon and the clock set for home port reads 3:00 P.M., ship time is 3 hours earlier than home time. Therefore, calculating 3 hours at 15° per hour puts the ship at 45° W longitude from home port.

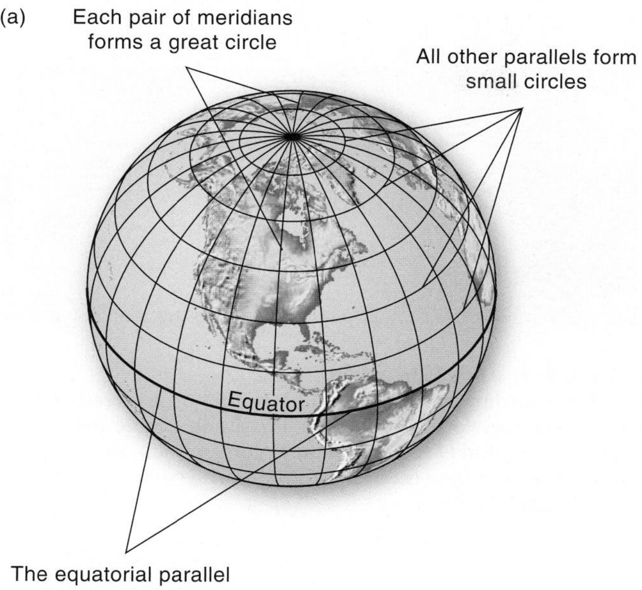

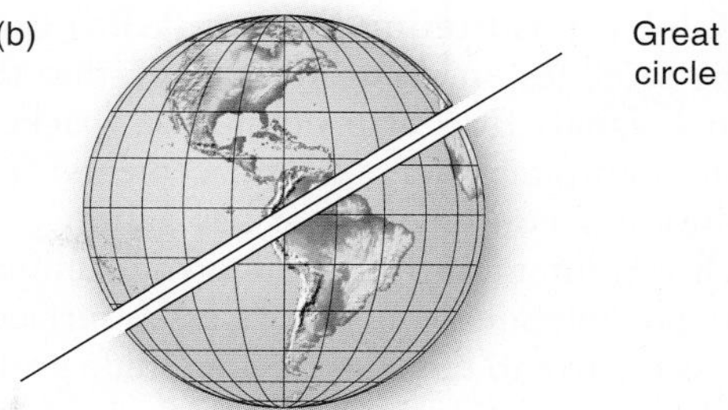

A plane intersecting the globe along a great circle divides the globe into equal halves and passes through its center

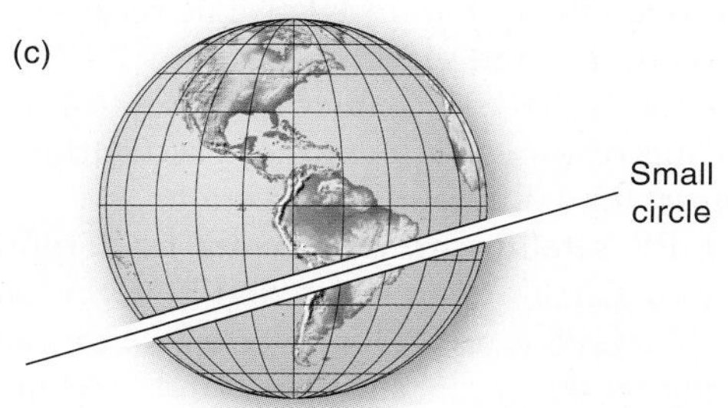

A plane that intersects the globe along a small circle splits the globe into unequal sections—this plane does not pass through the center of the globe

FIGURE 1.16 Great circles and small circles.
(a) Examples of great circles and small circles on Earth. (b) Any plane that divides Earth into equal halves intersects the globe along a great circle; this great circle is a full circumference of the globe and is the shortest distance between any two surface points. (c) Any plane surface that splits the globe into unequal portions intersects the globe along a small circle.

In contrast to meridians, only one parallel is a great circle—the *equatorial parallel*. All other parallels diminish in length toward the poles and, along with any other non-great circles that one might draw, constitute **small circles**. These circles have centers that do not coincide with Earth's center.

Figure 1.17 combines latitude and parallels with longitude and meridians to illustrate Earth's complete coordinate grid system. Note the dot that marks our measurement of 49° N and 60° E, a location in western Kazakstan. Next time you look at a world globe, follow the parallel and meridian that converge on your location.

CRITICAL THINKING 1.2

Where are you?

Select a location (for example, your campus, home, workplace, or a city) and determine the following: latitude, longitude, and elevation. Describe the resources you used to gather this geographic information, such as an atlas, web site, Google Earth™, or Global Positioning System measurement. Finally, using Table 1.1, find the nearest latitude and longitude value for the location you checked, and write down your calculation of the length of a degree of latitude and degree of longitude.

Global Positioning System

Today, using a handheld instrument that receives radio signals from satellites, you can accurately calibrate latitude, longitude, and elevation—and even know where your car is as you drive along, often with an electronic voice telling you where to turn.

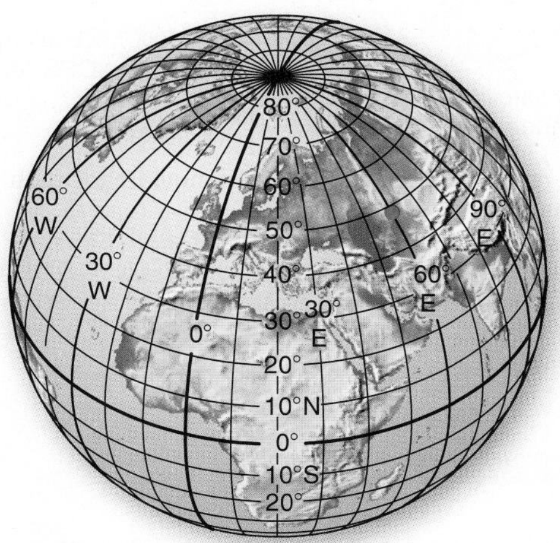

FIGURE 1.17 Earth's coordinate grid system.
Latitude and parallels, as well as longitude and meridians, allow us to locate all places on Earth precisely. The red dot is at 49° N latitude and 60° E longitude.

The **Global Positioning System (GPS)** comprises 24 orbiting satellites, in six orbital planes, that transmit navigational signals for Earth-bound use (backup GPS satellites are in orbital storage as replacements). Think of the satellites as a constellation of navigational beacons with which you interact to determine your unique location. Every possible square meter of Earth's surface has its own address relative to the latitude–longitude grid.

A GPS receiver, now built into cell phones, digital cameras, and camcorders, senses signals from three or more satellites and triangulates the receiver's location. This is accomplished by calculating the distance between each satellite and receiver using clocks built into each instrument, which time radio signals travelling at the speed of light between them (Figure 1.18). The receiver reports latitude, longitude, and elevation (Figure 1.19). Within *Differential GPS (DGPS)*, accuracy increases to 1 m to 3 m by comparing readings with another base station (reference receiver) for a differential correction.

The GPS satellite measurements help refine our knowledge of Earth's exact shape and how this shape is changing. The GPS is useful for diverse applications, such as navigating on the ocean, surveying land, tracking small changes in Earth's crust, managing the movement of fleets of trucks, mining and mapping resources, tracking wildlife migration and behavior, carrying out police and security work, and conducting environmental planning. Outdoor recreationists and sportspersons commonly carry handheld GPS devices that allow them to track their own location. Commercial airlines use the GPS to improve accuracy of routes flown and thus increase fuel efficiency.

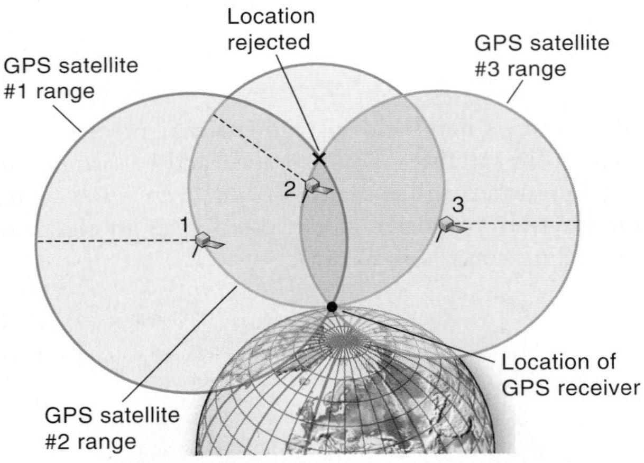

FIGURE 1.18 Using satellites to triangulate location through the GPS.
Imagine a ranging sphere around each of three GPS satellites. These spheres intersect at two points, one easily rejected because it is some distance above Earth and the other at the true location of the GPS receiver. The three spheres can intersect only at two points.

FIGURE 1.19 GPS in action.
Location coordinates on a GPS unit in operation on a research ship as it crosses Earth's equator at 0° latitude. Locate these latitude and longitude coordinates on a globe or map. How far from the nearest land is the ship? What is the name of this ocean? [Bobbé Christopherson.]

Scientific applications of GPS technology are extensive. Relative to earthquakes in southern California, the U.S. Geological Survey (USGS), Jet Propulsion Laboratory (JPL), National Aeronautics and Space Administration (NASA), and National Science Foundation (NSF), among other partners, operate the GPS Observation Office, which monitors a network of 250 continuously operating GPS seismic stations—the Southern California Integrated GPS Network (SCIGN, http://www.scign.org/). This system can record fault movement as small as 1 mm (0.04 in.).

Scientists used GPS technology to accurately determine the height of Mount Everest in the Himalayan Mountains (Figure 1.20). GPS measurements of Mount Kilimanjaro lowered its summit from 5895 m to 5892 m (from 19,340 ft to 19,330 ft).

In *precision agriculture*, farmers use GPS devices to determine crop yields on specific parts of their farms. A detailed plot map is made to guide the farmer on the addition of fertilizer, proper seed distribution, precise irrigation applications, or other work needed. A computer and GPS unit on board the farm equipment guides the

GEO REPORT 1.4 *GPS origins*

Originally devised in the 1970s by the U.S. Department of Defense for military purposes, the present GPS is commercially available worldwide. In 2000, the Pentagon shut down its Pentagon Selective Availability security control, making commercial resolution the same as military applications. Additional frequencies were added in 2003 and 2006, which increased accuracy significantly to less than 10 m (33 ft). For a GPS overview, see http://www.colorado.edu/geography/gcraft/notes/gps/gps_f.html.

FIGURE 1.20 GPS used to measure Everest's summit height. Installation of a Trimble 4800 GPS unit at Earth's highest benchmark, only 18 m (60 ft) below the 8850-m (29,035-ft) summit. Previously, the height was set at 8848 m. Scientists using GPS data accurately analyzed the rate the mountain range is moving due to tectonic forces. [Photo and GPS installation by adventurer/climber Wally Berg, May 20, 1998.]

work. This is the science of *variable-rate technology*, made possible by GPS technology.

The GPS is important to a locational, spatial geographic science because this precise technology reduces the need to maintain ground control points for location, mapping, and spatial analysis. Instead, geographers working in the field can determine their position accurately as they work. Boundaries and data points in a study area are easily determined and entered into a database, reducing the need for traditional surveys.

Prime Meridian and Standard Time

Coordination of international trade, airline schedules, business and agricultural activities, and daily living depends on a worldwide time system. Today, we take for granted standard time zones and an agreed-upon prime meridian, but such a standard is a relatively recent development.

Setting time was not a great problem in small European countries, most of which are less than 15° wide. But in North America, which spans more than 90° of longitude (the equivalent of six 15° time zones), the problem was serious. In 1870, railroad travelers going from Maine to San Francisco made 22 adjustments to their watches to stay consistent with local time.

In Canada, Sir Sanford Fleming directed the fight for standard time and for an international agreement on a prime meridian. His struggle led the United States and Canada to adopt a standard time in 1883. Twenty-seven countries attended the 1884 International Meridian Conference in Washington, DC, to set a global standard. Most participating nations chose the highly respected Royal Observatory at Greenwich, London, England, as the place for the prime meridian of 0° longitude for maps. Thus, a world standard was set—**Greenwich Mean Time (GMT)**—and a consistent Universal Time was established (see http://wwp.greenwichmeantime.com/).

The basis of time is that Earth revolves 360° every 24 hours, or 15° per hour (360° ÷ 24 = 15°) Thus, a time zone of 1 hour is established, spanning 7.5° on either side of a *central meridian*. Today, only three adjustments are needed in the continental United States—from Eastern Standard Time to Central, Mountain, and Pacific—and four changes across Canada.

Assuming it is 9:00 P.M. in Greenwich, then it is 4:00 P.M. in Baltimore (+5 hr), 3:00 P.M. in Oklahoma City (+6 hr), 2:00 P.M. in Salt Lake City (+7 hr), 1:00 P.M. in Seattle (+8 hr), noon in Anchorage (+9 hr), and 11:00 A.M. in Honolulu (+10 hr). To the east, it is midnight in Ar Riyāḑ, Saudi Arabia (–3 hr). The designation A.M. is for *ante meridiem*, "before noon," whereas P.M. is for *post meridiem*, "after noon." A 24-hour clock avoids the use of these designations: 3 P.M. is stated as 15:00 hours and 3 A.M. is 3:00 hours.

As you can see from the modern international time zones in Figure 1.21 on page 22, national boundaries and political considerations distort time boundaries. For example, China spans four time zones, but its government decided to keep the entire country operating at the same time. Thus, in some parts of China clocks are several hours off from what the Sun is doing. In the United States, parts of Florida and west Texas are in the same time zone.

GEO REPORT 1.5 *Magellan's crew loses a day!*

Early explorers had a problem before the date-line concept. For example, Magellan's crew returned from the first circumnavigation of Earth in 1522, confident from their ship's log that it was Wednesday, September 7. They were shocked when informed by local residents that it was actually Thursday, September 8. Without an International Date Line, they had no idea that they must advance a day somewhere when sailing around the world in a westward direction. Imagine the confusion of the arriving crew.

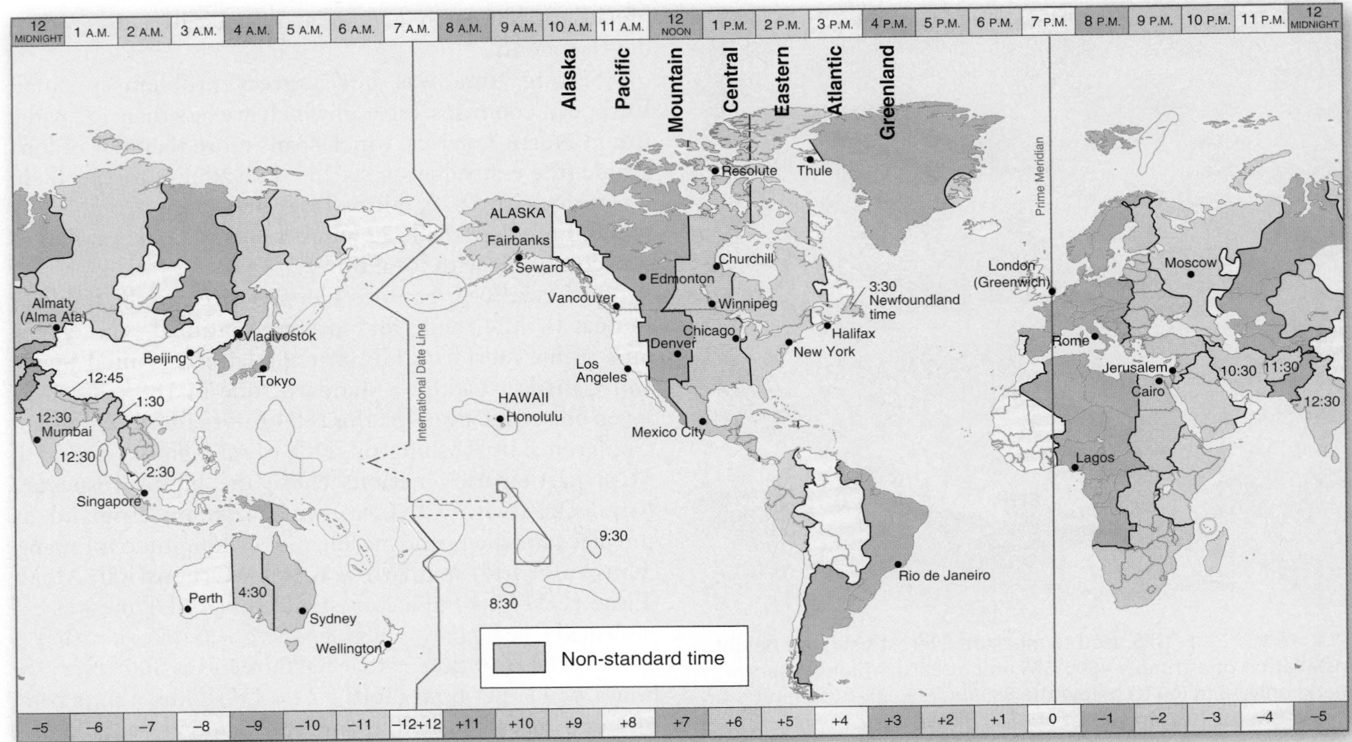

FIGURE 1.21 Modern international standard time zones. Numbers along the bottom of the map indicate how many hours each zone is earlier (plus sign) or later (minus sign) than Coordinated Universal Time (UTC) at the prime meridian. If it is 7 P.M. in Greenwich, determine the present time in Moscow, London, Halifax, Chicago, Winnipeg, Denver, Los Angeles, Fairbanks, Honolulu, Tokyo, and Singapore. [Adapted from Defense Mapping Agency. See http://aa.usno.navy.mil/faq/docs/world_tzones.html.]

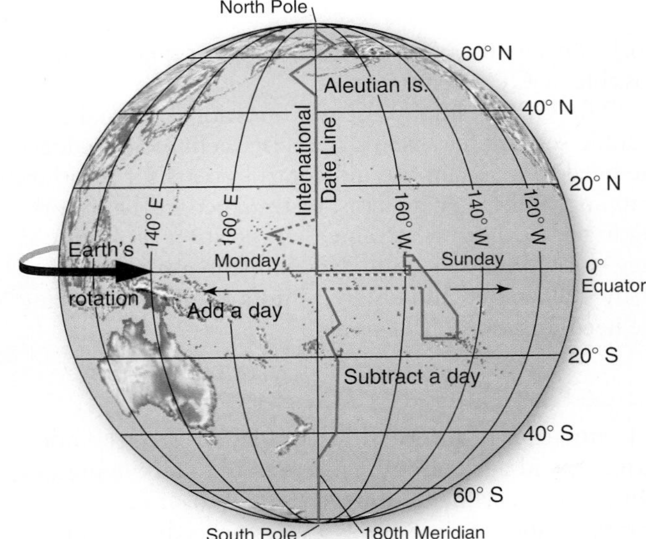

FIGURE 1.22 International Date Line. The International Date Line location is approximately along the 180th meridian (see the IDL location on Figure 1.21). Note the dotted lines on the map—island countries set their own time zones, but their political control extends only 3.5 nautical miles (4 mi) offshore. Officially, you gain 1 day crossing the IDL from east to west.

International Date Line An important corollary of the prime meridian is the 180° meridian on the opposite side of the planet. This meridian is the **International Date Line (IDL)**, which marks the place where each day officially begins (at 12:01 A.M.). From this "line," the new day sweeps westward. This *westward* movement of time is created by Earth's turning *eastward* on its axis. Locating the date line in the sparsely populated Pacific Ocean minimizes most local confusion.

At the IDL, the west side of the line is always one day ahead of the east side. No matter what time of day it is when the line is crossed, the calendar changes a day (Figure 1.22). Note in the illustration the departures from the IDL and the 180° meridian; this deviation is due to local administrative and political preferences.

Coordinated Universal Time For decades, Greenwich Mean Time (GMT) from the Royal Observatory's astronomical clocks was the world's Universal Time (UT) standard for accuracy. GMT was broadcast using radio time signals as early as 1910. Taking the initiative in 1912, the French government called a gathering of countries to better coordinate radio time signals. At this conference, GMT was made standard, and a new organization was set up to be the custodian of "exact" time—the International Bureau of Weights and Measures (BIPM, see http://www.bipm.org/). Progress was made in accurately measuring time with the invention of a quartz clock in 1939 and atomic clocks in the early 1950s.

Accurate time is not a matter of simply keeping track of Earth's rotation, since it is slowing down from the tug of tidal forces and rearrangement of water. Note that 150 million years ago a "day" was 22 hours long, and 150 million years in the future a "day" will be approaching 27 hours in length.

The **Coordinated Universal Time (UTC)*** time-signal system replaced GMT in 1972 and became the legal reference for official time in all countries. Although the prime meridian still runs through Greenwich, UTC is based on average time calculations from atomic clocks collected worldwide. You might still see official UTC referred to as GMT or Zulu time.

Time and Frequency Services of the National Institute for Standards and Technology (NIST), U.S. Department of Commerce, operates several of the most advanced clocks. For the time, call 303-499-7111 or 808-335-4363, or see http://nist.time.gov/ for UTC. In Canada, the Institute for Measurement Standards, National Research Council Canada, participates in determining UTC: English, 613-745-1578; French, 613-745-9426; for more see http://time5.nrc.ca/JavaClock/timeDisplayWE.shtml).

Daylight Saving Time In 70 countries, mainly in the temperate latitudes, time is set ahead 1 hour in the spring and set back 1 hour in the fall—a practice known as **daylight saving time**. The idea to extend daylight for early evening activities (at the expense of daylight in the morning), first proposed by Benjamin Franklin, was not adopted until World War I and again in World War II, when Great Britain, Australia, Germany, Canada, and the United States used the practice to save energy (1 less hour of artificial lighting needed).

In 1986 and again in 2007, the United States and Canada increased daylight saving time. Time "springs forward" 1 hour on the second Sunday in March and "falls back" 1 hour on the first Sunday in November, except in a few places that do not use daylight saving time (Hawai'i, Arizona, and Saskatchewan). In Europe, the last Sundays in March and October are used to begin and end the "summer-time period." (See http://webexhibits.org/daylightsaving/.)

*UTC is in use because agreement was not reached on whether to use the English word order, CUT, or the French order, TUC. UTC was the compromise and is recommended for all time-keeping applications; use of the term *GMT* is discouraged.

Maps, Scales, and Projections

The earliest known graphic maps date to 2300 B.C., when the Babylonians used clay tablets to record information about the region of the Tigris and Euphrates Rivers (the area of modern-day Iraq). Today, the making of maps and charts is a specialized science as well as an art, blending aspects of geography, engineering, mathematics, graphics, computer science, and artistic specialties. It is similar in ways to architecture, in which aesthetics and utility combine to produce a useful product.

A **map** is a generalized view of an area, usually some portion of Earth's surface, as seen from above and greatly reduced in size. **Cartography** is that part of geography that involves mapmaking. Maps are critical tools with which geographers depict spatial information and analyze spatial relationships.

We all use maps at some time to visualize our location in relation to other places, or maybe to plan a trip, or to coordinate commercial and economic activities. Have you found yourself looking at a map, planning real and imagined adventures to far-distant places? Maps are wonderful tools! Understanding a few basics about maps is essential to our study of physical geography.

The Scale of Maps

Architects, toy designers, and mapmakers have something in common: They all create scale models. They reduce real things and places to the more convenient scale of a drawing; a model car, train, or plane; a diagram; or a map. An architect renders a blueprint of a structure to guide the building contractors, selecting a scale so that a centimeter (or inch) on the drawing represents so many meters (or feet) on the proposed building. Often, the drawing is 1/50 or 1/100 real size.

The cartographer does the same thing in preparing a map. The ratio of the image on a map to the real world is **scale**; it relates a unit on the map to a similar unit on the ground. A 1:1 scale means that a centimeter on the map represents a centimeter on the ground, although this is an impractical map scale, since the map would be as large as the area mapped! A more appropriate scale for a local map is 1:24,000, in which 1 unit on the map represents 24,000 identical units on the ground.

Cartographers express map scale as a representative fraction, a graphic scale, or a written scale (Figure 1.23 on page 24). A *representative fraction* (*RF*, or *fractional scale*) is expressed with either a colon or a slash, as in 1:125,000 or 1/125,000. No actual units of measurement are mentioned because any unit is applicable as long as both parts of the fraction are in the same unit: 1 cm to 125,000 cm, 1 in. to 125,000 in., or even 1 arm length to 125,000 arm lengths.

A *graphic scale*, or *bar scale*, is a bar graph with units to allow measurement of distances on the map. An important advantage of a graphic scale is that, if the map is enlarged or reduced, the graphic scale enlarges or reduces along with the map. In contrast, written and fractional

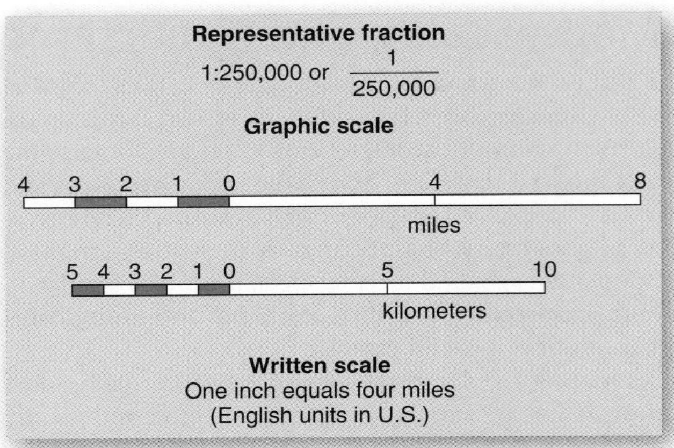

FIGURE 1.23 **Map scale.**
Three common expressions of map scale—representative fraction, graphic scale, and written scale.

scales become incorrect with enlargement or reduction. As an example, if you shrink a map from 1:24,000 to 1:63,360, the written scale "1 in. to 2000 ft" will no longer be correct. The new correct written scale is 1 in. to 5280 ft (1 mi).

Scales are *small*, *medium*, and *large*, depending on the ratio described. In relative terms, a scale of 1:24,000 is a large scale, whereas a scale of 1:50,000,000 is a small scale. The greater the denominator in a fractional scale (or the number on the right in a ratio expression), the smaller the scale and the more abstract the map is in relation to what is being mapped. Examples of selected representative fractions and written scales are listed in Table 1.2 for small-, medium-, and large-scale maps.

CRITICAL THINKING 1.3

Find and calculate some scales

Find a globe or map in the library or geography department and check the scale at which it was drawn. See if you can find examples of fractional, graphic, and written scales on wall maps, on highway maps, and in atlases. Find some examples of small- and large-scale maps, and note the different subject matter they portray.

Look at a world globe that is 61 cm (24 in.) in diameter. (Use different values depending on the globe you are using.) We know that Earth has an equatorial diameter of 12,756 km (7926 mi), so the scale of such a globe is the ratio of 61 cm to 12,756 km. To calculate the representative fraction for the globe in centimeters, divide Earth's actual diameter by the globe's diameter (12,756 km ÷ 61 cm). (*Hint*: 1 km = 1000 m, 1 m = 100 cm; therefore, Earth's diameter of 12,756 km represents 1,275,600,000 cm; and the globe's diameter is 61 cm.) In general, do you think a world globe is a small- or a large-scale map of Earth's surface?

Map Projections

A globe is not always a helpful map representation of Earth. When you go on a trip, you need more detailed information than a globe can provide. Consequently, to provide local detail, cartographers prepare large-scale *flat maps*, which are two-dimensional representations (scale models) of our three-dimensional Earth. Unfortunately, such conversion from three dimensions to two causes distortion.

A globe is the only true representation of *distance*, *direction*, *area*, *shape*, and *proximity*. A flat map distorts these properties. Therefore, in preparing a flat map, the cartographer must decide which characteristic to preserve, which to distort, and how much distortion is acceptable. To understand this problem, consider these important properties of a *globe*:

- Parallels always are parallel to each other, always are evenly spaced along meridians, and always decrease in length toward the poles.
- Meridians always converge at both poles and always are evenly spaced along any individual parallel.
- The distance between meridians always decreases toward poles, with the spacing between meridians at the 60th parallel equal to one-half the equatorial spacing.
- Parallels and meridians always cross each other at right angles.

The problem is that all these qualities cannot be reproduced on a flat surface. Simply taking a globe apart and laying it flat on a table illustrates the challenge faced by cartographers (Figure 1.24). You can see the empty spaces that open up between the sections, or gores, of the globe. This reduction of the spherical Earth to a flat surface is a **map projection**. Thus, no flat map projection of Earth can ever have all the features of a globe. Flat maps always possess some degree of distortion—much less for large-scale maps representing a few kilometers; much more for small-scale maps covering individual countries, continents, or the entire world.

TABLE 1.2 Sample Representative Fractions and Written Scales for Small-, Medium-, and Large-Scale Maps

System	Scale Size	Representative Fraction	Written System Scale
English	Small	1:3,168,000	1 in. = 50 mi
		1:1,000,000	1 in. = 16 mi
		1:250,000	1 in. = 4 mi
	Medium	1:125,000	1 in. = 2 mi
		1:63,360 (or 1:62,500)	1 in. = 1 mi
	Large	1:24,000	1 in. = 2000 ft
System		**Representative Fraction**	**Written System Scale**
Metric	Small	1:1,000,000	1 cm = 10.0 km
	Medium	1:25,000	1 cm = 0.25 km
	Large	1:10,000	1 cm = 0.10 km

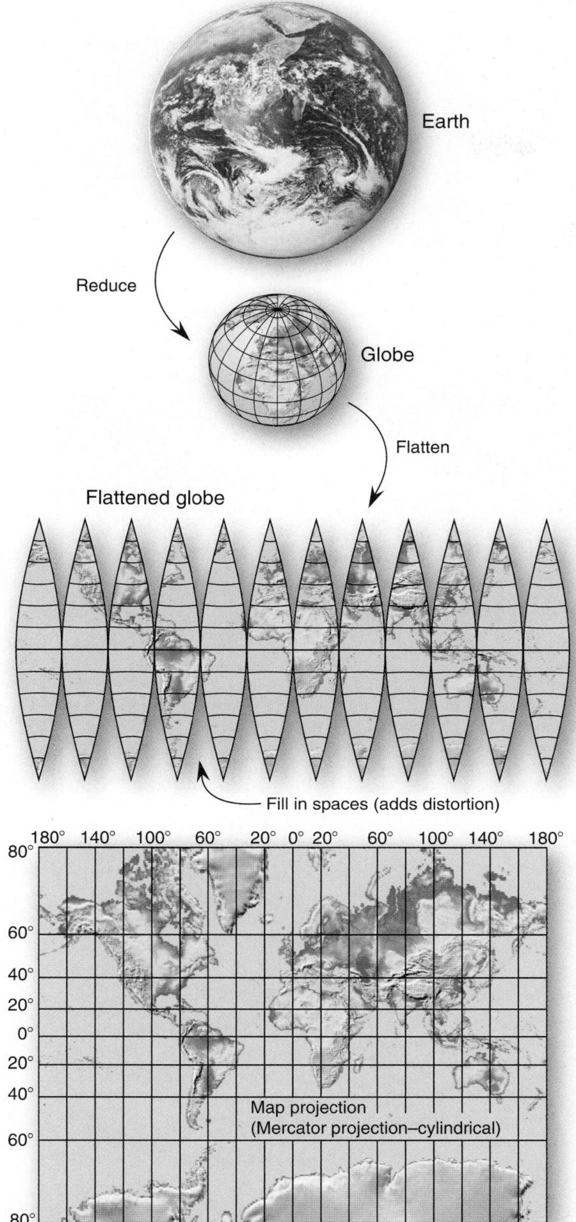

FIGURE 1.24 From globe to flat map. Conversion of the globe to a flat map projection requires decisions about which properties to preserve and the amount of distortion that is acceptable. [NASA astronaut photo from *Apollo 17*, 1972.]

Properties of Projections There are many map projections, four of which are shown in Figure 1.25 on page 26. *The best projection is always determined by its intended use.* The major decisions in selecting a map projection involve the properties of **equal area** (equivalence) and **true shape** (conformality). A decision for one property sacrifices the other, for they cannot be shown together on the same flat map.

If a cartographer selects equal area as the desired trait—for example, for a map showing the distribution of world climates—then true shape must be sacrificed by *stretching* and *shearing*, which allow parallels and meridians to cross at other than right angles. On an equal-area map, a coin covers the same amount of surface area no matter where you place it on the map. In contrast, if a cartographer selects the property of true shape, such as for a map used for navigational purposes, then equal area must be sacrificed, and the scale will actually change from one region of the map to another.

The Nature and Classes of Projections Despite the fact that modern cartographic technology uses mathematical constructions and computer-assisted graphics, the word *projection* is still used. The term comes from times past, when geographers actually projected the shadow of a wire-skeleton globe onto a geometric surface. The wires represented parallels, meridians, and outlines of the continents. A light source then cast a shadow pattern of latitude and longitude lines from the globe onto various geometric surfaces, such as a *cylinder*, *plane*, or *cone*.

Figure 1.25 illustrates the general classes of map projections and the perspectives from which they are generated. The classes shown include the cylindrical, planar (or azimuthal), and conic. Another class of projections, which cannot be derived from this physical-perspective approach, is the nonperspective oval shape. Still other projections derive from purely mathematical calculations.

With projections, the contact line or contact point between the wire globe and the projection surface—a *standard line* or *standard point*—is *the only place where all globe properties are preserved.* Thus, a *standard parallel* or *standard meridian* is a standard line true to scale along its entire length without any distortion. Areas away from this critical tangent line or point become increasingly distorted. Consequently, this line or point of accurate spatial properties should be centered by the cartographer on the area of interest.

The commonly used **Mercator projection** (from Gerardus Mercator, A.D. 1569) is a cylindrical projection (Figure 1.25a). The Mercator is a true-shape projection, with meridians appearing as equally spaced straight lines and parallels appearing as straight lines that are spaced closer together near the equator. The poles are infinitely stretched, with the 84th N parallel and 84th S parallel fixed at the same length as that of the equator. Note in Figures 1.24 and 1.25a that the Mercator projection is cut around the 80th parallel in each hemisphere because of the severe distortion at higher latitudes.

Unfortunately, Mercator classroom maps present false notions of the size (area) of midlatitude and poleward landmasses. A dramatic example on the Mercator projection is Greenland, which looks bigger than all of South America. In reality, Greenland is an island only one-eighth the size of South America and is actually 20% smaller than Argentina alone.

The advantage of the Mercator projection is that a line of constant direction, known as a **rhumb line**, or *loxodrome*, is straight and thus facilitates plotting directions between two points (Figure 1.26b). Thus, the

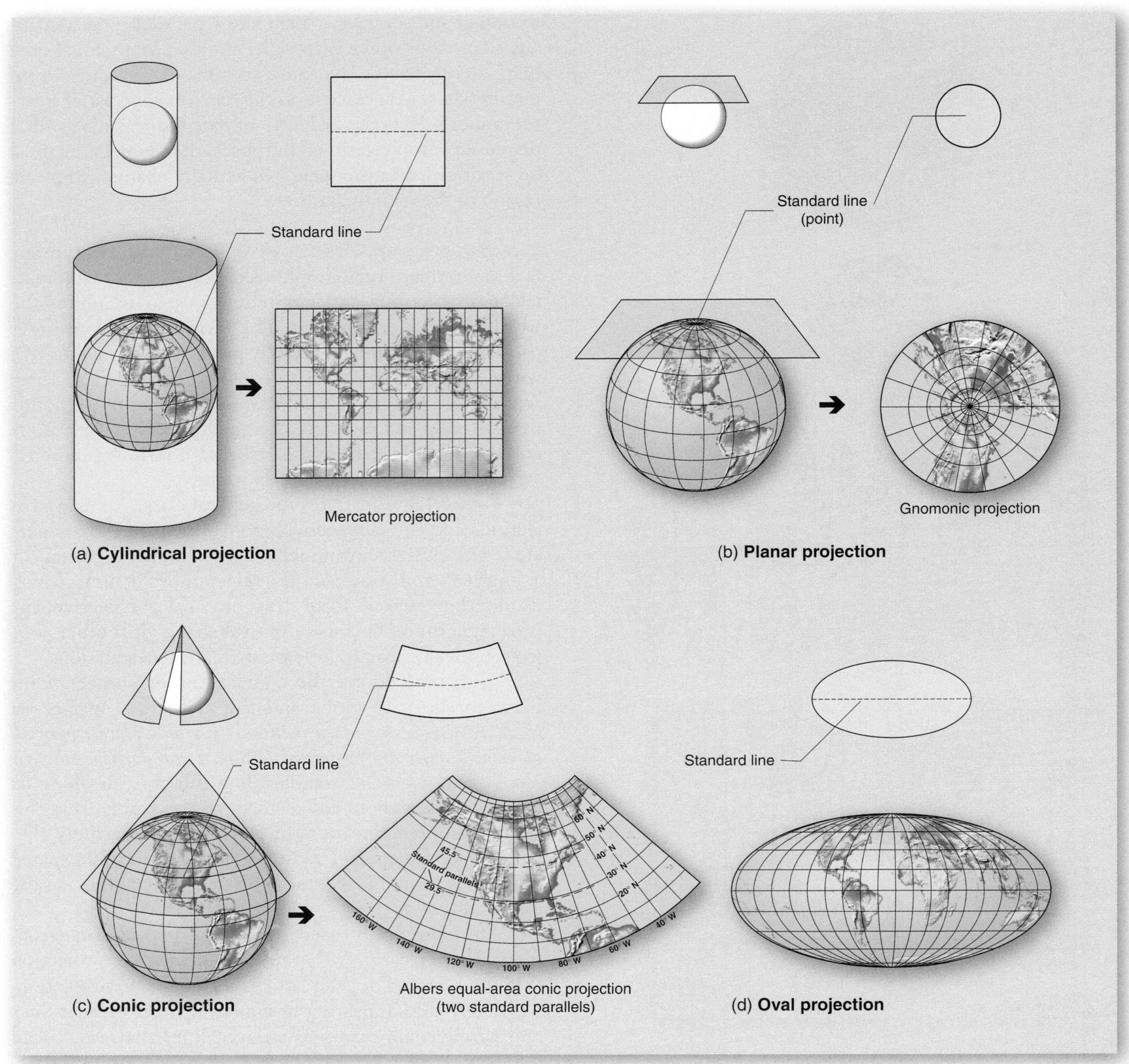

FIGURE 1.25 Classes of map projections.
Four general classes and perspectives of map projections—(a) cylindrical, (b) planar, (c) conic, and (d) oval projections.

Mercator projection is useful in navigation and is standard for nautical charts prepared by the National Ocean Service.

The *gnomonic*, or *planar*, *projection* in Figure 1.25b is generated by projecting a light source at the center of a globe onto a plane that is tangent to (touching) the globe's surface. The resulting severe distortion prevents showing a full hemisphere on one projection. However, a valuable feature is derived: *All great circle routes, which are the shortest distance between two points on Earth's surface, are projected as straight lines* (Figure 1.26a). The great circle routes plotted on a gnomonic projection then can be transferred to a true-direction projection, such as the Mercator, for determination of precise compass headings (Figure 1.26b).

For more information on maps used in this text and standard map symbols, turn to Appendix A, Maps in This Text and Topographic Maps. Topographic maps are essential tools of landscape analysis. Geographers, other scientists, travelers, and anyone visiting the outdoors may use topographic maps. Perhaps you have used a "topo" map in

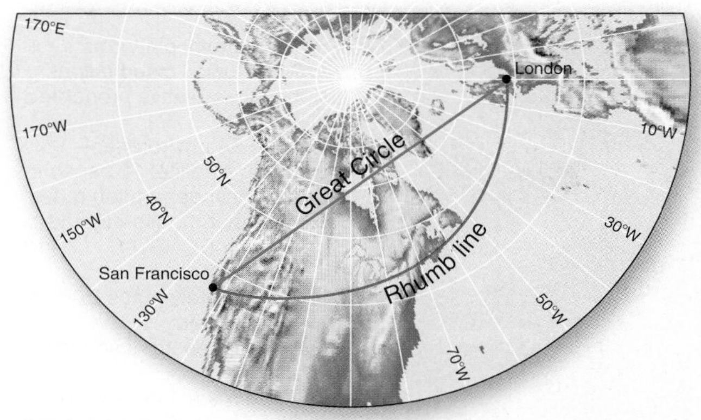

(a) Gnomonic projection

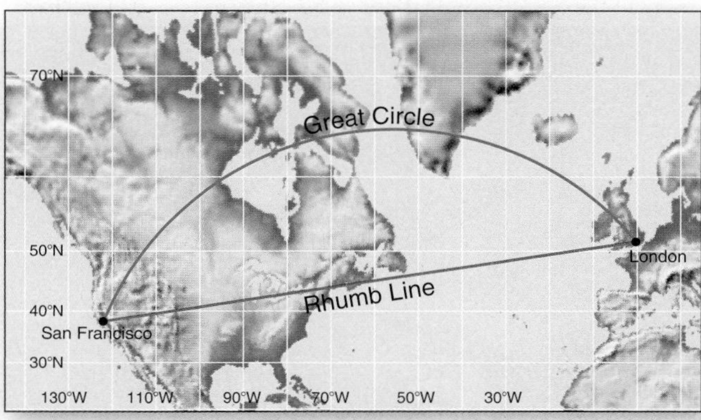

(b) Mercator projection (conformal, true shape)

FIGURE 1.26 Determining great circle routes.
A gnomonic projection (a) is used to determine the shortest distance—great circle route—between San Francisco and London because on this projection the arc of a great circle is a straight line. This great circle route is then plotted on a Mercator projection (b), which has true compass direction. Note that straight lines, or bearings, on a Mercator projection—rhumb lines—are not the shortest route.

planning a hike. USGS topographic maps appear in several chapters of this text.

Remote Sensing and GIS

Geographers probe, analyze, and map our home planet through remote sensing and geographic information systems (GIS). These technologies enhance our understanding of Earth. Geographers use remote-sensing data to study all the subjects in this text and the human activities that produce global change and impact Earth systems.

Remote Sensing

In this era of observations from orbit outside the atmosphere, from aircraft within it, and from remote submersibles in the oceans, scientists obtain a wide array of remotely sensed data (Figure 1.27 on page 28). Remote sensing is nothing new to humans; we do it with our eyes as we scan the environment, sensing the shape, size, and color of objects from a distance, registering energy from the visible-wavelength portion of the electromagnetic spectrum. Similarly, when a camera views the wavelengths for which its film or sensor is designed, it remotely senses energy that is reflected or emitted from a scene.

Our eyes and cameras are familiar means of obtaining **remote-sensing** information about a distant subject without having physical contact. Aerial photographs have been used for years to improve the accuracy of surface maps faster and more cheaply than can be done by on-site surveys. Deriving accurate measurements from photographs is the realm of **photogrammetry**, an important application of remote sensing.

Remote sensors on satellites, the International Space Station, and other craft sense a broader range of wavelengths than can our eyes. They can be designed to "see" wavelengths shorter than visible light (ultraviolet) and wavelengths longer than visible light (infrared and microwave radar).

Following launch, satellites can be set in several types of orbits: geostationary, polar, and Sun-synchronous, illustrated in Figure 1.28 on page 29. Figure 1.28a shows a geostationary (geosynchronous) orbit, typically at an altitude of 35,790 km (22,239 mi), where satellites pace Earth's rotation speed and therefore remain "parked" above a specific location, usually the equator, which provides a full hemisphere and is good for recording weather; an example is *GOES*.

Figure 1.28b shows a polar-orbiting satellite typically residing at an altitude between 200 km and 1000 km (124 mi–621 mi) and passing over every portion of the globe as Earth rotates beneath it, completing each orbit in 90 minutes. Such satellites are good for difficult regions such as

GEO REPORT 1.6 ***Mapping anniversary for the USGS***

December 2009 marked the 125th anniversary since Congress authorized the USGS to systematically map the country. The USGS National Mapping Program is complete for the lower 48 states, numbering 53,838 separate 7.5-minute quadrangles. *The National Map* (available at http://nationalmap.gov/), a collaborative effort by USGS and federal, state, and local agencies, provides downloadable digital topographic data for the entire United States. These data include aerial photographs, elevation, roads, geographic names, and land cover. The National Map is an important tool for scientists, teachers, students, and the public to access geographic information.

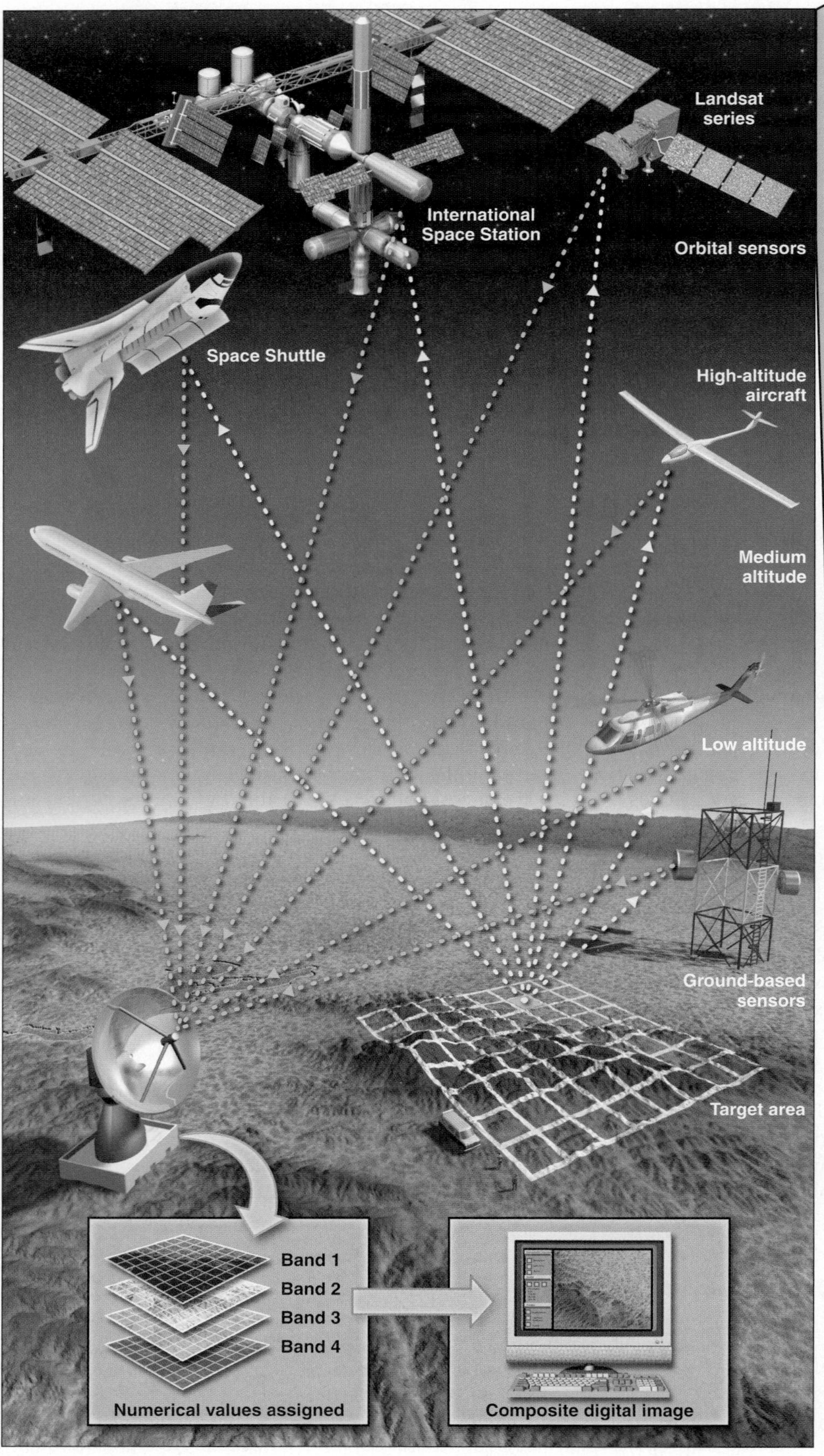

A sample of orbital platforms:

CloudSat: Studies cloud extent, distribution, radiative properties, and structure.

ENVISAT: ESA environment-monitoring satellite; 10 sensors, including next generation radar.

GOES: Weather monitoring and Forecasting; *GOES-11, -12, -13,* and *-14.*

GRACE: Accurately maps Earth's gravitational field.

JASON-1, -2: Measures sea-level heights.

Landsat: *Landsat-1* in 1972 to *Landsat-7* in 1999, provides millions of images for Earth systems science and global change.

NOAA: First in 1978 through *NOAA-15, -16, -17, -18,* and *-19* now in operation, global data gathering, short- and long-term weather forecasts.

RADARSAT-1, -2: Synthetic Aperture Radar in near-polar orbit, operated by Canadian Space Agency.

SciSat-1: Analyzes trace gases, thin clouds, atmospheric aerosols with Arctic focus.

SeaStar: Carries the *SeaWiFS* (Sea-viewing Wide Field-of-View instrument) to observe Earth's oceans and microscopic marine plants.

Terra and Aqua: Environmental change, error-free surface images, cloud properties, through five instrument packages.

TOMS-EP: Total Ozone Mapping Spectrometer, monitoring stratospheric ozone, similar instruments on *NIMBUS-7* and *Meteor-3.*

TOPEX-POSEIDON: Measures sea-level heights.

TRMM: Tropical Rainfall Measuring Mission, includes lightning detection, and global energy budget measurements.

For more info see:
http://www.nasa.gov/centers/goddard/missions/index.html

FIGURE 1.27 Remote-sensing technologies.
Remote-sensing technology measures and monitors Earth's systems from orbiting spacecraft, aircraft in the atmosphere, and ground-based sensors. Various wavelengths (bands) are collected from sensors. Computers process the data to produce digital images for analysis. A sample of remote-sensing platforms is listed along the side of the illustration. The Space Shuttle is shown in its inverted orbital flight mode. (Illustration is not to scale.)

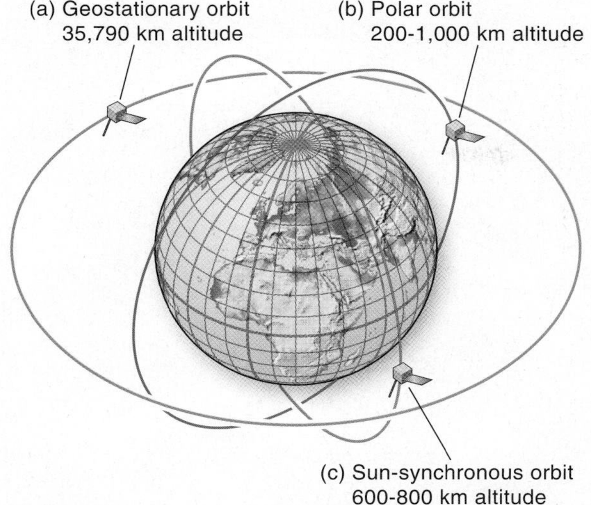

FIGURE 1.28 Three satellite orbital paths. (a) Geostationary (geosynchronous) orbit, typically 35,790-km altitude. (b) Polar-orbiting satellites typically are between 200-km and 1000-km altitude. (c) Sun-synchronous orbits shift their near-polar track from 600-km to 800-km altitude.

the poles, for environmental studies, and for spy surveillance; examples include *Terra* and *Aqua*.

Figure 1.28c demonstrates a Sun-synchronous orbit, shifting its 600 km to 800 km (373 mi to 497 mi) altitude, near-polar track, about 1° per day (365 days in a year, 360° in a circle), providing a surface below in constant sunlight for visible-wavelength images or in constant darkness for longwave radiation images.

Satellites do not take conventional-film photographs. Rather, they record *images* that are transmitted to Earth-based receivers in a way similar to television satellite transmissions or a digital camera. A scene is scanned and broken down into pixels (*pic*ture *el*ements), each identified by coordinates named *lines* (horizontal rows) and *samples* (vertical columns). For example, a grid of 6000 lines and 7000 samples forms 42,000,000 pixels, providing great detail. The large amount of data needed to produce a single image requires computer processing and data storage at ground stations.

Digital data are processed in many ways to enhance their utility: simulated natural color, "false" color to highlight a particular feature, enhanced contrast, signal filtering, and different levels of sampling and resolution. Active and passive are two types of remote-sensing systems.

Active Remote Sensing Active systems direct a beam of energy at a surface and analyze the energy reflected back. An example is *radar* (*ra*dio *d*etection *a*nd *r*anging). A radar transmitter emits short bursts of energy that have relatively long wavelengths (0.3 cm to 10 cm) toward the subject terrain, penetrating clouds and darkness. Energy reflected back to a radar receiver for analysis is known as *backscatter*. There are several important radar satellites in orbit at this time, such as *QuikSAT, RADARSAT–1* and *–2, SCISAT–1*, and *JERS–1* and *–2*.

A synthetic aperture radar system aboard an aircraft imaged the San Andreas fault near San Mateo, California. NASA and JPL developed systems to repeatedly capture detailed, three-dimensional images along this fault system to track movement and areas that are building up energy for future earthquakes (Figure 1.29). Images collected in a time series, allow scientists to compare images, pixel by pixel in computers to detect even the slightest horizontal or vertical movement between images; to diagnose accumulating strain.

Passive Remote Sensing Passive remote-sensing systems record energy radiated from a surface, particularly visible light and infrared. Our own eyes are passive remote sensors, as was the *Apollo 17* astronaut camera that made the photograph (film) of Earth on the back cover of this book from a distance of 37,000 km (23,000 mi).

Passive remote sensors on *Landsat* satellites provide a variety of data, as shown in images of the Appalachian Mountains and Mount St. Helens in Chapter 12, river deltas and New Orleans before and after Katrina in Chapter 14,

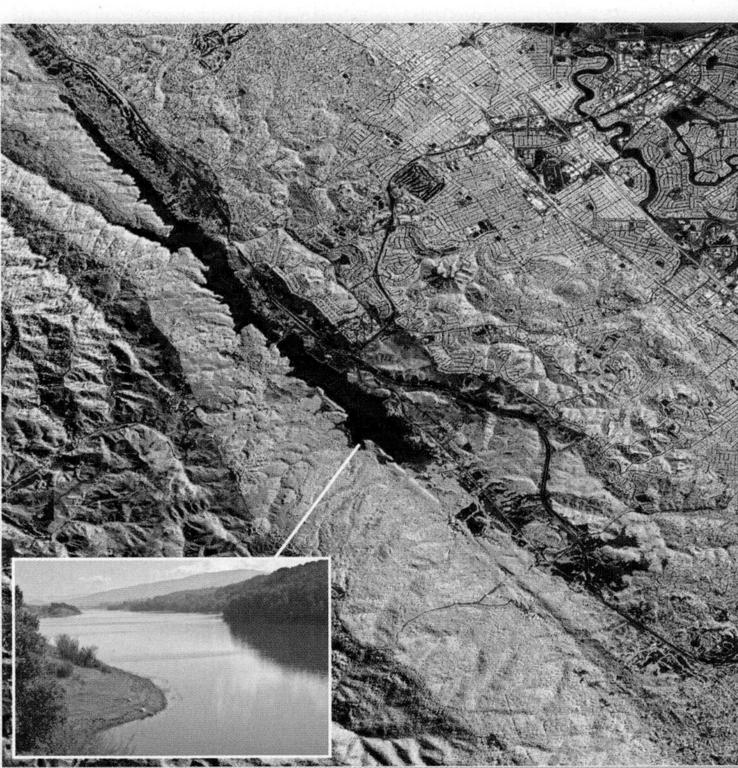

FIGURE 1.29 Radar image of active earthquake fault. Synthetic aperture radar image made from an aircraft repeatedly over time allows study of stress-accumulating portions of the fault. Discussion of this fault is in Chapter 12. Can you see the fault rift area running from southeast to northwest in this 2009 false-color image, in this area filled with a reservoir? [UACSAR image by JPL/NASA; photo by author.]

and the Vatnajökull ice cap in Iceland in Chapter 17. Two *Landsats* remain operational, *Landsat–5* and –7. See http://landsat.gsfc.nasa.gov/ and http://geo.arc.nasa.gov/sge/landsat/landsat.html for more information and additional links.

A NASA, National Oceanic and Atmospheric Administration (NOAA, see http://www.noaa.gov/), and United Kingdom joint venture, the *Polar Operational Environmental Satellites* (*POES*), complete nearly polar orbits 14 times a day, carrying the *advanced very high resolution radiometer* (*AVHRR*). These are in Sun-synchronous orbits aboard *NOAA–15* and *NOAA*–17 for morning hours and *NOAA–16*, *NOAA–18*, and *NOAA–19* for afternoon hours. See http://www.oso.noaa.gov/poesstatus/.

Key to NASA's Earth Observing System (EOS) is satellite *Terra*, which began beaming back data and images in 2000 (see http://terra.nasa.gov/), followed by another satellite in the series, *Aqua*. Five instrument packages observe Earth systems in detail. The image of Earth on the half-title page of this book includes a cloud snapshot from *GOES* added to *Terra* MODIS images.

The *Geostationary Operational Environmental Satellites*, known as *GOES*, became operational in 1994, providing the images you see on television weather reports. Geostationary satellites stay in semi-permanent positions at an altitude of 35,790 km (22,239 mi).

GOES–11 is now above 135° W longitude for western North America and eastern Pacific Ocean weather. *GOES–12* sits above 75° W longitude to monitor central and eastern North America and the western Atlantic. Think of these satellites as hovering over these meridians for continuous coverage, using visual wavelengths for daylight hours and infrared for nighttime views. For the future, *GOES–13* and *GOES–14* are parked in orbit ready when needed (Figure 1.30). See the Geostationary Satellite Server at http://www.goes.noaa.gov/. The *GOES* Project Science appears at http://rsd.gsfc.nasa.gov/goes/.

The USGS Earth Resources Observation and Science (EROS) Data Center, near Sioux Falls, South Dakota, is the national repository for geophysical and geospatial data—remote-sensing imagery, photos, maps, the Global Land Inventory System, NASA's EOS Gateway, and more. EROS is where we gather all the data and imagery in one place that will guide future generations about Earth systems operations and what conditions were like in this era (see http://eros.usgs.gov/).

For a sampling of global remote-sensing coverage from a variety of satellite platforms, go to the USGS Global Visualization Viewer, select the satellite and click on the map for a location, and then watch (http://glovis.usgs.gov/). For an update of all active and passive remote-sensing platforms now operational, see Remote Sensing Status Report in this chapter on the *Mastering Geography* web site. Note that Astronaut Mission Specialist Dr. Thomas Jones operated the radar and camera on Space Shuttle *Endeavour* to study the Kliuchevskoi volcano eruption. He is profiled in a Career Link along with these images on the *Mastering Geography* web site for this chapter.

FIGURE 1.30 ***GOES–14* first image.**
GEOS–14 made its first image July 27, 2009; the improved technology in this satellite produces 1-km (0.62-mi) resolution. Can you find the cloud-free southwestern United States? Thunderstorms along the Texas–Oklahoma border? There are two tropical waves visible in the eastern Pacific Ocean. [NASA.]

Geographic Information Systems

Techniques such as remote sensing acquire large volumes of spatial data that must be stored, processed, and retrieved in useful ways. A **geographic information system (GIS)** is a computer-based, data-processing tool for gathering, manipulating, and analyzing geographic information. Today's sophisticated computer systems allow the integration of geographic information from direct surveys (on-the-ground mapping) and remote sensing in complex ways never before possible.

Through a GIS, Earth and human phenomena are analyzed over time. This can include the study and forecasting of diseases such H1N1 flu and West Nile virus, or the population displacement caused by Hurricane Katrina, or the destruction from the Haiti and Chilean earthquakes in 2010, or interactive land-cover maps, to name a few examples (Figure 1.31b and c). Whereas printed maps are fixed at time of publication, GIS maps can be modified and evolve instantly.

The beginning component for any GIS is a map, with its associated coordinate system such as latitude–longitude. This establishes reference points against which to position data. The coordinate system is digitized, along with all areas, points, and lines. Increasingly, GPS units are now used to provide the basic mapping data needed for a GIS. Remotely sensed imagery and data are then added on the coordinate system. The resulting *digital elevation models* (*DEMs*) aid scientific analysis of topography, area–altitude distributions, slopes, and local stream-

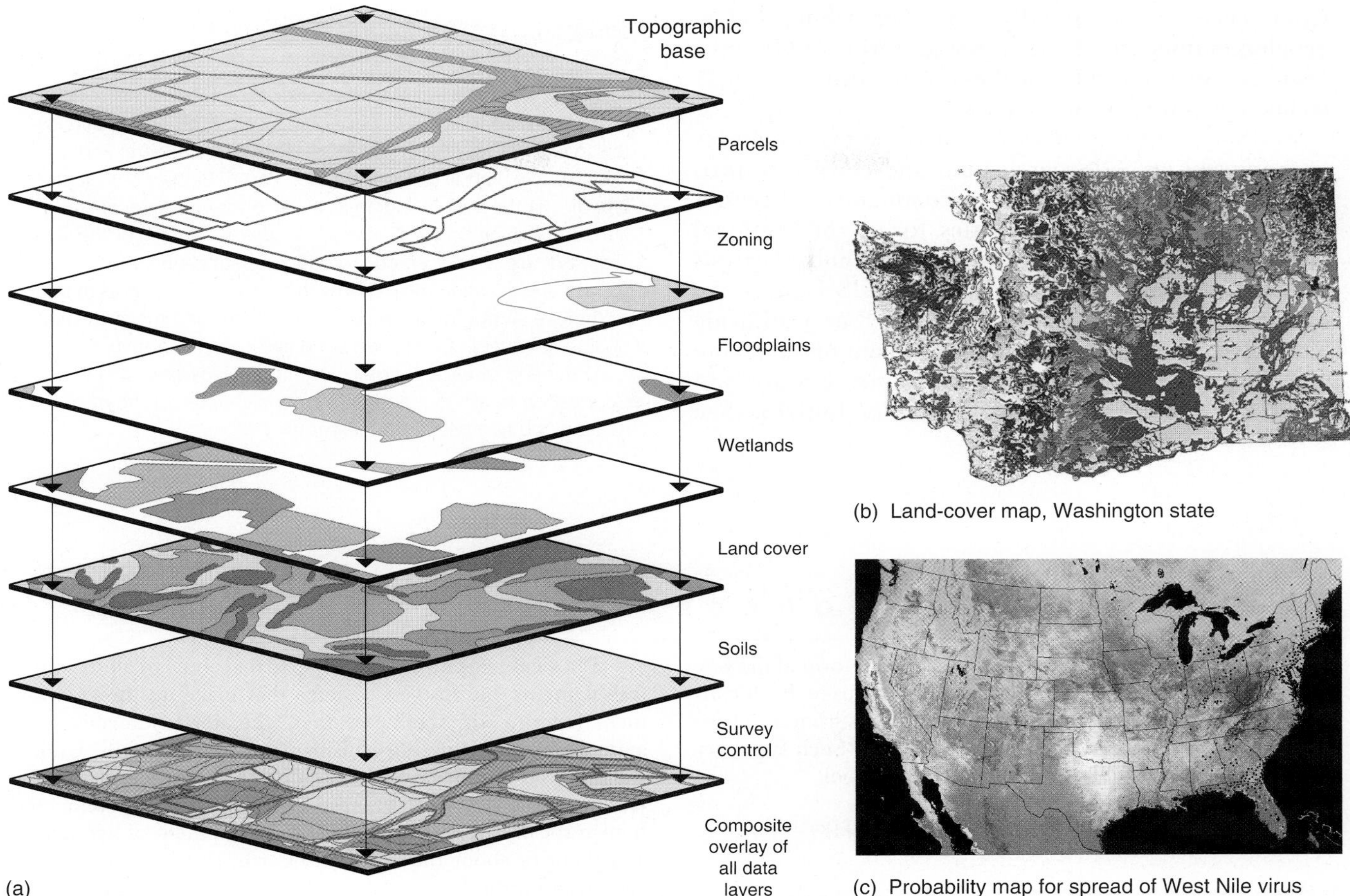

FIGURE 1.31 A geographic information system (GIS) model. (a) Layered spatial data in a GIS format. (b) This comprehensive land-cover map is an important component of a statewide GIS analysis for Washington State. The goal is to evaluate the protection of species and biodiversity. (c) A NASA GIS study uses satellite-derived temperature, humidity, and vegetation variables in a GIS overlay of environmental conditions for a probability map of the spread of West Nile virus. [(a) After USGS. (b) GIS map courtesy of Dr. Kelly M. Cassidy, *Gap Analysis of Washington State,* v. 5, Map 3, University of Washington. (c) NASA/GSFC.]

drainage characteristics, among many applications. The data then are available for computer manipulation, display, and GIS operations.

A GIS is capable of analyzing patterns and relationships within a single data plane, such as the floodplain or soil layer in Figure 1.31a. A GIS also can generate an overlay analysis where two or more data planes interact. When the layers are combined, the resulting synthesis—a *composite overlay*—is ready for use in analyzing complex problems. The utility of a GIS compared with that of a fixed map is the ability to manipulate the variables for analysis and to constantly change the map.

The applications of GIS are wide-ranging and cross many disciplines. Geographic information science (GISci) embodies the development, use, and application of GIS. An international conference series on GISci began in 2000 and continues biannually (see http://giscience.org/). These conferences bring together leading scientists from academia, industry, and government to explore emerging research across the disciplines and to seek applications in many fields. An important direction is toward more user-friendly GIS systems in the future.

Access to GIS is expanding with increased availability of numerous open-source GIS software packages. These are usually free, have on-line support systems, and are updated frequently (see http://opensourcegis.org/).

Programs for visualizing Earth, such as Google Earth™, provide three-dimensional viewing of the globe as well as geographic information in a program that can be downloaded from the Internet (http://earth.google.com/). Google Earth™ allows the user to "fly" anywhere on Earth and zoom in on landscapes and features of interest, using satellite imagery and aerial photography at varying resolutions. You can find where you live and your campus. Layers can be selected, as in a GIS model, by the user depending on the task at hand and the composite overlay displayed.

Career opportunities in GIS continue to increase. Regardless of your academic major, the ability to analyze data spatially is important. The annual ESRI (Environmental Systems Research Institute) International Users

Conference, whose attendance is approaching 15,000 people, continues to be the largest gathering of GIS professionals to share state-of-the-art applications of GIS technology (http://www.esri.com/).

GIS degree programs are available at many colleges and universities. GIS curriculum and certificate programs are now available at many community colleges. A consortium of three universities forms the National Center for Geographic Information and Analysis (NCGIA) for GIS education. These GIS centers are Department of Geography, University of California Santa Barbara, Santa Barbara; Department of Surveying and Engineering, University of Maine, Orono; and State University of New York–Buffalo, Buffalo. (See http://www.ncgia.ucsb.edu/.)

GEOSYSTEMS CONNECTION

Now that you have the essentials of geography and the *Geosystems* approach, we leave on a journey through the chapters in each of four Earth's spheres—Part I Atmosphere; Part II Hydrosphere; Part III Lithosphere; and Part IV Biosphere. Chapter 2 begins at the Sun, including its place in the Universe, operation, and energy flow to Earth. We look at the march of the seasons. We follow solar energy through the atmosphere to the surface and examine the patterns this creates. As outputs in the system we study global temperature patterns and the circulation of air and water in Earth's vast wind and ocean currents.

At the end of each chapter, you find a *Geosystems Connection* to act as a bridge from one chapter to the next, helping you to cross to the next topic.

KEY LEARNING CONCEPTS REVIEW

To assist you, here is a review in handy summary form of the Key Learning Concepts listed on this chapter's title page. Each concept review concludes with a list of the key terms from the chapter, their page numbers, and review questions. Such summary and review sections follow each chapter in the book.

■ ***Define*** **geography and physical geography in particular.**

Geography brings together disciplines from the physical and life sciences with those from the cultural and human sciences to attain a holistic view of Earth—physical geography is an essential aspect of the emerging **Earth systems sciences**. **Geography** is a science of method, a special way of analyzing phenomena over space; **spatial** refers to the nature and character of physical space. Geography integrates a wide range of subject matter, and geographic education recognizes five major themes: **location**, **region**, **human–Earth relationships**, **movement**, and **place**. Geography's method is **spatial analysis**, used to study the interdependence among geographic areas, natural systems, society, and cultural activities over space or area. **Process**—that is, a set of actions or mechanisms that operate in some special order—is central to geographic synthesis.

Physical geography applies spatial analysis to all the physical elements and process systems that make up the environment: energy, air, water, weather, climate, landforms, soils, animals, plants, microorganisms, and Earth itself. Understanding the complex relations among these elements is important to human survival because Earth's physical systems and human society are intertwined. The development of hypotheses and theories about the Universe, Earth, and life involves the **scientific method**.

Earth systems science (p. 3)
geography (p. 3)
spatial (p. 3)
location (p. 3)
region (p. 3)
human–Earth relationships (p. 3)
movement (p. 3)
place (p. 3)
spatial analysis (p. 3)
process (p. 3)
physical geography (p. 3)
scientific method (p. 3)

1. What is unique about the science of geography? On the basis of information in this chapter, define *physical geography* and review the geographic approach.
2. In general terms, how might a physical geographer analyze water pollution in the Great Lakes?
3. Assess your geographic literacy by examining atlases and maps. What types of maps have you used: Political? Physical? Topographic? Do you know what map projections they employed? Do you know the names and locations of the four oceans, seven continents, and most individual countries? Can you identify the new countries that have emerged since 1990?
4. Suggest a representative example for each of the five geographic themes and use that theme in a sentence.
5. Have you made decisions today that involve geographic concepts discussed within the five themes presented? Explain briefly.

- ***Describe*** systems analysis, open and closed systems, and feedback information, and ***relate*** these concepts to Earth systems.

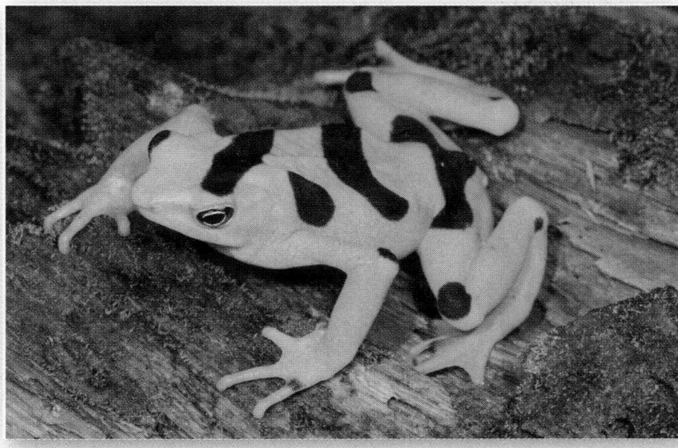

A **system** is any ordered, related set of things and their attributes, as distinct from their surrounding environment. Systems analysis is an important organizational and analytical tool used by geographers. Earth is an **open system** in terms of energy, receiving energy from the Sun, but it is essentially a **closed system** in terms of matter and physical resources.

As a system operates, "information" is returned to various points in the system via pathways of **feedback loops**. If the feedback information discourages change in the system, it is **negative feedback**. Further production of such feedback opposes system changes. Such negative feedback causes self-regulation in a natural system, stabilizing the system. If feedback information encourages change in the system, it is **positive feedback**. Further production of positive feedback stimulates system changes. Unchecked positive feedback in a system can create a runaway ("snowballing") condition. When the rates of inputs and outputs in the system are equal and the amounts of energy and matter in storage within the system are constant (or when they fluctuate around a stable average), the system is in **steady-state equilibrium**. A system that demonstrates a steady increase or decrease in system operations—a trend over time—is in **dynamic equilibrium**. A **threshold**, or tipping point, is the moment at which a system can no longer maintain its character, so it lurches to a new operational level, one that may not be compatible with previous conditions. Geographers often construct a simplified **model** of natural systems to better understand them.

Four immense open systems powerfully interact at Earth's surface: three **abiotic**, or nonliving, systems including **atmosphere**, **hydrosphere** (including the **cryosphere**), and **lithosphere** and a **biotic**, or living, system (**biosphere**, or **ecosphere**).

system (p. 5)
open system (p. 5)
closed system (p. 6)
feedback loops (p. 8)
negative feedback (p. 8)
positive feedback (p. 8)
steady-state equilibrium (p. 9)
dynamic equilibrium (p. 9)
threshold (p. 9)
model (p. 11)
abiotic (p. 11)
biotic (p. 11)
atmosphere (p. 11)
hydrosphere (p. 11)
cryosphere (p. 11)
lithosphere (p. 11)
biosphere (p. 13)
ecosphere (p. 13)

6. Define systems theory as an organizational strategy. What are open systems, closed systems, and negative feedback? When is a system in a steady-state equilibrium condition? What type of system (open or closed) is the human body? A lake? A wheat plant?
7. Describe Earth as a system in terms of both energy and matter; use simple diagrams to illustrate your description.
8. What are the three abiotic (nonliving) spheres that make up Earth's environment? Relate these to the biotic (living) sphere: the biosphere.

- ***Explain*** Earth's reference grid: latitude and longitude and latitudinal geographic zones and time.

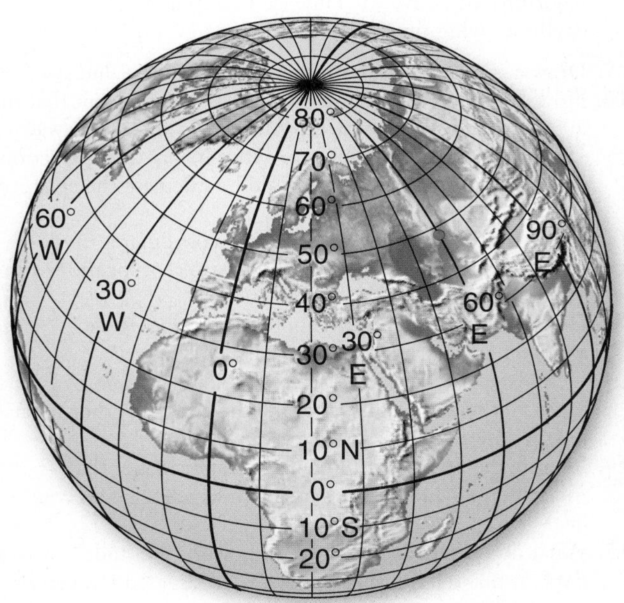

The science that studies Earth's shape and size is **geodesy**. Earth bulges slightly through the equator and is oblate (flattened) at the poles, producing a misshapen spheroid, a **geoid**. Absolute location on Earth is described with a specific reference grid of **parallels** of **latitude** (measuring distances north and south of the equator) and **meridians** of **longitude** (measuring distances east and west of a prime meridian). A historic breakthrough in navigation and timekeeping occurred with the establishment of an international **prime meridian** (0° through Greenwich, England) and the invention of precise chronometers that enabled accurate measurement of longitude. A **great circle** is any circle of Earth's circumference whose center coincides with the center of Earth. Great circle routes are the shortest distance between two points on Earth. **Small circles** are those whose centers do not coincide with Earth's center. Latitude, longitude, and elevation are accurately calibrated using a handheld **global positioning system (GPS)** instrument that reads radio signals from satellites.

The prime meridian provided the basis for **Greenwich Mean Time (GMT)**, the world's first universal time system. A corollary of the prime meridian is the 180° meridian, the **International Date Line**, which marks the place where each day officially begins. Today, **Coordinated Universal Time (UTC)** is the worldwide standard and the basis for international time zones. **Daylight saving time** is a seasonal change of clocks by 1 hour in summer months.

geodesy (p. 14)
geoid (p. 14)
latitude (p. 15)
parallel (p. 15)
longitude (p. 17)
meridian (p. 17)
prime meridian (p. 17)
Global Positioning System (GPS) (p. 20)
great circle (p. 18)
small circles (p. 19)
Greenwich Mean Time (GMT) (p. 21)
International Date Line (p. 22)
Coordinated Universal Time (UTC) (p. 23)
daylight saving time (p. 23)

9. Draw a simple sketch describing Earth's shape and size.
10. How did Eratosthenes use Sun angles to figure out that the 5000-stadia distance between Alexandria and Syene was 1/50 of Earth's circumference? Once he knew this fraction of Earth's circumference, how did he calculate the entirety of Earth's circumference?
11. What are the latitude and longitude coordinates (in degrees, minutes, and seconds) of your present location? Where did you find this information?
12. Define *latitude* and *parallel* and define *longitude* and *meridian* using a simple sketch with labels.
13. Define a *great circle*, *great circle routes*, and a *small circle*. In terms of these concepts, describe the equator, other parallels, and meridians.
14. Identify the various latitudinal geographic zones that roughly subdivide Earth's surface. In which zone do you live?
15. What does timekeeping have to do with longitude? Explain this relationship. How is Coordinated Universal Time (UTC) determined on Earth?
16. What and where is the prime meridian? How was the location originally selected? Describe the meridian that is opposite the prime meridian on Earth's surface.
17. What is the GPS and how does it assist you in finding location and elevation on Earth? Give a couple of examples where GPS technology was utilized to correct heights for some famous mountains.

■ ***Define*** cartography and mapping basics: map scale and map projections

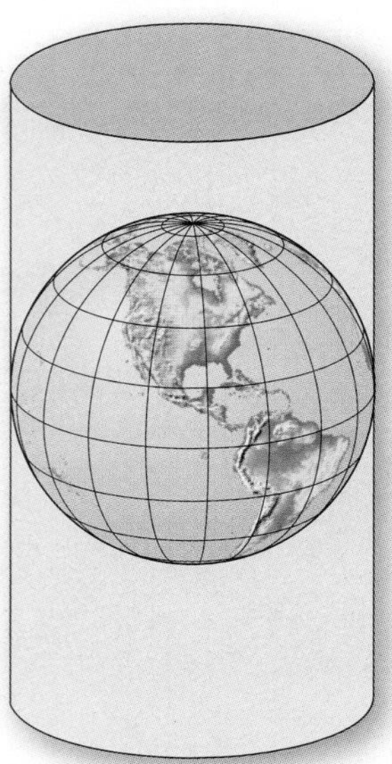

Cylindrical projection

A **map** is a generalized view of an area, usually some portion of Earth's surface, as seen from above and greatly reduced in size. The science and art of mapmaking is **cartography**. For the spatial portrayal of Earth's physical systems, geographers use maps. **Scale** is the ratio of the image on a map to the real world; it relates a unit on the map to an identical unit on the ground. Cartographers create a **map projection** for specific purposes, selecting the best compromise of projection for each application. Compromise is always necessary because Earth's round, three-dimensional surface cannot be exactly duplicated on a flat, two-dimensional map. **Equal area** (equivalence), **true shape** (conformality), true direction, and true distance are all considerations in selecting a projection. The **Mercator projection** is in the cylindrical class; it has true-shape qualities and straight lines that show constant direction. A **rhumb line** denotes constant direction and appears as a straight line on the Mercator.

map (p. 23)
cartography (p. 23)
scale (p. 23)
map projection (p. 24)
equal area (p. 25)
true shape (p. 25)
Mercator projection (p. 25)
rhumb line (p. 25)

18. Define cartography. Explain why it is an integrative discipline.
19. What is map scale? In what three ways may it be expressed on a map?

20. State whether the following ratios are large scale, medium scale, or small scale: 1:3,168,000; 1:24,000; 1:250,000.
21. Describe the differences between the characteristics of a globe and those that result when a flat map is prepared.
22. What type of map projection is used in Figure 1.14? In Figure 1.21? (See Appendix A.)

- ***Describe*** remote sensing, and ***explain*** the geographic information system (GIS); both as tools used in geographic analysis

Orbital and aerial **remote sensing** analyzes the operation of Earth's systems. Satellites do not take photographs, but record images that are transmitted to Earth-based receivers. Satellite images are recorded in digital form for later processing, enhancement, and generation. Aerial photographs have been used for years to improve the accuracy of surface maps. This is the realm of **photogrammetry**, an important application of remote sensing.

The mountain of data collected is put to work using **geographic information system (GIS)** technology. Computers process geographic information from direct surveys and remote sensing in complex ways never before possible. GIS methodology is an important step in better understanding Earth's systems and is a vital career opportunity for geographers.

The science of physical geography is in a unique position to synthesize the spatial, environmental, and human aspects of our increasingly complex relationship with our home planet—Earth.

remote sensing (p. 0)
photogrammetry (p. 0)
geographic information system (GIS) (p. 0)

23. What is remote sensing? What are you viewing when you observe a weather satellite image on TV or in the newspaper? Explain.
24. Describe the *Terra*, *Landsat*, *GOES*, and *NOAA* satellites, and detail them using several examples. Go to the NASA or ESA (European Space Agency) web site and view some satellite images.
25. If you were in charge of planning the development of a large tract of land, how would GIS methodologies assist you? How might planning and zoning be affected if a portion of the tract in the GIS is a floodplain or prime agricultural land?

MASTERING GEOGRAPHY

Visit www.masteringgeography.com for several figure animations, satellite loops, self-study quizzes, flashcards, visual glossary, case studies, career links, textbook reference maps, RSS Feeds, and an ebook version of *Geosystems*. Included are many geography web links to interesting Internet sources that support this chapter.

PART I

The Energy–Atmosphere System

In contrast to hot summers, winter snow covers the desert of northeastern Arizona. [Bobbé Christopherson.]

For more than 4.6 billion years, solar energy has traveled across interplanetary space to Earth, where a small portion of the solar output is intercepted. Our planet and our lives are powered by radiant energy from the star that is closest to Earth—the Sun. Because of Earth's curvature, the energy at the top of the atmosphere is unevenly distributed, creating imbalances from the equator to each pole—the equatorial region experiences energy surpluses; the polar regions experience energy deficits. Also, the annual pulse of seasonal change varies the distribution of energy during the year.

Earth's atmosphere acts as an efficient filter, absorbing most harmful radiation, charged particles, and space debris so that they do not reach Earth's surface. In the lower atmosphere the unevenness of daily energy receipt empowers atmospheric and surface energy budgets, giving rise to global patterns of temperature and circulation of wind and ocean currents. Each of us depends on many systems that are set into motion by energy from the Sun. These are the systems of Part I.

CHAPTER 2

Solar Energy to Earth and the Seasons

These boys and a donkey haul water on Fogo Island in the tropical West African country of Cape Verde. Note their shadow is cast almost directly beneath them. The inset photo captures the experience of having the Sun directly overhead—the subsolar point. [Bobbé Christopherson.]

KEY LEARNING CONCEPTS

After reading the chapter, you should be able to:

- ***Distinguish*** among galaxies, stars, and planets, and ***locate*** Earth.
- ***Examine*** the origin, formation, and development of Earth, and ***construct*** Earth's annual orbit about the Sun.
- ***Describe*** the Sun's operation, and ***explain*** the characteristics of the solar wind and the electromagnetic spectrum of radiant energy.
- ***Portray*** the intercepted solar energy and its uneven distribution at the top of the atmosphere.
- ***Define*** solar altitude, solar declination, and daylength, and ***describe*** the annual variability of each—Earth's seasonality.

Chasing the Subsolar Point

In this chapter we track the march of the seasons. The Sun's subsolar point, where direct overhead rays arrive at Earth's surface, migrates from December 21–22 along the Tropic of Capricorn at about 23.5° S latitude, northward to the equator on March 20–21, and onward to June 20–21 along the Tropic of Cancer at about 23.5° N. Then turning southward, the Sun's subsolar point returns to the equator on September 22–23, and on to the Southern Hemisphere in December (see Figure 2.15).

The chapter-opening photo shows two boys and a donkey hauling water in the West African island country of Cape Verde, around local noon. They were close to the subsolar point, the latitude where the Sun is highest in the sky and its rays are perpendicular to the surface. Outside of the tropics, the Sun is never directly overhead. For example, at 40° N latitude the noon Sun's altitude ranges from 26° above the horizon in December to 73° in June, and is never at 90°. On our expedition ship, we tracked our route and the Sun, to determine the closest encounter with the subsolar point, either on our ship or on an island in the Atlantic Ocean. Here is how we calculated this moment.

On most globes, globe designers place an *analemma* in the Southeast Pacific. The analemma is a figure-8 shaped chart where you can check any date, then trace to the *y-axis* and find the Sun's declination, which is the latitude of the subsolar point (Figure GN 2.1). Along the Tropic of Capricorn the subsolar point occurs on December 21–22, at the lower end of the analemma. Following the chart, you see that by March 20–21, the Sun's declination reaches the equator, and then moves on to the Tropic of Cancer in June. Use the chart to calculate the subsolar point location on your birthday.

The shape of the analemma as the Sun's declination migrates tropic-to-tropic is a result of Earth's axial tilt and elliptical orbit. Earth actually revolves in an elliptical orbit around the Sun, moving faster during December and January, and slower in June and July. This sets the *equation of time.*

An average day of 24 hours (86,400 seconds) sets *mean solar time.* However, *observed solar time* occurs each day at noon as the Sun crosses your meridian, and this sets the *apparent solar day.* You see on the chart that in October and November, *fast-Sun times* occur and the Sun arrives ahead of local noon (12:00), noted on the *x-axis* along the top of the chart. In February and March the Sun arrives later than local noon (12:00), causing *slow-Sun times.*

We chased the subsolar point as it moved from the Equator to the Tropic of Cancer between the March equinox and the June solstice. According to the analemma chart, the Sun's declination moves from about 1° N latitude to 15° N between March 23 and May 1. We were in these latitudes from April 25 to May 1.

So, when did we come closest to the subsolar point? The Sun was overhead (nearly a 90° angle, perpendicular to a flat surface) on May 1 when we were at 14.8° N, on Fogo Island, Cape Verde. The equation of time was two minutes fast. We measured the Sun's altitude and checked the analemma chart. We determined our location using a GPS receiver.

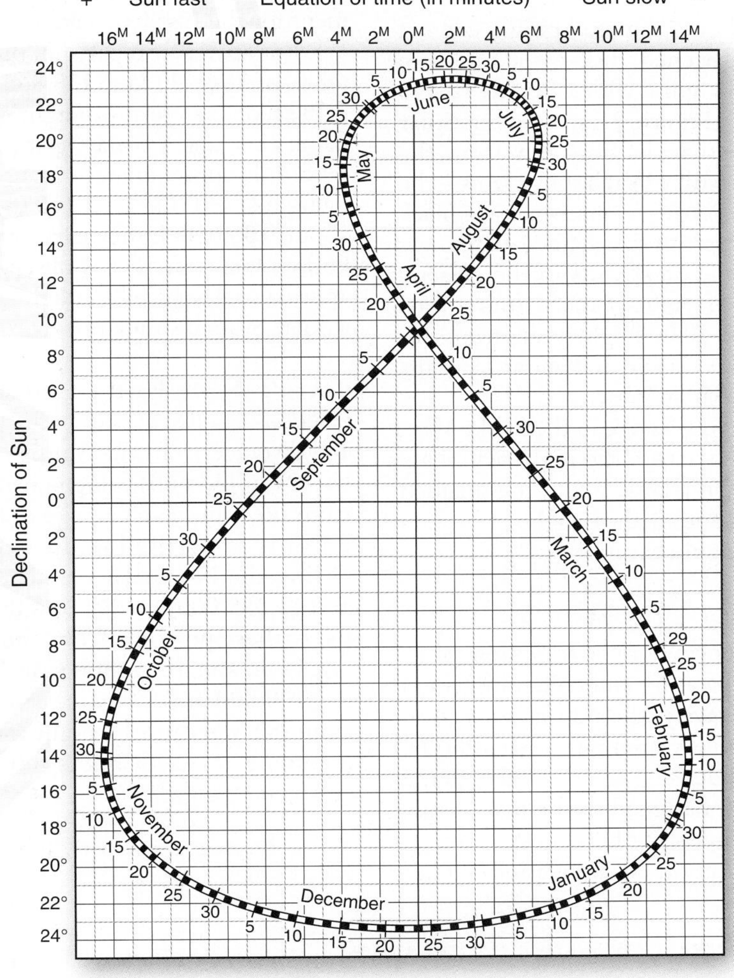

FIGURE GN 2.1 The analemma chart. Knowing a date, you can determine the Sun's declination and the equation of time. Begin your use of the analemma by looking up your birthday. Where is the subsolar point on that date?

The Universe is populated with at least 125 billion galaxies. One of these is our own Milky Way Galaxy, and it contains about 200 billion stars. Among these stars is an average yellow star, the Sun, although the dramatic *SOHO* satellite image in Figure 2.2 seems anything but average! Our Sun radiates energy in all directions and upon its family of orbiting planets. Of special interest to us is the solar energy that falls on the third planet, our immediate home.

In this chapter: Incoming solar energy arrives at the top of Earth's atmosphere, establishing the pattern of energy input that drives Earth's physical systems and influences our lives daily. This solar energy input to the atmosphere, plus Earth's tilt and rotation, produce daily, seasonal, and annual patterns of changing daylength and Sun angle. The Sun is the ultimate energy source for most life processes in our biosphere.

The Solar System, Sun, and Earth

Our Solar System is located on a remote, trailing edge of the **Milky Way Galaxy**, a flattened, disk-shaped mass in the form of a barred-spiral—a spiral with a slightly barred or elongated core of stars (Figure 2.1a, b). Our Solar System is embedded more than halfway out from the galactic center, in one of the Milky Way's spiral arms—the Orion Spur of the Sagittarius Arm. A super-massive black hole some 2 million solar masses in size, named *Sagittarius A** (pronounced "*Sagittarius A Star*"), sits in the galactic center. Our Solar System of eight planets, four dwarf planets, and asteroids is some 30,000 light-years from this black hole at the center of the Galaxy, and about 15 light-years above the plane of the Milky Way.

From our Earth-bound perspective in the Milky Way, the Galaxy appears to stretch across the night sky like a narrow band of hazy light. On a clear night, the unaided eye can see only a few thousand of these billions of stars gathered about us in our "neighborhood."

Solar System Formation and Structure

According to prevailing theory, our Solar System condensed from a large, slowly rotating and collapsing cloud of dust and gas, a *nebula*. **Gravity**, the mutual attraction exerted by the mass of an object upon all other objects, was the key force in this condensing solar nebula. As the nebular cloud organized and flattened into a disk shape, the early *protosun* grew in mass at the center, drawing more matter to it. Small accretion (accumulation) eddies swirled at varying distances from the center of the solar nebula; these were the *protoplanets*.

The **planetesimal hypothesis**, or *dust-cloud hypothesis*, explains how suns condense from nebular clouds with planetesimals forming in orbits about the central mass. Astronomers study this formation process in other parts of the Galaxy, where planets are observed orbiting distant stars. By spring 2010, astronomers had discovered more than 450 planets orbiting other stars.

Closer to home in our Solar System, some 165 moons (planetary satellites) are in orbit about six of the eight planets. As of 2010, the new satellite count for the four outer planets was: Jupiter, 63 moons; Saturn, 62 moons; Uranus, 27 moons; and Neptune, 13 moons.

Dimensions and Distances The **speed of light** is 300,000 kmps (kilometers per second), or 186,000 mps (miles per second)*—in other words, about 9.5 trillion kilometers per year, or nearly 6 trillion miles per year. This tremendous distance that light travels in a year is known as a *light-year*, and it is used as a unit of measurement for the vast Universe.

For spatial comparison, our Moon is an average distance of 384,400 km (238,866 mi) from Earth, or about 1.28 seconds in terms of light speed—for the *Apollo* astronauts this was a 3-day space voyage. Our entire Solar System is approximately 11 hours in diameter, measured by light speed (Figure 2.1c). In contrast, the Milky Way is about 100,000 light-years from side to side, and the known Universe that is observable from Earth stretches approximately 12 billion light-years in all directions. (See a Solar System simulator at http://space.jpl.nasa.gov/.)

Earth's Orbit Earth's orbit around the Sun is presently *elliptical*—a closed, oval path (Figure 2.1d). Earth's average distance from the Sun is approximately 150 million km (93 million mi), which means that light reaches Earth from the Sun in an average of 8 minutes and 20 seconds. Earth is at **perihelion**, its closest position to the Sun, during the Northern Hemisphere winter on January 3 at 147,255,000 km (91,500,000 mi). It is at **aphelion**, its farthest position from the Sun, during the Northern Hemisphere summer on July 4 at 152,083,000 km (94,500,000 mi). This seasonal difference in distance from the Sun causes a slight variation in the solar energy incoming to Earth but is not an immediate reason for seasonal change.

*In more precise numbers, light speed is 299,792 kmps, or 186,282 mps.

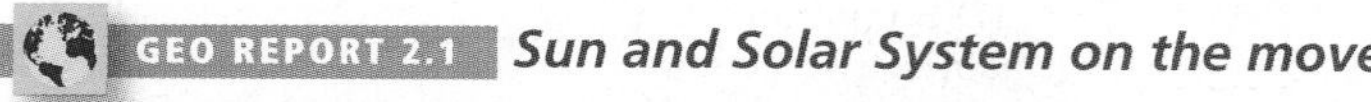

GEO REPORT 2.1 ***Sun and Solar System on the move***

The section above describes formation of the Sun and Earth, some 4.6 billion years ago. During this vast time span, the Sun, Earth and other planets completed 27 orbital trips around the Milky Way Galaxy. When you combine this travel distance with Earth's orbital revolution speed about the Sun of 107,280 kmph (66,660 mph) and Earth's equatorial rotation on its axis of 1675 kmph (1041 mph), you get an idea that "sitting still" is a relative term.

MG™
Animation
Nebular Hypothesis

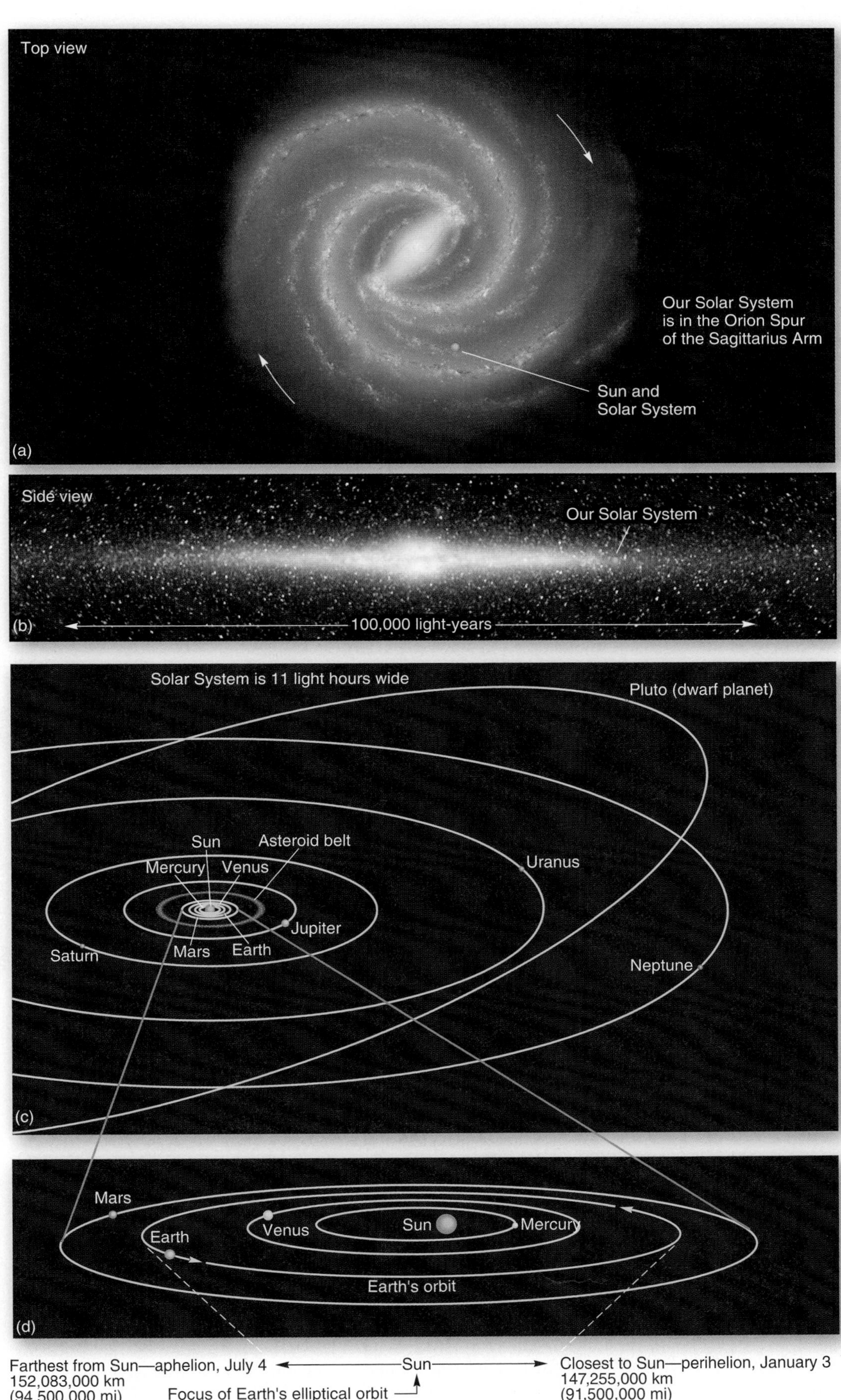

FIGURE 2.1
Milky Way Galaxy, Solar System, and Earth's orbit.
(a) The Milky Way Galaxy viewed from above in an artists conception, and (b) an image in cross-section side view. (c) All of the planets except dwarf Pluto have orbits closely aligned to the plane of the ecliptic. In 2006, Pluto was reclassified as part of the Kuiper Asteroid Belt. (d) The four inner terrestrial planets and the structure of Earth's elliptical orbit, illustrating perihelion (closest) and aphelion (farthest) positions during the year. Have you ever observed the Milky Way Galaxy in the night sky? [(a) and (b) illustration, courtesy of NASA/JPL.]

The structure of Earth's orbit is not a constant, but instead changes over long periods. As shown in Chapter 17, Figure 17.31, Earth's distance from the Sun varies more than 17.7 million km (11 million mi) during a 100,000-year cycle, placing it closer or farther at different periods in the cycle.

Solar Energy: From Sun to Earth

Our Sun is unique to us but is a commonplace star in the Galaxy. It is average in temperature, size, and color when compared with other stars, yet it is the ultimate energy source for most life processes in our biosphere.

The Sun captured about 99.9% of the matter from the original solar nebula. The remaining 0.1% of the matter formed all the planets, their satellites, asteroids, comets, and debris. Consequently, the dominant object in our region of space is the Sun. In the entire Solar System, it is the only object having the enormous mass needed to sustain a nuclear reaction in its core and produce radiant energy.

The solar mass produces tremendous pressure and high temperatures deep in its dense interior. Under these conditions, the Sun's abundant hydrogen atoms are forced together and pairs of hydrogen nuclei are joined in the process of **fusion**. In the fusion reaction, hydrogen nuclei form helium, the second-lightest element in nature, and enormous quantities of energy are liberated—literally, disappearing solar mass becomes energy.

A sunny day can seem so peaceful, certainly unlike the violence taking place on the Sun. The Sun's principal outputs consist of the *solar wind* and radiant energy in portions of the *electromagnetic spectrum*. Let us trace each of these emissions across space to Earth.

Solar Activity and Solar Wind

The Sun constantly emits clouds of electrically charged particles (principally, hydrogen nuclei and free electrons) that surge outward in all directions from the Sun's surface. This stream of energetic material travels more slowly than light—at about 50 million km (31 million mi) a day—taking approximately 3 days to reach Earth. The term **solar wind** was first applied to this phenomenon in 1958. The *Voyager* and *Pioneer* spacecraft launched in the 1970s are now far beyond our Solar System and have yet to escape the solar wind.

The Sun's most conspicuous features are large **sunspots**, caused by magnetic storms on the Sun. Individual sunspots may range in diameter from 10,000 to 50,000 km (6200 to 31,000 mi), with some growing as large as 160,000 km (100,000 mi), more than 12 times Earth's diameter. These surface disturbances produce flares and prominences (Figure 2.2a). In addition, outbursts of charged material, re-

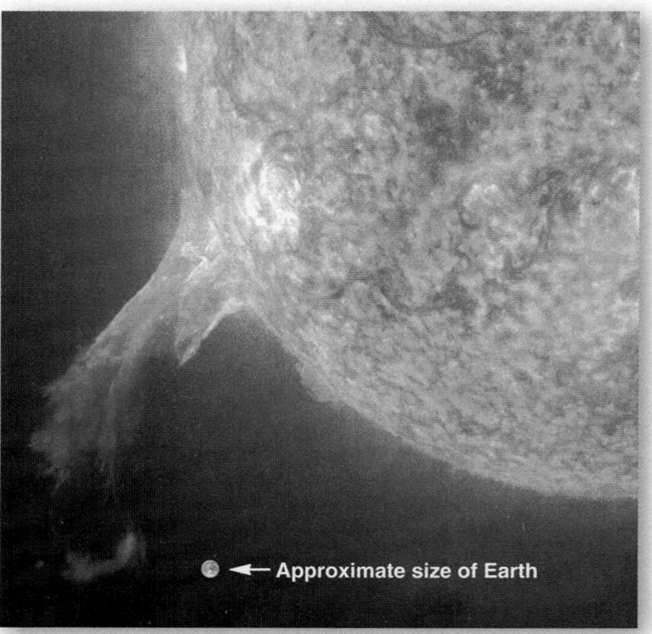

(a)

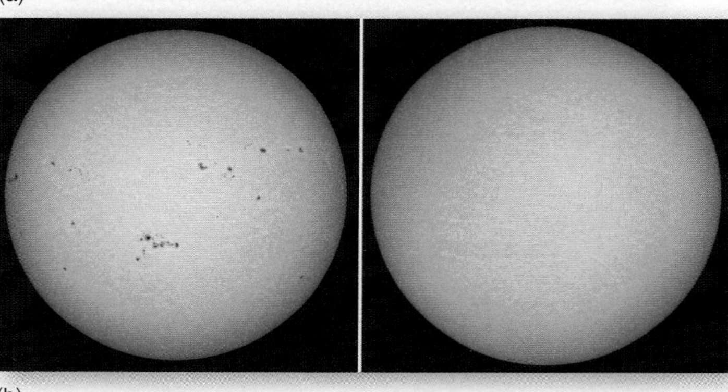

(b)

FIGURE 2.2 Image of the Sun and sunspots. (a) A dramatic Sun captured by the *SOHO* satellite. A prominence rises into the Sun's corona. Earth is shown for scale and is actually far smaller than an average sunspot. (b) The Sun at solar maximum in July 2000 and minimum in March 2009. Imaged by MDI (Michelson Doppler Imager) aboard satellite *SOHO*. [Images courtesy of *SOHO/EIT* Consortium (NASA and ESA); see http://sohowww.nascom.nasa.gov/.]

GEO REPORT 2.2 *Recent solar cycles*

In recent sunspot cycles, a solar minimum occurred in 1976 and a solar maximum in 1979, with more than 100 sunspots visible. Another minimum was reached in 1986, and an extremely active solar maximum followed in 1990–1991, with more than 200 sunspots visible at some time during the year. A sunspot minimum was in 1997, an intense maximum in 2000–2001 of more than 200. The forecast for the maximum in 2013 is that it will be less than the 2000 max. The present solar cycle began in 2008 and carries the name *Cycle 24*.

ferred to as *coronal mass ejections*, contribute to the solar wind flow of material into space.

A regular cycle exists for sunspot occurrences, averaging 11 years from maximum to maximum; however, the cycle may vary from 7 to 17 years (Figure 2.2b). A minimum in 2008 and a forecasted maximum in 2013 roughly maintains the average. (For more on the sunspot cycle, see http://solarscience.msfc.nasa.gov/SunspotCycle.shtml; and for the latest space weather, see http://www.spaceweather.com/.)

Solar Wind Effects The charged particles of the solar wind first interact with Earth's magnetic field as they approach Earth. The **magnetosphere** is a magnetic field surrounding Earth, generated by dynamo-like motions within our planet. The magnetosphere deflects the solar wind toward both of Earth's poles so that only a small portion of it enters the upper atmosphere.

Because the solar wind does not reach Earth's surface, research on this phenomenon must be conducted in space. In 1969, the *Apollo XI* astronauts exposed a piece of foil on the lunar surface as a solar wind experiment (Figure 2.3). When examined back on Earth, the exposed foil exhibited particle impacts that confirmed the character of the solar wind.

This interaction of the solar wind and the upper layers of Earth's atmosphere produces the remarkable **auroras** that occur toward both poles. These lighting effects are the *aurora borealis* (northern lights) and *aurora australis* (southern lights) in the upper atmosphere, 80–500 km (50–300 mi) above Earth's surface. They appear as folded sheets of green, yellow, blue, and red light that undulate across the skies of high latitudes poleward of 65°, as shown in Figure 2.4 During the 2001 solar maximum, auroras were visible as far south as Jamaica, Texas, and California.

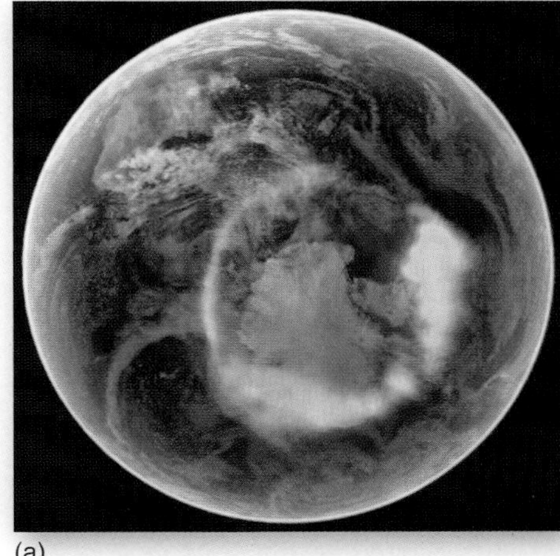

(a)

(b)

FIGURE 2.4 Auroras from an orbital perspective. (a) Aurora australis and (b) aurora borealis, as seen from orbit. [(a) Space Shuttle *Discovery* STS–114 and Image spacecraft, GSFC/NASA; (b) ISS astronaut Don Pettit/NASA.]

FIGURE 2.3 Astronaut and solar wind experiment. Without a protective atmosphere, the lunar surface receives solar wind charged particles and all the Sun's electromagnetic radiation. An *Apollo XI* astronaut in 1969 deploys a sheet of foil in the solar wind experiment. Earth-bound scientists analyzed the foil upon the astronauts' return. Why wouldn't this experiment work if we set it up on Earth's surface? [NASA.]

The solar wind disrupts certain radio broadcasts and some satellite transmissions and can cause overloads on Earth-based electrical systems. Astronauts working on the International Space Station in November 2003 had to take shelter in the shielded Service Module during a particularly strong outburst. Research continues as to possible links between the sunspot cycle and patterns of drought and periods of wetness—the sunspot–weather connection.

Our understanding of the solar wind is increasing dramatically as data are collected by a variety of satellites: the *SOHO* (*Solar and Heliospheric Observatory*), *FAST* (*Fast Auroral Snapshot*), *WIND*, *Ulysses*, *Genesis* with solar wind collectors, and the new *SDO* (*Solar*

Dynamics Observatory), among others. All satellite data are available on the Internet. (For auroral activity, see http://www.swpc.noaa.gov/pmap/; and for forecasts, see http://www.gi.alaska.edu/.)

Electromagnetic Spectrum of Radiant Energy

The key essential solar input to life is electromagnetic energy of various wavelengths. Solar radiation occupies a portion of the **electromagnetic spectrum** of radiant energy. This radiant energy travels at the speed of light to Earth. The total spectrum of this radiant energy is made up of different wavelengths. Figure 2.5 shows that a **wavelength** is the distance between corresponding points on any two successive waves. The number of waves passing a fixed point in 1 second is the *frequency*.

The Sun emits radiant energy composed of 8% ultraviolet, X-ray, and gamma-ray wavelengths; 47% visible light wavelengths; and 45% infrared wavelengths. A portion of the electromagnetic spectrum is illustrated in Figure 2.6, with wavelengths increasing from the top of the illustration to the bottom. Note the wavelengths at which various phenomena and human applications of energy occur.

An important physical law states that all objects radiate energy in wavelengths related to their individual surface temperatures: the hotter the object, the shorter the wavelengths emitted. This law holds true for the Sun and Earth. Figure 2.7 shows that the hot Sun radiates shorter wavelength energy, concentrated around 0.5 μm (micrometer).

The Sun's surface temperature is about 6000 K (6273°C, 11,459°F), and its emission curve shown in the figure is similar to that predicted for an idealized 6000 K surface, or *blackbody radiator* (shown in Figure 2.7). An ideal blackbody absorbs all the radiant energy that it receives and subsequently emits all that it receives. A hotter object like the Sun emits a much greater amount of energy per unit area of its surface than does a similar area of a cooler object like Earth. Shorter wavelength emissions are dominant at these higher temperatures.

Although cooler than the Sun, Earth radiates nearly all that it absorbs and acts as a blackbody (also shown in Figure 2.7). Because Earth is a cooler radiating body, longer wavelengths are emitted as radiation mostly in the infrared portion of the spectrum, centered around 10.0 μm. Some of the

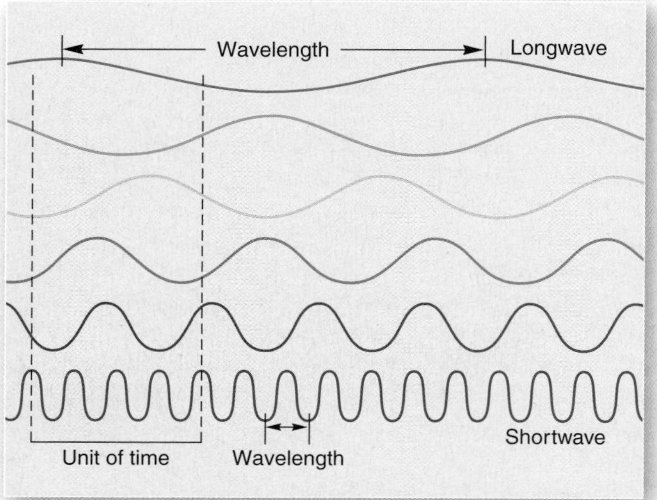

FIGURE 2.5 Wavelength and frequency.
Wavelength and frequency are two ways of describing electromagnetic wave motion. Short wavelengths are higher in frequency; long wavelengths are lower in frequency.

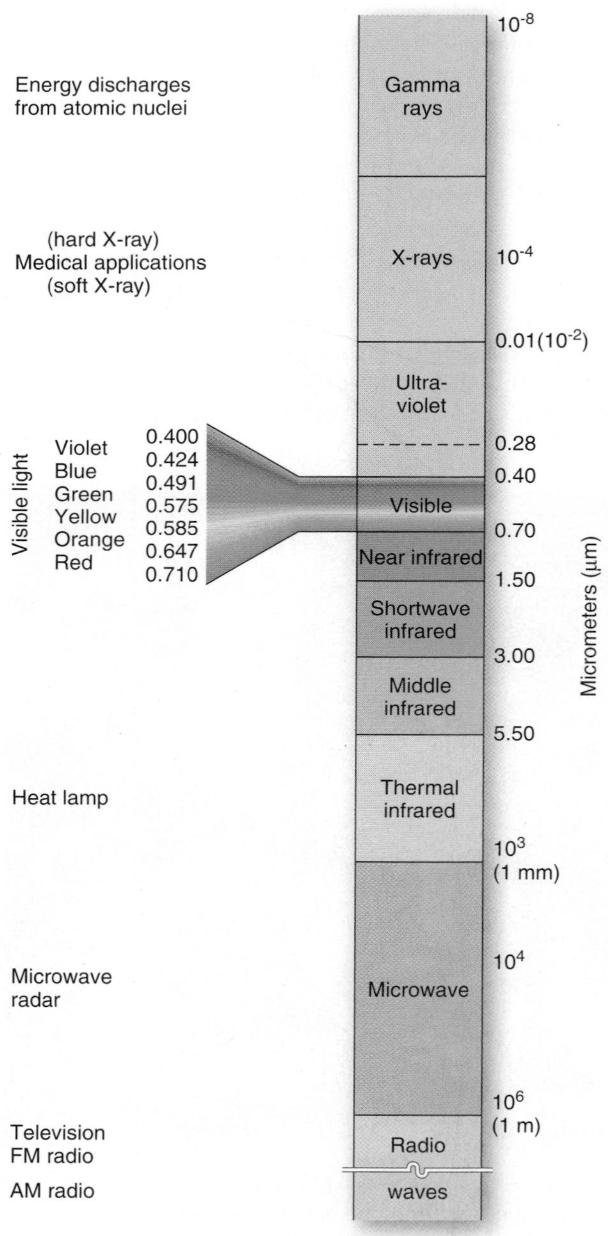

FIGURE 2.6 A portion of the electromagnetic spectrum of radiant energy.
Shorter wavelengths are at top; longer wavelengths toward the bottom.

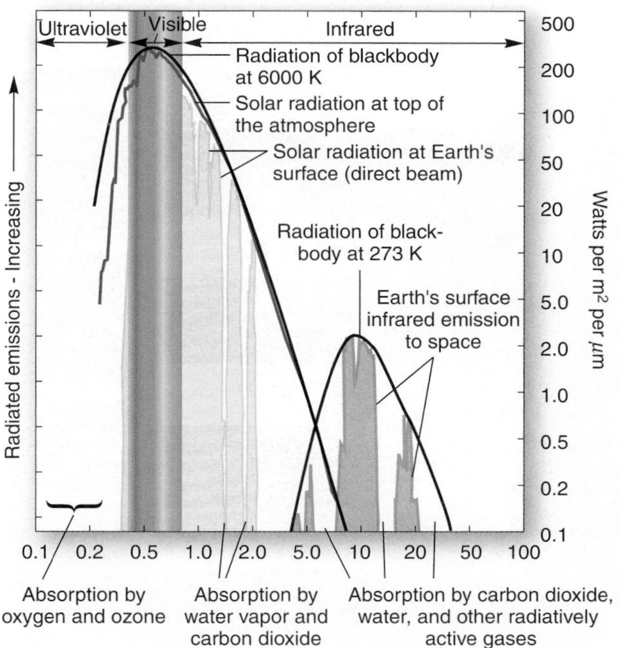

FIGURE 2.7 Solar and terrestrial energy distribution by wavelength.
A hotter Sun radiates shorter wavelengths, whereas a cooler Earth emits longer wavelengths. Dark lines represent ideal blackbody curves for the Sun and Earth. The dropouts in the plot lines for solar and terrestrial radiation represent absorption bands of water vapor, water, carbon dioxide, oxygen, ozone (O_3), and other gases. [Adapted from W. D. Sellers, *Physical Climatology* (Chicago: University of Chicago Press), p. 20. Used by permission.]

Animation Electromagnetic Spectrum and Plants

atmospheric gases vary in their response to radiation received, being transparent to some while absorbing others.

Figure 2.8 illustrates the flows of energy into and out of Earth systems. To summarize, the Sun's radiated energy is *shortwave radiation* that peaks in the short visible wavelengths, whereas Earth's radiated energy is *longwave radiation* concentrated in infrared wavelengths. In Chapter 4, we see that Earth, clouds, sky, ground, and things that are terrestrial radiate longer wavelengths in contrast to the Sun.

Incoming Energy at the Top of the Atmosphere

The region at the top of the atmosphere, approximately 480 km (300 mi) above Earth's surface, is the **thermopause** (see Figure 3.2). It is the outer boundary of Earth's energy system and provides a useful point at which to assess the arriving solar radiation before it is diminished by scattering and absorption in passage through the atmosphere.

Earth's distance from the Sun results in its interception of only one two-billionth of the Sun's total energy output. Nevertheless, this tiny fraction of energy from the Sun is an enormous amount of energy flowing into Earth's systems. Solar radiation that reaches a horizontal plane at Earth is **insolation**, derived from the words *in*coming *sol*ar radi*ation*. Insolation specifically applies to radiation arriving at Earth's atmosphere and surface. Insolation at the top of the atmosphere is expressed as the *solar constant.*

Solar Constant Knowing the amount of insolation incoming to Earth is important to climatologists and other scientists. The **solar constant** is the average insolation received at the thermopause when Earth is at its average distance from the Sun, a value of 1372 W/m² (watts per square meter).* As we follow insolation through the

*A *watt* is equal to 1 joule (a unit of energy) per second and is the standard unit of power in the International System of Units (SI). (See the conversion tables in Appendix C of this text for more information on measurement conversions.) In nonmetric *calorie* heat units, the solar constant is expressed as approximately 2 calories per square centimeter per minute, or 2 *langleys* per minute (a langley being 1 cal/cm²). A calorie is the amount of energy required to raise the temperature of 1 gram of water (at 15°C) 1 degree Celsius and is equal to 4.184 joules.

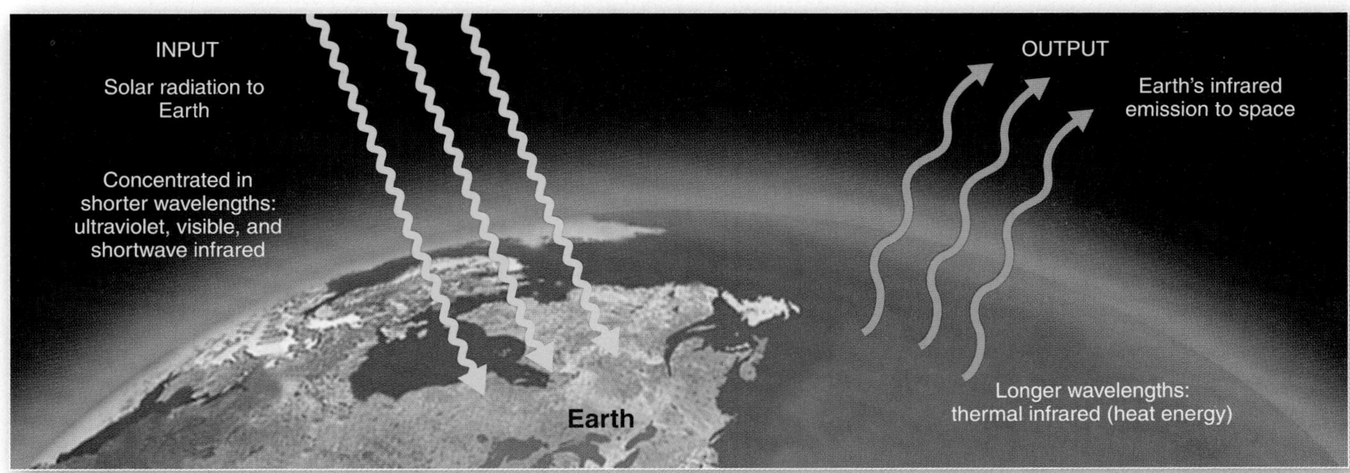

FIGURE 2.8 Earth's energy budget simplified.
Inputs of shorter wavelengths arrive at Earth from the Sun. Outputs of longer wavelengths of infrared radiate to space from Earth.

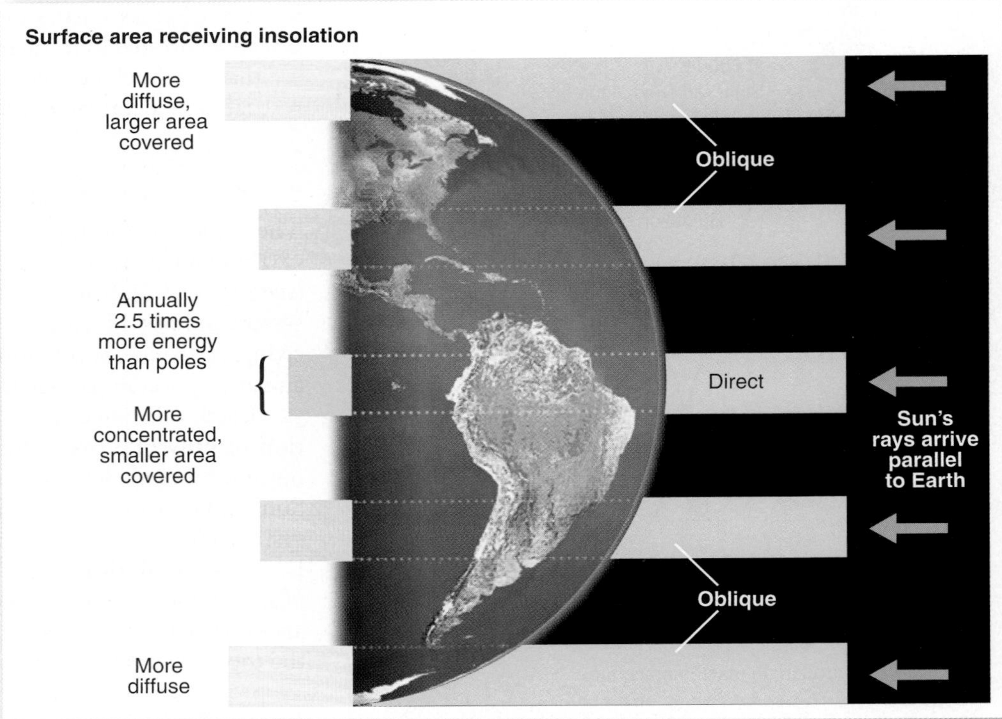

FIGURE 2.9 Insolation receipts and Earth's curved surface.
Solar insolation angles determine the concentration of energy receipts by latitude.

atmosphere to Earth's surface (Chapters 3 and 4), we see that the value of the solar constant is reduced by half or more through reflection, scattering, and absorption of shortwave radiation.

Uneven Distribution of Insolation Earth's curved surface presents a continually varying angle to the incoming parallel rays of insolation (Figure 2.9). Differences in the angle of solar rays at each latitude result in an uneven distribution of insolation and heating. The only point receiving insolation perpendicular to the surface (from directly overhead) is the **subsolar point**. During the year this point occurs at lower latitudes between the tropics (about 23.5° N and 23.5° S), where the energy received is more concentrated. All other places away from the subsolar point receive insolation at an angle less than 90° and thus experience more diffuse energy; this effect is pronounced at higher latitudes.

The thermopause above the equatorial region receives 2.5 times more insolation annually than the thermopause above the poles. Of lesser importance is the fact that the lower-angle solar rays toward the poles must pass through a greater thickness of atmosphere, resulting in greater losses of energy due to scattering, absorption, and reflection.

Figure 2.10 illustrates the daily variations throughout the year of energy at the top of the atmosphere for four

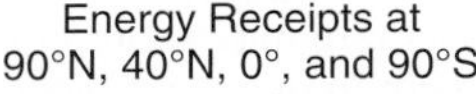

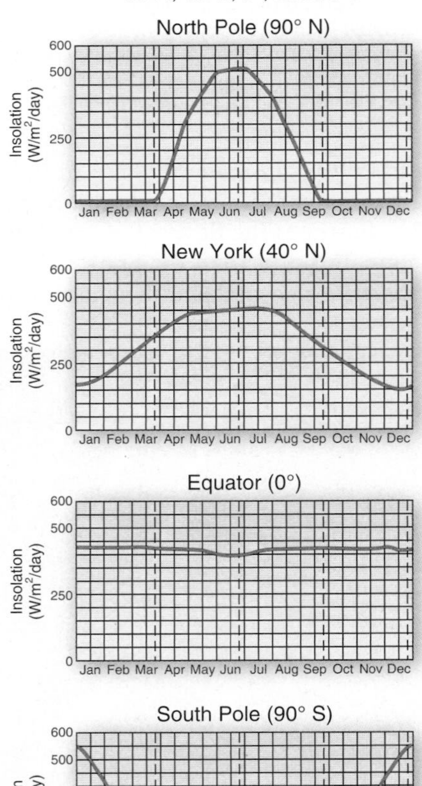

FIGURE 2.10 Daily insolation received at the top of the atmosphere.
The total daily insolation received at the top of the atmosphere is charted in watts per square meter per day for four locations (1 $W/m^2/day$ = 2.064 $cal/cm^2/day$).

locations in watts per square meter (W/m^2). The graphs show the seasonal changes in insolation from the equatorial regions northward and southward to the poles. In June, the North Pole receives slightly more than 500 W/m^2 per day, which is more than is ever received at 40° N latitude or at the equator. Such high values result from long daylengths at the poles in summer: 24 hours a day, compared with only 15 hours of daylight at 40° N latitude and 12 hours at the equator. However, at the poles the summertime Sun at noon is low in the sky, so a daylength twice that of the equator yields only about a 100 W/m^2 difference.

In December, the pattern reverses, as shown on the graph. Note that the top of the atmosphere at the South Pole receives even more insolation than the North Pole does in June (more than 550 W/m^2). This is a function of Earth's closer location to the Sun at perihelion (January 3 in Figure 2.1d). Along the equator, two maximum periods of approximately 430 W/m^2 occur at the spring and fall equinoxes, when the subsolar point is at the equator.

Global Net Radiation Earth Radiation Budget Experiment (ERBE) instruments aboard several satellites measured shortwave and longwave flows of energy at the top of the atmosphere. ERBE sensors collected the data used to develop the map in Figure 2.11. This map shows *net radiation*, or the balance between incoming shortwave and outgoing longwave radiation—energy inputs minus energy outputs.

Note the latitudinal energy imbalance in net radiation on the map—positive values in lower latitudes and negative values toward the poles. In middle and high latitudes, approximately poleward of 36° N and S latitudes, net radiation is negative. This occurs in these higher latitudes because Earth's climate system loses more energy to space than it gains from the Sun, as measured at the top of the atmosphere. In the lower atmosphere, these polar energy deficits are offset by flows of energy from tropical energy surpluses (as we will see in Chapters 4 and 6). The largest net radiation values, averaging 80 W/m^2, are above the tropical oceans along a narrow equatorial zone. Net radiation minimums are lowest over Antarctica.

Of interest is the −20 W/m^2 area over the Sahara region of North Africa. Here, typically clear skies—which permit great longwave radiation losses from Earth's surface—and light-colored reflective surfaces work together to reduce net radiation values at the thermopause. In other regions, clouds and atmospheric pollution in the lower atmosphere also affect net radiation patterns at the top of the atmosphere by reflecting more shortwave energy to space.

The atmosphere and ocean form a giant heat engine, driven by differences in energy from place to place

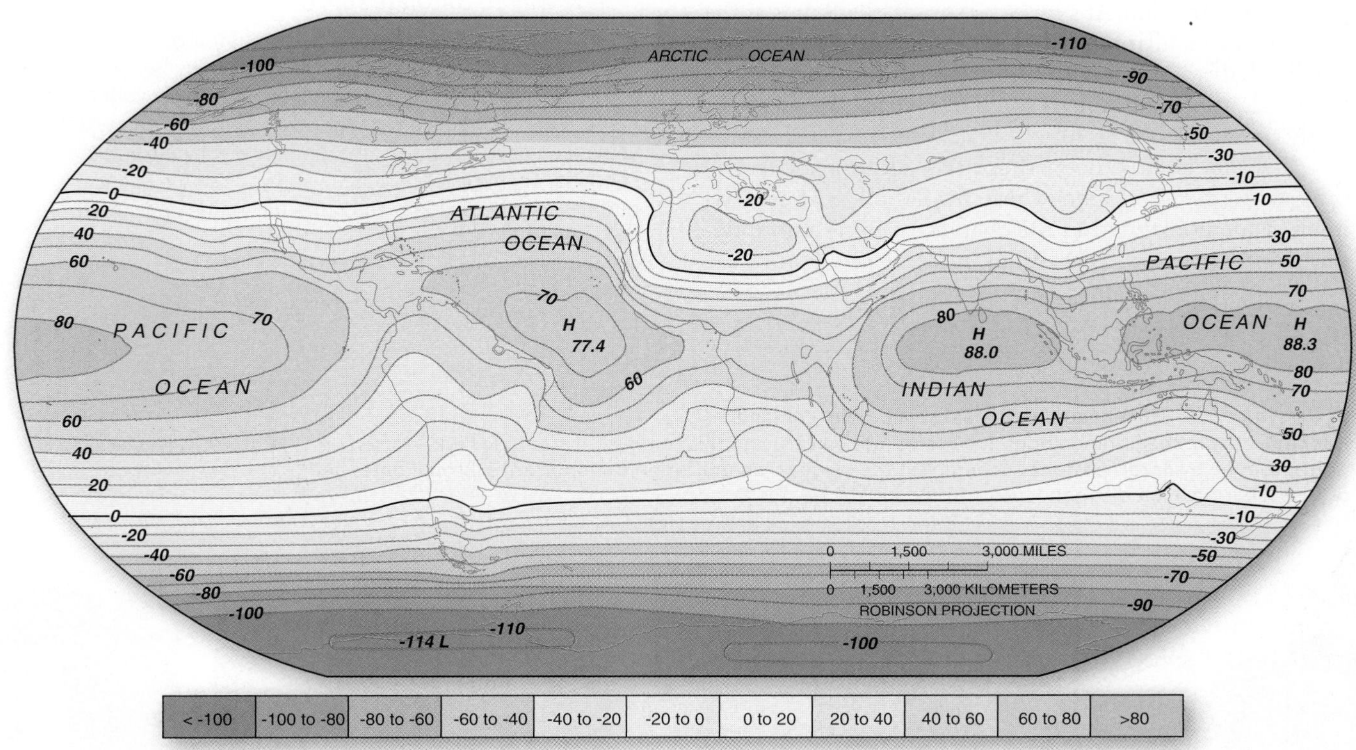

FIGURE 2.11 Daily net radiation patterns at the top of the atmosphere. Averaged daily net radiation flows measured at the top of the atmosphere by the Earth Radiation Budget Experiment (ERBE). Units are W/m^2. [Data for map courtesy of GSFC/NASA.]

which cause major circulations within the lower atmosphere and in the ocean. These circulations include global winds, ocean currents, and weather systems—subjects to follow in Chapters 6 and 8. As you go about your daily activities, let these dynamic natural systems remind you of the constant flow of solar energy through the environment.

Having examined the flow of solar energy to Earth and the top of the atmosphere, let us now look at how seasonal changes affect the distribution of insolation as Earth orbits the Sun during the year.

The Seasons

Earth's periodic rhythms of warmth and cold, dawn and daylight, twilight and night, have fascinated humans for centuries. In fact, many ancient societies demonstrated an intense awareness of seasonal change and formally commemorated these natural energy rhythms with festivals, monuments, ground markings, and calendars (Figure 2.12). Such seasonal monuments and calendar markings are found worldwide, including thousands of sites in North America, demonstrating an ancient awareness of seasons and astronomical relations.

Seasonality

Seasonality refers to both the seasonal variation of the Sun's position above the horizon and changing daylengths during the year. Seasonal variations are a response to changes in the Sun's **altitude**, or the angle between the horizon and the Sun. At sunrise or sunset, the Sun is at the horizon, so its altitude is 0°. If during the day, the Sun reaches halfway between the horizon and directly overhead, it is at 45° altitude. If the Sun reaches the point directly overhead, it is at 90° altitude.

The Sun is directly overhead (90° altitude, or *zenith*) only at the *subsolar point*, where insolation is at a maximum, as demonstrated in the chapter-opening photo described in *Geosystems Now*. At all other surface points, the Sun is at a lower altitude angle, producing more diffuse insolation.

The Sun's **declination** is the latitude of the subsolar point. Declination annually migrates through 47° of latitude, moving between the Tropic of Cancer and Tropic of Capricorn latitudes. Other than Hawai'i, which is between 19° N and 22° N, the subsolar point does not reach the continental United States or Canada; all other states and provinces are too far north.

Also, seasonality means changing **daylength**, or duration of exposure to insolation. Daylength varies during the year, depending on latitude. The equator always receives equal hours of day and night: If you live in Ecuador, Kenya, or Singapore, every day and night is 12 hours long, year-round. People living along 40° N latitude (Philadelphia, Denver, Madrid, Beijing), or 40° S latitude (Buenos Aires, Capetown, Melbourne), experience about 6 hours' difference in daylight between winter (9 hours) and summer (15 hours). At 50° N or S latitude (Winnipeg, Paris, Falkland, or Malvinas Islands), people experience almost 8 hours of annual daylength variation.

At the North and South poles, the range of daylength is extreme, with a 6-month period of no insolation, beginning with weeks of twilight, then darkness, then on to weeks of pre-dawn. Following sunrise, daylight covers a 6-month period of continuous 24-hour insolation—literally the poles experience one long day and one long night each year!

CRITICAL THINKING 2.1

A way to calculate sunrise and sunset

For a useful sunrise and sunset calculator for any location, go to http://www.srrb.noaa.gov/highlights/sunrise/sunrise.html, select a city near you or select "Enter lat/long" and enter your coordinates, enter your time difference offset from UTC and whether you are on daylight saving time, and enter the date you are checking. Then, click "Calculate" and you see the equation of time, solar declination, and times for sunrise and sunset. Give this a try.

FIGURE 2.12 Stonehenge on the Salisbury Plain, England.
Neolithic peoples beginning about 5000 years ago constructed this complex of standing rock—*sarsen* uprights, topped by *lintel* capstones. Inherent in the layout is an indication of important seasonal dates and precise lunar-cycle predictions. [Bobbé Christopherson.]

Reasons for Seasons

Seasons result from variations in the Sun's *altitude* above the horizon, the Sun's *declination* (latitude location of the subsolar point), and *daylength* during the year. These in turn are created by several physical factors that operate in concert: Earth's *revolution* in orbit around the Sun, its daily *rotation* on its axis, its *tilted axis*, the unchanging *orientation of its axis*, and its *sphericity* (Table 2.1). Of course, the essential ingredient is having a single source of radiant energy—the Sun. We now look at each of these factors individually. As we do, please note the distinction between revolution—Earth's travel around the Sun—and rotation—Earth's spinning on its axis (Figure 2.13).

Revolution Earth's orbital **revolution** about the Sun is shown in Figure 2.1d and 2.13. Earth's speed in orbit averages 107,280 kmph (66,660 mph). This speed, together with Earth's distance from the Sun, determines the time required for one revolution around the Sun and, therefore, the length of the year and duration of the seasons. Earth completes its annual revolution in 365.2422 days. This number is based on a *tropical year*, measured from equinox to equinox, or the elapsed time between two crossings of the equator by the Sun.

The Earth-to-Sun distance from aphelion to perihelion might seem a seasonal factor, but it is not significant. It varies about 3% (4.8 million km, or 3 million mi) during the year, amounting to a 50 W/m^2 difference between local polar summers. Remember that the Earth–Sun distance averages 150 million km (93 million mi).

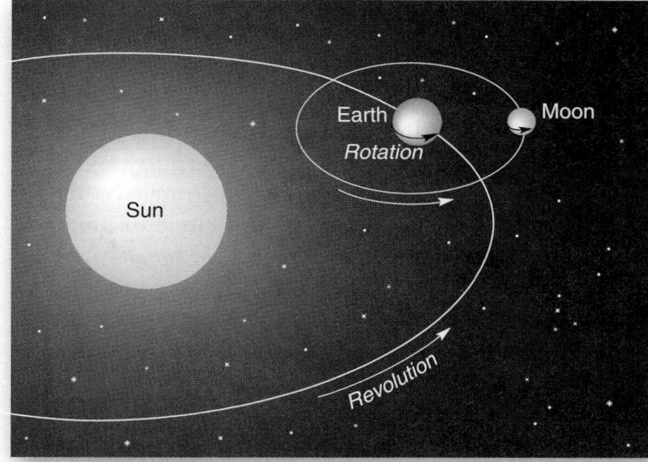

FIGURE 2.13 Earth's revolution and rotation.
Earth's *revolution* about the Sun and *rotation* on its axis, viewed from above Earth's orbit. Note the Moon's rotation on its axis and revolution are counterclockwise, as well.

TABLE 2.1 Five Reasons for Seasons

Factor	Description
Revolution	Orbit around the Sun; requires 365.24 days to complete at 107,280 kmph (66,660 mph)
Rotation	Earth turning on its axis; takes approximately 24 hours to complete
Tilt	Axis is aligned at about 23.5° angle from a perpendicular to the plane of the ecliptic (the plane of Earth's orbit)
Axial parallelism	Remains in a fixed alignment, with Polaris directly overhead at the North Pole throughout the year
Sphericity	Appears as an oblate spheroid to the Sun's parallel rays; the geoid

Rotation Earth's **rotation**, or turning on its axis, is a complex motion that averages slightly less than 24 hours in duration. Rotation determines daylength, creates the apparent deflection of winds and ocean currents, and produces the twice-daily rise and fall of the ocean tides in relation to the gravitational pull of the Sun and the Moon.

Earth rotates about its **axis**, an imaginary line extending through the planet from the geographic North Pole to the South Pole. When viewed from above the North Pole, Earth rotates counterclockwise around this axis. Viewed from above the equator, Earth rotates west to east, or eastward. This eastward rotation creates the Sun's *apparent* westward daily journey from sunrise in the east to sunset in the west. Of course, the Sun actually remains in a fixed position in the center of our Solar System.

Although every point on Earth takes the same 24 hours to complete one rotation, the linear velocity of rotation at any point on Earth's surface varies dramatically with latitude. The equator is 40,075 km (24,902 mi) long; therefore, rotational velocity at the equator must be approximately 1675 kmph (1041 mph) to cover that distance in one day. At 60° latitude, a parallel is only half the length of the equator, or 20,038 km (12,451 mi) long, so the rotational velocity there is 838 kmph (521 mph). At

 **GEO REPORT 2.3** *Why do we always see the same side of the Moon?*

Note in Figure 2.13 that the Moon both revolves around Earth and rotates on its axis in a counterclockwise direction when viewed from above Earth's North Pole. It does both these motions in the same amount of time. The Moon's speed in orbit varies slightly during the month, whereas its rotation speed is constant, so we see about 59% of the lunar surface during the month, or exactly 50% at any one moment—always the same side.

the poles, the velocity is 0. This variation in rotational velocity establishes the effect of the Coriolis force, discussed in Chapter 6. Table 2.2 lists the speed of rotation for several selected latitudes.

Earth's rotation produces the diurnal (daily) pattern of day and night. The dividing line between day and night is the **circle of illumination** (as illustrated in Figure 2.15). Because this day-night dividing circle of illumination intersects the equator, *daylength at the equator is always evenly divided*—12 hours of day and 12 hours of night. All other latitudes experience uneven daylength through the seasons, except for 2 days a year, on the equinoxes. (Any two great circles on a sphere bisect one another.)

A true day varies slightly from 24 hours, but by international agreement a day is defined as exactly 24 hours, or 86,400 seconds. This average, called *mean solar time*, eliminates predictable variations in rotation and revolution that cause the solar day to change slightly in length throughout the year.

Tilt of Earth's Axis To understand Earth's **axial tilt**, imagine a plane (a flat surface) that intersects Earth's elliptical orbit about the Sun, with half of the Sun and Earth above the plane and half below. Such a plane touching all points of Earth's orbit is the **plane of the ecliptic**. Earth's tilted axis remains fixed relative to this plane as Earth revolves around the Sun. The plane of the ecliptic is important to our discussion of Earth's seasons. Now, imagine a perpendicular (at a 90° angle) line passing through the plane. From this perpendicular, Earth's axis is tilted about 23.5°. It forms a 66.5° angle from the plane itself (Figure 2.14). The axis through Earth's two poles points just slightly off Polaris, which is named, appropriately, the *North Star*.

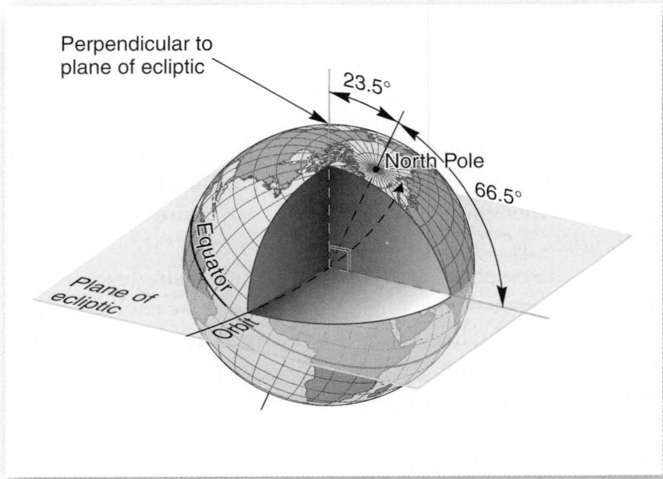

FIGURE 2.14 The plane of Earth's orbit—the ecliptic—and Earth's axial tilt.
Note on the illustration that the plane of the equator is inclined to the plane of the ecliptic at about 23.5°.

This text uses "about" with the tilt angle just described because Earth's axial tilt changes over a complex 41,000-year cycle (see Figure 17.31). The axial tilt ranges roughly between 22° and 24.5° from a perpendicular to the plane of the ecliptic. The present tilt is 23° 27′ (or 66° 33′ from the plane). In decimal numbers, 23° 27′ is approximately 23.45°. For convenience, this is rounded off to a 23.5° tilt (or 66.5° from the plane) in most usage. Scientific evidence shows that the angle of tilt is lessening in its 41,000-year cycle.

TABLE 2.2 Speed of Rotation at Selected Latitudes

Latitude	Speed kmph	Speed (mph)	Cities at Approximate Latitudes
90°	0	(0)	North Pole
60°	838	(521)	Seward, Alaska; Oslo Norway; Saint Petersburg, Russia
50°	1078	(670)	Chibougamau, Québec; Kyyîv (Kiev), Ukraine
40°	1284	(798)	Columbus, Ohio; Beijing, China; Valdivia, Chile
30°	1452	(902)	New Orleans, Louisiana; Pôrto Alegre, Brazil
0°	1675	(1041)	Pontianak, Indonesia; Quito, Ecuador

CRITICAL THINKING 2.2

Astronomical factors vary over long time frames

The variability of Earth's axial tilt, orbit about the Sun, and a wobble to the axis is described in Figure 17.31. Please refer to this figure and compare these changing conditions to the information in Table 2.1 and the related figures in this chapter.

What do you think the effect on Earth's seasons would be if the tilt of the axis was decreased? Or if the tilt was increased? You can take a ball or piece of round fruit and mark the poles and move the sphere around a light bulb, as if revolving it around the Sun. Note the alignment with no tilt and then with a 90° tilt to help complete your analysis. Or, what if Earth's orbit was more circular as opposed to its present elliptical shape? Earth's orbit actually does vary like this in a 100,000-year cycle.

GEO REPORT 2.4 ***Measuring Earth's rotation***

The complexity of Earth's rotation is measured by satellites in precise mathematical orbits: GPS (global positioning system), SLR (satellite-laser ranging), and VLBI (very-long baseline interferometry). All contribute to our knowledge of Earth's rotation. The International Earth Rotation Service at http://www.iers.org/ issues monthly and annual reports. Earth's rotation is gradually slowing, partially owing to the drag of lunar tidal forces. A "day" on Earth today is many hours longer than it was 4 billion years ago.

Hypothetically, if Earth were tilted on its side, with its axis parallel to the plane of the ecliptic, we would experience a maximum variation in seasons worldwide. In contrast, if Earth's axis were perpendicular to the plane of its orbit—that is, with no tilt—we would experience no seasonal changes, with something like a perpetual spring or fall season, and all latitudes would experience 12-hour days and nights. Do you agree?

Axial Parallelism Throughout our annual journey around the Sun, Earth's axis *maintains the same alignment* relative to the plane of the ecliptic and to Polaris and other stars. You can see this consistent alignment in Figure 2.15. If we compared the axis in different months, it would always appear parallel to itself, a condition known as **axial parallelism**.

Sphericity Although Earth is not a perfect sphere, as discussed in Chapter 1, we can refer to Earth's *sphericity* as a part of seasonality, for it produces the uneven receipt of insolation from pole to pole shown in Figures 2.9, 2.10, and 2.11.

All five reasons for seasons are summarized in Table 2.1: revolution, rotation, tilt, axial parallelism, and sphericity. Now, considering all these factors operating together, let's go through the march of the seasons.

Annual March of the Seasons

During the march of the seasons on Earth, daylength is the most obvious way of sensing changes in season at latitudes away from the equator. Daylength is the interval between **sunrise**, the moment when the disk of the Sun first appears above the horizon in the east, and **sunset**, that

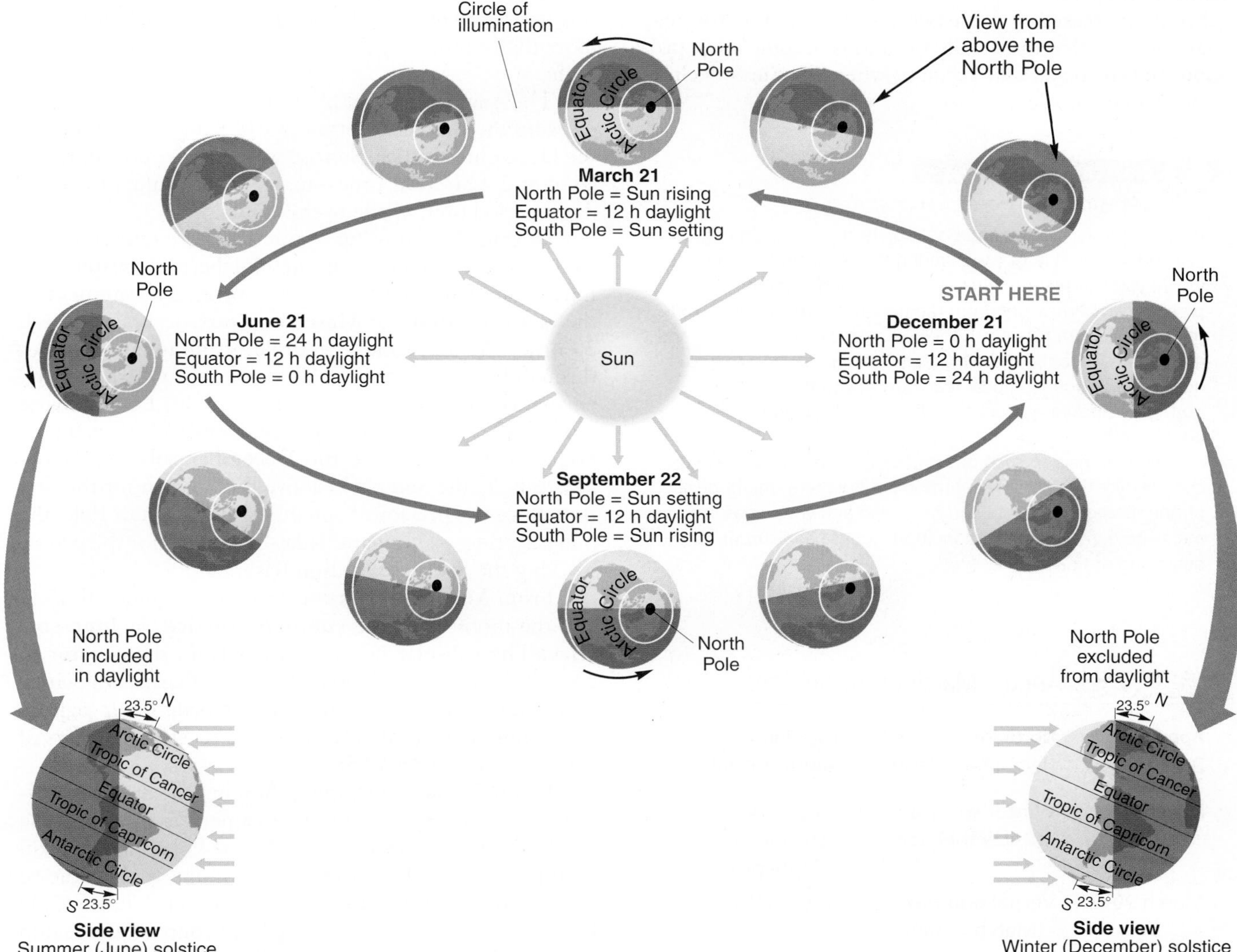

FIGURE 2.15 Annual march of the seasons.
Annual march of the seasons as Earth revolves about the Sun. Shading indicates the changing position of the circle of illumination. Note the hours of daylight for the equator and the poles. As you read, begin on the right side at December 21 and move counterclockwise. Given this illustration, turn to the back cover of this textbook and see whether you can determine the month during which the *Apollo* astronaut made the Earth photo. (The answer is on this book's copyright page.)

MG™ **Animation** Earth-Sun Rotations, Seasons

moment when it totally disappears below the horizon in the west.

The extremes of daylength occur in December and June. The times around December 21 and June 21 are *solstices*. Strictly speaking, the solstices are specific points in time at which the Sun's declination is at its position farthest north at the **Tropic of Cancer**, or south at the **Tropic of Capricorn**. "Tropic" is from *tropicus*, meaning a turn or change, so a tropic latitude is where the Sun's declination appears to stand still briefly (Sun stance, or *sol stice*); then it "turns" and heads toward the other tropic.

Table 2.3 presents the key seasonal anniversary dates during which the specific time of the equinoxes or solstices occur, their names, and the subsolar point location (declination). During the year, places on Earth outside of the equatorial region experience a continuous but gradual shift in daylength, a few minutes each day, and the Sun's altitude increases or decreases a small amount. You may have noticed that these daily variations become more pronounced in spring and autumn, when the Sun's declination changes at a faster rate.

CRITICAL THINKING 2.3

Measure and track seasonal change

Post a reminder in your notebook to check the position of sunrise and sunset at least twice during the semester from a place where you can see the horizon. If you have a magnetic compass, or can borrow one from the geography department, note the degrees from true north for sunrise and sunset. Be sure and ask your teacher about the *magnetic declination* of your location to adjust your compass readings. *Azimuth* is read from north in a clockwise direction—0° and 360° being the same point, north.

If you are new to such observations, the degree of location change over the span of months in a school term might be surprising. Based upon what you have read and your measurements here, make a prediction about sunrise and sunset for the beginning of finals week.

TABLE 2.3 Annual March of the Seasons

Approximate Date	Northern Hemisphere Name	Location of the Subsolar Point
December 21–22	Winter solstice (December solstice)	23.5° S latitude (Tropic of Capricorn)
March 20–21	Vernal equinox (March equinox)	0° (Equator)
June 20–21	Summer solstice (June solstice)	23.5° N latitude (Tropic of Cancer)
September 22–23	Autumnal equinox (September equinox)	0° (Equator)

Figure 2.15 demonstrates the annual march of the seasons and illustrates Earth's relationship to the Sun during the year. Let us begin with December, on the right side of the illustration. On December 21 or 22, at the moment of the **winter solstice** ("winter Sun stance"), or **December solstice**, the circle of illumination excludes the North Pole region from sunlight but includes the South Pole region. The subsolar point is about 23.5° S latitude, the Tropic of Capricorn parallel. The Northern Hemisphere is tilted away from these more direct rays of sunlight—our northern winter—thereby creating a lower angle for the incoming solar rays and thus a more diffuse pattern of insolation.

From about 66.5° N latitude to 90° N (the North Pole), the Sun remains below the horizon the entire day. This latitude (about 66.5° N) marks the **Arctic Circle**, the southernmost parallel (in the Northern Hemisphere) that experiences a 24-hour period of darkness. During this period, twilight and dawn provide some lighting for more than a month at the beginning and end of the Arctic night.

The Antarctic sunset photo in Figure 2.16a was made crossing the Bransfield Strait at 63° S, just a week before the December solstice. Sunrise was only a couple of hours later, at 2 A.M. local time—making the sunless twilight/dawn period only a little over 2 hours!

During the following 3 months, daylength and solar angles gradually increase in the Northern Hemisphere as Earth completes one-fourth of its orbit. The moment of the **vernal equinox**, or **March equinox**, occurs on March 20 or 21. At that time, the circle of illumination passes through both poles so that all locations on Earth experience a 12-hour day and a 12-hour night. People living around 40° N latitude (New York, Denver) have gained 3 hours of daylight since the December solstice. At the North Pole, the Sun peeks above the horizon for the first time since the previous September; at the South Pole, the Sun is setting—a dramatic 3-day "moment" for the people working the Scott–Amundson Base.

From March, the seasons move on to June 20 or 21 and the moment of the **summer solstice**, or **June solstice**. The subsolar point migrates from the equator to 23.5° N latitude, the Tropic of Cancer. Because the circle of illumination now includes the North Polar region, everything north of the Arctic Circle receives 24 hours of daylight—the *Midnight Sun*.

Figure 2.16b is a Midnight Sun in June at 80° N latitude in the Arctic Ocean. The camera is pointing due north, seeing the Sun across the North Pole, some 885 km (550 mi) distant. This seemed so strange at the time to have natural lighting that felt like about 3 P.M. (15:00 hours), yet have clocks at midnight. In contrast, the region from the **Antarctic Circle** to the South Pole (66.5°–90° S latitude) is in darkness. Those working in Antarctica call the June solstice *Midwinter's Day*.

September 22 or 23 is the moment in time of the **autumnal equinox**, or **September equinox**, when Earth's orientation is such that the circle of illumination again

(a)

(b)

FIGURE 2.16 The Midnight Sun and long days.
(a) Antarctic sunset, 11:30 P.M., approaching the continent of Antarctica at 63° S latitude in December. The low Sun angle produced incredible sunset colors for hours. Note a large iceberg and ice-covered mountains in the distance. (b) The Midnight Sun seen from a research ship in the Arctic Ocean at 80° N latitude in June. Can you find the December and June perspectives in Figure 2.15? [Bobbé Christopherson.]

passes through both poles so that all parts of the globe experience a 12-hour day and a 12-hour night. The subsolar point returns to the equator, with days growing shorter to the north and longer to the south. Researchers stationed at the South Pole see the disk of the Sun just rising, ending their 6 months of darkness. In the Northern Hemisphere, autumn arrives, a time of many colorful changes in the landscape, whereas in the Southern Hemisphere it is spring.

Dawn and Twilight *Dawn* is the period of diffused light that occurs before sunrise. The corresponding evening time after sunset is *twilight*. During both periods, light is scattered by molecules of atmospheric gases and reflected by dust and moisture in the atmosphere. The duration of both is a function of latitude, because the angle of the Sun's path above the horizon determines the thickness of the atmosphere through which the Sun's rays must pass. The illumination may be enhanced by the presence of pollution aerosols and suspended particles from volcanic eruptions or forest and grassland fires.

At the equator, where the Sun's rays are almost directly above the horizon throughout the year, dawn and twilight are limited to 30–45 minutes each. These times increase to 1–2 hours each at 40° latitude, and at 60° latitude they each range upward from 2.5 hours, with little true night in summer. The poles experience about 7 weeks of dawn and 7 weeks of twilight, leaving only 2.5 months of "night" during the 6 months when the Sun is completely below the horizon.

Seasonal Observations In the midlatitudes of the Northern Hemisphere, the position of sunrise on the horizon migrates from day to day, from the southeast in December to the northeast in June. Over the same period, the point of sunset migrates from the southwest to the northwest. The Sun's altitude at local noon at 40° N latitude increases from a 26° angle above the horizon at the winter (December) solstice to a 73° angle above the horizon at the summer (June) solstice—a range of 47° (Figure 2.17).

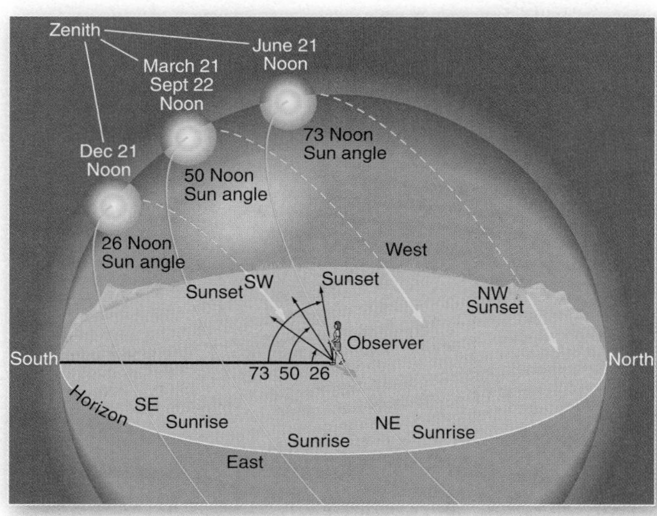

FIGURE 2.17 Seasonal observations—sunrise, noon, and sunset through the year.
Seasonal observations are at 40° N latitude for the December solstice, March equinox, June solstice, and September equinox. The Sun's altitude increases from 26° in December to 73° above the horizon in June—a difference of 47°. Note the changing position of sunrise and sunset along the horizon during the year.

CRITICAL THINKING 2.4

Measuring the Sun's changing altitude

Using the concepts in Figure 2.17 on page 19, use a protractor and stick or ruler to measure the angle of the Sun's altitude at noon (or 1 P.M., if in daylight saving time). Do not look at the Sun; rather, with your back to the Sun, align the stick so that it casts no shadow as you measure its rays against the protractor. Place this measurement in your notebook and affix a Post-It® in the textbook to remind you to repeat the measurement near the end of the semester. Compare and analyze seasonal change using your different measurements of the Sun's altitude.

Seasonal change is quite noticeable across the landscape away from the equator. Think back over the past year. What seasonal changes have you observed in vegetation, temperatures, and weather? Recently, the timing of the seasons is changing as global climates shift in middle and high latitudes. Spring and leafing out is occurring as much as 3 weeks earlier than expected from average conditions. Likewise, the timing of fall is happening later. Ecosystems are changing in response.

We found our place in the Universe, and the relations among Milky Way Galaxy, Sun, planets, and satellites. From the Sun the solar wind and electromagnetic spectrum of radiant energy flow across space to Earth, where we look at its distribution along the top of the atmosphere and how this changes in a seasonal rhythm.

Next we construct Earth's atmosphere and examine its composition, temperature, and functions. Electromagnetic energy cascades toward Earth's surface through the layers of the atmosphere, where harmful wavelengths are filtered out. Also, we examine human impacts on the atmosphere: the depletion of the ozone layer, acid deposition, and variable atmospheric ingredients, including human air pollution.

KEY LEARNING CONCEPTS REVIEW

- ***Distinguish*** among galaxies, stars, and planets, and ***locate*** Earth.

Our Solar System—the Sun and eight planets—is located on a remote, trailing edge of the **Milky Way Galaxy**, a flattened, disk-shaped mass estimated to contain up to 400 billion stars. **Gravity**, the mutual attracting force exerted by the mass of an object upon all other objects, is an organizing force in the Universe. The process of stars (like our Sun) condensing from nebular clouds with planetesimals (protoplanets) forming in orbits around their central masses is the **planetesimal hypothesis**.

Milky Way Galaxy (p. 40)
gravity (p. 40)
planetesimal hypothesis (p. 40)

1. Describe the Sun's status among stars in the Milky Way Galaxy. Describe the Sun's location, size, and relationship to its planets.
2. If you have seen the Milky Way at night, briefly describe it. Use specifics from the text in your description.
3. Briefly describe Earth's origin as part of the Solar System.
4. Compare the locations of the eight planets of the Solar System.

- ***Examine*** the origin, formation, and development of Earth, and ***construct*** Earth's annual orbit about the Sun.

The Solar System, planets, and Earth began to condense from a nebular cloud of dust, gas, debris, and icy comets approximately 4.6 billion years ago. Distances in space are so vast that the **speed of light** (300,000 kmps, or 186,000 mps, which is about 9.5 trillion km, or nearly 6 trillion mi, per year) is used to express distance.

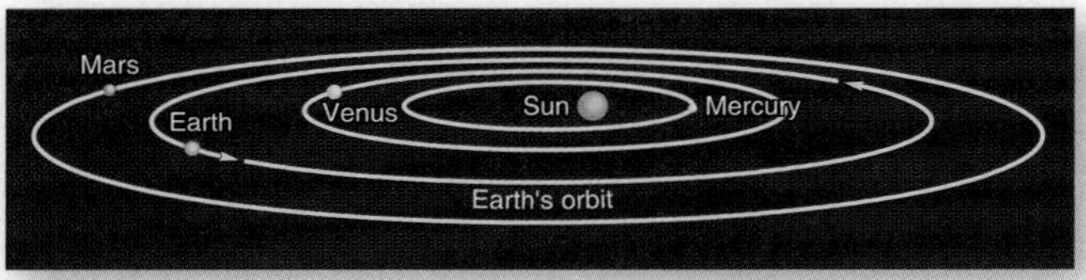

aphelion, July 4 perihelion, January 3

In its orbit, Earth is at **perihelion** (its closest position to the Sun) during our Northern Hemisphere winter (January 3 at 147,255,000 km, or 91,500,000 mi). It is at **aphelion** (its farthest position from the Sun) during our Northern Hemisphere summer (July 4 at 152,083,000 km, or 94,500,000 mi). Earth's average distance from the Sun is approximately 8 minutes and 20 seconds in terms of light speed.

speed of light (p. 40)
perihelion (p. 40)
aphelion (p. 40)

5. How far is Earth from the Sun in terms of light speed? In terms of kilometers and miles?
6. Briefly describe the relationship among these concepts: Universe, Milky Way Galaxy, Solar System, Sun, Earth, and Moon.
7. Diagram in a simple sketch Earth's orbit about the Sun. How much does it vary during the course of a year?

- ***Describe*** the Sun's operation, and ***explain*** the characteristics of the solar wind and the electromagnetic spectrum of radiant energy.

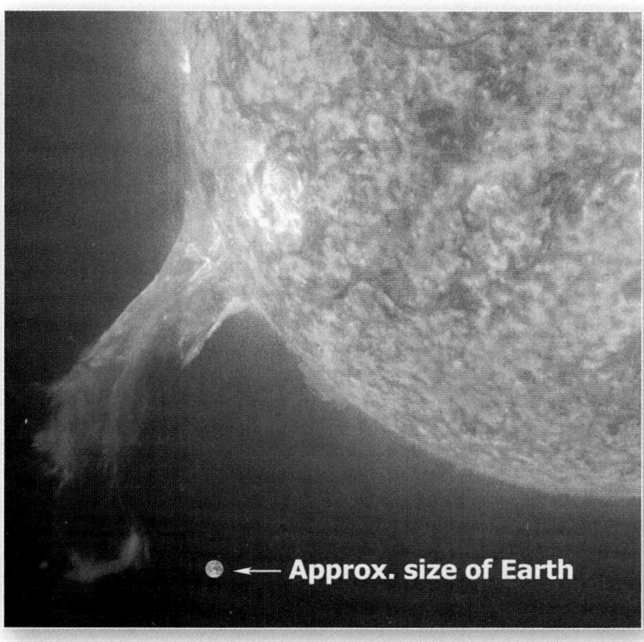

The **fusion** process—hydrogen nuclei forced together under tremendous temperature and pressure in the Sun's interior—generates incredible quantities of energy. Solar energy in the form of charged particles of **solar wind** travels out in all directions from magnetic disturbances on the Sun, such as from large **sunspots**. Solar wind is deflected by Earth's **magnetosphere**, producing various effects in the upper atmosphere, including spectacular **auroras**, the northern and southern lights, which surge across the skies at higher latitudes. Another effect of the solar wind in the atmosphere is its possible influence on weather.

The **electromagnetic spectrum** of radiant energy travels outward in all directions from the Sun. The total spectrum of this radiant energy is made up of different **wavelengths**—the distance between corresponding points on any two successive waves of radiant energy. Eventually, some of this radiant energy reaches Earth's surface.

fusion (p. 42)
solar wind (p. 42)
sunspots (p. 42)
magnetosphere (p. 43)
auroras (p. 43)
electromagnetic spectrum (p. 44)
wavelength (p. 44)

8. How does the Sun produce such tremendous quantities of energy?
9. What is the sunspot cycle? At what stage is the cycle in the year 2013?
10. Describe Earth's magnetosphere and its effects on the solar wind and the electromagnetic spectrum.
11. Summarize the presently known effects of the solar wind relative to Earth's environment.
12. Describe the various segments of the electromagnetic spectrum, from shortest to longest wavelength. What are the main wavelengths produced by the Sun? Which wavelengths does Earth radiate to space?

- ***Portray*** the intercepted solar energy and its uneven distribution at the top of the atmosphere.

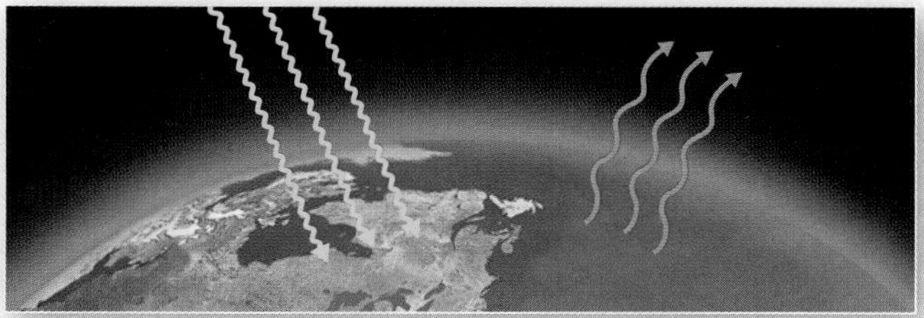

Electromagnetic radiation from the Sun passes through Earth's magnetic field to the top of the atmosphere—the **thermopause**, at approximately 500 km (300 mi) altitude. Incoming solar radiation that reaches a horizontal plane at Earth is **insolation**, a term specifically applied to radiation arriving at Earth's surface and atmosphere. The term **solar constant** describes insolation at the top of the atmosphere: the average insolation received at the thermopause when Earth is at its average distance from the Sun. The solar constant is measured as an average of 1372 W/m^2 (2.0 $cal/cm^2/min$; 2 langleys/min). The place receiving maximum insolation is the **subsolar point**, where solar rays are perpendicular to the surface (radiating from directly overhead). All other locations away from the subsolar point receive slanting rays and more diffuse energy.

thermopause (p. 45)
insolation (p. 45)
solar constant (p. 45)
subsolar point (p. 46)

13. What is the solar constant? Why is it important to know?
14. Study the graph for New York at 40° N latitude on Figure 2.10 and record the amount of energy in watts per square meter (W/m^2) per day in your notebook for each month throughout the year. Compare this with the amount at the North Pole and at the equator.
15. If Earth were flat and oriented at right angles to incoming solar radiation (insolation), what would be the latitudinal distribution of solar energy at the top of the atmosphere?

- ***Define*** solar altitude, solar declination, and daylength, and ***describe*** the annual variability of each—Earth's seasonality.

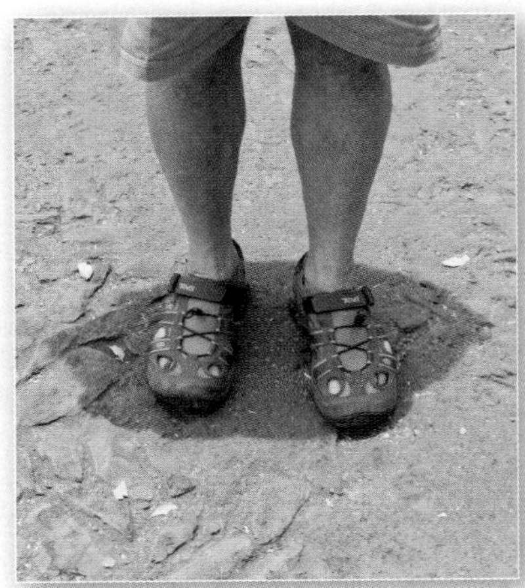

The angle between the Sun and the horizon is the Sun's **altitude**. The Sun's **declination** is the latitude of the subsolar point. Declination annually migrates through 47° of latitude, moving between the *Tropic of Cancer* at about 23.5° N (June) and the *Tropic of Capricorn* at about 23.5° S latitude (December). Seasonality means an annual change in the Sun's altitude and changing **daylength**, or duration of exposure.

Earth's distinct seasons are produced by interactions of **revolution** (annual orbit about the Sun) and **rotation** (Earth's turning on its **axis**). As Earth rotates, the boundary that divides daylight and darkness is the **circle of illumination**. Other reasons for seasons include **axial tilt** (at about 23.5° from a perpendicular to the **plane of the ecliptic**), **axial parallelism** (the parallel alignment of the axis throughout the year), and *sphericity*.

Earth rotates about its axis, an imaginary line extending through the planet from the geographic North Pole to the South Pole. In the Solar System, an imaginary plane touching all points of Earth's orbit is the *plane of the ecliptic*. Daylength is the interval between **sunrise**, the moment when the disk of the Sun first appears above the horizon in the east, and **sunset**, that moment when it totally disappears below the horizon in the west. The **Tropic of Cancer** parallel marks the farthest north the subsolar point migrates during the year, about 23.5° N latitude. The **Tropic of Capricorn** parallel marks the farthest south the subsolar point migrates during the year, about 23.5° S latitude.

On December 21 or 22, at the moment of the **winter solstice** ("winter Sun stance"), or **December solstice**, the circle of illumination excludes the North Pole but includes the South Pole. The subsolar point is at about 23.5° S latitude, the parallel called the Tropic of Capricorn. The Sun remains below the horizon the entire day. This latitude (about 66.5° N) marks the **Arctic Circle**, the southernmost parallel (in the Northern Hemisphere) that experiences a 24-hour period of darkness, or on June 21, a 24-hour period of daylight. The moment of the **vernal equinox**, or **March equinox**, occurs on March 20 or 21. At that time, the circle of illumination passes through both poles so that all locations on Earth experience a 12-hour day and a 12-hour night.

June 20 or 21 is the moment of the **summer solstice**, or **June solstice**. The subsolar point migrates from the equator to about 23.5° N latitude, the Tropic of Cancer. Because the circle of illumination now includes the North Polar region, everything north of the Arctic Circle receives 24 hours of daylight—the "Midnight Sun." In contrast, the area from the **Antarctic Circle** to the South Pole (66.5°–90° S latitude) is in darkness the entire 24 hours. September 22 or 23 is the time of the **autumnal equinox**, or **September equinox**, when Earth's orientation is such that the circle of illumination again passes through both poles so that all parts of the globe experience a 12-hour day and a 12-hour night.

altitude (p. 48)
declination (p. 48)
daylength (p. 48)
revolution (p. 49)
rotation (p. 49)
axis (p. 49)
circle of illumination (p. 50)
axial tilt (p. 50)
plane of the ecliptic (p. 50)
axial parallelism (p. 51)
sunrise (p. 51)
sunset (p. 51)
Tropic of Cancer (p. 52)
Tropic of Capricorn (p. 52)
winter solstice, December solstice (p. 52)
Arctic Circle (p. 52)
vernal equinox, March equinox (p. 52)
summer solstice, June solstice (p. 52)
Antarctic Circle (p. 52)
autumnal equinox, September equinox (p. 52)

16. Assess the 12-month Gregorian calendar, with its months of different lengths, and leap years, and its relation to the annual seasonal rhythms—the march of the seasons. What do you find?
17. The concept of seasonality refers to what specific phenomena? How do these two aspects of seasonality change during a year at 0° latitude? At 40°? At 90°?
18. Differentiate between the Sun's altitude and its declination at Earth's surface.
19. For the latitude at which you live, how does daylength vary during the year? How does the Sun's altitude vary? Does your local newspaper publish a weather calendar containing such information?
20. List the five physical factors that operate together to produce seasons.
21. Describe Earth's revolution and rotation and differentiate between them.
22. Define Earth's present tilt relative to its orbit about the Sun.
23. Describe seasonal conditions at each of the four key seasonal anniversary dates during the year. What are the solstices and equinoxes, and what is the Sun's declination at these times?

MASTERING GEOGRAPHY

Visit www.masteringgeography.com for several figure animations, satellite loops, self-study quizzes, flashcards, visual glossary, case studies, career links, textbook reference maps, RSS Feeds, and an ebook version of *Geosystems*. Included are many geography web links to Internet sources that support this chapter.

Fogo Island, in Cape (Capo) Verde, an island country in West Africa, in the Atlantic Ocean about 600 km (375 mi) from Africa. This is where the subsolar point was encountered in the *Geosystems Now* feature at the beginning of the chapter. The volcanic crater, Pico de Fogo, visible slightly east of image center, erupted in 1995. There are two small villages within the larger caldera. More on this in Chapter 12.

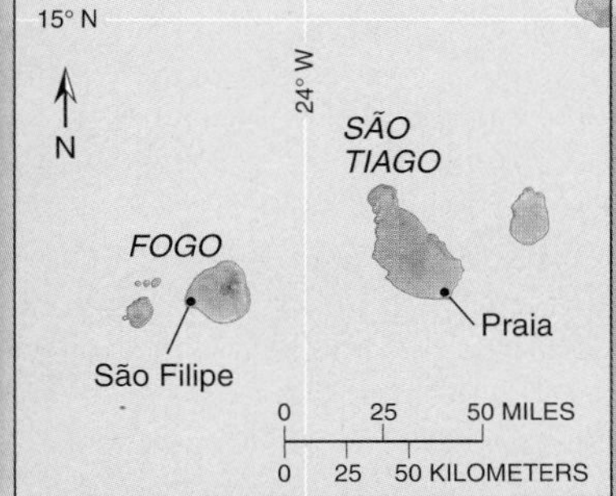

CHAPTER 3

Earth's Modern Atmosphere

Sunset at 30° S over the Atlantic Ocean. [Bobbé Christopherson.]

KEY LEARNING CONCEPTS

After reading the chapter, you should be able to:

- ***Construct*** a general model of the atmosphere based on three criteria—composition, temperature, and function—and ***diagram*** this model in a simple sketch.
- ***List*** the stable components of the modern atmosphere and their relative percentage contributions by volume, and ***describe*** each.
- ***Describe*** conditions within the stratosphere; specifically, ***review*** the function and status of the ozonosphere (ozone layer).
- ***Distinguish*** between natural and anthropogenic variable gases and materials in the lower atmosphere.
- ***Describe*** the sources and effects of carbon monoxide, nitrogen dioxide, and sulfur dioxide, and ***construct*** a simple diagram that illustrates photochemical reactions that produce ozone, peroxyacetyl nitrates, nitric acid, and sulfuric acid.

Humans Help Define the Atmosphere

The survival of humans in space depends on our knowledge of the upper atmosphere. Astronaut Mark Lee, on a spacewalk in 1994 (*STS–64*), was 241 km (150 mi) above the surface, in orbit (Figure GN 3.1). Beyond the protective shield of Earth's atmosphere, he was travelling at 28,165 kmph (17,500 mph), which is almost nine times faster than a high-speed rifle bullet. Where the Sun hit his spacesuit, temperatures climb to +120°C (+248°F), and in the shadows, temperatures drop to some −150°C (−238°F). The vacuum of space surround him. Radiation and solar wind impact his suit's surface. To survive in the upper atmosphere was an obvious challenge, one that relies on the ability of NASA spacesuits to duplicate the Earth's atmosphere.

What features must a spacesuit provide for human survival? The suit must give all the protection provided by the atmosphere, such as blocking radiation and particle impacts. The suit needs to protect the interior from thermal extremes.

FIGURE GN 3.1 **An untethered STS–64 astronaut as a satellite in Earth orbit on a working spacewalk.** [NASA.]

Earth's oxygen–carbon dioxide processing systems must be replicated in the suit, as must fluid delivery and waste management systems. The suit must maintain an internal air pressure against the space vacuum, and for pure oxygen, this is 4.7 psi (32.4 kPa; which roughly equals the pressure of only oxygen, water vapor, and carbon dioxide together at sea level). All 18,000 parts of the spacesuit work to duplicate what the atmosphere does for us on a daily basis. Think of this Chapter 3 as describing design features for such a suit.

In an earlier era before orbital flights, scientists didn't know how a human could survive in space or how to design an artificial atmosphere that is the modern spacesuit. In 1960, Air Force Captain Joseph Kittinger, Jr., stood at the opening of a small, unpressurized compartment, floating at an altitude of 31.3 km (19.5 mi), dangling from a helium-filled balloon. The air pressure is barely measurable—this altitude is the beginning of space in experimental aircraft testing. He then leaped into the stratospheric void and placed himself at tremendous personal risk for an experimental reentry into the atmosphere (Figure GN 3.2). He carried an instrument pack on his seat, his main chute, and pure oxygen for his breathing mask.

Initially frightening to him was that he heard nothing, no rushing sound, for there was not enough air to produce any sound. The fabric of his pressure suit did not flutter, for there was not enough air to create friction against the cloth. His speed was remarkable owing to the lack of air resistance in the stratosphere. Captain Kittinger quickly accelerated to 988 kmph (614 mph), near the speed of sound at sea level.

FIGURE GN 3.2 **A remotely triggered camera captures a stratospheric leap into history.** [Volkmar Wentzel/NGS Image Collection, *National Geographic Magazine*, Dec. 1960, p. 855. All rights reserved.]

His free fall took him through the stratosphere and its ozone layer, as the frictional drag of denser layers of atmospheric gases slowed his body. He dropped into the lower atmosphere, finally falling below airplane flying altitudes. His free fall lasted 4 minutes and 25 seconds to the opening of his main chute at 5500 m (18,000 ft). He safely drifted to Earth's surface. This remarkable 13-minute 35-second voyage through 99% of the atmospheric mass remains a record to this day.

The experiences of these two men illustrate the evolution of our understanding of survival in the upper atmosphere. From Joseph Kittinger's dangerous leap of discovery to the now routine spacewalks of astronauts such as Mark Lee, scientists have gained the ability to duplicate the atmosphere. This chapter describes our current knowledge and the design features for our spacesuit as we explore Earth's atmosphere.

Earth's atmosphere is a unique reservoir of gases, the product of 4.6 billion years of development. It protects us from hostile radiation and particles from the Sun and beyond. The atmosphere serves as an efficient filter. As shown in *Geosystems Now*, when humans venture away from the lower regions of the atmosphere, they must wear elaborate protective spacesuits that sustain them, providing services the atmosphere performs for us all the time.

In this chapter: We examine the modern atmosphere using the criteria of composition, temperature, and function. Our look at the atmosphere also includes the spatial aspects of both natural and human-produced air pollution. We all participate in the atmosphere with each breath we take, the energy we consume, the travelling we do, and the products we buy. Human activities cause stratospheric ozone losses and the blight of acid deposition on ecosystems. These topics are essential to physical geography, for we are influencing the atmospheric composition of the future.

Atmospheric Composition, Temperature, and Function

The modern atmosphere probably is the fourth general atmosphere in Earth's history, a gaseous mixture of ancient origin, the sum of all the exhalations and inhalations of life interacting on Earth throughout time. The principal substance of this atmosphere is air, the medium of life as well as a major industrial and chemical raw material. **Air** is a simple mixture of gases that is naturally odorless, colorless, tasteless, and formless, blended so thoroughly that it behaves as if it were a single gas (Figure 3.1).

As a practical matter, we consider the top of our atmosphere to be around 480 km (300 mi) above Earth's surface, the same altitude we used in Chapter 2 for measuring the solar constant and insolation received. Beyond that altitude is the **exosphere**, which means "outer sphere," where the rarefied, less dense atmosphere is nearly a vacuum. It contains scarce lightweight hydrogen and

FIGURE 3.1 Earthrise. Earthrise over the stark and lifeless lunar surface. [JPL/NASA. Quote from "The World's Biggest Membrane" in *The Lives of a Cell* by Lewis Thomas. Copyright © 1973 by the Massachusetts Medical Society.]

helium atoms, weakly bound by gravity as far as 32,000 km (20,000 mi) from Earth.

Atmospheric Profile

Think of Earth's modern atmosphere as a thin envelope of imperfectly shaped concentric "shells" or "spheres" that grade into one another, all bound to the planet by gravity. To study the atmosphere let us view it in layers, each with distinctive properties and purposes. Figure 3.2 charts the atmosphere in a vertical cross-section profile, or side view, and is a useful reference for the following discussion. To simplify we use three atmospheric criteria: see *composition*, *temperature*, and *function*, noted along the left side of Figure 3.2a.

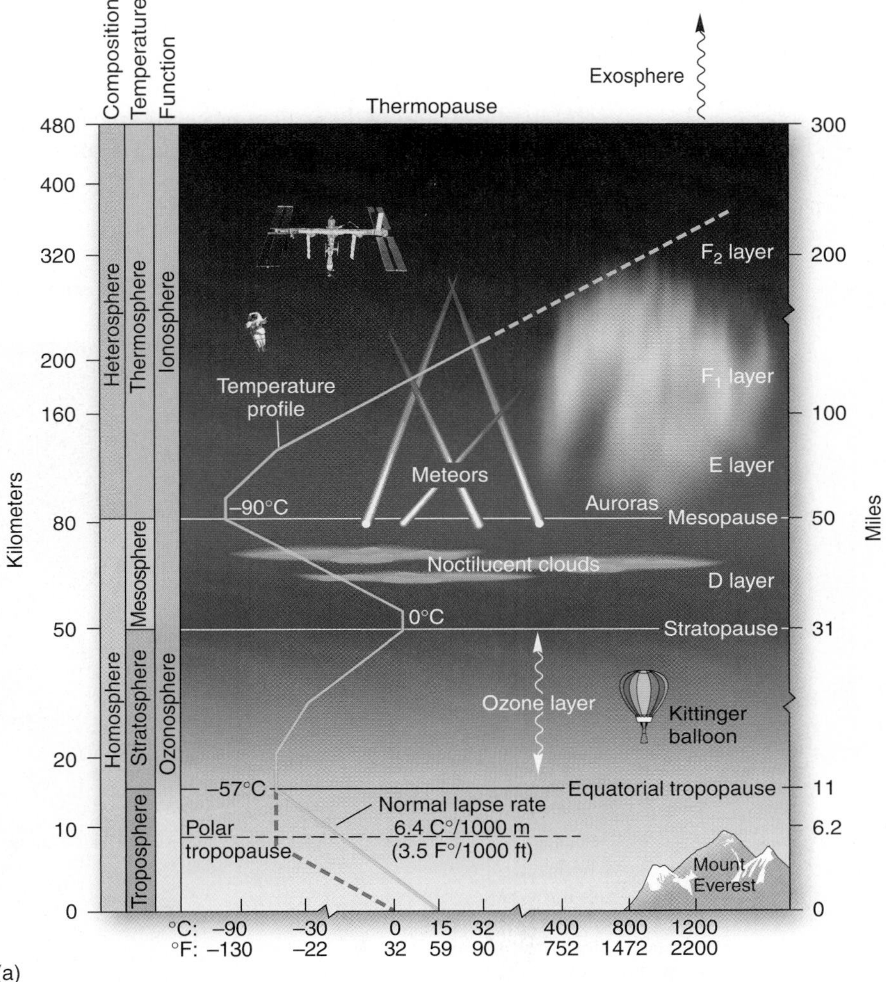

FIGURE 3.2 Profile of the modern atmosphere.
(a) The columns along the left side show the division of the atmosphere by composition, temperature, and function. The plot of temperature by altitude refers to the scale along the bottom axis. The small astronaut and balloon show the altitude achieved by astronaut Mark Lee and by Joseph Kittinger, discussed in *Geosystems Now*. (b) A sunset from orbit through various atmospheric layers. A silhouetted cumulonimbus thunderhead cloud tops out at the tropopause. [(b) NASA.]

Earth's atmosphere exerts its weight, pressing downward under the pull of gravity. Air molecules create air pressure through their motion, size, and number. Pressure is exerted on all surfaces in contact with the air. The weight (force over a unit area) of the atmosphere, or **air pressure**, pushes in on all of us. Fortunately, that same pressure also exists inside us, pushing outward; otherwise, we would be crushed by the mass of air around us.

The atmosphere exerts an average force of approximately 1kg/cm^2 (14.7 lb/in^2) at sea level. Gravity compresses air, making it denser near Earth's surface; it thins rapidly with increasing altitude (Figure 3.3a), and with reduced air pressure, less oxygen is carried in each breath a person takes.

Over half the total mass of the atmosphere is compressed below 5500 m (18,000 ft), 75% occurs below 10,700 m (35,100 ft), and 90% is below 16,000 m (52,500 ft). Only 0.1% of the atmosphere remains above an altitude of 50 km (31 mi), as shown in the pressure profile in Figure 3.3b (percentage column is farthest to the right).

At sea level, the atmosphere exerts a pressure of 1013.2 mb (*millibar*, or *mb*; a measure of force per square meter of surface area) or 29.92 in. of mercury (symbol, Hg), as measured by a barometer. In Canada and other countries, normal air pressure is expressed as 101.32 kPa (*kilopascal*; 1 kPa = 10 mb). More information on air pressure, the instruments that measure it, and the role it plays in generating winds is in Chapter 6.

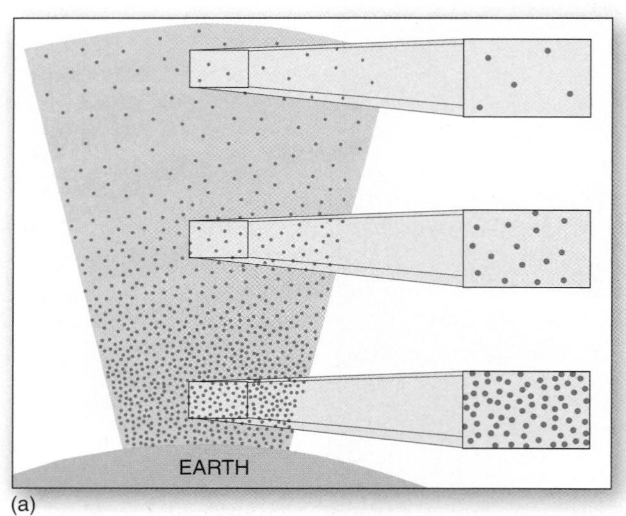

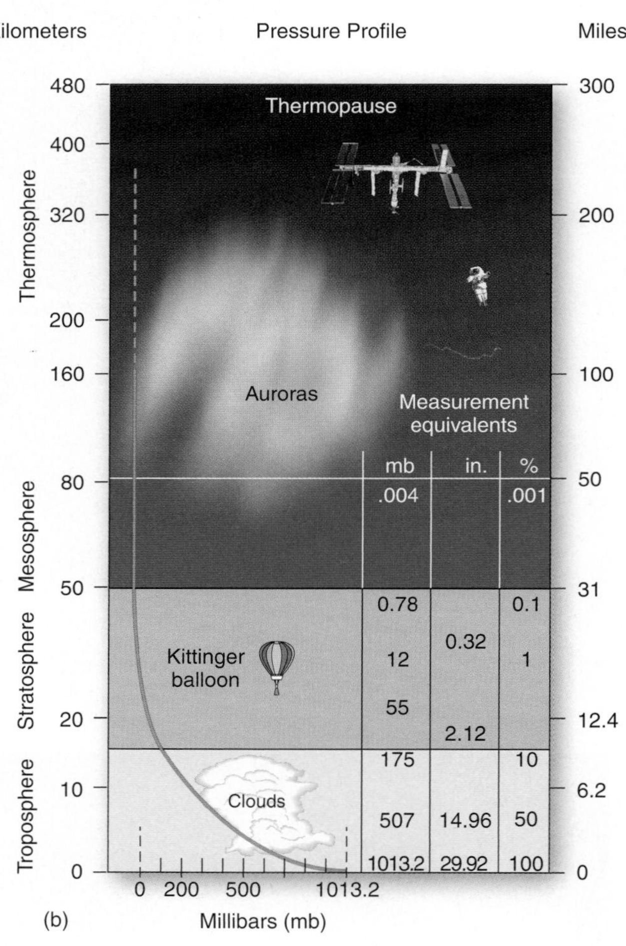

FIGURE 3.3 Density decreases with altitude. (a) The atmosphere rapidly decreases in density with altitude. Have you experienced pressure changes that you could feel on your eardrums? How high above sea level were you at the time? (b) In this pressure profile, note the rapid decrease in atmospheric pressure with altitude. The troposphere holds about 90% of the atmospheric mass (far-right % column).

GEO REPORT 3.1 *Outside the airplane*

Next time you fly somewhere consider a couple of things. Few people are aware that in routine air travel they are sitting above 80% of the total atmospheric volume and that the air pressure outside is only about 10% of surface air pressure. This means that only 20% of the atmospheric mass is above you. If your plane is at about 11,000 m (36,090 ft), think of Captain Kittinger's jump described in *Geosystems Now,* for he started another 20 km higher in altitude than your plane!

Atmospheric Composition Criterion

Using chemical *composition* as a criterion, the atmosphere divides into two broad regions, the *heterosphere* (80 to 480 km altitude) and the *homosphere* (Earth's surface to 80 km altitude), as shown along the left side of Figure 3.2a. As you read, note that the text follows the same path as incoming solar radiation that travels through the atmosphere to Earth's surface.

Heterosphere The **heterosphere** is the outer atmosphere in terms of composition. It begins in the exosphere and interplanetary space and extends downward to 80 km (50 mi) (see Figure 3.2). Less than 0.001% of the atmosphere's mass is in this rarefied heterosphere. The International Space Station (ISS) and most Space Shuttle missions orbit in the middle to upper heterosphere (note the ISS altitude in Figures 3.2 and 3.3).

As the prefix *hetero-* implies, this region is not uniform—its gases are not evenly mixed. Gases in the heterosphere occur in distinct layers sorted by gravity according to their atomic weight, with the lightest elements (hydrogen and helium) at the margins of outer space and the heavier elements (oxygen and nitrogen) dominant in the lower heterosphere. This distribution is quite different from the blended gases we breathe in the homosphere near Earth's surface.

Homosphere Below the heterosphere is the **homosphere**, extending from an altitude of 80 km (50 mi) to Earth's surface. Even though the atmosphere rapidly changes density in the homosphere—increasing pressure toward Earth's surface—the blend of gases is nearly uniform throughout. The only exceptions are the concentration of ozone (O_3) in the "ozone layer," from 19 to 50 km (12 to 31 mi), and the variations in water vapor, pollutants, and some trace chemicals in the lowest portion of the atmosphere.

The present blend of gases evolved approximately 500 million years ago. Table 3.1 lists by volume the stable ingredients that constitute dry, clean air in the homosphere. Air sampling occurs at the Mauna Loa Observatory, Hawai'i, operational since 1957. A marine inversion layer keeps volcanic emissions from nearby Kīlauea volcano and dust to a minimum.

The air of the homosphere is a vast reservoir of relatively inert *nitrogen*, originating principally from volcanic sources. Nitrogen is a key element of life, yet we exhale all the nitrogen we inhale. The explanation for this contradiction is that nitrogen integrates into our bodies not from the air we breathe, but through compounds in food. In the soil, nitrogen is tied up by nitrogen-fixing bacteria, and it returns to the atmosphere by denitrifying bacteria that remove nitrogen from organic materials. A complete discussion of the nitrogen cycle is in Chapter 19.

Oxygen, a by-product of photosynthesis, also is essential for life processes. Slight spatial variations occur in the percentage of oxygen in the atmosphere because of variations in photosynthetic rates with latitude, seasonal changes, and the lag time as atmospheric circulation slowly mixes the air. Although it forms about one-fifth of the atmosphere, oxygen forms compounds that compose about half of Earth's crust. Oxygen readily reacts with many elements to form these materials. Both nitrogen and

TABLE 3.1 Stable Components of the Modern Homosphere

Gas (Symbol)	Percentage by Volume	Parts per Million (ppm)
Nitrogen (N_2)	78.084	780,840
Oxygen (O_2)	20.946	209,460
Argon (Ar)	0.934	9,340
Carbon dioxide (CO_2)*	0.0393	393
Neon (Ne)	0.001818	18
Helium (He)	0.000525	5
Methane (CH_4)	0.00014	1.4
Krypton (Kr)	0.00010	1.0
Ozone (O_3)	Variable	
Nitrous oxide (N_2O)	Trace	
Hydrogen (H)	Trace	
Xenon (Xe)	Trace	

*May 2010 average CO2 measured at Mauna Loa, Hawai'i (see http://www.esrl.noaa.gov/gmd/ccgg/trends/.)

Mauna Loa Observatory, air sampling tower. [Bobbé Chrisotpherson.]

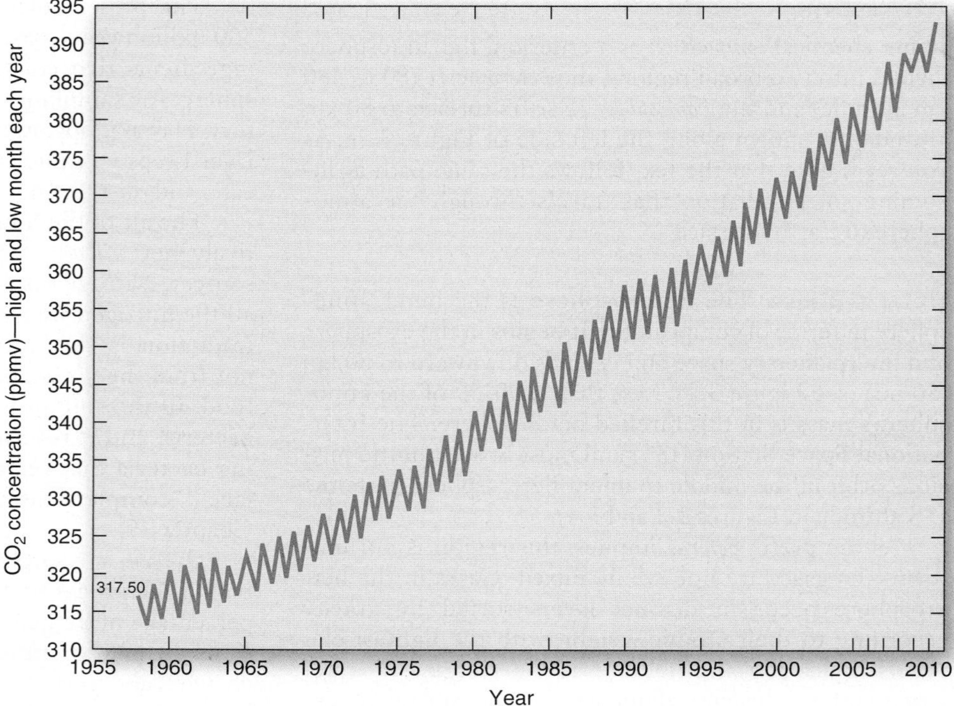

FIGURE 3.4 Carbon dioxide concentrations increase, 1958–2010. Fifty-two years of CO_2 data measured at the Mauna Loa Observatory through May 2010. The highest and lowest average months, usually May and October respectively, are plotted for each year on the graph in parts per million by volume (ppmv). [Data posted at ftp://ftp.cmdl.noaa.gov/ccg/co2/trends/co2_mm_mlo.txt.]

oxygen reserves in the atmosphere are so extensive that, at present, they far exceed human capabilities to disrupt or deplete them.

The gas *argon*, constituting less than 1% of the homosphere, is completely inert (an unreactive "noble" gas) and unusable in life processes. Argon is a residue from the radioactive decay of an isotope, or chemical form, of potassium, potassium-40 (symbolized ^{40}K). All the argon present in the modern atmosphere comes from slow accumulation over millions of years. Because industry has found uses for inert argon (in lightbulbs, welding, and some lasers), it is extracted or "mined" from the atmosphere, along with nitrogen and oxygen, for commercial, medical, and industrial uses.

Carbon dioxide *Carbon dioxide* (CO_2) is a natural byproduct of life processes. Although it is increasing rapidly, it is included with the homosphere's stable components in Table 3.1. Although its present percentage in the atmosphere is small at about 0.0393%, CO_2 is important to global temperatures.

Chapters 4, 5, and 10 discuss the role of carbon dioxide as an important greenhouse gas and its role in present global temperature increases. Carbon dioxide concentrations in the atmosphere since 1958 are graphed to 2010 in Figure 3.4. The annual fluctuation between summer and winter is evidence of seasons and the work of plant photosynthesis. Scientists agree that the increase is human-forced.

Over the past 200 years, the CO_2 percentage increased as a result of human activities, principally the burning of fossil fuels and deforestation. According to ice-core records, CO_2 is higher now than at any time in the past 800,000 years. This increase appears to be accelerating. From 1990 to 1999, CO_2 emissions rose at an average of 1.1% per year; compare this to the average emissions increase since 2000 of 3.1% per year—a 2 to 3 ppm per year increase. A distinct climatic threshold is approaching at 450 ppm, sometime in the decade of the 2020s. Beyond this tipping point, there would be irreversible ice-sheet and species losses. Chapter 10 discusses the implications of CO_2 increases for climate change.

GEO REPORT 3.2 *Atmospheric carbon dioxide accelerates*

Since the first edition of *Geosystems,* the global concentration of CO_2 in our atmosphere has increased more than 16%, from 354 ppm in 1992 to 393 ppm in 2010. For comparison, preindustrial levels of CO_2 were about 280 ppm. CO_2 today far exceeds the natural range of 180 to 300 ppm over the last 800,000 years. If this rate of increase continues the rest of the century, business as usual, then Table 3.1 could read 1400 ppm CO_2 in A.D. 2100, according to several forecasts.

Dr. James Hansen, a NASA scientist at Goddard Institute for Space Sciences, said in 2008:

> If humanity wishes to preserve a planet similar to that on which civilization developed, paleoclimatic evidence and ongoing climate change suggest that CO_2 will need to be reduced to at most 350 ppm [the average in 1987].

Atmospheric Temperature Criterion

Shifting to temperature as a criterion, the atmosphere has four distinct temperature zones—the *thermosphere, mesosphere, stratosphere,* and *troposphere* (labeled in Figure 3.2). Let's look at each of these, beginning with the highest in altitude.

Thermosphere The **thermosphere** ("heat sphere") roughly corresponds to the heterosphere (80 km out to 480 km, or 50–300 mi). The upper limit of the thermosphere is the **thermopause** (the suffix *-pause* means "to change"). During periods of a less active Sun, with fewer sunspots and coronal bursts, the thermopause may lower in altitude from the average 480 km (300 mi) to only 250 km (155 mi). During periods of a more active Sun, the outer atmosphere swells to an altitude of 550 km (340 mi), where it can create frictional drag on satellites in low orbit.

The temperature profile in Figure 3.2 (yellow curve) shows that temperatures rise sharply in the thermosphere, to 1200°C (2200°F) and higher. Despite such high temperatures, the thermosphere is not "hot" in the way you might expect. Temperature and heat are different concepts. The intense solar radiation in this portion of the atmosphere excites individual molecules (principally nitrogen and oxygen) to high levels of vibration. This **kinetic energy**, the energy of motion, is the vibrational energy that we measure as *temperature*.

However, the actual heat involved is small. The reason is that the density of molecules is so low. There is little actual **heat** produced or the flow of kinetic energy from one body to another because of a temperature difference between them. Heating in the atmosphere near Earth's surface is different because the greater number of molecules in the denser atmosphere transmits their kinetic energy as **sensible heat**, meaning that we can measure and feel it. (Density, temperature, and heat capacity determine the sensible heat of a substance.) More on this topic is in Chapters 4 and 5.

Mesosphere The **mesosphere** is the area from 50 to 80 km (30 to 50 mi) above Earth and is within the homosphere. As Figure 3.2 shows, the mesosphere's outer boundary, the *mesopause*, is the coldest portion of the atmosphere, averaging −90°C (−130°F), although that temperature may vary considerably (25–30 C°, or 45–54 F°). Note in Figure 3.3b the extremely low pressures (low density of molecules) in the mesosphere.

The mesosphere sometimes receives cosmic or meteoric dust particles, which act as nuclei around which fine ice crystals form. At high latitudes, an observer may see these bands of ice crystals glow in rare and unusual night **noctilucent clouds**, which are so high in altitude that they still catch sunlight after sunset. For reasons not clearly understood, these unique clouds are on the increase and may be seen in the midlatitudes. See, among several sites, http://lasp.colorado.edu/science/atmospheric/.

Stratosphere The **stratosphere** extends from 18 to 50 km (11 to 31 mi) from Earth's surface. Temperatures increase with altitude throughout the stratosphere, from −57°C (−70°F) at 18 km at the tropopause, warming to 0°C (32°F) at 50 km at the stratosphere's outer boundary, the *stratopause*. Stratospheric changes measured over the past 25 years show that chlorofluorocarbons are increasing and ozone concentrations decreasing, with greenhouse gases also on the increase. A noted stratospheric cooling is the response.

Troposphere The **troposphere** is the final layer encountered by incoming solar radiation as it surges through the atmosphere to the surface. It is the home of the biosphere, the atmospheric layer that supports life, and the region of principal weather activity.

Approximately 90% of the total mass of the atmosphere and the bulk of all water vapor, clouds, and air pollution are within the troposphere. An average temperature of −57°C (−70°F) defines the **tropopause**, the troposphere's upper limit, but its exact altitude varies with the season, latitude, and surface temperatures and pressures. Near the equator, because of intense heating from the surface, the tropopause occurs at 18 km (11 mi); in the middle latitudes, it occurs at an average of 12 km (8 mi); and at the North and South Poles, it averages only 8 km (5 mi) or less above Earth's surface. The marked warming with increasing altitude in the stratosphere above the tropopause causes the tropopause to act like a lid, generally preventing whatever is in the cooler (denser) air below from mixing into the warmer (less dense) stratosphere; visible in Figure 3.2b.

CRITICAL THINKING 3.1

Where is your tropopause?

On your next flight as the plane reaches its highest altitude, ask flight attendants to determine the temperature outside the plane. Some planes have video screens that display the outside air temperature. By definition, the tropopause is wherever the temperature −57°C (−70°F) occurs. Depending on the season of the year, altitude and temperature make an interesting comparison. Is the tropopause at a higher altitude in summer or winter?

Figure 3.5 illustrates the normal temperature profile within the troposphere during daytime. As the graph shows, temperatures decrease rapidly with increasing

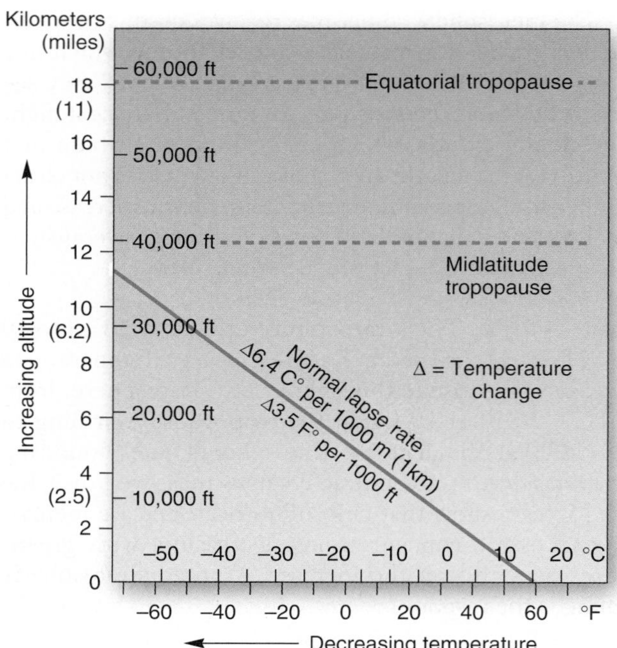

FIGURE 3.5 The temperature profile of the troposphere. Temperature decreases with increased altitude at a rate known as the *normal lapse rate*. Scientists use the *standard atmosphere* as a description of air temperature and pressure changes with altitude. Note the approximate locations of the equatorial and midlatitude tropopauses.

altitude at an average of 6.4 C° per km (3.5 F° per 1000 ft), a rate known as the **normal lapse rate**. This temperature plot also appears in the troposphere portion of Figure 3.2.

The normal lapse rate is an average. The actual lapse rate may vary considerably because of local weather conditions and is the **environmental lapse rate**. This variation in temperature gradient between normal and environmental rates in the lower troposphere is central to our discussion of weather processes in Chapters 7 and 8.

Atmospheric Function Criterion

Looking at our final atmospheric criterion of function, the atmosphere has two specific zones, the *ionosphere* and the *ozonosphere* (*ozone layer*), which remove most of the harmful wavelengths of incoming solar radiation and charged particles. Figure 3.6 depicts in a general way the absorption of radiation by the various functional layers of the atmosphere.

Ionosphere The outer functional layer, the **ionosphere**, extends throughout the thermosphere and into the mesosphere below (Figure 3.2). The ionosphere absorbs cosmic rays, gamma rays, X-rays, and shorter wavelengths of ultraviolet radiation, changing atoms to positively charged ions and giving the ionosphere its name. The glowing auroral lights discussed in Chapter 2 occur principally within the ionosphere.

Figure 3.2 shows the average daytime altitudes of four regions within the ionosphere, known as the D, E, F_1, and F_2 layers. These are important to broadcast communications, for they reflect certain radio wavelengths, including AM radio and other shortwave broadcasts, especially at night. However, during the day the ionosphere actively absorbs arriving radio signals, preventing distant reception. Normally unaffected are FM or television broadcast wavelengths, which pass through to space, requiring the use of communication satellites.

Ozonosphere That portion of the stratosphere that contains an increased level of ozone is the **ozonosphere**, or **ozone layer**. Ozone is a highly reactive oxygen molecule made up of three oxygen atoms (O_3) instead of the usual two atoms (O_2) that make up oxygen gas. Ozone absorbs certain wavelengths of ultraviolet (principally, all the UVC, 100–290 nm, and about 90% of the UVB,

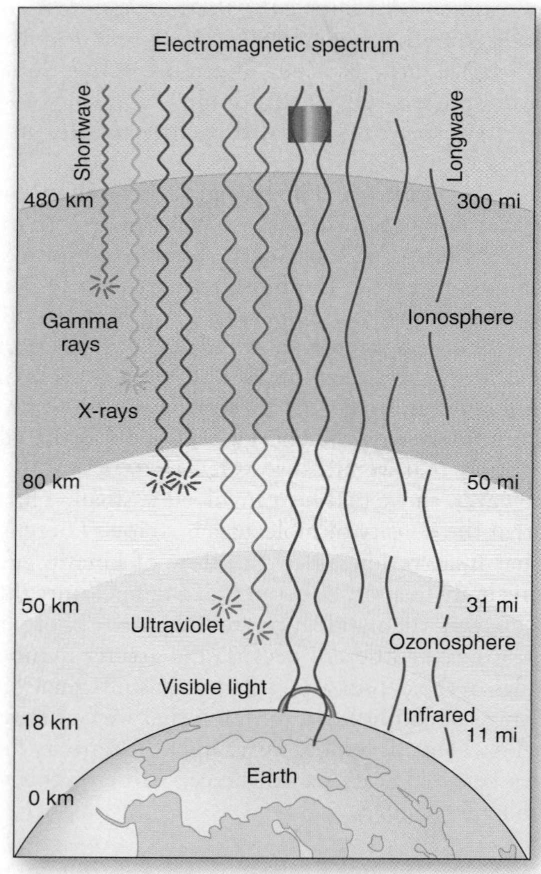

FIGURE 3.6 The atmosphere protects Earth's surface. As solar energy passes through the atmosphere, the shortest wavelengths are absorbed. Only a fraction of the ultraviolet radiation, and most of the visible light and infrared, reaches Earth's surface.

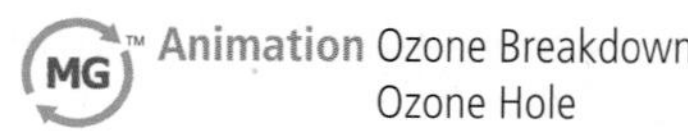

290–320 nm).* This absorbed energy subsequently radiates at longer wavelengths as infrared radiation. This process converts most harmful ultraviolet radiation, effectively "filtering" it and safeguarding life at Earth's surface. UVA, at 320–400 nm, makes up about 98% of all UV radiation that reaches Earth's surface.

The ozone layer is presumed to be relatively stable over the past several hundred million years (allowing for daily and seasonal fluctuations). Today, however, it is in a state of continuous change. Focus Study 3.1 presents an analysis of the crisis in this critical portion of our atmosphere.

CRITICAL THINKING 3.2

Finding your ozone column

To determine the total ozone column at your present location, go to "What was the total ozone column at your house?" at http://toms.gsfc.nasa.gov/teacher/ozone_overhead.html.

Select a point on the map or enter your latitude and longitude and the date you want to check. The ozone column is measured by the Ozone Monitoring Instrument (OMI) sensor aboard the *Aqua* satellite and is mainly sensitive to stratospheric ozone. (Note also the limitations listed on the extent of data availability.) If you check for several different dates, when do the lowest values occur? The highest values? Briefly explain and interpret the values you found.

UV Index Helps Save Your Skin Newspaper and television weather reports regularly include the *UV Index (UVI)* in daily forecasts, as reported by the National Weather Service (NWS) and the Environmental Protection Agency (EPA) since 1994. A revision in 2004 aligned the UVI with guidelines adopted by the World Health Organization (WHO) and World Meteorological Organization (WMO). Think of this as a status report on stratospheric functions.

The UVI, as shown in Table 3.2, forecasts the expected risk of overexposure to the Sun, using a scale from 1 to 11+. With stratospheric ozone levels at thinner-than-normal conditions, surface exposure to cancer-causing radiation increases. The public, especially babies and children, now is alerted to take extra precautions in the form of sunscreen, hats, protective clothing, and sunglasses. Remember, damage accumulates, and it may be decades before you experience the ill effects triggered by this summer's sunburn. See http://www.epa.gov/sunwise/uvindex.html for a UVI forecast for your location.

Variable Atmospheric Components

The troposphere contains natural and human-caused variable gases, particles, and other chemicals. The spatial aspects of these variables are important applied topics in physical geography and have important human-health implications.

Air pollution is not a new problem. Romans complained more than 2000 years ago about the foul air of their cities. Filling Roman air was the stench of open sewers, smoke from fires, and fumes from ceramic-making kilns and smelters (furnaces) that converted ores into metals. Throughout human civilization, cities have stressed the environment's natural ability to process and recycle waste. Historically, air pollution has occurred around population centers and is closely linked to human production and consumption of energy and resources.

Solutions require regional, national, and international strategies because the pollution sources often are distant from the observed impact—moving across political boundaries and crossing oceans. Regulations to curb human-caused air pollution have achieved great success, although much remains to be done. Before we discuss these topics, let's examine some natural pollution sources.

*Nanometer (nm) = one-billionth of a meter; 1 nm = 10^{-9} m. For comparison, a micrometer, or micron, (μm) = one-millionth of a meter; 1 μm = 10^{-6} m. A millimeter (mm) = one-thousandth of a meter; 1 mm = 10^{-3}.

TABLE 3.2 UV Index (EPA, NWS, WHO, WMO)

Exposure Risk Category	UVI Range	Comments
Low	less than 2	Low danger for average person. Wear sunglasses on bright days. Watch out for reflection off snow.
Moderate	3–5	Take covering precautions, such as sunglasses, sunscreen, hats, and protective clothing, and stay in shade during midday hours.
High	6–7	Use sunscreens with SPF of 15 or higher. Reduce time in the Sun between 11 A.M. and 4 P.M. Use protections mentioned above.
Very High	8–10	Minimize Sun exposure 10 A.M. to 4 P.M. Use sunscreens with SPF ratings of over 15. Use protections mentioned above.
Extreme	11+	Unprotected skin is at risk of burn. Sunscreen application every two hours if out-of-doors. Avoid direct Sun exposure during midday hours. Use protections mentioned above.

FOCUS STUDY 3.1

Stratospheric Ozone Losses: A Continuing Health Hazard

Consider:

- The stratospheric ozone loss above Antarctica during the months of Antarctic spring (September to November) has increased since 1979, with the record depletion in 2006 covering about 30 million km^2; subsequent years had similar losses, with 2008 recording the fifth largest loss. Recovery is proceeding more slowly than previously thought, despite the slowing rate of accumulation of offending chemicals.
- As stratospheric ozone flows into the depleted region from lower latitudes, ozone thins over southern South America, southern Africa, Australia, and New Zealand. In Ushuaia, Argentina, surface UVB intensity increases more than 225% compared to normal levels.
- An international scientific consensus confirmed the anthropogenic, or human-caused, disruption of the ozone layer. See *Scientific Assessment of Ozone Depletion,* by NASA, NOAA, UNEP (United Nations Environment Programme), and WMO.
- At Earth's opposite pole, a similar ozone depletion over the Arctic annually exceeds 30%, with a record 45% loss during spring 1997. The Canadian and U.S. governments report an "ultraviolet index" to help members of the public protect themselves (http://www.ec.gc.ca/ozone/).
- Increased ultraviolet radiation is affecting atmospheric chemistry, biological systems, oceanic phytoplankton (small photosynthetic organisms that form the basis of the ocean's primary food production), fisheries, crop yields, and human skin, eye tissues, and immunity.

More ultraviolet radiation than ever before is breaking through Earth's protective ozone layer. What is going on in the stratosphere everywhere and in the polar regions specifically? Why is this happening, and how are people and their governments responding? What effects does this have on you personally?

Monitoring Earth's Fragile Safety Screen

A sample from the ozone layer's densest part (at an altitude of 29 km, or 18 mi) contains only 1 part ozone per 4 million parts of air—compressed to surface pressure, the ozone layer would be only 3 mm thick. Yet this rarefied layer was in steady-state equilibrium for several hundred million years, absorbing intense ultraviolet radiation and permitting life to proceed safely on Earth.

The ozone layer has been monitored since the 1920s from ground stations. The *Total Ozone Mapping Spectrometer* (TOMS) began operations in 1978 aboard various satellites. Since 2004, the Ozone Monitoring Instrument (OMI) on board NASA's *Aura* satellite has tracked ozone depletion. Figure 3.1.1 shows the 2008 "ozone hole." (See http://ozonewatch.gsfc.nasa.gov/.)

Ozone Losses Explained

What is causing the decline in stratospheric ozone? In 1974, two atmospheric chemists, F. Sherwood Rowland and Mario Molina, hypothesized that some synthetic chemicals were releasing chlorine atoms that decompose ozone ("Stratospheric sink for chlorofluoromethanes . . . ," *Nature* 249 [1974]: 810). These **chlorofluorocarbons,** or **CFCs,** are synthetic molecules of chlorine, fluorine, and carbon.

CFCs are stable, or inert, under conditions at Earth's surface, and they possess remarkable heat properties. Both qualities made them valuable as propellants in aerosol sprays and as refrigerants. Also, some 45% of CFCs were solvents in the electronics industry and used as foaming agents. Being inert, CFC molecules do not dissolve in water and do not break down in biological processes. (In contrast, chlorine compounds derived from volcanic eruptions and ocean sprays are water soluble and rarely reach the stratosphere.)

Researchers Rowland and Molina hypothesized that stable CFC molecules slowly migrate into the stratosphere,

FIGURE 3.1.1 The Antarctic ozone hole. OMI image for September 12, 2008; the "hole," at nearly 27 million km^2 (10.4 million mi^2), ranks fifth in size in the satellite monitoring era; 2006 was the worst depletion year. Measurements dropped below 100 Dobson units in the region of blues and purples; greens, yellows, and reds denote more ozone. [OMI aboard *Aqua,* GSFC/NASA.]

where intense ultraviolet radiation splits them, freeing chlorine (Cl) atoms. This process produces a complex set of reactions that breaks up ozone molecules (O_3) and leaves oxygen gas molecules (O_2) in their place. The effect is severe, for single chlorine atoms decompose more than 100,000 ozone molecules. The long residence time of chlorine atoms in the ozone layer (40 to 100 years) is likely to produce long-term consequences through this century from the chlorine already in place. Tens of millions of tons of CFCs were sold worldwide since 1950 and subsequently released into the atmosphere.

Political Realities—An International Response and the Future

As the science became established, sales and production of CFCs declined until a March 1981 presidential order permitted the export and sales of banned products. Sales increased and hit a new peak in 1987 at 1.2 million metric tons (1.32 million tons), when an international agreement went into effect halting further sales growth.

Chemical manufacturers once claimed that no hard evidence existed to prove the ozone-depletion model, and they successfully delayed remedial action for 15 years. Today, with extensive scientific evidence and verification of losses, even the CFC manufacturers admit that the problem is serious.

The *Montreal Protocol on Substances That Deplete the Ozone Layer* (1987), as amended in 1990, 1992, 1995, 1997, and 1999, aims to reduce and eliminate CFC damage to the ozone layer. With 189 signatory countries, the protocol is regarded as probably the most successful international agreement in history (see http://ozone.unep.org/).

CFC sales continue their decline and all production of harmful CFCs ceased in 2010, although there is concern over some of the substitute compounds and a robust black market for banned CFCs, rivaling drug trafficking at some ports. If the protocol is fully enforced, scientists estimate that the stratosphere will return to more normal conditions in a century.

The world without this treaty might prove challenging. Imagine possible stratospheric global ozone losses of two-thirds, producing more cases of malignant melanoma (a skin cancer), and by 2100 a nonexistent ozone layer. We owe much to doctors Rowland and Molina.

For their work, Rowland, Molina, and another colleague, Paul Crutzen, received the 1995 Nobel Prize for Chemistry. In making the award, the Royal Swedish Academy of Sciences said, "By explaining the chemical mechanism that affects the thickness of the ozone layer, the three researchers have contributed to our salvation from a global environmental problem that could have catastrophic consequences."

Ozone Losses over the Poles

How do Northern Hemisphere CFCs become concentrated over the South Pole? Evidently, chlorine freed in the Northern Hemisphere midlatitudes concentrates over Antarctica through the work of atmospheric winds. Persistent cold temperatures over the South Pole and the presence of thin, icy clouds in the stratosphere promote development of the ozone hole, which forms in the Antarctic spring and usually peaks in September (Figure 3.1.2). Over the North Pole, conditions are more changeable, so the hole is smaller, although growing each year, during Arctic spring.

Polar stratospheric clouds (*PSCs*) are thin clouds that are important catalysts in the release of chlorine for ozone-depleting reactions. During the long, cold winter months, a tight circulation pattern forms over the Antarctic continent—the polar vortex. Chlorine that is freed from droplets in PSCs triggers the breakdown of the otherwise inert molecules and frees the chlorine for catalytic reactions.

Make a note to check some of the Internet data sources mentioned to see how our stratosphere is doing in years to come. What we learn here will be with us for the rest of this century.

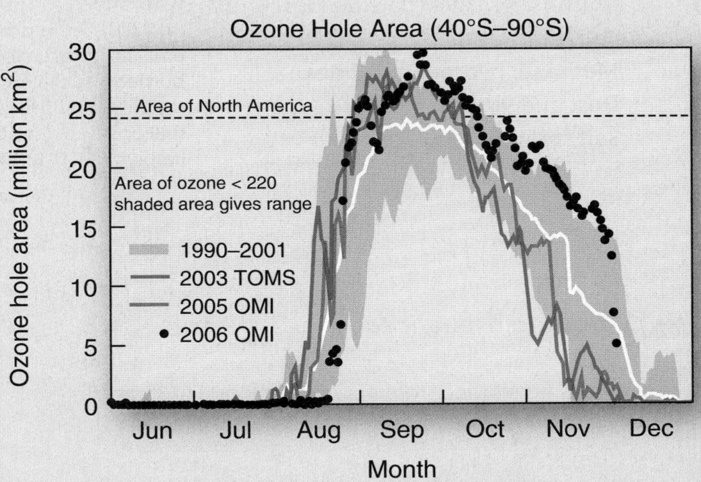

FIGURE 3.1.2 The timing and extent of ozone damage. The area of ozone depletion generally peaks in late September, reaching a record low size in 2006. [Data from GSFC/NASA.]

MG Satellite Southern Hemisphere Ozone 2002–2003

Natural Sources

Natural air pollution sources produce a greater quantity of pollutants—nitrogen oxides, carbon monoxide, hydrocarbons from plants and trees, and carbon dioxide—than do human-made sources. Table 3.3 lists some of these natural sources and the substances they contribute to the air. However, any attempt to dismiss the impact of human-made air pollution through a comparison with natural sources is irrelevant, for we evolved with and adapted to the natural ingredients in the air. We did not evolve in relation to the comparatively recent concentrations of *anthropogenic* (human-caused) contaminants in our metropolitan regions.

A dramatic natural source of pollution was the 1991 eruption of Mount Pinatubo in the Philippines (15° N 120° E), perhaps the twentieth century's second largest eruption. This event injected nearly 20 million tons of sulfur dioxide (SO_2) into the stratosphere. The spread of these emissions is shown in a sequence of satellite images in Figure 6.1, Chapter 6.

Devastating wildfires on several continents produce natural air pollution as each year's fire season worsens (Figure 3.7). Soot, ash, and gases darken skies and damage health in affected regions. Wind patterns spread the pollution from the fires to nearby cities, closing airports and forcing evacuations to avoid the health-related dangers. Wildfire smoke contains particulate matter (dust, smoke,

(a)

(b)

FIGURE 3.7 California wildfires fill the atmosphere with smoke.
(a) Wildfires in drought- and high-temperature-plagued portions of California, October 2007. More than 117,000 ha (50,000 acres) burned which continued through November. (b) Sierra Nevada mountain wildfires move along uncontrolled in 2008, consuming millions of acres of drought-damaged forest and chaparral.
[(a) *Terra* MODIS, NASA/GSFC; (b) Bobbé Christopherson.]

TABLE 3.3 Sources of Natural Variable Gases and Materials

Source	Contribution
Volcanoes	Sulfur oxides, particulates
Forest fires	Carbon monoxide and dioxide, nitrogen oxides, particulates
Plants	Hydrocarbons, pollens
Decaying plants	Methane, hydrogen sulfides
Soil	Dust and viruses
Ocean	Salt spray, particulates

GEO REPORT 3.3 *New atmospheric tools*

Studies of the atmosphere are benefitting from new technology. For example, small unpiloted aerial vehicles (UAVs) are lightweight aircraft guided by remote control. These instrumental drones can gather all types of data and carry various remote-sensing devices. In 2010, NASA's unpiloted Global Hawk completed its first scientific flight over the Pacific Ocean, flying for 14 hours on a preprogrammed flight path to altitudes above 18.3 km (60,000 ft) and carrying instruments to sample ozone depleting substances, aerosols, and other components of air quality in the upper troposphere and lower stratosphere. Aerosol-detecting satellites are in operation, such as Cloud-Aerosol Lidar and IR pathfinder Satellite (*CALIPSO*). Working since 2006, it measures atmospheric pollution and is discovering concentrations at 1 to 3 km altitudes.

soot, ash), nitrogen oxides, carbon monoxide, and volatile organic compounds.

Scientists established the connection between climate change and wildfire occurrence. This source of smoke is on the increase. A linkage exists between climate-change impacts and wildfires in the western United States:

> Higher spring and summer temperatures and earlier snowmelt are extending the wildfire season and increasing the intensity of wildfires in the western US . . . large wildfire activity increased suddenly and markedly in the mid-1980s, with higher large-wildfire frequency . . . longer wildfire durations, and longer wildfire seasons.*

These connections occur across the globe, as in drought-plagued Australia, where thousands of wildfires burned millions of hectares in recent years. Each year, worldwide, wildfire acreage breaks the record of the previous year (see http://www.usgs.gov/hazards/wildfires/).

Natural Factors That Affect Air Pollution

The problems resulting from both natural and human-made atmospheric contaminants are made worse by several important natural factors. Among these, wind, local and regional landscape characteristics, and temperature inversions in the troposphere dominate.

Winds Winds gather and move pollutants, sometimes reducing the concentration of pollution in one location while increasing it in another. Wind can produce dramatic episodes of dust movement, tracked by chemical analysis used by scientists to determine source areas. (Dust is defined as particles less than 62 μm, microns, or 0.0025 in.). Travelling on prevailing winds, dust from Africa contributes to the soils of South America and Europe, and Texas dust ends up across the Atlantic (Figure 3.8). Imagine at any one moment a billion tons of dust aloft in the atmospheric circulation, carried by the winds!

Winds make the atmosphere's condition an international issue. For example, prevailing winds transport air pollution from the United States to Canada, causing much complaint and negotiation between the two governments. Pollution from North America adds to European air problems. In Europe, the cross-boundary drift of pollution is an issue because of the proximity of countries, a factor in the formation of the European Union (EU).

Arctic haze is a term from the 1950s, when pilots noticed decreased visibility, either horizontally or at a slant angle from their aircraft, across the Arctic region. Haze is a concentration of microscopic particles and air pollution that diminishes air clarity. Since there is no heavy industry at these high latitudes and only sparse population, this seasonal haze is a remarkable attribute of industrialization in the Northern Hemisphere.

Also, an increase in wildfires across the Northern Hemisphere and agricultural burning in the midlatitudes contributes to Arctic haze. Simply, winds of atmospheric circulation transport pollution to sites far distant from points of origin. There is no comparable haze over the

(a)

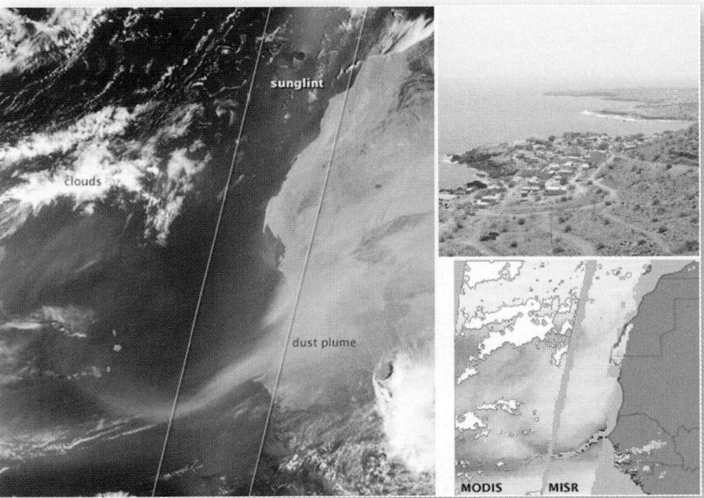

(b)

FIGURE 3.8 Natural variable dust in the atmosphere. (a) In 2009, dust in eastern Washington state from dryland agricultural fields is moved by winds up to 69 kmph (43 mph), closing some highways at the time. (b) Typical wind-carried dust moves off West Africa, across the Atlantic. The Cape Verde islands are usually in the path of African dust. Beneath such dust near Cape Verde at 15° N latitude, the ocean surface is difficult to distinguish from the sky on the horizon (inset photo). *Terra* MISR is able to gather data obscured by sunglint into the MODIS sensor. [(a) *Terra,* MODIS and (b) *Terra,* MODIS and MISR sensors; both NASA/GSFC; (inset) Bobbé Christopherson.]

*A. L. Westerling et al., "Warming and earlier spring increase western U.S. forest wildfire activity," *Science* 313, 5789 (Aug. 18, 2006): 940.

Antarctic continent. Based on this analysis, can you think of why Antarctica lacks such a condition?

Local and Regional Landscapes Local and regional landscapes are another important factor in air pollution. Surrounding mountains and hills can form barriers to air movement or can direct pollutants from one area to another. Some of the worst incidents result when local landscapes trap and concentrate air pollution.

Places with volcanic landscapes such as Iceland and Hawai'i have their own natural pollution. During periods of sustained volcanic activity at Kīlauea, some 2000 metric tons (2200 tons) of sulfur dioxide are produced a day. Concentrations are sometimes high enough to merit broadcast warnings about health concerns. The resulting acid rain and volcanic smog, called *vog* by Hawaiians (for *vo*lcanic smo*g*), cause losses to agriculture as well as other economic impacts.

Temperature Inversions Vertical temperature and atmospheric density in the troposphere also can worsen pollution conditions. A **temperature inversion** occurs when the normal temperature, which usually decreases with altitude (normal lapse rate), reverses trend and begins to increase at some point. This can happen at any elevation from ground level to several thousand meters.

Figure 3.9 compares a normal temperature profile with that of a temperature inversion. The normal profile (Figure 3.9a) permits warmer (less dense) air at the surface to rise, ventilating the valley and moderating surface pollution. But the warm-air inversion (Figure 3.9b) prevents the rise of cooler (denser) air beneath, halting the vertical mixing of pollutants with other atmospheric gases. Thus, instead of being carried away, pollutants are trapped under the *inversion layer*.

Inversions most often result from certain weather conditions, such as when the air near the ground is radiatively cooled on clear nights, or from topographic situations that produce cold-air drainage into valleys. In addition, the air above snow-covered surfaces or beneath subsiding air in a high-pressure system may cause a temperature inversion.

Anthropogenic Pollution

Anthropogenic, or human-caused, air pollution remains most prevalent in urbanized regions. Approximately 2% of annual deaths in the United States are attributable to air pollution. Comparable risks are identified in Canada, Europe, Mexico, and Asia, especially China and India.

As urban populations grow, human exposure to air pollution increases. By 2012, more than 50% of the world population will live in metropolitan regions, some one-third with unhealthful levels of air pollution. This represents a potentially massive public-health issue in this century.

Table 3.4 lists the names, chemical symbols, principal sources, and impacts of variable anthropogenic components in the air. The first seven pollutants in the table result from combustion of fossil fuels in transportation (specifically automobiles, including light trucks). For example, **carbon monoxide (CO)** results from incomplete

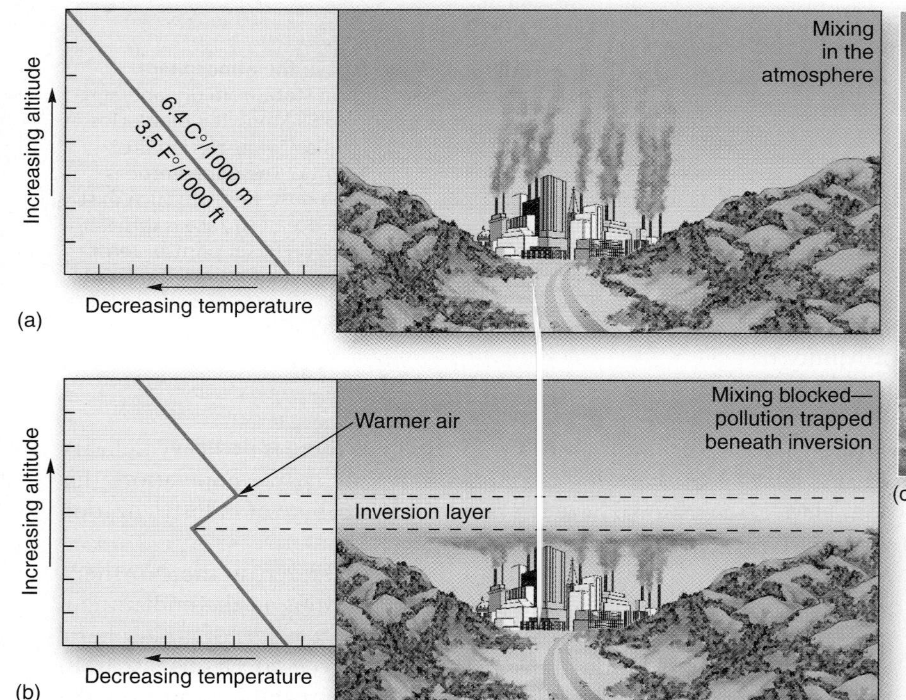

FIGURE 3.9 Normal and inverted temperature profiles.
(a) A normal temperature profile. (b) A temperature inversion in the lower atmosphere. Note how the warmer air layer prevents mixing of the denser (cooler) air below the inversion, thereby trapping pollution. (c) An inversion layer is visible in the morning hours over a valley. [(c) Bobbé Christopherson.]

TABLE 3.4 Anthropogenic Gases and Materials in the Lower Atmosphere

Name	Symbol	Sources	Description and Effects of Criteria Pollutants (Main sources listed for U.S.)
Carbon monoxide	CO	Incomplete combustion of fuels	Odorless, colorless, tasteless gas Toxicity is affinity for hemoglobin Displaces O_2 in bloodstream; 50 to 100 ppm cause headaches and vision and judgment losses *Source:* 78% transportation
Nitrogen oxides	NO_x (NO, NO_2)	High temperature/pressure combustion	Reddish-brown choking gas. Inflames respiratory system, destroys lung tissue Damages plants; 3 to 5 ppm is dangerous *Source:* 55% transportation
Volatile organic compounds	VOCs	Incomplete combustion of fossil fuels such as gasoline; cleaning and paint solvents	Prime agents of ozone formation *Sources:* 45% industrial 47% transportation
Ozone	O_3	Photochemical reactions	Highly reactive, unstable gas Oxidizes surfaces; dries rubber and elastic Damages plants at 0.01 to 0.09 ppm Agricultural losses at 0.1 ppm; 0.3 to 1.0 ppm irritates eyes, nose, throat
Peroxyacetyl nitrates	PANs	Photochemical reactions	Produced by NO + VOC photochemistry No human health effects Major damage to plants, forests, crops
Sulfur oxides	SO_x (SO_2, SO_3)	Combustion of sulfur-containing fuels	Colorless; irritating smell 0.1 to 1 ppm; impairs breathing, taste threshold Human asthma, bronchitis, emphysema Leads to acid deposition *Source:* 85% fuel combustion
Particulate matter	PM	Dust, dirt, soot, salt, metals, organics; agriculture, construction, roads	Complex mixture of solid and aerosol particles Dust, smoke, and haze affect visibility Various health effects: bronchitis, pulmonary function PM_{10} negative health effects established by researchers *Sources:* 42 % industrial 33% fuel combustion 25% transportation
Carbon dioxide	CO_2	Complete combustion, mainly from fossil-fuel consumption	Principal greenhouse gas Atmospheric concentration increasing; 64% of greenhouse warming effect
Methane	CH_4	Organic processes	Secondary greenhouse gas Atmospheric concentration increasing; 19% of greenhouse warming effect
Water vapor	H_2O vapor	Combustion processes, steam	See Chapter 7 for more on the role of water vapor in the atmosphere.

combustion, or when carbon in fuel does not burn completely. The main anthropogenic source of carbon monoxide is vehicle emissions.

Overall, automobiles contribute more than 60% of U.S. and 50% of Canadian human-caused air pollution. Reducing air pollution from the transportation sector involves known technologies and strategies, monetary savings for the consumer, and significant health benefits. These facts make the continuing delays in achieving better fuel efficiency all the more confusing.

Stationary sources, such as electric power plants and industrial plants that use fossil fuels, contribute the most sulfur oxides and particulates. For this reason, concentrations are focused in the Northern Hemisphere and

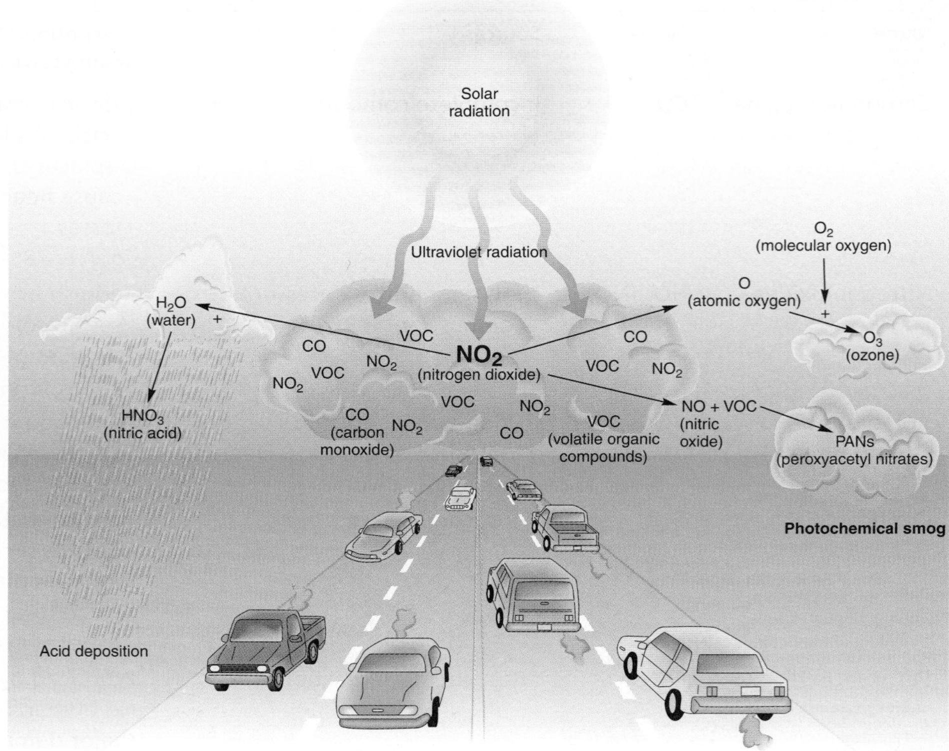

FIGURE 3.10 Photochemical reactions.
The interaction of automobile exhaust (NO_2, VOCs, CO) and ultraviolet radiation in sunlight causes photochemical reactions. To the left, note the formation of nitric acid and acid deposition.

the industrial, developed countries. The last three gases shown in Table 3.4 are discussed elsewhere in this text: Water vapor is examined with water and weather (Chapters 7 and 8); carbon dioxide and methane are discussed with greenhouse gases and climate (Chapters 4, 5, and 10).

Photochemical Smog Pollution Photochemical smog was not generally experienced until the advent of the automobile. Today, it is the major component of anthropogenic air pollution. **Photochemical smog** results from the interaction of sunlight and the combustion products in automobile exhaust (nitrogen oxides and VOCs). Although the term *smog*—a combination of the words *sm*oke and f*og*—is a misnomer, it is generally used to describe this pollution. Smog is responsible for the hazy sky and reduced sunlight in many of our cities.

The connection between automobile exhaust and smog was not determined until 1953 in Los Angeles, long after society had established its dependence upon cars and trucks. Despite this discovery, widespread mass transit declined, the railroads dwindled, and the polluting, inefficient individual automobile remains America's preferred transportation.

Figure 3.10 summarizes how car exhaust is converted into major air pollutants—PANs, ozone, and nitric acid. The high temperatures in automobile engines produce nitrogen dioxide (NO_2). Also, to a lesser extent nitrogen dioxide is emitted from power plants. Worldwide, the problem with **nitrogen dioxide** production is its concentration in metropolitan regions. North American urban areas may have from 10 to 100 times higher nitrogen dioxide concentrations than nonurban areas. Nitrogen dioxide interacts with water vapor to form nitric acid (HNO_3), a contributor to acid deposition by precipitation, the subject of Focus Study 3.2.

In the photochemical reaction in the illustration, ultraviolet radiation liberates atomic oxygen (O) and a ni-

GEO REPORT 3.4 ***Smoke gets in your eyes***

As your car idles at a downtown intersection or you walk through a parking garage, you may be exposed to between 50 ppm and 100 ppm of carbon monoxide, a colorless, odorless gas. Smoking a cigarette, CO levels reach 42,000 ppm at the burning ash. Secondhand smoke affects CO levels of anyone breathing nearby. What happens physiologically? CO combines with the oxygen-carrying hemoglobin of the blood, displacing the oxygen; thus, the hemoglobin no longer transports adequate oxygen to vital organs such as the heart and brain.

tric oxide (NO) molecule from the NO_2. The free oxygen atom combines with an oxygen molecule, O_2, to form the oxidant ozone, O_3. In addition, the nitric oxide (NO) molecule reacts with VOCs to produce *peroxyacetyl nitrates*. **Peroxyacetyl nitrates (PANs)** produce no known health effect in humans, but they are particularly damaging to plants, including both agricultural crops and forests.

Ozone is the primary ingredient in photochemical smog. (This is the same gas that is beneficial to us in the stratosphere in absorbing ultraviolet radiation.) The reactivity of ozone is a health threat, for it damages biological tissues. For several reasons, children are at greatest risk from ozone pollution—one in four children in U.S. cities is at risk of developing health problems from ozone pollution. This ratio is significant; it means that more than 12 million children are vulnerable in those metropolitan regions with the worst polluted air (Los Angeles, New York City, Atlanta, Houston, and Detroit). For more information and city rankings of the worst and the best, see http://www.lungusa.org(enter "city rankings" in the search box on the web site).

The **volatile organic compounds (VOCs)**, including hydrocarbons from gasoline, surface coatings, and combustion at electric utilities, are important factors in ozone formation. States such as California base their standards for control of ozone pollution on VOC emission controls—a scientifically accurate emphasis.

Industrial Smog and Sulfur Oxides Over the past 300 years, except in some developing countries, coal slowly replaced wood as the basic fuel used by society. The Industrial Revolution required high-grade energy to run machines. The changes involved conversion from *animate* energy (energy from animal sources, such as animal-powered farm equipment) to *inanimate* energy (energy from nonliving sources, such as coal, steam, and water). Air pollution associated with coal-burning industries is known as **industrial smog** (Figure 3.11). The term *smog* was coined by a London physician in A.D. 1900 to describe the combination of fog and smoke containing sulfur gases (sulfur is an impurity in fossil fuels).

Once in the atmosphere, **sulfur dioxide (SO_2)** reacts with oxygen (O) to form sulfur trioxide (SO_3), which is highly reactive and, in the presence of water or water vapor, forms **sulfate aerosols**, tiny particles about 0.1 to 1 μm in diameter. Sulfuric acid (H_2SO_4) can form even in moderately polluted air at normal temperatures. Coal-burning electric utilities and steel manufacturing are the main sources of sulfur dioxide. Sulfur dioxide–laden air is dangerous to health, corrodes metals, and deteriorates stone building materials at accelerated rates. Sulfuric acid deposition, added to nitric acid deposition, has increased in severity since it was first described in the 1970s. Focus Study 3.2 discusses this vital atmospheric issue and recent progress.

FIGURE 3.11 **Typical industrial smog.** Pollution generated by industry and coal-fired electrical generation differs from that produced by transportation. Industrial pollution has high concentrations of sulfur oxides, particulates, and carbon dioxide. This coal-fired power plant at Barentsburg, Svalbard, lacks scrubbers to reduce stack emissions. [Bobbé Christopherson.]

Particulates A diverse mixture of fine particles, both solid and aerosol, that impact human health is **particulate matter (PM)**. Haze, smoke, and dust are visible reminders of particulate material in the air we breathe. Such small particles in the air comprise **aerosols**. Remote sensing now provides a global portrait of such aerosols.

The effects of PM on human health vary with the size of the particle. For comparison, a human hair can range from 50 μm to 70 μm in diameter. $PM_{2.5}$, particulates that are 2.5 μm or less in diameter, pose the greatest health risk. These fine particles, such as combustion particles, organics, and metallic aerosols, can get into the lungs and bloodstream. Coarse particles (PM_{10}) are of less concern, although they can irritate a person's eyes, nose, and throat. New studies are implicating even smaller particles, known as *ultrafines*, which are $PM_{0.1}$. These are many times more potent than the larger $PM_{2.5}$ and PM_{10} particles because they can get into smaller channels in lung tissue and cause scarring, abnormal thickening, and damage called *fibrosis*.

Asthma prevalence has nearly doubled since 1980 in the United States, a major cause being motor vehicle–related air pollution—specifically, sulfur dioxide, ozone, and fine particulate matter.

We are now contributing significantly to the creation of the **anthropogenic atmosphere**, a tentative label for Earth's next atmosphere. The urban air we breathe today may be just a preview. What is the air quality like where you live, work, and go to college? Where could you go to find out its status?

FOCUS STUDY 3.2

Acid Deposition: Damaging to Ecosystems

Acid deposition is a major environmental problem in some areas of the United States, Canada, Europe, and Asia. Such deposition is most familiar as "acid rain," but it also occurs as "acid snow" and in dry form as dust or aerosols—tiny liquid droplets or solid particles. In addition, winds can carry the acid-producing chemicals many kilometers from their sources before they settle on the landscape, where they enter streams and lakes as runoff and groundwater flows.

Acid deposition is causally linked to serious problems: declining fish populations and fish kills in the northeastern United States, southeastern Canada, Sweden, and Norway; widespread forest damage in these same places and Germany; widespread changes in soil chemistry; and damage to buildings, sculptures, and historic artifacts. In New Hampshire's Hubbard Brook Experimental Forest (http://www.hubbardbrook.org/), a study covering 1960 to the present found half the nutrient calcium and magnesium base cations (see Chapter 18) were leached from the soil. Excess acids are the cause of the decline.

Despite scientific agreement about the problem, which the U.S. General Accounting Office calls a "combination of meteorological, chemical, and biological phenomena," corrective action was delayed by its complexity and politics.

The acidity of precipitation is measured on the pH scale, which expresses the relative abundance of free hydrogen ions (H+) in a solution. Free hydrogen ions in a solution are what make an acid corrosive, for they easily combine with other ions. The pH scale is logarithmic: Each whole number represents a tenfold change. A pH of 7.0 is neutral (neither acidic nor basic). Values less than 7.0 are increasingly *acidic*, and values greater than 7.0 are increasingly *basic*, or *alkaline*. A pH scale for soil acidity and alkalinity is portrayed graphically in Chapter 18.

Natural precipitation dissolves carbon dioxide from the atmosphere to form carbonic acid. This process releases hydrogen ions and produces an average pH reading for precipitation of pH 5.65. The normal range for precipitation is 5.3–6.0. Thus, normal precipitation is always slightly acidic.

Some anthropogenic gases are converted to acids in the atmosphere and then are removed by wet and dry deposition processes. Specifically, nitrogen and sulfur oxides released in the combustion of fossil fuels can produce nitric acid (HNO_3) and sulfuric acid (H_2SO_4) in the atmosphere.

Acid Precipitation Damage

Precipitation as acidic as pH 2.0 has fallen in the eastern United States, Scandinavia, and Europe. By comparison, vinegar and lemon juice register slightly less than 3.0. Aquatic plant and animal life perishes when lakes drop below pH 4.8.

More than 50,000 lakes and some 100,000 km (62,000 mi) of streams in the United States and Canada are at a pH level below normal (that is, below 5.3 pH), with several hundred lakes incapable of supporting any aquatic life. Acid deposition causes the release of aluminum and magnesium from clay minerals in the soil, and both of these are harmful to fish and plant communities.

Also, relatively harmless mercury deposits in lake-bottom sediments convert in acidified lake waters into highly toxic *methylmercury*, which is deadly to aquatic life. Local health advisories in 2 Canadian provinces and 22 U.S. states are regularly issued to warn those who fish of the methylmercury problem. Mercury atoms rapidly bond with carbon and move through biological systems as an *organometallic compound*.

Damage to forests results from the rearrangement of soil nutrients, the death of soil microorganisms, and an aluminum-induced calcium deficiency that is currently under investigation. The most advanced impact is seen in forests in Europe, especially in eastern Europe, principally because of its long history of burning coal and the density of industrial activity. In Germany and Poland, up to 50% of the forests are dead or damaged; in Switzerland, 30% are afflicted (Figure 3.2.1).

In the eastern United States, regional-scale decline in forest cover is significant, especially red spruce and sugar maples. In some maples, aluminum is collecting

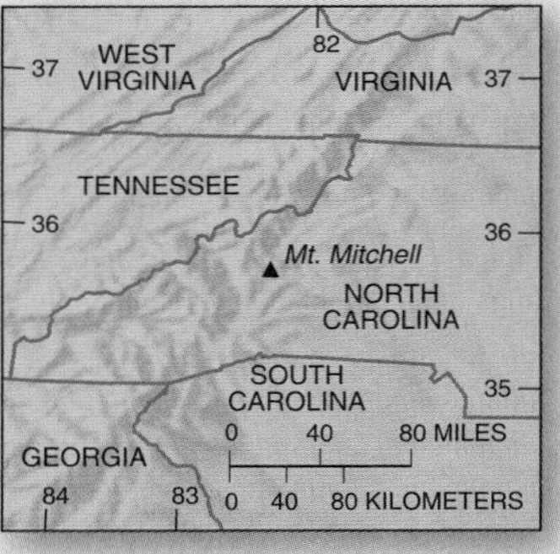

FIGURE 3.2.1 Acid deposition blight.
Stressed forests on Mount Mitchell in the Appalachians. [Will and Deni McIntyre/Photo Researchers.]

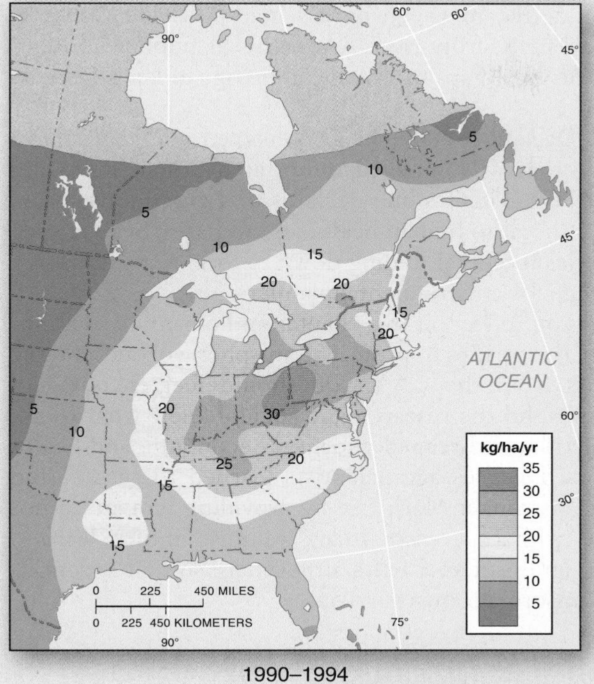

1990–1994

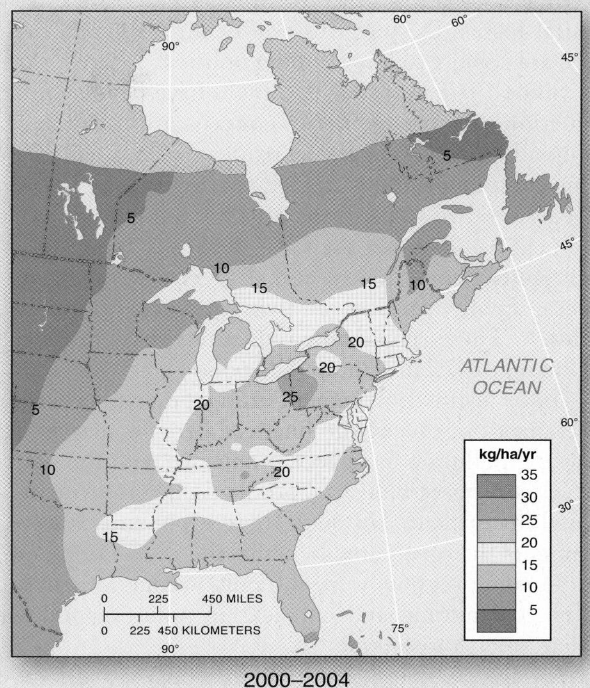

2000–2004

FIGURE 3.2.2 Improvement in sulfate wet deposition rate.
Spatial portrayal of annual sulfate (principally SO_4) wet deposition on the landscape, between two 5-year periods, in kilograms per hectare per year. The Canadian Environmental Protection Act, the Canada–United States Border Air Quality Strategy, and the U.S. Clean Air Act regulations together lowered levels of emissions that add acid to the environment. [Maps from Environment Canada, 2003. Reproduced with the permission of the Minister of Public Works and Government Services Canada.]

around rootlets; in spruce, acid fogs and rains leach calcium from needles directly. Affected trees are susceptible to winter cold, insects, and droughts. In red spruce, decline is evidenced by poor crown condition, reduced growth shown in tree ring analyses, and unusually high levels of tree mortality. For sugar maples, an indicator of forest damage is the reduction by almost half of the annual production of U.S. and Canadian maple sugar. Government estimates of damage in the United States, Canada, and Europe exceed $50 billion annually. Because wind and weather patterns are international, efforts at reducing acidic deposition also must be international in scope. Figure 3.2.2 maps the reduction in sulfate deposition between two 5-year periods, 1990–1994 and 2000–2004.

However, this progress is only a beginning. Research shows that power plants and other sources must cut sulfur dioxide emissions 80% beyond the Clean Air Act mandate to truly reverse damage trends.

Ten leading acid-deposition researchers reported in an extensive study in *BioScience*:

> Model calculations suggest that the greater the reduction in atmospheric sulfur deposition, the greater the magnitude and rate of chemical recovery. Less aggressive proposals for controls of sulfur emissions will result in slower chemical and biological recovery and in delays in regaining the services of a fully functional ecosystem. . . . North America and Europe are in the midst of a large-scale experiment. Sulfuric and nitric acids have acidified soils, lakes, and streams, thereby stressing or killing terrestrial and aquatic biota.*

Acid deposition is an issue of global spatial significance for which science is providing strong incentives for action. Reductions in troublesome emissions are closely tied to energy conservation and therefore directly related to production of greenhouse gases and global climate change concerns—thus, linking these environmental issues.

*C. T. Driscoll et al., "Acidic deposition in the northeastern United States: Sources and inputs, ecosystems effects, and management strategies," *BioScience* 51 (March 2001): 195.

Benefits of the Clean Air Act

The concentration of many air pollutants declined over the past several decades because of Clean Air Act (CAA) legislation (1970, 1977, 1990), resulting in the avoidance of trillions of dollars in health, economic, and environmental losses. Despite this reality, air pollution regulations are subject to a continuing political debate.

Since 1970 and the CAA, there have been significant reductions in atmospheric concentrations of carbon monoxide (−45%), nitrogen oxides (−22%), volatile organic compounds (−48%), PM_{10} particulates (−75%), sulfur oxides (−52%; see Focus Study 3.2, Figure 3.2.1), and lead (−98%). Prior to the CAA, lead was added to gasoline, emitted in exhaust, and dispersed over great distances, finally settling in living tissues, especially in children. These remarkable reductions show the successful linking of science and public policy.

To be justified, abatement (mitigation and prevention) costs must not exceed the financial benefits derived from reducing pollution damage. Compliance with the CAA affected patterns of industrial production, employment, and capital investment. Although these expenditures were investments that generated benefits, the dislocation and job loss in some regions were difficult—reductions in high-sulfur coal mining and cutbacks in polluting industries such as steel, for example.

In 1990, Congress requested the EPA to analyze the overall health, ecological, and economic benefits of the CAA compared with the costs of implementing the law. In response, the EPA's Office of Policy, Planning, and Evaluation performed an exhaustive cost-benefit analysis and published a report in 1997: *The Benefits of the Clean Air Act, 1970 to 1990.* The analysis provides a good lesson in cost-benefit analysis, valid to this day. The EPA report calculated the following:

- The *total direct cost* to implement all the federal, state, and local Clean Air Act rules from 1970 to 1990 was ***$523 billion*** (in 1990 dollars). This cost was borne by businesses, consumers, and government entities.
- The estimate of *direct economic benefits* from the Clean Air Act from 1970 to 1990 falls in a range from $5.6 to $49.4 trillion, with a central mean of ***$22.2 trillion***.
- Therefore, the estimated *net financial benefit* of the Clean Air Act is ***$21.7 trillion***!
- "The finding is overwhelming. The benefits far exceed the costs of the CAA," said EPA Administrator Richard Morgenstern in describing this 42-to-1 benefit-over-cost ratio.

The benefits to society, directly and indirectly, were widespread across the entire population, including improved health and environment, less lead to harm children, lowered cancer rates, less acid deposition, and an estimated 206,000 fewer deaths related to air pollution in 1990 alone. These benefits took place during a period in which the U.S. population grew by 22% and the economy expanded by 70%. The benefits continued between 1990 and 2010 as air quality continued to improve. Political efforts to weaken CAA regulations seem counterproductive to the progress made. Given a choice, which benefits from the CAA might the public choose to lose? Or would an informed public opt to keep benefits in place?

In December 2009, the EPA enacted an *Endangerment Finding* that will guide planning at national, state, and local levels. This finding declares that greenhouse gases "pose a threat to human health and welfare." Scaling such a significant federal finding down to local regulations is a challenge for environmental planners.

As you reflect on this chapter and our modern atmosphere, the treaties to protect stratospheric ozone, and the benefits from the CAA, you should feel encouraged. Scientists did the research and society knew what to do, took action, and reaped enormous economic and health benefits. Decades ago scientists learned to sustain and protect Astronaut Mark Lee by providing him an "atmosphere" with a spacesuit. Today, we must, in turn, learn to sustain and protect Earth's atmosphere so that it continues to ensure our own survival.

CRITICAL THINKING 3.3

Evaluating costs and benefits

In the scientific study *The Benefits of the Clean Air Act, 1970 to 1990,* the EPA determined that the Clean Air Act provided health, welfare, ecological, and economic benefits 42 times greater than its costs (estimated at $21.7 trillion in net benefits compared to $523 billion in costs). Do you think this is significant? Should this be part of the debate about future weakening or strengthening of the Clean Air Act? In your opinion, why is the public generally unaware of these details? What are the difficulties in informing the public?

Do you think a similar benefit pattern results from laws such as the Clean Water Act, or hazard zoning and planning, or action on global climate change and the Endangerment Finding by the EPA? Take a moment and brainstorm recommendations for action, education, and public awareness on these issues.

GEOSYSTEMS CONNECTION

We traversed the atmosphere from the thermosphere to Earth's surface. You learned the composition, temperature layers, and functions of the ionosphere and ozonosphere and the human impacts on the modern atmosphere. In the next chapter, we follow the flow of energy through the lower portions of the atmosphere as insolation makes its way to the surface. The Earth–atmosphere energy balance is established. Surface energy budgets are powered by this arrival of energy. We explore the possibilities of harvesting some of this solar energy.

KEY LEARNING CONCEPTS REVIEW

■ ***Construct*** a general model of the atmosphere based on three criteria—composition, temperature, and function—and ***diagram*** this model in a simple sketch.

Our modern atmosphere is a gaseous mixture so evenly mixed it behaves as if it were a single gas. The principal substance of this atmosphere is air—the medium of life. **Air** is naturally odorless, colorless, tasteless, and formless.

Above an altitude of 480 km (300 mi), the atmosphere is rarefied (nearly a vacuum) and is called the **exosphere,** which means "outer sphere." The weight (force over a unit area) of the atmosphere, exerted on all surfaces, is **air pressure**. It decreases rapidly with altitude.

By *composition*, we divide the atmosphere into the **heterosphere,** extending from 480 km (300 mi) to 80 km (50 mi), and the **homosphere,** extending from 80 km to Earth's surface. Within the heterosphere and using *temperature* as a criterion, we identify the **thermosphere**. Its upper limit, called the **thermopause,** is at an altitude of approximately 480 km. **Kinetic energy,** the energy of motion, is the vibrational energy that we measure and call temperature. The amount of heat actually produced in the thermosphere is very small. The density of the molecules is so low that little actual **heat,** the flow of kinetic energy from one body to another because of a temperature difference between them, is produced. Nearer Earth's surface, the greater number of molecules in the denser atmosphere transmits their kinetic energy as **sensible heat,** meaning that we can feel it.

The homosphere includes the **mesosphere, stratosphere,** and **troposphere,** as defined by temperature criteria. Within the mesosphere, cosmic or meteoric dust particles act as nuclei around which fine ice crystals form to produce rare and unusual night clouds, the **noctilucent clouds**.

The top of the troposphere is wherever a temperature of –57°C (–70°F) is recorded, a transition known as the **tropopause**. The normal temperature profile within the troposphere during the daytime decreases rapidly with increasing altitude at an average of 6.4 C° per km (3.5 F° per 1000 ft), a rate known as the **normal lapse rate**. The actual lapse rate at any particular time and place may deviate considerably because of local weather conditions and is called the **environmental lapse rate**.

We distinguish a region in the heterosphere by its *function*. The **ionosphere** absorbs cosmic rays, gamma rays, X-rays, and shorter wavelengths of ultraviolet radiation and converts them into kinetic energy. A functional region within the stratosphere is the **ozonosphere,** or **ozone layer,** which absorbs life-threatening ultraviolet radiation, subsequently raising the temperature of the stratosphere.

air (p. 60)
exosphere (p. 60)
air pressure (p. 62)
heterosphere (p. 63)
homosphere (p. 63)
thermosphere (p. 65)
thermopause (p. 65)
kinetic energy (p. 65)
heat (p. 65)
sensible heat (p. 65)
mesosphere (p. 65)
noctilucent clouds (p. 65)
stratosphere (p. 65)
troposphere (p. 65)
tropopause (p. 65)
normal lapse rate (p. 66)
environmental lapse rate (p. 66)
ionosphere (p. 66)
ozonosphere, ozone layer (p. 66)

1. What is air? Where did the components in Earth's present atmosphere originate?
2. In view of the cell analogy by Lewis Thomas, characterize the various functions the atmosphere performs that protect the surface environment.
3. What three distinct criteria are employed in dividing the atmosphere for study?
4. Describe the overall temperature profile of the atmosphere and list the four layers defined by temperature.
5. Describe the two divisions of the atmosphere on the basis of composition.
6. What are the two overall primary functional layers of the atmosphere and what does each do?

■ ***List*** the stable components of the modern atmosphere and their relative percentage contributions by volume, and ***describe*** each.

Even though the atmosphere's density decreases with increasing altitude in the homosphere, the blend (proportion) of gases is nearly uniform. This stable mixture of gases has evolved slowly.

The homosphere is a vast reservoir of relatively inert *nitrogen*, originating principally from volcanic sources and from bacterial action in the soil; *oxygen*, a by-product of photosynthesis; *argon*, constituting about 1% of the homosphere and completely inert; and *carbon dioxide*, a natural by-product of life processes and fuel combustion.

7. Name the four most prevalent stable gases in the homosphere. Where did each originate? Is the amount of any of these changing at this time?

■ ***Describe*** conditions within the stratosphere; specifically, ***review*** the function and status of the ozonosphere (ozone layer).

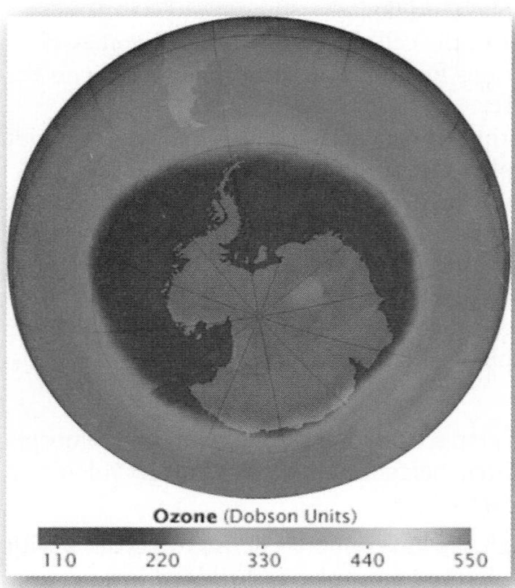

The overall reduction of the stratospheric ozonosphere, or ozone layer, during the past several decades represents a hazard for society and many natural systems and is caused by chemicals introduced into the atmosphere by humans. Since World War II, quantities of human-made **chlorofluorocarbons (CFCs)** and bromine-containing compounds have made their way into the stratosphere. The increased ultraviolet light at those altitudes breaks down these stable chemical compounds, thus freeing chlorine and bromine atoms. These atoms act as catalysts in reactions that destroy ozone molecules.

chlorofluorocarbons (CFCs) (p. 68)

8. Why is stratospheric ozone so important? Describe the effects created by increases in ultraviolet light reaching the surface.
9. Summarize the ozone predicament, and describe trends and any treaties to protect the ozone layer.
10. Evaluate Crutzen, Rowland, and Molina's use of the scientific method in investigating stratospheric ozone depletion and the public reaction to their findings.

■ ***Distinguish*** between natural and anthropogenic variable gases and materials in the lower atmosphere.

Within the troposphere, both natural and human-caused variable gases, particles, and other chemicals are part of the atmosphere. We coevolved with natural "pollution" and thus are adapted to it. But we are not adapted to cope with our own anthropogenic pollution. It constitutes a major health threat, particularly where people are concentrated in cities.

Vertical temperature and atmospheric density distribution in the troposphere can worsen pollution conditions. A **temperature inversion** occurs when the normal temperature decrease with altitude (normal lapse rate) reverses. In other words, temperature begins to increase at some altitude.

temperature inversion (p. 72)

11. Why are anthropogenic gases more significant to human health than those produced from natural sources?
12. In what ways does a temperature inversion worsen an air pollution episode? Why?

■ ***Describe*** the sources and effects of carbon monoxide, nitrogen dioxide, and sulfur dioxide and ***construct*** a simple diagram that illustrates photochemical reactions that produce ozone, peroxyacetyl nitrates, nitric acid, and sulfuric acid.

Odorless, colorless, and tasteless, **carbon monoxide (CO)** is produced by incomplete combustion (burning with limited oxygen) of fuels or other carbon-containing substances. Transportation is the major human-caused source for carbon monoxide. The toxicity of carbon monoxide is due to its affinity for blood hemoglobin, which is the oxygen-carrying pigment in red blood cells. In the presence of carbon monoxide, the oxygen is displaced and the blood becomes deoxygenated (Table 3.4).

Photochemical smog results from the interaction of sunlight and the products of automobile exhaust, the single largest contributor of pollution that produces smog. Car exhaust, containing *nitrogen dioxide* and *volatile organic compounds* (*VOCs*), in the presence of ultraviolet light in sunlight converts into the principal photochemical by-products—*ozone*, *peroxyacetyl nitrates* (*PANs*), and *nitric acid*.

Ozone (O_3) causes negative health effects, oxidizes surfaces, and kills or damages plants. **Peroxyacetyl nitrates (PANs)** produce no known health effects in humans, but are particularly damaging to plants, including both agricultural crops and forests. **Nitrogen dioxide (NO_2)** damages and inflames human respiratory systems, destroys lung tissue, and damages plants. Nitric oxides participate in reactions that produce nitric acid (HNO_3) in the atmosphere, forming both wet and dry acidic deposition. The **volatile organic compounds (VOCs),** including hydrocarbons from gasoline, surface coatings, and electric utility combustion, are important factors in ozone formation.

The distribution of human-produced **industrial smog** over North America, Europe, and Asia is related to transportation and electrical production. Such characteristic pollution contains **sulfur dioxide**. Sulfur dioxide in the atmosphere reacts to produce **sulfate aerosols,** which produce sulfuric acid (H_2SO_4) deposition, which can be dangerous to health and affect Earth's energy budget by scattering and reflecting solar energy. **Particulate matter (PM)** consists of dirt, dust, soot, and ash from industrial and natural sources. Small particles of dust, soot, and pollution suspended in the air are **aerosols**.

Increasing energy conservation and efficiency and reducing emissions are key to reducing air pollution. Earth's next atmosphere most accurately may be described as the **anthropogenic atmosphere** (human-influenced atmosphere).

carbon monoxide (CO) (p. 72)
photochemical smog (p. 74)
nitrogen dioxide (p. 74)
peroxyacetyl nitrates (PANs) (p. 75)
volatile organic compounds (VOCs) (p. 75)
industrial smog (p. 75)
sulfur dioxide (p. 75)
sulfate aerosols (p. 75)
particulate matter (PM) (p. 75)
aerosols (p. 75)
anthropogenic atmosphere (p. 75)

13. What is the difference between industrial smog and photochemical smog?
14. Describe the relationship between automobiles and the production of ozone and PANs in city air. What are the principal negative impacts of these gases?
15. How are sulfur impurities in fossil fuels related to the formation of acid in the atmosphere and acid deposition on the land?
16. In summary, what are the cost-benefit results from the first 20 years under Clean Air Act regulations?

MG MASTERING GEOGRAPHY

Visit www.masteringgeography.com for several figure animations, satellite loops, self-study quizzes, flashcards, visual glossary, case studies, career links, textbook reference maps, RSS Feeds, and an ebook version of *Geosystems*. Included are many geography web links to interesting Internet sources that support this chapter.

CHAPTER 4

Atmosphere and Surface Energy Balances

The phenomenal growth of the Phoenix metropolitan area between 1989 and 2009 is highlighted in these two *Landsat-5* images of southeastern Phoenix. The suburbs of Chandler to the lower right and Mesa just to the north represent urban sprawl with no central business core. South Mountain is to the lower left. This urban growth enhances the urban heat island effect on surface energy budgets. Because of the darker surfaces, impervious to water, Phoenix is about 7 C° to 11 C° warmer than surrounding rural regions. [*Landsat-5* TM images, NASA/GSFC.]

KEY LEARNING CONCEPTS

After reading the chapter, you should be able to:

- ***Identify*** the pathways of solar energy through the troposphere to Earth's surface: transmission, scattering, diffuse radiation, refraction, albedo (reflectivity), conduction, convection, and advection.
- ***Describe*** what happens to insolation when clouds are in the atmosphere, and ***analyze*** the effect of clouds and air pollution on solar radiation received at ground level.
- ***Review*** the energy pathways in the Earth–atmosphere system, the greenhouse effect, and the patterns of global net radiation.
- ***Plot*** the daily radiation curves for Earth's surface, and ***label*** the key aspects of incoming radiation, air temperature, and the daily temperature lag.
- ***Portray*** typical urban heat island conditions, and ***contrast*** the microclimatology of urban areas with that of surrounding rural environments.

Albedo Impacts, a Limit on Future Arctic Shipping?

For hundreds of years, explorers searched for the fabled "Northwest Passage"—a sea route connecting the Atlantic and Pacific oceans through the inland waterways of the Canadian Archipelago. The "Northeast Passage," or Northern Sea Route, is another Arctic shipping lane traversing the Russian coast and linking Europe and Asia. These northern routes eliminate the need to go through the Panama and Suez canals, respectively.

Arctic Ocean sea ice blocked these routes, making them unavailable for shipping—that is, until recently. In 2009, the first commercial navigation of the Northeast Passage by a pair of German container ships was successful owing to losses of Arctic ice associated with climate change.

The prospect of an ice-free Arctic Ocean is triggering commercial excitement for freighter and oil tanker traffic across these northern passages. With record losses of Arctic sea ice in 2007, the Northwest Passage achieved ice-free status and the Northeast Passage continued its ice-free summer status (Figure GN 4.1). The three years following had comparable ice losses caused by higher ocean and air temperatures across the region.

Given significant losses to multi-year ice, the National Snow and Ice Data Center (NSIDC) forecasted that summer ice-free conditions could happen for the Arctic Ocean before 2015. As the multi-year core ice fails, only the seasonal ice remains. This younger, thinner ice takes less energy to melt. Also, scientists report that strong Arctic winds blow ice southward, where it melts in warmer ocean waters, accounting for nearly a third of sea-ice loss.

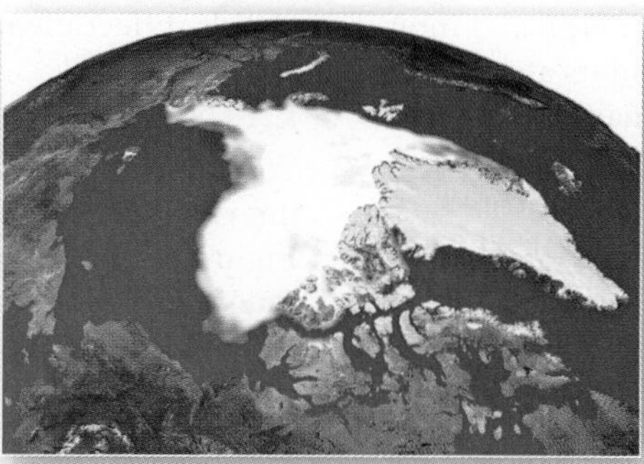

FIGURE GN 4.1 Arctic sea-ice extent. **Trace possible new shipping routes across the Northeast and Northwest passages on the 2007 satellite image.** [NASA/GSFC Image.]

The impacts of surface albedo also play a key role in sea ice melting. Day-to-day observations tell us that lighter surfaces tend to reflect sunlight and remain cooler, whereas darker surfaces tend to absorb sunlight and heat up. Snow- and ice-covered surfaces are great natural reflectors; in fact, sea ice reflects about 80%–95% of the solar energy it receives. The ocean surface is darker, reflecting only an average of about 10% of insolation. This percentage is albedo, or the reflective value of a surface. As the ice-covered area in the Arctic retreats, darker water or land receives direct sunlight and absorbs more heat, which adds to warming—a positive feedback.

Particulates in the atmosphere appear to enhance this albedo impact. Scientists now have evidence that soot and particulate aerosol fallout are significant causes of glacial snow and ice losses across the Himalayas. On the Greenland Ice Sheet, black soot, dust, and particulates accumulate with snow and harden into the ice, darkening the surface. As surface albedo decreases, ice absorbs more insolation, and increased melting causes soot to become even more concentrated and darker. The stack emissions from Arctic shipping produce similar impacts.

The 2009 *Arctic Marine Shipping Assessment* (Arctic Council, p. 140) stated that "ship stack emissions in and near the Arctic will increase." Shipping has already expanded by more than three weeks in the Arctic and two months in Hudson Bay and Manitoba since 2000 (Figure GN 4.2). The stack emissions from new freighter and tanker traffic will add soot and particulate material into the Arctic atmosphere. Fallout from these ship emissions could prove to be an albedo disaster, darkening remaining ice shelves and glaciers.

Whether to open the Northwest and Northeast passages to ship traffic is a question of surface albedo values and the principles we cover in Chapter 4. So far, these surface-energy budget considerations have not entered political or economic discussions.

FIGURE GN 4.2 Ship traffic and shipments in the Arctic Ocean. [Left–Canadian Coast Guard; right–German Shipowners Assn.]

Earth's biosphere pulses with flows of solar energy that sustain our lives and empower natural systems. Changing seasons, daily weather conditions, and the variety of Earth's climates remind us of this constant flow of energy cascading through the atmosphere. Earth's shifting seasonal rhythms are described in Chapter 2, and a profile of the atmosphere in Chapter 3. Energy and moisture exchanges between Earth's surface and its atmosphere are essential elements of weather and climate, discussed in later chapters.

In this chapter: We follow solar energy through the troposphere to Earth's surface. The *outputs* of reflected light and emitted longwave energy from the atmosphere and surface environment counter the *input* of insolation. Together, this input and output determine the net energy available to perform work. We examine surface energy budgets and analyze how net radiation is spent. Focus Study 4.1 on page 98 discusses solar energy, an applied aspect of surface energy and a renewable energy resource of great potential.

The chapter concludes with a look at the unique energy and moisture environment in our cities. The view across a hot parking lot of cars and pavement in summer is all too familiar—the air shimmers as heat energy radiates skyward. The climate of our urban areas differs measurably from that of surrounding rural areas, as demonstrated by the chapter-opening images.

Energy Essentials

When you look at a photograph of Earth taken from space, you clearly see the surface receipts of incoming insolation (see the photo of Earth on the back cover of this book). Land and water surfaces, clouds, and atmospheric gases and dust intercept solar energy. The flows of energy are in swirling weather patterns, powerful oceanic currents, and varied land-cover vegetation. Specific energy patterns differ for deserts, oceans, mountaintops, plains, rain forests, and ice-covered landscapes. In addition, the presence or absence of clouds may make a 75% difference in the amount of energy that reaches the surface because clouds reflect incoming energy.

Energy Pathways and Principles

Solar energy that heats Earth's atmosphere and surface is unevenly distributed by latitude and fluctuates seasonally. Figure 4.1 is a simplified flow diagram of shortwave and longwave radiation in the Earth–atmosphere system, discussed in the pages that follow. You will find it helpful to refer to this figure, and the more detailed energy-balance illustration in Figure 4.10 on page 92, as you read through the following section. We first look at some important pathways and principles for insolation as it passes through the atmosphere to Earth's surface.

Transmission refers to the passage of shortwave and longwave energy through either the atmosphere or water. Our budget of atmospheric energy comprises shortwave radiation *inputs* (ultraviolet light, visible light, and near-infrared wavelengths) and longwave radiation *outputs* (thermal infrared) that pass through the atmosphere by transmission.

Insolation Input

Insolation is the single energy input driving the Earth–atmosphere system. The world map in Figure 4.2 shows the distribution of average annual solar energy received at Earth's surface. It includes all the radiation that arrives at Earth's surface, both direct and diffuse (scattered by the atmosphere).

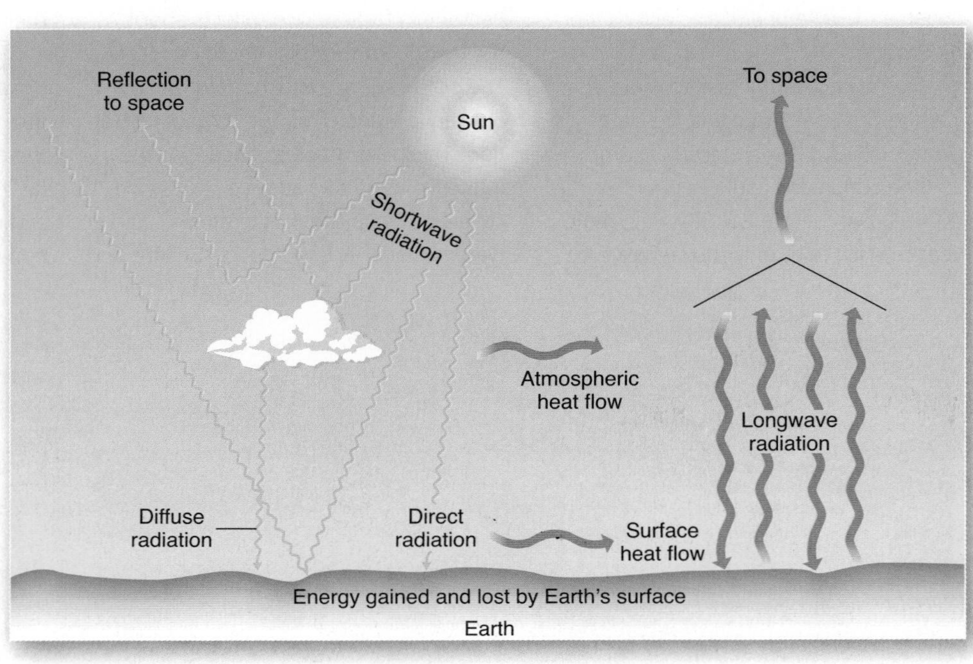

FIGURE 4.1 Simplified view of the Earth–atmosphere energy system.
Energy gained and lost by Earth's surface and atmosphere. Circuits include incoming shortwave insolation, reflected shortwave radiation, and outgoing longwave radiation.

Animation
Global Warming, Climate Change

Animation
Earth–Atmosphere Energy Balance

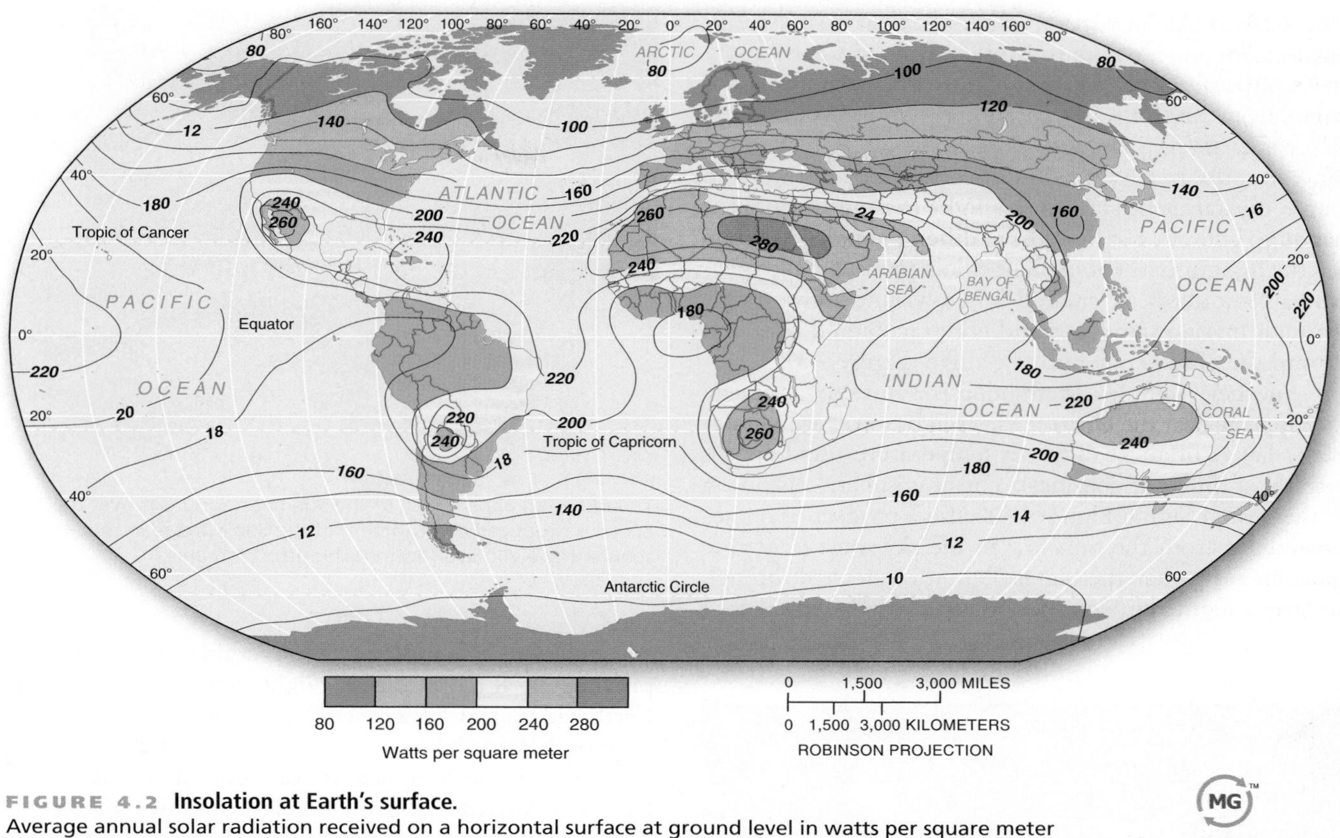

FIGURE 4.2 Insolation at Earth's surface.
Average annual solar radiation received on a horizontal surface at ground level in watts per square meter (100 W/m² = 75 kcal/cm²/year). [After M. I. Budyko, *The Heat Balance of the Earth's Surface* (Washington, DC: U.S. Department of Commerce, 1958), p. 99.]

MG™ Satellite
Global Shortwave Radiation

Several patterns are notable on the map. Insolation decreases poleward from about 25° latitude in both the Northern and the Southern hemispheres. Consistent daylength and high Sun altitude produce average annual values of 180–220 watts per square meter (W/m^2) throughout the equatorial and tropical latitudes. In general, greater insolation of 240–280 W/m^2 occurs in low-latitude deserts worldwide because of frequently cloudless skies. Note this energy pattern in the cloudless subtropical deserts in both hemispheres (for example, the Sonoran, Saharan, Arabian, Gobi, Atacama, Namib, Kalahari, and Australian deserts).

Scattering (Diffuse Radiation) Insolation encounters an increasing density of atmospheric gases as it travels toward the surface. Atmospheric gases, dust, cloud droplets, water vapor, and pollutants physically interact with insolation. The gas molecules redirect radiation, changing the direction of the light's movement without altering its wavelengths. **Scattering** describes this phenomenon and represents 7% of Earth's reflectivity, or albedo (see Figure 4.10).

Have you wondered why Earth's sky is blue? And why sunsets and sunrises are often red? These simple questions have an interesting explanation, based upon a principle known as Rayleigh scattering (named for English physicist Lord Rayleigh, 1881). This principle relates wavelength to the size of molecules or particles that cause the scattering.

The general rule is this: *The shorter the wavelength, the greater the scattering; the longer the wavelength, the lesser the scattering.* Small gas molecules in the air scatter shorter wavelengths of light. Thus, the shorter wavelengths of visible light—the blues and violets—scatter the most and dominate the lower atmosphere. And because there are more blue than violet wavelengths in sunlight, a blue sky prevails. A sky filled with smog and haze appears almost white because the larger particles associated with air pollution act to scatter all wavelengths of visible light.

The angle of the Sun's rays determines the thickness of the atmosphere through which they must pass to reach the surface. Therefore, direct rays (from overhead) experience less scattering and absorption than do low, oblique-angle rays, which must travel farther through the atmosphere. Insolation from the low-altitude Sun undergoes more scattering of shorter wavelengths, leaving only the residual oranges and reds to reach the observer at sunset or sunrise.

Some incoming insolation is diffused by clouds and atmosphere and transmits to Earth as **diffuse radiation**, the downward component of scattered light (labeled in Figure 4.10). Such diffuse light is multidirectional and thus casts shadowless light on the ground.

Refraction As insolation enters the atmosphere, it passes from one medium to another, from virtually empty space into atmospheric gases. This occurs when insolation passes from air into water, as well. The transition subjects the insolation to a change of speed, which also shifts its direction, the bending action of **refraction**. In the same way, a crystal or prism refracts light passing through it, bending different wavelengths to different angles, separating the light into its component colors to display the spectrum. A rainbow is created when visible light passes through myriad raindrops and is refracted and reflected toward the observer at a precise angle (Figure 4.3).

Another example of refraction is a **mirage**, an image that appears near the horizon where light waves are refracted by layers of air at different temperatures (and consequently of different densities). The atmospheric distortion of the setting Sun in Figure 4.4 is also a product of refraction—light from the Sun low in the sky must penetrate more air than when the Sun is high; it is refracted through air layers of different densities on its way to the observer.

FIGURE 4.3 A rainbow.
Raindrops—and in this photo, moisture droplets from the Niagara River—refract and reflect light to produce a primary rainbow. Note that in the primary rainbow the colors with the shortest wavelengths are on the inside and those with the longest wavelengths are on the outside. In the secondary bow, note that the color sequence is reversed because of an extra angle of reflection within each moisture droplet. [Bobbé Christopherson.]

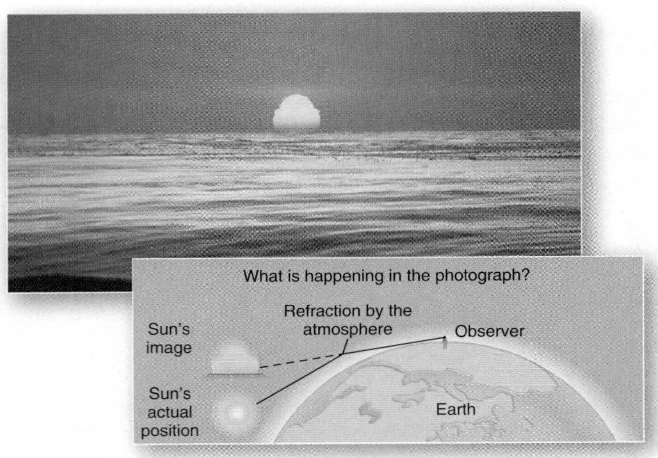

FIGURE 4.4 Sun refraction.
The distorted appearance of the Sun nearing sunset over the ocean is produced by refraction of the Sun's image in the atmosphere. Have you ever noticed this effect? [Author.]

An interesting function of refraction is that it adds approximately 8 minutes of daylight that we would lack if Earth had no atmosphere. We see the Sun's image about 4 minutes before the Sun actually peeks over the horizon. Similarly, as the Sun sets, its refracted image is visible from over the horizon for about 4 minutes afterward. These extra minutes vary with atmospheric temperature, moisture, and pollutants.

Albedo and Reflection A portion of arriving energy bounces directly back into space without being absorbed or performing any work—this is the **reflection** process. **Albedo** is the reflective quality, or intrinsic brightness, of a surface. It is an important control over the amount of insolation that is available for absorption by a surface. We state albedo as the percentage of insolation that is reflected (0% is total absorption; 100% is total reflectance). Examine the different surfaces and their albedo values in Figure 4.5.

In the visible wavelengths, darker colors have lower albedos, and lighter colors higher albedos. On water surfaces, the angle of the solar rays also affects albedo values: Lower angles produce a greater reflection than do higher angles. In addition, smooth surfaces increase albedo, whereas rougher surfaces reduce it.

GEO REPORT 4.1 *Earthshine*

Some of Earth's reflection of light to space shines on the Moon. We have all seen the effect of *earthshine* as a faint illumination of the dark portion of the Moon, especially in crescent phase, when the dark area of the lunar disk is barely visible. In contrast, the Moon reflects only 6% to 8% of the sunlight that hits its surface, producing the "moonlight" familiar to us all. Scientists from the New Jersey Institute of Technology and California Institute of Technology now measure earthshine at the Big Bear Solar Observatory (see http://www.bbso.njit.edu/). Earthshine studies began in 1925. Think about earthshine the next time you take a moonlit walk.

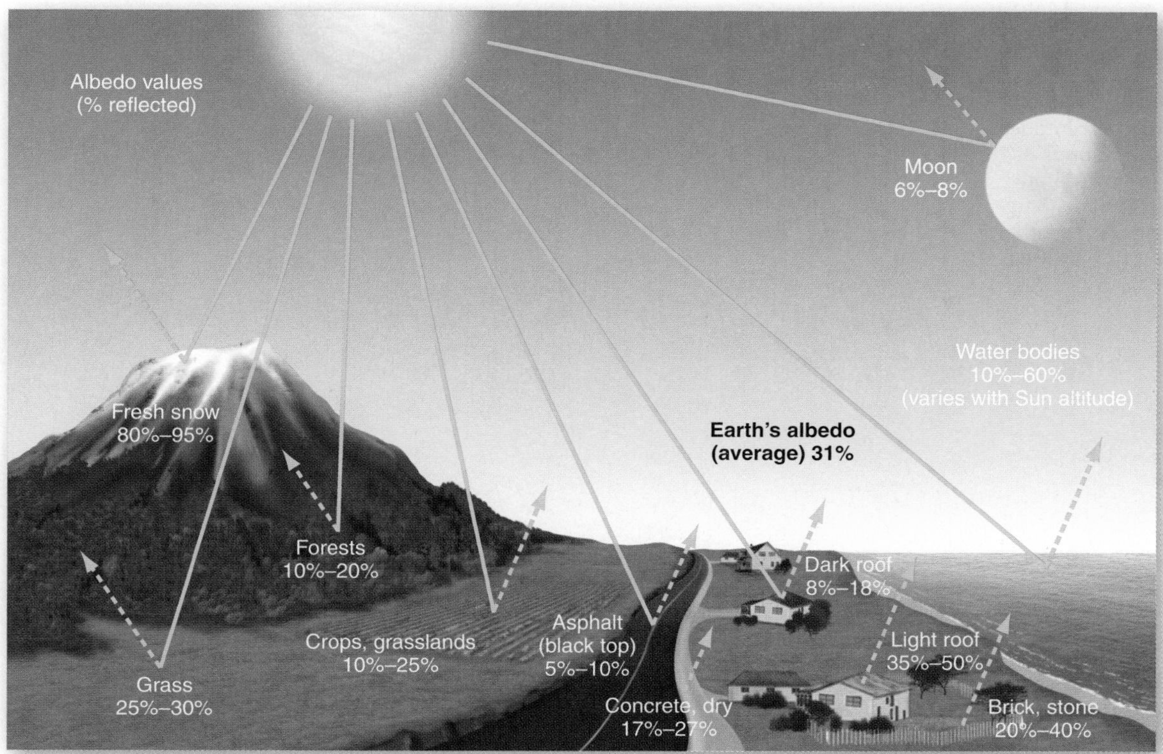

Satellite
Global Albedo Values, Global Shortwave Radiation

FIGURE 4.5 Various albedo values.
In general, light surfaces are more reflective than dark surfaces and thus have higher albedo values.

Specific locations experience highly variable albedo values during the year in response to changes in cloud and ground cover. Earth Radiation Budget (ERB) sensors aboard the *Nimbus*-7 satellite measured average albedos of 19%–38% for all surfaces between the tropics (23.5° N to 23.5° S), whereas albedos for the polar regions measured as high as 80% as a result of ice and snow. Tropical forests are characteristically low in albedo (15%), whereas generally cloudless deserts have higher albedos (35%).

Earth and its atmosphere reflect 31% of all insolation when averaged over a year. By comparison, a full Moon, which is bright enough to read by under clear skies, has only a 6%–8% albedo value. Thus, with *earthshine* being four times brighter than moonlight (four times the albedo), and with Earth four times greater in diameter than the Moon, it is no surprise that astronauts report how startling our planet looks from space.

Clouds, Aerosols, and the Atmosphere's Albedo

Clouds are an unpredictable factor in the tropospheric energy budget and therefore in the refining of climatic models. They reflect insolation and thus cool Earth's surface. The term **cloud-albedo forcing** refers to an increase in albedo caused by clouds. Yet clouds act as insulation, trapping longwave radiation from Earth and raising minimum temperatures. **Cloud-greenhouse forcing** is an increase in greenhouse warming caused by clouds. Figure 4.6 on page 88 illustrates the general effects of clouds on shortwave radiation and longwave radiation. Chapter 7 details more on clouds.

The eruption of Mount Pinatubo in the Philippines, which began explosively during June 1991, illustrates how Earth's internal processes affect the atmosphere. Approximately 15–20 megatons of sulfur dioxide droplets were injected into the stratosphere; winds rapidly spread these aerosols worldwide (see the images in Figure 6.1). As a result, atmospheric albedo increased worldwide and produced a temporary average cooling of 0.5°C (0.9°F).

Other mechanisms affect atmospheric albedo and therefore atmospheric and surface energy budgets. Industrialization is producing a haze of pollution, including sulfate aerosols, soot and fly ash, and black carbon, that is increasing the reflectivity of the atmosphere. Emissions of sulfur dioxide and subsequent chemical reactions in the atmosphere form *sulfate aerosols*. These aerosols act as an insolation-reflecting haze in clear-sky conditions.

Pollution causes both an atmospheric warming through absorption by the pollutants and a surface cooling through reduction in insolation reaching the surface. An overall term describes this decline in insolation making it to Earth's surface—**global dimming**. In determining present temperature increases associated with climate change, the dimming of sunlight is causing an underestimation of the actual amount of warming.

FIGURE 4.6 The effects of clouds on shortwave and longwave radiation.
(a) Clouds reflect and scatter shortwave radiation; a high percentage is returned to space. (b) Clouds absorb and radiate longwave radiation emitted by Earth; some longwave energy returns to space and some radiates toward the surface.

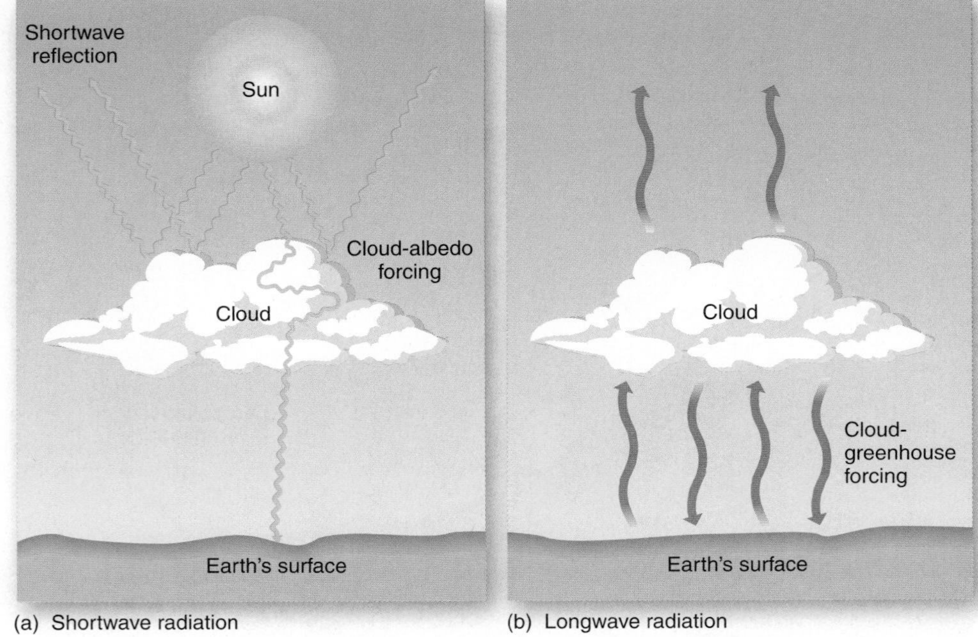

Images from the Clouds and the Earth Radiant Energy System (CERES) in Figure 4.7 show pollution-haze and soot effects on the atmospheric energy budget in the Indian Ocean region. The increased aerosols (1), containing black carbon and particulates, increase reflection off the atmosphere, thus raising atmospheric albedos (2). This causes an increase in atmospheric warming, as aerosols absorb energy (3), and an increase in surface cooling, as less insolation reaches Earth's surface (4). These aerosols reduce surface insolation by 10% and increase energy absorption in the atmosphere by 50%.

Such pollution effects in southern Asia influence the dynamics of the Asian monsoon through alteration of the

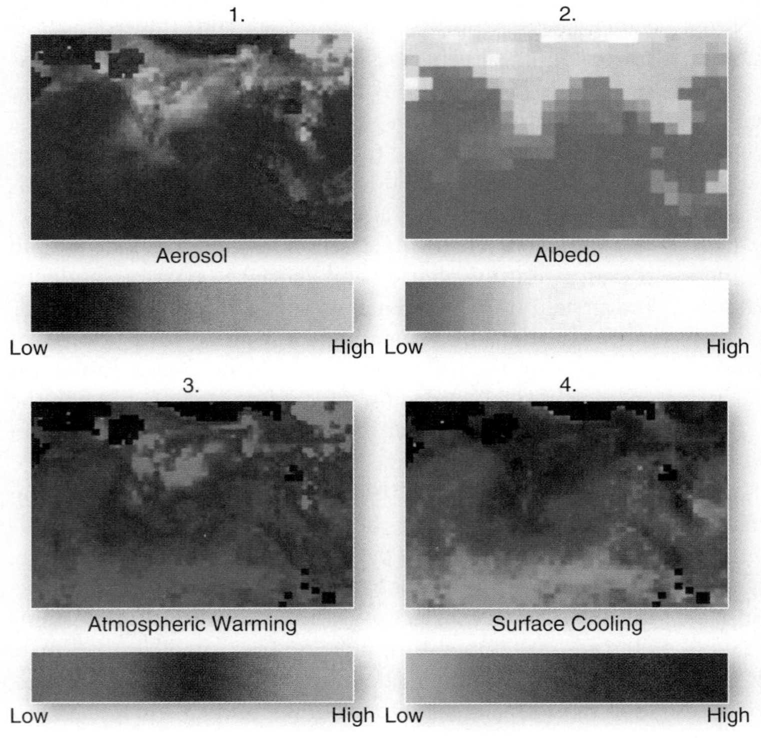

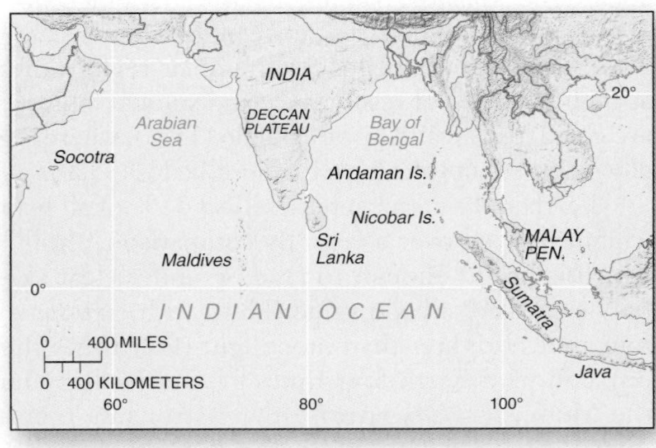

FIGURE 4.7 Aerosols impact Earth–atmosphere energy budgets.
Aerosols both reflect and absorb incoming insolation as identified in these four images from the CERES sensors aboard satellite *Terra,* made between January and March 2001 over southern Asia and the Indian Ocean. Note the color scale for each image rating from low to high values. [GSFC/NASA.]

region's Earth–atmosphere energy budget (see monsoons in Chapter 6). A weakening monsoonal flow will negatively impact regional water resources and agriculture. This research is part of the international, multiagency Indian Ocean Experiment (INDOEX; see http://www-indoex.ucsd.edu/).

Absorption The *assimilation* of radiation by molecules of matter and its conversion from one form of energy to another are **absorption**. Insolation, both direct and diffuse, that is not part of the 31% reflected from Earth's surface and atmosphere is absorbed. It is converted into either longwave radiation or chemical energy by plants in photosynthesis.

The temperature of the absorbing surface is raised in the process, and warmer surfaces radiate more total energy at shorter wavelengths—thus, *the hotter the surface, the shorter the wavelengths that are emitted.* Examples are an electric burner beginning to glow as it heats, or molten metal glowing at high temperatures. In addition to absorption by land and water surfaces (about 45% of incoming insolation), absorption occurs in atmospheric gases, dust, clouds, and stratospheric ozone (together about 24% of incoming insolation). Figure 4.10 summarizes the pathways of insolation and the flow of heat in the atmosphere and at the surface.

Conduction, Convection, and Advection There are several means of transferring heat energy in a system. **Conduction** is the molecule-to-molecule transfer of heat energy as it diffuses through a substance. As molecules warm, their vibration increases, causing collisions that produce motion in neighboring molecules, thus transferring heat from warmer to cooler materials.

Different materials (gases, liquids, and solids) conduct sensible heat directionally from areas of higher temperature to those of lower temperature. This heat flow transfers energy through matter at varying rates, depending on the conductivity of the material—Earth's land surface is a better conductor than air; moist air is a slightly better conductor than dry air.

Gases and liquids also transfer energy by movements of **convection** when the physical mixing involves a strong vertical motion. When horizontal motion dominates, the term **advection** applies. In the atmosphere or in bodies of water, warmer (less dense) masses tend to rise and cooler (denser) masses tend to sink, establishing patterns of convection. Sensible heat transports physically through the medium in this way.

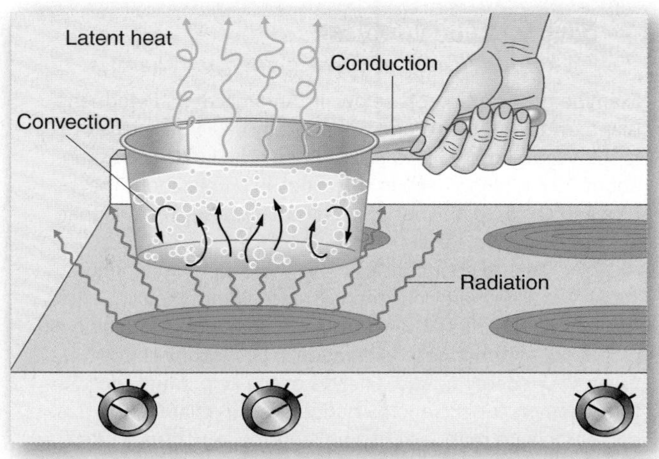

FIGURE 4.8 A pan of water on the stove illustrates heat transfer processes.
Infrared energy *radiates* from the burner to the saucepan and the air. Energy *conducts* through the molecules of the pan and the handle. The water physically mixes, carrying heat energy by *convection.*

You experience such energy flows in the kitchen: Energy is conducted through the handle of a pan, and boiling water bubbles in the saucepan in convective motions (Figure 4.8). Also in the kitchen, you may use a convection oven that has a fan to circulate heated air to uniformly cook food.

In physical geography, we find many examples of each physical transfer mechanism:

- *Conduction* includes surface energy budgets, temperature differences between land and water bodies and between darker and lighter surfaces, the heating of surfaces and overlying air, and soil temperatures;
- *Convection* is in atmospheric and oceanic circulation, air mass movements and weather systems, internal motions deep within Earth that produce a magnetic field, and movements in the crust; and
- *Advection* involves horizontal movement of winds from land to sea and back, fog that forms and moves to another area, and air mass movements from source regions.

Now, with these pathways and principles in mind, we put it all together to describe the energy budget of the lower atmosphere.

GEO REPORT 4.2 ***Monitoring Earth's radiant energy systems***

To better understand the role of natural clouds, as well as jet contrails and aerosols, NASA began operating sensors aboard satellites in 1997 with the Tropical Rainfall Measuring Mission (TRMM), followed by *Terra* and *Aqua* in 1999 and 2002, respectively. See CERES at http://science.larc.nasa.gov/; click on "Site Map" and then "Clouds and Aerosols" to review the various programs and spacecraft involved in this science.

CRITICAL THINKING 4.1

A kelp indicator of surface energy dynamics

Examine the photograph below. In Antarctica, on Petermann Island off the Graham Land Coast, we noticed this piece of kelp (a seaweed) dropped by a passing bird. The kelp was resting about 10 cm (4 in.) deep in the snow, in a hole about the shape of the kelp. In your opinion, what energy factors operated to make this scene?

Now, expand your conclusion to the issue of mining in Antarctica. There are coal deposits and other minerals in Antarctica. The international Antarctic Treaty blocks mining exploitation, although reconsideration is possible after 2048. Given your energy budget analysis of the kelp in the snow, information on surface energy budgets in this chapter, the dust and particulate output of mining, and the importance of a stable sea level, construct a case opposed to such mining activities based on these factors. Further, what factors can you think of that would favor such mining?

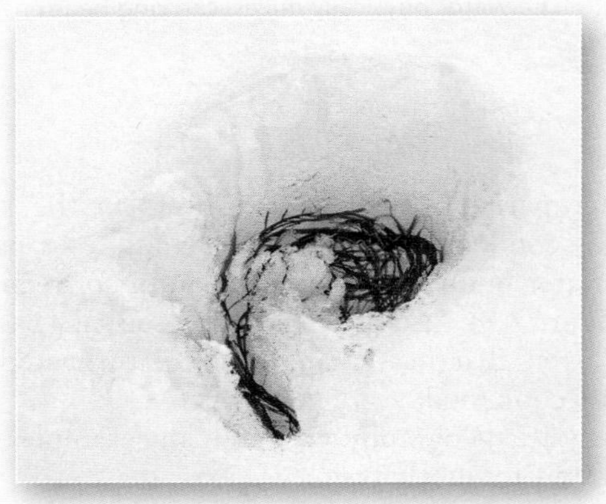

Energy Balance in the Troposphere

The Earth–atmosphere energy system budget naturally balances itself in a steady-state equilibrium. The atmosphere and surface eventually radiate longwave energy back to space, and this energy, together with reflected shortwave energy, equals the initial solar input—think of cash flows into and out of a checking account and the balance when cash deposits and withdrawals are equal.

Greenhouse gases in the atmosphere effectively delay longwave energy losses to space and act to warm the lower atmosphere. Let us examine this greenhouse effect and then develop an overall budget for the troposphere.

The Greenhouse Effect and Atmospheric Warming

Previously, we characterized Earth as a cool-body radiator, emitting energy in longer wavelengths from its surface and atmosphere toward space. However, some of this longwave radiation is absorbed by carbon dioxide, water vapor, methane, nitrous oxide, chlorofluorocarbons (CFCs), and other gases in the lower atmosphere and then emitted back toward Earth. This absorption and emission of energy are an important factor in warming the troposphere. The rough similarity between this process and the way a greenhouse operates gives the process its name—the **greenhouse effect**.

In a greenhouse, the glass is transparent to shortwave insolation, allowing light to pass through to the soil, plants, and materials inside, where absorption and conduction take place. The absorbed energy is then radiated as longwave radiation toward the glass, but the glass physically traps both the longer wavelengths and the warmed air inside the greenhouse. Thus, the glass acts as a one-way filter, allowing the shortwave energy in, but not allowing the longwave energy out except through conduction. You experience the same process in a car parked in direct sunlight.

Opening the greenhouse roof vent, or the car windows, allows the air inside to mix with the outside environment, thereby removing heated air physically from one place to another by convection. The interior of a car gets surprisingly hot, even on a day with mild temperatures outside. Many people place a removable sunscreen across the windshield to prevent shortwave energy from entering the car to begin the greenhouse process. Have you taken steps to reduce the insolation input into your car's interior or through windows at home?

Overall, the atmosphere behaves a bit differently. In the atmosphere, the greenhouse analogy does not fully apply because longwave radiation is not trapped as in a greenhouse. Rather, its passage to space is delayed as the longwave radiation is absorbed by certain gases, clouds, and dust in the atmosphere and is emitted to Earth's surface. According to scientific consensus, today's increasing carbon dioxide concentration is forcing more longwave radiation absorption in the lower atmosphere, thus producing a warming trend and changes in the Earth–atmosphere energy system.

Clouds and Earth's "Greenhouse"

Clouds affect the heating of the lower atmosphere in several ways, depending on cloud type. Not only is the percentage of cloud cover important, but also the cloud type, height, and thickness (water content and density) have an effect. High-altitude, ice-crystal clouds reflect insolation with albedos of about 50%, whereas thick, lower cloud cover reflects about 90% of incoming insolation.

To understand the actual effects on the atmosphere's energy budget, we must consider both transmission of shortwave and longwave radiation and cloud type. Figure 4.9a portrays the *cloud-greenhouse forcing* caused by high clouds (warming because their greenhouse effects exceed their albedo effects), and Figure 4.9b portrays the *cloud-albedo forcing* produced by lower, thicker clouds (cooling because albedo effects exceed greenhouse effects).

Jet contrails (condensation trails) produce high cirrus clouds stimulated by aircraft exhaust (Figure 4.9c and d)—sometimes called *false cirrus clouds*. As shown in Figure 4.9a, high-altitude ice-crystal clouds are efficient at reducing outgoing longwave radiation and therefore contribute to

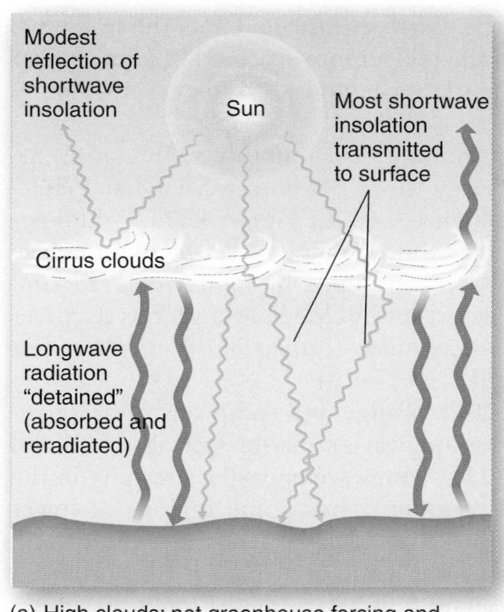

(a) High clouds: net greenhouse forcing and atmospheric warming

(b) Low clouds: net albedo forcing and atmospheric cooling

(c)

(d)

FIGURE 4.9 Energy effects of cloud types and contrails. (a) High, ice-crystal clouds (*cirrus*) produce a greater greenhouse forcing and a net warming of Earth. (b) Low, thick clouds (*stratus*) reflect most of the incoming insolation and produce a greater albedo forcing and a net cooling of Earth. (c) Jet aircraft exhaust triggers denser cirrus-cloud development. (d) Compare a newer, thinner contrail with one spreading to form false cirrus clouds. [(c) NASA/JSC; (d) Bobbé Christopherson.]

global heating. However, these contrail-seeded false cirrus clouds have a higher density and higher albedo than normal cirrus clouds, thus reducing some surface insolation.

The tragedy that struck the World Trade Center, and humanity, on September 11, 2001, inadvertently provided researchers with a chance to study these effects. Following 9/11 was a 3-day grounding of all commercial airline traffic and therefore no contrails across the United States. Weather data from 4000 stations over 30 years were compared with data during the 3-day flight shutdown. *Diurnal temperature range* (DTR)—the difference between daytime maximum and nighttime minimum temperatures—increased during the 3 days. Specifically, maximum temperatures were more responsive to the missing clouds than minimum—slightly higher afternoon and slightly lower nighttime temperature readings.

Later study of synoptic conditions (air pressure, humidity, and air mass data gathered at a specific time), analysis of Canadian air space, a check on conditions during adjacent 3-day periods, and a look at the few military flights that took place during the halt of air travel served to verify the initial findings and confirmed the energy-budget impacts of contrails.*

Earth–Atmosphere Energy Balance

If Earth's surface and its atmosphere are considered separately, neither exhibits a balanced radiation budget. The average annual energy distribution is positive (an energy surplus) for Earth's surface and negative (an energy deficit) for the atmosphere as it radiates energy to space. Considered together, these two equal each other, making it possible for us to construct an overall energy balance.

*See David J. Travis *et al.*, "Contrails reduce daily temperature range," *Nature* 418 (August 8, 2002): 601; and, D. J. Travis, A. M. Carleton, and R. G. Lauritsen, "Regional variations in U.S. diurnal temperature range for the 9/11–14/2001 aircraft groundings: Evidence of jet contrail influence on climate," *Journal of Climate* 17 (March 1, 2004): 1123–34.

Figure 4.10 summarizes the Earth–atmosphere radiation balance. It brings together all the elements discussed to this point in the chapter by following 100% of arriving insolation through the troposphere. The shortwave portion of the budget is on the left in the illustration; the longwave part of the budget is on the right.

Out of 100% of the solar energy arriving, Earth's average albedo is 31%. Absorption by atmospheric clouds, dust, and gases involves another 21% and accounts for the atmospheric heat input. Stratospheric ozone absorption and radiation account for another 3% of the atmospheric budget. About 45% of the incoming insolation transmits through to Earth's surface as direct and diffuse shortwave radiation.

The natural energy balance occurs through longwave energy transfers from the surface that are both *nonradiative* (physical motion) and *radiative*. Nonradiative transfers include convection, conduction, and the latent heat of evaporation (energy that is absorbed and dissipated by water as it evaporates and condenses). Radiative transfer is by longwave radiation among the surface, the atmosphere, and space, as illustrated to the right in Figure 4.10's depiction of the greenhouse effect.

To summarize, Earth eventually emits the longwave radiation part of the budget into space: 21% (atmosphere heating) + 45% (surface heating) + 3% (ozone emission) = 69%.

Figure 4.11 shows global monthly values in W/m^2 flux (meaning energy "flow") of both reflected and emitted radiation in Figure 4.10. In Figure 4.11a, lighter regions indicate where more sunlight is reflected into space than is absorbed—for example, by lighter-colored land surfaces such as deserts or by cloud cover such as that over tropical lands. Green and blue areas illustrate where less light was reflected.

In Figure 4.11b, orange and red pixels indicate regions where more longwave radiation was absorbed and emitted to space; less longwave energy is escaping in the blue and purple regions. Those blue regions of lower longwave emissions over tropical lands are due to tall, thick clouds along the equatorial region (the Amazon, equatorial Africa, and Indonesia); these clouds also cause higher shortwave reflection. Subtropical desert regions exhibit greater longwave radiation emissions owing to the presence of little cloud cover and greater radiative energy losses from surfaces that absorb a lot of energy.

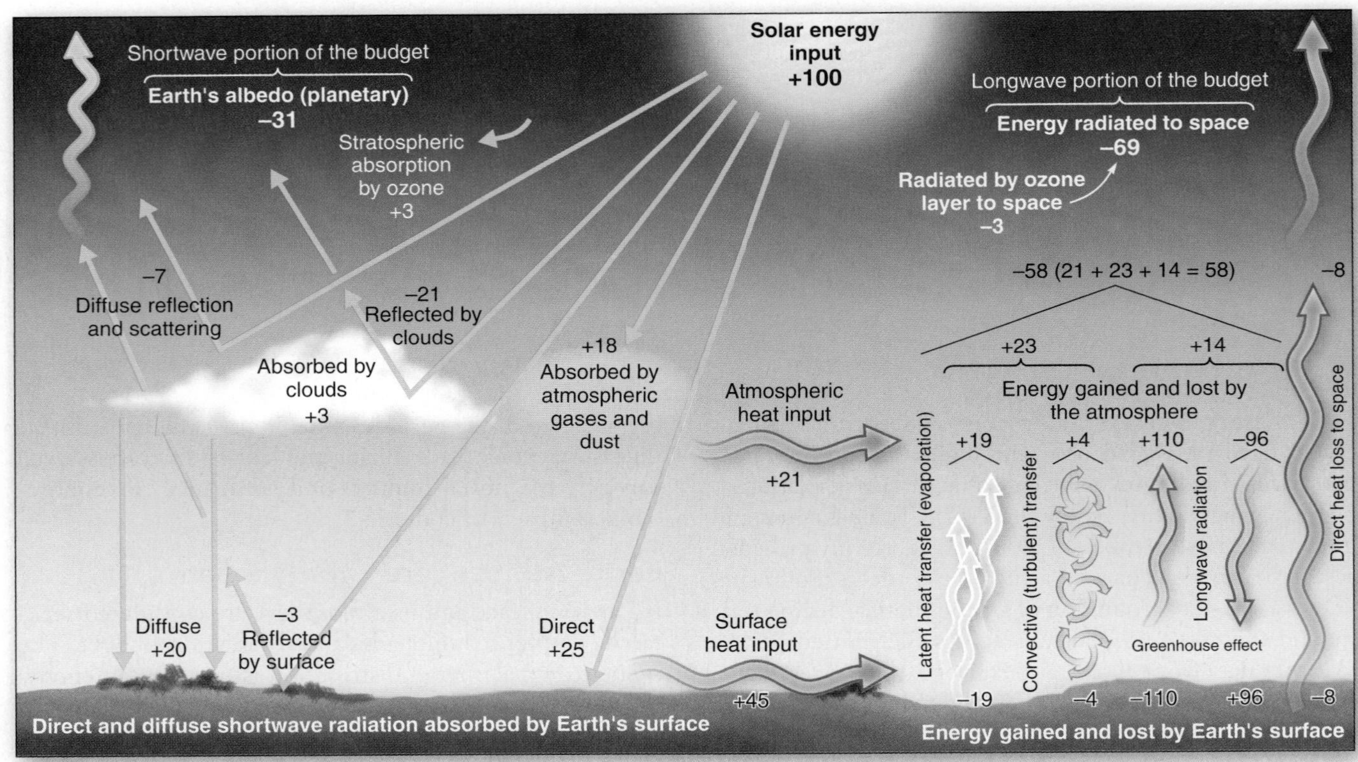

FIGURE 4.10 Detail of the Earth–atmosphere energy balance. Solar energy cascades through the lower atmosphere (left-hand portion of the illustration), where it is absorbed, reflected, and scattered. Clouds, atmosphere, and the surface reflect 31% of this insolation back to space. Atmospheric gases and dust and Earth's surface absorb energy and radiate longwave radiation (right-hand portion of the illustration). Over time, Earth emits, on average, 69% of incoming energy to space. When added to Earth's average albedo (31%, reflected energy), this equals the total energy input from the Sun (100%).

Animation
Global Warming, Climate Change

Animation
Earth–Atmosphere Energy Balance

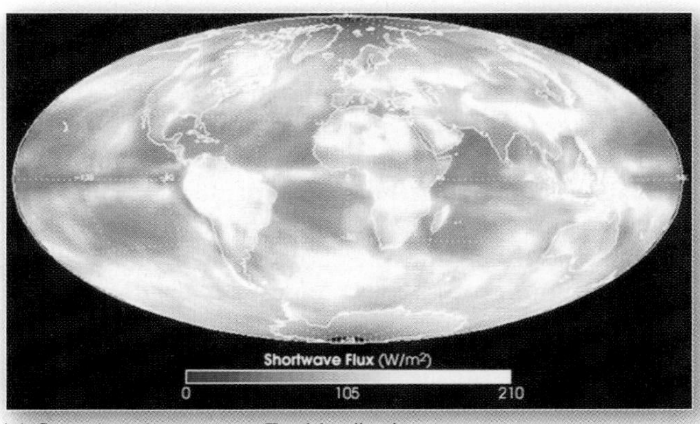

(a) Outgoing shortwave – Earth's albedo.

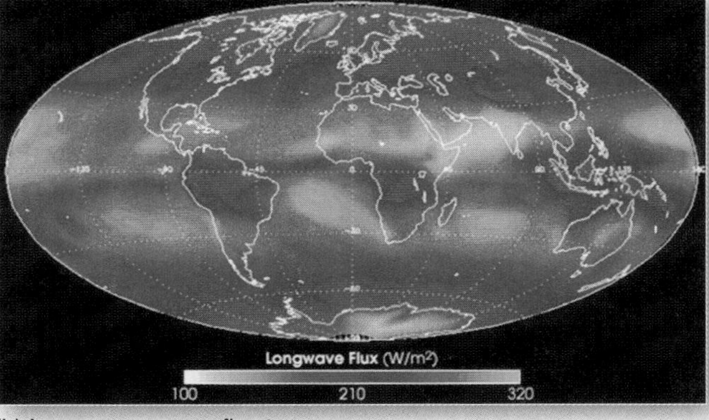

(b) Longwave energy flux to space.

FIGURE 4.11 Shortwave and longwave images show Earth's radiation budget components. The CERES sensors aboard *Terra* made these portraits, capturing (a) outgoing shortwave energy flux reflected from clouds, land, and water—Earth's albedo—and (b) longwave energy flux emitted by these surfaces back to space. The scale beneath each image displays values in W/m². [CERES, Langley Research Center, NASA.]

However, regionally and seasonally Earth absorbs more energy in the tropics and less in the polar regions, establishing the imbalance that drives global circulation patterns. Figure 4.12 summarizes the Earth–atmosphere energy balance for all shortwave and longwave energy by latitude:

- Between the tropics, the angle of incoming insolation is high and daylength is consistent, with little seasonal variation, so more energy is gained than lost—*energy surpluses dominate.*
- In the polar regions, the Sun is low in the sky, surfaces are light (ice and snow) and reflective, and for up to 6 months during the year no insolation is received, so more energy is lost than gained—*energy deficits prevail.*
- At around 36° latitude, a balance exists between energy gains and losses for the Earth–atmosphere system.

The imbalance of net radiation from the *tropical surpluses* and the *polar deficits* drives a vast global circulation of both energy and mass. The meridional (north–south) transfer agents are winds, ocean currents, dynamic weather systems, and related phenomena. Dramatic examples of such energy and mass transfers are tropical cyclones—hurricanes and typhoons. Forming in the tropics, these powerful storms mature and migrate to higher latitudes, carrying with them energy, water, and water vapor.

Given this Earth–atmosphere energy balance, we next focus on energy characteristics at Earth's surface and examine energy budgets along the ground level in Figure 4.10. The surface environment is the final stage in the Sun to Earth energy system.

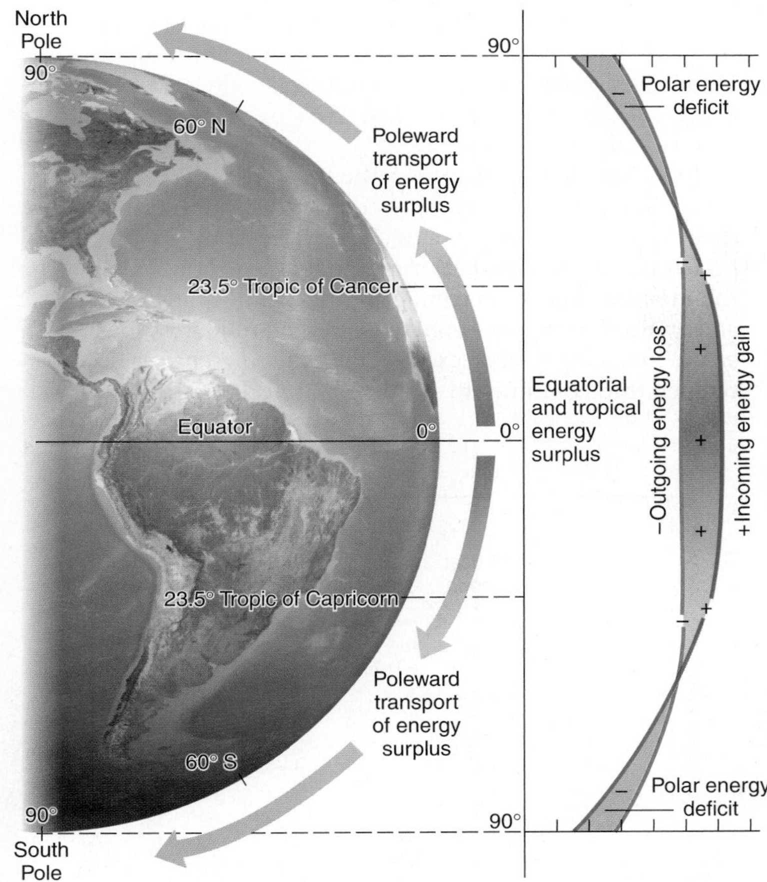

FIGURE 4.12 Energy budget by latitude. Earth's energy surpluses and deficits produce poleward transport of energy and mass in each hemisphere—atmospheric circulation and ocean currents. Outside of the tropics, atmospheric winds dominate the transport of energy toward each pole.

Energy Balance at Earth's Surface

Solar energy is the principal heat source at Earth's surface. Figure 4.10 illustrates the direct and diffuse shortwave radiation and longwave radiation arriving at the ground surface. These radiation patterns at Earth's surface are of great interest to geographers and should be of interest to everyone because these are the surface environments where we live.

Daily Radiation Patterns

Figure 4.13 shows the daily pattern of incoming shortwave energy absorbed and resulting air temperature. This graph represents idealized conditions for bare soil on a cloudless day in the middle latitudes. Incoming energy arrives during daylight, beginning at sunrise, peaking at noon, and ending at sunset.

The shape and height of this insolation curve vary with season and latitude. The highest trend for such a curve occurs at the time of the summer solstice (around June 21 in the Northern Hemisphere and December 21 in the Southern Hemisphere). The air temperature plot also responds to seasons and variations in insolation input. Within a 24-hour day, air temperature generally peaks between 3:00 and 4:00 P.M. and dips to its lowest point right at or slightly after sunrise.

The relationship between the insolation curve and the air temperature curve on the graph is interesting—they do not align; there is a *lag*. The warmest time of day occurs not at the moment of maximum insolation, but at the moment when a maximum of insolation is absorbed and emitted to the atmosphere from the ground. As long as the incoming energy exceeds the outgoing energy, air temperature continues to increase, not peaking until the incoming energy begins to diminish as the afternoon Sun's altitude decreases. In contrast, if you have ever gone camping in the mountains, you no doubt experienced the coldest time of day with a wake-up chill at sunrise!

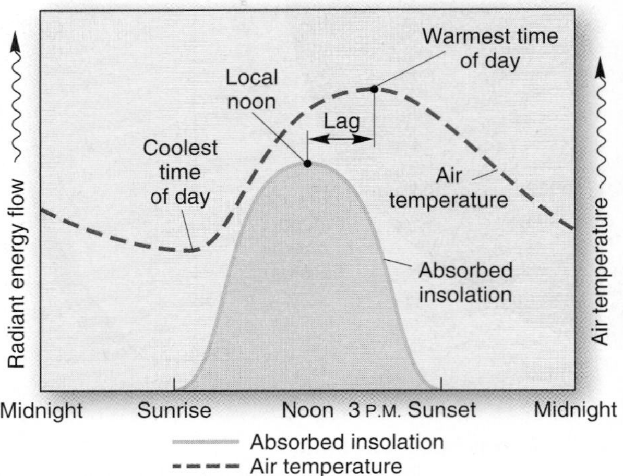

FIGURE 4.13 Daily radiation curves.
Sample radiation plot for a typical day shows the changes in insolation (solid line) and air temperature (dashed line). Comparing the curves demonstrates a lag between local noon (the insolation peak for the day) and the warmest time of day.

The annual pattern of insolation and air temperature exhibits a similar lag. For the Northern Hemisphere, January is usually the coldest month, occurring after the December solstice and the shortest days. Similarly, the warmest months of July and August occur after the June solstice and the longest days.

Simplified Surface Energy Balance

Energy and moisture are continually exchanged at the surface, creating worldwide "boundary-layer climates" of great variety. **Microclimatology** is the science of physical conditions at or near Earth's surface. The following discussion is more meaningful if you visualize an actual surface—perhaps a park, a front yard, or a place on campus.

The surface receives visible light and longwave radiation and it reflects light and longwave radiation according to the following simple scheme:

$$\underset{\text{(Insolation)}}{+SW\downarrow} \quad \underset{\text{(Reflection)}}{-SW\uparrow} \quad \underset{\text{(Infrared)}}{+LW\downarrow} \underset{\text{(Infrared)}}{-LW\uparrow} = \underset{\text{(Net Radiation)}}{\text{NET R}}$$

We use SW for shortwave and LW for longwave for simplicity. You may come across other symbols in the microclimatology literature, such as K for shortwave, L for longwave, and Q* for NET R (net radiation).

Figure 4.14 shows these components of a surface energy balance. The soil column shown continues to a depth at which energy exchange with surrounding materials or with the surface becomes negligible, usually less

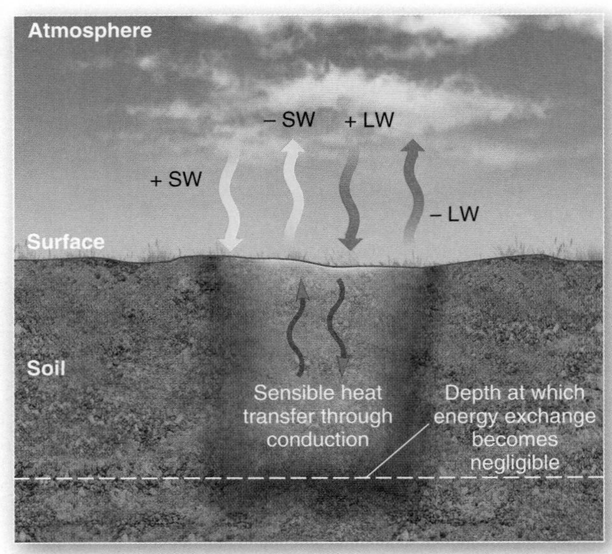

FIGURE 4.14 Surface energy budget.
Idealized input and output energy budget components for surface and a soil column. Sensible heat transfer in the soil is through conduction, predominantly downward during the day or in summer and toward the surface at night or in winter (SW = shortwave, LW = longwave).

than a meter. Sensible heat transfer in the soil is through conduction, predominantly downward during the day or in summer and toward the surface at night or in winter. Energy from the atmosphere that is moving toward the surface is a positive (a gain), and energy that is moving away from the surface, through sensible and latent heat transfers, is a negative (a loss) to the surface account.

Adding and subtracting the energy flow at the surface completes the calculation of **net radiation (NET R)**, or the balance of all radiation at Earth's surface. As the components of this simple equation vary with daylength through the seasons, cloudiness, and latitude, NET R varies. Figure 4.15 illustrates the components of a surface energy balance for a typical summer day at a midlatitude location, showing the daily change in the components of net radiation. The items plotted are direct and diffuse insolation (+SW↓), reflection by surface albedo value (−SW↑), and longwave radiation arriving (+LW↓) and leaving (−LW↑) from the surface.

At night, the net radiation value in Figure 4.15 becomes negative because the SW component ceases at sunset and the surface continues to lose longwave radiation to the atmosphere. The surface rarely reaches a zero net radiation value—a perfect balance—at any one moment. However, over time, Earth's surface naturally balances incoming and outgoing energies.

Net Radiation The net radiation (NET R of all wavelengths) available at Earth's surface is the final outcome of the entire energy-balance process discussed in this chapter.

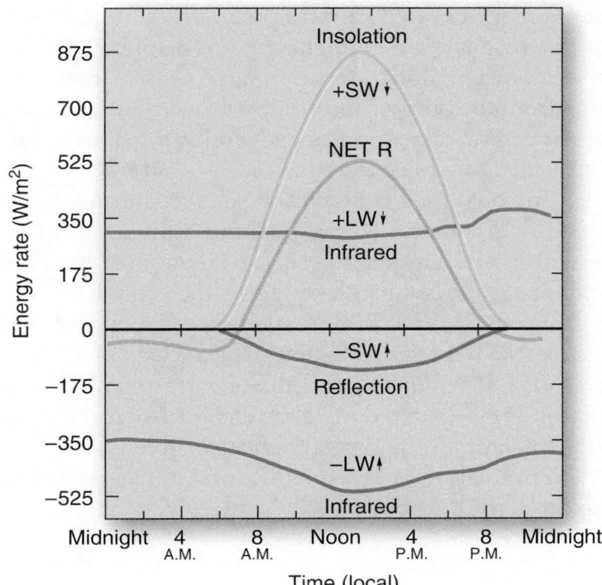

FIGURE 4.15 A day's radiation budget. Radiation budget on a typical July summer day at a midlatitude location (Matador in southern Saskatchewan, about 51° N). [Adapted by permission from T. R. Oke, *Boundary Layer Climates* (New York: Methuen & Co., 1978), p. 21.]

Figure 4.16 displays the global mean annual net radiation at ground level. The abrupt change in radiation balance from ocean to land surfaces is evident on the map. Note that all values are positive; negative values probably occur

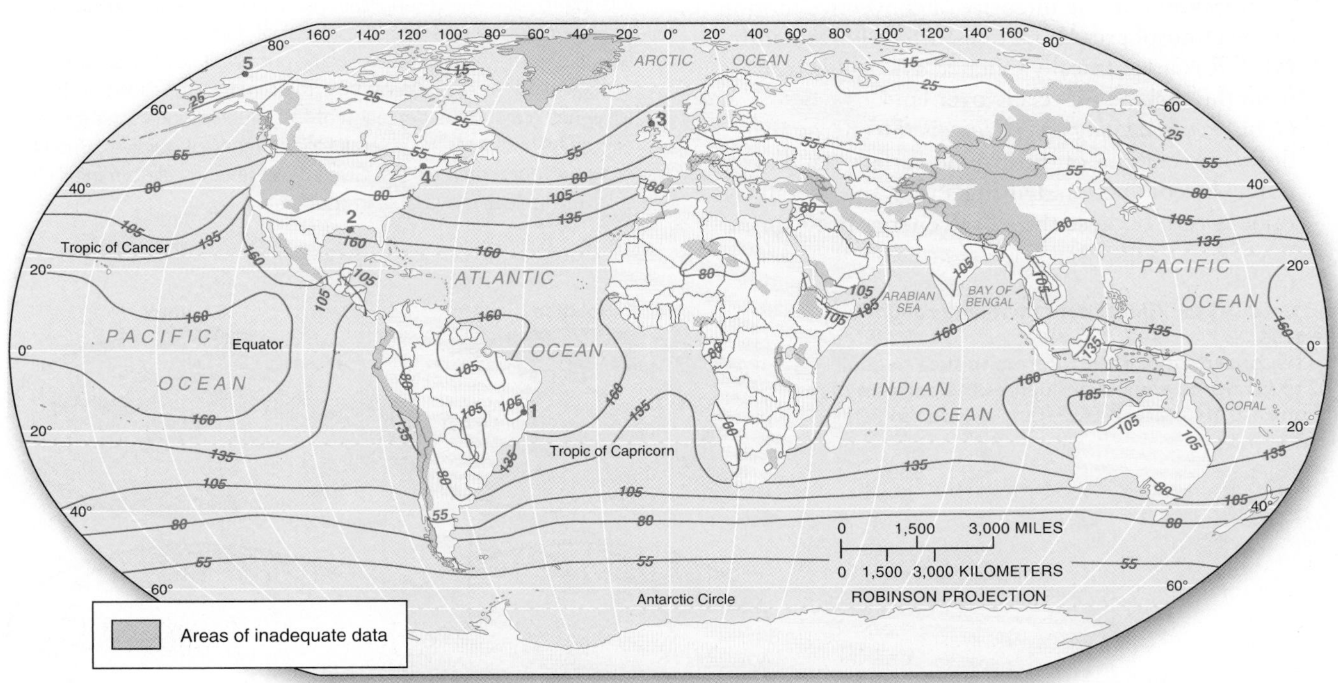

FIGURE 4.16 Global net radiation. Distribution of mean annual net radiation (NET R) at surface level in watts per square meter (100 W/m² = 75 kcal/cm²/year). Temperature graphs for the five cities noted on the map and in Figures 4.18 and 4.19 are graphed in Figure 5.4. [After M. I. Budyko, *The Heat Balance of the Earth's Surface* (Washington, DC: U.S. Department of Commerce, 1958), p. 106.]

1. Salvador, Brazil
2. New Orleans, Louisiana
3. Edinburgh, Scotland
4. Montreal, Canada
5. Barrow, Alaska (see Figure 5.4)

MG Satellite Global Net Radiation

only over ice-covered surfaces poleward of 70° latitude in both hemispheres. The highest net radiation, 185 W/m^2 per year, occurs north of the equator in the Arabian Sea. Aside from the obvious interruption caused by landmasses, the pattern of values appears generally zonal, or parallel, decreasing away from the equator.

A seasonal look is helpful to see the shifting patterns of NET R. Figure 4.17 shows the solstice months of December and June and the equinox months of March and September. The color scale gives you the NET R in W/m^2, all from the CERES sensors aboard *Terra* and *Aqua*. Positive NET R regions are toward the red color, with negative NET R regions toward the blue-green color. Notice tropical surpluses in reds near the equator, balanced yellows in the midlatitudes, and polar deficits in greens.

Net radiation is expended from a nonvegetated surface through three pathways:

- *LE*, or *latent heat of evaporation*, is the energy that is stored in water vapor as water evaporates. Water absorbs large quantities of this *latent heat* as it changes state to water vapor, thereby removing this heat energy from the surface. Conversely, this heat energy releases to the environment when water vapor changes state back to a liquid (discussed in Chapter 7). Latent heat is the dominant expenditure of Earth's entire NET R, especially over water surfaces.
- *H*, or *sensible heat*, is the back-and-forth transfer between air and surface in turbulent eddies through convection and conduction within materials. This activity depends on surface and boundary-layer temperatures and on the intensity of convective motion in the atmosphere. About one-fifth of Earth's entire NET R is mechanically radiated as sensible heat from the surface, especially over land.
- *G*, or *ground heating and cooling*, is the energy that flows into and out of the ground surface (land or water) by conduction. During a year, the overall *G* value is zero because the stored energy from spring and summer is equaled by losses in fall and winter. Another factor in ground heating is energy absorbed at the surface to melt snow or ice. In snow- or ice-covered landscapes, most available energy is in sensible and latent heat used in the melting and warming process.

On land, the highest annual values for latent heat of evaporation (*LE*) occur in the tropics and decrease toward the poles (Figure 4.18). Over the oceans, the highest *LE* values are over subtropical latitudes, where hot, dry air comes into contact with warm ocean water.

The values for sensible heat (*H*) are distributed differently, being highest in the subtropics (Figure 4.19). Here vast regions of subtropical deserts feature nearly waterless surfaces, cloudless skies, and almost vegetation-free landscapes. The bulk of NET R is expended as sensible heat in these dry regions. Moist and vegetated surfaces expend less in *H* and more in *LE*, as you can see by comparing the maps in Figures 4.18 and 4.19.

Understanding net radiation is essential to solar energy technologies that concentrate shortwave energy for use. Solar energy offers great potential worldwide and is presently the fastest-growing form of energy conversion by humans. Focus Study 4.1 on page 98 briefly reviews such direct application of surface energy budgets.

CRITICAL THINKING 4.2

Know more about applied solar technologies

On the Mastering Geography Web site for Chapter 4, as well as in this chapter of the text, you can find several listings of URLs relating to solar energy applications. Take some time and explore the Internet for a personal assessment of these necessary technologies (solar-thermal, solar-electric photovoltaic cells, solar-box cookers, and the like). As we near the climatic limitations of the fossil-fuel era and the depletion of fossil-fuel resources, renewable energy technologies are essential to the fabric of our lives. Briefly describe your search results. Given these findings, determine if solar technology is available in your area.

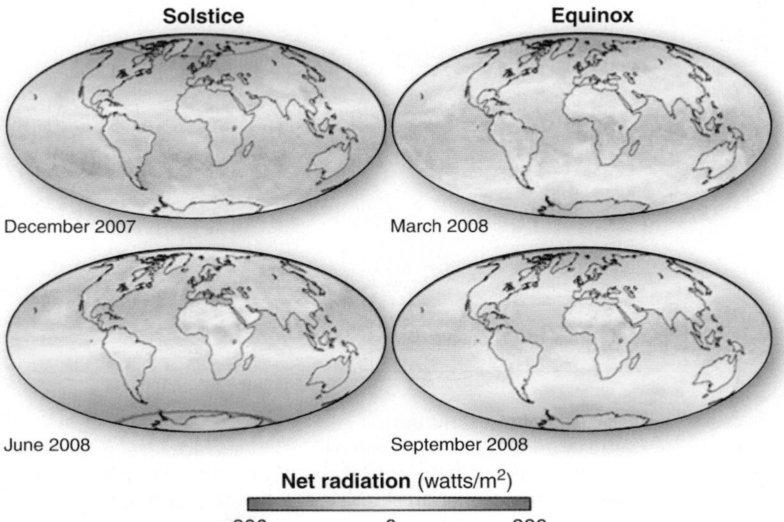

FIGURE 4.17 Global net radiation through the year. The seasonal rhythm of NET R from solstice to equinox to solstice to equinox influences the patterns of life on Earth's surface. [CERES, Langley Research Center, NASA.]

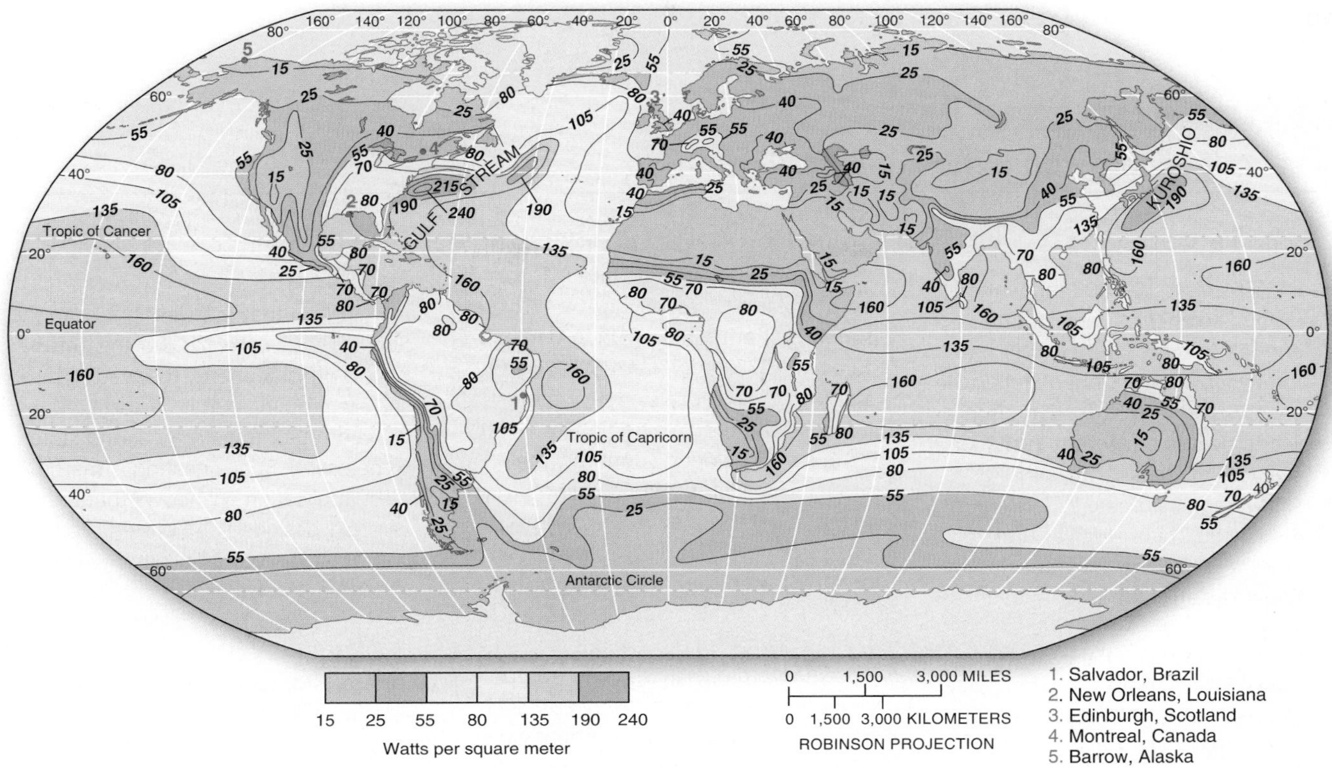

FIGURE 4.18 Global latent heat of evaporation.
Annual energy expenditure as the latent heat of evaporation (*LE*) at surface level in watts per square meter (100 W/m² = 75 kcal/cm²/year). Note the high values associated with high sea-surface temperatures in the area of the Gulf Stream and Kuroshio currents. [Adapted by permission from M. I. Budyko, *The Earth's Climate, Past and Future* (New York: Academic Press, 1982), p. 56.]

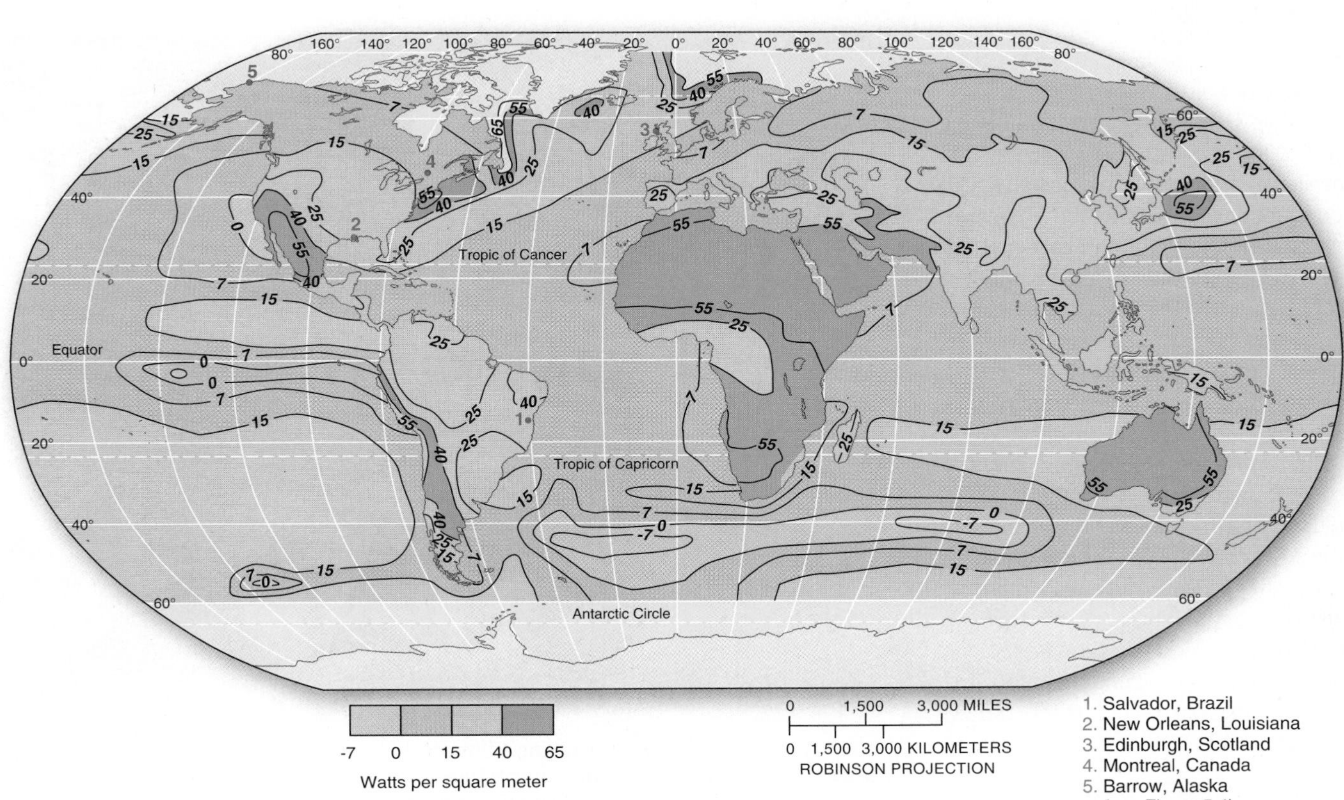

FIGURE 4.19 Global sensible heat.
Annual energy expenditure as sensible heat (*H*) at surface level in watts per square meter (100 W/m² = 75 kcal/cm²/year). [Adapted by permission from M. I. Budyko, *The Earth's Climate, Past and Future* (New York: Academic Press, 1982), p. 59.]

FOCUS STUDY 4.1

Solar Energy Collection and Concentration

Consider the following:

- Earth receives 100,000 terawatts (TW) of solar energy per hour, enough to meet world power needs for a year.* In the United States, the energy produced by fossil fuels in a year arrives in equivalent insolation every 35 minutes.
- An average commercial building in the United States receives 6 to 10 times more energy from the Sun hitting its exterior than is required to heat the inside.
- Photographs from the early 1900s show solar (flat-plate) water heaters used on rooftops. Today, installed solar water heating is making a comeback, serving 50 million households worldwide, more than 300,000 of these in the United States.
- Photovoltaic capacity is more than doubling every 2 years. Solar-electric power passed 20,000 megawatts installed in 2010.

Not only does insolation warm Earth's surface, but also it provides an inexhaustible supply of energy for humanity. Sunlight is direct and widely available, and solar power installations are decentralized, labor-intensive, and renewable. Although collected for centuries using various technologies, sunlight remains underutilized. The utility energy industry appears to prefer indirect, centralized, power-plant-specific, capital-intensive, and nonrenewable sources.

Rural villages in developing countries could benefit greatly from the simplest, most cost-effective solar application—the *solar cooker*. For example, people in Latin America and Africa walk many kilometers collecting fuel wood for cooking fires (Figure 4.1.1a), stripping nearby areas bare. Using solar cookers, villagers are able to cook meals and sanitize their drinking water without scavenging for wood (Figure 4.1.1b). These solar devices in Figure 4.1.1c are simple, yet efficient, reaching temperatures of 107°C–127°C (225°F–260°F). See http://solarcookers.org/ for more information.

In less-developed countries, the money for electrification (a *centralized* technology) is not available despite the push from more developed countries and energy corporations for large capital-intensive power projects. In reality, the pressing need is for *decentralized* energy sources, appropriate in scale to everyday needs, such as cooking, heating water, and pasteurizing. The net per capita (per person) cost for solar cookers is far less than for centralized electrical production, regardless of fuel source.

Collecting and Concentrating Solar Energy

Any surface that receives light from the Sun is a *solar collector.* But the diffuse nature of solar energy received at the surface requires that it be collected, concentrated, transformed, and stored to be useful. Space heating is the simplest application. Windows that are carefully designed and placed allow sunlight to shine into a building, where it is absorbed and converted into sensible heat—an everyday application of the greenhouse effect.

A *passive solar system* captures heat energy and stores it in a "thermal mass," such as water-filled tanks, adobe, tile, or concrete. An *active solar system* involves heating water or air in a collector and then pumping it through a plumbing system to a tank, where it can

(a)

(b)

(c)

FIGURE 4.1.1 The solar-cooking solution.
(a) Two women and a girl haul firewood many miles in Guatemala. (b) Women in East Africa carry home solar-box cookers that they made at a workshop. Construction is easy, using cardboard components. (c) These simple cookers collect insolation through transparent glass or plastic and trap longwave radiation in an enclosed box or cooking bag. [(a) and (b) Solar Cookers International, Sacramento, California; (c) Bobbé Christopherson.]

*Terawatt (10^{12} Watts) = 1 trillion W; gigawatt (10^9 Watts) = 1 billion W; megawatt (10^6 Watts) = 1 million W; kilowatt (10^3 Watts) = 1 thousand W.

provide hot water for direct use or for space heating.

Solar energy systems can generate heat energy of an appropriate scale for approximately half the present domestic applications in the United States, which includes space heating and water heating. In marginal climates, solar-assisted water and space heating is feasible as a backup; even in New England and the Northern Plains states, solar collection systems prove effective. Several techniques are being used to attain very high temperatures to heat water or other heat-storing fluids. Kramer Junction, California, about 225 km (140 mi) northeast of Los Angeles, in the Mojave Desert near Barstow, has the world's largest operating solar-electric generating system, with a capacity of 150 MW (megawatts; 150 million watts). Long troughs of computer-guided curved mirrors concentrate sunlight to create temperatures of 390°C (735°F) in vacuum-sealed tubes filled with synthetic oil. The heated oil heats water; the heated water produces steam that rotates turbines to generate cost-effective electricity. The facility converts 23% of the sunlight it receives into electricity during peak hours (Figure 4.1.2a), and operation and maintenance costs continue to decrease.

Electricity Directly from Sunlight

Photovoltaic cells (PVs) were first used to produce electricity in spacecraft in 1958. Familiar to us all are the solar cells in pocket calculators (hundreds of millions now in use). When light shines upon a semiconductor material in these cells, it stimulates a flow of electrons (an electrical current) in the cell.

The efficiency of these cells, often assembled in large arrays, has improved to the level that they are cost-competitive, especially if government policies and subsidies were to be balanced evenly among all energy sources. The residential installation in Figure 4.1.2b features 36 panels, generating 205 W each, or 7380 W total, at a 21.5% conversion efficiency. People are able to run their electric meters in reverse and supply electricity to the power grid.

The National Renewable Energy Laboratory (NREL, http://www.nrel.gov) and the National Center for Photovoltaics, http://www.nrel.gov/ncpv/, were established in 1974 to coordinate solar energy research, development, and testing in partnership with private industry. At NREL's Outdoor Test Facility in Golden, Colorado, testing is ongoing and includes development of solar cells that broke the 40% conversion efficiency barrier!

Rooftop photovoltaic electrical generation is now cheaper than power line construction to rural sites. PV roof systems provide power to hundreds of thousands of homes in Mexico, Indonesia, Philippines, South Africa, India, and Norway. (See the "Photovoltaic Home Page," at http://www.eere.energy.gov/solar/.)

Obvious drawbacks of both solar-heating and solar-electric systems are periods of cloudiness and night, which inhibit operations. Research is under way to enhance energy-storage and to improve battery technology.

The Promise of Solar Energy

Solar energy is a wise choice for the future. Solar is economically preferable to further development of our decreasing fossil-fuel reserves, further increases in oil imports and tanker and offshore oil-drilling spills, investment in foreign military incursions, or the addition of more troubled nuclear power, especially in a world with security issues.

Whether or not we follow the alternative path of solar energy is a matter of political control and not technological innovation. Much of the technology is ready for installation and is cost-effective when all direct and indirect costs are considered for the alternatives.

(a)

(b)

FIGURE 4.1.2 Solar thermal and photovoltaic energy production. (a) Kramer Junction solar-electric generating system in southern California. (b) A residence with a 7380-W rooftop solar photovoltaic array generates electricity. Excess electricity is fed to the grid for credits that offset 100% of the home's electric bill. [Bobbé Christopherson.]

Sample Stations A couple of real locations help complete our discussion. Variation in the expenditure of NET R among sensible heat (*H*, energy we can feel), latent heat (*LE*, energy for evaporation), and ground heating and cooling (*G*) produces the variety of environments we experience in nature. Let us examine the daily energy balance at two locations, El Mirage in California and Pitt Meadows in British Columbia.

El Mirage, at 35° N, is a hot desert location characterized by bare, dry soil with sparse vegetation (Figure 4.20a, b). Our sample is on a clear summer day, with a light wind in the late afternoon. The NET R value is lower than might be expected, considering the Sun's position close to zenith (June solstice) and the absence of clouds. But the income of energy at this site is countered by surfaces of higher albedo than forest or cropland and by hot soil surfaces that radiate longwave energy back to the atmosphere throughout the afternoon.

El Mirage has little or no energy expenditure for evaporation (*LE*). With little water and sparse vegetation, most of the available radiant energy dissipates through turbulent transfer of sensible heat (*H*), warming air and soil to high temperatures. Over a 24-hour period, *H* is 90% of NET R; the remaining 10% is for ground heating (*G*). The *G* component is greatest in the morning, when winds are light and turbulent transfers are

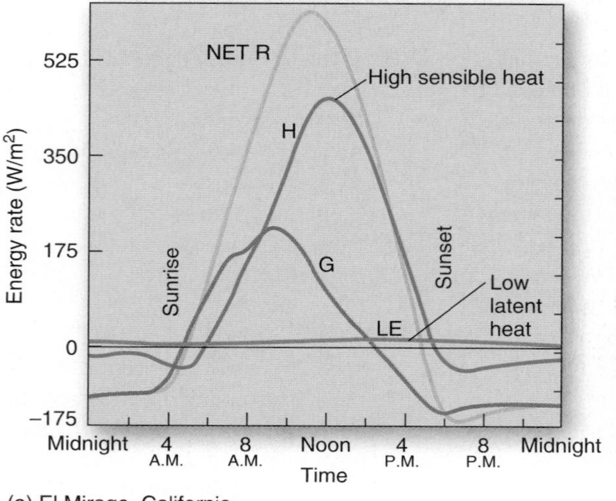
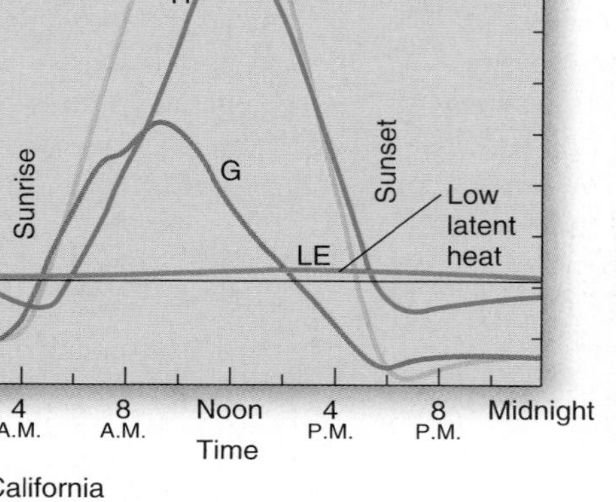

(a) El Mirage, California

(b) Mojave desert landscape

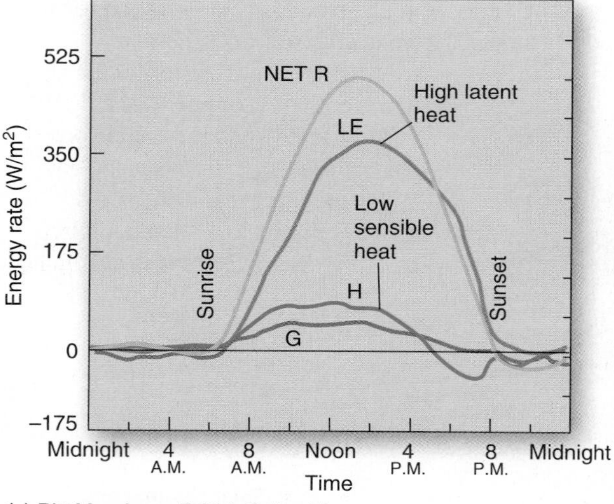

(c) Pitt Meadows, British Columbia

(d) Irrigated blueberry orchards

FIGURE 4.20 Radiation budget comparison for two stations.
(a) Graph of the daily net radiation expenditure for El Mirage, California, east of Los Angeles, at about 35° N; and (b) a typical desert landscape near the hot, dry site. (c) The daily net radiation expenditure for Pitt Meadows in southern British Columbia, at about 49° N; and (d) the irrigated blueberry orchards characteristic of agricultural activity in this moist environment of moderate temperatures. (*H* = turbulent sensible heat transfer; *LE* = latent heat of evaporation; *G* = ground heating and cooling.) [(a) Adapted from W. D. Sellers, *Physical Climatology*, © 1965 University of Chicago. (b) and (d) Bobbé Christopherson. (c) Adapted from T. R. Oke, *Boundary Layer Climates*, © 1978 Methuen & Co.]

lowest. In the afternoon, heated air rises off the hot ground, and convective heat expenditures are accelerated as winds increase.

Compare graphs for El Mirage (Figure 4.20a) and Pitt Meadows (Figure 4.20c). Pitt Meadows is midlatitude (49° N), vegetated, and moist, and its energy expenditures differ greatly from those at El Mirage. The Pitt Meadows landscape is able to retain much more of its energy because of lower albedo values (less reflection), the presence of more water and plants, and lower surface temperatures than those of El Mirage.

The energy-balance data for Pitt Meadows are from a cloudless summer day. Higher *LE* values result from the moist environment of rye grass and irrigated mixed-orchards for the sample area (Figure 4.20d), contributing to the more moderate sensible heat (*H*) levels during the day.

The Urban Environment

An urban landscape produces the temperatures most of you reading this book feel each day. Urban microclimates generally differ from those of nearby nonurban areas. In fact, the surface energy characteristics of urban areas are similar to those of desert locations. Because about 50% of the world's population lives in cities, urban microclimatology and other specific environmental effects related to cities are important topics for physical geographers.

The physical characteristics of urbanized regions produce an **urban heat island** that has, on average, both maximum and minimum temperatures higher than those of nearby rural settings. Table 4.1 lists five urban characteristics and the resulting temperature and moisture effects produced. Figure 4.21 on page 102 illustrates these traits. Every major city produces its own **dust dome** of airborne pollution, which can be blown from the city in elongated plumes; as noted in the table, such domes affect urban energy budgets.

Table 4.2 on page 102 compares climatic factors of rural and urban environments. The worldwide trend toward greater urbanization is placing more and more people on urban heat islands. The chapter-opening images of Phoenix, Arizona, and related studies show that these heat islands increase as population and urban sprawl

TABLE 4.1 Urban Physical Characteristics and Conditions

Urban Characteristics	Results and Conditions
Urban surfaces typically are metal, glass, asphalt, concrete, or stone, and their energy characteristics respond differently from natural surfaces.	Albedos of urban surfaces are lower, leading to higher net radiation values. Urban surfaces expend more energy as sensible heat than do nonurban areas (70% of the net radiation to *H*). Surfaces conduct up to three times more energy than wet, sandy soil and thus are warmer. During the day and evening, temperatures above urban surfaces are higher than those above natural areas.
Irregular geometric shapes in a city affect radiation patterns and winds.	Incoming insolation is caught in mazelike reflection and radiation "canyons." Delayed energy is conducted into surface materials, thus increasing temperatures. Buildings interrupt wind flows, diminishing heat loss through advective (horizontal) movement. Maximum heat island effects occur on calm, clear days and nights.
Human activity alters the heat characteristics of cities.	In summer, urban electricity production and use of fossil fuels release energy equivalent to 25%–50% of insolation. In winter, urban-generated sensible heat averages 250% greater than arriving insolation, reducing winter heating requirements.
Many urban surfaces are sealed (built on and paved), so water cannot reach the soil.	Central business district surfaces average 50% sealed, producing more water runoff than the suburbs that average 20% sealed surfaces. Urban areas respond as a desert landscape: A storm may cause a flash flood over the hard, sparsely vegetated surfaces, to be followed by dry conditions a few hours later.
Air pollution, including gases and aerosols, is greater in urban areas than in comparable natural settings; this increases possible convection and precipitation.	Pollution increases the atmosphere's reflectivity above a city, reducing insolation and absorbing infrared radiation, reradiating infrared downward. Increased particulates in pollution are condensation nuclei for water vapor, increasing cloud formation and precipitation. Urban-stimulated increases in precipitation may occur downwind from cities.

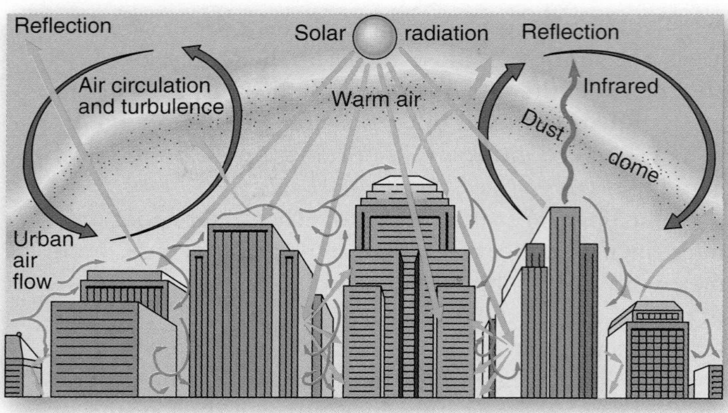

FIGURE 4.21 The urban environment. Insolation, wind movements, and dust dome in city environments.

increase. For an average city in North America, the sources of heating are modified surfaces, transportation, buildings, and industrial processes. For example, an average car (10 km/l or 25 mpg) produces enough heat to melt a 14-lb bag of ice for every mile driven (or 4.5 kg of ice per km driven).

Many known methods to make cities cooler await application. In addition to lowering urban temperatures, such strategies could reduce energy consumption and fossil-fuel use, thus reducing greenhouse gas emissions. An experiment in Phoenix lightened the color of rooftops on commercial buildings. The albedo of original roof gravel was 31%, whereas whitish gravel has an albedo of 72%. The resulting roof temperature was from 10 C° to 14 C° cooler, thus reducing energy costs. For structures with solar panel arrays, temperatures dropped dramatically with the roof under the shade of the panels.

Figure 4.22 illustrates a generalized cross section of a typical urban heat island, showing increasing temperatures toward the downtown central business district. Note that temperatures drop over areas of trees and parks. Sensible heat is lessened because of the latent heat of evaporation and plant effects such as transpiration and shade. Urban forests are important factors in cooling cities. NASA measurements in an outdoor mall parking lot found temperatures of 48°C (118°F), but a small planter with trees in the same lot was significantly cooler at 32°C (90°F)—16 C° (28 F°) lower in temperature! In Central Park in New York City, daytime temperatures average 5–10 C° (9–18 F°) cooler than in urban areas outside the park.

CRITICAL THINKING 4.3

Looking at your surface energy budget

Given what you now know about reflection, albedo, absorption, and net radiation expenditures, assess your wardrobe (fabrics and colors); house, apartment, or dorm (colors of exterior walls, especially south and west facing, or, in the Southern Hemisphere, north and west facing); roof (orientation relative to the Sun and roof color); automobile (color, use of sun shades); bicycle seat (color); and other aspects of your environment to determine a personal "Energy IQ." What grade do you give yourself? Can you roughly assess what this means in general terms of money costs? Begin a plan to change your energy budget. In Chapter 21 several sources lead you to Web sites to assess and reduce your carbon footprint, a direct link to these considerations of orientation, shade, color, and form of transportation. Be cool!

TABLE 4.2 Average Differences in Climatic Elements Between Urban and Rural Environments

Element	Urban Compared with Rural Environs	Element	Urban Compared with Rural Environs
Contaminants		**Precipitation (cont.)**	
Condensation nuclei	10 times more	Snowfall, downwind (lee) of city	10% more
Particulates	10 times more		
Gaseous admixtures	5–25 times more	Thunderstorms	10%–15% more
Radiation		**Temperature**	
Total on horizontal surface	0%–20% less	Annual mean	0.5–3 C° (0.9–5.4 F°) more
Ultraviolet, winter	30% less	Winter minima (average)	1.0–2 C° (1.8–3.6 F°) more
Ultraviolet, summer	5% less	Summer maxima	1.0–3 C° (1.8–3.0 F°) more
Sunshine duration	5%–15% less	Heating degree days	10% less
Cloudiness		**Relative Humidity**	
Clouds	5%–10% more	Annual mean	6% less
Fog, winter	100% more	Winter	2% less
Fog, summer	30% more	Summer	8% less
Precipitation		**Wind Speed**	
Amounts	5%–15% more	Annual mean	20%–30% less
Days with <5 mm (0.2 in.)	10% more	Extreme gusts	10%–20% less
Snowfall, inner city	5%–10% less	Calm	5%–20% more

Source: H. E. Landsberg, *The Urban Climate,* International Geophysics Series, vol. 28 (1981), p. 258. Reprinted by permission from Academic Press.

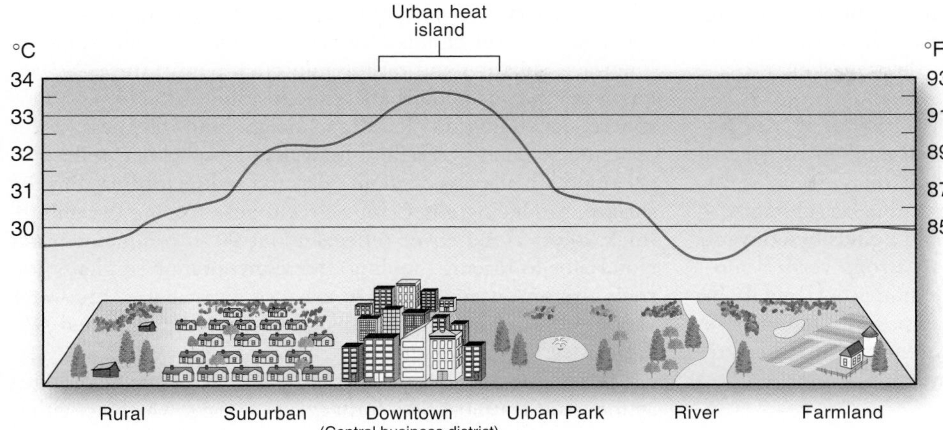

FIGURE 4.22 Typical urban heat island profile.
Cross section of a typical urban heat island from rural to downtown. Temperatures steeply rise in urban settings, plateau over the suburban built-up area, and peak where temperature is highest at the urban core. Note the cooling over the park area and river.

When you look across the urban landscape in which you live, what strategies do you think would reduce the urban heat island effect? Possibilities include lighter-colored surfaces for buildings, streets, and parking lots; more reflective roofs; and more trees, parks, and open space. With 60% of the global population living in cities by 2030 and with rising air and water temperatures from climate change, urban heat island issues will soon emerge as a critical concern. More information is available from the EPA, including strategies, publications, and hot-topic discussions, at http://www.epa.gov/hiri/.

We tracked the inputs of insolation, followed the cascade of energy through the atmosphere to Earth's surface, and saw the actions of movements and processes. We ended this journey from Sun to Earth analyzing the distribution of energy at the surface and the concepts of energy balance and net radiation available to do work. Now we shift to outputs in the energy–atmosphere system in Chapters 5 and 6—the results and consequences of global temperature patterns, and wind and ocean current circulations driven by these flows of energy.

KEY LEARNING CONCEPTS REVIEW

- ***Identify*** the pathways of solar energy through the troposphere to Earth's surface: transmission, scattering, diffuse radiation, refraction, albedo (reflectivity), conduction, convection, and advection.

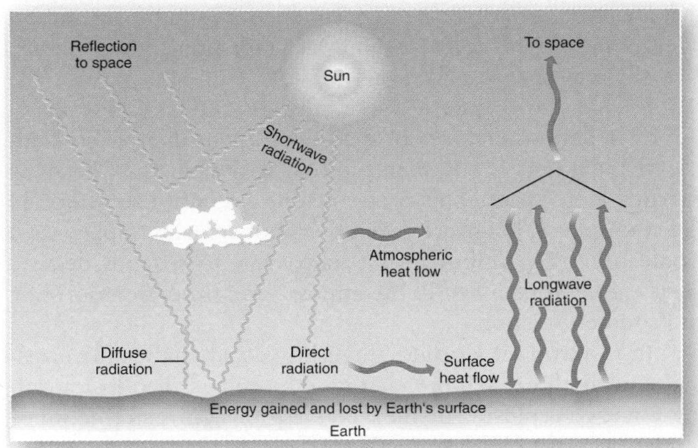

Radiant energy from the Sun that cascades through complex circuits to the surface powers Earth's biosphere. Our budget of atmospheric energy comprises shortwave radiation *inputs* (ultraviolet light, visible light, and near-infrared wavelengths) and longwave radiation *outputs* (thermal infrared).

Transmission refers to the passage of shortwave and longwave energy through either the atmosphere or water. The gas molecules redirect radiation, changing the direction of the light's movement *without altering its wavelengths.* This **scattering** represents 7% of Earth's reflectivity, or albedo. Dust particles, pollutants, ice, cloud droplets, and water vapor produce further scattering. Some incoming insolation is diffused by clouds and atmosphere and is transmitted to Earth as **diffuse radiation**, the downward component of scattered light. The speed of insolation entering the atmosphere changes as it passes from one medium to another; the change of speed causes a bending action called **refraction. Mirage** is a refraction effect when an image appears near the horizon where light waves are refracted by layers of air at different temperatures (and consequently of different densities).

A portion of arriving energy bounces directly back into space without being converted into heat or performing any work. **Albedo** is the reflective quality (intrinsic brightness) of a surface; this returned energy is **reflection.** Albedo is an important control over the amount of insolation that is available for absorption by a surface. We state albedo as the percentage of insolation that is reflected. Earth and its atmosphere reflect 31% of all insolation when averaged over a year.

An increase in albedo and reflection of shortwave radiation caused by clouds is **cloud-albedo forcing.** Also, clouds can act as insulation, thus trapping longwave radiation and raising minimum temperatures. An increase in greenhouse warming caused by clouds is **cloud-greenhouse forcing**. **Global dimming** describes the decline in sunlight reaching Earth's surface owing to pollution, aerosols, and clouds and is perhaps masking the actual degree of global warming occurring.

Absorption is the assimilation of radiation by molecules of a substance and its conversion from one form to another—for example, visible light to infrared radiation. **Conduction** is the molecule-to-molecule transfer of energy as it diffuses through a substance.

Energy also is transferred in gases and liquids by **convection** (when the physical mixing involves a strong vertical motion) or **advection** (when the dominant motion is horizontal). In the atmosphere or bodies of water, warmer portions tend to rise (they are less dense) and cooler portions tend to sink (they are more dense), establishing patterns of convection.

transmission (p. 84)
scattering (p. 85)
diffuse radiation (p. 85)
refraction (p. 86)
mirage (p. 86)
reflection (p. 86)
albedo (p. 86)
cloud-albedo forcing (p. 87)
cloud-greenhouse forcing (p. 87)
global dimming (p. 87)
absorption (p. 89)
conduction (p. 89)
convection (p. 89)
advection (p. 89)

1. Diagram a simple energy balance for the troposphere. Label each shortwave and longwave component and the directional aspects of related flows.
2. What would you expect the sky color to be at an altitude of 50 km (30 mi)? Why? What factors explain the lower atmosphere's blue color?
3. Define *refraction*. How is it related to daylength? To a rainbow? To the beautiful colors of a sunset?
4. List several types of surfaces and their albedo values. Explain the differences among these surfaces. What determines the reflectivity of a surface?
5. Using Figure 4.5, explain the differences in albedo values for various surfaces. Based upon albedo alone, which surface is cooler? Which is warmer? Why do you think this is?
6. Define the concepts of transmission, absorption, diffuse radiation, conduction, and convection.

■ ***Describe*** what happens to insolation when clouds are in the atmosphere, and ***analyze*** the effect of clouds and air pollution on solar radiation received at ground level.

Clouds reflect *insolation*, thus cooling Earth's surface—*cloud-albedo forcing*. Yet clouds also act as insulation, thus trapping longwave radiation and raising minimum temperatures—*cloud-greenhouse forcing*. Clouds affect the heating of the lower atmosphere, depending on cloud type, height, and thickness (water content and density). High-altitude, ice-crystal clouds reflect insolation with albedos of about 50% and redirect infrared to the surface, producing a net cloud-greenhouse forcing (warming); thick, lower cloud cover reflects about 90%, producing a net cloud-albedo forcing (cooling). **Jet contrails,** or condensation trails, are produced by aircraft exhaust, particulates, and water vapor and can form high cirrus clouds, sometimes called *false cirrus clouds*.

Emissions of sulfur dioxide and the subsequent chemical reactions in the atmosphere form *sulfate aerosols*, which act either as an insolation-reflecting haze in clear-sky conditions or as a stimulus to condensation in clouds that increases reflectivity.

jet contrails (p. 90)

7. What role do clouds play in the Earth–atmosphere radiation balance? Is cloud type important? Compare high, thin cirrus clouds and lower, thick stratus clouds.
8. Jet contrails affect the Earth–atmosphere balance in what ways? Describe the recent scientific findings.
9. In what way does the presence of sulfate aerosols affect solar radiation received at ground level? How does it affect cloud formation?

■ ***Review*** the energy pathways in the Earth–atmosphere system, the greenhouse effect, and the patterns of global net radiation.

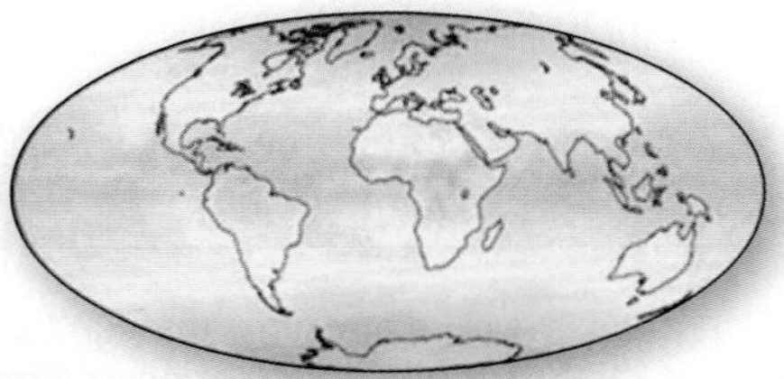

The Earth–atmosphere energy system naturally balances itself in a steady-state equilibrium. It does so through energy transfers that are *nonradiative* (convection, conduction, and the latent heat of evaporation) and *radiative* (longwave radiation among the surface, the atmosphere, and space).

Carbon dioxide, water vapor, methane, CFCs (chloroflurocarbons), and other gases in the lower atmosphere absorb infrared radiation that is then emitted to Earth, thus delaying energy loss to space. This process is the **greenhouse effect**. In the atmosphere, longwave radiation is not actually trapped, as it would be in a greenhouse, but its passage to space is delayed (heat energy is detained in the atmosphere) through absorption and counterradiation.

In the tropical latitudes, high insolation angle and consistent daylength cause more energy to be gained than lost, forming energy surpluses. In the polar regions, an extremely low insolation angle, highly reflective surfaces, and up to 6 months of no insolation annually cause more energy to be lost, forming energy deficits. This imbalance of net radiation from tropical surpluses to polar deficits drives a vast global circulation of both energy and mass.

Surface energy balances are used to summarize the energy expenditure for any location. Surface energy measurements are used as an analytical tool of **microclimatology**. Adding and subtracting the energy flow at the surface lets us calculate **net radiation (NET R)**, or the balance of all radiation at Earth's surface available to do work—shortwave (SW) and longwave (LW).

greenhouse effect (p. 90)
microclimatology (p. 94)
net radiation (NET R) (p. 95)

10. What are the similarities and differences between an actual greenhouse and the gaseous atmospheric greenhouse? Why is Earth's greenhouse effect changing?
11. In terms of surface energy balance, explain the term *net radiation (NET R)*.
12. What are the expenditure pathways for surface net radiation? What kind of work is accomplished?
13. Generalize the pattern of global net radiation. How might this pattern drive the atmospheric weather machine? (See Figures 4.16 and 4.17.)
14. What is the role played by latent heat in surface energy budgets?
15. In terms of energy expenditures for latent heat of evaporation, describe the annual pattern as mapped in Figure 4.18.
16. Compare the daily surface energy balances of El Mirage, California, and Pitt Meadows, British Columbia. Explain the differences.

■ ***Plot*** the daily radiation curves for Earth's surface, and ***label*** the key aspects of incoming radiation, air temperature, and the daily temperature lag.

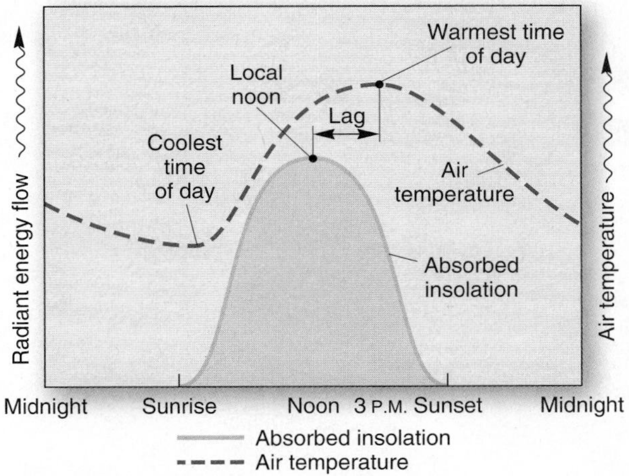

The greatest insolation input occurs at the time of the summer solstice in each hemisphere. Air temperature responds to seasons and variations in insolation input. Within a 24-hour day, air temperature peaks between 3:00 and 4:00 P.M. and dips to its lowest point right at or slightly after sunrise.

Air temperature lags behind each day's peak insolation. The warmest time of day occurs not at the moment of maximum insolation, but at the moment when a maximum of insolation is absorbed.

17. Why is there a temperature lag between the highest Sun altitude and the warmest time of day? Relate your answer to the insolation and temperature patterns during the day.

■ ***Portray*** typical urban heat island conditions, and ***contrast*** the microclimatology of urban areas with that of surrounding rural environments.

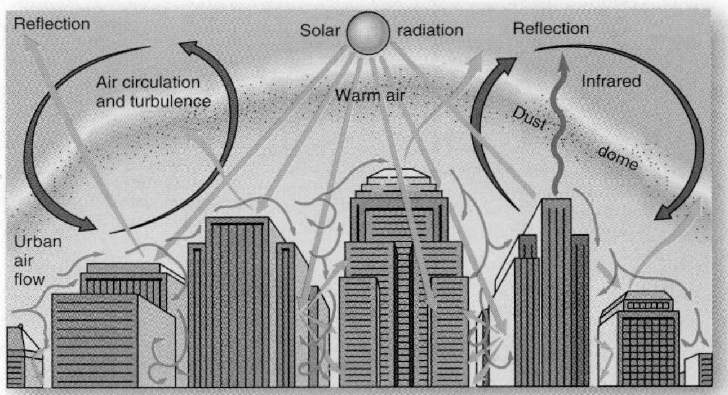

A growing percentage of Earth's people live in cities and experience a unique set of altered microclimatic effects: increased conduction, lower albedos, higher NET R values, increased water runoff, complex radiation and reflection patterns, anthropogenic heating, and the gases, dusts, and aerosols of urban pollution. Urban surfaces of metal, glass, asphalt, concrete, and stone conduct up to three times more energy than does wet, sandy soil and thus are warmed, producing an **urban heat island**. Air pollution, including gases and aerosols, is greater in urban areas than in rural ones. Every major city produces its own **dust dome** of airborne pollution.

urban heat island (p. 101)
dust dome (p. 101)

18. What is the basis for the urban heat island concept? Describe the climatic effects attributable to urban as compared with nonurban environments.
19. Which of the items in Table 4.2 have you yourself experienced? Explain.
20. Assess the potential for solar energy applications in our society. What are some negatives? What are some positives?

MG™ **MASTERING GEOGRAPHY**

Visit www.masteringgeography.com for several figure animations, satellite loops, self-study quizzes, flashcards, visual glossary, case studies, career links, textbook reference maps, geography links, RSS Feeds, and an ebook version of *Geosystems*. Included are many geography web links to interesting Internet sources that support this chapter.

CHAPTER 5

Global Temperatures

Global temperature change is affecting the wild sheep abandoned on this Scottish Isle. Stone cleits sheltered the residents of Hirta for thousands of years. Note more rock structures along the hillside in the background. [Bobbé Christopherson.]

KEY LEARNING CONCEPTS

After reading the chapter, you should be able to:

- ***Define*** the concepts of temperature, kinetic energy, and sensible heat, and ***distinguish*** among Kelvin, Celsius, and Fahrenheit temperature scales and how they are measured.
- ***List*** and ***review*** the principal controls and influences that produce global temperature patterns.
- ***Review*** the factors that produce different marine effects and continental effects as they influence temperatures, and ***utilize*** several pairs of cities to illustrate these differences.
- ***Interpret*** the pattern of Earth's temperatures from their portrayal on January and July temperature maps and on a map of annual temperature ranges.
- ***Contrast*** wind chill and heat index, and ***determine*** human response to these apparent temperature effects.

GEOSYSTEMS NOW

Temperature Change Affects St. Kilda's Sheep

On the windswept island of Hirta in Scotland's Outer Hebrides, global temperature change is affecting ecosystems. The wild Soay sheep have shrunk in body size over the past quarter century, a trend that has puzzled scientists. Now, however, the explanation appears clearly linked to milder winters and longer summers associated with recent climate change.

Some 180 km (110 mi) west of the Scottish mainland, the St. Kilda Archipelago comprises three islands, Hirta, Soay, and Dun, centered at 57.75° N latitude (Figure GN 5.1). Climate conditions are moderate for the latitude because of warm ocean currents from the Gulf Stream (see Figure 5.10 on page 114). The marine west coast climate is similar to that of Ireland and Britain, but with a cooler summer. January average temperatures are 5.6°C (42.1°F), with July rising to an average of 11.8°C (53.2°F). Snow is a rare occurrence. Precipitation is evenly distributed, an average year receiving 1400 mm (55 in.). Domestic sheep were introduced to these islands about 2000 years ago and evolved in isolation as a small, primitive breed now known as Soay sheep.

St. Kilda's principal island, Hirta, was continuously occupied over the last 3000 to 4000 years. The people of Hirta lived simply and stored food in sod-roofed stone shelters known as *cleits*. More traditional walled houses of stone did not appear until the 19th century. In 1930, the last 30 residents of Hirta were evacuated as their small society failed, and the Soay sheep remained, unmanaged to this day (Figure GN 5.2).

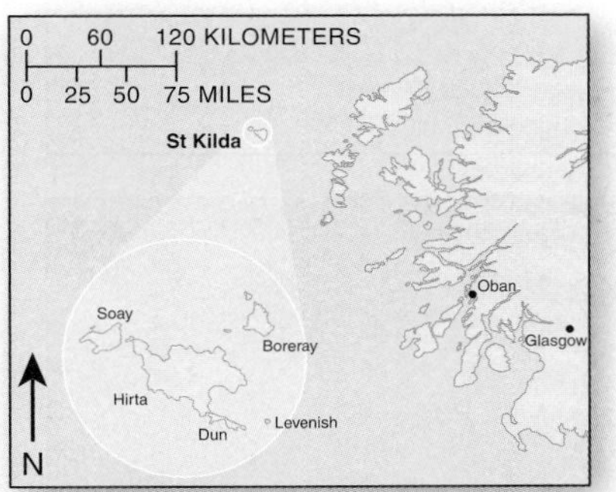

FIGURE GN 5.1 St. Kilda Archipelago, Scotland.

FIGURE GN 5.2 Wild Soay sheep on Hirta Island. [Bobbé Christopherson.]

Hirta's sheep provide scientists an ideal opportunity to study an isolated population with no significant competitors or predators. According to evolutionary theory, wild sheep such as the Soay should gradually increase in size over many generations because larger, stronger sheep are more likely to survive winter and reproduce in the spring. This trend follows the principle of natural selection. Remarkably, scientists are seeing the opposite trend as global temperature change affects these sheep.

Present climate change is extending summer as spring comes a few weeks earlier and fall a few weeks later across the globe. Also, winters are milder and shorter in length. Scientists began studying the impact of changing climatic conditions on Hirta's wild Soay sheep in 1985 and published their results in 2009. Surprisingly, scientists found that these sheep were not getting larger in size and that concepts of adaptive evolution could not explain what was happening.

Scientists found, first, that females are giving birth at younger ages, when they can only physically produce offspring that are smaller in size than they were at birth. This "young mum effect" explains why sheep size is not increasing. But why then are sheep shrinking? Since 1985, female Soay sheep of all ages dropped 5% in body mass, with leg length and body weight decreasing. In addition, lambs are not growing as quickly. The explanation ties to warming temperatures: Longer summers increase the availability of grass feed, and milder winters mean that lambs do not need to put on as much weight in their first months, allowing even slower-growing individuals to survive. Smaller lambs that previously would not survive winter are surviving; thus, smaller individuals are becoming more common in the population.

This study indicates that local factors, such as temperature change, can cause rapid *phenotypic change* as a species' genetic makeup interacts with the environment. Such factors can override the pressures of natural selection and evolution. For Chapter 5 and our discussion of global temperatures, we must keep in mind the interconnections among all life and Earth's physical systems as global temperatures increase. (*See* A. Oxgul, et al., "The dynamics of Phenotypic Change and the Shrinking Sheep of St. Kilda," *Science*, July 24, 2009, v. 325, no. 5939: 464–67; and "Supporting Online Material," by study authors, *Science*, July 2, 2009.)

What is the temperature now—both indoors and outdoors—as you read these words? How is it measured, and what does the value mean? How is air temperature influencing your plans for the day? Our bodies subjectively sense temperature and judge comfort and react to changing temperature. Air temperature plays a remarkable role in our lives, both at the micro level and at the macro and global levels.

Global temperatures have risen an average of 0.17 C° (0.3 F°) per decade since 1970, and this rate is accelerating. In Figure 5.1, surface temperature anomalies by decade are compared to the standard 1951–1980 base period in the record. Ice-core data and other proxy measures indicate that present temperatures are higher than at any time during the past 125,000 years. This warming trend in air temperature is the subject of much scientific, geographic, and political action.

Temperatures of global land, air, and water set records in the first part of 2010. This trend is in response to carbon dioxide emissions, which have been increasing at 3.0% per year since 2000. As mentioned in Chapter 3, present CO_2 levels in the atmosphere are higher than at any time in the past 800,000 years.

In this chapter: Temperature affects societies, decision making, and resources consumed across the globe. Understanding temperature concepts helps us begin our study of related Earth energy, weather, and climate systems. We look at the principal temperature controls of latitude, altitude and elevation, cloud cover, and land–water heating differences as they interact to produce Earth's temperature patterns. We end by examining the effect of temperature on the human body and discussing current temperature trends associated with global warming.

Temperature Concepts and Measurement

Heat and temperature are not the same. *Heat* is a form of energy that flows from one system or object to another because the two are at different temperatures. **Temperature** is a measure of the average kinetic energy (motion) of individual molecules in matter. We feel the effect of temperature as the *sensible heat* transfer from warmer objects to cooler objects. For instance, when you jump into a cool lake, you can sense the heat transfer from your skin to the water as kinetic energy leaves your body and flows to the water.

Temperature and heat are related because changes in temperature are caused by the absorption or emission (gain or loss) of heat energy. The term *heat energy* is frequently used to describe energy that is added to or removed from a system or substance.

Temperature Scales

The temperature at which atomic and molecular motion in matter completely stops is *absolute zero*, or *0° absolute temperature*. Its value on the different temperature-measuring scales is −273° Celsius (C), −459.4° Fahrenheit (F), and 0 Kelvin (K). Figure 5.2 compares these three scales. Formulas for converting among Celsius, SI (Système International), and English units are in Appendix C.

The Fahrenheit scale places the melting point of ice at 32°F, with 180 subdivisions to the boiling point of water at 212°F. The scale is named for its developer, Daniel G. Fahrenheit, a German physicist (1686–1736). Note that there is only one melting point for ice, but there are many freezing points for water, ranging from 32°F down to −40°F, depending on its purity, its volume, and certain conditions in the atmosphere.

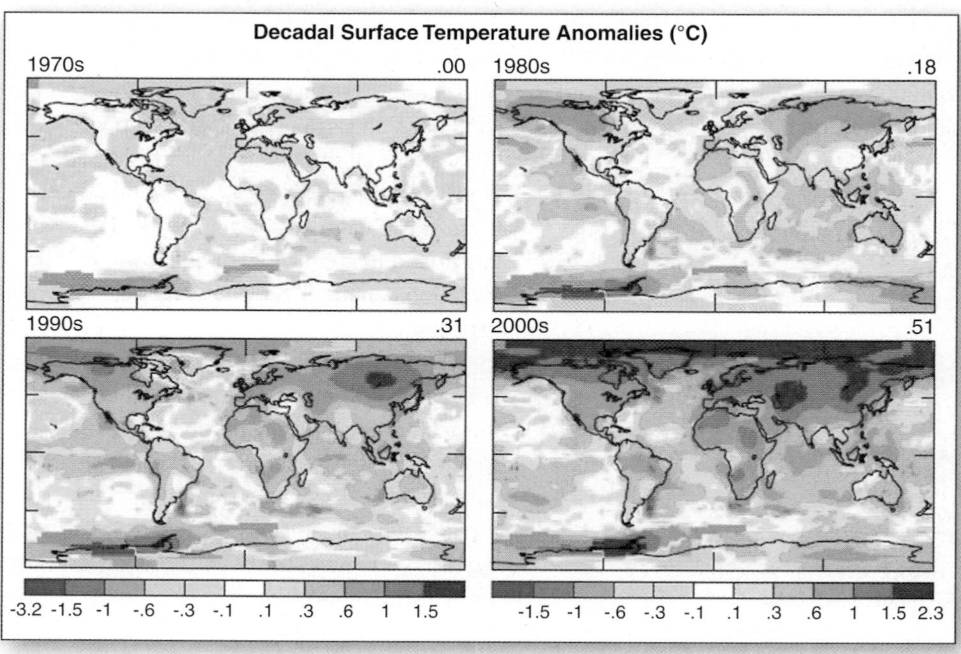

FIGURE 5.1 Surface temperature anomalies by decade. Increasing temperature patterns are apparent in this decadal look at temperature anomalies for the past four decades, as compared to the 1951–1980 baseline. [GISS/NASA.]

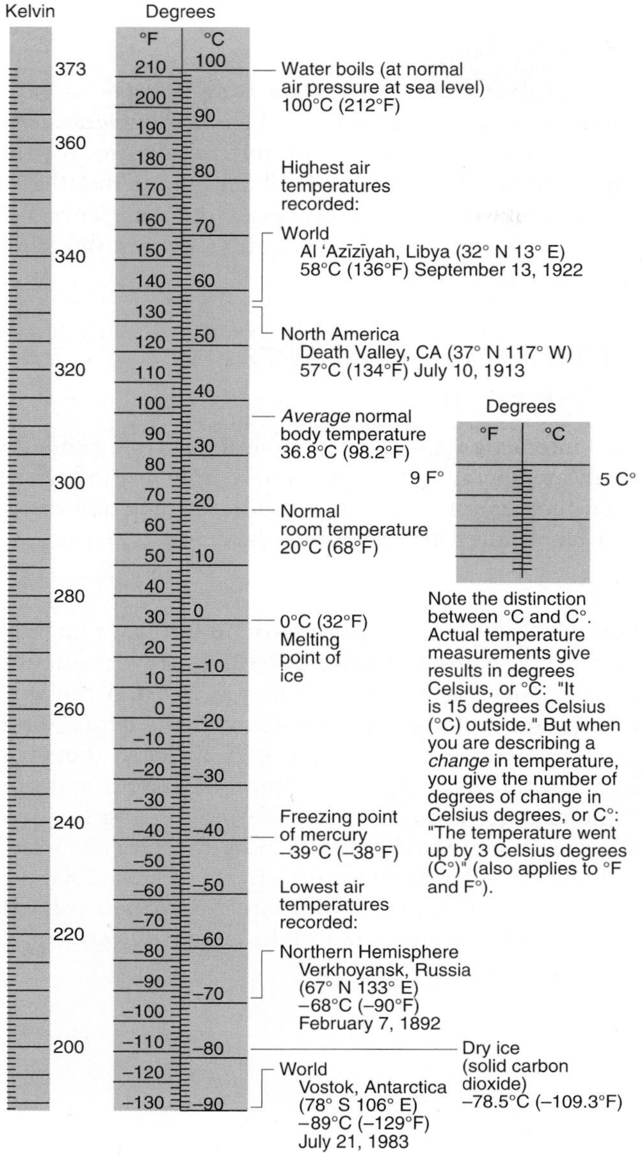

Kelvin | Degrees °F °C

— Water boils (at normal air pressure at sea level) 100°C (212°F)

Highest air temperatures recorded:

World
Al 'Azīzīyah, Libya (32° N 13° E)
58°C (136°F) September 13, 1922

North America
Death Valley, CA (37° N 117° W)
57°C (134°F) July 10, 1913

— *Average* normal body temperature 36.8°C (98.2°F)

— Normal room temperature 20°C (68°F)

Degrees °F °C — 9 F° — 5 C°

— 0°C (32°F) Melting point of ice

Note the distinction between °C and C°. Actual temperature measurements give results in degrees Celsius, or °C: "It is 15 degrees Celsius (°C) outside." But when you are describing a *change* in temperature, you give the number of degrees of change in Celsius degrees, or C°: "The temperature went up by 3 Celsius degrees (C°)" (also applies to °F and F°).

— Freezing point of mercury −39°C (−38°F)

Lowest air temperatures recorded:

Northern Hemisphere
Verkhoyansk, Russia
(67° N 133° E)
−68°C (−90°F)
February 7, 1892

— Dry ice (solid carbon dioxide) −78.5°C (−109.3°F)

World
Vostok, Antarctica
(78° S 106° E)
−89°C (−129°F)
July 21, 1983

FIGURE 5.2 Temperature scales.
Scales for expressing temperature in Kelvin (K) and degrees Celsius (°C) and Fahrenheit (°F), with details of what we experience. Note the distinction between temperature and units of temperature expression and the placement of the degree symbol.

About a year after the adoption of the Fahrenheit scale, Swedish astronomer Anders Celsius (1701–1744) developed the Celsius scale (formerly centigrade). He placed the melting point of ice at 0°C and the boiling temperature of water at sea level at 100°C, dividing his scale into 100 degrees using a decimal system.

British physicist Lord Kelvin (born William Thomson, 1824–1907) proposed the Kelvin scale in 1848. Science uses this scale because it starts at absolute zero temperature and thus readings are proportional to the actual kinetic energy in a material. The Kelvin scale's melting point for ice is 273 K, and its boiling point of water is 373 K.

Most countries use the Celsius scale to express temperature. The United States remains the only major country still using the Fahrenheit scale. The continuing pressure from the scientific community, not to mention the rest of the world, makes adoption of Celsius and SI units inevitable in the United States. This textbook presents Celsius (with Fahrenheit equivalents in parentheses) throughout to help bridge this transitional era in the United States.

Measuring Temperature

A *mercury thermometer* or *alcohol thermometer* is a sealed glass tube that measures outdoor temperatures. Fahrenheit invented both the alcohol and the mercury thermometers. Cold climates demand alcohol thermometers because alcohol freezes at −112°C (−170°F), whereas mercury freezes at −39°C (−38.2°F). The principle of these thermometers is simple: When fluids are heated, they expand; upon cooling, they contract. A thermometer stores fluid in a small reservoir at one end and is marked with calibrations to measure the expansion or contraction of the fluid, which reflects the temperature of the thermometer's environment.

Thermometers for standardized official readings are placed outdoors in small shelters that are white (for high albedo) and louvered (for ventilation) to avoid overheating of the instruments. They are placed at least 1.2–1.8 m (4–6 ft) above a surface, usually on turf; in the United States, 1.2 m (4 ft) is standard. Official temperature measurements are made in the shade to prevent the effect of direct insolation.

The instrument shelter in Figure 5.3 on page 110 contains a *thermistor*, which measures temperature by sensing the electrical resistance of a semiconducting material. Temperature is sensed because resistance changes at 4% per C°, which is reported electronically to the weather station.

Temperature readings are taken daily, sometimes hourly, at more than 16,000 weather stations worldwide. Some stations with recording equipment also report the duration of temperatures, rates of rise or fall, and temperature variation over time throughout the day and night. One of the goals of the Global Climate Observing System (GCOS, http://www.wmo.ch/pages/prog/gcos/) is to establish a reference network of one station per 250,000 km² (95,800 mi²). (See the World Meteorological Organization at http://www.wmo.ch/.)

Getting the measurement right

The Goddard Institute for Space Studies, NASA, constantly refines the methodology of gathering global temperature readings. The temperature data used for the maps in Figure 5.1 were gathered from 6300 stations. Corrections are calibrated for city settings where the urban heat island affects temperatures. Population data and the pattern of nightlights captured by satellites further refine the data base. For more discussion, see http://data.giss.nasa.gov/.

FIGURE 5.3 **Instrument shelter.**
This standard thermometer shelter is white and louvered, installed above a turf surface. These are replacing the traditional Cotton-region, or Stephenson screen, shelters for temperature measurement. [Bobbé Christopherson.]

The *daily mean temperature* is an average of daily minimum–maximum readings. The *monthly mean temperature* is the total of daily mean temperatures for the month divided by the number of days in the month. An *annual temperature range* expresses the difference between the lowest and highest monthly mean temperatures for a given year.

Principal Temperature Controls

The interaction of several physical controls produces Earth's temperature patterns. These principal influences upon temperatures include latitude, altitude and elevation, cloud cover, and land–water heating differences.

Latitude

Insolation is the single most important influence on temperature variations. Figure 2.9 shows how insolation intensity decreases as one moves away from the subsolar point—a point that migrates annually between the Tropic of Cancer and Tropic of Capricorn (between 23.5° N and 23.5° S). In addition, daylength and Sun angle change throughout the year, increasing seasonal effect with increasing latitude. The five cities graphed in Figure 5.4 demonstrate the effects of latitudinal position on temperature. From equator to poles, Earth ranges from continually warm, to seasonally variable, to continually cold.

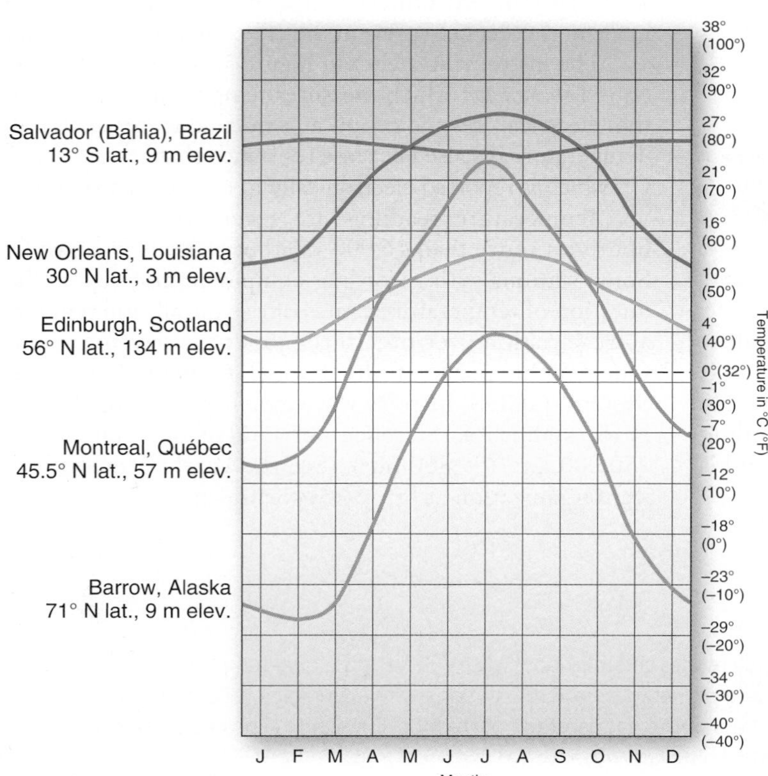

Located on maps in:
Figure 1.14
Figure 4.16 – net radiation
Figure 4.18 – latent heat
Figure 4.19 – sensible heat

FIGURE 5.4 **Latitude affects temperatures.**
A comparison of five cities from near the equator to north of the Arctic Circle demonstrates changing seasonality and increasing differences between average minimum and maximum temperatures as one moves to higher latitudes.

Altitude/Elevation

Within the troposphere, temperatures decrease with increasing altitude above Earth's surface. (Recall that the *normal lapse rate* of temperature change with altitude is 6.4 C°/1000 m, or 3.5 F°/1000 ft; see Figure 3.5.) The density of the atmosphere also diminishes with increasing altitude. In fact, the density of the atmosphere at an elevation of 5500 m (18,000 ft) is about half that at sea level. As the atmosphere thins, there is a loss in its ability to absorb and radiate sensible heat. Thus, worldwide, mountainous areas experience lower temperatures than do regions nearer sea level, even at similar latitudes.

The consequences are that, at high elevations, average air temperatures are lower, nighttime cooling is greater, and the temperature range between day and night is greater than at low elevations. The temperature difference between areas of sunlight and shadow is greater at elevation than at sea level. You may have felt temperatures decrease noticeably in the shadows and shortly after sunset when you were in the mountains. Surfaces both gain energy rapidly from and lose energy rapidly to the thinner atmosphere.

The snow line seen in mountain areas indicates where winter snowfall exceeds the amount of snow lost through summer melting and evaporation. The snow line's location is a function both of elevation and of latitude, and local microclimatic conditions of exposure. Permanent ice fields and glaciers exist on equatorial mountain summits in the Andes and East Africa. In equatorial mountains, the snow line occurs at approximately 5000 m (16,400 ft). With increasing latitude, snow lines gradually lower in elevation from 2700 m (8850 ft) in the midlatitudes to lower than 900 m (2950 ft) in southern Greenland.

Two cities in Bolivia illustrate the interaction of the two temperature controls, latitude and elevation. Figure 5.5 displays temperature data for the cities of Concepción and La Paz, which are near the same latitude (about 16° S). Note the elevation, average annual temperature, and precipitation for each location at the bottom of the figure.

The hot, humid climate of Concepción at its much lower elevation stands in marked contrast to the cool, dry climate of highland La Paz. People living around La Paz are able to grow wheat, barley, and potatoes—crops characteristic of the cooler midlatitudes—despite the fact that

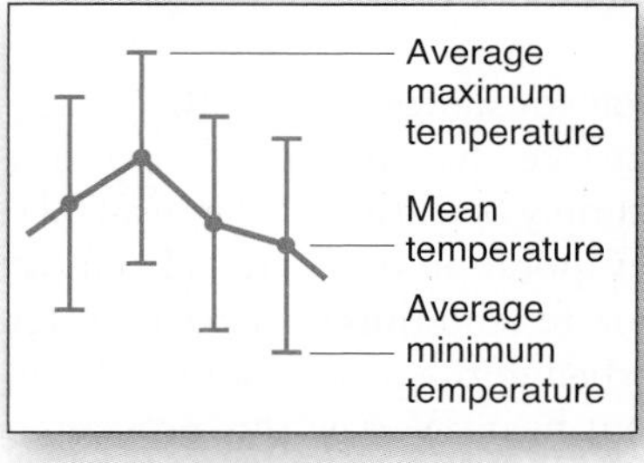

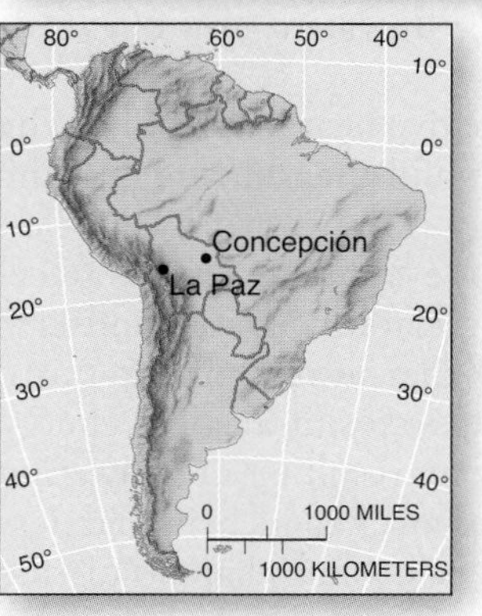

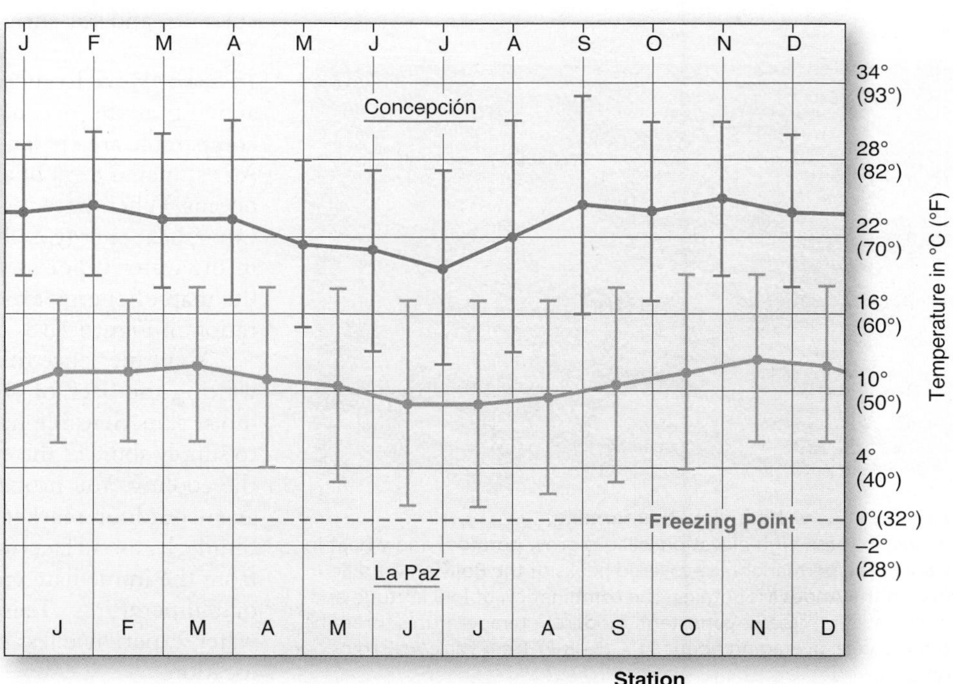

	Concepción, Bolivia	La Paz, Bolivia
Latitude/longitude	16° 15′ S 62° 03′ W	16° 30′ S 68° 10′ W
Elevation	**490 m (1608 ft)**	**4103 m (13,461 ft)**
Avg. ann. temperature	23°C (73.4°F)	11°C (51.8°F)
Ann. temperature range	6.5 C° (11.7 F°)	3.5 C° (6.3 F°)
Ann. precipitation	121.2 cm (47.7 in.)	55.5 cm (21.9 in.)
Population	10,000	810,300 (Administrative division 1.6 million)

FIGURE 5.5 Effects of latitude and elevation. Comparison of temperature patterns in two Bolivian cities.

La Paz is 4103 m (13,461 ft) above sea level (Figure 5.6). (For comparison, the summit of Pikes Peak in Colorado is at 4301 m, or 14,111 ft, and that of Mount Rainier, Washington, is at 4392 m, or 14,410 ft.)

The combination of elevation and low-latitude location guarantees La Paz nearly constant daylength and moderate temperatures, averaging about 11°C (51.8°F) through the year. Such moderate temperature and moisture conditions lead to the formation of more fertile soils than those found in the warmer, wetter climate of Concepción, with an annual average temperature of 24°C (75°F).

Cloud Cover

Orbiting satellites reveal that approximately 50% of Earth is cloud covered at any given moment. Clouds moderate temperature, and their effect varies with cloud type, height, and density. Cloud moisture reflects, absorbs, and liberates large amounts of energy released upon condensation and cloud formation.

At night, clouds act as insulation and radiate longwave energy, preventing rapid energy loss. During the day, clouds reflect insolation as a result of their high albedo values. In general, they lower daily maximum temperatures and raise nighttime minimum temperatures. In the last chapter, Figures 4.6 and 4.9 portray cloud-albedo forcing and cloud-greenhouse forcing as they relate to the presence of clouds and cloud types. Clouds also reduce latitudinal and seasonal temperature differences.

FIGURE 5.6 High-elevation farming.
People in these high-elevation villages grow potatoes and wheat in view of the permanent ice-covered peaks of the Bolivian Cordillera Real in the Andes Mountains. The combination of low latitude and high elevation creates consistent, moderate temperatures throughout the year, averaging about 11°C (51.8°F). [Seux Paule/AGE Fotostock America, Inc.]

Clouds are the most variable factor influencing Earth's radiation budget, making them the subject of simulations in computer models of atmospheric behavior. The International Satellite Cloud Climatology Project (http://isccp.giss.nasa.gov/), part of the World Climate Research Programme, is presently in the midst of such research. Clouds and the Earth Radiant Energy System (CERES, http://science.larc.nasa.gov/ceres/) sensors aboard the *TRMM*, *Terra*, and *Aqua* satellites are assessing cloud effects on longwave, shortwave, and net radiation patterns as never before possible.

Land–Water Heating Differences

Another major control over temperature is the different ways in which land and water respond to insolation. Earth presents these two surfaces in an irregular arrangement of continents and oceans. Each absorbs and stores energy differently than the other and therefore contributes to the global pattern of temperature. Water bodies tend to produce moderate temperature patterns, whereas inland in continental interiors more extreme temperatures occur.

The physical nature of land (rock and soil) and water (oceans, seas, and lakes) is the reason for **land–water heating differences**—land heats and cools faster than water. Figure 5.7 visually summarizes the following discussion of the land–water temperature controls: evaporation, transparency, specific heat, movement, ocean currents, and sea-surface temperatures.

Evaporation Evaporation consumes more of the energy arriving at the ocean's surface than is expended over a comparable area of land, simply because water is available. An estimated 84% of all evaporation on Earth is from the oceans. When water evaporates and thus changes to water vapor, heat energy is absorbed in the process and is stored in the water vapor as latent heat. We saw this process in the map of energy expended for the latent heat of evaporation in Figure 4.18.

You experience this evaporative heat loss (cooling) by wetting the back of your hand and then blowing on the moist skin. Sensible heat energy is drawn from your skin to supply some of the energy for evaporation, and you feel the cooling. You probably appreciate the water "misters" at an outdoor market or café that lower temperatures. Similarly, as surface water evaporates, it absorbs energy from the immediate environment, resulting in a lowering of temperatures. Temperatures over land, with far less water, experience less evaporative cooling than do marine locations.

GEO REPORT 5.2 ***Referring to heights above sea level***

Altitude and elevation are different. Note that *altitude* refers to airborne objects or heights above Earth's surface, whereas *elevation* usually refers to the height of a point on Earth's surface above some plane of reference. For example, we ask, "At what altitude did the jet fly?" and, "The ski resort is at what elevation?"

FIGURE 5.7 Land–water heating differences.
The differential heating of land and water produces contrasting marine (more moderate) and continental (more extreme) temperature regimes.

Transparency The transmission of light obviously differs between soil and water: Solid ground is opaque; water is transparent. Consequently, light striking a soil surface does not penetrate, but is absorbed, heating the ground surface. That energy is accumulated during times of exposure and is rapidly lost at night or in shadows.

Figure 5.8 shows the profile of diurnal (daily) temperatures for a column of soil and the atmosphere above it at a midlatitude location. You can see that maximum and minimum temperatures generally are experienced right at ground level. Below the surface, even at shallow depths, temperatures remain about the same throughout the day. You encounter this at a beach, where surface sand may be painfully hot to your feet, but as you dig in your toes and feel the sand a few centimeters below the surface, it is cooler, offering relief.

In contrast, when light reaches a body of water, it penetrates the surface because of water's **transparency**—water is clear and light passes through it to an average depth of 60 m (200 ft) in the ocean (Figure 5.9). This illuminated zone is the *photic layer* and occurs in some ocean waters to depths of 300 m (1000 ft). This characteristic of water results in the distribution of available heat energy over a much greater depth and volume, forming a larger energy reservoir than that of land surfaces.

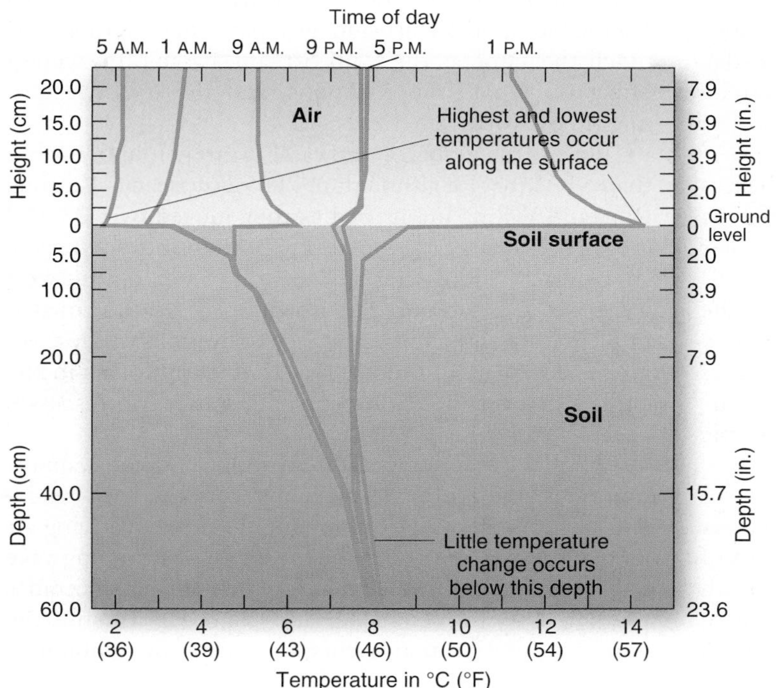

FIGURE 5.8 Land is opaque.
Profile measurements of air and soil temperatures in Seabrook, New Jersey. Note that little temperature change occurs throughout the day at depth, whereas daily extremes of temperature register along the surface, where insolation is absorbed. [Adapted from Mather, *Climatology: Fundamentals and Applications* (New York: McGraw-Hill, 1974).]

FIGURE 5.9 **Ocean is transparent.**
The transparency of the ocean permits insolation to penetrate to greater depths than land, increasing the volume of water that absorbs energy. Sunlight empowers myriad forms of life covering the substrate, including a sea urchin at 3.0 m (9.8 ft) depth. [Bobbé Christopherson.]

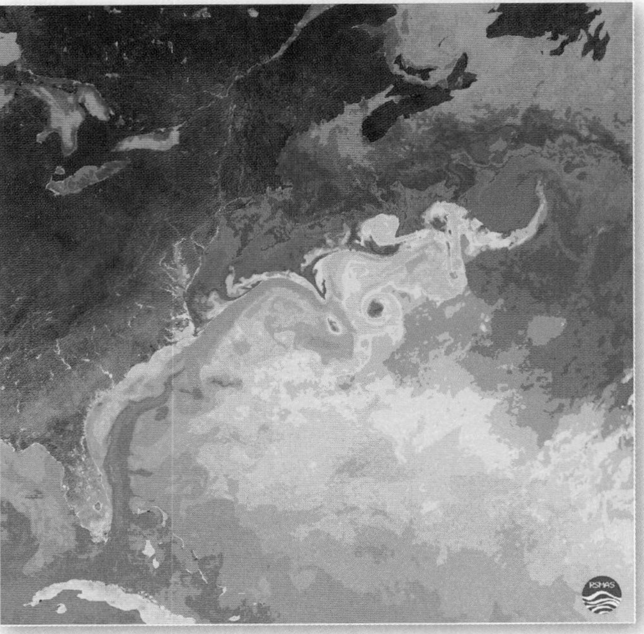

FIGURE 5.10 **The Gulf Stream.**
Satellite instruments sensitive to thermal infrared wavelengths imaged the Gulf Stream. Temperature differences are noted by computer-enhanced false colors: reds/oranges = 25°–29°C (76°–84°F), yellows/greens = 17°–24°C (63°–75°F); blues = 10°–16°C (50°–61°F); and purples = 2°–9°C (36°–48°F). [Imagery by RSMAS, University of Miami.]

Specific Heat When comparing equal volumes of water and land, water requires far more energy to increase its temperature than does land. In other words, *water can hold more heat than can soil or rock*, and therefore water is said to have a higher **specific heat**, the heat capacity of a substance. On average, the specific heat of water is about four times that of soil. A given volume of water represents a more substantial energy reservoir than does an equal volume of soil or rock and therefore heats and cools more slowly. For this reason, day-to-day temperatures near a substantial body of water tend to be moderated.

Movement Land is a rigid, solid material, whereas water is a fluid and is capable of movement. Differing temperatures and currents result in a mixing of cooler and warmer waters, and that mixing spreads the available energy over an even greater volume than if the water were still. Surface water and deeper waters mix, redistributing energy. Both ocean and land surfaces radiate longwave radiation at night, but land loses its energy more rapidly than does the moving oceanic energy reservoir.

Ocean Currents and Sea-Surface Temperatures The **Gulf Stream** moves northward off the east coast of North America, carrying warm water far into the North Atlantic (Figure 5.10). As a result, the southern third of Iceland experiences much milder temperatures than would be expected for a latitude of 65° N, just south of the Arctic Circle (66.5°). In Reykjavík, on the southwestern coast of Iceland, monthly temperatures average above freezing during all months of the year. Similarly, the Gulf Stream moderates temperatures in coastal Scandinavia and northwestern Europe. In the western Pacific Ocean, the Kuroshio, or Japan Current, functions much the same as the Gulf Stream, having a warming effect on Japan, the Aleutians, and the northwestern margin of North America.

Across the globe, ocean water is rarely found warmer than 31°C (88°F), although in 2005 Hurricanes Katrina, Rita, and Wilma intensified as they moved over 33.3°C (92°F) sea-surface temperatures in the Gulf of Mexico. In contrast, along midlatitude and subtropical west coasts, cool ocean currents flowing toward the equator influence air temperatures. When conditions in these regions are warm and moist, fog frequently forms in the chilled air over the cooler currents. Chapter 6 discusses ocean currents.

Higher ocean temperatures produce higher evaporation rates, and more energy is lost from the ocean as latent heat. As the water vapor content of the overlying air mass increases, the ability of the air to absorb longwave radiation also increases. Therefore, the air mass becomes warmer. The warmer the air and the ocean become, the more evaporation that occurs, increasing the amount of water vapor entering the air mass. More water vapor leads to cloud formation, which reflects insolation and produces lower temperatures. Lower temperatures of air and ocean reduce evaporation rates and the ability of the air mass to absorb water vapor—an interesting *negative feedback* mechanism.

The Physical Oceanography Distributed Active Archive Center (PODAAC, http://podaac.jpl.nasa.gov/) is responsible for storage and distribution of data relevant to the physical state of the ocean. Sea-level height, currents, and ocean temperatures are measured as part of the effort to understand oceans and climate interactions.

Sea-surface temperature (SST) satellite data correlate well with actual sea-surface measurements. Figure 5.11 displays SST satellite data for January 2010 and July 2009 from satellites in the NOAA/NASA Pathfinder AVHRR. The Western Pacific Warm Pool is the deep red color in the southwestern Pacific Ocean with temperatures above 30°C (86°F). This region has the highest average ocean temperatures in the world. Note the seasonal change in ocean temperatures and the cool currents off the west coasts of North and South America, Europe, and Africa.

Mean annual SSTs increased steadily from 1982 through 2010 to record-high levels—2010 breaking records for both ocean and land temperatures. Increasing warmth is measured at depths to 1000 m (3200 ft), and even slight increases were found in the temperature of deep bottom water first reported in 2004. Scientists suggest that the ocean's ability to absorb excess heat energy from the atmosphere may be nearing its capacity.

Summary of Marine Effects versus Continental Effects As noted, Figure 5.7 summarizes the operation of the land–water temperature controls presented:

MG Satellite Global Sea-surface Temperatures

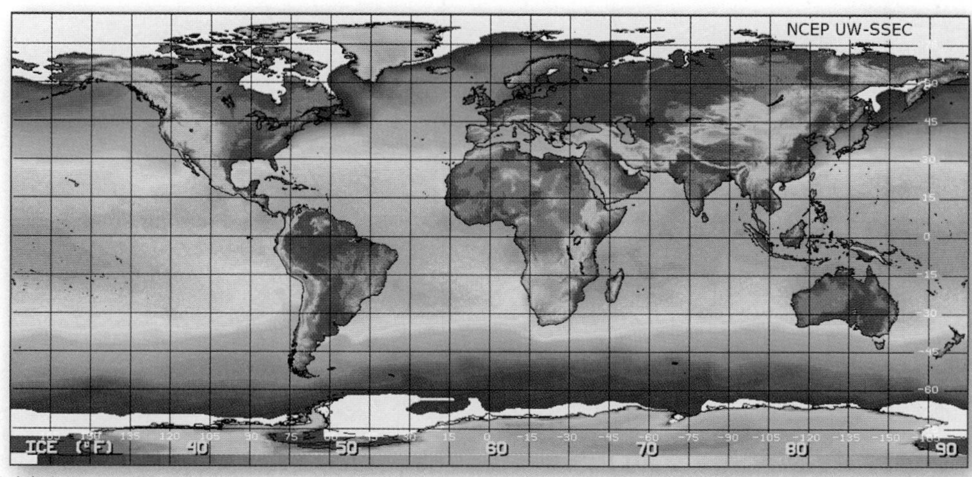

(a) January 23, 2010

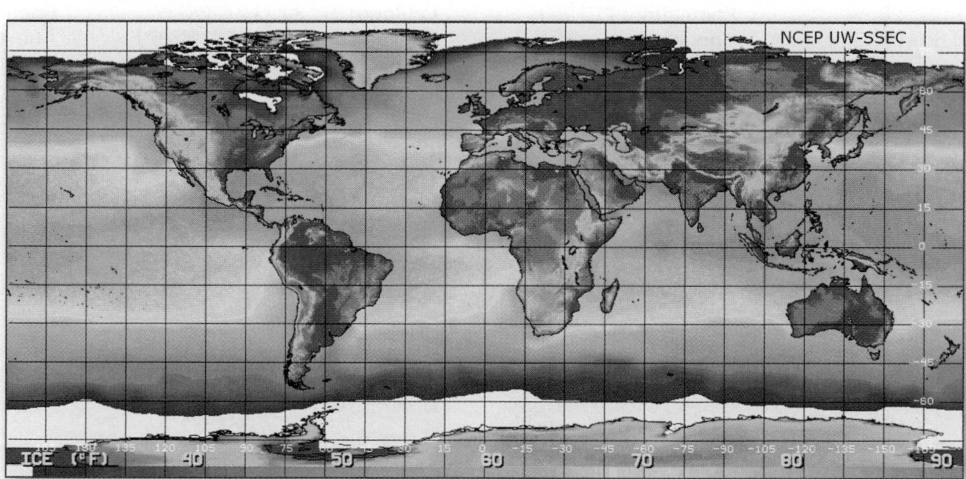

(b) July 23, 2009

FIGURE 5.11 Sea-surface temperatures.
Average annual sea-surface temperatures for (a) January 23, 2010, and (b) July 23, 2009. [Satellite data courtesy of Space Science and Engineering Center, University of Wisconsin, Madison.]

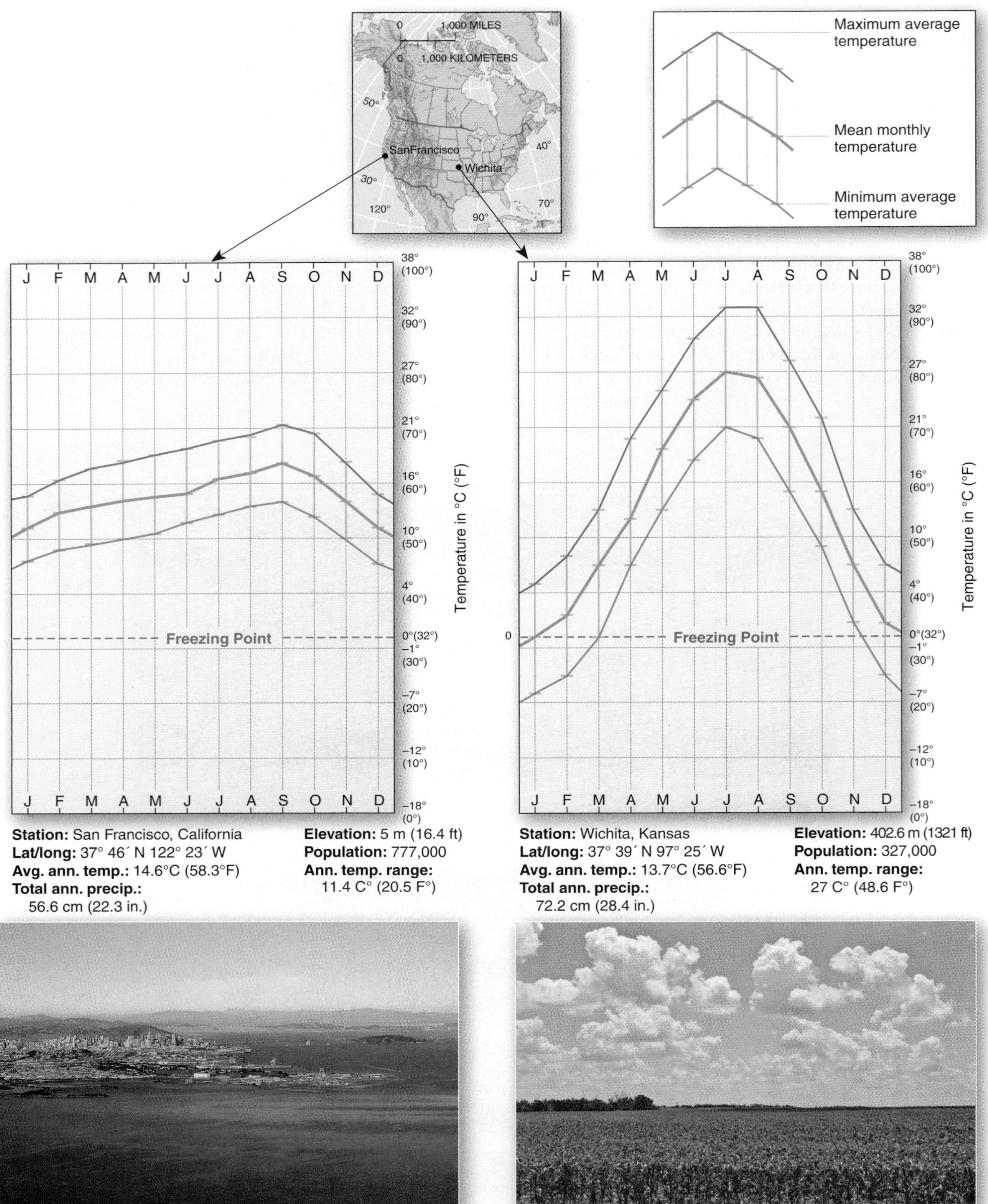

FIGURE 5.12 Marine and continental cities—United States.
Compare temperatures in coastal San Francisco, California, with those of continental Wichita, Kansas. [Photos by Bobbé Christopherson.]

evaporation, transparency, specific heat, movement, ocean currents, and sea-surface temperatures. The term **marine effect**, or *maritime*, describes locations that exhibit the moderating influences of the ocean, usually along coastlines or on islands. **Continental effect**, a condition of *continentality*, refers to areas less affected by the sea and therefore having a greater range between maximum and minimum temperatures on both a daily and a yearly basis.

San Francisco, California, and Wichita, Kansas, provide us an example of marine and continental conditions (Figure 5.12). Both cities are at approximately 37° 40′ N latitude. The cooling waters of the Pacific Ocean and San Francisco Bay surround San Francisco on three sides. Summer fog helps delay the warmest summer month in San Francisco until September. In 100 years of weather records, marine effects moderated temperatures, so that only a few days a year have summer maximums that exceed 32.2°C (90°F). Winter minimums rarely drop below freezing.

In contrast, the continental location of Wichita is susceptible to freezes from late October to mid-April, with daily variations slightly increased by its elevation, experiencing −30°C (−22°F) as a record low. West of Wichita, winters increase in severity with increasing distance from the moderating influences of invading air masses from the Gulf of Mexico. Wichita's temperature reaches 32.2°C (90°F) or higher more than 65 days each year, with 46°C (114°F) as a record high. Temperatures exceeded 38°C (100°F) for 23 nonconsecutive days in 2003.

Earth's Temperature Patterns

Earth's temperature patterns result from the combined effect of the controlling factors in our discussion. Let us now look at maps of mean air temperatures worldwide and for the polar regions for January (Figures 5.13 and 5.14 on pages 118 and 119) and July (Figures 5.16 and 5.17 on pages 121 and 122). To complete the analysis, Figure 5.18 on page 122 presents the temperature range differences between the January and July maps, or the differences between averages of the coolest and warmest months.

Maps are for January and July instead of the solstice months of December and June because a lag occurs between insolation received and maximum or minimum temperatures experienced, as explained in Chapter 4. The U.S. National Climate Data Center provided the data for these maps. Some ship reports go back to 1850 and land reports to 1890, although the bulk of the record is representative of conditions since 1950.

The lines on temperature maps are known as *isotherms*. An **isotherm** is an isoline—a line along which there is a constant value—that connects points of equal temperature and portrays the temperature pattern, just as a contour line on a topographic map illustrates points of equal elevation. Isotherms help with the spatial analysis of temperatures.

CRITICAL THINKING 5.1

Begin a record and physical geography profile

With each temperature map (Figures 5.13, 5.16, and 5.18), begin by finding your own city or town and noting the temperatures indicated by the isotherms for January and July and the annual temperature range. Record the information from these maps in your notebook. Remember that the small scale of these maps permits only generalizations about actual temperatures at specific sites, allowing you to make only a rough approximation for your location.

As you work through the different maps throughout this text, note atmospheric pressure and winds, annual precipitation, climate type, landforms, soil orders, vegetation, and terrestrial biomes. Add this information to your profile in your notebook. By the end of the course, you will have a complete physical geography profile for your regional environment.

January Temperature Maps

Figure 5.13 maps January's mean temperatures for the world. In the Southern Hemisphere, the higher Sun altitude causes longer days and summer weather conditions; in the Northern Hemisphere, the lower Sun angle causes the short days of winter. Isotherms generally are zonal, trending east–west, parallel to the equator, and they appear to be interrupted by the presence of landmasses. Isotherms mark the general decrease in insolation and net radiation with distance from the equator.

The **thermal equator** is an isotherm connecting all points of highest mean temperature, roughly 27°C (80°F); it trends southward into the interior of South America and Africa, indicating higher temperatures over landmasses. In the Northern Hemisphere, isotherms shift equatorward as cold air chills the continental interiors. The oceans, in contrast, are more moderate, with warmer conditions extending farther north than over land at comparable latitudes.

As an example, in Figure 5.13, follow along 50° N latitude (the 50th parallel) and compare isotherms: 3°C to 6°C in the North Pacific and 3°C to 9°C in the North Atlantic, as contrasted to −18°C in the interior of North America and −24°C to −30°C in central Asia. Also, note the orientation of isotherms over areas with mountain ranges and how they illustrate the cooling effects of elevation. Check the South American Andes as an example of these elevation effects.

The north polar and south polar regions (Figure 5.14a and b on page 119) are featured on two maps. Remember, the north polar region is an ocean surrounded by land, in contrast to the south polar region, which is the Antarctic continent surrounded by ocean. On the northern map (Figure 5.14a), the island of Greenland has a summit elevation of 3240 m (10,630 ft) on Earth's second largest ice sheet, the highest elevation north of the Arctic Circle. Two-thirds of the island is north of the Arctic Circle. Its north shore is only 800 km (500 mi) from the North Pole. This combination of high latitude and interior high elevation on the ice sheet produces cold midwinter temperatures.

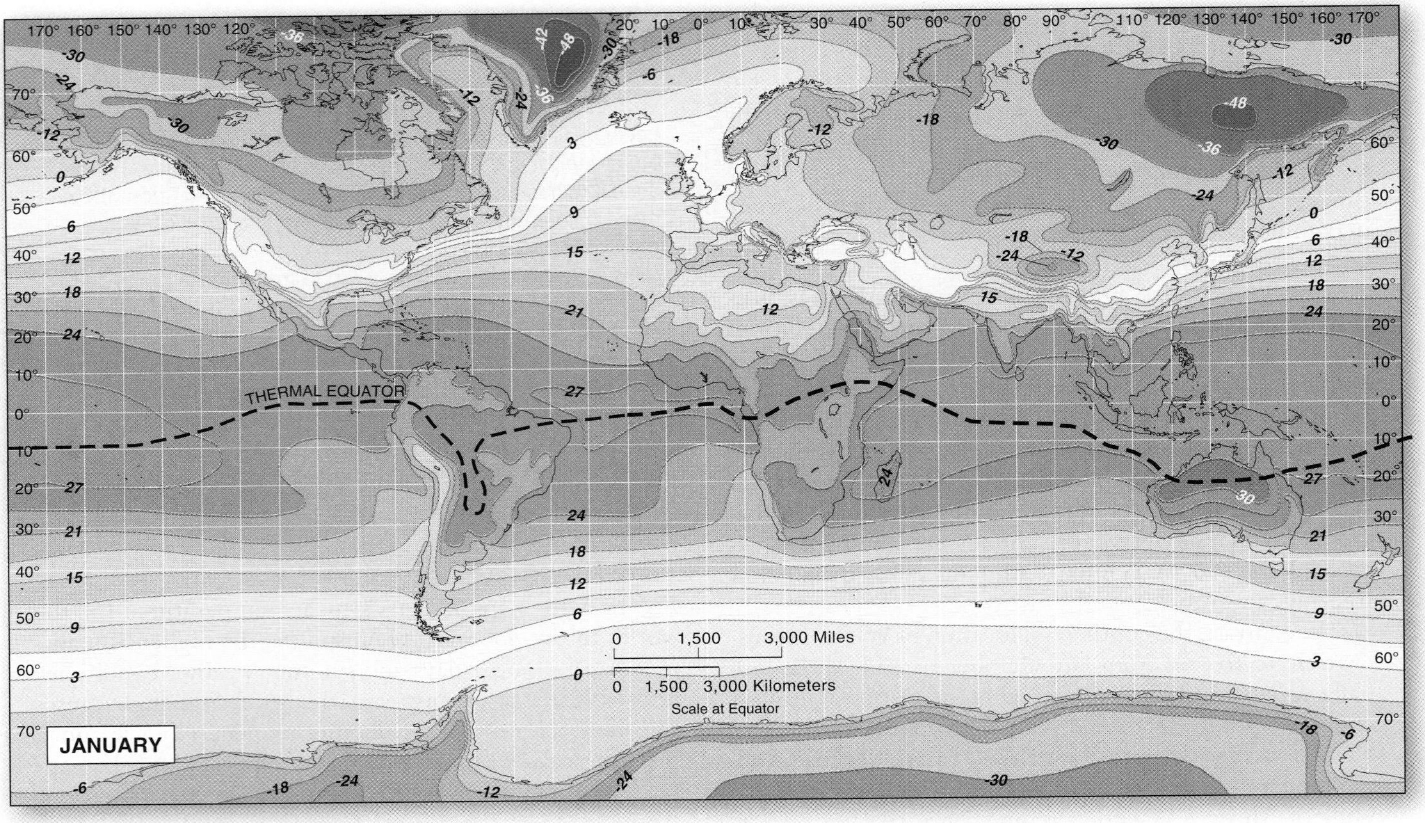

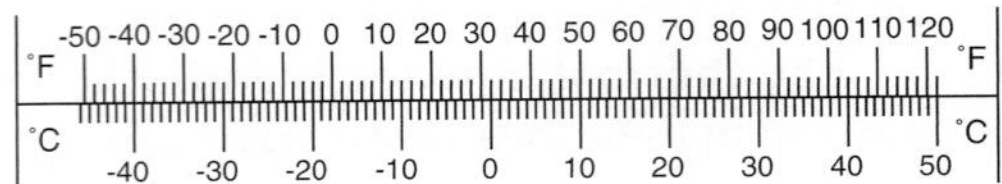

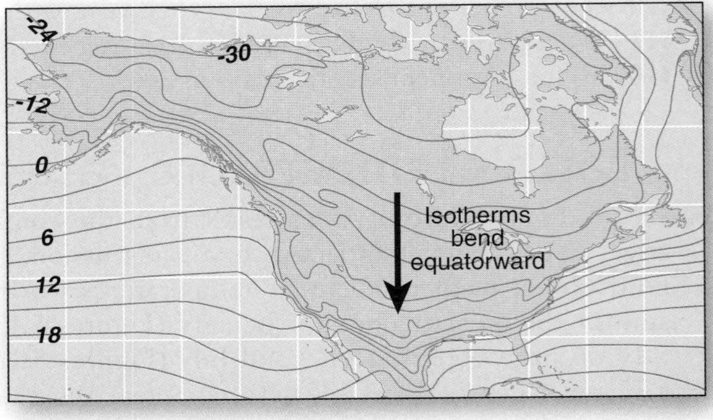

FIGURE 5.13 Global mean temperatures for January. Temperatures are in Celsius (convertible to Fahrenheit by means of the scale) as taken from separate air-temperature databases for ocean and land. Note the inset map of North America and the equatorward-trending isotherms in the interior. Compare with Figure 5.16. [Adapted by author and redrawn from National Climatic Data Center, *Monthly Climatic Data for the World,* 47 (January 1994), and WMO and NOAA.]

MG Satellite Global Surface Temperatures, Land and Ocean

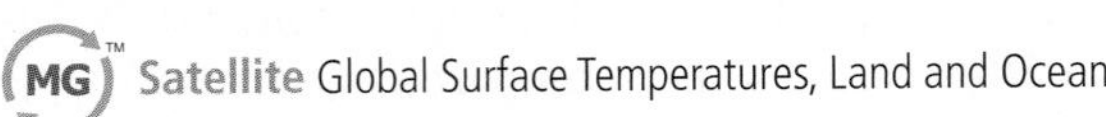

In Antarctica (Figure 5.14b), it is "summer" in December and January. This is Earth's coldest and highest average elevation landmass. January average temperatures for the three scientific bases noted on the map are McMurdo Station, at the coast on Ross Island, −3°C; the Amundsen–Scott Station at the South Pole (elevation 2835 m, 9301 ft), −28°C; and the Russian Vostok Station at a more continental location (elevation 3420 m, 11,220 ft), −32°C; respectively, 26.6°F, −18.4°F, and −25.6°F.

For a continental region other than Antarctica, Russia—and specifically, northeastern Siberia—is the coldest area on the map. The intense cold results from

GEO REPORT 5.3 *Polar regions show greatest warming*

Climate change is affecting the higher latitudes at a pace exceeding the middle and lower latitudes. Since 1978, warming has increased in the Arctic region to a rate of 1.2 C° (2.2 F°) per decade, which means the last 20 years warmed at nearly seven times the rate of the last 100 years. Since 1970, nearly 60% of the Arctic sea ice has disappeared in response to increasing air and ocean temperatures, with record low levels of ice from 2007 through 2010. Similar warming trends are impacting the Antarctic Peninsula and the West Antarctic Ice Sheet, as ice shelves collapse and retreat along the coast.

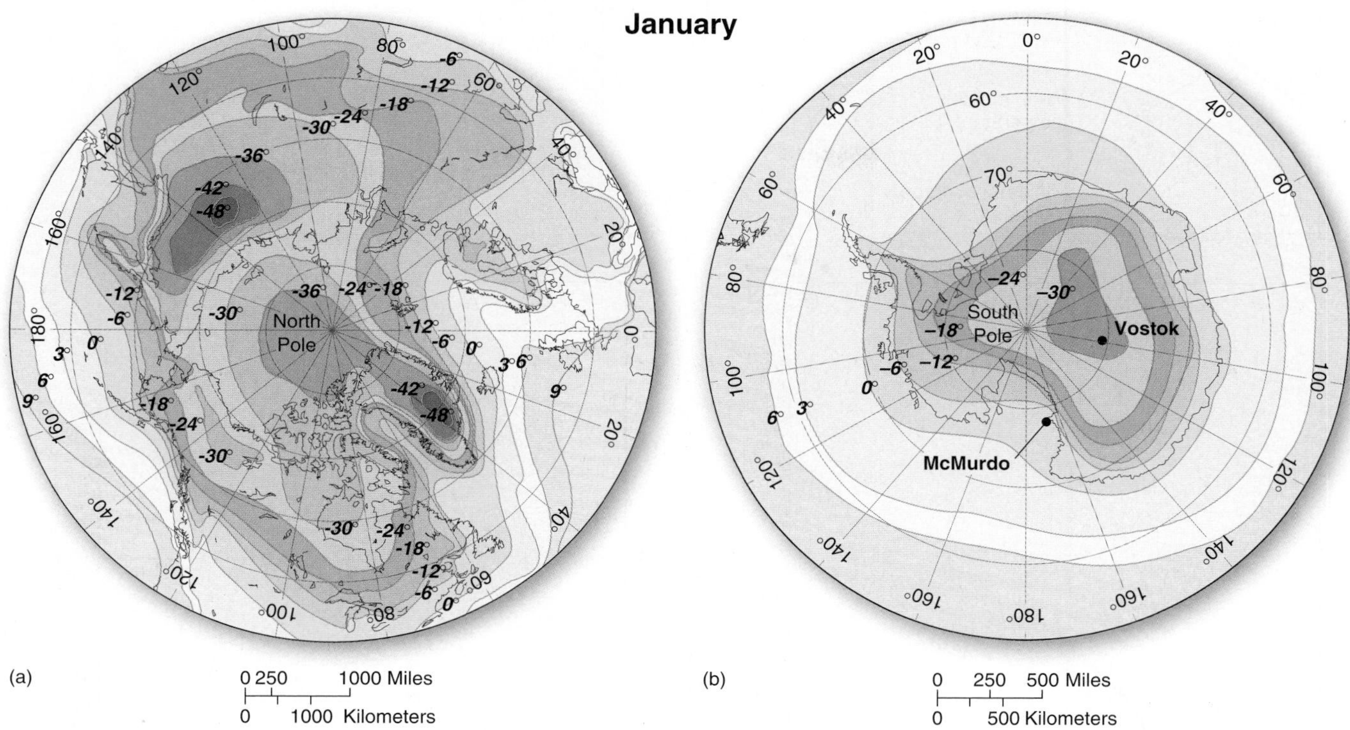

FIGURE 5.14 January mean temperatures for polar regions.
January temperatures in (a) North Polar region and (b) South Polar region. Use temperature conversions in Figure 5.13. Note that each map is at a different scale. [Author-prepared maps using same sources as Figure 5.13.]

winter conditions of consistently clear, dry, calm air; small insolation input; and an inland location far from moderating maritime effects. Prevailing global winds prevent moderating effects from the Pacific Ocean to the east. Verkhoyansk, Russia (located within the −48°C isotherm on the map), actually recorded a minimum temperature of −68°C (−90°F) and experiences a daily average of −50.5°C (−58.9°F) for January. Verkhoyansk (Figure 5.15) has 7 months of temperatures below freezing, including at least 4 months below −34°C (−30°F).

In contrast, Verkhoyansk was hit with a maximum temperature of +37°C (+98°F) in July—an incredible 105 C° (189 F°) minimum–maximum range! People do live and work in Verkhoyansk, which has a population of 1400; the town has been occupied continuously since 1638 and is today a secondary mining district.

Trondheim, Norway, is near the latitude of Verkhoyansk and at a similar elevation. But Trondheim's coastal location moderates its annual temperature regime (Figure 5.15). January minimum and maximum temperatures range between −17°C and +8°C (+1.4°F and +46°F), and July minimum and maximum temperatures range between +5°C and +27°C (+41°F and +81°F). The lowest minimum and highest maximum temperatures ever recorded in Trondheim are −30°C and +35°C (−22°F and +95°F)—quite a difference from the extremes at Verkhoyansk.

July Temperature Maps

Average July temperatures are presented in Figure 5.16 on page 121 for the world. The longer days of summer and higher Sun altitude are in the Northern Hemisphere. Winter dominates the Southern Hemisphere, although it is milder than winters north of the equator because continental landmasses are smaller and the dominant oceans and seas store and release more energy. The *thermal equator* shifts northward with the high summer Sun and reaches the Persian Gulf–Pakistan–Iran area. The Persian Gulf is the site of the highest recorded sea-surface temperature of 36°C (96°F), difficult to imagine for a large water body.

During July in the Northern Hemisphere, isotherms shift poleward over land as higher temperatures dominate continental interiors. July temperatures in Verkhoyansk average more than 13°C (56°F), which represents a 63 C° (113 F°) seasonal range between January and July averages. The Verkhoyansk region of Siberia is probably Earth's most dramatic example of temperature continentality.

The hottest places on Earth occur in Northern Hemisphere deserts during July. The reasons are simple: clear skies, strong surface heating, virtually no surface water, and few plants. Prime examples are portions of the Sonoran Desert of North America and the Sahara of Africa. The most extreme shade temperature ever recorded, higher than 58°C (136°F), occurred on September 13, 1922, at Al 'Azīzīyah, Libya.

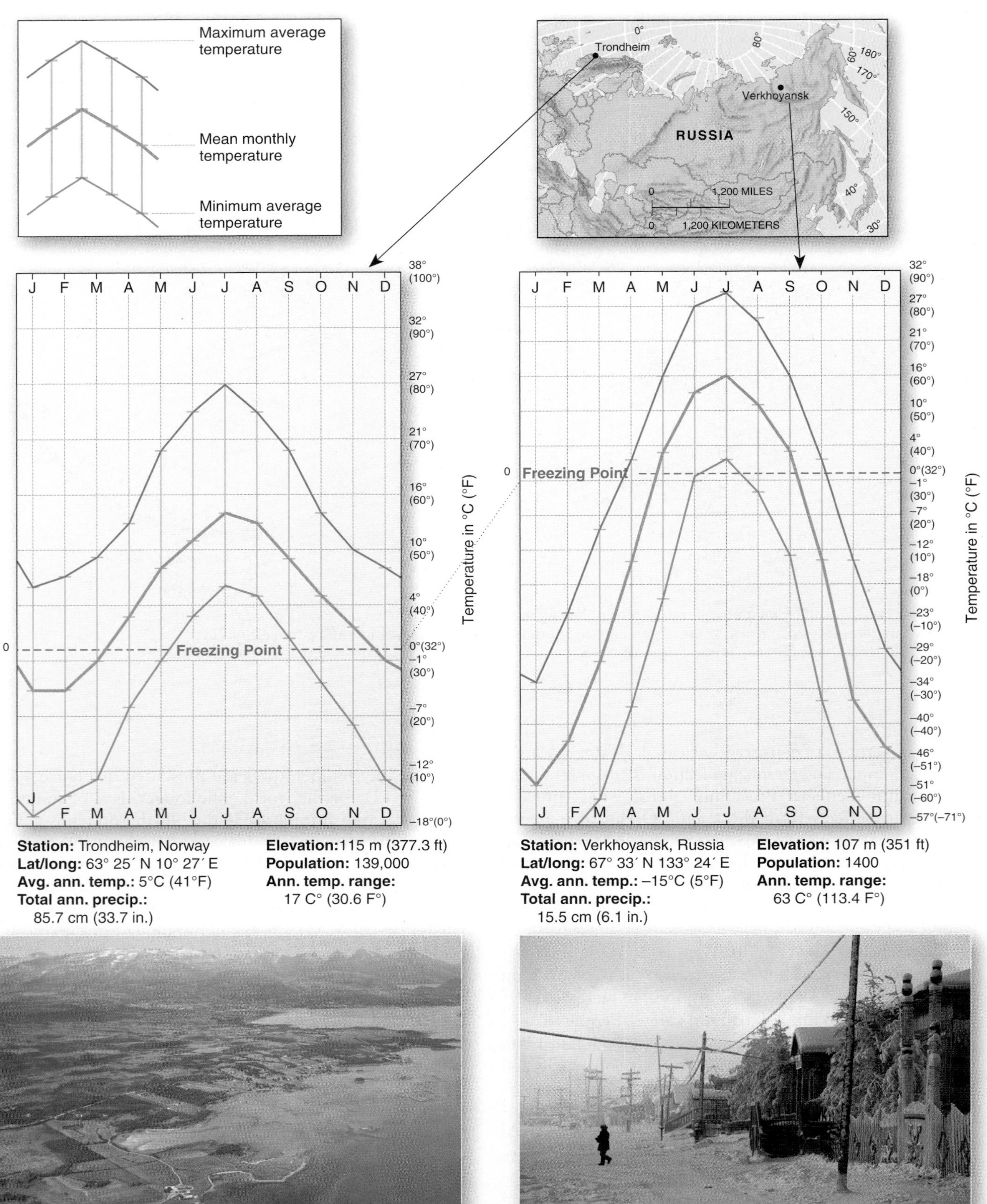

FIGURE 5.15 Marine and continental cities—Eurasia.
Compare temperatures in coastal Trondheim, Norway, and continental Siberian Russia. Note that the freezing levels on the two graphs are positioned differently to accommodate the contrasting data. [Coastal Norway by Bobbé Christopherson; Verkhoyansk by Bryan and Cherry Alexander/Arcticphoto.]

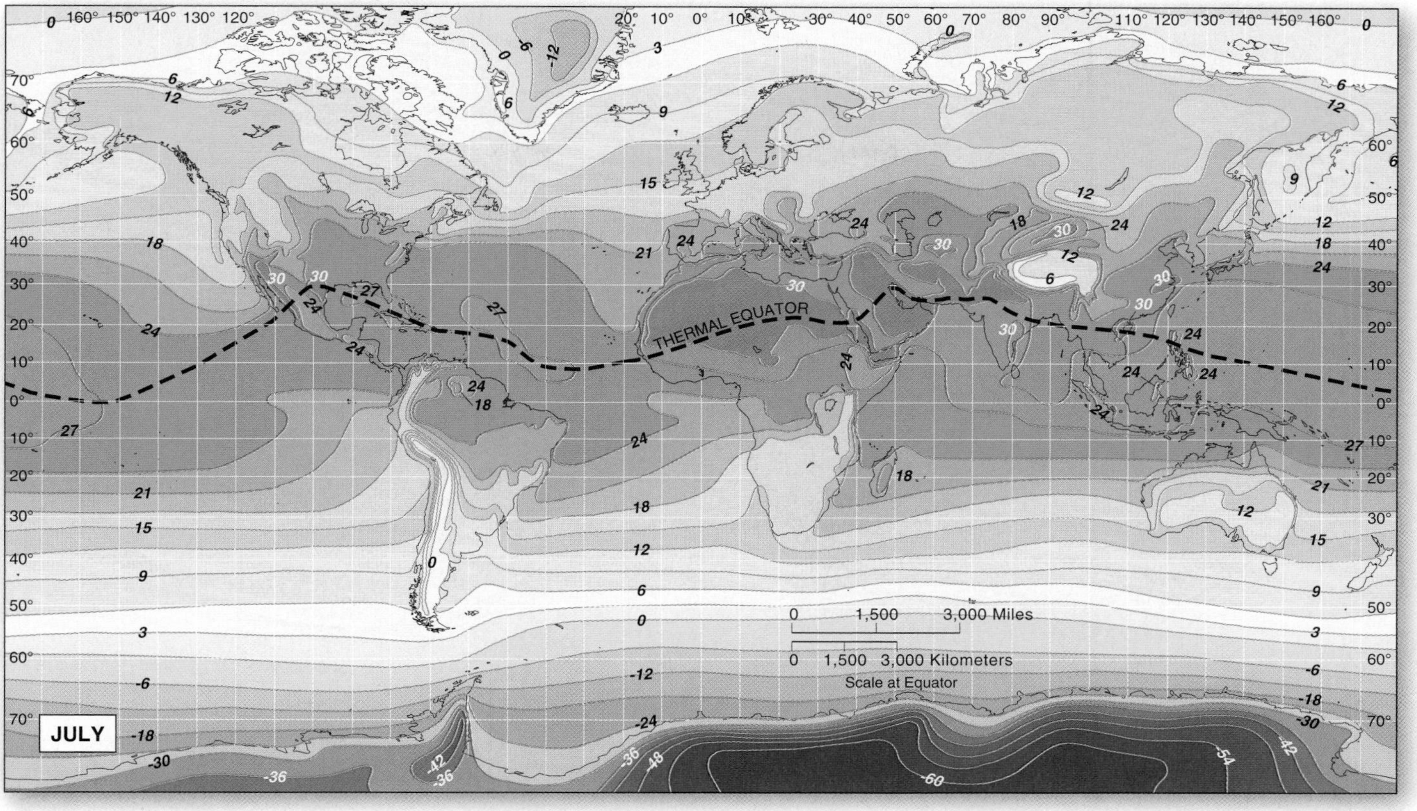

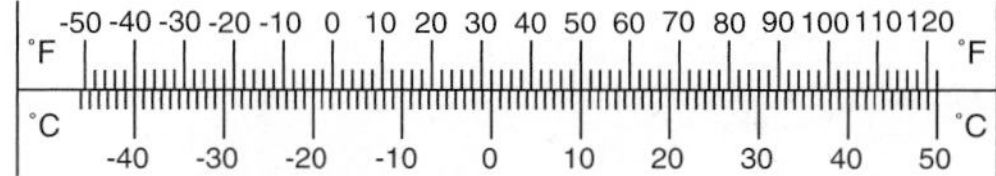

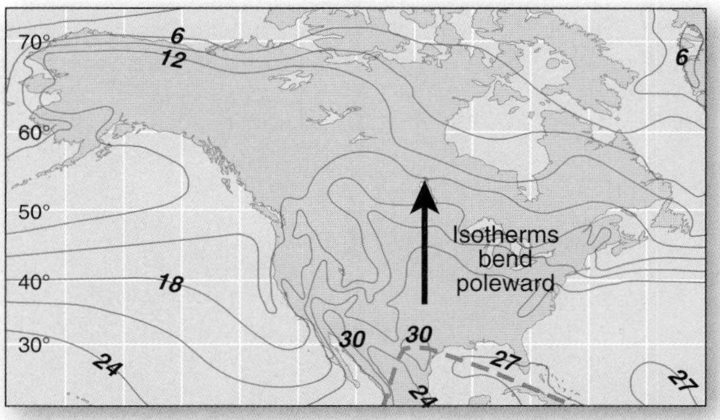

FIGURE 5.16 Global mean temperatures for July. Temperatures are in Celsius (convertible to Fahrenheit by means of the scale) as taken from separate air-temperature databases for ocean and land. Note the inset map of North America and the poleward-trending isotherms in the interior. Compare with Figure 5.13. [Adapted by author and redrawn from National Climatic Data Center, *Monthly Climatic Data for the World,* 47 (July 1994), and WMO and NOAA.]

Satellite Global Surface Temperatures, Land and Ocean

The highest maximum and annual average temperatures in North America occurred in Death Valley, California, where the Greenland Ranch Station reached 57°C (134°F) in 1913. The station is at 37° N and is −54.3 m, or −178 ft (the minus indicates meters or feet below sea level). Chapter 15 discusses such hot, arid lands.

The north polar and south polar regions (Figure 5.17a and b) are featured on two maps. July is "summer" in the Arctic Ocean (Figure 5.17a) and brings a thinning of multi-year and seasonal pack ice. See the Midnight Sun photo in Figure 2.16b to see what continuous daylight is like. Open cracks and leads stretched all the way to the North Pole over the past several years.

In July, nights in Antarctica are 24 hours long. This lack of insolation results in the lowest natural temperature reported on Earth, a frigid −89.2°C (−128.6°F) recorded on July 21, 1983, at the Russian Vostok Station, Antarctica (noted on Figure 5.17b). Such a temperature is 11 C° (19.8 F°) colder than the freezing point of dry ice (solid carbon dioxide)! If the concentration of carbon dioxide were large enough, such a cold temperature would theoretically freeze tiny carbon dioxide dry-ice particles out of the sky.

The average July temperature around Vostok is −68°C (−90.4°F). For average temperature comparisons, the Amundsen–Scott Station at the South Pole experiences −60°C (−76°F), and McMurdo Station has a −26°C (−14.8°F) reading. Note that the coldest temperatures in Antarctica are usually in August, not July, just before the equinox sunrise in September at the end of the long polar night.

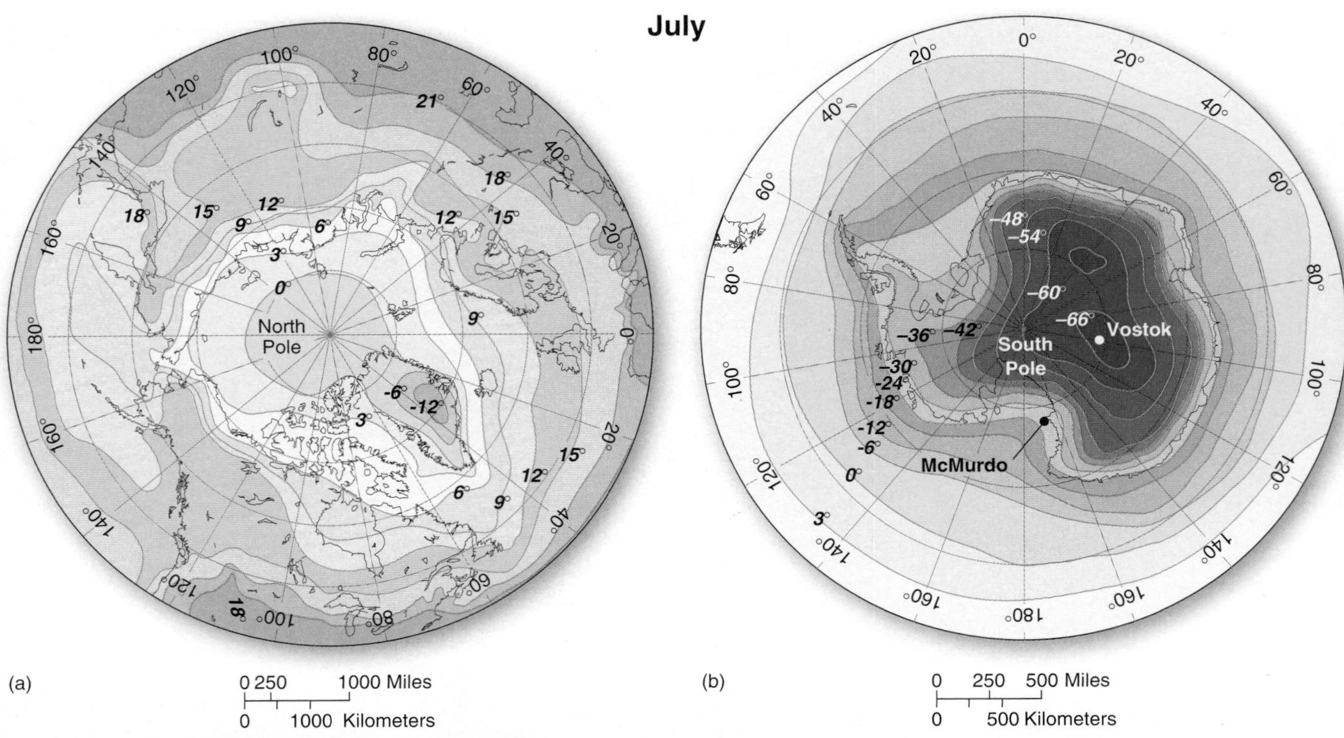

FIGURE 5.17 July mean temperatures for polar regions.
July temperatures in (a) north polar region and (b) south polar region. Use temperature conversions in Figure 5.16. Note that each map is at a different scale. [Author-prepared maps using same sources as Figure 5.16.]

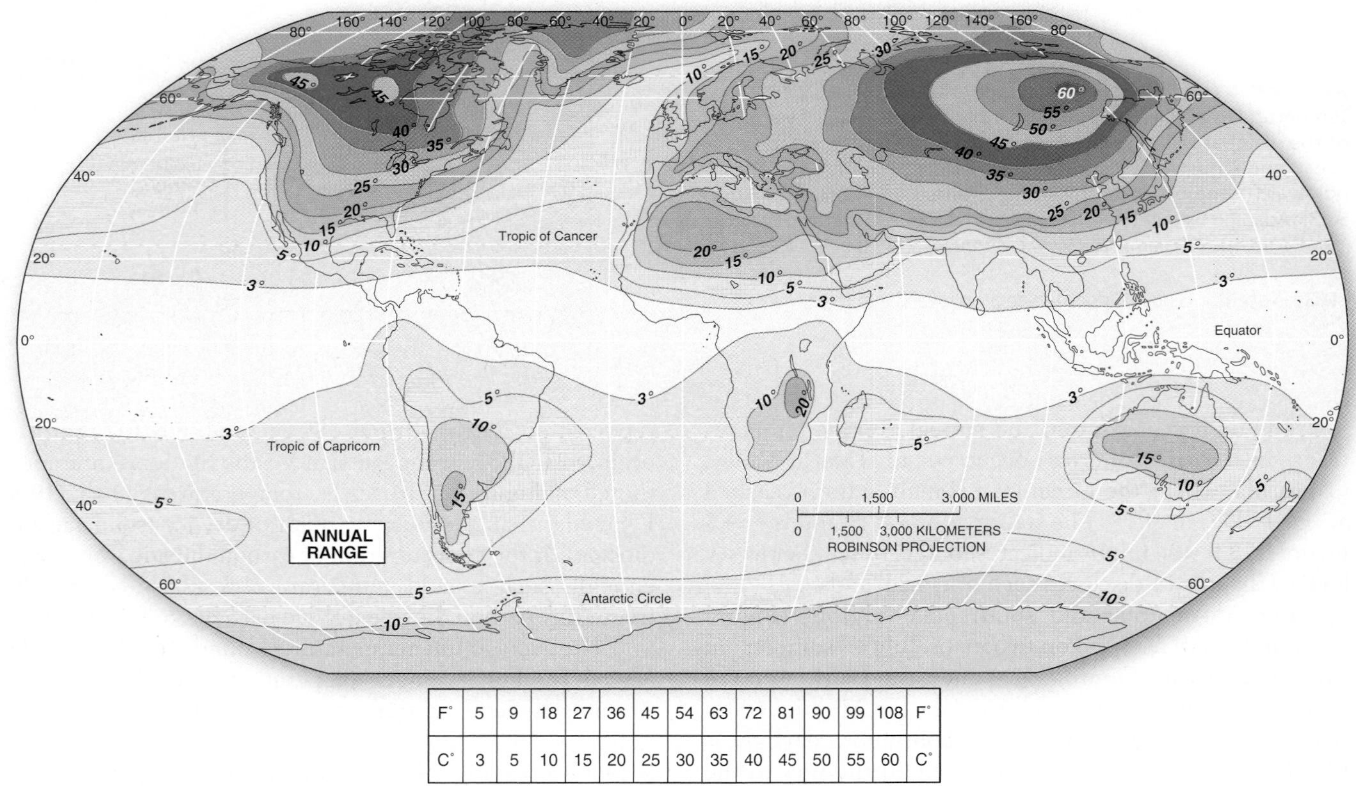

F°	5	9	18	27	36	45	54	63	72	81	90	99	108	F°
C°	3	5	10	15	20	25	30	35	40	45	50	55	60	C°

FIGURE 5.18 Global annual temperature ranges.
The annual range of global temperatures in Celsius degrees, C°, with conversions to Fahrenheit degrees, F°, shown on scale. The mapped data show the difference between average January and July temperature maps.

Annual Temperature Range Map

The temperature range map helps identify regions that experience the greatest annual extremes and, in contrast, the most moderate temperature regimes, as demonstrated in Figure 5.18. As you might expect, the largest temperature ranges occur at subpolar locations within the continental interiors of North America and Asia, where average ranges of 64 C° (115 F°) are recorded (dark brown area on the map). The Southern Hemisphere, in contrast, has little seasonal variation in mean temperatures, owing to the lack of large landmasses and the vast expanses of water to moderate temperature extremes.

Southern Hemisphere temperature patterns are generally maritime, and Northern Hemisphere patterns feature continentality, although interior regions in the Southern Hemisphere experience some continentality effects. For example, in January (Figure 5.13), Australia is dominated by isotherms of 20°–30°C (68°–86°F), whereas in July (Figure 5.16), Australia is crossed by the 12°C (54°F) isotherm. The Northern Hemisphere, with greater land area overall, registers a slightly higher average surface temperature than does the Southern Hemisphere.

Imagine living in some of these regions and the degree to which you would need to adjust your personal wardrobe and make other comfort adaptations. See Focus Study 5.1 for more on air temperature and the human body. Summer heat-wave deaths remain a powerful reminder of the role of temperatures in our lives.

Record Temperatures and Greenhouse Warming

Humans are experiencing greater temperature-related challenges owing to complex changes now under way in the lower atmosphere. Scientists agree that human activities, principally the burning of fossil fuels, are increasing greenhouse gases that absorb longwave radiation, delaying losses of heat energy to space. We are experiencing human-forced climate change, as Earth's natural greenhouse is enhanced.

The Dome C ice core taken from the East Antarctic Plateau shows that the levels of carbon dioxide, methane, and nitrous oxides—all greenhouse gases—are higher in the present decade than in any other decade in the entire 800,000-year record! These are remarkable times in which we live. The last 15 years feature the warmest years in the climate record.

The Intergovernmental Panel on Climate Change (IPCC), through major reports beginning in 1990 and its *Fourth Assessment Report* (*AR4*) in 2007, confirms that global warming is occurring. The IPCC *AR4* reached a consensus, stating, "Warming of the climate system is unequivocal, as is now evident from observations of increases in global average air and ocean temperatures, widespread melting of snow and ice, and rising global average sea level."

The U.S. National Research Council of the National Academy of Sciences (NRC/NAS) released three reports, totaling 825 pages, in fall 2010. In the volume *Advancing the Science of Climate Change*, the NRC/NAS stated on page 2:

> Climate change is occurring, is caused largely by human activities, and poses significant risks for—and in many cases is already affecting—a broad range of human and natural systems. . . . Earth is warming. . . . Most of the warming over the last several decades can be attributed to human activities that release carbon dioxide. . . . Global warming is closely associated with a broad spectrum of other climate changes, such as increases in the frequency of intense rainfall, decreases in snow cover and sea ice, more frequent and intense heat waves, rising sea levels, and widespread ocean acidification. . . . [T]hese changes pose risks for a wide range of human and environmental systems.

The Association of American Geographers (AAG) at the 2006 Chicago meeting passed the AAG Resolution Requesting Action on Climate Change. This resolution places the AAG in line with other professional organizations and national academies of sciences that are now on record as supporting action to slow rates of climate change.

In July 2010 the National Climate Data Center, NOAA, issued *The State of the Climate* report. In part NCDC summarized their study:

> A comprehensive review of key climate indicators confirms the world is warming and the past decade was the warmest on record. More than 300 scientists from 48 countries analyzed data on 37 climate indicators, including sea ice, glaciers and air temperatures. . . . A warmer climate means higher sea level, humidity and temperatures in the air and ocean. A warmer climate also means less snow cover, melting Arctic sea ice and shrinking glaciers. . . . Recent studies show the world's oceans are heating up as they absorb most of the extra heat being added to the climate system from the build-up of heat-trapping gases. . . . Continued temperature increases will threaten many aspects of our society, including coastal cities and infrastructure, water supply and agriculture. People have spent thousands of years building society for one climate and now a new one is being created – one that's warmer and more extreme.

All these indicators are happening as detailed in this report.

Animation Global Warming, Climate Change

Notebook GISS Surface Temperature Analysis, 1891 to 2006

FOCUS STUDY 5.1

Air Temperature and the Human Body

We humans are capable of sensing small changes in temperature in the environment around us. Our perception of temperature is described by the terms *apparent temperature* and *sensible temperature*. This perception of temperature varies among individuals and cultures. Through complex mechanisms, our bodies maintain an average internal temperature ranging within a degree of 36.8°C (98.2°F), slightly lower in the morning or in cold weather and slightly higher at emotional times or during exercise and work.*

*The traditional value for "normal" body temperature, 37°C (98.6°F), was set in 1868 using old methods of measurement. According to Dr. Philip Mackowiak of the University of Maryland School of Medicine, a more accurate modern assessment places normal at 36.8°C (98.2°F), with a range of 2.7 C° (4.8 F°) for the human population (*Journal of the American Medical Association*, September 23–30, 1992).

Taken together, the water vapor content of air, wind speed, and air temperature affect each individual's sense of comfort. High temperatures, high humidity, and low winds produce the most heat discomfort, whereas low humidity and strong winds enhance cooling rates. Although modern heating and cooling systems, if available and affordable, can reduce the impact of uncomfortable temperatures indoors, the danger to human life from excessive heat or cold persists outdoors. When changes occur in the surrounding air, the human body reacts in various ways to maintain its core temperature and to protect the brain at all cost.

Wind-Chill and Heat Indexes

The *wind-chill index* is important to those who experience winters with freezing temperatures. The wind-chill factor indicates the enhanced rate at which body heat is lost to the air. As wind speeds increase, heat loss from the skin increases. The National Weather Service (NWS, http://www.nws.noaa.gov/om/windchill/) and the Meteorological Services of Canada (MSC) developed the Wind Chill Temperature (WCT) Index in Figure 5.1.1.

Using the chart, if the air temperature is −7°C (20°F) and the wind is blowing at 32 kmph (20 mph), skin temperatures will be at −16°C (4°F). The lower wind-chill values present a serious freezing hazard to exposed flesh. Imagine the wind chill experienced by a downhill ski racer going 130 kmph (80 mph)—frostbite is a definite possibility during a 2-minute run. The wind chill does omit consideration of sunlight intensity, a person's physical activity, and the use of protective clothing, such as a windbreaker, that prevents wind access to your skin.

The *heat index* (HI) indicates the human body's reaction to air tempera-

FIGURE 5.1.1 Wind Chill Temperature Index for various temperatures and wind speeds.

Actual Air Temperature in °C (°F)

Wind speed, kmph (mph)	4° (40°)	−1° (30°)	−7° (20°)	−12° (10°)	−18° (0°)	−23° (−10°)	−29° (−20°)	−34° (−30°)	−40° (−40°)
Calm									
8 (5)	2° (36°)	−4° (25°)	−11° (13°)	−17° (1°)	−24° (−11°)	−30° (−22°)	−37° (−34°)	−43° (−46°)	−49° (−57°)
16 (10)	1° (34°)	−6° (21°)	−13° (9°)	−20° (−4°)	−27° (−16°)	−33° (−28°)	−41° (−41°)	−47° (−53°)	−54° (−66°)
24 (15)	0° (32°)	−7° (19°)	−14° (6°)	−22° (−7°)	−28° (−19°)	−36° (−32°)	−43° (−45°)	−50° (−58°)	−57° (−71°)
32 (20)	−1° (30°)	−8° (17°)	−16° (4°)	−23° (−9°)	−30° (−22°)	−37° (−35°)	−44° (−48°)	−52° (−61°)	−59° (−74°)
40 (25)	−2° (29°)	−9° (16°)	−16° (3°)	−24° (−11°)	−31° (−24°)	−38° (−37°)	−46° (−51°)	−53° (−64°)	−61° (−78°)
48 (30)	−2° (28°)	−9° (15°)	−17° (−1°)	−24° (−12°)	−32° (−26°)	−39° (−39°)	−47° (−53°)	−55° (−67°)	−62° (−80°)
56 (35)	−2° (28°)	−10° (14°)	−18° (0°)	−26° (−14°)	−33° (−27°)	−41° (−41°)	−48° (−55°)	−56° (−69°)	−63° (−82°)
64 (40)	−3° (27°)	−11° (13°)	−18° (−1°)	−26° (−15°)	−34° (−29°)	−42° (−43°)	−49° (−57°)	−57° (−71°)	−64° (−84°)
72 (45)	−3° (26°)	−11° (12°)	−19° (−2°)	−27° (−16°)	−34° (−30°)	−42° (−44°)	−50° (−58°)	−58° (−72°)	−66° (−86°)
80 (50)	−3° (26°)	−11° (12°)	−19° (−3°)	−27° (−17°)	−35° (−31°)	−43° (−45°)	−51° (−60°)	−59° (−74°)	−67° (−88°)

Frostbite times: 30 min. 10 min. 5 min.

ture and water vapor. The water vapor in air is expressed as relative humidity, a concept presented in Chapter 7. For now, we simply can assume that the amount of water vapor in the air affects the evaporation rate of perspiration from the skin because the more water vapor in the air (the higher the humidity), the less water from perspiration the air can absorb through evaporation, thus reducing natural evaporative cooling. The heat index indicates how the air feels to an average person—in other words, its apparent temperature.

Figure 5.1.2 is an abbreviated version of the heat index used by the NWS and now included in its daily weather summaries during appropriate months. The table beneath the graph describes the effects of heat-index categories on higher-risk groups. A combination of high temperature and high humidity can severely reduce the body's natural ability to regulate internal temperature. The NWS provides a heat-index forecast warning map (see http://www.nws.noaa.gov/om/heat/index.shtml); a heat-index calculator is at http://www.crh.noaa.gov/jkl/?n=heat_index_calculator.

Summer Heat-Wave Deaths

Heat waves are a major cause of weather-related deaths, causing the loss of more than 8000 lives in the United States between 1979 and 1999. In the NRC/NAS report cited earlier in this chapter, the National Academy of Sciences stated, "Hot days, hot nights, and heat waves have become more frequent in recent decades, and their frequency, intensity, and duration are projected to increase. . . ."

Europe was paralyzed in the summer of 2003 with high heat-index conditions when temperatures topped 40°C (104°F) in June, July, and August. Official estimates across the continent placed the death count at more than 20,000.

For nearly a week during July 1995, Chicago's heat-index values went to category I in dwellings that lacked air conditioning. Chicago had never experienced a 48-hour period during which temperatures did not go below 31.5°C (89°F)! With high pressure (hot, stable air) dominating from the Midwest to the Atlantic and moist air from the Gulf of Mexico, all the ingredients for disaster were present.

This combination of high temperatures and water vapor produced stifling conditions, particularly affecting the sick and elderly. As record temperatures occurred at Midway Airport (41°C;106°F), some apartments without air conditioning exceeded indoor heat-index temperatures of 54°C (130°F). In Chicago, 700 people died. Nearly 1000 people died overall in the Midwest and East from these conditions.

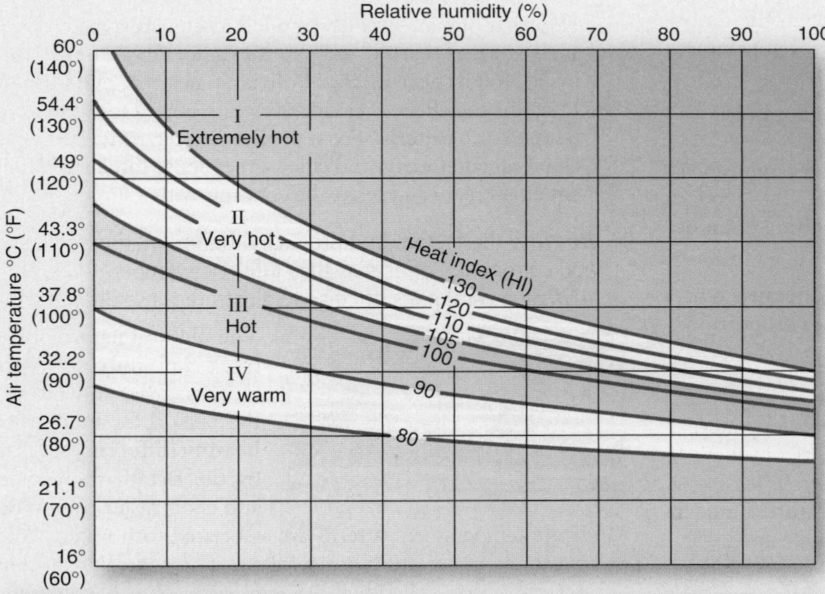

Level of concern	Category	Heat Index Apparent Temperature	General Effect of Heat Index on People in High-Risk Groups
Extreme danger	I	54°C (130°F) or higher	Heat/sunstroke highly likely with continued exposure
Danger	II	41° – 54°C (105° – 130°F)	Sunstroke, heat cramps, or heat exhaustion likely and heatstroke possible with prolonged exposure and/or physical activity
Extreme caution	III	32° – 41°C (90° – 105°F)	Sunstroke, heat cramps, and heat exhaustion possible with prolonged exposure and/or physical activity
Caution	IV	27° – 32°C (80° – 90°F)	Fatigue possible with prolonged exposure and/or physical activity

FIGURE 5.1.2 Heat index for various temperatures and relative humidity levels.

CRITICAL THINKING 5.2

What's your temperature comfort level?

Have you ever experienced discomfort from low-temperature or high-temperature stress? Where were you at the time of each experience? Make copies of the *wind-chill* and *heat-index* charts given in Focus Study 5.1 for later reference. On an appropriate day, check air temperature and wind speed, and determine their combined effect on skin chilling. In contrast, on an appropriate day, check air temperature and relative humidity values, and determine their combined effect on your personal comfort. The *Mastering Geography* website presents several interesting temperature links.

Global temperature patterns are a significant output of the energy–atmosphere system. We examined the complex interactions of several factors that produce these patterns and studied maps of their distributions. In the next chapter, we shift to another output in this energy–atmosphere system, that of global circulation of winds and ocean currents. We look at the forces that interact to produce these movements of air and water. Such events as the Gulf of Mexico oil disaster that began in April 2010 make this knowledge important as we see how ocean circulation spreads the toxic debris for thousands of miles.

KEY LEARNING CONCEPTS REVIEW

■ *Define* the concepts of temperature, kinetic energy, and sensible heat, and *distinguish* among Kelvin, Celsius, and Fahrenheit temperature scales and how they are measured.

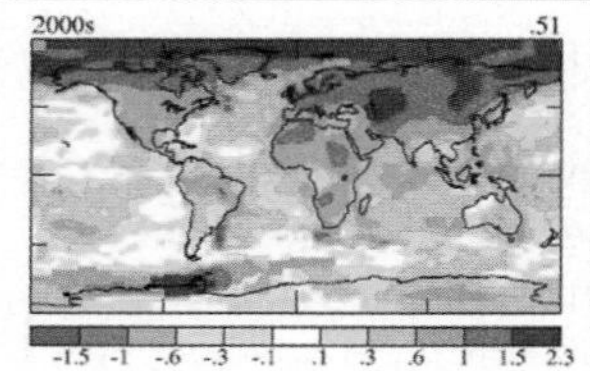

Temperature is a measure of the average kinetic energy (motion) of individual molecules in matter. We feel the effect of temperature as the sensible heat transfer from warmer objects to cooler objects when these objects are touching. Temperature scales include

- Kelvin scale: 100 units between the melting point of ice (273 K) and the boiling point of water (373 K).
- Celsius scale: 100 degrees between the melting point of ice (0°C) and the boiling point of water (100°C).
- Fahrenheit scale: 180 degrees between the melting point of ice (32°F) and the boiling point of water (212°F).

The Kelvin scale is used in scientific research because temperature readings start at absolute zero and thus are proportional to the actual kinetic energy in a material.

temperature (p. 108)

1. Distinguish between sensible heat and sensible temperature.
2. What does air temperature indicate about energy in the atmosphere?
3. Compare the three scales that express temperature. Find "normal" human body temperature on each scale and record the three values in your notes.
4. What is your source of daily temperature information? Describe the highest temperature and the lowest temperature you have experienced. From what we discussed in this chapter, can you identify the factors that contributed to these temperatures?

■ *List* and *review* the principal controls and influences that produce global temperature patterns.

Principal controls and influences upon temperature patterns include *latitude* (the distance north or south of the equator), *altitude* and *elevation*, *cloud cover* (reflection, absorption, and radiation of energy), and *land–water heating differences* (the nature of evaporation, transparency, specific heat, movement, ocean currents, and sea-surface temperatures).

5. Explain the effect of altitude/elevation on air temperature. Why is air at higher altitude/elevation lower in temperature? Why does it feel cooler standing in shadows at higher elevation than at lower elevation?
6. What noticeable effect does air density have on the absorption and radiation of energy? What role does elevation play in that process?
7. How is it possible to grow moderate-climate-type crops such as wheat, barley, and potatoes at an elevation of 4103 m (13,460 ft) near La Paz, Bolivia, so near the equator?
8. Describe the effect of cloud cover with regard to Earth's temperature patterns. From the last chapter, review the cloud-albedo forcing and cloud-greenhouse forcing of different cloud types and relate the concepts with a simple sketch.

■ *Review* the factors that produce different marine effects and continental effects as they influence temperatures, and *utilize* several pairs of cities to illustrate these differences.

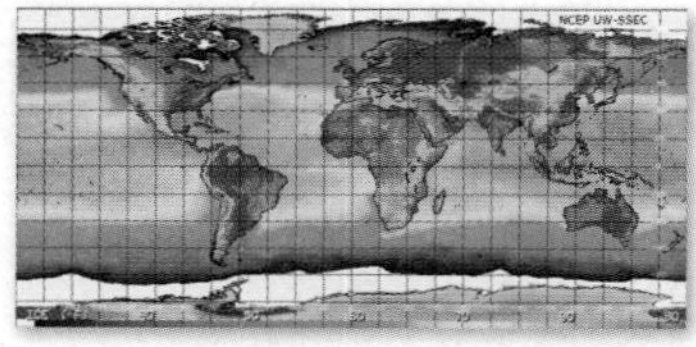

The physical nature of land (rock and soil) and water (oceans, seas, and lakes) is the reason for **land–water heating differences**, caused by the fact that land heats and cools faster than water. Moderate temperature patterns are associated with water bodies, and extreme temperatures occur inland. The controls that cause differences between land and water surfaces are *evaporation*, *transparency*, *specific heat*, *movement*, *ocean currents*, and *sea-surface temperatures*.

Light penetrates water because of its **transparency**. Water is clear and light passes through it to an average depth of 60 m (200 ft) in the ocean. This penetration distributes available heat energy over a much greater volume than could occur on opaque land; thus, a larger energy reservoir is formed. When equal volumes of water and land are compared, water requires far more energy to increase its temperature than does land. In other words, water can hold more energy than can soil or rock, so water has a higher **specific heat**, the heat capacity of a substance, averaging about four times that of soil.

Ocean currents affect temperature. An example of the effect of ocean currents is the **Gulf Stream**, which moves northward off the east coast of North America, carrying warm water far into the North Atlantic. As a result, the southern third of Iceland experiences much milder temperatures than would be expected for a latitude of 65° N, just below the Arctic Circle (66.5°).

Marine effect, or maritime, describes locations that exhibit the moderating influences of the ocean, usually along coastlines or on islands. **Continental effect** refers to the condition of areas that are less affected by the sea and therefore have a greater range between maximum and minimum temperatures diurnally and yearly.

land–water heating differences (p. 112)
transparency (p. 113)
specific heat (p. 114)
Gulf Stream (p. 114)
marine effect (p. 117)
continental effect (p. 117)

9. List the physical aspects of land and water that produce their different responses to heating from absorption of insolation. What is the specific effect of transparency in a medium?
10. What is specific heat? Compare the specific heat of water and soil.
11. Describe the pattern of sea-surface temperatures (SSTs) as determined by satellite remote sensing. Where is the warmest ocean region on Earth?
12. What effect does sea-surface temperature have on air temperature? Describe the negative feedback mechanism created by higher sea-surface temperatures and evaporation rates.
13. Differentiate between marine and continental temperatures. Give geographic examples of each from the text discussion of the United States, Norway, and Russia.

■ ***Interpret*** the pattern of Earth's temperatures from their portrayal on January and July temperature maps and on a map of annual temperature ranges.

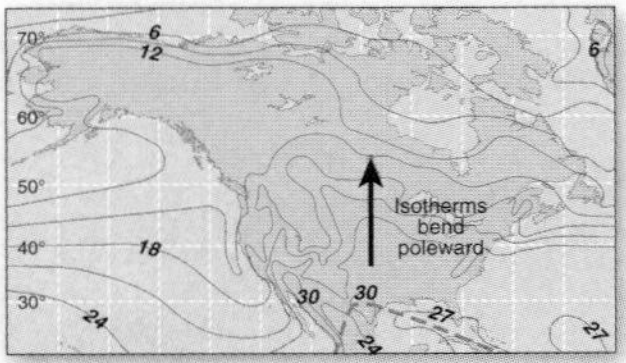

Maps for January and July instead of the solstice months of December and June are used for temperature comparison because of the natural lag that occurs between insolation received and maximum or minimum temperatures experienced. Each line on these temperature maps is an **isotherm**, an isoline that connects points of equal temperature. Isotherms portray temperature patterns.

Isotherms generally are zonal, trending east–west, parallel to the equator. They mark the general decrease in insolation and net radiation with distance from the equator. The **thermal equator** (isoline connecting all points of highest mean temperature) trends southward in January and shifts northward with the high summer Sun in July. In January, it extends farther south into the interior of South America and Africa, indicating higher temperatures over landmasses.

In the Northern Hemisphere in January, isotherms shift equatorward as cold air chills the continental interiors. The coldest area on the map is in Russia—specifically, northeastern Siberia. The intense cold experienced there results from winter conditions of consistently clear, dry, calm air; small insolation input; and an inland location far from any moderating maritime effects.

isotherm (p. 117)
thermal equator (p. 117)

14. What is the thermal equator? Describe its location in January and in July. Explain why it shifts position annually.
15. Observe trends in the pattern of isolines over North America, and compare the January average temperature map with the July map. Why do the patterns shift locations?
16. Describe and explain the extreme temperature range experienced in north-central Siberia between January and July.
17. Where are the hottest places on Earth? Are they near the equator or elsewhere? Explain. Where is the coldest place on Earth?
18. From the maps in Figures 5.13, 5.16, and 5.18, determine the average temperature values and annual range of temperatures for your present location.
19. Compare the maps in Figures 5.13 and 5.16: (a) Describe what you find in central Greenland between January and July; (b) look at the south polar region, and describe seasonal changes between the two maps. Characterize conditions along the Antarctic Peninsula between January and July (around 60° W longitude).

■ ***Contrast*** wind chill and heat index, and ***determine*** human response to these apparent temperature effects.

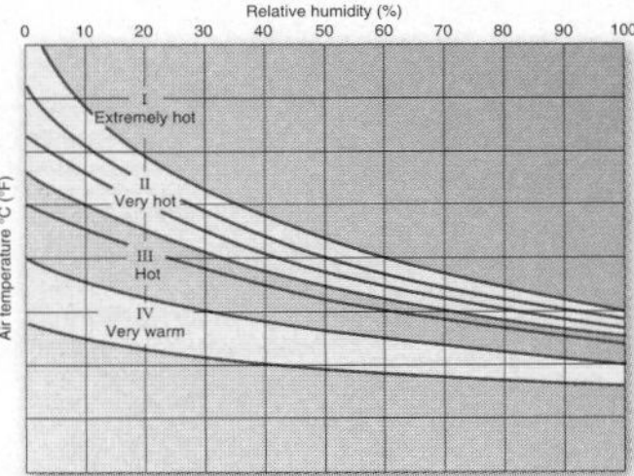

The *wind-chill factor* indicates the enhanced rate at which body heat is lost to the air. As wind speeds increase, heat loss from the skin increases. The *heat index* (HI) indicates the human body's reaction to air temperature and water vapor. The level of humidity in the air affects our natural ability to cool through evaporation from skin.

20. What is the wind-chill temperature on a day with an air temperature of −12°C (10°F) and a wind speed of 32 kmph (20 mph)?
21. On a day when temperature reaches 37.8°C (100°F), how does a relative humidity reading of 50% affect apparent temperature?

MASTERING GEOGRAPHY

Visit www.masteringgeography.com for several figure animations, satellite loops, self-study quizzes, flashcards, visual glossary, case studies, career links, textbook reference maps, RSS Feeds, and an ebook version of *Geosystems*. Included are many geography web links to interesting Internet sources that support this chapter.

CHAPTER 6

Atmospheric and Oceanic Circulations

Six wind turbines operate on Deadwood Plain, St. Helena (15° 55′ S, 5° 43′ W), a remote island in the South Atlantic Ocean. These six turbines provide about 15% of the island's electrical energy needs, avoiding the importation of hundreds of thousands of barrels of oil. Note the wind-sculpted trees shaped by the prevailing trade winds. [Bobbé Christopherson.]

KEY LEARNING CONCEPTS

After reading the chapter, you should be able to:

- ***Define*** the concept of air pressure, and ***describe*** instruments used to measure air pressure.
- ***Define*** wind, and ***explain*** how wind is measured, how wind direction is determined, and how winds are named.
- ***Explain*** the four driving forces within the atmosphere—gravity, pressure gradient force, Coriolis force, and friction force—and ***locate*** the primary high- and low-pressure areas and principal winds.
- ***Describe*** upper-air circulation, and ***define*** the jet streams.
- ***Summarize*** several multi-year oscillations of air temperature, air pressure, and circulation in the Arctic, Atlantic, and Pacific oceans.
- ***Explain*** several types of local winds, and the regional monsoons.
- ***Discern*** the basic pattern of Earth's major surface ocean currents and deep thermohaline circulation.

GEOSYSTEMS NOW

Ocean Currents Bring Invasive Species

As human civilization pumps and throws more refuse and chemicals into the oceans, the garbage travels in currents that sweep across the globe. A dramatic episode of such ocean-current transport happened in the South Atlantic in 2006. Let us begin this story here in Chapter 6 and continue with the ecological impacts in *Geosystems Now,* Chapter 20.

High-pressure systems guide the winds and ocean currents in the major oceanic basins. These systems flow clockwise in the Northern Hemisphere and counterclockwise in the Southern Hemisphere. In the South Atlantic, this counterclockwise flow drives the South Equatorial Current westward and the West Wind Drift eastward (Figure GN 6.1). Along the African coast, the cool Benguela Current flows northward, while the warm Brazil Current moves south along the coast of South America. This is the circulation gyre of the South Atlantic.

Along the southeastern portion of this gyre is a remote island group, the Tristan da Cunha Archipelago comprising four islands, some 2775 km (1725 mi) from Africa and 3355 km (2085 mi) from South America (see location in Figure GN 6.1). Only 298 people live on Tristan in the village of Edinburgh (Figure GN 6.2).

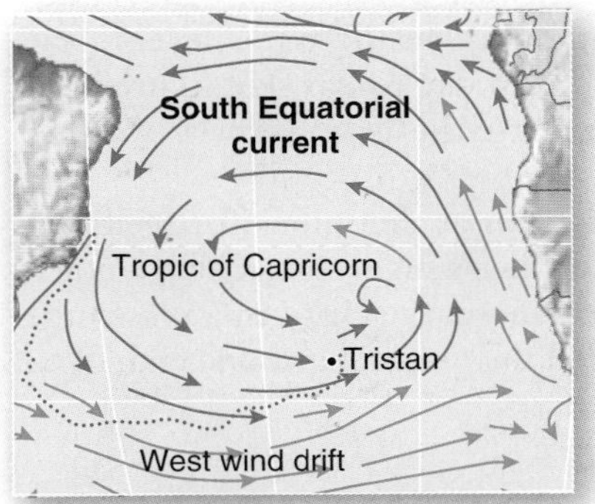

FIGURE GN 6.1 Principal ocean currents of the South Atlantic. Tristan da Cunha is at approximately 37° 4′ S and 12° 19′ W. Note the probable route of the oil-drilling platform.

FIGURE GN 6.2 Village of Edinburgh, Tristan da Cunha. [Bobbé Christopherson]

This unique small society does subsistence farming, collectively grows and exports potatoes, and depends on rich marine life that they carefully manage. The Tristan rock lobster (a crawfish) is harvested and quick frozen at a community factory and exported around the world. Although Tristan lacks an airport and has no pier or port, several ships a year transport these commodities. In 2006, ocean currents brought to this island a grim reminder of the outside world.

Oil-drilling platforms are like small ships that can be towed to a desired site for drilling undersea wells. Petrobras, Brazil's state-controlled oil company, had such a platform, the *Petrobras XXI*. The oil company towed the 80-m × 67-m × 34-m high (264-ft × 220-ft × 113-ft high) platform from Macae, Brazil, on March 5, 2006—destination Singapore. The company chartered a tug, the *Mighty Deliverer,* to do the job. This tug was actually built for inland work as a "pusher" tug, like ones you might see on the Mississippi River or in the Great Lakes. Yet, it was contracted to haul this drilling rig across the Southern Ocean, easily the most treacherous seas on Earth. The *Deliverer* headed south, encountering rough seas after a couple of weeks. Conditions forced the tug's crew to cut the platform loose on April 30, and after a few days, it was lost.

The *Petrobras XXI* was missing in rough seas, caught in the currents of the West Wind Drift and westerlies. Sometime in late May, after almost a month adrift, the oil-drilling platform ran aground on Tristan in Trypot Bay, where it was discovered by Tristanian fishermen on June 7 (Figure GN 6.3).

FIGURE GN 6.3 *Petrobras XXI* aground in Trypot Bay, Tristan. [Photo by biologist Sue Scott. All rights reserved.]

The owners of the drilling platform had not cleaned it prior to towing, which is the usual practice. Clean rigs move through the water with less friction, require less fuel, and result in lower labor costs. Consequently, *Petrobras XXI* carried some 62 non-native species of marine life, including free-swimming silver porgy and blenny fish that tagged along with the rig as it drifted in the currents.

What the marine scientists found when they surveyed the grounded rig was an unprecedented species invasion of Tristan. Exclusive photos by one of the scientist divers, interviewed for this report, follow in Chapter 20. Here in Chapter 6, we learn about wind and ocean circulation, the forces that brought the *Petrobas XXI* to Tristan.

After 635 years of dormancy, Mount Pinatubo in the Philippines erupted in 1991 (Figure 6.1a). This event had tremendous impact, lofting 15–20 million tons of ash, dust, and sulfur dioxide (SO_2) into the atmosphere. As the sulfur dioxide rose into the stratosphere, it quickly formed sulfuric acid (H_2SO_4) aerosols, which concentrated at an altitude of 16–25 km (10–15.5 mi). This debris increased atmospheric albedo about 1.5%, giving scientists an estimate of the aerosol volume generated by the eruption (see Figure 6.1).

The AVHRR instrument aboard *NOAA-11* monitored the reflected solar radiation from Mount Pinatubo's aerosols as global winds swept them around Earth. Figure 6.1b–e shows images made at three-week intervals that clearly track the spread of the debris worldwide. Some 60 days after the eruption (the last satellite image in the sequence), the aerosol cloud spanned about 42% of the globe, from 20° S to 30° N. For almost two years, colorful sunrises and sunsets and a small temporary lowering of average temperatures followed. The eruption provided a unique insight into the dynamics of atmospheric circulation.

Today, technology permits a depth of analysis unknown in the past—satellites now track the atmospheric effects from dust storms, forest fires, industrial haze, warfare, and the dispersal of volcanic explosions. As an example, millions of tons of dust from African soils cross the Atlantic each year, borne by the winds of atmospheric circulation, visible in Figure 6.1b.

Global winds are certainly an important reason why the United States, the former Soviet Union, and Great Britain signed the 1963 Limited Test Ban Treaty. That treaty banned aboveground testing of nuclear weapons because atmospheric circulation spread radioactive contamination worldwide. Such agreements illustrate how the fluid movement of the atmosphere links humanity more than perhaps any other natural or cultural factor. Our atmosphere makes the world a spatially bound society—one person's or country's exhalation is another's inhalation.

In this chapter: We begin with a discussion of wind essentials, consisting of air pressure and its measurement and a description of wind. The driving forces that produce surface winds are pressure gradient, Coriolis, and friction. We examine the circulation of Earth's atmosphere and the patterns of global winds, including principal pressure systems. We consider Earth's wind-driven oceanic currents and explain multi-year oscillations in atmospheric flows, local winds, and the powerful monsoons. The energy driving all this movement comes from one source: the Sun.

Wind Essentials

The primary circulation of winds across Earth has fascinated travelers, sailors, and scientists for centuries, although only in the modern era is a true picture emerging of global winds. Driven by the imbalance between equatorial energy surpluses and polar energy deficits, Earth's atmospheric circulation transfers both energy and mass on a grand scale, determining Earth's weather patterns and the flow of ocean currents. Scientists at the National Center for Atmospheric Research determined that the atmosphere is the dominant medium for redistributing energy from about 35° latitude to the poles in each hemisphere, whereas ocean currents redistribute more heat in a zone straddling the equator between the 17th parallels in each hemisphere. Atmospheric circulation also spreads air pollutants, whether natural or human-caused, worldwide, far from their point of origin.

Atmospheric circulation is categorized at three levels: primary circulation, consisting of general worldwide circulation; secondary circulation of migratory high-pressure and low-pressure systems; and tertiary circulation, which includes local winds and temporal weather patterns. Winds that move principally north or south along meridians of longitude are meridional flows. Winds moving east or west along parallels of latitude are zonal flows.

Air Pressure and Its Measurement

Air pressure, including its measurement and expression, is key to understanding wind. The molecules that constitute air create **air pressure** through their motion, size, and number—the factors that determine the temperature and density of the air. Pressure is exerted on all surfaces in contact with air.

In 1643, Galileo's pupil Evangelista Torricelli was working on a mine-drainage problem. His work led him to discover a method for measuring air pressure (Figure 6.2a). He knew that pumps in the mine were able to "pull" water upward about 10 m (33 ft), but no higher. He did not know why. Careful observation led him to discover that this was caused not by weak pumps, but by the atmosphere itself. Torricelli noted that the water level in the vertical pipe fluctuated from day to day. He figured out that air pressure, the weight of the air, varies with weather conditions.

To simulate the problem at the mine, Torricelli devised an instrument, at Galileo's suggestion, using a much denser fluid than water—mercury (Hg)—in a glass tube 1 m high. Torricelli sealed the glass tube at one end, filled

GEO REPORT 6.1 *Blowing in the wind*

Dust originating in Africa sometimes increases the iron content of the waters off Florida, promoting the toxic algal blooms (*Karenia brevis*) known as "Red Tides." In the Amazon, soil samples bear the dust print of these former African soils that crossed the Atlantic. Active research on such dust is part of the U.S. Navy's Aerosol Analysis and Prediction System (NAAPS); see links at http://www.nrlmry.navy.mil/aerosol/.

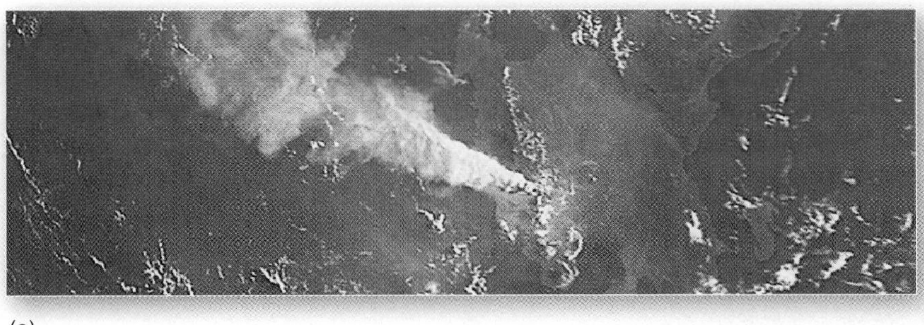

(a)

Luzon Island, Philippines

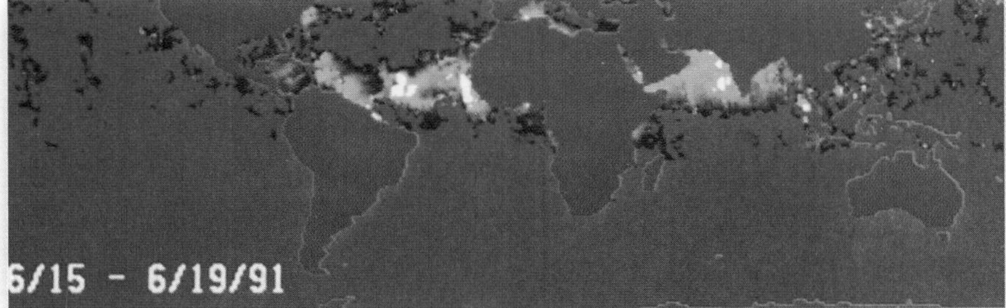

(b)

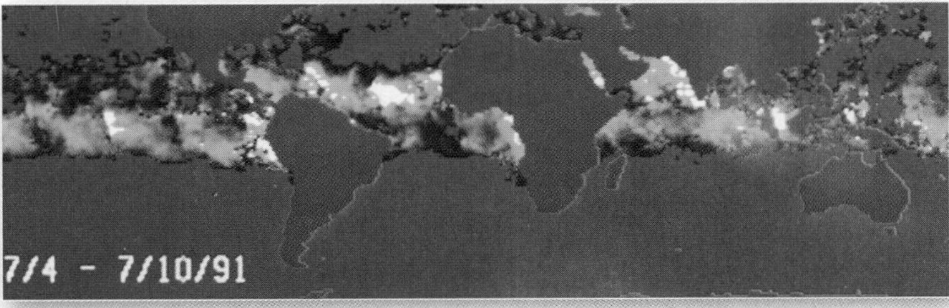

(c)

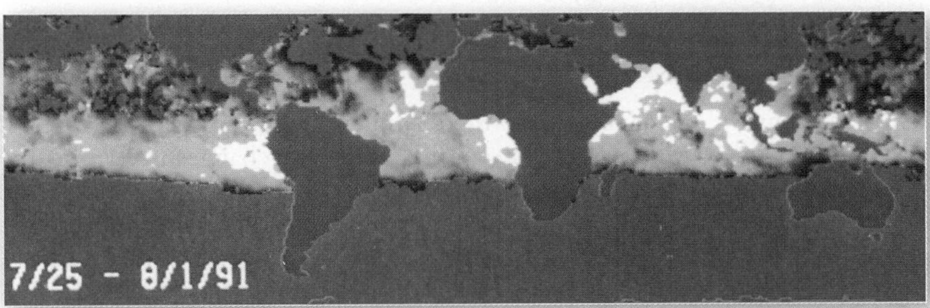

(d)

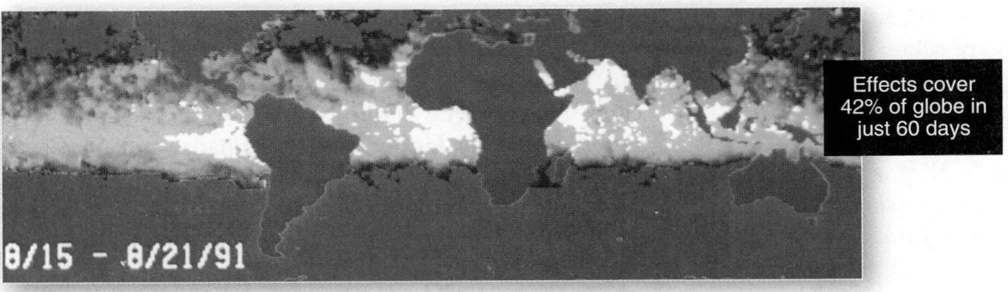

(e)

FIGURE 6.1 Volcanic eruption effects spread worldwide by winds. (a) Mount Pinatubo eruption, June 15, 1991. False-color images (b–e) of aerosols from Mount Pinatubo, smoke from fires, and dust storms, all swept about the globe by the general atmospheric circulation. The false color shows aerosol optical thickness (AOT): White is densest, dull yellow indicates medium values, and brown areas have the lowest aerosol concentration. In (b), note the dust moving westward from Africa, smoke from Kuwaiti oil well fires set during the first Persian Gulf War, smoke from forest fires in Siberia, and haze off the East Coast of the United States. [(a) AVHRR satellite image, Data available from U.S. Geological Survey, EROS Data Center, Sioux Falls, SD; (b–e)AVHRR satellite AOT images NESDIS/NOAA.]

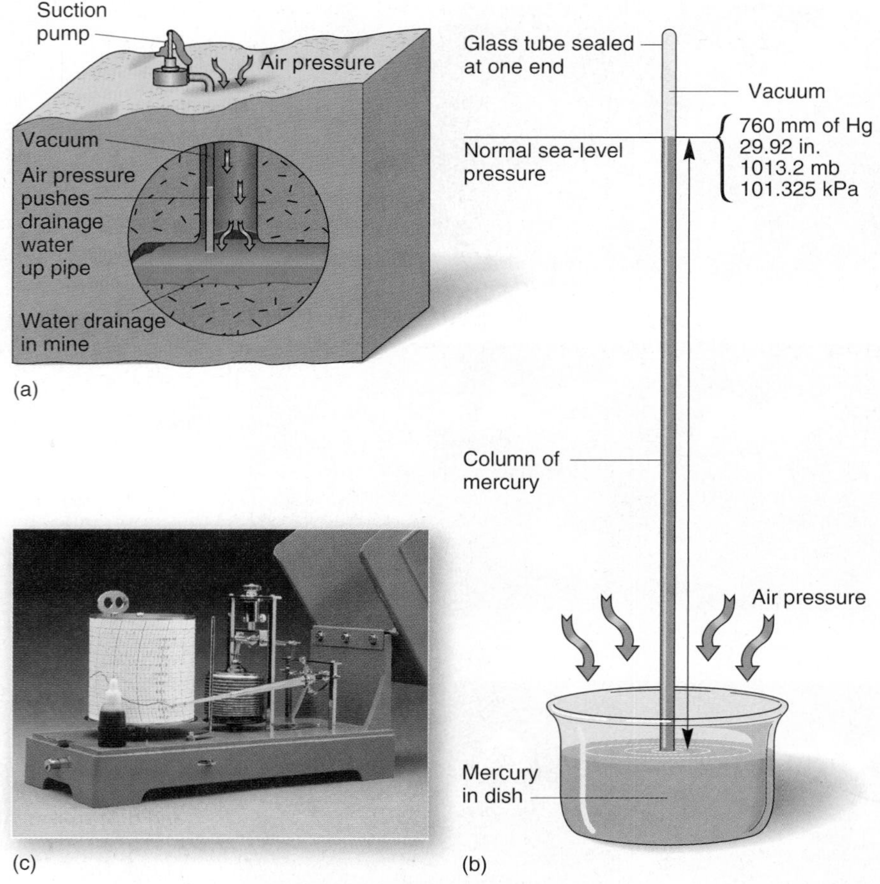

FIGURE 6.2 **Developing the barometer.**
(a) Torricelli developed the barometer to measure air pressure while trying to solve a mine-drainage problem. Two types of instruments are used to measure atmospheric pressure: (b) a mercury barometer (idealized sketch) and (c) an aneroid barometer. Have you used a barometer? If so, what type is it? Have you tried to reset it using a local weather information source? [(c) Qualimetrics, Inc.]

it with mercury, and inverted it into a dish of mercury (Figure 6.2b). He determined that the average height of the column of mercury in the tube was 760 mm (29.92 in.) and that it did vary day to day as the weather changed. He concluded that the mass of surrounding air was exerting pressure on the mercury in the dish that counterbalanced the column of mercury.

Using similar instruments, scientists set a standard of normal sea-level pressure at 1013.2 mb (millibar, which expresses force per square meter of surface area) or 29.92 in. of mercury (Hg). In Canada and other countries, normal air pressure is expressed as 101.32 kPa (kilopascal; 1 kPa = 10 mb).

Any instrument that measures air pressure is a barometer (from the Greek *baros*, meaning "weight"). Torricelli developed a **mercury barometer**. A more compact barometer design, which works without a meter-long tube of mercury, is the **aneroid barometer**, shown in Figure 6.2c. Aneroid means "using no liquid." The aneroid barometer principle is simple: Imagine a small chamber, partially emptied of air, which is sealed and connected to a mechanism attached to a needle on a dial. As air pressure increases, it presses on the chamber; as air pressure decreases, it relieves pressure on the chamber—changes in air pressure move the needle. An aircraft altimeter is a type of aneroid barometer. It accurately measures altitude because air pressure diminishes with altitude above sea level. For accuracy, the altimeter must be adjusted for temperature changes.

Figure 6.3 illustrates comparative scales in millibars and inches used to express air pressure and its relative force. The normal range of Earth's atmospheric pressure from strong high pressure to deep low pressure is about 1050 to 980 mb (31.00 to 29.00 in.). The figure also indicates the extreme highest and lowest pressures ever recorded in the United States, Canada, and on Earth. Hurricane Wilma (2005) is now the U.S. record holder. For comparison, note hurricanes Gilbert (1988) and Rita and Katrina (2005) and their lowest central pressures.

Wind: Description and Measurement

Simply stated, **wind** is generally the horizontal motion of air across Earth's surface. Turbulence adds wind updrafts and downdrafts and a vertical component to this definition. Differences in air pressure (density) between one lo-

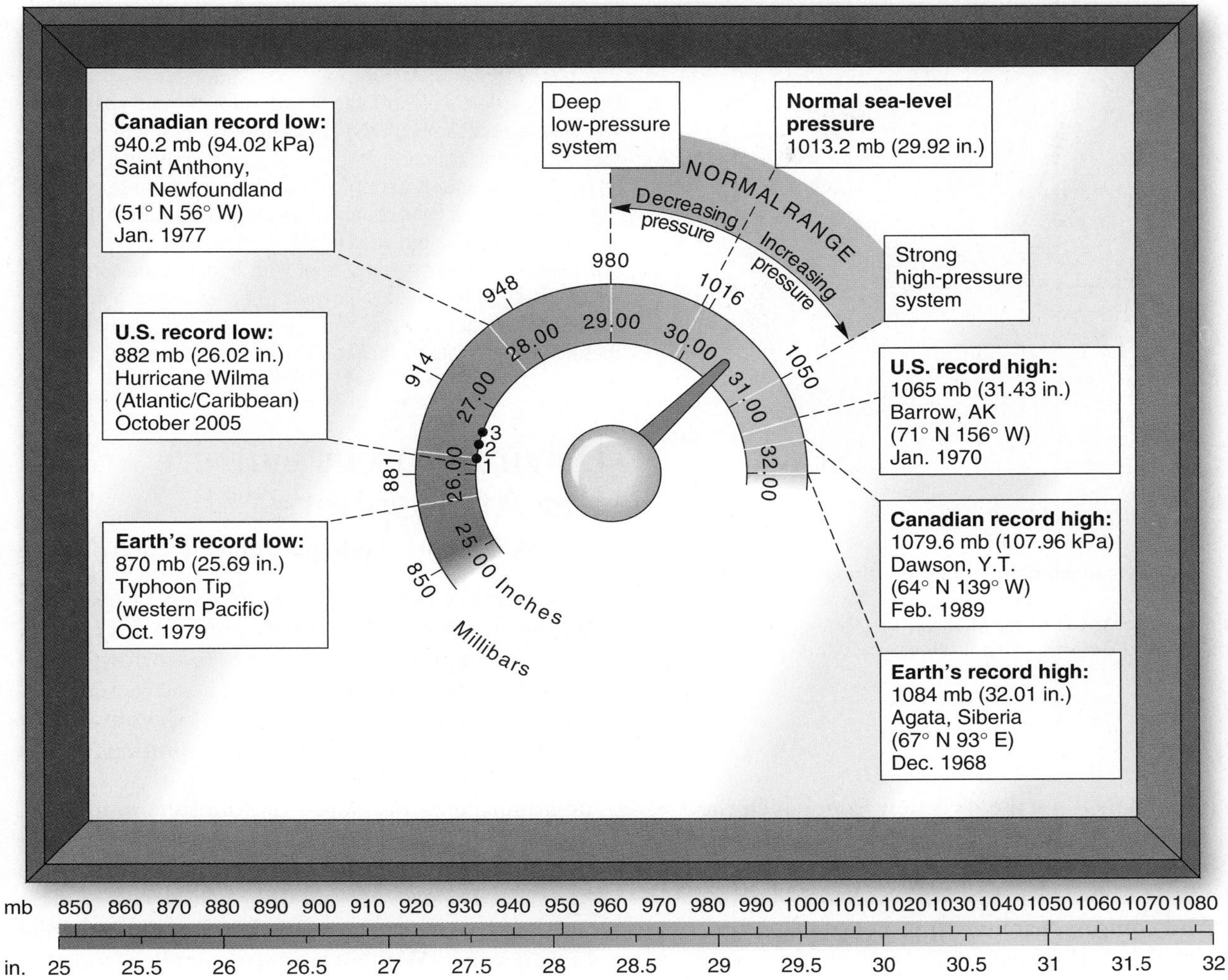

Canadian record low:
940.2 mb (94.02 kPa)
Saint Anthony, Newfoundland
(51° N 56° W)
Jan. 1977

U.S. record low:
882 mb (26.02 in.)
Hurricane Wilma
(Atlantic/Caribbean)
October 2005

Earth's record low:
870 mb (25.69 in.)
Typhoon Tip
(western Pacific)
Oct. 1979

Deep low-pressure system

Normal sea-level pressure
1013.2 mb (29.92 in.)

NORMAL RANGE
Decreasing pressure
Increasing pressure

Strong high-pressure system

U.S. record high:
1065 mb (31.43 in.)
Barrow, AK
(71° N 156° W)
Jan. 1970

Canadian record high:
1079.6 mb (107.96 kPa)
Dawson, Y.T.
(64° N 139° W)
Feb. 1989

Earth's record high:
1084 mb (32.01 in.)
Agata, Siberia
(67° N 93° E)
Dec. 1968

850 881 914 948 980 1016 1050 Millibars
25.00 26.00 27.00 28.00 29.00 30.00 31.00 32.00 Inches

mb 850 860 870 880 890 900 910 920 930 940 950 960 970 980 990 1000 1010 1020 1030 1040 1050 1060 1070 1080

in. 25 25.5 26 26.5 27 27.5 28 28.5 29 29.5 30 30.5 31 31.5 32

1. Hurricane Gilbert	2. Hurricane Rita	3. Hurricane Katrina
888 mb (26.23 in.)	897 mb (26.46 in.)	902 mb (26.61 in.)
Sept. 1988	Sept. 2005	August 2005

Some convenient conversions:
1.0 in. (Hg) = 33.87 mb = 25.40 mm (Hg) = 0.49 lb/in.^{2}
1.0 mb = 0.0295 in. (Hg) = 0.75 mm (Hg) = 0.0145 lb/in.^{2}

FIGURE 6.3 Air pressure readings and conversions.
Scales express barometric air pressure in millibars and inches, with average air pressure values and recorded pressure extremes. Canadian values include kilopascal equivalents, 10 mb = 1 kPa. The intensity of the 2005 hurricane season is evident; note the positions of Katrina, Rita, and Wilma on the pressure dial.

cation and another produce wind. Wind's two principal properties are speed and direction, and instruments measure each. An **anemometer** measures wind speed in kilometers per hour (kmph), miles per hour (mph), meters per second (mps), or knots. (A knot is a nautical mile per hour, covering 1 minute of Earth's arc in an hour, equivalent to 1.85 kmph, or 1.15 mph.) A **wind vane** determines wind direction; the standard measurement is taken 10 m (33 ft) above the ground to reduce the effects of local topography on wind direction (Figure 6.4).

FIGURE 6.4 Wind vane and anemometer.
Instruments used to measure wind direction (wind vane, left) and wind speed (anemometer, right) at a weather station installation. [NOAA Photo Library.]

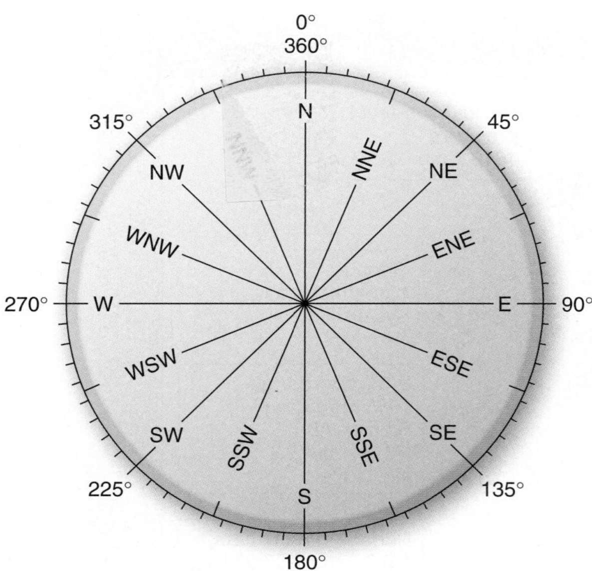

FIGURE 6.5 Sixteen wind directions identified on a wind compass.
Winds are named for the direction from which they originate. For example, a wind from the west is a westerly wind.

Winds are named for the direction from which they originate. For example, a wind from the west is a westerly wind (it blows eastward); a wind out of the south is a southerly wind (it blows northward). Figure 6.5 illustrates a simple wind compass, naming 16 principal wind directions used by meteorologists.

The traditional Beaufort wind scale is a descriptive scale useful in visually estimating wind speed (Table 6.1). Admiral Beaufort of the British Navy introduced his wind scale in 1806. In 1926, G. C. Simpson expanded Beaufort's scale to include wind speeds on land. The National Weather Service (old Weather Bureau) standardized the scale in 1955 (see http://www.hpc.ncep.noaa.gov/html/beaufort.shtml). The scale is still referenced on ocean charts, enabling estimation of wind speed without instruments. Although most ships use sophisticated equipment to perform such measurements, you find the Beaufort scale posted on the bridge of many ships.

CRITICAL THINKING 6.1

Measure the wind

Use the Beaufort scale in Table 6.1 to estimate wind speed as you walk across campus on a day with wind. Estimate wind speed and wind direction at least twice during the day, and record them in your notebook. If possible, check these against the reports from a local source of weather information. For wind direction, moisten your finger, hold it in the air, and sense evaporative cooling on one side of your finger to indicate from which direction the wind is blowing. What changes in wind speed and direction do you notice over several days? How did these changes relate to the weather you experienced?

Driving Forces Within the Atmosphere

Four forces determine both speed and direction of winds:

- The pressure that Earth's *gravitational force* exerts on the atmosphere is virtually uniform. Gravity equally compresses the atmosphere worldwide, with the density decreasing as altitude increases. The gravitational force counteracts the outward centrifugal force acting on Earth's spinning surface and atmosphere. Without gravity, there would be no atmospheric pressure—or atmosphere, for that matter.
- **Pressure gradient force** drives air from areas of higher barometric pressure (more dense air) to areas of lower barometric pressure (less dense air), thereby causing winds. Without a pressure gradient force, there would be no wind.
- **Coriolis force**, a deflective force, makes wind that travels in a straight path appear to be deflected in relation to Earth's rotating surface. The Coriolis force deflects wind to the right in the Northern Hemisphere and to the left in the Southern Hemisphere. Without Coriolis force, winds would move along straight paths between high- and low-pressure areas.
- **Friction force** drags on the wind as it moves across surfaces; it decreases with height above the surface. Without friction, winds would simply move in paths parallel to isobars and at high rates of speed.

GEO REPORT 6.2 *A force is with us*

Note that we call Coriolis a *force*. This label is appropriate because, as the physicist Sir Isaac Newton (1643–1727) stated, when something is accelerating over a space, a force is in operation (mass times acceleration). This apparent force (in classical mechanics, an inertial force) acts as an effect on moving objects. It is named for Gaspard Coriolis (1792–1843), a French mathematician and researcher of applied mechanics, who first described the force in 1831. For deeper insight into the physics of this phenomenon, consider an analysis by Anders Persson, "How do we understand the Coriolis force?" in the *Bulletin of the American Meteorological Society* 79, no. 7 (July 1998): 1373–85.

TABLE 6.1 Beaufort Wind Scale

Wind Speed			Beaufort	Wind	Beaufort Wind Scale	
kmph	mph	knots	Number	Description	Observed Effects at Sea	Observed Effects on Land
< 1	< 1	< 1	0	Calm	Glassy calm, like a mirror	Calm, no movement of leaves
1–5	1–3	1–3	1	Light air	Small ripples; wavelet scales; no foam on crests	Slight leaf movement; smoke drifts; wind vanes still
6–11	4–7	4–6	2	Light breeze	Small wavelets; glassy look to crests, which do not break	Leaves rustling; wind felt; wind vanes moving
12–19	8–12	7–10	3	Gentle breeze	Large wavelets; dispersed whitecaps as crests break	Leaves and twigs in motion; small flags and banners extended
20–29	13–18	11–16	4	Moderate breeze	Small, longer waves; numerous whitecaps	Small branches moving; raising dust, paper, litter, and dry leaves
30–38	19–24	17–21	5	Fresh breeze	Moderate, pronounced waves; many whitecaps; some spray	Small trees and branches swaying; wavelets forming on inland waterways
39–49	25–31	22–27	6	Strong breeze	Large waves, white foam crests everywhere; some spray	Large branches swaying; overhead wires whistling; difficult to control an umbrella
50–61	32–38	28–33	7	Moderate (near) gale	Sea mounding up; foam and sea spray blown in streaks in the direction of the wind	Entire trees moving; difficult to walk into wind
62–74	39–46	34–40	8	Fresh gale (or gale)	Moderately high waves of greater length; breaking crests forming sea spray; well-marked foam streaks	Small branches breaking; difficult to walk; moving automobiles drifting and veering
75–87	47–54	41–47	9	Strong gale	High waves; wave crests tumbling and the sea beginning to roll; visibility reduced by blowing spray	Roof shingles blown away; slight damage to structures; broken branches littering the ground
88–101	55–63	48–55	10	Whole gale (or storm)	Very high waves and heavy, rolling seas; white appearance to foam-covered sea; overhanging waves; visibility reduced	Uprooted and broken trees; structural damage; considerable destruction; seldom occurring
102–116	64–73	56–63	11	Storm (or violent storm)	White foam covering a breaking sea of exceptionally high waves; small- and medium-sized ships lost from view in wave troughs; wave crests frothy	Widespread damage to structures and trees, a rare occurrence
> 117	> 74	> 64	12–17	Hurricane	Driving foam and spray filling the air; white sea; visibility poor to nonexistent	Severe to catastrophic damage; devastation to affected society

All four of these forces operate on moving air and ocean currents at Earth's surface and influence global wind circulation patterns. The following sections describe the actions of the pressure gradient, Coriolis, and friction forces. (The gravitational force operates uniformly worldwide.)

Pressure Gradient Force

High- and low-pressure areas exist in the atmosphere principally because Earth's surface is unequally heated. For example, cold, dense air at the poles exerts greater pressure than warm, less dense air along the equator. These pressure differences establish a pressure gradient force.

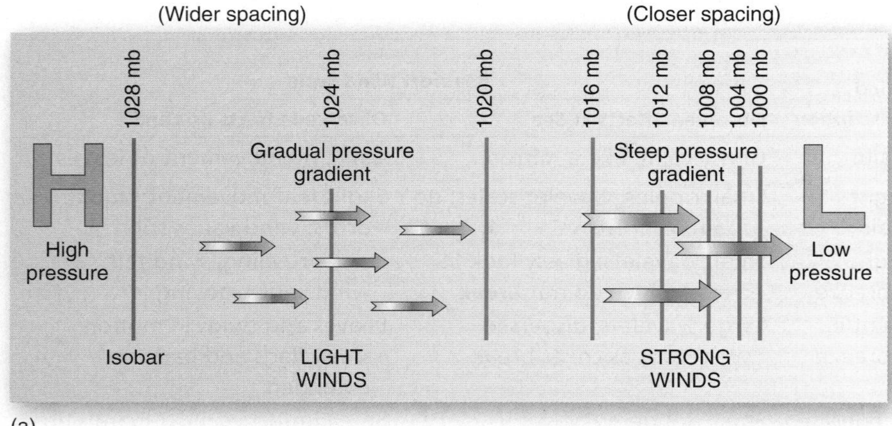

(a)

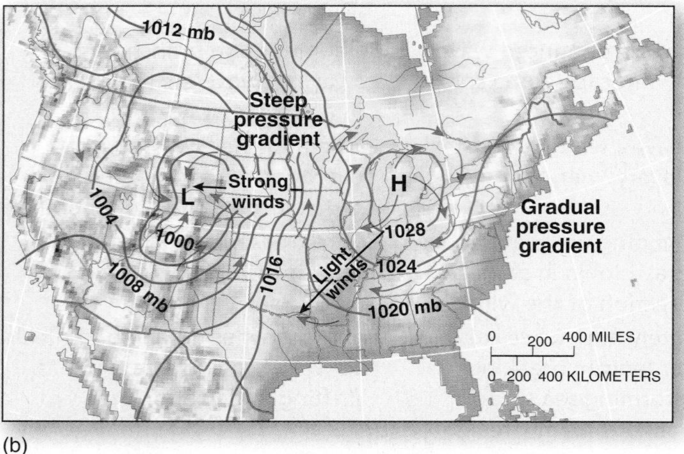

(b)

FIGURE 6.6 Pressure gradient determines wind speed. Pressure gradient (a) portrayed on a weather map (b). See the relationship between pressure gradient and wind strength. Here we see surface winds spiraling clockwise out of a high-pressure system and spiraling counterclockwise into a low-pressure system.

An **isobar** is an isoline (a line along which there is a constant value) plotted on a weather map to connect points of equal pressure. A pattern of isobars on a weather map provides a portrait of the pressure gradient between an area of higher pressure and one of lower pressure. The spacing between isobars indicates the intensity of the pressure difference, or pressure gradient.

Just as closer contour lines on a topographic map indicate a steeper slope on land, so closer isobars denote steepness in the pressure gradient. In Figure 6.6a, note the spacing of the isobars (green lines). A steep gradient causes faster air movement from a high-pressure area to a low-pressure area. Isobars spaced wider apart from one another mark a more gradual pressure gradient, one that creates a slower airflow. Along a horizontal surface, the pressure gradient force alone acts at right angles to the isobars, so wind blows across isobars at right angles. Note the location of steep ("strong winds") and gradual ("light winds") pressure gradients and their relationship to wind intensity on the map in Figure 6.6b.

Figure 6.7 illustrates the forces that direct winds. Figure 6.7a shows the pressure gradient force acting alone. A field of subsiding, or sinking, air develops in a high-pressure area. Air descends in a high-pressure area and diverges outward at the surface in all directions. In contrast, in a low-pressure area, as air rises, it pulls air from all directions, converging into the area of lower pressure at the surface.

Coriolis Force

You might expect surface winds to move in a straight line from areas of higher pressure to areas of lower pressure. On a nonrotating Earth, they would. But on our rotating planet, the Coriolis force deflects anything that flies or flows across Earth's surface—wind, an airplane, or ocean currents—from a straight path. In simplest terms, this force is an effect of Earth's rotation. Earth's rotational speed varies with latitude, increasing from 0 kmph at the poles (surface is at Earth's axis) to 1675 kmph (1041 mph) at the equator (surface is farthest from Earth's axis). Table 2.2 lists rotational speeds by latitude.

The deflection occurs regardless of the direction in which the object is moving. Because Earth rotates eastward, objects appear to curve to the right in the Northern Hemisphere and to the left in the Southern

GEO REPORT 6.3 *Coriolis: Not a force on sinks or toilets*

A common misconception about the Coriolis force is that it affects water draining out of a sink, tub, or toilet. Moving water or air must cover some distance across space and time before the Coriolis force noticeably deflects it. Long-range artillery shells and guided missiles do exhibit small amounts of deflection that must be corrected for accuracy. But water movements down a drain are too small in spatial extent to be noticeably affected by this force.

(a) Pressure gradient

Top view and side view of air movement in an idealized high-pressure area and low-pressure area on a nonrotating Earth.

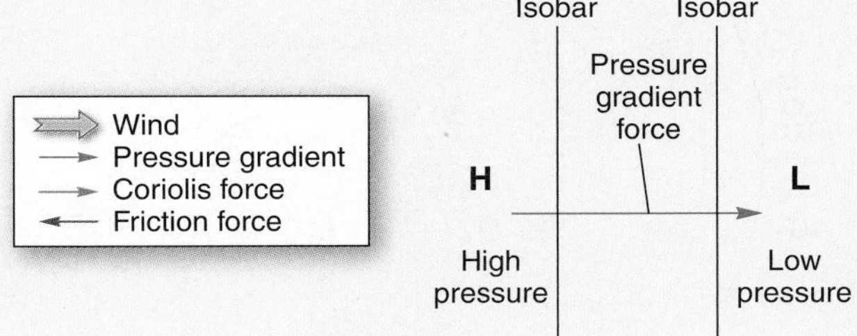

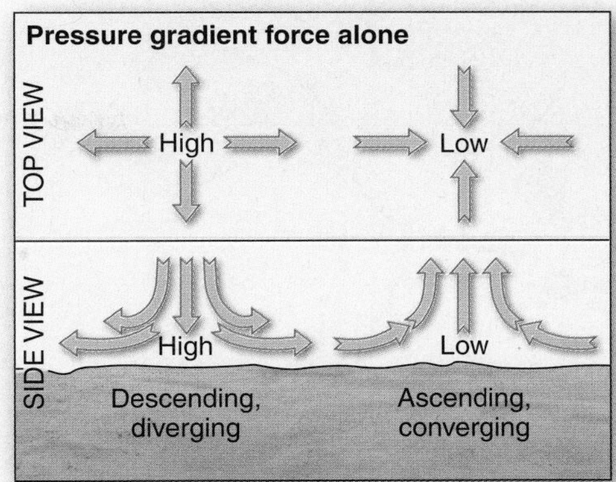

(b) Pressure gradient + Coriolis forces (upper level winds)

Earth's rotation adds the Coriolis force and a "twist" to air movements. High-pressure and low-pressure areas develop a rotary motion, and wind flowing between highs and lows flows parallel to isobars.

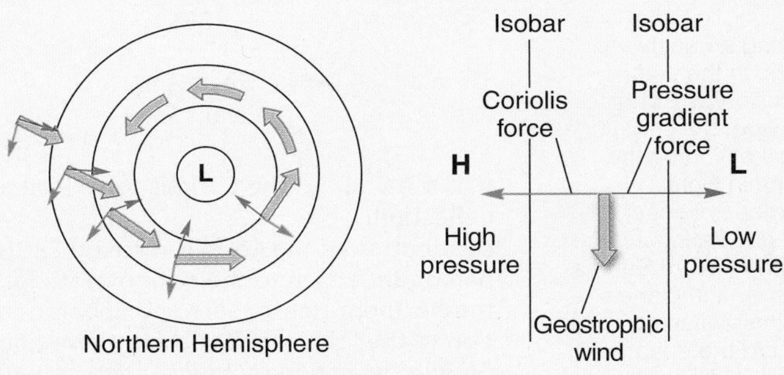

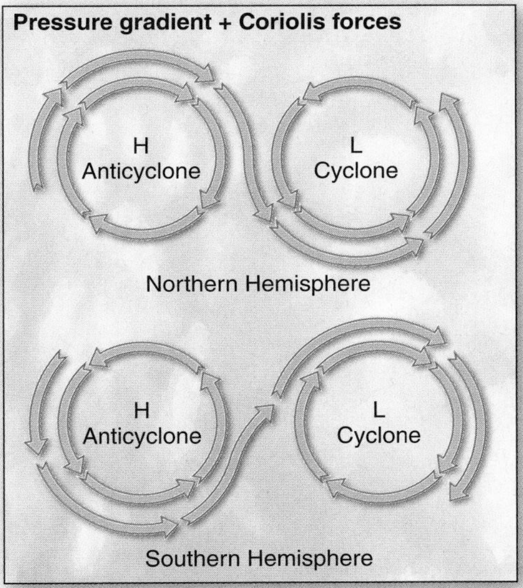

(c) Pressure gradient + Coriolis + friction forces (surface winds)

Surface friction adds a countering force to Coriolis, producing winds that spiral out of a high-pressure area and into a low-pressure area. Surface winds cross isobars at an angle. Air flows into low-pressure cyclones and turns to the left, because of deflection to the right.

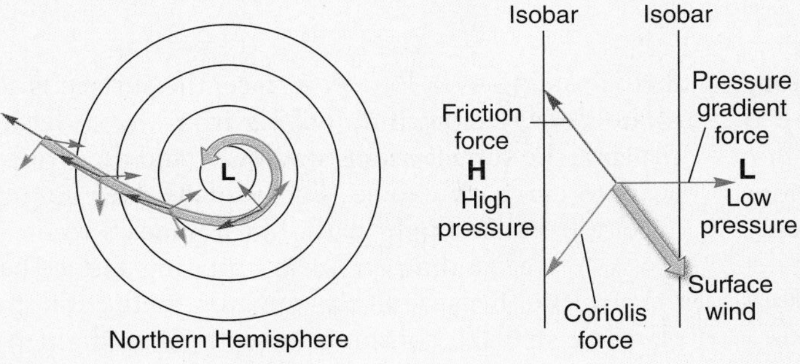

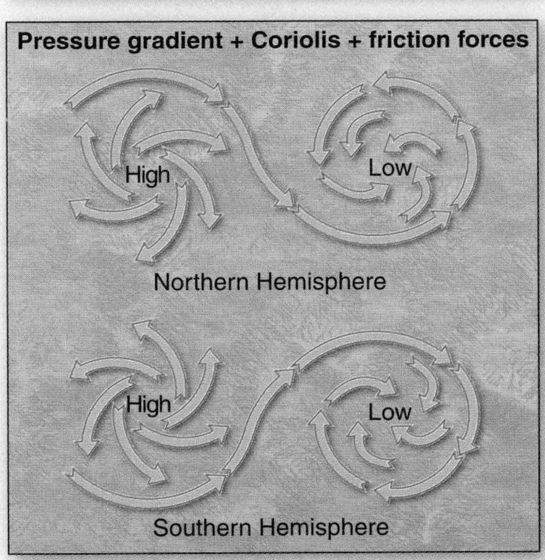

FIGURE 6.7 Three physical forces that produce winds.
Three physical forces interact to produce wind patterns at the surface and aloft: (a) pressure gradient force; (b) Coriolis force, which counters the pressure gradient force, producing a geostrophic wind flow in the upper atmosphere; and (c) friction force, which, combined with the other two forces, produces characteristic surface winds.

Animation
Wind Pattern Development

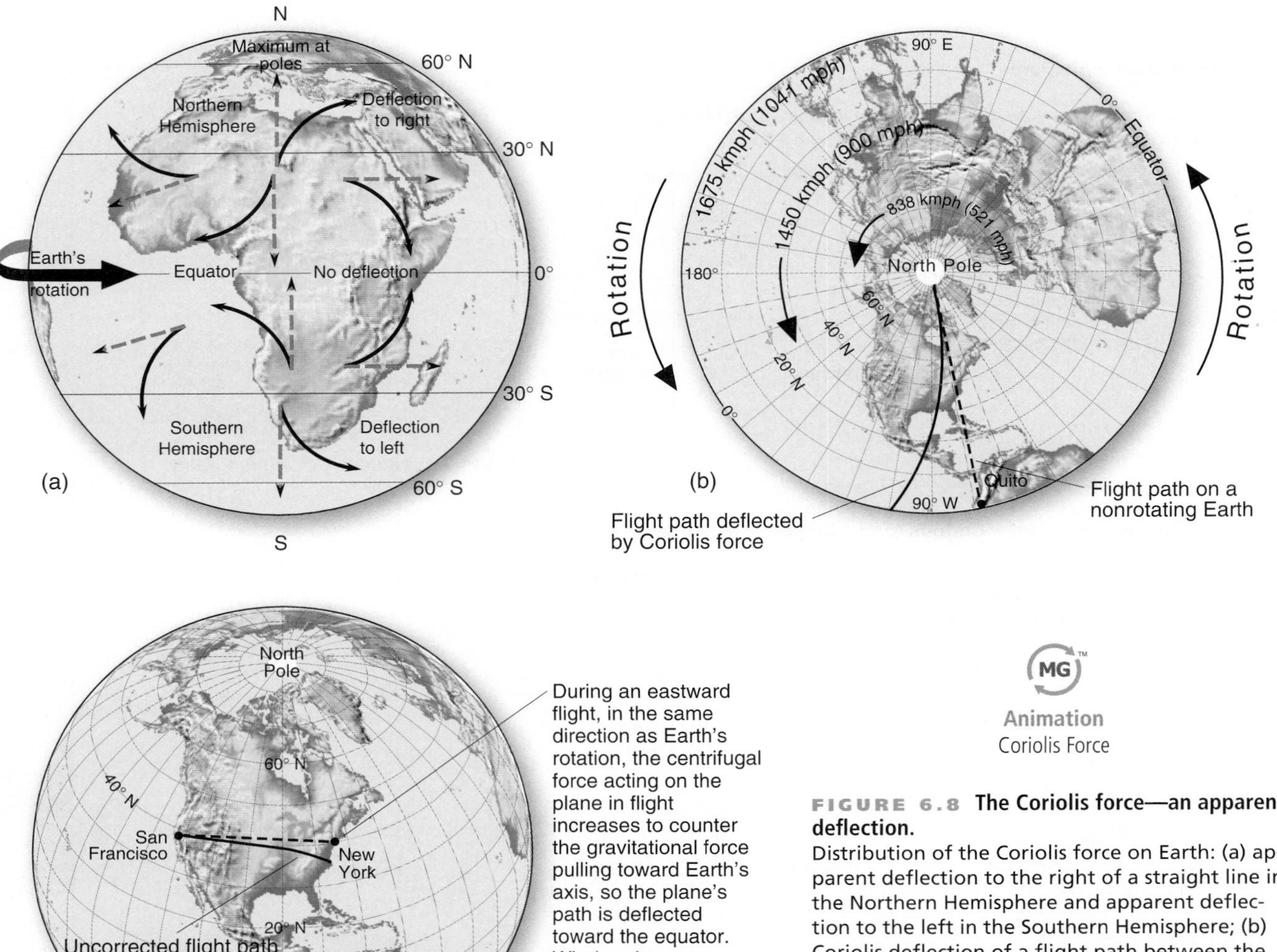

FIGURE 6.8 The Coriolis force—an apparent deflection.
Distribution of the Coriolis force on Earth: (a) apparent deflection to the right of a straight line in the Northern Hemisphere and apparent deflection to the left in the Southern Hemisphere; (b) Coriolis deflection of a flight path between the North Pole and Quito, Ecuador, which is on the equator; (c) deflection of a flight path between San Francisco and New York. Deflection from a straight path occurs regardless of the direction of movement.

Hemisphere (Figure 6.8a). The Coriolis force is zero along the equator, increases to half the maximum deflection at 30° N and 30° S latitude, and reaches maximum deflection flowing away from the poles. The effect of the Coriolis force increases as the speed of the moving object increases; thus, the faster the wind speed, the greater its apparent deflection. The Coriolis force does not normally affect small-scale motions that cover insignificant distance and time.

We work through a simple explanation here focused on the observed phenomena. If you choose, you can use physics and identify the variables that interplay to produce these effects. Mathematical formulas explain how the conservation of angular and linear momentum, as the distance to Earth's axis changes with latitudinal location, operates to produce the deflection, as well as the interplay of the opposing gravitational and centrifugal forces.

With this in mind, we observe the Coriolis force from different viewpoints. From the viewpoint of an airplane that is passing over Earth's surface, the surface is seen to rotate slowly below. But, looking from the surface at the airplane, the surface seems stationary, and the airplane appears to curve off course. The airplane does not actually deviate from a straight path, but it appears to do so because we are standing on Earth's rotating surface beneath the airplane. Because of this apparent deflection, the airplane must make constant corrections in flight path to maintain its "straight" heading relative to a rotating Earth.

As an example of the effect of this force, see Figure 6.8b. A pilot leaves the North Pole and flies due south toward Quito, Ecuador. If Earth were not rotating, the aircraft would simply travel along a meridian of longitude and arrive at Quito. But Earth is rotating eastward beneath the aircraft's flight path. If the pilot does not allow for this increase in rotational speed, the plane will reach the equator over the ocean along an apparently curved path, far to the west of the intended destination. Likewise, flying northward on a return flight, the plane maintains its eastward speed as it makes its way northward over slower rotational speeds of Earth. Without correction, the plane ends up to the east of the pole, in a right-hand deflection. Pilots must correct for this Coriolis deflection in their navigational calculations.

This effect occurs regardless of the direction of the moving object. A flight from California to New York provides an example in Figure 6.8c. The Coriolis deflection occurs as the airplane flies to New York because Earth continues to rotate eastward; this increases the centrifugal force acting on the plane (Earth rotation speed + plane speed), so the plane moves toward the equator in a right-hand deflection (moving farther from Earth's axis at a lower latitude). Unless the pilot corrects for this deflective force, the flight would end up somewhere in North Carolina. Likewise, flying westward on a return flight opposite Earth's rotation direction, which decreases the centrifugal force (Earth rotation speed – plane speed), the plane moves away from the equator, in a right-hand deflection (moving closer to Earth's axis at a higher latitude).

Coriolis Force and Winds How does the Coriolis force affect wind? As air rises from the surface through the lowest levels of the atmosphere, it leaves the drag of surface friction behind and increases speed. This increases the Coriolis force, spiraling the winds to the right in the Northern Hemisphere or to the left in the Southern Hemisphere, generally producing upper-air westerly winds from the subtropics to the poles. In the upper troposphere, the Coriolis force just balances the pressure gradient force. Consequently, the winds between higher-pressure and lower-pressure areas in the upper troposphere flow parallel to the isobars.

Figure 6.7b illustrates the combined effect of the pressure gradient force and the Coriolis force on air currents aloft. Together, they produce winds that do not flow directly from high to low, but that flow around the pressure areas, remaining parallel to the isobars. Such winds are **geostrophic winds** and are characteristic of upper tropospheric circulation. (The suffix *-strophic* means "to turn.") Note the inset illustration showing the effects of the pressure gradient and Coriolis forces that produce a geostrophic flow of air. Geostrophic winds produce the characteristic pattern shown on the upper-air weather map just ahead in Figure 6.14.

Friction Force

Figure 6.7c adds the effect of friction to the Coriolis and pressure gradient forces on wind movements; combining all three forces produces the wind patterns we see along Earth's surface. The effect of surface friction extends to a height of about 500 m (around 1600 ft) and varies with surface texture, wind speed, time of day and year, and atmospheric conditions. In general, rougher surfaces produce more friction.

Near the surface, friction disrupts the equilibrium established in geostrophic wind flows between the pressure gradient and Coriolis forces—note the inset illustration in Figure 6.7c. Because surface friction decreases wind speed, it reduces the effect of the Coriolis force and causes winds to move across isobars at an angle. The reason the balance of forces causes the winds to curve to the left into low-pressure cyclones as they are deflected to the right is shown.

In Figure 6.7c, you can see that the Northern Hemisphere winds spiral out from a high-pressure area *clockwise* to form an **anticyclone** and spiral into a low-pressure area *counterclockwise* to form a **cyclone**. (In the Southern Hemisphere these circulation patterns are reversed, with winds flowing counterclockwise out of anticyclonic high-pressure cells and clockwise into cyclonic low-pressure cells.)

Atmospheric Patterns of Motion

With these forces and motions in mind, we are ready to build a general model of total atmospheric circulation. The warmer, less-dense air along the equator rises, creating low pressure at the surface, and the colder, more-dense air at the poles sinks, creating high pressure. If Earth did not rotate, the result would be a simple wind flow from the poles to the equator, a meridional flow caused solely by pressure gradient.

However, Earth does rotate, creating a more complex flow system. On a rotating Earth, the poles-to-equator flow is predominantly zonal (latitudinal), both at the surface and aloft. Winds are westerly (eastward-moving) in the middle and high latitudes and easterly (westward-moving) in the low latitudes toward the equator in both hemispheres. Earth's global circulation system transfers thermal energy, air, and water masses from equatorial energy surpluses to polar energy deficits, using waves, streams, and eddies on a planetary scale.

Primary High-Pressure and Low-Pressure Areas

The following discussion of Earth's pressure and wind patterns refers often to Figure 6.9, isobaric maps showing average surface barometric pressure in January and July. Indirectly, these maps indicate prevailing surface winds, which are suggested by the isobars.

The primary high- and low-pressure areas of Earth's general circulation appear on these maps as cells or uneven belts of similar pressure that stretch across the face of the planet, interrupted by landmasses. Between these areas flow the primary winds. Secondary highs and lows, from a few hundred to a few thousand kilometers in diameter and hundreds to thousands of meters high, form within these primary pressure areas. The secondary systems seasonally migrate to produce changing weather patterns in the regions over which they pass.

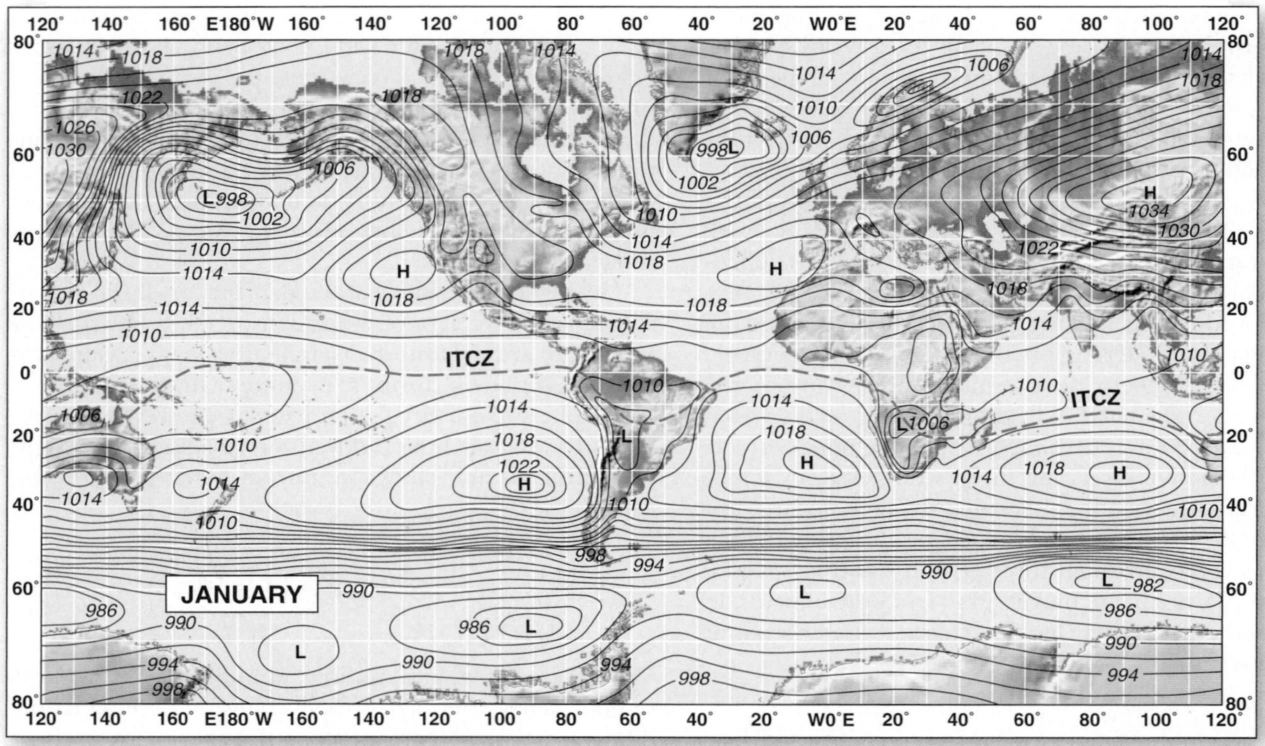

(a)

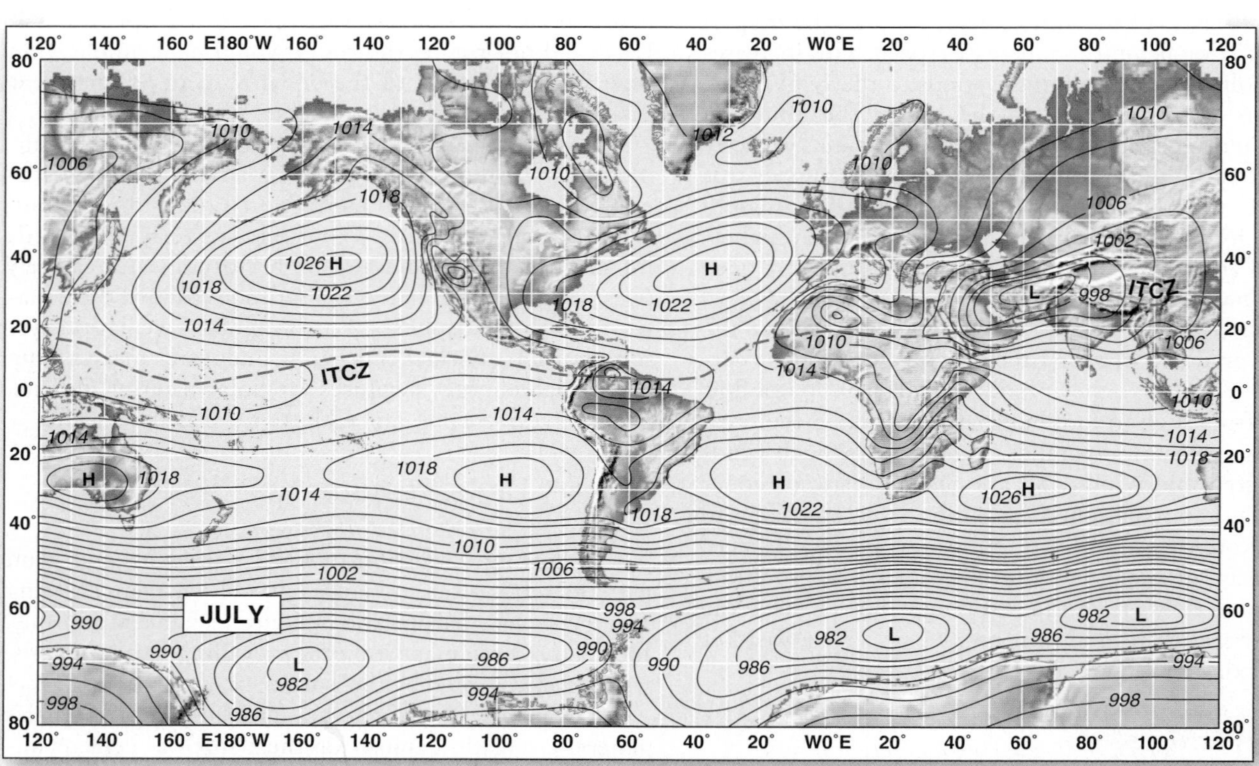

(b)

FIGURE 6.9 Global barometric pressures for January and July. Maps portray average surface barometric pressure (millibars) for (a) January and (b) July. The dashed line marks the general location of the intertropical convergence zone (ITCZ). Compare specific regions for January and July—for instance, the North Pacific, the North Atlantic, and the central Asian landmass. [Adapted by author and redrawn from National Climatic Data Center, *Monthly Climatic Data for the World*, 46 (January and July 1993), and WMO and NOAA.]

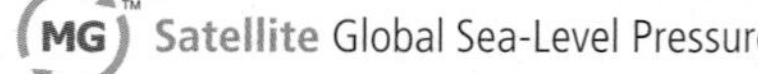

TABLE 6.2 Four Hemispheric Pressure Areas

Name	Cause	Location	Air Temperature/ Moisture
Polar high-pressure cells	Thermal	90° N, 90° S	Cold/dry
Subpolar low-pressure cells	Dynamic	60° N, 60° S	Cool/wet
Subtropical high-pressure cells	Dynamic	20°–35° N, 20°–35° S	Hot/dry
Equatorial low-pressure trough	Thermal	10° N to 10° S	Warm/wet

Four broad pressure areas cover the Northern Hemisphere, and a similar set exists in the Southern Hemisphere. In each hemisphere, two of the pressure areas are stimulated by thermal (temperature) factors. These are the **equatorial low-pressure trough** (marked by the ITCZ line on the maps) and the weak **polar high-pressure cells** at the North and South poles (not shown, as the maps are cut off at 80° N and 80° S). The other two pressure areas—the **subtropical high-pressure cells** (H) and **subpolar low-pressure cells** (L)—are formed by dynamic (mechanical) factors. Table 6.2 summarizes the characteristics of these pressure areas. We now examine each principal pressure region.

Equatorial Low-Pressure Trough—ITCZ: Warm and Rainy Constant high Sun altitude and consistent daylength (12 hours a day, year-round) make large amounts of energy available in this region throughout the year. The warming creates lighter, less-dense, ascending air, with surface winds converging along the entire extent of the low-pressure trough. This converging air is extremely moist and full of latent heat energy. As it rises, the air expands and cools, producing condensation; consequently, rainfall is heavy throughout this zone. Vertical cloud columns frequently reach the tropopause, in thunderous strength and intensity.

The combination of heating and convergence forces air aloft and forms the **intertropical convergence zone (ITCZ)**. The ITCZ is identified by bands of clouds associated with the convergence of winds along the equator and is noted on the January and July pressure maps (see Figure 6.9; the maps show the ITCZ as a dashed line). In January, the zone crosses northern Australia and dips southward in eastern Africa and South America. Note, using precipitation patterns, this ITCZ location on the satellite images in Figure 6.10. During summer, a marked wet season accompanies the shifting ITCZ over various regions.

Figure 6.10 features satellite images showing the equatorial low-pressure trough of the ITCZ using associated precipitation for January and July. A broken band of precipitation accumulation (monthly averages) reveals this low-pressure system and its position slightly north of the equator in July. Note the position of the ITCZ in Figure 6.9 and compare this with the precipitation pattern captured by the TRMM (Tropical Rainfall Measuring Mission) sensors in Figure 6.10. The equatorial low-pressure trough is an elongated, undulating narrow band of low pressure (converging, ascending airflow), cloudiness, and precipitation that nearly encircles the planet, consistently over the oceans and interrupted over land surfaces.

Figure 6.11 on page 142 shows two views of Earth's general circulation. The winds converging on the equatorial low-pressure trough are known generally as the

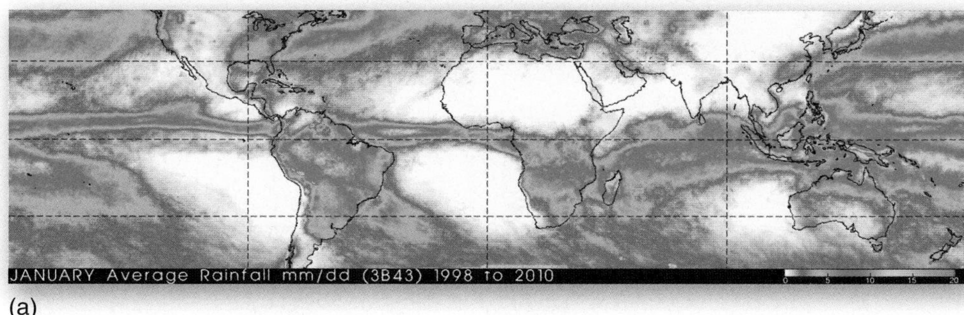

(a)

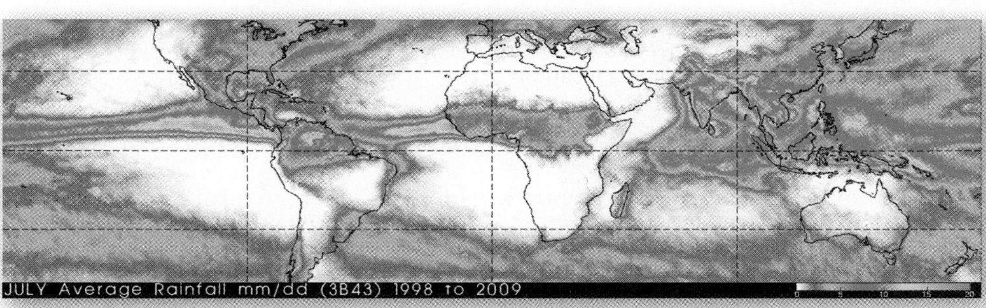

(b)

FIGURE 6.10 Precipitation accumulation portrays equatorial and subtropical circulation patterns. (a) January average rainfall 1998–2010. An interrupted band of average precipitation in millimeters per day for 31 days along the intertropical convergence zone (ITCZ). (b) July average rainfall 1998–2010. Note to the north and south of the ITCZ in the subtropics several high-pressure systems and clear skies (lacking coloration). [TRMM Microwave Imager on several geosynchronous satellites, courtesy of GSFC/NASA.]

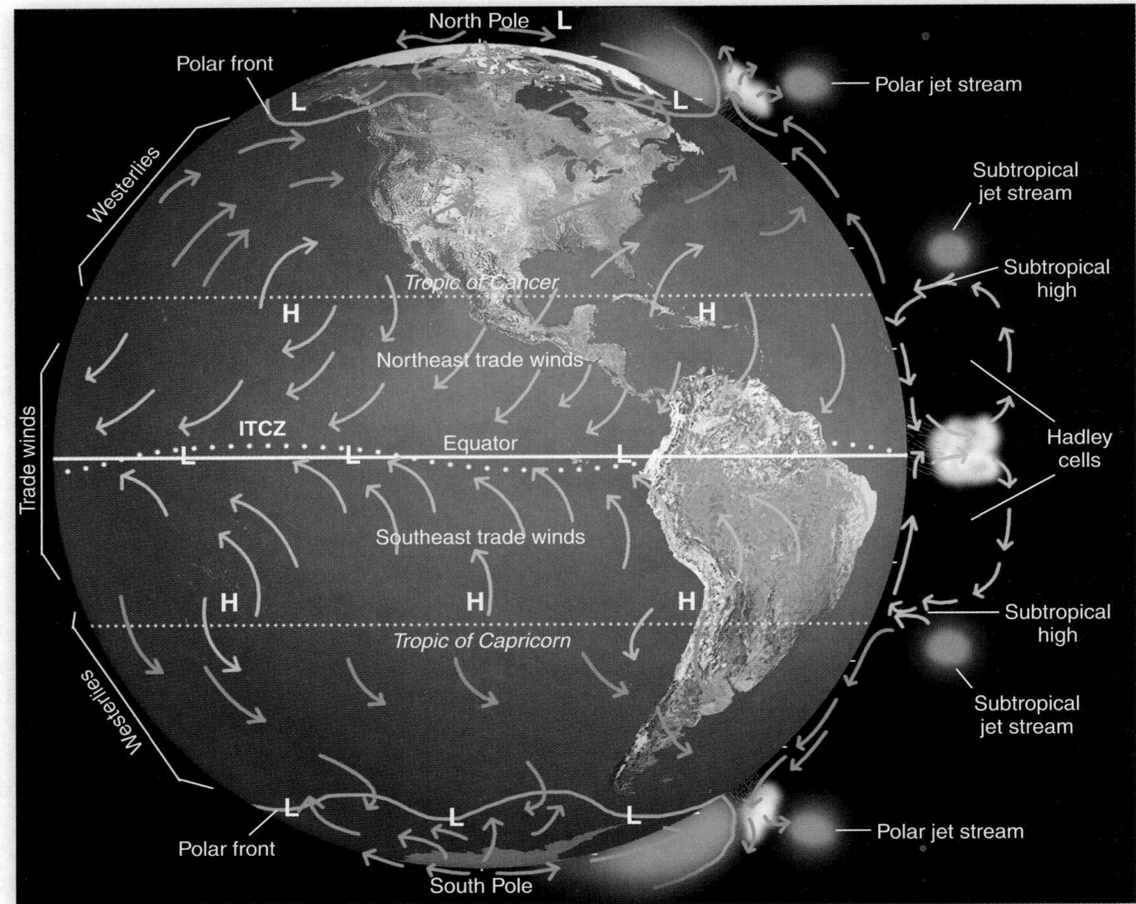

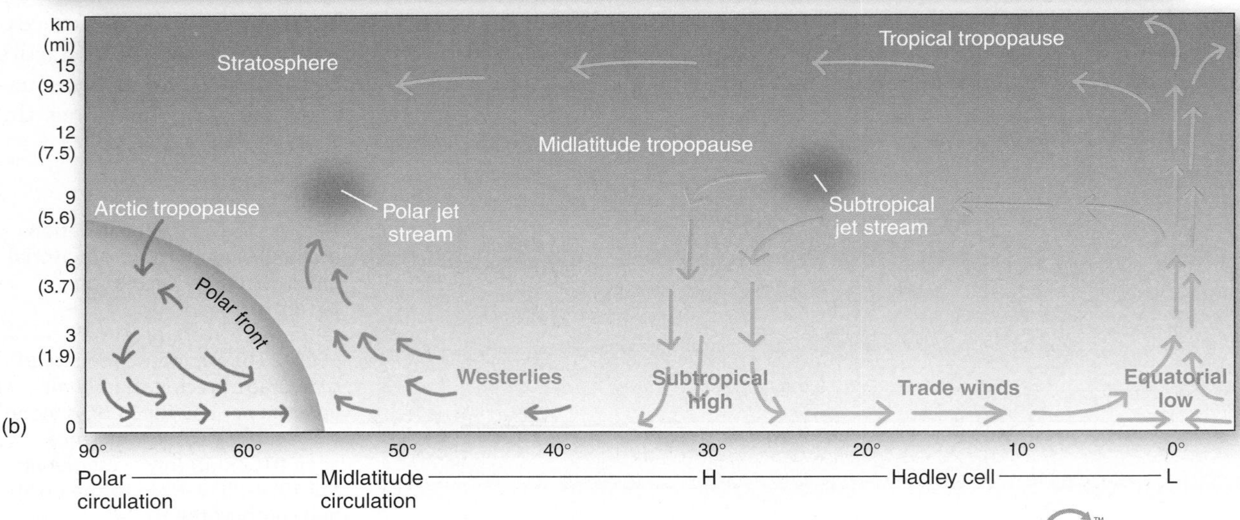

FIGURE 6.11 General atmospheric circulation model. Two views of the general atmospheric circulation: (a) general circulation schematic; (b) equator-to-pole cross section of the Northern Hemisphere. Both views show Hadley cells, subtropical highs, the polar front, the subpolar low-pressure cells, and approximate locations of the subtropical and polar jet streams.

MG Animation Global Wind Circulation, Hadley Cells

MG Satellite Global Infrared

trade winds, or *trades*. *Northeast trade winds* blow in the Northern Hemisphere and *southeast trade winds* in the Southern Hemisphere. These are labeled in Figure 6.11a. The trade winds were named during the era of sailing ships that carried trade across the seas.

The trade winds pick up large quantities of moisture as they return through the Hadley circulation cell for another cycle of uplift and condensation (shown in cross section in Figure 6.11b). These *Hadley cells* in each hemisphere, named for the eighteenth-century English scientist who described the trade winds, denote the circuit completed by winds rising along the ITCZ. Air moves northward and southward into the subtropics, descending to the surface and returning to the ITCZ as the trade winds. During the year in each hemisphere, this circulation pattern appears most vertically symmetrical near the equinoxes.

Within the ITCZ, winds are calm or mildly variable because of the even pressure gradient and the vertical ascent of air. These equatorial calms are the doldrums, a name formed from an older English word meaning "foolish," because of the difficulty sailing ships encountered when attempting to move through this zone. The rising air from the equatorial low-pressure area spirals upward into a geostrophic flow to the north and south. These upper-air winds turn eastward, flowing from west to east, beginning at about 20° N and 20° S, and form descending masses of air and high-pressure systems in the subtropical latitudes.

Subtropical High-Pressure Cells: Hot and Dry Between 20° and 35° latitude in both hemispheres, a broad high-pressure zone of hot, dry air is evident across the globe (see Figures 6.9 and 6.10). Clear, frequently cloudless skies over the Sahara and the Arabian deserts and portions of the Indian Ocean dominate these regions. Can you identify these desert regions on the TRMM satellite image in Figure 6.10? How about on the photo of Earth on the text's back cover?

The dynamic cause of these subtropical anticyclones is too complex to detail here, but they generally form as air above the subtropics is mechanically pushed downward and heats by compression on its descent to the surface (shown in Figure 6.11). Warmer air has a greater capacity to absorb water vapor than does cooler air, making this descending warm air relatively dry (large water-vapor capacity, low water-vapor content). The air is also dry because heavy precipitation along the equatorial portion of the circulation removes moisture.

Surface air diverging from the subtropical high-pressure cells generates Earth's principal surface winds: the westerlies and the trade winds. The **westerlies** are the dominant surface winds from the subtropics to high latitudes. They diminish somewhat in summer and are stronger in winter in both hemispheres.

As you examine the global pressure maps in Figure 6.9, you find several high-pressure areas. In the Northern Hemisphere, the Atlantic subtropical high-pressure cell is the **Bermuda high** (in the western Atlantic) or the **Azores high** (when it migrates to the eastern Atlantic in winter). The area in the Atlantic under subtropical high pressure features clear, warm waters and large quantities of Sargassum (a seaweed) that gives the area its name—the Sargasso Sea. The **Pacific high**, or *Hawaiian high*, dominates the Pacific in July, retreating southward in January. In the Southern Hemisphere, three large high-pressure centers dominate the Pacific, Atlantic, and Indian oceans, especially in January, and tend to move along parallels of latitude in shifting zonal positions.

The entire high-pressure system migrates with the summer high Sun, fluctuating about 5°–10° in latitude. The eastern sides of these anticyclonic systems are drier and more stable (exhibit less convective activity) and feature cooler ocean currents than do the western sides. The drier eastern sides of these systems and dry-summer conditions influence climate along subtropical and midlatitude west coasts (Figure 6.12). In fact, Earth's major deserts generally occur within the subtropical belt and extend to the west coast of each continent except Antarctica. In Figures 6.12 and Figure 6.20, note that the desert regions of Africa come right to the shore in both hemispheres, with the cool, southward-flowing *Canaries Current* offshore in the north and the cool, northward-flowing *Benguela Current* offshore in the south.

Because the subtropical belts are near 25° N and 25° S latitude, these areas sometimes are known as the *calms of Cancer* and the *calms of Capricorn*. These zones of windless, hot, dry desert air, so deadly in the era of sailing ships, earned the name *horse latitudes*. Although the term's true origin is uncertain, it is popularly attributed to becalmed and stranded sailing crews of past centuries who destroyed the horses on board, not wanting to share food or water with the livestock.

Subpolar Low-Pressure Cells: Cool and Moist In January, two low-pressure cyclonic cells exist over the oceans around 60° N latitude, near their namesakes: the North Pacific **Aleutian low** and the North Atlantic **Icelandic low** (see Figure 6.9a). Both cells are dominant in winter and weaken or disappear in summer with the

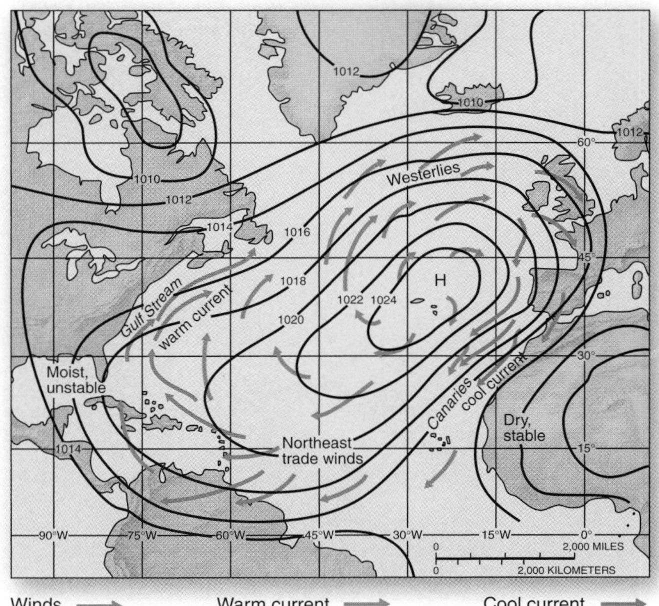

FIGURE 6.12 Subtropical high-pressure system in the Atlantic.
Characteristic circulation in the Northern Hemisphere. Note deserts extending to the shores of Africa with offshore cool currents, whereas the southeastern United States is moist and humid, with offshore warm currents.

strengthening of high-pressure systems in the subtropics. The area of contrast between cold air from higher latitudes and warm air from lower latitudes forms the **polar front**, where masses of air with different characteristics battle. This front encircles Earth, focused in these low-pressure areas.

Figure 6.11 illustrates this confrontation between warm, moist air from the westerlies and cold, dry air from the polar and Arctic regions. Cooling and condensation in the lifted air occurs because of upward displacement of the warm air. Low-pressure cyclonic storms migrate out of the Aleutian and Icelandic frontal areas and may produce precipitation in North America and Europe, respectively. Northwestern sections of North America and Europe generally are cool and moist as a result of the passage of these cyclonic systems onshore—consider the weather in British Columbia, Washington, Oregon, Ireland, and the United Kingdom.

In the Southern Hemisphere, a discontinuous belt of subpolar low-pressure systems surrounds Antarctica. The spiraling cloud patterns produced by these cyclonic systems are visible on the spacecraft image in Figure 6.13. Severe cyclonic storms can cross Antarctica, producing strong winds and new snowfall. How many cyclonic systems can you identify around Antarctica on the image?

Polar High-Pressure Cells: Frigid and Dry Polar high-pressure cells are weak. The polar atmospheric mass is small, receiving little energy from the Sun to put it into motion. Variable winds, cold and dry, move away from the polar region in an anticyclonic direction. They descend and diverge clockwise in the Northern Hemisphere (counterclockwise in the Southern Hemisphere) and form weak, variable winds of the **polar easterlies**.

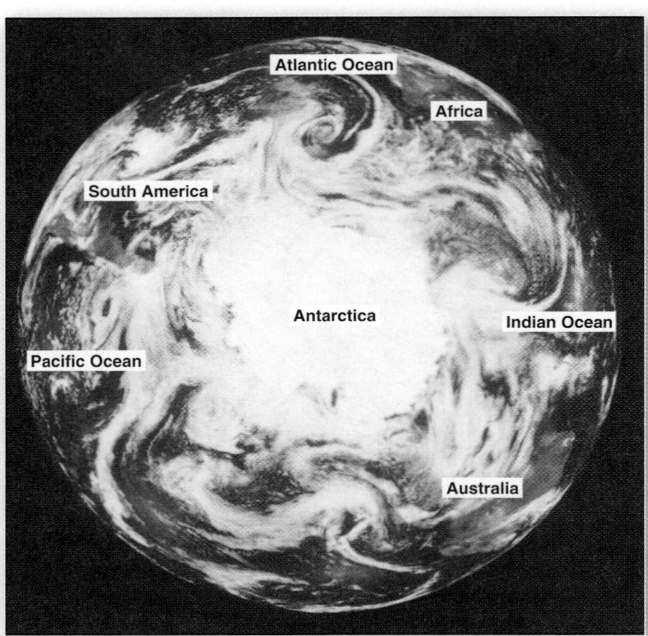

FIGURE 6.13 Clouds portray subpolar and polar circulation patterns.
Centered on Antarctica, this image shows a series of subpolar low-pressure cyclones in the Southern Hemisphere. Antarctica is fully illuminated by a midsummer Sun as the continent approaches the December solstice. The Galileo spacecraft made this image during its December 1990 flyby of Earth. [Image courtesy of ESA and the late Dr. W. Reid Thompson, Cornell University.]

Of the two polar regions, the **Antarctic high** is stronger and more persistent, forming over the Antarctic landmass. Less pronounced is a polar high-pressure cell over the Arctic Ocean. When it does form, it tends to locate over the colder northern continental areas in winter (Canadian and Siberian highs) rather than directly over the relatively warmer Arctic Ocean.

Upper Atmospheric Circulation

Circulation in the middle and upper troposphere is an important component of the atmosphere's general circulation. Just as sea level is a reference datum for evaluating air pressure at the surface, we use a pressure level such as 500 mb as a **constant isobaric surface** for a pressure-reference datum in the upper atmosphere.

On upper-air pressure maps, we plot the height above sea level at which an air pressure of 500 mb occurs. The isobaric chart and illustration for an April day in Figure 6.14a and b illustrate such an undulating isobaric surface, upon which all points have the same pressure. In contrast, on surface weather maps we plot different pressures at the fixed elevation of sea level—a *constant-height surface*.

We use this 500-mb level to analyze upper-air winds and the possible support they provide for surface weather conditions. Similar to surface maps, closer spacing of the isobars indicates faster winds; wider spacing indicates slower winds. On this constant-isobaric-pressure surface, altitude variations from the reference datum are ridges for high pressure (with isobars on the map bending poleward) and troughs for low pressure (with isobars on the map bending equatorward). Looking at the figure, can you identify such ridges and troughs in the isobaric surface?

The pattern of ridges and troughs in the upper-air wind flow is important in sustaining surface cyclonic (low-pressure) and anticylonic (high-pressure) circulation. Near ridges in the isobaric surface, winds slow and converge (pile up), whereas winds near the area of maximum wind speeds along the troughs in the isobaric surface accelerate and diverge (spread out). Note the wind-speed indicators and labels in Figure 6.14a near the ridge (over Alberta, Saskatchewan, Montana, and Wyoming); now compare these with the wind-speed indicators around the trough (over Kentucky, West Virginia, the New England states, and the Maritimes). Also, note the wind relationships off the Pacific Coast.

Now look at Figure 6.14c. Divergence in the upper-air flow is important to cyclonic circulation at the surface because it creates an outflow of air aloft that stimulates an inflow of air into the low-pressure cyclone (like what happens when you open a chimney damper to create an upward draft). Convergence aloft, in contrast, drives descending airflows and divergent winds at the surface, moving out from high-pressure anticyclones.

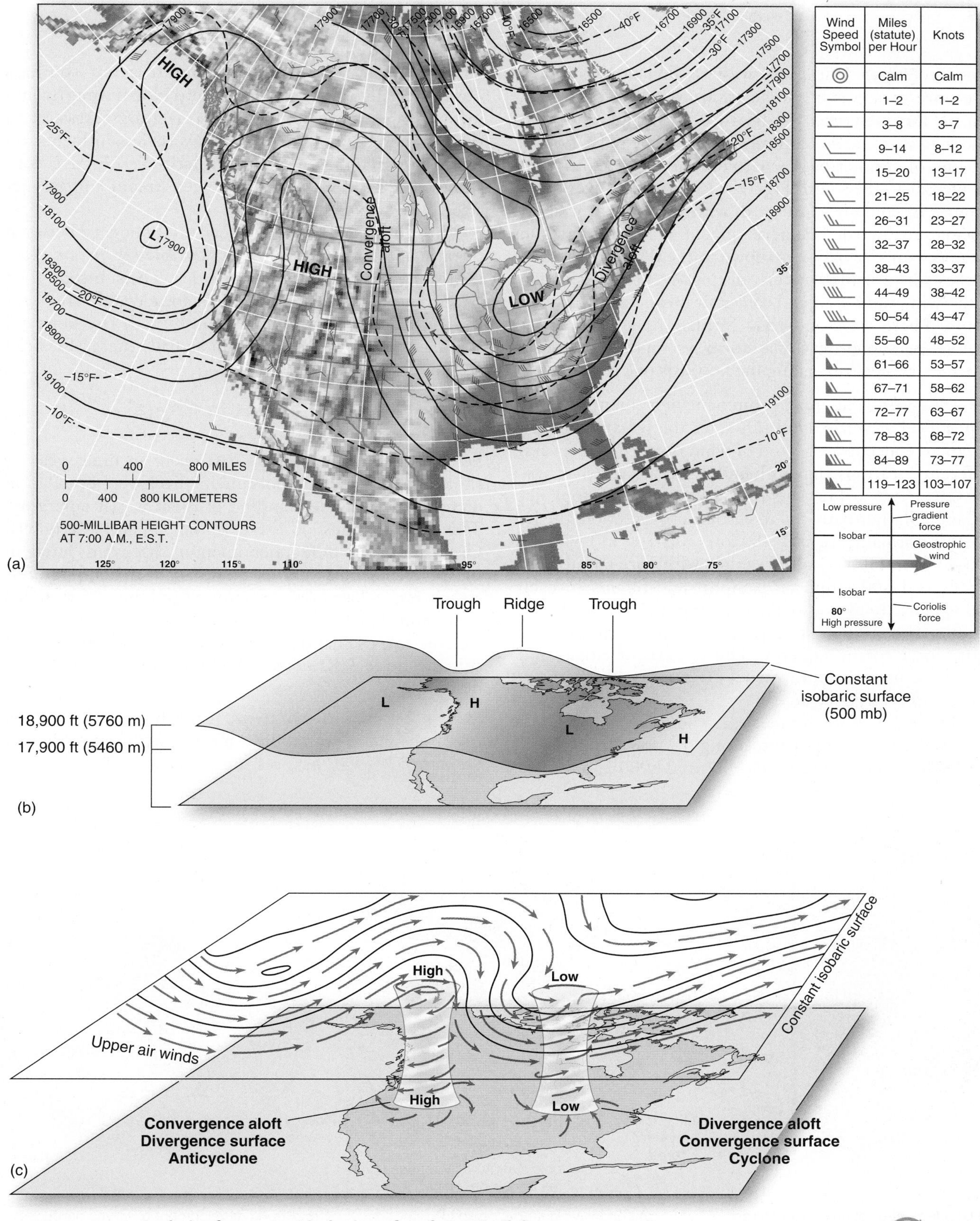

Wind Speed Symbol	Miles (statute) per Hour	Knots
◎	Calm	Calm
	1–2	1–2
	3–8	3–7
	9–14	8–12
	15–20	13–17
	21–25	18–22
	26–31	23–27
	32–37	28–32
	38–43	33–37
	44–49	38–42
	50–54	43–47
	55–60	48–52
	61–66	53–57
	67–71	58–62
	72–77	63–67
	78–83	68–72
	84–89	73–77
	119–123	103–107

FIGURE 6.14 Analysis of a constant isobaric surface for an April day.
(a) Contours show elevation (in feet) at which 500-mb pressure occurs—a constant isobaric surface. The pattern of contours reveals a 500-mb isobaric surface and geostrophic wind patterns in the troposphere ranging from 16,500 to 19,100 ft in elevation. (b) Note on the map and in the sketch beneath the chart the "ridge" of high pressure over the Intermountain West, at an altitude of 5760 m, and the "trough" of low pressure over the Great Lakes region and off the Pacific Coast, at an altitude of 5460 m. (c) Note areas of convergence aloft (corresponding to surface divergence) and divergence aloft (corresponding to surface convergence).

MG™
Animation
Cyclones and Anticyclones

Rossby Waves Within the westerly flow of geostrophic winds are great waving undulations, the **Rossby waves**, named for meteorologist Carl G. Rossby, who first described them mathematically in 1938. The polar front is the line of conflict between colder air to the north and warmer air to the south (Figure 6.15). Rossby waves bring tongues of cold air southward, with warmer tropical air moving northward. The development of Rossby waves in the upper-air circulation is shown in the three-part figure. As these disturbances mature, distinct cyclonic circulation forms, with warmer air and colder air mixing along distinct fronts. These wave-and-eddy formations and upper-air divergences support cyclonic storm systems at the surface. Rossby waves develop along the flow axis of a jet stream.

Jet Streams The most prominent movement in these upper-level westerly wind flows are the **jet streams**, an irregular, concentrated band of wind occurring at several different locations that influences surface weather systems. (Figure 6.11a shows the location of four jet streams.) Rather flattened in vertical cross section, the jet streams normally are 160–480 km (100–300 mi) wide by 900–2150 m (3000–7000 ft) thick, with core speeds that can exceed 300 kmph (190 mph). Jet streams in each hemisphere tend to weaken during the hemisphere's summer and strengthen during its winter as the streams shift closer to the equator. The pattern of ridges and troughs causes variation in jet-stream speeds (convergence and divergence).

The *polar jet stream* meanders between 30° and 70° N latitude, at the tropopause along the polar front, at altitudes between 7600 and 10,700 m (24,900 and 35,100 ft). The polar jet stream can migrate as far south as Texas, steering colder air masses into North America and influencing surface storm paths travelling eastward. In the summer, the polar jet stream exerts less influence on storms by staying over higher latitudes. Figure 6.16 shows a stylized view of a polar jet stream and two jet streams over North America.

In subtropical latitudes, near the boundary between tropical and midlatitude air, the *subtropical jet stream* flows near the tropopause (see Figure 6.11). The subtropical jet stream meanders from 20° to 50° latitude and may occur over North America simultaneously with the polar jet stream—sometimes the two will actually merge for brief episodes.

Multi-year Oscillations in Global Circulation

Several system fluctuations that occur in multi-year or shorter periods are important in the global circulation picture, yet an understanding of them is just emerging. The most famous of these is the El Niño–Southern Oscillation (ENSO) phenomenon discussed later in Chapter 10. Multi-year oscillations affect temperatures, air masses, and air pressure patterns and thus affect global winds and climates. Here we describe three of these hemisphere-scale oscillations in the briefest terms. Please consult the three URLs listed for illustrations and further descriptions.

North Atlantic Oscillation A north–south fluctuation of atmospheric variability marks the *North Atlantic Oscillation (NAO)*, as pressure differences between the

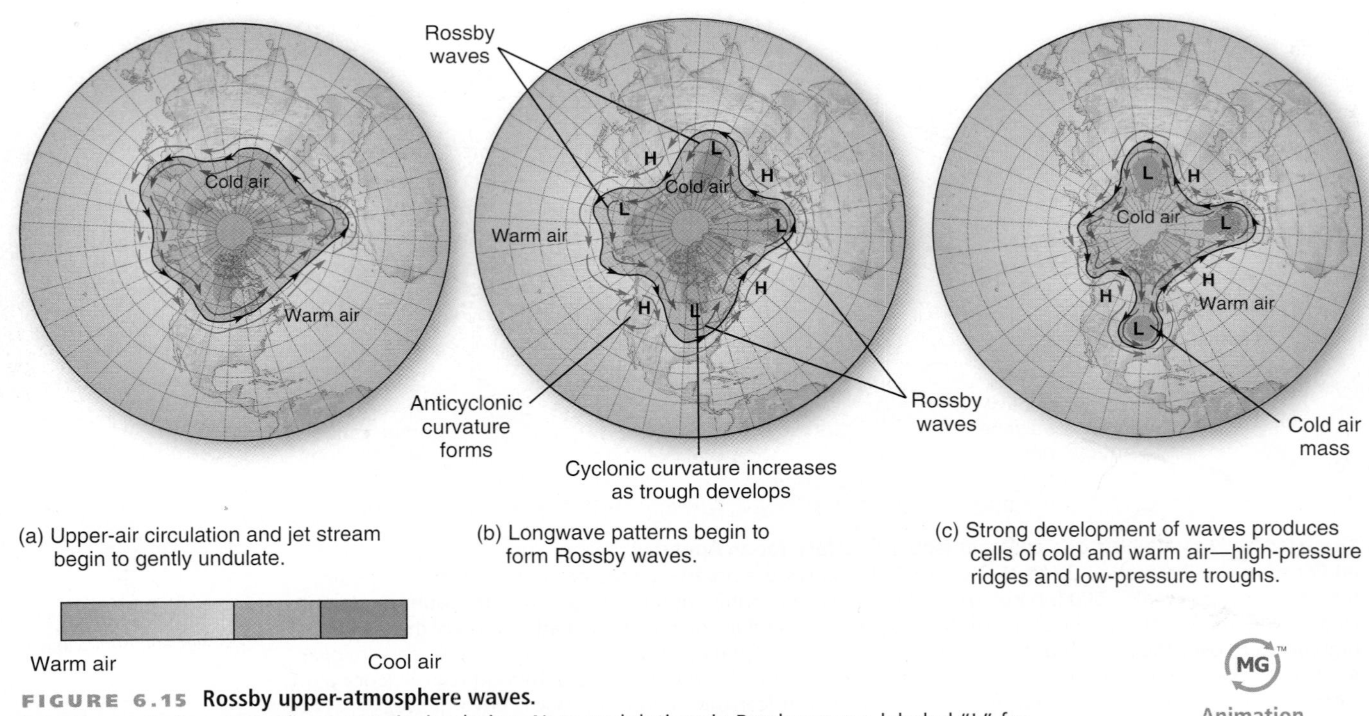

FIGURE 6.15 Rossby upper-atmosphere waves.
Development of waves in the upper-air circulation. Note undulations in Rossby waves, labeled "L" for troughs and "H" for ridges.

MG Animation
Jet Stream, Rossby Waves

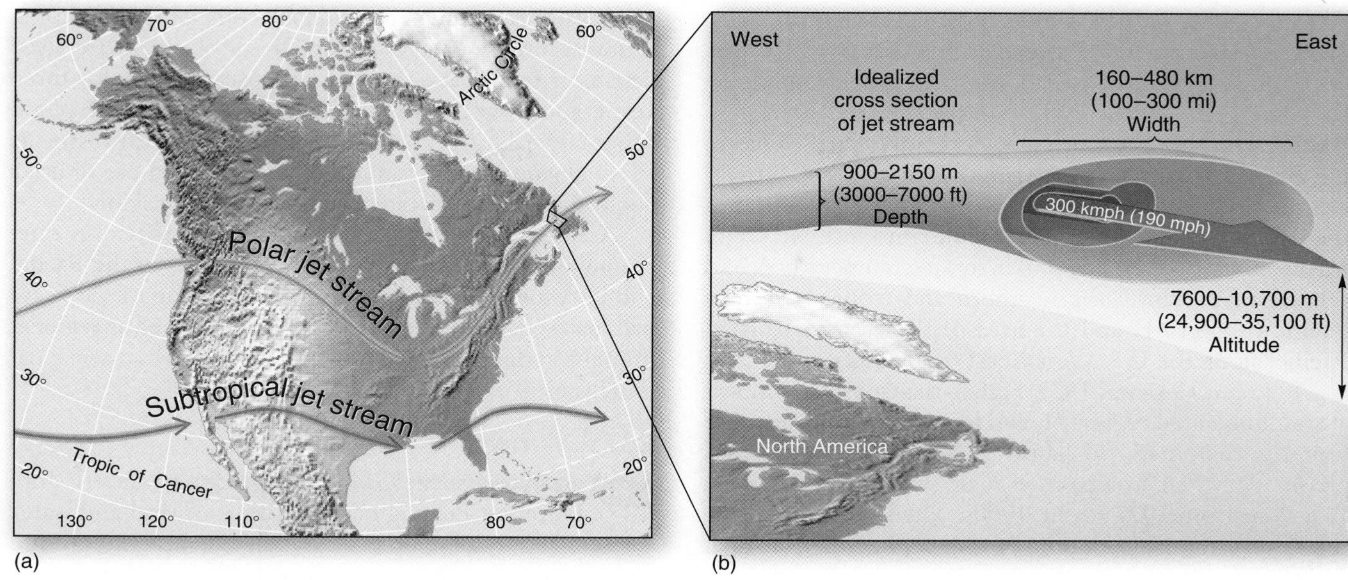

FIGURE 6.16 Jet streams.
(a) Average locations of the two jet streams over North America. (b) Stylized portrait of a polar jet stream.

Icelandic low and the Azores high in the Atlantic alternate from a weak to a strong pressure gradient. The *NAO Index* is in its *positive phase* when the Icelandic low-pressure system is lower than normal and the Azores (west of Portugal) high-pressure cell is higher than normal. Strong westerly winds and jet stream cross the eastern Atlantic. Northerly winds blow in the Labrador Sea (west of Greenland), and southerly winds move through the Norwegian Sea (east of Greenland). In the eastern United States, winters tend to be less severe in contrast to the strong, warm, wet storms hitting northern Europe; however, the Mediterranean region is dry.

In its *negative phase*, the NAO features a weaker pressure gradient than normal between the Azores and Iceland and reduced westerlies and jet stream. Storm tracks shift southward in Europe, bringing moist conditions to the Mediterranean and cold, dry winters to northern Europe. The eastern United States experiences cold, snowy winters as Arctic air masses plunge to lower latitudes.

With unpredictable flips between positive and negative phases, sometimes changing from week to week, a trend to more positive than negative phases has emerged since 1960. From 1980 to 2008, the NAO was averaging a more positive phase condition; however, through 2009 and into early 2010, the NAO was more into a negative phase through the winter (see http://www.ldeo.columbia.edu/NAO/).

Arctic Oscillation Variable fluctuations between middle- and high-latitude air mass conditions over the Northern Hemisphere produce the *Arctic Oscillation (AO)*. The AO is associated with the NAO, especially in winter, and the two phases of the *AO Index* correlate with the *NAO Index*. In the *positive phase* (warm phase) of the AO Index (positive NAO), the pressure gradient is affected by lower pressure than normal over the North Pole region and relatively higher pressures at lower latitudes. This sets up stronger westerly winds and a consistently strong jet stream and flow of warmer Atlantic water currents into the Arctic Ocean. Cold air masses do not migrate as far south in winter, whereas winters are colder than normal in Greenland.

In the *negative phase* (cold phase) of the AO Index (negative NAO), the pattern reverses. Higher-than-normal pressure is over the polar region and relatively lower pressure in the central Atlantic. The weaker zonal wind flow allows winter cold air masses into the eastern United States, northern Europe, and Asia, and sea ice in the Arctic Ocean becomes a bit thicker. Greenland is warmer than normal. The Northern Hemisphere winter of 2009–2010 featured unusual flows of cold air into the midlatitudes. In December

 **GEO REPORT 6.4** ***Icelandic ash caught in the jet stream***

Although smaller than either the Mount St. Helens 1980 eruption or the Mount Pinatubo 1991 eruption, the 2010 eruption of Eyjafjallajökull volcano in Iceland injected about a tenth of a cubic kilometer of volcanic debris into the jet-stream flow. The ash cloud from Iceland was swept toward the European mainland and the United Kingdom. Aircraft cannot risk ingesting volcanic ash into jet engines; therefore, airports were shut down, and thousands of flights cancelled. People's attention was focused on the guiding jet stream as it impacted their flight schedules and lives.

2009, the *Arctic Index* was in its most negative phase since 1970; it was even more negative in February 2010 (see http://nsidc.org/arcticmet/patterns/arctic_oscillation.html).

Pacific Decadal Oscillation Across the Pacific Ocean, the *Pacific Decadal Oscillation (PDO)*, which lasts 20 to 30 years, is longer-lived than the 2- to 12-year variation in the ENSO. The PDO term came into use in 1996. Involved are two regions of sea-surface temperatures and related air pressure: the northern and tropical western Pacific (region #1) and the area of the eastern tropical Pacific, along the U.S. West Coast (region #2).

Between 1947 and 1977, higher-than-normal temperatures dominated region #1, and lower temperatures were found in region #2; this is the PDO *negative phase* (cool phase). A switch to a *positive phase* (warm phase) in the PDO ran from 1977 to the 1990s, when lower temperatures were found in region #1 and higher-than-normal temperatures dominated region #2, coinciding with a time of more intense ENSO events. In 1999, a *negative phase* began for four years, then a mild positive phase for three years, and now the PDO has been in a negative phase since 2008. Unfortunately for the U.S. Southwest, this PDO negative phase can mean a decade or more of drier conditions for the already drought-plagued region.

Causes of the PDO and its cyclic variability over time are unknown. Scientists monitor conditions in the Pacific and look for patterns. A better understanding of the PDO will help scientists predict ENSO events as well as regional drought cycles (see http://topex-www.jpl.nasa.gov/science/pdo.html).

Local Winds

Land and sea breezes occur on most coastlines (Figure 6.17). Different heating characteristics of land and water surfaces create these breezes. Land gains heat energy and

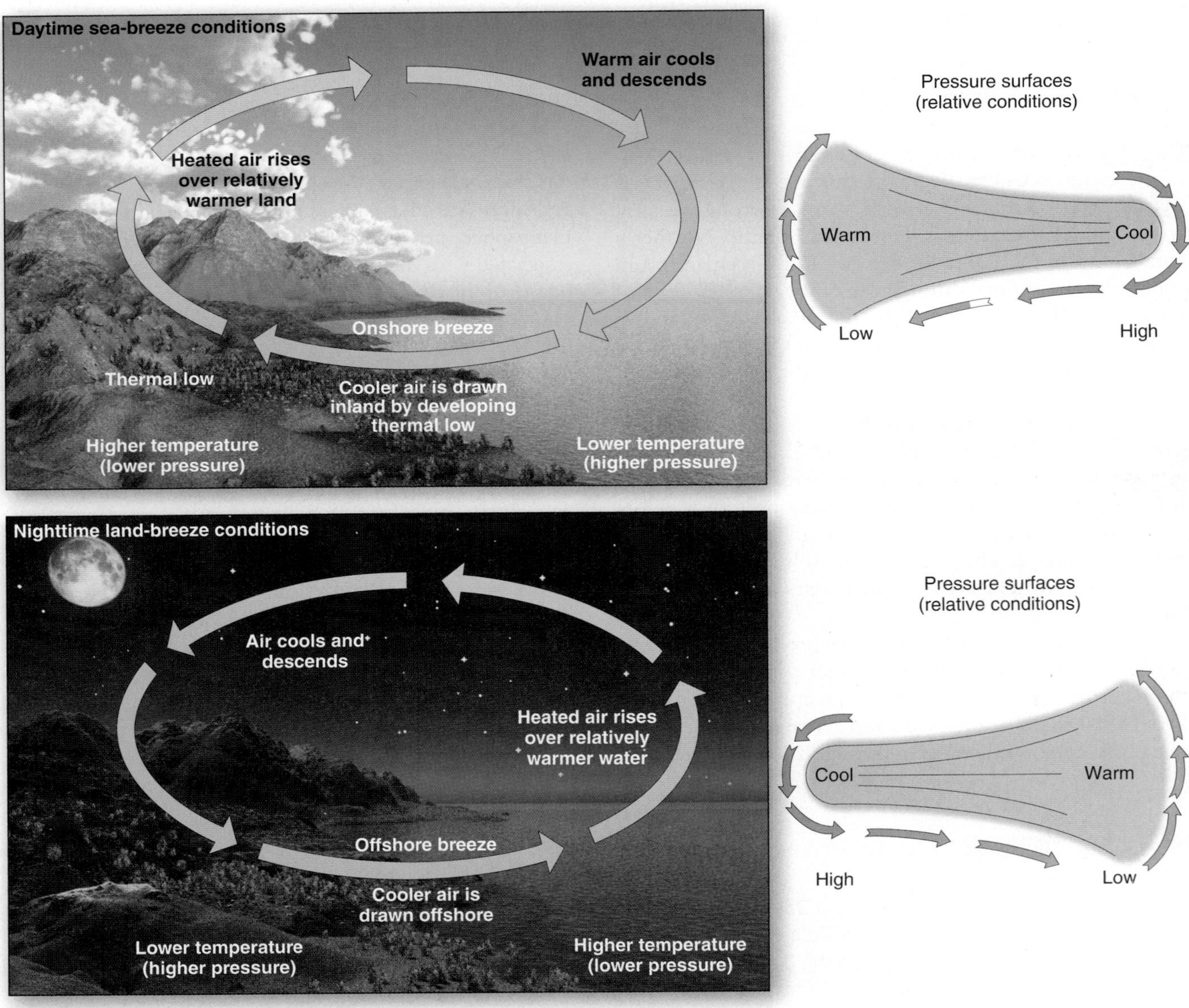

FIGURE 6.17 Land and sea breezes characteristic of day and night.

warms faster than the water offshore during the day. Because warm air is less dense, it rises and triggers an onshore flow of cooler marine air to replace the rising warm air—the flow is usually strongest in the afternoon. At night, inland areas cool (radiate heat energy) faster than offshore waters. As a result, the cooler air over the land subsides and flows offshore over the warmer water, where the air is lifted. This night pattern reverses the process that developed during the day.

Mountain and valley breezes result when mountain air cools rapidly at night, and valley air gains heat energy rapidly during the day (Figure 6.18). Thus, warm air rises upslope during the day, particularly in the afternoon; at night, cooler air subsides downslope into the valleys.

Santa Ana winds result from a pressure gradient generated when high pressure builds over the Great Basin of the western United States. A strong, dry wind flows out across the desert to southern California coastal areas. Compression heats the air as it flows from higher to lower elevations, and with increasing speed, it moves through constricting valleys to the southwest. These winds bring dust, dryness, and heat to populated areas near the coast and create dangerous wildfire conditions.

Katabatic winds, or gravity drainage winds, are of larger regional scale and are usually stronger than local winds, under certain conditions. An elevated plateau or highland is essential to their formation, where layers of air at the surface cool, become denser, and flow downslope. Such gravity winds are not specifically related to the pressure gradient. The ferocious winds that can blow off the ice sheets of Antarctica and Greenland are katabatic in nature.

Regionally, wind represents a significant source of renewable energy of increasing importance. Focus Study 6.1 briefly explores the wind-power resource.

(text continued on page 152)

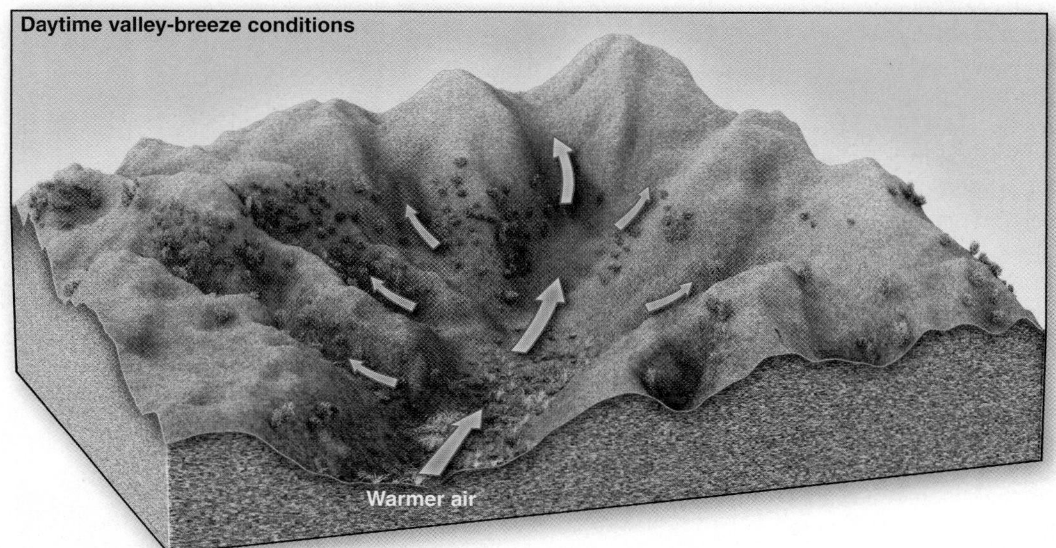

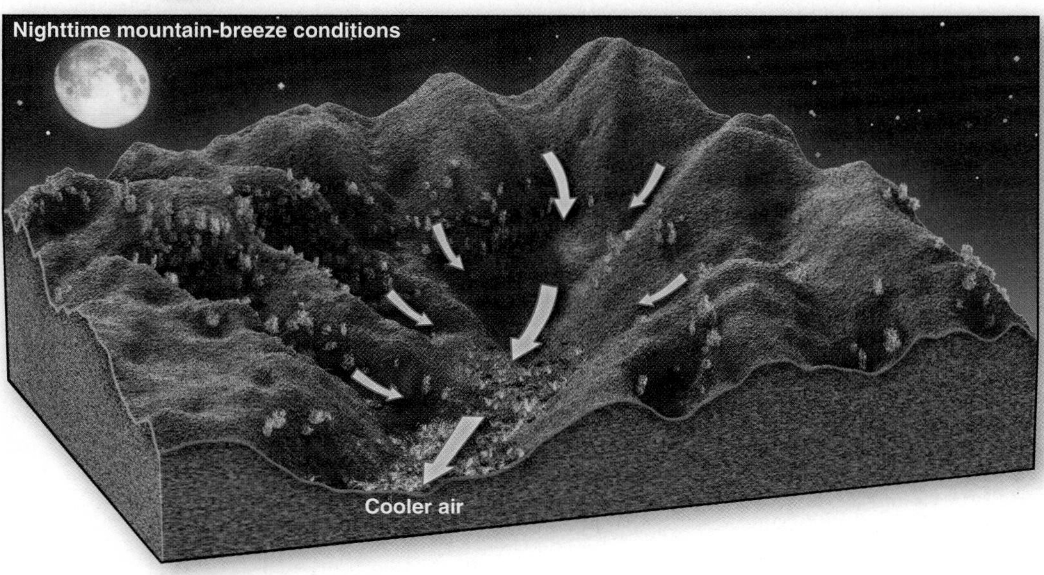

FIGURE 6.18 Pattern of mountain and valley breezes day and night.

FOCUS STUDY 6.1

Wind Power: An Energy Resource for the Present and Future

The principles of wind power are ancient, but the technology is modern and the benefits are substantial. Scientists estimate that wind as a resource could potentially produce many times more energy than is currently in demand on a global scale. Yet, despite the available technology, wind-power development continues to be slowed mainly by the changing politics of renewable energy.

The Nature of Wind Energy

Power generation from wind depends on site-specific characteristics of the wind resource. Favorable settings for consistent wind are (1) along coastlines influenced by trade winds and westerly winds; (2) where mountain passes constrict air flow and interior valleys develop thermal low-pressure areas, thus drawing air across the landscape; or (3) where localized winds occur, such as an expanse of relatively flat prairies, or in areas with katabatic or monsoonal winds. Many developing countries are generally located in areas blessed by such steady winds, such as the trade winds across the tropics.

Where winds are sufficient, electricity is generated by groups of wind turbines, in wind farms, or by individual installations (see the chapter-opening photo and Figure 6.1.1). If winds are reliable less than 25%–30% of the time, only small-scale use of wind power is economically feasible.

The potential of wind power in the United States is enormous. In the Midwest, power from the winds of North and South Dakota and Texas alone could meet all U.S. electrical needs. In the California Coast Ranges, land–sea breezes blow between the Pacific and Central Valley, peaking in intensity from April to October, which happens to match peak electrical demands for air conditioning during the hot summer months.

On the eastern shore of Lake Erie sits the closed Bethlehem Steel mill, contaminated with industrial waste until the site was redeveloped with eight 2.5-MW wind-power turbines in 2007 (Figure 6.1.2). This former "brownfield" site now supplies the equivalent of 75% of the electricity for Lackawanna, New York. Twelve turbines will soon be added, making this a 50-MW electrical generation facility, the largest urban installation in the country. An additional proposal for 500 MW of wind power from some 167 turbines is in the works for an installation offshore in Lake Erie. This

(a)

(b)

FIGURE 6.1.1 Wind farms: Kansas and Ascension Island
(a) The Gray County Wind Farm, 40 km (25 mi) west of Dodge City, near Montezuma, Kansas, generates 112 MW of electricity. All the towers and roads combined occupy only 6 acres of land, allowing the balance of 12,000 acres to stay in agricultural production. (b) Wind turbines on Ascension Island (at about 8° S 14° W in the Atlantic Ocean): British installation (left), U.S. installation (right). Four of the U.S. turbines offset the need for the importation of 300,000 barrels of oil per year. [Bobbé Christopherson.]

former steel town is using wind power to lift itself out of an economic depression with the slogan "Turning the Rust Belt into the Wind Belt."

While most U.S. wind power is land-based, offshore wind development has high potential. The proposed Cape Wind Farm near Cape Cod, Massachusetts, was recently approved as the nation's first offshore project. Proponents hope that despite the additional expense, offshore production will increase, especially along the eastern seaboard, where population centers are close together. At least 12 offshore projects are currently under consideration, most of them on the East Coast.

Land-based wind-power development is enhanced by the income it brings. Farmers in Iowa and Minnesota receive $2000 in annual income from electrical production by a leased turbine and $20,000 a year from electrical production by an owned turbine—requiring only one-quarter acre to site the wind machine. The Midwest is on the brink of an economic boom if this wind-energy potential is developed and transmission line capacity installed.

Wind-Power Status and Benefits

Wind-generated energy resources are the fastest-growing energy technology—capacity has risen worldwide in a continuing trend of doubling every three years. Total world capacity exceeded 159,213 MW (megawatts; 159.2 GW or gigawatts) by the end of 2009 from installations in 80 countries. Globally, the wind-power industry employs more than 550,000 people with global investments passing the U.S.$63 billion mark in 2009. Forecasts state that employment will top the million-worker mark in 2012. Average-size wind farms produce more than 100 MW.

In the United States, installed wind capacity exceeded 35,000 MW through 2009, an increase of 300% in 3 years. Installations are operating in 37 states, with the most capacity in Texas, Iowa, California, Washington, Oregon, and Illinois. This puts U.S. installed wind capacity highest on a global level, followed by China and Germany.

The European Wind Energy Association announced installed capacity exceeding 74,767 MW at the end of 2009, a 23% increase in 1 year. Germany has the most, followed by Spain, Italy, France, and the United Kingdom. The European Union has a goal of 20% of all energy from renewable sources by 2020.

The economic and social benefits from using wind resources are numerous. With all costs considered, wind energy is cost-competitive and actually cheaper than oil, coal, natural gas, and nuclear power. Wind power is renewable and does not cause adverse human health effects or environmental degradation. Relative to concerns about turbines hurting birds, a survey found that wind turbines ranked behind other causes of bird mortality when compared to communication towers, pesticides, high-tension lines, domestic cats, and collisions with buildings and windows.

To put numbers in meaningful perspective, every 10,000 MW of wind-generation capacity reduces carbon dioxide emissions by 33 million metric tons if it replaces coal or by 21 million metric tons if it replaces mixed fossil fuels. As an example, if countries rally and create a proposed $600 billion industry by installing 1,250,000 MW of wind capacity by 2020, that would supply 12% of global electrical needs.

Whether or not governments and transnational energy corporations support full-scale implementation of renewable resources such as wind, the energy realities of the near future leave no alternative, and global political realities affirm this urgency. By the middle of this century, wind-generated electricity could be routine, along with other renewable energy sources, conservation, and energy efficiency.

FIGURE 6.1.2 Wind turbines at a former industrial site. Known as "The Steel Wind," eight turbines deliver 20 MW of wind power in Lackawanna, New York, on the eastern shore of Lake Erie, south of Buffalo. The former industrial site is to expand to 50 MW. [Bobbé Christopherson.]

Worldwide, a variety of terrain produces local winds, which bear many local names. The *mistral* of the Rhône Valley in southern France can cause frost damage to vineyards as the cold north winds move over the region to the Gulf of Lions and the Mediterranean Sea. The frequently stronger *bora*, driven by the cold air of winter high-pressure systems inland, flows across the Adriatic Coast to the west and south. In Alaska, such winds are called the *taku*.

CRITICAL THINKING 6.2

Construct your own wind-power assessment report

Go to http://www.awea.org/ and http://www.ewea.org/, the Web sites of the American Wind Energy Association and the European Wind Energy Association, respectively. Sample the materials presented as you assess the potential for wind-generated electricity. What are your thoughts concerning this resource—its potential, the reasons for delays, and the competitive economics presented? Propose a brief action plan for the future of this resource.

Monsoonal Winds

Some regional wind systems seasonally change direction. Intense, seasonally shifting wind systems occur in the tropics over Southeast Asia, Indonesia, India, northern Australia, and equatorial Africa. A mild version of such monsoonal-type flow affects the extreme southwestern United States. These winds involve an annual cycle of returning precipitation with the summer Sun. Note the change in precipitation visible on the TRMM images in Figure 6.10; compare January and July.

The Arabic word for season, *mausim*, or **monsoon**, is the source of the name. Specific monsoonal weather, associated climate types, and vegetation regions are discussed in Chapters 8, 10, and 20. The location and size of the Asian landmass and its proximity to the Indian Ocean drive the monsoons of southern and eastern Asia (Figure 6.19). Also important to the generation of monsoonal flows are wind and pressure patterns in the upper-air circulation.

The extreme temperature range from summer to winter over the Asian landmass is due to its continentality (isolation from the modifying effects of the ocean). An intense high-pressure anticyclone dominates this continental landmass in winter (see Figure 6.9a and Figure 6.19a), whereas the equatorial low-pressure trough (ITCZ) dominates the central area of the Indian Ocean. This pressure gradient produces cold, dry winds from the Asian interior over the Himalayas and across India. Average temperatures range between 15°C and 20°C (between 60°F and 68°F) at lower elevations. These winds desiccate, or dry out, the landscape, giving way to hot weather from March through May.

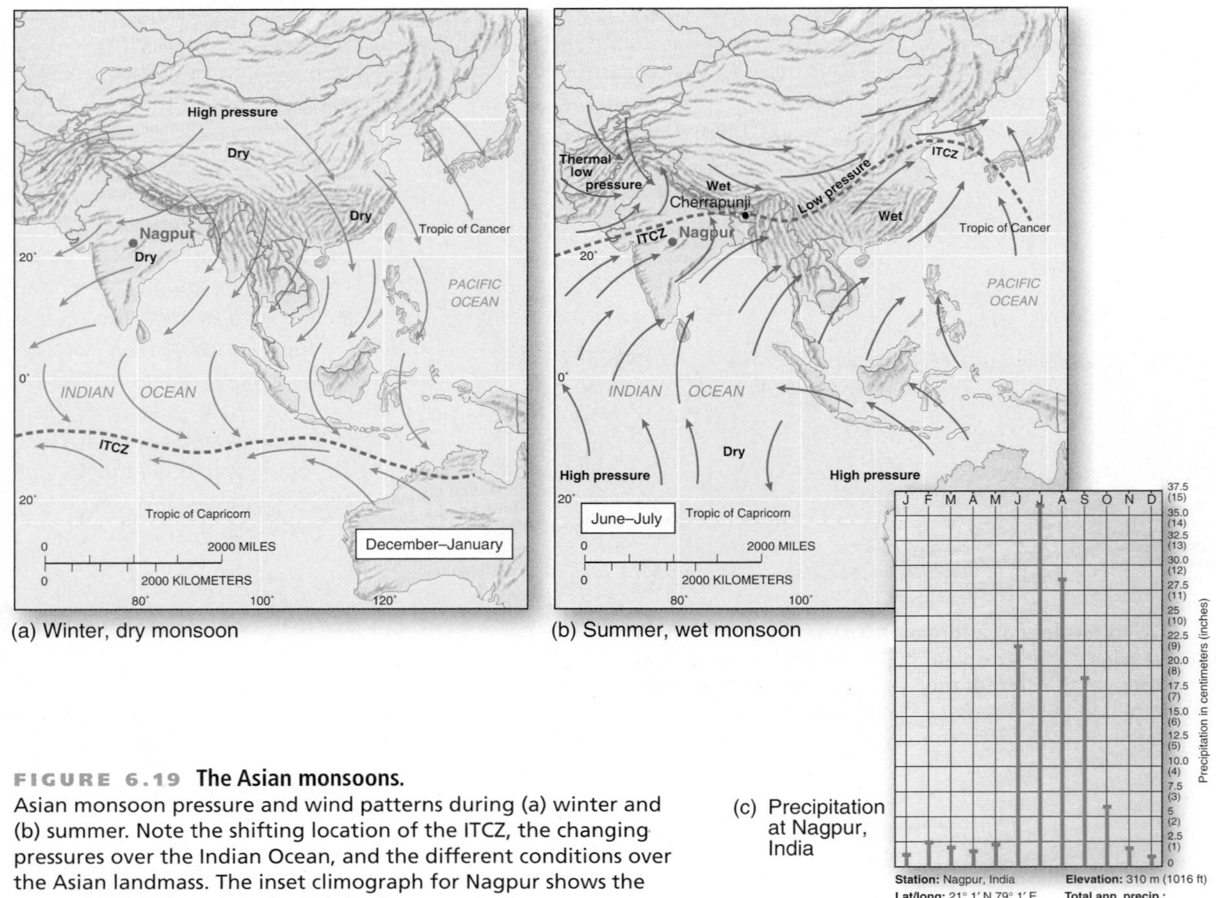

(a) Winter, dry monsoon

(b) Summer, wet monsoon

(c) Precipitation at Nagpur, India

FIGURE 6.19 The Asian monsoons.
Asian monsoon pressure and wind patterns during (a) winter and (b) summer. Note the shifting location of the ITCZ, the changing pressures over the Indian Ocean, and the different conditions over the Asian landmass. The inset climograph for Nagpur shows the severe contrast in seasonal precipitation.

During the June–September wet period, the subsolar point (direct overhead rays of sunlight) shifts northward to the Tropic of Cancer, near the mouths of the Indus and Ganges rivers. The ITCZ shifts northward over southern Asia, and the Asian continental interior develops a thermal low pressure, associated with high average temperatures (remember the summer warmth in Verkhoyansk, Siberia, from Chapter 5). Meanwhile, subtropical high pressure dominates the Indian Ocean, with a surface temperature of 30°C (86°F). As a result of this reversed pressure gradient, hot subtropical air sweeps over the warm ocean, producing extremely high evaporation rates (Figure 6.19b).

By the time this air mass and the convergence zone reach India, the air is laden with moisture in thunderous, dark clouds. When the monsoonal rains arrive from June to September, they are welcome relief from the dust, heat, and parched land of Asia's springtime. Likewise, the annual monsoon is an integral part of Indian music, poetry, and life. In the Himalayas, the monsoon brings snowfall. The photo of the GPS unit in Figure 1.20 was made near the summit of Mount Everest on May 20, 1998, at the end of the spring climbing season in Nepal just before the onset of the summer monsoon.

World-record rainfalls of the wet monsoon drench India. Cherrapunji, India, received both the second highest average annual rainfall (1143 cm, or 450 in.) and the highest single-year rainfall (2647 cm, or 1042 in.) on Earth.

The fact that the monsoons of southern Asia involve vast global pressure systems leads us to ask whether future climate change might affect present monsoonal patterns. Concern increases following events such as the monsoonal deluge on July 27, 2005, that drowned Mumbai (Bombay)—in only a few hours, 94.2 cm (37.1 in.) of rain fell, causing widespread flooding. Again in August 2007, extensive flooding from an intense monsoon occurred across India, and in 2010 Pakistan was devastated by record-breaking monsoon rains.

Scientists forecast higher rainfall if consideration is given to increases in carbon dioxide and its greenhouse warming alone. However, when consideration is given to increases in aerosols—principally sulfur compounds and black carbon—then a drop in precipitation of 7%–14% appears in the model. Air pollution reduces surface heating and therefore decreases the pressure differences at the heart of monsoonal flows. Considering that 70% of the annual precipitation for the entire region comes during the wet monsoon, any such changes would force difficult societal adjustments to reduced water resources. (See the satellite images in Figure 4.7 and review the text discussion.)

Oceanic Currents

The driving force for ocean currents is the frictional drag of the winds, thus linking the atmospheric and oceanic systems. Also important in shaping these currents is the interplay of the Coriolis force, density differences caused by temperature and salinity, the configuration of the continents and ocean floor, and astronomical forces (the tides).

Surface Currents

Figure 6.20 portrays the general patterns of major ocean currents. Because ocean currents flow over distance and through time, the Coriolis force deflects them. However, their pattern of deflection is not as tightly circular as that of the atmosphere. Compare this ocean-current map with the map showing Earth's pressure systems (see Figure 6.9), and you can see that ocean currents are driven by the circulation around subtropical high-pressure cells in

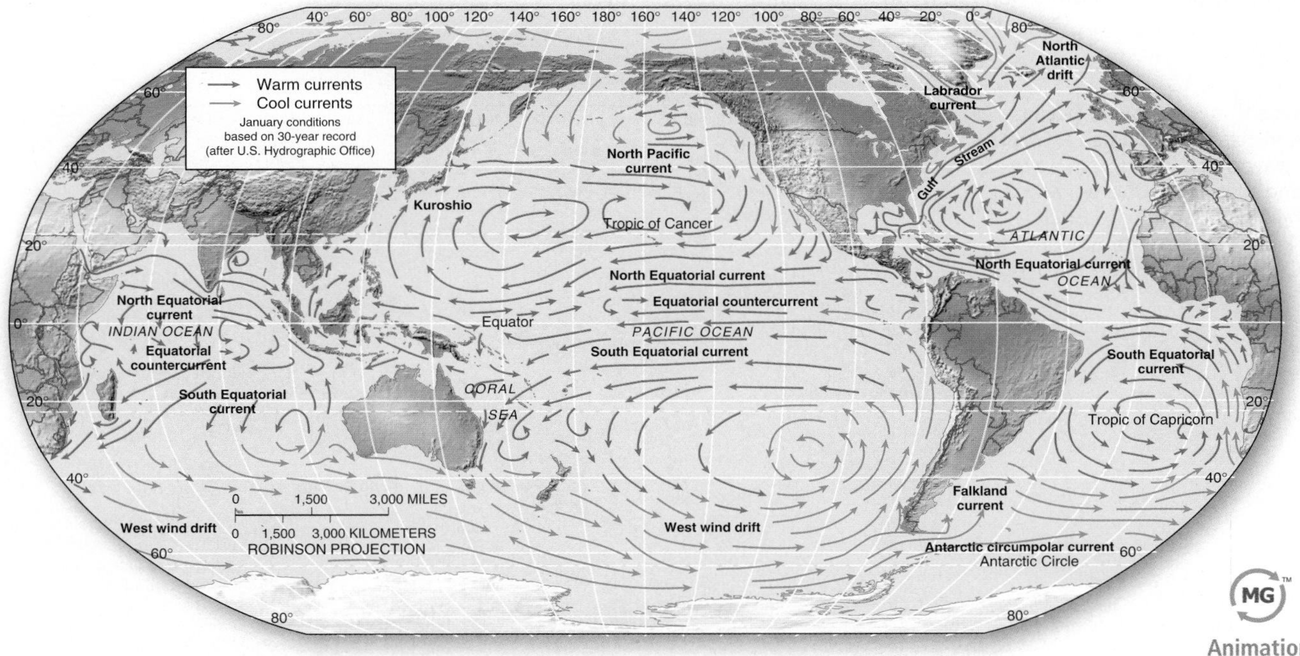

FIGURE 6.20 Major ocean currents. [After the U.S. Naval Oceanographic Office.]

MG Animation Ocean Circulation

both hemispheres. These circulation systems are known as *gyres* and generally appear to be offset toward the western side of each ocean basin. Remember, in the Northern Hemisphere, winds and ocean currents move clockwise about high-pressure cells; in the Southern Hemisphere, circulation is counterclockwise, evident on the map. You saw in *Geosystems Now* how these currents guided an oil-drilling platform to Tristan da Cunha, where it ran aground.

In sailing days, the Spanish galleons would leave San Blas and Acapulco, Mexico (16.5° N), and sail southwest, catching the northeast trade winds across the Pacific to Manila (14° N) in the Philippines. Goods would be traded and new commodities obtained. Leaving, the ships would move northward to catch the westerlies in the midlatitudes to be blown across the ocean to the shores of present-day Alaska or British Columbia and sail along the coast. The galleons would sail south along North America, fighting the rain, frequent fog, and right-hand Coriolis deflection pushing them away from the coast. The journey would end back in Mexico with cargo from the Orient. Thus, the history of the *Manila Galleons* is a lesson in oceanic currents around the Pacific gyre.

Examples of Gyre Circulation In 1992, a child at Dana Point, California (33.5° N), a small seaside community south of Los Angeles, placed a letter in a glass juice bottle and tossed it into the waves. Thoughts of distant lands and fabled characters filled the child's imagination as the bottle disappeared. The vast circulation around the Pacific high, clockwise-circulating gyre now took command (Figure 6.21).

Three years passed before ocean currents carried the message in a bottle to the coral reefs and white sands of Mogmog, a small island in Micronesia (7° N). A 7-year-old child there had found a pen pal from afar. Imagine the journey of that note from California—travelling through storms and calms, clear moonlit nights and typhoons, as it floated on ancient currents as the galleons once had.

In January 1994, a powerful storm ravaged a container ship from Hong Kong loaded with toys and other goods. One of the containers on board split apart in the wind off the coast of Japan, dumping nearly 30,000 rubber ducks, turtles, and frogs into the North Pacific. Westerly winds and the North Pacific current swept this floating cargo at up to 29 km (18 mi) a day across the ocean to the coast of Alaska, Canada, Oregon, and California. Other toys, still adrift, went through the Bering Sea and into the Arctic Ocean (this route is indicated by the dashed line in Figure 6.21).

On August 19, 2006, Tropical Storm Ioke formed about 1285 km (800 mi) south of Hawai'i. The storm's track moved across the Pacific Ocean over Wake Island and Johnston Atoll, becoming the strongest Super Typhoon in recorded history—a category 5, as we discuss in Chapter 8. Turning northward east of Japan and the Kamchatka Peninsula of Russia, the storm moved into higher latitudes. Typhoon Ioke remnants eventually crossed over the Aleutian Islands, reaching 55° N as an extratropical depression. The storm roughly followed the path around the Pacific gyre and along the track of the rubber duckies. Scientists are using incidents such as the message in a bottle and rubber duckies and such remarkable storm tracks to learn more about wind systems and ocean currents.

Equatorial Currents In Figure 6.20, you can see that trade winds drive the ocean surface waters westward in a concentrated channel along the equator. These equatorial currents are kept near the equator by the Coriolis force, which diminishes to zero at the equator. As these surface currents approach the western margins of the oceans, the

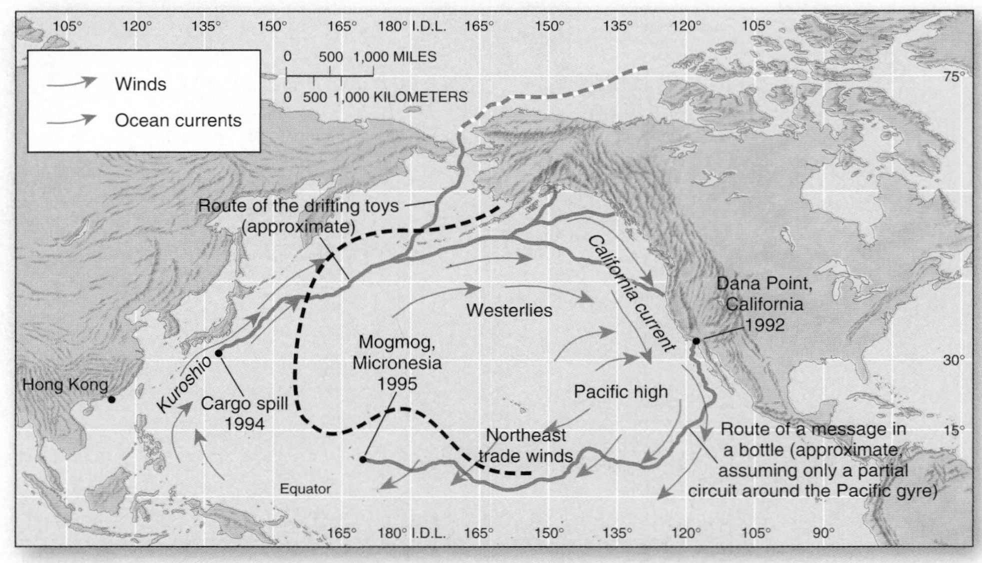

FIGURE 6.21 Pacific Ocean currents transport human artifacts. The travels of a message in a bottle and toy rubber duckies, and the path of Typhoon Ioke (dashed black line), in currents moving clockwise around the Pacific gyre.

water actually piles up against the eastern shores of the continents. The average height of this pileup is 15 cm (6 in.). This phenomenon is the **western intensification**.

The piled-up ocean water then goes where it can, spilling northward and southward in strong currents, flowing in tight channels along the eastern shorelines. In the Northern Hemisphere, the *Gulf Stream* and the *Kuroshio* (a current east of Japan) move forcefully northward as a result of western intensification. Their speed and depth increase with the constriction of the area they occupy. The warm, deep, clear water of the ribbon-like Gulf Stream (Figure 5.10) usually is 50–80 km (30–50 mi) wide and 1.5–2.0 km (0.9–1.2 mi) deep, moving at 3–10 kmph (1.8–6.2 mph). In 24 hours, ocean water can move 70–240 km (40–150 mi) in the Gulf Stream.

Upwelling and Downwelling Flows Where surface water is swept away from a coast, either by surface divergence (induced by the Coriolis force) or by offshore winds, an **upwelling current** occurs. This cool water generally is nutrient-rich and rises from great depths to replace the vacating water. Such cold upwelling currents exist off the Pacific coasts of North and South America and the subtropical and midlatitude west coast of Africa. These areas are some of Earth's prime fishing regions.

In other regions with an accumulation of water—such as at the western end of an equatorial current, or in the Labrador Sea, or along the margins of Antarctica—the excess water gravitates downward in a **downwelling current**. These are the deep currents that flow vertically and along the ocean floor and travel the full extent of the ocean basins, carrying heat energy and salinity.

Thermohaline Circulation—The Deep Currents

Differences in temperatures and salinity produce density differences important to the flow of deep currents. This is Earth's **thermohaline circulation**, and it is different from wind-driven surface currents. Travelling at slower speeds than surface currents, the thermohaline circulation hauls larger volumes of water. Figure 16.4 illustrates the ocean's physical structure and profiles of temperature, salinity, and dissolved gases; note the temperature and salinity differences with depth in this illustration.

To picture such a deep, salty current, imagine a continuous channel of water beginning with cold water downwelling in the North Atlantic and along Antarctica and upwelling in the Indian Ocean and North Pacific (Figure 6.22). Here it warms and then is carried in surface currents back to the North Atlantic. A complete circuit of the current may require 1000 years from downwelling, to deep cold current covering thousands of kilometers, to its reemergence at the surface elsewhere. Even deeper Antarctic bottom water flows northward in the Atlantic Basin beneath these currents.

These current systems appear to play a profound role in global climate; in turn, global warming has the potential to disrupt the downwelling in the North Atlantic and the thermohaline circulation. There is a freshening of ocean surface waters in both polar regions, contrasted

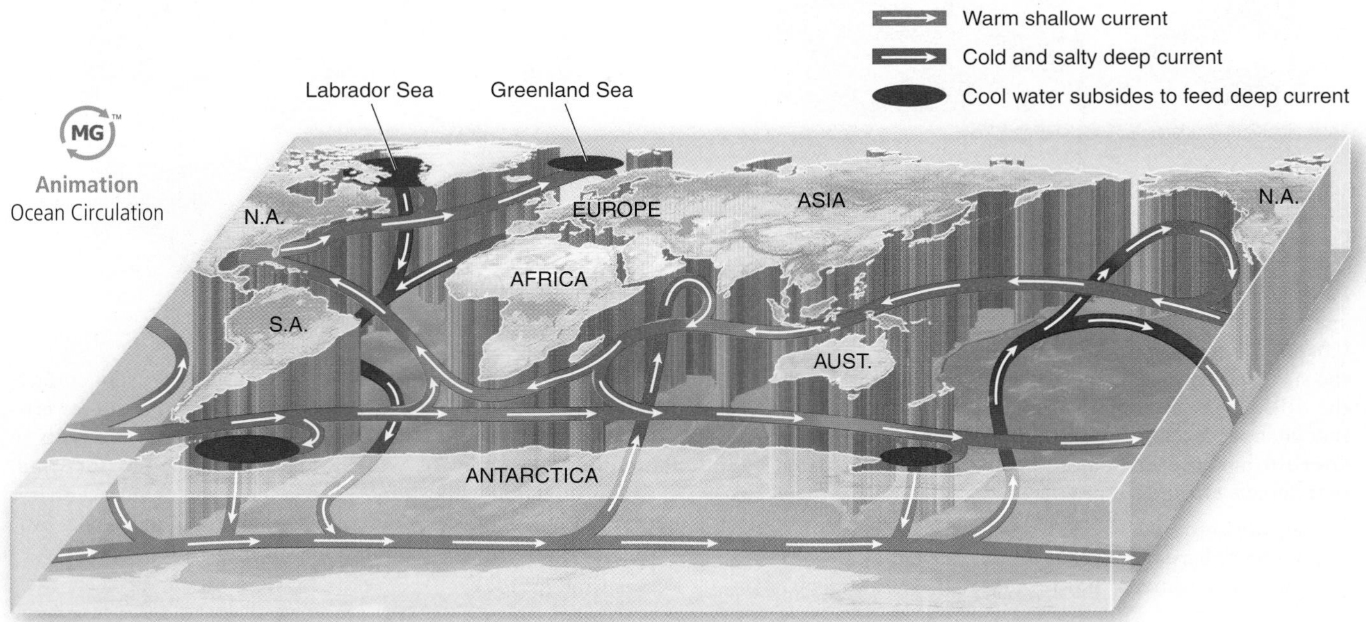

FIGURE 6.22 Deep-ocean thermohaline circulation.
This global circulation is a vast conveyor belt of water, drawing heat energy from warm shallow currents and transporting it for release in the depths of the ocean basins in cold, deep, salty currents. Four blue areas at high latitudes are where surface water cools, sinks, and feeds the deep circulation.

with large increases in salinity in the surface waters at lower latitudes. Increased rates of glacial, sea-ice, and ice-sheet melting produce these fresh, lower-density surface waters that ride on top of the denser saline water. The concern is that such changes in ocean temperature and salinity could dampen the rate of the North Atlantic Deep Water downwelling in the North Atlantic.

This is a vital research frontier because Earth's hydrologic cycle is still little understood. Research groups such as Scripps Institution of Oceanography (http://sio.ucsd.edu/), Woods Hole Oceanographic Institution (http://www.whoi.edu/), Institute for the Study of Earth, Oceans, and Space (http://www.eos.sr.unh.edu/), and Bedford Institute of Oceanography (http://www.bio.gc.ca/), to name only a few of many, are places to periodically check for updated information.

With this chapter, you have completed the inputs–actions–outputs through the chapters of PART I, *The Energy–Atmosphere System*. Along the way, we have examined the many ways these Earth systems impact our lives and the many ways human society is impacting these same systems. Given this foundation, we move on to PART II, *The Water, Weather, and Climate Systems*. Water and ice cover 71% of Earth's surface by area, making Earth the water planet. This water is driven by the pattern of energy we have been studying to produce weather, water resources, and climate.

KEY LEARNING CONCEPTS REVIEW

- ***Define*** **the concept of air pressure, and** ***describe*** **instruments used to measure air pressure.**

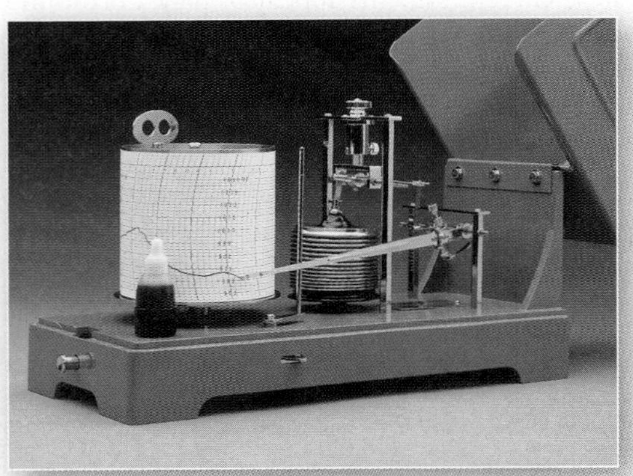

The weight (created by motion, size, and number of molecules) of the atmosphere is **air pressure,** which exerts an average force of approximately 1 kg/cm^2 (14.7 lb/in.^{2}), from Chapter 3. A **mercury barometer** measures air pressure at the surface (mercury in a tube—closed at one end and open at the other, with the open end placed in a vessel of mercury—that changes level in response to pressure changes), as does an **aneroid barometer** (a closed cell, partially evacuated of air, that detects changes in pressure).

air pressure (p. 130)
mercury barometer (p. 132)
aneroid barometer (p. 132)

1. How does air exert pressure? Describe the basic instrument used to measure air pressure. Compare the operation of two different types of instruments discussed.
2. What is normal sea-level pressure in millimeters? Millibars? Inches? Kilopascals?

- ***Define*** **wind, and** ***explain*** **how wind is measured, how wind direction is determined, and how winds are named.**

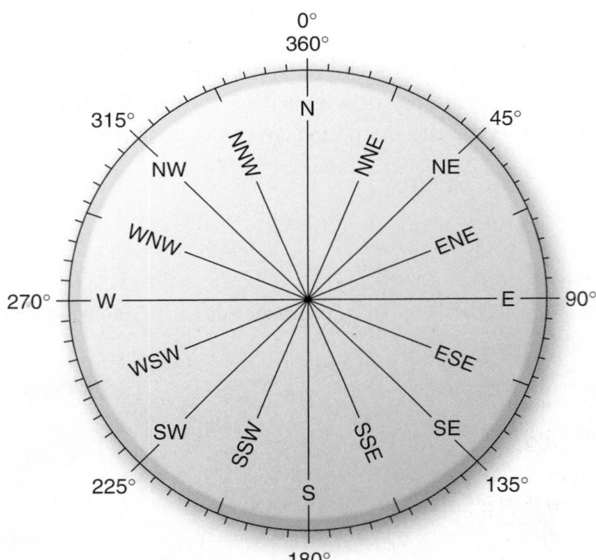

Wind is the horizontal movement of air across Earth's surface; turbulence adds wind updrafts and downdrafts, and thus a vertical component to the definition. Its speed is measured with an **anemometer** (a device with cups that are pushed by the wind) and its direction with a **wind vane** (a flat blade or surface that is directed by the wind). A descriptive scale useful in visually estimating wind speed is the traditional Beaufort wind scale.

wind (p. 132)
anemometer (p. 133)
wind vane (p. 133)

3. What is a possible explanation for the beautiful sunrises and sunsets during the summer of 1992 in North America? Relate your answer to global circulation.

4. Explain this statement: "The atmosphere socializes humanity, making the world a spatially linked society." Illustrate your answer with some examples.
5. Define wind. How is it measured? How is its direction determined?
6. Distinguish among primary, secondary, and tertiary general classifications of global atmospheric circulation.
7. What is the purpose of the Beaufort wind scale? Characterize winds given Beaufort numbers of 4, 8, and 12, and describe specific effects over both water and land.

■ ***Explain*** the four driving forces within the atmosphere—gravity, pressure gradient force, Coriolis force, and friction force—and ***locate*** the primary high- and low-pressure areas and principal winds.

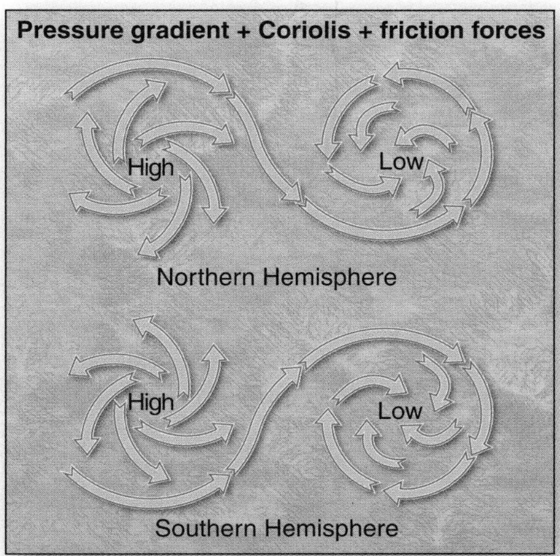

The pressure that Earth's gravitational force exerts on the atmosphere is virtually uniform worldwide. Winds are driven by the **pressure gradient force,** as air moves from areas of high pressure to areas of low pressure. Winds are deflected by the **Coriolis force,** an apparent deflection in the path of winds or ocean currents caused by the rotation of Earth and surface distance from Earth's axis as the gravitational force and centrifugal force work in opposition, deflecting objects to the right in the Northern Hemisphere and to the left in the Southern Hemisphere. Finally, winds are dragged by the **friction force,** in which Earth's varied surfaces exert a drag on wind movements in opposition to the pressure gradient. Maps portray air pressure patterns using the **isobar**—an isoline that connects points of equal pressure. A combination of the pressure gradient and Coriolis force alone produces **geostrophic winds,** which move parallel to isobars, characteristic of winds above the surface frictional layer.

Winds descend and diverge, spiraling outward to form an **anticyclone** (clockwise in the Northern Hemisphere), and they converge and ascend, spiraling upward to form a **cyclone** (counterclockwise in the Northern Hemisphere). The pattern of high and low pressures on Earth in generalized belts in each hemisphere produces the distribution of specific wind systems. These primary pressure regions are the **equatorial low-pressure trough,** the weak **polar high-pressure cells** (at both the North and the South poles), the **subtropical high-pressure cells,** and the **subpolar low-pressure cells**.

All along the equator winds converge into the equatorial low, creating the **intertropical convergence zone (ITCZ)**. Air rises along the equator and descends in the subtropics in each hemisphere. The winds returning to the ITCZ from the northeast in the Northern Hemisphere and from the southeast in the Southern Hemisphere produce the **trade winds**.

Winds flowing out of the subtropics to higher latitudes produce the **westerlies** in each hemisphere. The subtropical high-pressure cells on Earth, generally between 20° and 35° in each hemisphere, are variously named the **Bermuda high, Azores high**, and **Pacific high**.

Along the polar front and the series of low-pressure cells, the **Aleutian low** and **Icelandic low** dominate the North Pacific and Atlantic, respectively. This region of contrast between colder air toward the poles and warmer air toward the equator is the **polar front**. The weak and variable **polar easterlies** diverge from the polar high-pressure cells, particularly the **Antarctic high**.

pressure gradient force (p. 134)
Coriolis force (p. 134)
friction force (p. 134)
isobar (p. 136)
geostrophic winds (p. 139)
anticyclone (p. 139)
cyclone (p. 139)
equatorial low-pressure trough (p. 141)
polar high-pressure cells (p. 141)
subtropical high-pressure cells (p. 141)
subpolar low-pressure cells (p. 141)
intertropical convergence zone (ITCZ) (p. 141)
trade winds (p. 142)
westerlies (p. 143)
Bermuda high (p. 143)
Azores high (p. 143)
Pacific high (p. 143)
Aleutian low (p. 143)
Icelandic low (p. 143)
polar front (p. 144)
polar easterlies (p. 144)
Antarctic high (p. 144)

8. What does an isobaric map of surface air pressure portray? Contrast pressures over North America for January and July.
9. Describe the effect of the Coriolis force. Explain how it apparently deflects atmospheric and oceanic circulations.
10. What are geostrophic winds, and where are they encountered in the atmosphere?
11. Describe the horizontal and vertical air motions in a high-pressure anticyclone and in a low-pressure cyclone.
12. Construct a simple diagram of Earth's general circulation; begin by labelling the four principal pressure belts or zones, and then add arrows between these pressure systems to denote the three principal wind systems.
13. How is the intertropical convergence zone (ITCZ) related to the equatorial low-pressure trough? How does it appear on satellite images of accumulated precipitation for January and July in Figure 6.10?
14. Characterize the belt of subtropical high pressure on Earth: Name the specific cells. Describe the generation of westerlies and trade winds. Discuss sailing conditions.
15. What is the relation among the Aleutian low, the Icelandic low, and migratory low-pressure cyclonic storms in North America? In Europe?

■ ***Describe*** upper-air circulation and ***define*** the jet streams.

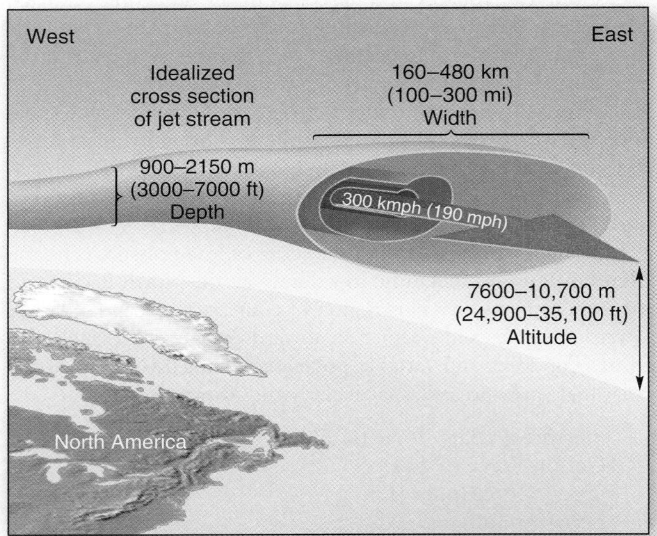

A **constant isobaric surface,** or surface along which the same pressure, such as 500 mb, occurs regardless of altitude, describes air pressure in the middle and upper troposphere. The height of this surface above the ground forms ridges and troughs that support the development of and help sustain surface pressure systems; surface lows are sustained by divergence aloft, and surface highs are sustained by convergence aloft.

Vast, flowing, longwave undulations in these upper-air westerlies form wave motions, the **Rossby waves**. Prominent streams of high-speed westerly winds in the upper-level troposphere are the **jet streams**. Depending on their latitudinal position in either hemisphere, they are termed the polar jet stream or the subtropical jet stream.

constant isobaric surface (p. 144)
Rossby waves (p. 146)
jet streams (p. 146)

16. What is the relation between wind speed and the spacing of isobars?
17. How is the constant isobaric surface (ridges and troughs) related to surface pressure systems? To divergence aloft and surface lows? To convergence aloft and surface highs?
18. Relate the jet-stream phenomenon to general upper-air circulation. How is the presence of this circulation related to airline schedules for the trip from New York to San Francisco and for the return trip to New York?

■ ***Summarize*** several multi-year oscillations of air temperature, air pressure, and circulation in the Arctic, Atlantic, and Pacific oceans.

Several system fluctuations that occur in multi-year or shorter periods are important in the global circulation picture. The most famous of these is the *El Niño–Southern Oscillation (ENSO)* phenomenon. A north–south fluctuation of atmospheric variability marks the *North Atlantic Oscillation (NAO)* as pressure differences between the Icelandic low and the Azores high in the Atlantic alternate between weaker and stronger pressure gradients. The *Arctic Oscillation (AO)* is the variable fluctuation between middle- and high-latitude air mass conditions over the Northern Hemisphere. The AO is associated with the NAO, especially in

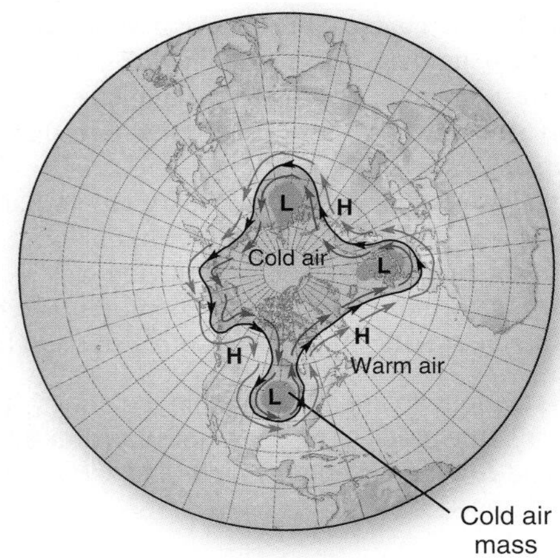

winter, and the two phases of the AO Index correlate to the two-phased NAO Index. During the winter of 2009–2010, the AO Index was at the lowest negative phase since 1970.

Across the Pacific Ocean, the *Pacific Decadal Oscillation (PDO)* involves variability between two regions of sea-surface temperatures and related air pressure: the northern and tropical western Pacific (region #1) and the area of the eastern tropical Pacific, along the West Coast (region #2). The PDO switches between positive and negative phases in 20- to 30-year cycles.

19. What phases are identified for the AO and NAO Indexes? What winter weather conditions generally affect the eastern United States during each phase? What happened during the 2009–2010 winter season in the Northern Hemisphere?
20. What is the apparent relation between the PDO and the strength of El Niño events? Between PDO phases and the intensity of drought in the southwestern United States?

■ ***Explain*** several types of local winds and regional monsoons.

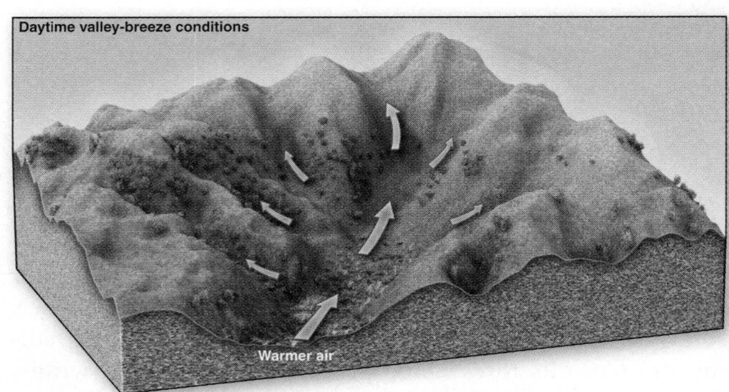

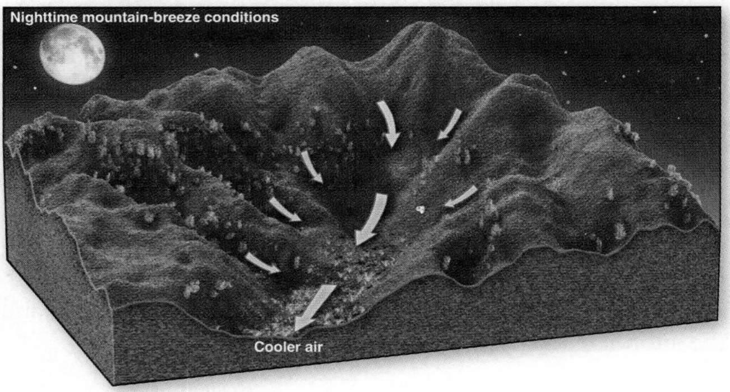

Different heating characteristics of land and water surfaces create **land–sea breezes**. Temperature differences during the day and evening between valleys and mountain summits cause **mountain–valley breezes**. **Katabatic winds,** or gravity drainage winds, are of larger regional scale and are usually stronger than mountain–valley breezes, under certain conditions. An elevated plateau or highland is essential, where layers of air at the surface cool, become denser, and flow downslope.

Intense, seasonally shifting wind systems occur in the tropics over Southeast Asia, Indonesia, India, northern Australia, equatorial Africa, and southern Arizona. These winds involve an annual cycle of returning precipitation with the summer Sun and carry the Arabic word for season, *mausim*, or **monsoon**. The location and size of the Asian landmass and its proximity to the Indian Ocean and the seasonally shifting ITCZ drive the monsoons of southern and eastern Asia.

land–sea breezes (p. 148)
mountain–valley breezes (p. 149)
katabatic winds (p. 149)
monsoon (p. 152)

21. People living along coastlines generally experience variations in winds from day to night. Explain the factors that produce these changing wind patterns.
22. The arrangement of mountains and nearby valleys produces local wind patterns. Explain the day and night winds that might develop.
23. Describe the seasonal pressure patterns that produce the Asian monsoonal wind and precipitation patterns. Contrast January and July conditions.
24. This chapter presents wind power as well developed and cost effective, given the information presented and your additional critical thinking work, what conclusions or observations have you made?

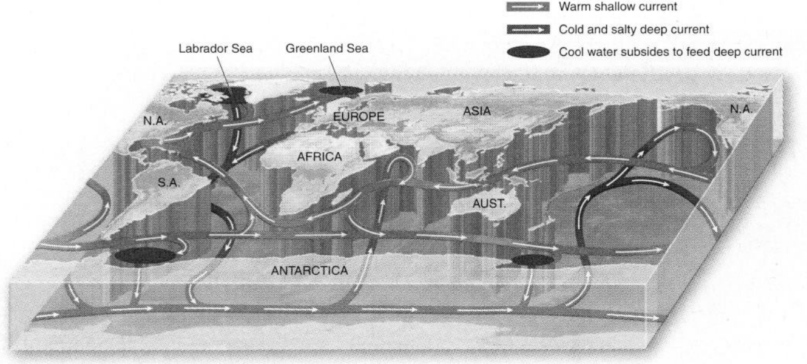

■ ***Discern*** the basic pattern of Earth's major surface ocean currents and deep thermohaline circulation.

Ocean currents are primarily caused by the frictional drag of wind and occur worldwide at varying intensities, temperatures, and speeds, both along the surface and at great depths in the oceanic basins. The circulation around subtropical high-pressure cells in both hemispheres is notable on the ocean circulation map—these gyres are usually offset toward the western side of each ocean basin.

The trade winds converge along the ITCZ and push enormous quantities of water that pile up along the eastern shore of continents in a process known as the **western intensification**. Where surface water is swept away from a coast, either by surface divergence (induced by the Coriolis force) or by offshore winds, an **upwelling current** occurs. This cool water generally is nutrient-rich and rises from great depths to replace the vacating water. In other portions of the sea where there is an accumulation of water, the excess water gravitates downward in a **downwelling current**. These currents generate important mixing currents that flow along the ocean floor and travel the full extent of the ocean basins, carrying heat energy and salinity.

Differences in temperatures and salinity produce density differences important to the flow of deep, sometimes vertical, currents; this is Earth's **thermohaline circulation**. Travelling at slower speeds than wind-driven surface currents, the thermohaline circulation hauls larger volumes of water. There is scientific concern that increased surface temperatures in the ocean and atmosphere, coupled with climate-related changes in salinity, can alter the rate of thermohaline circulation in the oceans.

western intensification (p. 155)
upwelling current (p. 155)
downwelling current (p. 155)
thermohaline circulation (p. 155)

25. Define the western intensification. How is it related to the Gulf Stream and the Kuroshio Current?
26. Define the western intensification. How is it related to the Gulf Stream and the Kuroshio Current?
27. Where on Earth are upwelling currents experienced? What is the nature of these currents? Where are the four areas of downwelling that feed these dense bottom currents?
28. What is meant by deep-ocean thermohaline circulation? At what rates do these currents flow? How might this circulation be related to the Gulf Stream in the western Atlantic Ocean?
29. Relative to Question 28, what relation do these deep currents have to global warming and possible climate change?

MASTERING GEOGRAPHY

Visit www.masteringgeography.com for several figure animations, study quizzes, flashcards, visual glossary, case studies, career links, textbook reference maps, geography links, RSS Feeds, and an ebook version of *Geosystems*. Included are many geography web links to interesting Internet sources that support this chapter.

PART II

The Water, Weather, and Climate Systems

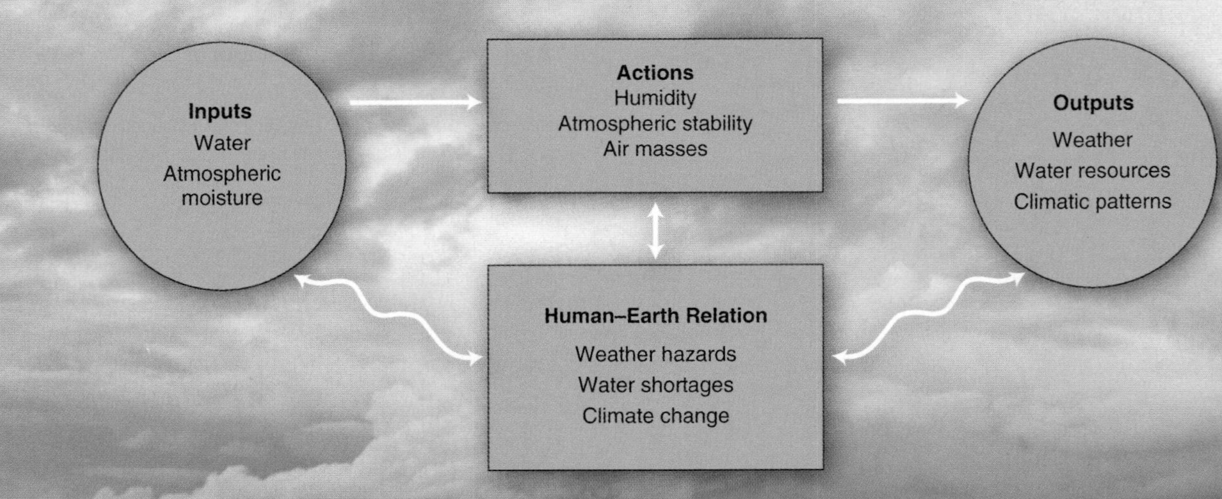

Dramatic cumulonimbus clouds forming and heavy downpour, evidence of large quantities of heat energy. [Bobbé Christopherson.]

Earth is the water planet. Chapter 7 describes the remarkable qualities and properties water possesses, where it occurs, and how it is distributed. We see the daily dynamics of the atmosphere—the powerful interaction of moisture and energy, the resulting stability and instability, and the variety of cloud forms—all important to understanding weather. Chapter 8 examines weather and its causes. Topics include the interaction of air masses, the daily weather map, and analysis of violent phenomena in thunderstorms, tornadoes, and hurricanes, and the recent trends for each of these.

Chapter 9 explains water circulation on Earth in the hydrologic cycle—an output of the water–weather system. We examine the water-budget concept, which is useful in understanding soil-moisture and water-resource relationships on global, regional, and local scales. Potable (drinking) water is emerging as the critical global political issue this century. In Chapter 10, we see the spatial implications over time of the energy–atmosphere and water–weather systems and the output of Earth's climatic patterns. In this way, Chapter 10 interconnects all the system elements from Chapters 2 through 9. PART II closes with a discussion of global climate-change science, a look at present conditions, and a forecast of future climate trends.

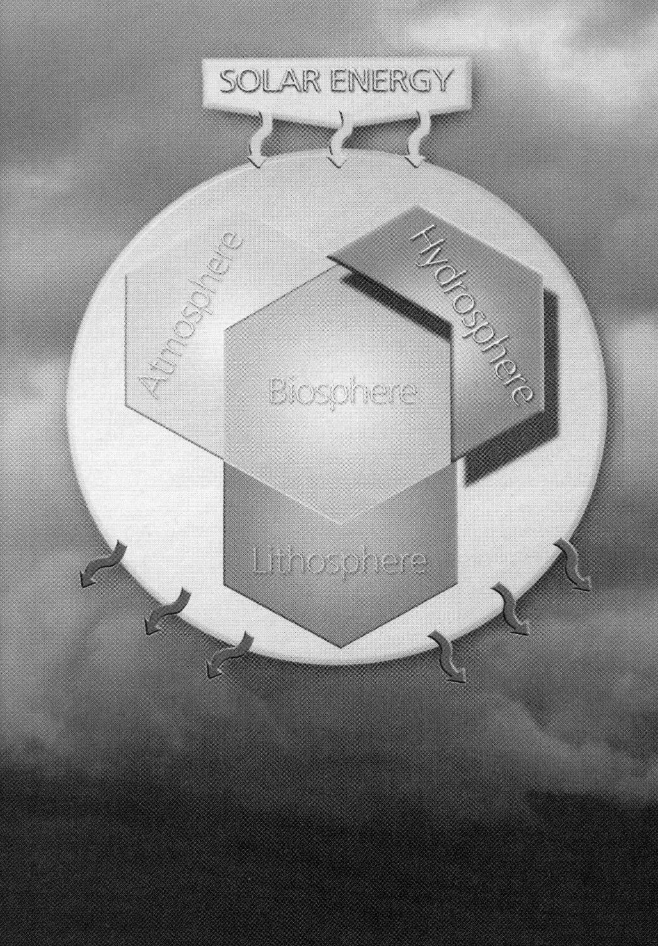

CHAPTER 7

Water and Atmospheric Moisture

A marine layer fog contributes important moisture to this pasture and trees on the island of St. Helena in the middle of the Atlantic Ocean. [Bobbé Christopherson.]

KEY LEARNING CONCEPTS

After reading the chapter, you should be able to:

- ***Describe*** the origin of Earth's waters, ***define*** the quantity of water that exists today, and ***list*** the locations of Earth's freshwater supply.
- ***Describe*** the heat properties of water, and ***identify*** the traits of its three phases: solid, liquid, and gas.
- ***Define*** humidity and the expressions of the relative humidity concept, and ***explain*** dew-point temperature and saturated conditions in the atmosphere.
- ***Define*** atmospheric stability, and ***relate*** it to a parcel of air that is ascending or descending.
- ***Illustrate*** three atmospheric conditions—unstable, conditionally unstable, and stable—with a simple graph that relates the environmental lapse rate (ELR) to the dry adiabatic rate (DAR) and moist adiabatic rate (MAR).
- ***Identify*** the requirements for cloud formation, and ***explain*** the major cloud classes and types, including fog.

GEOSYSTEMS NOW

Earth's Lakes Provide an Important Warming Signal

All lakes combined contain only 0.33% of freshwater on Earth, yet lakes are significant regionally for water, food, livelihoods, and transport. Most lakes, including the seven major lakes on Earth listed in Table 7.1, are presently being impacted by climate change.

Increasing air temperatures are having a range of effects on lakes throughout the world, from rising lake levels in response to melting of glacial ice, to falling lake levels associated with drought and high evaporation rates. Such warming is damaging to lake ecology and health. Changes in the thermal structure of a lake tend to block the normal mixing between deep and surface waters.

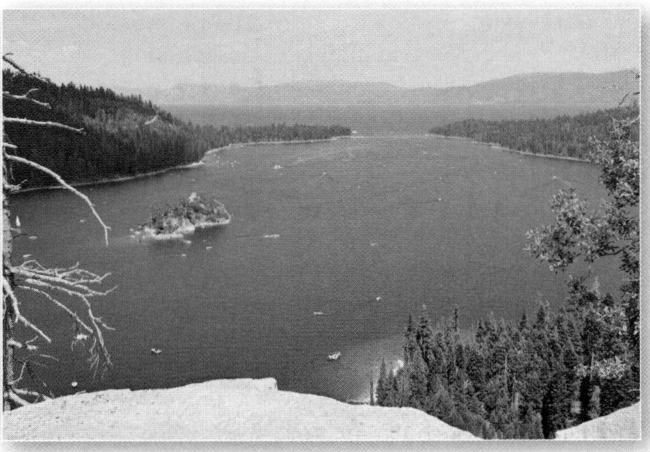

FIGURE GN 7.1 Recent warming is affecting ecosystems such as Lake Tahoe. [Bobbé Christopherson.]

Normally, stratification occurs in the summer in most lakes, with warmer water at the surface and cooler water at depth. This leads to a depletion of nutrients in the shallows and loss of dissolved oxygen at depth by late summer. In the fall, as the surface cools, water sinks, creating a turnover that replenishes surface nutrients and dissolved oxygen levels throughout the lake. In some lakes, wind assists such mixing.

As regional temperatures warm, the summerlike stratification appears earlier in the year and persists later in the fall, and mixing slows or stops. Cold-water species become stressed and invasive warmer-water species flourish.

Lake Tahoe in the Sierra Nevada mountain range along the California–Nevada border is warming at twice the rate of the surrounding region, at about 1.3° C (2.3° F) per decade (Figure GN 7.1). Some 7300 measurements collected over the past 35 years confirm the trend. The rate of warming is highest in the upper 10 m (33 ft), causing increased thermal stability and resistance to mixing. Invasive large-mouth bass, carp, and Asian clam are on the rise, while cold-water species decline. Such temperature response is seen in the declining snowpack in the surrounding mountains, forecasted to drop some 80% this century.

Lake Tanganyika in East Africa is surrounded by an estimated 10 million people, with most depending on its fish stocks, especially freshwater sardines, for food. Present water temperatures are unprecedented, reaching 26°C (79°F), highest in a 1500-year climate record exposed by lake sediment cores. Mixing of surface and deep waters is necessary to replenish nutrients in the upper 200 m (656 ft) of the lake, where the sardines reside. The surface warming and fixed stratification even resist mixing by wind, and scientists fear that fish stocks will continue to decline.

The largest and deepest lake on Earth, Lake Baykal in Russia, is also one of the richest in biological diversity for marine life. Here the effects of warming temperatures are focused around melting ice. Some 60 years of research show that regional temperatures today are higher, especially through winter, with increases in precipitation in response to the warmer air. These changes shorten the amount of time the lake is frozen over and have led to a decrease in the thickness and transparency of the ice.

Microscopic algae are the basis of Lake Baykal's food web and have adapted to the lake's ice cover for growth and reproduction. Present changes in ice thickness and transparency affect the spring algal blooms and decrease growth rates, thus decreasing food for crustaceans that feed on the algae, which decreases food available for fish as the effect moves up the food chain. With earlier spring melting, Lake Baykal's seal population suffers, as this unique freshwater species mates and gives birth on the ice. Climate change is affecting Lake Baykal in other ways, including increased nutrient inputs and industrial pollution from melting permafrost and increased regional precipitation.

These examples illustrate some of the ecosystem impacts of climate change on Earth's lakes. Many of these topics involve "Water and Atmospheric Moisture," Chapter 7.

Water is critical to our daily lives and is an extraordinary compound in nature. In the Solar System, water occurs in significant quantities only on our planet—water covers 71% of Earth by area. Several satellite probes detected ice beneath the poles of the Moon. Elsewhere, three orbiting spacecraft and two robot landers imaged and detected flowing water early in Martian history, and one craft landing in Martian high latitudes found itself amongst ice-caused patterned ground. Scientists are finding evidence on Mars of existing subsurface water, polar water and ice, and modern erosional features caused by water. Farther from the Sun, icy expanses can be seen on two of Jupiter's moons, Europa and Calisto—exciting discoveries about water in our Solar System.

Pure water is colorless, odorless, and tasteless; yet, because it is a solvent (dissolves solids), pure water rarely occurs in nature. Water weighs 1 g/cm^3 (gram per cubic centimeter), or 1 kg/L (kilogram per liter). In the English system, water weighs 62.3 lb/ft^3, or 8.337 lb/gal.

Water constitutes nearly 70% of our bodies by weight and is the major ingredient in plants, animals, and our food. A human can survive 50 to 60 days without food, but only 3 or 5 days without water. The water we use must be adequate, both in quantity and in quality, for its many tasks—everything from personal hygiene to vast national water projects. Water is the medium of life.

In this chapter: We examine water on Earth and the dynamics of atmospheric moisture and stability—the essentials of weather. The key questions answered include these: What is the origin of water? How much water is there? Where is water located? An irony exists with this most common compound because it possesses such uncommon physical characteristics. Water's unique heat properties and existence in all three states in nature are critical to powering Earth's weather systems. Condensation of water vapor and atmospheric conditions of stability and instability are key to cloud formation. We end the chapter with clouds, our beautiful indicators of atmospheric conditions.

Water on Earth

Earth's hydrosphere contains about 1.36 billion cubic kilometers of water (specifically, 1,359,208,000 km^3, or 326,074,000 mi^3). According to scientific evidence, much of Earth's water originated from icy comets and hydrogen- and oxygen-laden debris that were part of the planetesimals that coalesced to form the planet. In 2007, the orbiting Spitzer Space Telescope observed for the first time the presence of water vapor and ice as planets form in a system 1000 light-years from Earth. Such discoveries prove water to be abundant throughout the Universe. As a planet forms, water from within migrates to its surface and outgasses.

Outgassing is a continuing process by which water and water vapor emerge from layers deep within and below the crust, 25 km (15.5 mi) or more below Earth's surface. Figure 7.1 presents two such areas, among many sites worldwide, in Iceland and Wyoming.

In the early atmosphere, massive quantities of outgassed water vapor condensed and then fell to Earth in torrential rains. For water to remain on Earth's surface, land temperatures had to drop below the boiling point of 100°C (212°F), something that occurred about 3.8 billion years ago. The lowest places across the face of Earth then began to fill with water—first ponds, then lakes and seas, and eventually ocean-sized bodies of water. Massive flows of water washed over the landscape, carrying both dissolved and solid materials to these early seas and oceans. Outgassing of water has continued ever since and is visible in volcanic eruptions, geysers, and seepage to the surface.

Worldwide Equilibrium

Today, water is the most common compound on the surface of Earth, having attained the present volume approximately 2 billion years ago. This quantity has remained relatively constant, even though water is continuously being lost from the system. Water is lost when it dissociates into hydrogen and oxygen and the hydrogen escapes Earth's gravity to space or when it breaks down and forms

(a)

(b)

FIGURE 7.1 Water outgassing from the crust.
Outgassing of water from Earth's crust in geothermal areas: (a) in southern Iceland west of where the Eyjafjallajökull volcano erupted in 2010; and (b) thermal features, travertine deposits, and hot mineral pools visible from the trail in Yellowstone National Park, Wyoming. [(a) Bobbé Christopherson; (b) author.]

new compounds with other elements. As pristine water not previously at the surface emerges from within Earth's crust, it replaces lost water in the system. The net result of these water inputs and outputs is that Earth's hydrosphere is in a steady-state equilibrium in terms of quantity.

Despite this overall net balance in water quantity, worldwide changes in sea level do occur. The concept **eustasy** describes the global sea-level condition. Eustatic changes relate to the water volume in the oceans and not changes in the overall quantity of planetary water. Some of these changes result when the amount of water stored in glaciers and ice sheets varies; these are **glacio-eustatic** factors (see Chapter 17). In cooler times, as more water is bound up in glaciers (on mountains worldwide and in high latitudes) and in ice sheets (Greenland and Antarctica), sea level lowers. In warmer times, less water is stored as ice, so sea level rises.

Some 18,000 years ago, during the most recent ice age, sea level was more than 100 m (330 ft) lower than it is today; 40,000 years ago it was 150 m (about 500 ft) lower. Over the past 100 years, mean sea level has risen by 20–40 cm (8–16 in.) and is still rising worldwide at an accelerating pace as higher temperatures melt more ice and cause ocean water to thermally expand.

Isostasy refers to actual vertical physical movement in landmasses, such as continental uplift or subsidence. Such landscape changes cause apparent changes in sea level relative to coastal environments. Chapter 11 discusses such land elevation changes.

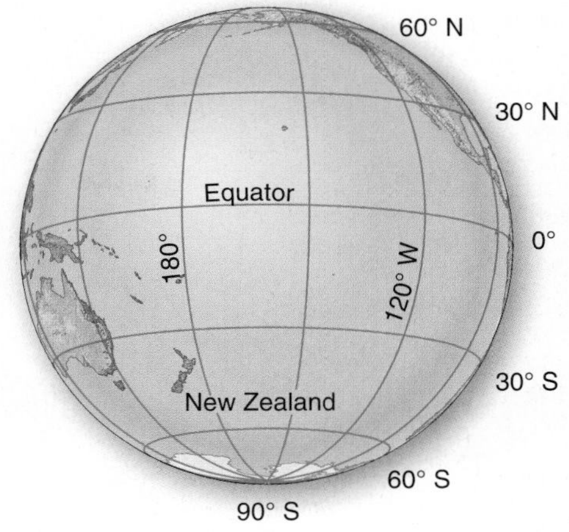

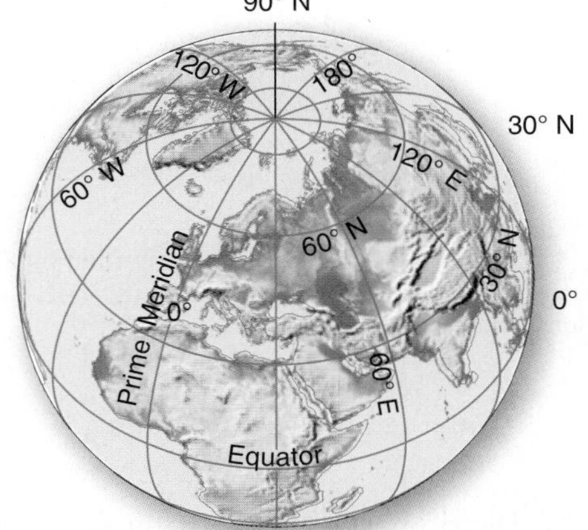

FIGURE 7.2 Land and water hemispheres. Two perspectives that roughly illustrate Earth's ocean hemisphere and land hemisphere.

Distribution of Earth's Water Today

From a geographic point of view, ocean and land surfaces are distributed unevenly. If you examine a globe, it is obvious that most of Earth's continental land is in the Northern Hemisphere, whereas water dominates the Southern Hemisphere. In fact, when you look at Earth from certain angles, it appears to have an *oceanic hemisphere* and a *land hemisphere* (Figure 7.2).

The present location of all of Earth's liquid and frozen water—whether fresh or saline, surface or underground—is indicated in Figure 7.3 on page 166. The oceans contain 97.22% of all water (Figure 7.3a). A table in the figure lists and details the four oceans—Pacific, Atlantic, Indian, and Arctic. The extreme southern portions of the Pacific, Atlantic, and Indian Oceans that surround the Antarctic continent are collectively the "Southern Ocean." Fourth in size among the five oceans, the Southern Ocean lacks precise boundaries, so we meld it into its three parent oceans in the table.

Only 2.78% of all of Earth's water is freshwater (nonoceanic). The middle pie chart, along with Table 7.1, details this freshwater portion—surface water and subsurface water (Figure 7.3b). Ice sheets and glaciers are the greatest single repository of surface freshwater; they contain 77.14% of all of Earth's freshwater. Adding subsurface groundwater to frozen surface water accounts for 99.36% of all freshwater.

The remaining freshwater, which resides in lakes, rivers, and streams, actually represents less than 1% of all water (Figure 7.3c). All the world's freshwater lakes total only 125,000 km^3 (30,000 mi^3), with 80% of this volume in just 40 of the largest lakes and about 50% contained in just 7 lakes (listed with Table 7.1).

The greatest single volume of lakewater resides in 25-million-year-old Lake Baykal in Siberian Russia. It contains almost as much water as all five North American Great Lakes combined. Africa's Lake Tanganyika contains the next largest volume, followed by the five Great Lakes. Overall, 70% of lake water is in North America, Africa, and Asia, with about a fourth of lakewater worldwide in small lakes too numerous to count. More than 3 million lakes exist in Alaska alone, and Canada has over 750 km^2 of lake surface.

Not connected to the ocean are saline lakes and salty inland seas. They usually exist in regions of interior river drainage (no outlet to the ocean), which allows salts resulting from evaporation over time to become concentrated. They contain 104,000 km^3 (25,000 mi^3) of water. Examples of such lakes include Utah's Great Salt Lake (Figure 7.4), California's Mono Lake, Southwest Asia's Caspian and Aral Seas, and the Dead Sea between Israel and Jordan.

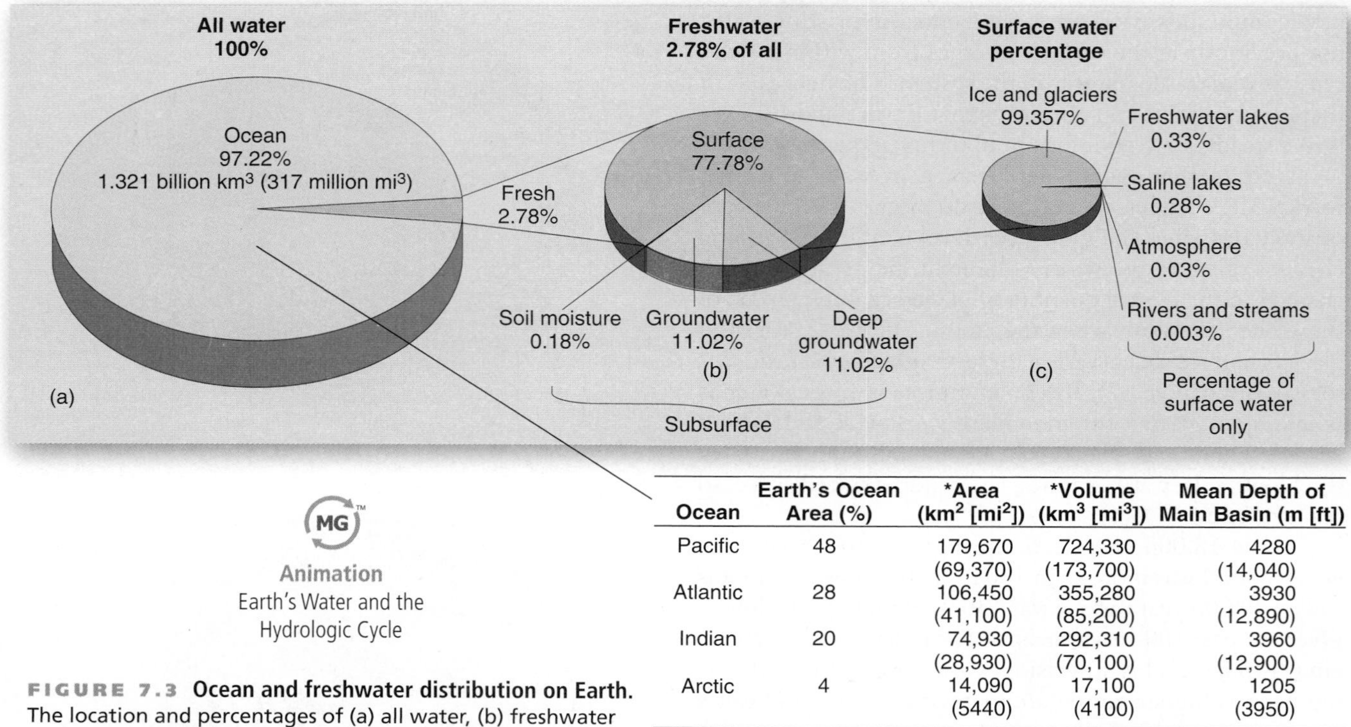

Ocean	Earth's Ocean Area (%)	*Area (km² [mi²])	*Volume (km³ [mi³])	Mean Depth of Main Basin (m [ft])
Pacific	48	179,670 (69,370)	724,330 (173,700)	4280 (14,040)
Atlantic	28	106,450 (41,100)	355,280 (85,200)	3930 (12,890)
Indian	20	74,930 (28,930)	292,310 (70,100)	3960 (12,900)
Arctic	4	14,090 (5440)	17,100 (4100)	1205 (3950)

*Data in thousands (000): includes all marginal seas.

MG
Animation
Earth's Water and the Hydrologic Cycle

FIGURE 7.3 Ocean and freshwater distribution on Earth. The location and percentages of (a) all water, (b) freshwater including subsurface water, and (c) surface water.

TABLE 7.1 Distribution of Freshwater on Earth

Location	Amount (km³ [mi³])		Percentage of Freshwater	Percentage of Total Water
Surface Water				
Ice sheets and glaciers	29,180,000	(7,000,000)	77.14	2.146
Freshwater lakes*	125,000	(30,000)	0.33	0.009
Saline lakes and inland seas	104,000	(25,000)	0.28	0.008
Atmosphere	13,000	(3,100)	0.03	0.001
Rivers and streams	1,250	(300)	0.003	0.0001
Total surface water	29,423,250	(7,058,400)	77.78	2.164
Subsurface Water				
Groundwater—surface to 762 m (2500 ft) depth	4,170,000	(1,000,000)	11.02	0.306
Groundwater—762 to 3962 m (2500 to 13,000 ft) depth	4,170,000	(1,000,000)	11.02	0.306
Soil moisture storage	67,000	(16,000)	0.18	0.005
Total subsurface water	8,407,000	(2,016,000)	22.22	0.617
Total Freshwater (rounded)	**37,800,000**	(9,070,000)	**100.00%**	2.78%

*Major Freshwater Lakes	Volume (km³ [mi³])		Surface Area (km² [mi²])		Depth (m [ft])	
Baykal (Russia)	22,000	(5,280)	31,500	(12,160)	1,620	(5,315)
Tanganyika (Africa)	18,750	(4,500)	39,900	(15,405)	1,470	(4,923)
Superior (U.S./Canada)	12,500	(3,000)	83,290	(32,150)	397	(1,301)
Michigan (U.S.)	4,920	(1,180)	58,030	(22,400)	281	(922)
Huron (U.S./Canada)	3,545	(850)	60,620	(23,400)	229	(751)
Ontario (U.S./Canada)	1,640	(395)	19,570	(7550)	237	(777)
Erie (U.S./Canada)	485	(115)	25,670	(9910)	64	(210)

FIGURE 7.4 Great Salt Lake, a saline lake.
The Great Salt Lake has no outlets other than evaporation and is a remnant of paleoloake Bonneville. At its maximum, some 18,000 years ago, Lake Bonneville extended across eastern Nevada, northern Utah, and southern Idaho—the basin is outlined by ancient shorelines, which scientists actively study. [Bobbé Christopherson.]

Think of all the moisture in the atmosphere and Earth's thousands of flowing rivers and streams. Combined, they amount to only 14,250 km^3 (3400 mi^3), or only 0.033% of freshwater, or 0.0011% of all water! Yet this small amount is dynamic. A water molecule travelling through atmospheric and surface water paths moves through the entire hydrologic cycle (ocean–atmosphere–precipitation–runoff) in less than two weeks. Contrast this to a water molecule in deep-ocean circulation, groundwater, or a glacier; moving slowly, it takes thousands of years to move through the system.

Unique Properties of Water

Earth's distance from the Sun places it within a most remarkable temperate zone when compared with the locations of the other planets. This temperate location allows all three states of water—ice, liquid, and vapor—to occur naturally on Earth.

Even though water is the most common compound on Earth's surface, it exhibits most uncommon properties. Two atoms of hydrogen and one of oxygen, which readily bond, make up each water molecule (suggested in Figure 7.5, upper left). Once hydrogen and oxygen atoms join in a covalent, or double, bond, they are difficult to separate, thereby producing a water molecule that remains stable in Earth's environment. This water molecule is a versatile solvent and possesses extraordinary heat characteristics.

The nature of the hydrogen–oxygen bond gives the hydrogen side of a water molecule a positive charge and the oxygen side a negative charge. As a result of this *polarity*, water molecules attract each other: The positive (hydrogen) side of a water molecule attracts the negative (oxygen) side of another. This bonding between water molecules is *hydrogen bonding* (see Figure 7.5, upper right).

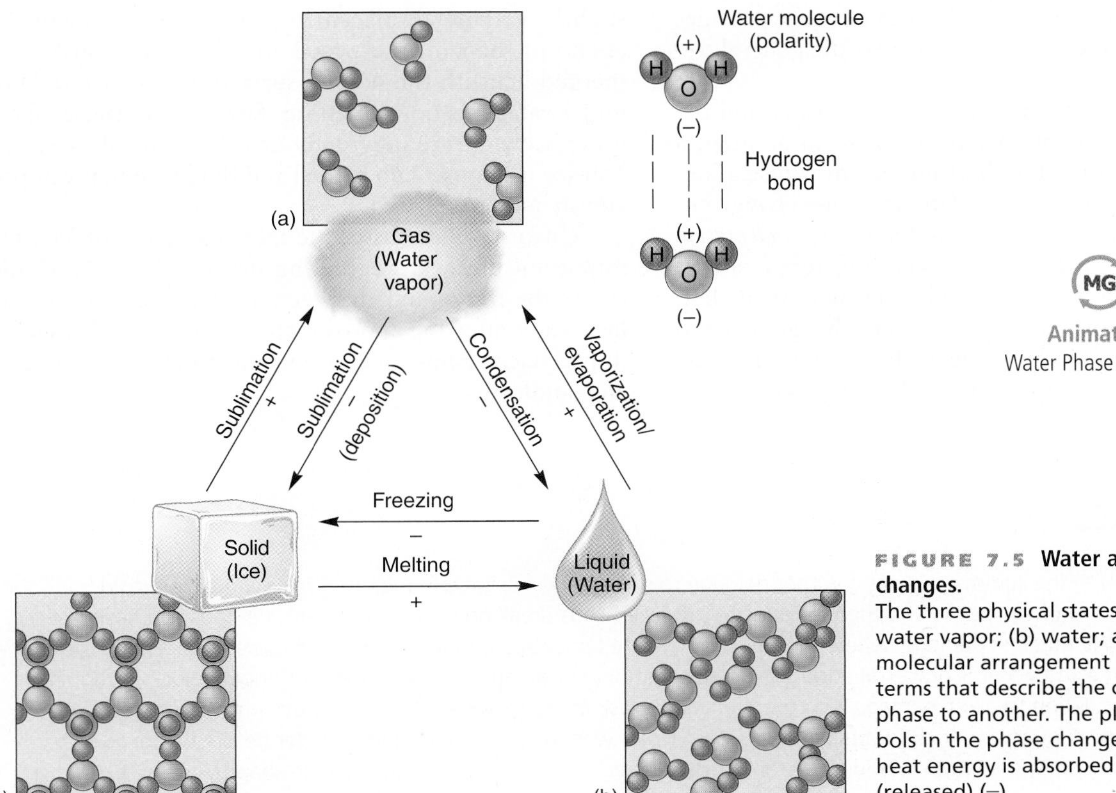

MG Animation
Water Phase Changes

FIGURE 7.5 Water and water's phase changes.
The three physical states of water: (a) gas, or water vapor; (b) water; and (c) ice. Note the molecular arrangement in each state and the terms that describe the changes from one phase to another. The plus and minus symbols in the phase changes denote whether heat energy is absorbed (+) or liberated (released) (–).

The polarity of water molecules also explains why water "acts wet," dissolving many substances. Because of this solvent ability, pure water is rare in nature; something is usually dissolved in it.

The effects of hydrogen bonding in water are observable in everyday life. Hydrogen bonding creates the *surface tension* that allows you to float a steel needle on the surface of water, even though steel is much denser than water. This *surface tension* allows you to slightly overfill a glass with water; webs of millions of hydrogen bonds hold the water slightly above the rim.

Hydrogen bonding is the cause of *capillarity*, which you observe when you "dry" something with a paper towel. The towel draws water through its fibers because hydrogen bonds make each molecule pull on its neighbor. In chemistry laboratory classes, students observe the curved *meniscus*, or surface of the water, which forms in a cylinder or a test tube because hydrogen bonding allows the water to slightly "climb" the glass sides. *Capillary action* is an important component of soil-moisture processes, discussed in Chapters 9 and 18. Without hydrogen bonding to hold molecules together in water and ice, water would be a gas at normal surface temperatures.

Heat Properties

For water to change from one state to another (solid, liquid, or vapor), *heat energy must be absorbed or liberated (released)*. To cause a change of state, the amount of heat energy must be sufficient to affect the hydrogen bonds between molecules. This relation between water and heat energy is important to atmospheric processes. In fact, the heat exchanged between physical states of water provides more than 30% of the energy that powers the general circulation of the atmosphere.

Figure 7.5 presents the three states of water and the terms describing a change from one state to another, a **phase change**. Along the bottom of the illustration, *melting* and *freezing* describe the familiar phase change between solid and liquid. At the right, the terms *condensation* and *evaporation* (or *vaporization* at boiling temperature) apply to the change between liquid and vapor. At the left, the term **sublimation** refers to the direct change of ice to water vapor or water vapor to ice; when water vapor attaches directly to an ice crystal, it is referred to as *deposition*. The deposition of water vapor to ice may form *frost* on surfaces.

Ice, the Solid Phase As water cools, it behaves like most compounds and contracts in volume. However, it reaches its greatest density not as ice, but as water at 4°C (39°F). Below that temperature, water behaves differently from other compounds. It begins to expand as more hydrogen bonds form among the slower-moving molecules, creating the hexagonal (six-sided) structures shown in Figure 7.5c and Figure 7.6a. This six-sided preference applies to ice crystals of all shapes: plates, columns, needles, and dendrites (branching or treelike forms). Ice crystals demonstrate a unique interaction of chaos (all ice crystals are different) and the determinism of physical principles (all have a six-sided structure).

This expansion of water and ice that begins at 4°C continues to a temperature of –29°C (–20°F)—up to a 9% increase in volume is possible (Figure 7.6c). This expansion is important in the weathering of rocks, highway and pavement damage, and burst water pipes. For more on ice crystals and snowflakes, see http://www.its.caltech.edu/∼atomic/snowcrystals/ or http://emu.arsusda.gov/snowsite/default.html.

The expansion in volume that accompanies the freezing process results in a decrease in density (the same number of molecules occupies a larger space). Specifically, pure ice has 0.91 times the density of water, so it floats. Without this change in density, much of Earth's freshwater would be bound in masses of ice on the ocean floor.

In nature, this density varies slightly given the age of the ice and air content. Therefore, icebergs float with slightly varying displacements that average about 1/7 (14%) of their mass exposed and about 6/7 (86%) submerged beneath the ocean's surface (Figure 7.6d). With underwater portions melting faster than those above water, icebergs are inherently unstable and will overturn. You see in Figure 7.6b ruffled and fluted ice that was previously underwater.

Chapter 13 discusses the freezing action of ice as an important physical weathering process. Chapter 17 discusses the freeze-and-thaw action of surface and subsurface water affecting approximately 30% of Earth's surface in periglacial landscapes, producing a variety of processes and landforms.

GEO REPORT 7.1 *Breaking roads and pipes*

Road crews are busy in the summer repairing winter damage to streets and freeways in regions where winters are cold. Rainwater seeps into roadway cracks and then expands as it freezes, thus breaking up the pavement. Perhaps you have noticed that bridges suffer the greatest damage, where cold air can circulate beneath a bridge and produce more freeze–thaw cycles. The expansion of freezing water is powerful enough to crack plumbing or an automobile radiator or engine block. Wrapping water pipes with insulation to avoid damage is a common winter task in many places. Historically, this physical property of water was put to use in quarrying rock for building materials. Holes were drilled and filled with water before winter so that, when cold weather arrived, the water would freeze and expand, cracking the rock into manageable shapes.

FIGURE 7.6 The uniqueness of ice forms.
(a) Computer-enhanced photo reveals ice-crystal patterns, dictated by the internal structure between water molecules. (b) Iceberg arch in the Antarctic Sound near 63.5° S. (c) Ice needles form in the freezing process of a meltwater stream. (d) A bergy bit (small iceberg) off the coast of Antarctica illustrates the density–buoyancy state of floating ice. [(a) Photo enhancement © Scott Camazine/Photo Researchers, Inc., after W. A. Bentley; (b), (c), and (d) Bobbé Christopherson.]

CRITICAL THINKING 7.1

Iceberg analysis

Examine the iceberg photographs in Figure 7.6b and Figure 7.6d. These photos show Antarctic waters. Determine what caused the thermal notch above the ocean surface around the iceberg. Why do you think the iceberg appears to be riding higher in the water as it melts? Place an ice cube in a clear glass almost filled with water. Then approximate the amount of ice above the water's surface compared to the amount below water. Pure ice is 0.91 the density of water, whereas ice actually contains air bubbles, so icebergs are about 0.86 the density of water. How do your measurements compare with this range of density?

Water, the Liquid Phase As a liquid, water assumes the shape of its container and is a noncompressible fluid. For ice to change to water, heat energy must increase the motion of the water molecules to break some of the hydrogen bonds (Figure 7.5b). Despite the fact that there is no change in sensible temperature between ice at 0°C (32°F) and water at 0°C, 80 calories* of heat energy must be absorbed for the phase change of 1 g of ice to melt to 1 g of water (Figure 7.7, upper left). Heat energy involved in the phase change is **latent heat** and is hidden within the structure of water. It becomes liberated whenever the phase reverses and a gram of water freezes. The *latent heat of freezing* and the *latent heat of melting* each involves 80 calories.

To raise the temperature of 1 g of water at 0°C (32°F) to boiling at 100°C (212°F), we must add 100 cal, gaining an increase of 1 C° (1.8 F°) for each calorie added. No phase change is involved in this temperature gain.

*Remember, from Chapter 2, that a calorie (cal) is the amount of energy required to raise the temperature of 1 g of water (at 15°C) 1 degree Celsius and is equal to 4.184 joules.

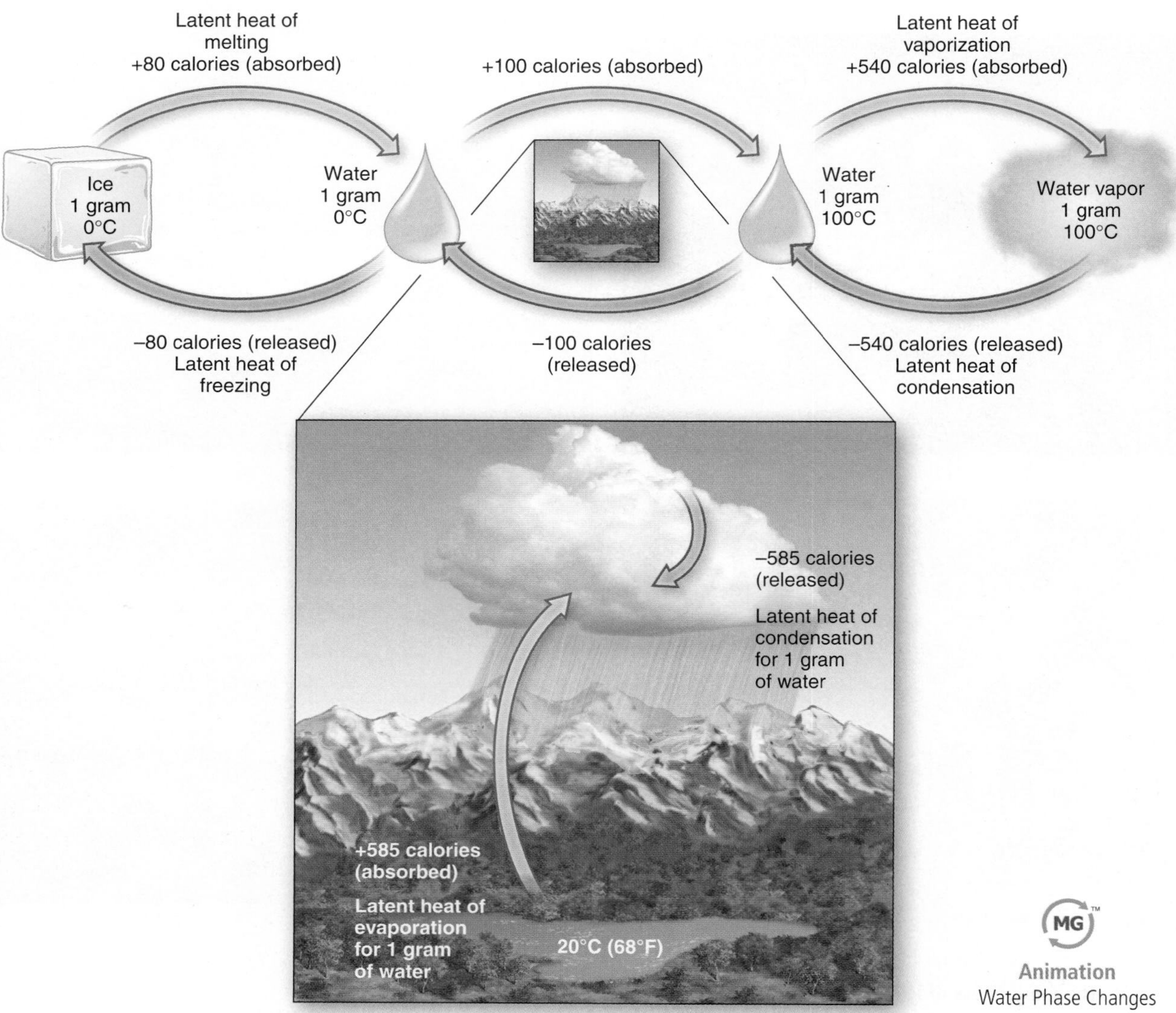

MG Animation Water Phase Changes

FIGURE 7.7 Water's heat-energy characteristics.
The phase changes of water absorb or release a lot of latent heat energy. To transform 1 g of ice at 0°C to 1 g of water vapor at 100°C requires 720 cal: 80 + 100 + 540. The landscape illustrates phase changes between water (lake at 20°C) and water vapor under typical conditions in the environment.

Water Vapor, the Gas Phase Water vapor is an invisible and compressible gas in which each molecule moves independently of the others (Figure 7.5a). The phase change from liquid to vapor at boiling temperature, under normal sea-level pressure, requires the addition of 540 cal for each gram, the **latent heat of vaporization** (Figure 7.7). When water vapor condenses to a liquid, each gram gives up its hidden 540 cal as the **latent heat of condensation**. Perhaps you have felt the liberation of the latent heat of condensation on your skin from steam when you drained steamed vegetables or pasta or filled a hot teakettle.

In summary, taking 1 g of ice at 0°C and changing its phase to water and then to water vapor at 100°C—from a

GEO REPORT 7.2 *Iceberg hazards and the end of a trip*

A major hazard to ships in higher latitudes is posed by floating ice. Since ice has approximately 0.86 the density of water, an iceberg sits with approximately 6/7 of its mass below water level. The irregular edges of submarine ice can impact the side of a passing ship. Hitting an iceberg buckled plates and substandard rivets along the side of the RMS *Titanic* on its maiden voyage in 1912, causing it to sink and perhaps triggering a brief lapse in society's faith in technology.

solid to a liquid to a gas—*absorbs* 720 cal (80 cal + 100 cal + 540 cal). Reversing the process, or changing the phase of 1 g of water vapor at 100°C to water and then to ice at 0°C, *liberates* 720 cal into the surrounding environment. On the *Mastering Geography* Web site is an excellent animation illustrating these concepts.

Heat Properties of Water in Nature

In a lake or stream or in soil water, at 20°C (68°F), every gram of water that breaks away from the surface through evaporation must absorb from the environment approximately 585 cal as the *latent heat of evaporation* (see the natural scene in Figure 7.7). This is slightly more energy than would be required if the water were at a higher temperature such as boiling (540 cal). You can feel this absorption of latent heat as evaporative cooling on your skin when it is wet. This latent heat exchange is the dominant cooling process in Earth's energy budget.

The process reverses when air cools and water vapor condenses back into the liquid state, forming moisture droplets and thus liberating 585 cal for every gram of water as the *latent heat of condensation*. When you realize that a small, puffy, fair-weather cumulus cloud holds 500–1000 tons of moisture droplets, think of the tremendous latent heat released when water vapor condenses to droplets.

The **latent heat of sublimation** absorbs 680 cal as a gram of ice transforms into vapor. Water vapor freezing directly to ice releases a comparable amount of energy.

Humidity

Humidity refers to water vapor in the air. The capacity of air for water vapor is primarily a function of temperature—the temperatures of both the air and the water vapor, which are usually the same. Humidity can be expressed in several ways, discussed in this section.

We are all aware of humidity in the air, for its relationship to air temperature determines our sense of comfort. North Americans spend billions of dollars a year to adjust humidity, either with air conditioning (extracting water vapor and cooling) or with air humidifying (adding water vapor). We discussed the relation between humidity and temperature and the heat index in Chapter 5. To determine the energy available for powering weather, one has to know the water-vapor *content* of air and relate this in a ratio to the air's *saturation equilibrium* at a given temperature.

Relative Humidity

After air temperature and barometric pressure, relative humidity is the most common piece of information given in local weather broadcasts. **Relative humidity** is a ratio (expressed as a percentage) of the amount of water vapor that is actually in the air compared to the maximum water vapor possible in the air at a given temperature.

Relative humidity varies because of water vapor or temperature changes in the air. The formula to calculate the relative humidity ratio and express it as a percentage places actual water vapor in the air as the numerator and water vapor possible in the air at that temperature as the denominator:

$$\text{Relative humidity} = \frac{\text{Actual water vapor in the air}}{\text{Maximum water vapor possible in the air at that temperature}} \times 100$$

Warmer air increases the evaporation rate from water surfaces, whereas cooler air tends to increase the condensation rate of water vapor to water surfaces. Because there is a maximum amount of water vapor that can exist in a volume of air at a given temperature, the rates of evaporation and condensation can reach equilibrium at some point; the air is then saturated with humidity. Relative humidity tells us how near the air is to saturation and is an expression of an ongoing process of water molecules moving between air and moist surfaces.

In Figure 7.8 at 5 P.M., the evaporation rate exceeds condensation at the higher temperatures of this time of day and relative humidity is at 20%. At 11 A.M., the evaporation rate still exceeds condensation, though not by as much, since daytime temperatures are not as high, so the same volume of water vapor now occupies 50% of the maximum possible capacity. At 5 A.M., in the cooler morning air, saturation equilibrium exists, and any further cooling or addition of water vapor produces net condensation. When the air is saturated with maximum water vapor for its temperature, the relative humidity percentage is 100%.

Saturation As stated, air is at **saturation**, at 100% relative humidity, when the rate of evaporation and the rate of condensation—the net transfer of water molecules—reach equilibrium. Saturation indicates that any further addition of water vapor or any decrease in temperature that reduces the evaporation rate results in active condensation (clouds, fog, or precipitation).

The temperature at which a given mass of air becomes saturated and net condensation begins to form

GEO REPORT 7.3 *Katrina had the power*

Meteorologists estimated that the moisture in Hurricane Katrina (2005) weighed more than 30 trillion tons at its maximum power and mass. With 585 cal released for every gram as the latent heat of condensation, a weather event such as a hurricane involves a staggering amount of energy. Do the quick math in English units (28 g to an ounce, 16 ounces to the pound, 2000 pounds to the ton, times 30 trillion tons).

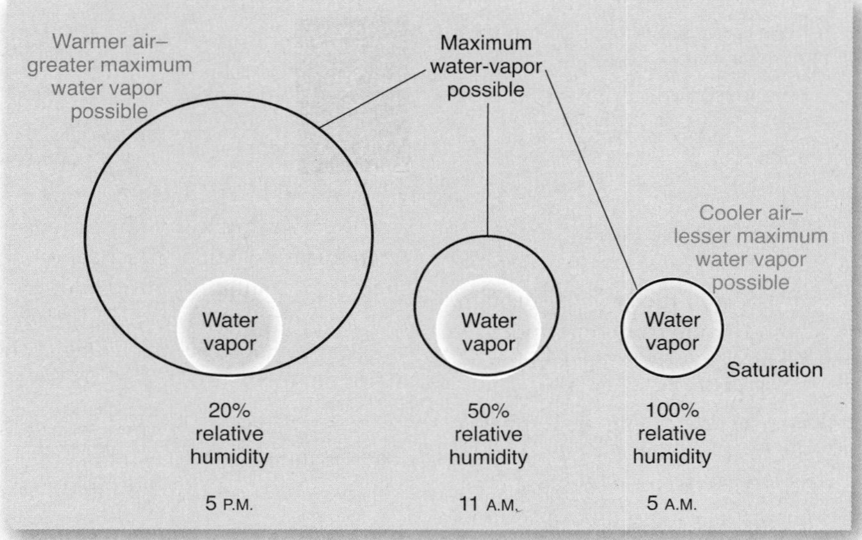

FIGURE 7.8 Water vapor, temperature, and relative humidity. The maximum water vapor possible in warm air is greater (net evaporation) than that possible in cold air (net condensation), so relative humidity changes with temperature, even though in this example the actual water vapor present in the air stays the same during the day.

water droplets is the **dew-point temperature**. *The air is saturated when the dew-point temperature and the air temperature are the same*. When temperatures are below freezing, the term *frost point* is sometimes used.

A cold drink in a glass provides a common example of these conditions (Figure 7.9). The water droplets that form on the outside of the glass condense from the air because the air layer next to the glass is chilled to below its dew-point temperature and thus becomes saturated. Figure 7.9 gives an additional example of saturated air and active condensation above a wet rock surface in the cooler air layer near the rock. As you walk to classes on some cool mornings, you perhaps notice damp lawns or dew on windshields, an indication of dew-point conditions.

Satellites using infrared sensors now routinely sense water vapor in the lower atmosphere. Water vapor absorbs long wavelengths (infrared), making it possible to distinguish areas of relatively high water vapor from areas of low water vapor. Figure 7.10 includes a global image composite by sensors of the "water-vapor channel." This ability is important to forecasting because it shows the moisture available to weather systems and therefore the available latent heat energy and precipitation potential.

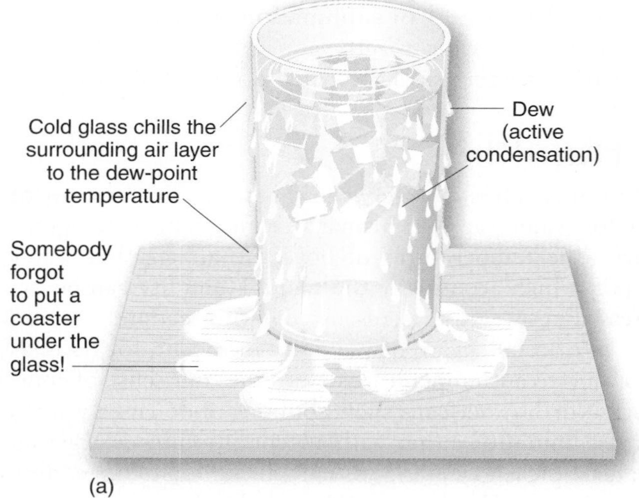

FIGURE 7.9 Dew-point temperature examples.
(a) The low temperature of the glass chills the surrounding air layer to the dew-point temperature and saturation. Thus, water vapor condenses out of the air and onto the glass as dew.
(b) Cold air above the rain-soaked rocks is at the dew point and is saturated. Water evaporates from the rock into the air and condenses in a changing veil of clouds. [Author photo.]

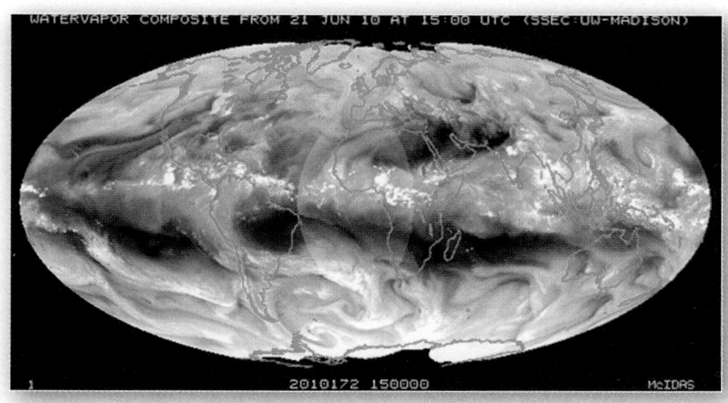

FIGURE 7.10 Composite infrared image from *GOES* (United States), *Meteosat* (European Space Agency), and *MTSAT* (Japan) of water vapor in the atmosphere.
In this June 21, 2010, image, higher water-vapor content is lighter and lower water-vapor content is darker. [Satellite data courtesy of Space Science and Engineering Center, University of Wisconsin, Madison.]

Daily and Seasonal Relative Humidity Patterns An inverse relation occurs during a typical day between air temperature and relative humidity—as temperature rises, relative humidity falls (Figure 7.11). Relative humidity is highest at dawn, when air temperature is lowest. If you park outdoors, you know about the wetness of the dew that condenses on your car or bicycle overnight. In your own experience, you probably have noticed this pattern—the morning dew on windows, cars, and lawns evaporates by late morning as net evaporation increases with air temperature.

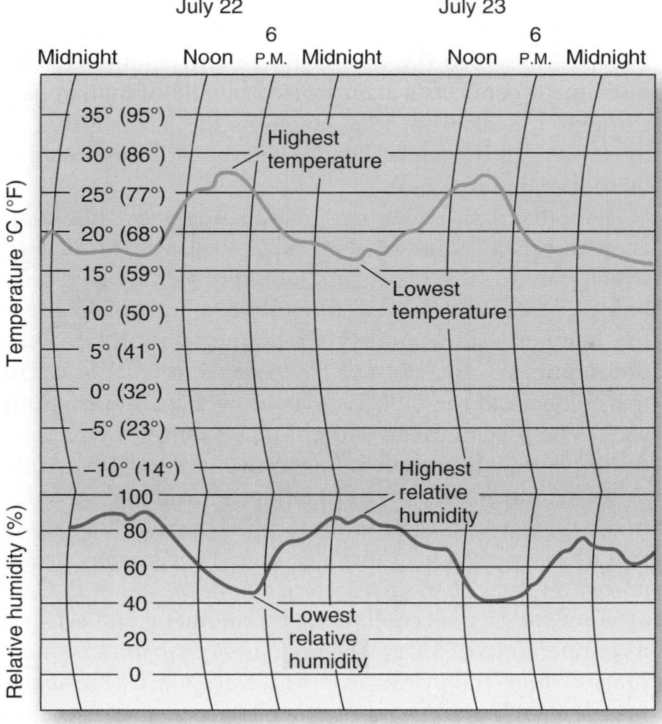

FIGURE 7.11 Daily relative humidity patterns.
Typical daily variations demonstrate temperature and relative humidity relations.

Relative humidity is lowest in the late afternoon, when higher temperatures increase the rate of evaporation. As shown in Figure 7.8, the actual water vapor present in the air may remain the same throughout the day. However, relative humidity changes because the temperature, and therefore the rate of evaporation, varies from morning to afternoon. Seasonally, January readings are higher than July readings because air temperatures are lower overall in winter. Similar relative humidity records at most weather stations demonstrate the same relation among season, temperature, and relative humidity.

Expressions of Humidity

There are several ways to express humidity and relative humidity. Each has its own utility and application. Two measures involve vapor pressure and specific humidity.

Vapor Pressure As free water molecules evaporate from surfaces into the atmosphere, they become water vapor. Now part of the air, water-vapor molecules exert a portion of the air pressure along with nitrogen and oxygen molecules. The share of air pressure that is made up of water-vapor molecules is **vapor pressure**, expressed in millibars (mb).

As explained earlier, *saturation* is reached when the movement of water molecules between surface and air is in equilibrium. Air that contains as much water vapor as possible at a given temperature is at *saturation vapor pressure*. Any temperature increase or decrease will change the saturation vapor pressure.

Figure 7.12 on page 174 graphs the saturation vapor pressure at various air temperatures. The graph illustrates that, for every temperature increase of 10 C° (18 F°), the saturation vapor pressure in air nearly doubles. This relation explains why warm tropical air over the ocean can contain so much water vapor, thus providing much latent heat to power tropical storms. It also explains why cold air is "dry" and why cold air toward the poles does not produce a lot of precipitation (it contains too little water vapor, even though it is near the dew-point temperature).

As marked on the graph, air at 20°C (68°F) has a saturation vapor pressure of 24 mb; that is, the air is saturated if the water-vapor portion of the air pressure also is at 24 mb. Thus, if the water vapor actually present is exerting a vapor pressure of only 12 mb in 20°C air, the relative humidity is 50% ($12 \text{ mb} \div 24 \text{ mb} = 0.50 \times 100 = 50\%$). The inset in Figure 7.12 compares saturation vapor pressure over water and over ice surfaces at subfreezing temperatures. You can see that saturation vapor pressure is greater above a water surface than over an ice surface—that is, it takes more water-vapor molecules to saturate air above water than it does above ice. This fact is important to condensation processes and rain-droplet formation, both of which are described later in this chapter.

Specific Humidity A useful humidity measure is one that remains constant as temperature and pressure change. **Specific humidity** is the mass of water vapor (in grams) per mass of air (in kilograms) at any specified temperature.

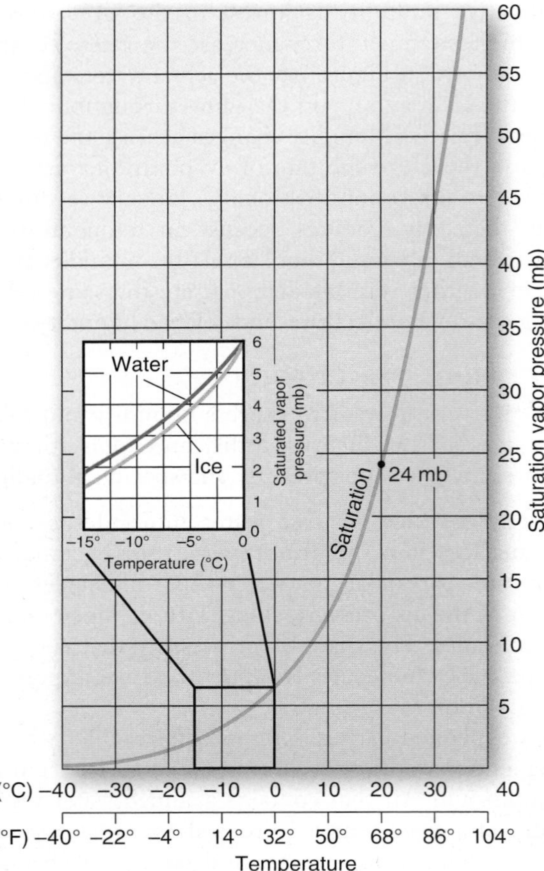

FIGURE 7.12 Saturation vapor pressure at various temperatures.
Saturation vapor pressure is the maximum possible water vapor, as measured by the pressure it exerts (mb). Inset compares saturation vapor pressures over water surfaces with those over surfaces at subfreezing temperatures. Note the 24-mb label for the text discussion.

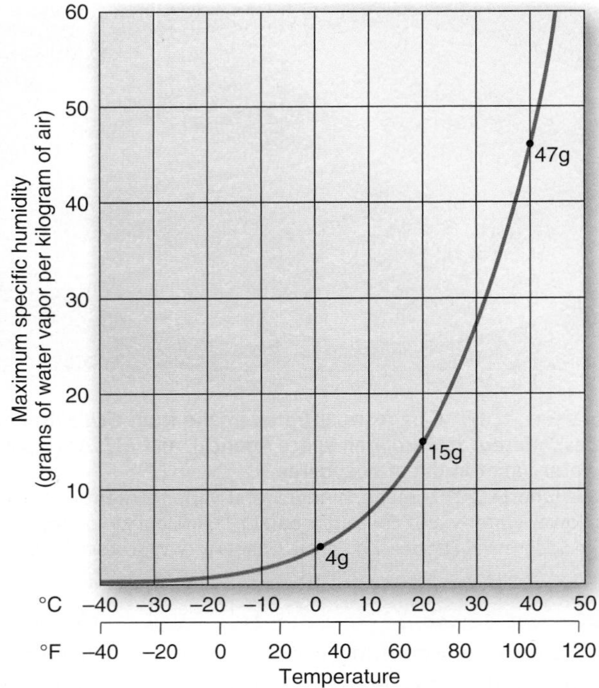

FIGURE 7.13 Maximum specific humidity at various temperatures.
Maximum specific humidity is the maximum possible water vapor in a mass of water vapor per unit mass of air (g/kg). Note the 47-g, 15-g, and 4-g labels for the text discussion.

Because it is measured in mass, specific humidity is not affected by changes in temperature or pressure, as when an air parcel rises to higher elevations. Specific humidity stays constant despite volume changes.*

The maximum mass of water vapor possible in a kilogram of air at any specified temperature is the *maximum specific humidity*, plotted in Figure 7.13. Noted on the graph is that a kilogram of air could hold a maximum specific humidity of 47 g of water vapor at 40°C (104°F), 15 g at 20°C (68°F), and about 4 g at 0°C (32°F). Therefore, if a kilogram of air at 40°C has a specific humidity of 12 g, its relative humidity is 25.5% (12 g ÷ 47 g = 0.255 × 100 × 25.5%). Specific humidity is useful in describing the moisture content of large air masses that are interacting in a weather system and provides information necessary for weather forecasting.

*Another similar measure used in relative humidity that approximates specific humidity is the *mixing ratio*—that is, the ratio of the mass of water vapor (grams) per mass of dry air (kilograms), as in g/kg.

Instruments for Measuring Humidity Various instruments measure relative humidity. The **hair hygrometer** uses the principle that human hair changes as much as 4% in length between 0% and 100% relative humidity. The instrument connects a standardized bundle of human hair through a mechanism to a gauge. As the hair absorbs or loses water in the air, it changes length, indicating relative humidity (Figure 7.14a).

Another instrument used to measure relative humidity is a **sling psychrometer**. Figure 7.14b shows this device, which has two thermometers mounted side by side on a holder. One is the *dry-bulb thermometer*; it simply records the ambient (surrounding) air temperature. The other thermometer is the *wet-bulb thermometer*; it is set lower in the holder, and the bulb is covered by a moistened cloth wick. The psychrometer is then spun ("slung") by its handle or placed where a fan forces air over the wet-bulb wick.

The rate at which water evaporates from the wick depends on the relative saturation of the surrounding air. If the air is dry, water evaporates quickly, *absorbing specific heat as the latent heat of evaporation from the wet-bulb thermometer and its wick*, cooling the thermometer and causing its temperature to lower (the *wet-bulb depression*). In conditions of high humidity, little water evaporates from the wick; in low humidity, more water evaporates. After being spun a minute or two, a comparison of temperature on each bulb and a check on a relative humidity (psychrometric) chart give the relative humidity reading.

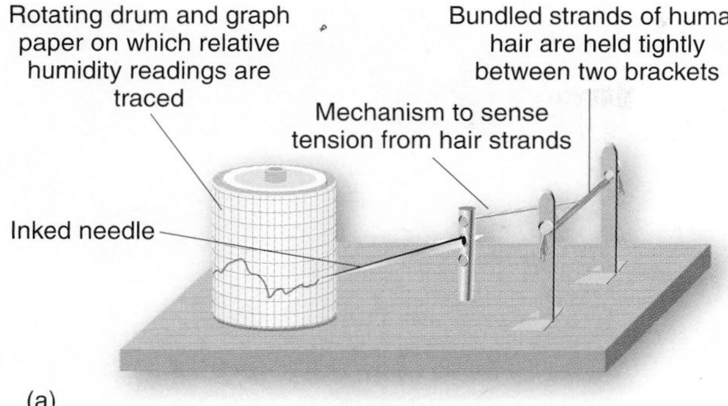

(a)

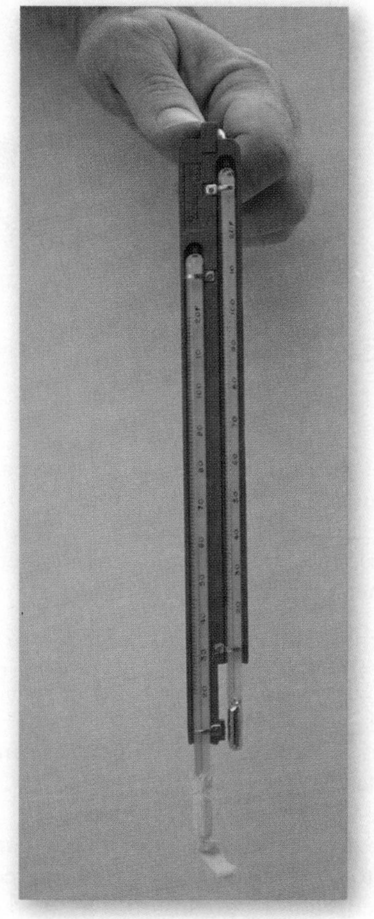
(b)

FIGURE 7.14 Instruments that measure relative humidity. (a) The principle of a hair hygrometer. (b) Sling psychrometer with wet and dry bulbs. [Bobbé Christopherson.]

CRITICAL THINKING 7.2

Using relative humidity and dew-point maps

Please refer to our discussion of relative humidity, air temperature, dew-point temperature, and saturation, on the *Mastering Geography* Web site. View the dew-point temperature map in relation to the map of air temperatures. As you compare and contrast the maps, describe in general terms what relative humidity conditions you find in the Northwest? How about along the Gulf Coast? How does this contrast to the Southeast and to the Southwest? Remember, the closer the air temperature is to the dew-point temperature, then the closer the air is to saturation and the possibility of condensation.

Atmospheric Stability

Meteorologists use the term *parcel* to describe a body of air that has specific temperature and humidity characteristics. Think of an air parcel as a volume of air, perhaps 300 m (1000 ft) in diameter or more. The temperature of the volume of air determines the density of the air parcel. Warm air produces a lower density in a given volume of air; cold air produces a higher density.

Two opposing forces work on a parcel of air: an upward *buoyant force* and a downward *gravitational force*. A parcel of lower density than the surrounding air rises (is more buoyant); a rising parcel expands as external pressure decreases. In contrast, a parcel of higher density descends (is less buoyant); a falling parcel compresses as external pressure increases. Figure 7.15 illustrates these relationships.

An indication of weather conditions is the relative stability of air parcels in the atmosphere. **Stability** refers to the tendency of an air parcel, with its water-vapor cargo, either to remain in place or to change vertical position by ascending (rising) or descending (falling). An air parcel is *stable* if it resists displacement upward or, when disturbed, tends to return to its starting place. An air parcel is *unstable* if it continues to rise until it reaches an altitude where the surrounding air has a density and temperature similar to its own.

To visualize this, imagine a hot-air balloon launch. The air-filled balloon sits on the ground with the same air temperature inside as in the surrounding environment, like a stable air parcel. As the burner ignites, the balloon fills with hot (less dense) air and rises buoyantly (Figure 7.16 on page 176), like an unstable parcel of air. In the photos, why do you think these launches are taking place in the early morning? The concepts of stable and unstable air lead us to the specific temperature characteristics that produce these conditions.

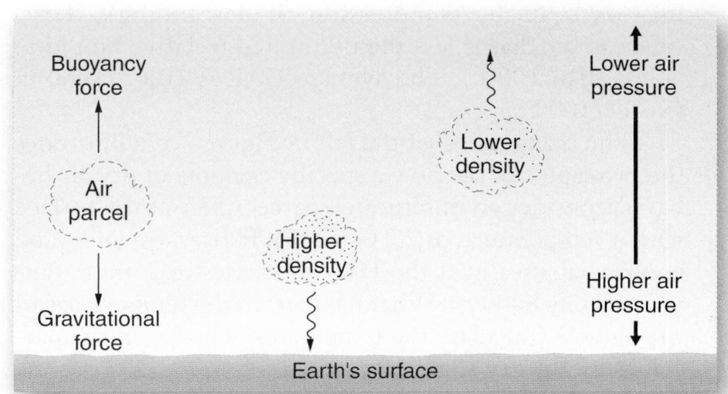

FIGURE 7.15 The forces acting on an air parcel. Buoyancy and gravitational forces work on an air parcel. Different densities produce rising or falling parcels in response to imbalance in these forces.

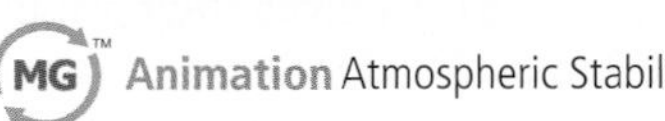

Animation Atmospheric Stability

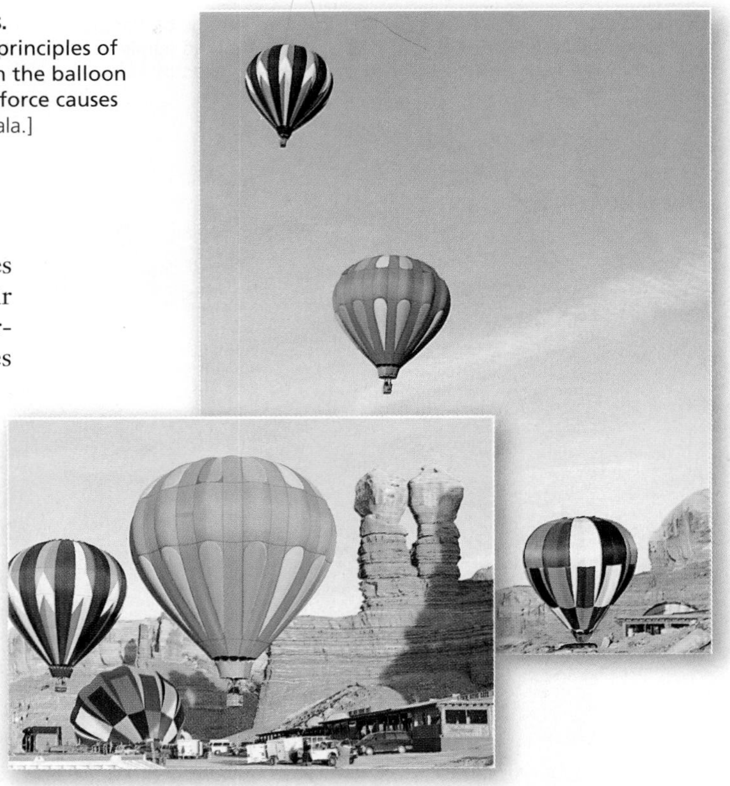

FIGURE 7.16 **Principles of air stability and balloon launches.** Hot-air balloons being launched in southern Utah illustrate the principles of stability. As the temperature inside a balloon increases, the air in the balloon becomes less dense than the surrounding air, and the buoyancy force causes the balloon to rise, acting like a warm-air parcel. [Steven K. Huhtala.]

Adiabatic Processes

Determining the degree of stability or instability requires measuring two temperatures: the temperature inside an air parcel and the temperature of the air surrounding the parcel. The contrast of these two temperatures determines stability. Such temperature measurements are made daily with *radiosonde* instrument packages carried aloft by helium-filled balloons at thousands of weather stations.

The *normal lapse rate*, introduced in Chapter 3, is the average decrease in temperature with increasing altitude, a value of 6.4 C°/1000 m (3.5 F°/1000 ft). This rate of temperature change is for still, calm air, and it can vary greatly under different weather conditions. Consequently, the *environmental lapse rate* (ELR) is the actual lapse rate at a particular place and time. It can vary by several degrees per thousand meters.

Adiabatic describes the warming and cooling rates for a parcel of expanding or compressing air. *An ascending parcel of air tends to cool by expansion*, responding to the reduced pressure at higher altitudes. In contrast, *descending air tends to heat by compression*. *Diabatic* means occurring with an exchange of heat; *adiabatic* means occurring without a loss or gain of heat—that is, without any heat exchange between the surrounding environment and the vertically moving parcel of air. Adiabatic temperatures are measured with one of two specific rates, depending on moisture conditions in the parcel: dry adiabatic rate (DAR) and moist adiabatic rate (MAR).

Dry Adiabatic Rate The **dry adiabatic rate (DAR)** is the rate at which "dry" air cools by expansion (if ascending) or heats by compression (if descending). "Dry" refers to air that is less than saturated (relative humidity is less than 100%). The average DAR is 10 C°/1000 m (5.5 F°/1000 ft).

The rising parcel at the left in Figure 7.17a illustrates the principle. To see how a specific example of dry air behaves, consider an unsaturated parcel of air at the surface with a temperature of 27°C (81°F). It rises, expands, and cools adiabatically at the DAR as it rises to 2500 m (approximately 8000 ft). What happens to the temperature of the parcel? Calculate the temperature change in the parcel, using the dry adiabatic rate:

$$(10\ \text{C}°/1000\ \text{m}) \times 2500\ \text{m} = 25\ \text{C}° \text{ of total cooling}$$
$$(5.5\ \text{F}°/1000\ \text{ft}) \times 8000\ \text{ft} = 44\ \text{F}° \text{ of total cooling}$$

Subtracting the 25 C° (44 F°) of adiabatic cooling from the starting temperature of 27°C (81°F) gives the temperature in the air parcel at 2500 m as 2°C (36°F).

In Figure 7.17b, assume that an unsaturated air parcel with a temperature of –20°C at 3000 m (24°F at 9800 ft) descends to the surface, heating adiabatically. Using the dry adiabatic lapse rate, we determine the temperature of the air parcel when it arrives at the surface:

$$(10\ \text{C}°/1000\ \text{m}) \times 3000\ \text{m} = 30\ \text{C}° \text{ of total warming}$$
$$(5.5\ \text{F}°/1000\ \text{ft}) \times 9800\ \text{ft} = 54\ \text{F}° \text{ of total warming}$$

Adding the 30 C° of adiabatic warming to the starting temperature of –20°C gives the temperature in the air parcel at the surface as 10°C (50°F).

Moist Adiabatic Rate The **moist adiabatic rate (MAR)** is the rate at which an ascending air parcel that is moist, or saturated, cools by expansion. The average MAR is 6 C°/1000 m (3.3 F°/1000 ft). This is roughly 4 C° (2 F°) less than the dry adiabatic rate. From this average, the MAR varies with moisture content and temperature and can range from 4 C° to 10 C° per 1000 m (2 F° to 5.5 F° per 1000 ft). (Note that a descending parcel of saturated air warms at the MAR as well because the evaporation of liquid droplets, absorbing sensible heat, offsets the rate of compressional warming.)

The cause of this variability, and the reason that the MAR is lower than the DAR, is the latent heat of condensation. As water vapor condenses in the saturated air, latent heat is liberated, becoming sensible heat, thus decreasing the adiabatic rate. The release of latent heat may vary with temperature and water-vapor content. The MAR is much lower than the DAR in warm air, whereas the two rates are more similar in cold air.

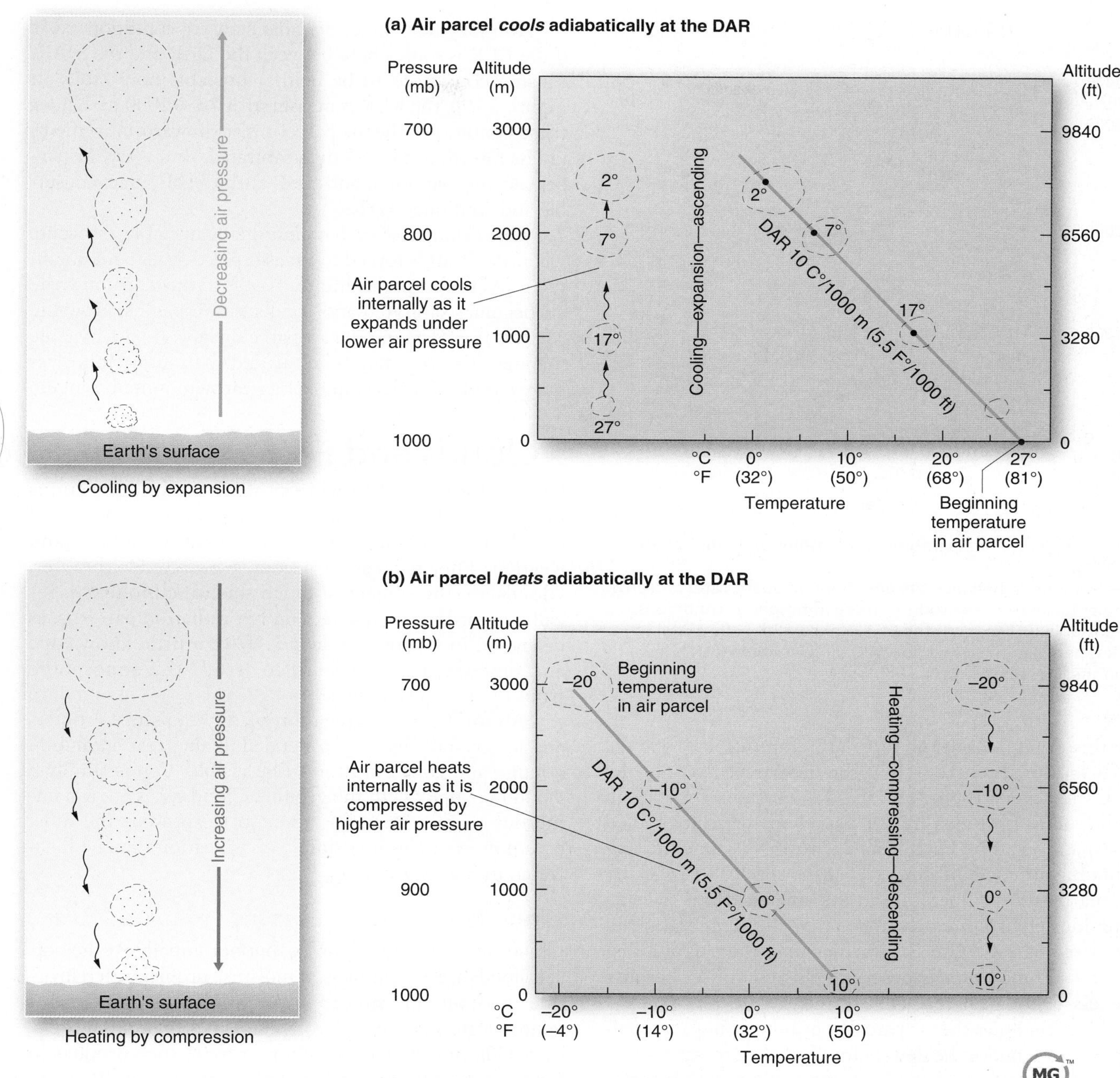

FIGURE 7.17 Vertically moving air experiences temperature changes—adiabatic cooling and heating. (a) A rising air parcel that is less than saturated cools adiabatically by expansion at the DAR. (b) A descending air parcel that is less than saturated heats adiabatically by compression at the DAR.

MG Animation Atmospheric Stability

Stable and Unstable Atmospheric Conditions

Now we bring this discussion of adiabatic processes to determine atmospheric stability. The relation among the DAR, MAR, and ELR at a given time and place determines the stability of the atmosphere over an area. You see examples of three stability relationships in Figure 7.18 on page 178.

Temperature relationships in the atmosphere produce the three different conditions: unstable, conditionally unstable, and stable. For the sake of illustration, the three examples in Figure 7.19 on page 179 begin with an air parcel at the surface at 25°C (77°F). In each example, compare the temperatures of the air parcel and the surrounding environment. Assume that a lifting mechanism, such as surface heating, a mountain range, or weather fronts, is present to get the parcel started (we examine lifting mechanisms in Chapter 8).

Given unstable conditions in Figure 7.19a, the air parcel continues to rise through the atmosphere because it is warmer (less dense and more buoyant) than the surrounding environment. Note the environmental lapse rate

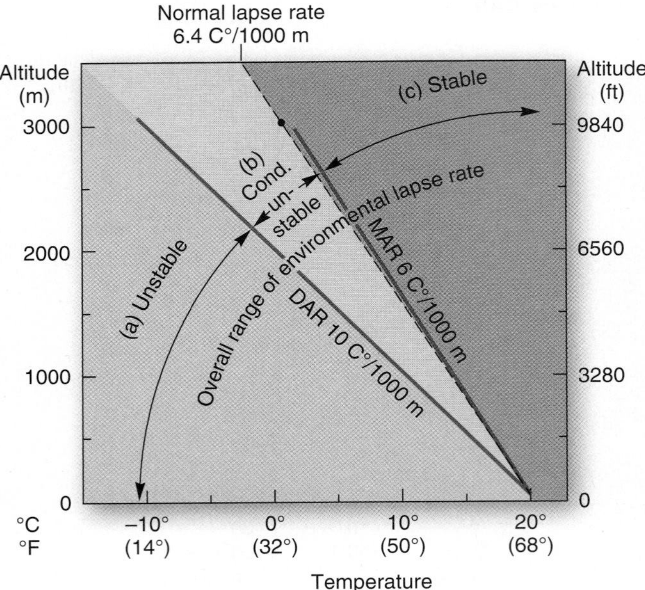

FIGURE 7.18 Temperature relationships and atmospheric stability.
The relationship between dry and moist adiabatic rates and environmental lapse rates produces three atmospheric conditions: (a) unstable (ELR exceeds the DAR), (b) conditionally unstable (ELR is between the DAR and MAR), and (c) stable (ELR is less than the DAR and MAR).

for this example is set at 12 C°/1000 m (6.6 F°/1000 ft). That is, the air surrounding the air parcel is cooler by 12 C° for every 1000-m increase in altitude. By 1000 m (about 3300 ft), the rising air parcel adiabatically cooled by expansion at the DAR from 25° to 15°C, while the surrounding air cooled from 25°C at the surface to 13°C. By comparing the temperature in the air parcel and the surrounding environment, you see that the temperature in the parcel is 2 C° (3.6 F°) warmer than the surrounding air at 1000 m. *Unstable* describes this condition because the less-dense air parcel will continue to lift.

Eventually, as the air parcel continues rising and cooling, it may achieve the dew-point temperature, saturation, and active condensation. This point of saturation forms the *lifting condensation level* that you see in the sky as the flat bottoms of clouds.

The example in Figure 7.19c shows stable conditions resulting when the ELR is only 5 C°/1000 m (3 F°/1000 ft). An ELR of 5 C°/1000 m is less than both the DAR and the MAR, a condition in which the air parcel has a lower temperature (is more dense and less buoyant) than the surrounding environment. The relatively cooler air parcel tends to settle back to its original position—it is *stable*. The denser air parcel resists lifting, unless forced by updrafts or a barrier, and the sky remains generally cloud-free. If clouds form, they tend to be stratiform (flat clouds) or cirroform (wispy), lacking vertical development. In regions experiencing air pollution, stable conditions in the atmosphere worsen the pollution by slowing exchanges in the surface air.

You may be wondering what stability condition exists if the ELR is somewhere between the DAR and the MAR, causing conditions to be neither unstable nor stable. In Figure 7.19b, the ELR is measured at 7 C°/1000 m. Under these conditions, the air parcel resists upward movement, unless forced, if it is less than saturated. But, if the air parcel becomes saturated and cools at the MAR, it acts unstable and continues to rise.

One example of such conditionally unstable air occurs when stable air is forced to lift as it passes over a mountain range. As the air parcel lifts and cools to the dew point, the air becomes saturated and condensation begins. Now the MAR is in effect, and the air parcel behaves in an unstable manner. The sky may be clear and without a cloud, yet huge clouds may develop over a nearby mountain range.

Clouds and Fog

Clouds are more than whimsical, beautiful decorations in the sky; they are fundamental indicators of overall conditions, including stability, moisture content, and weather. They form as air becomes saturated with water. Clouds are the subject of much scientific inquiry, especially regarding their effect on net radiation patterns, as discussed in Chapters 4 and 5. With a little knowledge and practice, you can learn to "read" the atmosphere from its signature clouds.

A **cloud** is an aggregation of tiny moisture droplets and ice crystals that are suspended in air, great enough in volume and concentration to be visible. Fog is simply a cloud in contact with the ground. Cloud types are too numerous to fully describe here, so we present the most common examples in a simple classification scheme illustrated in Figure 7.21.

Cloud Formation Processes

Clouds may contain raindrops, but not initially. At the outset, clouds are a great mass of moisture droplets, each invisible without magnification. A **moisture droplet** is approximately 20 μm (micrometers) in diameter (0.002 cm, or 0.0008 in.). It takes a million or more such droplets to form an average raindrop with a diameter of 2000 μm (0.2 cm, or 0.078 in.), as shown in Figure 7.20.

As an air parcel rises, it may cool to the dew-point temperature and 100% relative humidity. (Under certain conditions, condensation may occur at slightly less or more than 100% relative humidity.) More lifting of the air parcel cools it further, producing condensation of water vapor into water. Water does not just condense among the air molecules. Condensation requires **cloud-condensation nuclei**, microscopic particles that always are present in the atmosphere.

Continental air masses average 10 billion cloud-condensation nuclei per cubic meter. Ordinary dust, soot, and ash from volcanoes and forest fires and particles from burned fuel, such as sulfate aerosols, typically provide these nuclei. The air over cities contains great concentrations of such nuclei. In maritime air masses, nuclei average

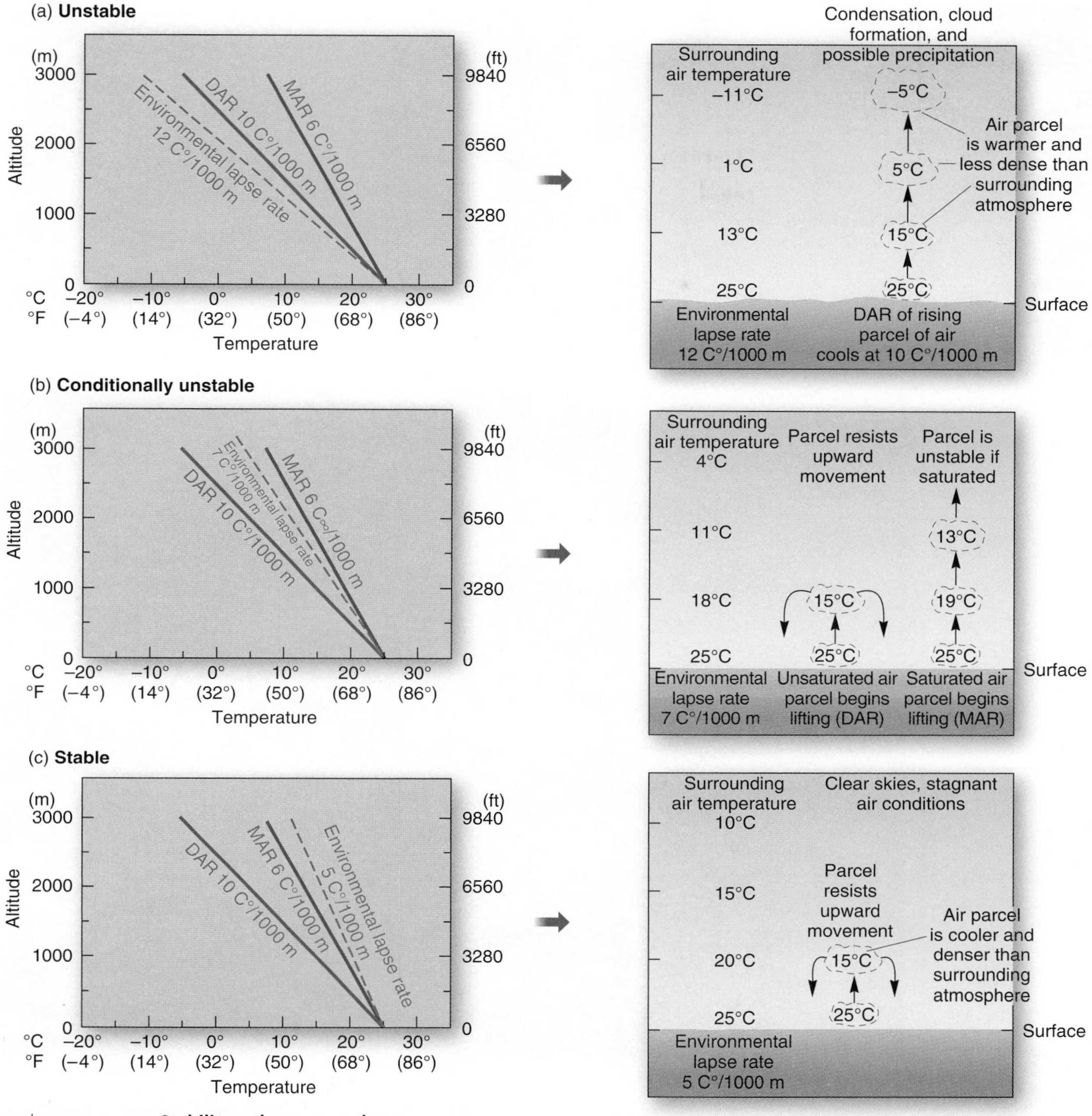

FIGURE 7.19 Stability—three examples.
Specific examples of (a) unstable, (b) conditionally unstable, and (c) stable conditions in the lower atmosphere. Note the response to these three conditions in the air parcel on the right side of each diagram.

1 billion per cubic meter, including sea salts derived from ocean sprays that supply some of the needed nuclei. The lower atmosphere never lacks cloud-condensation nuclei.

Given the presence of saturated air, the availability of cloud-condensation nuclei, and the presence of cooling (lifting) mechanisms in the atmosphere, condensation occurs. Two principal processes account for the majority of the world's raindrops and snowflakes: the *collision–coalescence process* (for warmer clouds, involving falling coalescing

(text continued on page 182)

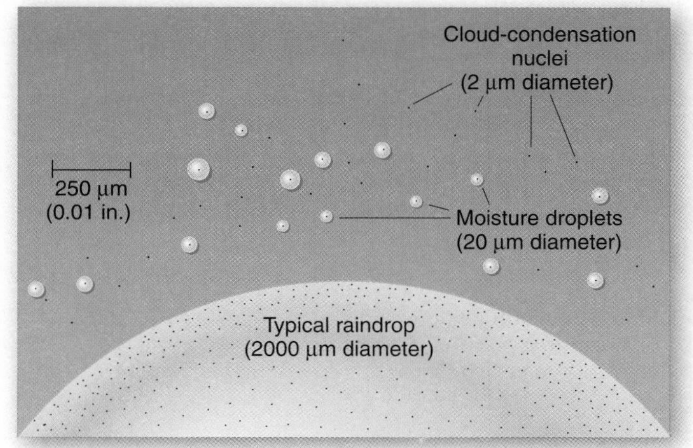

FIGURE 7.20 Moisture droplets and raindrops.
Cloud-condensation nuclei, moisture droplets, and a raindrop enlarged many times—compared at roughly the same scale.

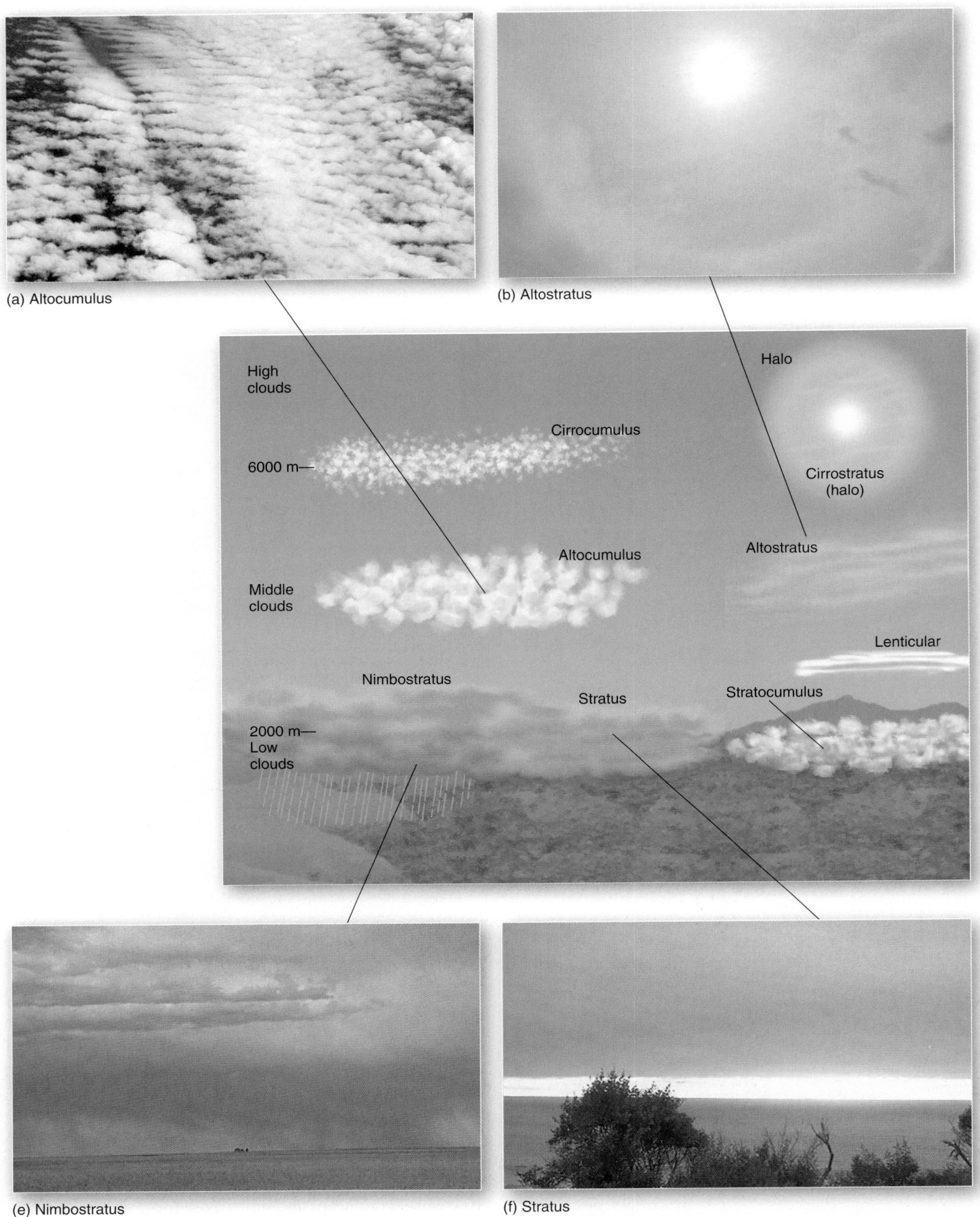

FIGURE 7.21 Principal cloud types.
Principal cloud types, classified by form (cirroform, stratiform, and cumuliform) and altitude (low, middle, high, and vertically developed across altitude): (a) altocumulus (aerial), (b) altostratus, (c) cirrus, (d) cirrostratus, (e) nimbostratus, (f) stratus, (g) cumulus, and (h) cumulonimbus. [(a), (b), (c), (g), and (h) by Bobbé Christopherson; (d), (e), and (f) by author.]

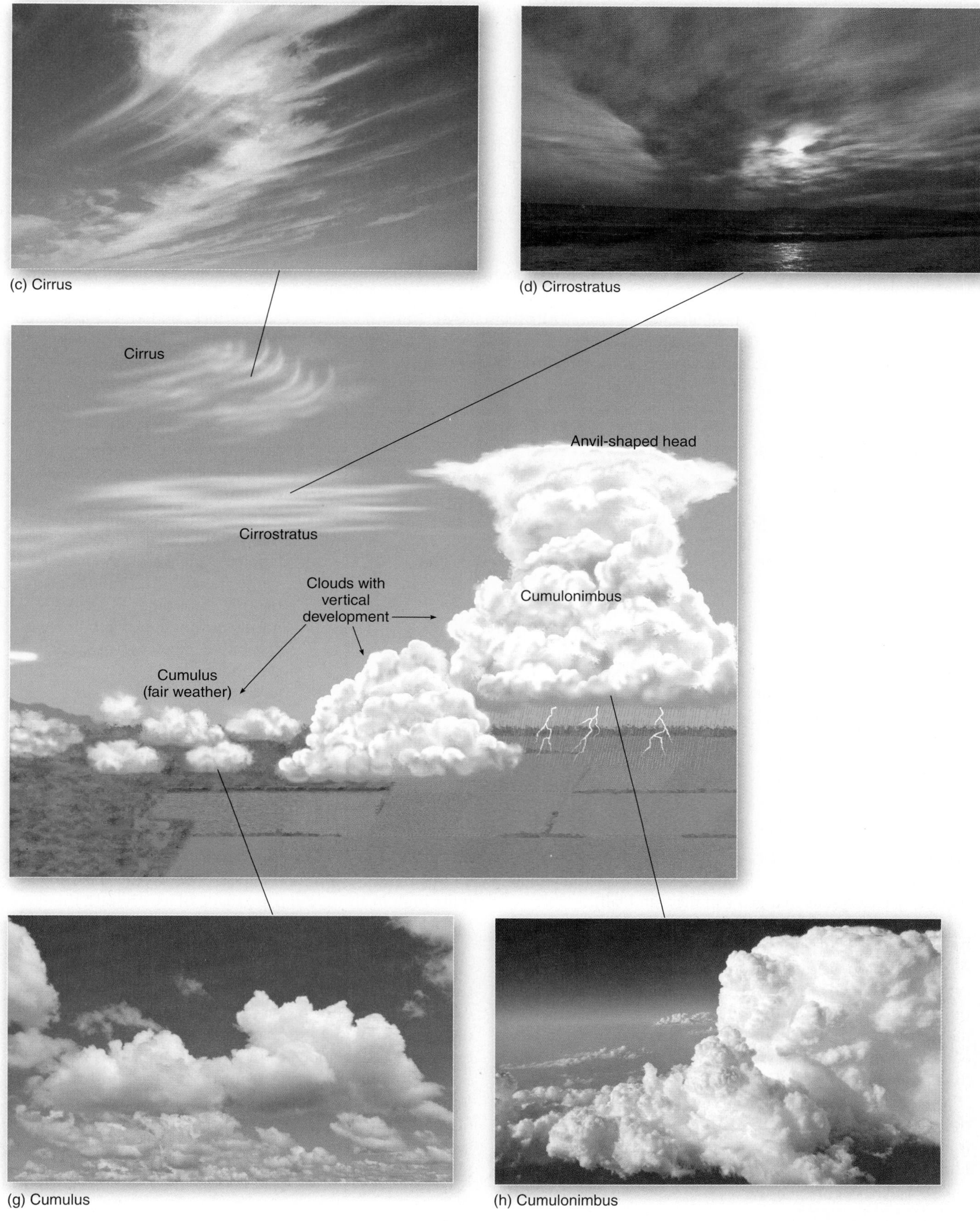

(c) Cirrus

(d) Cirrostratus

(g) Cumulus

(h) Cumulonimbus

TABLE 7.2 Cloud Classes and Types

Class	Altitude/Composition at Midlatitudes	Type	Symbol	Description
Low clouds (C_L)	Up to 2000 m (6500 ft)	Stratus (St)		Uniform, featureless, gray, like high fog
	Water	Stratocumulus (Sc)		Soft, gray, globular masses in lines, groups, or waves, heavy rolls, irregular overcast patterns
		Nimbostratus (Ns)		Gray, dark, low, with drizzling rain
Middle clouds (C_M)	2000–6000 m (6500–20,000 ft)	Altostratus (As)		Thin to thick, no halos, Sun's outline just visible, gray day
	Ice and water	Altocumulus (Ac)		Patches of cotton balls, dappled, arranged in lines or groups, rippling waves, the lenticular clouds associated with mountains
High clouds (C_H)	6000–13,000 m (20,000–43,000 ft)	Cirrus (Ci)		Mares' tails, wispy, feathery, hairlike, delicate fibers, streaks, or plumes
	Ice	Cirrostratus (Cs)		Veil of fused sheets of ice crystals, milky, with Sun and Moon halos
		Cirrocumulus (Cc)		Dappled, "mackerel sky," small white flakes, tufts, in lines or groups, sometimes in ripples
Vertically developed clouds	Near surface to 13,000 m (43,000 ft)	Cumulus (Cu)		Sharply outlined, puffy, billowy, flat-based, swelling tops, fair weather
	Water below, ice above	Cumulonimbus (Cb)		Dense, heavy, massive, dark thunderstorms, hard showers, explosive top, great vertical development, towering, cirrus-topped plume blown into anvil-shaped head

droplets) and the *Bergeron ice-crystal process* (involving supercooled water droplets, which evaporate and are absorbed by ice crystals that grow in mass and fall).

Cloud Types and Identification

As noted, a cloud is a collection of moisture droplets and ice crystals suspended in air in sufficient volume and concentration to be visible. In 1803, English biologist and amateur meteorologist Luke Howard, in his article "On the Modification of Clouds," established a classification system for clouds and coined Latin names for them that we still use. *Altitude* and *shape* are key to cloud classification. Clouds occur in three basic forms—flat, puffy, and wispy—and in four primary altitude classes and ten basic cloud types. Horizontally developed clouds—flat and layered—are *stratiform* clouds. Vertically developed clouds—puffy and globular—are *cumuliform* clouds. Wispy clouds usually are quite high in altitude and are made of ice crystals; these are *cirroform*.

These three basic forms occur in four altitudinal classes: low, middle, high, and those vertically developed through the troposphere. Table 7.2 presents the basic cloud classes and types, and cloud symbols. Figure 7.21 on pages 180-181 illustrates the general appearance of each type and includes representative photographs.

Low clouds, ranging from the surface up to 2000 m (6500 ft) in the middle latitudes, are simply *stratus* or *cumulus* (Latin for "layer" and "heap," respectively). **Stratus** clouds appear dull, gray, and featureless. When they yield precipitation, they become **nimbostratus** (*nimbo-* denotes

 **GEO REPORT 7.4** *Luke Howard knew the signs for clouds*

In 1803, when Luke Howard published his article and gave a series of lectures on his cloud classification system, he not only designated all the Latin names, but also designed the symbols used for each cloud type. These symbols, shown in Table 7.2, are still in full use to this day.

"stormy" or "rainy"), and their showers typically fall as drizzling rain (Figure 7.21e).

Cumulus clouds appear bright and puffy, like cotton balls. When they do not cover the sky, they float by in infinitely varied shapes. Vertically developed cumulus clouds are in a separate class in Table 7.2 because further vertical development can produce cumulus clouds that extend beyond low altitudes into middle and high altitudes (illustrated at the far right in Figure 7.21 and Figure 7.21h).

Sometimes near the end of the day **stratocumulus** may fill the sky in patches of lumpy, grayish, low-level clouds. Near sunset, these spreading puffy stratiform remnants may catch and filter the Sun's rays, sometimes indicating clearing weather.

The prefix *alto-* (meaning "high") denotes middle-level clouds. They are made of water droplets, and, when cold enough, these droplets can be mixed with ice crystals. **Altocumulus** clouds, in particular, represent a broad category that occurs in many different styles: patchy rows, wave patterns, a "mackerel sky," or lens-shaped (lenticular) clouds.

Ice crystals in thin concentrations compose clouds occurring above 6000 m (20,000 ft). These wispy filaments, usually white except when colored by sunrise or sunset, are **cirrus** clouds (Latin for "curl of hair"), sometimes dubbed mares' tails. Cirrus clouds look as though an artist took a brush and added delicate feathery strokes high in the sky. Cirrus clouds can indicate an oncoming storm, especially if they thicken and lower in elevation. The prefix *cirro-*, as in *cirrostratus* and *cirrocumulus*, indicates other high clouds that form a thin veil or puffy appearance, respectively.

A cumulus cloud can develop into a towering giant called **cumulonimbus** (again, *-nimbus* in Latin denotes "rain storm" or "thundercloud"; Figure 7.22). Such clouds

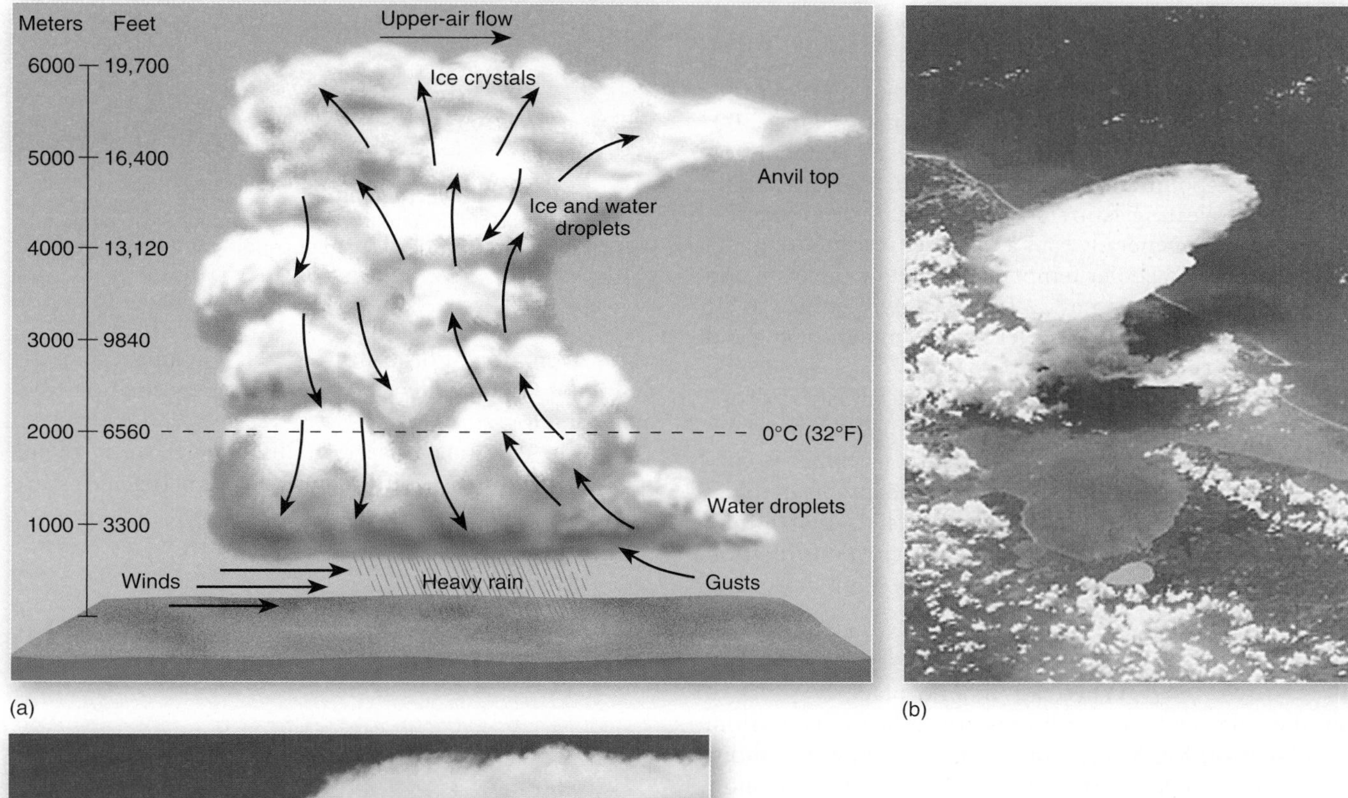

FIGURE 7.22 Cumulonimbus thunderhead.
(a) Structure and form of a cumulonimbus cloud. Violent updrafts and downdrafts mark the circulation within the cloud. Blustery wind gusts occur along the ground. (b) Space Shuttle astronauts capture a dramatic cumulonimbus thunderhead as it moves over Galveston Bay, Texas. (c) Few acts of nature can match the sheer power released by an intense thunderstorm and related turbulence for air travelers, here rising above the plane's altitude of 10,000 m (32,808 ft). [(b) NASA; (c) Bobbé Christopherson.]

are *thunderheads* because of their shape and associated lightning and thunder. Note the surface wind gusts, updrafts and downdrafts, heavy rain, and ice crystals present at the top of the rising cloud column. High-altitude winds may then shear the top of the cloud into the characteristic anvil shape of the mature thunderhead.

FIGURE 7.23 Advection fog.
San Francisco's Golden Gate Bridge shrouded by an invading advection fog characteristic of summer conditions along a western coast. [Author.]

CRITICAL THINKING 7.3

Be a cloud and weather observer

Using Figure 7.21, begin to observe clouds on a regular basis. See if you can relate the cloud type to particular weather conditions at the time of observation. You may want to keep a log in your notebook during this physical geography course. Seeing clouds in this way will help you to understand weather and make studying Chapter 8 easier. See if by the end of this course you can have each type checked off in Table 7.2.

Fog

By international definition, **fog** is a cloud layer on the ground, with visibility restricted to less than 1 km (3300 ft). The presence of fog tells us that the air temperature and the dew-point temperature at ground level are nearly identical, indicating saturated conditions. A temperature-inversion layer generally caps a fog layer (warmer temperatures above and cooler temperatures below the inversion altitude), with as much as 22 C° (40 F°) difference in air temperature between the cooler ground under the fog and the warmer, sunny skies above.

Almost all fog is warm—that is, its moisture droplets are above freezing. Supercooled fog, which occurs when the moisture droplets are below freezing, is special because it can be dispersed by means of artificial seeding with ice crystals or other crystals that mimic ice, following the principles of the ice-crystal formation process described earlier. Let's briefly look at several types of fog.

Advection Fog When air in one place *migrates* to another place where conditions are right for saturation, an **advection fog** forms. For example, when warm, moist air overlays cooler ocean currents, lake surfaces, or snow masses, the layer of migrating air directly above the surface becomes chilled to the dew point and fog develops. Off all subtropical west coasts in the world, summer fog forms in the manner just described (Figure 7.23).

Desert organisms have adapted remarkably to the presence of coastal fog along western coastlines in subtropical latitudes. For example, sand beetles in the Namib Desert in extreme southwestern Africa harvest water from the fog. They hold up their wings so condensation collects and runs down to their mouths. As the day's heat arrives, they burrow into the sand, only to emerge the next night or morning when the advection fog brings in more water for harvesting. Throughout history, people harvested water from fog. For centuries, coastal villages in the deserts of Oman collected water drips deposited on trees by coastal fogs.

In the Atacama Desert of Chile and Peru, residents stretch large nets to intercept advection fog; moisture

FIGURE 7.24 Fog harvesting.
In the mountains inland from Chungungo, Chile, polypropylene mesh stretched between two posts captures advection fog for local drinking-water supplies. [Robert S. Schemenauer.]

condenses on the netting and drips into trays, flowing through pipes to a 100,000-L (26,000-gal.) reservoir. Large sheets of plastic mesh along a ridge of the El Tofo Mountains harvest water from advection fog (Figure 7.24). Chungungo, Chile, receives 10,000 L (2,600 gal.) of water from 80 fog-harvesting collectors in a project developed by Canadian (International Development Research Center) and Chilean interests and made operational in 1993. At least 30 countries across the globe experience conditions suitable for this water resource technology. (See http://www.oas.org/dsd/publications/unit/oea59e/ch33. htm.)

Another type of advection fog forms when cold air lies over the warm water of a lake, an ocean surface, or even a swimming pool. This wispy **evaporation fog**, or *steam fog*, may form as water molecules evaporate from the water surface into the cold overlying air, effectively humidifying the air to saturation, followed by condensation to form fog (Figure 7.25). When evaporation fog happens at sea, it is a shipping hazard called *sea smoke*.

Advection fog also forms when moist air flows to higher elevations along a hill or mountain. This upslope lifting leads to adiabatic cooling by expansion as the air rises. The resulting **upslope fog** forms a stratus cloud at the condensation level of saturation. Along the Appalachians and the eastern slopes of the Rockies such fog is common in winter and spring. Another advection fog associated with topography is **valley fog**. Because cool air is denser than warm air, it settles in low-lying areas, producing a fog in the chilled, saturated layer near the ground in the valley (Figure 7.26).

Radiation Fog When radiative cooling of a surface chills the air layer directly above that surface to the dew-point temperature, creating saturated conditions, a **radiation fog** forms. This fog occurs over moist ground, especially on clear nights; it does not occur over water because water does not cool appreciably overnight. Slight movements of air deliver even more moisture to the cooled area for more fog formation of greater depth (Figure 7.27).

FIGURE 7.26 **Valley fog.**
Cold air settles in the valleys of the Appalachian Mountains, chilling the air to the dew point and forming a valley fog. [Author photo.]

FIGURE 7.25 **Evaporation fog.**
Evaporation fog, or sea smoke, highlights dawn on a cold morning at Donner Lake, California. Later that morning as air temperatures rose, what do you think happened to the evaporation fog? [Bobbé Christopherson.]

FIGURE 7.27 **Winter radiation fog.**
Radiation fog in the southern Great Valley of California. This fog is locally known as a ***tule fog*** (pronounced "toolee") because of its association with the tule (bulrush) plants that line the low-elevation islands and marshes of the Sacramento River and San Joaquin River delta regions. [*Terra* MODIS image, NASA/NOAA.]

FIGURE 7.28 Two kinds of fog.
The cold air above the river creates evaporation fog along portions of the channel, whereas radiation fog is forming over the farmland. [Bobbé Christopherson.]

In Figure 7.28 can you tell what two kinds of fog are pictured? The river water is warmer than the cold overlying air, producing an evaporation fog, especially beyond the bend in the river. The moist farmlands have radiatively cooled overnight, chilling the air along the surface to the dew point, resulting in active condensation. You see wisps of radiation fog as light air movements flow from right to left in the photo.

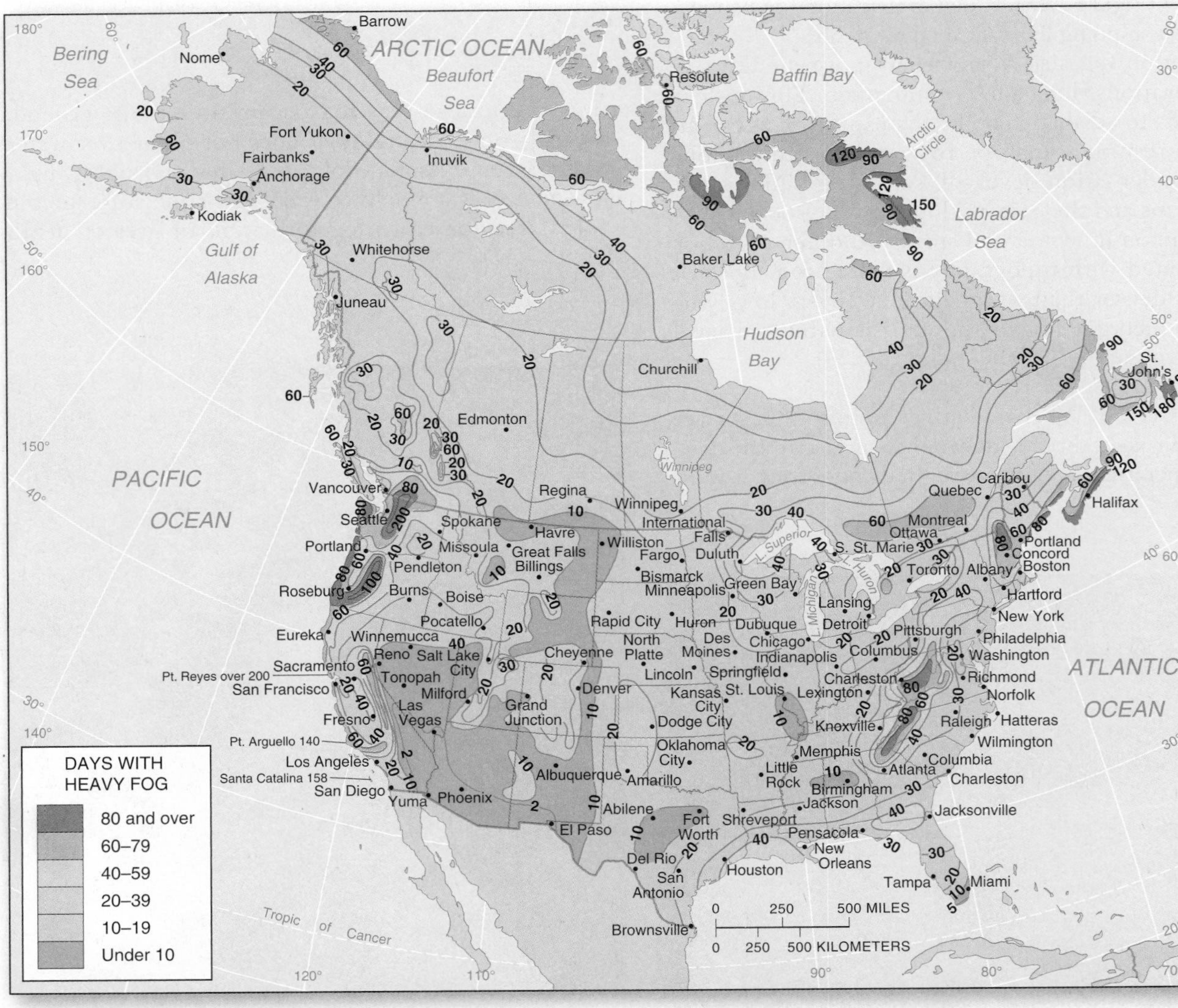

FIGURE 7.29 Mean annual number of days with heavy fog in the United States and Canada.
The foggiest spot in the United States is the mouth of the Columbia River where it enters the Pacific Ocean at Cape Disappointment, Washington. One of the foggiest places in the world is Newfoundland's Avalon Peninsula, specifically Argentia and Belle Isle, which regularly exceed 200 days of fog each year. [Data courtesy of NWS; *Climatic Atlas of Canada*, Atmospheric Environment Service Canada; and *The Climates of Canada*, Environment Canada, 1990.]

Every year the media carry stories of multicar pileups on stretches of highway where vehicles drive at high speed in foggy conditions. These crash scenes can involve dozens of cars and trucks. Fog is a hazard to drivers, pilots, sailors, pedestrians, and cyclists, even though its conditions of formation are quite predictable. The spatial aspects of fog occurrence should be a planning element for any proposed airport, harbor facility, or highway. The prevalence of fog throughout the United States and Canada is shown in Figure 7.29.

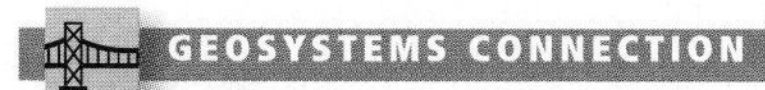

We explored the origins of water, where it is located, and its remarkable physical properties. The role of water in the atmosphere and humidity concepts formed the basis for examining stability and instability conditions, which lead to clouds and fog formation. With this foundation we move to Chapter 8 and identification of air masses and the conditions that force these to lift, cool, condense, and produce weather phenomena. Also, we look at violent weather events, such as thunderstorms, strong winds, tornadoes, and tropical cyclones.

KEY LEARNING CONCEPTS REVIEW

- ***Describe*** the origin of Earth's waters, ***define*** the quantity of water that exists today, and *list* the locations of Earth's freshwater supply.

The next time it rains where you live, pause and reflect on the journey each of those water molecules made. Water molecules came from within Earth over a period of billions of years, in the **outgassing** process. Thus began endless cycling of water through the hydrologic system of evaporation–condensation–precipitation. Water covers about 71% of Earth. Approximately 97% of it is salty seawater, and the remaining 3% is freshwater—most of it frozen.

The present volume of water on Earth is estimated at 1.36 billion km^3 (326 million mi^3), an amount achieved roughly 2 billion years ago. This overall steady-state equilibrium might seem in conflict with the many changes in sea level that have occurred over Earth's history, but is not. **Eustasy** refers to worldwide changes in sea level and relates to the change in volume of water in the oceans. The amount of water stored in glaciers and ice sheets explains these changes as **glacio-eustatic** factors. At present, sea level is rising because of increases in the temperature of the oceans and the record melting of glacial ice.

outgassing (p. 164)
eustasy (p. 165)
glacio-eustatic (p. 165)

1. Approximately where and when did Earth's water originate?
2. If the quantity of water on Earth has been quite constant in volume for at least 2 billion years, how can sea level have fluctuated? Explain.
3. Describe the locations of Earth's water, both oceanic and fresh. What is the largest repository of freshwater at this time? In what ways is this distribution of water significant to modern society?
4. Why would climate change be a concern, given this distribution of water?
5. Why might you describe Earth as the water planet? Explain.

- ***Describe*** the heat properties of water, and ***identify*** the traits of its three phases: solid, liquid, and gas.

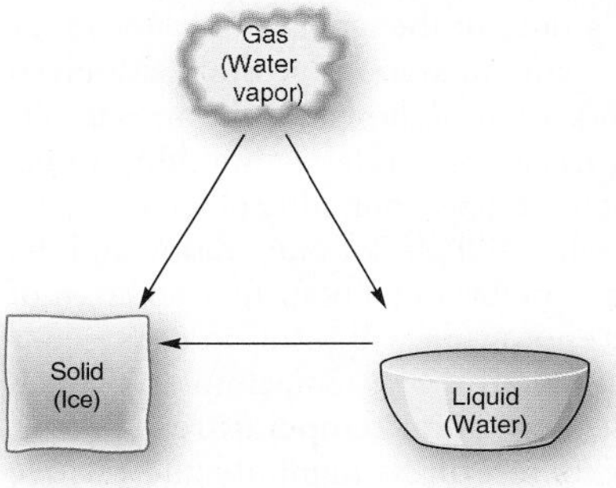

Water is the most common compound on the surface of Earth, and it possesses unusual solvent and heat characteristics. Part of Earth's uniqueness is that its water exists naturally in all three states—solid, liquid, and gas—owing to Earth's temperate position relative to the Sun. A change from one state to another is a **phase change**. The change from solid to vapor is **sublimation**; from liquid to solid, freezing; from solid to liquid, melting; from vapor to liquid, condensation; and from liquid to vapor, vaporization or evaporation.

The heat energy required for water to change phase is **latent heat** because, once absorbed, it is hidden within the structure of the water, ice, or water vapor. For 1 g of water to become 1 g of water vapor at boiling requires the addition of 540 cal, or the **latent heat of vaporization**. When this 1 g of water vapor condenses, the same amount of heat energy is liberated, as 540 cal of the **latent heat of condensation**. The **latent heat of sublimation** is the energy exchanged in the phase change from ice to vapor and vapor to ice. Weather is powered by the tremendous amount of latent heat energy involved in the phase changes among the three states of water.

phase change (p. 168)
sublimation (p. 168)
latent heat (p. 169)
latent heat of vaporization (p. 170)
latent heat of condensation (p. 170)
latent heat of sublimation (p. 171)

6. Describe the three states of matter as they apply to ice, water, and water vapor.
7. What happens to the physical structure of water as it cools below 4°C (39°F)? What are some visible indications of these physical changes?

8. What is latent heat? How is it involved in the phase changes of water?
9. Take 1 g of water at 0°C and follow it through to 1 g of water vapor at 100°C, describing what happens along the way. What amounts of energy are involved in the changes that take place?
10. Describe the physical structure of water and ice as it applies to Figure 7.6. Explain.

■ ***Define*** humidity and the expressions of the relative humidity concept, and ***explain*** dew-point temperature and saturated conditions in the atmosphere.

The amount of water vapor in the atmosphere is **humidity**. The maximum water vapor possible in air is principally a function of the temperature of the air and of the water vapor (usually these temperatures are the same). Warmer air produces higher net evaporation rates and maximum possible water vapor, whereas cooler air can produce net condensation rates and lower possible water vapor.

Relative humidity is a ratio of the amount of water vapor actually in the air to the maximum amount possible at a given temperature. Relative humidity tells us how near the air is to saturation. Relatively dry air has a lower relative humidity value; relatively moist air has a higher relative humidity percentage. Air is said to be at **saturation** when the rate of evaporation and the rate of condensation reach equilibrium; any further addition of water vapor or temperature lowering will result in active condensation (100% relative humidity). The temperature at which air achieves saturation is the **dew-point temperature**.

Among the various ways to express humidity and relative humidity are vapor pressure and specific humidity. **Vapor pressure** is that portion of the atmospheric pressure produced by the presence of water vapor. A comparison of vapor pressure with the saturation vapor pressure at any moment yields a relative humidity percentage. **Specific humidity** is the mass of water vapor (in grams) per mass of air (in kilograms) at any specified temperature. Because it is measured as a mass, specific humidity does not change as temperature or pressure changes, making it a valuable measurement in weather forecasting. A comparison of specific humidity with the maximum specific humidity at any moment produces a relative humidity percentage.

Two instruments measure relative humidity, and indirectly the actual humidity content of the air; they are the **hair hygrometer** and the **sling psychrometer**.

humidity (p. 171)
relative humidity (p. 171)
saturation (p. 171)
dew-point temperature (p. 172)
vapor pressure (p. 173)
specific humidity (p. 173)
hair hygrometer (p. 174)
sling psychrometer (p. 174)

11. What is humidity? How is it related to the energy present in the atmosphere? To our personal comfort and how we perceive apparent temperatures?
12. Define relative humidity. What does the concept represent? What is meant by the terms *saturation* and *dew-point temperature*?
13. Using different measures of humidity in the air than those used in the chapter, derive relative humidity values (vapor pressure/saturation vapor pressure; specific humidity/ maximum specific humidity) in Figures 7.12 and 7.13.
14. How do the two instruments described in this chapter measure relative humidity?
15. How does the daily distribution of relative humidity compare with the daily trend in air temperature?

■ ***Define*** atmospheric stability, and ***relate*** it to a parcel of air that is ascending or descending.

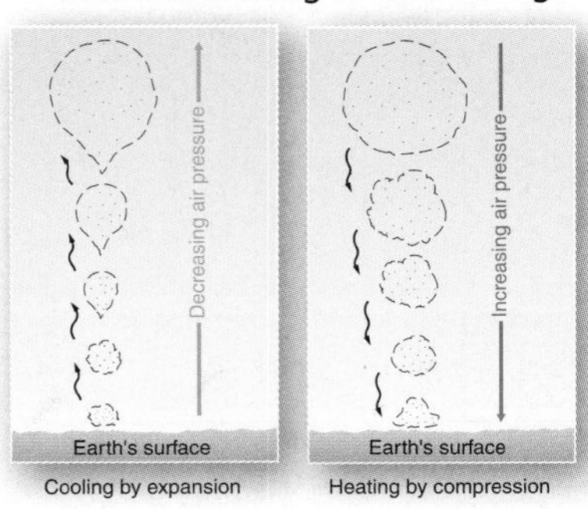

Meteorologists use the term *parcel* to describe a body of air that has specific temperature and humidity characteristics. Think of a parcel of air as a volume of air, perhaps 300 m (1000 ft) in diameter. The temperature of the volume of air determines the density of the air parcel. Warm air has a lower density in a given volume of air; cold air has a higher density.

Stability refers to the tendency of an air parcel, with its water-vapor cargo, either to remain in place or to change vertical position by ascending (rising) or descending (falling). An air parcel is *stable* if it resists displacement upward or, when disturbed, it tends to return to its starting place. An air parcel is *unstable* if it continues to rise until it reaches an altitude where the surrounding air has a density (air temperature) similar to its own.

stability (p. 175)

16. Differentiate between stability and instability relative to a parcel of air rising vertically in the atmosphere.
17. What are the forces acting on a vertically moving parcel of air? How are they affected by the density of the air parcel?

■ ***Illustrate*** three atmospheric conditions—unstable, conditionally unstable, and stable—with a simple graph that relates the environmental lapse rate (ELR) to the dry adiabatic rate (DAR) and moist adiabatic rate (MAR).

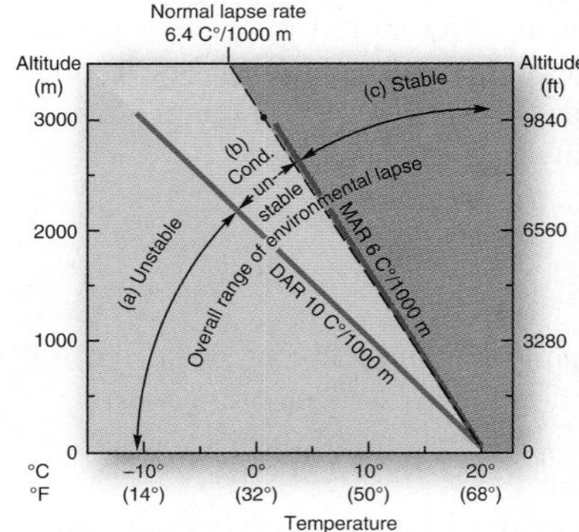

An ascending (rising) parcel of air cools by expansion, responding to the reduced air pressure at higher altitudes. A descending (falling) parcel heats by compression. These temperature changes internal to a moving air parcel are explained by physical laws that govern the behavior of gases. Temperature changes in both ascending and descending air parcels occur without any significant heat exchange between the surrounding environment and the vertically moving parcel of air. The warming and cooling rates for a parcel of expanding or compressing air are termed **adiabatic**.

The **dry adiabatic rate (DAR)** is the rate at which "dry" air cools by expansion (if ascending) or heats by compression (if descending). The term *dry* is used when air is less than saturated (relative humidity is less than 100%). The DAR is 10 C°/1000 m (5.5 F°/1000 ft). The **moist adiabatic rate (MAR)** is the average rate at which ascending air that is moist (saturated) cools by expansion or descending air warms by compression. The average MAR is 6 C°/1000 m (3.3 F°/1000 ft). This is roughly 4 C° (2 F°) less than the dry rate. The MAR, however, varies with moisture content and temperature and can range from 4 to 10 C° per 1000 m (2 to 5.5 F° per 1000 ft).

A simple comparison of the DAR and MAR in a vertically moving parcel of air with that of the ELR in the surrounding air determines the atmosphere's stability—whether it is unstable (air parcel continues lifting), stable (air parcel resists vertical displacement), or conditionally unstable (air parcel behaves as though unstable if the MAR is in operation and stable otherwise).

adiabatic (p. 176)
dry adiabatic rate (DAR) (p. 176)
moist adiabatic rate (MAR) (p. 176)

18. How do the adiabatic rates of heating or cooling in a vertically displaced air parcel differ from the normal lapse rate and environmental lapse rate?
19. Why is there a difference between the dry adiabatic rate (DAR) and the moist adiabatic rate (MAR)?
20. What atmospheric temperature and moisture conditions would you expect on a day when the weather is unstable? When it is stable? Relate in your answer what you would experience if you were outside watching.
21. Use the "Atmospheric Stability" animation in the Mastering Geography™ Study Area. Try different temperature settings on the sliders to produce stable and unstable conditions.

■ ***Identify*** **the requirements for cloud formation, and** ***explain*** **the major cloud classes and types, including fog.**

A **cloud** is an aggregation of tiny moisture droplets and ice crystals suspended in the air. Clouds are a constant reminder of the powerful heat-exchange system in the environment. **Moisture droplets** in a cloud form when saturated air and the presence of **cloud-condensation nuclei** lead to *condensation*. Raindrops are formed from moisture droplets through either the *collision–coalescence process* or the *Bergeron ice-crystal process*.

Low clouds, ranging from the surface up to 2000 m (6500 ft) in the middle latitudes, are **stratus** (flat clouds, in layers) or **cumulus** (puffy clouds, in heaps). When stratus clouds yield precipitation, they are **nimbostratus**. Sometimes near the end of the day, lumpy, grayish, low-level clouds called **stratocumulus** may fill the sky in patches. Middle-level clouds are denoted by the prefix *alto-*. **Altocumulus** clouds, in particular, represent a broad category that occurs in many different styles. Clouds at high altitude, principally composed of ice crystals, are called **cirrus**. A cumulus cloud can develop into a towering giant **cumulonimbus** cloud (*-nimbus* in Latin denotes "rain storm" or "thundercloud"). Such clouds are called *thunderheads* because of their shape and their associated lightning, thunder, surface wind gusts, updrafts and downdrafts, heavy rain, and hail.

Fog is a cloud that occurs at ground level. **Advection fog** forms when air in one place migrates to another place where conditions exist that can cause saturation—for example, when warm, moist air moves over cooler ocean currents. Another type of advection fog forms when cold air flows over the warm water of a lake, ocean surface, or swimming pool. This **evaporation fog**, or steam fog, may form as the water molecules evaporate from the water surface into the cold overlying air. **Upslope fog** is produced when moist air is forced to higher elevations along a hill or mountain. Another fog caused by topography is **valley fog**, formed because cool, denser air settles in low-lying areas, producing fog in the chilled, saturated layer near the ground. Radiative cooling of a surface that chills the air layer directly above the surface to the dew-point temperature creates saturated conditions and a **radiation fog**.

cloud (p. 178)
moisture droplet (p. 178)
cloud-condensation nuclei (p. 178)
stratus (p. 182)
nimbostratus (p. 182)
cumulus (p. 183)
stratocumulus (p. 183)
altocumulus (p. 183)
cirrus (p. 183)
cumulonimbus (p. 183)
fog (p. 184)
advection fog (p. 184)
evaporation fog (p. 185)
upslope fog (p. 185)
valley fog (p. 185)
radiation fog (p. 185)

22. Specifically, what is a cloud? Describe the droplets that form a cloud.
23. Explain the condensation process: What are the requirements? What two principal processes are discussed in this chapter?
24. What are the basic forms of clouds? Using Table 7.2, describe how the basic cloud forms vary with altitude.
25. Explain how clouds might be used as indicators of the conditions of the atmosphere and of expected weather.
26. What type of cloud is fog? List and define the principal types of fog.
27. Describe the occurrence of fog in the United States and Canada. Where are the regions of highest incidence?

MASTERING GEOGRAPHY

Visit www.masteringgeography.com for several figure animations, satellite loops, self-study quizzes, flashcards, visual glossary, case studies, career links, textbook reference maps, RSS Feeds, and an ebook version of *Geosystems*. Included are many geography web links to interesting Internet sources that support this chapter.

CHAPTER 8

Weather

Standing in the exact path where 4 days earlier a tornado tore through many residential blocks in Wadena, Minnesota. The milewide EF-4 tornado leveled neighborhoods. Advanced warnings led to no loss of life in Wadena. On June 17, 2010, central and southern Minnesota was hit by 18 tornadoes, 3 of which were rated at EF-4 (267–322 kmph; 166–200 mph). Minnesota averages 26 tornadoes a year. [Bobbé Christopherson.]

KEY LEARNING CONCEPTS

After reading the chapter, you should be able to:

- ***Describe*** air masses that affect North America, and ***relate*** their qualities to source regions.
- ***Identify*** and ***describe*** four types of atmospheric lifting mechanisms, and give an example of each.
- ***Analyze*** the pattern of orographic precipitation, and ***describe*** the link between this pattern and global topography.
- ***Describe*** the life cycle of a midlatitude cyclonic storm system, and ***relate*** this to its portrayal on weather maps.
- ***List*** the measurable elements that contribute to modern weather forecasting, and ***describe*** the technology and methods employed.
- ***Analyze*** various forms of violent weather and their characteristics, and ***review*** several examples of each.

On the Front Lines of Intense Weather

How do you deal with the televised images of intense weather, tornadoes, or floods? Imagine what people face on the front lines of these events. Improved warnings have reduced loss of life, yet higher storm intensities and frequencies have caused huge increases in property damage.

Think of Greensburg, Kansas, southeast of Dodge City, with its 1600 residents. For tourists, the town features the world's largest hand-dug water well. Nearby stood the old water tower with the town's name inscribed before the tornado hit—here is small town, Midwest, USA.

On the evening of May 4, 2007, a frontal system moved through. As tornado warning sirens and NOAA weather radio signaled the alert, people went to their basements and storm shelters. Just before 10 P.M., an EF-5–intensity tornado roared in from the southwest. This was the first designated "5" since the Fujita Scale revision in February 2007, described in this chapter.

The monster was 2.7 km (1.7 mi) wide, and it attacked Greensburg. It shredded all that was familiar into a chaotic mess in less than 10 minutes. After the storm, people came out into the now-silent evening, lights were out, some called for help, yet nothing familiar was left. Flashlights came on, as did headlights when an owner could find keys and a surviving car. Sunrise brought shock; all the tree-lined streets were barren, trees pruned and broken by the winds. Eleven people died that night.

Heroism was all about, the National Guard mobilized, a federal disaster area was declared, the media arrived, and in a few days politicians began their tours. Residents appeared on TV, interviewed for their anecdotes. Disaster assistance processing was set up, forms were filled out, and weeks of delays began. In the void of factual information, rumors spread quickly.

In a couple of weeks, the outsiders left; streets and lots were cleared of debris, piled into a hastily prepared dump. All that was familiar was in the pile north of town. The months rolled by with no media coverage, just the daily task of dealing with shattered reality. A year later, when we asked people about that night, they readily told their stories. How do people on disaster front lines physically and mentally deal with such changes in their lives?

FIGURE GN 8.1 Most of Greensburg, Kansas, was erased by an EF-5 tornado in 2007. A new water tower stands a year later where the old one fell. [Bobbé Christopherson.]

We drive along the Bolivar Peninsula on the Gulf Coast of Texas one year after the September 13, 2008, 2 A.M. landfall of Hurricane Ike. We pass 35 km (21 mi) of leveled coastal neighborhoods. Ike's winds were hurricane force for more than 177 km (110 mi) along the coast. Debris protrudes from the barrier-island sand, power is still out in spots, and few trees are left standing. Near the western end of the island, toward the Galveston ferry dock, large piles of former homes and businesses stand above the sand. Several piles are burning, filling the air with foul smoke—and this is a year since the hurricane.

Bolivar, like Hatteras and many other barrier islands, uses septic sewage systems, which become disrupted by storm surge. In 2009, residents in Bolivar were wrestling with whether to attach to a sewer line for the first time. One developer is constructing small houses on 8.5-m (28-ft) stilts, perched in ominous preparation for the next hurricane.

We can easily expand this narrative to thoughts of New Orleans years after Katrina, Rita, Gustav, and Cindy, and now the BP oil-spill disaster; or the floods in Missouri, North Dakota, Oklahoma City, parts of Texas, Milwaukee, Chicago, and elsewhere; or the Wadena tornado in the chapter-opening photo. Again the question: How do people on the front lines cope with such disaster? Contemplate the human dimension as you work through the weather chapter.

FIGURE GN 8.2 Hurricane Ike swept away all homes and the pavement in this neighborhood. Many kilometers of coastline were destroyed on Bolivar Peninsula, Texas. [Bobbé Christopherson.]

Water has a leading role in the vast drama played out daily on Earth's stage. It affects the stability of air masses and their interactions and produces powerful and beautiful special effects in the lower atmosphere. Air masses come into conflict; they move and shift, dominating now one region and then another, varying in strength and characteristics. Think of the weather as a play, North America as the stage, and the air masses as actors of varying ability.

In this chapter: We follow huge air masses across North America, observe powerful lifting mechanisms in the atmosphere, revisit the concepts of stable and unstable conditions, and examine migrating cyclonic systems with attendant cold and warm fronts. We conclude with a portrait of violent and dramatic weather so often in the news in recent years.

Water, with its ability to absorb and release vast quantities of heat energy, drives this daily drama in the atmosphere. The spatial implications of these weather phenomena and their relationship to human activities strongly link meteorology and weather forecasting to physical geography and this chapter.

Weather Essentials

Weather is the short-term, day-to-day condition of the atmosphere, contrasted with *climate*, which is the long-term average (over decades) of weather conditions and extremes in a region. Weather is, at the same time, both a "snapshot" of atmospheric conditions and a technical status report of the Earth–atmosphere heat-energy budget. Important elements that contribute to weather are temperature, air pressure, relative humidity, wind speed and direction, and seasonal factors such as insolation receipt related to daylength and Sun angle.

The cost of weather-related destruction (due to droughts, floods, hail, tornadoes, derechos, tropical systems, storm surges, blizzards and ice storms, and wildfires) has risen more than 500% since 1975, making this a subject of great interest. NOAA's National Climate Data Center publication *Billion Dollar Climate and Weather Disasters, 1980–2009* (http://www.ncdc.noaa.gov/img/reports/billion/billionz-2009.pdf) summarizes 96 events surpassing $1 billion in cost.

We turn for weather forecasts to the National Weather Service (NWS) in the United States (http://www.nws.noaa.gov/) or the Canadian Meteorological Centre, a branch of the Meteorological Service of Canada (MSC) (http://www.msc-smc.ec.gc.ca/cmc/index_e.html), to see current satellite images and to hear weather analysis. Internationally, the World Meteorological Organization coordinates weather information (see http://www.wmo. ch/). Many sources of weather information and related topics are found on our *Mastering Geography* Web site.

Meteorology is the scientific study of the atmosphere. (*Meteor* means "heavenly" or "of the atmosphere.") Meteorologists study the atmosphere's physical characteristics and motions; related chemical, physical, and geologic processes; the complex linkages of atmospheric systems; and weather forecasting. Massive computers handle the volumes of data from surface, aircraft, and orbital platforms used to accurately forecast near-term weather and to study trends in long-term weather, climatology, and climatic change.

An essential part of understanding weather is Doppler radar (Figure 8.1). Using backscatter from two radar pulses,

(a)

(b)

FIGURE 8.1 NWS weather installation.
(a) A radar antenna is within the dome structure at the Indianapolis International Airport. (b) WSR-88D coverage in the United States. Other installations are maintained in Japan, Guam, South Korea, and the Azores. [(a) Bobbé Christopherson. (b) NWS Radar Operations Center, Norman, Oklahoma.]

it detects the direction of moisture droplets toward or away from the radar, indicating wind direction and speed. This information is critical to providing accurate severe storm warnings. As part of the Next Generation Weather Radar (NEXRAD) program, 159 WSR-88D (*W*eather *S*urveillance *R*adar) Doppler radar systems are operational through the NWS, Federal Aviation Administration, and Department of Defense.

Air Masses

Each area of Earth's surface imparts its temperature and moisture characteristics to overlying air. The effect of the surface on the air creates regional air masses with a homogenous mix of temperature, humidity, and stability. These masses of air interact to produce weather patterns. Such a distinctive body of air is an **air mass**, and it initially reflects the characteristics of its *source region* and sometimes extends through the lower half of the troposphere. For example, weather forecasters speak of a "cold Canadian air mass" and "moist tropical air mass."

Air Masses Affecting North America

We classify air masses generally according to the moisture and temperature characteristics of their source regions:

1. *Moisture*—designated **m** for maritime (wet) and **c** for continental (dry).
2. *Temperature* (latitude factor)—designated **A** for arctic, **P** for polar, **T** for tropical, **E** for equatorial, and **AA** for Antarctic.

The principal air masses that affect North America in winter and summer are mapped in Figure 8.2.

Continental polar (cP) air masses form only in the Northern Hemisphere and are most developed in winter and cold-weather conditions. These cP air masses are major players in middle- and high-latitude weather. The cold, dense cP air displaces moist, warm air in its path, producing lifting, cooling, and condensation. An area covered by cP air in winter experiences cold, stable air; clear skies; high pressure; and anticyclonic wind flow—all visible on the weather map in Figure 8.3. The Southern Hemisphere lacks the necessary continental landmasses at high latitudes to create such a cP air mass.

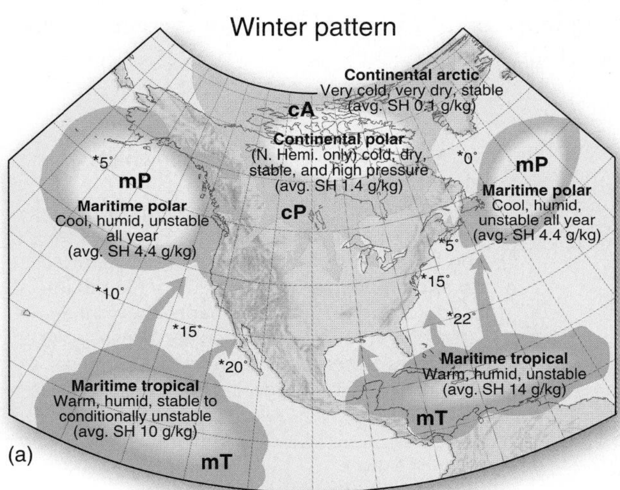

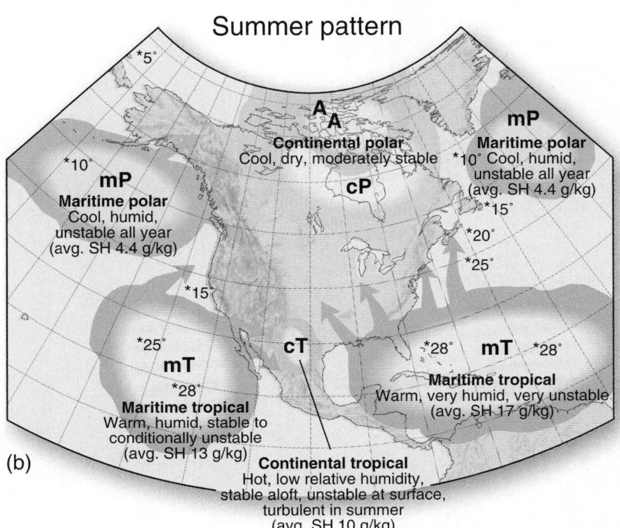

FIGURE 8.2 Principal air masses.
Air masses and their source regions influence North America during (a) winter and (b) summer. (*Sea-surface temperature in °C; SH = specific humidity.)

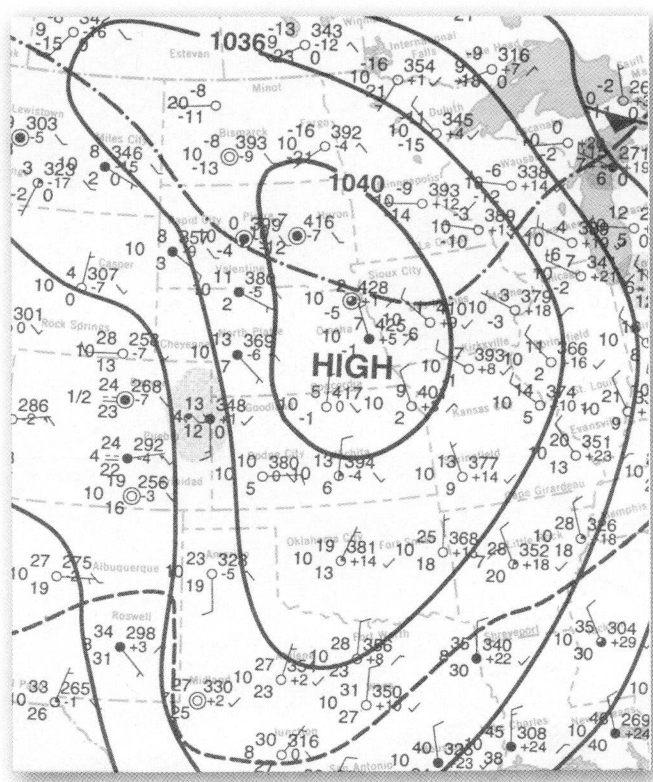

FIGURE 8.3 Winter high-pressure system.
A cP air mass dominates the Midwest with a central pressure of 1042.8 mb (30.76 in.), air temperature of –17°C (2°F), and dewpoint temperature of –21°C (–5°F), with clear, calm, stable conditions. Note the pattern of isobars portraying the cP air mass. The dotted lines are the –18°C (0°F) and 0°C (32°F) isotherms. [NWS, NOAA.]

Maritime polar (mP) air masses in the Northern Hemisphere exist over the northern oceans. Within them, cool, moist, unstable conditions prevail throughout the year. The Aleutian and Icelandic subpolar low-pressure cells reside within these mP air masses, especially in their well-developed winter pattern (see the January isobaric pressure map in Figure 6.9a).

Two *maritime tropical* (mT) air masses—the mT Gulf/Atlantic and the mT Pacific—influence North America. The humidity experienced in the East and Midwest is created by the mT Gulf/Atlantic air mass, which is particularly unstable and active from late spring to early fall. In contrast, the mT Pacific is stable to conditionally unstable and generally lower in moisture content and available energy. As a result, the western United States, influenced by this weaker Pacific air mass, receives lower average precipitation than the rest of the country. Please review Figure 6.12 and the discussion of subtropical high-pressure cells.

Air Mass Modification

As air masses migrate from source regions, their temperature and moisture characteristics modify and slowly take on the characteristics of the land over which they pass. For example, an mT Gulf/Atlantic air mass may carry humidity to Chicago and on to Winnipeg, but gradually loses its initial characteristics of high humidity and warmth with each day's passage northward.

Similarly, below-freezing temperatures occasionally reach into southern Texas and Florida, brought by an invading winter cP air mass from the north. However, that air mass warms from the −50°C (−58°F) of its winter source region in central Canada, especially after it leaves areas covered by snow.

Modification of cP air as it moves south and east produces snowbelts that lie to the east of each of the Great Lakes. As below-freezing cP air passes over the warmer Great Lakes, it absorbs heat energy and moisture from the lake surfaces and becomes *humidified*. This enhancement produces heavy lake-effect snowfall downwind into Ontario, Québec, Michigan, northern Pennsylvania, and New York—some areas receiving in excess of 250 cm (100 in.) in average snowfall a year (Figure 8.4).

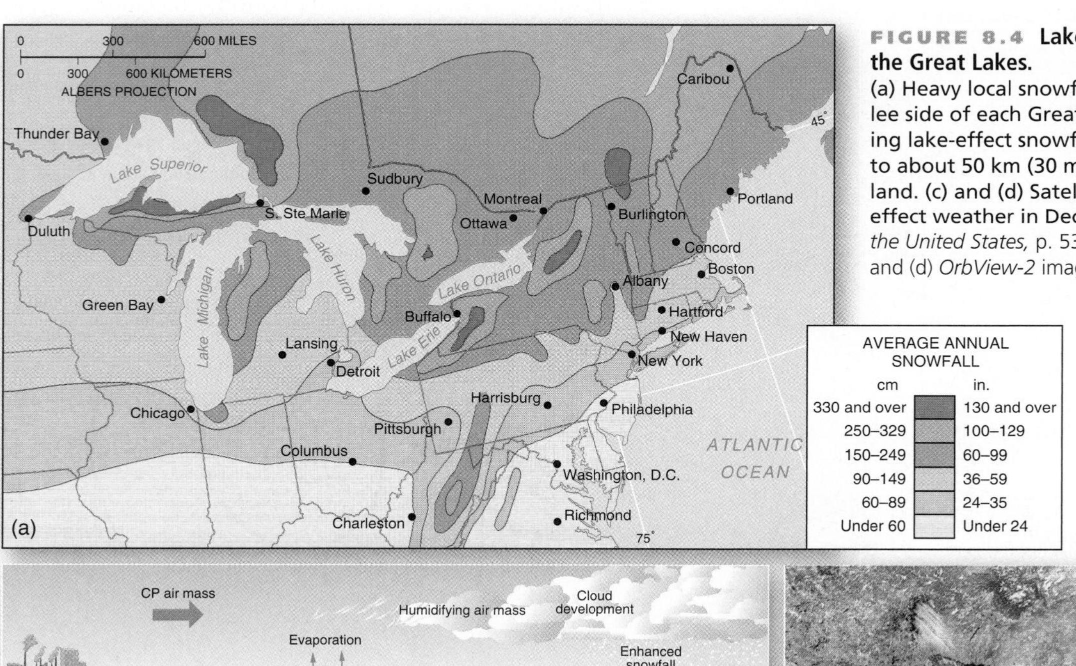

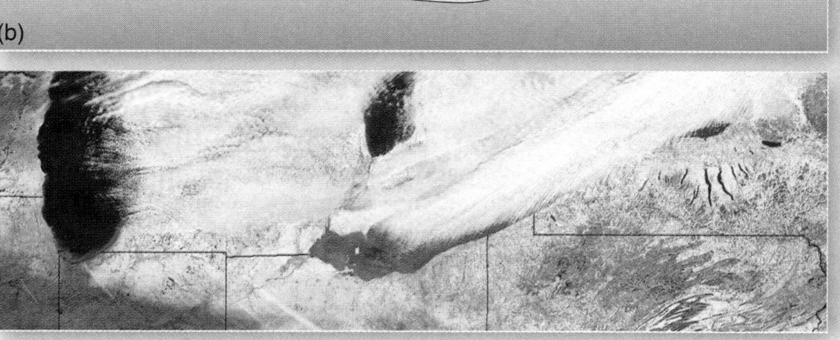

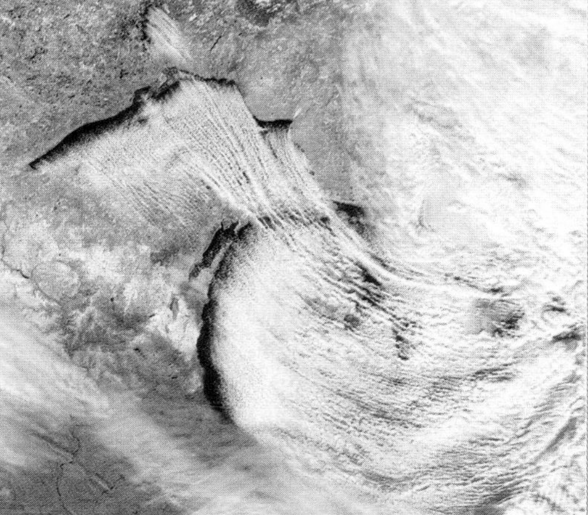

(c) (d)

FIGURE 8.4 Lake-effect snowbelts of the Great Lakes. (a) Heavy local snowfall is associated with the lee side of each Great Lake. (b) Processes causing lake-effect snowfall are generally limited to about 50 km (30 mi) to 100 km (60 mi) inland. (c) and (d) Satellite images show lake-effect weather in December. [(a) *Climatic Atlas of the United States,* p. 53. (c) *Terra* MODIS image and (d) *OrbView-2* image; both NASA/GSFC.]

Severity correlates to a low-pressure system positioned north of the Great Lakes, with counterclockwise winds pushing air across the lakes. With global warming, regional temperatures are increasing, which leads forecasters to predict enhanced lake-effect snowfall over the next several decades, since warmer air can absorb more water vapor. Forecast models show that later in the century, snowfall will decrease as temperatures rise, although rainfall totals will continue to increase over the leeside region. Temperature and evaporation rates are increasing for all of these lakes.

Atmospheric Lifting Mechanisms

Lifting air masses cool adiabatically (by expansion) to reach the dew-point temperature. Moisture in the saturated air then can condense and form clouds, and perhaps precipitation. Four principal lifting mechanisms operate in the atmosphere:

- Convergent lifting—air flows toward an area of low pressure
- Convectional lifting—air is stimulated by local surface heating
- Orographic lifting—air is forced over a barrier such as a mountain range
- Frontal lifting—air is displaced upward along the leading edges of contrasting air masses

Descriptions of all four mechanisms follow (see also Figure 8.5).

Convergent Lifting

Air flowing from different directions into the same low-pressure area is converging, displacing air upward in **convergent lifting**. All along the equatorial region, the

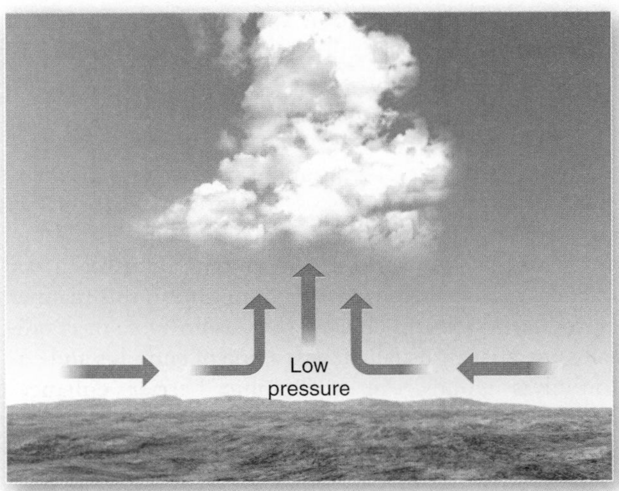

(a) Convergent lifting

(b) Convectional lifting

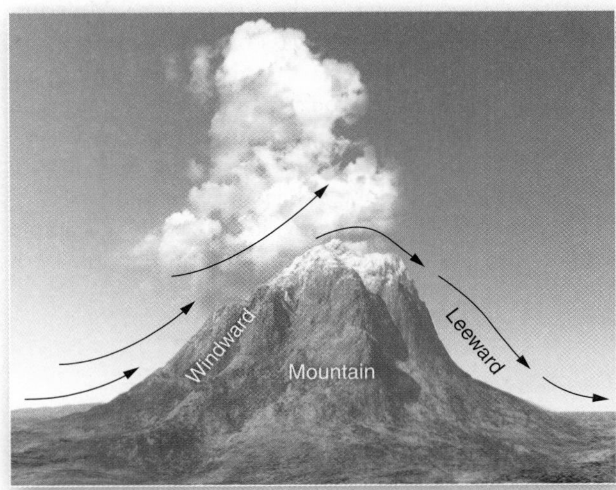

(c) Orographic lifting

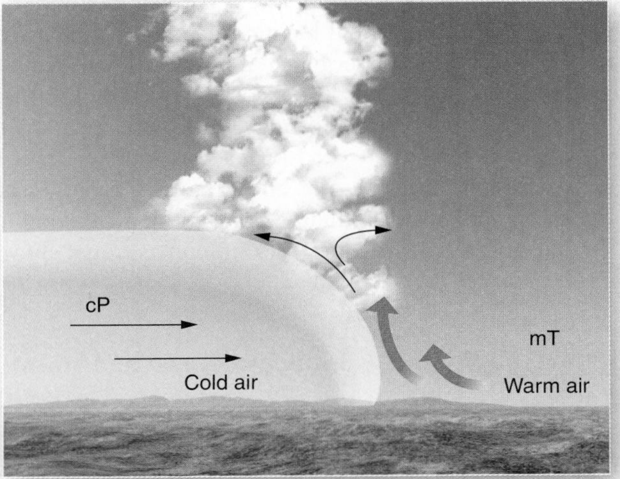

(d) Frontal lifting, cold front example

FIGURE 8.5 Four atmospheric lifting mechanisms.

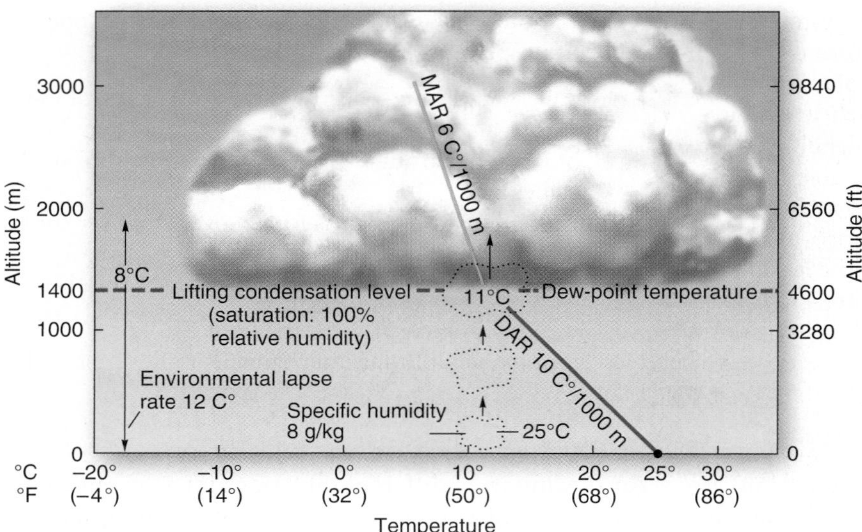

FIGURE 8.6 Convectional activity in unstable conditions.
Work through these unstable atmospheric conditions, with an environmental lapse rate of 12 C°/1000 m. Specific humidity of the air parcel is 8 g/kg and the beginning temperature is 25°C. Looking back to Figure 7.13, you find that air with a specific humidity of 8 g/kg must be cooled to 11°C to achieve the dew-point temperature. Here the dew point is reached after 14 C° of adiabatic cooling at 1400 m (4600 ft). Note that the DAR (dry adiabatic rate) is used when the air parcel is less than saturated, changing to the MAR (moist adiabatic rate) above the lifting-condensation level at 1400 m.

southeast and northeast trade winds converge, forming the intertropical convergence zone (ITCZ) and areas of extensive uplift, towering cumulonimbus cloud development, and high average annual precipitation (see Figure 6.10).

Convectional Lifting

When an air mass passes from a maritime source region to a warmer continental region, heating from the warmer land causes lifting and convection in the air mass. Other sources of surface heating might include an urban heat island or an area of dark soil in a plowed field—the warmer surfaces produce **convectional lifting**. If conditions are unstable, initial lifting continues and clouds develop. Figure 8.5b illustrates convectional action stimulated by local heating, with unstable conditions present in the atmosphere. The rising parcel of air continues its ascent because it is warmer and therefore less dense than the surrounding environment (Figure 8.6).

Florida's precipitation generally illustrates both convergent and convectional lifting mechanisms. Heating of the land produces convergence of onshore winds from the Atlantic and the Gulf of Mexico. As an example of local heating and convectional lifting, Figure 8.7 depicts a day on which the landmass of Florida was warmer than the surrounding Gulf of Mexico and Atlantic Ocean. Because the Sun's radiation gradually heats the land throughout the day and warms the air above it, convectional showers tend to form in the afternoon and early evening. Thus, Florida has the highest frequency of days with thunderstorms in the United States.

Orographic Lifting

The physical presence of a mountain acts as a topographic barrier to migrating air masses. **Orographic lifting** (*oro* means "mountain") occurs when air is forcibly lifted upslope as it is pushed against a mountain. The lifting air cools adiabatically. Stable air forced upward in this manner may produce stratiform clouds, whereas unstable or conditionally unstable air usually forms a line of cumulus and cumulonimbus clouds. An orographic barrier enhances convectional activity and causes additional lifting during the passage of weather fronts and cyclonic systems, thereby extracting more moisture from passing air masses.

Figure 8.8a (p. 198) illustrates the operation of orographic lifting under unstable conditions. The wetter intercepting slope is the *windward slope*, as opposed to the drier far-side slope, known as the *leeward slope*. Moisture condenses from the lifting air mass on the windward side of the mountain; on the leeward side, the descending air mass heats by compression, and any remaining water in the air evaporates (Figure 8.8b). Thus, air beginning its ascent up a mountain can be warm and moist, but finishing its descent on the leeward slope, it becomes hot and dry.

The state of Washington provides an excellent example of this concept, as shown in Figures 8.8c and 8.9 (p. 199).

Snow avalanche of a storm

In October 2006, a lake-effect storm dubbed the "surprise storm" dropped 0.6 m (2 ft) of wet snow in the Buffalo, New York area. This was the earliest date for such a storm in 137 years of records and remarkable because the ratio of snow to water content was 6:1. An "avalanche of a storm," as the media described it, hit upstate New York in February 2007 when 2 m (6.5 ft) of snow fell in 1 day, and over 3.7 m (12 ft) fell over 7 days.

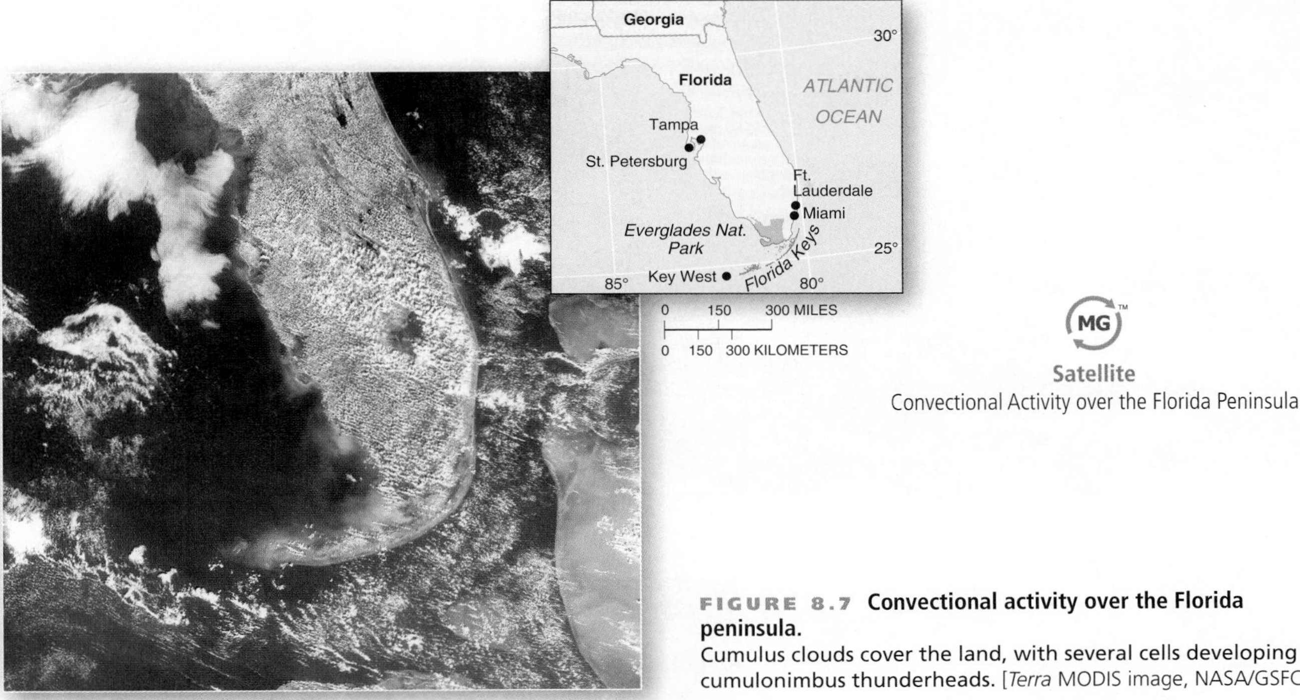

MG
Satellite
Convectional Activity over the Florida Peninsula

FIGURE 8.7 Convectional activity over the Florida peninsula.
Cumulus clouds cover the land, with several cells developing into cumulonimbus thunderheads. [*Terra* MODIS image, NASA/GSFC.]

The Olympic Mountains and Cascade Mountains orographically lift invading mP air masses from the North Pacific Ocean, squeezing out annual precipitation of more than 500 cm and 400 cm, respectively (200 in. and 160 in.).

The Quinault Ranger and Rainier Paradise weather stations demonstrate windward-slope rainfall. The cities of Sequim, in the Puget Trough, and Yakima, in the Columbia Basin, are in the rain shadow on the leeward side of these mountain ranges and are characteristically low in annual rainfall. Find these stations on the landscape profile and on the precipitation map in Figure 8.9.

In North America, **chinook winds** (called *föhn* or *foehn* winds in Europe) are the warm, downslope air flows characteristic of the leeward side of mountains. Such winds can bring a 20 C° (36 F°) jump in temperature and greatly reduce relative humidity.

The term **rain shadow** is applied to dry regions leeward of mountains. East of the Cascade Range, Sierra Nevada, and Rocky Mountains, such rain-shadow patterns predominate. In fact, the precipitation pattern of windward and leeward slopes persists worldwide, as confirmed by the precipitation maps for North America (Figure 9.5) and the world (Figure 10.2).

Frontal Lifting (Cold and Warm Fronts)

The leading edge of an advancing air mass is its *front*. Vilhelm Bjerknes (1862–1951) first applied the term while working with a team of meteorologists in Norway during World War I. Weather systems seemed to them to be migrating air mass "armies" doing battle along fronts. A front is a place of atmospheric discontinuity, a narrow zone forming a line of conflict between two air masses of different temperature, pressure, humidity, wind direction and speed, and cloud development. The leading edge of a cold air mass is a **cold front**, whereas the leading edge of a warm air mass is a **warm front** (Figures 8.10 and 8.11, p. 200).

GEO REPORT 8.2 *Mountains cause record rains*

Mount Waialeale, on the island of Kaua'i, Hawai'i, rises 1569 m (5147 ft) above sea level. On its windward slope, rainfall averaged 1234 cm (486 in., or 40.5 ft) a year for the years 1941–1992. In contrast, the rain-shadow side of Kaua'i receives only 50 cm (20 in.) of rain annually. If no islands existed at this location, this portion of the Pacific Ocean would receive only an average 63.5 cm (25 in.) of precipitation a year. (These statistics are from established weather stations with a consistent record of weather data; several stations claim higher rainfall values, but do not have dependable measurement records.)

Cherrapunji, India, is 1313 m (4309 ft) above sea level at 25° N latitude, in the Assam Hills south of the Himalayas. Summer monsoons pour in from the Indian Ocean and the Bay of Bengal, producing 930 cm (366 in., or 30.5 ft) of rainfall in one month. Not surprisingly, Cherrapunji is the all-time precipitation record holder for a single year, 2647 cm (1042 in., or 86.8 ft), and for every other time interval from 15 days to 2 years. The average annual precipitation there is 1143 cm (450 in., 37.5 ft), placing it second only to Mount Waialeale.

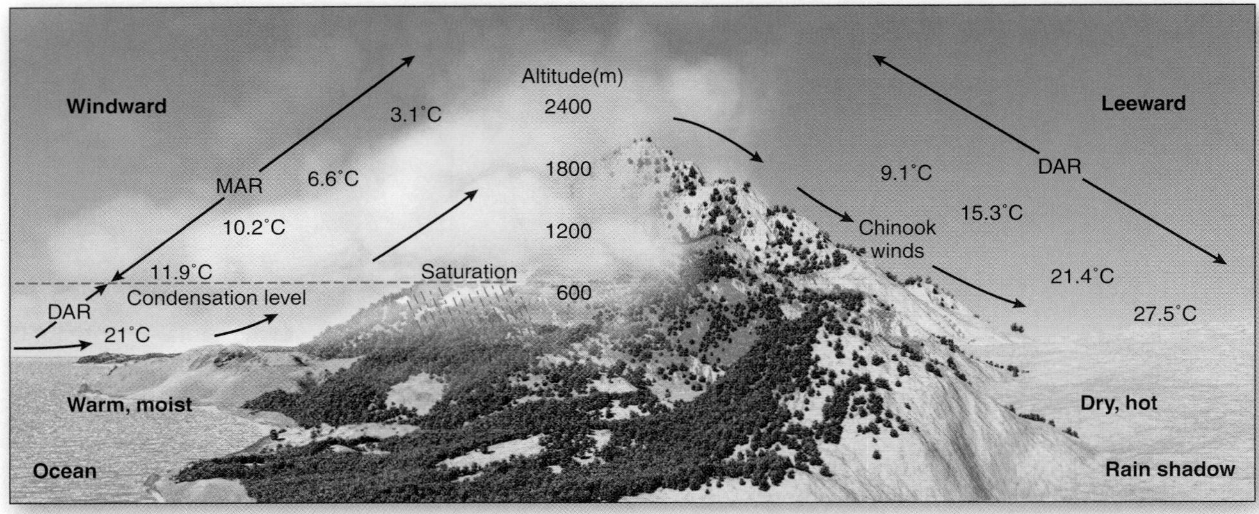

(a)

(b) (c)

FIGURE 8.8 Orographic precipitation, unstable conditions assumed.
(a) Prevailing winds force warm, moist air upward against a mountain range, producing adiabatic cooling, eventual saturation and net condensation, cloud formation, and precipitation. On the leeward slope, as the "dried" air descends, compressional heating warms it and net evaporation dominates, creating the hot, relatively dry rain shadow. (b) Rain shadow produced by descending, warming air contrasts with the clouds of the windward side. Dust is stirred up by leeward downslope winds. (c) The wetter windward slopes are in contrast to the drier leeward landscapes in Washington—check the map in Figure 8.9. [(b) Author. (c) *Terra* MODIS image, NASA/GSFC.]

Cold Front On weather maps, such as the example in Critical Thinking 8.1, a cold front is a line with triangular spikes that point in the direction of frontal movement along an advancing cP or mP air mass. The steep face of the cold air mass suggests its ground-hugging nature, caused by its greater density and uniform physical character compared to the warmer air mass it displaces (Figure 8.10a).

Warm, moist air in advance of the cold front lifts upward abruptly and experiences the same adiabatic rates of cooling and factors of stability or instability that pertain to all lifting air parcels. A day or two ahead of the cold front's passage, high cirrus clouds appear, telling observers that a lifting mechanism is on the way.

A wind shift, temperature drop, and lowering barometric pressure mark a cold front's advance due to lifting

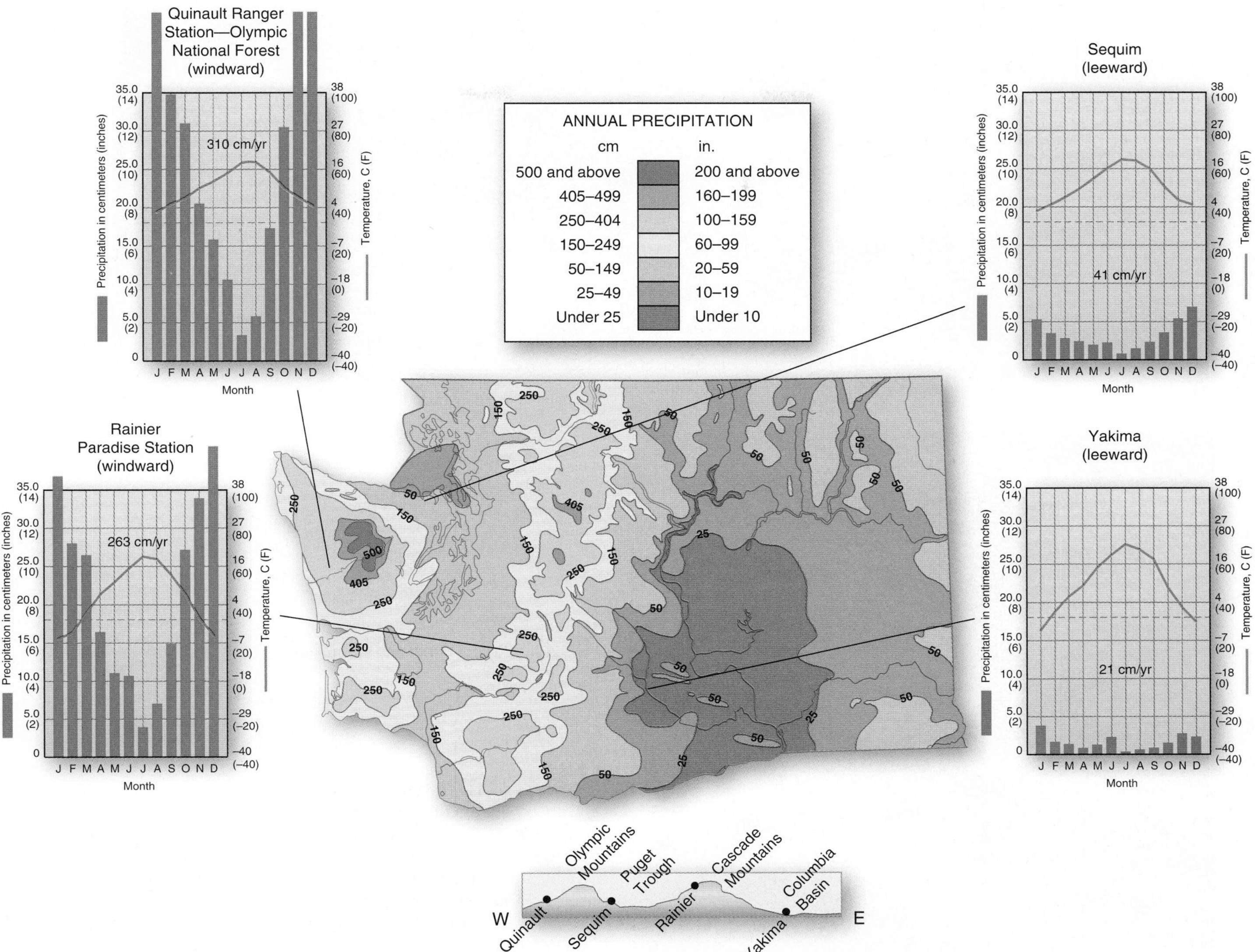

FIGURE 8.9 Orographic patterns in Washington State.
Four stations in Washington provide examples of orographic effects: windward precipitation and leeward rain shadows. Isohyets (isolines of equal precipitation amounts) on the map indicate rainfall (in inches). [Data from J. W. Scott and others, *Washington: A Centennial Atlas* (Bellingham, WA: Center for Pacific Northwest Studies, Western Washington University, 1989), p. 3.]

along the front. As the line of most intense lifting passes, usually just ahead of the front itself, air pressure drops to a local low. Clouds may build along the cold front into characteristic cumulonimbus form and may appear as an advancing wall of clouds. Precipitation usually is heavy, containing large droplets, and can be accompanied by hail, lightning, and thunder.

The aftermath of a cold front's passage usually brings northerly winds in the Northern Hemisphere and southerly winds in the Southern Hemisphere as anticyclonic high-pressure advances. Temperatures are lower and air pressure rises in response to the cooler, more dense air; cloud cover breaks and clears.

The particular shape and size of the North American landmass and its latitudinal position present conditions where cP and mT air masses are best developed and have the most direct access to each other. The resulting contrast can lead to dramatic weather, particularly in late spring, with sizable temperature differences from one side of a cold front to the other.

A fast-advancing cold front can cause violent lifting and create a zone right along or slightly ahead of the front, a **squall line**. Along a squall line, such as the one in the Gulf of Mexico shown in Figure 8.10b, wind patterns are turbulent and wildly changing, and precipitation is intense. The well-defined frontal clouds in the photograph rise abruptly, with new thunderstorms forming along the front. Tornadoes also may develop along such a squall line.

Warm Front A line with semicircles facing in the direction of frontal movement denotes a warm front on weather

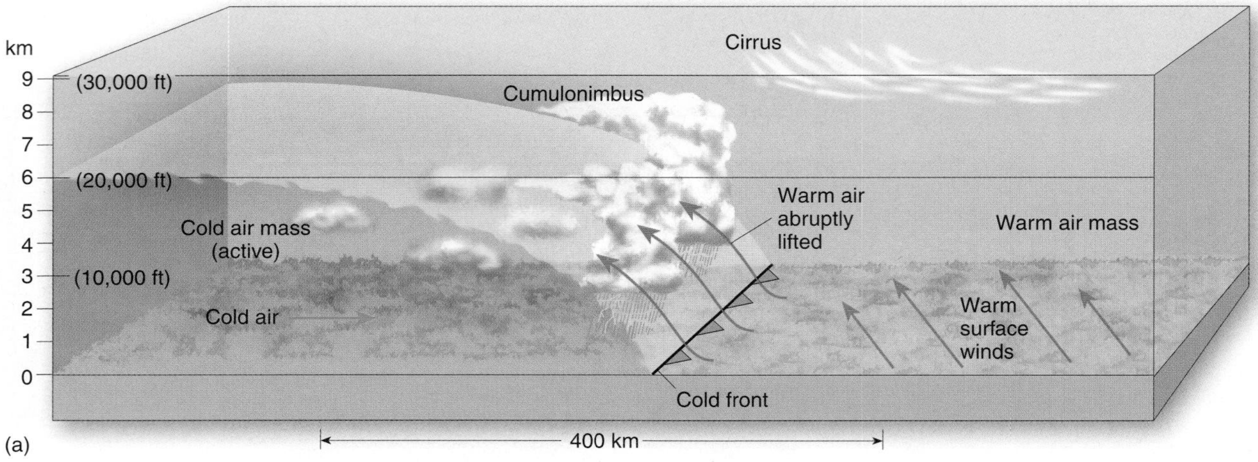

FIGURE 8.10 A typical cold front.
(a) Denser, advancing cold air forces warm, moist air to lift abruptly. As the air is lifted, it cools by expansion at the DAR, cooling to the dew-point temperature as it rises to a level of condensation and cloud formation. (b) A sharp line of cumulonimbus clouds near the Texas coast and Gulf of Mexico marks a cold front and squall line. The cloud formation rises to 17,000 m (56,000 ft). The passage of such a frontal system over land often produces strong winds, cumulonimbus clouds, large raindrops, heavy showers, lightning and thunder, hail, and the possibility of tornadoes. [NASA.]

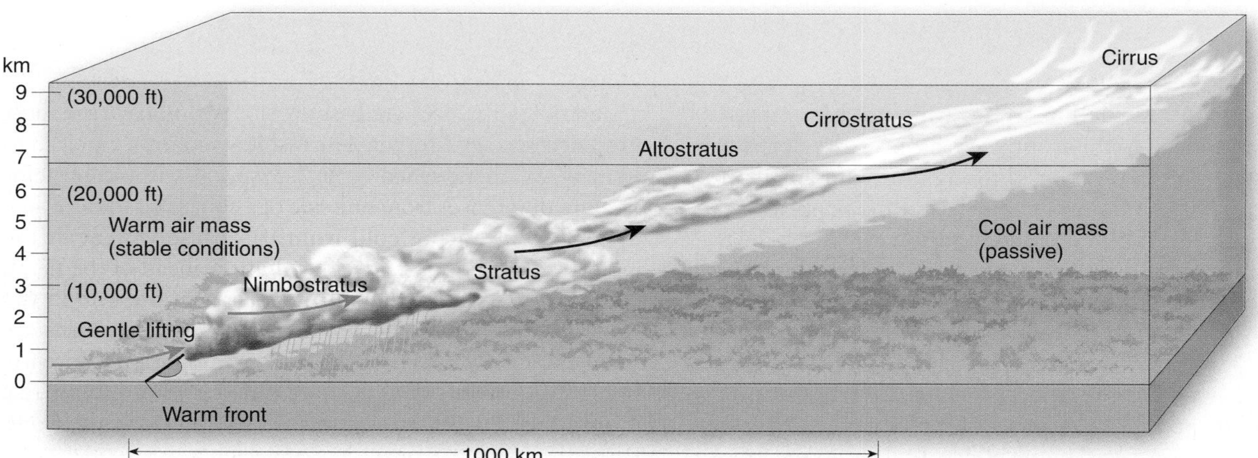

FIGURE 8.11 A typical warm front.
Note the sequence of cloud development as the warm front approaches. Warm air slides upward over a wedge of cooler, passive air near the ground. Gentle lifting of the warm, moist air produces nimbostratus and stratus clouds and drizzly rain showers, in contrast to the more dramatic cold-front precipitation.

maps (see the map in Figure 8.12). The leading edge of an advancing warm air mass is unable to displace cooler, passive air, which is denser along the surface. Instead, the warm air tends to push the cooler, underlying air into a characteristic wedge shape, with the warmer air sliding up over the cooler air. Thus, in the cooler-air region a temperature inversion is present, sometimes causing poor air drainage and stagnation.

Figure 8.11 illustrates a typical warm front in which mT air is gently lifted, leading to stratiform cloud development and characteristic nimbostratus clouds as well as drizzly precipitation. A warm front presents a progression of cloud development to an observer: High cirrus and cirrostratus clouds announce the advancing frontal system; then the clouds lower and thicken to altostratus; and finally the clouds lower and thicken to stratus within several hundred kilometers of the front.

Midlatitude Cyclonic Systems

The conflict between contrasting air masses can develop a **midlatitude cyclone**, or **wave cyclone**. This migrating low-pressure center with converging, ascending air spirals inward counterclockwise in the Northern Hemisphere and inward clockwise in the Southern Hemisphere. Because of the undulating nature of frontal boundaries and the steering flow of the jet streams, the term *wave* is appropriate. The combination of the *pressure gradient force*, *Coriolis force*, and *surface friction* generates this cyclonic motion (see discussion in Chapter 6).

Before World War I, weather maps displayed only pressure and wind patterns. Vilhelm Bjerknes added the concept of fronts, and his son Jacob Bjerknes contributed the concept of migrating centers of cyclonic low-pressure systems.

Wave cyclones dominate weather patterns in the middle and higher latitudes of both the Northern and the Southern hemispheres and act as a catalyst for air mass conflict. Such a midlatitude cyclone can initiate along the polar front, particularly in the region of the Icelandic and Aleutian subpolar low-pressure cells in the Northern Hemisphere. The intense high-speed winds of the jet streams guide cyclonic systems along their tracks (see Figures 6.15 and 6.16).

Life Cycle of a Midlatitude Cyclone

Figure 8.12 shows the birth, maturity, and death of a typical midlatitude cyclone in several stages, along with an idealized weather map. On the average, a midlatitude cyclonic system takes 3–10 days to progress through these stages from the area where it develops to the area where it finally dissolves. However, chaos rules and every day's weather map departs from the ideal in some manner.

Cyclogenesis The atmospheric process in which low-pressure wave cyclones develop and strengthen is **cyclogenesis**. This process usually begins along the polar front, where cold and warm air masses converge and are drawn into conflict.

The polar front is a discontinuity of temperature, moisture, and winds that establishes potentially unstable conditions. For a wave cyclone to form along the polar front, a compensating area of divergence aloft matches a surface point of air convergence. Even a slight disturbance along the polar front, perhaps a small change in the path of the jet stream, can initiate the converging, ascending flow of air and thus a surface low-pressure system (illustrated in Figure 8.12a).

In addition to the polar front, certain other areas are associated with wave cyclone development and intensification: the eastern slope of the Rockies and other north–south mountain barriers, the Gulf Coast, and the east coasts of North America and Asia.

Open Stage To the east of the developing low-pressure center, warm air begins to move northward along an advancing front, while cold air advances southward to the west of the center. See this organization in Figure 8.12b and around the center of low pressure, located over western Nebraska, in the accompanying map. The growing circulation system then vents into upper-level winds. As the midlatitude cyclone matures, the counterclockwise flow draws the cold air mass from the north and west and the warm air mass from the south. In the cross section, you can see the profiles of both a cold front and a warm front and each air mass segment. Compare these with the illustrations in Figures 8.10 and 8.11.

On the map in Figure 8.12, Denver has just experienced the passage of a cold front. You can see on the map that before the front passed, winds were from the southwest, but now the winds have shifted and are from the northwest. Temperature and humidity went from the warm, moist mT air mass to colder conditions as the cP air mass moves over Denver. Meanwhile, Wichita, Kansas, experienced the passage of a warm front and now is in the midst of the warm-air segment of the cyclone.

Occluded Stage Remember the relation between air temperature and density of an air mass. The colder cP air mass is denser than the warmer mT air mass. This cooler, more unified air mass acts like a bulldozer blade and therefore moves faster than the warm front. Cold fronts can travel at an average 40 kmph (25 mph), whereas warm fronts average roughly half that, at 16–24 kmph (10–15 mph). Thus, a cold front often overtakes the cyclonic warm front, wedging beneath it, producing an **occluded front** (*occlude* means "to close") in Figure 8.12c.

On the idealized weather map in Figure 8.12, an occluded front stretches south from the center of low pressure in Virginia to the border between the Carolinas. Precipitation may be moderate to heavy initially and then taper off as the warmer air wedge is lifted higher by the advancing cold air mass. Note the still active warm front in the extreme Southeast and the flow of mT air. What conditions do you observe in Tallahassee, Florida? What

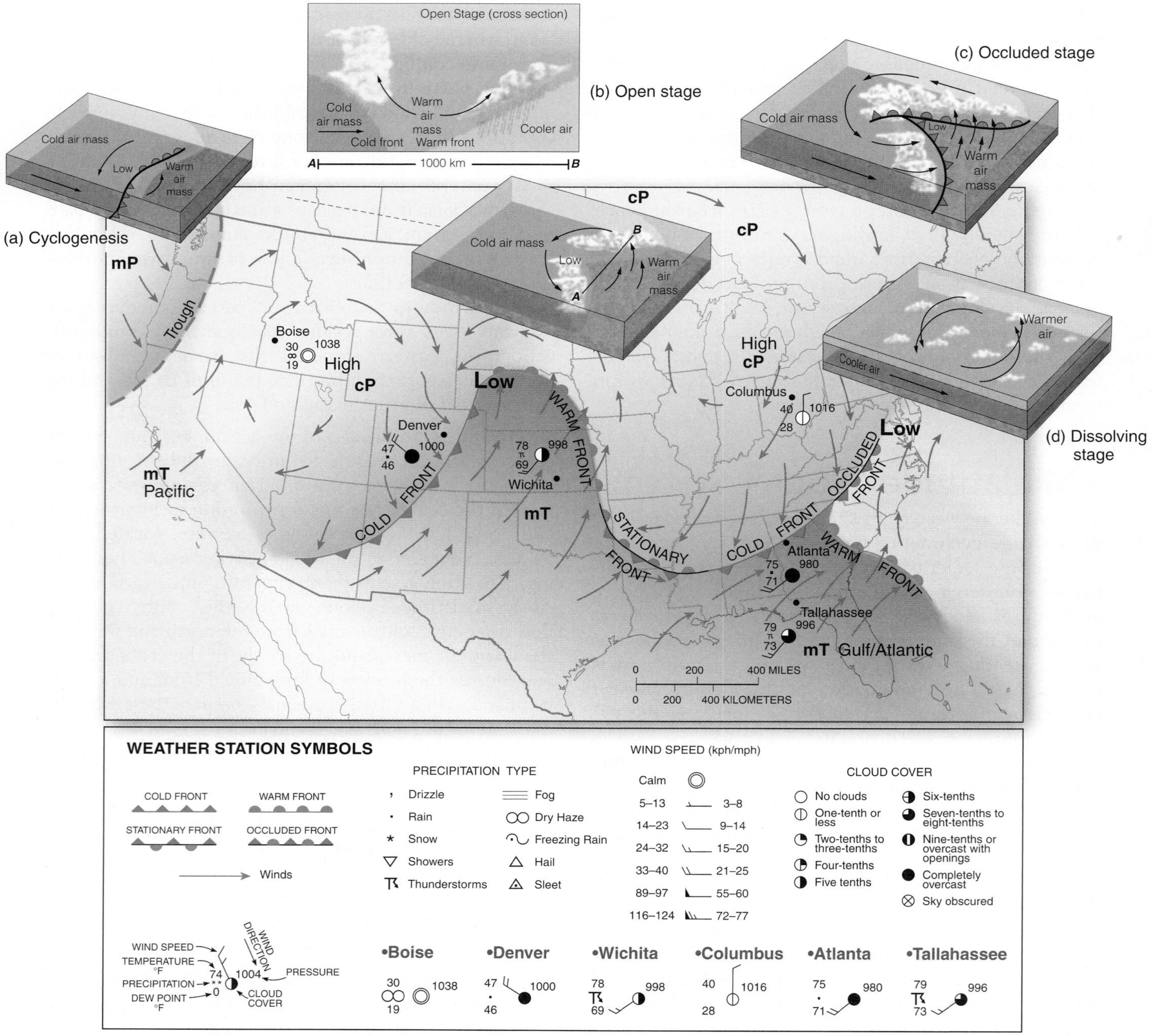

FIGURE 8.12 Idealized stages of a midlatitude wave cyclone. Standard weather station symbols and current conditions for six cities appear below the map. The block diagrams show (a) cyclogenesis, where surface convergence and lifting begin; (b) the open stage; (c) the occluded stage; and (d) the dissolving stage, reached at the end of the storm track as the cyclone spins down, no longer energized by the latent heat from condensing moisture.

Animation
Midlatitude Cyclones

might it be like there in the next 12 hours if the cold front passes south of the city?

When there is a stalemate between cooler and warmer air masses where air flow on either side is almost parallel to the front, although in opposite directions, a **stationary front** results. Some gentle lifting might produce light to moderate precipitation. Eventually, the stationary front will begin to move, as one of the air masses assumes dominance, evolving into a warm or a cold front.

Dissolving Stage The final, dissolving, stage of the midlatitude cyclone occurs when its lifting mechanism is completely cut off from the warm air mass, which was its source of energy and moisture. Remnants of the cyclonic system then dissipate in the atmosphere, perhaps after passage across the country (Figure 8.12d).

Storm Tracks Cyclonic storms can be 1600 km (1000 mi) wide. These wave cyclones and their attendant air masses move across the continent along **storm tracks**, which shift in latitude with the Sun and the seasons. Typical storm tracks that cross North America are farther northward in summer and farther southward in winter (Figure 8.13a). Noted on the map are some of the regional names for cyclonic source regions. As the storm tracks begin to shift northward in the spring, cP and mT air masses are in their clearest conflict. This is the time of strongest frontal activity, featuring thunderstorms and tornadoes. Storm tracks follow the path of upper-air winds, which direct storm systems across the continent.

A map of actual storm tracks for a typical March in Figure 8.13b demonstrates several areas of cyclogenesis: the northwest over the Pacific Ocean, the Gulf of Mexico (Gulf Lows), the eastern seaboard (Nor'easter, Hatteras Low), and the Arctic. Cyclonic circulation also frequently develops on the lee side of mountain ranges, as along the Rockies from Alberta (Alberta Clipper) south to Colorado (Colorado Low). By crossing the mountains, such systems gain access to the moisture-laden, energy-rich mT air masses from the Gulf of Mexico and are strengthened.

Analysis of Daily Weather Maps—Forecasting

Synoptic analysis is the evaluation of weather data collected at a specific time. Building a database of wind, pressure, temperature, and moisture conditions is key to *numerical* (computer-based) *weather prediction* and the development of weather-forecasting models. Development of numerical models is a great challenge because the atmosphere

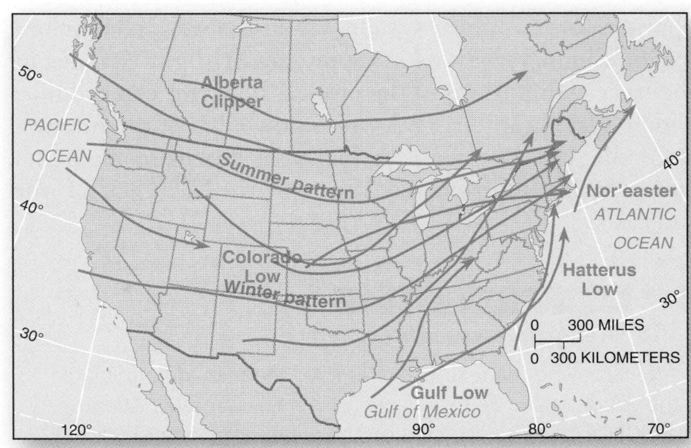

(a) Average storm tracks

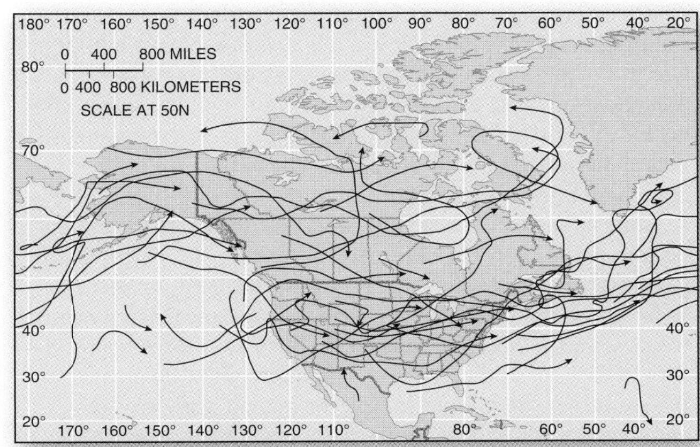

(b) Actual storm tracks for a typical March

FIGURE 8.13 Typical and actual storm tracks. (a) Cyclonic storm tracks over North America vary seasonally. The tracks indicate several locations of cyclogenesis; note the regional names. (b) Actual cyclonic tracks during an average March over North America. [(b) NOAA/NESDIS/NCDC.]

(a) (b)

FIGURE 8.14 ASOS weather instruments and AWIPS workstation display. [(a) Bobbé Christopherson. (b) NWS, ESRL.]

operates as a nonlinear system, tending toward chaotic behavior. Slight variations in input data or slight changes in the basic assumptions of the model's behavior can produce widely varying forecasts. As our knowledge of the interactions that produce weather increases and as instruments and software improve, so, too, will the accuracy of our forecasts.

CRITICAL THINKING 8.1

Analyzing a weather map

After studying the text, test your knowledge by determining the conditions in Boise, Denver, Wichita, Columbus, Atlanta, and Tallahassee as depicted on the weather map in Figure 8.12.

Next, look at the weather map showing a classic open stage in Figure 8.1.1 for March 31, 2007, 7 A.M. EST, along with a *GOES-12* satellite image of water vapor at that time. Using the legend in Figure 8.12 for weather map symbols, briefly analyze this map: Find the center of low pressure, note the counterclockwise winds, compare temperatures on either side of the cold front, note air temperatures and dew-point temperatures, determine why the fronts are placed where they are on the map, and locate the center of high pressure.

You can see that the air and dew-point temperatures in New Orleans are 68°F and 63°F, respectively; yet, in St. George, Utah (southwest corner of state), you see 27°F and 22°F, respectively. To become saturated, the warm, moist air in New Orleans needs to cool to 63°F, whereas the dry, cold air in St. George needs to cool to 22°F for saturation. In the far north, find Churchill on the western shore of Hudson Bay. The air and dew-point temperatures are 6°F and 3°F, respectively, and the state of the sky is clear. If you were there, what would you experience at this time? Why would you be reaching for the lip balm?

Describe the pattern of air masses on this weather map. How are these air masses interacting, and what kind of frontal activity do you see? What does the pattern of isobars tell you about high- and low-pressure areas?

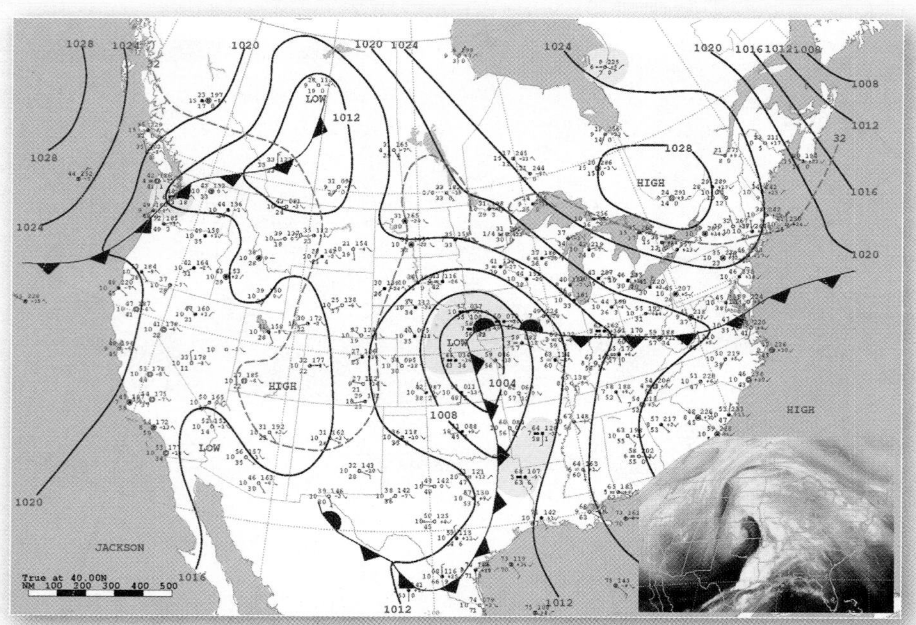

FIGURE 8.1.1 Weather map and water vapor image, March 31, 2007. [Map courtesy of Hydrometeorological Prediction Center, NCEP, NWS.]

Weather data necessary for the preparation of a synoptic map and forecast include:

- Barometric pressure (sea level and altimeter setting)
- Pressure tendency (steady, rising, falling)
- Surface air temperature
- Dew-point temperature
- Wind speed, direction, and character (gusts, squalls)
- Type and movement of clouds
- Current weather
- State of the sky (current sky conditions)
- Visibility; vision obstruction (fog, haze)
- Precipitation since last observation

For many related Internet links to weather maps, current forecasts, satellite images, and the latest radar, go to Chapter 8 on the *Mastering Geography* Web site. NOAA's Earth System Research Laboratory (ESRL, http://www.esrl.noaa.gov/) in Boulder, Colorado, uses a number of forecasting tools, including a network of 35 wind profilers using radar from the surface to high altitudes; a high-performance supercomputing facility and state-of-the-art data center for 3-D weather models and a range of other computations; a ground-based GPS meteorology project; and a system for improved international cooperation and weather data sharing.

The Automated Surface Observing System (ASOS) is the primary network observing surface weather in the United States (Figure 8.14a, p. 203). An ASOS instrument array includes rain gauge (tipping bucket type), temperature/dew-point sensor, barometer, present weather identifier, wind-speed indicator, direction sensor, cloud-height indicator, freezing-rain sensor, thunderstorm sensor, and visibility sensor, among other items.

NOAA's Advanced Weather Interactive Processing System (AWIPS) integrates data from a number of sources in order to improve the accuracy of forecasts and warnings. An AWIPS workstation displays data across three monitors—for example, virtual 3-D images of pressure, water vapor, humidity, Doppler radar, lightning strikes in real time, and wind profiles (Figure 8.14b, p. 203).

Preparing a weather report and forecast requires analysis of daily weather maps and the use of standard weather symbols, such as those in Figure 8.12. Although the actual pattern of cyclonic passage over North America is widely varied in shape and duration, you can apply this general model, along with your understanding of warm and cold fronts, to reading the daily weather map. A weather map is a useful tool, especially when weather turns dramatic in violent episodes.

Violent Weather

Weather provides a continuous reminder of the flow of energy across the latitudes that at times can set into motion destructive, violent weather conditions. We focus in this chapter on thunderstorms, derechos, tornadoes, and hurricanes; coverage of floods appears in Chapter 14 and coastal hazards in Chapter 16.

Weather is often in the news. Weather-related destruction has risen more than 500% over the past three decades as population has increased in areas prone to violent weather and as climate change intensifies weather anomalies. Weather-related losses exceeded $10 billion per year through the 1990s, eclipsing the previous annual average of less than $2 billion. Each year of the 21st century has far surpassed the past decade. Hurricane Katrina and the 2005 season caused $130 billion in damages and Hurricanes Gustav and Ike (2008) $100 billion in damages! Government research and monitoring of violent weather is centered at NOAA's National Severe Storms Laboratory and Storm Prediction Center—see http://www.nssl.noaa.gov/ and http://www.spc.noaa.gov/, among other agencies; consult these sites for each of the topics that follow.

Violent weather includes *ice storms* of **sleet** (freezing rain, ice glaze, and ice pellets), blizzards, and low temperatures. Sleet is caused when precipitation falls through a below-freezing layer of air near the ground. During the ice storm of January 1998, a large region in Canada and the United States with 700,000 residents was without power for weeks. Imagine a coating of ice on everything, weighing on power lines and tree limbs. More than 80 hours of freezing rain and drizzle hit the region; this is more than double typical ice-storm duration. In Montreal, over 100 mm (4 in.) of ice accumulated. Hypothermia claimed 25 lives.

Thunderstorms

Tremendous energy is liberated by the condensation of large quantities of water vapor. This process locally heats the air, causing violent updrafts and downdrafts as rising parcels of air pull surrounding air into the column and as the frictional drag of raindrops pulls air toward the ground. Giant cumulonimbus clouds can create dramatic weather moments—squall lines of heavy precipitation, lightning, thunder, hail, blustery winds, and tornadoes. Thunderstorms may develop within an air mass, in a line along a front (particularly a cold front), or where mountain slopes cause orographic lifting.

Thousands of thunderstorms occur on Earth at any given moment. Equatorial regions and the ITCZ experience many of them, exemplified by the city of Kampala, Uganda, in East Africa (north of Lake Victoria), which sits virtually on the equator and averages a record 242 days a year with thunderstorms. Figure 8.15 shows the annual distribution of days with thunderstorms across the United States and Canada. You can see that, in North America, most thunderstorms occur in areas dominated by mT air masses.

Atmospheric Turbulence Most airplane flights experience at least some turbulence—the encountering of air of different densities or air layers moving at different speeds and directions. This is a natural state of the atmosphere, and passengers are asked to keep seat belts fastened when in their seats to avoid injury, even if the seat-belt light is turned off.

Thunderstorms can produce severe turbulence in the form of downbursts, which are exceptionally strong downdrafts. Downbursts are classified by size: A *macroburst* is at least 4.0 km (2.5 mi) wide and in excess of 210 kmph (130 mph); a *microburst* is smaller in size and speed. A microburst causes rapid changes in wind speed and direction, and it causes the dreaded *wind shear* that can bring down aircraft. Such turbulence events are short-lived and hard to detect, although ESRL, among others, is making progress in developing forecasting methods.

Lightning and Thunder An estimated 8 million lightning strikes occur each day on Earth. **Lightning** refers to flashes of light caused by enormous electrical discharges—tens of millions to hundreds of millions of volts—that briefly superheat the air to temperatures of 15,000°–30,000°C (27,000°–54,000°F). A buildup of electrical energy polarity between areas within a cumulonimbus cloud or between the cloud and the ground creates lightning. The violent expansion of this abruptly heated air sends shock waves through the atmosphere as the sonic bang of **thunder**.

Lightning poses a hazard to aircraft, people, animals, trees, and structures. Certain precautions are mandatory when a lightning discharge threatens because lightning causes nearly 200 deaths and thousands of injuries each year in the United States and Canada. When lightning is imminent, the NWS issues *severe storm warnings* and cautions people to remain indoors.

NASA's Lightning Imaging Sensor (LIS) aboard the *Tropical Rainfall Measuring Mission* (*TRMM*) satellite monitors lightning and other phenomena. The LIS can image lightning strikes day or night, within clouds or between cloud and ground. The sensor's data show that about 90%

You can feel the warning

If the hair on your head or neck begins to stand on end, get indoors, or if you are in the open, get low to the ground, crouched on both feet and not lying down, in a low spot, if possible. Your hair is telling you that a lightning charge is building in that area. The place *not* to seek shelter is beneath a tree, for trees are good conductors of electricity and often are hit by lightning.

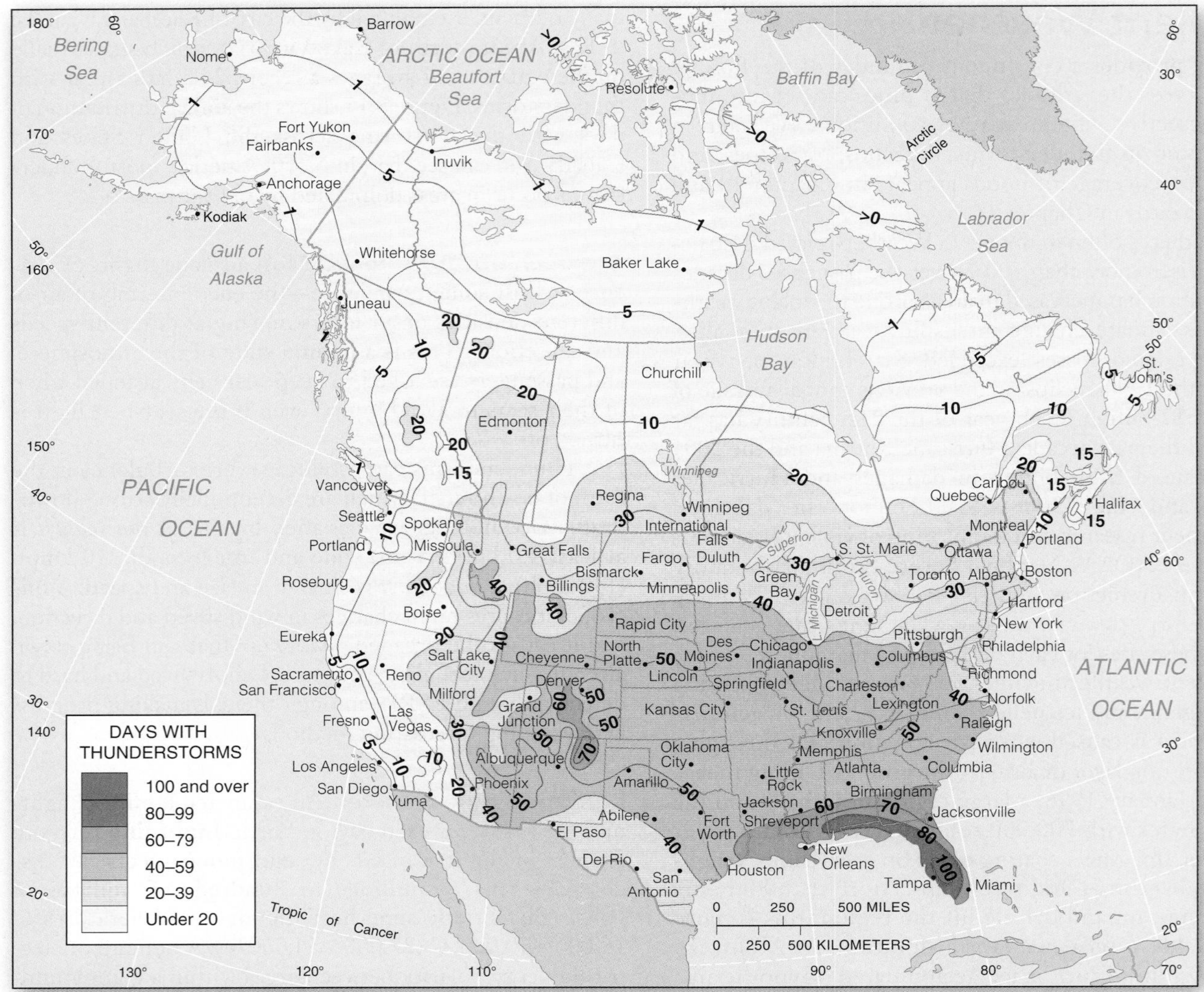

FIGURE 8.15 Thunderstorm occurrence.
Average annual number of days experiencing thunderstorms. [Data courtesy of NWS; Map Series 3; *Climatic Atlas of Canada,* Atmospheric Environment Service, Canada.]

of all strikes occur over land in response to increased convection over relatively warmer continental surfaces. Strikes shift with expected seasonal shifts of the high Sun, as shown in Figure 8.16a and b (see http://thunder.msfc.nasa.gov/lis/).

Hail Ice pellets of **hail** generally form within a cumulonimbus cloud. Raindrops circulate repeatedly above and below the freezing level in the cloud, adding layers of ice until the circulation in the cloud can no longer support their weight. Hail may also grow from the addition of moisture on a snow pellet.

Pea-sized (0.25 in. diameter) hail is common, although hail the size of quarters (1.00 in.), golf balls (1.75 in.), hen's eggs (2.00 in.), baseballs (2.75 in.), and softballs (4.50 in.) is known to happen. Baseball-sized hail fell a half dozen times in 2010 in the United States alone (Figure 8.17). For larger hail to form, the frozen pellets must stay aloft for longer periods. The largest authenticated hailstone in the world fell from a thunderstorm supercell in Aurora, Nebraska, June 22, 2003; it measured 17.8 cm (7.0 in.) in diameter and 47.6 cm (18.75 in.) in circumference.

Hail is common in the United States and Canada, although somewhat infrequent at any given place. Hail occurs perhaps every 1 or 2 years in the highest-frequency areas. Annual hail damage in the United States tops $800 million. The pattern of hail occurrence across the United States and Canada is similar to that of thunderstorms shown in Figure 8.15. Days with hail are on the increase since 2000.

Derechos

Although tornadoes and hurricanes grab headlines, straight-line winds associated with thunderstorms and bands of showers cause significant damage and crop losses. These strong linear winds in excess of 26 m/s (58 mph) are known as **derechos**, and in Canada sometimes called *plow winds*. Wind outbursts from convective storms can

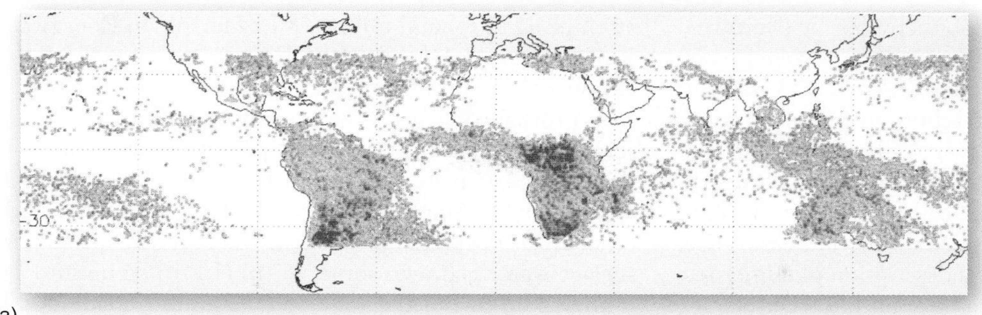

(a)

Winter (Dec. 1999, Jan. and Feb. 2000)

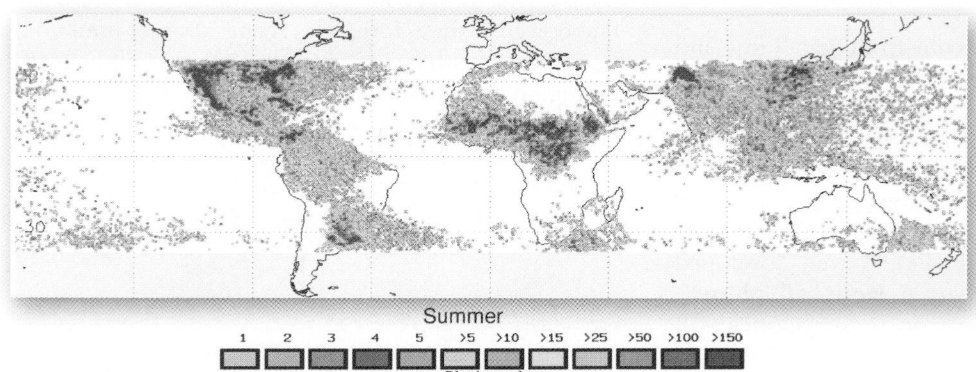

(b)

Summer (June, July, August 2000)

(c)

FIGURE 8.16 Seasonal images show global lightning. A composite of 3 months' worth of data derived from NASA's Lightning Imaging Sensor (LIS) records of all lightning strikes between 35° N and 35° S latitudes during (a) winter (December 1999–February 2000) and (b) summer (June, July, August 2000). The LIS is aboard the *Tropical Rainfall Measuring Mission* satellite launched in 1997. (c) Multiple lightning strikes in southern Arizona captured in a time-lapse photo. [(a) and (b) *TRMM* images courtesy of NASA's Global Hydrology and Climate Center, MSC. (c) Kieth Kent/Photo Researchers.]

produce groups of downburst clusters from a thunderstorm system. These derechos tend to blast in linear paths fanning out along curved-wind fronts over a wide swath of land. The name, coined by physicist G. Hinrichs in 1888, derives from a Spanish word meaning "direct" or "straight ahead."

A derecho in 1998 in eastern Wisconsin exceeded 57 m/s (128 mph). In August 2007, a series of derechos across northern Illinois reached this same intensity. Researchers identified 377 derechos between 1986 and 2003, an average of about 21 per year. Derechos pose

FIGURE 8.17 Hailstones and ruler for measurement. [Weatherstock/Photolibrary.]

distinct hazards to summer outdoor activities by overturning boats, hurling flying objects, and causing broken trees and limbs. Their highest frequency (about 70%) is from May to August in the region stretching from Iowa, across Illinois, and into the Ohio River Valley in the upper Midwest. By September and on through to April, areas of activity migrate southward to eastern Texas through Alabama. Reported wind events are on the increase since 2000. For more information and a description of numerous derechos, see http://www.spc.noaa.gov/misc/AbtDerechos/derechofacts.htm.

Tornadoes

The updrafts associated with a cold-front squall line and cumulonimbus cloud development appear on satellite images as pulsing bubbles of clouds. According to one hypothesis, a spinning, cyclonic, rising column of mid-troposphere-level air forms a mesocyclone. Ranging up to 10 km (6 mi) in diameter, a **mesocyclone** rotates vertically within a *supercell cloud* (the parent cloud) to a height of thousands of meters (Figure 8.18a). A well-developed mesocyclone will produce heavy rain, large hail, blustery winds, and lightning; some mature mesocyclones will generate tornado activity.

As more moisture-laden air is drawn up into the circulation of a mesocyclone, more energy is liberated, and the rotation of air increases speed. The narrower the mesocyclone, the faster the spin of converging parcels of air being sucked into the rotation. The swirl of the mesocyclone itself is visible, as are smaller, dark gray **funnel clouds** that pulse from the bottom side of the parent cloud. The potential of this stage of development is the lowering of a funnel cloud to Earth—a **tornado** (Figure 8.18b).

A tornado can range from a few meters to a few hundred meters in diameter and can last anywhere from a few moments to tens of minutes. Tornadoes have been increasing in average speed and duration over the last two decades. Greensburg, Kansas, was essentially erased from the map by an EF-5 tornado in May 2007. The statistics are unprecedented: Winds exceeded 330 kmph (205 mph) in the 2.7-km-wide (1.7-mi-wide) funnel that was on the ground along a 35-km (22-mi) track. A photograph taken 1 year later (Figure 8.18c) shows block after block of empty lots in what was a residential district. Greensburg is attempting to rebuild following Leadership in Energy and Environmental Design (LEED) "Green" standards for energy conservation and efficiency.

A tornado's scoured path is visible on the ground in a series of semicircular "suction" prints and, of course, in the trail of damage, such as in the image of a tornado's path through Maryland (Figure 8.18d). When tornado circulation occurs over water, a **waterspout** forms, and surface water is drawn some 3–5 m (10–16 ft) up into the funnel. The rest of the waterspout funnel is made visible by the rapid condensation of water vapor.

The chapter-opening photo shows and *Geosystems Now* discusses the tornado in Wadena, Minnesota, in some detail. The power of these tornadoes seems abstract until you see the impacts firsthand. Your author along with his nature-photographer wife visited Wadena four days after the EF–4 destroyed 260 homes and many community buildings, a swimming pool, and the high school (Figure 8.19). Multiply the destruction times the near doubling of average occurrences and an increase in EF–3, –4, and –5 events in the United States, and it does take one's breath away.

Tornado Measurement and Totals Pressures inside a tornado usually are about 10% less than those in the surrounding air. The inrushing convergence created by such a horizontal pressure gradient causes high wind speeds. The late Theodore Fujita, a noted meteorologist from the University of Chicago, designed the Fujita Scale (1971), which classifies tornadoes according to wind speed as indicated by related property damage. A refinement in the scale, adopted in February 2007, is the *Enhanced Fujita Scale*, or *EF Scale* (Table 8.1; http://www.tornadoproject.com/fscale/fscale.htm). The revision filled the need to better assess damage, correlate wind speed to damage caused, and account for structural construction quality. To assist with wind estimates, the EF Scale contains Damage Indicators relating to types of structures and vegetation affected, along with Degrees of Damage ratings, both of which you can view at the URL listed in the table note.

North America experiences more tornadoes than anywhere on Earth because its latitudinal position and topography permit contrasting air masses to confront one another. Tornadoes have struck all 50 states and all the

GEO REPORT 8.4 *Tornadoes on a day in May*

During May 2003, 543 tornadoes struck in some 20 states. The first 12 days of the month featured 354 tornadoes. This is far above the average for May—and more tornadoes for any single month since 1950. The results caused more than 40 deaths and hundreds of millions of dollars in property damage.

Eighty-four tornadoes hit on Sunday, May 4, in a path across eight states. Tornado damage ratings indicated four EF-4s and eight EF-3s struck in the Midwest. Five tornadoes struck in or near Kansas City. One of the Kansas City EF-4 tornadoes achieved a funnel-base width of 450 m (1500 ft). Some damage appeared to be from downbursts and derecho linear-wind effects. (For an animation of satellite images, see http://www.osei.noaa.gov/Events/Severe/US_Midwest/2003/SVRusMW125_G12.avi.)

MG Animation
Tornado Wind Patterns

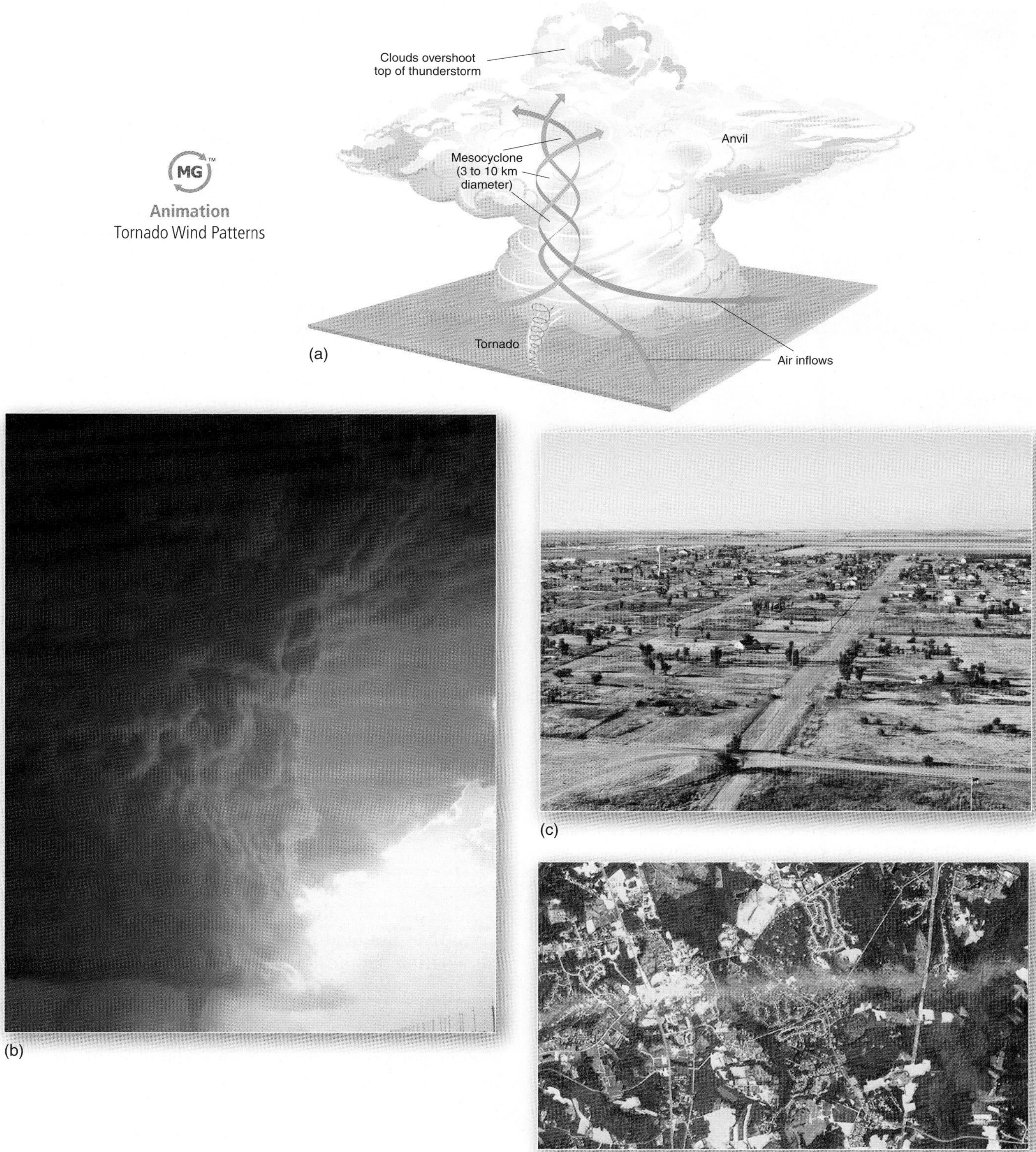

FIGURE 8.18 Mesocyclone and tornado formation.
(a) Strong wind aloft establishes spinning; as updraft from thunderstorm development tilts the rotating air, a mesocyclone forms as a rotating updraft within the thunderstorm. If one forms, a tornado will descend from the lower portion of the mesocyclone. (b) A supercell tornado descends from the cloud base near Spearman, Texas. Strong hail is falling to the left of the tornado. (c) 2007 EF-5 tornado leveled whole neighborhoods in Greensburg, Kansas; photograph taken in 2008. (d) A tornado leaves its mark, as this *Landsat-7* image captures. The damage path tracks for 39 km (24 mi) where an EF-4 tornado tore through La Plata, Maryland. More than 900 homes and 200 businesses were destroyed. [(b) Photo by Howard Bluestein, all rights reserved. (c) Bobbé Christopherson. (d) *Landsat-7* image, USGS/EROS.]

(a)

(b)

FIGURE 8.19 June 17, 2010, Wadena, Minnesota. (a) The power of 322-kmph (200-mph) wind bends steel poles. (b) Sheet metal from the destroyed community center blew through the house to the left, where a man survived, clutching his dog, both laying together on his living room floor. [Bobbé Christopherson.]

TABLE 8.1 The Enhanced Fujita Scale

EF-number	3-Second Gust Wind Speed; Damage Specs
EF-0 Gale	105–137 kmph; 65–85 mph: *light damage*: branches broken, chimneys damaged.
EF-1 Weak	138–177 kmph; 86–110 mph: *moderate damage*: beginning of hurricane wind-speed designation, roof coverings peeled off, mobile homes pushed off foundations.
EF-2 Strong	178–217 kmph; 111–135 mph: *considerable damage*: roofs torn off frame houses, large trees uprooted or snapped, box cars pushed over, small missiles generated.
EF-3 Severe	218–266 kmph; 136–165 mph: *severe damage*: roofs torn off well-constructed houses, trains overturned, trees uprooted, cars thrown.
EF-4 Devastating	267–322 kmph; 166–200 mph: *devastating damage*: well-built houses leveled, cars thrown, large missiles generated.
EF-5 Incredible	More than 322 kmph; >200 mph: *incredible damage*: house lifted and carried distance to disintegration, car-sized missiles fly farther than 100 m, bark removed from trees.

Note: See http://www.wind.ttu.edu/EFScale.pdf for details.

Canadian provinces and territories. May and June are the peak months, as you can see in Figure 8.20, based on 51 years of records. Note the comparison with statistics for the 7-year period from 2003 to 2009 and the annual trend for tornado occurrence on the graph.

In the United States, 53,757 tornadoes were recorded in the years from 1950 through June 2010, causing over 5000 deaths, or about 85 deaths per year. Of note is the fact that these tornadoes resulted in more than 80,000 injuries and property damage of over $28 billion for this period of record. The yearly average of $500 million in damage is rising each year.

The long-term annual average number of tornadoes before 1990 was 787. Interestingly, after 1990 the average per year rose to over 1000. In 1998, the annual average reached 1270 tornadoes; the peak year was 2004, with 1820 tornadoes; 1156 tornadoes were reported in 2009.

Importantly, the number of EF-4 and EF-5 tornadoes is on the increase, reaching more than a dozen each season. Note on the graph in Figure 8.20 that the number of tornadoes nearly doubled in March and more than doubled in September, meaning that access to maritime tropical air masses is occurring earlier in spring and later in fall. This suggests climate change is forcing this trend.

Canada experiences an average of 80 observed tornadoes per year, although unpopulated rural areas go unreported. University of Leeds researchers said that in the United Kingdom observers report 60 to 80 a year. Other continents experience a small number of tornadoes each year.

The Storm Prediction Center in Kansas City, Missouri, provides short-term forecasting for thunderstorms and tornadoes to the public and to NWS field offices. Warning times of from 12 to 30 minutes are possible with current technology.

Reasons for the annual increase in tornado frequency in North America range from global climate change and more intense thunderstorm activity to better reporting through Doppler radar and a larger population with video

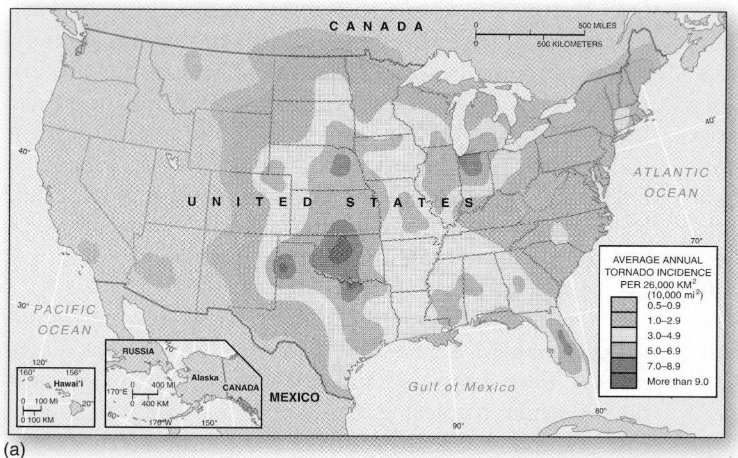

(a)

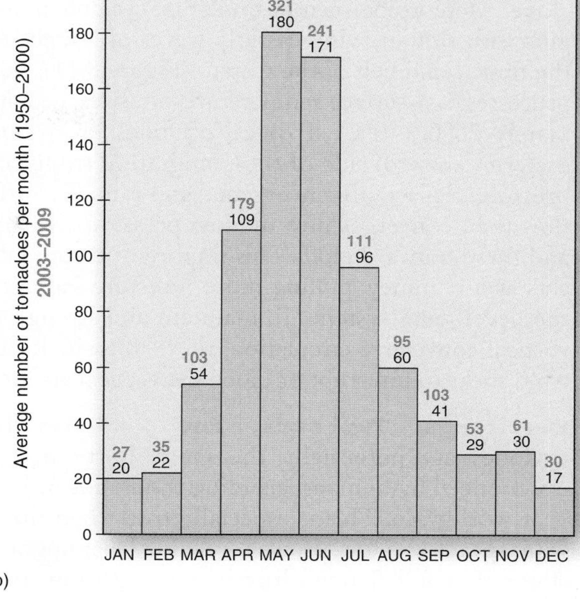

(b)

FIGURE 8.20 Trends in tornado occurrence in the United States. (a) Average number of tornadoes per 26,000 km² (10,000 mi²). (b) Average number of tornadoes per month from 1950 to 2000. Remarkably, the 7 years from 2003 to 2009 have seen a significant increase in the average number of tornadoes per month. [Data courtesy of the Storm Prediction Center, NWS, and NOAA sources.]

recorders. Researchers are pursuing many questions with an enthusiasm for the science and an awe of the subject, as summarized by Howard Bluestein, prominent tornado researcher and meteorologist:

> Our quest for discovery has not taken away from the respect we have for the awesome power the tornado harbors, nor the thrill for viewing the violent motions in the tornado or the beauty of the storm. We eagerly await the next act in the atmospheric play starring the tornado.*

Tropical Cyclones

A powerful manifestation of the Earth–atmosphere energy budget is the **tropical cyclone**, which originates entirely within tropical air masses. The tropics extend from the Tropic of Cancer at 23.5° N to the Tropic of Capricorn at 23.5° S, containing the equatorial zone between 10° N and 10° S. Approximately 80 tropical cyclones occur annually worldwide. Some 45 per year are powerful enough to be classified as hurricanes, typhoons, and cyclones (different regional names for the same type of tropical storm)—30% of these occur in the western North Pacific. Table 8.2 presents the criteria for classification of tropical cyclones based on wind speed.

Cyclonic systems forming in the tropics are quite different from midlatitude cyclones because the air of the tropics is essentially homogeneous, with no fronts or conflicting air masses of differing temperatures. In addition, the warm air and warm seas ensure abundant water vapor and thus the necessary latent heat to fuel these storms. Tropical cyclones convert heat energy from the ocean into mechanical energy in the wind—the warmer the ocean and atmosphere, the more intense the conversion and powerful the storm.

*H. Bluestein, *Tornado Alley, Monster Storms of the Great Plains* (New York: Oxford University Press, 1999), p. 162.

TABLE 8.2 Tropical Cyclone Classification

Designation	Winds	Features
Tropical disturbance	Variable, low	Definite area of surface low pressure; patches of clouds
Tropical depression	Up to 34 knots or 63 kmph (39 mph)	Gale force, organizing circulation; light to moderate rain
Tropical storm	35–63 knots or 63–118 kmph (39–73 mph)	Closed isobars; definite circular organization; heavy rain; assigned a name
Hurricane (Atlantic and East Pacific) Typhoon (West Pacific) Cyclone (Indian Ocean, Australia)	Greater than 64 knots or 119 kmph (74 mph)	Circular, closed isobars; heavy rain, storm surges; tornadoes in right-front quadrant

What mechanism triggers the start of a tropical cyclone? Meteorologists now think that cyclonic motion begins with slow-moving easterly waves of low pressure in the trade-wind belt of the tropics (Figure 8.21). For these processes, sea-surface temperatures must exceed approximately 26°C (79°F). Tropical cyclones form along the eastern (leeward) side of these migrating troughs of low pressure, a place of convergence and rainfall. Surface air flow then converges into the low-pressure area, ascends, and flows outward aloft. This important divergence aloft acts as a chimney, pulling more moisture-laden air into the developing system. To maintain and strengthen this vertical convective circulation, there must be little or no wind shear to interrupt or block the vertical air flow.

Hurricanes, Typhoons, and Cyclones Tropical cyclones are potentially the most destructive storms experienced by humans, claiming thousands of lives each year worldwide. This is especially true when they attain wind speeds and low-pressure readings that upgrade their status to a full-fledged **hurricane**, **typhoon**, or *cyclone* (>65 knots, >119 kmph, >74 mph). Across the globe, such storms bear these three different names, among others: *hurricane* around North America, *typhoon* in the western Pacific (Japan, Philippines), and *cyclone* in Indonesia, Bangladesh, and India. Worldwide, about 10% of all tropical disturbances have the right ingredients to become hurricanes or typhoons. For coverage and reporting, see the National Hurricane Center at http://www.nhc.noaa.gov/ or the Joint Typhoon Warning Center at http://www.usno.navy.mil/JTWC.

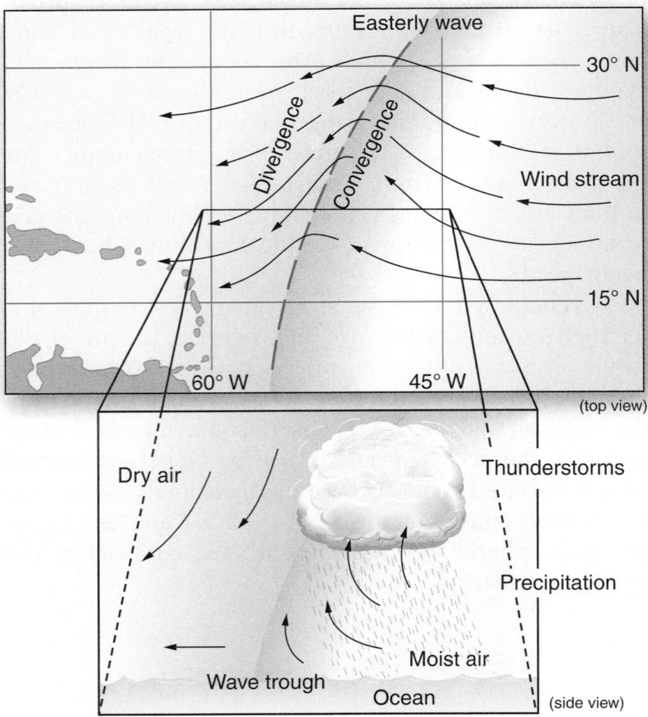

FIGURE 8.21 Easterly wave in the tropics. A low-pressure center develops along an easterly (westward-moving) wave. Moist air rises in an area of convergence at the surface to the east of the wave trough. Wind flows bend and converge before the trough and diverge downwind from the trough.

The map in Figure 8.22a shows areas in which tropical cyclones form and some of their characteristic storm tracks. This map also indicates the range of months during which tropical cyclones are most likely to appear. For example, storms that strike the southeastern United States do so mostly between August and October. Officially, tropical cyclone season for the Atlantic is from June 1 to November 30 each year. Figure 8.22d shows the pattern of actual tracks and intensities for tropical cyclones between 1856 and 2006.

Relative to North and Central America, tropical depressions (low-pressure areas) intensify into tropical storms as they cross the Atlantic. A rule of thumb has emerged: If tropical storms mature early along their track, they tend to curve northward toward the North Atlantic and miss the United States. The critical position is approximately 40° W longitude. If a tropical storm matures after it reaches the longitude of the Dominican Republic (70° W), then it has a higher probability of hitting the United States.

In contradiction to these average conditions, in October 2005, the remnants of Tropical Storm Vince became the first Atlantic tropical cyclone to strike Spain (Figure 8.22b). In the Southern Hemisphere, no hurricane was ever observed turning from the equator into the south Atlantic (note on the storm track map). Yet in March 2004, Hurricane Catarina made landfall in Brazil south of the resorts at Laguna, in the state of Santa Catarina. NASA's *Terra* satellite in Figure 8.22c clearly shows an organized hurricane, with characteristic central eye and rain bands. Compare the circulation patterns of Hurricane Gilbert (see Figure 8.23a) to this satellite image of Catarina. What do you see? Note the Coriolis force in action, counterclockwise in the Northern Hemisphere and clockwise in the Southern Hemisphere.

Then in 2006, the first tropical storm to become a category 5 super typhoon developed in the Central Pacific, some 1285 km (800 mi) south of Hawai'i. Super Typhoon Ioke continued for almost three weeks, with final remnants causing large-scale erosion along Alaskan shores. And in 2007, Super Cyclone Gonu became the strongest tropical cyclone on record occurring in the

GEO REPORT 8.5 Tropical cyclone peak

For the North Atlantic region from 1871 to 2009, 1229 tropical cyclones (storms and hurricanes) developed. The peak day in this Atlantic region (assuming a 9-day moving average calculation) is September 10, with 73 cyclones for the 138-year period.

FIGURE 8.22 Worldwide pattern of the most intense tropical cyclones. (a) Typical tropical storm tracks with principal months of occurrence and regional names. Super Typhoon Ioke (2006) and Super Cyclone Gonu (2007) are noted. (b) The remnants of Tropical Storm Vince hit Spain in October 2005, the first storm to take such a track from the tropics. (c) Hurricane Catarina approaches Brazil's southeastern coast, March 27, 2004—a unique occurrence in recorded history. (d) Global tropical cyclone tracks from 1856 to 2006. Can you find Hurricane Catarina on the map? [(b) October 11, 2005, *Meteosat–7* image, University of Ulm. (c) *Terra* MODIS image, NASA/GSFC. (d) R. Rohde/NASA/GSFC.]

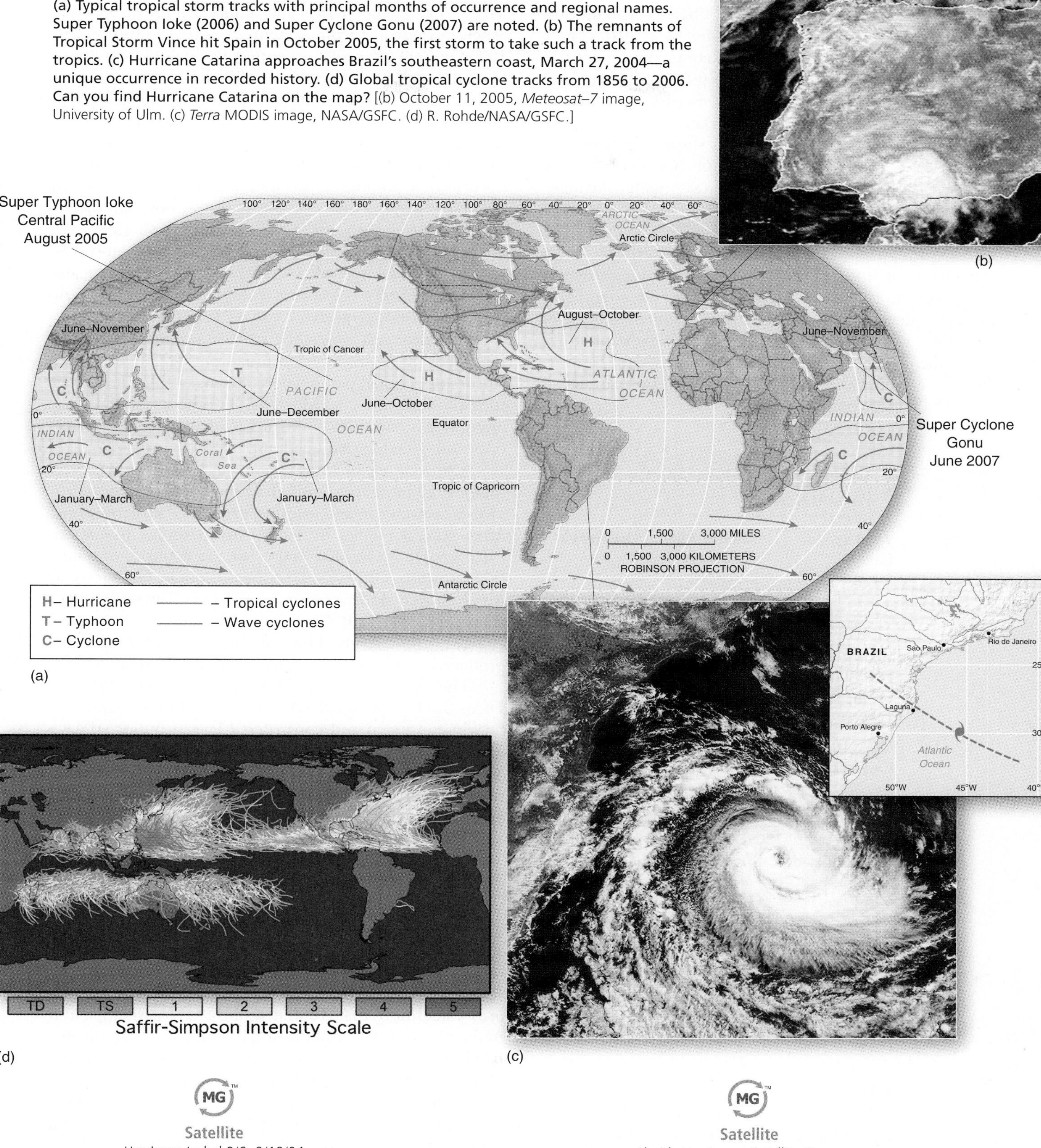

MG Satellite
Hurricane Isabel 9/6–9/19/04

MG Satellite
Florida Hurricanes Satellite Coverage

Arabian Sea, eventually hitting Oman and the Arabian Peninsula. Another tropical cyclone, Phet, took an unusual path into the Arabian Peninsula during June 2010. These storms apparently are signaling changes under way in tropical meteorology relative to the increased occurrence and intensity of cyclonic systems, as related to higher oceanic and atmospheric temperatures.

When you hear meteorologists speak of a "category 4" hurricane, they are using the *Saffir–Simpson Hurricane Damage Potential Scale* to estimate possible damage from hurricane-force winds. Table 8.3 presents this scale, which ranks hurricanes and typhoons in five categories using wind speeds and central pressure criteria, from category 1 for smaller storms to category 5 for extremely dangerous storms. Damage depends on the degree of property development at a storm's landfall site, how prepared citizens are for the blow, the height of storm surge, the presence of embedded tornadoes, and the overall intensity of the storm. Related central pressure levels are included for comparison.

TABLE 8.3 Saffir–Simpson Hurricane Damage Potential Scale

Category	Wind Speed [Central Pressure (mb)]	Notable Atlantic Examples (landfall rating)
1	64–82 knots 119–154 kmph (74–95 mph) [> 980 mb]	
2	83–95 knots 155–178 kmph (96–110 mph) [965–979 mb]	1954 Hazel; 1999 Floyd; 2003 Isabel (was a cat. 5), Juan; 2004 Francis; 2008 Dolly; 2010 Alex
3	96–113 knots 179–210 kmph (111–130 mph) [945–964 mb]	1985 Elena; 1991 Bob; 1995 Roxanne, Marilyn; 1998 Bonnie; 2003 Kate; 2004 Ivan (was a cat. 5), Jeanne; 2005 Dennis; Rita and Wilma (were cat. 5); 2007 Henrietta; 2008 Gustav, Ike (were cat. 4)
4	114–135 knots 211–250 kmph (131–155 mph) [920–944 mb]	1979 Frederic; 1985 Gloria; 1995 Felix, Luis, Opal; 1998 Georges; 2004 Charley; 2005 Emily, Katrina (were cat. 5)
5	> 135 knots > 250 kmph (> 155 mph) [< 920 mb]	1935 No. 2; 1938 No. 4; 1960 Donna; 1961 Carla; 1969 Camille; 1971 Edith; 1977 Anita; 1979 David; 1980 Allen; 1988 Gilbert, Mitch; 1989 Hugo; 1992 Andrew; 2004 Ivan; 2007 Dean, Felix

Physical Structure Fully organized tropical cyclones have an intriguing physical appearance (Figure 8.23). They range in diameter from a compact 160 km (100 mi), to 1000 km (600 mi), to some western Pacific super typhoons that attain 1300–1600 km (800–1000 mi). Vertically, these storms dominate the full height of the troposphere.

The inward-spiraling clouds form dense rain bands, with a central area designated the *eye*. Around the eye swirls a thunderstorm cloud called the *eyewall*, which is the area of most intense precipitation. The eye remains an enigma, for in the midst of devastating winds and torrential rains the eye is quiet and warm with even a glimpse of blue sky or stars possible. The structure of the rain bands, central eye, and eyewall is clearly visible for Hurricane Gilbert in Figure 8.23a (top view), 8.23b (oblique view from an artist's perspective), and 8.23c (side-radar view).

The strongest winds of a tropical cyclone are usually recorded in its right-front quadrant (relative to the storm's directional path), where dozens of fully developed tornadoes may be embedded at the time of landfall. For example, Hurricane Camille in 1969 had up to 100 tornadoes embedded in that quadrant.

A tropical cyclone moves along at 16–40 kmph (10–25 mph). When it makes **landfall**, or moves ashore, the storm pushes seawater inland, causing dangerous **storm surges** often several meters in depth. Storm surges frequently catch people by surprise and cause the majority of hurricane drownings. In 1998, Hurricanes Bonnie, Georges, and Mitch moved slowly onshore, producing damaging storm surges along North Carolina, Mississippi, and Central America, respectively. The slow progress of these storms produced extensive flooding from the sustained rains.

The storm surge from Hurricane Katrina in 2005 (see Focus Study 8.1, Figure 8.1.1) caused failure in numerous poorly conceived and constructed levees and canals; we discuss this in Chapter 14 under floodplain management. The flood disaster that befell New Orleans after Katrina resulted more from human engineering and construction errors than from the storm itself, which was actually downgraded to a category 3 hurricane when it made landfall.

Sample Devastating Tropical Cyclones The tropical cyclone that struck Bangladesh in 1970 killed an estimated 300,000 people, and the one in 1991 claimed over 200,000. In the United States, death tolls are much lower, but still significant. The Galveston, Texas, hurricane of 1900 killed 6000; Hurricane Katrina and engineering failures killed more than 1000 in Louisiana, Mississippi, and Alabama in 2005. Other storms resulting in significant fatalities include Hurricane Audry (1957), 400; Hurricane Gilbert (1988), 318; Hurricane Camille (1969), 256; and Hurricane Agnes (1972), 117. In 1998, Hurricanes Bonnie and Georges damaged property and took lives. Hurricane Mitch (October 26–November 4, 1998) was the deadliest Atlantic hurricane in two centuries, killing more than 12,000 people in Central America. In 2007, for the first time in history in a single Atlantic season, two category 5 hurricanes simultaneously made landfall, Hurricane Dean (Yucatán) and Hurricane Felix (Honduras).

(a)

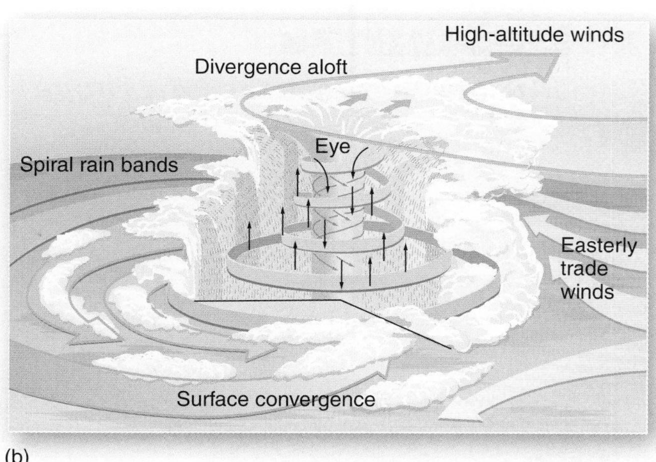

(b)

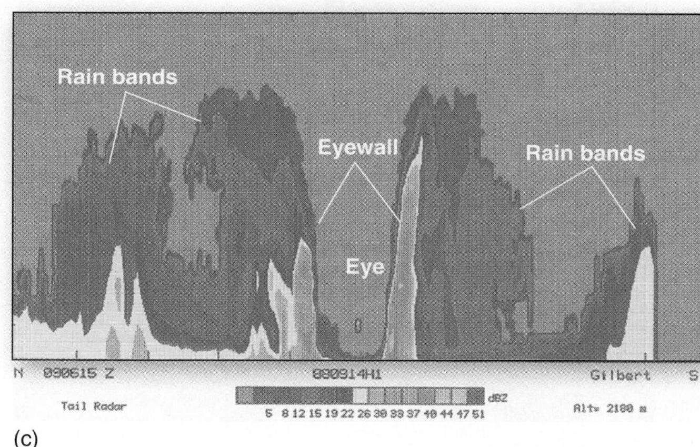

(c)

FIGURE 8.23 Profile of a hurricane.
For the Western Hemisphere, Hurricane Gilbert attained the record size (1600 km, or 1000 mi, in diameter) and second lowest barometric pressure (888 mb, or 26.22 in.). Gilbert's sustained winds reached 298 kmph (185 mph), with peaks exceeding 320 kmph (200 mph). Hurricane Gilbert, September 13, 1988: (a) *GOES-7* satellite image. (b) A stylized portrait of a mature hurricane, drawn from an oblique perspective (cutaway view shows the eye, rain bands, and wind-flow patterns). (c) *SLAR* (side-looking airborne radar) image from an aircraft flying through the center of the storm. Rain bands of greater cloud density are false-colored in yellows and reds. The clear sky in the central eye is dramatically portrayed. [(a) and (c) NOAA and the NHC.]

In September 2003, Hurricane Isabel struck the Outer Banks and Cape Hatteras of North Carolina, causing extensive flooding and wind damage to these fragile barrier islands as well as 36 deaths and about $2 billion in damages. Hurricane Charley brought category 4 winds to Florida's Gulf Coast in 2004, causing approximately $12 billion in damage and taking about two dozen lives (Figure 8.24a, p. 218).

The 2008 season was notable for seven landfalls featuring four major storms, including Hurricanes Gustav hitting Louisiana and Ike hitting Galveston, the southeastern Texas coast, and southern Louisiana. Hurricane Ike reached category 4 status in the Caribbean, crossed Cuba, and maintained category 2 as it grew in overall size. On September 13, it made landfall on Galveston Island, with the worst storm surge and damage striking to the east on the Bolivar Peninsula, where 80% to 95% of homes were destroyed (Figure 8.24b, p. 218); see the photograph and discussion in *Geosystems Now* for this chapter and more coverage of barrier islands in Chapter 16.

The tragedy of Hurricane Andrew, which struck Florida in 1992, is that the storm destroyed or seriously damaged 70,000 homes and left 200,000 people homeless between Miami and the Florida Keys. The Everglades, the coral reefs north of Key Largo, some 10,000 acres of mangrove wetlands, and coastal southern pine forests all had significant damage from Andrew. Scientific assessments judge the Everglades to be quite resilient because the region's ecosystems evolved naturally with periodic hurricanes over the millennia. In fact, urbanization, agriculture, pollution, and water diversion pose a greater ongoing threat to the Everglades than do hurricanes.

Thus, tropical cyclones usually damage human structures more than they damage natural systems. This recurrent, yet avoidable, cycle—construction, devastation, reconstruction, devastation—was reinforced ironically in *Fortune* magazine more than 40 years ago after the destruction caused by Hurricane Camille:

> Before long the beachfront is expected to bristle with new motels, apartments, houses, condominiums,

FOCUS STUDY 8.1

Atlantic Hurricanes and the Future

Forecasters at the National Hurricane Center (NHC) in Miami were busy throughout the 1995–2009 hurricane seasons. This was the most active 15-year period in the history of the NHC, with 207 named tropical storms, including 111 hurricanes (48 of which were intense, meaning category 3 or higher). This was a record level of activity and individual storm intensity despite the reduced number of tropical cyclones during the 1997 El Niño season.

Statistically, the damage caused by tropical cyclones is increasing substantially as more and more development occurs along susceptible coastlines, whereas loss of life is decreasing in most parts of the world owing to better forecasting of these storms. The journal *Science* offered this "Perspective":

> The risk of human losses is likely to remain low . . . because of a well-established warning and rescue system and ongoing improvements in hurricane prediction. A main concern is the risks of high damage costs (up to $100 billion in a single event) because of ongoing population increases in coastal areas and increasing investment in buildings and extensive infrastructure in general.*

*L. Bengtsson, "Hurricane threats," *Science* 293 (July 20, 2001): 441.

Notebook
Record-breaking
2005 Atlantic hurricane season

(a)

(b)

(c)

FIGURE 8.1.1 Katrina surge clears neighborhoods and destroys highway bridge. (a) Hurricane Katrina, August 28, 2005, 11 A.M. CDT, category 5 strength. (b) NOAA aerial photo shows the power of wind and storm surge as entire Mississippi Gulf Coast neighborhoods are obliterated, such as here in Long Beach, with the debris carried by storm surge a kilometer from shore. (c) Remnants of the U.S. 90 bridge and causeway out of Bay St. Louis, Mississippi. [(a) *Terra* image, NASA/GSFC. (b) NOAA. (c) Randall Christopherson.]

Satellite
2005: 27 Storms,
Arlene to Zeta

As an example, the 2005 Atlantic hurricane season broke several records, including the most named tropical storms in a single year, totaling 27 (the average is 10); the most hurricanes, totaling 15 (the average is 5); and the highest number of intense hurricanes, category 3 or higher, totaling 7 (the average is 2). The 2005 season was the first time that three category 5 storms (Katrina, Rita, and Wilma) occurred in the Gulf of Mexico and the first time that three category 3 or above hurricanes made U.S. landfall in subsequent years (2004 and 2005). The 2005 season also recorded the greatest damage total in one year, at more than $130 billion. Hurricane Wilma is now the most intense storm (lowest central pressure) in recorded history for the Atlantic. On our *Mastering Geosystems* Web site, a map and table details all the storms in this remarkable season of tropical cyclones and hurricanes, along with a satellite-image movie of the entire 2005 season.

The Gulf Coast and Atlantic Seaboard sustained hits and near misses in 2005 from Hurricanes Cindy, Dennis, Katrina, Ophelia, Rita, Wilma, and Gamma; and, when added to Charley, Francis, Ivan, and Jeanne in 2004 and to Dolly, Fay, Gustav, and Ike in 2008, the United States sustained more than $250 billion in damages and a loss of more than 2500 lives.

In the aftermath of Hurricane Ike in 2008, government delays held up water, ice, and food supplies for the 2 million evacuees until 4 or 5 days after landfall, thus repeating mistakes made after Katrina (Figure 8.24b). As we learn through hearings and investigations of the post-storm failure to respond and assist these regions and as we see continuing mistakes in planning and engineering, the case for an integrated scientific approach is clearly made. We discuss related issues for this region, including engineering, floodplain management, and levee considerations, in Chapter 14.

Prediction and the Future

The Tropical Prediction Center, at the NHC in Miami, Florida, posts tropical cyclone forecasts on its web site (http://www.nhc.noaa.gov/). Their analysis of weather records for the period 1900–2006 disclosed a significant causal relationship between Atlantic tropical storms, including hurricanes that make landfall along the U.S. Gulf and East Coasts, and several meteorological variables, such as sea-surface temperatures, the presence of upper tropospheric crosswinds that cause shearing of vertical tropical storm circulation, and conditions in the Pacific Ocean, among other factors.

Subsequent research published in 2005 correlates longer storm lifetimes and greater storm intensity with rising sea-surface temperatures,* a link that was further reinforced in 2008:

> Atlantic tropical cyclones are getting stronger on average, with a 30-year trend that has been related to an increase in ocean temperatures. . . . The results presented here are conclusive in showing significant increasing trends in the satellite-derived lifetime-maximum wind speeds of the strongest tropical cyclones globally, and are quantitatively consistent with the heat-engine theory of cyclone intensity. Thus, as seas warm, the ocean has more energy that can be converted to tropical cyclone wind.†

Hurricanes Katrina, Rita, and Wilma in 2005 and Dean and Felix in 2007 all recorded remarkable central pressure drops as they passed over record sea-surface temperatures (see the satellite image of Katrina in Figure 8.1.1a).

*K. Emanuel, "Increasing destructiveness of tropical cyclones over the past 30 years," *Nature* 436 (August 4, 2005): 686–688.

†J. B. Elsner, J. P. Kossin, and T. H. Jagger, "The increasing intensity of the strongest tropical cyclones," *Nature* 455 (September 4, 2008): 92–95.

Wilma experienced a 100-mb central pressure drop in a little over 24 hours, October 19, 2005. The record sea-surface temperatures that are being recorded across the globe produce warmer water to depth, and the natural mixing that such storms trigger only brings up more heat to fuel the circulation. Many scientists now suggest that future warming with climate change, when considered alongside increased coastal population settlement, will produce unwanted, yet substantial, hurricane-related damage losses (Figure 8.1.1b and c).

Ironically, the relatively mild hurricane seasons from 1971 to 1994 encouraged weak zoning and rapid development of vulnerable coastal lowlands. New buildings, apartments, and government offices are opening right next to still visible rubble and bare foundation pads. Unfortunately, public or private decision makers practice thoughtful hazard planning infrequently, whether along coastal lowlands, river floodplains, or earthquake fault zones. The consequence is that our entire society bears the financial cost of planning failure, not to mention those victims who directly shoulder the physical, emotional, and economic hardship of the event.

No matter how accurate storm forecasts become, coastal and lowland property damage will continue to increase until better hazard zoning and development restrictions are in place. The property insurance industry appears to be taking action to promote these improvements. It is requiring tougher building standards to obtain coverage—or, in some cases, it is refusing to insure property along vulnerable coastal lowlands. Given the increased intensity and "power dissipation" of these storms and a rise in sea level, the public, politicians, and business interests must respond to somehow mitigate this hazardous predicament. Given what we have experienced and what we know of Earth systems, did society learn anything? The Atlantic hurricane season begins each June 1.

(a)

(b)

FIGURE 8.24 Damage from Hurricanes Charley and Ike.
(a) Category 4 Hurricane Charley in 2004 devastated Punta Gorda, near Port Charlotte, Florida; U.S. damage totaled at $15 billion. (b) Category 2 Hurricane Ike, which was 965 km (600 mi) in diameter, devastated Bolivar Peninsula and the Texas coast, with damage hitting $40 billion. [Bobbé Christopherson.]

> and office buildings. Gulf Coast businessmen, incurably optimistic, doubt there will ever be another hurricane like Camille, and even if there is, they vow, the Gulf Coast will rebuild bigger and better after that.*

**Fortune*, October 1969, p. 62.

Sadly, Hurricane Katrina obliterated the same Gulf Coast towns in 2005—Waveland, Bay Saint Louis, Pass Christian, Long Beach, Gulfport, among others—that Camille washed away in 1969. The below-sea-level sections of New Orleans may never recover their former status; however the public is told that "the Gulf Coast will rebuild bigger and better after that."

CRITICAL THINKING 8.2

Hazard perception and planning: What seems to be missing?

Relative to coastal devastation from tropical weather, the following statement appears in the text: "This recurrent, yet avoidable, cycle—construction, devastation, reconstruction, devastation. . . ." This describes the ever-increasing dollar losses to property from tropical storms and hurricanes at a time when improved forecasts have resulted in a significant reduction in loss of life. Given rising sea levels along coastlines and the doubling of total power dissipation in these tropical storms since 1970, in your opinion, what is the solution to halting this cycle of destruction and increasing losses? How would you implement your ideas?

GEOSYSTEMS CONNECTION

Weather is the expression of energy, water, water vapor, and the atmosphere at any given moment. The patterns of precipitation produced across the globe form the input to water resource analysis. In the next chapter, we use the water balance model and examine inputs and outputs of the water budget. A budget can be established for any time frame or area, from a house plant to a front lawn, to a region, or to a country. Issues of water quality and quantity and the availability of potable water loom as major issues for the global society.

KEY LEARNING CONCEPTS REVIEW

Weather is the short-term condition of the atmosphere; **meteorology** is the scientific study of the atmosphere. The spatial implications of atmospheric phenomena and their relationship to human activities strongly link meteorology to physical geography.

weather (p. 192)
meteorology (p. 192)

■ ***Describe*** air masses that affect North America, and ***relate*** their qualities to source regions.

Specific conditions of humidity, stability, and cloud coverage occur in a regional, homogenous **air mass**. The longer an air mass remains stationary over a region, the more definite its physical attributes become. The homogeneity of temperature and humidity in an air mass sometimes extends through the lower half of the troposphere. Air masses are categorized by their moisture content—**m** for maritime (wetter) and **c** for continental (drier)—and

their temperature, a function of latitude—designated **A** (arctic), **P** (polar), **T** (tropical), **E** (equatorial), and **AA** (Antarctic).

air mass (p. 193)

1. How does a source region influence the type of air mass that forms over it? Give specific examples of each basic classification.
2. Of all the air masses, which are of greatest significance to the United States and Canada? What happens to them as they migrate to locations different from their source regions? Give an example of air mass modification.

■ ***Identify*** and ***describe*** four types of atmospheric lifting mechanisms, and give an example of each.

Air masses can rise through **convergent lifting** (air flows conflict, forcing some of the air to lift); **convectional lifting** (air passing over warm surfaces gains buoyancy); **orographic lifting** (air passes over a topographic barrier); and *frontal lifting*. In North America, **chinook winds** (called *föhn* or *foehn* winds in Europe) are the warm, downslope air flows characteristic of the leeward side of mountains. Orographic lifting creates wetter windward slopes and drier leeward slopes situated in the **rain shadow** of the mountain. Conflicting air masses at a front produce a **cold front** (and sometimes a zone of strong wind and rain) or a **warm front**. A zone right along or slightly ahead of the front, called a **squall line**, is characterized by turbulent and wildly changing wind patterns and intense precipitation.

convergent lifting (p. 195)
convectional lifting (p. 196)
orographic lifting (p. 196)
chinook winds (p. 197)
rain shadow (p. 197)
cold front (p. 197)
warm front (p. 197)
squall line (p. 199)

3. Explain why it is necessary for an air mass to be lifted if there is to be saturation, condensation, and precipitation.
4. What are the four principal lifting mechanisms that cause air masses to ascend, cool, condense, form clouds, and perhaps produce precipitation? Briefly describe each.
5. Differentiate between the structure of a cold front and that of a warm front.

■ ***Analyze*** the pattern of orographic precipitation, and ***describe*** the link between this pattern and global topography.

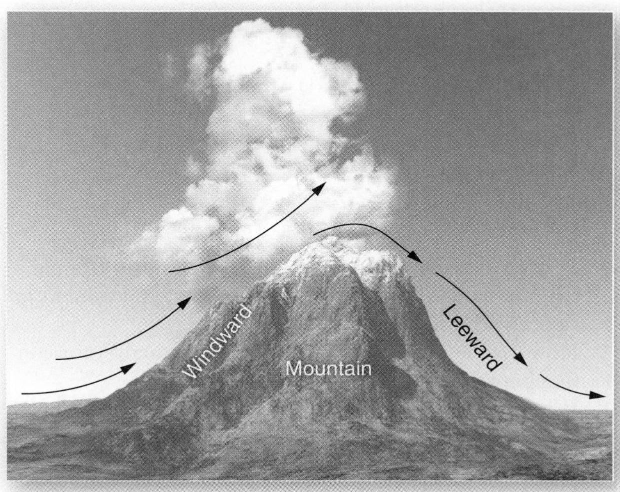

The physical presence of a mountain acts as a topographic barrier to migrating air masses. *Orographic lifting* (*oro-* means "mountain") occurs when air is forcibly lifted upslope as it is pushed against a mountain. It cools adiabatically. An orographic barrier enhances convectional activity. The precipitation pattern of windward and leeward slopes persists worldwide.

6. When an air mass passes across a mountain range, many things happen to it. Describe each aspect of a moist air mass crossing a mountain. What is the pattern of precipitation that results?
7. Explain how the distribution of precipitation in the state of Washington is influenced by the principles of orographic lifting.

■ ***Describe*** the life cycle of a midlatitude cyclonic storm system, and ***relate*** this to its portrayal on weather maps.

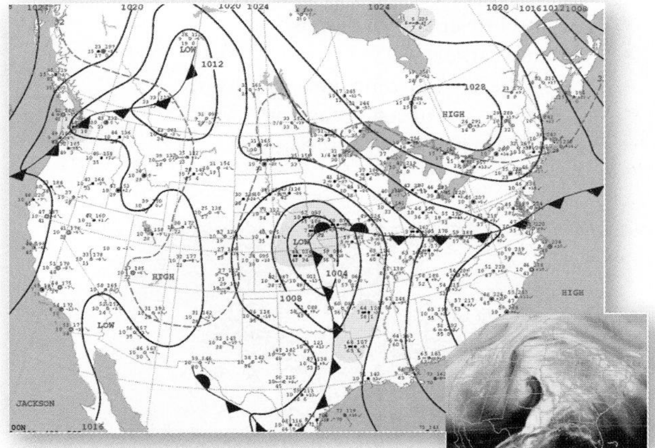

A **midlatitude cyclone,** or **wave cyclone,** is a vast low-pressure system that migrates across the continent, pulling air masses into conflict along fronts. **Cyclogenesis**, the birth of the low-pressure circulation, can occur off the west coast of North America, along the polar front, along the lee slopes of the Rockies, in the Gulf of Mexico, and along the East Coast. A midlatitude cyclone can be thought of as having a life cycle of birth, maturity, old age, and dissolution. An **occluded front** is produced when a cold front overtakes a warm front in the maturing cyclone. Sometimes a **stationary front** develops between conflicting air masses, where air flow is parallel to the front on both sides. These systems are guided by the jet streams of the upper troposphere along seasonally shifting **storm tracks**.

midlatitude cyclone (p. 201)
wave cyclone (p. 201)
cyclogenesis (p. 201)
occluded front (p. 201)
stationary front (p. 202)
storm tracks (p. 203)

8. Differentiate between a cold front and a warm front as types of frontal lifting, and describe what you would experience with each one.
9. How does a midlatitude cyclone act as a catalyst for conflict between air masses?
10. What is meant by cyclogenesis? In what areas does it occur and why? What is the role of upper-tropospheric circulation in the formation of a surface low?
11. Diagram a midlatitude cyclonic storm during its open stage. Label each of the components in your illustration, and add arrows to indicate wind patterns in the system.

■ ***List*** **the measurable elements that contribute to modern weather forecasting, and** ***describe*** **the technology and methods employed.**

Synoptic analysis involves the collection of weather data at a specific time, as shown on the synoptic weather maps inserted in Figures 8.3, 8.12, and CT 8.1.1. The standard weather station symbols used on these maps are identified in the legend to Figure 8.12. Building a database is key to computer-based weather prediction and the development of weather-forecasting models, including the following:

- Barometric pressure (sea level and altimeter setting)
- Pressure tendency (steady, rising, falling)
- Surface air temperature
- Dew-point temperature
- Wind speed, direction, and character (gusts, squalls)
- Type and movement of clouds
- Current weather
- State of the sky (current sky conditions)
- Visibility; vision obstruction (fog, haze)
- Precipitation since last observation

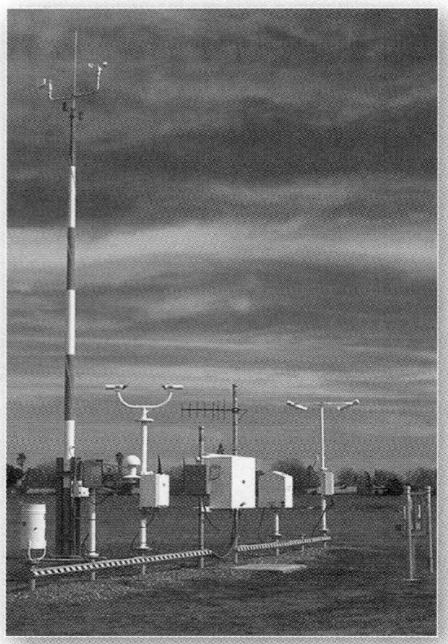

Technology employed includes Doppler radar, wind profilers, AWIPS workstations, ASOS instrument arrays, various satellite platforms, and GPS.

12. What is your principal source of weather data, information, and forecasts? Where does your source obtain its data? Have you used the Internet and World Wide Web to obtain weather information? What benefits do you see?

■ ***Analyze*** **various forms of violent weather and their characteristics, and** ***review*** **several examples of each.**

The violent power of some weather phenomena poses a hazard to society. Severe ice storms involve **sleet** (freezing rain, ice glaze, and ice pellets), snow blizzards, and crippling ice coatings on roads, power lines, and crops. Thunderstorms produce **lightning** (electrical discharges in the atmosphere), **thunder** (sonic bangs produced by the rapid expansion of air after intense heating by lightning), and **hail** (ice pellets formed within cumulonimbus clouds).

Strong linear winds in excess of 26 m/s (58 mph), known as **derechos**, are associated with thunderstorms and bands of showers crossing a region. Straight-line winds can cause significant damage and crop losses. A spinning, cyclonic column rising to mid-troposphere level—a **mesocyclone**—is sometimes visible as the swirling mass of a cumulonimbus cloud, especially in a *supercell* system. Dark gray **funnel clouds** pulse from the bottom side of the parent cloud. A **tornado** is formed when the funnel connects with Earth's surface. A **waterspout** forms when a tornado circulation occurs over water.

Within tropical air masses, large low-pressure centers can form along easterly wave troughs. Under the right conditions, a tropical cyclone is produced. Depending on central pressure, a **tropical cyclone** becomes a **hurricane**, **typhoon**, or *cyclone* when winds exceed 65 knots (119 kmph, 74 mph). As forecasting of weather-related hazards improves, loss of life decreases, although property damage continues to increase. Great damage occurs to occupied coastal lands when hurricanes make **landfall** and when winds drive ocean water inland in **storm surges**.

sleet (p. 205)
lightning (p. 205)
thunder (p. 205)
hail (p. 206)
derechos (p. 206)

mesocyclone (p. 208)
funnel clouds (p. 208)
tornado (p. 208)
waterspout (p. 208)
tropical cyclone (p. 211)
hurricane (p. 212)
typhoon (p. 212)
landfall (p. 214)
storm surges (p. 214)

13. What constitutes a thunderstorm? What type of cloud is involved? What type of air mass would you expect in an area of thunderstorms in North America?
14. Lightning and thunder are powerful phenomena in nature. Briefly describe how they develop.
15. Describe the formation process of a mesocyclone. How is this development associated with that of a tornado?
16. Evaluate the pattern of tornado activity in the United States. What generalizations can you make about the distribution and timing of tornadoes? Do you perceive a trend in tornado occurrences in the United States? Explain.
17. What are the different classifications for tropical cyclones? List the various names used worldwide for hurricanes. Have any hurricanes ever occurred in the south Atlantic?
18. What factors contributed to the incredible damage cost of Hurricane Andrew? Why have such damage figures increased even though loss of life has decreased over the past 30 years?
19. What have scientists said to explain the intensity of tropical cyclones since 1970? Put into your own words the quotations in this chapter from scientific journals related to changes in storm intensity.
20. Relative to improving weather forecasting, what are some of the technological innovations discussed in this chapter?

MASTERING GEOGRAPHY

Visit www.masteringgeography.com for several figure animations, satellite loops, self-study quizzes, flashcards, visual glossary, case studies, career links, textbook reference maps, RSS Feeds, and an ebook version of *Geosystems*. Included are many geography web links to interesting Internet sources that support this chapter.

(a)

(b)

Bolivar Peninsula, TX, before and after Hurricane Ike. [USGS photo comparisons.]

CHAPTER 9

Water Resources

The western United States, particularly the Southwest, is experiencing a prolonged drought. Hoover Dam on the Arizona–Nevada border backs up Lake Mead on the Colorado River. The maximum reservoir elevation of 374.6 m (1229 ft) is shown in the July 1983 photo of Lake Mead. Twenty-six years later, in March 2009, Lake Mead was at 331 m (1086 ft), or about 43.6 m (143 ft) lower than 1983. More on this in Focus Study 15.1 in Chapter 15. [(1983) Author. (2009) Bobbé Christopherson.]

KEY LEARNING CONCEPTS

After reading the chapter, you should be able to:

- ***Illustrate*** the hydrologic cycle with a simple sketch, and ***label*** it with definitions for each water pathway.
- ***Relate*** the importance of the water-budget concept to your understanding of the hydrologic cycle, water resources, and soil moisture for a specific location.
- ***Construct*** the water-balance equation as a way of accounting for the expenditures of water supply, and ***define*** each of the components in the equation and its specific operation.
- ***Describe*** the nature of groundwater, and ***define*** the elements of the groundwater environment.
- ***Identify*** critical aspects of freshwater supplies for the future, and ***cite*** specific water-resource issues and potential remedies for any shortfalls.

GEOSYSTEMS NOW

Water Budgets, Climate Change, and the Southwest

The Colorado River system begins in the Rockies of Colorado and flows almost 2317 km (1440 mi) through the southwestern United States. This region includes the four fastest-growing states in terms of population: Nevada, Arizona, Colorado, and Utah. The rapidly expanding urban areas of Las Vegas, Phoenix, Tucson, Denver, San Diego, and Albuquerque all depend on Colorado River water, which supplies approximately 30 million people in seven states. Ending as a trickle of water in the sand, the river now disappears kilometers short of its former mouth in the Gulf of California (Figure GN 9.1).

Reservoirs store water to offset variable Colorado River flows. The largest are Lake Mead, formed by Hoover Dam, and Lake Powell, formed by Glen Canyon Dam, which together account for over 80% of total system storage. However, Lake Mead and Lake Powell are in water-budget trouble, since water allocations already exceed the long-term average flows in the system and have for years. Thus, an impending water supply crisis is emerging. A convergence of factors in the American Southwest is reducing water inflows to the Colorado River and increasing outflow demands, revealing that current water-management practices are not sustainable.

Drought conditions affecting the Colorado River system began in 2000 and result primarily from climate change, which causes higher temperatures and evaporation rates as well as reduced mountain snowpack and earlier spring melt. An added factor is the shifting of the subtropical high-pressure system to higher latitudes, thus expanding the region of permanent drought. According to regional tree-ring records, nine droughts have affected the American Southwest since A.D. 1226. However, the present drought is the first to occur in the Southwest in the presence of an increasing human demand for water and shifting global pressure systems.

Effects of prolonged drought are evident throughout the Colorado River system. Water inflow to Lake Powell is running at less than one-third of long-term averages—with the low year of 2002 at just 25%. System reservoirs are at record low levels. The production of hydroelectric power has fallen to 50% of capacity. With the system in an official declared "drought condition," as total system storage was at 58% of capacity and Lake Mead at 41% of capacity by mid-2010.

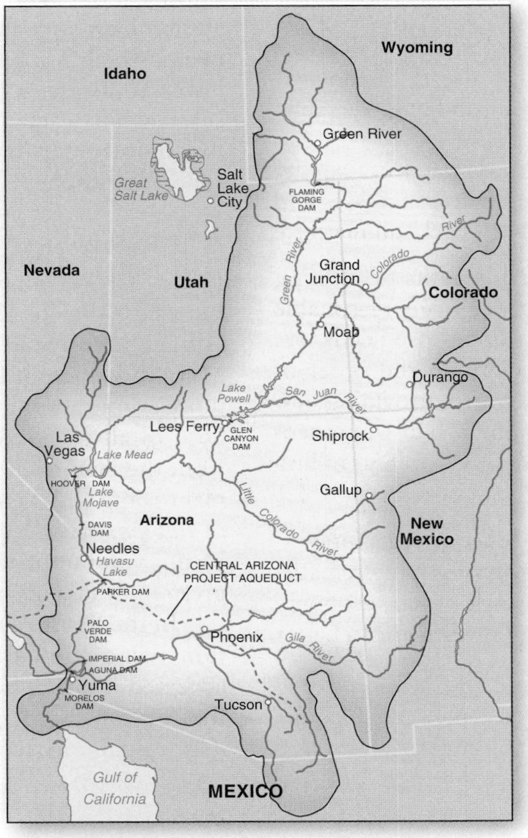

Recent research shows that reservoir storage is at risk as average streamflows decrease and water use increases throughout the region. In 2008, a team of scientists forecasted: (1) a 50% probability that water storage in Lake Mead could be fully depleted by 2021, without enough water to turn turbines by 2017, and (2) have a 10% chance of no "live storage" (no water releases through gravitational flow) at Lake Mead and Lake Powell by 2013. (See http://scrippsnews.ucsd.edu/Releases/?releaseID=876; also see T. P. Barnett and W. Pierce, "When will Lake Mead go dry?" *Water Resource Research* 44, WR3201 [March 29, 2008].)

Engineers estimate that it could take 13 normal winters of snowpack in the Rockies and basinwide precipitation to "reset" the system. Such a recovery is much slower than previous drought recoveries because of increasing evaporation associated with warming temperatures across the Southwest as well as from increasing human water use. Present probabilities for the period after 2016 put the chances of surplus discharge occurring in the Colorado River system in any given year at only 20%.

Limiting or halting further metropolitan construction and population growth in the Southwest are not part of present planning. Despite conditions along the Colorado River, we still seem to be "supply" focused. A detailed focus study on this river system is in Chapter 15. Here in Chapter 9, we look at water resources as a product of supply and demand in the water-balance model.

FIGURE GN 9.1 The Colorado River Basin. The region is divided between an Upper Basin and a Lower Basin. The river from 1990 to 2009 averaged only 11.20 maf (million acre feet) of discharge per year. In its water budget, inputs do not equal outputs and the reservoirs are declining.

The physical reality of life is defined by water. Our lives are bathed and infused with water. Our own bodies are about 70% water, as are plants and animals. We use water to cook, bathe, wash clothing, and dilute our wastes. We water small gardens and vast agricultural tracts. Most industrial processes would be impossible without water. It is the essence of our existence and is therefore the most critical resource supplied by Earth systems—the essential resource for life.

Water is not always naturally available where and when we want it. Consequently, we rearrange surface-water resources to suit our needs. We drill wells, build cisterns and reservoirs, and dam and divert streams to redirect water either spatially (geographically, from one area to another) or temporally (over time, from one part of the calendar to another). All of this activity constitutes water-resource management.

Fortunately, water is a renewable resource, constantly cycling through the environment. Even so, some 1.1 billion people lack safe drinking water. People in 80 countries face impending water shortages, either in quantity or in quality, or both. Approximately 2.4 billion people lack adequate sanitary facilities—80% of these in Africa and 13% in Asia. This translates to approximately 2 million deaths a year due to lack of water and 5 million deaths a year from waterborne infections and disease. An investment in safe drinking water, sanitation, and hygiene could decrease these numbers.

During the first half of this century, water availability per person will drop by 74% as population increases and adequate water decreases. Author Peter Gleick summarized:

> The overall economic and social benefits of meeting basic water requirements far outweigh any reasonable assessment of the costs of providing those needs. . . . Water-related diseases cost society on the order of $125 billion per year. . . . Yet the cost of providing new infrastructure needs for all major urban water sectors has been estimated at around $25 to $50 billion per year.*

Thus, the role for geographic science with its integrative, spatial approach is key in analyzing a systematic, comprehensive global portrait, a world-water economy approach, to more effectively allocate investment to better mitigate the coming crisis (see the World Water Assessment Program, UNESCO, http://www.unesco.org/water/wwap/).

In this chapter: We begin with the hydrologic cycle, which gives us a model for understanding the global water balance. We look at water resources using a water-budget approach—similar in many ways to a money budget—in which we examine water "receipts" and "expenses" at specific locations. This budget approach can be applied at any scale, from a small garden, to a farm, to a regional landscape such as the Colorado River in *Geosystems Now*.

About half the U.S. population draws freshwater from groundwater resources, yet groundwater is tied to surface-water supplies for recharge. This chapter concludes by considering the quantity and quality of the water we withdraw and consume for irrigation, industrial, and municipal uses—our specific water supply. Water supplies of adequate quantity and quality loom as the most important resource issue for many parts of the world in this century.

*P. Gleick, *The World's Water*, a biennial report (Washington, DC: Island Press, 2008; http://www.worldwater.org/).

The Hydrologic Cycle

Vast currents of water, water vapor, ice, and energy are flowing about us continuously in an elaborate, open, global plumbing system. Together, they form the **hydrologic cycle**, which has operated for billions of years, from the lower atmosphere to several kilometers beneath Earth's surface. The cycle involves the circulation and transformation of water throughout Earth's atmosphere, hydrosphere, lithosphere, and biosphere.

A Hydrologic Cycle Model

Figure 9.1 is a simplified model of this complex system. Let us use the ocean as a starting point for our discussion, although we could jump into the model at any point. More than 97% of Earth's water is in the ocean, and here most evaporation and precipitation occur. We can trace 86% of all evaporation to the ocean. The other 14% is from the land, including water moving from the soil into plant roots and passing through their leaves (the process of *transpiration*, described later in this chapter). Estimates of the volume of water involved in these pathways are given in thousands of km^3.

In the figure, you can see that of the 86% of evaporation rising from the ocean, 66% combines with 12% advected (moving horizontally) from the land to produce the 78% of all precipitation that falls back into the ocean. The remaining 20% of moisture evaporated from the

GEO REPORT 9.1 *The water we use*

On an individual level in the United States, urban dwellers directly use an average of 680 L (180 gal) of water per person per day, whereas rural populations average only 265 L (70 gal) or less per person per day. Take a moment and speculate why there is a difference. However, overall individual water use is actually much higher than these figures because of indirect use, such as consuming food and beverages made using water-intensive practices. See http://www.worldwater.org/data.html, *World's Water 2008–2009,* Table 19, for more on these indirect uses of water.

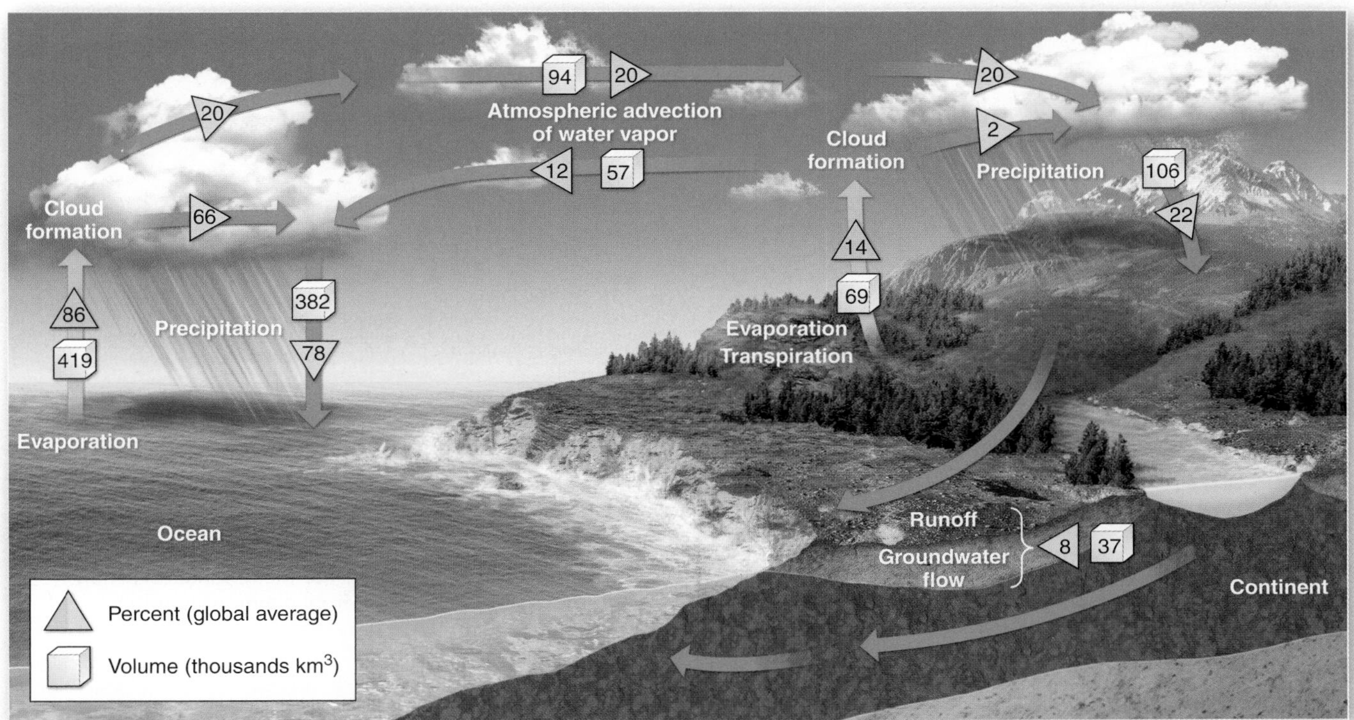

FIGURE 9.1 The hydrologic cycle model.
Water travels endlessly through the hydrosphere, atmosphere, lithosphere, and biosphere. The triangles show global average values as percentages. Note that all evaporation (86% + 14% = 100%) equals all precipitation (78% + 22% = 100%), and advection in the atmosphere is balanced by surface and subsurface water runoff when all of Earth is considered (1 km^3 × 0.24 = 1 mi^3).

Animation
Earth's Water and the Hydrologic Cycle

ocean, plus 2% of land-derived moisture, produces the 22% of all precipitation that falls over land. Clearly, the bulk of continental precipitation comes from the oceanic portion of the cycle. Regionally, the different parts of the cycle will vary, creating imbalances and, depending on climate, surpluses in one region and shortages in another.

Surface Water

Precipitation that reaches Earth's surface follows two basic pathways: It either flows overland or soaks into the soil. Along the way, **interception** occurs when precipitation strikes vegetation or other ground cover. Intercepted water that drains across plant leaves and down their stems to the ground is *stem flow* and can be an important moisture route to the ground. Precipitation that falls directly to the ground, coupled with drips from vegetation (excluding stem flow), constitutes *throughfall*. Water soaks into the subsurface through **infiltration**, or penetration of the soil surface. It further permeates soil or rock through the downward movement of **percolation**. These concepts are shown in Figure 9.2.

The atmospheric advection of water vapor from sea to land and land to sea at the top of Figure 9.1 appears to be unbalanced: 20% (94,000 km^3) moving inland, but only 12% (57,000 km^3) moving out to sea. However, this exchange is balanced by the 8% (37,000 km^3) that is land runoff, which flows to the sea. Most of this runoff—about 95%—comes from surface waters that wash across land as overland flow and streamflow. Only 5% of runoff is slow-moving subsurface groundwater. These percentages indicate that the small amount of water in rivers and streams

GEO REPORT 9.2 *How is water measured?*

In most of the United States, hydrologists measure streamflow in cubic feet per second (ft^3/s); Canadians use cubic meters per second (m^3/s). In the eastern United States, and for large-scale assessments, water managers use millions of gallons a day (MGD), billions of gallons a day (BGD), or billions of liters per day (BLD).

In the western United States, where irrigated agriculture is so important, the measure frequently used is acre-feet per year. One acre-foot is an acre of water, 1 ft deep, equivalent to 325,872 gal (43,560 ft^3, or 1234 m^3, or 1,233,429 L). An acre is an area that is about 208 ft on a side and is 0.4047 hectares. For global measurements, 1 km^3 = 1 billion m^3 = 810 million acre-feet; 1000 m^3 = 264,200 gal = 0.81 acre-feet. For smaller measures, 1 m^3 = 1000 L = 264.2 gal.

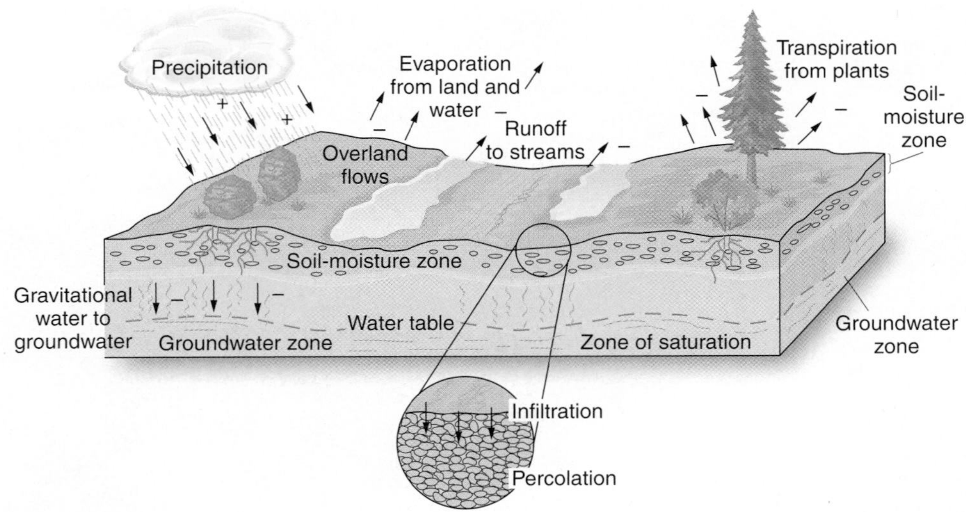

FIGURE 9.2 The soil-moisture environment.
The principal pathways for precipitation include interception by plants; throughfall to the ground; collection on the surface, forming overland flow to streams; transpiration and evaporation from plants; evaporation from land and water; and gravitational water moving to subsurface groundwater.

is dynamic; in contrast, the large quantity of subsurface water is sluggish and represents only a small portion of total runoff.

The residence time for a water molecule in any part of the hydrologic cycle determines its relative importance in affecting Earth's climates. The short time spent by water in transit through the atmosphere (an average of 10 days) plays a role in temporary fluctuations in regional weather patterns. Long residence times, such as the 3000–10,000 years in deep-ocean circulation, groundwater aquifers, and glacial ice, act to moderate temperature and climatic changes. These slower parts of the cycle work as a "system memory"; the long periods over which heat energy is stored and released can buffer the effects of system changes.

Soil-Water-Budget Concept

An effective method for assessing water resources is to establish a **soil-water budget** for any area of Earth's surface—a continent, country, region, field, or front yard. Key is measuring the precipitation "supply" input and its distribution to satisfy the "demand" outputs of plants, evaporation, and soil-moisture storage in the area considered. Such a budget can examine any time frame, from minutes to years.

Think of a soil-water budget as a money budget: Precipitation income must be balanced against expenditures for evaporation, transpiration, and runoff. Soil-moisture storage acts as a savings account, accepting deposits and withdrawals of water. Sometimes all expenditure demands are met, and any extra water results in a surplus. At other times, precipitation and soil-moisture income are inadequate to meet demands, and a deficit, or water shortage, results.

Geographer Charles W. Thornthwaite (1899–1963) pioneered applied water-resource analysis and worked with others to develop a water-balance methodology. They applied water-balance concepts to solving problems, especially related to irrigation, which requires accurate quantity and timing of water application to maximize crop yields. Thornthwaite also developed methods for estimating evaporation and transpiration. He recognized the important relation between water supply and water demand as an essential climatic element. In fact, his initial use of these techniques was to develop a climatic classification system.

The Soil-Water-Balance Equation

To understand Thornthwaite's water-balance methodology and "accounting" or "bookkeeping" procedures, we must first understand some terms and concepts. Figure 9.2 illustrates the essential aspects of a soil-water budget. Precipitation (mostly rain and snow) provides the moisture input. The object, as with a money budget, is to account for the ways in which this supply is distributed: actual water taken by evaporation and plant transpiration, extra water that exits in streams and subsurface groundwater, and recharge or utilization of soil-moisture storage.

Figure 9.3 organizes the water-balance components into an equation. As in all equations, the two sides must balance; that is, the precipitation receipt (left side) must be fully accounted for by expenditures (right side). Follow this water-balance equation as you read the following paragraphs. To help you learn these concepts, we use letter abbreviations (such as PRECIP) for the components.

Precipitation (PRECIP) Input The moisture supply to Earth's surface is **precipitation** (PRECIP, or P). It arrives in several different forms depending upon temperature and moisture supply. On the *Mastering Geography* web site, Table 9.1 summarizes the different types of precipitation.

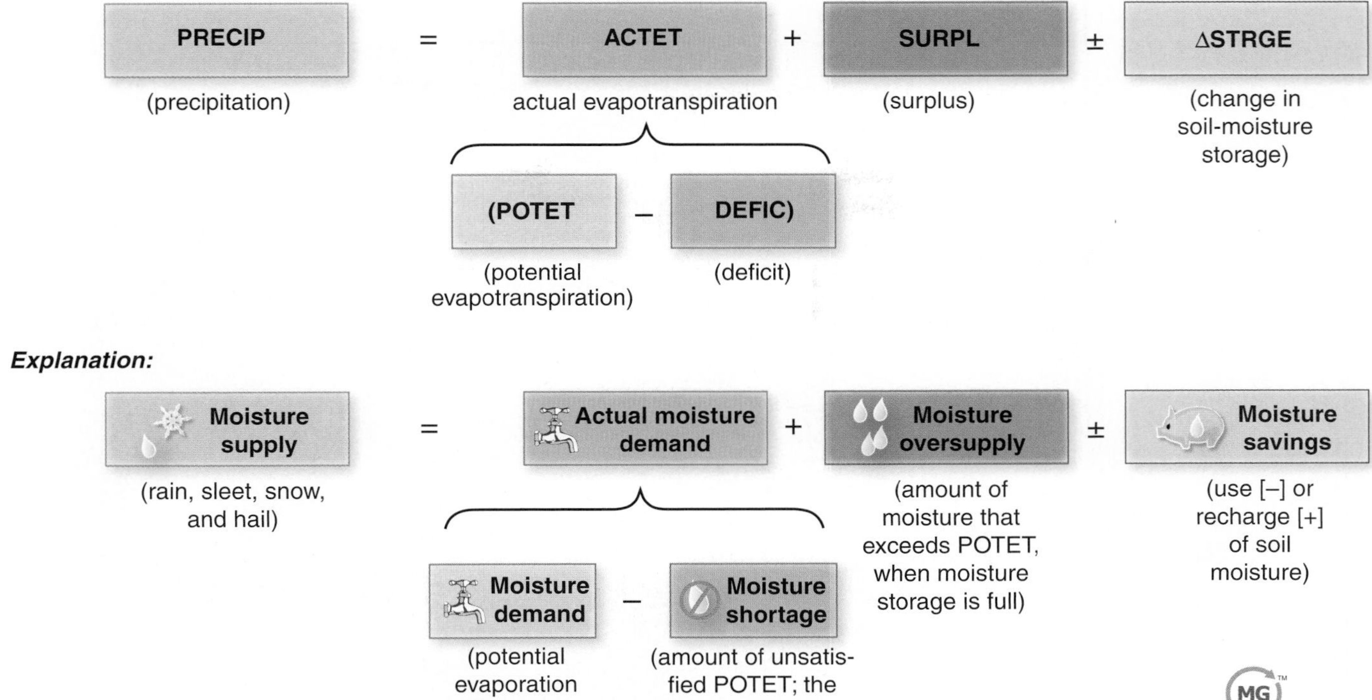

MG Satellite Global Water Balance Components

FIGURE 9.3 The water-balance equation explained.
The outputs (components to the right of the equal sign) are an accounting of expenditures from the precipitation moisture-supply input (to the left).

As you read through this table, recall which types of precipitation you have experienced—rain, sleet, snow, and hail being most common.

MG Notebook Chapter 9, Table 9.1, Types of Precipitation

Precipitation is measured with the **rain gauge**, essentially a large measuring cup that collects rainfall and snowfall so the water can be measured by depth, weight, or volume (Figure 9.4). Wind can cause an undercatch because the drops or snowflakes are not falling vertically. For example, a wind of 37 kmph (23 mph) produces an undercatch as great as 40%, meaning that an actual 1-in. rainfall might gauge at only 0.6 in. The *wind shield* you see above the gauge's opening reduces this error by catching raindrops that arrive at an angle. According to the World Meteorological Organization (http://www.wmo.ch/), more than 40,000 weather-monitoring stations are operating worldwide, with more than 100,000 places measuring precipitation.

Figure 9.5 maps precipitation patterns for the United States and Canada. Note the patterns of wetness and dryness as you remember our discussion of air masses and lifting mechanisms in Chapter 8. Chapter 10 presents a world precipitation map in Figure 10.2. PRECIP is the principal input to the water-balance equation in Figure 9.3. All the components discussed next are outputs or expenditures of this water.

FIGURE 9.4 A rain gauge.
A funnel guides water into a bucket that is sitting on an electronic weighing device. The gauge minimizes evaporation, which would cause low readings. The wind shield around the top of the gauge minimizes the undercatch produced by wind. [Bobbé Christopherson.]

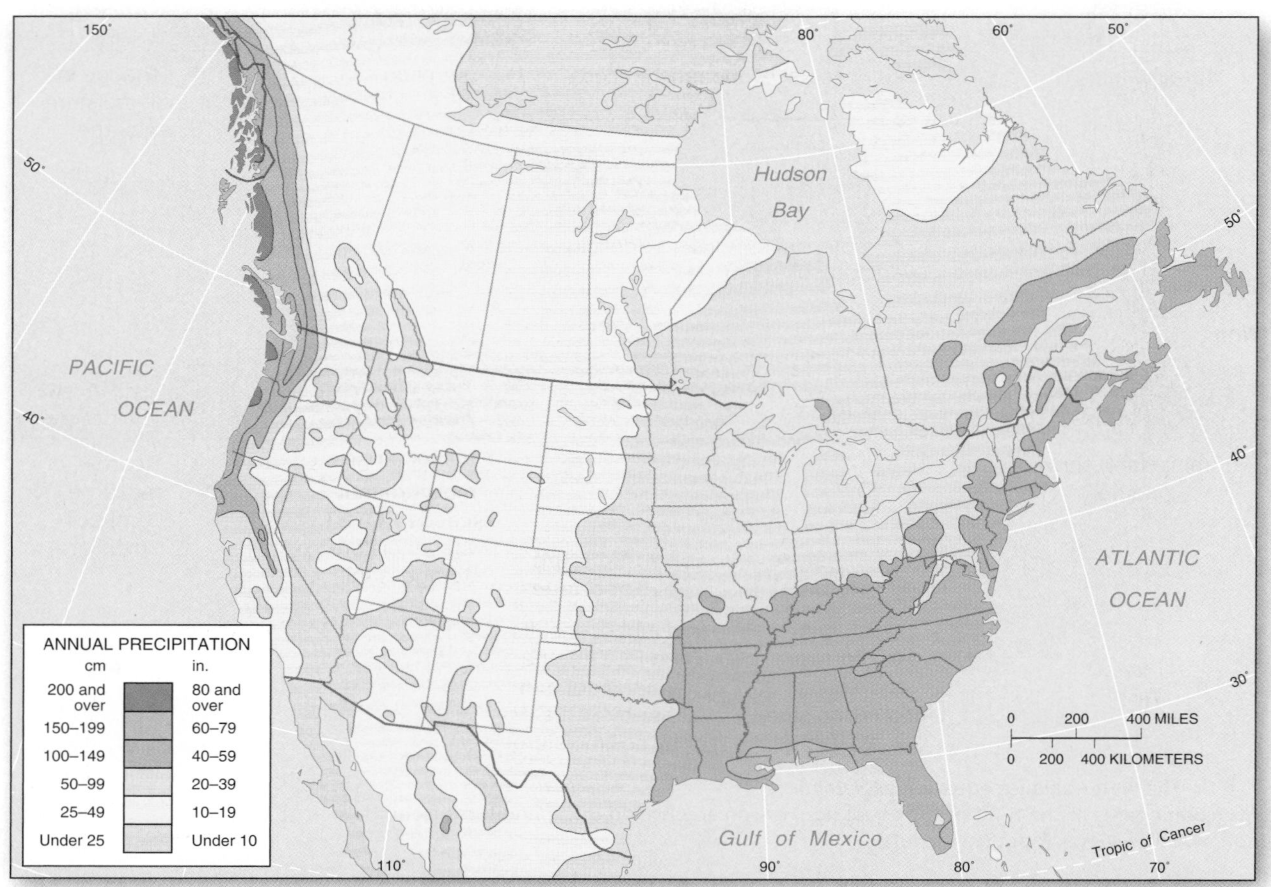

FIGURE 9.5 Precipitation (PRECIP) in North America—the water supply. [Adapted from NWS, U.S. Department of Agriculture, and Environment Canada.]

MG Satellite Global Water Balance Components

Actual Evapotranspiration (ACTET) As discussed in Chapters 5 and 7, **evaporation** is the net movement of free water molecules away from a wet surface into air that is less than saturated. **Transpiration** is a cooling mechanism in plants. When a plant transpires, it moves water through small openings called stomata in the underside of its leaves. The water evaporates, cooling the plant, much as perspiration cools humans. Transpiration is partially regulated by the plants themselves. Control cells around the stomata conserve or release water. Transpired quantities can be significant: On a hot day, a single tree can transpire hundreds of liters of water; a forest, millions of liters. Evaporation and transpiration are combined into one term—**evapotranspiration**—and 14% of evapotranspiration occurs from land and plants. Now, let's examine ways to estimate evapotranspiration rates.

Potential Evapotranspiration (POTET) Evapotranspiration is an actual expenditure of water. In contrast, **potential evapotranspiration** (POTET, or PET, or PE) is the amount of water that would evaporate and transpire under optimum moisture conditions when adequate precipitation and adequate soil-moisture supply are present.

Filling a bowl with water and letting the water evaporate illustrates this concept: When the bowl becomes dry, is there still an evaporation demand? The demand remains, of course, regardless of whether the bowl is dry. If you always kept water in the bowl, the amount of water that would evaporate or transpire is the POTET, or the optimum demand, given a constant supply of water. If you let the bowl dry out, the amount of POTET demand that went unmet is the shortage, or deficit (DEFIC). Note that in the water-balance equation, when we subtract the deficit from the potential evapotranspiration, we derive what actually happened—ACTET.

Figure 9.6 presents POTET values derived using Thornthwaite's approach for the United States and Canada. Note that higher values occur in the South, with highest readings in the Southwest, where higher average air temperature and lower relative humidity exist. Lower POTET values are found at higher latitudes and elevations, which have lower average temperatures.

Compare this POTET (demand) map with the PRECIP (supply) map in Figure 9.5. The relation between the two determines the remaining components of the water-balance equation in Figure 9.3. From the two maps, can you identify regions where PRECIP is greater than POTET (for example, the eastern United States)? Or where POTET is greater than PRECIP (for example, the southwestern United States)? Where you live, is the water

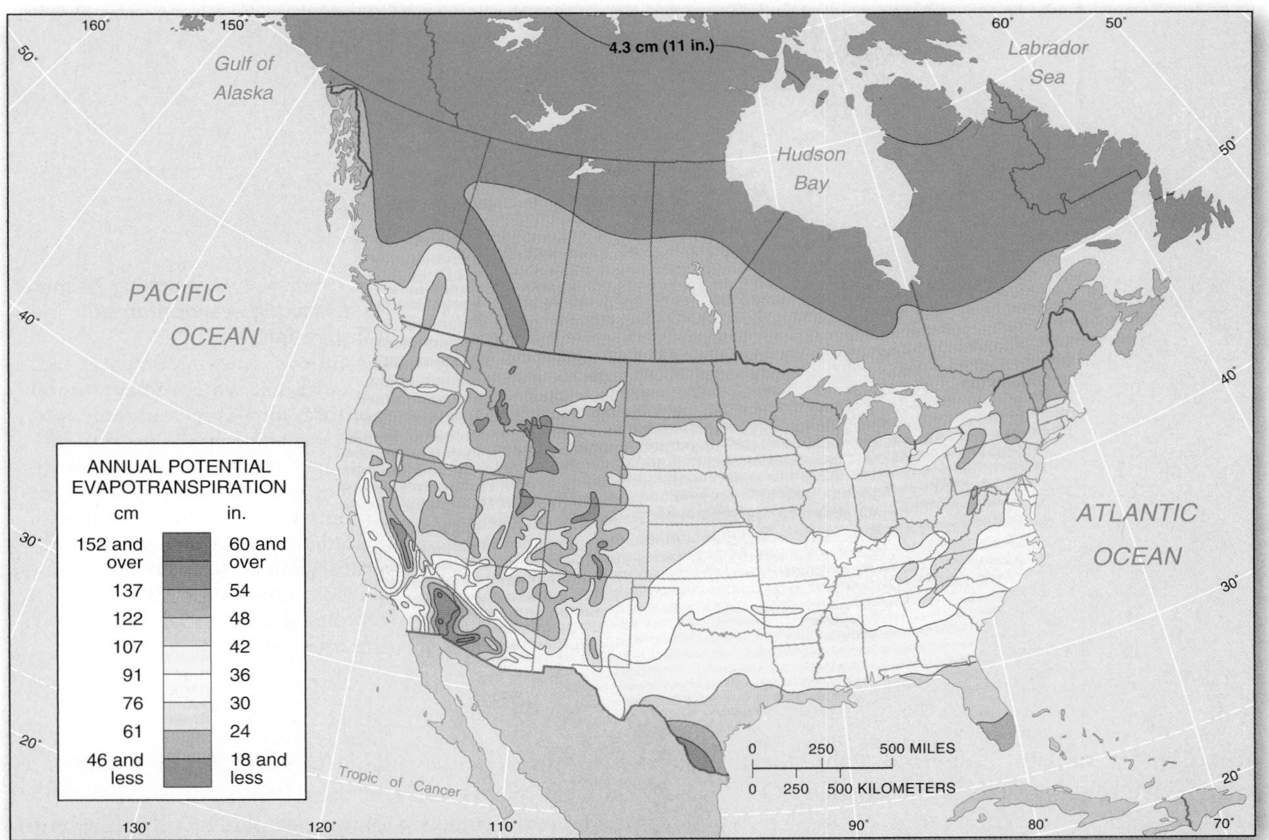

FIGURE 9.6 Potential evapotranspiration (POTET) for the United States and Canada—the water demand.
[From C. W. Thornthwaite, "An approach toward a rational classification of climate," *Geographical Review* 38 (1948): 64, © American Geographical Society. Canadian data adapted from M. Sanderson, "The climates of Canada according to the new Thornthwaite classification," *Scientific Agriculture* 28 (1948): 501–517.]

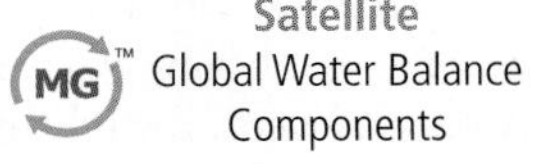

demand usually met by the precipitation supply? Or does your area experience a natural shortage? How might you find out?

Determining POTET Although exact measurement is difficult, one method of measuring POTET employs an **evaporation pan**, or *evaporimeter*. As evaporation occurs, water in measured amounts is automatically replaced in the pan, equaling the amount that evaporated. Mesh screens over the pan protect against overmeasurement due to wind, which accelerates evaporation.

A more elaborate measurement device is a **lysimeter**. A tank, approximately a cubic meter or larger in size, is buried in a field with its upper surface left open. The lysimeter isolates a representative volume of soil, subsoil, and plant cover, thus allowing measurement of the moisture moving through the sampled area. A weighing lysimeter has this embedded tank resting on a weighing scale (Figure 9.7). A rain gauge next to the lysimeter measures the precipitation input.

The number of lysimeters and evaporation pans is somewhat limited across North America, but they still provide a database from which to develop ways of estimating POTET. Several methods of estimating POTET on the basis of meteorological data are used widely and are easily implemented for regional applications. Thornthwaite developed one of these methods.

He discovered that, if you know monthly mean air temperature and daylength, you can approximate POTET. These data are readily available, so calculating POTET is easy and fairly accurate for most midlatitude locations. (Recall that daylength is a function of a station's latitude.) His method can work with data from hourly, daily, or annual time frames or with historical data to re-create water conditions in past environments.

Thornthwaite's method works better in some climates than in others. It ignores warm or cool air movements across a surface, sublimation and surface water retention at subfreezing temperatures, poor drainage, and frozen soil (*permafrost*, Chapter 17), although there is allowance for snow accumulations. Nevertheless, we use his water-balance method here because of its great utility as a teaching tool and its overall ease of application and acceptable results for geographic studies.

Deficit (DEFIC) The POTET demand can be satisfied in three ways: by PRECIP, by moisture stored in the soil, or through artificial irrigation. If these three sources are

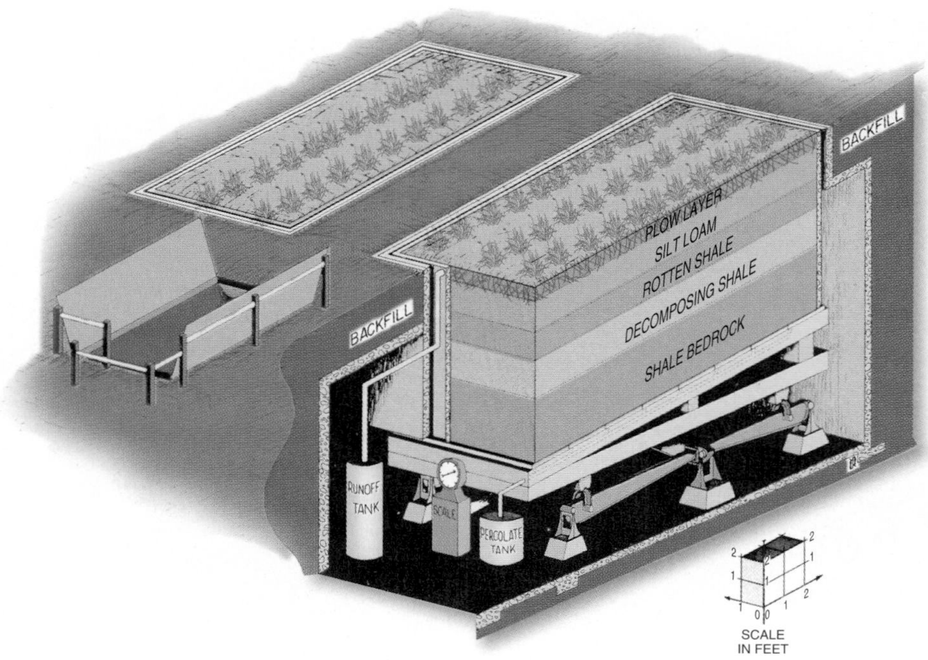

FIGURE 9.7 A weighing lysimeter for measuring evaporation and transpiration.
The various pathways of water are tracked: Some water remains as soil moisture, some is incorporated into plant tissues, some drains from the bottom of the lysimeter, and the remainder is credited to evapotranspiration. Given natural conditions, the lysimeter measures actual evapotranspiration. [Adapted from illustration courtesy of Lloyd Owens, Agricultural Research Service, USDA, Coshocton, Ohio.]

inadequate to meet moisture demands, the location experiences a moisture shortage. This unsatisfied POTET is **deficit** (DEFIC).

By subtracting DEFIC from POTET, we determine the **actual evapotranspiration**, or ACTET, that takes place (see Figure 9.3). Under ideal conditions for plants, potential and actual amounts of evapotranspiration are about the same, so plants do not experience a water shortage; droughts result from deficit conditions.

Surplus (SURPL) If POTET is satisfied and the soil is full of moisture, then additional water input becomes **surplus** (SURPL). This excess water might sit on the surface in puddles, ponds, and lakes, or it might flow across the surface toward stream channels or percolate through the soil underground. The **overland flow** combines with precipitation and groundwater flows into river channels to make up the **total runoff**. Because surplus water generates most streamflow, or runoff, the water-balance approach is useful for indirectly estimating streamflow.

Soil-Moisture Storage (Δ STRGE) A "savings account" of water that receives recharge "deposits" and provides for utilization "withdrawals" is **soil-moisture storage** (Δ STRGE). This is the volume of water stored in the soil that is accessible to plant roots. The delta symbol, **Δ**, means that this component includes both *recharge* and *utilization* (use) of soil moisture; the Δ in math means "change." Soil moisture comprises two categories of water—hygroscopic and capillary—but only capillary water is accessible to plants (Figure 9.8a).

Hygroscopic water is inaccessible to plants because it is a molecule-thin layer that is tightly bound to each soil particle by the hydrogen bonding of water molecules (Figure 9.8a, left). Hygroscopic water exists even in the desert, but it is unavailable to meet POTET demands. Relative to plants, soil is at the **wilting point** when all that remains is this inaccessible water; plants wilt and eventually die after a prolonged period of such moisture stress.

Capillary water is generally accessible to plant roots because it is held against the pull of gravity in the soil by hydrogen bonds between water molecules, which cause surface tension, and by hydrogen bonding between water molecules and the soil. Most capillary water that remains in the soil is **available water** in soil-moisture storage. After some water drains from the larger pore spaces, the amount of available water remaining for plants is termed **field capacity**, or storage capacity. It is removable to meet POTET demands through the action of plant roots and surface evaporation (Figure 9.8a, center). Field capacity is specific to each soil type, and the amount can be determined by soil surveys.

When soil becomes saturated after a precipitation event, any water surplus in the soil body becomes **gravitational water**. It percolates from the shallower capillary zone to the deeper groundwater zone (Figure 9.8a, right).

Figure 9.8b shows the relation of soil texture to soil-moisture content. Different plant species send roots to different depths and therefore reach different amounts of soil moisture. A soil blend that maximizes available water is best for plants. On the basis of Figure 9.8b, can you determine the soil texture with the greatest quantity of available water?

As **soil-moisture utilization** removes soil water, the plants must exert greater effort to extract the amount of moisture they need. As a result, even though a small amount of water may remain in the soil, plants may be unable to exert enough pressure to use it. The resulting *unsatisfied demand* is a *deficit* (DEFIC). Avoiding a deficit and reducing plant growth inefficiencies are the goals of

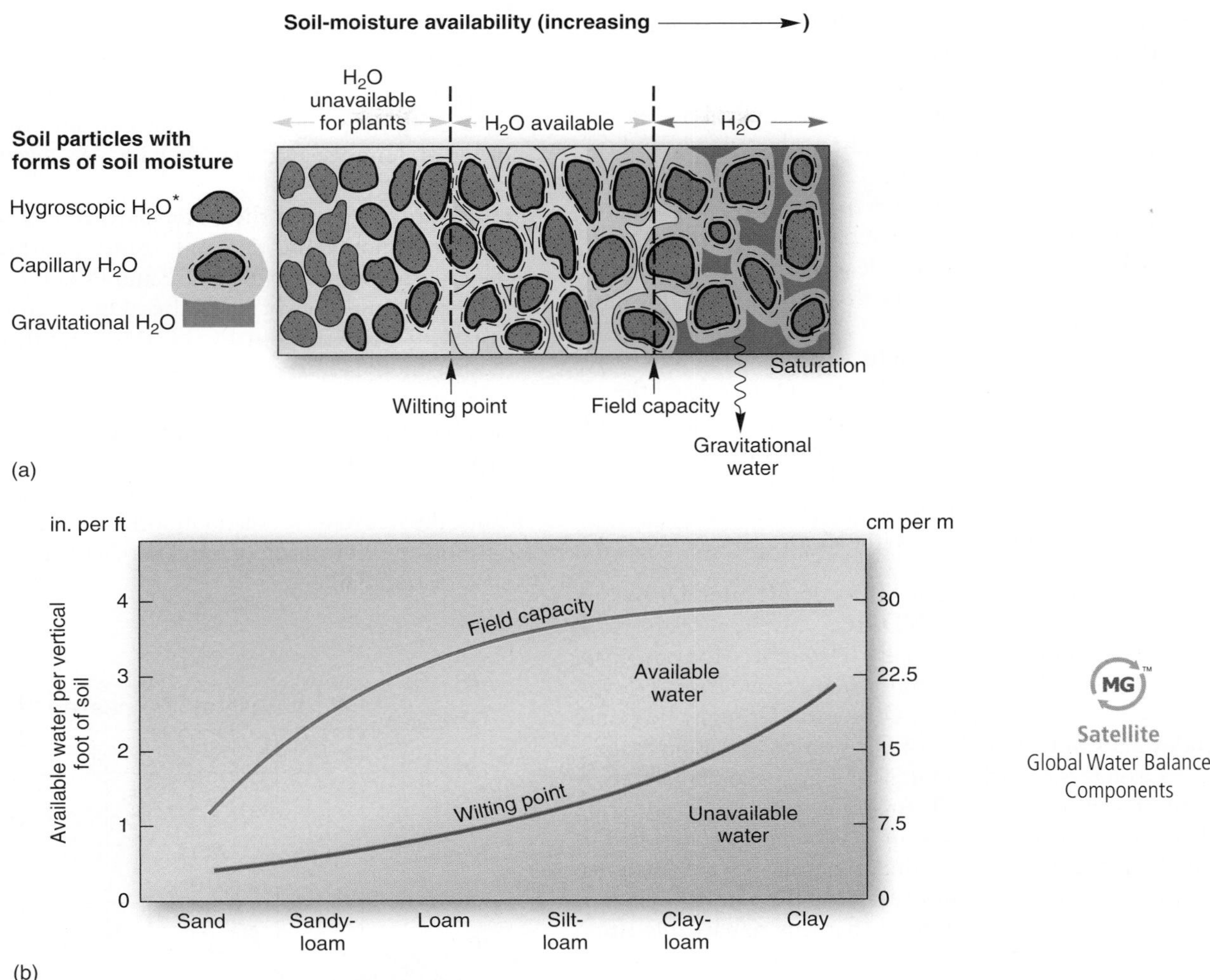

FIGURE 9.8 Types of soil moisture and its availability.
(a) Hygroscopic and gravitational water is unavailable to plants; only capillary water that is not bound hygroscopically to soil is available. (b) The relation between soil-moisture availability and soil texture determines the distance between the two curves that show field capacity and wilting point. A loam soil (one-third each of sand, silt, and clay) has roughly the most available water per vertical foot of soil exposed to plant roots. [(a) After D. Steila, *The Geography of Soils,* © 1976, p. 45, © Pearson Prentice Hall, Inc. (b) After U.S. Department of Agriculture, *1955 Yearbook of Agriculture—Water,* p. 120.]

MG Satellite Global Water Balance Components

irrigation, for the harder plants must work to get water, the smaller their yield and growth will be.

Whether from natural precipitation or artificial irrigation, water infiltrates the soil and replenishes available water, the process of **soil-moisture recharge**. The texture and the structure of the soil dictate available pore spaces, or **porosity**. The property of the soil that determines the rate of soil-moisture recharge is its **permeability**. Permeability depends on particle sizes and the shape and packing of soil grains.

Water infiltration is rapid in the first minutes of precipitation and slows as the upper soil layers become saturated, even though the deeper soil is still dry. Agricultural practices, such as plowing and adding sand or manure to loosen soil structure, can improve both soil permeability and the depth to which moisture can efficiently penetrate to recharge soil-moisture storage. You may have found yourself working to improve soil permeability with a house plant or garden—that is, working the soil to increase the rate of soil-moisture recharge.

Drought

Drought might seem a simple thing to define: Less precipitation and higher temperatures make for drier conditions over an extended period of time. However, scientists and resource managers use four technical definitions for **drought**. The key is to consider not only precipitation, temperature, and soil moisture content, but also the water-resource demand as it relates to the lack of water.

- *Meteorological drought* is defined by the degree of dryness, as compared to a regional average, and the duration of dry conditions. This definition is region-specific, since it relates to atmospheric conditions that change from area to area.

- *Agricultural drought* occurs when shortages of precipitation and soil moisture affect crop yields. Losses can be significant and are running in the tens of billions of dollars each year in the United States, although agricultural drought evolves slowly and gets little media coverage.
- *Hydrological drought* relates to the effects of precipitation shortages (both rain and snow) on water supply, such as when streamflow decreases, reservoir levels drop, mountain snowpack declines, and groundwater mining increases.
- *Socioeconomic drought* results when reduced water supply causes the demand for goods to exceed the supply, such as when hydroelectric power production declines with reservoir depletion. This is a more comprehensive measure that considers loss of life, water rationing, wildfire events, and other widespread impacts of water shortfalls.

The National Drought Mitigation Center, University of Nebraska–Lincoln (NDMC; http://www.drought.unl.edu/), publishes a weekly Drought Monitor map, released each Thursday, and a newsletter, *DroughtScape*. The NDMC site includes a Vegetation Drought Response Index map and a Number of Days Since Last Rain map.

Droughts of various intensities are occurring on every continent through 2010. Australia is in a decade-long drought, its worst drought in 110 years. The first half of 2010 was the driest in that time span for western Australia. Regarding the U.S. Southwest drought, under way since early 2000, scientists reported that this remarkable situation links directly to global climate change:

> The projected future climate of intensified aridity in the Southwest is caused by . . . a poleward expansion of the subtropical dry zones. The drying of subtropical land areas . . . imminent or already underway is unlike any climate state we have seen in the instrumental record. It is also distinct from the multi-decadal mega-droughts that afflicted the American Southwest during Medieval times. . . . The most severe future droughts will still occur during persistent La Niña events, but they will be worse than any since the Medieval period, because the La Niña conditions will be perturbing a base state that is drier than any state experienced recently.*

In other words, previous droughts in the region related to sea-surface temperature changes in the distant tropical Pacific Ocean will still occur, but they will be worsened by climate change and the expansion farther into the region of the hot, dry, subtropical high-pressure system and the summertime continental tropical (cT) air mass. The spatial implications of such semipermanent drought in a region of explosive population growth and urbanization are serious and tie in directly to the need for water-resource planning.

*Richard Seager et al., "Model projections of an imminent transition to a more arid climate in southwestern North America," *Science* 316, no. 5828 (May 25, 2007): 1184.

Sample Water Budgets

Using all of these concepts, we graph the water-balance components for several representative cities. Let's begin by looking at Kingsport, in the extreme northeastern corner of Tennessee at 36.6° N 82.5° W, elevation 390 m (1280 ft).

Figure 9.9 graphs the water-balance components for long-term supply and demand, using monthly averages. The monthly values for PRECIP and POTET smooth the actual daily and hourly variability. On the graph, a comparison of PRECIP and POTET by month determines whether there is a *net supply* or a *net demand* for water. The cooler time from October to May shows a net supply (blue line), but the warm days from June to September create a net water demand. If we assume a soil-moisture storage capacity of 100 mm (4.0 in.), typical of shallow-rooted plants, the net water-demand months are satisfied through soil-moisture utilization (green area), with a small summer soil-moisture deficit.

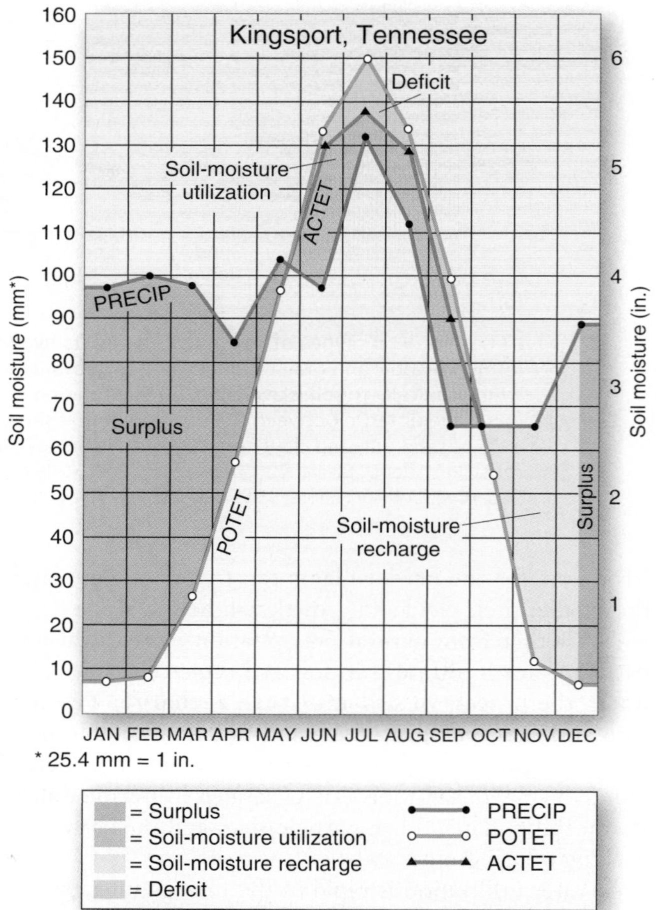

FIGURE 9.9 Sample water budget for Kingsport, Tennessee. Compare the average plots for precipitation inputs and potential evapotranspiration outputs to determine the condition of the soil-moisture environment. A typical pattern of spring surplus, summer soil-moisture utilization, a small summer deficit, autumn soil-moisture recharge, and ending surplus highlights the year.

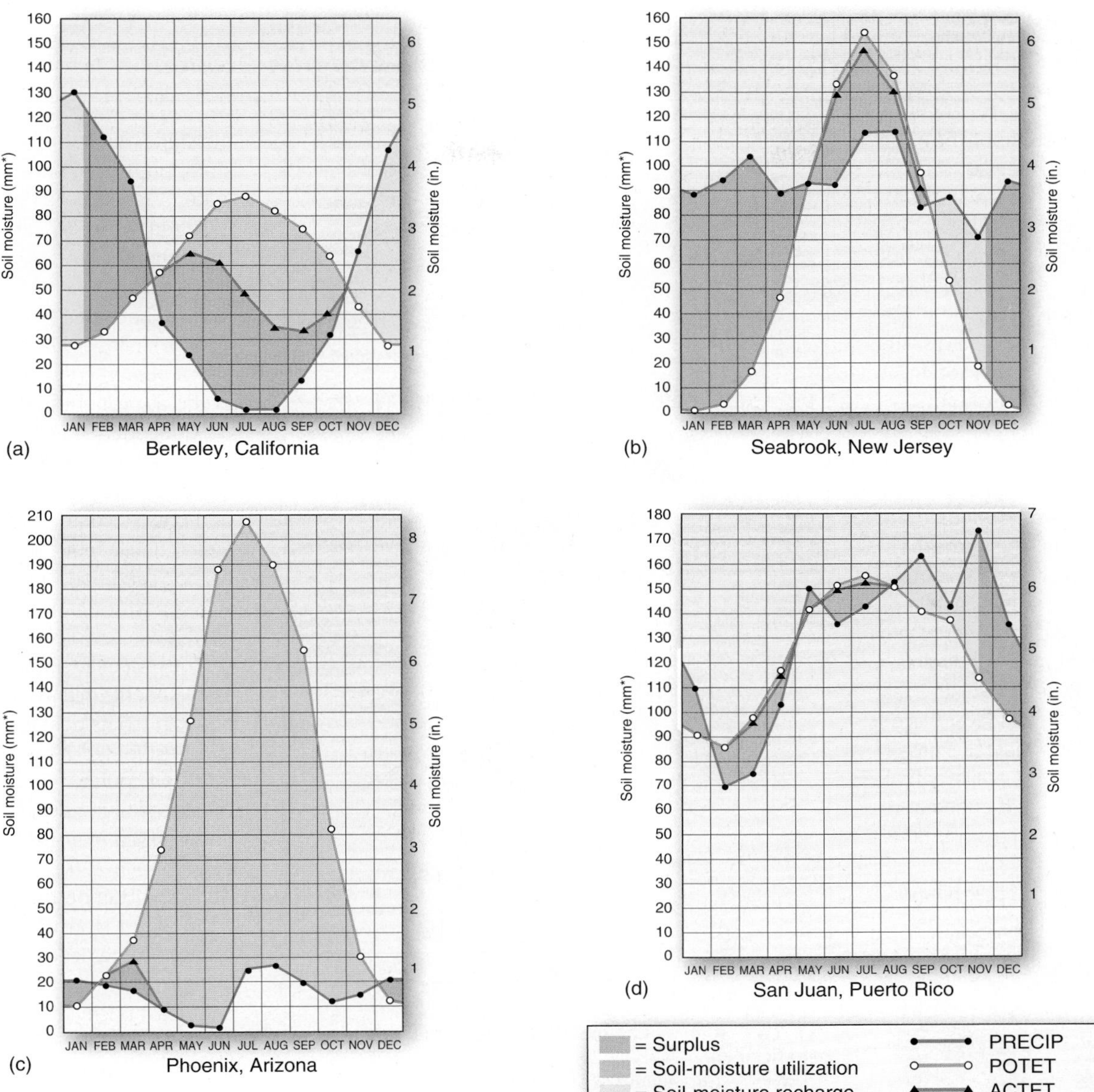

FIGURE 9.10 **Sample water budgets for selected U.S. stations.**

Obviously, not all stations experience the surplus moisture patterns of this humid continental region. Different climatic regimes experience different relations among water-balance components. Figure 9.10 presents water-balance graphs for several other cities in the United States. Among these examples, compare the summer minimum precipitation of Berkeley, California, with the summer maximum precipitation of Seabrook, New Jersey, and the precipitation receipts of the coastal city of San Juan, Puerto Rico, with those of the desert city of Phoenix, Arizona.

An interesting application of the water-balance approach to water resources is analyzing an event such as Hurricane Camille (1969), which generally had positive effects on regional water resources, but caused severe coastal damage, see Focus Study 9.1.

CRITICAL THINKING 9.1

Your local water budget

Select your campus, yard, or perhaps a house plant and apply the water-balance concepts. Where does the water supply originate—its source? Estimate the ultimate water supply and demand for the area you selected. For a general idea of PRECIP and POTET, find your locale on Figures 9.5 and 9.6. Consider the seasonal timing of this supply and demand, and estimate water needs and how they vary as components of the water-budget change.

Water Budget and Water Resources

Water distribution is uneven over space and time. Because we require a steady supply, we build large-scale management projects intended to redistribute the water

FOCUS STUDY 9.1

Hurricane Camille, 1969: Water-Balance Analysis and the End of a Drought

Hurricane Camille was one of the most devastating hurricanes of the 20th century. Hurricanes Ivan in 2004; Rita, Katrina, and Wilma in 2005; Humberto in 2007; and Gustav in 2008 brought back memories of Camille to Gulf Coast residents. Ironically, Camille was significant not only for the disaster it brought (256 dead, $1.5 billion in damage), but also for the drought it ended.

Figure 9.1.1 shows Camille inland from the Gulf Coast north of Biloxi, Mississippi. Hurricane-force winds sharply diminished after the hurricane made landfall, leaving a vast rainstorm that travelled from the Gulf Coast through Mississippi, western Tennessee, and Kentucky and into central Virginia. Severe flooding drowned the Gulf Coast near landfall and the James River Basin of Virginia, where torrential rains produced record floods. But Camille actually had a beneficial aspect: It ended a year-long drought along major portions of its storm track.

Figure 9.1.2a maps the precipitation from Camille and portrays the storm's track from landfall in Mississippi to the coast of Virginia and Delaware. By comparing actual water budgets along Camille's track with water budgets for the same three days with Camille's rainfall totals removed artificially, the moisture impact of the storm on the region's water budget could be analyzed. Figure 9.1.2b maps the moisture shortages that were avoided because of Camille's rains—termed *deficit abatement*. Think of this as drought that did not continue because Camille's rains occurred. Over vast portions of the affected area, Camille reduced dry-soil conditions, restored pastures, and filled low reservoirs.

Thus, Camille's monetary benefits inland outweighed its damage by an estimated 2-to-1 ratio. (Of course, the tragic loss of life does not fit into a financial equation.) In 2007, Hurricane Humberto, a category 1 storm, brought some needed relief across the Southeast once it moved inland. Hurricanes should be viewed as normal and natural meteorological events that not only have terrible destructive potential, mainly to coastal lowlands, but also contribute to the precipitation regimes of the southern and eastern United States. According to weather records, about one-third of all hurricanes making landfall in the United States provide beneficial precipitation to local water budgets.

FIGURE 9.1.1 ***ESSA-9* satellite image of Hurricane Camille.** Hurricane Camille made landfall on the night of August 17, 1969, and continued inland on August 18. The storm was progressing northward through Mississippi in the image. [NOAA.]

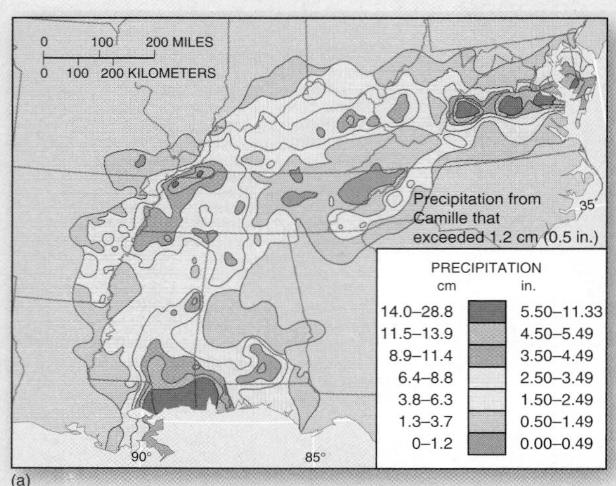

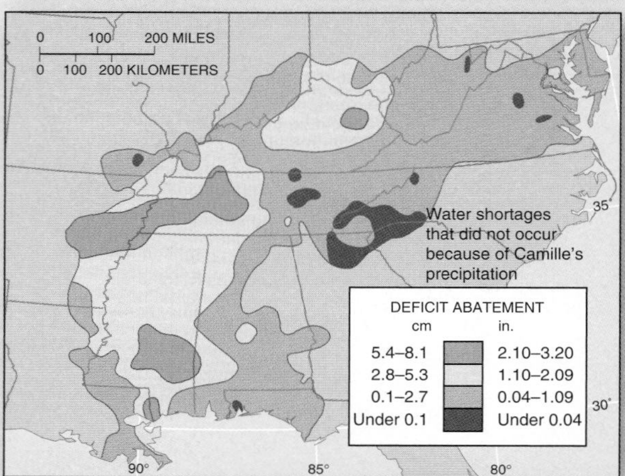

FIGURE 9.1.2 Camille affects water budgets in a beneficial way. A water-resource view of Hurricane Camille's impact on local water budgets: (a) Precipitation attributable to the storm (moisture supply). (b) Resulting deficit abatement (avoided moisture shortages) attributable to Camille. [Data and maps by Robert Christopherson.]

resource either geographically, by moving water from one place to another, or over time, by storing water from time of receipt until it is needed. In this way, deficits are reduced, surpluses are held for later release, and water availability is improved to satisfy evapotranspiration and society's demands. The water balance permits analysis of the contribution that streams make to water resources.

Streams may be *perennial* (constantly flowing; *perennial* is Latin for "through the year") or *intermittent*. In either case, the total runoff that moves through them comes from surplus surface water runoff, subsurface throughflow, and

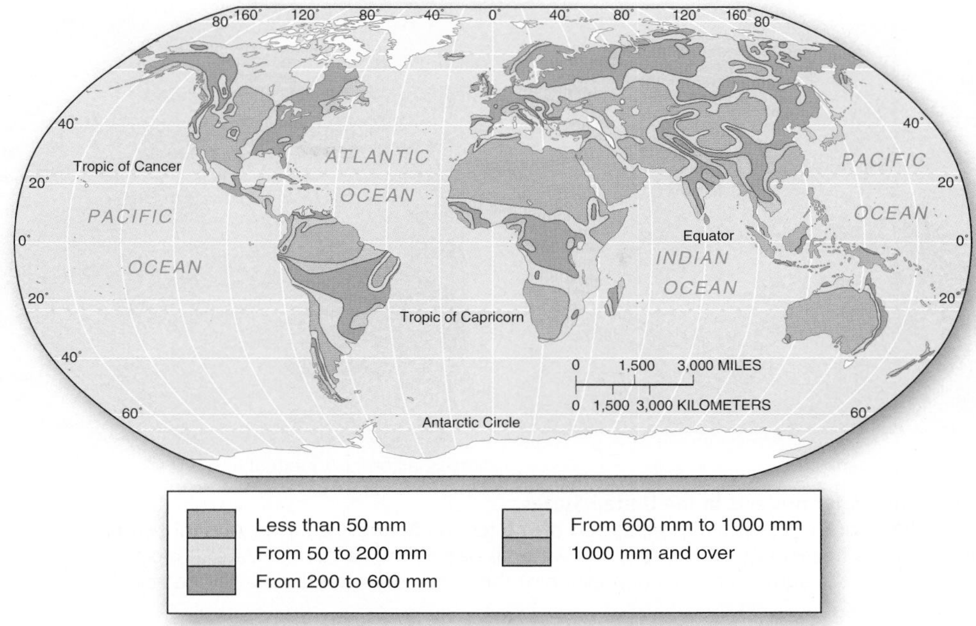

FIGURE 9.11 **Annual global river runoff.**
Distribution of runoff is closely correlated with climatic region, as expected. But it is poorly correlated with human population distribution and density.

groundwater. Figure 9.11 maps annual global river runoff for the world. The highest runoff amounts are along the equator within the tropics, reflecting the continual rainfall along the Intertropical Convergence Zone (ITCZ). Southeast Asia also experiences high runoff, as do northwest coastal mountains in the Northern Hemisphere. In countries having great seasonal fluctuations in runoff, groundwater becomes an important reserve water supply. Regions of lower runoff coincide with Earth's subtropical deserts, rain-shadow areas, and continental interiors, particularly in Asia.

The new Three Gorges Dam controls the treacherous Yangtze River in China (Figure 9.12). At a length of 2.3 km (1.5 mi) and a height of 185 m (607 ft), it is the largest dam and related construction in the world in overall size. Forced relocation of entire cities and more than 1 million people made room for the 600-km-long (370-mi-long) reservoir. Controversies surrounded the immense scale of environmental, historical, and cultural losses. Gained is flood control, water resources for redistribution, and power production. Electrical production capacity is 22,000 MW.

In the United States, several large water-management projects already operate: the Bonneville Power Administration (BPA) in Washington State, the Tennessee Valley Authority in the Southeast, the California Water Project, and the Central Arizona Project. BPA comprises 31 hydroelectric dams and a transmission grid and began operation in 1937 (Figure 9.13a).

FIGURE 9.12 **Major dam in China.**
Satellite views in 2000 (construction) and 2006 (completion) of the $25 billion Three Gorges Dam on the Yangtze River, flowing from upper left to right. For scale, the main portion of the dam is 2.3 km in length. This is the world's largest individual construction project. The channel to the north of the dam is a system of locks for shipping. [*Terra* ASTER images, NASA/GSFC/MITI/ERSDAC/JAROS.]

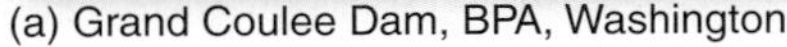

(a) Grand Coulee Dam, BPA, Washington

(b) California Aqueduct

FIGURE 9.13 Two major water projects in the United States.
(a) Grand Coulee Dam holds back the Columbia River, as part of the federal Bonneville Power Administration, in Washington State. (b) The California Aqueduct transports water as part of the California Water Project. The Palmdate Reservoir beyond the aqueduct is in a depression of the San Andreas fault zone. [(a) Author. (b) Bobbé Christopherson.]

Literally, the California Water Project rearranges the water budget in the state, holding back winter runoff for release in summer and pumping water from the north to the south. Completed in 1971, this 1207-km-long (750-mi-long) "river" flows from the Sacramento River delta to the Los Angeles region, servicing irrigated agriculture in the San Joaquin Valley along the way (Figure 9.13b).

The best sites for multipurpose hydroelectric projects are already taken, so battles among conflicting interests and analysis of negative environmental impacts invariably accompany new project proposals. Chapter 14 discusses river management through such construction in further detail.

Groundwater Resources

Groundwater is an important part of the hydrologic cycle, although it lies beneath the surface, beyond the soil-moisture root zone. It is tied to surface supplies through pores in soil and rock. Groundwater is the largest potential freshwater source in the hydrologic cycle—larger than all surface lakes and streams combined. Between Earth's land surface and a depth of 4 km (13,000 ft) worldwide, some 8,340,000 km^3 (2,000,000 mi^3) of water resides, a volume comparable to 70 times all the freshwater lakes in the world. Groundwater is generally free of sediment, color, and disease organisms, although polluted groundwater conditions are considered irreversible. Pollution threatens groundwater quality, and overconsumption depletes groundwater volume in quantities beyond natural replenishment rates.

Remember: *Groundwater is not an independent source of water, for it is tied to surface supplies for recharge.* An important consideration in many regions is that groundwater accumulation occurred over millions of years, so care must be taken not to exceed this long-term buildup with excessive short-term demands.

About 50% of the U.S. population derives a portion of its freshwater from groundwater sources. In some states, such as Nebraska, groundwater supplies 85% of water needs, with that figure as high as 100% in rural areas. Between 1950 and 2000, annual groundwater withdrawal in the United States and Canada increased more than 150%. Figure 9.14 shows potential groundwater resources in both countries.

Groundwater Profile and Movement

Figure 9.15 brings together many groundwater elements in a single illustration and is the basis for the following discussion. Groundwater begins as surplus water, which percolates downward as gravitational water from the zone of capillary water. This excess surface water moves through the **zone of aeration**, where soil and rock are less than saturated (some pore spaces contain air).

The *porosity* of any rock layer depends on the arrangement, size, and shape of its individual particles; the nature of any cementing elements between them; and their degree of compaction. Subsurface rocks are either *permeable* or *impermeable*. Their permeability depends on whether they conduct water readily (higher permeability) or tend to obstruct its flow (lower permeability).

Eventually, the water reaches an area where subsurface water accumulates, the **zone of saturation**. Here the pores are completely filled with water. Like a hard sponge made of sand, gravel, and rock, the saturation zone stores water in its countless pores and voids. An **aquifer** is a rock layer that is permeable to groundwater flow adequate for wells and springs. An **aquiclude** (also called an *aquitard*) is a body of rock that does not conduct water in usable

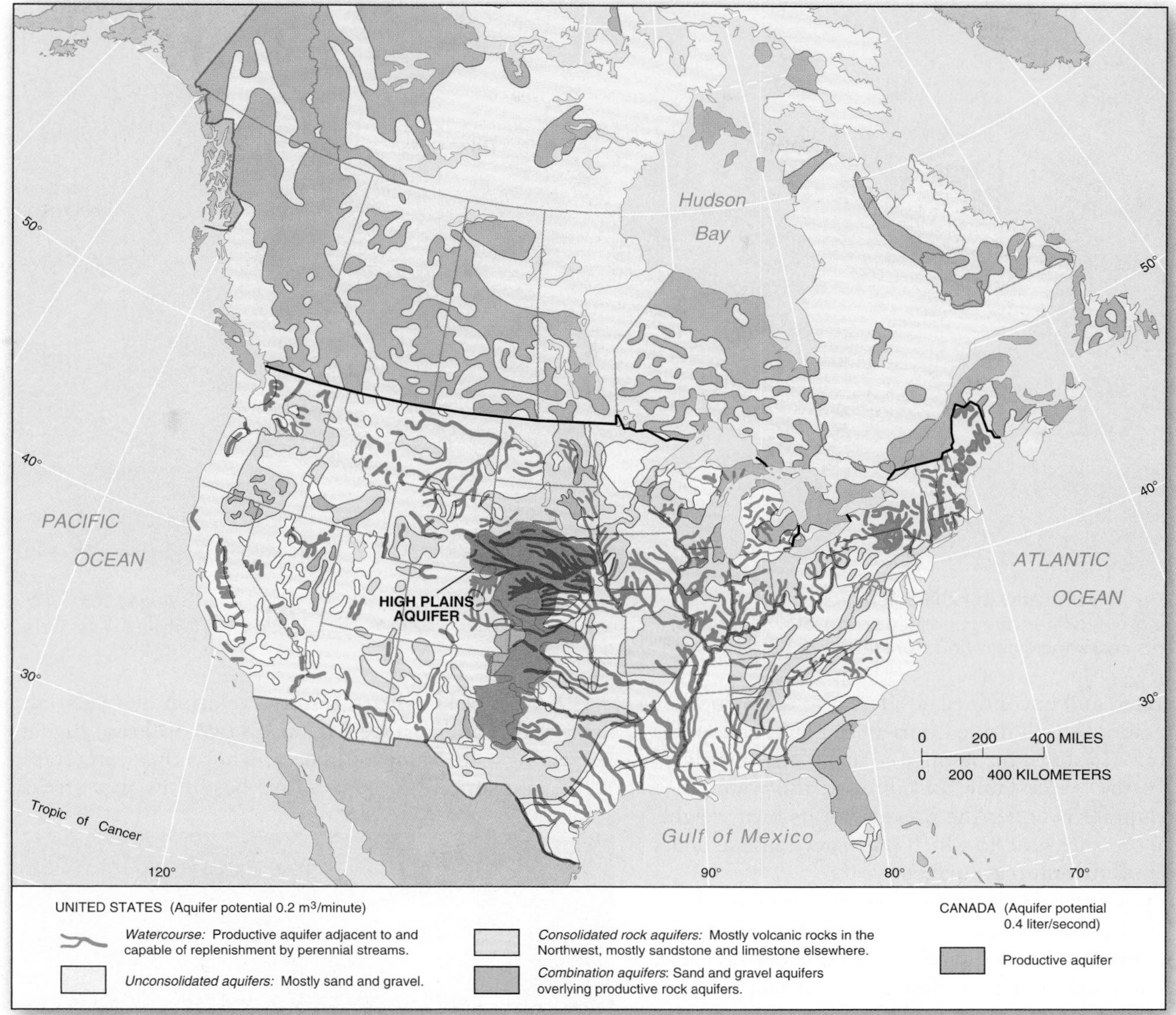

FIGURE 9.14 Groundwater resource potential for the United States and Canada. Highlighted areas of the United States are underlain by productive aquifers capable of yielding freshwater to wells at 0.2 m³/min or more (for Canada, 0.4 L/s). [Courtesy of Water Resources Council for the United States and the *Inquiry on Federal Water Policy* for Canada.]

amounts. The saturated zone may include the saturated portion of the aquifer and a part of the underlying *aquiclude*, even though the latter has such low permeability that its water is inaccessible.

The upper limit of the water that collects in the zone of saturation is the **water table**. It is the surface where the zone of saturation, a narrow capillary fringe layer, and the zone of aeration are in transition (all across Figure 9.15). The slope of the water table, which generally follows the contours of the land surface, controls groundwater movement.

Aquifers, Wells, and Springs

For many people, adequate water supply depends on a good aquifer beneath them that is accessible with a well or exposed on a hillside where water emerges as a spring. Is any of the water you consume from a well or spring?

Confined and Unconfined Aquifers Whether an aquifer is confined or unconfined affects its behavior. A **confined aquifer** is bounded above and below by impermeable layers of rock or sediment. An **unconfined aquifer** has a permeable layer on top and an impermeable one beneath (see Figure 9.15).

Confined and unconfined aquifers differ in the size of their recharge area, which is the ground surface where water enters an aquifer to recharge it. For an unconfined aquifer, the **aquifer recharge area** generally extends above the entire aquifer; the water simply percolates down to the water table. But in a confined aquifer, the recharge area is far more restricted, as you can see in Figure 9.15. Pollution of this limited recharge area causes groundwater contamination; note in the illustration the improperly located disposal pond on the aquifer recharge area, contaminating wells to the right.

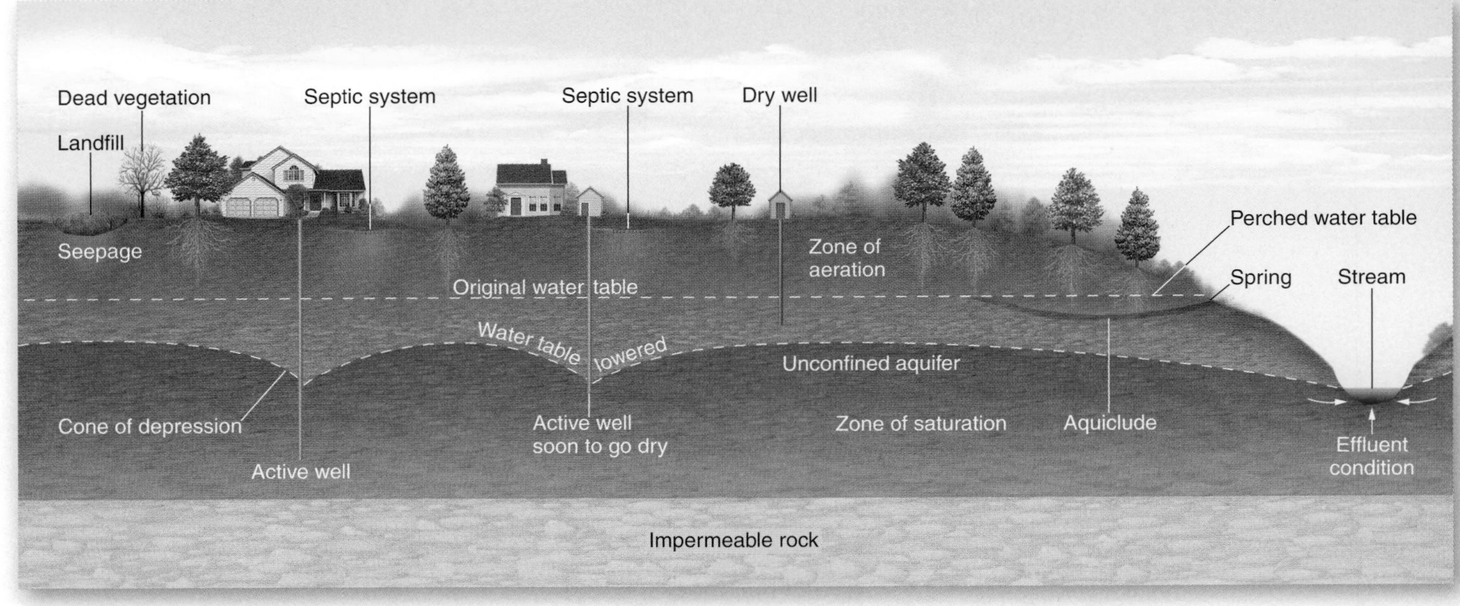

FIGURE 9.15 Groundwater characteristics.
Subsurface groundwater characteristics, processes, water flows, and human interactions. Follow the concepts across the illustration from left to right as you read along in the text.

Confined and unconfined aquifers also differ in their water pressure. A well drilled into an unconfined aquifer (see Figure 9.15) must be pumped to make the water level rise above the water table. In contrast, the water in a confined aquifer is under the pressure of its own weight, creating a pressure level to which the water can rise on its own, the **potentiometric surface**.

The potentiometric surface actually can be above ground level (right side in figure). Under this condition, **artesian water**, or groundwater confined under pressure, may rise in a well and even flow at the surface without pumping if the top of the well is lower than the potentiometric surface. (These wells are called *artesian* for the Artois area in France, where they are common.) In other wells, however, pressure may be inadequate, and the artesian water must be pumped the remaining distance to the surface.

Wells, Springs, and Streamflows The slope of the water table, which broadly follows the contour of the land surface, controls groundwater movement (see Figure 9.15). Groundwater tends to move toward areas of lower pressure and elevation.

Water wells can work only if they penetrate the water table. Too shallow a well will be a "dry well"; too deep a well will punch through the aquifer and into the impermeable layer below, also yielding little water. This is why water wells should be drilled in consultation with a *hydrogeologist*.

Where the water table intersects the surface, it creates springs (Figure 9.16). Such an intersection also occurs in lakes and riverbeds. Ultimately, groundwater may enter stream channels to flow as surface water (stream near center in Figure 9.15). In fact, during dry periods, the water table may sustain river flows.

Figure 9.17 illustrates the relation between the water table and surface streams in two different climatic settings. In humid climates, where the water table generally supplies a continuous base flow to a stream

FIGURE 9.16 An active spring.
Springs provide evidence of groundwater emerging at the surface. [Author.]

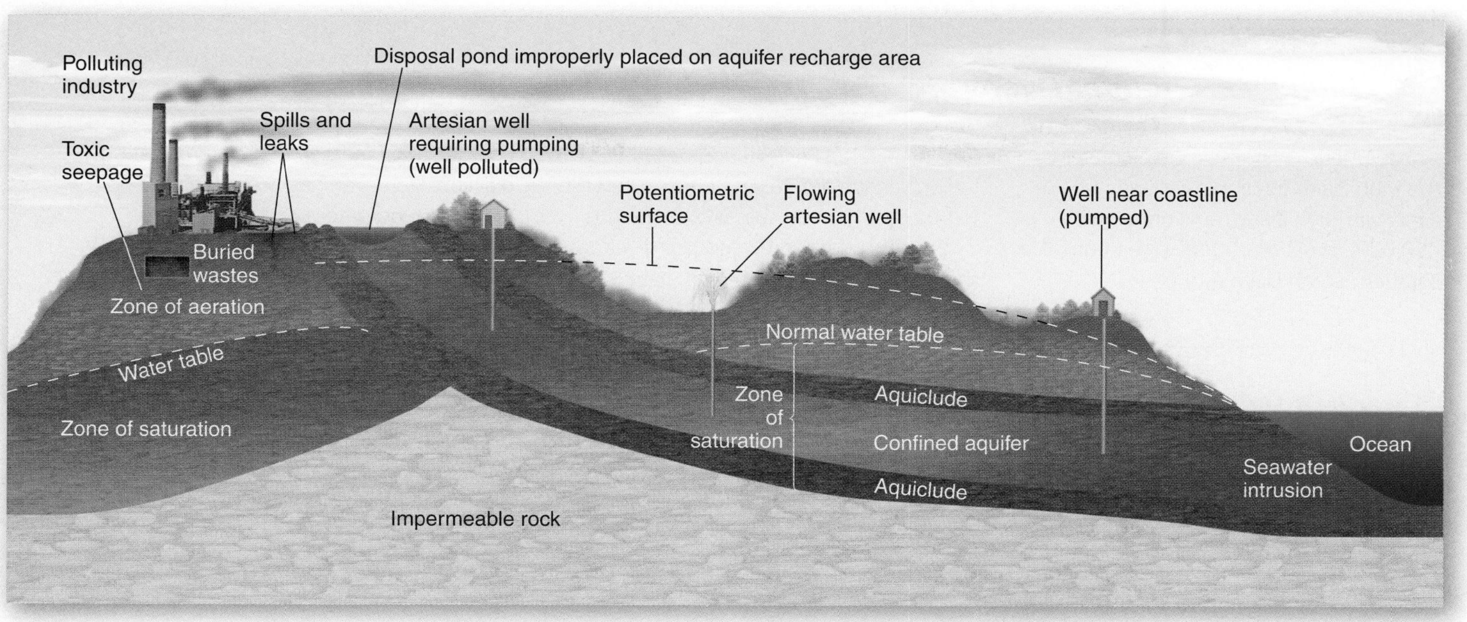

MG Animation Forming a Cone of Depression

and is higher than the stream channel, the stream is *effluent* because it receives the water flowing out (effluent) from the surrounding ground. The Mississippi River and countless other streams are examples. In drier climates, with lower water tables, water from *influent* streams flows into the adjacent ground, sustaining vegetation along the stream. The Colorado and Rio Grande rivers of the American West are examples of influent streams.

The Arkansas River no longer flows across western and central Kansas. As you see in Focus Study 9.2, "High Plains Aquifer Overdraft," groundwater mining is prevalent throughout this region. When the bottom of the stream bed is no longer in contact with the declining water table, streamflow seeps into the aquifer (Figure 9.17c). These influent conditions produce dry river, stream, and creek beds across the region.

Overuse of Groundwater

As water is pumped from a well, the surrounding water table within an unconfined aquifer may experience **drawdown**, or become lowered. Drawdown occurs if the pumping rate exceeds the replenishment flow of water into the aquifer or the horizontal flow around the well. The resultant lowering of the water table around the well is a **cone of depression** (see Figure 9.15, left).

(text continued on page 242)

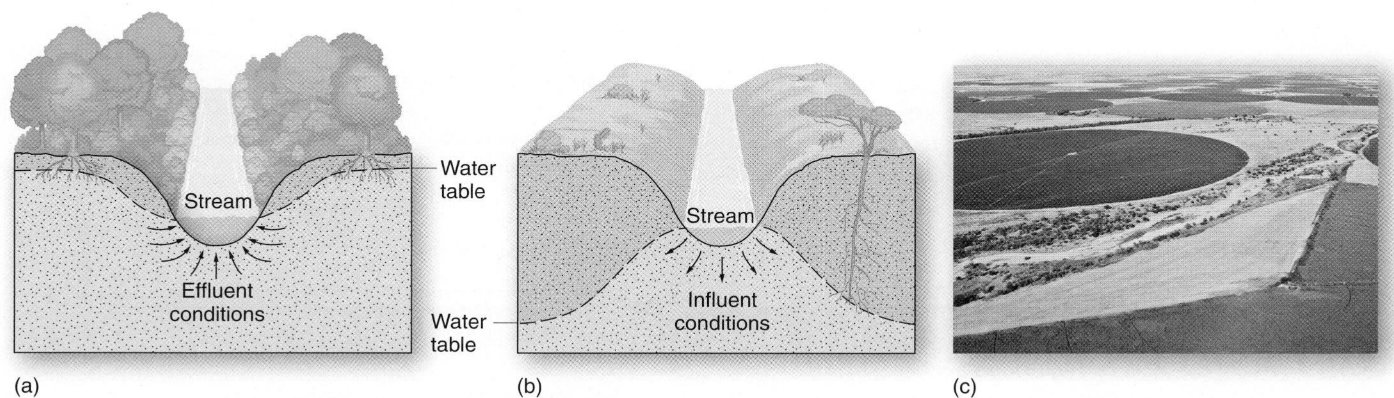

FIGURE 9.17 Groundwater interaction with streamflow.
(a) Effluent stream base flow is partially supplied by a high water table, characteristic of humid regions. (b) Influent stream supplies a lower water table, characteristic of drier regions. The water table may drop below the stream channel, effectively drying it out. (c) The former Arkansas River is dry as water flows into subsurface groundwater aquifers. The river channel appears to be used as an off-road course in the photo. [Bobbé Christopherson.]

FOCUS STUDY 9.2

High Plains Aquifer Overdraft

North America's largest known aquifer is the High Plains aquifer. It lies beneath the American High Plains, an eight-state, 450,600-km² (174,000-mi²) area from southern South Dakota to Texas (Figure 9.2.1a). Precipitation over the region varies from 30 cm in the southwest to 60 cm in the northeast (12 to 24 in.).

For several hundred thousand years, the aquifer's sand and gravel were charged with meltwaters from retreating glaciers. However, heavy mining of High Plains groundwater began 100 years ago, intensifying after World War II with the introduction of center-pivot

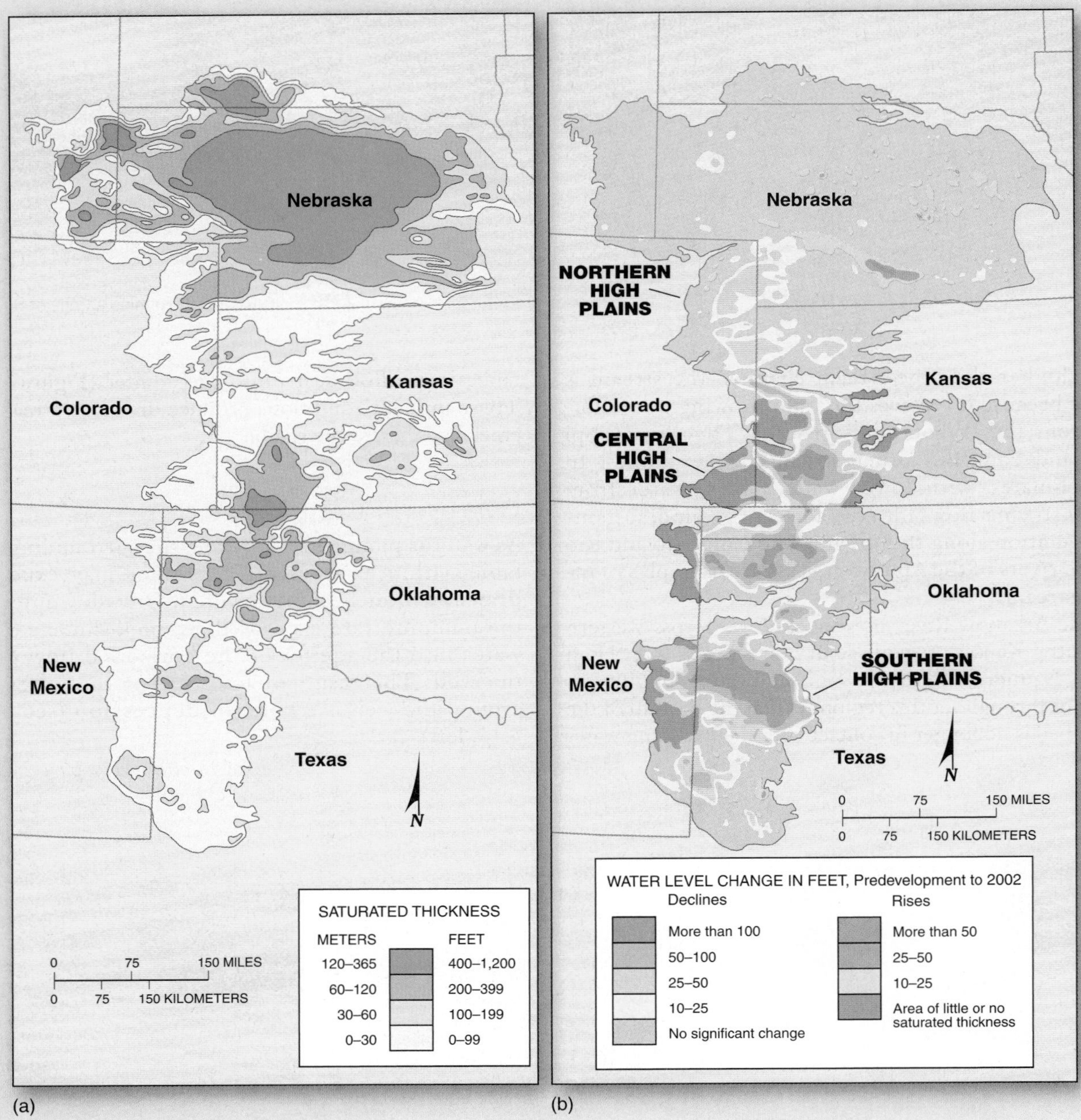

Nebraska
Kansas
Colorado
Oklahoma
New Mexico
Texas
N
0 75 150 MILES
0 75 150 KILOMETERS

SATURATED THICKNESS

METERS	FEET
120–365	400–1,200
60–120	200–399
30–60	100–199
0–30	0–99

(a)

Nebraska
NORTHERN HIGH PLAINS
Kansas
Colorado
CENTRAL HIGH PLAINS
Oklahoma
New Mexico
SOUTHERN HIGH PLAINS
Texas
N
0 75 150 MILES
0 75 150 KILOMETERS

WATER LEVEL CHANGE IN FEET, Predevelopment to 2002

Declines	Rises
More than 100	More than 50
50–100	25–50
25–50	10–25
10–25	Area of little or no saturated thickness
No significant change	

(b)

FIGURE 9.2.1 High Plains aquifer.
(a) Average saturated thickness. (b) Water level changes, using 1950 to represent "predevelopment." The color scale in the legend indicates widespread declines and a few areas of water-level rise. [Maps: (a) After D. E. Kromm and S. E. White, "Interstate groundwater management preference differences: The High Plains region," *Journal of Geography* 86, no. 1 (January–February 1987): 5. (b) Adapted from "Water-Level Changes in the High Plains Aquifer, Predevelopment to 2002, 1980 to 2002, and 2001 to 2002," by V. L. McGuire, USGS Fact Sheet 2004-3026, Fig. 1.]

irrigation (Figure 9.2.2). These large, circular devices provide vital water to wheat, sorghums, cotton, corn, and about 40% of the grain fed to cattle in the United States. The USGS began monitoring this groundwater mining from more than 7000 wells in 1988.

A 2004 USGS map of the water-level changes from predevelopment (about 1950) to 2002 appears in Figure 9.2.1b. Wells were measured in the winter or early spring when water levels were recovered from the previous irrigation season. Declining water levels are most severe in northern Texas, where the saturated thickness of the aquifer is least, through the Oklahoma panhandle and into western Kansas. Rising water levels are noted in portions of south-central Nebraska and a portion of Texas due to recharge from surface irrigation, a period of above-normal precipitation years, and downward percolation from canals and reservoirs. (See http://co.water.usgs.gov/nawqa/hpgw/HPGW_home.html.)

The High Plains aquifer irrigates about one-fifth of all U.S. cropland: 120,000 wells provide water for 5.7 million hectares (14 million acres). This is down from the peak of 170,000 wells in 1978. In 1980, water was pumped from the aquifer at the rate of 26 billion cubic meters (21 million acre-feet) a year, an increase of more than 300% since 1950. By the turn of this century, withdrawals had decreased 10% due to declining well yields and increasing pumping costs.

During the past five decades, the water table in the aquifer dropped more than 30 m (100 ft), and throughout the 1980s, it averaged a 2-m (6-ft) drop each year. The USGS estimates that recovery of the High Plains aquifer (those portions that have not collapsed) would take at least 1000 years if groundwater mining stopped today!

Obviously, billions of dollars of agricultural activity cannot be abruptly halted, but neither can profligate water mining continue. This issue raises tough questions: How do we best manage cropland? Can extensive irrigation continue? Can the region continue to meet the demand to produce commodities for export? Should we continue high-volume farming of certain crops that are in chronic oversupply? Should we rethink federal policy on crop subsidies and price supports? What changes will biofuels production force? What would be the impact on farmers and rural communities of any changes to the existing system?

Present irrigation practices, if continued, will deplete about half of the High Plains aquifer resource (and two-thirds of the Texas portion) by 2020. Add to this the approximate 10% loss of soil moisture due to increased evapotranspiration demand caused by climatic warming for this region by 2050, as forecast by computer models, and we have a portrait of a major regional water problem and a challenge for society.

FIGURE 9.2.2 Center-pivot irrigation.
Myriad center-pivot irrigation systems water crops in the western Great Plains. A growing season for corn requires from 10 to 20 revolutions of the sprinkler arm, depending on the weather. The arm delivers about 3 cm (1.18 in.) of water per revolution (rainfall equivalent). [Bobbé Christopherson.]

Overpumping Aquifers frequently are pumped beyond their flow and recharge capacities, a condition known as **groundwater mining**. In the United States today, large tracts experience chronic groundwater overdrafts in the Midwest, West, lower Mississippi Valley, and Florida and in the intensely farmed Palouse region of eastern Washington State. In many places, the water table or artesian water level has declined more than 12 m (40 ft). Groundwater mining is of special concern in the massive High Plains aquifer, which has declined over 30 m (100 ft) in some areas.

About half of India's irrigation water needs and half of its industrial and urban water needs are met by the groundwater reserve. In rural areas, groundwater supplies 80% of domestic water from some 3 million hand-pumped bore holes. And in approximately 20% of India's agricultural districts, groundwater mining through more than 17 million wells (motorized dug wells and tube wells) is beyond recharge rates.

In the Middle East, groundwater overuse is even more severe. The groundwater resource beneath Saudi Arabia accumulated over tens of thousands of years, forming "fossil aquifers," but today's increasing withdrawals are not being naturally recharged due to the desert climate—in essence, groundwater has become a *nonrenewable* resource. Some researchers suggest that groundwater in the region will be depleted in a decade, although worsening water-quality problems will no doubt arise before then. Imagine having the hottest issue in the Middle East become water, not oil!

> Like any renewable resource, groundwater can be tapped indefinitely as long as the rate of extraction does not exceed the rate of replenishment. But just like a bank account, a groundwater reserve will dwindle if withdrawals exceed deposits. Few governments have established and enforced rules and regulations to insure that groundwater sources are exploited at a sustainable rate. . . . No government has yet adequately tackled the issue of groundwater depletion, but it is at least getting more attention.*

Collapsing Aquifers A possible effect of water removal from an aquifer is that the aquifer, which is a layer of rock or sediment, will lose its internal support. Water in the pore spaces between rock grains is not compressible, so it adds structural strength to the rock. If the water is removed through overpumping, air infiltrates the pores. Air is readily compressible, and the tremendous weight of overlying rock may crush the aquifer. On the surface, the visible result may be land subsidence, cracked house foundations, and changes in drainage.

Houston, Texas, provides an example. The removal of groundwater and crude oil caused land within an 80-km (50-mi) radius of Houston to subside more than 3 m (10 ft) over the years. Yet another example is the Fresno area of California's San Joaquin Valley. After years of intensive pumping of groundwater for irrigation, land levels have dropped almost 10 m (33 ft) because of a combination of water removal and soil compaction from agricultural activity.

Surface mining (also called strip mining) for coal in eastern Texas, northeastern Louisiana, Wyoming, Arizona, and other locales also destroys aquifers. Aquifer collapse is an increasing problem in West Virginia, Kentucky, and Virginia, where low-sulfur coal is mined to meet current air pollution requirements for coal-burning electric power plants.

Unfortunately, collapsed aquifers may not be rechargeable, even if surplus gravitational water becomes available, because pore spaces may be permanently collapsed. The water-well fields that serve the Tampa Bay–St. Petersburg, Florida, area are a case in point. Pond and lake levels, swamps, and wetlands in the area are declining and land surfaces subsiding as the groundwater drawdown for export to the cities increases.

An additional problem arises when aquifers are overpumped near the ocean or coastal seas. Along a coastline, fresh groundwater and salty seawater establish a natural interface, or *contact surface*, with the less-dense freshwater flowing on top. But excessive withdrawal of freshwater can cause this interface to migrate inland. As a result, wells near the shore may become contaminated with saltwater, and the aquifer may become useless as a freshwater source. Figure 9.15 illustrates this seawater intrusion (far-right side). Pumping freshwater back into the aquifer may halt contamination by seawater, but once contaminated, the aquifer is difficult to reclaim.

Pollution of Groundwater

When surface water is polluted, groundwater inevitably becomes contaminated because it is recharged from surface-water supplies. Surface water flows rapidly and

*L. Brown et al. and the Worldwatch Institute, *Vital Signs: The Environmental Trends That Are Shaping Our Future* (New York: W. W. Norton & Co., 2000), pp. 122, 123.

GEO REPORT 9.3 Water pipelines in the Middle East

One water pipeline proposal brings water overland from Turkey to the Middle East through two branches. The Gulf water line would run southeast through Jordan and Saudi Arabia, with extensions into Kuwait, Abu Dhabi, and Oman. The other branch would run south through Syria, Jordan, and Saudi Arabia to the cities of Makkah (Mecca) and Jeddah. This pipeline would cover some 1500 km (930 mi), the distance from New York City to St. Louis. (See the links listed in Middle East Water Information Network at http://www.columbia.edu/cu/lweb/indiv/mideast/cuvlm/water.html.)

flushes pollution downstream, but slow-moving groundwater, once contaminated, remains polluted virtually forever.

Pollution can enter groundwater from many sources: industrial injection wells (wastes pumped into the ground), septic tank outflows, seepage from hazardous-waste disposal sites, industrial toxic waste, agricultural residues (pesticides, herbicides, fertilizers), and urban solid-waste landfills. As an example, leakage from some 10,000 underground gasoline storage tanks at U.S. gasoline stations is suspected. Some cancer-causing additives to gasoline, such as the oxygenate MTBE (methyl tertiary butyl ether), are contaminating thousands of local water supplies.

For management purposes, about 35% of pollution is *point source*, such as a gasoline tank or septic tank; 65% is *nonpoint source*, coming from a broad area, such as an agricultural field or urban runoff. Regardless of the spatial nature of the source, pollution can spread over a great distance, as illustrated in Figure 9.15. Since the nature of aquifers makes them inaccessible, the extent of groundwater pollution is underestimated.

In the face of government inaction, and even attempts to reverse some protection laws, serious groundwater contamination continues nationwide, adding validity to a quote from more than 25 years ago:

> One characteristic is the practical irreversibility of groundwater pollution, causing the cost of clean-up to be prohibitively high. . . . It is a questionable ethical practice to impose the potential risks associated with groundwater contamination on future generations when steps can be taken today to prevent further contamination.*

Our Water Supply

Human thirst for adequate water supplies, both in quantity and in quality, will be a major issue in this century. Internationally, increases in per capita water use are double the rate of population growth. Since we are so dependent on water, it seems that humans should cluster where good water is plentiful. But accessible water supplies are not well correlated with population distribution or the regions where population growth is greatest.

Table 9.1 shows estimated world water supplies, world population in 2009, estimated population figures for 2025, a population-change forecast between 2009 and 2050 at present growth rates, and a comparison of population and global runoff percentages for each continent. Also note 2006 data on carbon dioxide emissions per person for each region. The table gives an idea of the unevenness of Earth's water supply, which is tied to climatic variability, and water demand, which is tied to level of development, affluence, and per capita consumption.

For example, North America's mean annual discharge is 5960 km^3 and Asia's is 13,200 km^3. However, North America has only 6.6% of the world's population, whereas Asia has 60.5%, with a population doubling time less than half of North America's. In northern China, 550 million people living in approximately 500 cities lack adequate water supplies. For comparison, note that the 1990 floods cost China $10 billion, whereas water shortages are running at more than $35 billion a year in costs to the Chinese economy. An analyst for the World Bank stated that China's water shortages pose a more serious threat than floods during this century, yet in news coverage, we hear only of floods when they happen.

CRITICAL THINKING 9.2

Water on your campus

How much do you know about the water on your campus? If the campus has its own wells, how is the quality tested? Who on campus is in charge of supervising these wells? Lastly, what is your subjective assessment of the water you use: taste, smell, hardness, clarity? Write a brief critique following a taste test. Compare these perceptions with others in your class.

In Africa, 56 countries draw from a varied water-resource base; these countries share more than 50 river and lake watersheds. Increasing population growth and

*J. Tripp, "Groundwater protection strategies," in *Groundwater Pollution: Environmental and Legal Problems* (Washington, DC: American Association for the Advancement of Science, 1984), p. 137. Reprinted by permission.

GEO REPORT 9.4 ***The water it takes for food and necessities***

Simply providing the variety of food we enjoy requires voluminous water. For example, 77 g (2.7 oz) of broccoli requires 42 L (11 gal) of water to grow and process; producing 250 ml (8 oz) of milk requires 182 L (48 gal) of water; producing 28 g (1 oz) of cheese requires 212 L (56 gal); producing 1 egg requires 238 L (63 gal); and producing a 113-g (4-oz) beef patty requires 2314 L (616 gal). And then there are the toilets, the majority of which still flush approximately 4 gal of water. Imagine the spatial complexity of servicing the desert city of Las Vegas, with 150,000 hotel rooms times the number of toilet flushes per day; of providing all support services for 38 million visitors per year, plus 38 golf courses. One hotel confirms that it washes 14,000 pillowcases a day.

TABLE 9.1 Estimate of Available Global Water Supply Compared by Region and Population

Region (2009 population in millions)	Land Area in Thousands of km² (mi²)	Share of Mean Annual Discharge in km³/yr (BGD)	Global Stream Runoff (%)	Projected Population, 2025 (millions)	Population Change (+) 2009–2050 (%)	Metric Tons of CO_2, Emissions per Person, 2006
Africa (999)	30,600 (11,800)	4,220 (3,060)	10.8	1,385	100	0.9
Asia (4117)	44,600 (17,200)	13,200 (9,540)	34.0	4,858	33	3.0 China: 4.3
Australia–Oceania (36)	8,420 (3,250)	1,960 (1,420)	5.1	45	60	Australia: 19.0 New Zealand: 8.9
Europe (738)	9,770 (3,770)	3,150 (2,280)	8.1	736	–5	8.4
North America (451) (Canada, Mexico, U.S.)	22,100 (8,510)	5,960 (4,310)	15.3	518	38	18.4 Mexico: 4.0
Central and South America (470)	17,800 (6,880)	10,400 (7,510)	26.7	545	40	3.1
Global (6810) (excluding Antarctica)	134,000 (51,600)	38,900 (28,100)	—	8,087	38	More developed: 11.5 Less developed: 2.4

Note: BGD, billion gallons per day. Population data from *2009 World Population Data Sheet* (Washington, DC: Population Reference Bureau, 2009). Numbers are rounded. Metric tons × 1.1 for short (U.S.) tons.

percentage of urban dwellers cause water demands to rise and lead to an overexploitation of the resource. Twelve African countries experience water stress (less than 1700 m³ per person/year) and 14 have water scarcity (less than 1000 m³ per person/year).

Water resources are different from other resources in that there is *no alternative substance to water*. Water-resource stress related to quantity and/or quality shortfalls will dominate future political agendas. Water shortages increase the probability for international conflict, endanger public health, reduce agricultural productivity, and damage life-supporting ecological systems. For the 21st century a transition to a new approach is required, away from large, centralized structures toward decentralized, community-based strategies, focused on demand management and efficient technologies.

Water Supply in the United States

The U.S. water supply derives from surface and groundwater sources that are fed by an average daily precipitation of 4200 BGD (billion gallons a day). That sum is based on an average annual precipitation value of 76.2 cm (30 in.) divided evenly among the 48 contiguous states (excluding Alaska and Hawai'i).

The 4200 billion gallons of average daily precipitation mentioned are unevenly distributed across the country and unevenly distributed throughout the year. For example, New England's water supply is so abundant that only about 1% of available water is consumed each year. The same is true in Canada, where the resource greatly exceeds that in the United States.

Figure 9.18 shows how the U.S. supply of 4200 BGD is distributed. Viewed daily, the national water budget has two general outputs: 71% actual evapotranspiration (ACTET) and 29% surplus (SURPL). The 71% actual evapotranspiration involves 2970 BGD of the daily supply. It passes through nonirrigated land, including farm crops and pasture, forest and browse (young leaves, twigs, and shoots), and nonagricultural vegetation. Eventually, it returns to the atmosphere to continue its journey through the hydrologic cycle. The remaining 29% surplus is what we directly use.

Note that groundwater is considered part of surface water because it is linked to surface supplies. In 2005, about 67% of fresh groundwater withdrawals were for irrigation and 18% for public supply water needs. Just six states accounted for 50% of withdrawals: California, Texas, Nebraska, Arkansas, Idaho, and Florida (where more than half of the public supply water is from groundwater).

Instream, Nonconsumptive, and Consumptive Uses

The surplus 1230 BGD is runoff, available for withdrawal, consumption, and various instream uses.

- *Instream* uses are those that use stream water in the channel, without removing it: examples include navigation, wildlife and ecosystem preservation, waste dilution and removal, hydroelectric power production, fishing, and recreation.
- *Nonconsumptive uses*, sometimes called **withdrawal**, or *offstream use*, remove water from the supply, use it, and then return it to the same supply. Nonconsumptive water is used by industry, agriculture, and municipalities and in steam–electric power generation. A portion of the water withdrawn is consumed.

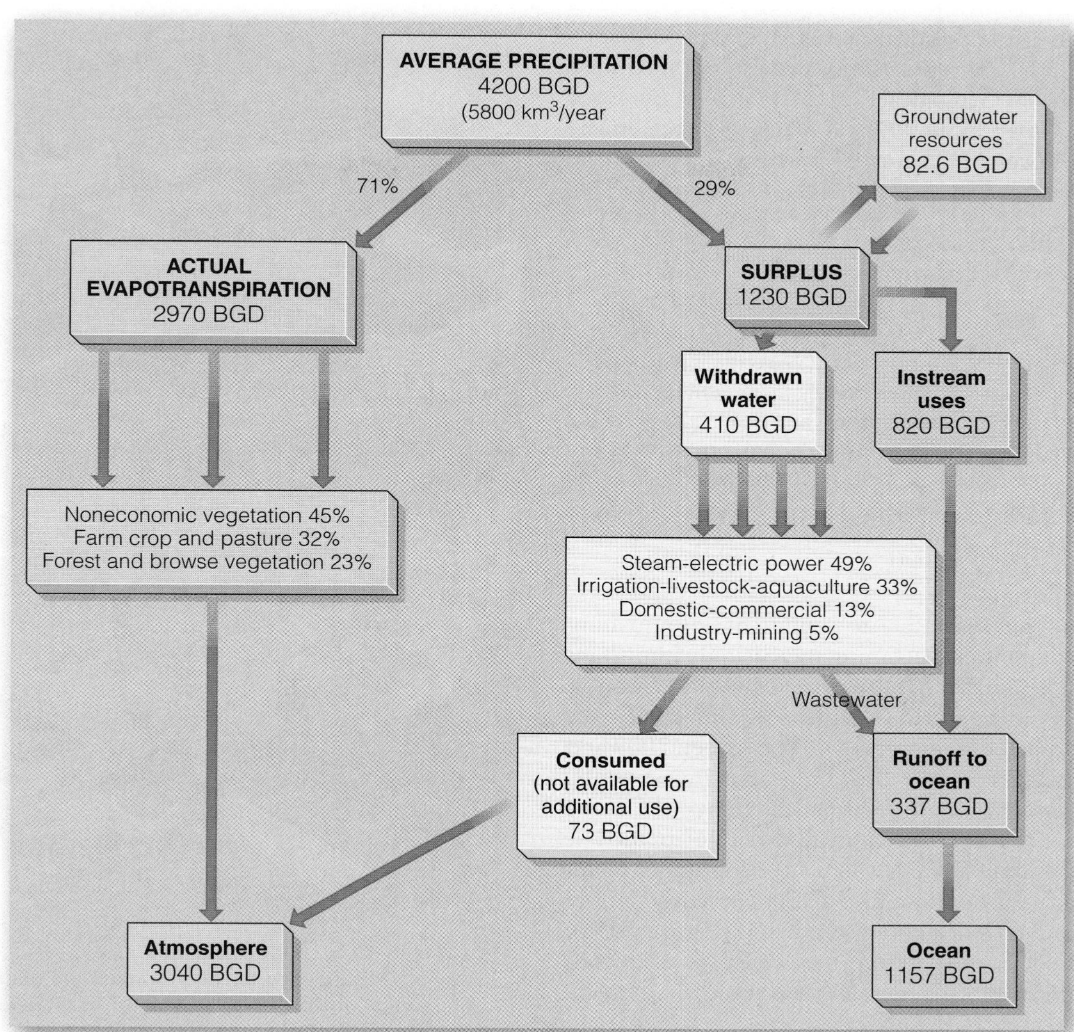

FIGURE 9.18 **U.S. water budget.**
Daily water budget for the contiguous 48 states in billions of gallons a day (BGD). [Data from Joan Kenny, Nancy Barber, Susan Hutson, Kristin Linsey, John Lovelace, and Molly Maupin, *Estimated Use of Water in the United States in 2005* (Denver, CO: USGS Circular 1344, 2005)—the latest year for which data are available.]

- **Consumptive uses** remove water from a stream, but do not return it, so it is not available for a second or third use. Some consumptive examples include water that evaporates or is vaporized in steam–electric plants.

When water returns to the system, water quality usually is altered—water is contaminated chemically with pollutants, or waste, or thermally with heat energy. In Figure 9.18, this portion of the budget is returned to runoff and eventually the ocean. This wastewater represents an opportunity to extend the resource through reuse.

Contaminated or not, returned water becomes a part of all water systems downstream. An example is New Orleans, the last city to withdraw municipal water from the Mississippi River. New Orleans receives diluted and mixed contaminants added throughout the entire Missouri–Ohio–Mississippi River drainage. This includes the effluent from chemical plants, runoff from millions of acres of farm fields treated with fertilizer and pesticides, treated and untreated sewage, oil spills, gasoline leaks, wastewater from thousands of industries, runoff from urban streets and storm drains, and turbidity from countless construction sites and from mining, farming, and logging activities. Abnormally high cancer rates among citizens living along the Mississippi River between Baton Rouge and New Orleans have led to the ominous label "Cancer Alley" for the region. The infamous "muck" that covered the region when the levees broke in 2005 contained some of this in its brew.

The estimated U.S. withdrawal of water for 2005 was 410 BGD, almost 3 times the usage rate in 1940, double that in 1950, but a decrease of 5% since the peak year of 1980. A shift in emphasis from supply-side to demand-side planning, pricing, and management produced these efficiencies. Total per capita use was 1367 gal/day in 2005.

The four main uses of withdrawn water in the United States in 2005 were steam–electric power (49%, about 29% met with saline ocean and coastal water), industry–mining

(5%), irrigation–livestock–aquaculture (33%), and domestic–commercial (13%). For studies of water use in the United States and the U.S. Geological Survey's (USGS's) *Estimated Use of Water in the U.S. in 2005* (available as a free download), see http://water.usgs.gov/public/watuse/.

CRITICAL THINKING 9.3

That next glass of water

You have no doubt had several glasses of water between when you awoke this morning and the time you are reading these words. Where did this water originate? Obtain the name of the water company or agency, determine whether it is using surface or groundwater to meet demands, and check how the water is metered and billed. If your state or province requires water-quality reporting, obtain a copy from the water supplier of this analysis of your tap water.

Desalination

Desalination of seawater to augment diminishing groundwater supplies is becoming increasingly important as a freshwater source. More than 12,000 desalination plants are now in operation worldwide, and plant construction is expected to increase 140% between 2009 and 2017, amounting to some $60 billion in investments. Approximately 50% of all desalination plants are in the Middle East. Desalination is an alternative to further groundwater mining and saltwater intrusion under Saudi Arabia; in fact, 30 desalination plants currently supply 70% of the country's drinking water needs (Figure 9.19a). In the United States, along the coast of southern California and in Florida, desalination is slowly increasing in use.

The Tampa–St. Petersburg area in Florida took a giant step forward with the opening of its own Tampa Bay desalination plant, which produces 25 MGD, about 10% of the region's drinking water supply, making it the largest in the United States. This output is about 12% of the production at the Jabal plant in Saudi Arabia. The seawater intake is from the Big Bend Power Station's cooling water discharge. The Tampa Bay plant produces drinking water through a series of filters to remove algae and particles and a reverse-osmosis process where the water is forced through semipermeable membranes (Figure 9.19b).

In Sydney, Australia, a reverse-osmosis desalination plant opened in 2010. The $17 billion Kurnell plant supplies about 15% of the region's water needs. Some 67 wind turbines supply energy needs for the operation.

(a) Saudi Arabia

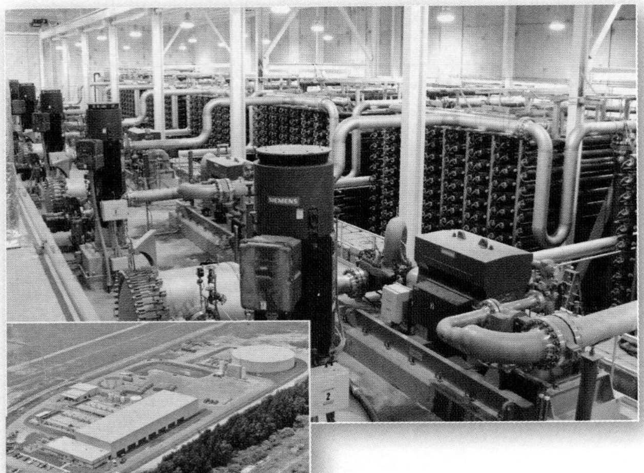

(b) Tampa Bay Florida

FIGURE 9.19 Water desalination.
(a) Freshwater is supplied to Saudi Arabia from the Jabal water desalination plant along the Red Sea. (b) The Tampa Bay desalination plant provides 25 MGD to the region's water resources through filtration and reverse osmosis processes.
[(a) Liaison Agency. (b) Tampa Bay Water.]

Future Considerations

When precipitation is budgeted, the limits of the water resource become apparent. How can we satisfy the growing demand for water? Water availability per person declines as population increases, and individual demand increases with economic development, affluence, and technology. Thus, world population growth since 1970 reduced per capita water supplies by a third. Also, pollution limits the water-resource base, so that even before *quantity* constraints are felt, *quality* problems may limit the health and growth of a region.

The international aspect of the water-resource problem is illustrated by the 200 major river drainage basins in the world. One hundred and forty-five countries possess territory within an international river drainage basin—truly a global commons.

Working Group II, in its report entitled *Impacts, Adaptation, and Vulnerability*—part of the Intergovernmental Panel on Climate Change's *Fourth Assessment Report* (2007)—concluded that global climate change and greenhouse warming will cause added water problems:

> By mid-century, annual average river runoff and water availability are projected to increase 10–40% at high latitudes and in some wet tropical areas, and decrease by 10–30% over some dry regions at mid-latitudes and in the dry tropics, some of which are presently water-stressed areas. . . . Drought-affected

areas will likely increase in extent. . . . In the course of the century, water supplies stored in glaciers and snow cover are projected to decline, reducing water availability in regions supplied by meltwater from major mountain ranges, where more than one-sixth of the world population currently lives.

Clearly, cooperation is needed, yet we continue toward a water crisis without a concept of a *world water economy* as a frame of reference. The geospatial question is this: When will more international coordination begin, and which country or group of countries will lead the way to sustain future water resources? The soil-water-budget approach detailed in this chapter is a place to start.

GEOSYSTEMS CONNECTION

In this chapter, we looked at water resources through a water-balance approach. Such a systems view, considering both water supply and water demand, is the best method to understand the water resource. The ongoing drought in the western United States brings the need for such a water-budget strategy to the forefront. With water resources as the ultimate output of the water–weather system, we shift to climate. The examination of world climates in Chapter 10 completes our journey from Chapter 2, which began with solar energy flowing through the atmosphere to Earth's surface, which then moved to the outputs of temperature and wind and ocean circulation, and finally shifted to water and weather. The synthesis in the next chapter brings us to the end of Part II.

KEY LEARNING CONCEPTS REVIEW

- ***Illustrate*** **the hydrologic cycle with a simple sketch, and** ***label*** **it with definitions for each water pathway.**

The **hydrologic cycle** is a model of Earth's water system, which has operated for billions of years from the lower atmosphere to several kilometers beneath Earth's surface. **Interception** occurs when precipitation strikes vegetation or other ground cover. Water soaks into the subsurface through **infiltration**, or penetration of the soil surface. It further permeates soil or rock through vertical movement called **percolation**.

hydrologic cycle (p. 224)
interception (p. 225)
infiltration (p. 225)
percolation (p. 225)

1. Sketch and explain a simplified model of the complex flows of water on Earth—the hydrologic cycle.
2. What are the possible routes that a raindrop may take on its way to and into the soil surface?
3. Compare precipitation and evaporation volumes from the ocean with those over land. Describe advection flows of moisture and countering surface and subsurface runoff.

- ***Relate*** **the importance of the water-budget concept to your understanding of the hydrologic cycle, water resources, and soil moisture for a specific location.**

A **soil-water budget** can be established for any area of Earth's surface by measuring the precipitation input and the output of various water demands in the area considered. Understanding both the supply of the water resource and the natural demands on the resource is essential to sustainable human interaction with the hydrologic cycle. Streams represent only a tiny fraction of all water (1250 km^3, or 300 mi^3), the smallest volume of any of the freshwater categories. Yet streams represent four-fifths of all the water we use. Groundwater is the largest potential freshwater source in the hydrologic cycle and is tied to surface supplies.

soil-water budget (p. 226)

4. How might an understanding of the hydrologic cycle in a particular locale or a soil-moisture budget of a site assist you in assessing water resources? Give some specific examples.

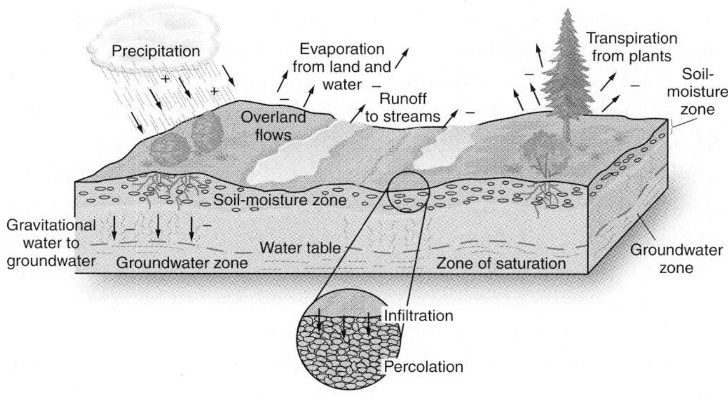

- ***Construct*** **the water-balance equation as a way of accounting for the expenditures of water supply, and** ***define*** **each of the components in the equation and its specific operation.**

The moisture supply to Earth's surface is **precipitation** (PRECIP, or P), arriving as rain, sleet, snow, and hail. Precipitation is measured with the **rain gauge**. **Evaporation** is the net movement of free water molecules away from a wet surface into air. **Transpiration** is the movement of water through plants and back into the

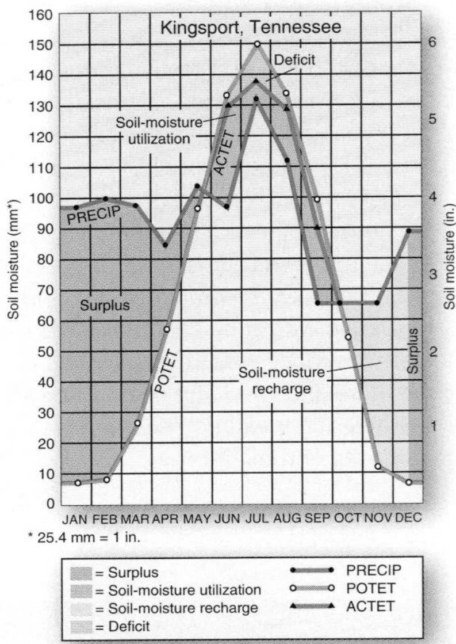

atmosphere; it is a cooling mechanism for plants. Evaporation and transpiration are combined into one term—**evapotranspiration**. The ultimate demand for moisture is **potential evapotranspiration** (POTET, or PE), the amount of water that *would* evaporate and transpire under optimum moisture conditions (adequate precipitation and adequate soil moisture). Evapotranspiration is measured with an **evaporation pan** (*evaporimeter*) or the more elaborate **lysimeter.**

Unsatisfied POTET is a **deficit** (DEFIC). By subtracting DEFIC from POTET, we determine **actual evapotranspiration**, or ACTET. Ideally, POTET and ACTET are about the same, so that plants have sufficient water. If POTET is satisfied and the soil is full of moisture, then additional water input becomes **surplus** (SURPL), which may puddle on the surface, flow across the surface toward stream channels, or percolate underground through the soil. The **overland flow** to streams includes precipitation and groundwater flows into river channels to make up the **total runoff** from the area.

A "savings account" of water that receives recharge deposits and provides for utilization withdrawals as water-balance conditions change is the **soil-moisture storage**, ΔSTRGE. This is the volume of water stored in the soil that is accessible to plant roots. In soil, **hygroscopic water** is inaccessible because it is a molecule-thin layer that is tightly bound to each soil particle by hydrogen bonding. As available water is utilized, soil reaches the **wilting point** (all that remains is unextractable water). **Capillary water** is generally accessible to plant roots because it is held in the soil by surface tension and hydrogen bonding between water and soil. Almost all capillary water that remains in the soil is **available water** in soil-moisture storage. After water drains from the larger pore spaces, the available water remaining for plants is termed **field capacity**, or storage capacity. When soil is saturated after a precipitation event, surplus water in the soil becomes **gravitational water** and percolates to groundwater. As **soil-moisture utilization** removes soil water, the plants work harder to extract the same amount of moisture, whereas **soil-moisture recharge** is the rate at which needed moisture enters the soil.

The texture and the structure of the soil dictate available pore spaces, or **porosity**. The soil's **permeability** is the degree to which water can flow through it. Permeability depends on particle sizes and the shape and packing of soil grains.

Drought does not have a simple water-budget definition; rather, it can occur in at least four forms: *meteorological drought*, *agricultural drought*, *hydrologic drought*, and/or *socioeconomic drought*.

precipitation (p. 226)
rain gauge (p. 227)
evaporation (p. 228)
transpiration (p. 228)
evapotranspiration (p. 228)
potential evapotranspiration (p. 228)
evaporation pan (p. 229)
lysimeter (p. 229)
deficit (p. 230)
actual evapotranspiration (p. 230)
surplus (p. 230)
overland flow (p. 230)
total runoff (p. 230)
soil-moisture storage (p. 230)
hygroscopic water (p. 230)
wilting point (p. 230)
capillary water (p. 230)
available water (p. 230)
field capacity (p. 230)
gravitational water (p. 230)
soil-moisture utilization (p. 230)
soil-moisture recharge (p. 231)
porosity (p. 231)
permeability (p. 231)
drought (p. 231)

5. Explain the meaning of this statement: "The soil-water budget is an assessment of the hydrologic cycle at a specific site."
6. What are the components of the water-balance equation? Construct the equation, and place each term's definition below its abbreviation in the equation.
7. Using the annual water-balance chart for Kingsport, Tennessee, work the concepts through the water-balance equation as you interpret the relation of POTET to PRECIP.
8. Explain how to derive actual evapotranspiration (ACTET) in the water-balance equation.
9. What is potential evapotranspiration (POTET)? How do we go about estimating this potential rate? What factors did Thornthwaite use to determine this value?
10. Explain the operation of soil-moisture storage, soil-moisture utilization, and soil-moisture recharge. Include discussion of the field capacity, capillary water, and wilting point concepts.
11. In the case of silt-loam soil from Figure 9.8, roughly what is the available water capacity? How is this value derived?
12. Describe the four types of drought. Why do you think media coverage of drought seems lacking?

■ ***Describe*** the nature of groundwater, and ***define*** the elements of the groundwater environment.

Groundwater is a part of the hydrologic cycle, but it lies beneath the surface beyond the soil-moisture root zone. Groundwater does not exist independently because its replenishment is tied to surface surpluses. Excess surface water moves through the **zone of aeration**, where soil and rock are less than saturated. Eventually, the water reaches the **zone of saturation**, where the pores are completely filled with water.

The *permeability* of subsurface rocks depends on whether they conduct water readily (higher permeability) or tend to obstruct its

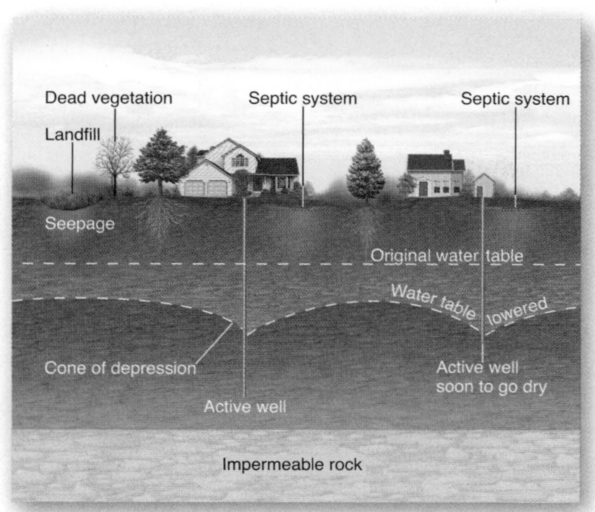

flow (lower permeability). They can even be impermeable. An **aquifer** is a rock layer that is permeable to groundwater flow in usable amounts. An **aquiclude** (aquitard) is a body of rock that does not conduct water in usable amounts.

The upper limit of the water that collects in the zone of saturation is the **water table**; it is the contact surface between the zones of saturation and aeration. A **confined aquifer** is bounded above and below by impermeable layers of rock or sediment. An **unconfined aquifer** has a permeable layer on top and an impermeable one beneath. The **aquifer recharge area** extends over an entire unconfined aquifer. Water in a confined aquifer is under the pressure of its own weight, creating a pressure level to which the water can rise on its own, the **potentiometric surface**, which can be above ground level. Groundwater confined under pressure is **artesian water**; it may rise up in wells and even flow out at the surface without pumping if the head of the well is below the potentiometric surface.

As water is pumped from a well, the surrounding water table within an unconfined aquifer will experience **drawdown**, or become lower, if the rate of pumping exceeds the horizontal flow of water in the aquifer around the well. This excessive pumping causes a **cone of depression**. Aquifers frequently are pumped beyond their flow and recharge capacities, a condition known as **groundwater mining**.

groundwater (p. 236)
zone of aeration (p. 236)
zone of saturation (p. 236)
aquifer (p. 236)
aquiclude (p. 236)
water table (p. 237)
confined aquifer (p. 237)
unconfined aquifer (p. 237)
aquifer recharge area (p. 237)
potentiometric surface (p. 238)
artesian water (p. 238)
drawdown (p. 239)
cone of depression (p. 239)
groundwater mining (p. 242)

13. Are groundwater resources independent of surface supplies, or are the two interrelated? Explain your answer.
14. Make a simple sketch of the subsurface environment, labelling zones of aeration and saturation and the water table in an unconfined aquifer. Then add a confined aquifer to the sketch.
15. At what point does groundwater utilization become groundwater mining? Use the High Plains aquifer example to explain your answer.
16. What is the nature of groundwater pollution? Can contaminated groundwater be cleaned up easily? Explain.

- ***Identify*** critical aspects of freshwater supplies for the future, and ***cite*** specific water-resource issues and potential remedies for any shortfalls.

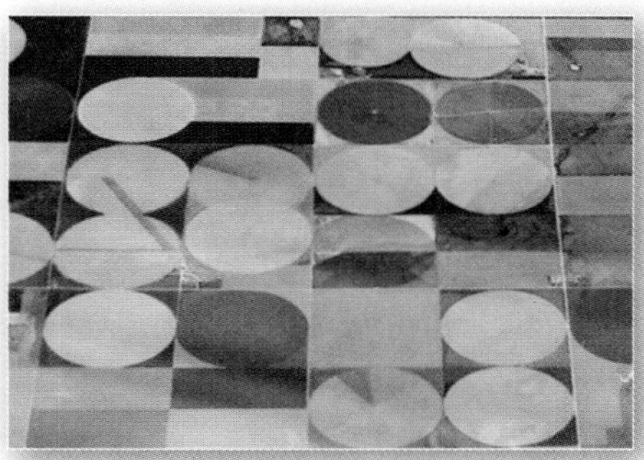

Nonconsumptive uses, or water **withdrawal**, or *offstream use* remove water from the supply, use it, and then return it to the stream. **Consumptive uses** remove water from a stream, but do not return it, so the water is not available for a second or third use. Americans in the 48 contiguous states withdraw approximately one-third of the available surplus runoff for irrigation, industry, and municipal uses. **Desalination** as a water resource involves the removal of organics, debris, and salinity from seawater through distillation or reverse osmosis. This processing yields potable water for domestic uses.

withdrawal (p. 244)
consumptive uses (p. 245)
desalination (p. 246)

17. Describe the principal pathways involved in the water budget of the contiguous 48 states. What is the difference between withdrawal and consumptive use of water resources? Compare these with instream uses.
18. Briefly assess the status of world water resources. What challenges exist in meeting the future needs of an expanding population and growing economies?
19. If wars in the 21st century are predicted to be about water availability in quantity and quality, what action could we take to understand the issues and avoid the conflicts?

MASTERING GEOGRAPHY

Visit www.masteringgeography.com for several figure animations, satellite loops, self-study quizzes, flashcards, visual glossary, case studies, career links, textbook reference maps, RSS Feeds, and an ebook version of *Geosystems*. Included are many geography web links to interesting Internet sources that support this chapter.

CHAPTER 10

Climate Systems and Climate Change

Across southern Puerto Rico in the rain shadow of the Cordillera Central, drier conditions produce tropical savanna climate conditions. The acacia trees in the foreground are suggestive of similar savanna landscapes in Africa and coastal Southeast Asia. The area in the photo receives about 69 cm (27 in.) of rainfall, with 5 months receiving less than 6 cm (2.4 in.). [Bobbé Christopherson.]

KEY LEARNING CONCEPTS

After reading the chapter, you should be able to:

- ***Define*** climate and climatology, and ***explain*** the difference between climate and weather.
- ***Review*** the role of temperature, precipitation, air pressure, and air mass patterns used to establish climatic regions.
- ***Review*** the development of climate classification systems, and ***compare*** genetic and empirical systems as ways of classifying climate.
- ***Describe*** the principal climate classification categories other than deserts, and ***locate*** these regions on a world map.
- ***Explain*** the precipitation and moisture efficiency criteria used to determine the arid and semiarid climates, and ***locate*** them on a world map.
- ***Explain*** the causes and potential consequences of climate change, and ***outline*** future climate patterns from forecasts.

GEOSYSTEMS NOW

A Look at Puerto Rico's Climate at a Larger Scale

Examining the small-scale maps in this text, you see generalizations such as on the world climates map in Figure 10.6, where the map scale is 1 inch equals 1650 miles. On a larger-scale map of an island such as Puerto Rico we can see more climatic detail and response to local elevation and orientation to trade winds. Similarly, on the map of terrestrial biomes in Figure 20.4, Puerto Rico is designated as rain forest. A larger-scale map shows three biomes—rain forest, seasonal forest and scrub, and savanna, as you can see in the photos in Figure GN 10.1.

Puerto Rico is centered between the Greater and Lesser Antilles island chains, separating the Atlantic Ocean and the Caribbean Sea, a key location in the history of the entire region. Puerto Rico is within important shipping lanes to the Panama Canal, a situation of historical importance. The capital, San Juan, has an outstanding natural harbor. The 18th parallel runs along Puerto Rico's south coast, positioning it to receive trade winds from the northeast. The island is nearly rectangular, comprising 9104 km^2 (3515 mi^2) and stretching 65 km (40 mi) north and south and 180 km (112 mi) east and west. Puerto Rico is an official territory of the United States, sharing the same currency and president and with nonvoting representation in Congress.

The mountains of the Cordillera Central, rising to 1338 m (4389 ft), divides Puerto Rico lengthwise, and the Sierra de Luquillo in the east, rising to 650 m (2133 ft), is home to the tropical rain forest of El Yunque National Park, administered by the National Park Service. The moisture-intercepting northern and eastern slopes are in contrast to the rain shadow southern slopes and flatlands. Across Puerto Rico, all months average above 18°C (64.4°F), qualifying it as tropical.

On the map, you see the extent of tropical rain forest climates, where precipitation exceeds 6 cm (2.4 in.) every month. The mountains of the Cordillera Central add an orographic component to the climate, increasing annual precipitation. In the leeward south, we find a tropical monsoon climate where precipitation is less than 6 cm for 3 to 5 months during the year. Ponce and the south coast have a tropical savanna climate, where rainfall for 6 months is less than 6 cm (2.4 in.), and features an acacia–savanna landscape, as shown in the chapter-opening photo and in Figure GN 10.1. With this closer look in mind, we shift our perspective to examine climate categories across the globe.

San Juan, 150.9 cm

El Yunque National Park, 508 cm

Pico del Este, 434.5 cm

Near Ponce, 69.7 cm

North of Adjuntas, 198.3 cm

FIGURE GN 10.1
Climates, representative photos, and rainfall data.
What do you see in the photos that indicate differences in annual rainfall? Annual rain fall totals in centimeters (1 cm = 0.3937 in.; 1 in. = 2.54 cm).
[All photos by Bobbé Christopherson.]

Earth experiences an almost infinite variety of *weather* at any given time and place. But if we consider the weather over many years, including its variability and extremes, a pattern emerges that constitutes **climate**. Think of climatic patterns as dynamic rather than static, owing to the fact that we are witnessing climate change. Climate is more than a consideration of simple averages of temperature and precipitation.

Today, climatologists know that intriguing global-scale linkages exist in the Earth–atmosphere–ocean system. For instance, strong monsoonal rains in West Africa are correlated with the development of intense Atlantic hurricanes. One year an El Niño in the Pacific was linked to rains in the American West, floods in Louisiana and Northern Europe, and a weak Atlantic hurricane season. Yet the persistent La Niña in 2007 and 2010 strengthened drought conditions in the American West. The spring wildflower bloom in Death Valley provides visible evidence of the resultant heavy rains during El Niño conditions (Figure 10.1a), as compared to normal desert conditions. The departure from climate norms, such as the El Niño/La Niña phenomenon, is the subject of Focus Study 10.1 (pages 254–255).

Climatologists, among other scientists, are analyzing global climate change—record-breaking global average temperatures, glacial ice melt, drying soil-moisture conditions, changing crop yields, the spread of infectious disease, changing distributions of plants and animals, declining coral reef health and fisheries, and the thawing of high-latitude lands and seas. Climatologists are concerned about observed changes in the global climate, as these are occurring at a pace not evidenced in the records of the past millennia. Climate and natural vegetation shifts during the next 50 years could exceed the total of all changes since the peak of the last ice-age episode, some 18,000 years ago. Geographer/scientist Jack Williams summarized the issue:

> By the end of the 21st century, large portions of the Earth's surface may experience climates not found at present, and some 20th-century climates may disappear. . . . Novel climates are projected to develop primarily in the tropics and subtropics. . . . Disappearing climates increase the likelihood of species extinctions and community disruption for species endemic to particular climatic regimes, with the largest impacts projected for poleward and tropical montane regions.*

We need to realize that the climate map and climate designations we study in this chapter are not fixed, but are on the move as temperature and precipitation relationships shift.

In this chapter: Climates are so diverse that no two places on Earth's surface experience exactly the same climatic conditions; in fact, Earth is a vast collection of microclimates. However, broad similarities among local climates permit their grouping into climatic regions.

Many of the physical elements of the environment, studied in the first nine chapters of this text, link together to explain climates. Here we survey the patterns of climate using a series of sample cities and towns. *Geosystems* uses a simplified classification system based on physical factors that help uncover the "why" question—why climates are in certain locations. Though imperfect, this method is easily understood and is based on a widely used classification system devised by climatologist Wladimir

*J. W. Williams, S. T. Jackson, and J. E. Kutzbach, "Projected distributions of novel and disappearing climates by 2100 AD," *Proceedings of the National Academy of Sciences* 104 (April 3, 2007): 5739.

(a) 1998, during El Niño

(b) 2002, during La Niña

FIGURE 10.1 El Niño's impact on the desert.
Death Valley, in southeastern California, during spring in (a) full bloom following record rains triggered by El Niño and (b) the same scene in its stark desert grandeur during a La Niña a few years later. This dramatic effect is caused by changes in the distant tropics of the Pacific Ocean described in Focus Study 10.1. [Bobbé Christopherson.]

Köppen (pronounced KUR-pen). For reference, Appendix B details the Köppen climate classification system and all its criteria.

Climatologists use powerful computer models to simulate changing complex interactions in the atmosphere, hydrosphere, lithosphere, and biosphere. This chapter concludes with a discussion of climate change and its vital implications for society.

Earth's Climate System and Its Classification

Climatology, the study of climate and its variability, analyzes long-term weather patterns over time and space and the controls that produce Earth's diverse climatic conditions. One type of climatic analysis locates areas of similar weather statistics and groups them into **climatic regions**. Observed patterns grouped into regions are at the core of climate classification.

The climate where you live may be humid with distinct seasons, or dry with consistent warmth, or moist and cool—almost any combination is possible. Some places have rainfall totaling more than 20 cm (8 in.) each month, with monthly average temperatures remaining above 27°C (80°F) year-round. Other places may be rainless for a decade at a time. A climate may have temperatures that average above freezing every month, yet still threaten severe frost problems for agriculture. Students reading *Geosystems* in Singapore experience precipitation every month, ranging from 13.1 to 30.6 cm (5.1 to 12.0 in.), or 228.1 cm (89.8 in.) during an average year, whereas students at the university in Karachi, Pakistan, measure only 20.4 cm (8 in.) of rain over an entire year.

Climates greatly influence *ecosystems*, the natural, self-regulating communities formed by plants and animals in their nonliving environment. On land, the basic climatic regions determine to a large extent the location of the world's major ecosystems. These regions, called *biomes*, include forest, grassland, savanna, tundra, and desert. Plant, soil, and animal communities are associated with these biomes. Because climate cycles through periodic change, it is never really stable; therefore, ecosystems should be thought of as being in a constant state of adaptation and response.

The present global climatic warming trend is producing changes in plant and animal distributions. Figure 10.2 presents a schematic view of Earth's climate system, showing both internal and external processes and linkages that influence climate and thus regulate such changes.

(*text continued on page 256*)

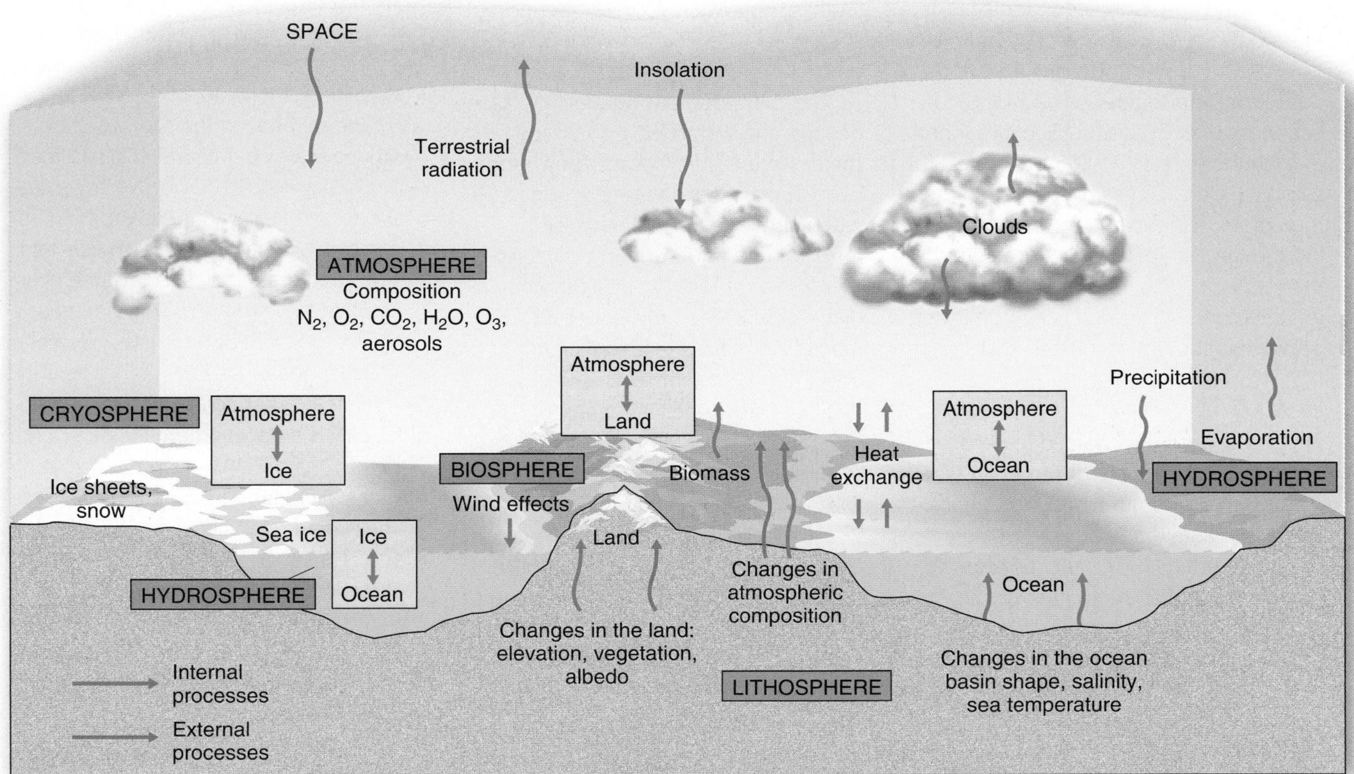

FIGURE 10.2 A schematic of Earth's climate system.
Imagine you are hired to write a computer program that simulates Earth's climates. *Internal processes* that influence climate involve the atmosphere, hydrosphere (streams and oceans), cryosphere (polar ice masses and glaciers), biosphere, and lithosphere (land)—all energized by insolation. *External processes,* principally from human activity, affect this climatic balance and force climate change. [After J. Houghton, *The Global Climate* (Cambridge, England: Cambridge University Press, 1984); and the Global Atmospheric Research Program (coordinated by WMO).]

 FOCUS STUDY 10.1

The El Niño Phenomenon—Global Linkages

Climate is the consistent behavior of weather over time, but average weather conditions also include extremes that depart from normal. The El Niño–Southern Oscillation (ENSO) in the Pacific Ocean forces the greatest interannual variability of temperature and precipitation on a global scale. The two strongest ENSO events in 120 years hit in 1982–1983 and 1997–1998. Peruvians coined the name El Niño ("the boy child") because these episodes seem to occur around the traditional December celebration time of Christ's birth. Actually, El Niños can occur as early as spring and summer and persist through the year.

Revisit Figure 6.20 and see that the northward-flowing Peru Current dominates the region off South America's west coast. These cold waters move toward the equator and join the westward movement of the South Equatorial Current.

The Peru Current is part of the normal counterclockwise circulation of winds and surface ocean currents around the subtropical high-pressure cell dominating the eastern Pacific in the Southern Hemisphere. As a result, a location such as Guayaquil, Ecuador, normally receives 91.4 cm (36 in.) of precipitation each year under dominant high pressure, whereas islands in the Indonesian archipelago receive more than 254 cm (100 in.) under dominant low pressure. This normal alignment of pressure is shown in Figure 10.1.1a.

What Is ENSO?

Occasionally, for unexplained reasons, pressure patterns and surface ocean temperatures shift from their usual locations. Higher pressure than normal develops over the western Pacific and lower pressure develops over the eastern Pacific. Trade winds normally moving from east to west weaken and can be reduced or even replaced by an eastward (west-to-east) flow. The shifting of atmospheric pressure and wind patterns across the Pacific is the *Southern Oscillation.* Chapter 6 discussed the Pacific Decadal Oscillation (PDO) and its interrelation with ENSO.

Sea-surface temperatures increase, sometimes more than 8 C° (14 F°) above normal in the central and eastern Pacific during an ENSO, replacing the normally cold, upwelling, nutrient-rich water along Peru's coastline. Such ocean-surface warming, the "warm pool," may extend to the International Date Line. This surface pool of warm water is known as El Niño. Thus, the designation ENSO—El Niño–Southern Oscillation—is derived. This condition is shown in Figure 10.1.1b in illustration and satellite image.

The expected interval for recurrence is 3 to 5 years, but the interval may range from 2 to 12 years. The frequency and intensity of ENSO events increased through the 20th century, a topic of extensive scientific research looking for a link to global climate change. Recent studies suggest ENSO might be more responsive to global change than previously thought.

The thermocline (boundary of colder, deep-ocean water) lowers in depth in the eastern Pacific Ocean. The change in wind direction and warmer surface water slow the normal upwelling currents that control nutrient availability. This loss of nutrients affects the phytoplankton and food chain, depriving many fish, marine mammals, and predator birds of nourishment.

Scientists at the National Oceanographic and Atmospheric Administration (NOAA) speculate that ENSO events have occurred more than a dozen times since the 14th century. The latest El Niño subsided in May 2010, followed by a brief neutral period and the start of La Niña conditions in July 2010. See the satellite images in Figure 10.1.1b–j.

La Niña—El Niño's Cousin

When surface waters in the central and eastern Pacific cool to below normal by 0.4 C° (0.7 F°) or more, the condition is dubbed La Niña, Spanish for "the girl." This is a weaker condition and less consistent than El Niño. There is no correlation in the strength or weakness of each. For instance, following the record 1997–1998 ENSO event, the subsequent La Niña was not as strong as predicted and shared the Pacific with lingering warm water (Figure 10.1.1c, d, h, and j).

Global Effects Related to ENSO and La Niña

Effects related to ENSO and La Niña occur worldwide: droughts in South Africa, southern India, Australia, and the Philippines; strong hurricanes in the Pacific, including Tahiti and French Polynesia; and flooding in the southwestern United States and mountain states, Bolivia, Cuba, Ecuador, and Peru. In India, every drought for more than 400 years seems linked to ENSO events. The Atlantic hurricane season weakens during El Niño years and strengthens during La Niñas.

Precipitation in the southwestern United States is greater in El Niño than in La Niña years. The Pacific Northwest is wetter with La Niña than with El Niño. Discovery of these truly Earthwide relations and spatial impacts is at the heart of physical geography. The climate of one location is related to climates elsewhere. "It is fascinating that what happens in one area can affect the whole world. . . . Scientists are trying to make order out of chaos," says NOAA scientist Alan Strong. (For ENSO monitoring and forecasts, see the National Climate Prediction Center at http://www.cpc.ncep.noaa.gov/, the Jet Propulsion Laboratory at http://sealevel.jpl.nasa.gov/science/el-nino.html, or NOAA's El Niño Theme Page at http://www.pmel.noaa.gov/toga-tao/el-nino/nino-home.html.)

FIGURE 10.1.1 Normal, El Niño, and La Niña changes in the Pacific.
(a) Normal patterns in the Pacific. (b) El Niño wind and weather patterns across the Pacific Ocean and satellite image for November 1997 (white and red colors indicate warmer surface water—a warm pool). (c) Image of La Niña conditions in transition in the Pacific in October 12, 1998 (purple and blue colors for cooler surface water—a cool pool). (d) through (j) track La Niña and El Niño conditions from 2000 to 2010. [(a) and (b) adapted and author corrected from C. S. Ramage, "El Niño." © 1986 by *Scientific American, Inc.* (b)–(e) *TOPEX/Poseidon* and (f)–(j) *Jason–1* images courtesy of Jet Propulsion Laboratory, NASA.]

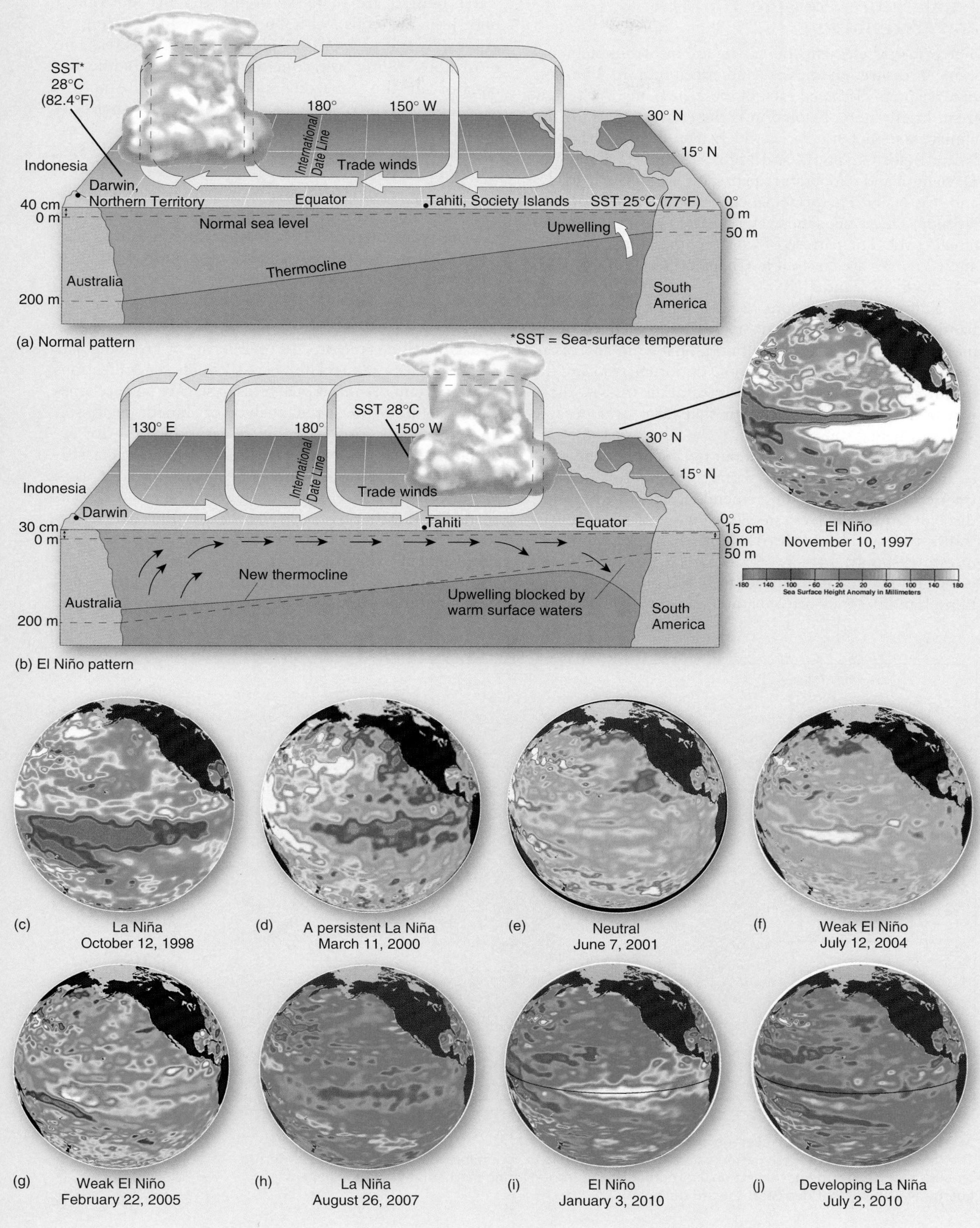
SST*
28°C
(82.4°F)
180°
150° W
30° N
15° N
Indonesia
International Date Line
Trade winds
Darwin,
Northern Territory
40 cm
0 m
Equator
Tahiti, Society Islands
SST 25°C (77°F)
0°
0 m
Normal sea level
Upwelling
50 m
Thermocline
Australia
South America
200 m
(a) Normal pattern
*SST = Sea-surface temperature
SST 28°C
130° E
180°
150° W
30° N
15° N
Indonesia
International Date Line
Trade winds
Darwin
30 cm
0 m
Tahiti
Equator
0°
15 cm
0 m
50 m
New thermocline
Upwelling blocked by
warm surface waters
Australia
South America
200 m
(b) El Niño pattern
El Niño
November 10, 1997
-180 -140 -100 -60 -20 20 60 100 140 180
Sea Surface Height Anomaly in Millimeters
(c) La Niña
October 12, 1998
(d) A persistent La Niña
March 11, 2000
(e) Neutral
June 7, 2001
(f) Weak El Niño
July 12, 2004
(g) Weak El Niño
February 22, 2005
(h) La Niña
August 26, 2007
(i) El Niño
January 3, 2010
(j) Developing La Niña
July 2, 2010

Climate Components: Insolation, Temperature, Pressure, Air Masses, and Precipitation

The principal elements of climate are insolation, temperature, pressure, air masses, and precipitation. The first nine chapters discussed each of these elements. We review them briefly here. Insolation is the energy input for the climate system, but it varies widely over Earth's surface by latitude (see Chapter 2 and Figures 2.9, 2.10, and 2.11). Daylength and temperature patterns vary diurnally (daily) and seasonally. The principal controls of temperature are latitude, elevation, land-water heating differences, and cloud cover. The pattern of world temperatures and their annual ranges are covered in Chapter 5 (see Figures 5.13, 5.14, 5.16, 5.17, and 5.18).

Temperature variations result from a coupling of dynamic forces in the atmosphere to Earth's pattern of atmospheric pressure and resulting global wind and ocean systems (see Figures 6.9 and 6.11). Important, too, are the location and physical characteristics of air masses, those vast bodies of homogeneous air that form over oceanic and continental source regions.

Moisture, the remaining input to climate, is represented by precipitation in all its forms. The hydrologic cycle transfers moisture, with its tremendous latent heat energy, through Earth's climate system (see Figure 9.1). Figure 10.3 shows the worldwide distribution of precipitation, our moisture supply. In contrast average temperatures and daylength help us approximate POTET (potential evapotranspiration), a measure of natural moisture demand.

Most of Earth's desert regions, areas of permanent water deficit, are in lands dominated by subtropical high-pressure cells, with bordering lands grading to grasslands and to forests as precipitation increases. The most consistently wet climates on Earth straddle the equator in the Amazon region of South America, the Congo region of Africa, and Indonesia and Southeast Asia, all of which are influenced by equatorial low pressure and the intertropical convergence zone (ITCZ; see Figure 6.10).

Simply relating the two principal climatic components—temperature and precipitation—reveals general climate types (Figure 10.4). Temperature and precipitation patterns, plus other weather factors, provide the key to climate classification.

Classification of Climatic Regions

Classification is the process of ordering or grouping data or phenomena into related classes. Such generalizations are important organizational tools in science and are especially useful for the spatial analysis of climatic regions. A climate classification based on *causative* factors—for example, the interaction of air masses—is a **genetic classification**. This approach explores the "why" question as to the mix of climatic ingredients in certain locations. A climate classification based on *statistics* or other data determined by measurement of observed effects is an **empirical classification**.

Climate classifications based on temperature and precipitation are examples of the empirical approach. One

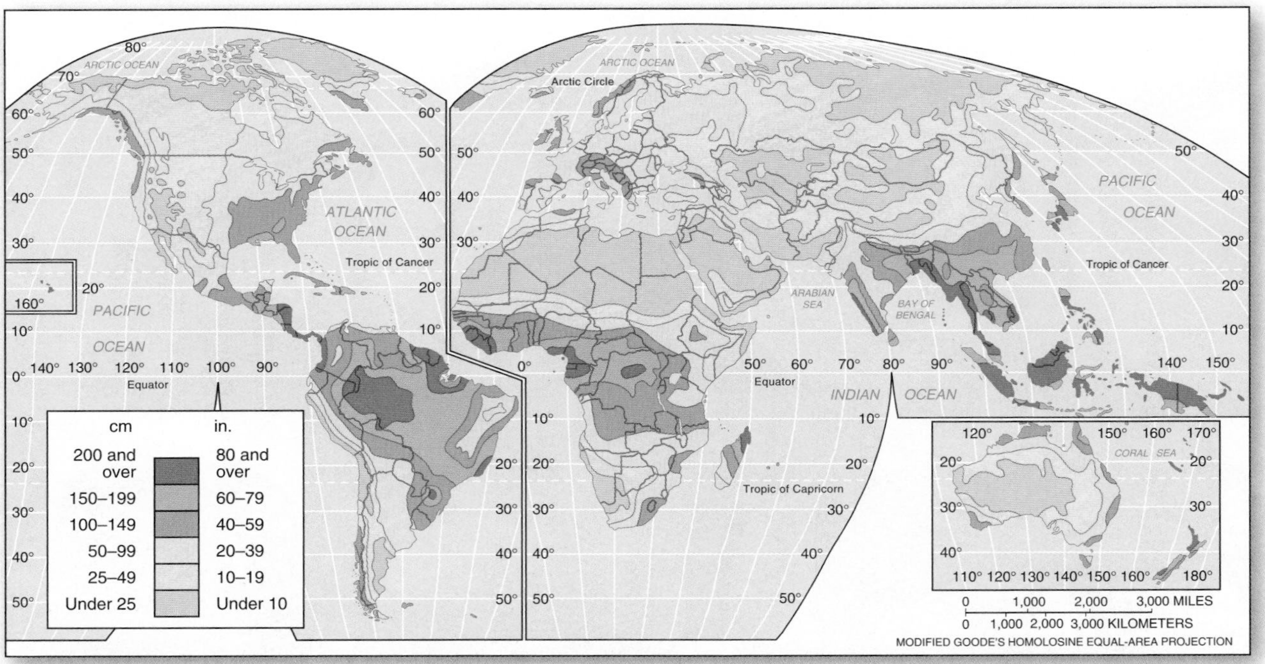

FIGURE 10.3 Worldwide average annual precipitation.
The causes that produce these patterns should be recognizable to you: temperature and pressure patterns; air mass types; convergent, convectional, orographic, and frontal lifting mechanisms; and the general energy availability that decreases toward the poles.

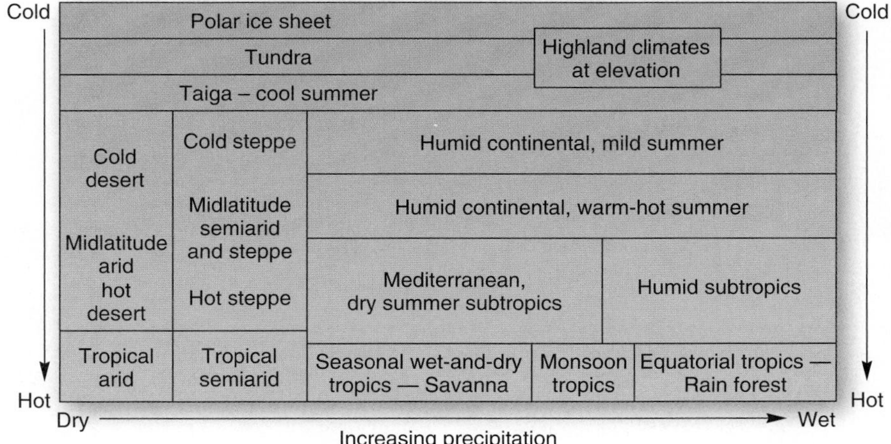

FIGURE 10.4 Climatic relationships.
A temperature and precipitation schematic reveals climatic relationships. Based on general knowledge of your college location, can you identify its approximate location on the schematic diagram? Now locate the region of your birthplace.

empirical classification system, published by C. W. Thornthwaite, identified moisture regions using aspects of the water-balance approach (discussed in Chapter 9) and vegetation types. Another empirical system is the widely recognized Köppen climate classification, designed by Wladimir Köppen (1846–1940), a German climatologist and botanist. His classification work began with an article on heat zones in 1884 and continued throughout his career. The first wall map showing world climates, coauthored with his student Rudolph Geiger, was introduced in 1928 and soon was widely adopted. Köppen continued to refine it until his death. In Appendix B, you will find a description of his system and the detailed criteria he used to distinguish climatic regions and their boundaries.

Classification Categories The basis of any classification system is the choice of criteria or causative factors used to draw lines between categories. Some climate elements used include average monthly temperatures, average monthly precipitation, total annual precipitation, air mass characteristics, ocean currents and sea-surface temperatures, moisture efficiency, insolation, and net radiation, among others. As we devise spatial categories and boundaries, we must remember that boundaries really are transition zones of gradual change. The trends and overall patterns of boundary lines are more important than their precise placement, especially with the small scales generally used for world maps.

For the *Geosystems* classification, we focus on temperature and precipitation measurements and, for the desert areas, moisture efficiency. Keep in mind these are measurable results produced by the climatic elements listed in the last paragraph. Figure 10.5 portrays the six basic climate categories and their regional types that provide us with a structure for our discussion in this chapter:

- Tropical (tropical latitudes)
  - rain forest (rainy all year)
  - monsoon (6 to 12 months rainy)
  - savanna (less than 6 months rainy)
- Mesothermal (midlatitudes, mild winters)
  - humid subtropical (hot summers)
  - marine west coast (warm to cool summers)
  - Mediterranean (dry summers)
- Microthermal (mid and high latitudes, cold winters)
  - humid continental (hot to warm, mild summers)
  - subarctic (cool summers to very cold winters)
- Polar (high latitudes and polar regions)
  - tundra (high latitude or high elevation)
  - ice caps and ice sheets (perpetually frozen)
  - polar marine

GEO REPORT 10.1 *Ancient thinking on climate classifications*

The ancient Greeks simplified their view of world climates into three zones: The "torrid zone" referred to warmer areas south of the Mediterranean; the "frigid zone" was to the north; and the area where they lived was labelled the "temperate zone," which they considered the optimum climate. They believed that travel too close to the equator or too far north would surely end in death. But the world is a diverse place, and Earth's myriad climatic variations are more complex than these simple views.

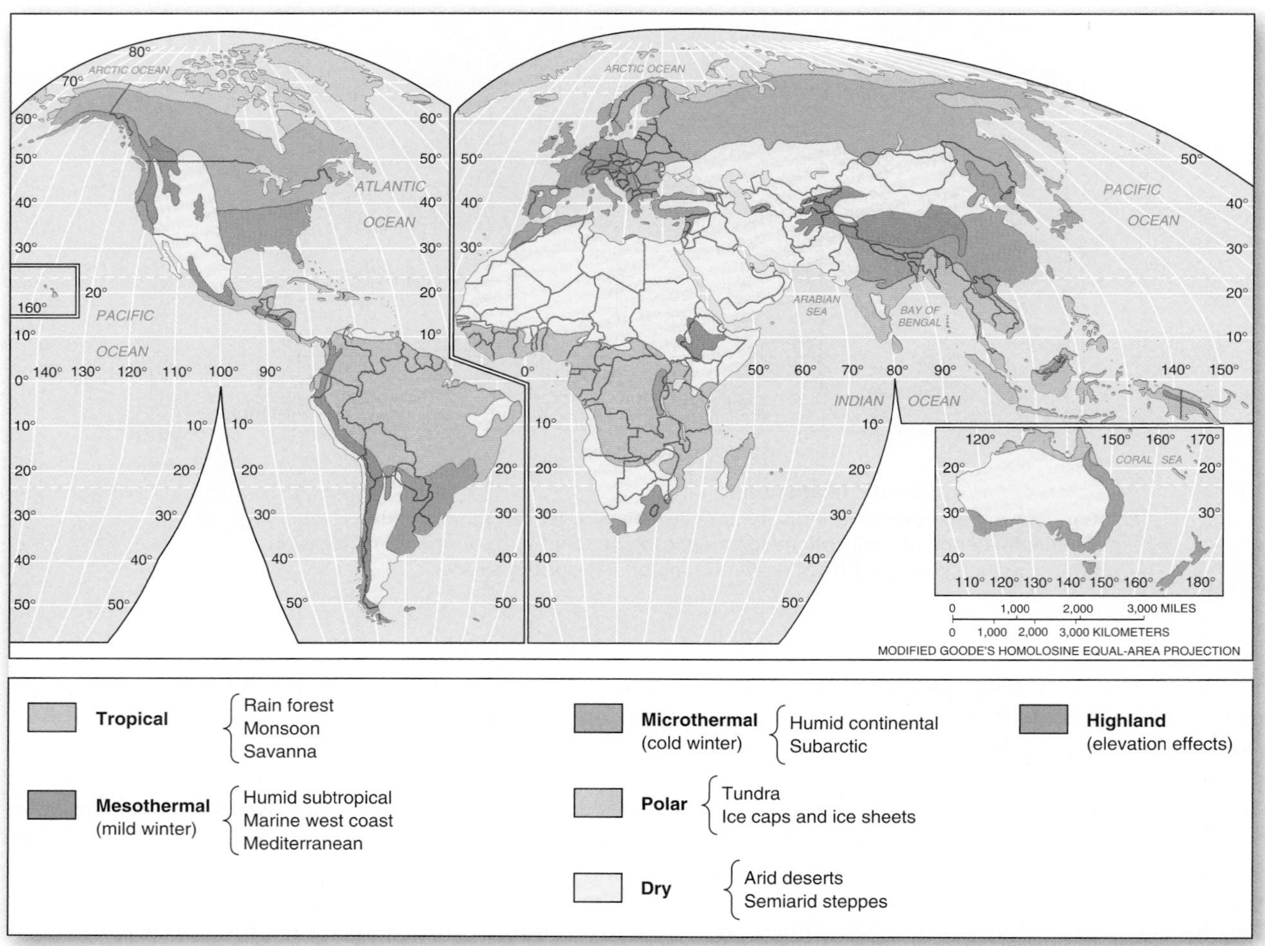

FIGURE 10.5 Climate regions generalized.
Six general climate categories, keyed to the legend and coloration in the Figure 10.6 map. Relative to the question asked about your campus and birthplace in the caption to Figure 10.4, locate these two places on this climate map.

- Highland (high elevations at all latitudes; highlands have lower temperatures)

Only one climate category is based on moisture efficiency as well as temperature:

- Dry (permanent moisture deficits)
  - arid deserts (tropical, subtropical hot and midlatitude cold)
  - semiarid steppes (tropical, subtropical hot and midlatitude cold)

Global Climate Patterns Adding detail to Figure 10.5, we develop the world climate map presented in Figure 10.6. The following sections describe specific climates, organized around each of the main climate categories listed previously. An opening box at the

GEO REPORT 10.2 *Boundary considerations and changes under way*

The boundary between mesothermal and microthermal climates is sometimes placed along the isotherm where the coldest month is −3°C (26.6°F) or lower. That might be an accurate criterion for Europe, but for conditions in North America, the 0°C (32°F) isotherm is considered more appropriate. The difference between the 0°C and −3°C isotherms covers an area about the width of the state of Ohio. In Figure 10.6, you see the 0°C boundary used.

Climate change means these statistical boundaries are shifting. The Intergovernmental Panel on Climate Change (IPCC; discussed later in this chapter) predicts a 150- to 550-km (90- to 350-mi) range of possible poleward shift of climatic patterns in the midlatitudes during this century. Such change in climate would place Ohio, Indiana, and Illinois within climate regimes now experienced in Arkansas and Oklahoma. As you examine North America in Figure 10.6, use the graphic scale to get an idea of the magnitude of these potential shifts.

beginning of each climate section gives (1) a simple description of the climate category and causal elements that are in operation and (2) a world map showing distribution and the featured representative cities. The names of the climates appear in italics in the chapter.

Selected cities represent each climate category, with their information on a **climograph**, a graph that shows monthly temperature and precipitation for a station. Also included are the location coordinates, average annual temperature, total annual precipitation, elevation, local population, annual temperature range, annual hours of sunshine (if available, as an indication of cloudiness), and location map for each city. Along the top of each climograph are the dominant weather features that are influential in that climate.

Discussions of soils, vegetation, and major terrestrial biomes that fully integrate these global climate patterns appear in Part IV of this text. Table 20.1 synthesizes all this information and enhances your understanding of this chapter, so please place a tab on that page and refer to it as you read.

CRITICAL THINKING 10.1

Finding your climate

The caption text for Figures 10.4 and 10.5 asked that you locate the climate conditions for your campus and your birthplace. Now, mark these locations with a dot on Figures 10.4 through 10.6. For these places, obtain monthly precipitation and temperature data. Briefly describe the information sources you used: library, Internet, teacher, and phone calls to state and provincial climatologists. Now, refer to Appendix B to refine your assessment of climate for the two locations. Briefly show how you worked through the Köppen climate criteria given in the appendix that established the climate classification for your two cities.

Tropical Climates (tropical latitudes)

Tropical climates, the most extensive, occupy about 36% of Earth's surface, including both ocean and land areas. The tropical climates straddle the equator from about 20° N to 20° S, roughly between the Tropics of Cancer and Capricorn—thus, the name. Tropical climates stretch northward to the tip of Florida and south-central Mexico, central India, and Southeast Asia, and southward to northern Australia, Madagascar, central Africa, and southern Brazil. These climates truly are winterless. Important causal elements include

- Consistent daylength and insolation input, which produce consistently warm temperatures;
- Effects of the intertropical convergence zone (ITCZ), which brings rains as it shifts seasonally with the high Sun; and
- Warm ocean temperatures and unstable maritime air masses.

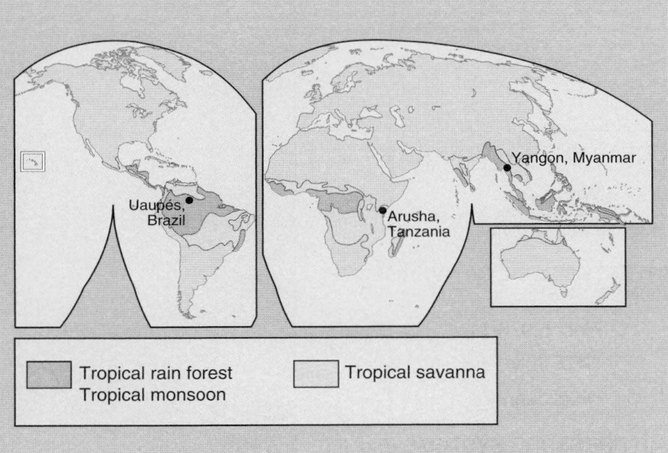

Tropical climates have three distinct regimes: *tropical rain forest* (ITCZ present all year), *tropical monsoon* (ITCZ present 6 to 12 months), and *tropical savanna* (ITCZ present less than 6 months).

Tropical Rain Forest Climates

Tropical rain forest climates are constantly moist and warm. Convectional thunderstorms, triggered by local heating and trade-wind convergence, peak each day from mid-afternoon to late evening inland and earlier in the day where marine influence is strong along coastlines. Precipitation follows the migrating ITCZ (Chapter 6). The ITCZ shifts northward and southward with the summer Sun throughout the year, but it influences *tropical rain forest* regions all year long. Not surprisingly, water surpluses are enormous, creating the world's greatest stream discharges in the Amazon and Congo river basins.

High rainfall sustains lush evergreen broadleaf tree growth, producing Earth's equatorial and tropical rain forests. The leaf canopy is so dense that little light diffuses to the forest floor, leaving the ground surface dim and sparse in plant cover. Dense surface vegetation occurs along riverbanks, where light is abundant. (Widespread deforestation of Earth's rain forest is detailed in Chapter 20.)

High temperature promotes energetic bacterial action in the soil, so that organic material is quickly consumed. Heavy precipitation washes away certain minerals and nutrients. The resulting soils are somewhat sterile and can support intensive agriculture only if supplemented by fertilizer.

(*text continued on page 262*)

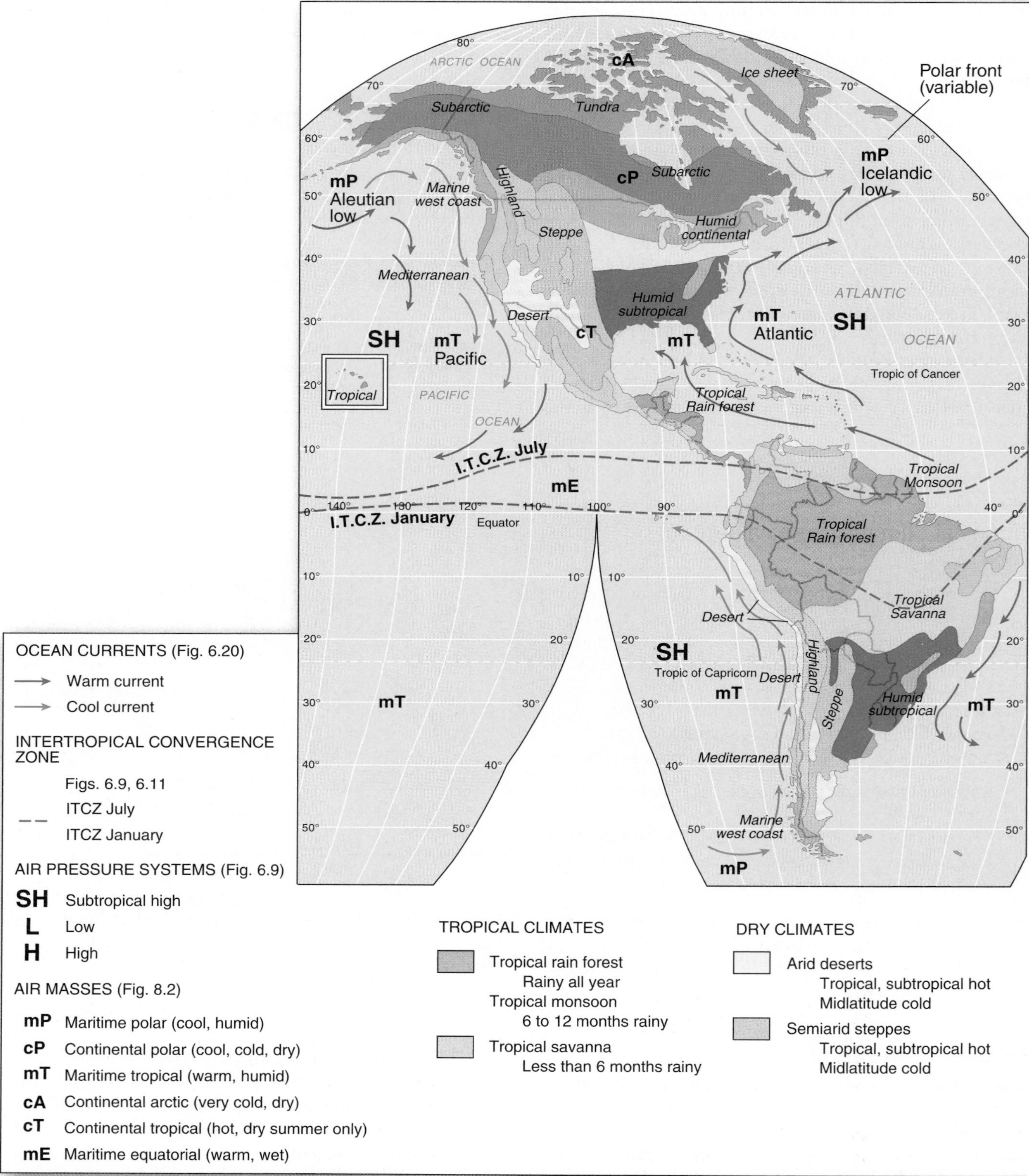

FIGURE 10.6 World climate classification.
Annotated on this map are selected air masses, near-shore ocean currents, pressure systems, and the January and July locations of the ITCZ. Use the colors in the legend to locate various climate types; some labels of the climate names appear in italics on the map to guide you.

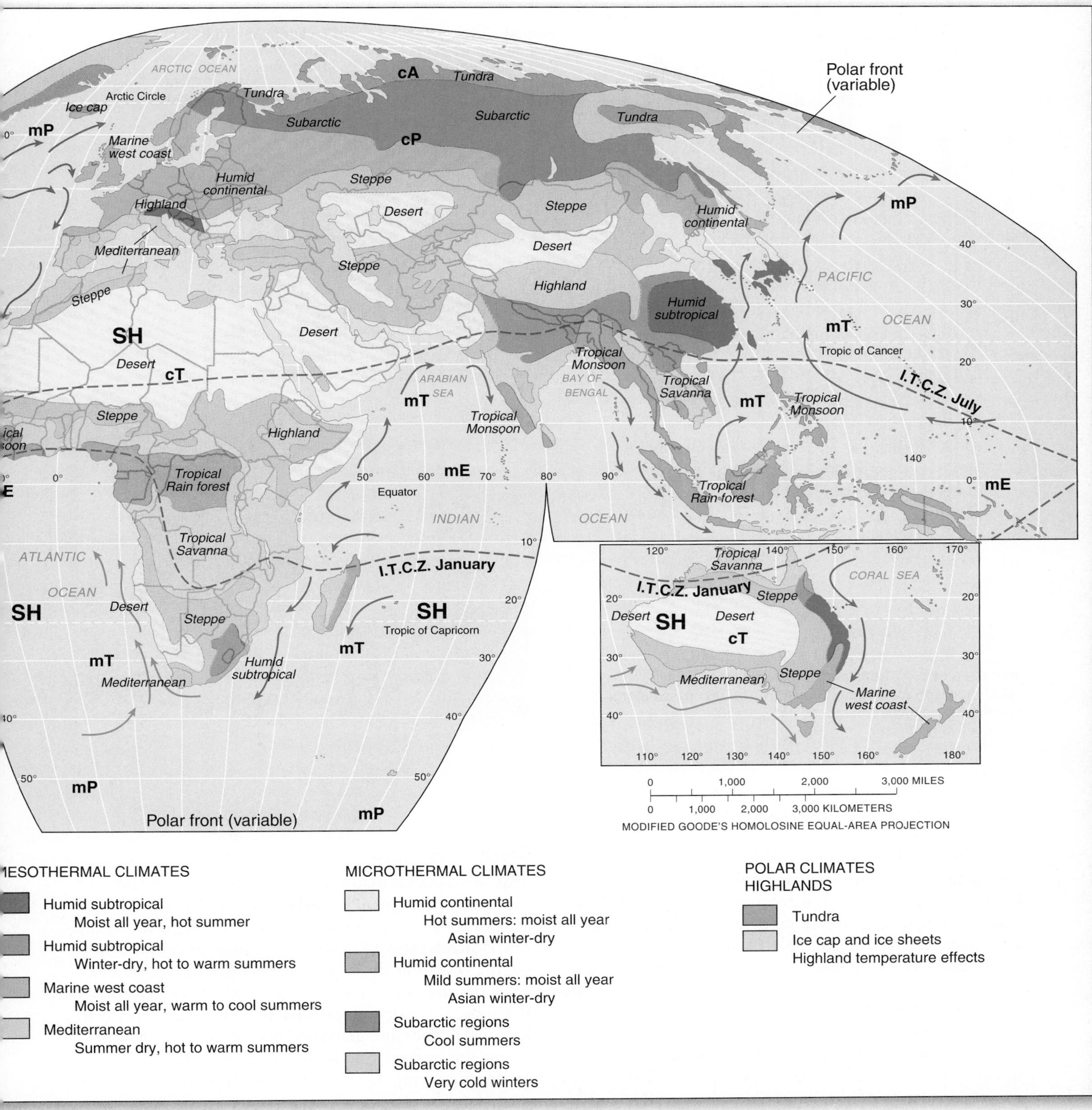

ARCTIC OCEAN
cA
Tundra
Polar front (variable)
Arctic Circle
Ice cap
Tundra
Subarctic
Subarctic
Tundra
mP
Marine west coast
cP
Humid continental
Steppe
Highland
Desert
Steppe
Humid continental
mP
Desert
Mediterranean
Steppe
Highland
PACIFIC
Steppe
Humid subtropical
SH
Desert
mT
OCEAN
Desert
cT
Tropical Monsoon
Tropic of Cancer
ARABIAN SEA
BAY OF BENGAL
Tropical Savanna
I.T.C.Z. July
mT
mT
Tropical Monsoon
Steppe
Tropical Monsoon
Highland
Tropical Rain forest
mE
Equator
Tropical Rain forest
mE
INDIAN
OCEAN
Tropical Savanna
ATLANTIC
OCEAN
I.T.C.Z. January
SH
Tropical Savanna
CORAL SEA
I.T.C.Z. January
Steppe
Desert
SH
Desert
Steppe
SH
Tropic of Capricorn
Desert
cT
mT
Humid subtropical
mT
Mediterranean
Mediterranean
Steppe
Marine west coast
mP
Polar front (variable)
mP
0 1,000 2,000 3,000 MILES
0 1,000 2,000 3,000 KILOMETERS
MODIFIED GOODE'S HOMOLOSINE EQUAL-AREA PROJECTION
MESOTHERMAL CLIMATES
Humid subtropical
Moist all year, hot summer
Humid subtropical
Winter-dry, hot to warm summers
Marine west coast
Moist all year, warm to cool summers
Mediterranean
Summer dry, hot to warm summers
MICROTHERMAL CLIMATES
Humid continental
Hot summers: moist all year
Asian winter-dry
Humid continental
Mild summers: moist all year
Asian winter-dry
Subarctic regions
Cool summers
Subarctic regions
Very cold winters
POLAR CLIMATES
HIGHLANDS
Tundra
Ice cap and ice sheets
Highland temperature effects

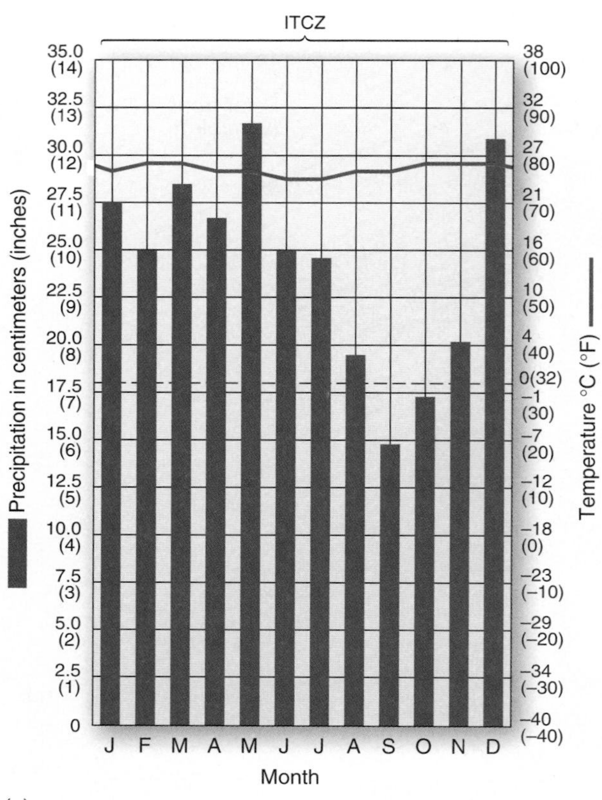

Station: Uaupés, Brazil
Lat/long: 0° 08′ S 67° 05′ W
Avg. Ann. Temp.: 25°C (77°F)
Total Ann. Precip.: 291.7 cm (114.8 in.)
Elevation: 86 m (282.2 ft)
Population: 10,000
Ann. Temp. Range: 2 C° (3.6 F°)
Ann. Hr of Sunshine: 2018

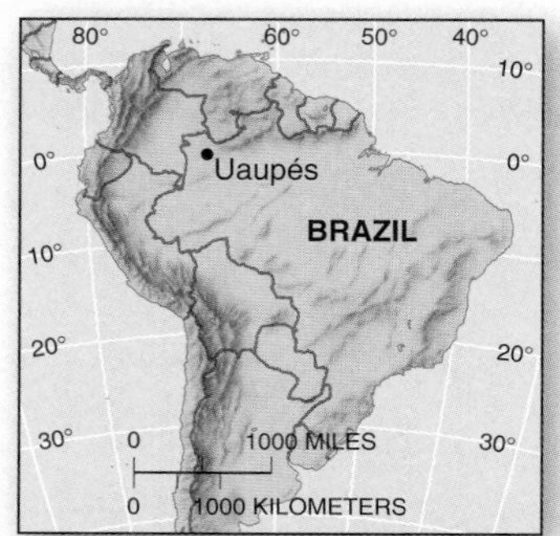

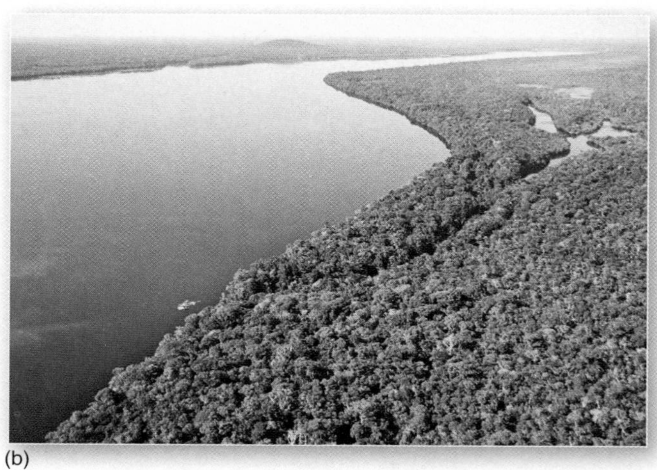

FIGURE 10.7 Tropical rain forest climate. (a) Climograph for Uaupés, Brazil. (b) The rain forest along the Rio Negro River, Amazonas state, Brazil. [Gerard & Margi Moss/peterarnold.com]

Uaupés, Brazil (Figure 10.7), is characteristic of *tropical rain forest*. On the climograph, you can see that the lowest-precipitation month receives nearly 15 cm (6 in.) and the annual temperature range is barely 2 C° (3.6 F°). In all such climates, the diurnal (day-to-night) temperature range exceeds the annual average minimum–maximum (coolest to warmest) range: Day–night temperatures can range more than 11 C° (20 F°), more than five times the annual monthly average range.

The only interruption in the distribution of *tropical rain forest* climates across the equatorial region is in the highlands of the South American Andes and in East Africa (see Figure 10.6). There, higher elevations produce lower temperatures; Mount Kilimanjaro is less than 4° south of the equator, but at 5895 m (19,340 ft) it has permanent glacial ice on its summit (although this ice is nearly gone due to warming air temperatures). Such mountainous sites fall within the highland climate category.

Tropical Monsoon Climates

Tropical monsoon climates feature a dry season that lasts 1 or more months. Rainfall brought by the ITCZ falls in these areas from 6 to 12 months of the year. (Remember, the ITCZ affects the *tropical rain forest* climate region throughout the year.) The dry season occurs when the convergence zone is not overhead. Yangon, Myanmar (formerly Rangoon, Burma), is an example of this climate type, as illustrated by the climograph and photograph in Figure 10.8. Mountains prevent cold air masses from central Asia from getting into Yangon, resulting in its high average annual temperatures.

About 480 km (300 mi) north in another coastal city, Sittwe (Akyab), Myanmar, on the Bay of Bengal, annual precipitation rises to 515 cm (203 in.) compared to Yangon's 269 cm (106 in.). Therefore, Yangon is a drier *tropical monsoon* area than farther north along the coast, but it still receives more than the 250-cm criteria in use for this classification.

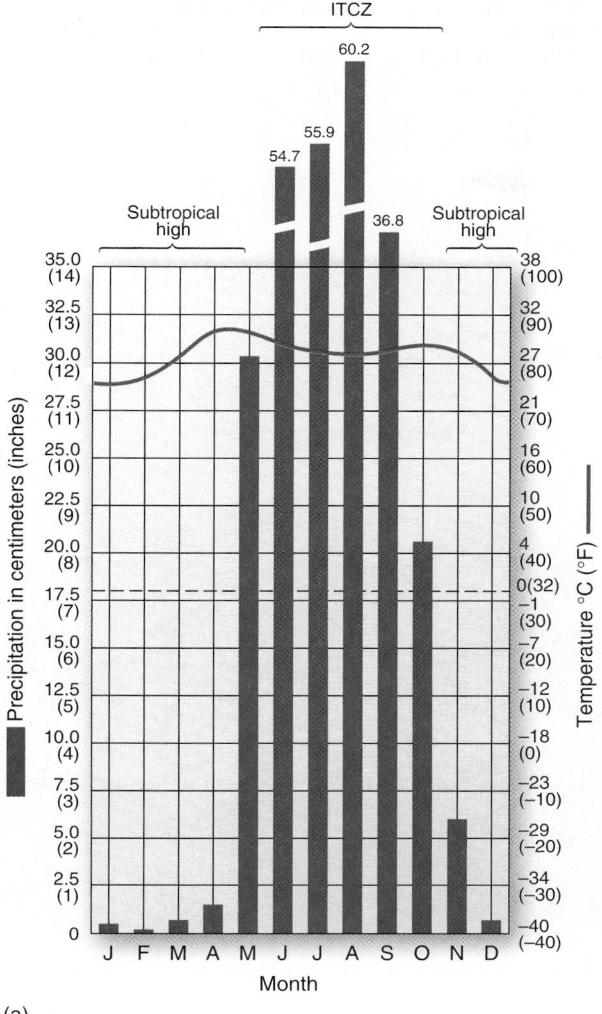

(a)

Station: Yangon, Myanmar*
Lat/long: 16° 47′ N 96° 10′ E
Avg. Ann. Temp.: 27.3°C (81.1°F)
Total Ann. Precip.: 268.8 cm (105.8 in.)
Elevation: 23 m (76 ft)
Population: 6,000,000
Ann. Temp. Range: 5.5 C° (9.9 F°)

*(Formerly Rangoon, Burma)

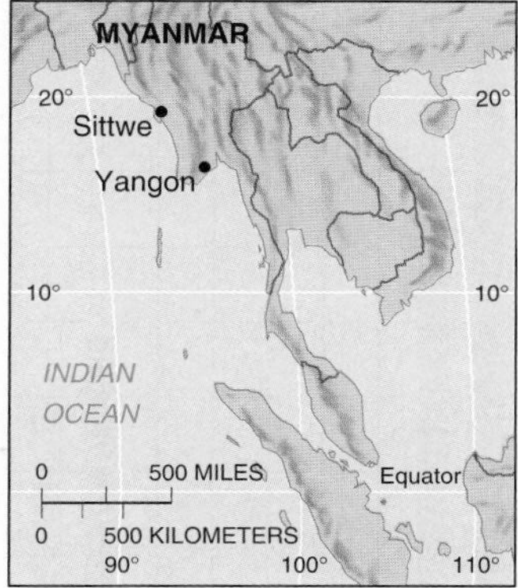

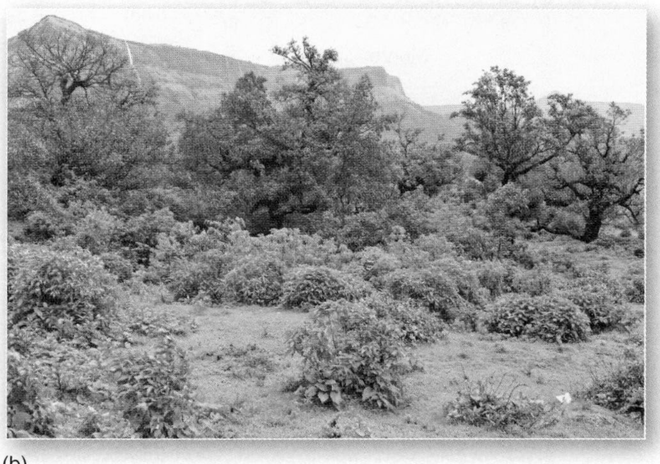

(b)

FIGURE 10.8 Tropical monsoon climate.
(a) Climograph for Yangon, Myanmar (formerly Rangoon, Burma); city of Sittwe also noted on map. (b) Mixed monsoonal forest and scrub characteristic of the region here in eastern India. [shaileshnanal/Shuttershock.]

Tropical monsoon climates lie principally along coastal areas within the *tropical rain forest* climatic realm and experience seasonal variation of wind and precipitation. Evergreen trees grade into thorn forests on the drier margins near the adjoining savanna climates.

Tropical Savanna Climates

Tropical savanna climates exist poleward of the *tropical rain forest* climates. The chapter-opening photo from Puerto Rico is such a region. The ITCZ reaches these climate regions for about 6 months or less of the year as it migrates with the summer Sun. Summers are wetter than winters because convectional rains accompany the shifting ITCZ when it is overhead. This produces a notable dry condition when the ITCZ is farthest away and high pressure dominates. Thus, POTET (natural moisture demand) exceeds PRECIP (natural moisture supply) in winter, causing water-budget deficits.

Temperatures vary more in *tropical savanna* climates than in *tropical rain forest* regions. The *tropical savanna* regime can have two temperature maximums during the year because the Sun's direct rays are overhead twice—before and after the summer solstice in each hemisphere as the Sun moves between the equator and the tropics. Dominant grasslands with scattered trees, drought resistant to cope with the highly variable precipitation, characterize the *tropical savanna* regions.

Arusha, Tanzania, is a *tropical savanna* city (Figure 10.9). This metropolitan area of more than 1,368,000 people is east of the famous Serengeti Plains savanna and Olduvai Gorge, site of human origins, and north of Tarangire National Park. Temperatures are consistent with tropical climates, despite the elevation (1387 m) of the station. Note the marked dryness from June to October, which defines changing dominant pressure systems rather than annual changes in temperature. This region is near the transition to the drier *desert hot steppe* climates to the northeast.

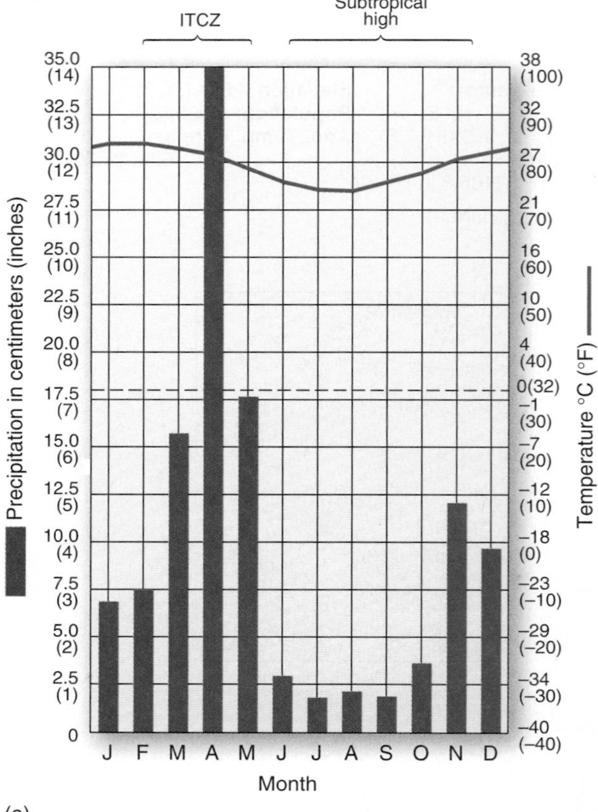

(a)

Station: Arusha, Tanzania
Lat/long: 3° 24′ S 36° 42′ E
Avg. Ann. Temp.: 26.5°C (79.7°F)
Total Ann. Precip.: 119 cm (46.9 in.)
Elevation: 1387 m (4550 ft)
Population: 1,368,000
Ann. Temp. Range: 4.1 C° (7.4 F°)
Ann. Hr of Sunshine: 2600

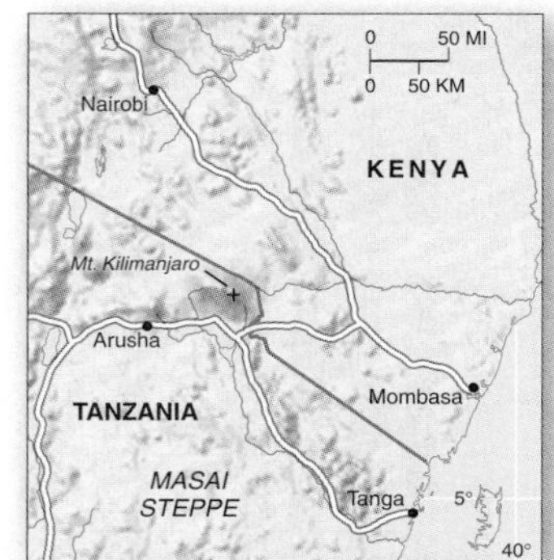

(b)

FIGURE 10.9 Tropical savanna climate. (a) Climograph for Arusha, Tanzania; note the intense dry period. (b) Characteristic landscape in the Ngorongoro Conservation Area, Tanzania, near Arusha, with plants adapted to seasonally dry water budgets. [Blaine Harrington III/Corbis.]

Mesothermal Climates (midlatitudes, mild winters)

Mesothermal, meaning "middle temperature," describes these warm and temperate climates, where true seasonality begins and seasonal contrasts in vegetation, soil, and human lifestyle adaptations are evident. More than half the world's population resides in these mesothermal climates. These climates occupy about 27% of Earth's land and sea surface, the second largest percentage behind the tropical climates. Together, the tropical and mesothermal climates dominate over half of Earth's oceans and about one-third of its land area.

The mesothermal climates, and nearby portions of the microthermal climates (cold winters), are regions of great weather variability, for these are the latitudes of greatest air mass interaction. Causal elements include

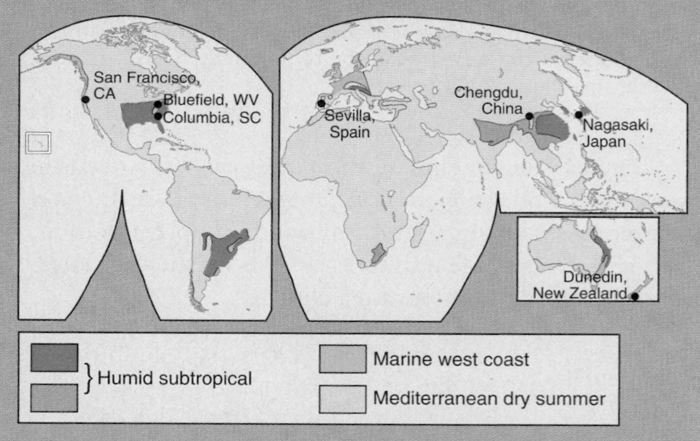

- Shifting of maritime and continental air masses, as they are guided by upper-air westerly winds and undulating Rossby waves and jet streams;
- Migration of cyclonic (low-pressure) and anticyclonic (high-pressure) systems, bringing changeable weather conditions and air mass conflicts;
- Effects of sea-surface temperatures on air mass strength: Cooler temperatures along west coasts weaken air masses, and warmer temperatures along east coasts strengthen air masses; and
- Latitudinal effects on insolation and temperature, as summers transition from hot to warm to cool moving poleward from the tropics.

Mesothermal climates are humid, except where subtropical high pressure produces dry-summer conditions. Their four distinct regimes based on precipitation variability are *humid subtropical hot-summer* (moist all year), *humid subtropical winter-dry* (hot to warm summers, in Asia), *marine west coast* (warm to cool summers, moist all year), and *Mediterranean dry-summer* (warm to hot summers).

Humid Subtropical Climates

Humid subtropical hot-summer climates either are moist all year or have a pronounced winter-dry period, as occurs in eastern and southern Asia. Maritime tropical air masses generated over warm waters off eastern coasts influence *humid subtropical hot-summer* climates during summer. This warm, moist, unstable air produces convectional showers over land. In fall, winter, and spring, maritime tropical and continental polar air masses interact, generating frontal activity and frequent midlatitude cyclonic storms. These two mechanisms produce year-round precipitation. Overall, precipitation averages 100–200 cm (40–80 in.) a year.

Nagasaki, Japan (Figure 10.10), is characteristic of an Asian *humid subtropical hot-summer* station, whereas Columbia, South Carolina, is characteristic of the North American *humid subtropical hot-summer* climate found across the southeastern United States (Figure 10.11). Unlike the winter precipitation of *humid subtropical hot-summer* cities in the United States (Atlanta, Memphis, Norfolk, New Orleans, and Columbia), Nagasaki's winter precipitation is less because of the effects of the Asian monsoon. However, the lower precipitation of winter is not quite dry enough to change its category to a *humid subtropical winter-dry* climate. In comparison to higher rainfall amounts in Nagasaki (196 cm, 77 in.), Columbia's precipitation totals 126.5 cm (49.8 in.; Figure 10.11b); compare this to Atlanta's 122 cm (48 in.) annually.

Humid subtropical winter-dry climates are related to the winter-dry, seasonal pulse of the monsoons. They extend poleward from *tropical savanna* climates and have a summer month that receives 10 times more precipitation than their driest winter month.

A representative station in Asia is Chengdu, China. Figure 10.12 (p. 267) demonstrates the strong correlation between precipitation and the high-summer Sun. In May 2008, this city and the surrounding region were hit with a magnitude 7.9 earthquake, with many strong aftershocks, causing widespread damage, killing 85,000 people, and injuring nearly 400,000 people (Figure 10.12b). This was the worst earthquake to strike China since the 1976 Tangshan quake that killed 250,000 people, discussed in Chapter 12.

The habitability of the *humid subtropical hot-summer* and *humid subtropical winter-dry* climates and their ability to sustain populations are borne out by the concentration of people in north-central India, the bulk of China's 1.3 billion people, and the many who live in climatically similar portions of the United States.

The intense summer rains of the Asian monsoon can cause problems, as they did during many years of the last decade, producing floods in India and Bangladesh. The

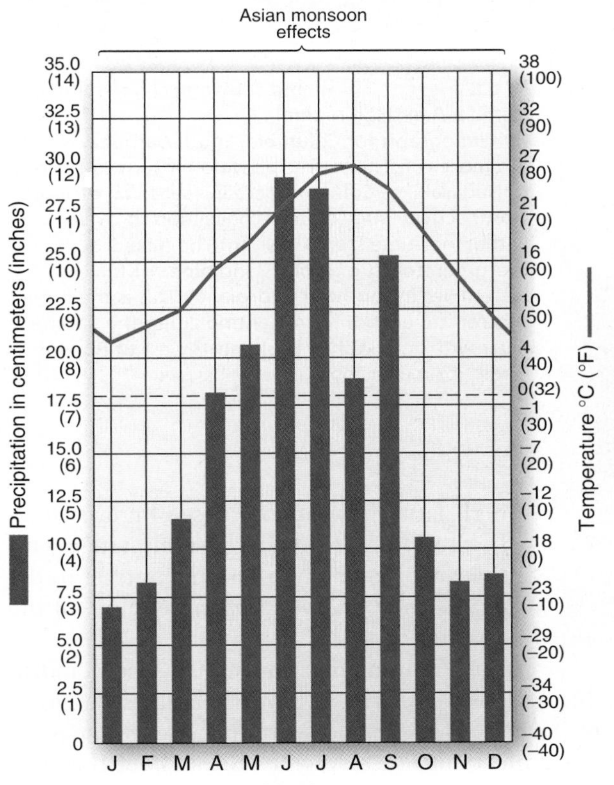

Station: Nagasaki, Japan
Lat/long: 32° 44′ N 129° 52′ E
Avg. Ann. Temp.: 16°C (60.8°F)
Total Ann. Precip.: 195.7 cm (77 in.)
Elevation: 27 m (88.6 ft)
Population: 1,585,000
Ann. Temp. Range: 21 C° (37.8 F°)
Ann. Hr of Sunshine: 2131

(a)

(b)

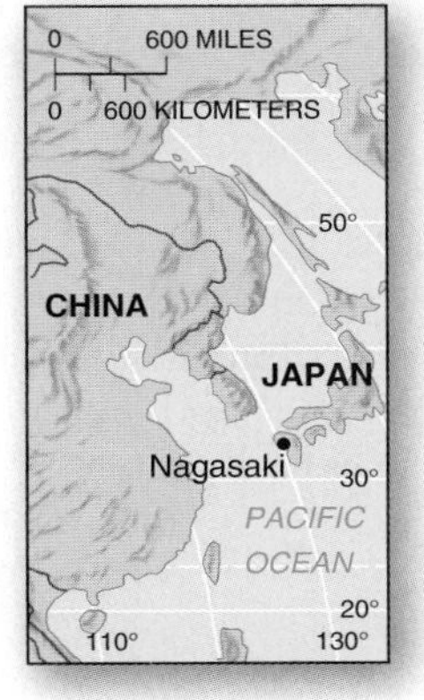

FIGURE 10.10 Humid subtropical hot-summer climate, Asian region. (a) Climograph for Nagasaki, Japan. (b) Landscape on Kitakyujukuri Island, near Nagasaki and Sasebo, Japan, in spring season. [JTB Photo/photolibrary.com.]

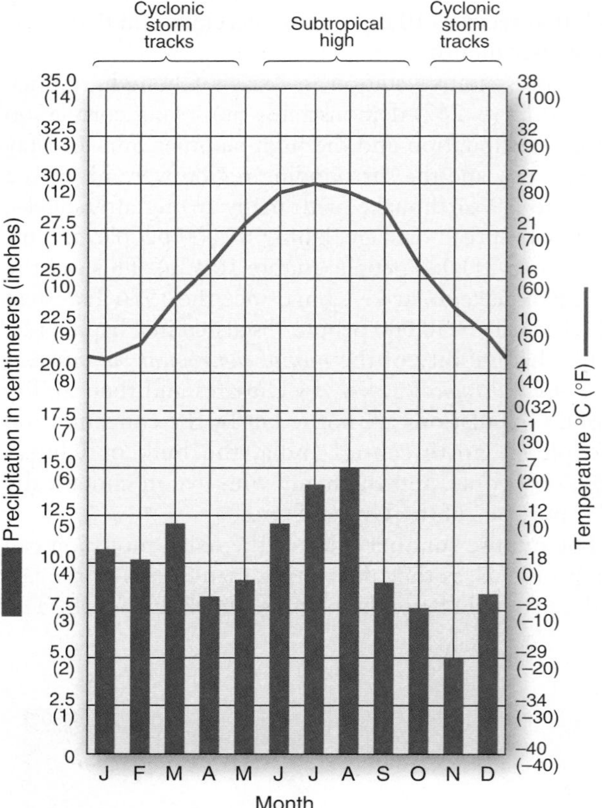

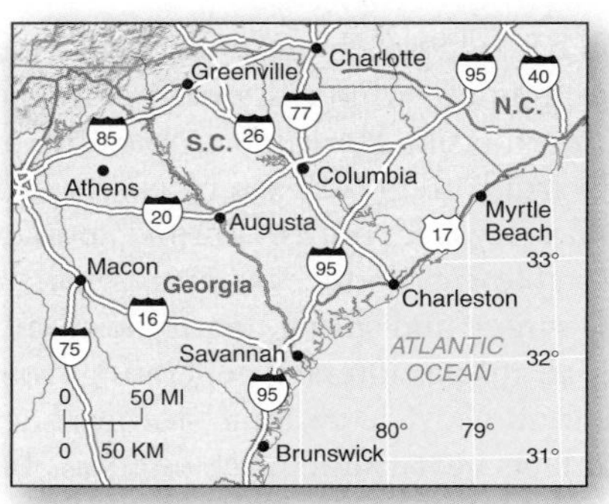

(b)

FIGURE 10.11 Humid subtropical hot-summer climate, American region.
(a) Climograph for Columbia, South Carolina. Note the more consistent precipitation pattern compared to Nagasaki, as Columbia receives seasonal cyclonic storm activity and summer convection showers within maritime tropical air. (b) The mixed evergreen forest of cypress and pine, with water lilies in southern Georgia. (c) This is the humid subtropical portion of Argentina along the Paraná River with a makeshift dock, northwest of Buenos Aires. [(b) and (c) Bobbé Christopherson.]

(c)

monsoonal winter-dry climates hold several precipitation records, as discussed in Chapter 8. Sometimes dramatic thunderstorms and tornadoes are notable in the southeastern United States, with tornado occurrences each year breaking previous records.

Marine West Coast Climates

Marine west coast climates, featuring mild winters and cool summers, dominate Europe and other middle-to-high-latitude west coasts (see Figure 10.6). In the United States, these climates, with their cooler summers, are in contrast to the *humid subtropical hot-summer* climate of the Southeast.

Maritime polar air masses—cool, moist, unstable—dominate *marine west coast* climates. Weather systems forming along the polar front and maritime polar air masses move into these regions throughout the year, making weather quite unpredictable. Coastal fog, annually totaling 30 to 60 days, is a part of the moderating marine influence. Frosts are possible and tend to shorten the growing season.

Marine west coast climates are unusually mild for their latitude. They extend along the coastal margins of the Aleutian Islands in the North Pacific, cover the southern

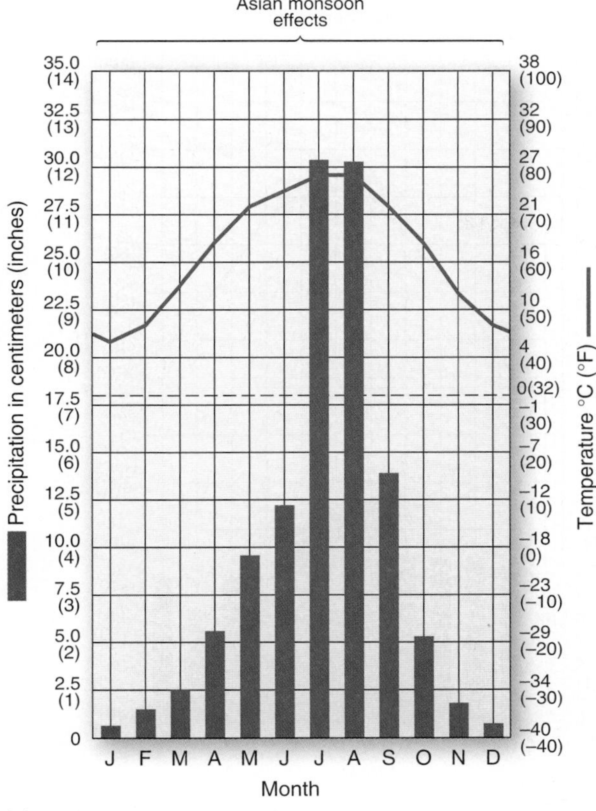

Station: Chengdu, China
Lat/long: 30° 40′ N 104° 04′ E
Avg. Ann. Temp.: 17°C (62.6°F)
Total Ann. Precip.: 114.6 cm (45.1 in.)
Elevation: 498 m (1633.9 ft)
Population: 2,500,000
Ann. Temp. Range: 20 C° (36 F°)
Ann. Hr of Sunshine: 1058

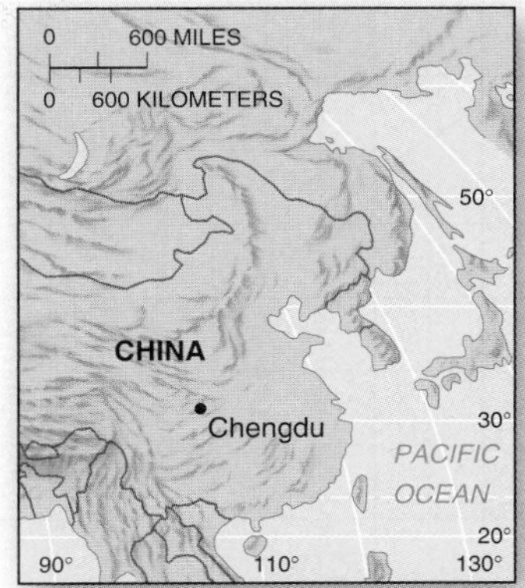

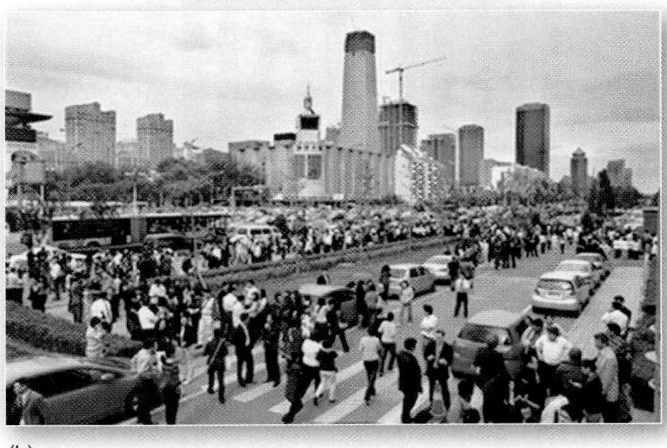

FIGURE 10.12 Humid subtropical winter-dry climate. (a) Climograph for Chengdu, China. Note the summer-wet monsoonal precipitation. (b) People flee high-rise buildings following a major earthquake in Chengdu. [Photo courtesy of *China Daily*.]

third of Iceland in the North Atlantic, and coastal Scandinavia, and dominate the British Isles. Many of us might find it hard to imagine that such high-latitude locations can have average monthly temperatures above freezing throughout the year.

Unlike the extensive influence of *marine west coast* regions in Europe, mountains restrict this climate to coastal environs in Canada, Alaska, Chile, and Australia. The temperate rain forest of Vancouver Island is representative of these moist and cool conditions (Figure 10.13).

The climograph for Dunedin, New Zealand, demonstrates the moderate temperature patterns and the annual temperature range for a *marine west coast* city in the Southern Hemisphere (Figure 10.14).

An interesting anomaly occurs in the eastern United States. In portions of the Appalachian highlands, increased elevation lowers summer temperatures in the surrounding *humid subtropical hot-summer* climate, producing a *marine west coast* cooler summer. The climograph for Bluefield, West Virginia (Figure 10.15), reveals *marine west coast* temperature and precipitation patterns, despite its location in the East. Vegetation similarities between the Appalachians and the Pacific Northwest have enticed many emigrants from the East to settle in these climatically familiar environments in the Northwest.

Mediterranean Dry-Summer Climates

The *Mediterranean dry-summer* climate designation specifies that at least 70% of annual precipitation occurs during the winter months. This is in contrast to the majority

(*text continued on page 270*)

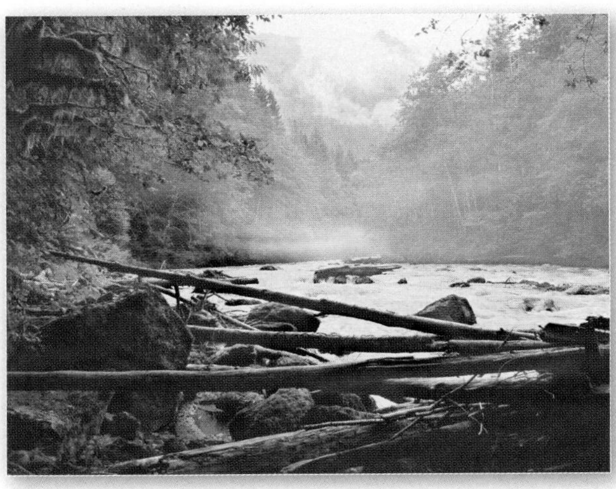

FIGURE 10.13 A marine west coast climate. Temperate rain forest along the Little Qualicum River, central Vancouver Island, adjacent to the Macmillan Provincial Park, British Columbia. [Bobbé Christopherson.]

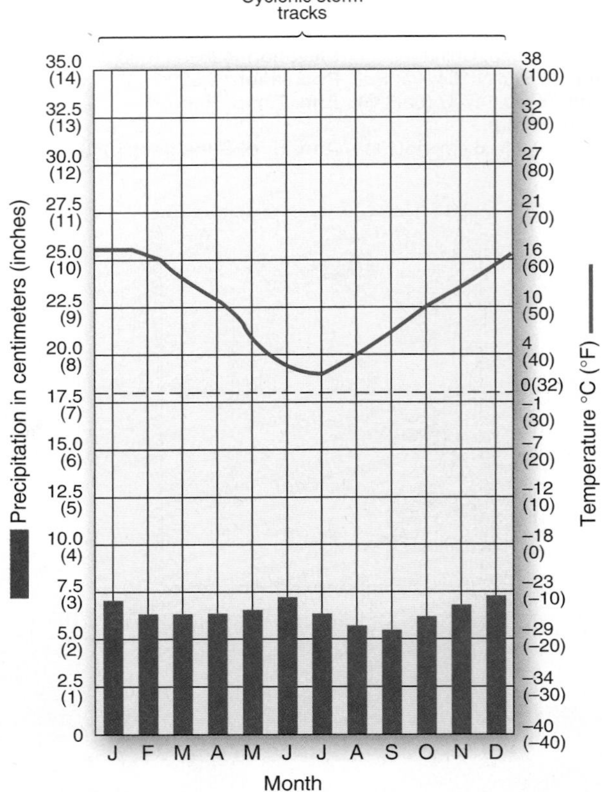

Station: Dunedin, New Zealand
Lat/long: 45° 54′ S 170° 31′ E
Avg. Ann. Temp.: 10.2°C (50.3°F)
Total Ann. Precip.: 78.7 cm (31.0 in.)
Elevation: 1.5 m (5 ft)
Population: 120,000
Ann. Temp. Range: 14.2 C° (25.5 F°)

(a)

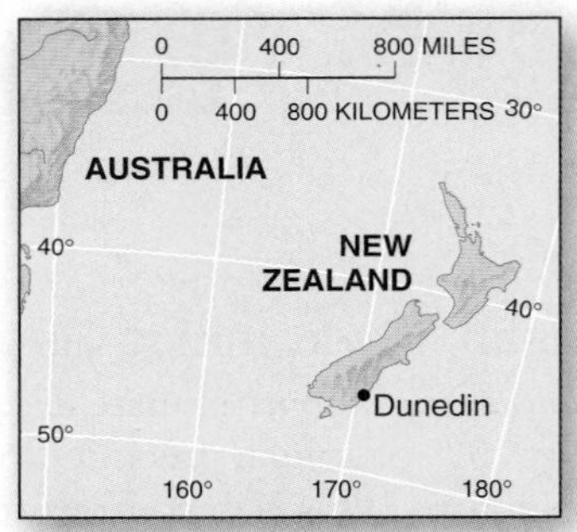

(b)

FIGURE 10.14 A Southern Hemisphere marine west coast climate.
(a) Climograph for Dunedin, New Zealand. (b) Meadow, forest, and mountains on South Island, New Zealand.
[Photo by Brian Enting/Photo Researchers, Inc.]

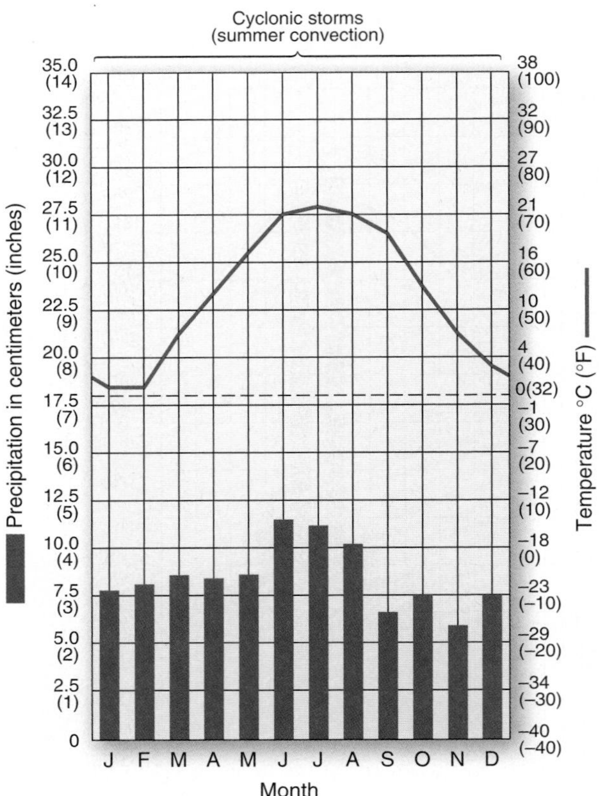

Station: Bluefield, West Virginia
Lat/long: 37° 16′ N 81° 13′ W
Avg. Ann. Temp.: 12°C (53.6°F)
Total Ann. Precip.: 101.9 cm (40.1 in.)
Elevation: 780 m (2559 ft)
Population: 11,000
Ann. Temp. Range: 21 C° (37.8 F°)

(a)

(b)

FIGURE 10.15 Marine west coast climate in the Appalachians of the eastern United States.
(a) Climograph for Bluefield, West Virginia.
(b) Characteristic mixed forest in Appalachia in winter.
[Author.]

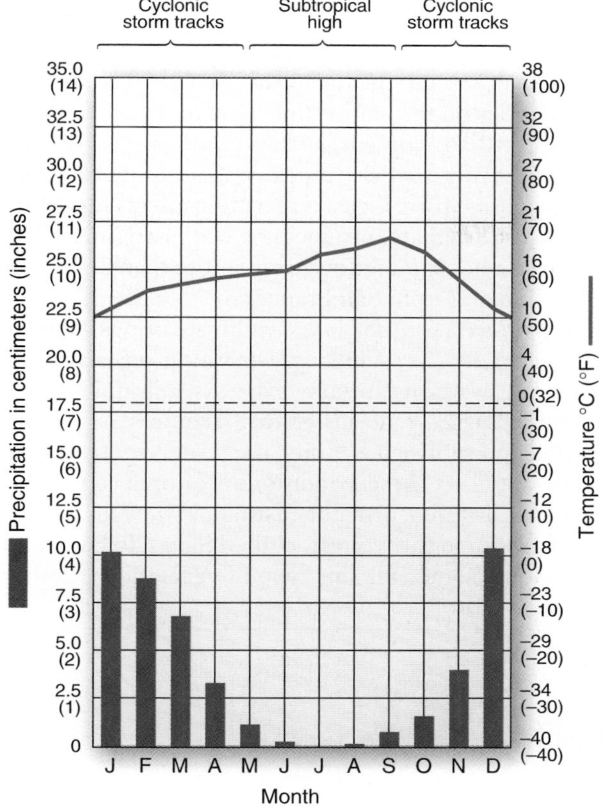

Station: San Francisco, California
Lat/long: 37° 37′ N 122° 23′ W
Avg. Ann. Temp.: 14.6°C (57.2°F)
Total Ann. Precip.: 56.6 cm (22.3 in.)
Elevation: 5 m (16.4 ft)
Population: 777,000
Ann. Temp. Range: 11.4 C° (20.5 F°)
Ann. Hr of Sunshine: 2975

(a)

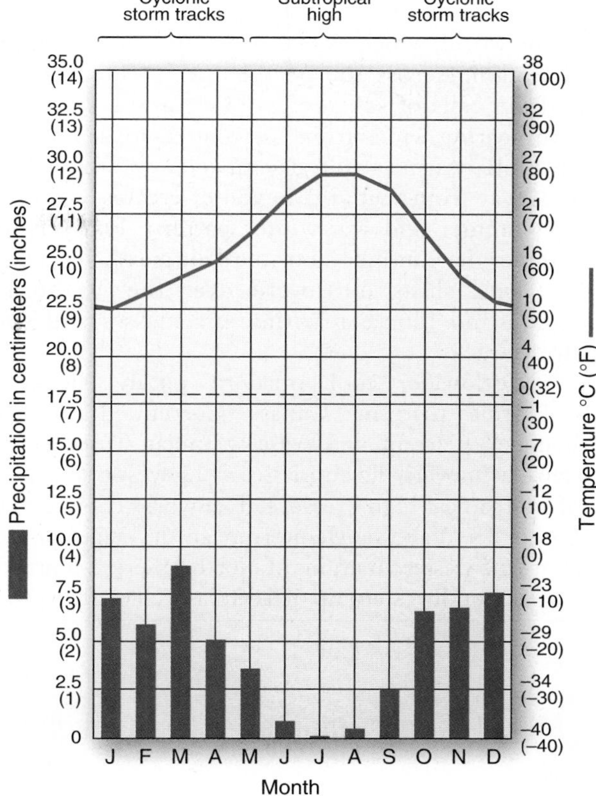

Station: Sevilla, Spain
Lat/long: 37° 22′ N 6° W
Avg. Ann. Temp.: 18°C (64.4°F)
Total Ann. Precip.: 55.9 cm (22 in.)
Elevation: 13 m (42.6 ft)
Population: 1,764,000
Ann. Temp. Range: 16 C° (28.8 F°)
Ann. Hr of Sunshine: 2862

(b)

(c)

(d)

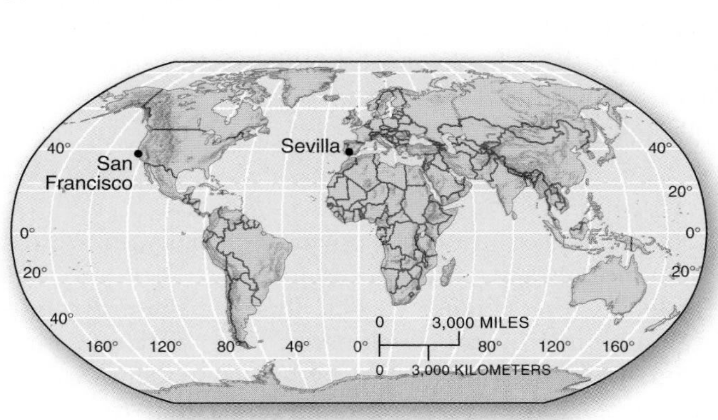

FIGURE 10.16 Mediterranean climates, California and Spain.
Climographs for (a) San Francisco, California, with its cooler dry summer, and (b) Sevilla, Spain, with its hotter dry summer. (c) Central California Mediterranean landscape of oak savanna. (d) Near Sevilla, Spain, the El Penon mountains in the distance. [(c) Bobbé Christopherson. (d) Michael Thornton/Design Pics/Corbis.]

of the world, which experiences summer-maximum precipitation. Across the planet during summer months, shifting cells of subtropical high pressure block moisture-bearing winds from adjacent regions. This shifting of stable, warm to hot, dry air over an area in summer and away from that area in winter creates a pronounced dry-summer and wet-winter pattern. For example, in summer the continental tropical air mass over the Sahara in Africa shifts northward over the Mediterranean region and blocks maritime air masses and cyclonic storm tracks.

Worldwide, cool offshore ocean currents (the California Current, Canary Current, Peru Current, Benguela Current, and West Australian Current) produce stability in overlying air masses along west coasts, poleward of subtropical high pressure. The world climate map (see Figure 10.6) shows Mediterranean dry-summer climates along the western margins of North America, central Chile, and the southwestern tip of Africa as well as across southern Australia and the Mediterranean Basin—the climate's namesake region. Examine the offshore currents along each of these regions on the world climate map.

Figure 10.16 compares the climographs of the *Mediterranean dry-summer* cities of San Francisco and Sevilla (Seville), Spain. Coastal maritime effects moderate San Francisco's climate, producing a cooler summer. The transition to a hot summer occurs no more than 24–32 km (15–20 mi) inland from San Francisco.

The *Mediterranean dry-summer* climate brings summer water-balance deficits. Winter precipitation recharges soil moisture, but water use usually exhausts soil moisture by late spring. Large-scale agriculture requires irrigation, although some subtropical fruits, nuts, and vegetables are uniquely suited to these conditions. Natural vegetation features a hard-leafed, drought-resistant variety known locally as *chaparral* in the western United States. (Chapter 20 discusses local names for this type of vegetation in other parts of the world.)

Microthermal Climates (mid and high latitudes, cold winters)

Humid microthermal climates have a winter season with some summer warmth. Here the term *microthermal* means cool temperate to cold. Approximately 21% of Earth's land surface is influenced by these climates, equaling about 7% of Earth's total surface.

These climates occur poleward of the mesothermal climates and experience great temperature ranges related to continentality and air mass conflicts. Temperatures decrease with increasing latitude and toward the interior of continental landmasses and result in intensely cold winters. In contrast to moist-all-year regions (northern tier across the United States and Canada, eastern Europe through the Ural Mountains) is the winter-dry association with the Asian dry monsoon and cold air masses across the region.

In Figure 10.6, note the absence of microthermal climates in the Southern Hemisphere. Because the Southern Hemisphere lacks substantial landmasses, microthermal climates develop only in highlands. Important causal elements include

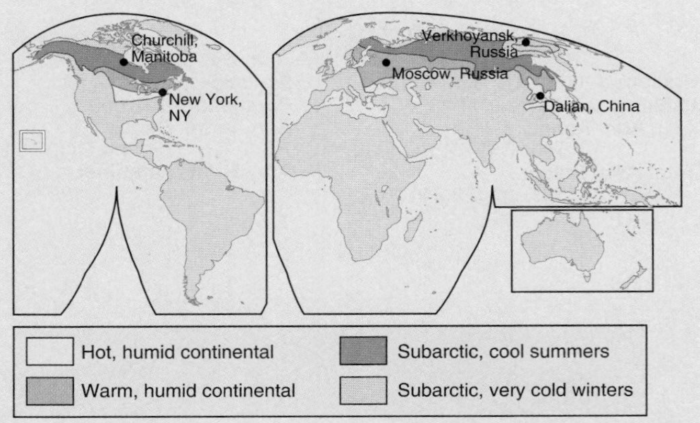

- Increasing seasonality (daylength and Sun altitude) and greater temperature ranges (daily and annually);
- Upper-air westerly winds and undulating Rossby waves, which bring warmer air northward and colder air southward for cyclonic activity, and convectional thunderstorms from mT air masses in summer;
- An Asian winter-dry pattern associated with continental high pressure and related air masses, increasing east of the Ural Mountains to the Pacific Ocean and eastern Asia;
- Hot summers cooling northward, with insolation and net radiation declines from the mesothermal climates, and short spring and fall seasons surrounding winters that are cold to very cold; and
- Continental interiors serving as source regions for intense continental polar air masses that dominate winter, blocking cyclonic storms.

Microthermal climates have four distinct regimes based on increasing cold with latitude and precipitation variability: *humid continental hot-summer* (Chicago, New York); *humid continental mild-summer* (Duluth, Toronto, Moscow); *subarctic cool-summer* (Churchill, Manitoba); and the formidable extremes of frigid *subarctic with very cold winters* (Verkhoyansk and northern Siberia).

Humid Continental Hot-Summer Climates

Humid continental hot-summer climates are differentiated by their annual precipitation distribution. In the summer, maritime tropical air masses influence both *humid continental moist-all-year* and *humid continental winter-dry* climates. In North America, frequent weather activity is possible between conflicting air masses—maritime tropical and continental polar—especially in winter. The climographs and photos for New York City and Dalian, China (Figure 10.17), illustrate these two hot-summer microthermal climates. The Dalian, China, climograph demonstrates a dry-winter tendency caused by the intruding cold continental air that forces dry monsoon conditions.

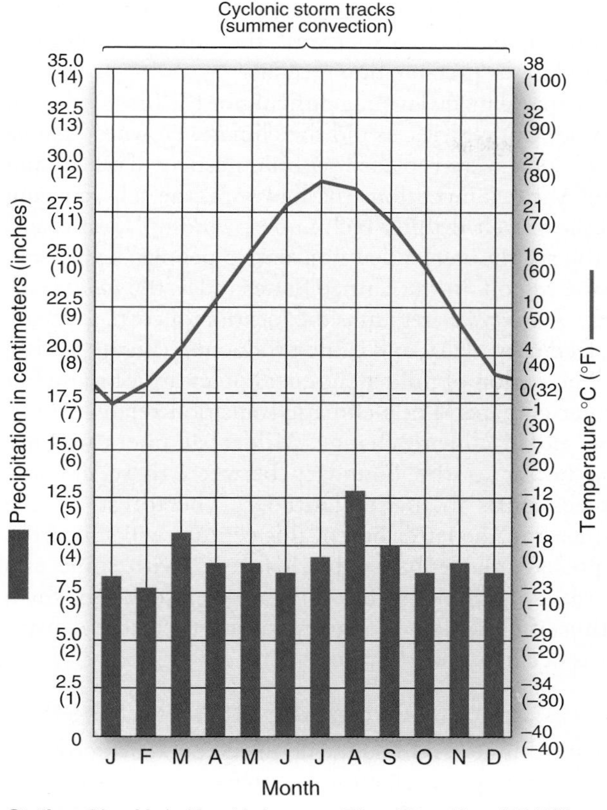

Station: New York, New York
Lat/long: 40° 46′ N 74° 01′ W
Avg. Ann. Temp.: 13°C (55.4°F)
Total Ann. Precip.: 112.3 cm (44.2 in.)
Elevation: 16 m (52.5 ft)
Population: 8,092,000
Ann. Temp. Range: 24 C° (43.2 F°)
Ann. Hr of Sunshine: 2564

(a)

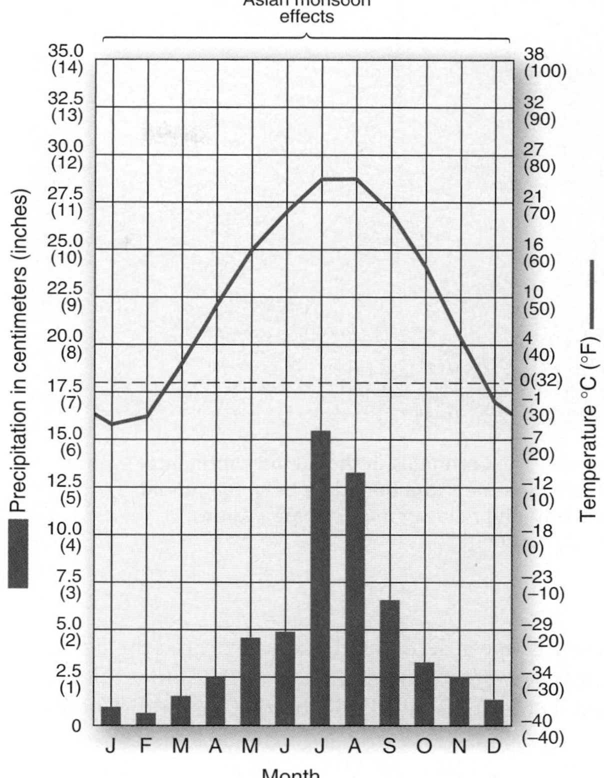

Station: Dalian, China
Lat/long: 38° 54′ N 121° 54′ E
Avg. Ann. Temp.: 10°C (50°F)
Total Ann. Precip.: 57.8 cm (22.8 in.)
Elevation: 96 m (314.9 ft)
Population: 5,550,000
Ann. Temp. Range: 29 C° (52.2 F°)
Ann. Hr of Sunshine: 2762

(b)

(c)

(d)

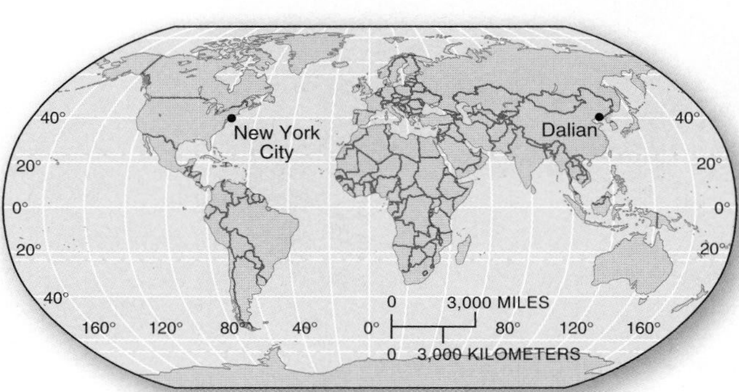

FIGURE 10.17 Humid continental hot-summer climates, New York and China.
Climographs for (a) New York City (humid continental hot-summer climate that is moist all year) and (b) Dalian, China (humid continental hot-summer climate that is dry in winter). (c) Belvedere Castle, 1872, in New York's Central Park; location of weather station from 1919 to 1960. (d) Dalian, China, cityscape and park in summer. [(c) Bobbé Christopherson. (d) Courtesy of the Paul Louis collection.]

FIGURE 10.18 Cornfields in the humid continental prairie. East of Minneapolis near the boundary between humid continental hot- and mild-summer climate regions. [Bobbé Christopherson.]

Before European settlement, forests covered the *humid continental hot-summer* climatic region of the United States as far west as the Indiana–Illinois border. Beyond that approximate line, tall-grass prairies extended westward to about the 98th meridian (98° W in central Kansas) and the approximate location of the 51-cm (20-in.) *isohyet* (a line of equal precipitation). Farther west, the short-grass prairies reflected lower precipitation receipts.

Deep sod made farming difficult for the first settlers of the American prairies, as did the climate. However, native grasses soon were replaced with domesticated wheat and barley. Various inventions (barbed wire, the self-scouring steel plow, well-drilling techniques, windmills, railroads, and the six-shooter) aided non-native peoples' expansion into the region. In the United States today, the *humid continental hot-summer* region is the location of corn, soybean, hog, feed crop, dairy, and cattle production (Figure 10.18).

The region of milder summers poleward of these hot summer climates is in a climatic transition related to ongoing global climate change. Milder summers will become hotter as the boundary between these climatic subtypes shifts to higher latitudes. The forests of the Northeast in the latter half of this century will shift from the present maple–beech–birch forests, with some elm and ash, to a dominant oak–hickory–hemlock mix. Such shifting of climatic boundaries and their related ecosystems is occurring worldwide.

Humid Continental Mild-Summer Climates

Soils are thinner and less fertile in the cooler microthermal climates, yet agricultural activity is important and

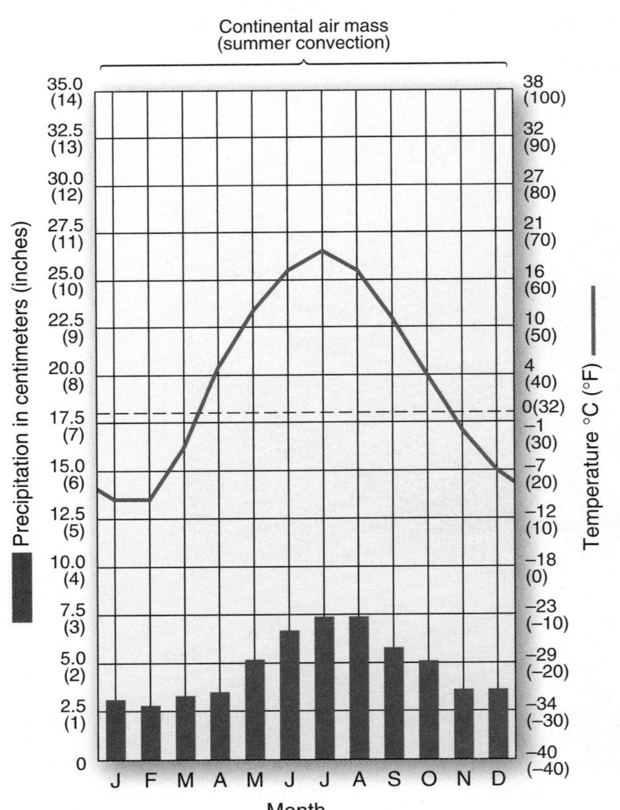

Station: Moscow, Russia
Lat/long: 55° 45′ N 37° 34′ E
Avg. Ann. Temp.: 4°C (39.2°F)
Total Ann. Precip.: 57.5 cm (22.6 in.)
Elevation: 156 m (511.8 ft)
Population: 11,460,000
Ann. Temp. Range: 29 C° (52.2 F°)
Ann. Hr of Sunshine: 1597

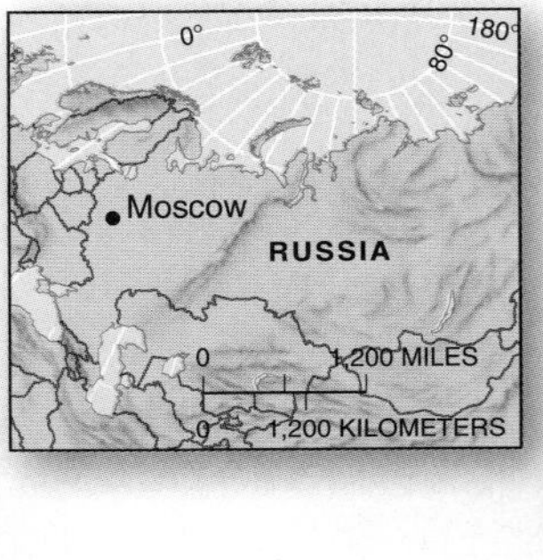

(b)

(c)

FIGURE 10.19 Humid continental mild-summer climate.
(a) Climograph for Moscow, Russia. (b) Landscape between Moscow and St. Petersburg along the Volga River. (c) Winter scene of pathway through mixed forest near Brunswick, Maine, in the North American region with this climate type. [(b) Dave G. Houser/Corbis; (c) Bobbé Christopherson.]

includes dairy, poultry, flax, sunflower, sugar beet, wheat, and potato production. Frost-free periods range from fewer than 90 days in the north to as many as 225 days in the south. Overall, precipitation is less than in the hot-summer regions to the south; however, notably heavier snowfall is important to soil-moisture recharge when it melts. Various snow-capturing strategies are in use, including fences and tall stubble left standing after harvest in fields to create snowdrifts and thus more moisture retention on the soil.

Characteristic cities are Duluth, Minnesota, and Saint Petersburg, Russia. Figure 10.19 presents a climograph for Moscow, which is at 55° N, or about the same latitude as the southern shore of Hudson Bay in Canada.

The dry-winter aspect of the mild-summer climate occurs only in Asia, in a far-eastern area poleward of the winter-dry mesothermal climates. A representative *humid continental mild-summer* climate along Russia's east coast is Vladivostok, usually one of only two ice-free ports in that country.

Subarctic Climates

Farther poleward, seasonal change becomes greater. The short growing season is more intense during long summer days. The *subarctic* climates include vast stretches of Alaska, Canada, and northern Scandinavia with their cool summers and Siberian Russia with its very cold winters (Figure 10.20). Discoveries of minerals and petroleum reserves and the Arctic Ocean sea-ice losses have led to new

Continental air mass

Precipitation in centimeters (inches): 35.0 (14), 32.5 (13), 30.0 (12), 27.5 (11), 25.0 (10), 22.5 (9), 20.0 (8), 17.5 (7), 15.0 (6), 12.5 (5), 10.0 (4), 7.5 (3), 5.0 (2), 2.5 (1), 0

Temperature °C (°F): 38 (100), 32 (90), 27 (80), 21 (70), 16 (60), 10 (50), 4 (40), 0(32), −1 (30), −7 (20), −12 (10), −18 (0), −23 (−10), −29 (−20), −34 (−30), −40 (−40)

J F M A M J J A S O N D

Month

(a)

Station: Churchill, Manitoba
Lat/long: 58° 45′ N 94° 04′ W
Avg. Ann. Temp.: −7°C (19.4°F)
Total Ann. Precip.: 44.3 cm (17.4 in.)
Elevation: 35 m (114.8 ft)
Population: 1400
Ann. Temp. Range: 40 C° (72 F°)
Ann. Hr of Sunshine: 1732

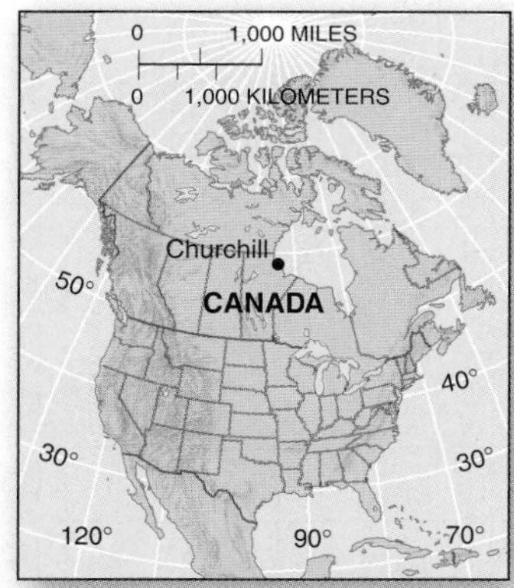

(b)

(c)

FIGURE 10.20 Subarctic cool-summer climate. (a) Climograph for Churchill, Manitoba. (b) Winter scene in Churchill showing the port facilities on Hudson Bay. (c) Polar bear mom and older cub taking protective shelter near Hudson Bay, waiting for the ice. [Bobbé Christopherson.]

interest in portions of these regions, such as the port expansion in Churchill (see Figure 10.20b).

Areas that receive 25 cm (10 in.) or more of precipitation a year on the northern continental margins and are covered by the so-called snow forests of fir, spruce, larch, and birch are the *boreal forests* of Canada and the *taiga* of Russia. These forests are in transition to the more open northern woodlands and to the tundra region of the far north. Forests thin out to the north when the warmest month drops below an average temperature of 10°C (50°F). During the decades ahead, the boreal forests will shift northward into the tundra in response to climate change with higher temperatures.

Soils are thin in these lands once scoured by glaciers. Precipitation and potential evapotranspiration both are low, so soils are generally moist and either partially or totally frozen beneath the surface, a phenomenon known as *permafrost* (discussed in Chapter 17).

The Churchill, Manitoba, climograph (Figure 10.20) shows average monthly temperatures below freezing for 7 months of the year, during which time light snow cover and frozen ground persist. High pressure dominates Churchill during its cold winter—this is the source region for the continental polar air mass. Churchill is representative of the *subarctic cool-summer* climate, with an annual temperature range of 40 C° (72 F°) and low precipitation of 44.3 cm (17.4 in.).

The *subarctic* climates that feature a dry and very cold winter occur only within Russia. The intense cold of Siberia and north-central and eastern Asia is difficult to comprehend, for these areas experience an average temperature lower than freezing for 7 months; minimum temperatures of below –68°C (–90°F) were recorded there, as described in Chapter 5. Yet summer maximum temperatures in these same areas can exceed 37°C (98°F).

An example of this extreme *subarctic climate with very cold winters* is Verkhoyansk, Siberia (Figure 10.21). For 4 months of the year, average temperatures fall below –34°C (–30°F). Verkhoyansk has probably the world's greatest annual temperature range from winter to summer: a remarkable 63 C° (113.4 F°). Winters feature brittle metals and plastics, triple-thick windowpanes, and temperatures that render straight antifreeze a solid.

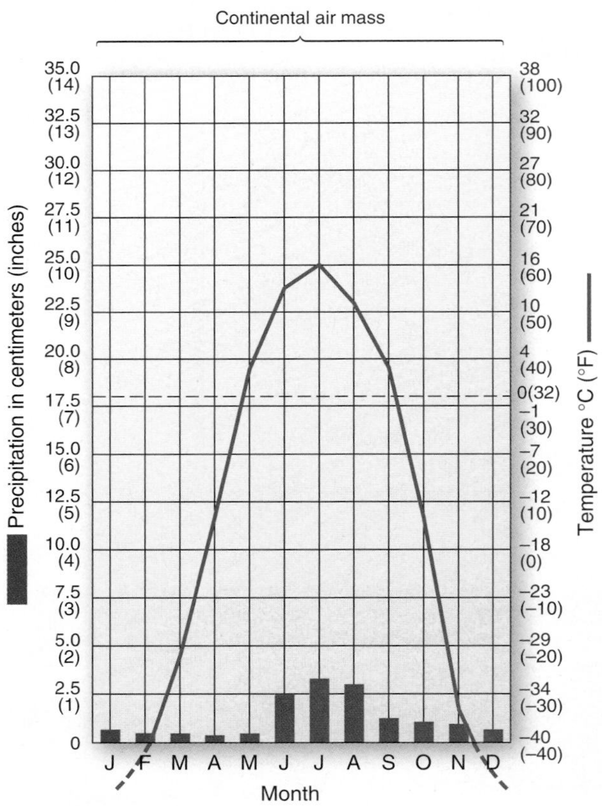

(a)

Station: Verkhoyansk, Russia
Lat/long: 67° 35′ N 133° 27′ E
Avg. Ann. Temp.: –15°C (5°F)
Total Ann. Precip.: 15.5 cm (6.1 in.)
Elevation: 137 m (449.5 ft)
Population: 1500
Ann. Temp. Range: 63 C° (113.4 F°)

(b)

FIGURE 10.21 Extreme subarctic cold winter climate.
(a) Climograph for Verkhoyansk, Russia. (b) A summer scene in this unique continental location; the top layers of permafrost thaw, creating many ponds. [Dean Conger/Corbis.]

Polar and Highland Climates

The polar climates have no true summer like that in lower latitudes. Dominating the South Pole is the Antarctic continent, surrounded by the Southern Ocean, whereas the North Pole region is the Arctic Ocean, surrounded by the continents of North America and Eurasia. Poleward of the Arctic and Antarctic circles, daylength increases in summer until daylight becomes continuous, yet average monthly temperatures never rise above 10°C (50°F). These temperature conditions do not allow tree growth. See the polar region temperature maps for January and July in Figure 5.14 and Figure 5.17. Principal causal elements in these frozen and barren regions are the following:

- Low Sun altitude even during the long summer days, which is the principal climatic factor;
- Extremes of daylength between winter and summer, which determine the amount of insolation received;
- Extremely low humidity, producing low precipitation amounts—Earth's frozen deserts; and
- Surface albedo impacts, as light-colored surfaces of ice and snow reflect substantial energy away from the ground, thus reducing net radiation.

Polar climates have three regimes: *tundra* (high latitude or high elevation); *ice caps* and *ice sheets* (perpetually frozen); and *polar marine* (oceanic association, slight moderation of extreme cold).

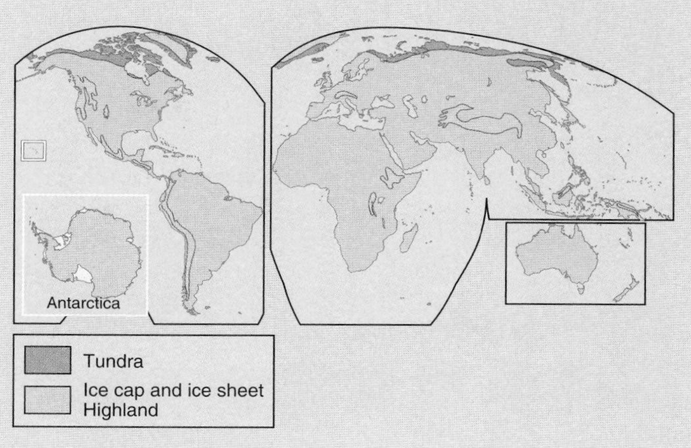

Also in this climate category we include highland climates. Even at low latitudes, the effects of elevation can produce tundra and polar conditions. Glaciers on tropical mountain summits attest to the cooling effects of altitude. Highland climates on the map follow the pattern of Earth's mountain ranges.

Tundra Climate

In a *tundra* climate, land is under continuous snow cover for 8–10 months, with the warmest month above 0°C, yet never warming above 10°C (50°F). Because of elevation, the summit of Mount Washington in New Hampshire (1914 m, or 6280 ft) statistically qualifies as a highland *tundra* climate on a small scale. On a larger scale, approximately 410,500 km^2 (158,475 mi^2) of Greenland are ice-free, an area of tundra and rock about the size of California.

In spring when the snow melts, numerous plants appear—stunted sedges, mosses, flowering plants, and lichens—and persist through the short summer (Figure 10.22a). Some of the dwarf willows (7.5 cm-, or 3-in., tall) can exceed 300 years in age. Much of the area experiences permafrost and ground ice conditions; these are Earth's periglacial regions (see Chapter 17).

Global warming is bringing dramatic changes to the tundra. In parts of Canada and Alaska, registered temperatures as much as 5 C° (9 F°) above average are a regular occurrence, setting many records. As organic peat deposits in the tundra thaw, vast stores of carbon are released to the atmosphere, further adding to the greenhouse gas problem (see *Geosystems Now* for Chapter 18.). Temperatures in the Arctic are warming at a rate twice that of the global average increase.

Tundra climates are strictly in the Northern Hemisphere, except for elevated mountain locations in the Southern Hemisphere and a portion of the Antarctic Peninsula.

Ice-Cap and Ice-Sheet Climate

Earth's two ice sheets cover the Antarctic continent and most of the island of Greenland. Thus, most of Antarctica and central Greenland fall within the *ice-sheet* climate, as does the North Pole, with all months averaging below freezing. These regions are dominated by dry, frigid air masses, with vast expanses that never warm above freezing. In fact, winter minimum temperatures in central Antarctica (July) frequently drop below the temperature of solid carbon dioxide or "dry ice" (–78°C, or –109°F).

The area of the North Pole is actually a sea covered by ice, whereas Antarctica is a substantial continental landmass. Antarctica is constantly snow-covered, but receives less than 8 cm (3 in.) of precipitation each year. However, Antarctic ice has accumulated to several kilometers deep and is the largest repository of freshwater on Earth.

Ice caps are smaller in extent than ice sheets, roughly less than 50,000 km^2 (19,300 mi^2), yet they completely bury the landscape like an ice sheet. The Vatnajökull Ice Cap in southeastern Iceland is an example.

Figure 10.23 shows two scenes of the repositories of multiyear ice on the Antarctic and Greenland ice sheets. This ice contains a vast historical record of Earth's atmosphere. Within it, evidence of thousands of past volcanic eruptions from all over the world and ancient combinations of atmospheric gases lies trapped in frozen bubbles. Chapter 17 presents analysis of ice cores taken from

(a)

(b)

FIGURE 10.22 Greenland tundra and a research settlement.
(a) Late September in East Greenland with fall colors and muskoxen. (b) Utilidors (raised shelters that carry water and sewage) and housing in the scientific settlement of Ny-Ålesund, Spitsbergen (Svalbard). [Bobbé Christopherson.]

Greenland and the latest one taken from Antarctica, which pushed the climate record to 800,000 years before the present.

Polar Marine Climate

Polar marine stations are more moderate than other polar climates in winter, with no month below –7°C (20°F), yet they are not as warm as *tundra* climates. Because of marine influences, annual temperature ranges are low. This climate exists along the Bering Sea, on the southern tip of Greenland, and in northern Iceland, Norway, and the Southern Hemisphere, generally over oceans between 50° S and 60° S. Macquarie Island at 54° S in the Southern Ocean, south of New Zealand, is *polar marine*.

South Georgia Island is *polar marine*, made famous as the place where Earnest Shackelton sought rescue help for him and his men in 1916 after their failed Antarctic expedition (Figure 10.24). Although the island is in the Southern Ocean and part of Antarctica, the annual temperature range is only 8.5 C° (15.3 F°) between the seasons (average January is 7°C and July is –1.5°C; 44.6°F and 29.3°F), with 7 months averaging slightly above freezing. Although cold at between 0°C and 4°C, the ocean temperatures moderate the climate despite its location at 54° S latitude. Average precipitation is 150 cm (59 in.), and it can snow during any month.

(a)

(b)

FIGURE 10.23 Earth's ice sheets—Antarctica and Greenland.
These are Earth's frozen freshwater reservoirs. (a) In the Antarctic Sound, between Bransfield Strait and the Weddell Sea, mountains rise from the mist behind a flank of tabular icebergs. (b) Glaciers move toward the ocean into Davis Strait from West Greenland and the ice sheet on the horizon. [Bobbé Christopherson.]

FIGURE 10.24 **South Georgia Island, a polar marine climate.** Here is the abandoned whaling station at Grytviken, South Georgia, processing more than 50,000 whales between 1904 and 1964, about a third of all whale processed on South Georgia among seven factories. The whales of the Southern Ocean were driven almost to extinction. [Bobbé Christopherson.]

Dry Climates (permanent moisture deficits)

For understanding the dry climates, we consider moisture efficiency along with temperature. These dry regions occupy more than 35% of Earth's land area and are by far the most extensive climate over land. Sparse vegetation leaves the landscape bare; water demand exceeds the precipitation water supply throughout, creating permanent water deficits. The extent of these deficits distinguishes two types of dry climatic regions: *arid deserts*, where precipitation supply is roughly less than one-half of the natural moisture demand; and *semiarid steppes*, where the precipitation supply is roughly more than one-half of natural moisture demand. (See specific annual and daily desert temperature regimes, including the highest record temperatures and surface energy budgets in Chapters 4 and 5; desert landscapes in Chapter 15; and desert environments in Chapter 20.)

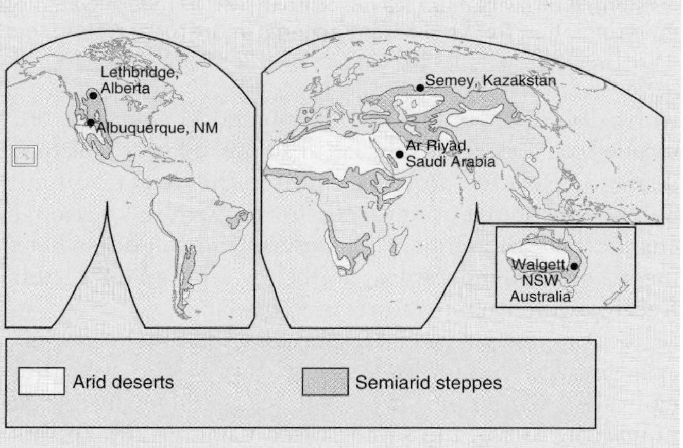

Important causal elements in these dry lands include

- The dominant presence of dry, subsiding air in subtropical high-pressure systems;
- Location in the rain shadow (or leeward side) of mountains, where dry air subsides after moisture is intercepted on the windward slopes;
- Location in continental interiors, particularly central Asia, which are far from moisture-bearing air masses; and
- Shifting subtropical high-pressure systems, which produce semiarid steppes around the periphery of arid deserts.

Dry climates are distributed by latitude and the amount of moisture deficits in four distinct regimes: arid climates include *tropical, subtropical hot desert*, and *midlatitude cold desert*; semiarid climates include *tropical, subtropical hot steppe*, and *midlatitude cold steppe*.

Characteristics of Dry Climates

The world climate map in Figure 10.6 reveals the pattern of Earth's dry climates, which cover broad regions between 15° and 30° N and S latitude. In these areas, subtropical high-pressure cells predominate, with subsiding, stable air and low relative humidity (Figure 10.25). Under generally cloudless skies, these subtropical deserts extend to western continental margins, where cool, stabilizing ocean currents operate offshore and summer advection fog forms. The Atacama Desert of Chile, the Namib Desert of Namibia, the Western Sahara of Morocco, and the Australian Desert each lie adjacent to such a coastline.

Orographic lifting intercepts moisture-bearing weather systems to create rain shadows along mountain ranges that extend these dry regions into higher latitudes. Note these rain shadows in North and South America on the climate map. The isolated interior of Asia, far distant from any moisture-bearing air masses, also falls within the dry climate classification.

Dry climates are subdivided into deserts and steppes, which are distinguished by the amount of moisture deficit; deserts have greater moisture deficits than do steppes, but both have permanent water shortages. Important is whether precipitation falls principally in the winter with a dry summer, falls principally in the summer

(a)

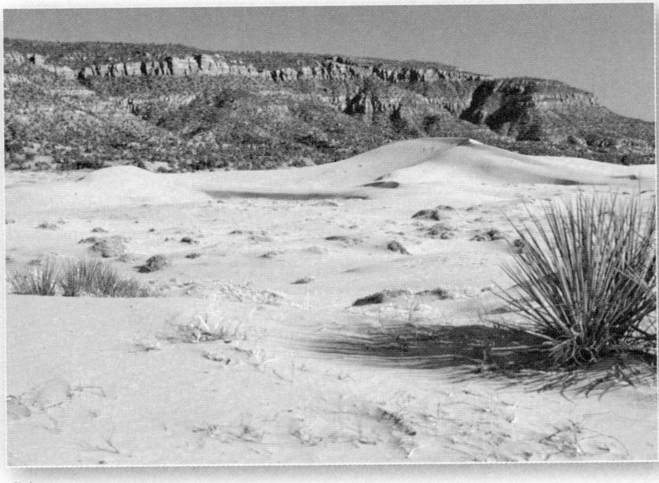

(b)

FIGURE 10.25 Desert landscapes.
(a) Desert vegetation such as these plants in the Anza-Borrego Desert is typically *xerophytic*: drought-resistant, with waxy, hard leaves, or no leaves, to reduce water loss by transpiration. (b) Sand dunes get their coloration from the parent material in the rocky cliffs in southern Utah. [Bobbé Christopherson.]

with a dry winter, or is evenly distributed. Winter rains are most effective because they fall at a time of lower moisture demand. Relative to temperature, the lower-latitude deserts and steppes tend to be hotter with less seasonal change than the midlatitude deserts and steppes, where mean annual temperatures are below 18°C (64.4°F) and freezing winter temperatures are possible.

A **steppe** is a regional term referring to the vast semiarid grassland biome of Eastern Europe and Asia (the equivalent biome in North America is called shortgrass prairie; in Africa, the savanna; see Chapter 20). In this chapter, we use steppe in a climatic context; a steppe climate is considered too dry to support forest, but too moist to be a desert.

Tropical, Subtropical Hot Desert Climates

Tropical, subtropical hot desert climates are Earth's true tropical and subtropical deserts and feature annual average temperatures above 18°C (64.4°F). They generally are found on the western sides of continents, although Egypt, Somalia, and Saudi Arabia also fall within this classification. Rainfall is from local summer convectional showers. Some regions receive almost no rainfall, whereas others may receive up to 35 cm (14 in.) of precipitation a year. A representative *subtropical hot desert* city is Ar Riyāḑ (Riyadh), Saudi Arabia (Figure 10.26).

Along the Sahara's southern margin is a drought-tortured region, the Sahel. Human populations suffered great hardship as desert conditions gradually expanded over their homelands. The arid environment sets the stage for a rugged lifestyle and subsistence economies, such as on volcanic Fogo Island in Cape Verde, a West African country (Figure 10.26c). *Geosystems Now* in Chapter 15 presents the process of desertification (expanding desert conditions).

Death Valley, California (Figure 10.1), features a *subtropical hot desert* climate with an average annual temperature of 24.4°C (76°F). July and August average temperatures are 46°C and 45°C (115°F, 113°F), respectively. Temperatures over 50°C (122°F) are not uncommon.

In recent years, the media focused on the Iraq war and related political events. What seemed overlooked in many reports was the fact that the air temperatures in Baghdad (located on the map in Figure 10.26) were actually higher than those in Death Valley during some days in July and August. Soldiers and civilians experience temperatures of 50°C (122°F) and higher in the city. Such readings broke records for Baghdad in 2007. In January, averages for Death Valley (11°C; 52°F) and Baghdad (9.4°C; 49°F) are comparable. Death Valley is drier, with 5.9 cm of precipitation, compared to 14 cm in Baghdad (2.33 in.; 5.5 in.), both low amounts. Baghdad's May to September period is remarkable, with zero precipitation, as it is dominated by an intense subtropical high-pressure system. Keep these hot desert climates in mind as you follow events in the news.

Midlatitude Cold Desert Climates

Midlatitude cold desert climates cover only a small area: the countries along the southern border of Russia, the Gobi Desert, and Mongolia in Asia; the central third of Nevada and areas of the American Southwest, particularly at high elevations; and Patagonia in Argentina. Because of lower temperatures and lower moisture-demand values, rainfall must be low for a station to qualify as a *midlatitude cold desert* climate; consequently, total annual average rainfall is only about 15 cm (6 in.).

A representative station is Albuquerque, New Mexico, with 20.7 cm (8.1 in.) of precipitation and an annual average temperature of 14°C (57.2°F) (Figure 10.27). Note the precipitation increase from summer convectional showers on the climograph. A characteristic expanse of *midlatitude cold desert* stretches across central Nevada, over

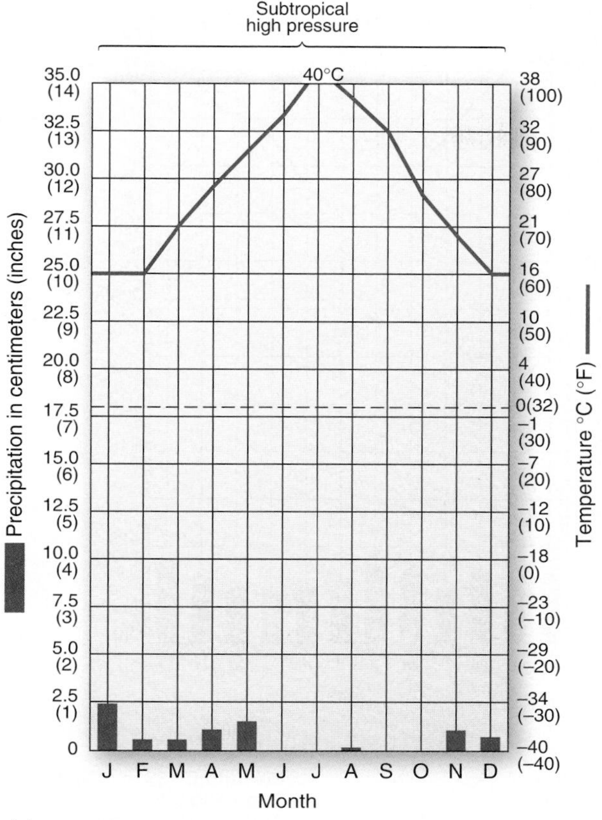

Station: Ar Riyāḍ (Riyadh), Saudi Arabia
Lat/long: 24° 42′ N 46° 43′ E
Avg. Ann. Temp.: 26°C (78.8°F)
Total Ann. Precip.: 8.2 cm (3.2 in.)
Elevation: 609 m (1998 ft)
Population: 5,024,000
Ann. Temp. Range: 24 C° (43.2 F°)

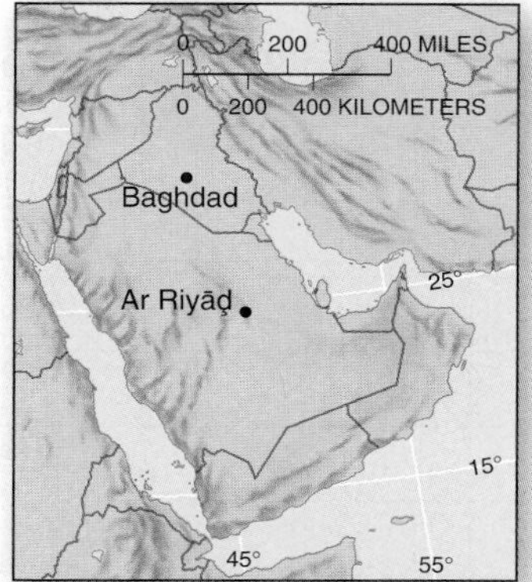

(b)

(c)

FIGURE 10.26 Tropical, subtropical hot desert climate. (a) Climograph for Ar Riyāḍ (Riyadh), Saudi Arabia. (b) The Arabian desert landscape of Red Sands near Ar Riyāḍ. (c) The desert landscape on Fogo Island, Cape Verde, viewed from a higher elevation with enough moisture to support scattered trees. [(b) Andreas Wolf/agefotostock; (c) Bobbé Christopherson.]

the region of the Utah–Arizona border, and into northern New Mexico (Figure 10.27b).

Tropical, Subtropical Hot Steppe Climates

Tropical, subtropical hot steppe climates generally exist around the periphery of hot deserts, where shifting subtropical high-pressure cells create a distinct summer-dry and winter-wet pattern. Average annual precipitation in *subtropical hot steppe* areas is usually below 60 cm (23.6 in.). Walgett, in interior New South Wales, Australia, provides a Southern Hemisphere example of this climate (Figure 10.28). This climate is seen around the Sahara's periphery and in the Iran, Afghanistan, Turkmenistan, and Kazakhstan region.

Midlatitude Cold Steppe Climates

The *midlatitude cold steppe* climates occur poleward of about 30° latitude and the *midlatitude cold desert* climates. Such midlatitude steppes are not generally found in the Southern Hemisphere. As with other dry climate regions, rainfall in the steppes is widely variable and undependable, ranging from 20 to 40 cm (7.9 to 15.7 in.). Not all rainfall is convectional, for cyclonic storm tracks penetrate the continents; however, most storms produce little precipitation.

Figure 10.29 presents a comparison between Asian and North American *midlatitude cold steppe*. Consider Semey (Semipalatinsk) in Kazakstan, with its greater temperature range and more evenly distributed precipitation, and Lethbridge, Alberta, with its lesser temperature range and summer maximum convectional precipitation.

(*text continued on page 282*)

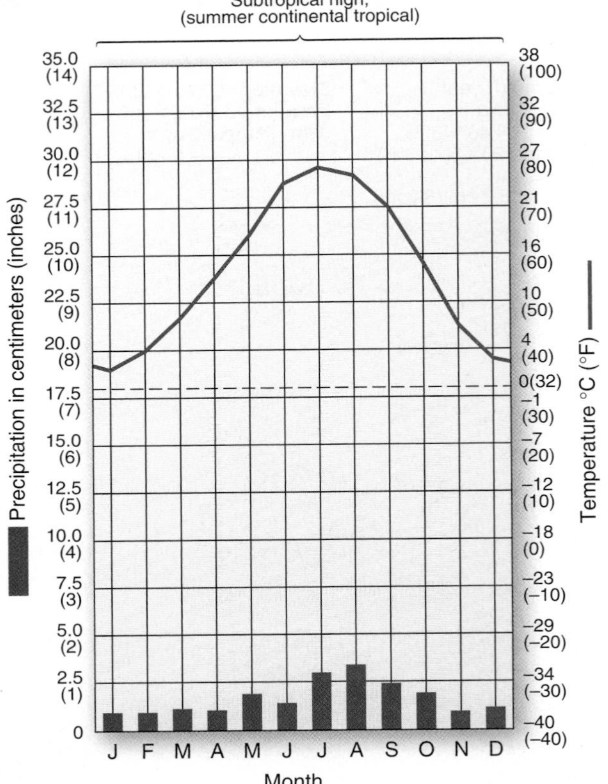

(a)

Station: Albuquerque, New Mexico
Lat/long: 35° 03′ N 106° 37′ W
Avg. Ann. Temp.: 14°C (57.2°F)
Total Ann. Precip.: 20.7 cm (8.1 in.)
Elevation: 1620 m (5315 ft)
Population: 522,000
Ann. Temp. Range: 24 C° (43.2 F°)
Ann. Hr of Sunshine: 3420

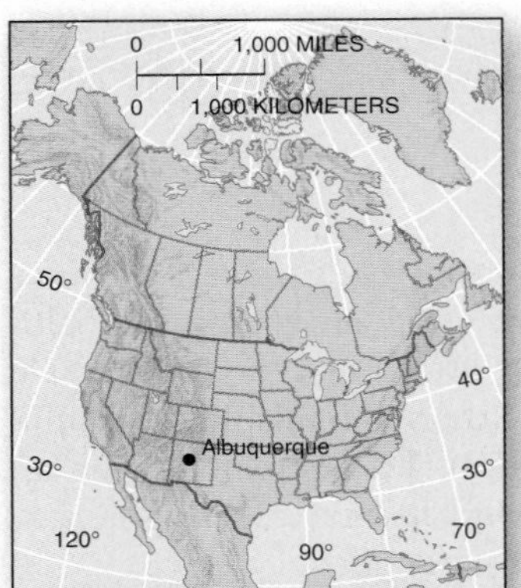

(b)

FIGURE 10.27 Midlatitude cold desert climate.
(a) Climograph for Albuquerque, New Mexico. (b) A winter scene near the New Mexico–Arizona border. [Bobbé Christopherson.]

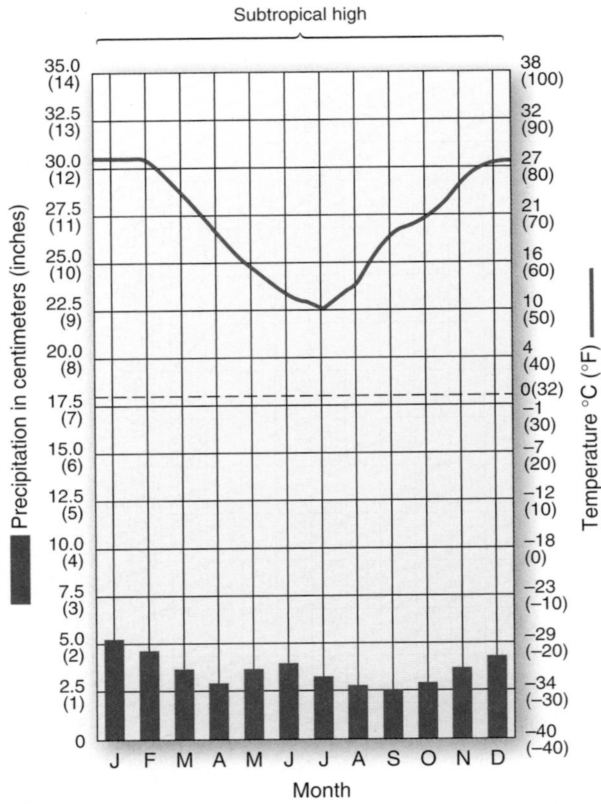

(a)

Station: Walgett, New South Wales, Australia
Lat/long: 30° S 148° 07′ E
Avg. Ann. Temp.: 20°C (68°F)
Total Ann. Precip.: 45.0 cm (17.7 in.)
Elevation: 133 m (436 ft)
Population: 8200
Ann. Temp. Range: 17 C° (31 F°)

(b)

FIGURE 10.28 Tropical, subtropical hot steppe climate.
(a) Climograph for Walgett, New South Wales, Australia. (b) Vast plains characteristic of north-central New South Wales. [Otto Rogge/Stock Market.]

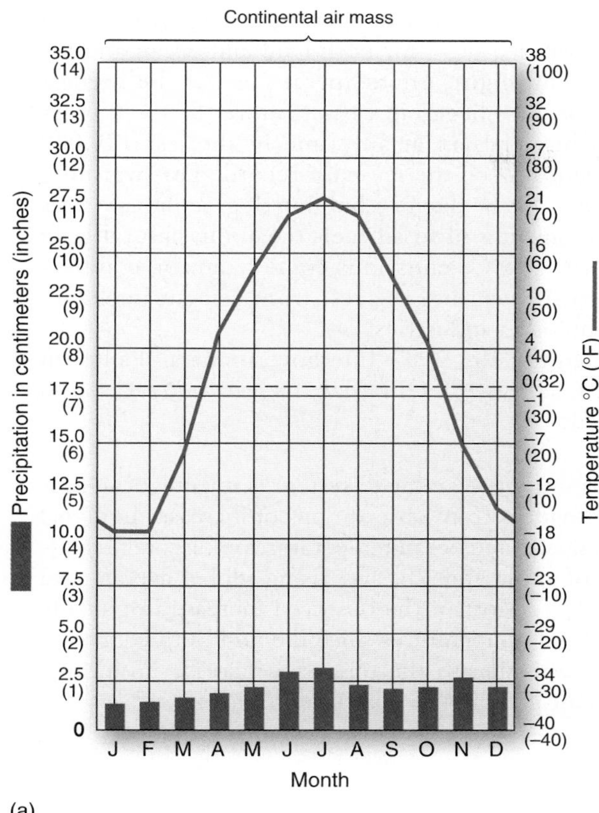

(a)

Station: Semey (Semipalatinsk), Kazakhstan
Lat/long: 50° 21′ N 80° 15′ E
Avg. Ann. Temp.: 3°C (37.4°F)
Total Ann. Precip.: 26.4 cm (10.4 in.)
Elevation: 206 m (675.9 ft)
Population: 270,500
Ann. Temp. Range: 39 C° (70.2 F°)

(b)

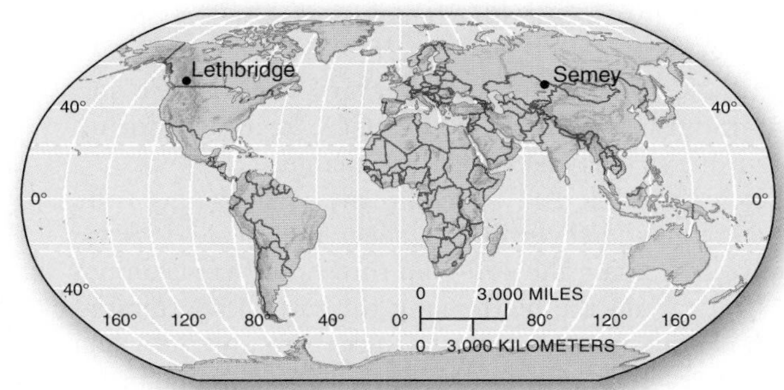

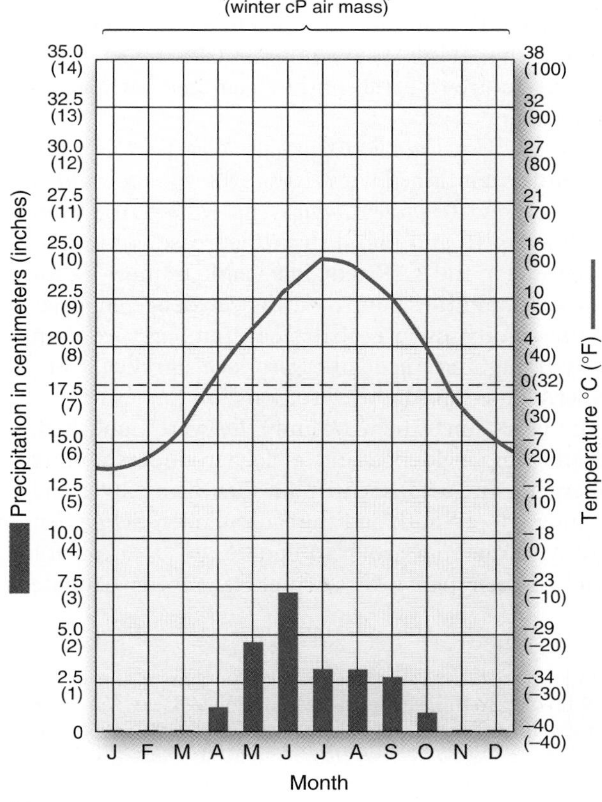

(c)

Station: Lethbridge, Alberta
Lat/long: 49° 42′ N 110° 50′ W
Avg. Ann. Temp.: 2.9°C (37.3°F)
Total Ann. Precip.: 25.8 cm (10.2 in.)
Elevation: 910 m (2985 ft)
Population: 73,000
Ann. Temp. Range: 24.3 C° (43.7 F°)

(d)

FIGURE 10.29 **Midlatitude cold steppe climates, Kazakstan and Canada.**
(a) Climograph for Semey (Semipalatinsk) in Kazakhstan. (b) Aerial view of Semey and the Ertis River.
(c) Climograph for Lethbridge in Alberta. (d) Grain elevators highlight an Alberta, Canada, landscape. [(b) Dinara Sagatova; (d) Design Pics RF/Getty Images.]

Global Climate Change

Let us begin this important section of *Geosystems* with a bottom line. The most dangerous greenhouse gas, as we learned in Chapter 3 and Figure 3.4, is carbon dioxide. The annual emissions of CO_2 to the atmosphere has been increasing at about 3.0% per year since 2000. The measurements of CO_2 in the lower atmosphere, beginning in 1958, are made at the Mauna Loa station in Hawai'i, and are worth reviewing:

- May 2005 = 382 ppm
- May 2006 = 385 ppm
- May 2007 = 387 ppm
- May 2008 = 389 ppm
- May 2009 = 390 ppm
- May 2010 = 393 ppm

Over the past 800,000 years, CO_2 levels in the atmosphere ranged between 100 ppm and 300 ppm—never over 300. You can check the data yourself and see what the level of CO_2 is during the month when you are working on this chapter (**ftp://ftp.cmdl.noaa.gov/ccg/co2/trends/co2_mm_mlo.txt**). May and October are the highest and lowest months, respectively, for CO_2 readings during the year due to seasonal vegetation changes.

Scientists think that we are already in the danger zone and that to avoid irreversible ice sheet and species loss, we must avoid crossing the 450 ppm threshold for maximum CO_2. You can do the math using the rate of accelerated growth in emissions for the past 10 years to find out when we might reach this threshold. Further, to obtain climate stability, scientists tell us that we should implement policies to return the level of CO_2 to 350 ppm.* This is the bottom line based on actual levels of the most dangerous greenhouse gas.

Significant climatic change has occurred on Earth in the past and most certainly will occur in the future. Society can do nothing about long-term climatic influences that cycle Earth through swings from ice ages to warmer periods. However, our global society must address short-term changes that are influencing global temperatures within the life span of present generations. This is especially true since these changes are due to human activities—an anthropogenic forcing of climate. Article 2 of the United Nations Framework Convention on Climate Change (UNFCCC), signed in 1992, commits countries to stabilize greenhouse gases at levels that will "prevent dangerous anthropogenic interference with the climate system." This agreement was completed during the copyright year of *Geosystems*, first edition.

Record-high global temperatures have dominated the past two decades—for both land and ocean and for both day and night. Air temperatures are the highest since recordings began in earnest more than 140 years ago and higher than at any time in the last 120,000 years, according to the ice-core record. This warming trend is *very likely** due to a buildup of greenhouse gases from the burning of fossil fuels (coal, oil, natural gas); about 94% of CO_2 emissions come from the combustion of fossil fuels and the rest from cement manufacturing and other industries.

Remember the 2007 Intergovernmental Panel on Climate Change (IPCC) *Fourth Assessment Report* (*AR4*) concluded:

> Warming of the climate system is unequivocal, as is now evident from observations of increases in global average air and ocean temperatures, widespread melting of snow and ice, and rising global average sea level. . . . Most of the observed increase in globally averaged temperatures since the mid-20th century is *very likely* due to the observed increase in anthropogenic greenhouse gas concentrations.†

"Reasons for Concern"

In the IPCC's *Third Assessment Report* (*TAR*), published in 2001, "Reasons for Concern" were detailed in Chapter 19 of the volume entitled *Vulnerability to Climate Change and Reasons for Concern: A Synthesis* (pp. 911–967), prepared by Working Group II. Scientists grouped together categories of concern with specific observations and impact analyses, proposals for future research, and indications that specific impacts were underway.

In the IPCC's *Fourth Assessment Report* (*AR4*), the IPCC scientists continued with *Assessing Key Vulnerabilities and the Risk from Climate Change* (pp. 779–810), from Working Group II, and found that "the 'reasons for concern' identified in the TAR remain a viable framework for assessing vulnerabilities." Further, that evidence suggests that smaller increases in global mean temperature than previously thought are needed to produce significant impacts described within the five "reasons for concern."

In the years since the *TAR* and *AR4* were published, more confirming evidence and a deeper understanding have emerged. The lead scientist authors of the 2001 and 2007 chapters, Joel Smith and the late Stephen Schneider, respectively, gathered a group to update the "Reasons for Concern" in their paper "Assessing Dangerous Climate

*Among many sources, see J. Hansen et al., "Target atmospheric CO_2: Where should humanity aim?" *Open Atmospheric Science Journal* 2 (March 2008): 217–231; and K. Zickfeld, M. Eby, H. D. Matthews, and A. J. Weaver, "Setting cumulative emissions targets to reduce the risk of dangerous climate change," *Proceedings of the National Academy of Sciences* 106, 38 (September 22, 2009): 16129–16134.

*As a standard scientific reference on climate change, the IPCC uses the following to indicate levels of confidence: *Virtually certain* > 99% probability of occurrence; *Extremely likely* > 95%; *Very likely* > 90%; *Likely* > 66%; *More likely than not* > 50%; *Unlikely* < 33%; *Very unlikely* < 10%; and *Extremely unlikely* < 5%.

†IPCC, Working Group I, *Fourth Assessment Report: Climate Change 2007: The Physical Science Basis* (Geneva, Switzerland: IPCC Secretariat, 2007), pp. 5, 10.

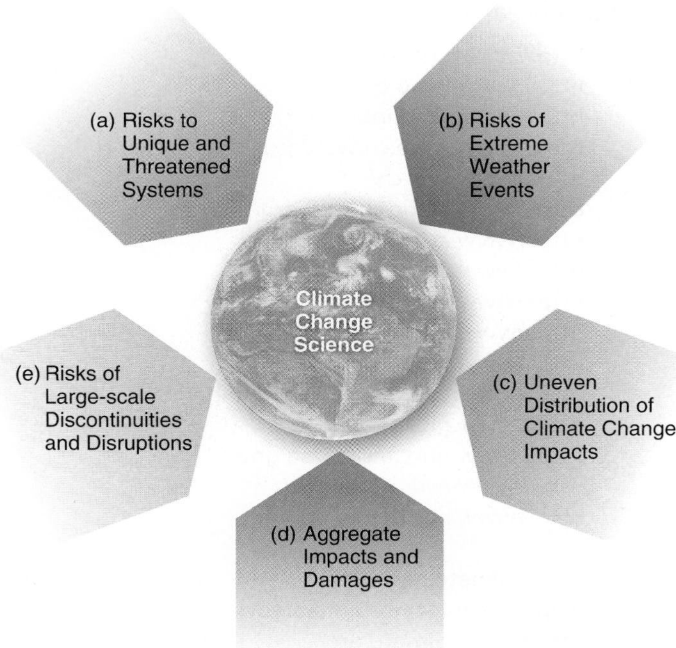

FIGURE 10.30 Summary of IPCC's five "Reasons for Concern." [Based on sources presented in the text.]

Change Through an Update of the IPCC 'Reasons for Concern'" [*Proceedings of the National Academy of Sciences* 106, 11 (March 17, 2009): 4133–4137]. The following summary is drawn from all three publications.

Figure 10.30 summarizes the five "Reasons for Concern" as they each inform climate change science. As you explore climate change, perhaps this framework can assist you in organizing your approach. The five IPCC "Reasons for Concern" are

A. **Risks to Unique and Threatened Systems**—Risks include irreversible losses to physical systems (e.g., glaciers, coral reefs, lakes, barrier islands, small islands) and biological systems (e.g., biodiversity hotspots, mangroves, ecosystem migrations, species extinctions), among others.

B. **Risks of Extreme Weather Events**—Risks include extreme event magnitude and frequency increases; intense precipitation and floods; tropical cyclone and tornado intensity increases; heat waves; droughts; and wildfires, among others.

C. **Uneven Distribution of Climate Change Impacts**—Impacts are greatest in low-latitude and less-developed countries and for circumpolar Arctic indigenous peoples; disparity occurs among economic classes of people; range and seasonality of disease vectors and health impacts tend to impact lower-latitude countries.

D. **Aggregate Impacts and Damages**—Effects are combined to assess total monetary damage and lives changed or lost; few developing countries or poor have the financial, technical, or institutional base to support adaptation.

E. **Risks of Large-Scale Discontinuities and Disruptions**—Abrupt and dramatic system responses occur at thresholds in systems which fluctuate above and below an average, yet show a trend over time—a dynamic equilibrium. As conditions change, the system can no longer maintain that operational level and spasms to a new equilibrium; that moment is a "tipping point." Resulting disruptions include the melting of mountain glaciers, Greenland and West Antarctic ice sheet losses, ocean acidification, permafrost melt and CO_2 liberation, and rapid sea-level rise, among others.

The fact that climate change is occurring is an accepted consensus among mainstream scientists. Determining the accelerating rate of this change is essential. In geologic time, changes in the environment occurred over millions of years, whereas elements of climate change today are happening in the span of decades.

The 800,000-year climate record from the Antarctic Dome-C ice core, discussed in Chapter 17, revealed that atmospheric CO_2 has never changed 30 ppm upward or downward during a period of less than 1000 years. Yet today, we see that atmospheric CO_2 rose 30 ppm in only the last 12 years. The rate of change affects the amount of time available for physical and biological systems to adapt. The accelerating rate of change is a key factor in all five "Reasons for Concern."

The International Scientific Conference on Climate Change met in March 2009 in Copenhagen to update the

GEO REPORT 10.3 *The UNFCCC process and annual meetings*

The UNFCCC was a product of the 1992 Earth Summit in Rio de Janeiro. The leading body of the Convention is the Conference of the Parties (COP), maintained by the 192 countries that ratified the FCCC Treaty. Meetings were then held in Berlin (*COP-1*, 1995) and Geneva (*COP-2*, 1996). These meetings set the stage for *COP-3* in Kyoto, Japan, in December 1997, where 10,000 participants adopted the *Kyoto Protocol* by consensus. Nineteen national academies of science endorsed the Kyoto Protocol, and 84 countries are signatories, including the United States after delays. For updates on the status of the Kyoto Protocol, see http://unfccc.int/2860.php. The 2010 gathering, *COP-16*, is in Mexico City.

TABLE 10.1 Sources of Information for Global Climate Change Research

United Nations Environment Programme	http://www.unep.ch/
World Meteorological Organization	http://www.wmo.ch/
World Climate Research Programme	http://wcrp.wmo.int/
Global Climate Observing System	http://www.wmo.ch/web/gcos/gcoshome.html
Intergovernmental Panel on Climate Change	http://www.ipcc.ch/
Arctic Council	http://www.arctic-council.org/
Arctic Climate Impact Assessment	http://www.acia.uaf.edu/
Arctic Monitoring and Assessment Program	http://www.amap.no/
Conservation of Flora and Fauna	http://www.caff.is/
International Arctic Science Committee	http://www.iasc.no/
U.S. Global Change Research Program	http://www.globalchange.gov.
U.S. Environmental Protection Agency	http://www.epa.gov/climatechange/
Goddard Institute for Space Studies	http://www.giss.nasa.gov/
Global Hydrology and Climate Center	http://www.ghcc.msfc.nasa.gov
National Climate Data Center	http://www.ncdc.noaa.gov/
National Environmental Satellite, Data, and Information Service	http://www.nesdis.noaa.gov/
NASA Global Climate Change	http://climate.nasa.gov/
NWS Climate Prediction Center	http://www.cpc.ncep.noaa.gov/
Climate Institute	http://www.climate.org/
Pew Center on Global Climate Change	http://www.pewclimate.org/
National Ice Center	http://www.natice.noaa.gov/
National Snow and Ice Data Center	http://nsidc.org/
National Center for Atmospheric Research	http://www.ncar.ucar.edu/
Environment Canada	http://www.ec.gc.ca/climate/
British Antarctic Survey	http://www.antarctica.ac.uk/
International Union for the Conservation of Nature and Natural Resources	http://www.iucn.org/
World.Org 100 Top Climate Change Sites	http://www.world.org/weo/clima

AR4. The conclusions from these meetings reinforce the "Reasons for Concern" in part:

- "The worst-case IPCC projections, or even worse, are being realized,"
- "Emissions are soaring, projections of sea-level rise are higher than expected,"
- "The acceleration of Greenland's and Antarctica's ice losses of mass are very clearly in evidence in satellite images,"
- "Inaction is inexcusable, weaker emissions targets for 2020 increases the risk of crossing numerous tipping points."*

*Summarized in E. Kintisch, "Global Warming: Projections of Climate Change Go from Bad to Worse, Scientists Report," *Science* 323, 5921 (March 20, 2009): 1546–1547.

Table 10.1 presents a sampling of major scientific organizations and web sites dealing with climate change topics and reports. Our *Mastering Geography* Web site also links to several more sites.

Climate Change Measurements

In the field of **paleoclimatology**, the science that studies past climates (discussed in Chapter 17), scientists use a number of proxy, or substitute, indicators for climate reconstruction. Data from ice cores, lake and ocean sediments, coral reefs, ancient pollen, and tree-ring density, among others, point to the present time as the warmest in the last 120,000 years. These proxy indicators also suggest that the increase in temperature during the 20th century is *very likely* the largest in any century over the past 1000 years (Figure 10.31). Earth is

GEO REPORT 10.4 *The IPCC and the Nobel Peace Prize*

The IPCC, formed in 1988, is an organization operating under the United Nations Environment Programme (UNEP) and the World Meteorological Organization (WMO), and is the scientific organization coordinating global climate-change research, climate forecasts, and policy formulation. In 2007, the IPCC shared the Nobel Peace Prize with former Vice President Albert Gore for their two decades of work raising understanding and awareness about global climate-change science.

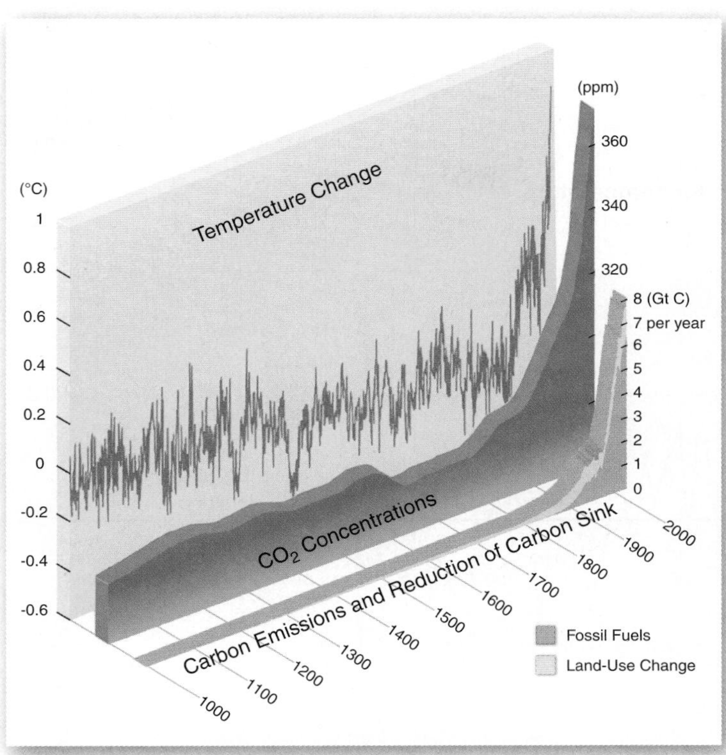

FIGURE 10.31 A thousand years of record and the covariance of carbon dioxide and temperature. Carbon emissions, CO_2 concentrations, and temperature change correlate over the past 1000 years. This graph was prepared before the record-setting temperatures of 2005. By May 2010, CO_2 levels were at 393 ppm. Earth systems are in record territory. [Illustration from ACIA, *Impacts of a Warming Arctic* (London: Cambridge University Press, 2004), p. 2. Used by permission of ACIA.]

less than 1 C° (1.8 F°) from equaling the highest average temperature of the past 125,000 years.

Figure 10.32a plots observed annual land-surface air temperatures and 5-year mean temperatures from 1880 through 2009. The record year for warmth for land-surface temperatures was 2005, with 2007, 2009, and 1998 the second, third, and fourth warmest. Through September 2010, all temperature records for land and ocean were set. For the Southern Hemisphere, 2009 was the warmest in the modern record. The decade of 2001–2010 was the warmest decade in the entire record.

The map in Figure 10.32b uses the same base period as the graph (1951–1980) to give you an idea of temperature anomalies across the globe from January to June 2010. Refer back to the temperature anomaly maps for several decades in Figure 1.1. Note the Arctic region through Hudson Bay, Canada, where new records for land and water temperatures and sea-ice melt were set. Anomalies exceeding 3.5 C° (6.3 F°) were measured in the Canadian Arctic and the Antarctic Peninsula. On the *Mastering Geography* web site for this text, there is a movie of temperature anomalies from 1881 to 2006 in which you can see the warming patterns over this time span.

MG™ **Satellite** GISS Surface Temperature Analysis Movie, 1891–2006.

Scientists are determining the difference between *forced fluctuations* (human-caused) and *unforced fluctuations* (natural) as a key to predicting future climate trends. Figure 10.33 shows a comparison of climate models, which include natural forcing (shaded blue area) and anthropogenic forcing (shaded pink area) for the globe. The black line on the graph plots observations from 1906 to 2005. Clearly, only those coupled climate models that include human factors accurately simulate increasing

GEO REPORT 10.5 *Professional consensus on climate change*

The American Association for the Advancement of Science (AAAS) reported in "The Scientific Consensus on Climate Change," by Naomi Oreskes [*Science* 306, 5702 (December 3, 2004): 1686], the results of a survey of all 928 climate-change papers published between 1993 and 2003 and concluded: "Remarkably, none of the papers disagreed with the consensus position."

The Association of American Geographers (AAG; 11,000 members) passed the *AAG Resolution Requesting Action on Climate Change* in 2006. This resolution places the AAG in line with other professional organizations and many national academies of sciences, now on record for action to slow rates of climate change. The American Meteorological Society (AMS; 13,000 members) published its *AMS Statement on Climate Change* in 2007. The American Physical Society (APS; 46,000 members) adopted its *APS National Policy on Climate Change* in 2007, declaring that "human causes of climate change are well understood and beyond dispute."

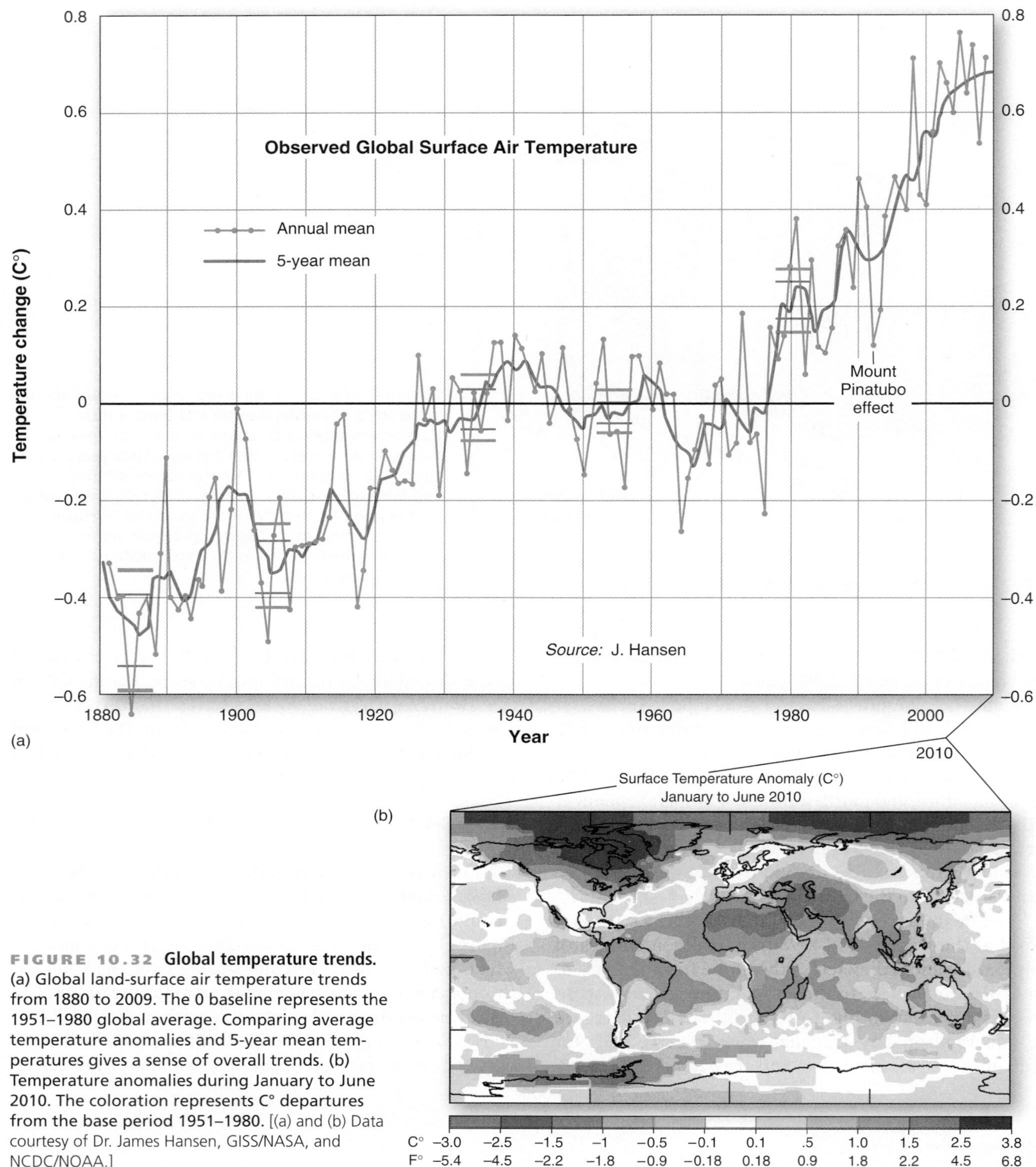

FIGURE 10.32 Global temperature trends. (a) Global land-surface air temperature trends from 1880 to 2009. The 0 baseline represents the 1951–1980 global average. Comparing average temperature anomalies and 5-year mean temperatures gives a sense of overall trends. (b) Temperature anomalies during January to June 2010. The coloration represents C° departures from the base period 1951–1980. [(a) and (b) Data courtesy of Dr. James Hansen, GISS/NASA, and NCDC/NOAA.]

temperatures. Solar variability and volcanic output alone do not explain the increase.

Because the enhanced greenhouse gases are anthropogenic in origin, various management strategies are possible to reduce human-forced climatic changes. Let's begin by examining the problem at its roots.

Greenhouse Gases Carbon dioxide and water vapor are the principal radiatively active gases causing Earth's natural greenhouse effect. *Radiatively active* gases include atmospheric gases, such as carbon dioxide (CO_2), methane (CH_4), nitrous oxide (N_2O), chlorofluorocarbons (CFCs), and water vapor, which absorb and radiate longwave

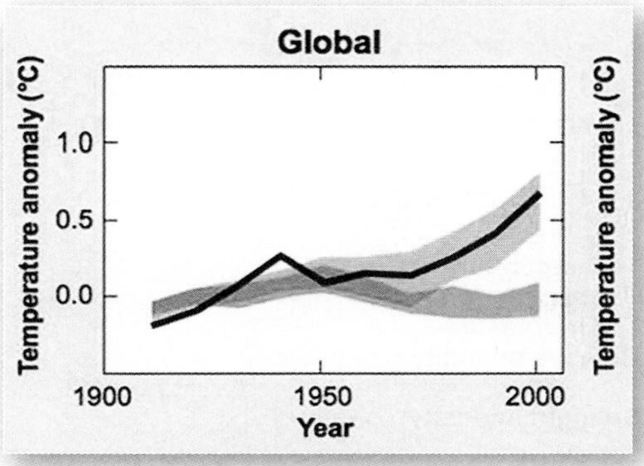

FIGURE 10.33 Explaining global temperature changes. Computer models accurately track observed temperature change (black line) when they factor in human-forced influences on climate (pink shading). Solar activity and volcanoes do not explain the increases (blue shading). [IPCC, Working Group I, *Fourth Assessment Report: Climate Change 2007: The Physical Science Basis* (Geneva, Switzerland: IPCC Secretariat, February 2007), Fig. SPM-4, p. 11.]

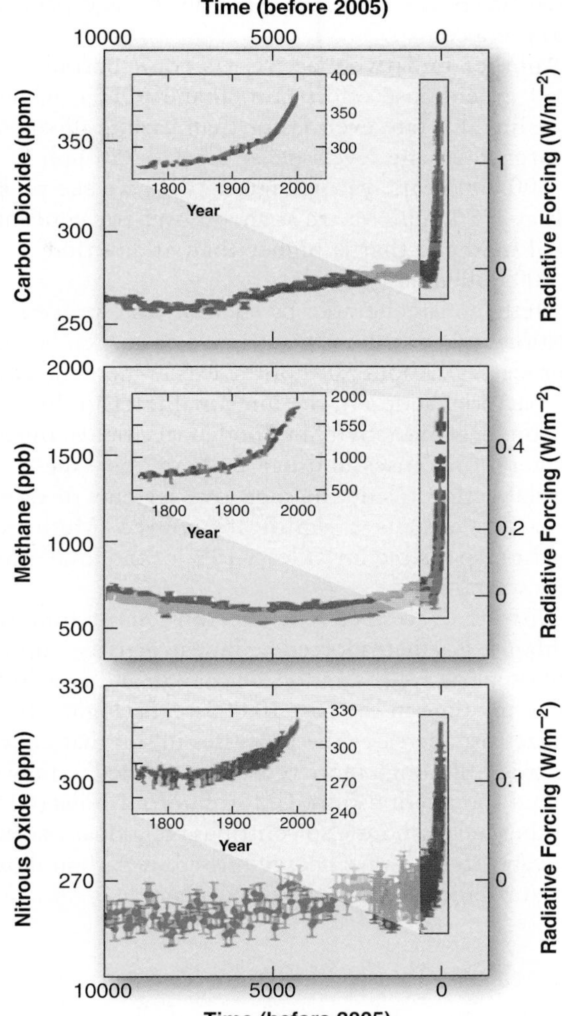

FIGURE 10.34 Greenhouse gas changes over the last 10,000 years.
Ice-core and modern data show the trends in carbon dioxide, methane, and nitrous oxide over the last 10,000 years. We are living in unique territory on these graphs. To update, during May 2010, CO_2 passed the 393 ppm level, CH_4 was at 1780 ppb, and N_2O approached 330 ppb. [IPCC, Working Group I, *Fourth Assessment Report: Climate Change 2007: The Physical Science Basis* (Geneva, Switzerland: IPCC Secretariat, February 2007), Fig. SPM-1, p. 3.]

CRITICAL THINKING 10.2

The forcing element of fuel economy

Transportation is the major sector producing carbon dioxide emissions. In effect since 1975 in the United States, corporate average fuel economy (CAFÉ) requirements for automobiles are presently set at 27.7 mpg (8.55 L/100 km). For light trucks (including SUVs), these requirements were set at 22.2 mpg in 2007 (10.6 L/100 km) and changed to 23.5 in 2010. In Figure 10.2.1, examine the higher standards in all the other countries sampled.

A 10-mile-per-gallon increase in efficiency could save more than a million barrels of oil per day, or more than half what we import from the Persian Gulf states, making a 35 mpg goal for 2020 quite appealing. Increased numbers of drivers and driving will negate some of these savings, making the solution more complex than just having more-fuel-efficient cars and trucks. Study the graph, and explain the relative inefficiency of U.S. vehicles. Why are U.S. fuel standards about half those of the European Union and so far below the standard enforced in China? Why is having higher CAFÉ standards such a hard sell, given the price of fuel and all the problems involved with importing oil (conflicts, oil spills, balance of trade deficits) and drilling for oil? How do we communicate the upfront savings in money for drivers and the improved air quality that higher standards would bring?

FIGURE 10.2.1 Fuel economy standards for selected countries.
[Adapted from IPCC, *Climate Change 2007: Mitigation and Climate Change* (Geneva, Switzerland: IPCC Secretariat, February 2007), Fig. 5.18, p. 373.]

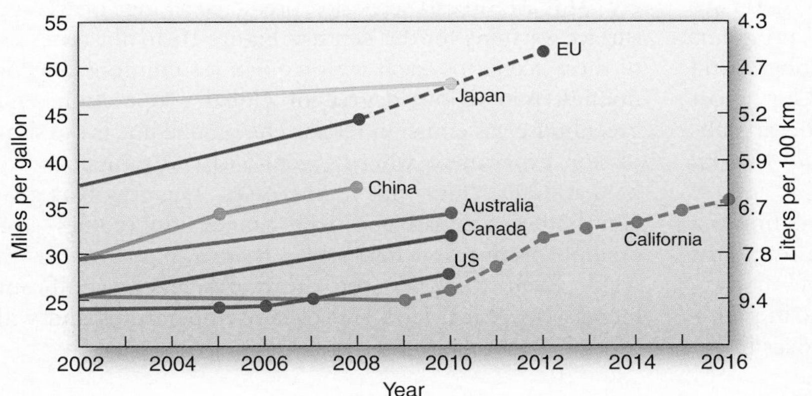

energy. Figure 10.34 plots changes in three of these greenhouse gases over the past 10,000 years.

Over the past several years, China took the lead for carbon dioxide emissions, with the United States a close second. China, with 19.5% of global population, produces 4.6 tons per person, whereas the United States, with 4.5% of world population, is at 19.8 tons of CO_2 emissions per person. Clearly, per capita CO_2 emissions in the

United States are far above what an average person in China produces.

Another radiatively active gas contributing to the overall greenhouse effect is methane (CH_4), which is increasing at a rate even faster than carbon dioxide. In ice cores, methane levels never topped 750 ppb in the past 800,000 years, yet in Figure 10.34 we see present levels at 1780 ppb. We are at an atmospheric concentration of methane that is higher than at any time in the past 800 millennia.

Methane is generated by such organic processes as digestion and rotting in the absence of oxygen (anaerobic processes). About 50% of the excess methane comes from bacterial action in the intestinal tracts of livestock and from organic activity in flooded rice fields. Burning of vegetation causes another 20% of the excess, and bacterial action inside the digestive systems of termite populations also is a significant source. Methane is thought responsible for at least 19% of the total atmospheric warming.

Nitrous oxide (N_2O) is the third most important greenhouse gas that is forced by human activity—up 17% in atmospheric concentration since 1750 and now higher than at any time in the past 10,000 years (Figure 10.34). Fertilizer use increases the processes in soil that emit nitrous oxide, although more research is needed to fully understand the relationships. Chlorofluorocarbons (CFCs) and other halocarbons also contribute to global warming. CFCs absorb longwave energy missed by carbon dioxide and water vapor in the lower troposphere.

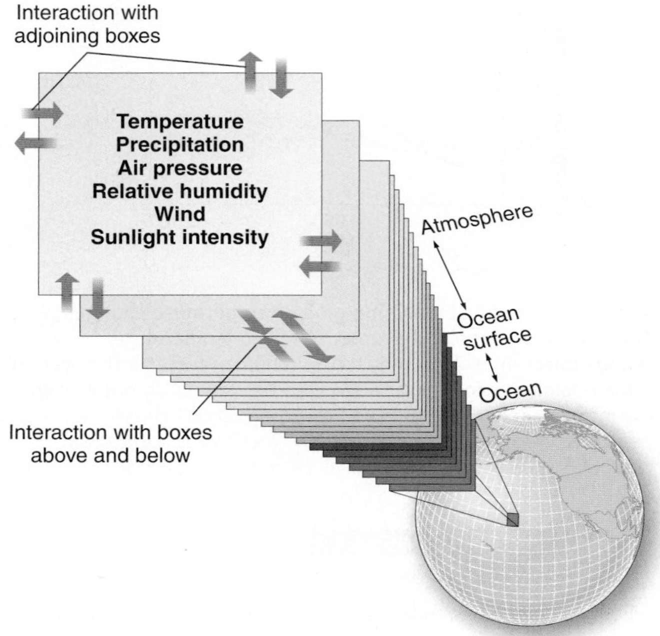

FIGURE 10.35 A general circulation model scheme. Temperature, precipitation, air pressure, relative humidity, wind, and sunlight intensity are sampled in myriad grid boxes. In the ocean, sampling is limited, but temperature, salinity, and ocean current data are considered. The interactions within a grid layer, and between layers on all six sides, are modeled in a general circulation model program.

Climate Models and Future Temperatures

In addition to actual measurements of climatic elements and the paleoclimatic records, scientists use computer models to run climate-forecasting programs. The scientific challenge in understanding climate change is to sense climatic trends in what is essentially a nonlinear, chaotic natural system. Imagine the tremendous task of building a computer model of all climatic components and programming these linkages (shown in Figure 10.2) over different time frames and at various scales.

Using mathematical models originally established for forecasting weather, scientists developed a complex computer climate model known as a **general circulation model (GCM)**. At least a dozen established GCMs are now in operation around the world. Submodel programs for the atmosphere, ocean, land surface, cryosphere, and biosphere operate within the GCM. The most sophisticated models couple atmosphere and ocean submodels and are known as *Atmosphere–Ocean General Circulation Models (AOGCMs)*.

The first step in describing a climate is defining a manageable portion of Earth's climatic system for study. Climatologists create dimensional "grid boxes" that extend from beneath the ocean to the tropopause, in multiple layers (Figure 10.35). Resolution of these boxes in the atmosphere has greatly improved, and they are now sized at 180 km (112 mi) in the horizontal and 1 km (0.6 mi) in the vertical; in the ocean, the boxes use the same horizontal and a vertical resolution of about 200 to 400 m (650–1300 ft). Analysts deal not only with the climatic components within each grid layer, but also with the interaction among the layers on all sides.

A comparative benchmark among the operational GCMs is *climatic sensitivity* to doubling of carbon dioxide levels in the atmosphere. GCMs do not predict specific temperatures, but they do offer various scenarios of global warming. GCM-generated maps correlate well with the observed global warming patterns experienced since 1990.

The 2007 IPCC *Fourth Assessment Report*, using a variety of GCM forecast scenarios, predicted a range of average surface warming for this century. Figure 10.36 illustrates six of these scenarios, each with its own assumptions of economics, population, degree of global cooperation, and greenhouse gas emission levels. The orange line is the simulation experiment where greenhouse gas emissions are held at 2000 values with no increases. The gray bars give you the best estimate and likely ranges of outcomes—for example, from a "low forecast" in B1 to a "high forecast" in A1Fl. Even the "B" scenario represents a significant increase in global land and ocean temperatures and will produce consequences.

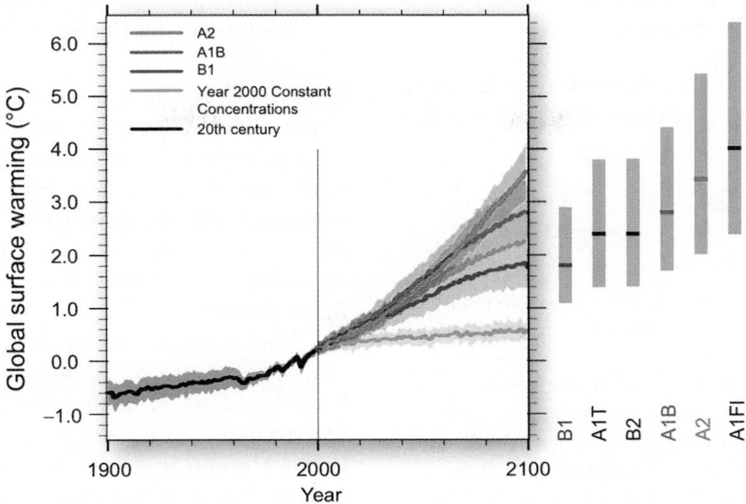

FIGURE 10.36 Model scenarios for surface warming.
Ranging from business as usual, lack of international cooperation, and continuation of the use of fossil fuels (A2FI) to a low-emission scenario (B1), each scenario spells climatic catastrophe to some degree. Holding emissions at the 2000 concentration (orange line plot) results in the least temperature impacts. [Graph from IPCC, Working Group I, *Fourth Assessment Report: Climate Change 2007: The Physical Science Basis* (Geneva, Switzerland: IPCC Secretariat, February 2007), Fig. SPM-5, p. 14.]

Changes in Sea Level Sea-level rise is occurring, and the rate accelerating. The rate in the late 20th century was about three times that of the late 19th century. During the last century, sea level rose 10–20 cm (4–8 in.), a rate 10 times greater than the average rate during the last 3000 years. Sea level is rising because of melting ice sheets and glaciers that add water and because of thermal expansion that occurs as the ocean absorbs heat energy and warms. There is more on this subject in Chapters 16 and 17.

The 2007 IPCC forecast scenarios for global mean sea-level rise this century are already being exceeded. Unfortunately, new measurements of Greenland's ice-loss acceleration and Antarctic ice losses were not within the time frame for the *AR4* report. Scientists are considering at least a 1.4-m (4.6-ft) "high case" for estimates of sea-level rise sometime this century as more realistic, given Greenland's present losses coupled with mountain glacial ice losses worldwide. Remember that a 0.3-m rise in sea level would produce a shoreline retreat of 30 m (98 ft) on average. Here is the concern:

> The data now available raise concerns that the climate system, in particular sea level, may be responding more quickly than climate models indicate. . . . The rate of sea-level rise for the past 20 years is 25% faster than the rate of rise in any 20-year period in the preceding 115 years. . . . Since 1990 the observed sea-level has been rising faster than the rise projected by models.*

*S. Rahmstorf et al., "Recent climate observations compared to projections," *Science* 316, 5825 (May 4, 2007): 709.

These increases would continue beyond 2100 even if greenhouse gas concentrations were stabilized.

A quick survey of world coastlines shows that even a moderate rise could bring change of unparalleled proportions. At stake are the river deltas, lowland coastal farming valleys, and low-lying mainland areas, all contending with high water, high tides, and higher storm surges. Particularly tragic social and economic consequences will affect small island states, which are unable to adjust within their present country boundaries—disruption of biological systems, loss of biodiversity, reduction in water resources, and evacuation of residents are among the impacts.

The future may bring both internal and international migration of affected human populations, spread over decades, as people move away from coastal flooding caused by the sea-level rise—a 1-m rise will displace 130 million people. Presently, there is no body of world law that covers "environmental refugees." In Chapter 16, maps present what coastlines will experience in the event of such a 1-m rise in sea level.

"No Regrets"

Reading through all this climate-change science must seem pretty "heavy." Instead, think of this information as empowering and as motivation to take action—personally, locally, regionally, nationally, and globally.

The IPCC declared that "no regrets" opportunities to reduce carbon dioxide emissions are available in most countries and are defined as follows:

> No regrets options are by definition greenhouse gas emissions reduction options that have negative net costs, because they generate direct and indirect benefits that are large enough to offset the costs of implementing the options.

Benefits that equal or exceed their cost to society include reduced energy cost, improved air quality and health, reduction in tanker spills and oil imports, and deployment of renewable and sustainable energy sources, with the growth of new related business opportunities, among others. This holds true without even considering the benefits of slowing the rate of climate change.

For Europe, scientists determined that carbon emissions could be reduced to less than half the 1990 level by 2030, at a negative cost. A study by the U.S. Department of Energy's national laboratories found that we could meet the Kyoto carbon emission reduction targets with negative overall costs (cash benefit savings) ranging from –\$7 to –\$34 billion a year. Of course, one key to "no regrets" is untapped energy-efficiency and conservation potentials.

The most comprehensive economic analysis is *The Economics of Climate Change: The Stern Review*, prepared by Nicholas Stern for the British government (Cambridge, England: Cambridge University Press, 2007; 692 pages). This report concluded as follows:

- There is still time to avoid the worst impacts of climate change, if we take strong action now.
- Climate change could have very serious impacts on growth and development.
- The costs of stabilizing the climate are significant but manageable; delay would be dangerous and much more costly.
- Action on climate change is required across all countries, and it need not cap the aspirations of growth in rich and poor countries.
- A range of options exists to cut emissions; strong, deliberate policy action is required to motivate their take-up.
- Climate change demands an international response, based on shared understanding of long-term goals and agreement on frameworks for action.

II CRITICAL THINKING 10.3

Thinking through an action plan to reduce human climate forcing

Many external factors force climate, as you can see in Figure 10.3.1. Let us work with the chart to see how the variables influence climate change decisions.

The estimates of radiative forcing (RF) in Watts-per-square-meter units are given on the x-axis (horizontal axis). In the far-right column, LOSU refers to "level of scientific understanding." Those bars that extend to the right of the "0" value (in red, orange, and yellow) indicate warming, or *positive forcing*, such as "long-lived greenhouse gases." Bars that go to the left of the "0" value (in blue) indicate cooling, or *negative forcing*, such as cloud-albedo effect. Each bar also includes an estimate of the uncertainty range.

Assume you are a policy maker with a goal of reducing the rate of climate change—that is, reducing positive radiative forcing of the climate system. What strategies do you suggest to alter the extent of the columns and adjust the mix of elements that cause temperature increases? Assign priorities to each suggested strategy to denote the most- to least-effective strategies in moderating climate change. Brainstorm and discuss your strategies with others.

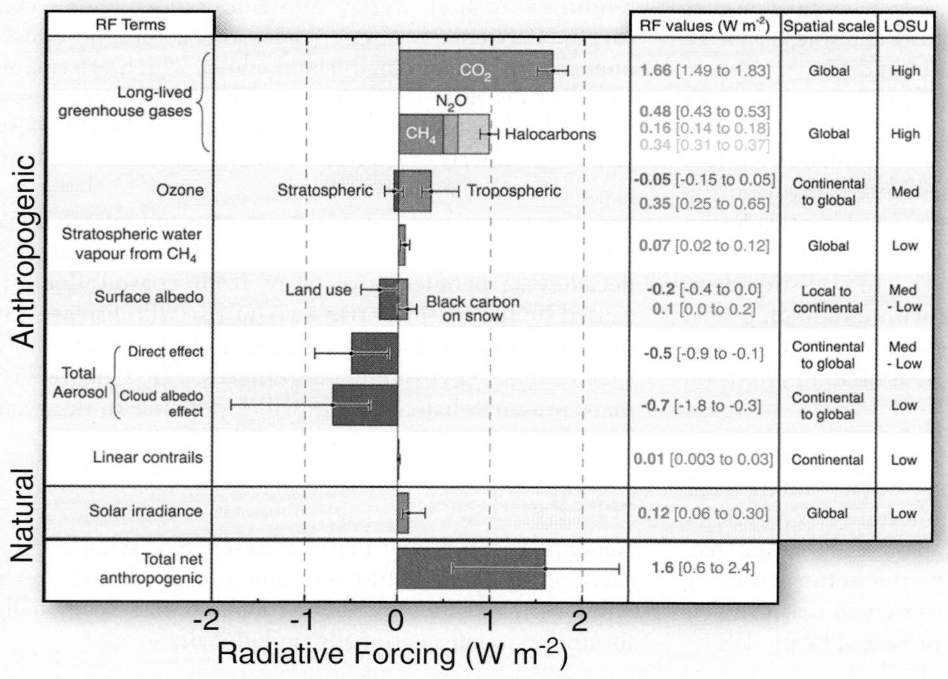

FIGURE 10.3.1 Analysis of radiative forcing of temperatures. [IPCC, Working Group I, *Fourth Assessment Report: Climate Change 2007: The Physical Science Basis* (Geneva, Switzerland: IPCC Secretariat, February 2007), Figure SPM-2, p. 4, and Figure 2.20, p. 203.]

Twenty-two years ago climatologists Richard Houghton and George Woodwell described the present climatic condition:

> The world is warming. Climatic zones are shifting. Glaciers are melting. Sea level is rising. These are not hypothetical events from a science fiction movie; these changes and others are already taking place, and we expect them to accelerate over the next years as the amounts of carbon dioxide, methane, and other trace gases accumulating in the atmosphere through human activities increase.*

Your author knows this quote well because it has been in every edition of *Geosystems* since the first edition in 1992.

*R. Houghton and G. Woodwell, "Global climate change," *Scientific American* (April 1989): 36.

Chapter 10 is a synthesis of systems in Chapters 2 through 9 of *Geosystems* and marks the end of Part II. Climate classification categories give us a portrait of energy and moisture and of solar insolation and surface energy budgets and provide a baseline that we can use to track changing climatic boundaries, given climate change. Next, we move to Part III, where Earth and the atmosphere interface—where the systems of each meet. We begin with the endogenic system in Chapters 11 and 12, where energy wells up from within Earth—with such power that crustal plates are dragged and pushed around, sometimes in dramatic episodes. We study rock-forming processes and the nature of deformation by folding, faulting, and displacement, and we spend time with earthquakes and volcanic eruptions.

KEY LEARNING CONCEPTS REVIEW

■ ***Define*** climate and climatology, and ***explain*** the difference between climate and weather.

Climate is dynamic, not static. **Climate** is a synthesis of weather phenomena at many scales, from planetary to local, in contrast to weather, which is the condition of the atmosphere at any given time and place. Earth experiences a wide variety of climatic conditions that can be grouped by general similarities into climatic regions. **Climatology** is the study of climate and attempts to discern similar weather statistics and identify **climatic regions**.

climate (p. 252)
climatology (p. 253)
climatic regions (p. 253)

1. Define climate and compare it with weather. What is climatology?
2. Explain how a climatic region synthesizes climate statistics.
3. How does the El Niño phenomenon produce the largest interannual variability in climate? What are some of the changes and effects that occur worldwide?

■ ***Review*** the role of temperature, precipitation, air pressure, and air mass patterns used to establish climatic regions.

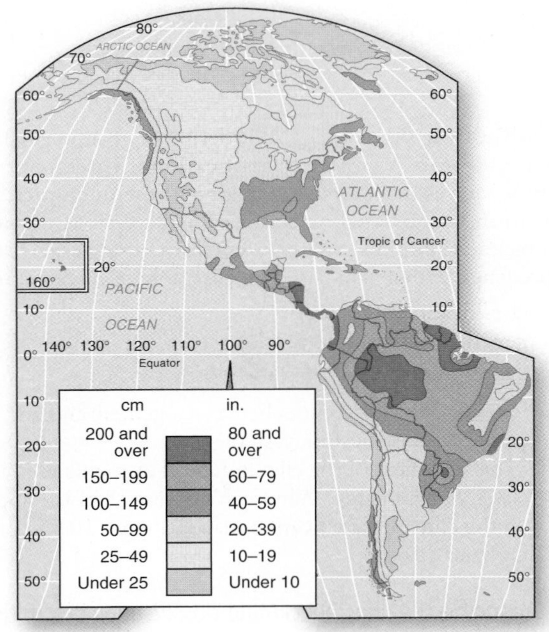

Climatic inputs include insolation (pattern of solar energy in the Earth–atmosphere environment), temperature (sensible heat energy content of the air), precipitation (rain, sleet, snow, and hail—the supply of moisture), air pressure (varying patterns of atmospheric density), and air masses (regional-sized homogeneous units of air). Climate is the basic element in ecosystems, the natural, self-regulating communities of plants and animals that thrive in specific environments.

4. How do radiation receipts, temperature, air-pressure inputs, and precipitation patterns interact to produce climate types? Give an example from a humid environment and one from an arid environment.
5. Evaluate the relationships among a climatic region, an ecosystem, and a biome.

- ***Review*** the development of climate classification systems, and ***compare*** genetic and empirical systems as ways of classifying climate.

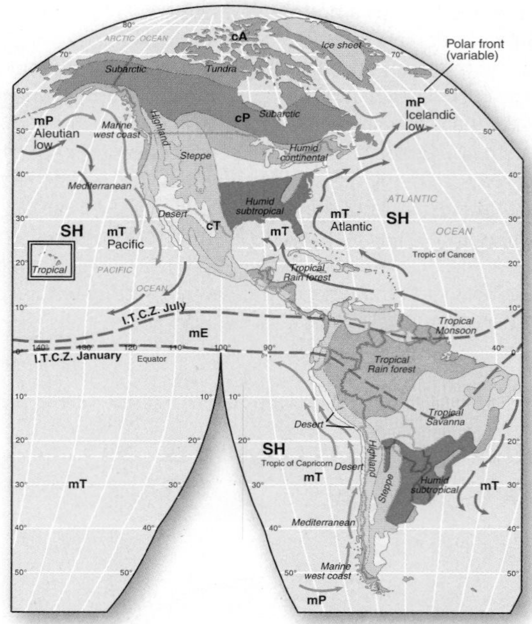

Classification is the process of ordering or grouping data in related categories. A **genetic classification** is based on causative factors, such as the interaction of air masses. An **empirical classification** is one based on statistical data, such as temperature or precipitation. This text analyzes climate using aspects of both approaches, with a map based on climatological elements.

classification (p. 256)
genetic classification (p. 256)
empirical classification (p. 256)

6. What are the differences between a genetic and an empirical classification system?
7. What are some of the climatological elements used in classifying climates? Why each of these? Use the approach on the climate classification map in Figure 10.6 in preparing your answer.

- ***Describe*** the principal climate classification categories other than deserts, and ***locate*** these regions on a world map.

Keep in mind temperature and precipitation are measurable results produced by interacting elements of weather and climate. These data are plotted on a **climograph** to display the characteristics of the climate.

There are six basic climate categories. Temperature and precipitation considerations form the basis of five climate categories and their regional types:

- Tropical (tropical latitudes)
  - rain forest (rainy all year)
  - monsoon (6 to 12 months rainy)
  - savanna (less than 6 months rainy)
- Mesothermal (midlatitudes, mild winters)
  - humid subtropical (hot summers)
  - marine west coast (warm to cool summers)
  - Mediterranean (dry summers)
- Microthermal (mid and high latitudes, cold winters)
  - humid continental (hot to warm, mild summers)
  - subarctic (cool summers to very cold winters)
- Polar (high latitudes and polar regions)
  - tundra (high latitude or high elevation)
  - ice caps and ice sheets (perpetually frozen)
  - polar marine
- Highland (high elevations at all latitudes; highlands have lower temperatures)

Only one climate category is based on moisture efficiency as well as temperature:

- Dry (permanent moisture deficits)
  - arid deserts (tropical, subtropical hot and midlatitude cold)
  - semiarid steppes (tropical, subtropical hot and midlatitude cold)

climograph (p. 259)

8. List and discuss each of the principal climate categories. In which one of these general types do you live? Which category is the only type associated with the annual distribution and amount of precipitation?
9. What is a climograph, and how is it used to display climatic information?
10. Which of the major climate types occupies the most land and ocean area on Earth?
11. Characterize the tropical climates in terms of temperature, moisture, and location.
12. Using Africa's tropical climates as an example, characterize the climates produced by the seasonal shifting of the ITCZ with the high Sun.
13. Mesothermal climates (midlatitudes, mild winters) occupy the second-largest portion of Earth's entire surface. Describe their temperature, moisture, and precipitation characteristics.
14. Explain the distribution of the *humid subtropical hot-summer* and *Mediterranean dry-summer* climates at similar latitudes and the difference in precipitation patterns between the two types. Describe the difference in vegetation associated with these two climate types.

15. Which climates are characteristic of the Asian monsoon region?
16. Explain how a *marine west coast* climate can occur in the Appalachian region of the eastern United States.
17. What role do offshore ocean currents play in the distribution of the *marine west coast* climates? What type of fog is formed in these regions?
18. Discuss the climatic conditions for the coldest places on Earth outside the poles.

■ ***Explain*** the precipitation and moisture efficiency criteria used to determine the arid and semiarid climates, and ***locate*** them on a world map.

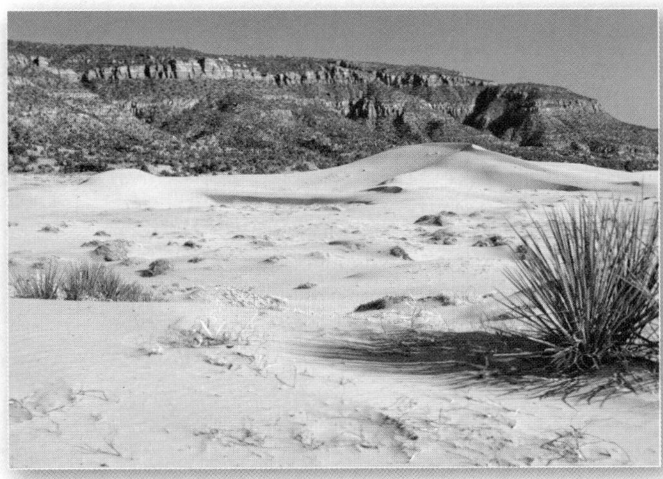

The dry climates are described by precipitation rather than temperature.

Major subdivisions are *arid deserts* in tropical and midlatitude areas (precipitation—natural water supply—less than one-half of natural water demand) and *semiarid steppes* in tropical and midlatitude areas (precipitation more than one-half of natural water demand). A **steppe** is a regional term referring to the vast semiarid grassland biome of Eastern Europe and Asia (the equivalent biome in North America is called shortgrass prairie; in Africa, the savanna).

steppe (p. 278)

19. In general terms, what are the differences among the four desert classifications? How are moisture and temperature distributions used to differentiate these subtypes?
20. Relative to the distribution of arid and semiarid climates, describe at least three locations where they occur across the globe and the reasons for their presence in these locations.

■ ***Explain*** the causes and potential consequences of climate change, and ***outline*** future climate patterns from forecasts.

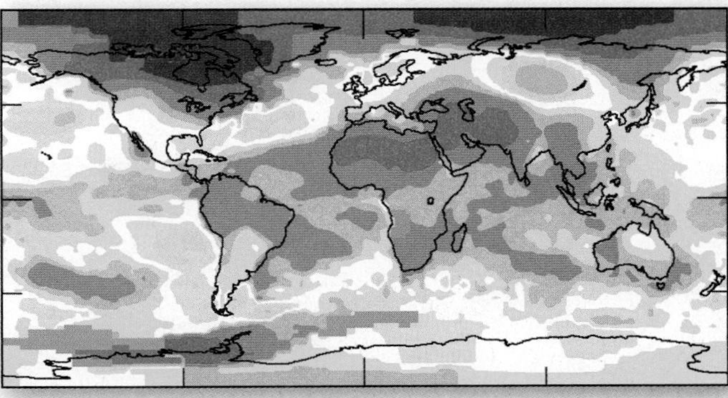

Various activities of present-day society are producing climatic changes, particularly a global warming trend. The highest average annual temperatures experienced since the advent of instrumental measurements have dominated the last 25 years. There is a scientific consensus that climate change is related to the anthropogenic impacts that are enhancing the greenhouse effect.

The 2007 *Fourth Assessment Report* affirms this consensus. The IPCC has predicted surface temperature increases in response to a doubling of carbon dioxide. Natural climatic variability over the span of Earth's history is the subject of **paleoclimatology**. A **general circulation model (GCM)** forecasts climate patterns and is providing greater capability and accuracy than in the past. People and their political institutions can use GCM forecasts to formulate policies aimed at reducing unwanted climate change.

paleoclimatology (p. 284)
general circulation model (GCM) (p. 288)

21. Explain climate forecasts. How do general circulation models (GCMs) produce such forecasts?
22. Describe the potential climatic effects of global warming on polar and high-latitude regions. What are the implications of these climatic changes for persons living at lower latitudes?
23. How is climatic change affecting agricultural and food production? Natural environments? Forests? The possible spread of disease?
24. What are the actions being taken at present to delay the effects of global climate change? What is the Kyoto Protocol? What is the current status of the U.S. government action on the protocol?

MASTERING GEOGRAPHY

Visit www.masteringgeography.com for several figure animations, satellite loops, self-study quizzes, flashcards, visual glossary, case studies, career links, textbook reference maps, RSS Feeds, and an ebook version of *Geosystems*. Included are many geography web links to interesting Internet sources that support this chapter.

PART III

The Earth–Atmosphere Interface

Inputs
Earth
Endogenic systems
Exogenic systems

Actions
Rock and mineral formation
Tectonic processes
Weathering
Erosion, transport, deposition

Outputs
Crustal deformation
Orogenesis and vulcanism
Landforms: karst, fluvial, eolian, coastal, glacial

Human–Earth Relation
Hazard perception
Humans as geomorphic agent
Fluvial impacts; Desertification; Sea-level rise; Ice-melt

The American Southwest is a region of dramatic landscapes. Just east of Monument Valley, near the Arizona–Utah border, are spectacular buttes and mesas where weathering and erosion removed massive amounts of rock over millions of years, leaving towering monuments. Note the sand dunes in the background. The center of the scene is at 36° 54′ N 109° 59′ W. [Bobbé Christopherson.]

Earth is a dynamic planet whose surface is shaped by active physical agents of change. Two broad systems—endogenic and exogenic—organize these agents in Part III. The **endogenic system** (Chapters 11 and 12) encompasses internal processes that produce flows of heat and material from deep below Earth's crust. Radioactive decay principally powers these processes. The materials involved constitute the solid realm of Earth. Earth's surface responds by moving, warping, and breaking, sometimes in dramatic episodes of earthquakes and volcanic eruptions, constructing the crust.

At the same time, the **exogenic system** (Chapters 13 through 17) involves external processes that set into motion air, water, and ice, all powered by solar energy—this is the fluid realm of Earth's environment. These media carve, shape, and reduce the landscape. One such process, weathering, breaks up and dissolves the crust. Erosion picks up these materials; transports them in rivers, winds, coastal waves, and flowing glaciers; and deposits them along the way. Thus, Earth's surface is the interface between two vast open systems: one that builds the landscape and topographic relief and one that tears it down into sedimentary plains.

CHAPTER 11

The Dynamic Planet

Pico de Fogo volcano, in the Cape Verde Islands in the Atlantic Ocean, rises to 2830 m (9285 ft). An eruption from the small vent cone on the slopes of Pico (stained with sulfur compounds) happened in 1995. This forced the evacuation of the village Chã das Caldeiras, located in the crater of the larger volcano Chã, which last erupted in 1637. Farming, including some vineyards and coffee, is done in the volcanic soils; homes are built with blocks of basaltic rock. [Bobbé Christopherson.]

KEY LEARNING CONCEPTS

After reading the chapter, you should be able to:

- ***Distinguish*** between the endogenic and exogenic systems, ***determine*** the driving force for each, and ***explain*** the pace at which these systems operate.
- ***Diagram*** Earth's interior in cross section, and ***describe*** each distinct layer.
- ***Illustrate*** the geologic cycle, and ***relate*** the rock cycle and rock types to endogenic and exogenic processes.
- ***Describe*** Pangaea and its breakup, and ***relate*** several physical proofs that crustal drifting is continuing today.
- ***Portray*** the pattern of Earth's major plates, and ***relate*** this pattern to the occurrence of earthquakes, volcanic activity, and hot spots.

GEOSYSTEMS NOW

Earth's Migrating Magnetic Poles

In this chapter, we journey to the center of Earth and study Earth's core, divided between a solid inner core and a dense, but fluid outer core. Earth's magnetic field is principally generated by motions in this outer core. In Chapter 1, we discussed the north geographic pole—the axial pole centered where the meridians of longitude converge. This is *true north* and is a fixed point. However, Earth's magnetic field is focused at a different set of poles, which move at a measurable rate.

When we use a compass to find directions, the needle points to the *North Magnetic Pole* (NMP). At this pole, the magnetic field is directed vertically downward, the magnetic dip, relative to Earth's surface. This is the observed magnetic pole, located and mapped by magnetic surveys.

Consider North at the top of a circle and azimuth as the degrees of arc clockwise around the circle from North. This makes due East 90°, due South 180°, due West 270°, and returning to North at 360° or 0°. The compass indicates the NMP, which is not the same as the geographic pole, or true north. Based on the location where you are measuring with the compass there could be an angular distance in degrees between the NMP and true north, this is the magnetic declination.

In 2008, north of San Francisco, California, there was a west declination of 15°. However, at the junction of the Iowa, Missouri, Illinois borders there was a 0° declination, in other words the NMP and true north were aligned. Whereas, for someone in Boston there was a 15° east declination in 2008. To get true north from your compass reading, you add or subtract the declination accordingly. Isogonal maps are published that plot lines of equal magnetic declination. Most topographic maps give you this information in a corner legend but it is only accurate for the date the map was printed. The NMP is on the move.

As the magnetic field of Earth changes, the NMP slowly drifts across the Canadian Arctic. Scientists have traced a long history of change in the location of the NMP.

As a surface expression of Earth's magnetic field, the NMP moved 1100 km (685 mi) just this past century. The Geological Survey of Canada (GSC) tracks the motion of the NMP to determine the pole's location (see http://gsc.nrcan.gc.ca/geomag/nmp/northpole_e.php). Surveys are a collaborative effort between the GSC and the Bureau de Recherches Géologiques et Minières in France. Presently the NMP is moving northwest toward Siberia at approximately 50 km (31 mi) per year, and since 2007 the rate of movement has increased to 55–60 km (34–37 mi) per year. The observed positions for 1831–2012 are mapped in Figure GN 11.1. Each day the actual magnetic pole migrates in a small oval pattern around the average locations given on the map.

The NMP is near 85° N by 147° W in 2012 in the Canadian Arctic. It's antipode, or opposite pole, is the *South Magnetic Pole* (SMP), which lies off the coast of Wilkes Land, Antarctica. The SMP moves separately from the NMP and is presently headed northwest at just 5 km (3.1 mi) per year. This chapter discusses the changing intensity of Earth's magnetic field and how the field periodically reverses polarity.

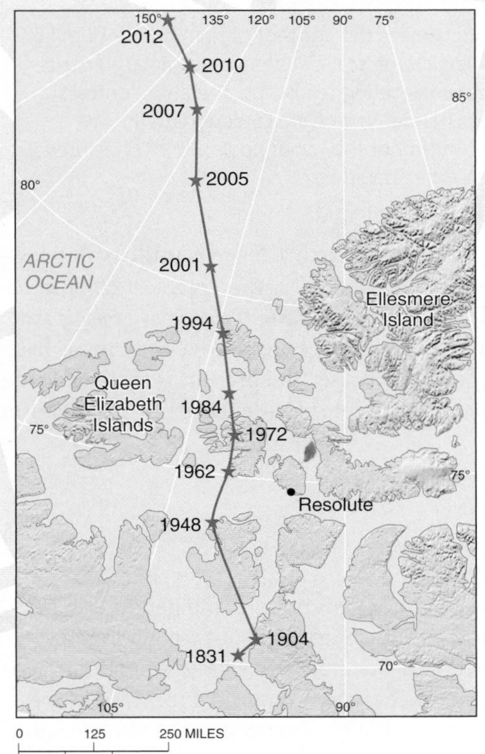

FIGURE GN 11.1 North Magnetic Pole Movement, 1831 to 2012.

TABLE GN 11.1 Approximate North Magnetic Pole coordinates 2001 to 2015

Year	Latidude (°N)	Longitude (°W)
2001	81.3	110.8
2002	81.6	111.6
2003	82.0	112.4
2004	82.3	113.4
2005	82.7	114.4
2006	83.9	119.9
2007	84.4	121.7
2008	84.2	124.9
2009	84.9	131.0
2010	85.0	132.6
2011	85.1	134.0
2012	85.9	147.0
2015 prediction	86.1	153.0

Discoveries have revolutionized our understanding of Earth systems science and how continents and oceans came to be arranged as they are. A new era of understanding is emerging, combining various sciences within the study of physical geography. One task of physical geography is to explain the spatial implications of all this new information.

In this chapter: Earth's interior is organized as a core surrounded by roughly concentric shells of material. It is unevenly heated by the radioactive decay of unstable elements. A rock cycle produces three classes of rocks through igneous, sedimentary, and metamorphic processes. Internal processes coupled with the rock cycle, hydrologic cycle, and tectonic cycle result in a varied crustal surface, featuring irregular fractures, extensive mountain ranges both on land and on the ocean floor, drifting continental and oceanic crust, and frequent earthquakes and volcanic events. All of this movement of material results from *endogenic* forces within Earth—the subject of this chapter.

The Pace of Change

The **geologic time scale** is a summary timeline of all Earth history, shown in Figure 11.1. It reflects currently accepted names of time intervals for each segment of Earth's history, from vast *eons* through briefer *eras*, *periods*, and *epochs*. Present thinking places Earth's age at about 4.6 billion years, with the Moon about 30 million years younger, formed when a Mars-sized object struck the early Earth.

The time scale depicts two important kinds of time: *relative* (what happened in what order) and *absolute* (actual number of years before the present). Also on the geologic time scale, see the labels denoting the six major extinctions of life forms in Earth history. Rather than being complete extinctions, think of some of these spasms in planetary life as significant *depletions*. These range from 440 million years ago (m.y.a.) to the ongoing present-day episode caused by modern civilization.

Relative time is the sequence of events, based on the relative positions of rock strata above or below each other. Relative time is based on the important general principle of *superposition*, which states that *rock and sediment always are arranged with the youngest beds "superposed" toward the top of a rock formation and the oldest at the base, if they have not been disturbed*. The study of these sequences is *stratigraphy*. Thus, relative time places the Precambrian at the bottom (beginning) of the time scale and the Holocene (today) at the top. Important time clues—namely, *fossils*, the remains of ancient plants and animals—lie embedded within these strata. Since approximately 4.0 billion years ago, life has left its evolving imprint in the rocks.

Scientific methods such as radiometric dating determine *absolute time*, the actual "millions of years ago" shown on the time scale. These absolute ages permit scientists to actively update geologic time, refining the time-scale sequence and lending greater accuracy to relative dating sequences. For more on the geologic time scale, see http://www.ucmp.berkeley.edu/exhibit/geology.html.

Holocene is the name given to the youngest epoch in the geologic time scale, consisting of the last 11,500 years since the retreat of the continental glaciers. Because of the impact of human society on planetary systems, discussion is under way about designating a new name for the human epoch. A proposal is to call this time the *Anthropocene*.

However, when would such a time begin? Some think the beginning of the Industrial Revolution in the 18th century is the starting point, whereas others think an appropriate time is some 5000 years ago, when Asians began flooding rice fields that released methane gas into the atmosphere. Or perhaps 8000 years ago is a better chronology with the spread of domesticated crops, widespread clearing of forests, and the human use of fire to modify environments. Human impacts on the atmosphere began at the roots of civilization.

CRITICAL THINKING 11.1

Thoughts about an "Anthropocene" Epoch

Take a moment to explore this concept of naming our human epoch in the geologic time scale. Determine some arguments for such a change considering landscape alteration, deforestation, and climate change. If we were to designate the late Holocene as the Anthropocene, what do you think the criteria for the beginning date should be?

A guiding principle of Earth science is **uniformitarianism**, which assumes that *the same physical processes active in the environment today have been operating throughout geologic time*. For example, if streams carve valleys now, they must have done so 500 m.y.a. The phrase "the present is the key to the past" describes this principle. Evidence from exploration

GEO REPORT 11.1 ***Radioactive dating and Earth time***

Radiometric dating of Earth materials involves the steady decay of certain atoms. An atom contains protons and neutrons in its nucleus. Certain forms of atoms called isotopes have unstable nuclei; that is, the protons and neutrons do not remain together indefinitely. As particles break away and the nucleus disintegrates, radiation is emitted, and the atom decays into a different element—this process is *radioactivity*. This provides the steady time clock needed to measure the age of ancient rocks, since the decay rates for different isotopes are determined precisely. Scientists compare the amount of original isotope in the sample with the amount of decayed end product in the sample to determine the date the rock formed.

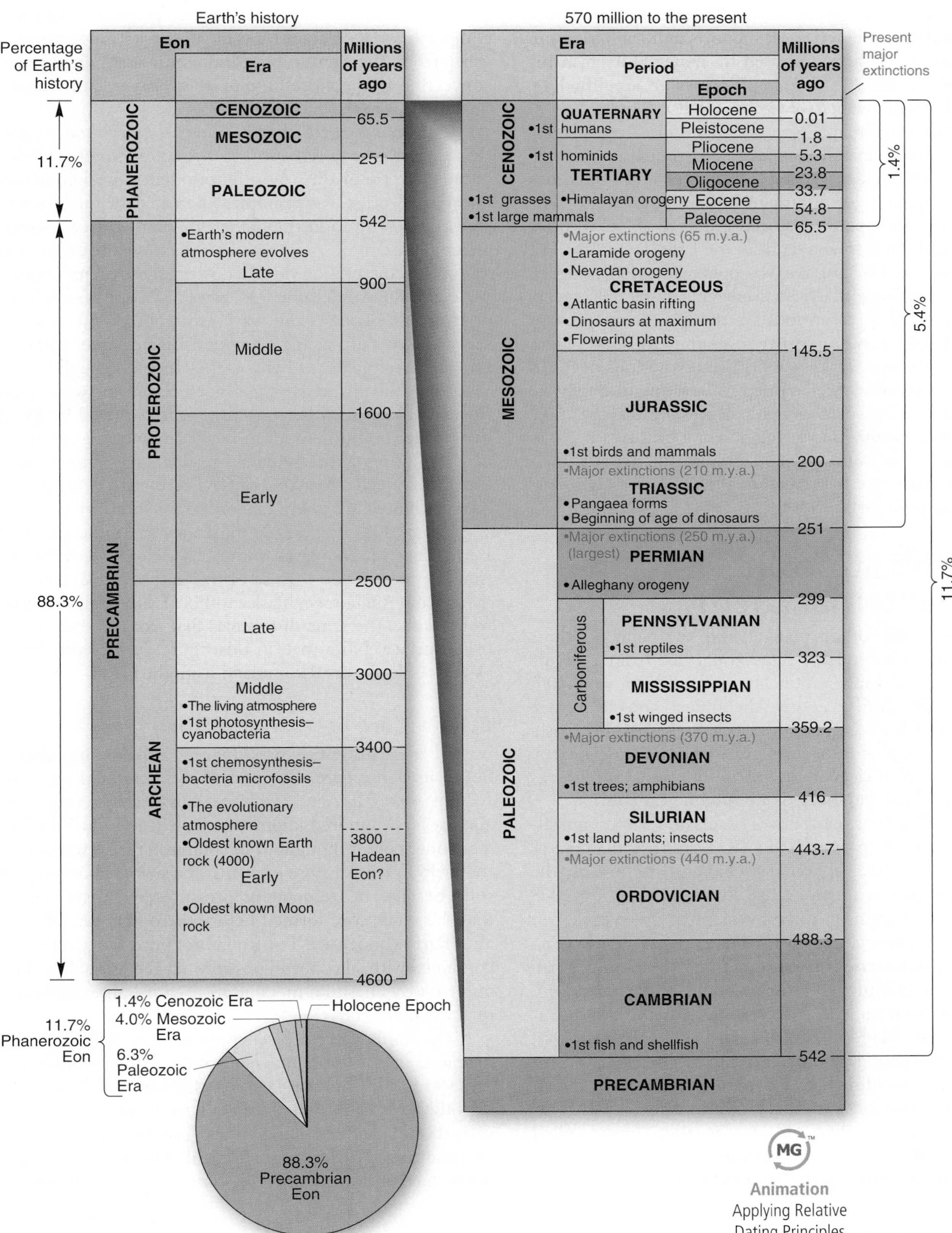

FIGURE 11.1 Geologic time scale, showing highlights of Earth's history.
The six major extinctions or depletions of life forms are shown in red. The sixth extinction episode is under way at the present time. In the column to the left, note that 88.3% of geologic time occurred during the Precambrian Eon. Dates appear in m.y.a. (million years ago). [Data and update from Geological Society of America and *Nature* 429 (May 13, 2004): 124–125.]

and from the landscape record of volcanic eruptions, earthquakes, and exogenic processes supports uniformitarianism. The concept was first proposed by James Hutton in his *Theory of the Earth* (1795) and was later amplified by Charles Lyell in *Principles of Geology* (1830).

Within the principle of uniformitarianism, catastrophic events such as massive landslides, earthquakes, volcanic episodes, extraterrestrial asteroid impacts, and sometimes recurring episodes of mountain building punctuate geologic time. Vast superfloods occurred as ice-blocked lakes tore through their frozen dams in the later days of the last ice age, such as glacial Lake Missoula, which left evidence across a broad region. These localized catastrophic events occur as small interruptions in the generally uniform processes that shape the slowly evolving landscape. Here the *punctuated equilibrium* concept (interruptions occur in the flow of events, systems jump to new operation levels) studied in the life sciences and paleontology might apply to aspects of Earth's long developmental history.

We start our journey deep within the planet. A knowledge of Earth's internal structure and energy is key to understanding the surface.

Earth's Structure and Internal Energy

Along with the other planets and the Sun, Earth is thought to have condensed and congealed from a nebula of dust, gas, and icy comets about 4.6 billion years ago (Chapter 2). Until recently, the oldest known surface rocks on Earth were in northwestern Canada. The Acasta Gneiss is a 3.96-billion-year-old formation; the nearby Slave Province rocks date to 4.03 billion years old. Rocks from Greenland date back to 3.8 billion years old.

Detrital zircons (particles of preexisting zirconium silica oxides that formed rock) in Western Australia date to between 4.2 and 4.4 billion years old and are possibly the oldest materials in Earth's crust. Recent research in 2008 found mineral grains in the Nuvvuagittuq greenstone belt of northern Quebec, Canada, that date to 4.28 billion years old. These discoveries tell us something significant: Earth was forming continental crust at least 4 billion years ago, during the Archean Eon. Although not an official designation, note the Hadean Eon in Figure 11.1, the proposed name for the period before the Archean.

As Earth solidified, gravity sorted materials by density. Heavier substances such as iron gravitated slowly to its center, and lighter elements such as silica slowly welled upward to the surface and became concentrated in the crust. Consequently, Earth's interior is sorted into roughly concentric layers, each one distinct in either chemical composition or temperature. Heat energy migrates outward from the center by conduction and by physical convection in the more fluid or plastic layers in the mantle and nearer the surface.

Our knowledge of Earth's *internal differentiation* into these layers is acquired entirely through indirect evidence because we are unable to drill more than a few kilometers into Earth's crust. For example, when an earthquake sends shock waves through the planet, the cooler areas, which generally are more rigid, transmit these **seismic waves** at a higher velocity than do the hotter areas, where seismic waves are slowed to lower velocity. Plastic zones simply do not transmit some seismic waves; they absorb them. Some seismic waves are reflected as densities change, whereas others are refracted, or bent, as they travel through Earth.

Thus, the distinctive ways in which seismic waves pass through Earth and the time they take to travel between two surface points help seismologists deduce the structure of Earth. This is the science of *seismic tomography*, as if Earth is subjected to a kind of CAT scan with every earthquake. Figure 11.2 is a model of Earth's interior.

Figure 11.3 (page 302) illustrates the dimensions of Earth's interior compared with surface distances in North America to give you a sense of size and scale. An airplane flying from Anchorage, Alaska, to Fort Lauderdale, Florida, would travel the same distance as that from Earth's center to its surface. Note that, in this figure, Earth's thin crust extends only about 30 km inland from the coast.

Earth's Core and Magnetism

A third of Earth's entire mass, but only a sixth of its volume, lies in its dense core. The **core** is differentiated into two regions—*inner core* and *outer core*—divided by a transition zone several hundred kilometers wide (see Figure 11.2b). The inner core is thought to be solid iron that is well above the melting temperature of iron at the surface, but remains solid because of tremendous pressure. Scientists believe that the inner core formed before the outer core, shortly after Earth condensed. The iron is not pure, but probably is combined with silicon, and possibly oxygen and sulfur. The outer core is molten, metallic iron with lighter densities than the inner core.

The fluid outer core generates at least 90% of Earth's magnetic field and the magnetosphere that surrounds and protects Earth from solar wind and cosmic radiation. One hypothesis explains that circulation in the outer core converts thermal and gravitational energy into magnetic energy, producing Earth's magnetic field. The locations of the north and south magnetic poles are surface expressions of

How much does our planet weigh? A revised estimate of Earth's mass, or weight, calculated in 2000 set its weight at 5.972 sextillion metric tons (5972 followed by 18 zeros).

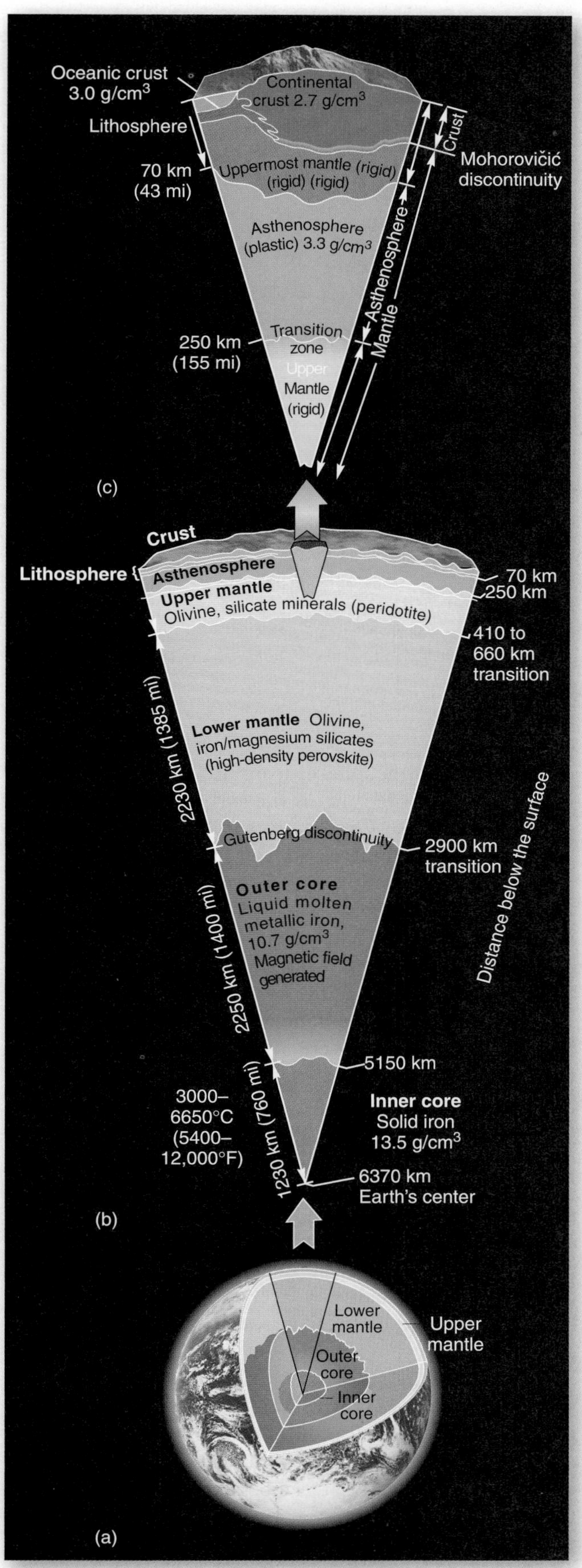

FIGURE 11.2 Earth in cross section.
(a) Cutaway showing Earth's interior. (b) Earth's interior in cross section, from the inner core to the crust. (c) Detail of the structure of the lithosphere and its relation to the asthenosphere. (For comparison to the densities noted, the density of water is 1.0 g/cm³, and mercury, a liquid metal, is 13.0 g/cm³.) [Photo from NASA.]

Earth's magnetic field and migrate as plotted on the map in *Geosystems Now* for this chapter.

An intriguing feature of Earth's magnetic field is that its polarity sometimes fades to zero and then returns to full strength, with the NMP and SMP reversed. In the process, the field does not blink on and off, but instead diminishes slowly to low intensity, perhaps 25% strength, and then rapidly regains full power. This **geomagnetic reversal** has taken place nine times during the past 4 million years and hundreds of times over Earth's history. During the transition interval of low strength, Earth's surface receives higher levels of cosmic radiation and solar particles, but note that past reversals do not correlate with species extinctions. The evolution of life has weathered many of these transitions.

The average period of a magnetic reversal is about 500,000 years; several hundred documented occurrences varied from as short as 20,000 to 30,000 years to 50 million years. Transition periods of low intensity last from 1000 to 10,000 years. The last reversal was 790,000 years ago, preceded by a transition ranging from 2000 years in length along the equator to 10,000 years in the midlatitudes. Given present rates of magnetic field decay over the last 150 years, we are perhaps 1000 years away from entering the next phase of field changes, although there is no expected pattern to use to forecast the timing. An apparent trend in recent geologic time is toward more frequent reversals. The obvious question for us is, Why does this happen?

The reasons for these magnetic reversals are unknown. However, the spatial patterns they create at Earth's surface are a key tool in understanding the evolution of landmasses and the movements of the continents. When new iron-bearing rocks solidify from molten material (lava) at Earth's surface, the small magnetic particles in the rocks align according to the orientation of the magnetic poles at that time. As the rocks cool and solidify, this alignment locks in place. When Earth is without polarity in its magnetic field, a random pattern of magnetism in crustal rock results.

All across Earth, rocks of the same age bear an identical record of magnetic reversals in the form of measurable magnetic "stripes" in any magnetic material they contain, such as iron particles. These stripes illustrate global patterns of changing magnetism. Matching segments allow scientists to reassemble past continental arrangements. Later in this chapter, we see the importance of these magnetic reversals.

Earth's Mantle

Figure 11.2b shows a transition zone several hundred kilometers wide at an average depth of about 2900 km (1800 mi), dividing Earth's outer core from its mantle. This zone is a *discontinuity*, or a place where physical

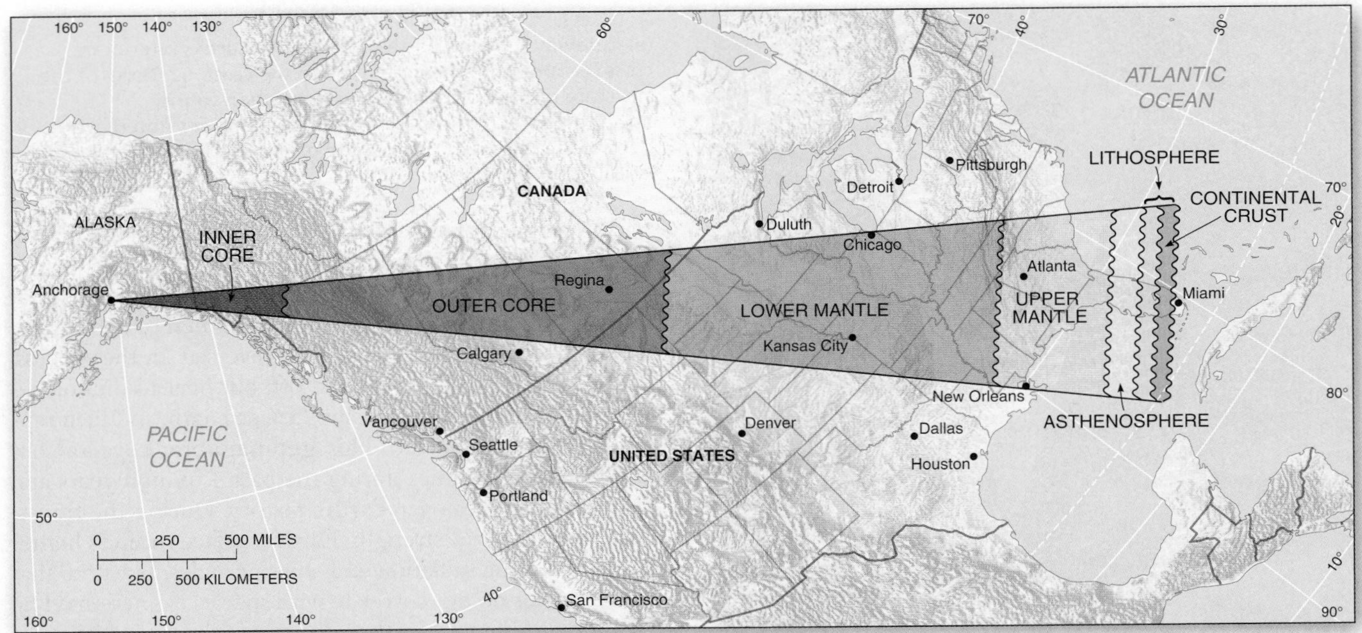

FIGURE 11.3 Distances from core to crust.
The distance from Anchorage, Alaska, to Fort Lauderdale, Florida, is the same distance as that from Earth's center to its outer crust. The continental crust thickness extends only an equivalent distance to the outskirts of Fort Lauderdale and its suburbs, some 30 km (18.6 mi) from the coast.

differences occur between adjoining regions in Earth's interior. Scientists at the California Institute of Technology analyzed more than 25,000 earthquakes and determined that this transition area, called the *Gutenberg discontinuity*, is uneven, with ragged peak-and-valley-like formations. Experts think that this zone of rough texture and low velocity creates some of the motion in Earth's mantle.

The lower and upper **mantle** represents about 80% of Earth's total volume. The mantle is rich in oxides of iron and magnesium and silicates (FeO, MgO, and SiO_2), which are dense and tightly packed at depth, grading to lesser densities toward the surface. A broad transition zone of several hundred kilometers, centered between 410 and 660 km (255–410 mi) below the surface, separates the upper mantle from the lower mantle. The entire mantle experiences a gradual temperature increase with depth and a stiffening due to increased pressures. The denser lower mantle is thought to contain a mixture of iron, magnesium, and silicates, with some calcium and aluminum. One hypothesis states that a tremendous amount of water is bound up with minerals in the mantle.

The upper mantle divides into three fairly distinct layers: upper mantle, asthenosphere, and uppermost mantle, shown in Figure 11.2c. The uppermost mantle is a high-velocity zone just below the crust where seismic waves transmit through a rigid, cooler layer. This uppermost mantle, along with the crust, makes up the **lithosphere**, approximately 45–70 km (28–43 mi) thick.

Below the lithosphere, from about 70 km down to 250 km (43 mi to 155 mi), is the **asthenosphere**, or plastic layer (from the Greek *asthenos*, meaning "weak"). It contains pockets of increased heat from radioactive decay and is susceptible to slow convective currents in these hotter, less dense materials.

Because of its dynamic condition, the asthenosphere is the least rigid region of the mantle, with densities averaging 3.3 g/cm^3 (or 3300 kg/m^3). About 10% of the asthenosphere is molten in uneven patterns and hot spots. The resulting slow movement in this zone disturbs the overlying crust and creates tectonic activity—the folding, faulting, and general deformation of surface rocks. The depth affected by convection currents is the subject of much scientific research. One body of evidence states that mixing occurs throughout the entire mantle, upwelling from great depths at the core–mantle boundary—sometimes small blobs and other times "megablobs" of mantle convect toward and away from the crust. Another view states that mixing in the mantle is layered, segregated above and below the 660-km boundary. Presently, evidence indicates some truth in both positions.

As an example, there are hot spots on Earth, such as those under Hawai'i, Easter Island, and Tahiti, that appear to be at the top of tall plumes of rising mantle rock anchored deep in the lower mantle, where the upward flow of warmer, less-dense material begins. The plume beneath Iceland appears to begin at the 660-km transition. Yet there are other surface hot spots that sit atop smaller plumes going down some 200 km (124 mi).

Earth's Lithosphere and Crust

The lithosphere includes the **crust** and uppermost mantle to about 70 km (43 mi) in depth (Figure 11.2c). An important internal boundary between the crust and the high-velocity portion of the uppermost mantle is another discontinuity, called the **Mohorovičić discontinuity**, or **Moho** for short. It is named for the Yugoslavian seismologist who determined that seismic waves change at this depth due to sharp contrasts of materials and densities.

Figure 11.2c illustrates the relation of the crust to the rest of the lithosphere and the asthenosphere below. Crustal areas beneath mountain masses extend deep, perhaps to 50–60 km (31–37 mi), whereas the crust beneath continental interiors averages about 30 km (19 mi) in thickness. Oceanic crust averages only 5 km (3 mi). The crust is only a fraction of Earth's overall mass. Drilling through the crust and Moho discontinuity (the crust–mantle boundary) into the uppermost mantle remains an elusive scientific goal.

The longest-lasting deep-drilling attempt is on the northern Kola Peninsula near Zapolyarny, Russia, 250 km north of the Arctic Circle—the *Kola Borehole (KSDB)*. Twenty years of high-technology drilling (1970–1989) produced a hole 12.23 km deep (7.6 mi, or 40,128 ft), purely for exploration and science. Crystalline rock 1.4 billion years old at 180°C (356°F) was reached. The site has other active boreholes, and the fifth is under way. (See an analysis log at http://www.icdp-online.de/.) A record holder for depth for a gas well is in Oklahoma; it was stopped at 9750 m (32,000 ft) when the drill bit ran into molten sulfur.

Oceanic crust is thinner than continental crust and is the object of several drilling attempts. The Integrated Ocean Drilling Program (IODP), a cooperative effort among the governments of Japan, the United States, and the European Union, replaced the former Ocean Drilling Program (ODP) effort directed by Texas A & M University. Over the past 40 years, about 1700 bore holes were completed, yielding 160 km of sediment and rock cores and more than 35,000 samples for researchers. Much of this science windfall helped put together the puzzle of geologic time and plate tectonics. See http://www.oceandrilling.org/ for the IODP home page information.

Japan's *Chikyu*, the largest deep-ocean drilling ship, began work in 2007. The ship is the first capable of high-latitude drilling in the Arctic (Figure 11.4). The *Chikyu* is "riserless" in that it uses seawater as the primary drilling fluid. It can drill to 7 km, more than three times the capability of the drilling ship *JOIDES Resolution*, operated by the United States as its Scientific Ocean Drilling Vessel. The *JOIDES*, workhorse of the former ODP, has now been converted to riserless status as well (see http://joidesresolution.org/).

The composition and texture of continental and oceanic crusts are quite different, and this difference is a key to the concept of drifting continents. The oceanic crust is denser than continental crust. In collisions, the denser oceanic material plunges beneath the lighter, more buoyant continental crust.

FIGURE 11.4 Ocean drilling ship.
The new *Chikyu* research ship began drilling core samples from the ocean floor in 2007. [IODP.]

- *Continental crust* is essentially **granite**; it is crystalline and high in silica, aluminum, potassium, calcium, and sodium. (Sometimes continental crust is called *sial*, shorthand for *si*lica and *al*uminum.) Continental crust is relatively low in density, averaging 2.7 g/cm^3 (or 2700 kg/m^3). Compare this with other densities given in Figure 11.2.
- *Oceanic crust* is **basalt**; it is granular and high in silica, magnesium, and iron. (Sometimes oceanic crust is called *sima*, shorthand for *si*lica and *ma*gnesium.) It is denser than continental crust, averaging 3.0 g/cm^3 (or 3000 kg/m^3).

Buoyancy is the principle that something less dense, such as wood, floats in something denser, such as water. The principles of buoyancy and balance were combined in the 1800s into the important principle of **isostasy**, which explains certain vertical movements of Earth's crust.

Think of Earth's crust as floating on the denser layers beneath, much as a boat floats on water. Where the load is greater, owing to glaciers, sediment, or mountains, the crust tends to sink, or ride lower in the asthenosphere. Without that load (for example, when a glacier melts), the crust rides higher, in a recovery uplift known as *isostatic rebound*. Thus, the entire crust is in a constant state of compensating adjustment, or isostasy, slowly rising and sinking in response to its burdens, as it is pushed, and dragged, and pulled over the asthenosphere (Figure 11.5). An interesting development in Alaska relates this concept of isostatic rebound and climate change.

The retreat of glacial ice following the last ice-age cycle unloaded much weight off the crust in Alaska. Researchers from the Geophysical Institute at the University of Alaska–Fairbanks, using an array of global

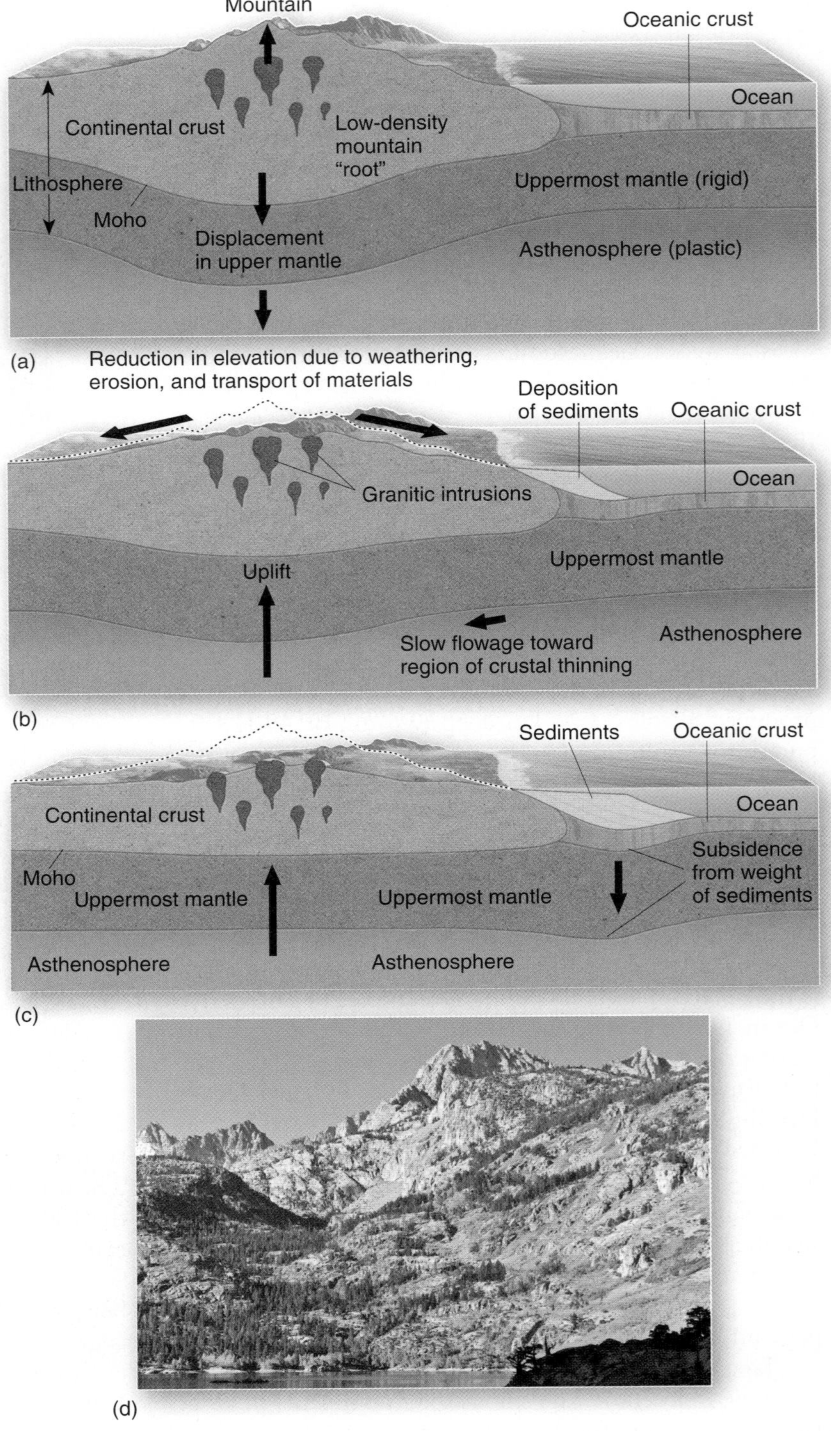

FIGURE 11.5 Isostatic adjustment of the crust.
Earth's entire crust is in a constant state of compensating adjustment, as suggested by these three sequential stages: (a) The mountain mass slowly sinks, displacing mantle material. (b) Because of the loss of mass through erosion and transportation, the crust isostatically adjusts upward, and sediments accumulate in the ocean. (c) As the continent thins and rises, the heavy sediment load offshore begins to deform the lithosphere beneath the ocean. (d) The melting of ice from the last ice age and losses of overlying sediments are thought to produce an ongoing isostatic uplift of portions of the Sierra Nevada batholith. [Bobbé Christopherson.]

positioning system (GPS) receivers, measured isostatic rebound of the crust following glacial ice losses. They expected to find a slowed rate of crustal rebound in southeastern Alaska as compared to the more rapid response in the distant past when the ice first retreated.

Instead, they detected some of the most rapid vertical motion on Earth, averaging about 36 mm (1.42 in.) per year. Scientists attribute this isostatic rebound to the loss of modern glaciers in the region, especially the areas from Yakutat Bay and the Saint Elias Mountains in the north, through Glacier Bay, Juneau, and on south through the inland passage. This rapid rebound is attributable to glacial melt and retreat over the past 150 years and correlates with record warmth across Alaska.

Earth's crust, an irregular, brittle layer, is the outermost shell, residing restlessly on a dynamic and diverse interior. Let us examine the processes at work on this crust and the variety of rock types that compose the landscape.

The Geologic Cycle

Earth's crust is in an ongoing state of change, being formed, deformed, moved, and broken down by physical, chemical, and biological processes. While the endogenic (internal) system is at work building landforms, the exogenic (external) system is busily wearing them down. This vast give-and-take at the Earth–atmosphere–ocean interface is the **geologic cycle**. It is fueled from two sources—Earth's internal heat and solar energy from space—influenced by the ever-present leveling force of Earth's gravity.

Figure 11.6 illustrates the geologic cycle, combining many of the elements presented in this text. As you can see in the figure, the geologic cycle is composed of three

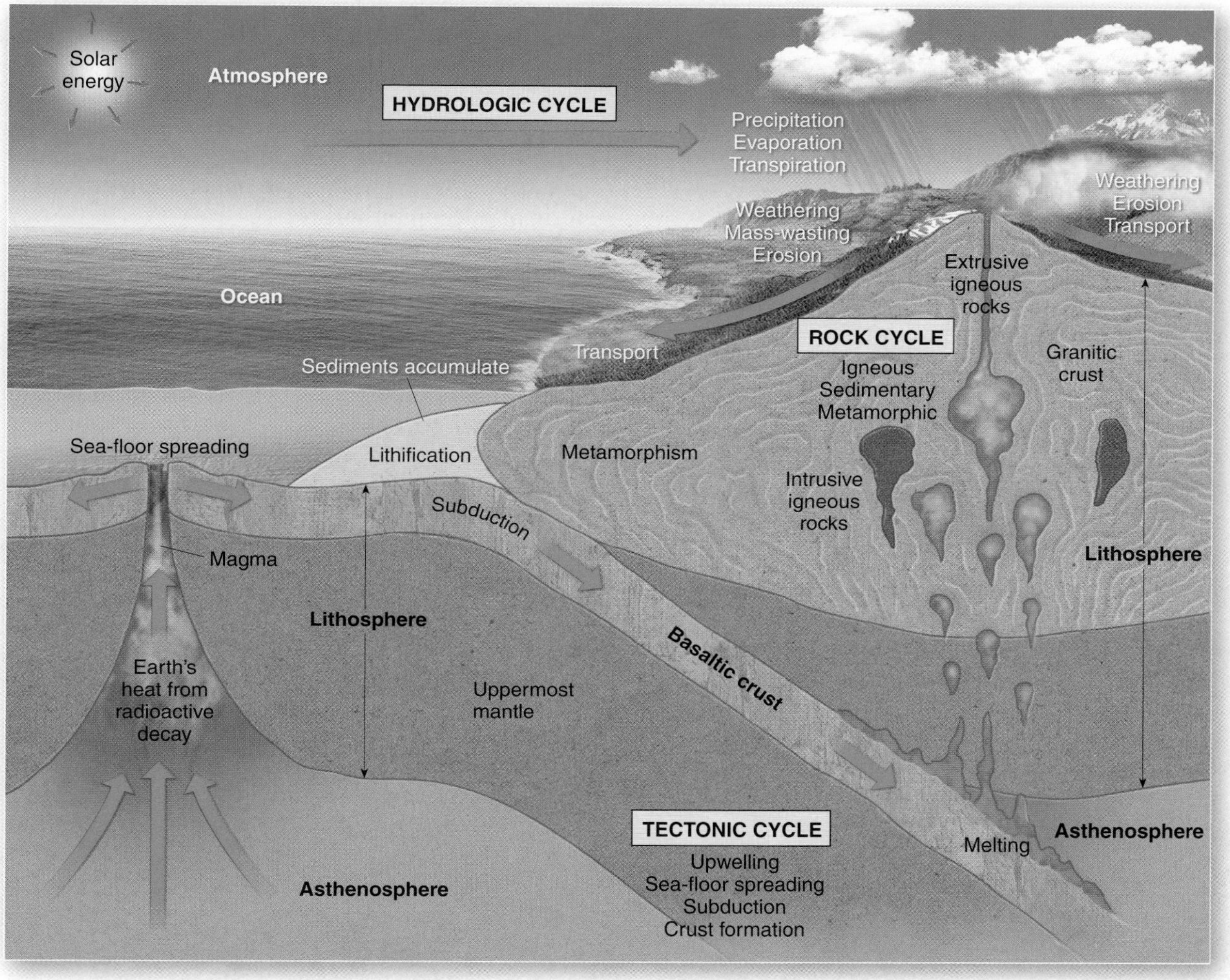

(a)

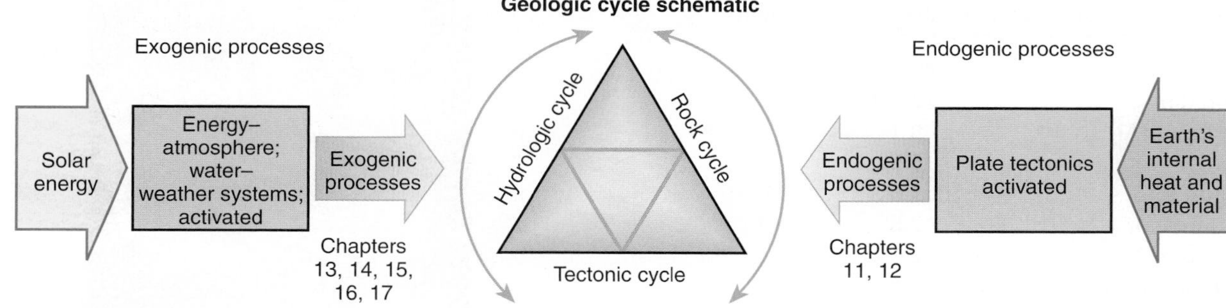

(b)

FIGURE 11.6 The geologic cycle.
(a) The geologic cycle is a model showing the interactive relation among the hydrologic cycle, rock cycle, and tectonic cycle. (b) Earth's surface is where two dynamic systems—the endogenic (internal) and exogenic (external)—interact.

Animation
Convection in a Lava Lamp

principal cycles—hydrologic, rock, and tectonic. The *hydrologic cycle*, along the top of Figure 11.6, through erosion, transportation, and deposition, processes Earth materials with the chemical and physical action of water, ice, and wind. The *rock cycle* produces the three basic rock types found in the crust—igneous, metamorphic, and sedimentary. The *tectonic cycle* brings heat energy and new material to the surface and recycles material, creating movement and deformation of the crust.

The Rock Cycle

Only eight natural elements compose 98.5% of Earth's crust, and only two of these—oxygen and silicon—account for 74.3% (Table 11.1). Oxygen is the most reactive gas in the lower atmosphere, readily combining with other elements. For this reason, the percentage of oxygen is higher in the crust than the 21% concentration in the atmosphere. The internal differentiation process, in which less dense elements migrate toward the surface, explains the relatively large percentages of lightweight elements such as silicon and aluminum in the crust.

Minerals and Rocks Earth's elements combine to form minerals. A **mineral** is an inorganic, or nonliving, natural compound having a specific chemical formula and usually possessing a crystalline structure. The combination of elements and the crystal structure give each mineral its characteristic hardness, color, density, and other properties. For example, the common mineral *quartz* is silicon dioxide, SiO_2, and has a distinctive six-sided crystal. Note that mercury, a liquid metal, is an exception to this definition, for it has no crystalline structure; also note that water is not a mineral, but ice does fit the definition of a mineral.

Of the more than 4200 minerals, about 30 comprise the *rock-forming minerals* most commonly encountered. *Mineralogy* is the study of the composition, properties, and classification of minerals. (See http://webmineral.com/).

One of the most widespread mineral families on Earth is the *silicates* because silicon and oxygen are so common and because they readily combine with each other and with other elements. Roughly 95% of Earth's crust is made up of silicates. This mineral family includes quartz, feldspar, clay minerals, and numerous gemstones. *Oxides* are another group of minerals in which oxygen combines with metallic elements, such as iron, to form hematite, Fe_2O_3. Also, there are the *sulfides* and *sulfates* groups, in which sulfur compounds combine with metallic elements to form pyrite (FeS_2) and anhydrite ($CaSO_4$), respectively.

Another important mineral family is the *carbonate* group, which features carbon in combination with oxygen and other elements such as calcium, magnesium, and potassium. An example is the mineral calcite ($CaCO_3$), a form of calcium carbonate.

A **rock** is an assemblage of minerals bound together (such as granite, a rock containing three minerals), or a mass of a single mineral (such as rock salt), or undifferentiated material (such as the noncrystalline glassy obsidian), or even solid organic material (such as coal). Thousands of different rocks have been identified. All can be sorted into one of three kinds, depending on the processes that formed them: *igneous* (melted), *sedimentary* (from settling out), and *metamorphic* (altered). Figure 11.7 illustrates these three processes and the interrelations among them that constitute the **rock cycle**. Let us examine each rock-forming process.

Igneous Processes

An **igneous rock** is one that solidifies and crystallizes from a molten state. Familiar examples are granite, basalt, and rhyolite. Igneous rocks form from **magma**, which is molten rock beneath the surface (hence the name *igneous*, which means "fire-formed" in Latin). Magma is fluid, highly gaseous, and under tremendous pressure. Either it

TABLE 11.1 Common Elements in Earth's Crust

Element	Percentage of Earth's Crust by Weight
Oxygen (O)	46.6
Silicon (Si)	27.7
Aluminum (Al)	8.1
Iron (Fe)	5.0
Calcium (Ca)	3.6
Sodium (Na)	2.8
Potassium (K)	2.6
Magnesium (Mg)	2.1
All others	1.5
Total	100.00

Note: A quartz crystal (SiO_2) consists of Earth's two most abundant elements, silicon (Si) and oxygen (O). Inset photo from *Laboratory Manual in Physical Geology*, 3rd ed., R. M. Busch, ed. © 1993 by Macmillan Publishing Co.

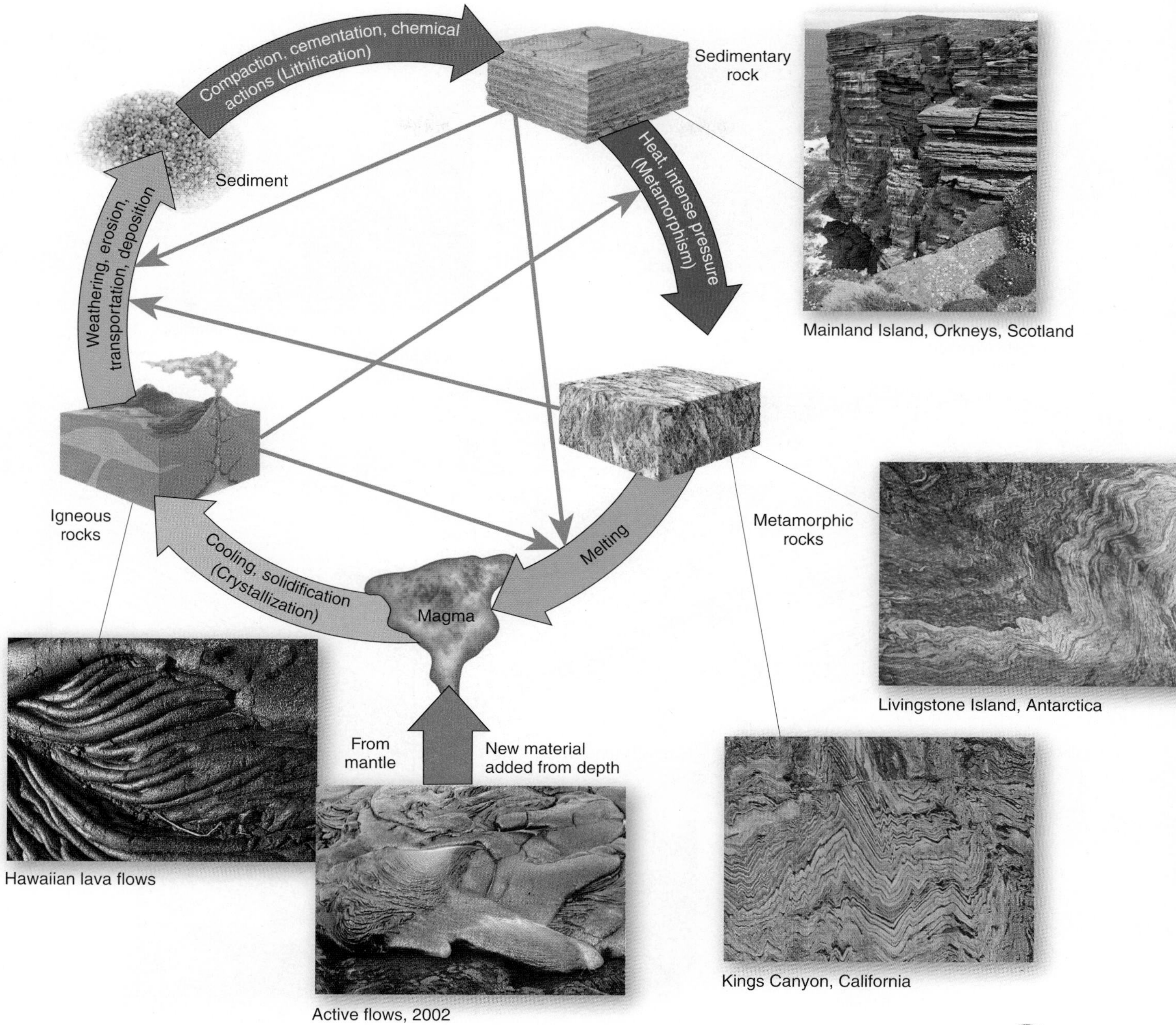

FIGURE 11.7 The rock cycle.
A rock-cycle schematic demonstrates the relation among igneous, sedimentary, and metamorphic processes. The arrows indicate that each rock type can enter the cycle at various points and be transformed into other rock types. [Photos by Bobbé Christopherson.]

MG Animation The Rock Cycle

intrudes into crustal rocks, cools, and hardens, or it *extrudes* onto the surface as **lava**.

The cooling history of an igneous rock—how fast it cooled and how steadily its temperature dropped—determines its crystalline physical characteristics, or crystallization. Igneous rocks range from coarse-grained (slower cooling, with more time for larger crystals to form) to fine-grained or glassy (faster cooling).

Igneous rocks make up approximately 90% of Earth's crust, although sedimentary rocks (sandstone, shale, limestone), soil, or oceans frequently cover them. Figure 11.8 illustrates the variety of occurrences of igneous rocks, both on and beneath Earth's surface.

Intrusive and Extrusive Igneous Rocks Intrusive igneous rock that cools slowly in the crust forms a **pluton**, a general term for any intrusive igneous rock body, regardless of size or shape, that invaded layers of crustal rocks. The Roman god of the underworld, Pluto, is the namesake. The largest pluton form is a **batholith**, defined as an irregular-shaped mass with a surface greater than 100 km^2 (40 mi^2) (Figure 11.9a, page 309). Batholiths form the mass

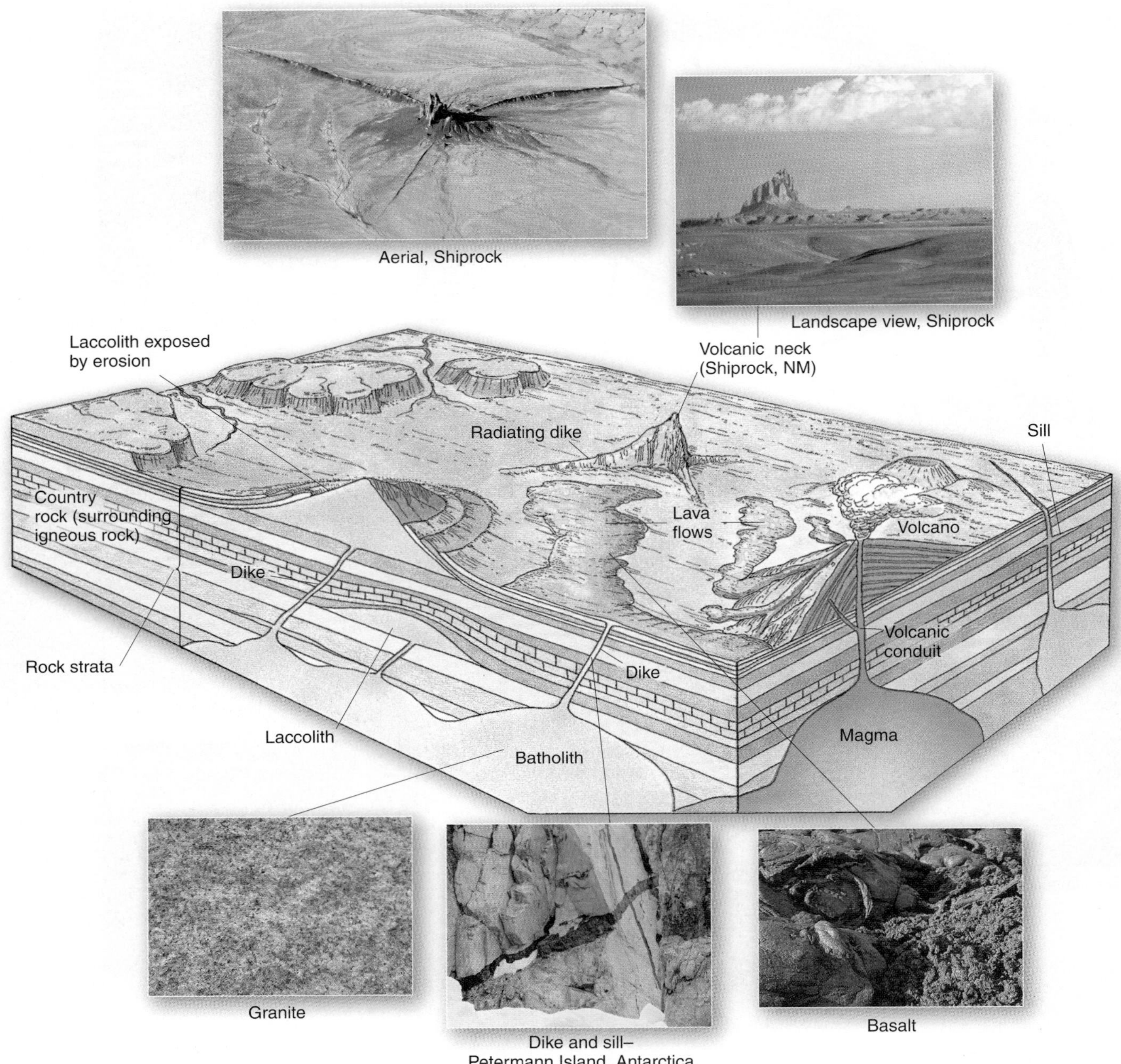

FIGURE 11.8 Igneous rock types and landforms.
The variety of igneous rocks, both intrusive (below the surface) and extrusive (on the surface), and associated landforms. Inset photographs show samples of granite (intrusive), basalt (extrusive), and other igneous forms. [Photos of rock samples and aerial by Bobbé Christopherson; surface photo of Shiprock volcanic neck by author.]

of many large mountain ranges—for example, the Sierra Nevada batholith in California (see Figure 11.5), the Idaho batholith, and the Coast Range batholith of British Columbia and Washington State.

Smaller plutons include the magma conduits of ancient volcanoes that have cooled and hardened. Those that form parallel to layers of sedimentary rock are *sills*; those that cross layers of the rock they invade are *dikes*. You see these two forms in Figure 11.8 at Petermann Island, Antarctica. Magma also can bulge between rock strata and produce a lens-shaped body, a *laccolith*, a type of sill. In addition, magma conduits themselves may solidify in roughly cylindrical forms that stand starkly above the landscape when finally exposed by weathering

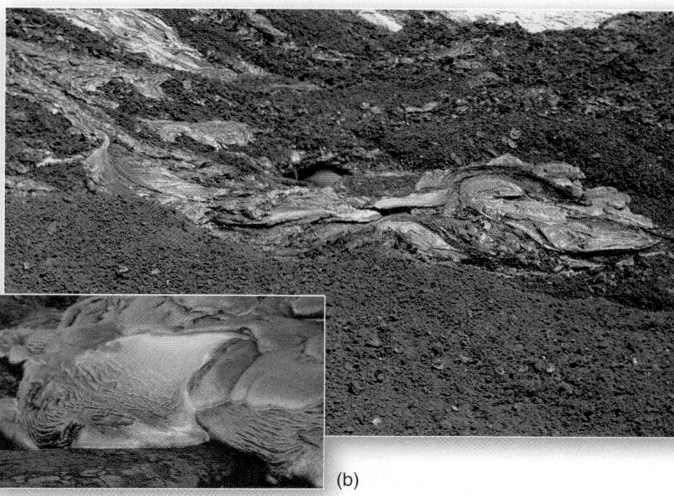

FIGURE 11.9 Intrusive and extrusive rocks.
(a) Exposed intrusive granite, part of a batholith. (b) Basaltic lava flows on Hawai'i. The glowing opening is a skylight into an active lava tube where molten lava flows; the shiny surface is where lava recently flowed out of the skylight. [Bobbé Christopherson.]

and erosion. Shiprock *volcanic neck* in New Mexico is such a feature, rising 518 m (1700 ft) above the surrounding plain, as suggested by the art in Figure 11.8 and shown in the inset photos; note the radiating dikes in the aerial photo. Weathering action of air, water, and ice can expose all of these intrusive forms.

Volcanic eruptions and flows produce extrusive igneous rock, such as lava that cools and forms basalt (Figure 11.9b). Chapter 12 presents volcanism and treats this in detail.

Classifying Igneous Rocks Mineral composition and texture usually classify igneous rocks (Table 11.2). The two broad categories are

1. *Felsic* igneous rocks—derived both in composition and in name from *fel*dspar and *si*li*ca*. Felsic minerals are generally high in silica, aluminum, potassium, and sodium and have low melting points. Rocks formed from felsic minerals generally are lighter in color and less dense than mafic mineral rocks.
2. *Mafic* igneous rocks—derived both in composition and in name from *ma*gnesium and *f*err*ic* (Latin for iron). Mafic minerals are low in silica, are high in magnesium and iron, and have high melting points. Rocks formed from mafic minerals are darker in color and of greater density than felsic mineral rocks.

The same magma that produces coarse-grained granite (when it slowly cools beneath the surface) can form fine-grained *rhyolite* (when it cools above the surface—see inset photo in Table 11.2). If it cools quickly, magma having silica content comparable to granite and rhyolite can form the dark, smoky, glassy-textured rock called *obsidian*, or volcanic glass (see inset photo in Table 11.2). Another glassy rock, *pumice*, forms when escaping gases bubble a frothy texture into the lava. Pumice is full of small holes, is light in weight, and is low enough in density to float in water (see inset photo in Table 11.2).

On the mafic side, basalt is the most common fine-grained extrusive igneous rock. It makes up the bulk of the ocean floor, accounting for 71% of Earth's surface. It appears in lava flows such as those on the big island of Hawai'i (Figure 11.9b). An intrusive counterpart to basalt, formed by slow cooling of the parent magma, is *gabbro*.

Sedimentary Processes

Solar energy and gravity drive the process of *sedimentation*, with water as the principal transporting medium. Existing rock is disintegrated and dissolved by weathering, picked up and moved by erosion and transportation, and deposited along river, beach, and ocean sites, where burial initiates the rock-forming process. The formation of **sedimentary rock** involves **lithification** processes of cementation, compaction, and a hardening of sediments.

Most sedimentary rocks derive from fragments of existing rock or organic materials. Bits and pieces of former rocks—principally quartz, feldspar, and clay minerals—erode and then are mechanically transported by water, ice (glacial action), wind, and gravity. They are transported from "higher-energy" sites, where the carrying medium has the energy to pick up and move them, to "lower-energy" sites, where the material is dumped.

TABLE 11.2 Igneous Rock Minerals

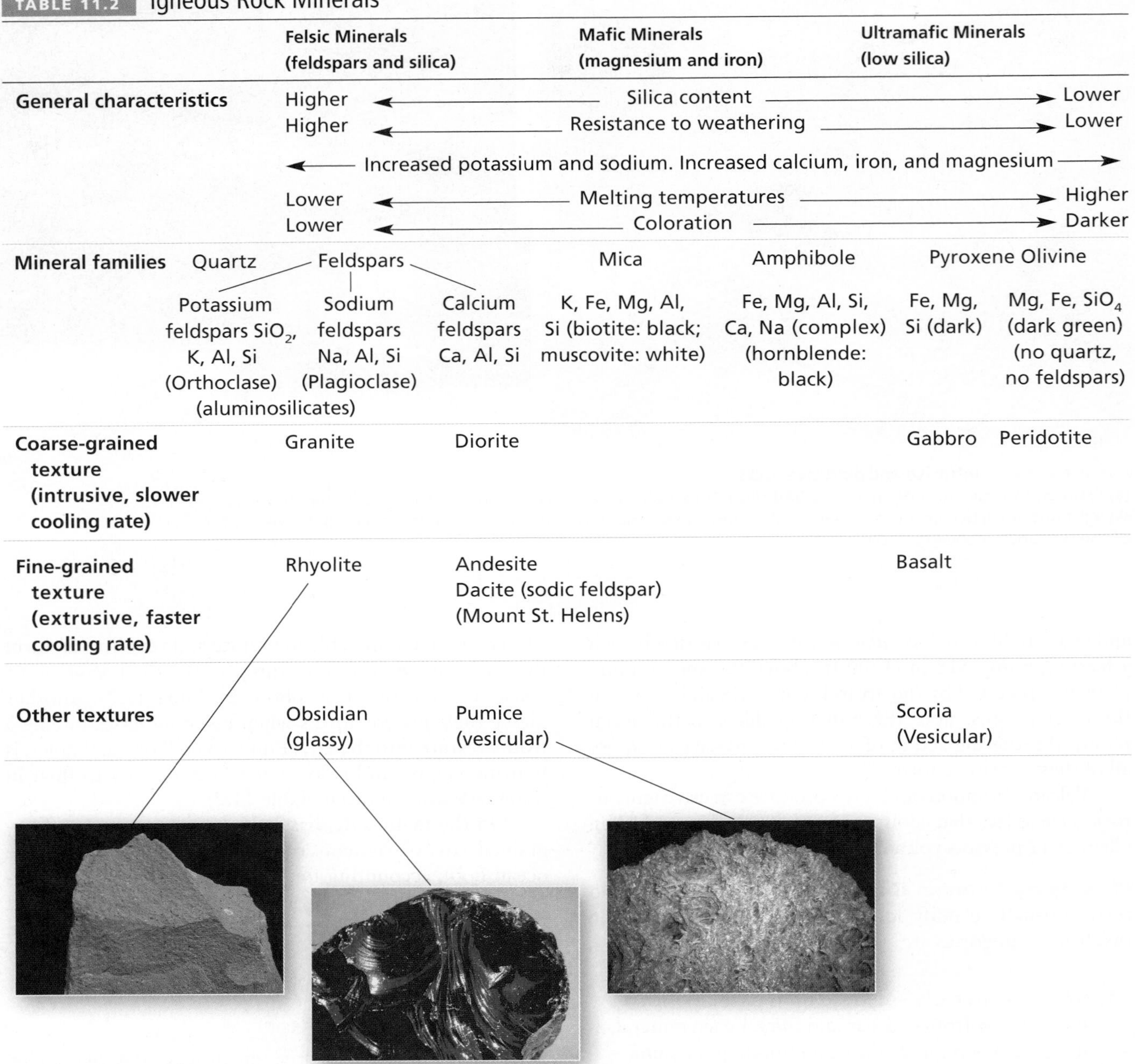

	Felsic Minerals (feldspars and silica)			Mafic Minerals (magnesium and iron)		Ultramafic Minerals (low silica)	
General characteristics	Higher ← Silica content → Lower						
	Higher ← Resistance to weathering → Lower						
	← Increased potassium and sodium. Increased calcium, iron, and magnesium →						
	Lower ← Melting temperatures → Higher						
	Lower ← Coloration → Darker						
Mineral families	Quartz	Feldspars		Mica	Amphibole	Pyroxene	Olivine
	Potassium feldspars SiO_2, K, Al, Si (Orthoclase) (aluminosilicates)	Sodium feldspars Na, Al, Si (Plagioclase)	Calcium feldspars Ca, Al, Si	K, Fe, Mg, Al, Si (biotite: black; muscovite: white)	Fe, Mg, Al, Si, Ca, Na (complex) (hornblende: black)	Fe, Mg, Si (dark)	Mg, Fe, SiO_4 (dark green) (no quartz, no feldspars)
Coarse-grained texture (intrusive, slower cooling rate)	Granite		Diorite			Gabbro	Peridotite
Fine-grained texture (extrusive, faster cooling rate)	Rhyolite		Andesite Dacite (sodic feldspar) (Mount St. Helens)			Basalt	
Other textures	Obsidian (glassy)		Pumice (vesicular)			Scoria (Vesicular)	

Source: Photos from R. M. Busch, ed., *Laboratory Manual in Physical Geology,* 3rd ed., © 1993 by Macmillan Publishing Co.

The common sedimentary rocks are *sandstone* (cemented sand), *shale* (compacted mud), *limestone* (lithified bones and shells or calcium carbonate that precipitated in ocean and lake waters), and *coal* (ancient plant remains that became compacted into rock).

Various cements, depending on availability, fuse rock particles together. Lime, or calcium carbonate ($CaCO_3$), is the most common, followed by iron oxides (Fe_2O_3) and silica (SiO_2). Drying (dehydration), heating, or chemical reactions can also unite particles.

Some minerals, such as calcium carbonate, dissolve into solution and form sedimentary deposits by precipitating from those solutions to form rock. This is an important process in both the oceanic and karst (weathered limestone; see Chapter 13) environments.

The layered strata of sedimentary rocks form an important record of past ages. **Stratigraphy** is the study of the sequence (superposition), thickness, and spatial distribution of strata. These sequences yield clues to the age and origin of the rocks. Figure 11.10a shows a sandstone sedimentary rock in a desert landscape. Note the multiple layers in the formation and how differently they resist weathering processes. The climatic history of this rock's formation environment is disclosed in the stratigraphy.

(a)

(b)

(c) Martian sedimentary formations

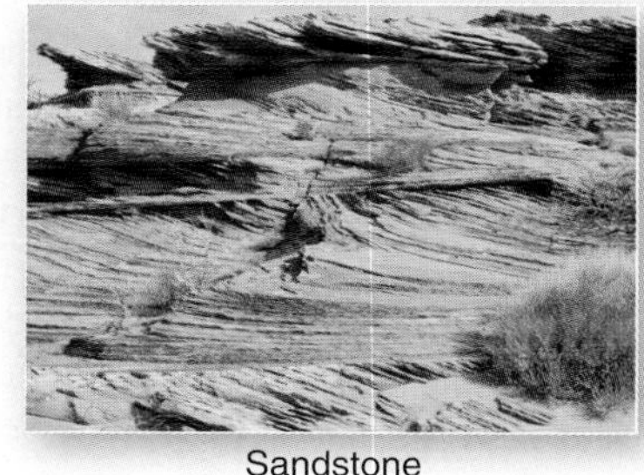
Sandstone

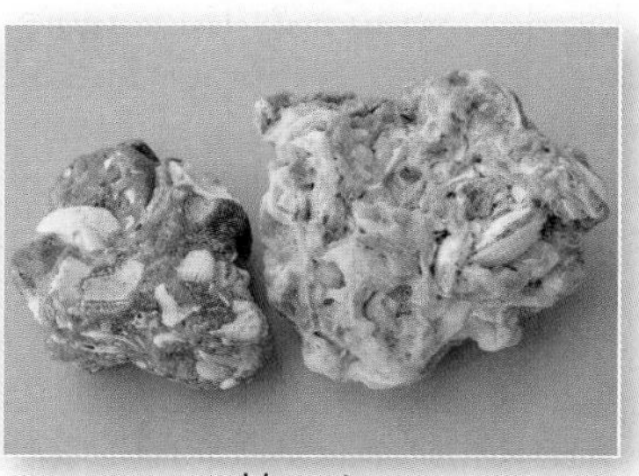
Limestone

FIGURE 11.10 **Sedimentary rock types.**
(a) Sandstone formation with sedimentary strata subjected to differential weathering; note weaker underlying siltstone, leaving a balanced rock above. (b) A limestone landscape, formed by chemical sedimentary processes, in south-central Indiana; inset samples have whole shells and some clasts (parts) cemented together. (c) Ancient sediment deposition is evident in the Martian western Arabia Terra (8° N and 7° W). [(a) and (b) Bobbé Christopherson. (c) *Mars Global Surveyor,* NASA/JPL/Malin Space Science Systems.]

TABLE 11.3 Clast Sizes and Related Sedimentary Rock Form

Unconsolidated Sediment	Grain Size	Rock Form
Boulders, cobbles	>80 mm	Conglomerate (breccia, if pieces are angular)
Pebbles, gravel	>2 mm	Conglomerate
Coarse sand	0.5–2.0 mm	Sandstone
Medium-to-fine sand	0.062–0.5 mm	Sandstone
Silt	0.0039–0.062 mm	Siltstone (mudstone)
Clay	<0.0039 mm	Shale

Figure 11.10c shows sedimentary rock strata on Mars that imply water deposition of sediments sometime in the distant Martian past.

The two primary sources of sedimentary rocks are *clastic sediments,* formed from the mechanically transported fragments of older rock, and *chemical sediments* from the dissolved minerals in solution, some having organic origins.

Clastic Sedimentary Rocks Weathered and fragmented rocks that are further worn in transport provide *clastic sediments.* Table 11.3 lists the range of *clast* sizes—everything from boulders to microscopic clay particles—and the form they take as *lithified rock.* Common sedimentary rocks from clastic sediments include siltstone or mudstone (silt-sized particles), shale (clay to silt-sized particles), and sandstone (sand-sized particles ranging from 0.06 to 2 mm in diameter).

Chemical Sedimentary Rocks *Chemical sedimentary* rocks are formed not from physical pieces of broken rock, but instead from dissolved minerals, transported in solution and chemically precipitated from solution (they are essentially nonclastic). The most common chemical sedimentary rock is **limestone**, which is lithified calcium carbonate, derived from inorganic and organic sources. A similar form is *dolomite*, which is lithified calcium–magnesium carbonate, $CaMg(CO_3)_2$. Limestone from marine organic origins is most common; it is biochemical—derived from shell and bone produced by biological activity (see inset photo in Figure 11.10b). Once formed, these rocks are vulnerable to chemical weathering, which produces unique landforms, as discussed in the weathering section of Chapter 13.

Chemical sediments from inorganic sources are deposited when water evaporates and leaves behind a residue of salts. These *evaporites* may exist as common salts, such as gypsum or sodium chloride (table salt). They often appear as flat, layered deposits across a dry landscape. The pair of photographs in Figure 11.11 dramatically demonstrates this process; one was made in Death Valley National Park one day after a record 2.57-cm (1.01-in.) rainfall, and the other photo was made 1 month later at the exact same spot.

Chemical deposition also occurs in the water of natural hot springs from chemical reactions between minerals and oxygen. This is a sedimentary process related to *hydrothermal activity*. An example is the massive deposit of travertine, a form of calcium carbonate, at Mammoth Hot Springs in Yellowstone National Park (Figure 11.12a).

Hydrothermal activity also occurs on the ocean floor along the rift valleys formed by sea-floor spreading. These "black smokers" belch dark clouds of hydrogen sulfides, minerals, and metals that hot water (in excess of 380°C, or 716°F) leached from the basalt. In contact with seawater, metals and minerals precipitate into deposits (Figure 11.12b). Along the Mid-Atlantic Ridge, one region of hydrothermal activity has 30- to 60-m (98- to 198-ft) towering-chimney vents of calcium carbonate that are at least 30,000 years old and still active.

Metamorphic Processes

Any rock, either igneous or sedimentary, may be transformed into a **metamorphic rock** by going through profound physical or chemical changes under pressure and increased temperature. (The name *metamorphic* comes from a Greek word meaning "to change form.") Metamorphic rocks generally are more compact than the original rock and therefore are harder and more resistant to weathering and erosion (Figure 11.13).

Several conditions can cause metamorphism. Most common is when subsurface rock is subjected to high temperatures and high compressional stresses over millions of years. Igneous rocks become compressed during collisions between slabs of Earth's crust (see the discussion of plate tectonics later in this chapter). Sometimes rocks simply are crushed under great weight when a crustal area is thrust beneath other crust. In another setting, igneous rocks may be sheared and stressed along earthquake fault zones, causing metamorphism.

Metamorphic rocks comprise the ancient roots of mountains. Exposed at the bottom of the inner gorge of the Grand Canyon in Arizona, a Precambrian (Archean) metamorphic rock, the Vishnu Schist, is a remnant of such a mountain root (Figure 11.13b). Despite the fact that the schist is harder than steel, the Colorado River has cut down into the uplifted Colorado Plateau, exposing and eroding these ancient rocks.

Another metamorphic condition occurs when sediments collect in broad depressions in Earth's

(a)

(b)

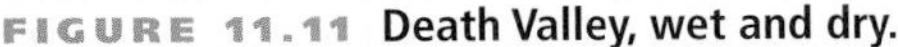
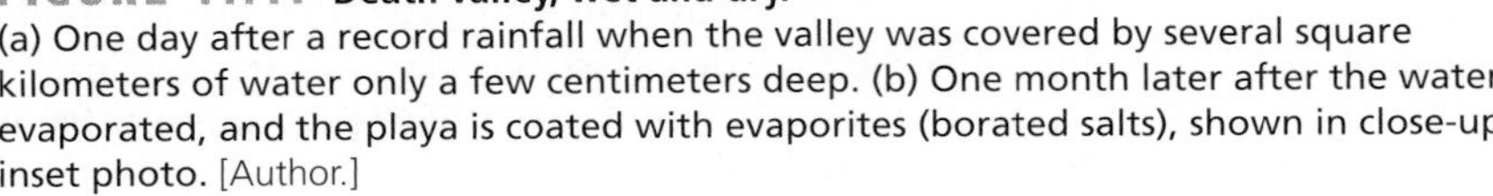

FIGURE 11.11 **Death Valley, wet and dry.**
(a) One day after a record rainfall when the valley was covered by several square kilometers of water only a few centimeters deep. (b) One month later after the water evaporated, and the playa is coated with evaporites (borated salts), shown in close-up inset photo. [Author.]

(a)

(b)

FIGURE 11.12 Hydrothermal deposits.
(a) Mammoth Hot Springs, Yellowstone National Park, is an example of a hydrothermal deposit principally composed of travertine ($CaCO_3$), deposited as a chemical precipitate from heated spring water as it evaporated. (b) Hydrothermal vents with black smokers rise along the Mid-Atlantic Ridge. [(a) Author. (b) Scripps Institution of Oceanography/University of California, San Diego.]

(a)

(b)

FIGURE 11.13 Metamorphic rocks.
(a) A metamorphic rock outcrop in Greenland, the Amitsoq Gneiss, at 3.8 billion years old, one of the oldest rock formations on Earth. (b) Precambrian Vishnu Schist makes up these rock cliffs in the inner gorge of the Grand Canyon in northern Arizona. These metamorphic rocks lie beneath many layers of sedimentary rocks deep in the continent. [(a) Photo by Kevin Schafer/Photolibrary/Peter Arnold, Inc. (b) Author.]

Animation Foliation (Metamorphic Rock)

crust and, because of their own weight, create enough pressure in the bottommost layers to transform the sediments into metamorphic rock, a process of *regional metamorphism*. Also, molten magma rising within the crust may "cook" adjacent rock, a process of *contact metamorphism*.

Metamorphic rocks may be changed both physically and chemically from the original rocks. Table 11.4 lists some metamorphic rock types, their parent rocks, and their resultant textures. If the mineral structure demonstrates a particular alignment after metamorphism, the rock is *foliated*, and some minerals may appear as wavy striations (streaks or lines) in the new rock. The four inset photographs demonstrate foliated and nonfoliated textures. Which form do you see—foliated or nonfoliated—in Figure 11.13a or in the far-right inset photo of metamorphic rock from Livingston Island in Figure 11.7?

On the Isle of Lewis (Outer Hebrides), northwest of the Scottish coast, people constructed the Standing Stones of Calanais (Callanish) beginning approximately 5000 years ago (Table 11.4, far-right inset). These Neolithic people used the 3.1-billion-year-old metamorphic Lewisian gneiss and arranged the monument with the foliations aligned vertically. The standing stone in the photo is about 3.5 m (11.5 ft) tall.

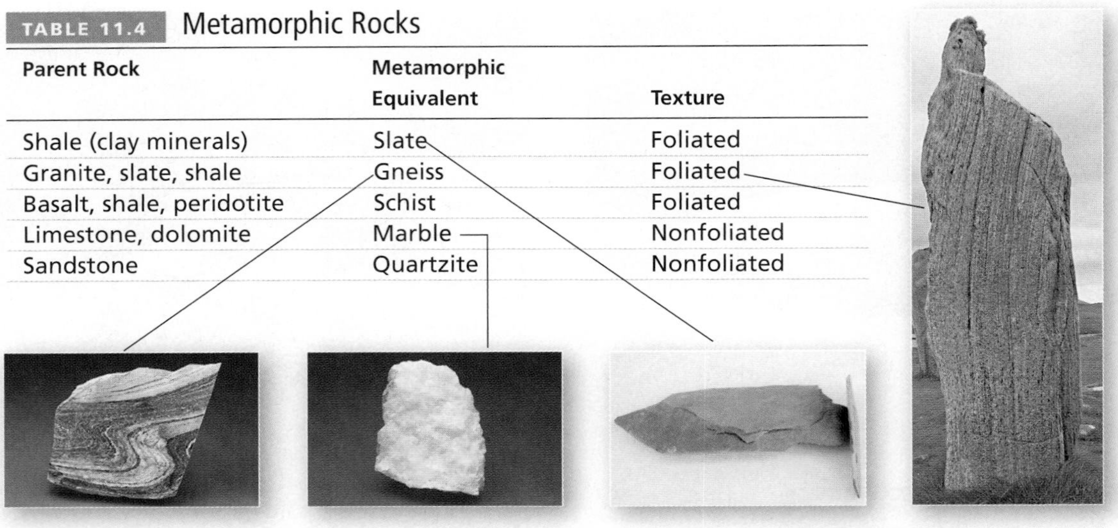

TABLE 11.4 Metamorphic Rocks

Parent Rock	Metamorphic Equivalent	Texture
Shale (clay minerals)	Slate	Foliated
Granite, slate, shale	Gneiss	Foliated
Basalt, shale, peridotite	Schist	Foliated
Limestone, dolomite	Marble	Nonfoliated
Sandstone	Quartzite	Nonfoliated

You see how the rock-forming processes yield the igneous, sedimentary, and metamorphic materials of Earth's crust. Next, we look at how vast *tectonic processes* press, push, and drag portions of the crust in large-scale movements, causing the continents to *drift*.

Plate Tectonics

Have you ever looked at a world map and noticed that a few of the continental landmasses appear to have matching shapes like pieces of a jigsaw puzzle—particularly South America and Africa? The incredible reality is that the continental pieces once did fit together! Continental landmasses not only migrated to their present locations, but also continue to move at speeds up to 6 cm (2.4 in.) per year. We say that the continents are *adrift* because convection currents in the asthenosphere and upper mantle are dragging them around. The key point is that the arrangement of continents and oceans we see today is not permanent, but is in a continuing state of change.

A continent, such as North America, is actually a collage of crustal pieces that migrated from elsewhere to form the landscape we know. Through a historical chronology, let us trace the discoveries that led to the plate tectonics theory—a major revolution in Earth systems science.

A Brief History

As early mapping gained accuracy, some observers noticed the symmetry among continents, particularly the fit between South America and Africa. Abraham Ortelius (1527–1598), a geographer, noted the apparent fit of some continental coastlines in his *Thesaurus Geographicus* (1596). In 1620, English philosopher Sir Francis Bacon noted gross similarities between the edges of Africa and South America (although he did not suggest that they had drifted apart). Benjamin Franklin wrote in 1780 that the crust must be a shell that can break and shift by movements of fluid below. Others wrote—unscientifically—about such apparent relationships, but it was not until much later that a valid explanation came forward.

In 1912, German geophysicist and meteorologist Alfred Wegener publicly presented in a lecture his idea that Earth's landmasses migrate. His book *Origin of the Continents and Oceans* appeared in 1915. Today, Wegener is regarded as the father of this concept, which he first called **continental drift**. Scientists at the time were unreceptive to Wegener's revolutionary proposal. Thus, a great debate began, lasting almost 50 years until a scientific consensus developed that Wegener was right after all.

Wegener thought that all landmasses formed one supercontinent approximately 225 m.y.a., during the Triassic Period. **Pangaea** is the name he gave this one landmass, meaning "all Earth" (see Figure 11.17b). Although his initial model kept the landmasses together too long and his proposal included an incorrect driving mechanism for the moving continents, Wegener's arrangement of Pangaea and its breakup was correct. Pangaea was only the latest of earlier supercontinent arrangements over the span of Earth history.

To come up with his Pangaea fit, Wegener studied the geologic record represented in the rock strata, the fossil record, and the climatic record for the continents. He concluded that South America and Africa correlated in many complex ways. He decided that the large midlatitude coal deposits, which date to the Permian and Carboniferous Periods (251–354 m.y.a.), exist because these regions once were nearer the equator and therefore covered by lush vegetation that became coal.

As modern scientific capabilities built the case for continental drift, the 1950s and 1960s saw a revival of interest in Wegener's concepts and, finally, confirmation. Aided by an avalanche of discoveries, the plate tectonics theory today is universally accepted as an accurate model of the way Earth's surface evolves.

Tectonic, from the Greek *tektonikùs*, meaning "building" or "construction," refers to changes in the configuration of Earth's crust as a result of internal forces. **Plate tectonics** include the processes of upwelling of magma; lithospheric plate movements; sea-floor spreading and lithospheric subduction; earthquakes; volcanic activity; and lithospheric deformation such as warping, folding, and faulting.

Sea-Floor Spreading and Production of New Crust

The key to establishing the theory of continental drift was a better understanding of the seafloor. The seafloor has a remarkable feature: an interconnected worldwide mountain chain, forming a ridge some 64,000 km (40,000 mi) in extent and averaging more than 1000 km (600 mi) in width. A striking view of this great undersea mountain chain opens Chapter 12. How did this global mountain chain get there?

In the early 1960s, geophysicists Harry H. Hess and Robert S. Dietz proposed **sea-floor spreading** as the mechanism that builds this mountain chain and drives continental movement. Hess said that these submarine mountain ranges were **mid-ocean ridges** and the direct result of upwelling flows of magma from hot areas in the upper mantle and asthenosphere and perhaps from the deeper lower mantle. These are sites of intense hydrothermal activity, as mentioned earlier.

When mantle convection brings magma up to the crust, the crust fractures, and the magma extrudes onto the seafloor and cools to form new seafloor. This process builds the mid-ocean ridges and spreads the seafloor laterally, rifting as the two sides move apart. The concept is illustrated in Figure 11.14, which shows how the ocean floor is rifted and scarred along mid-ocean ridges and includes a remote-sensing image of the mid-ocean ridge and a map of the system in the Atlantic.

Figure 11.14d shows a portion of the mid-ocean ridge system that surfaces at Thingvellir, Iceland, forming these rifts—the North American plate along one side and the European plate on the other. Iceland's first Althing (parliament) met here in A.D. 930; leaders used the acoustics of the rift walls to be heard by the gathering. Only in Iceland does a portion of Earth's 64,000-km (40,000-mi) mid-ocean ridge system surface.

As new crust generates and the seafloor spreads, magnetic particles in the lava orient with the magnetic field in force at the time the lava cools and hardens. The particles become locked in this alignment as new seafloor forms, creating a kind of magnetic tape recording in the seafloor. The continually forming oceanic crust records each magnetic reversal and reorientation of Earth's polarity.

Figure 11.15 illustrates just such a recording from the Mid-Atlantic Ridge south of Iceland. The colors denote the alternating magnetic polarization preserved in the minerals of the oceanic crust. Note the mirror images that develop on either side of the rift as a result of the nearly symmetrical spreading of the seafloor; the relative ages of the rocks increase with distance from the ridge. These periodic reversals of Earth's magnetic field are a valuable clue to understanding sea-floor spreading, helping scientists fit together pieces of Earth's crust.

These sea-floor recordings of Earth's magnetic-field reversals and other measurements allowed the age of the seafloor to be determined. The complex harmony of the two concepts of plate tectonics and sea-floor spreading thus became clearer. The youngest crust anywhere on Earth is at the spreading centers of the mid-ocean ridges, and with increasing distance from these centers, the crust gets steadily older (Figure 11.16). The oldest seafloor is in the western Pacific near Japan, dating to the Jurassic Period. Note on the map in the figure the distance between this basin and its spreading center in the South Pacific, west of South America.

Overall, the seafloor is relatively young; nowhere is it more than 208 million years old—remarkable when you remember that Earth's age is 4.6 billion years. The reason is that oceanic crust is short-lived—the oldest sections, farthest from the mid-ocean ridges, are slowly plunging beneath continental crust along Earth's deep oceanic trenches. The discovery that the seafloor is young demolished earlier thinking that the oldest rocks would be found there.

Subduction of the Lithosphere

In contrast to the upwelling zones along the mid-ocean ridges are the areas of descending lithosphere elsewhere. On the left side of Figure 11.14a, note how one plate of the crust is diving, or being dragged, beneath another into the mantle. Recall that the basaltic ocean crust has a density of 3.0 g/cm^3, whereas continental crust averages a lighter 2.7 g/cm^3. As a result, when continental crust and oceanic crust slowly collide, the denser ocean floor will grind beneath the lighter continental crust, thus forming a **subduction zone**, as shown in the figure.

The subducting slab of crust exerts a gravitational pull on the rest of the plate—an important driving force in plate motion. This plate motion can trigger through shear traction along the base of the lithosphere, a flow in the mantle material beneath.

The world's deep ocean trenches coincide with these subduction zones and are the lowest features on Earth's surface. Deepest is the Mariana Trench near Guam, which descends below sea level to –11,030 m (–36,198 ft); next deepest is the Tonga Trench, also in the Pacific, which drops to –10,882 m (35,702 ft). For comparison, in the Atlantic Ocean, the Puerto Rico Trench drops to –8605 m (–28,224 ft), and, in the Indian Ocean, the Java Trench drops to –7125 m (–23,376 ft).

The subducted portion of lithosphere travels down into the asthenosphere, where it remelts and eventually is recycled as magma, rising again toward the surface through deep fissures and cracks in crustal rock. Volcanic mountains such as the Andes in South America and the

(text continued on page 320)

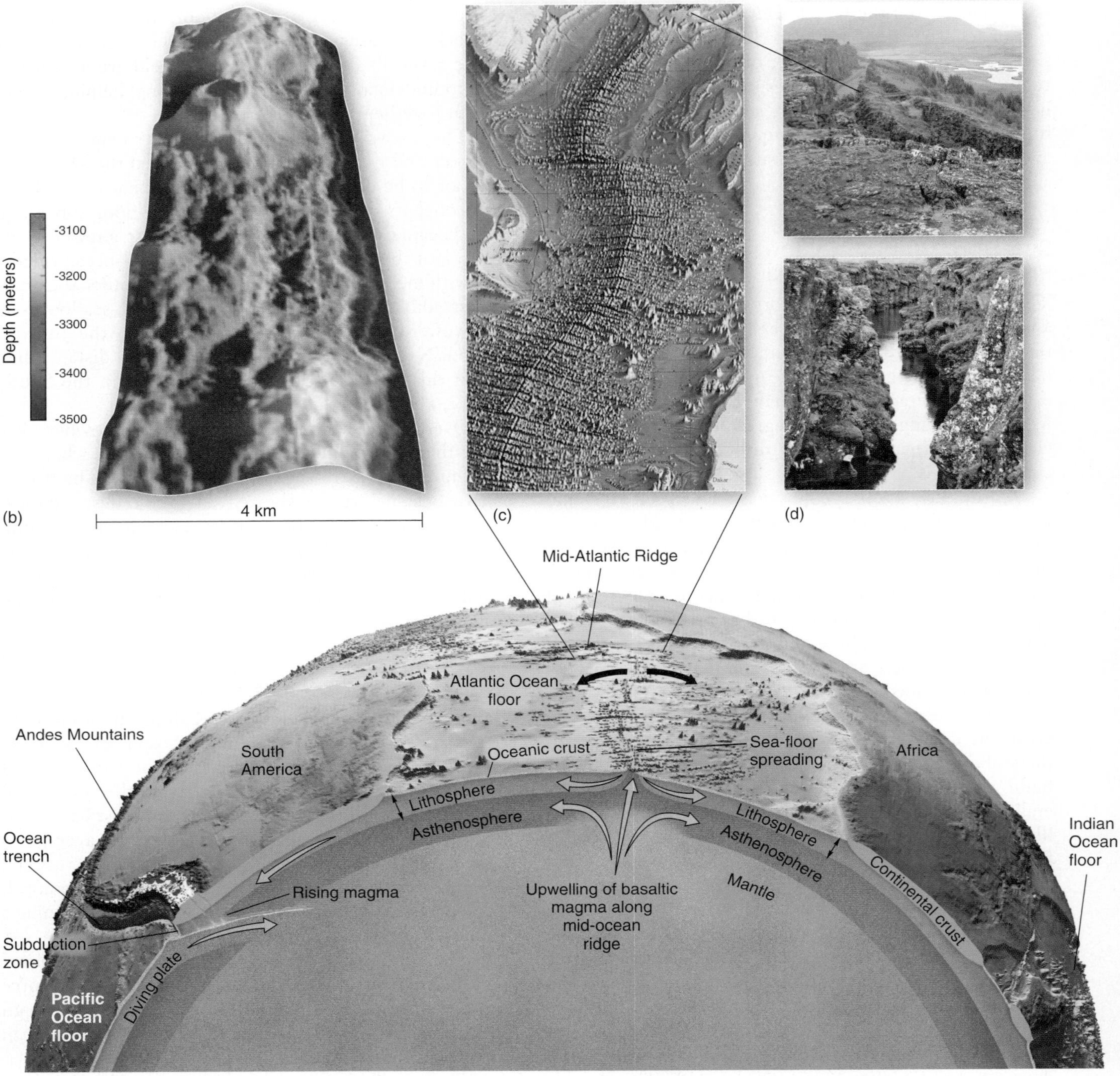

FIGURE 11.14 Crustal movements.
(a) Sea-floor spreading, upwelling currents, subduction, and plate movements, shown in cross section. Arrows indicate the direction of the spreading. (b) A 4-km-wide image of the Mid-Atlantic Ridge showing linear faults, a volcanic crater, a rift valley, and ridges. The image was taken by the TOBI (towed ocean-bottom instrument) at approximately 29° N latitude. (c) Detail from ocean-floor map. (d) The mid-Atlantic rift surfaces through Iceland, where the North American and European plates of Earth's crust meet.
[(b) Image courtesy of D. K. Smith, Woods Hole Oceanographic Institute. (c) Office of Navel Research. (d) Bobbé Christopherson.]

Animation
Sea-Floor Spreading, Subduction

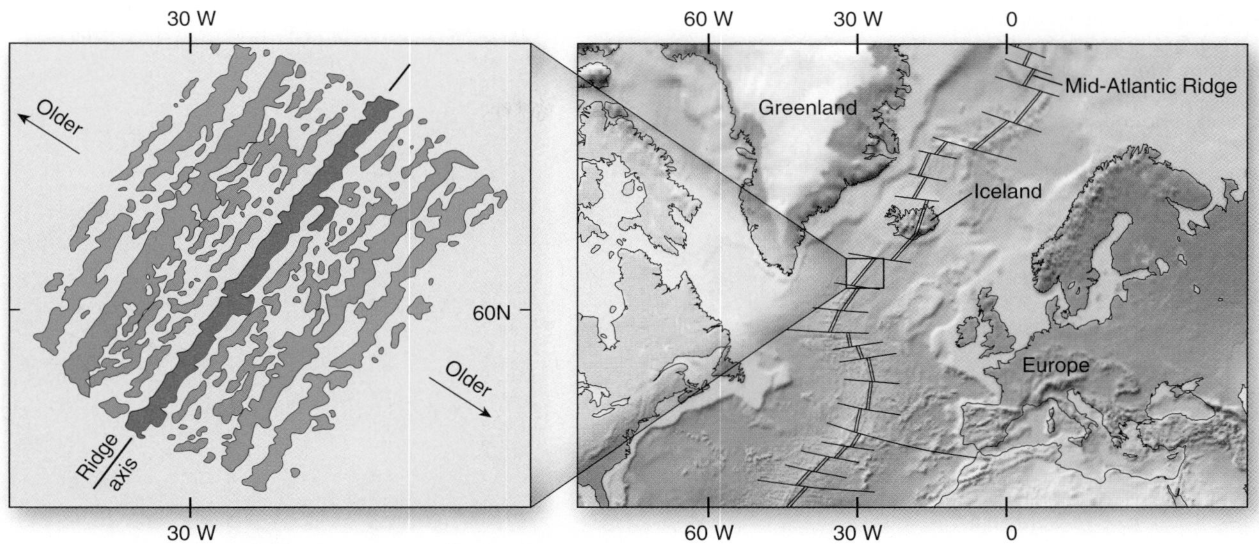

FIGURE 11.15 Magnetic reversals recorded in the seafloor.
Magnetic reversals recorded in the seafloor south of Iceland along the Mid-Atlantic Ridge. [Adapted from J. R. Heirtzler, S. Le Pichon, and J. G. Baron, *Deep-Sea Research* 13, © 1966, Pergamon Press, p. 247.]

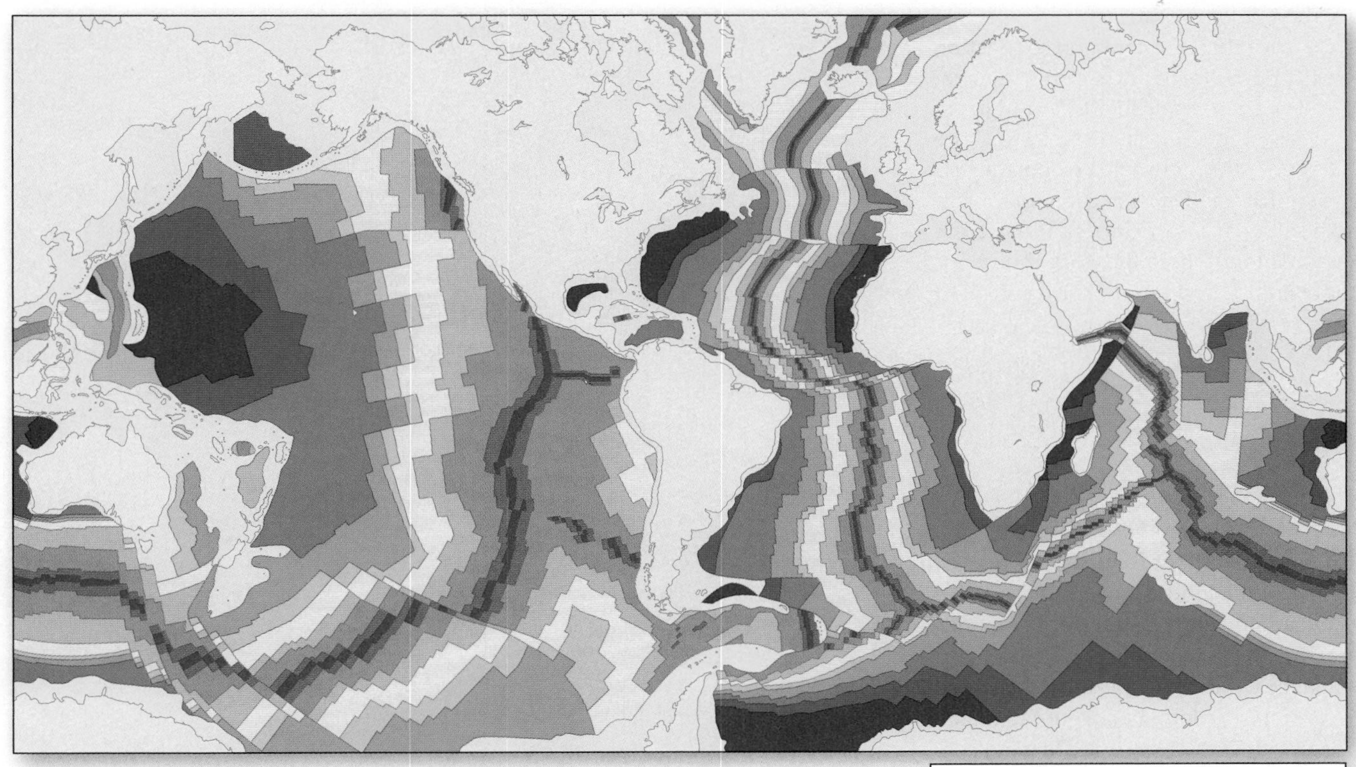

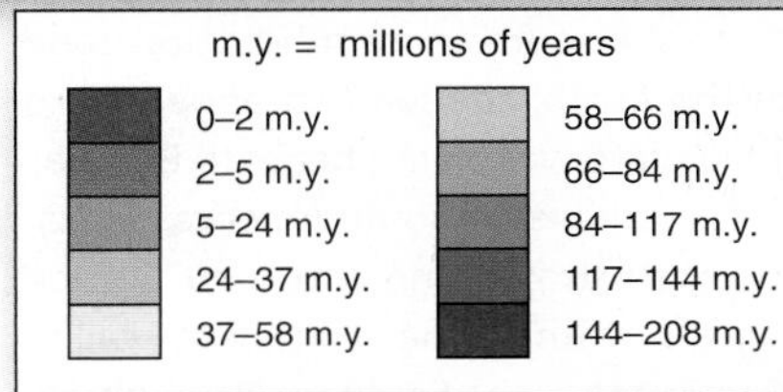

FIGURE 11.16 Relative age of the oceanic crust.
Compare the dark-red color near the East Pacific rise in the eastern Pacific Ocean with the same dark-red color along the Mid-Atlantic Ridge. What does the difference in width tell you about the rates of plate motion in the two locations? [Adapted from *The Bedrock Geology of the World* by R. L. Larson, *et al.*, © 1985, W. H. Freeman and Company.]

FIGURE 11.17 Continents adrift, from 465 m.y.a. to the present. Observe the formation and breakup of Pangaea and the types of motions occurring at plate boundaries. [(a) From R. K. Bambach, "Before Pangaea: The geography of the Paleozoic world," *American Scientist* 68 (1980): 26–38, reprinted by permission. (b–e) From R. S. Dietz and J. C. Holden, *Journal of Geophysical Research* 75, no. 26 (September 10, 1970): 4939–4956, © The American Geophysical Union. (f) Remote-sensing image courtesy of S. Tighe, University of Rhode Island, and R. Detrick, Woods Hole National Oceanic and Atmospheric Administration]

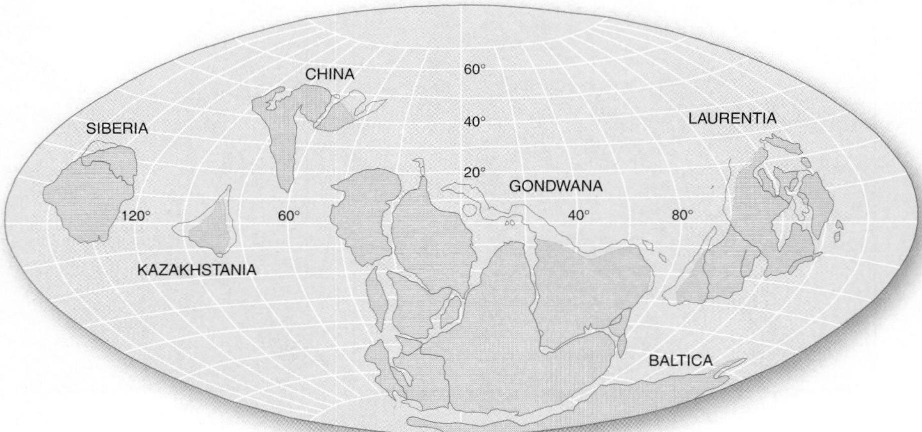

(a) 465 million years ago

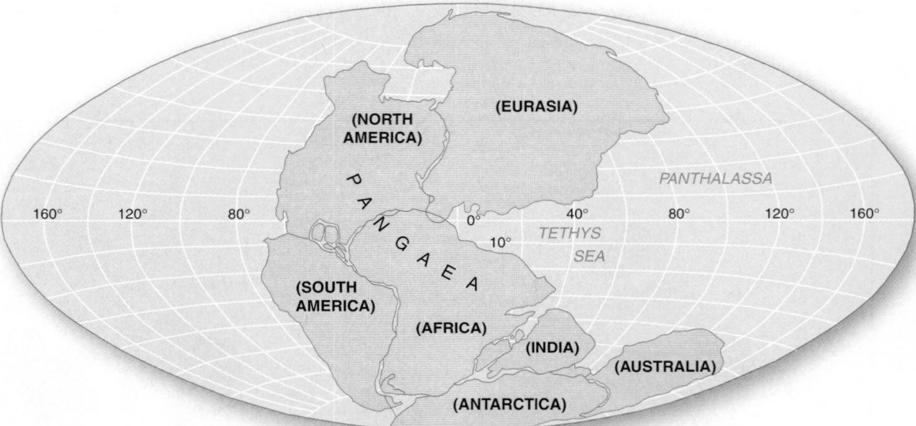

(b) 225 million years ago

Pangaea—all Earth. Panthalassa ("all seas") became the Pacific Ocean, and the Tethys Sea (partly enclosed by the African and the Eurasian plates) became the Mediterranean Sea; trapped portions of former ocean became the present-day Caspian Sea. The Atlantic Ocean did not exist.

Africa shared a common connection with both North and South America. Today, the Appalachian Mountains in the eastern United States and the Lesser-Atlas Mountains of northwestern Africa reflect this common ancestry; they are, in fact, portions of the same mountain range, torn thousands of kilometers apart.

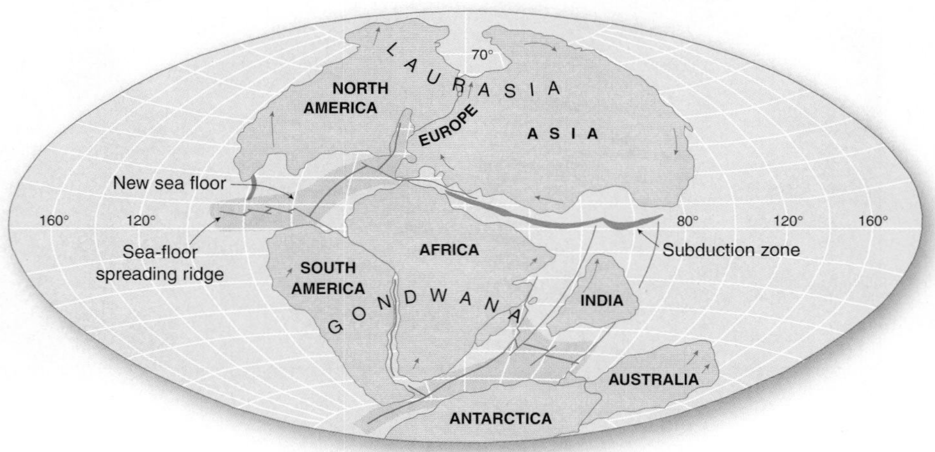

(c) 135 million years ago

New seafloor that formed since the last map is highlighted (gray tint). An active spreading center rifted North America away from landmasses to the east, shaping the coast of Labrador. India was farther along in its journey to collide with Eurasia, with a spreading center to the south and a subduction zone to the north; the leading edge of the India plate was diving beneath Eurasia.

The outlines of South America, Africa, India, Australia, and southern Europe appear within the continent called *Gondwana* in the Southern Hemisphere. North America, Europe, and Asia make up *Laurasia,* the northern portion.

Animation Motions at Plate Boundaries

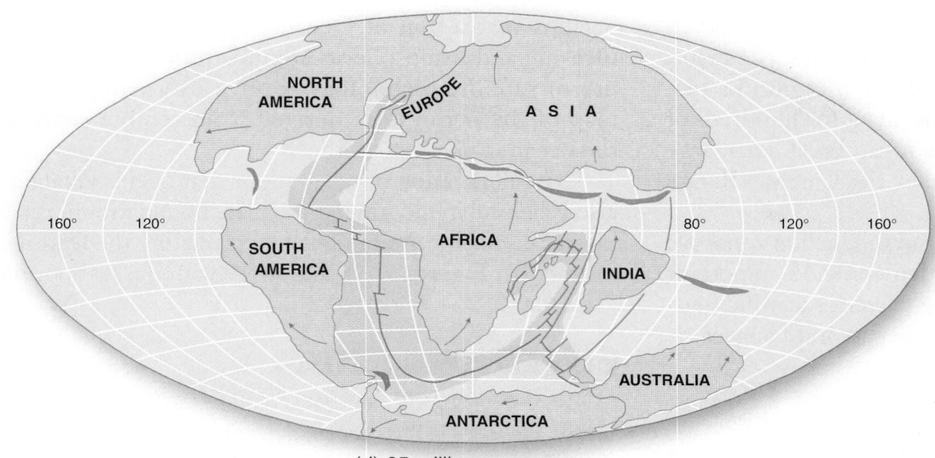

NORTH AMERICA
EUROPE
A S I A
160°
120°
80°
120°
160°
SOUTH AMERICA
AFRICA
INDIA
AUSTRALIA
ANTARCTICA
(d) 65 million years ago

Along the Mid-Atlantic Ridge, the seafloor spread some 3000 km (almost 1900 mi) in 70 million years. Africa moved northward about 10° in latitude, leaving Madagascar split from the mainland and opening up the Gulf of Aden. The rifting along what would be the Red Sea began. The India plate moved three-fourths of the way to Asia, as Asia continued to rotate clockwise. Of all the major plates, India travelled the farthest—almost 10,000 km.

MG Notebook Plate Boundaries

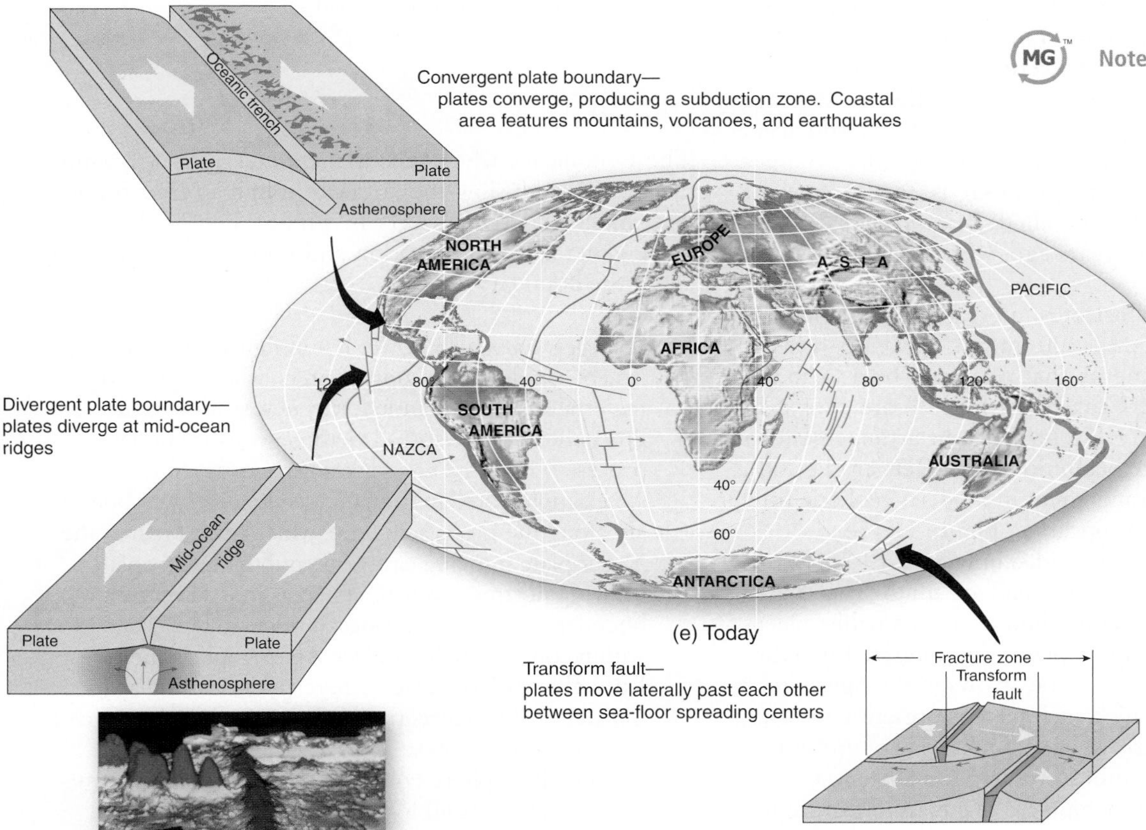
Oceanic trench
Plate
Plate
Asthenosphere
Convergent plate boundary—
plates converge, producing a subduction zone. Coastal area features mountains, volcanoes, and earthquakes
NORTH AMERICA
EUROPE
A S I A
PACIFIC
AFRICA
120°
80°
40°
0°
40°
80°
120°
160°
SOUTH AMERICA
NAZCA
AUSTRALIA
40°
60°
ANTARCTICA
Divergent plate boundary—
plates diverge at mid-ocean ridges
Mid-ocean ridge
Plate
Plate
Asthenosphere
Transform fault—
plates move laterally past each other between sea-floor spreading centers
Fracture zone
Transform fault
(e) Today

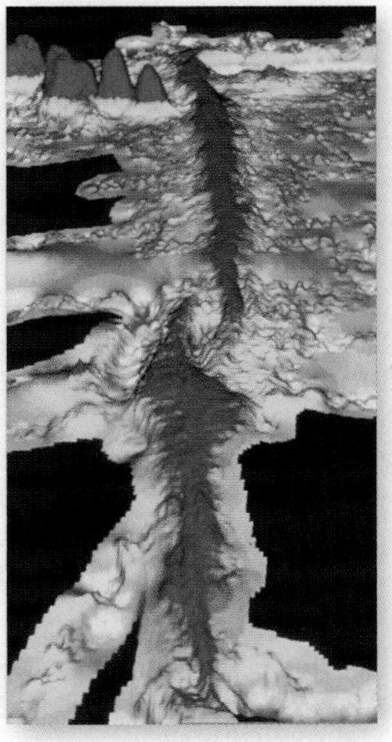
(f) East-Pacific rise, sea-floor spreading ridge (center of image at 9° N)

From 65 m.y.a. until the present, more than half of the ocean floor was renewed. The northern reaches of the India plate underthrust the southern mass of Asia through subduction, forming the Himalayas in the upheaval created by the collision. Plate motions continue to this day.

MG Animation Correlating Processes and Plate Boundaries

MG Animation Forming a Divergent Boundary

Cascade Range from northern California to the Canadian border form inland of these subduction zones as a result of rising plumes of magma, as suggested in Figure 11.14a. Sometimes the diving plate remains intact for hundreds of kilometers, whereas at other times it can break into large pieces, thought to be the case under the Cascade Range with its distribution of volcanoes.

The fact that spreading ridges and subduction zones are areas of earthquake and volcanic activity provides important proof of plate tectonics. Now, using current scientific findings, let us go back and reconstruct the past and Pangaea.

The Formation and Breakup of Pangaea

The supercontinent of Pangaea and its subsequent breakup into today's continents represent only the last 225 million years of Earth's 4.6 billion years, or only the most recent 1/23 of Earth's existence. During the remaining 22/23 of geologic time, other things were happening. The landmasses as we know them were unrecognizable.

Figure 11.17a (page 318) begins with the pre-Pangaea arrangement of 465 m.y.a. (during the middle Ordovician Period). Researchers are discovering other, older arrangements of continents. One proposal is for an ancient supercontinent, tentatively called *Rodinia*, that formed approximately a billion years ago, with breakup at about 700 million.

Figure 11.17b illustrates an updated version of Wegener's Pangaea, 225–200 m.y.a. (Triassic–Jurassic periods). Plate movements that occurred by 135 m.y.a. (the beginning of the Cretaceous Period) are shown in Figure 11.17c. Figure 11.17d presents the arrangement 65 m.y.a. (shortly after the beginning of the Tertiary Period). Finally, the present arrangement in modern geologic time (the late Cenozoic Era) appears in Figure 11.17e.

Earth's present crust is divided into at least 14 plates, of which about half are major and half are minor in terms of area (Figure 11.18). Literally hundreds of smaller pieces and perhaps dozens of microplates that migrated together make up these broad plates. The arrows in the figure indicate the direction in which each plate is presently moving, and the length of the arrows suggests the rate of movement during the past 20 million years. On the *Mastering Geography* site, you can manipulate this map to feature the speed arrows, or the plates, with or without the landmasses.

Compare Figure 11.18 with the image of the ocean floor in Figure 11.19. In the image, satellite radar-altimeter measurements determined the sea-surface height to a remarkable accuracy of 0.03 m, or 1 in. The sea-surface elevation is far from uniform; it is higher or lower in direct response to the mountains, plains, and trenches of the ocean floor beneath it. This sea-floor topography causes slight differences in Earth's gravity.

For example, a massive mountain on the ocean floor exerts a high gravitational field and attracts water to it, producing higher sea level above the mountain; over a trench, there is less gravity, resulting in a drop in sea level. A mountain 2000 m (6500 ft) high that is 20 km (12 mi) across its base creates a 2-m rise in the sea surface. Scripps Institution of Oceanography and the National Oceanic and Atmospheric Administration developed this use of satellite technology to give us a comprehensive map of the ocean *floor*, derived from remote sensing of the ocean *surface*!

The illustration of the seafloor that begins Chapter 12 also is helpful in identifying the plate boundaries in the following discussion. You might try correlating the features illustrated in Figures 11.18 and 11.19 with the seafloor map.

CRITICAL THINKING 11.2

Tracking your location since Pangaea

Using the maps in this chapter, determine your present location relative to Earth's crustal plates. Now, using Figure 11.17b, approximately identify where your present location was 225 m.y.a.; express it in a rough estimate using the equator and the longitudes noted on the map.

Plate Boundaries

The boundaries where plates meet clearly are dynamic places, although slow-moving within human time frames. The block diagram inserts in Figure 11.17e show the three general types of motion and interaction that occur along the boundary areas:

- *Divergent boundaries* (lower left in figure) are characteristic of sea-floor spreading centers, where upwelling material from the mantle forms new seafloor and lithospheric plates spread apart—a constructional process. The spreading makes these zones of tension. An example noted in the figure is the divergent boundary along the East Pacific rise, which gives birth to the Nazca plate (moving eastward) and the Pacific plate (moving northwestward). Whereas most divergent boundaries occur at mid-ocean ridges, a few occur within continents themselves. An example is the Great Rift Valley of East Africa, where crust is rifting apart.
- *Convergent boundaries* (upper left) are characteristic of collision zones, where areas of continental and oceanic lithosphere collide. These are zones of compression and crustal loss—a destructional process. Examples include the subduction zone off the west coast of South and Central America and the area along the Japan and Aleutian trenches. Along the western edge of South America, the Nazca plate collides with and is subducted beneath the South American plate. This convergence creates the Andes Mountains chain and related volcanoes. The collision of India and Asia is another example of a convergent boundary, mentioned with Figure 11.17.
- *Transform boundaries* (lower right) occur where plates slide laterally past one another at right angles to a sea-floor spreading center, neither diverging nor converging, and usually with no volcanic eruptions. These are the right-angle fractures stretching across the mid-ocean ridge system worldwide.

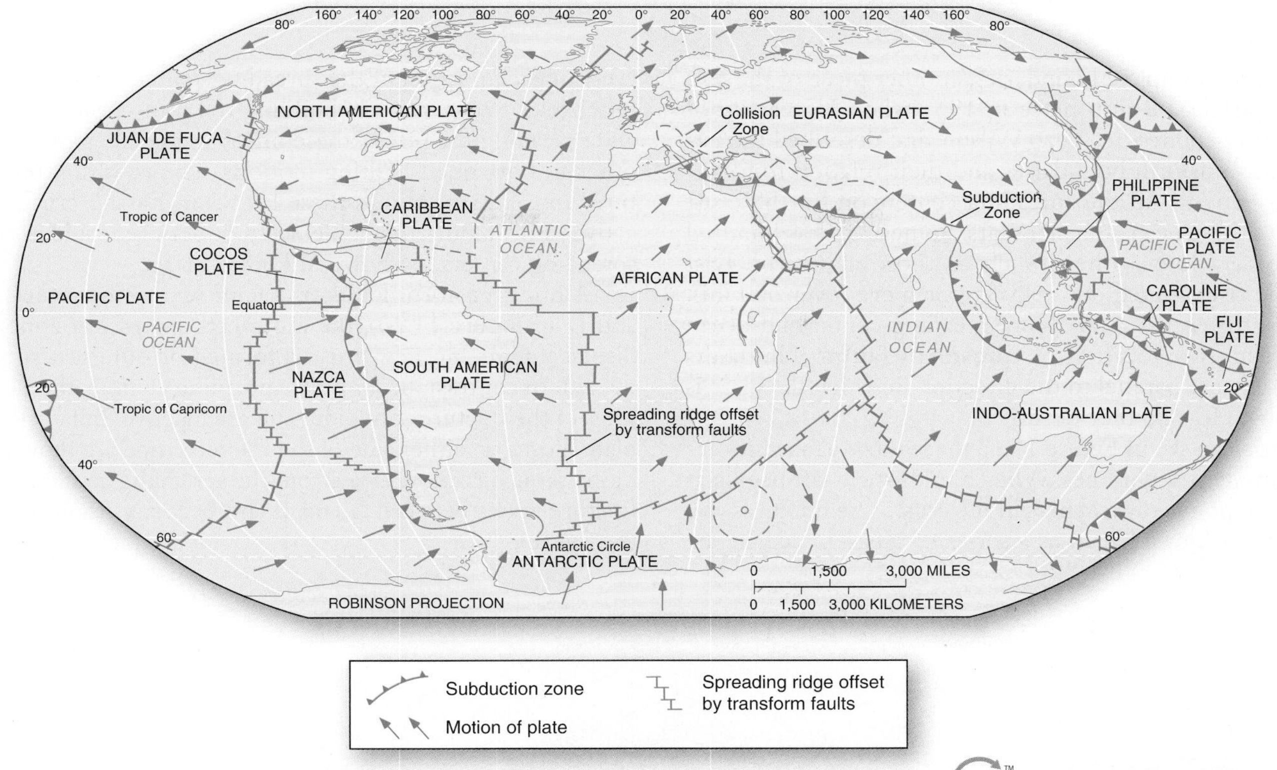

FIGURE 11.18 Earth's major lithospheric plates and their movements.
Each arrow represents 20 million years of movement. The longer arrows indicate that the Pacific and Nazca plates are moving more rapidly than are the Atlantic plates. Compare the length of these arrows with those light-gray areas on Figure 11.17. [Adapted from U.S. Geodynamics Committee, National Academy of Sciences and National Academy of Engineering.]

Animation India Collision with Asia

Notebook Plate Boundaries

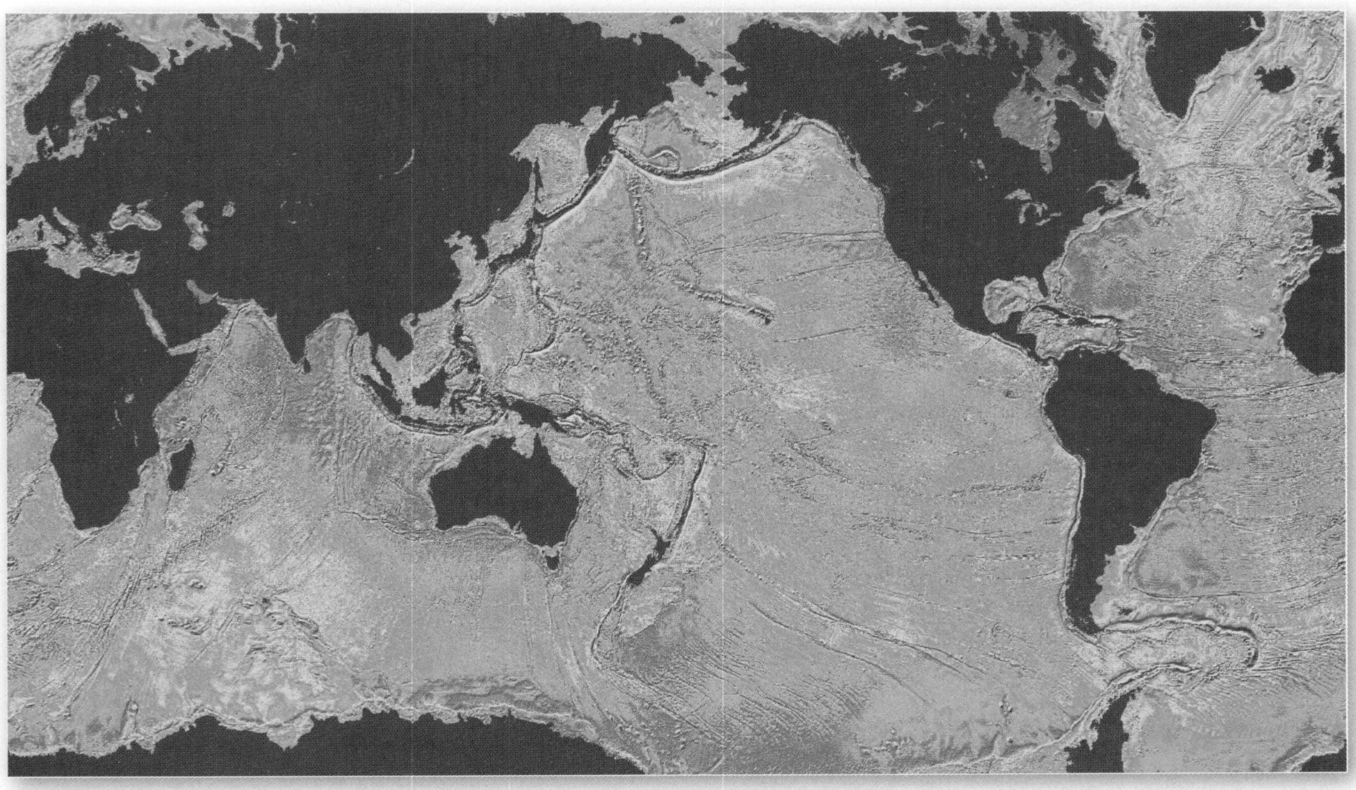

FIGURE 11.19 The ocean floor revealed.
A global gravity anomaly map derived from *Geosat* and *ERS-1* altimeter data. The radar altimeters measured sea-surface heights. Variation in sea-surface elevation is a direct indication of the topography of the ocean floor. [Image courtesy of D. T. Sandwell, Scripps Institution of Oceanography. All rights reserved, 1995.]

Related to plate boundaries, another piece of the tectonic puzzle fell into place in 1965 when University of Toronto geophysicist Tuzo Wilson first described the nature of transform boundaries and their relation to earthquake activity. All the spreading centers on Earth's crust feature these perpendicular scars. Some are a few hundred kilometers long; others, such as those along the East Pacific rise, stretch out 1000 km or more (over 600 mi). Across the entire ocean floor, spreading-center mid-ocean ridges are the location of **transform faults**. The faults generally are parallel to the direction in which the plate is moving. How do they occur?

You can see in Figure 11.18 that mid-ocean ridges are not simple straight lines. When a mid-ocean rift begins, it opens at points of weakness in the crust. The fractures you see in the figure began as a series of offset breaks in the crust in each portion of the spreading center. As new material rises to the surface, building the mid-ocean ridges and spreading the plates, these offset areas slide past each other in horizontal faulting motions. The resulting fracture zone, which follows these breaks in Earth's crust, is active only along the fault section between ridges of spreading centers, as shown in Figure 11.20.

Along transform faults (the fault section between C and D in Figure 11.20), the motion is one of horizontal displacement—no new crust is formed or old crust subducted. In contrast, beyond the spreading centers, the two sides of the fracture zones join and are inactive. In fact, the plate pieces on either side of the fracture zone are moving in the same direction, away from the spreading center (the fracture zones between A and B and between E and F in the figure).

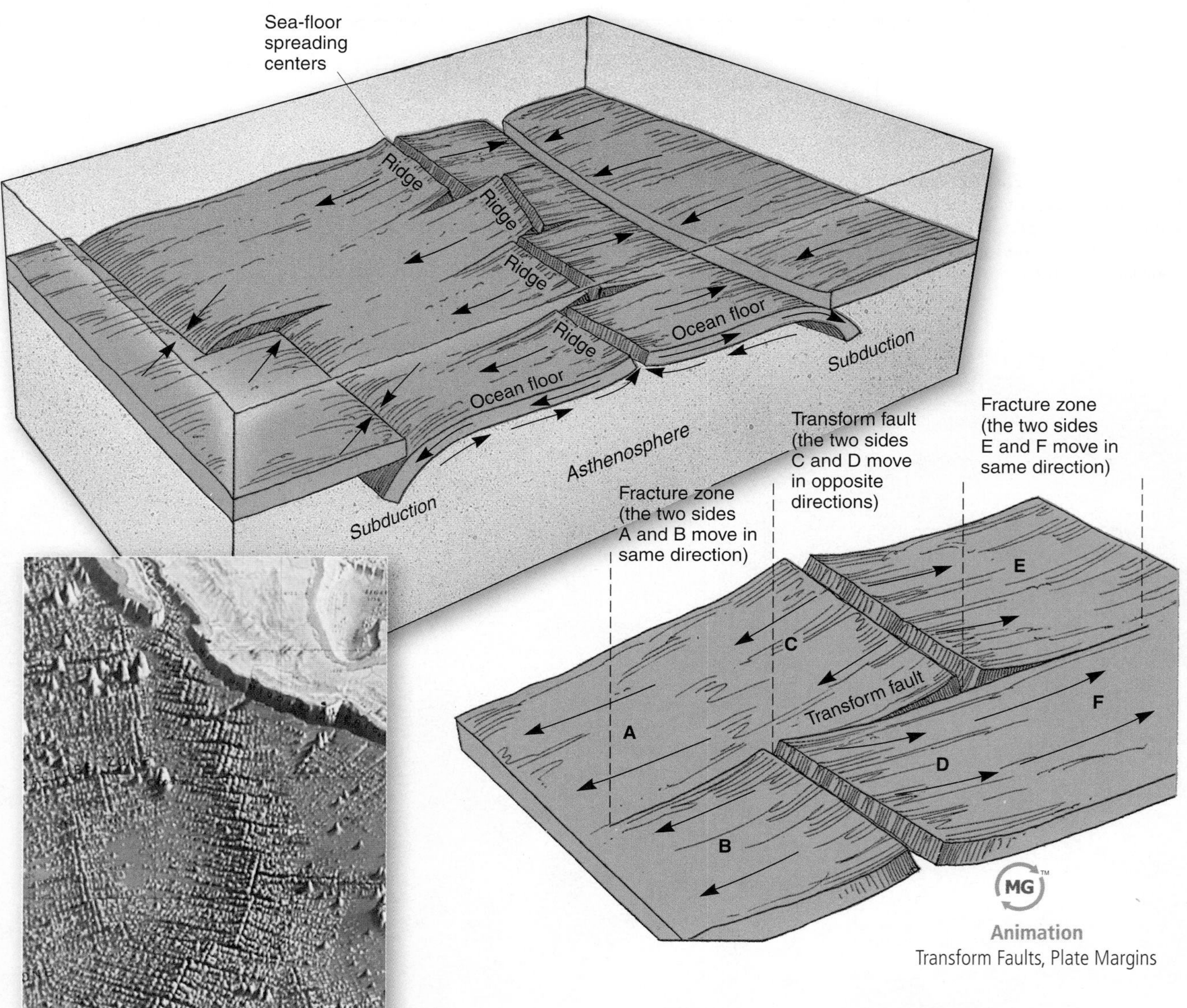

FIGURE 11.20 Transform faults.
Appearing along fracture zones, a transform fault is only that section between spreading centers where adjacent plates move in opposite directions—between C and D. [Adapted from B. Isacks, J. Oliver, and L. R. Sykes, *Journal of Geophysical Research* 73 (1968): 5855–5899, © The American Geophysical Union.]

The name *transform* was assigned because of this apparent transformation in the direction of the fault movement. The famous San Andreas fault system in California, where continental crust has overridden a transform system, relates to this type of motion. The fault that triggered the 1995 Kobe earthquake in Japan also has this type of horizontal motion, as does the North Anatolian fault in Turkey and its 1999 and later quakes.

Earthquake and Volcanic Activity

Plate boundaries are the primary location of earthquake and volcanic activity, and the correlation of these phenomena is an important aspect of plate tectonics. In 2010, the massive earthquakes that hit Haiti and Chile and the volcanic eruption in Iceland focused world attention on these principles. The next chapter discusses earthquakes and volcanic activity in more detail.

Figure 11.21 maps earthquake zones, volcanic sites, hot spots, and plate motion. The "ring of fire" surrounding the Pacific Basin, named for the frequent incidence of volcanoes, is evident. The subducting edge of the Pacific plate thrusts deep into the crust and mantle and produces molten material that makes its way back toward the surface. The upwelling magma forms active volcanoes along the Pacific Rim. Such processes occur similarly at plate boundaries throughout the world.

Hot Spots

A dramatic aspect of Earth's internal dynamics is the estimated 50 to 100 **hot spots** across Earth's surface (shown in Figure 11.22, page 326). These are individual sites of upwelling material arriving at the surface in tall plumes from the mantle or rising from near the surface that produce thermal effects in groundwater and crust. Some of these sites can be developed for geothermal power (Focus Study 11.1). Hot spots occur beneath both oceanic and continental crust. Some hot spots are anchored deep in the stiff lower mantle, tending to remain fixed relative to migrating plates; others appear to be above plumes that move themselves, or shift with plate motion. Thus, the area of a plate that is above a hot spot is locally heated for the brief geologic time it is there (a few hundred thousand or million years).

The Pacific plate moved across a hot, upward-erupting plume over the last 80 million years, creating a string of volcanic islands stretching northwestward away from the hot spot. This hot spot produced, and continues to form, the Hawaiian–Emperor Islands chain (Figure 11.22). Thus, the ages of the islands and seamounts (submarine mountains that do not reach the surface) in the chain increase northwestward from the island of Hawai'i, as you can see from the ages marked in the figure. The oldest island in the Hawaiian part of the chain is Kaua'i, approximately

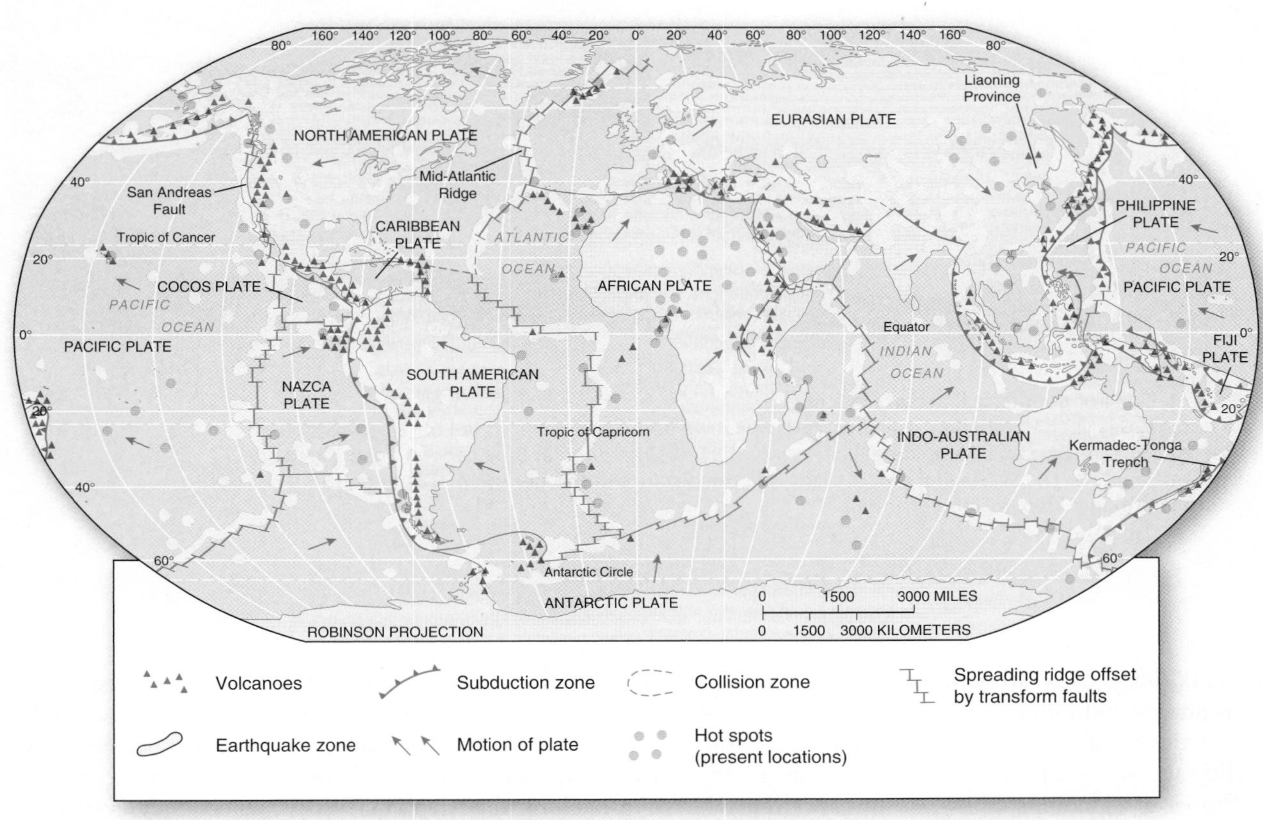

FIGURE 11.21 Earthquake and volcanic activity locations.
Earthquake and volcanic activity in relation to major tectonic plate boundaries and principal hot spots. [Earthquake, volcano, and hot spot data adapted from U.S. Geological Survey.]

FOCUS STUDY 11.1

Heat from Earth—Geothermal Energy and Power

A tremendous amount of endogenic energy flows from Earth's interior toward the surface. Temperatures at the base of Earth's crust range from 200°C to 1000°C (392°F to 1800°F). Convection and conduction transport this geothermal energy in enormous quantities from the mantle to the crust, yet geothermal power development is limited in extent to certain locations (Figure 11.1.1).

What Is Geothermal Energy?

Geothermal energy is produced when pockets of magma and hot portions of the crust heat groundwater. This energy is transmitted to the surface by heated water or steam accessed through the drilling of wells. For an effective *underground thermal reservoir* to form, an aquifer must have high porosity and high permeability that allows heated water to move freely through connecting pore spaces. Ideally, this aquifer should heat groundwater from 180°C to 350°C (355°F to 600°F) and be accessible to drilling within 3 km (1.9 mi) of the surface, although 6- to 7-km (3.7- to 4.3-mi) depths are workable.

Geothermal energy literally refers to heat from Earth's interior, whereas *geothermal power* relates to specific applied strategies of geothermal electric or geothermal direct applications. *Geothermal electric* uses steam, or hot water that flashes to steam, to drive a turbine-generator. In *geothermal direct*, hot water is used to heat and cool buildings using a heat-exchange system; to heat greenhouses, soil, and swimming pools; and for manufacturing processes and aquaculture, among many other uses.

Geothermal Power Production

The first geothermal electricity-generating station—built in Larderello, Italy, in 1904—has operated continuously since 1913 with an installed capacity of 360 MWe (megawatts electric). Today, geothermal applications are in use in 30 countries, through some 200 plants, producing an output of 8000 MWe and direct heat of 12,000 MWt (megawatts thermal). In Reykjavík, Iceland, the majority of space heating (87%) and some of the electrical production (18%) is geothermal. The Svartsengi power plant (40 MWe) on the Reykjavík Peninsula is an example of such a modern facility (Figure 11.1.2). In the same area, a large array of wells collects heat from hot water and pumps it to the city for geothermal direct applications.

(a)

(b)

FIGURE 11.1.1 Surface geothermal activity in Iceland.
(a) This is the original Geysir, in Haukadalur, Iceland, from which we derive the name for all other geysers on Earth; historical accounts of Geysir's activity date back to A.D. 1294. (b) The popular Blue Lagoon Spa is filled with hot geothermal waters. The Svartsengi power plant is in the distance in this geothermal field on the Reykjavík Peninsula, Iceland. [Bobbé Christopherson.]

5 million years old; it is weathered, eroded, and deeply etched with canyons and valleys.

To the northwest of this active hot spot in Hawai'i, the island of Midway rises as a part of the same system. From there, the Emperor seamounts follow northwestward until they reach about 40 million years in age. At that point, this linear island chain shifts direction northward. This bend in the chain is now thought to be from both movement in the plume and a possible change in the plate motion itself. The old thinking that the plumes remained fixed relative to the migrating plate has been revised. At the northernmost extreme, the seamounts that formed about 80 m.y.a. are now approaching the Aleutian Trench, where they eventually will be subducted beneath the Eurasian plate. Perhaps some 80 million years hence, the Hawaiian Islands, too, will slowly disappear into the trench. Aloha!

(text continued on page 327)

FIGURE 11.1.2 Svartsengi geothermal power plant. On the Reykjavík Peninsula, southwest of Iceland's capital, the Svartsengi power plant produces electricity and provides hot water for homes and businesses. The plant's name means "Black Meadow," referring to the extensive lava beds throughout this region of Iceland. [Bobbé Christopherson.]

In the Paris Basin, France, some 25,000 residences are heated using geothermal direct 70°C (158°F) water. In the Philippines, almost 30% of total electrical production is generated with geothermal energy. The top six countries for installed geothermal electrical generation are the United States, Philippines, Italy, Mexico, Indonesia, and Japan. Worldwide operational capacity is 8933 MWe of a potential estimated to be in excess of 80,000 MWe and hundreds of thousands of MWt. The Geysers Geothermal Field in northern California (so named despite the lack of any geysers in the area) began production in 1960, increased to a peak in 1989 of 1967 MWe, and lowered to a present capacity of 1070 MWe. By the mid-1990s, the field contained some 600 wells, with an average well depth of 2500 m (8200 ft).

Geothermal power capacity of some 2800 MWe is operating in California, Hawai'i, Nevada, and Utah, with a new 185-MWe plant proposed in the desert near Brawley, California. Figure 11.1.3 maps low-, medium-, and high-temperature known geothermal resource areas (KGRAs) in the United States. About 300 cities are within 8 km (5 mi) of a KGRA. The Department of Energy has identified about 9000 potential development sites. For more information, see http://www1.eere.energy.gov/geothermal/, http://geothermal.marin.org/, http://www.geothermal.org/, or http://smu.edu/geothermal/.

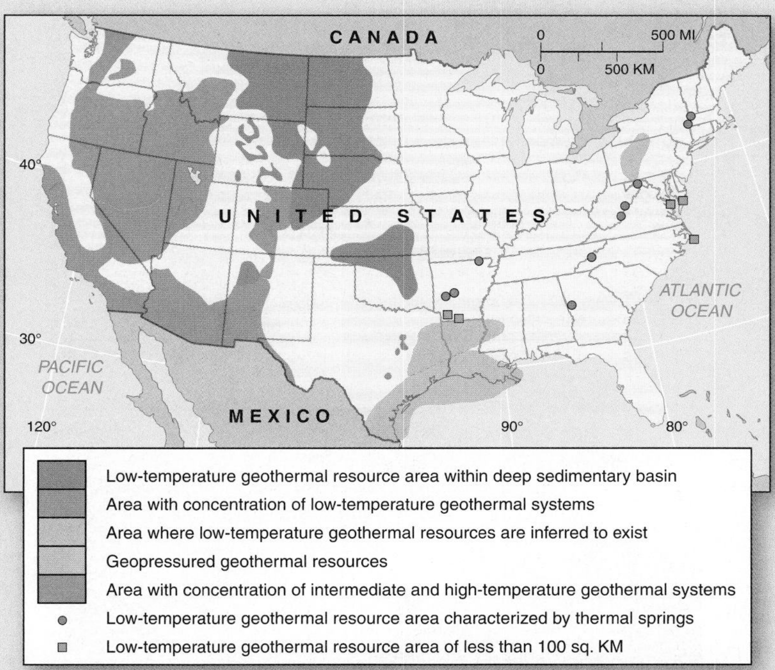

CANADA
UNITED STATES
MEXICO
PACIFIC OCEAN
ATLANTIC OCEAN
0 500 MI
0 500 KM
40°
30°
120°
90°
80°

- Low-temperature geothermal resource area within deep sedimentary basin
- Area with concentration of low-temperature geothermal systems
- Area where low-temperature geothermal resources are inferred to exist
- Geopressured geothermal resources
- Area with concentration of intermediate and high-temperature geothermal systems
- Low-temperature geothermal resource area characterized by thermal springs
- Low-temperature geothermal resource area of less than 100 sq. KM

FIGURE 11.1.3 Known geothermal resource areas. Geothermal resources in the conterminous United States. [Map courtesy W. A. Duffield *et al., Tapping the Earth's Natural Heat,* USGS Circular 1125, 1994, p. 35.]

GEO REPORT 11.3 *A lot of basalt in Hawai'i*

A mass of basalt forms the island of Hawai'i. From seafloor to summits, it contains about 40,000 km^3 (9600 mi^3) of basalt, which is enough to cover the states of Massachusetts, Connecticut, and Rhode Island to a depth of 1 km.

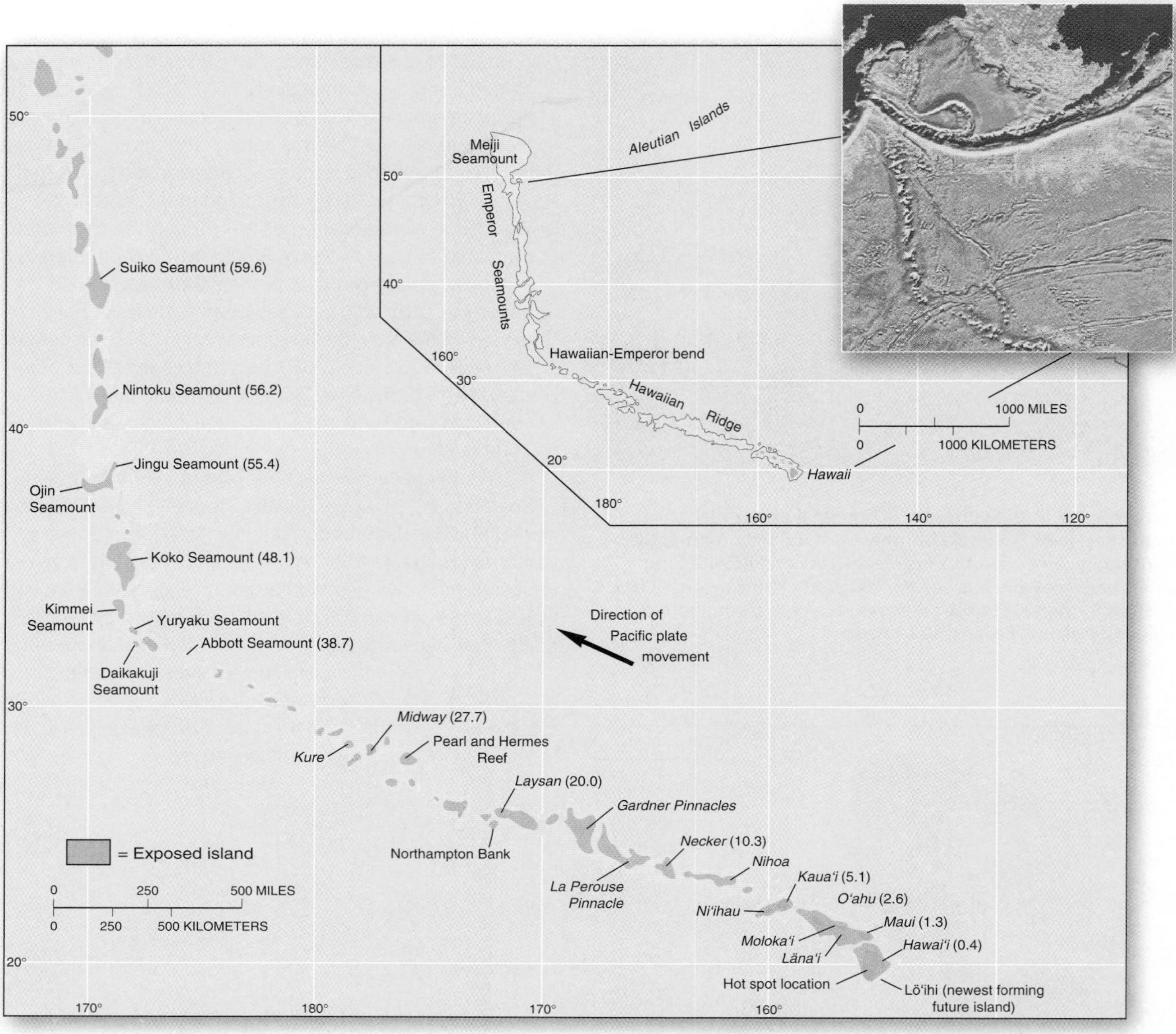

Aleutian Islands
Meiji Seamount
Emperor Seamounts
Hawaiian-Emperor bend
Hawaiian Ridge
Hawaii
0 1000 MILES
0 1000 KILOMETERS
Suiko Seamount (59.6)
Nintoku Seamount (56.2)
Jingu Seamount (55.4)
Ojin Seamount
Koko Seamount (48.1)
Kimmei Seamount
Yuryaku Seamount
Abbott Seamount (38.7)
Daikakuji Seamount
Direction of Pacific plate movement
Midway (27.7)
Kure
Pearl and Hermes Reef
Laysan (20.0)
Gardner Pinnacles
Northampton Bank
Necker (10.3)
Nihoa
La Perouse Pinnacle
Kaua'i (5.1)
O'ahu (2.6)
Ni'ihau
Maui (1.3)
Moloka'i
Lāna'i
Hawai'i (0.4)
Hot spot location
Lō'ihi (newest forming future island)
= Exposed island
0 250 500 MILES
0 250 500 KILOMETERS
50° 40° 30° 20° 160° 180° 170° 160° 140° 120°

(a)

Animation Hot Spot Volcano Tracks

(b)

FIGURE 11.22 Hot spot tracks across the North Pacific.
Hawai'i and the linear volcanic chain of islands known as the Emperor Seamounts. (a) The islands and seamounts in the chain are progressively younger toward the southeast. Ages, in m.y.a., are shown in parentheses. Note that Midway Island is 27.7 million years old, meaning that the site was over the plume 27.7 m.y.a. (b) Lo'ihi is forming 975 m (3200 ft) beneath the Pacific Ocean off South Point; presently an undersea volcano (seamount), it is growing into the next Hawaiian island. [(a) After D. A. Clague, "Petrology and K–Ar (Potassium–Argon) ages of dredged volcanic rocks from the western Hawaiian ridge and the southern Emperor seamount chain," *Geological Society of America Bulletin* 86 (1975): 991; inset from global gravity anomaly map image, Scripps Institution of Oceanography. (b) Bobbé Christopherson.]

The big island of Hawai'i, the newest, actually took less than 1 million years to build to its present stature. The island is a huge mound of lava, melded from several sea-floor fissures through five volcanoes, rising from the seafloor 5800 m (19,000 ft) to the ocean surface. From sea level, its highest peak, Mauna Kea, rises to an elevation of 4205 m (13,796 ft). This total height of almost 10,000 m (32,800 ft) represents the highest mountain on Earth if measured from the seafloor.

The youngest island in the Hawaiian chain is still a seamount. It rises 3350 m (11,000 ft) from its base, but is still 975 m (3200 ft) beneath the ocean surface. Even though this new island will not experience the tropical Sun for about 10,000 years, it is already named Lō'ihi (noted on the map).

CRITICAL THINKING 11.3

How fast is the Pacific plate moving?

Relative to the motion of the Pacific plate shown in Figure 11.22 (note the map's graphic scale in the lower-left corner), the island of Midway formed 27.7 m.y.a. over the hot spot that is active under the southeast coast of the big island of Hawai'i today. Given the scale of the map, roughly determine the average annual speed of the Pacific plate in centimeters per year for Midway to have travelled this distance.

Iceland is a result of an active hot spot sitting astride a mid-ocean ridge—visible on the different maps and images of the seafloor (see Figure 11.19). It is an excellent example of a segment of mid-ocean ridge rising above sea level. This hot spot has generated enough material to form Iceland, and eruptions continue from deep in the mantle, as happened in 2010. As a result, Iceland is still growing in area and volume. This is further evidence that, indeed, Earth is a dynamic planet.

GEOSYSTEMS CONNECTION

Beginning with the internal structure of Earth, we tracked the flow of energy toward the surface. Owing to these internal dynamics, the crust is forced about, and gravity pulls on heavier elements. Plate tectonics is the science studying these continent-sized migrating pieces of crust that can move on a collision course with other plates. Earth's present surface map is the result of the vast forces and motions. In Chapter 12, we look at the surface expressions of all this energy and matter in motion: the stress and strain of folding, faulting, and deformation; the building of mountains; and the sometimes dramatic activity of earthquakes and volcanoes.

KEY LEARNING CONCEPTS REVIEW

■ ***Distinguish*** between the endogenic and exogenic systems, ***determine*** the driving force for each, and ***explain*** the pace at which these systems operate.

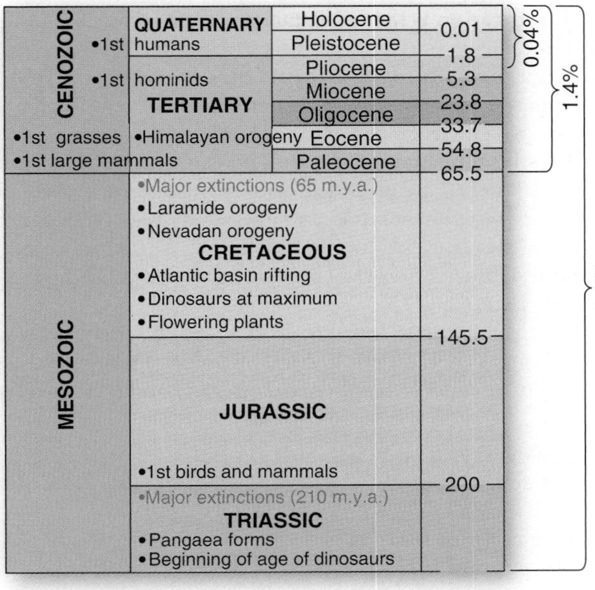

Era	Period / events	Epoch	m.y.a.	
CENOZOIC	QUATERNARY •1st humans	Holocene	0.01	0.04%
		Pleistocene	1.8	
	•1st hominids	Pliocene	5.3	1.4%
	TERTIARY	Miocene	23.8	
		Oligocene	33.7	
	•1st grasses •Himalayan orogeny	Eocene	54.8	
	•1st large mammals	Paleocene	65.5	
MESOZOIC	CRETACEOUS •Major extinctions (65 m.y.a.) •Laramide orogeny •Nevadan orogeny •Atlantic basin rifting •Dinosaurs at maximum •Flowering plants		145.5	5.4%
	JURASSIC •1st birds and mammals		200	
	TRIASSIC •Major extinctions (210 m.y.a.) •Pangaea forms •Beginning of age of dinosaurs			

The Earth–atmosphere interface is where the **endogenic system** (internal), powered by heat energy from within the planet, interacts with the **exogenic system** (external), powered by insolation and influenced by gravity. These systems work together to produce Earth's diverse landscape. The **geologic time scale** is an effective device for organizing the vast span of geologic time. It depicts the sequence of Earth's events (relative time) and the approximate actual dates (absolute time).

The most fundamental principle of Earth science is **uniformitarianism**. Uniformitarianism assumes that the *same physical processes active in the environment today have been operating throughout geologic time*, although dramatic episodes, such as massive landslides or volcanic eruptions, can interrupt the flow of events, leading to a state of *punctuated equilibrium*.

endogenic system (p. 295)
exogenic system (p. 295)
geologic time scale (p. 298)
uniformitarianism (p. 298)

1. To what extent is Earth's crust active at this time in its history?
2. Define the endogenic and the exogenic systems. Describe the driving forces that energize these systems.
3. How is the geologic time scale organized? What is the basis for the time scale in relative and absolute terms? What era, period, and epoch are we living in today?
4. Contrast uniformitarianism and punctuated equilibrium models for Earth's development.

■ ***Diagram*** Earth's interior in cross section, and ***describe*** each distinct layer.

We have learned about Earth's interior from indirect evidence—the way its various layers transmit **seismic waves**. The **core** is

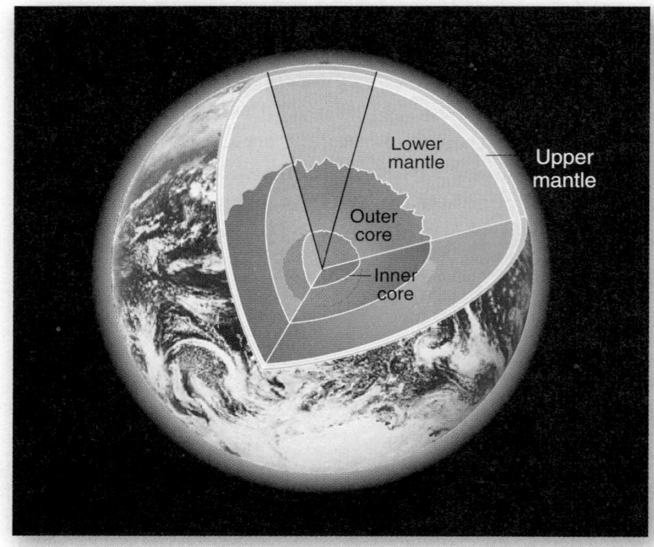

differentiated into an inner core and an outer core, divided by a transition zone. Earth's magnetic field is generated almost entirely within the outer core. Polarity reversals in Earth's magnetism are recorded in cooling magma that contains iron minerals. The patterns of **geomagnetic reversal** frozen in rock help scientists piece together the story of Earth's mobile crust.

Beyond Earth's core lies the **mantle,** differentiated into lower mantle and upper mantle. It experiences a gradual temperature increase with depth and a stiffening due to increased pressures. The upper mantle is divided into three fairly distinct layers. The uppermost mantle, along with the crust, makes up the **lithosphere**. Below the lithosphere is the **asthenosphere,** or *plastic layer*. It contains pockets of increased heat from radioactive decay and is susceptible to slow convective currents in these hotter materials. An important internal boundary between the **crust** and the high-velocity portion of the uppermost mantle is the **Mohorovičić discontinuity,** or **Moho**. *Continental crust* is basically **granite**; it is crystalline and high in silica, aluminum, potassium, calcium, and sodium. *Oceanic crust* is **basalt**; it is granular and high in silica, magnesium, and iron. The principles of buoyancy and balance produce the important principle of **isostasy**. Isostasy explains certain vertical movements of Earth's crust, such as isostatic rebound when the weight of ice is removed.

seismic waves (p. 300)
core (p. 300)
geomagnetic reversal (p. 301)
mantle (p. 302)
lithosphere (p. 302)
asthenosphere (p. 302)
crust (p. 303)
Mohorovičić discontinuity (Moho) (p. 303)
granite (p. 303)
basalt (p. 303)
isostasy (p. 303)

5. Make a simple sketch of Earth's interior, label each layer, and list the physical characteristics, temperature, composition, and range of size of each on your drawing.
6. How does Earth generate its magnetic field? Is the magnetic field constant, or does it change? Explain the implications of your answer.
7. Describe the asthenosphere. Why is it also known as the plastic layer? What are the consequences of its convection currents?
8. What is a discontinuity? Describe the principal discontinuities within Earth.
9. Define isostasy and isostatic rebound, and explain the crustal equilibrium concept.
10. Diagram the uppermost mantle and crust. Label the density of the layers in grams per cubic centimeter. What two types of crust were described in the text in terms of rock composition?

- ***Illustrate*** the geologic cycle, and ***relate*** the rock cycle and rock types to endogenic and exogenic processes.

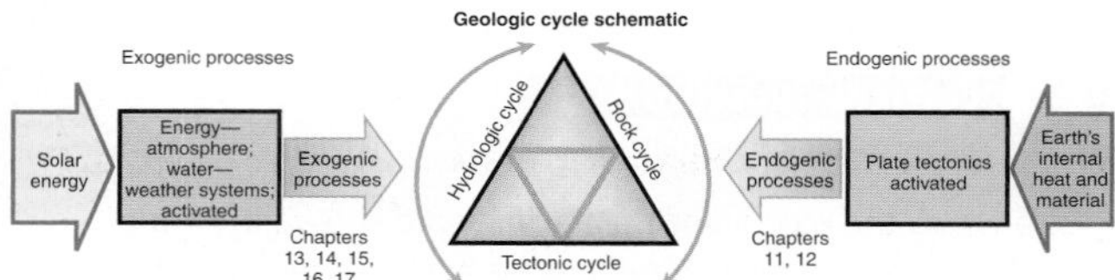

The **geologic cycle** is a model of the internal and external interactions that shape the crust. A **mineral** is an inorganic natural compound having a specific chemical formula and possessing a crystalline structure. A **rock** is an assemblage of minerals bound together (such as granite, a rock containing three minerals), or it may be a mass of a single mineral (such as rock salt).

The geologic cycle comprises three cycles: the *hydrologic cycle*, the *tectonic cycle*, and the **rock cycle**. The rock cycle describes the three principal rock-forming processes and the rocks they produce. **Igneous rocks** form from **magma,** which is molten rock beneath the surface. Magma is fluid, highly gaseous, and under tremendous pressure. Either it *intrudes* into crustal rocks, cools, and hardens, or it *extrudes* onto the surface as **lava**. Intrusive igneous rock that cools slowly in the crust forms a **pluton**. The largest pluton form is a **batholith**.

Lithification is the cementation, compaction, and hardening of sediments into **sedimentary rocks**. These layered strata form important records of past ages. **Stratigraphy** is the study of the sequence (superposition), thickness, and spatial distribution of strata, which yield clues to the age and origin of the rocks.

Clastic sedimentary rocks are derived from the fragments of weathered rocks. Chemical sedimentary rocks are dissolved minerals, transported in solution and chemically precipitated out of solution (they are essentially nonclastic). The most common chemical sedimentary rock is **limestone,** which is lithified calcium carbonate, $CaCO_3$.

Any rock, either igneous or sedimentary, may be transformed into a **metamorphic rock** by going through profound physical or chemical changes under pressure and increased temperature.

geologic cycle (p. 305)
mineral (p. 306)
rock (p. 306)
rock cycle (p. 306)
igneous rock (p. 306)
magma (p. 306)
lava (p. 307)
pluton (p. 307)
batholith (p. 307)
sedimentary rock (p. 309)

lithification (p. 309)
stratigraphy (p. 310)
limestone (p. 312)
metamorphic rock (p. 312)

11. Illustrate the geologic cycle, and define each component: rock cycle, tectonic cycle, and hydrologic cycle.
12. What is a mineral? A mineral family? Name the most common minerals on Earth. What is a rock?
13. Describe igneous processes. What is the difference between intrusive and extrusive types of igneous rocks?
14. Characterize felsic and mafic minerals. Give examples of both coarse- and fine-grained textures.
15. Briefly describe sedimentary processes and lithification. Describe the sources and particle sizes of sedimentary rocks.
16. What is metamorphism, and how are metamorphic rocks produced? Name some original parent rocks and their metamorphic equivalents.

■ ***Describe*** Pangaea and its breakup, and ***relate*** several physical proofs that crustal drifting is continuing today.

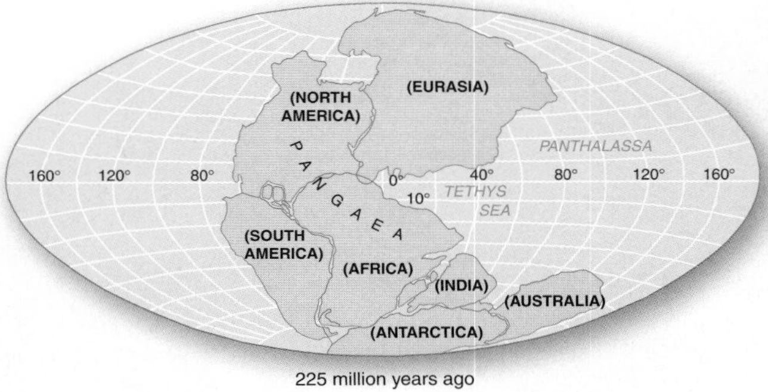

The present configuration of the ocean basins and continents is the result of *tectonic processes* involving Earth's interior dynamics and crust. Alfred Wegener coined the phrase **continental drift** to describe his idea that the crust is moved by vast forces within the planet. **Pangaea** was the name he gave to a single assemblage of continental crust some 225 m.y.a. that subsequently broke apart. Earth's lithosphere is fractured into huge slabs or plates, each moving in response to gravitational pull and to flowing currents in the mantle that create frictional drag on the plate. The all-encompassing theory of **plate tectonics** includes **sea-floor spreading** along **mid-ocean ridges** and denser oceanic crust diving beneath lighter continental crust along **subduction zones**.

continental drift (p. 314)
Pangaea (p. 314)
plate tectonics (p. 315)
sea-floor spreading (p. 315)
mid-ocean ridge (p. 315)
subduction zone (p. 315)

17. Briefly review the history of the theory of continental drift, sea-floor spreading, and the all-inclusive plate tectonics theory. What was Alfred Wegener's role?
18. Define upwelling, and describe related features on the ocean floor. Define subduction, and explain the process.
19. What was Pangaea? What happened to it during the past 225 million years?
20. Characterize the three types of plate boundaries and the actions associated with each type.

■ ***Portray*** the pattern of Earth's major plates, and ***relate*** this pattern to the occurrence of earthquakes, volcanic activity, and hot spots.

Occurrences of often-damaging earthquakes and volcanoes are correlated with plate boundaries. Three types of plate boundaries form: divergent, convergent, and transform. Along the offset portions of mid-ocean ridges, horizontal motions produce **transform faults**.

As many as 50 to 100 **hot spots** exist across Earth's surface, where plumes of magma—some anchored in the lower mantle, others originating from shallow sources in the upper mantle—generate a flow upward. **Geothermal energy** literally refers to heat from Earth's interior, whereas *geothermal power* relates to specific applied strategies of geothermal electric or geothermal direct applications.

transform faults (p. 322)
hot spots (p. 323)
geothermal energy (p. 324)

21. What is the relation between plate boundaries and volcanic and earthquake activity?
22. What is the nature of motion along a transform fault? Name a famous example of such a fault.

MASTERING GEOGRAPHY

Visit www.masteringgeography.com for several figure animations, satellite loops, self-study quizzes, flashcards, visual glossary, case studies, career links, textbook reference maps, RSS Feeds, and an ebook version of *Geosystems*. Included are many geography web links to interesting Internet sources that support this chapter.

CHAPTER 12

Tectonics, Earthquakes, and Volcanism

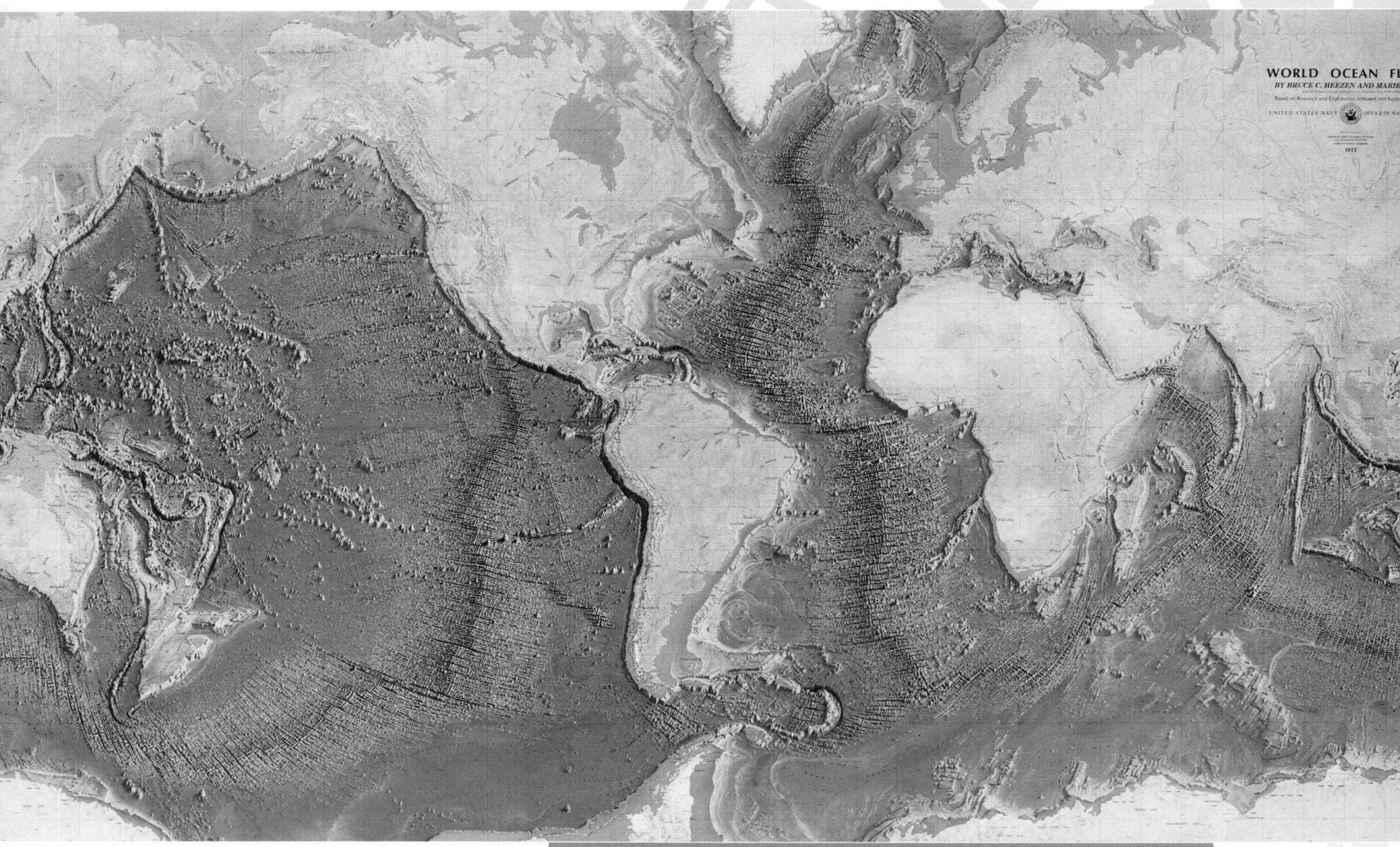

The scarred ocean floor is clearly visible: sea-floor spreading centers marked by oceanic ridges that stretch over 64,000 km (40,000 mi), subduction zones indicated by deep oceanic trenches, and transform faults slicing across oceanic ridges. [*World Ocean Floor* by Bruce C. Heezen and Marie Tharp, 1977, Office of Naval Research.]

KEY LEARNING CONCEPTS

After reading the chapter, you should be able to:

- ***Describe*** first, second, and third orders of relief, and ***relate*** examples of each from Earth's major topographic regions.
- ***Describe*** the several origins of continental crust, and ***define*** displaced terranes.
- ***Explain*** compressional processes and folding, and ***describe*** four principal types of faults and their characteristic landforms.
- ***Relate*** the three types of plate collisions associated with orogenesis, and ***identify*** specific examples of each.
- ***Explain*** the nature of earthquakes, including their characteristics, measurement, fault mechanics, and forecasting methods.
- ***Distinguish*** between an effusive and an explosive volcanic eruption, and ***describe*** related landforms, using specific examples.

GEOSYSTEMS NOW

The San Jacinto Fault Connection

The San Jacinto fault is an active earthquake fault in southern California. It is part of the San Andreas fault system, which is made up of hundreds of horizontally moving faults. These faults are produced by the relative motion of the northwestward-moving Pacific plate against the southeastward-moving North American plate; some 5 cm (2 in.) per year of motion builds stress and accumulated strain. The San Jacinto fault runs from north of the Mexican border to the "Inland Empire" of San Bernardino, somewhat parallel to the Elsinore fault, which continues on into Orange County, and the Whittier fault, which runs into Los Angeles. Numerous other related faults run through the region with this same alignment (Figure GN 12.1).

From the small inland communities in the Mohave Desert to the sprawling coastal population centers of metropolitan Los Angeles, the people of southern California live with earthquakes. One overriding question for everyone is, When will the "Big One" occur? The U.S. Geological Survey (USGS) regards the San Jacinto fault as capable of an M 7.5 quake ("M" is the abbreviation for *moment magnitude* and refers to an earthquake rating scale discussed in this chapter). The San Jacinto fault's relation to the Elsinore, Whittier, and other fault lines in the San Andreas system indicates the potential for major earthquakes across the metropolitan southern California landscape. An M 7.5 would release almost 30 times more energy than that produced by the 1994 Northridge earthquake—thus, a major quake compared to a strong quake. Remember, the Northridge quake remains the most costly in terms of property damage in U.S. history.

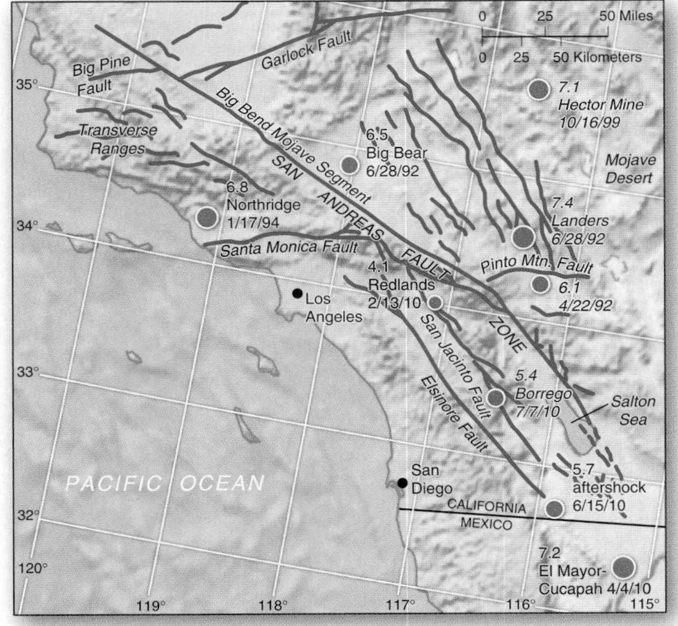

FIGURE GN 12.1 Southern California seismic map.

Scientific evidence regarding the nature of faulting and crustal movement suggests that a major earthquake may be preceded by a number of small earthquakes in an area. Recent evidence points to a "wavelike pattern" of small quakes as rupturing spreads along the fault plane. The latest earthquake activity along the San Jacinto and Elsinore faults may be following this pattern.

Records show that at least five quakes have occurred along the San Jacinto fault in the last 50 years: an M 5.8 and M 6.5 in 1968, an M 5.3 in 1980, and an M 5.0 in 2005. Then an M 4.1 hit February 13, 2010, about 40 km (25 mi) south of Palm Springs. The largest of the sequence was an M 7.2 Sierra El Mayor quake April 4, 2010, some 65 km south-southeast of El Centro, in Baja, California. An aftershock of M 5.7 occurred on June 15.

The April quake transferred strain to the San Jacinto fault, with large clusters of aftershocks toward the north in the fault structure. Continuing the pattern, an M 5.4 quake hit along the San Jacinto fault on July 7, 2010, the Borrego quake, with aftershocks again spread to the northwest. If strain is moving northward along the system, then there should be much concern. A network of 100 global positioning system (GPS) stations monitors any crustal change across the region that might indicate an upcoming earthquake. More on earthquake forecasting is ahead in this chapter.

As said so many times in this text, such a situation adds to the importance of Earth systems science and education of the public to ensure proper planning, zoning, and preparation. Zoning could block new construction in particularly vulnerable areas and promote the retrofitting of poorly built infrastructure. A great quake along the San Jacinto and related faults should not be a surprise, but should be anticipated with adaptation and mitigation actions in place.

Earth's endogenic systems produce flows of heat and material toward the surface to form crust. Ongoing processes produce continental landscapes and oceanic sea-floor crust, sometimes in dramatic episodes. Earth's physical systems move to the front pages whenever an earthquake strikes or after one of the nearly 50 volcanic eruptions a year that threaten a city.

The setting was the impoverished country of Haiti, January 12, 2010, just before 5 P.M., when an M 7.0 earthquake struck. The epicenter was 15 km (10 mi) southwest of the capital. Seven minutes later an aftershock of M 6.0 continued the tragedy. More than 222,000 people died, with 300,000 injuries. Almost 300,000 houses were destroyed or damaged (Figure 12.1a). Total damage exceeded the country's $14 billion gross domestic product (GDP).

The fault plane solution—that is, the orientation of the fault plane and the direction of slip motion—determined that the Caribbean plate moved eastward relative to the North American plate's westward shift, in a horizontal strike-slip motion along a 50-km section of the Enriquillo–Plaintain Garden fault. Having this fault segment snap added more stress to the eastern segment of the fault, which is some 600 km in length. The entire fault system is accumulating elastic energy at the same rate as the section that broke in the quake. Port-Au-Prince was destroyed in 1751, 1770, and 2010.

Six weeks later the Nazca plate, moving eastward beneath the westward-moving South American plate, suddenly shifted, generating an M 8.8 quake southwest of Santiago, near Maule, Chile (Figure 12.1b). This is near Concepcion and the epicenter of the greatest earthquake of the twentieth century, the M 9.6 that hit in 1960. Some of the fault-slip zone from that quake matched areas for this February 27, 2010, quake. Damage exceeded $30 billion, compared to Chile's GDP of $45 billion. The deaths of 521 people and the injuries to 12,000 others marked the human dimension. At least two tsunami were generated, arriving in Hilo, Hawai'i, about 15 hours later, each only a meter in height.

The fault plane solution here was a subduction zone earthquake. This area of subduction of the Nazca plate beneath the South American plate is moving at a relative speed of 7 m (23 ft) per century. Damage across Chile was reduced by strict building codes that went into effect in 1985.

Volcanic eruptions pose threats to the lives and property of those who live in proximity to the hazard. Earth systems science is providing analysis and warnings to affected populations as never before. For instance, in April 2010 the Eyjafjallajökull volcano in southern Iceland erupted (see Figure 12.22). Previously, this mountain had erupted in 1821 and 1823. The ash cloud rose to 10,660 m (35,000 ft) and quickly

(a) Port-Au-Prince, Haiti

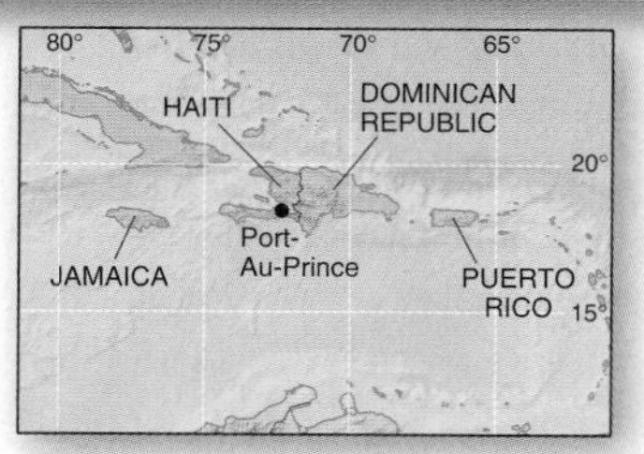

(b) Santiago, Chile

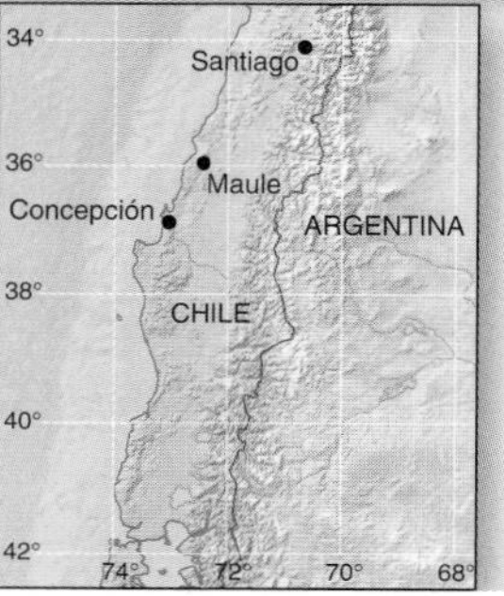

FIGURE 12.1 Earthquakes strike Haiti and Chile.
(a) Haiti was devastated by an M 7.0 quake in January 2010, destroying homes in downtown Port-Au-Prince, Haiti. (b) In February 2010, an M 8.8 quake struck southwest of Santiago, Chile. This devastated apartment building in Santiago, Chile, reflects lateral land movement of 1 to 3 m (3.3 to 9.8 ft) across the region. [(a) Cameron Davidson/Corbis. (b) Ian Salas/eps/Corbis.]

dispersed toward Europe and commercial airline corridors. Airspace was closed for five days, cancelling more than 100,000 flights. When flights resumed, routes were changed to avoid lingering ash. Because this eruption was essentially horizontal in the lower atmosphere, there was not a climate signal in Earth's atmosphere, as occurred from Mount Pinatubo in 1991 and its vertical blast.

In this chapter: We examine processes that construct Earth's surface and create world structural regions. Tectonic processes deform, recycle, and reshape Earth's crust. These processes occur sometimes in dramatic episodes, but most often in slow, deliberate motions that build the landscape. Continental crust has been forming throughout most of Earth's 4.6-billion-year existence.

The arrangement of continents and oceans, the origin of mountain ranges, the topography of the land and the seafloor, and the locations of earthquake and volcanic activity are all evidence of our dynamic Earth. Principal seismic and volcanic zones occur along plate boundaries and hot-spot locations, thus linking plate tectonics to the potential devastation of earthquakes and to the local threats of major volcanic eruptions and their attendant global climatic impact.

We begin our look at tectonics and volcanism on the ocean floor, hidden from direct view. The map that opens this chapter is a striking representation of the concepts learned in the previous chapter, laying the foundation for this and subsequent chapters. Be sure to take the quick tour presented in Critical Thinking 12.1 for a closer look. Try to correlate this sea-floor illustration with the maps of crustal plates and plate boundaries shown in Figure 11.18 and the gravity anomaly image in Figure 11.19.

CRITICAL THINKING 12.1

Ocean floor tectonics tour

Follow the ocean ridge and spreading center known as the East Pacific Rise northward as it trends beneath the west coast of the North American plate, disappearing under earthquake-prone California. Locate the continents, the offshore-submerged continental shelves, and the expanse of the sediment-covered abyssal plain on the map.

On the floor of the Indian Ocean, you see the wide track along which the Indian plate travelled northward to its collision with the Eurasian plate. Vast deposits of sediment cover the Indian Ocean floor, south of the Ganges River and to the east of India. Sediments derived from the Himalayan Range blanket the floor of the Bay of Bengal (south of Bangladesh) to a depth of 20 km (12.4 mi). These sediments result from centuries of soil erosion in the land of the monsoons.

In the area of the Hawaiian Islands, a chain of islands and seamounts marks the hot-spot track on the Pacific plate from Hawai'i to the Aleutians. Visible as dark trenches are subduction zones south and east of Alaska and Japan and along the western coast of South and Central America. Follow the Mid-Atlantic Ridge spreading center the full length of the Atlantic to where it passes through Iceland, sitting astride the ridge. Take time with this map and note the many concepts discussed in Chapters 11 and 12.

Earth's Surface Relief Features

Relief refers to vertical elevation differences in the landscape. Examples include the low relief of Nebraska and Saskatchewan, the medium relief of foothills along mountain ranges, and the high relief of the Rockies and Himalayas. The undulating form of Earth's surface, including its relief, is **topography,** portrayed so effectively on topographic maps—the lay of the land.

The relief and topography of Earth's landforms played a vital role in human history: High mountain passes both protected and isolated societies, ridges and valleys dictated transportation routes, and vast plains necessitated developing faster methods of communication and travel. Earth's topography has stimulated human invention and spurred adaptation.

Crustal Orders of Relief and Hypsometry

Modern computer capabilities and tools such as the GPS for determining location and elevation assist in studying Earth's relief and topography. Scientists put elevation data in digital form for computer manipulation and display in *digital elevation models* (*DEMs*).

An example is the digitized shaded-relief map of the United States prepared by the USGS. The entire map is made up of 12 million spot elevations, with less than a kilometer between any two. On the USGS web site, "A Tapestry of Time and Terrain," you can see a U.S. map that combines topography and geology; this composite map offers a detailed color illustration of land surfaces and the ages of underlying formations (see http://tapestry.usgs.gov/).

For convenience of description, geographers group the landscape's topography into three *orders of relief.* These orders classify landscapes by scale, from vast ocean basins and continents down to local hills and valleys. The first order of relief is the coarsest level of landforms, consisting of the continents and oceans. **Continental landmasses** are those portions of crust that reside above or near sea level, including the undersea continental shelves along the coastlines. **Ocean basins,** featured in the chapter-opening map illustration, are entirely below sea level. Approximately 71% of Earth is covered by water.

The second order of relief is the intermediate level of landforms, for both continental and ocean-basin features. Continental features in the second order of relief include mountain masses, plains, and lowlands. A few examples are the Alps, Canadian and American Rockies, west Siberian lowland, and Tibetan Plateau. The great rock "shields" that form the heart of each continental mass are of this second order. In the ocean basins, the second order of relief includes continental rises, slopes, abyssal plains, mid-ocean ridges, submarine canyons, and oceanic trenches (subduction zones)—all visible in the sea-floor illustration that opens this chapter.

The third and most detailed order of relief includes individual mountains, cliffs, valleys, hills, and other

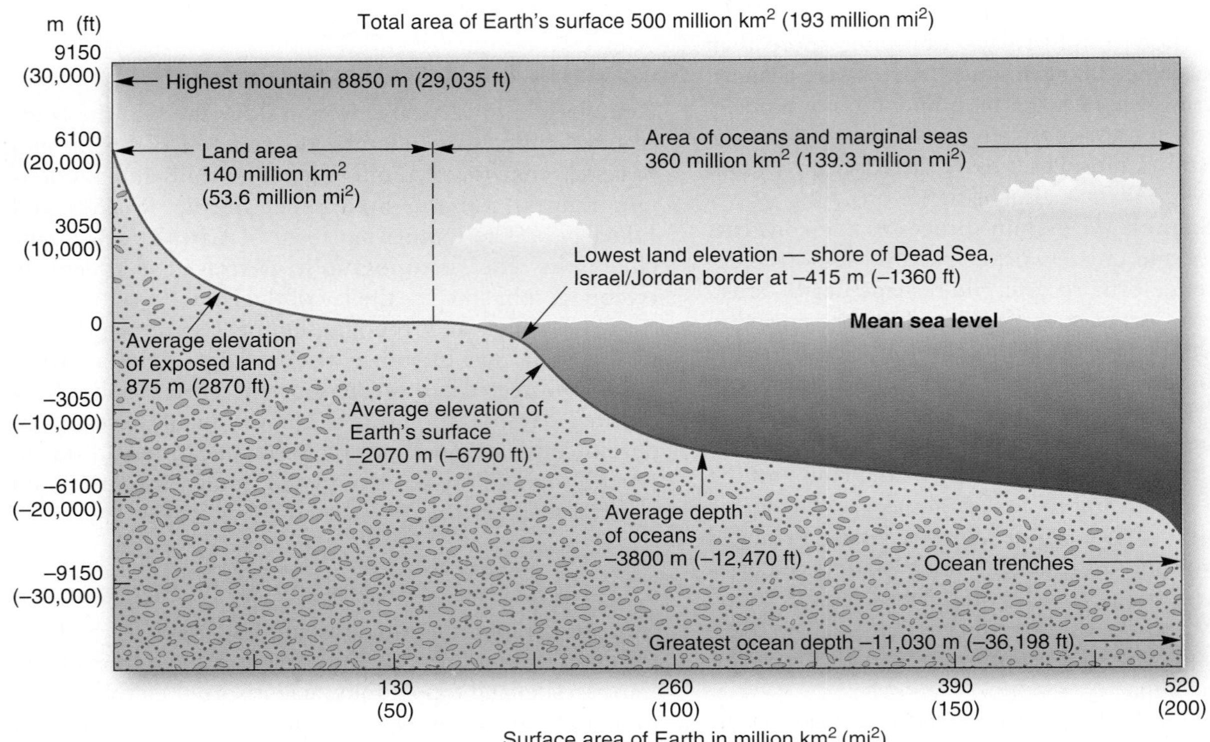

FIGURE 12.2 Earth's hypsometry.
Hypsographic curve of Earth's surface area and elevation as related to mean sea level. From the highest point above sea level (Mount Everest) to the deepest oceanic trench (Mariana Trench), Earth's overall relief is almost 20 km (12.5 mi).

landforms of smaller scale. These features are identifiable as local landscapes.

Figure 12.2 is a *hypsographic curve* (from the Greek *hypsos*, meaning "height") that shows the distribution of Earth's surface by area and elevation in relation to sea level. Relative to Earth's diameter of 12,756 km (7926 mi), the surface is of low relief—only about 20 km (12.5 mi) from highest peak to lowest oceanic trench. For perspective, Mount Everest is 8.8 km (5.5 mi) above sea level, and the Mariana Trench is 11 km (6.8 mi) below sea level.

The average elevation of Earth's solid surface is actually under water: –2070 m (–6790 ft) below mean sea level. The average elevation for exposed land is only 1875 m (2870 ft). For the ocean depths, average elevation is –3800 m (–12,470 ft). From this description you can see that, on the average, the oceans are much deeper than continental regions are high. Overall, the underwater ocean basins, ocean floor, and submarine mountain ranges form Earth's largest "landscape."

Earth's Topographic Regions

The three orders of relief can be further generalized into six topographic regions: plains, high tablelands, hills and low tablelands, mountains, widely spaced mountains, and depressions (Figure 12.3). An arbitrary elevation or descriptive limit that is in common use defines each type of topography (see the map legend).

Four of the continents possess extensive *plains*, areas with local relief of less than 100 m (325 ft) and slope angles

GEO REPORT 12.1 *Mount Everest at new heights*

The announcement came November 11, 1999, in Washington, DC: Mount Everest had a new revised elevation, determined by direct GPS placement on the mountain's icy summit. This was a continuation of a measurement effort begun in 1995 by the late Bradford Washburn, renowned mountain photographer and explorer.

Everest's new elevation of 8850 m (29,035 ft) is close to the previous official measure of 8848 m (29,028 ft) set in 1954 by the Survey of India. Also measured was the fact that the Himalayas are moving northeastward, as the mountain range is being driven farther into Asia by plate tectonics—the continuing collision of Indian and Asian landmasses.

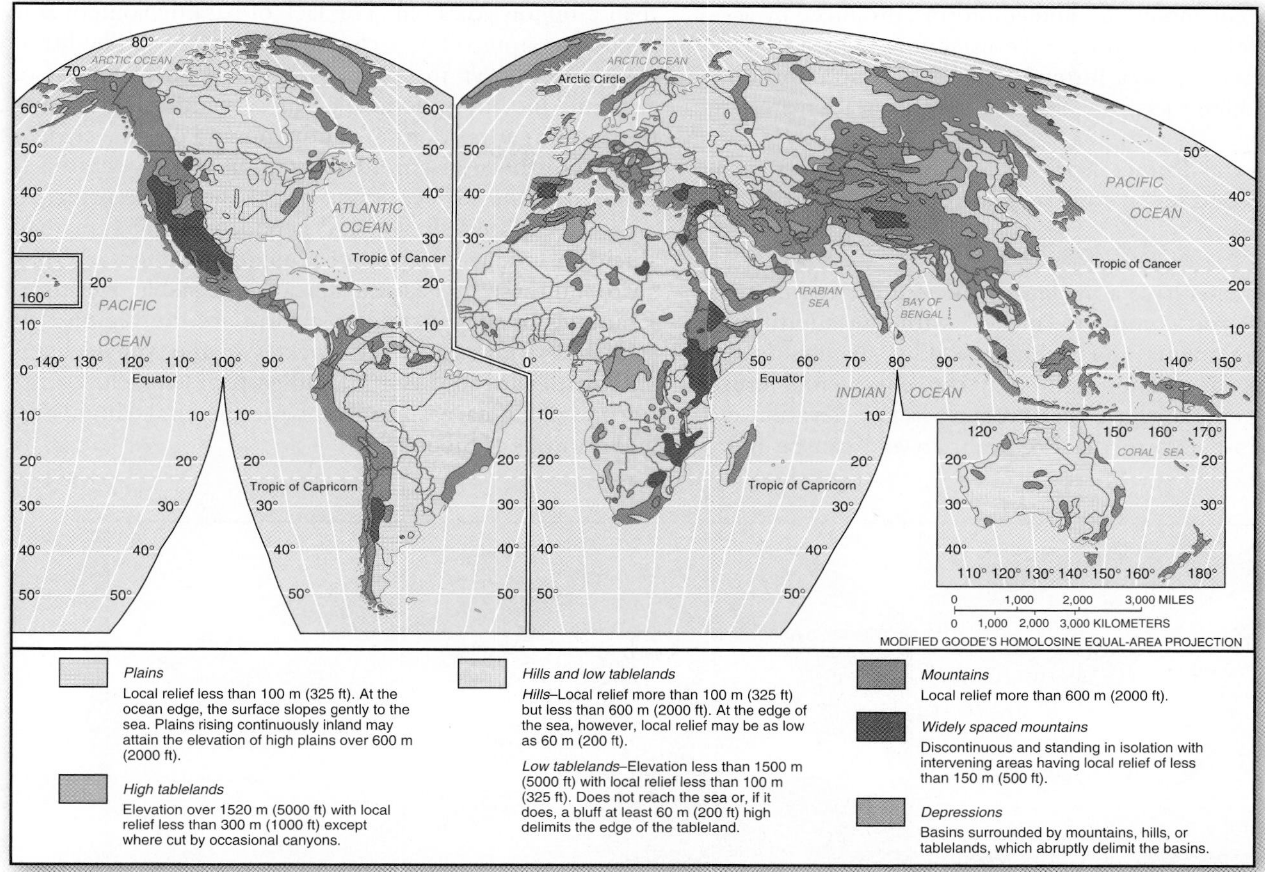

FIGURE 12.3 Earth's topographic regions.
Compare the map and legend to the world physical map inside the back cover of this text. [After R. E. Murphy, "Landforms of the world," *Annals of the Association of American Geographers* 58, 1 (March 1968). Adapted by permission.]

of 5° or less. Some plains have high elevations of more than 600 m (2000 ft); in the United States, the high plains attain elevations above 1220 m (4000 ft). The Colorado Plateau, Greenland, and Antarctica are notable *high tablelands* (the latter two composed of ice), with elevations exceeding 1520 m (5000 ft). *Hills* and *low tablelands* dominate Africa.

Mountain ranges, characterized by local relief exceeding 600 m (2000 ft), occur on each continent. Earth's relief and topography are undergoing constant change as a result of processes that form crust.

CRITICAL THINKING 12.2

Scaling topographic regions to your area

Using the topographic regions map in Figure 12.3, assess the region within 100 km (62 mi) and 1000 km (620 mi) of your campus. Describe the topographic character and the variety of relief within these two regional scales. You may want to consult local maps and atlases in your analysis. Do you perceive that this type of topographic region influences lifestyles? Economic activities? Transportation? Was it influential in the history of the region?

Crustal Formation Processes

How did Earth's continental crust form? What gave rise to the three orders of relief just discussed? Earth's surface is a battleground of opposing processes: On the one hand, tectonic activity, driven by our planet's internal energy, builds crust; on the other hand, the exogenic processes of weathering and erosion, powered by the Sun through the actions of air, water, waves, and ice, tear down crust. In the PART III opening photo, imagine the uplift of the Colorado Plateau being countered by weathering and erosion, resulting in the buttes, mesas, and scenic vistas.

Tectonic activity generally is slow, requiring millions of years. Endogenic processes result in gradual uplift and new landforms, with major mountain building occurring along plate boundaries. These uplifted crustal regions are quite varied, but we think of them in three general categories, all discussed in this chapter:

- Residual mountains and stable continental cratons, formed from inactive remnants of ancient tectonic activity;

- Tectonic mountains and landforms, produced by active folding, faulting, and crustal movements; and
- Volcanic features, formed by the surface accumulation of molten rock from eruptions of subsurface materials.

Thus, several distinct processes operate in concert to produce the continental crust we see around us.

Continental Shields

All continents have a nucleus of ancient crystalline rock on which the continent "grows" with the addition of crustal fragments and sediments. This nucleus is the *craton*, or heartland region, of the continental crust. Cratons generally have been eroded to a low elevation and relief. Most date to the Precambrian and can be more than 2 billion years old. The lack of basaltic components in these cratons offers a clue to their stability. The lithosphere is thicker in cratonic regions, which consist of crust and the lithospheric uppermost mantle, than it is beneath younger portions of continents and oceanic crust.

A **continental shield** is a region where a craton is exposed at the surface. Figure 12.4 shows the principal areas of exposed shields and a photo of the Canadian shield in Québec. Layers of younger sedimentary rock surround these shields and appear quite stable over time. Examples of such a stable platform are the region that stretches from east of the Rockies to the Appalachians and northward into central and eastern Canada, a large portion of China, eastern Europe to the Ural Mountains, and portions of Siberia.

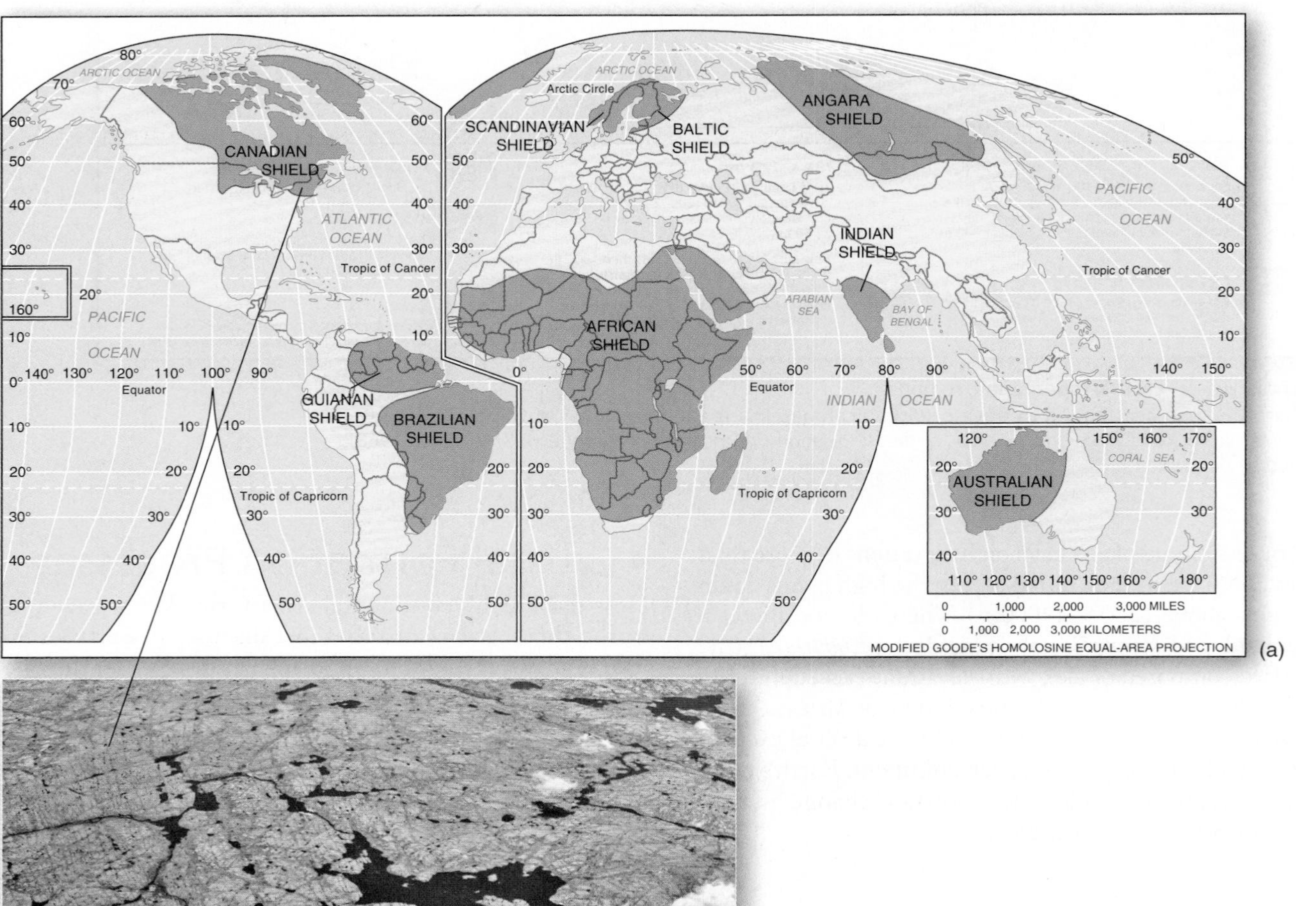

FIGURE 12.4 Continental shields.
(a) Portions of major continental shields exposed by erosion. Adjacent portions of these shields remain covered by younger sedimentary layers. (b) Canadian shield landscape in northern Québec, stable for hundreds of millions of years, stripped by past glaciations and marked by intrusive igneous dikes (magmatic intrusions). Can you see evidence in the photo of directional flow of the continental glaciers? [(a) After R. E. Murphy, "Landforms of the world," *Annals of the Association of American Geographers* 58, 1 (March 1968). Adapted by permission. (b) Bobbé Christopherson.]

Building Continental Crust and Terranes

The formation of continental crust is complex and takes hundreds of millions of years. It involves the entire sequence of sea-floor spreading and formation of oceanic crust, its later subduction and remelting, and its subsequent rise as new magma, all summarized in Figure 12.5.

To understand this process, study Figure 12.5 and the inset photos, and recheck Figure 11.14. Begin with the magma that originates in the asthenosphere and wells up along the mid-ocean ridges. Basaltic magma is formed from minerals in the upper mantle that are rich in iron and magnesium. Such magma contains less than 50% silica and has a low-viscosity (thin) texture—it tends to flow. This mafic material rises to erupt at spreading centers and cools to form new basaltic seafloor, which spreads outward to collide with continental crust along its far edges. This denser oceanic crust plunges beneath the lighter continental crust, into the mantle, where it remelts. The new magma then rises and cools, forming more continental crust, in the form of intrusive granitic igneous rock.

As the subducting oceanic plate works its way under a continental plate, it takes with it trapped seawater and sediment from eroded continental crust. The remelting incorporates the seawater, sediments, and surrounding crust into the mixture. As a result, the magma, generally called a *melt*, that migrates upward from a subducted plate contains 50%–75% silica and aluminum (called *andesitic* or *silicic*, depending on silica content). The melt has a high-viscosity (thick) texture—it tends to block and plug conduits to the surface.

Bodies of such silica-rich magma may reach the surface in explosive volcanic eruptions, or they may stop short and become subsurface intrusive bodies in the crust, cooling slowly to form granitic crystalline plutons such as batholiths (see Figures 11.8 and 11.9a). Note that this composition is quite different from the magma that rises directly from the asthenosphere at sea-floor spreading centers. In these processes of crustal formation, you can literally follow the cycling of materials through the tectonic cycle.

Each of Earth's major lithospheric plates actually is a collage of many crustal pieces acquired from a variety of sources. Crustal fragments of ocean floor, curving chains (or arcs) of volcanic islands, and other pieces of continental crust all have been forced against the edges of continental shields and platforms. These slowly migrating crustal pieces, which have become attached or accreted to

FIGURE 12.5 Crustal formation.
Material from the asthenosphere upwells along sea-floor spreading centers. Basaltic ocean floor is subducted beneath lighter continental crust, where it melts, along with its cargo of sediments, water, and minerals. This melting generates magma, which makes its way up through the crust to form igneous intrusions and extrusive eruptions. The dacite volcanic bomb (lighter rock on right, higher in silica) is from Mount St. Helens, and the basalt rock (darker rock on left, lower in silica) is from Hawai'i. What does this coloration tell you about the rocks' history and composition? Find these rock types in Table 11.2. [Bobbé Christopherson.]

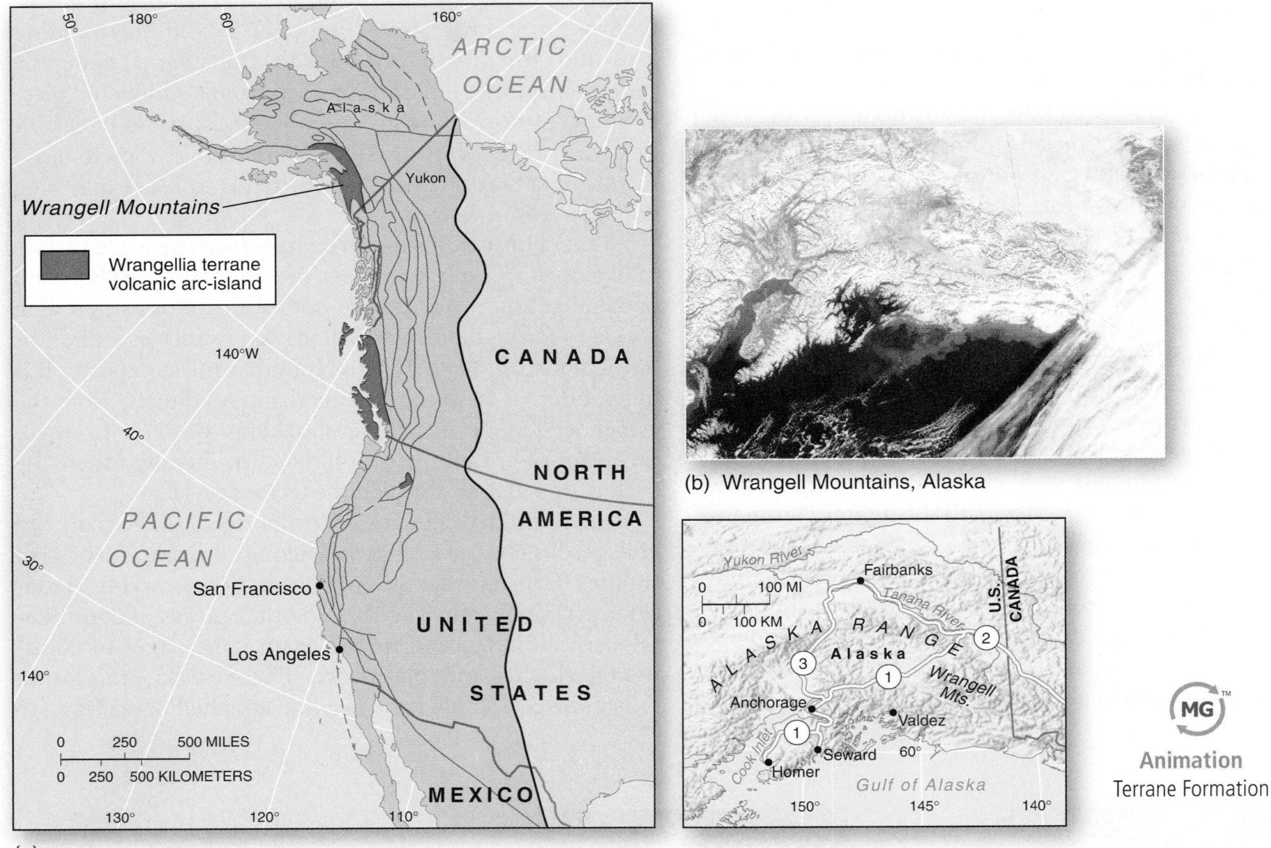

FIGURE 12.6 North American terranes.
(a) Wrangellia terranes occur in four segments, highlighted in red among the outlines of other terranes along the western margin of North America (light blue shaded area). (b) Snow-covered Wrangell Mountains north of the Chugach Mountains in this November 7, 2001, image in east-central Alaska and across the Canadian border. Note Mount McKinley (Denali) at 6194 m (20,322 ft), the highest elevation in North America, in the upper-left part of the image, north of Cook Inlet (highways numbers on locator map). [(a) Based on data from USGS. (b) *Terra* MODIS image, NASA/GSFC.]

the plates, are **terranes** (not to be confused with *terrain*, which refers to the topography of a tract of land). These displaced terranes, sometimes called *microplate* or *exotic terranes*, have histories different from those of the continents that capture them. They are usually framed by fault-zone fractures and differ in rock composition and structure from their new continental homes.

In the region surrounding the Pacific, accreted terranes are particularly prevalent. At least 25% of the growth of western North America can be attributed to the accretion of at least 50 terranes since the early Jurassic Period (190 million years ago). A good example is the Wrangell Mountains, which lie just east of Prince William Sound and the city of Valdez, Alaska. The *Wrangellia terranes*—a former volcanic island arc and associated marine sediments from near the equator—migrated approximately 10,000 km (6200 mi) to form the Wrangell Mountains and three other distinct formations along the western margin of the continent (Figure 12.6).

The Appalachian Mountains, extending from Alabama to the Maritime Provinces of Canada, possess bits of land once attached to ancient Europe, Africa, South America, Antarctica, and various oceanic islands. The discovery of terranes, made only in the 1980s, demonstrates one of the ways continents are assembled.

Crustal Deformation Processes

Rocks, whether igneous, sedimentary, or metamorphic, are subjected to powerful stress by tectonic forces, gravity, and the weight of overlying rocks. Figure 12.7 shows three types of stress: *tension* (stretching), *compression* (shortening), and *shear* (twisting or tearing).

Strain is how rocks respond to stress. Strain is expressed in rocks by *folding* (bending) or *faulting* (breaking). Think of stress as a force and the resulting strain as the deformation in the rock. Whether a rock bends or breaks depends on several factors, including composition and how much pressure is on the rock. An important quality is whether the rock is *brittle* or *ductile*. The patterns created by these processes are clearly visible in the landforms we see today, especially in mountain areas.

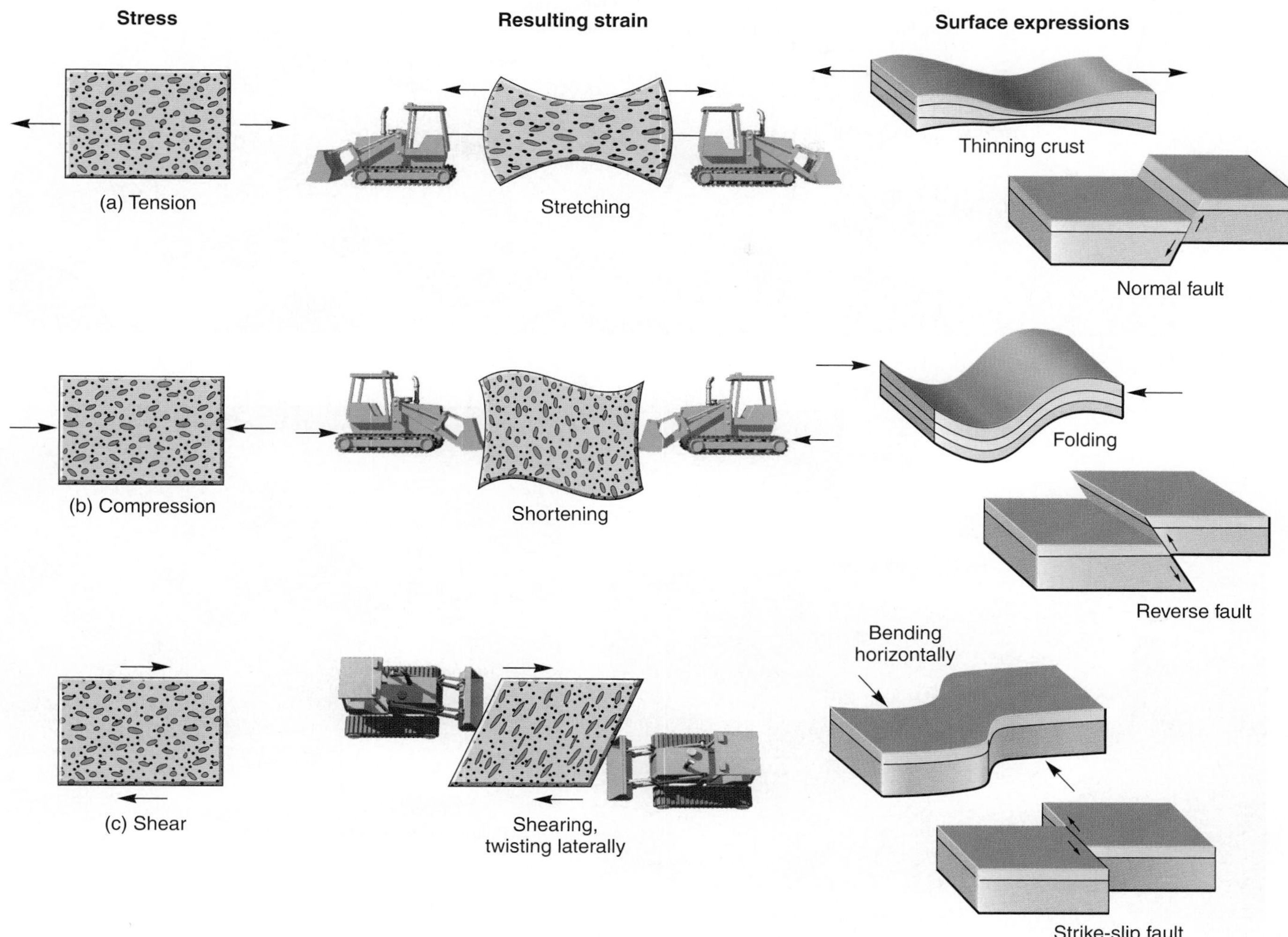

FIGURE 12.7 Three kinds of stress and strain and the resulting surface expressions. The bulldozers represent the stress, or force, on the rock strata. (a) Tension stress produces a stretching and thinning of the crust and a *normal fault.* (b) Compression produces shortening and folding and *reverse faulting* of the crust. In a back-and-forth horizontal motion, (c) shear stress produces a bending of the crust and, on breaking, a *strike-slip fault.*

Figure 12.7 illustrates each type of stress and the resulting strain and surface expressions that develop.

Folding and Broad Warping

When rock strata that are layered and flat are subjected to compressional forces, they become deformed (Figure 12.8). Convergent plate boundaries intensely compress rocks, deforming them in a process known as **folding**. As an analogy, if we take sections of thick fabric, stack them flat on a table, and then slowly push on opposite ends of the stack, the cloth layers will bend and rumple into folds similar to those shown in Figure 12.8a.

If we then draw a line down the center axis of a resulting ridge and a line down the center axis of a resulting trough, we see how the names of the folds are assigned. Along the *ridge* of a fold, layers *slope downward away from the axis,* forming an **anticline**. In the *trough* of a fold, however, layers *slope downward toward the axis,* resulting in a **syncline**.

If the axis of either type of fold is not "level" (horizontal, or parallel to Earth's surface), the fold axis then *plunges,* inclined (dipped down) at an angle. Knowledge of how folds are angled to Earth's surface and where they are located is important for the petroleum industry. For example, petroleum geologists know that oil and natural gas collect in the upper portions of anticlinal folds in permeable rock layers such as sandstone. This is the "anticlinal theory of petroleum accumulation" attributed to I. C. White, founder of the West Virginia Geological Survey.

Figure 12.8a further illustrates the landforms that result from the effects of weathering and other exogenic processes in folded landscapes:

- A residual *synclinal ridge* may form within a syncline because different rock strata offer greater resistance to weathering processes. An actual synclinal ridge is exposed dramatically in an interstate highway road cut in Figure 12.8b.
- Compressional forces often push folds far enough that they actually overturn upon their own strata (*overturned anticline*).

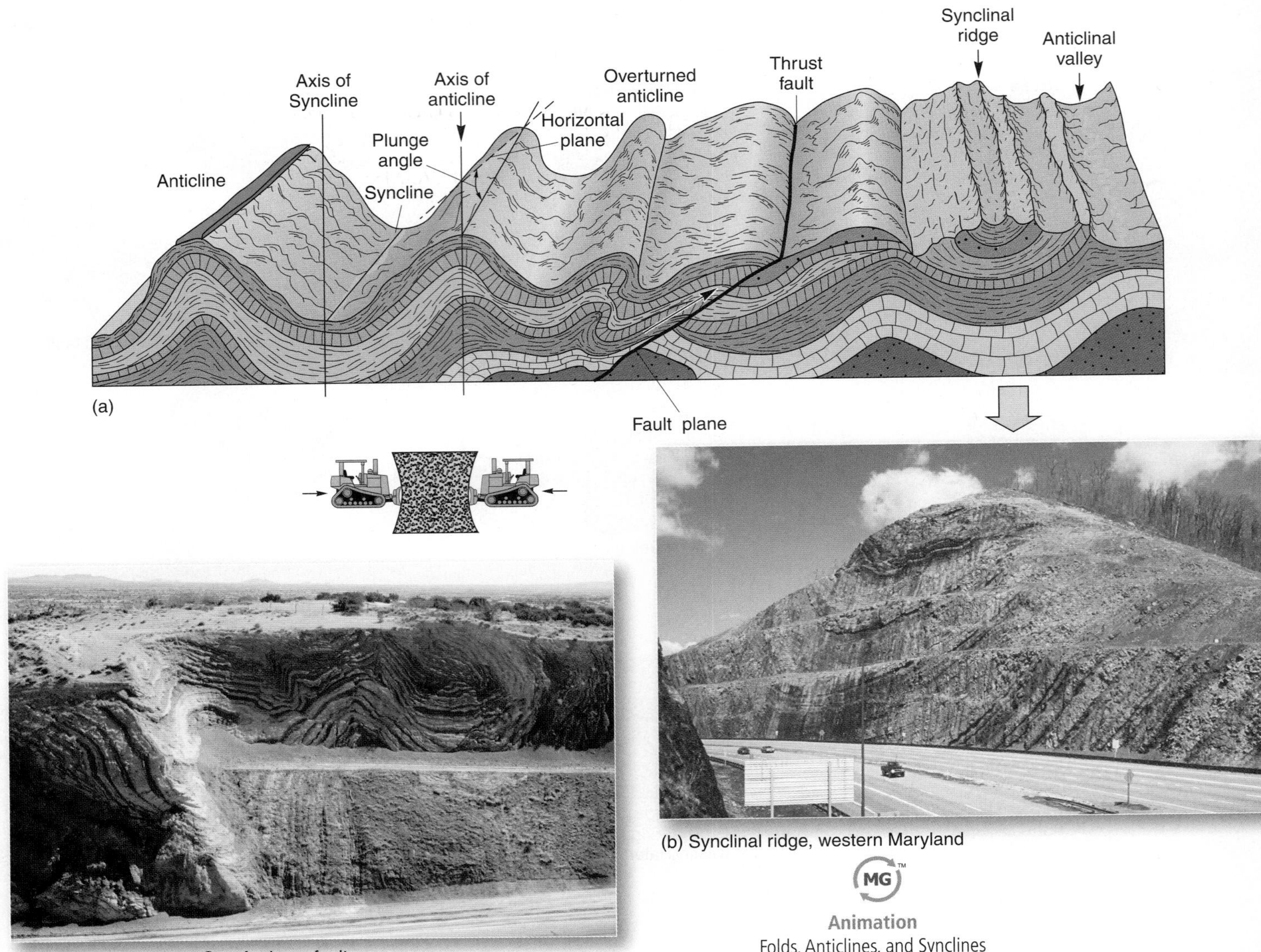

(b) Synclinal ridge, western Maryland

MG
Animation
Folds, Anticlines, and Synclines

(c) Folded strata along San Andreas fault

FIGURE 12.8 Folded landscapes.
(a) Folded landscape and the basic types of fold structures. (b) A road cut exposes a synclinal ridge in western Maryland. This syncline is a natural outdoor classroom, and the state of Maryland built an interpretive center with a walkway above the highway. (c) Folding, squeezing, and uplift from strong compressional and shearing forces show well in a road cut along the San Andreas fault zone, beside the Antelope Valley Freeway. [(b) Mike Boroff/Photri-Microstock. (c) Bobbé Christopherson.]

- Further stress eventually fractures the rock strata along distinct lines, and some overturned folds are thrust upward, causing a considerable shortening of the original strata (*thrust fault*). Areas of intense stress, compressional folding, and faulting are visible in a road cut where the San Andreas fault in southern California passes beneath the Antelope Valley Freeway (Figure 12.8c).

The Canadian Rocky Mountains, the Appalachian Mountains, and areas of the Middle East illustrate the complexity of folded landscapes. Satellites allow us to view many of these structures from an orbital perspective, as in Figure 12.9, north of the Persian Gulf in the Zagros Mountains of Iran. This area was a dispersed terrane that separated from the Eurasian plate. However, the collision produced by the northward push of the Arabian block is now shoving this terrane back into Eurasia and forming an active margin known as the Zagros crush zone, a continuing collision more than 400 km (250 mi) wide. In the satellite image, anticlines form the parallel ridges; active weathering and erosion processes are exposing the underlying strata.

In addition to the rumpling of rock strata just discussed, broad warping actions affect Earth's continental crust. These actions produce similar up-and-down bending of strata, but the bends are far greater in extent than those produced by folding. Warping forces include mantle convection, isostatic adjustment such as that caused by the weight of previous ice loads across northern Canada, or crustal swelling above an underlying hot spot. Warping features can be small, individual, foldlike structures called *basins* and *domes* (Figure 12.10a, b). They can also range up to regional features the size of the Ozark Mountain complex in Arkansas and Missouri, the Colorado Plateau in the

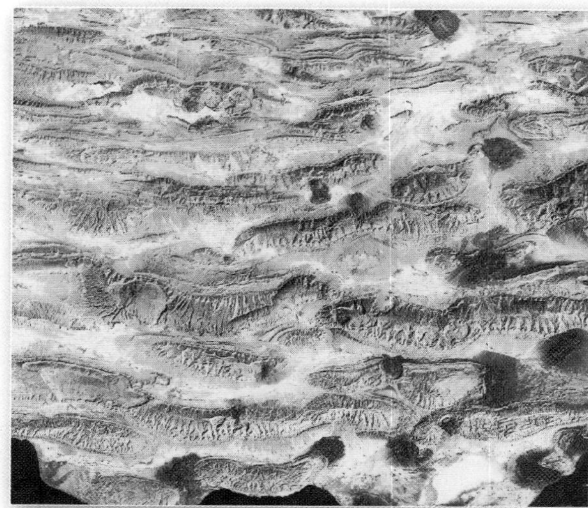

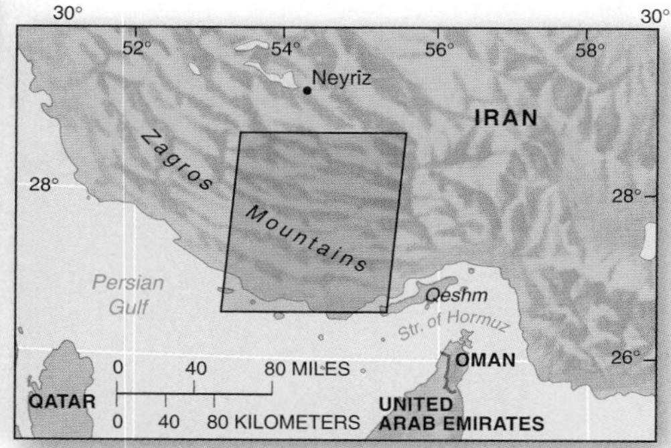

FIGURE 12.9
Folding in the Zagros crush zone, Iran.
The Zagros Mountains are a product of the Zagros crush zone between the Arabian and Eurasian plates, where the northward push of the Arabian block is causing the folded mountains shown. [NASA.]

West, the Richat Dome in Mauritania, or the Black Hills of South Dakota (Figure 12.10c, d).

Faulting

A freshly poured concrete sidewalk is smooth and strong. But stress the sidewalk by driving heavy equipment over it, and the resulting strain might cause a fracture. Pieces on either side of the fracture may move up, down, or horizontally depending on the direction of stress. Similarly, when rock strata are stressed beyond their ability to remain a solid unit, they express the strain as a fracture. Rocks on either side of the fracture displace relative to the other side in a process known as **faulting**. Thus, *fault zones* are areas where fractures in the rock demonstrate crustal movement. At the moment of fracture, a sharp release of energy occurs, producing an **earthquake** or *quake*.

The fracture surface along which the two sides of a fault move is the *fault plane*. The names of the three basic types of faults, illustrated in Figure 12.11, are based on the tilt and orientation of the fault plane. A normal fault forms when rocks are pulled apart by tensional stress. A thrust or reverse fault results when rocks are forced together by compressional stress. A strike-slip fault forms when rocks are torn by lateral-shearing stress.

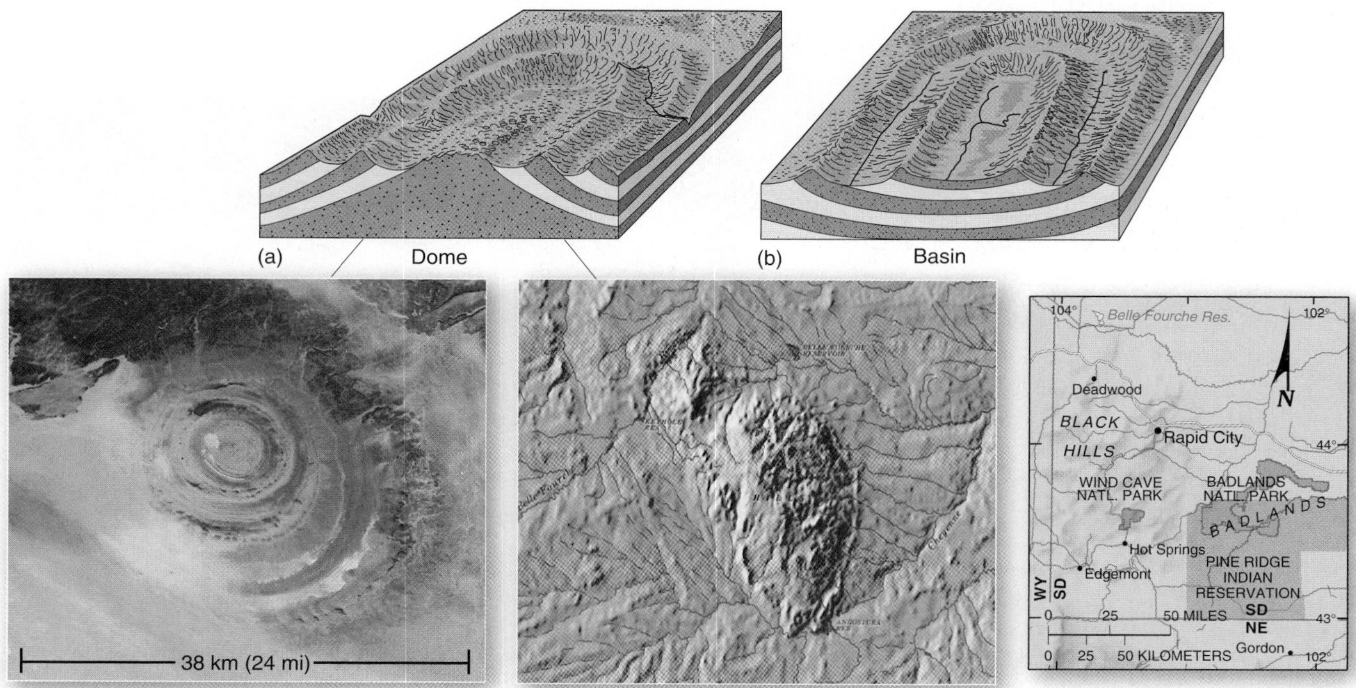

FIGURE 12.10 Domes and basins.
(a) An upwarped dome. (b) A structural basin. (c) The Richat Dome structure of Mauritania. (d) The Black Hills of South Dakota are a dome structure, shown here on a digitized relief map. [(c) *Terra* image, NASA/Mark Marten/Science Source/Photo Researchers, Inc. (d) From USGS DEM I-2206, 1992.]

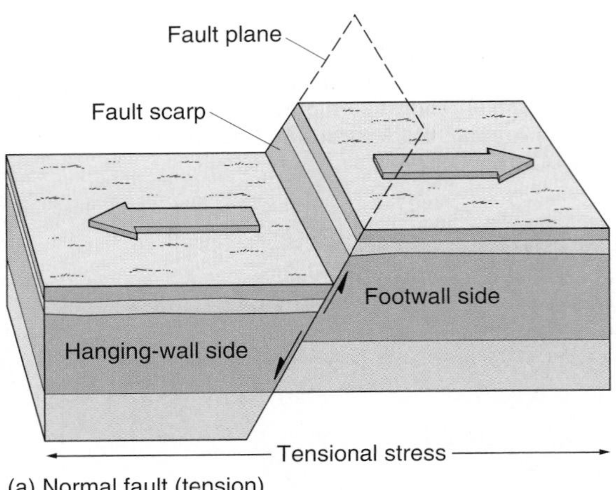

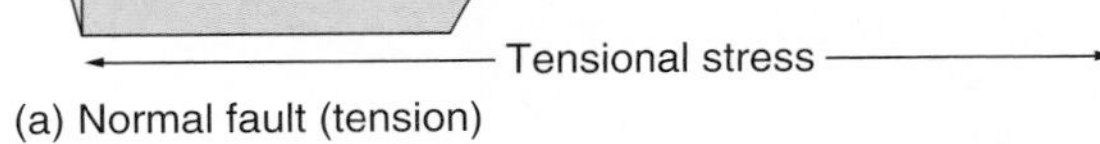

(a) Normal fault (tension)

(a)

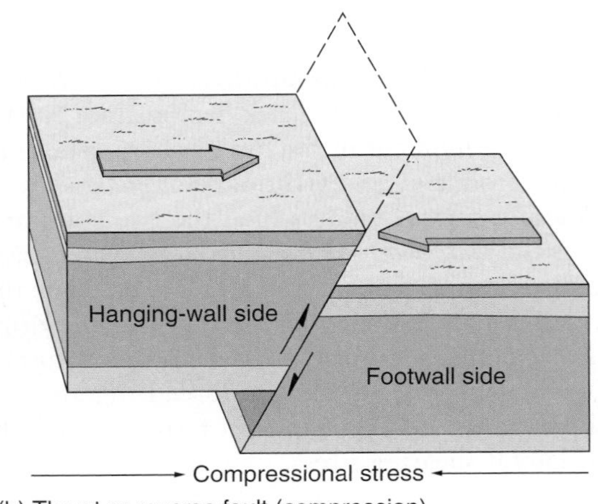

(b) Thrust or reverse fault (compression)

(b)

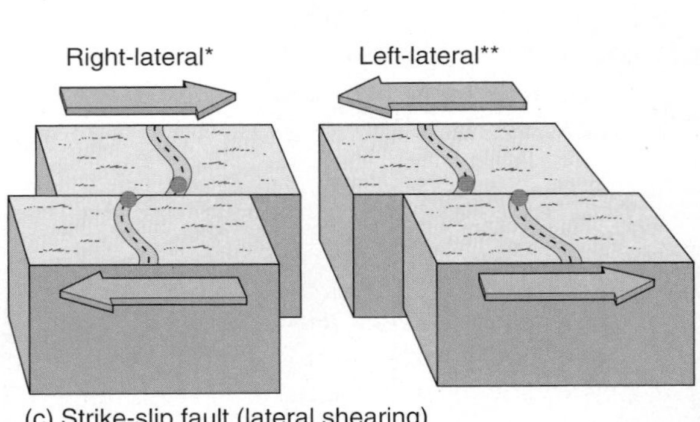

(c) Strike-slip fault (lateral shearing)

* Viewed from either dot on each road, movement to opposite side is *to the right.*
** Viewed from either dot on each road, movement to opposite side is *to the left.*

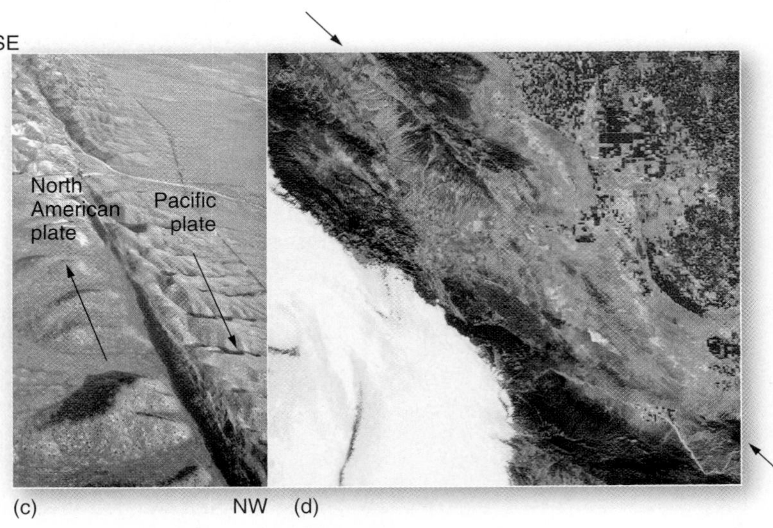

(c) (d)

FIGURE 12.11 Types of faults.
(a) A normal fault produced by tension in the crust, visible along the edge of mountain ranges in California and along the Wasatch Front in Utah. (b) A thrust, or reverse fault, produced by compression in the crust, visible in these offset strata in coal seams and volcanic ash in British Columbia. (c) A strike-slip fault produced by lateral shearing, clearly seen looking southeast along the San Andreas fault rift zone, with the Carrizo Plain–Pacific plate to the right and the North American plate to the left; arrows show relative motion. (d) Can you find the San Andreas fault on the satellite image? Look for a linear rift stretching from southeast to northwest, at the edge of the coastal mountains and the San Joaquin Valley. Use the arrows as guides. [(a) and (d) Bobbé Christopherson. (b) Fletcher and Baylis/Photo Researchers, Inc. (c) Kevin Schafer/Photolibrary/Peter Arnold, Inc. (d) *Terra* MISR image NASA/GSFC/JPL.]

Animation
Fault Types, Transform Faults, Plate Margins

When forces pull rocks apart, the tension causes a **normal fault,** or *tension fault*. When the break occurs, rock on one side moves vertically along an inclined fault plane (Figure 12.11a). The downward-shifting side is the *hanging wall*; it drops relative to the *footwall block*. The exposed fault plane sometimes is visible along the base of faulted mountains, where individual ridges are truncated by the movements of the fault and appear as triangular facets at the ends of the ridges. A cliff formed by faulting is commonly called a *fault scarp*, or *escarpment*.

Compressional forces associated with converging plates force rocks to move *upward* along the fault plane. This is a **reverse fault,** or *compression fault* (Figure 12.11b). On the surface, it appears similar to a normal fault, although more collapse and landslides may occur from the hanging wall component. In England, when miners worked along a reverse fault, they would stand on the lower side (footwall) and hang their lanterns on the upper side (hanging wall), giving rise to these terms.

If the fault plane forms a low angle relative to the horizontal, the fault is termed a **thrust fault,** or *overthrust fault*, indicating that the overlying block has shifted far over the underlying block (see Figure 12.8, "Thrust fault"). Place your hands palms-down on your desk, with fingertips together, and slide one hand up over the other—this is the motion of a low-angle thrust fault, with one side pushing over the other.

In the Alps, several such overthrusts result from compressional forces of the ongoing collision between the African and Eurasian plates. Beneath the Los Angeles Basin, overthrust faults produce a high risk of earthquakes and caused many quakes in the twentieth century, including the $30 billion 1994 Northridge earthquake. These blind (unknown until they rupture) thrust faults beneath the Los Angeles region remain a major earthquake threat.

If movement along a fault plane is horizontal, such as produced along a transform fault, it forms a **strike-slip fault** (Figure 12.11c; refer to Figure 11.20 for review). The movement is *right-lateral* or *left-lateral* depending on the motion perceived when you observe movement on one side of the fault relative to the other side.

Although strike-slip faults do not produce cliffs (scarps), as do the other types of faults, they can create linear rift valleys. This is the case with the San Andreas fault system of California. The rift valley is clearly visible in Figure 12.11c, where the edges of the North American and Pacific plates are grinding past one another as a result of transform-fault movement.

The evolution of the San Andreas fault system is shown in Figure 12.12. Note in the figure how the East Pacific rise developed as a spreading center with associated transform faults (1), while the North American plate was progressing westward after the breakup of Pangaea. Forces then shifted the transform faults toward a northwest–southeast alignment along a weaving axis (2). Finally, the western margin of North America overrode those shifting transform faults (3). Note the convergence rate is a rapid 4 cm (1.6 in.) per year.

In *relative* terms, the motion along this series of transform faults is right-lateral, whereas in *absolute* terms, the North American plate is still moving westward. Consequently, the San Andreas system is a series of faults that are *transform* (associated with a former spreading center), *strike-slip* (horizontal in motion), and *right-lateral* (one side is moving to the right relative to the other side).

Faults in Concert Combinations of faults can produce distinctive landscapes (Figure 12.13a). In the U.S. interior west, the *Basin and Range Province* (featuring a roughly parallel series of faulted mountains and valleys) experienced tensional forces caused by uplifting and thinning of the crust (illustrated later in Chapter 15, Figure 15.22). This movement cracked the surface to form aligned pairs of normal faults and a distinctive landscape.

The term **horst** applies to upward-faulted blocks; **graben** refers to downward-faulted blocks. One example of a horst and graben landscape is the Great Rift Valley of East Africa (associated with crustal spreading); it extends northward to the Red Sea, which fills the rift formed by parallel normal faults (Figure 12.13b). Another example is the Rhine graben, through which the Rhine River flows in Europe.

Orogenesis (Mountain Building)

Orogenesis literally means the birth of mountains (*oros* comes from the Greek for "mountain"). An *orogeny* is a mountain-building episode, occurring over millions of years. It can occur through large-scale deformation and uplift of the crust. An orogeny may include the capture of migrating exotic terranes, addition by accretion to the continental margins, or the intrusion of granitic magmas to form plutons. The net result of this accumulating material is a thickening of the crust. The granite plutons often become exposed by erosion following uplift. Uplift is the final act of an orogenic cycle of mountain building. Earth's major chains of folded and faulted mountains, called *orogens*, are remarkably well correlated with the plate tectonics model.

No orogeny is a simple event; many involve previous developmental stages dating back far into Earth's past, and the processes are ongoing today. Major mountain ranges, and the latest related orogenies that caused them, include the following:

- Rocky Mountains of North America—*Laramide orogeny*, 40–80 million years ago (m.y.a.); third mountain-building episode, beginning 170 m.y.a., including the *Sevier orogeny;*
- Sierra Nevada and Klamath Mountains of California— *Nevadan orogeny*, with faulting 29–35 m.y.a. and older batholithic intrusion emplacements dating back 80 to 180 million years;
- Appalachian Mountains and the Ridge and Valley Province (nearly parallel ridges and valleys) of the

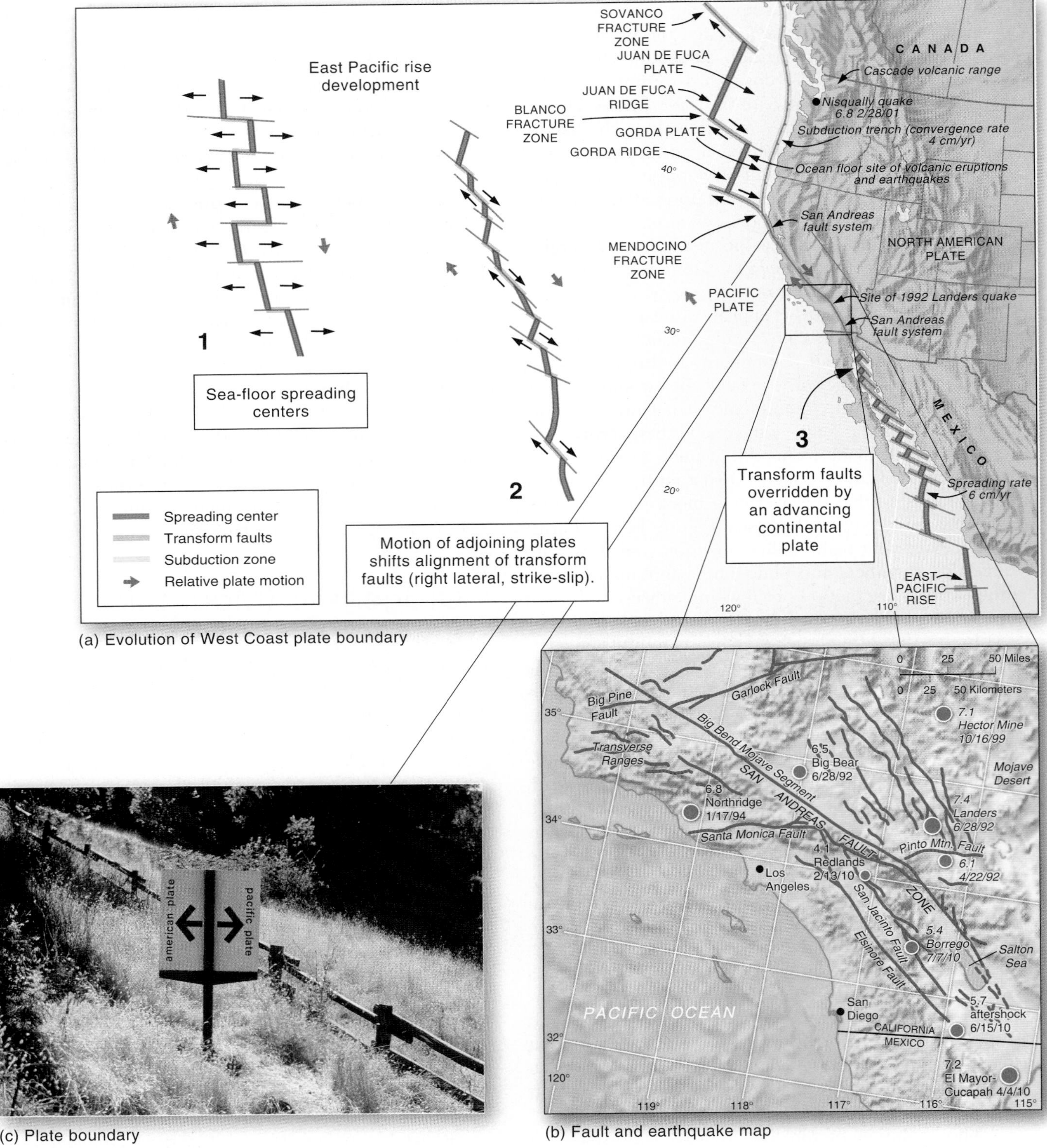

(c) Plate boundary

(b) Fault and earthquake map

FIGURE 12.12 San Andreas fault formation.
(a) Formation of the San Andreas fault system as a series of transform faults in three successive stages.
(b) Enlargement shows the portion of southern California where the 1992 Landers, 1994 Northridge (Reseda), 1999 Hector Mine, and 2010 Redlands, Borrego, and El Mayor earthquakes occurred. Note the January 2001 M 6.8 Nisqually quake, in Washington State, related to the subduction zone offshore on map (a). (c) Trail sign near the epicenter of the 1906 San Francisco earthquake roughly marks the Pacific–North American plate boundary, Marin County, Point Reyes National Seashore, California. [(c) Author.]

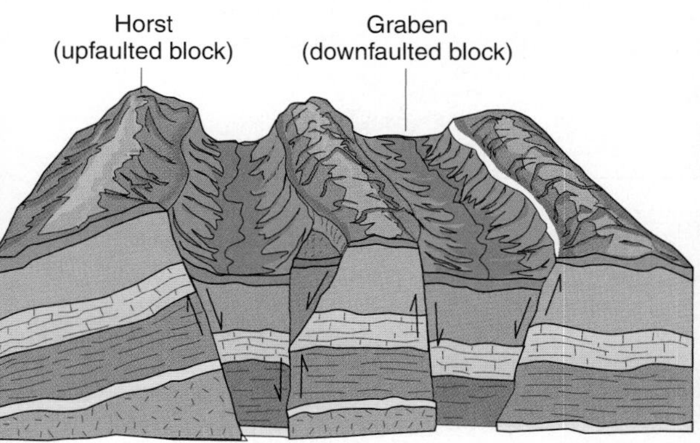

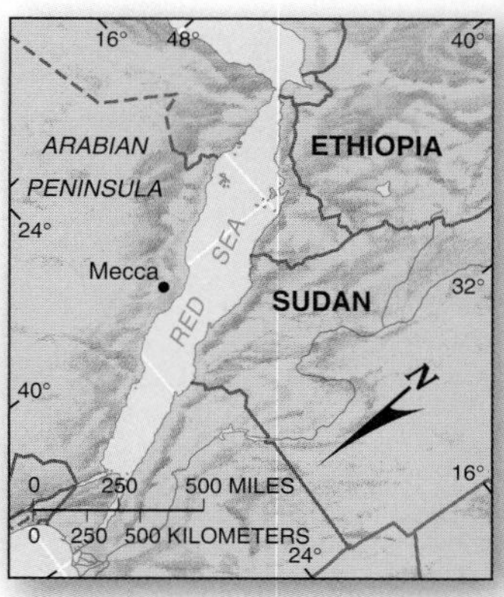

FIGURE 12.13 Faulted landscapes.
(a) Pairs of faults produce a horst-and-graben landscape. (b) The Red Sea occupies a down-dropped block that is part of the rift system through East Africa, a graben. [(b) NASA *Gemini*.]

eastern United States and Canadian Maritime Provinces—*Alleghany orogeny*, 250–300 m.y.a., associated with the collision of Africa and North America, preceded by at least two earlier orogenies; in Europe, this is contemporary with the *Hercynian orogeny*;

- Alps of Europe—*Alpine orogeny*, 2 to 66 m.y.a., principally in the Tertiary Period and continuing to the present across southern Europe and the Mediterranean, with many earlier episodes (see Figure 12.14); and
- Himalayas of Asia—*Himalayan orogeny*, 45–54 m.y.a., beginning with the collision of the India plate and Eurasia plate and continuing to the present.

Types of Orogenies

Figure 12.15 illustrates three types of convergent plate collisions that cause orogenesis along plate margins:

(a) **Oceanic plate–continental plate collision orogenesis.** This type of convergence is now occurring along the Pacific coast of the Americas and has formed the Andes, the Sierra of Central America, the Rockies, and other western mountains. We see folded sedimentary formations, with intrusions of magma forming granitic plutons at the heart of these mountains. Their buildup has been augmented by capturing displaced terranes, accreted during

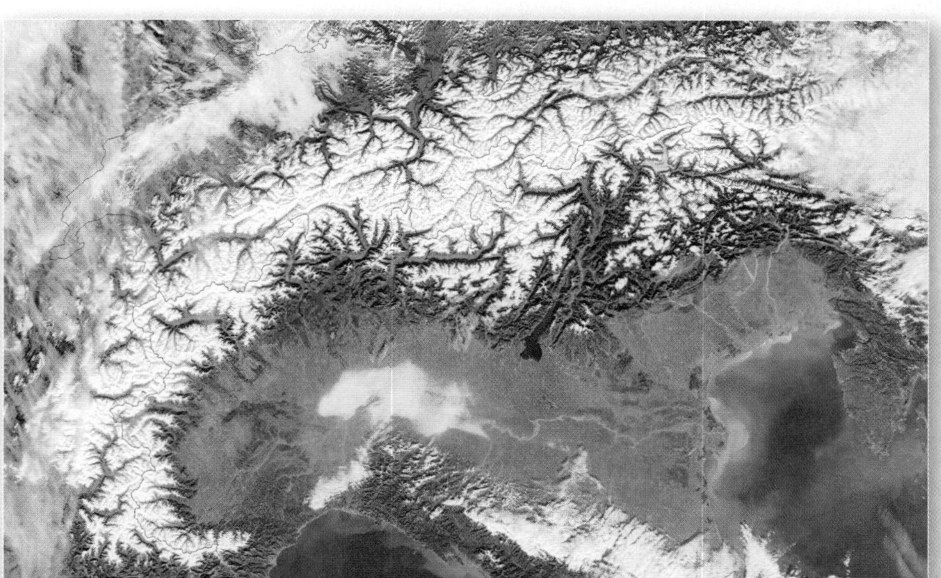

FIGURE 12.14 European Alps. Western (France), central (Italy), and eastern (Austria) segments make up the crescent shape of the Alps. Complex overturned faults and crustal shortening due to compressional forces occur along convergent plates. The Alps are some 1200 km (750 mi) in length, occupying 207,000 km^2 (80,000 mi^2). Note the snow coverage in this December image. [*Terra* MODIS image, NASA/GSFC.]

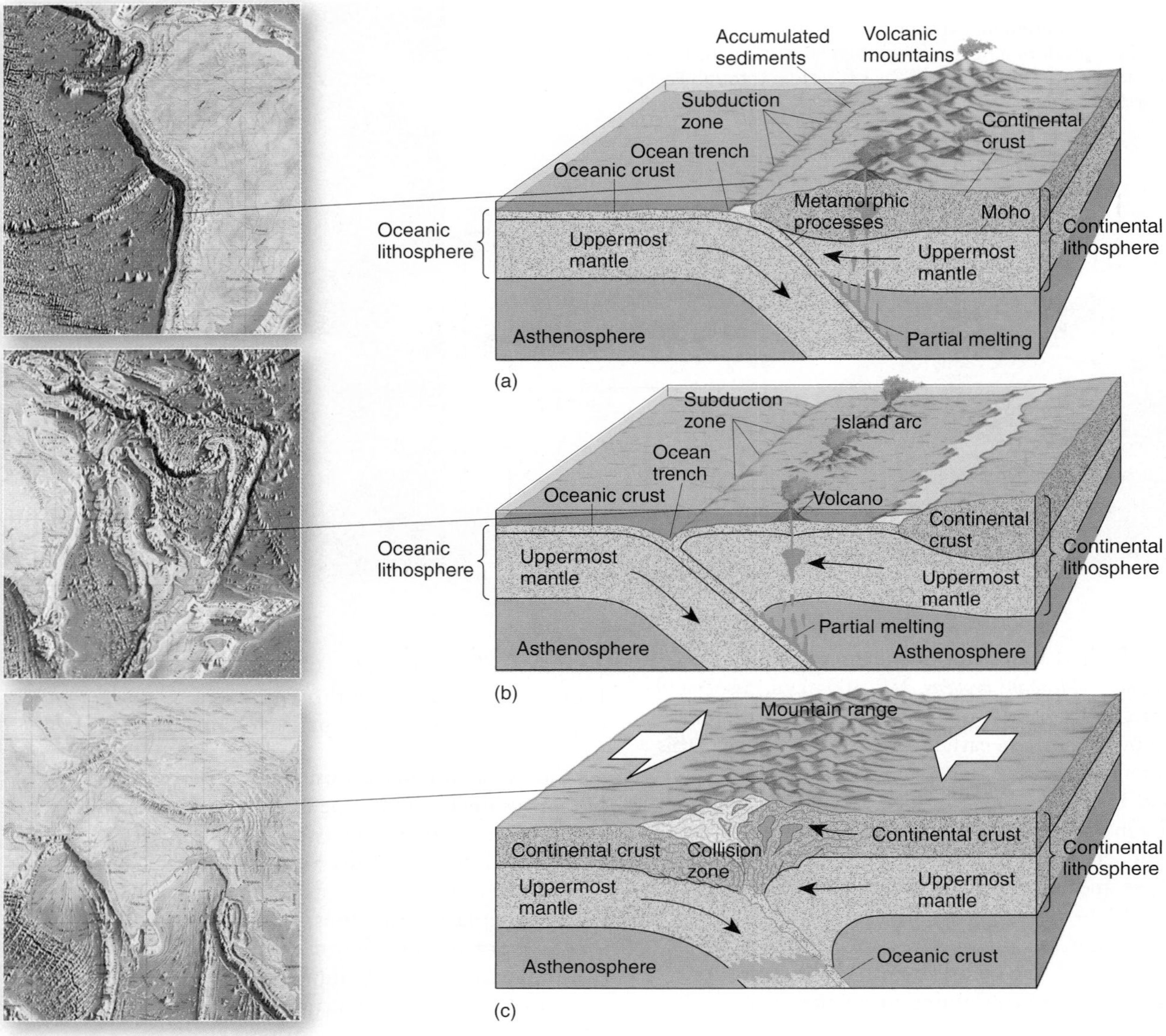

FIGURE 12.15 Three types of plate convergence.
(a) Oceanic plate–continental plate (example: Nazca plate–South American plate collision and subduction). (b) Oceanic plate–oceanic plate (example: New Hebrides Trench near Vanuatu, 16° S, 168° E). (c) Continental plate–continental plate (example: Indian plate and Eurasian landmass collision and resulting Himalayan Mountains). Can you identify more of these plate convergence areas on the chapter-opening map?

Animation
Plate Boundaries

collision with the continental mass. Also, note the associated volcanic activity inland from the subduction zone, the eruption of plateau (flood) basalts (see Figures 12.26 and 12.31), and the development of composite volcanoes (see Figure 12.32).

(b) **Oceanic plate–oceanic plate collision orogenesis.** Such collisions can produce either simple volcanic island arcs or more complex arcs, such as Indonesia and Japan, which include deformation and metamorphism of rocks and granitic intrusions. These processes formed the chains of island arcs and volcanoes that continue from the southwestern Pacific to the western Pacific, the Philippines, the Kurils, and on through portions of the Aleutians.

Both collision types—(a) oceanic plate–continental plate and (b) oceanic plate–oceanic plate—are active around the Pacific Rim. Both are thermal in nature because the diving plate melts and migrates back toward the surface as molten rock. The region of active volcanoes and earthquakes around the Pacific is known as the **circum-Pacific belt** or, more popularly, the **ring of fire**.

(c) **Continental plate–continental plate collision orogenesis.** Here the orogenesis is mechanical; large masses of continental crust are subjected to intense folding, overthrusting, faulting, and uplifting. The converging plates crush and deform both marine sediments and basaltic oceanic crust.

> The formation of the European Alps is a result of such compression forces and includes considerable crustal shortening, forming great overturned folds, called *nappes*.

As mentioned earlier, the collision of India with the Eurasian landmass produced the Himalayan Mountains. That collision is estimated to have shortened the overall continental crust by as much as 1000 km (about 600 mi) and to have produced telescoping sequences of thrust faults at depths of 40 km (25 mi). The Himalayas feature the tallest above-sea-level mountains on Earth, including Mount Everest at 8850 m elevation (29,035 ft) and all 10 of Earth's highest peaks (measured from sea level).

The disruption created by this plate collision has reached far under China, and frequent earthquakes there signal the continuation of this rapid-paced collision, as India moves northeastward at about 2 m (6.6 ft) a year. As evidence of this ongoing strain, the January 2001 quake in Gujarat, India, was along a shallow, east–west-tending thrust fault that gave way under the pressure of the northward-pushing Indian plate. More than 1 million buildings were destroyed or damaged during this event along the 2500-km (1553-mi) fault zone that marks this plunging Indian subcontinent. Again, a 40-km (25-mi) segment snapped in October 2005, when an M 7.6 quake hit the Kashmir region of Pakistan, killing more than 83,000 people.

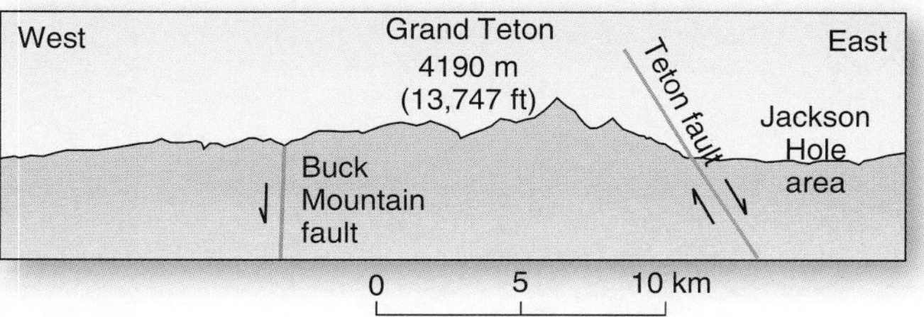

FIGURE 12.16 Tilted-fault block.
The Teton Range in Wyoming is an example of a tilted-fault block, a range of scenic beauty featuring a 2130-m (7,000-ft) rugged relief between Jackson Hole and the summits. The Grand Teton is the highest peak in the range at 4190 m (13,747 ft). [Author.]

The Grand Tetons and the Sierra Nevada

The Sierra Nevada of California and the Grand Tetons of Wyoming are examples of recent stages of mountain building. Each is a *tilted-fault-block* mountain range, in which a normal fault on one side of the range has produced a tilted mountain landscape of dramatic relief (Figure 12.16). Magma intruded into those blocks, slowly cooling to form granitic cores of coarsely crystalline rock. After tremendous tectonic uplift and the removal of overlying material through weathering, erosion, and transport, those granitic masses are now exposed in each mountain range. In some areas, these batholiths were previously covered by more than 7500 m (25,000 ft) of overlying material.

Recent research in the Sierra Nevada disclosed that some of the uplift was isostatic, caused by the erosion of overburden and the loss of melting ice mass following the last ice age some 18,000 years ago. The accumulation of sediments in the adjoining valley to the west depressed the crust, thus further enhancing relief in the landscape.

The Appalachian Mountains

The eroded, fold-and-thrust belt of the eastern United States and southeastern Canada has origins dating to the formation of Pangaea and the collision of Africa and North America (250–300 million years old). The Appalachians are in contrast to the higher mountains of western North America (35–80 million years old). As noted, the complexity of the *Alleghany orogeny* derives from at least two earlier orogenic cycles of uplift and the accretion of several captured terranes.

Linked during the Pangaea collision and now separated by an ocean, the mountain ranges on either side of the Atlantic Ocean share both structure and composition. In fact, the Lesser (or Anti-) Atlas Mountains of Mauritania and northwestern Africa were connected to the Appalachians at some time in the past, but the mountains embedded in the African plate rafted apart from the Appalachians.

The Appalachian Mountain region comprises several landscape subregions (refer to the locator map in Figure 12.17): the Ridge and Valley Province (elongated sequences of folded sedimentary rock); the Blue Ridge Province (principally of crystalline rock, highest where North Carolina, Virginia, and Tennessee converge); the Piedmont (hilly to gentle terrain along most of the eastern and southern margins of the mountains); the east coastal plain (from gentle hills to flat plains that extend to the coast); and the Canadian maritime provinces.

Figure 12.17 displays the linear folds of the Appalachian system. Note how *water gaps* form where rivers cut through dissected ridges. These important breaks in the rugged ridges greatly influenced migration, settlement patterns, and the diffusion of cultural traits in the 1700s. The initial flow of people, goods, and ideas was

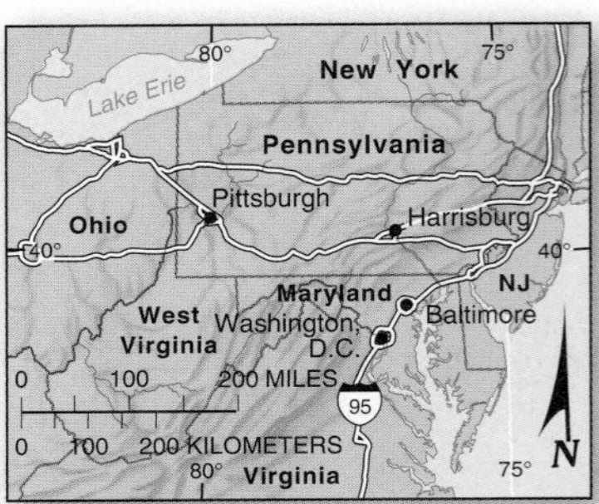

(b) Little and Line Mountains

(c) Water gaps

(a) Appalachian Mountains

FIGURE 12.17 The Appalachian Mountains.
(a) Appalachian Mountain region of Pennsylvania south through Maryland, Virginia, and West Virginia in a *Terra* image from October 11, 2000, showing true fall colors. The highly folded nature of this entire region is clearly visible in the satellite image. (b) The folded Little and Line mountains end at the Susquehanna River, along the left edge of this aerial photo, following a light March snowfall. (c) Note how the water gap directs transportation corridors. [(a) *Terra* MODIS image, NASA/GSFC. (b) and (c) Bobbé Christopherson.]

guided by this topography. The Susquehanna River in east-central Pennsylvania is one watercourse that interrupts the ridges and valleys of the Appalachians.

World Structural Regions

Examine the first two maps in this chapter (the chapter opener and Figure 12.3) and note two vast alpine systems on the continents. In the Western Hemisphere, the *Cordilleran system* stretches from Tierra del Fuego at the southern tip of South America to the massive peaks of Alaska, including the relatively young Rocky and Andes mountains along the western margins of the North and South American plates. In the Eastern Hemisphere, the *Eurasian-Himalayan system* stretches from the European Alps across Asia to the Pacific Ocean and contains younger and older components.

These mountain systems also are shown on the structural region map as the *Alpine system* (Figure 12.18). The map defines seven fundamental structural regions that possess distinctive types of landscapes, grouped because of their shared physical characteristics. Looking at the distribution of these regions helps summarize the three rock-forming processes (igneous, sedimentary, and metamorphic), plate tectonics, landform origins and construction, and overall orogenesis.

As you examine the map, identify the continental shields at the heart of each landmass. Continental platforms composed of younger sedimentary deposits surround these areas. Various mountain chains, rifted regions, and isolated volcanic areas are noted on the map. On the continent of Australia, you see older mountain sequences to the east, sedimentary layers covering basement rocks west of these ranges, and portions of the original Gondwana, an ancient landscape in the central and western region. Remember that Gondwana was a landmass that included Antarctica, Australia, South America, Africa, and the southern portion of India; it broke away from Pangaea some 200 million years ago.

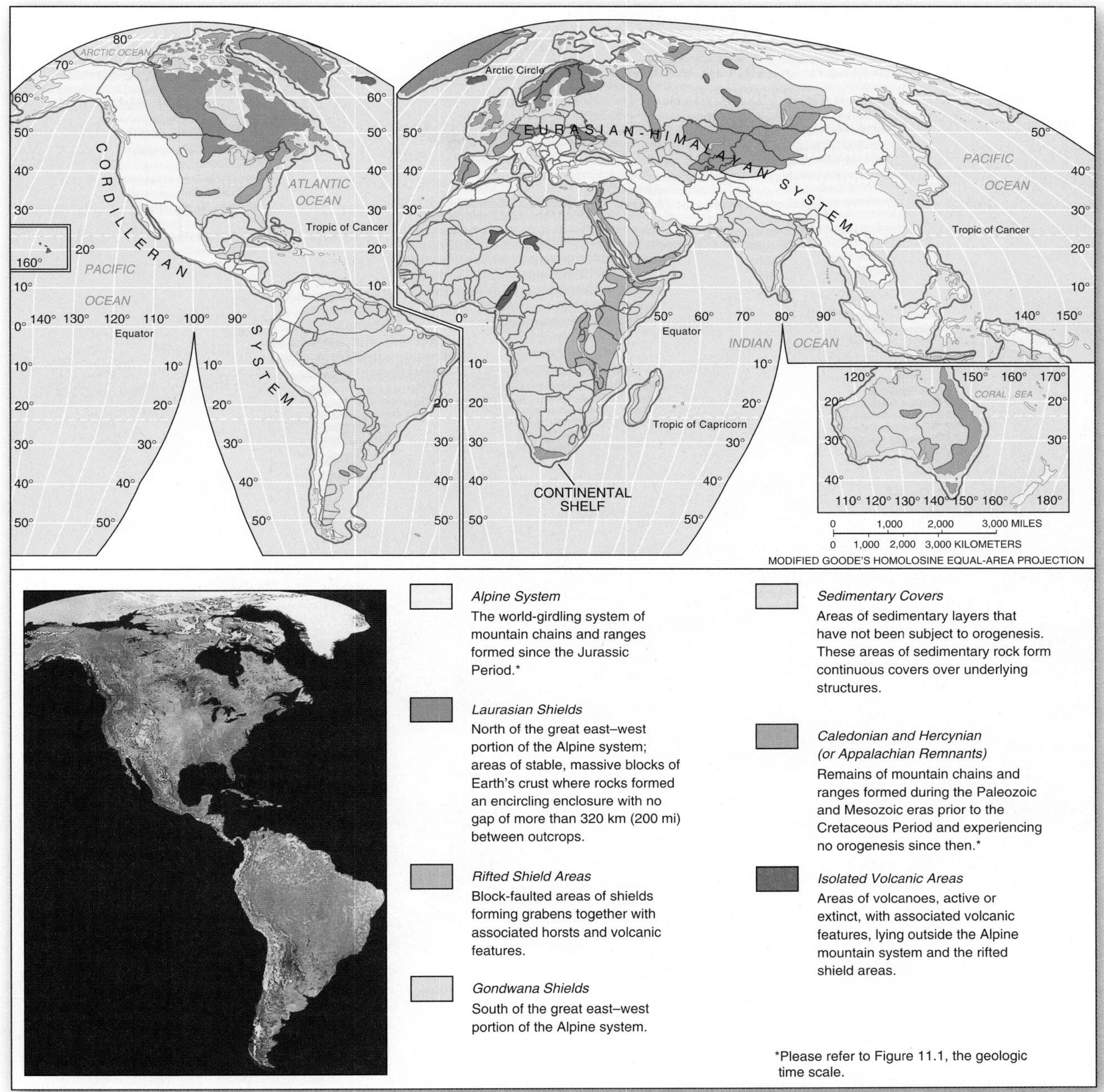

FIGURE 12.18 World structural regions and major mountain systems.
Some of the regions appear larger than the structures themselves because each region includes related landforms adjacent to the central feature. (Correlate aspects of this map with Figures 12.3 and 12.4.) Structural regions in the Western Hemisphere are visible on the composite false-color *Landsat* image inset (vegetation is portrayed in red). [After R. E. Murphy, "Landforms of the world," *Annals of the Association of American Geographers* 58, 1 (March 1968). Adapted by permission. Inset image EROS Data Center.]

Earthquakes

Crustal plates do not glide smoothly past one another. Instead, tremendous friction exists along plate boundaries. The *stress* (a force) of plate motion builds *strain* (a deformation) in the rocks until friction is overcome and the sides along plate boundaries or fault lines suddenly break loose. The two sides of the fault plane then lurch into new positions, moving from centimeters to several meters, and release enormous amounts of seismic energy into the surrounding crust. This energy radiates throughout the planet, diminishing with distance, but sufficient enough to register on instruments worldwide.

In 2007, southern Sumatra, Indonesia, was hit with M 8.4 and M 7.8 quakes, near the same plate boundary where four other quakes greater than M 7.9 had occurred since 1998. At this boundary, the Australian plate moves northeast, subducting beneath the Sunda plate on which Sumatra rides, at a relative speed of 6.6 cm/year

(2.6 in./year). This is where the M 9.1 Sumatra–Andaman earthquake struck December 26, 2004, triggering the devastating Indian Ocean tsunami (more on this in Chapter 16); the quake and seismic wave took 228,000 lives. For complete listings of these and other earthquakes, go to http://earthquake.usgs.gov/earthquakes/.

What are the mechanisms that produce such dramatic tectonic events? Why do most quakes strike in total surprise, such as in Haiti and Chile in 2010?

Focus, Epicenter, Foreshock, and Aftershock

The subsurface area along a fault plane, where the motion of seismic waves is initiated, is the *focus*, or hypocenter, of an earthquake (see the labels in Figure 12.19). The area at the surface directly above the focus is the *epicenter*. Shock waves produced by an earthquake radiate outward through the crust from the focus and epicenter. Some of the seismic waves are conducted throughout the planet to

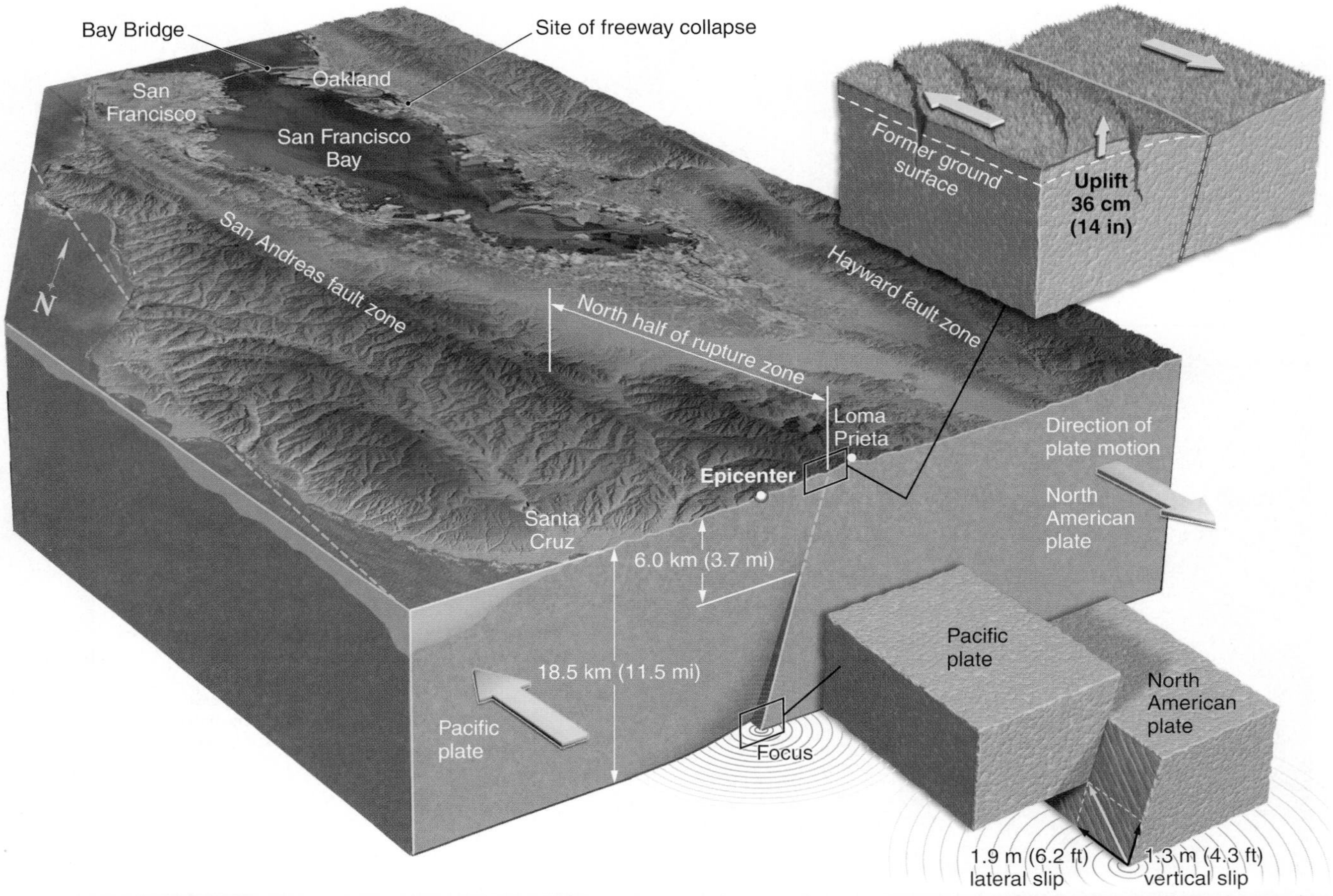

FIGURE 12.19 Anatomy of an earthquake.
(a) The fault-plane solution for the 1989 Loma Prieta, California, earthquake shows the lateral and vertical (thrust) movements occurring at depth. There was no surface expression of this fault plane. (b) In only 15 seconds, more than 2 km (1.2 mi) of the Route 880 Cypress Freeway collapsed. (c) Section failure in the San Francisco–Oakland Bay Bridge. [(a) After USGS. (b) and (c) Courtesy of California Department of Transportation.]

Animation
P- and S-Waves, Seismology

distant instruments. Scientists explore Earth's interior using these seismic wave patterns and the nature of their transmission through the layers of the planet.

An *aftershock* may occur after the main shock, sharing the same general area of the epicenter; some aftershocks rival the main tremor in magnitude. A *foreshock*, which precedes the main shock, also is possible. The pattern of foreshocks is now regarded as an important consideration in the forecasting effort. Before the 1992 Landers earthquake in southern California, at least two dozen foreshocks occurred along the soon-to-fail portion of the fault.

Unlike previous earthquakes—such as San Francisco in 1906, when the plates shifted a maximum of 6.4 m (21 ft) relative to each other—no fault plane or rifting was evident at the surface in the 1989 Loma Prieta, California, area. Instead, the fault plane suggested in Figure 12.19 shows the two plates moving horizontally approximately 2 m (6 ft) past each other deep below the surface, with the Pacific plate thrusting 1.3 m (4.3 ft) upward. This vertical motion is unusual for the San Andreas fault and indicates that this portion of the San Andreas system is more complex than previously thought. Damage totaled $8 billion, 14,000 people were displaced from their homes, 4000 were injured, and 67 were killed.

Earthquake Intensity and Magnitude

Tectonic earthquakes are those quakes associated with faulting. A worldwide network of more than 4000 **seismograph** instruments records vibrations transmitted as waves of energy throughout Earth's interior and in the crust. Using this device and actual observations, scientists rate earthquakes on two kinds of scales: a *qualitative* damage intensity scale and a *quantitative* magnitude-of-energy-released scale.

A damage *intensity scale* is useful in classifying and describing damage to terrain and structures after an earthquake. Earthquake damage intensity is rated on the arbitrary *Mercalli scale*, a Roman-numeral scale from I to XII, representing "barely felt" to "catastrophic total destruction." It was designed in 1902 and modified in 1931. Table 12.1 shows this scale and the number of quakes in each category that are expected each year. The strongest earthquake thought to have struck was the M 9.6 in Chile in 1960, producing Mercalli XII damage. The fault-line break in the crust extended over 1000 km (629 mi) in that quake.

In 1935, Charles Richter designed a system to estimate earthquake magnitude. In this method, a seismograph located at least 100 km (62 mi) from the epicenter of the quake records the amplitude of seismic waves. The measurement is then charted on the **Richter scale**. The relation of magnitude to energy released is still a useful feature of his scale.

The Richter scale is logarithmic: Each whole number on it represents a 10-fold increase in the measured wave amplitude. Translated into energy, each whole number signifies a 31.5-fold increase in energy released. Thus, a magnitude of 3.0 on the Richter scale represents 31.5 times more energy than a 2.0 and 992 times more energy than a 1.0. The Richter scale has since become more quantitative. Revision was needed because the scale did not properly measure or differentiate between quakes of high intensity. Seismologists wanted to know more about what they call the *seismic moment* to understand a broader range of possible motions during an earthquake.

The **moment magnitude scale,** in use since 1993, is more accurate for large earthquakes than is Richter's *amplitude magnitude* scale. Moment magnitude considers the amount of fault slippage produced by the earthquake, the size of the surface (or subsurface) area that ruptured, and the nature of the materials that faulted, including how resistant they were to failure. The new scale considers extreme ground acceleration (movement upward), which the Richter amplitude magnitude method underestimated.

A reassessment of past quakes has increased the rating of some and decreased that of others. As an example, the 1964 earthquake at Prince William Sound in Alaska had an amplitude magnitude of 8.6, but on the moment magnitude scale, it increases to an M 9.2. Note that Table 12.2 includes

TABLE 12.1 Magnitude, Intensity, and Frequency of Earthquakes

Description	Effects on Populated Areas	Moment Magnitude Scale	Modified Mercalli Scale	Number per Year*
Great	Damage nearly total	8.0 and higher	XII	1
Major	Great damage	7–7.9	X–XI	17
Strong	Considerable-to-serious damage to buildings; railroad tracks bent	6–6.9	VIII–IX	134
Moderate	Felt-by-all, with slight building damage	5–5.9	V–VII	1,319
Light	Felt-by-some to felt-by-many	4–4.9	III–IV	13,000 (estimated)
Minor	Slight, some feel it	3–3.9	I–II	130,000 (estimated)
Very minor	Not felt, but recorded	2–2.9	None to I	1,300,000 (estimated)

*Based on observations since 1990.

Source: USGS, Earthquake Information Center.

TABLE 12.2 A Sampling of Significant Earthquakes*

Year	Date	Location	Number of Deaths	Mercalli Intensity	Moment Magnitude (Richter)
1556	Jan. 23	Shaanxi Province, China	830,000	**	**
1737	Oct. 11	Calcutta, India	300,000	**	**
1812	Feb. 7	New Madrid, Missouri	Several	XI–XII	**
1857	Jan. 9	Fort Tejon, California	**	X–XI	**
1870	Oct. 21	Montreal to Québec, Canada	**	IX	**
1886	Aug. 31	Charleston, South Carolina	**	IX	6.7
1906	Apr. 18	San Francisco, California	3,000	XI	7.7 (8.25)
1923	Sept. 1	Kwanto, Japan	143,000	XII	7.9 (8.2)
1939	Dec. 27	Erzincan, Turkey	40,000	XII	7.6 (8.0)
1960	May 22	Southern Chile	5,700	XII	9.6 (8.6)
1964	Mar. 28	Southern Alaska	131	X–XII	9.2 (8.6)
1970	May 31	Northern Peru	66,000	**	7.9 (7.8)
1971	Feb. 9	San Fernando, California	65	VII–IX	6.7 (6.5)
1972	Dec. 23	Managua, Nicaragua	5,000	X–XII	6.2 (6.2)
1976	July 28	Tangshan, China	250,000	XI–XII	7.4 (7.6)
1978	Sept. 16	Iran	25,000	X–XII	7.8 (7.7)
1985	Sept. 19	Mexico City, Mexico	7,000	IX–XII	8.1 (8.1)
1988	Dec. 7	Armenia–Turkey border	30,000	XII	6.8 (6.9)
1989	Oct. 17	Loma Prieta (near Santa Cruz, California)	66	VII–IX	7.0 (7.1)
1991	Oct. 20	Uttar Pradesh, India	1,700	IX–XI	6.2 (6.1)
1994	Jan. 17	Northridge (Reseda), California	66	VII–IX	6.8
1995	Jan. 17	Kobe, Japan	5,500	XII	6.9
1996	Feb. 17	Indonesia	110	X	8.1
1997	Feb. 28	Armenia–Azerbaijan	1,100	XII	6.1
1997	May 10	Northern Iran	1,600	XII	7.3
1998	May 30	Afghanistan–Tajikistan	4,000	XII	6.9
1998	July 17	Papua New Guinea	2,200	X	7.1
1999	Jan. 26	Armenia, Colombia	1,000	VIII–IX	6.0
1999	Aug. 17	Izmit, Turkey	17,100	VIII–XI	7.4
1999	Sept. 7	Athens, Greece	150	VI–VIII	5.9
1999	Sept. 20	Chi-Chi, Taiwan	2,500	VI–X	7.6
1999	Sept. 30	Oaxaca, Mexico	33	VI	7.5
1999	Oct. 16	Hector Mine, California	0	**	7.1
1999	Nov. 12	Düzce, Turkey	700	VI–X	7.2
2001	Jan. 26	Gujarat state, India	19,998	X–XII	7.7
2002	Nov. 3	Near Denali National Park, Alaska	1	X	7.9
2003	Dec. 26	Bam, Iran	30,000	X–XII	6.9
2004	Dec. 26	Northern Sumatra (west coast)	170,000+***	X–XII	9.3
2005	Oct. 8	Kashmir region, Pakistan	83,000+	X–XII	7.6
2008	May 12	Chengdu, China	87,000+	X–XII	7.9
2010	Jan. 12	Port-Au-Prince, Haiti	223,000	X–XII	7.0
2010	Feb. 27	Maule, Chile	521	IX–XI	8.8

* Note that earthquakes are not on an increase; rather, the list reflects a greater sample for recent years.

** Data not available.

*** Estimated deaths in 11 countries from the quake and resulting tsunami that travelled across the Indian Ocean.

MG™ **Animation** Seismograph, How it Works

GEO REPORT 12.2 *Large earthquakes affect the entire Earth system*

Scientific evidence is mounting that Earth's largest earthquake events have a global influence. The 2004 Sumatra–Andaman earthquake appears to have affected fault strengths and global seismicity as changes in measured stress and strain increased. Related to this was an unusual increase in M 8.0 and higher quakes after 2004.

An M 7.9 quake near Denali National Park, Alaska, was the largest on Earth in 2002. This quake triggered far-reaching effects: more than 200 small earthquakes in Yellowstone, seismic crustal waves that were felt in Texas, and altered water tables and well water in Pennsylvania. Earth indeed behaves as a vast tectonic system.

more earthquakes for the period following 1960 than for the years before 1960. This is not because earthquake frequency has increased; rather, it reflects an effort to include recent events that affected increased population densities in vulnerable areas.

The National Earthquake Information Center in Golden, Colorado, reports earthquake epicenters. For references, see http://earthquake.usgs.gov/regional/neic/ and the National Geophysical Data Center at http://www.ngdc.noaa.gov/hazard/.

The Nature of Faulting

We earlier described types of faults and faulting motions. The specific mechanics of how a fault breaks remain under study, but **elastic-rebound theory** describes the basic process. Generally, two sides along a fault appear to be locked by friction, resisting any movement despite the powerful forces acting on the adjoining pieces of crust. Stress continues to build strain along the *fault plane* surfaces, storing elastic energy like a wound-up spring. When the strain buildup finally exceeds the frictional lock, both sides of the fault abruptly move to a condition of less strain, releasing a burst of mechanical energy.

Think of the fault plane as a surface with irregularities that act as sticking points, preventing movement, similar to two pieces of wood held together by drops of glue of different sizes rather than an even coating of glue. Research scientists at the USGS and the University of California identify these small areas of high strain as *asperities*. They are the points that break and release the sides of the fault.

If the fracture along the fault line is isolated to a small asperity break, the quake will be small in magnitude. Clearly, as some asperities break (perhaps recorded as small foreshocks), the strain increases on surrounding asperities that remain intact. Thus, small earthquakes in an area may be precursors to a major quake. However, if the break involves the release of strain along several asperities, the quake will be greater in extent and will involve the shifting of massive amounts of crust. The latest evidence points to a wavelike pattern, as rupturing spreads along the fault plane, rather than the entire fault surface giving way at once.

The buildup and release of stress and strain accumulates as the two sides of the fault attempt to move laterally past one another, yet remain locked (Figure 12.20). Stress forces continue to build strain until it overwhelms the frictional sticking points and the two sides snap violently and at great speed into new positions—an earthquake hits.

Earthquake Prediction and Planning

The map in Figure 12.21 plots the earthquake hazards for the United States. These occurrences give some indication of relative risk by region. The challenge is to discover how to predict the *specific time and place* for a quake in the short term in these regions of past earthquake experience.

Actual implementation of an action plan to reduce death, injury, and property damage from earthquakes is difficult. The political environment adds complexity, since studies show that an accurate earthquake prediction may threaten a region's economy. Research examining the potential socioeconomic impact of earthquake prediction on an urban community shows surprising negative economic

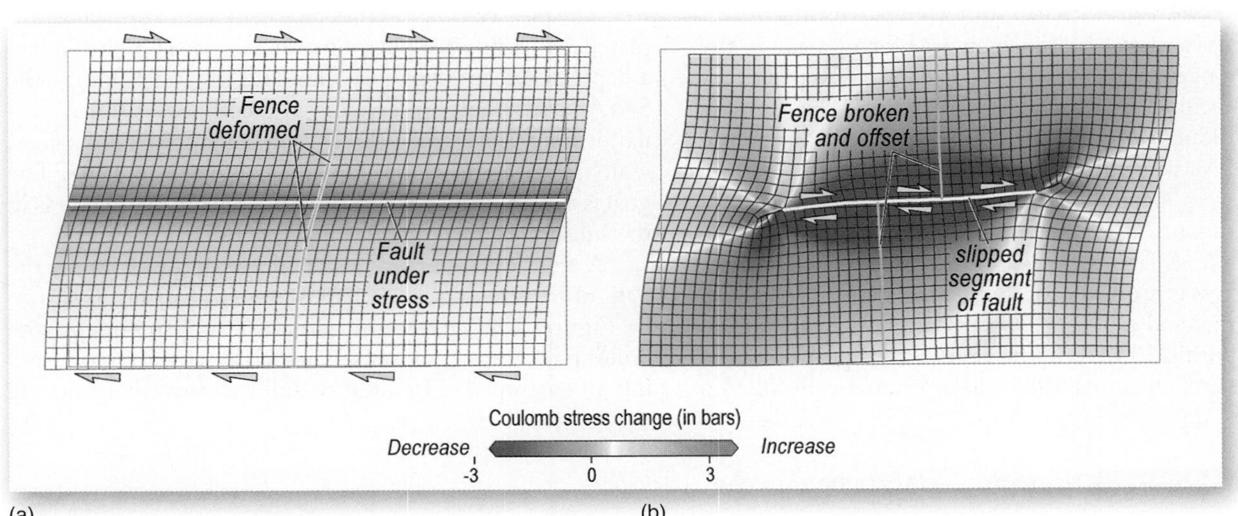

FIGURE 12.20 Buildup and release of stress and strain along a fault system. The elastic-rebound mechanism applies to strike-slip faults such as the San Andreas system of faults in California, North Anatolian fault in Turkey, Enriquillo fault in Haiti, and Nojima fault zone near Kobe, Japan. Note the fence that crosses the fault line. This stress continues to build to the break point, the two sides snap into new positions, strain is released, and the fence no longer forms a continuous property line. The segments of the fault that remain locked now have added strain. A coulomb is a unit of shear-stress failure overcoming friction and resistance to movement; the color scale ranges from blue (decreased stress) to red (increased stress). [Illustrations courtesy of Serkan Bozkurt, USGS.]

Animation
Elastic Rebound

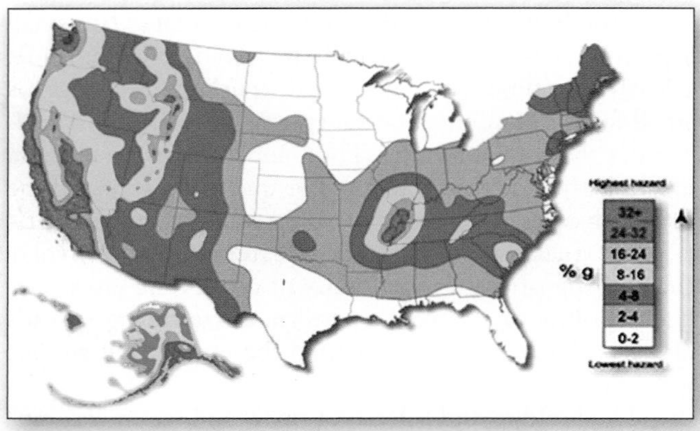

FIGURE 12.21 Earthquake hazard map for the United States.
Earthquake ground accelerations having a 10% probability of being topped in the next 50 years. Acceleration is expressed as a percentage of 1 g (gravity), with red being the highest shaking. Active seismic regions include the West Coast, the Wasatch Front of Utah northward into Canada, the central Mississippi Valley, the southern Appalachians, and portions of South Carolina, upstate New York, and Ontario. [USGS; see http://eqhazmaps.usgs.gov/.]

impacts in the period before the quake hits. Imagine a chamber of commerce, bank, real estate agent, tax assessor, or politician who would welcome an earthquake prediction and such negative publicity for their city. These factors work against effective planning and preparation.

How do you forecast an earthquake? One approach is to examine the history of each plate boundary and determine the frequency of past earthquakes, the study of *paleoseismology*. Paleoseismologists construct maps that provide an estimate of expected earthquake activity based on past performance. An area that is quiet and overdue for an earthquake is a *seismic gap*; such an area forms a gap in the earthquake occurrence record and is therefore a place that possesses accumulated strain. The area along the Aleutian Trench subduction zone had three such gaps until the great 1964 Alaskan earthquake filled one of them.

The areas around San Francisco and northeast and southeast of Los Angeles represent other such gaps where the San Andreas fault system appears to be locked by friction and stress is producing accumulating strain. The USGS in 1988 predicted a 30% chance of an M 6.5 earthquake occurring within 30 years in the Loma Prieta portion of the fault system. The actual 1989 quake dramatically filled a portion of the seismic gap in the central California region. The 1995 Nojima fault quake in the area of Kobe, Japan, occurred in another gap.

A second approach to forecasting is to observe and measure phenomena that might precede an earthquake. *Dilitancy* refers to the slight increase in volume of rock produced by small cracks that form under stress and accumulated strain. The affected region may tilt and swell in response to strain. Tiltmeters measure these changes in suspect areas. Another indicator of dilitancy is an increase in radon (a naturally occurring, slightly radioactive gas) dissolved in groundwater. At present, earthquake hazard zones have thousands of radon monitors taking samples in test wells; increases in gas levels, however, have come before quakes and yet also happened when there was no subsequent quake.

In the absence of effective forecasting weeks, months, or years ahead, earthquake warning systems can be useful. Following the 1989 Loma Prieta earthquake near San Francisco, freeway-repair workers were linked to a broadcast system that gave them 20-second alerts of aftershocks from distant epicenters. In southern California, the USGS is coordinating a Southern California Seismographic Network (SCSN; http://www.scsn.org/) to correlate 350 instruments for immediate earthquake location analysis and disaster coordination. Also, for the southern California region, see http://earthquake.usgs.gov/regional/sca/. A seismic network in operation in Mexico City is providing 70-second warnings of arriving seismic-wave energy from distant epicenters.

The science behind earthquake forecasting is informed by a range of data that is integrated to predict future events. The most intensely studied seismic area in the world is along the San Andreas fault outside Parkfield, California. The area features tiltmeter instruments, gas-monitoring wells, *lidar* laser-ranging measurements, and a newly completed drill hole where instruments are placed in the fault at a depth between 2 km and 3 km (1.2 mi and 1.9 mi)—the San Andreas Fault Observatory at Depth (SAFOD). A similar drilling project has been completed in the Nojima fault, which caused the Kobe, Japan, earthquake in 1995. The goal is to sample material at the seismogenic depth that is in the subsurface area of fault slippage.

A synthetic-aperture radar image of a portion of the San Andreas fault zone in San Mateo, California, is shown in Figure 1.29. Even the slightest movements in the buildup to a quake are sensed, which is critical to forecasting an earthquake. In addition, all the movement during

GEO REPORT 12.3 *Northridge classes held in the outdoors*

Students and faculty at California State University–Northridge, in the San Fernando Valley of southern California, were near the epicenter of the most devastating earthquake in U.S. history in terms of property damage—$30 billion in destruction across the region. The quake was caused by one of the many low-angle thrust faults that underlie the Los Angeles region. The January 17, 1994, Northridge (Reseda) earthquake, an M 6.8, and more than 10,000 aftershocks caused approximately $350 million damage to campus buildings two weeks before the beginning of spring classes. The semester began just three weeks late in 450 temporary trailers. Geography professors taught some class sessions outdoors for a brief time. Amazingly, graduation was still held in May.

the quake and the settling down of aftershocks following a quake are tracked and measured.

In the San Francisco Bay Area, the present forecast predicts a 62% chance of an M 6.7, or higher, striking between 2000 and 2030 along the San Andreas system of faults. Added to this is a 14% chance of a significant earthquake occurring along some yet unknown fault. Adding to the risk is the extent of landfill in the Bay Area, where about half the original bay is now filled and occupied with buildings. In an earthquake, such fill fails in a process of liquefaction, where shaking brings water to the surface and liquefies the soil.

Someday accurate earthquake forecasting may be a reality, but several questions remain: How will humans respond to a forecast? Can a major metropolitan region be evacuated for short periods of time? Can a city relocate after a disaster to an area of lower risk?

Long-range planning is a complex subject. After the Loma Prieta earthquake in 1989, the National Research Council concluded in a report:

> One of the most jarring lessons from these quakes may be that earthquake professionals have long known many of the things that could have been done to reduce devastation. . . . The cost-effectiveness of mitigation and the importance of closing the knowledge gap among researchers, building professionals, government officials, and the public are only two of the many lessons from the Loma Prieta that need immediate action.

A valid and applicable generalization is that *humans and their institutions seem unable or unwilling to perceive hazards in a familiar environment*. In other words, we tend to feel secure in our homes and communities, even if they are sitting on a quiet fault zone. Such an axiom of human behavior certainly helps explain why large populations continue to live and work in earthquake-prone settings. Similar questions also can be raised about populations in areas vulnerable to floods, droughts, hurricanes, and coastal storm surges and in settlements on barrier islands. See the Natural Hazards Center at the University of Colorado at http://www.colorado.edu/hazards/index.html, home of the publication *Natural Hazards Observer*.

Volcanism

Volcanic eruptions across the globe remind us of Earth's internal energy. Ongoing eruption activity matches plate tectonic activity, as shown on the map in Figure 11.21. A few recent eruptions, among many, include Mount Merapi (Indonesia), Eyjafjallajökull (Iceland), Kīlauea (Hawai'i), Soufriere Hills on Montserrat (Lesser Antilles), Mount Etna (Sicily, Italy), Nyamuragira (Democratic Republic of the Congo), several on the Kamchatka Peninsula (Russia), Rabaul (Papua New Guinea), Grímsvötn (Iceland), Axial Summit and the Jackson Segment (sea-floor spreading center off the coast of Oregon), and Shishaldin (Unimak Island, Alaska; see Figure 12.32).

The Eyjafjallajökull eruption on April 13, 2010, was notable because its ash forced the 5-day closure of European airspace, grounding some 100,000 flights (Figure 12.22). Iceland's Meteorological Office monitored the effusive eruption with its 56-station seismic network, implemented with risk assessment and evacuation plans in 1999. Such subglacial eruptions do produce floodwaters and sometimes explosive reactions between meltwater and lava. No large cities were threatened. The region near the volcano is mainly rural farm and pasture land (see Figure 7.1a).

Over 1300 identifiable volcanic cones and mountains exist on Earth, although fewer than 600 are active (have had at least one eruption in recorded history). In an average year, about 50 volcanoes erupt worldwide, varying from modest activity to major explosions. An index to the world's volcanoes is at http://vulcan.wr.usgs.gov/Volcanoes/framework.html, and the National Museum of Natural History Global Volcanism Program lists information for more than 8500 eruptions at http://www.nmnh.si.edu/gvp/. For more volcanic eruption listings, check http://www.geo.mtu.edu/volcanoes/links/observatories.html for links to volcano observatories across the globe.

Eruptions in remote locations and at depths on the seafloor go largely unnoticed, but the occasional eruption of great magnitude near a population center makes headlines. North America has about 70 volcanoes (mostly inactive) along the western margin of the continent. Mount St. Helens in Washington State is a famous active example, and over 1 million visitors a year travel to Mount St. Helens Volcanic National Monument to see volcanism for themselves.

Volcanic Features

A **volcano** forms at the end of a central vent or pipe that rises from the asthenosphere and upper mantle through the crust into a volcanic mountain. A **crater,** or circular surface depression, usually forms at or near the summit.

FIGURE 12.22 **The Eyjafjallajökull eruption Iceland, April 2010.** [Rakal Osk Sigurdardottir/Nordi.]

Animation
Tectonic Settings and Volcanic Activity

(a) Aa

(b) Pahoehoe

FIGURE 12.23 Two types of basaltic lava—Hawaiian examples, close up. (a) A rough, sharp-edged lava, aa is said to get its name from the sounds people make if they attempt to walk on it. The gold-colored strands are clumps of natural spun glass formed by the lava as it blew out into the air—called *Pele's hair*. (b) Pahoehoe forms ropy chords in twisted folds. [Bobbé Christopherson.]

Magma rises and collects in a magma chamber deep below the volcano until conditions are right for an eruption. This subsurface magma emits tremendous heat; in some areas, it boils groundwater, as seen in the thermal springs and geysers of Yellowstone National Park and elsewhere in the world—geothermal energy.

Lava (molten rock), gases, and **pyroclastics,** or *tephra* (pulverized rock and clastic materials of various sizes ejected violently during an eruption), pass through the vent to the surface and build the volcanic landform. Flowing basaltic lava takes two principal forms, both named in Hawaiian terms (Figure 12.23). One form, called **aa,** is rough and jagged with sharp edges. This texture happens because the lava loses trapped gases, flows slowly, and develops a thick skin that cracks into the jagged surface. Another principal lava texture that is more fluid than aa is **pahoehoe**. As this lava flows, it forms a thin crust that develops folds and appears "ropy," like coiled, twisted rope. Both forms can come from the same eruption, and sometimes pahoehoe will become aa as the flow progresses. Other types of basaltic magma are described later in this section.

The fact that lava can occur in many different textures and forms accounts for the varied behavior of volcanoes and the different landforms they build. In this section, we look at five volcanic landforms and their origins: cinder cones, calderas, shield volcanoes, plateau basalts, and composite volcanoes.

A **cinder cone** is a small, cone-shaped hill usually less than 450 m (1500 ft) high, with a truncated top formed from cinders that accumulate during moderately explosive eruptions. Cinder cones are made of pyroclastic material and *scoria* (cindery rock, full of air bubbles). For example, about 100 cinder cones and craters are found on the Ascension Islands, located about 8° S latitude in the Atlantic Ocean (Figure 12.24).

A **caldera** (Spanish for "kettle") is a large, basin-shaped depression. It forms when summit material on a volcanic mountain collapses inward after an eruption or other loss of magma. A caldera may fill with rainwater, as it did in beautiful Crater Lake in southern Oregon.

The Long Valley caldera (Figure 12.25), near the California–Nevada border, was formed by a powerful volcanic eruption 760,000 years ago. The oval caldera is about 15 by 30 km (9 by 19 mi) long in a north–south direction at an elevation of 2000 m (6500 ft). About 1200 tons of carbon dioxide are coming up through the soil in the old caldera each day, killing about 170 acres (69 ha) of forest in six different areas. The source is active and moving magma at a depth of some 3 km (1.86 mi). These gas emissions signal volcanic activity and are useful indicators of potential eruptions. For updates, see http://lvo.wr.usgs.gov/.

On Fogo Island in the Cape Verde Islands (15° N 24.5° W), the Cha caldera opens to the sea on the eastern side. The best farmland is in the caldera, so the people choose to live and work there. Eruptions began in 1995 along the flank of Pico do Fogo, pictured in the Chapter 11 opening photo, destroying farmland and forcing the evacuation of 5000 people. Previously, eruptions occurred between A.D. 1500 and 1750, followed by six more in the nineteenth century, each destroying houses and crops. The population returned to work the volcanic soils and to develop the beginnings of an ecotourism business when

FIGURE 12.24 Many cinder cones on Ascension Island. Ascension Island is a massive composite volcano that rises over 3000 m (9842 ft) from the ocean floor, with 858 m above sea level. It is classified as inactive. [Bobbé Christopherson.]

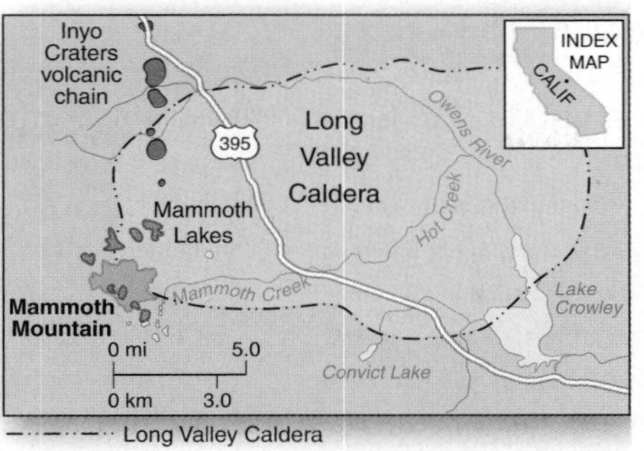

FIGURE 12.25 Forest death from CO_2 emission, Long Valley caldera, California.
A warning sign signals the danger from CO_2 gassing up through soils from subsurface magma and related volcanic activity. The hazard has killed one person to date. [Bobbé Christopherson.]

the eruption sequence subsided. Grape vines planted in basaltic soil are pictured in Figure 18.23c.

Animation Formation of Crater Lake

Location and Types of Volcanic Activity

The location of volcanic mountains on Earth is a function of plate tectonics and hot-spot activity. Volcanic activity occurs in three settings (with representative examples):

1. Along subduction boundaries at continental plate–oceanic plate convergence (such as Mount St. Helens and Kliuchevskoi, Siberia) or oceanic plate–oceanic plate convergence (Philippines and Japan).
2. Along sea-floor spreading centers on the ocean floor (Iceland on the Mid-Atlantic Ridge and off the coast of Oregon and Washington) and along areas of rifting on continental plates (the rift zone in East Africa).
3. At hot spots, where individual plumes of magma rise to the crust (such as Hawai'i and Yellowstone National Park).

Figure 12.26 illustrates these three settings for volcanic activity, which you can compare with the active volcano sites and plate boundaries shown in Figure 11.21.

During a single eruption, a volcano may behave in several different ways. The primary factors in determining an eruption type are (1) the magma's chemistry, which is related to its source, and (2) the magma's viscosity. Viscosity is the magma's resistance to flow ("thickness"), ranging from low viscosity (very fluid) to high viscosity (thick and flowing slowly). We consider two types of eruptions—effusive and explosive—and the characteristic landforms they build.

Effusive Eruptions

Effusive eruptions are the relatively gentle ones that produce enormous volumes of lava annually on the seafloor

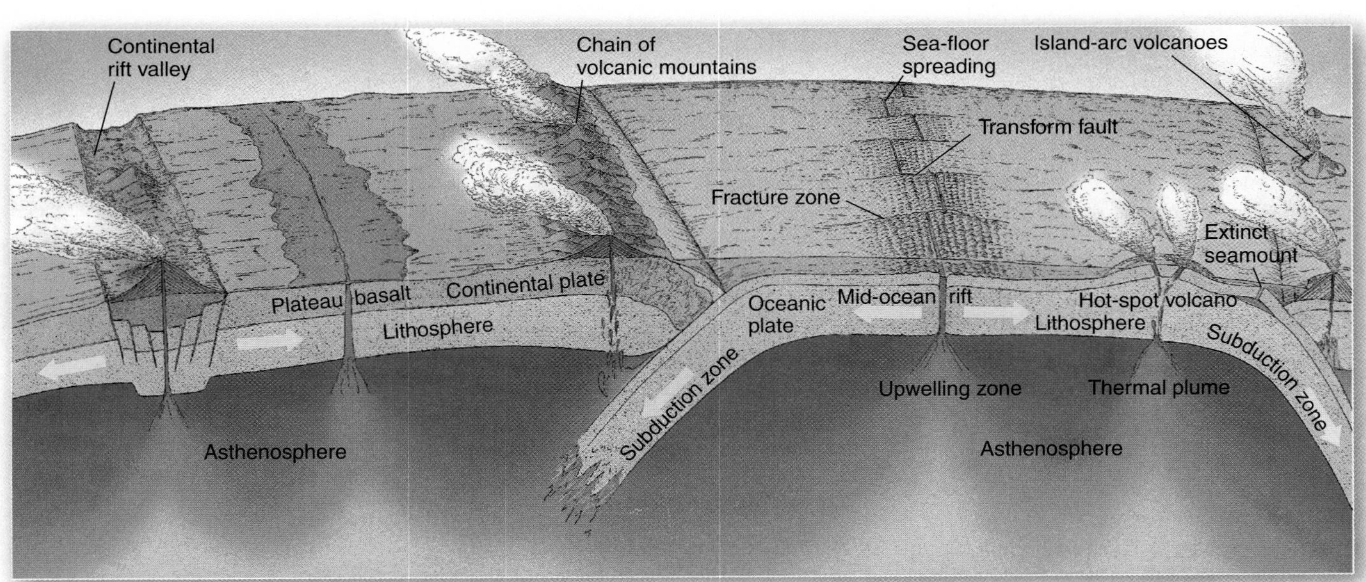

FIGURE 12.26 Tectonic settings of volcanic activity.
Magma rises and lava erupts from rifts, through crust above subduction zones, and where thermal plumes at hot spots break through the crust. [Adapted from U.S. Geological Survey, *The Dynamic Planet* (Washington, DC: Government Printing Office, 1989).]

Animation Forming Types of Volcanoes

and in places such as Hawai'i and Iceland. These direct eruptions from the asthenosphere and upper mantle produce a low-viscosity magma that is fluid and cools to form a dark, basaltic rock low in silica (less than 50% silica) and rich in iron and magnesium). Gases readily escape from this magma because of its low viscosity, adding to a gentle **effusive eruption** that pours out on the surface, with relatively small explosions and few pyroclastics. However, dramatic fountains of basaltic lava sometimes shoot upward, powered by jets of rapidly expanding gases.

An effusive eruption may come from a single vent or from the flank of a volcano, through a side vent. If such vents form a linear opening, they are *fissures*; these sometimes erupt in a dramatic *curtain of fire* (sheets of molten rock spraying into the air). Rift zones capable of erupting tend to converge on the central crater, or vent, as they do in Hawai'i. The interior of such a crater, often a sunken caldera, may fill with low-viscosity magma during an eruption, forming a molten lake, which then may overflow lava downslope in dramatic rivers and falls of molten rock.

On the island of Hawai'i, the continuing Kīlauea eruption is the longest in recorded history, active since January 3, 1983 (Figure 12.27). Although this eruption is located on the slopes of the massive Mauna Loa volcano, scientists have determined that Kīlauea has its own magma plumbing system down some 60 km into Earth, allowing it to produce more lava than any other vent in recorded history. To date, the eruption has produced 3.1 km^3 (0.7 mi^3) of lava, covering 117 km^2 (45 mi^2). During 1989–1990, lava flows from Kīlauea actually consumed several visitor buildings in the Hawai'i Volcanoes National Park. Beginning on Mother's Day 2002 (May 12), an aggressive eruption sequence intensified, flowing over the Chain of Craters Road and continuing for several years to send streams of lava to the ocean (Figure 12.27b). Fourteen km (7 mi) of this former road are now covered by lava flows, making National Park Service rangers keep their end-of-road checkpoint for tourists movable.

Pu'u O'o is the active crater on Kīlauea and is presently continuing to erupt. In 2007, a portion of Pu'u O'o crater's west side collapsed, filled with lava, and collapsed again (Figure 12.28)—see http://hvo.wr.usgs.gov/.

In 2008, a vent blasted open a pit crater in the floor of the Halema'uma'u Crater within Kīlauea. This is within full view of the USGS Hawaiian Volcano Observatory and a dedicated web cam with hourly updated views (http://volcanoes.usgs.gov/hvo/cams/KIcam/). The vent in the crater floor is directly below the former tourist overlook and parking area on the rim of Halema'uma'u (Figure 12.29). Since 2008, the vent has cycled through lava filling and draining events and through debris collapses blocking the vent and then clearing. This activity is producing enormous quantities of sulfur dioxide, which contributes to the volcanic smog problem, *vog*.

A typical mountain landform built from effusive eruptions is gently sloped, gradually rising from the surrounding landscape to a summit crater. The shape is similar in outline to a shield of armor lying face up on the ground and therefore is called a **shield volcano**. The shield shape and size of Mauna Loa in Hawai'i are distinctive when compared with Mount Rainier in Washington, which is a different type of volcano (explained shortly) and the largest in the Cascade Range (Figure 12.30).

(a)

(b)

FIGURE 12.27 Kīlauea landscape.
(a) Aerial photo of the newest land on the planet produced by the massive flows of basaltic lavas from the Kīlauea volcano, Hawai'i Volcano National Park. In the distance, ocean water quenches the intense 1200°C (2200°F) lava on contact, producing steam and hydrochloric acid mist. This entire area is slowly slipping toward the sea. (b) Dramatic Hawaiian active lava flows form the Kamoamoa bench as it builds into the sea. The glowing lava is clearly visible through the steam. Even at the safe distance of this aerial photo, the radiated heat from the lava flow is incredible to feel. [Bobbé Christopherson.]

(a)

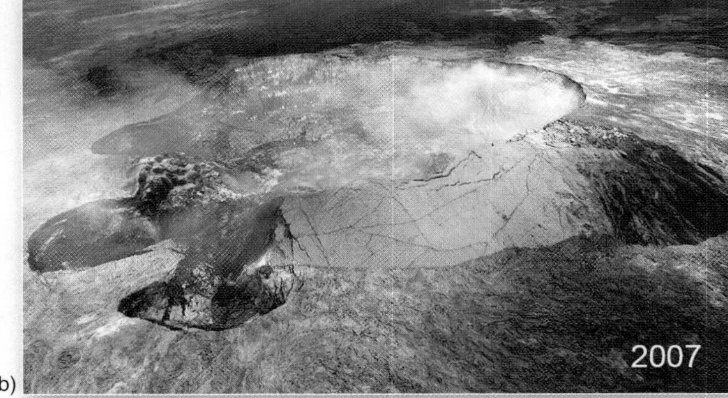

(b)

FIGURE 12.28 Scientists monitor the active Pu'u O'o, at Kīlauea, Hawai'i.
(a) The cracked and faulted crater of Pu'u O'o in 1999. (b) Note the changes 8 years later. Portions of the crater wall collapsed in summer 2007, refilled with lava, and collapsed again. Active lava from the crater continues to make its way to the coast. [(a) Bobbé Christopherson, 1999. (b) USGS Hawaiian Volcano Observatory, 2007.]

(a)

(b)

FIGURE 12.29 Halema'uma'u crater eruptions from 2008 to the present.
Halema'uma'u crater is 100 m (330 ft) deep, roughly circular, and about 800 m (2625 ft) in diameter. A pit crater formed and began eruptions in 2008. The writer Mark Twain witnessed and described eruptions at this same site in his travels in 1866. [USGS 2009 photos.]

GEO REPORT 12.4 *Slow slip events across Kīlauea's south flank*

The entire south flank of the Kīlauea volcano is in a long-term motion toward the ocean, along a low-angle fault at a rate of about 7 cm per year—see the area in Figure 12.27a. GPS units, tiltmeters, and small earthquakes mark the slow movement. In February 2010, in one 36-hour period, a slow slip event of 3 cm occurred, accompanied by a flurry of small earthquakes. A bench collapse or failure of this new basaltic landscape into the sea is possible, but the eventual outcome of these slow slip events is unknown.

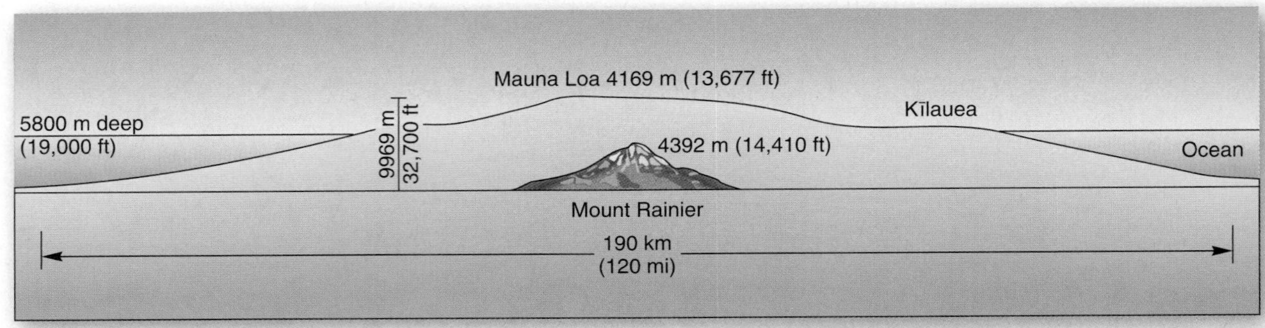

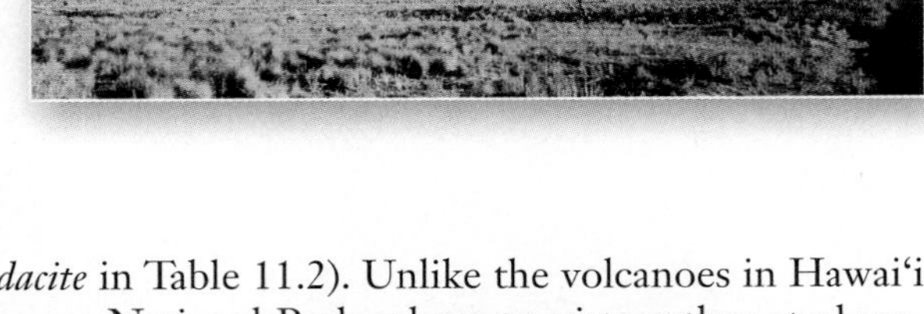

FIGURE 12.30 Shield and composite volcanoes compared. Comparison of Mauna Loa in Hawai'i, a shield volcano, and Mount Rainier in Washington State, a composite volcano. Their strikingly different profiles signify their different tectonic origins. Note the gently sloped shield shape in the inset photo. [After U.S. Geological Survey, *Eruption of Hawaiian Volcanoes* (Washington, DC: Government Printing Office, 1986). Inset photo by Bobbé Christopherson.]

The height of the Mauna Loa shield is the result of successive eruptions, flowing one on top of another. Mauna Loa is one of five shield volcanoes that make up the island of Hawai'i. At least 1 million years were needed to accumulate its mass. Mauna Kea is slightly taller, but Mauna Loa is the most massive single mountain on Earth.

In another volcanic setting, effusive eruptions send material out through hot spots or elongated fissures, such as a continental rift valley, forming extensive sheets of basaltic lava on the surface (see Figure 12.26). The Columbian Plateau of the northwestern United States, some 2–3 km thick, is the result of the eruption of **plateau basalts,** or *flood basalts*.

More than double the size of the Columbian Plateau is the Deccan Traps, which dominates west-central India. *Trap* is Dutch for "staircase," referring to the typical step-like eroded lavas. The Siberian Traps is more than twice the area of that in India and is exceeded only by the Ontong Java Plateau, which covers an area of the seafloor in the Pacific (Figure 12.31). These regions are sometimes referred to as *plateau basalt provinces.* Of these extensive igneous provinces, no presently active sites come close in size to the largest of the extinct igneous provinces, some of which formed more than 200 million years ago.

Explosive Eruptions

Volcanic activity inland from subduction zones produces the infamous explosive volcanoes. Magma produced by the melting of subducted oceanic plate and other materials is thicker (more viscous) than magma that forms effusive volcanoes. It is 50%–75% silica and high in aluminum; consequently, it tends to block the magma conduit inside the volcano. The blockage traps and compresses gases, causing pressure to build for a possible **explosive eruption**.

From this magma, a lighter dacitic rock forms at the surface, as illustrated in the comparison in Figure 12.5 (see *dacite* in Table 11.2). Unlike the volcanoes in Hawai'i Volcanoes National Park, where tourists gather at observation platforms or hike out across the cooling lava flows to watch the relatively calm effusive eruptions, these explosive volcanoes do not invite close inspection and can explode with little warning.

The term **composite volcano** describes these explosively formed mountains. (They are sometimes called *stratovolcanoes* because they are built up in alternating layers of ash, rock, and lava, but shield volcanoes also can exhibit a stratified structure, so *composite* is the preferred term.) Composite volcanoes tend to have steep sides, are more conical in shape than shield volcanoes, and therefore are also known as *composite cones.* If a single summit vent erupts repeatedly, a remarkable symmetry may develop as the mountain grows in size, as demonstrated by Mount Orizaba in Mexico, Mount Shishaldin in Alaska (Figure 12.32), Mount Fuji in Japan, Mount Mayon in the Philippines, and the pre-1980-eruption shape of Mount St. Helens in Washington.

Probably the most studied and photographed composite volcano on Earth is Mount St. Helens, located 70 km (45 mi) northeast of Portland, Oregon, and 130 km (80 mi) south of the Tacoma–Seattle area. Mount St. Helens is the youngest and most active of the Cascade Range of volcanoes, which form a line from Mount Lassen in California to Mount Meager in British Columbia. The Cascade Range is the product of the Juan de Fuca seafloor spreading center off the coast of northern California, Oregon, Washington, and British Columbia and the plate subduction that occurs offshore, as identified in the upper right of Figure 12.12a. Focus Study 12.1 details the highly publicized 1980 eruption of Mount St. Helens that so drastically altered the landscape (Figure 12.33).

Owing to its chemical makeup and high viscosity, rising magma in a composite volcano forms a plug near the surface. The blockage causes tremendous pressure to

(text continued on page 364)

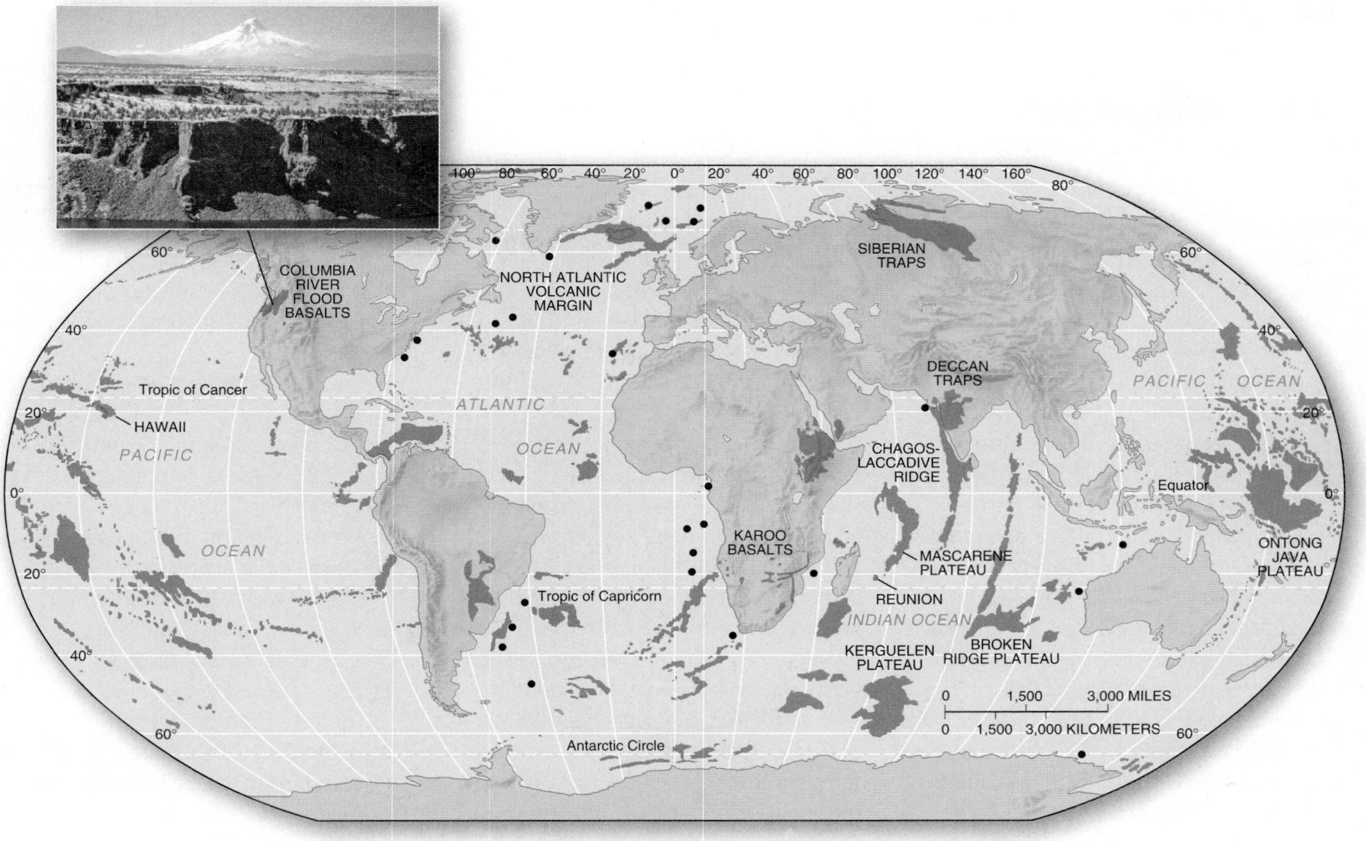

FIGURE 12.31 Earth's extensive igneous provinces.
(a) Plateau basalts form large igneous provinces across the globe. Many are ancient and inactive. The largest known is the Ontong Java Plateau, which covers nearly 2 million km^2 (0.8 million mi^2) on the Pacific Ocean floor. Black dots are areas of ancient volcanism along continental margins. (b) In the foreground are plateau (flood) basalts characteristic of the Columbia Plateau in Oregon. Mount Hood, in the background to the west of the plateau, is a composite volcano. [(a) After M. F. Coffin and O. Eldholm, "Large igneous provinces," *Scientific American* (October 1993): 42–43. © Scientific American, Inc. (b) Author.]

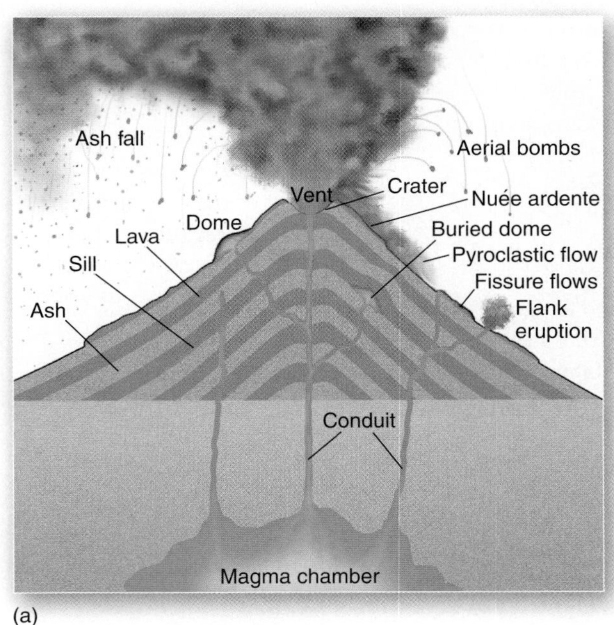

(a)

(b)

FIGURE 12.32 A composite volcano.
(a) A typical composite volcano with its cone-shaped form in an explosive eruption. (b) Mount Shishaldin, Unimak Island in the eastern Aleutian Islands, Alaska, is a 2857-m (9372-ft) composite volcano. Here shown during a 1995 eruption; it was active through 2001. [(b) Photo courtesy of AeroMap U.S., Inc., Anchorage, Alaska.]

FOCUS STUDY 12.1

The 1980 Eruption of Mount St. Helens

Quiet since 1857, Mount St. Helens awoke in March 1980, with a sharp earthquake registering an M 4.1. The first eruptive outburst occurred a week later, beginning with an M 4.5 quake and continuing with a thick black plume of ash and the development of a small summit crater. Ten days later, after scientists hurriedly placed instruments around the volcano, the first volcanic earthquake, called a *harmonic tremor,* registered. Harmonic tremors are slow, steady vibrations, unlike the sharp releases of energy associated with tectonic earthquakes and faulting. These vibrations indicated that magma was on the move within the mountain.

Also developing was a massive bulge on Mount St. Helens' north side, indicating the direction of the magma flow within the volcano. A bulge represents the greatest risk from a composite volcano, for it could signal a potential lateral burst through the bulge and across the landscape.

Early on Sunday, May 18, the area north of the mountain was rocked by an M 5.0 quake, the strongest to date. The mountain, with its distended 245-m (800-ft) bulge, was shaken, but nothing happened. Then a second quake, at M 5.1, hit a few minutes later, loosening the bulge and launching the eruption. David Johnston, a volcanologist with the USGS, was only 8 km (5 mi) from the mountain, servicing instruments, when he saw the eruption begin. He radioed headquarters in Vancouver, Washington, saying "Vancouver, Vancouver, this is it!" He perished in the eruption. For a continuously updated live picture of Mount St. Helens from Johnston's approximate observation point and other specifics, go to http://www.fs.fed.us/gpnf/volcanocams/msh/. The camera is mounted below the roofline at the Johnston Ridge Observatory, which is open to the public.

As the contents of the mountain exploded, a surge of hot gas (about 300°C, or 570°F), steam-filled ash, pyroclastics, and a *nuée ardente* (rapidly moving, very hot, explosive ash and incandescent gases) moved northward, hugging the ground and traveling at speeds up to 400 kmph (250 mph) for a distance of 28 km (17 mi).

The slumping north face of the mountain produced the greatest landslide witnessed in recorded history; about 2.75 km^3 (0.67 mi^3) of rock, ice, and trapped air, all fluidized with steam, surged at speeds approaching 250 kmph (155 mph). Landslide materials traveled for 21 km (13 mi) into the valley, blanketing the forest, covering a lake, and filling the rivers below. A series of photographs, taken at 10-second intervals from the east looking west, records this sequence (Figure 12.1.1). The eruption continued with intensity for 9 hours, first clearing out old rock from the throat of the volcano and then blasting new material for days.

As destructive as such eruptions are, they also are constructive, for this is the way in which a volcano eventually builds its height. Before the eruption, Mount St. Helens was 2950 m (9677 ft) tall; the eruption blew away 418 m (1370 ft). Today, Mount St. Helens is building a lava dome within its crater. The thick lava rapidly and repeatedly plugs and breaks in a series of lesser dome eruptions that may continue for several decades. The several lava domes are more than 300 m (1000 ft) high, so a new mountain is being born from the eruption of the old.

The lava dome is covered with a dozen survey benchmarks and several tiltmeters to monitor the status of the volcano. Swarms of minor earthquakes along the north flank of the mountain began in November 2001, marking this mountain as still unstable. Dome eruptions of varying intensities occurred through 2007. Intensive scientific research and monitoring has paid off, as every eruption since 1980 was successfully forecasted from days to as long as 3 weeks in advance, with the exception of one small eruption in 1984.

An ever-resilient ecosystem is recovering, as shown in the dramatic comparison photos from 1983 and 1999 in Figure 19.23. More than three decades have passed, yet strong interest in the area continues; scientists and more than 1 million tourists visit the Mount St. Helens Volcanic National Monument each year. For more information, see the Cascades Volcano Observatory at http://vulcan.wr.usgs.gov/.

The attention drawn to Mount St. Helens over these three decades is deserved, especially when you think of the status of volcanic science 30 years ago, before cell phones, laptops, and GPS. A recent paper summarizes this well:

> The eruption remains a seminal historical event—studying it and its aftermath revolutionized the way scientists approach the field of volcanology. Not only was the eruption spectacular, but also it occurred in daytime, at an accessible volcano, in a country with the resources to transform the disaster into scientific opportunity, amid a transformation in digital technology. Lives lost and the impact of the eruption on people and infrastructure downstream and downwind made it imperative for scientists to investigate events and work with communities to lessen losses from future eruptions.*

*J. W. Vallance et al., "Mount St. Helens: A 30-year legacy of volcanism," *EOS, Transactions* (American Geophysical Union), May 11, 2010, p. 169.

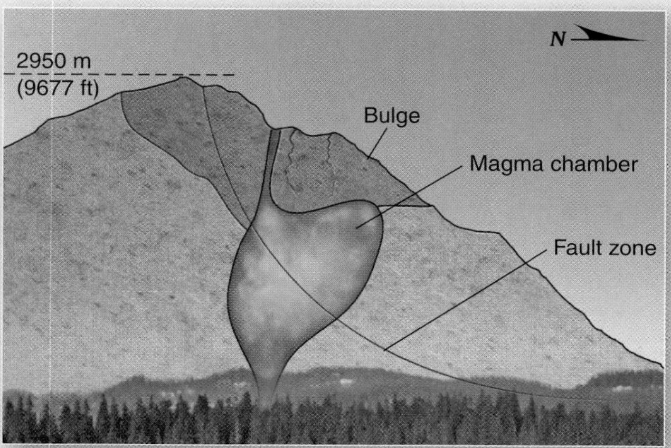

FIGURE 12.1.1 The Mount St. Helens eruption sequence and corresponding schematics. [Photo sequence by Keith Ronnholm.]

Animation Debris Avalanche and eruption of Mount St. Helens

(a)

(b)

(c)

FIGURE 12.33 Mount St. Helens before and after the eruption—days and years later. (a) Mount St. Helens prior to the 1980 eruption. (b) The devastated and scorched land shortly after the eruption in 1980. The scorched earth and tree blow-down area covered some 38,950 hectares (95,000 acres). (c) A landscape in recovery as life moves back in and takes hold to establish new ecosystems in 1999 along the Toutle River and debris flow. [(a) Pat and Tom Lesson/Photo Researchers. (b) Krafft-Explorer/Photo Researchers, Inc. (c) Bobbé Christopherson.]

build, keeping the trapped gases compressed and liquefied. When the blockage can no longer hold back this pressurized inferno, an explosion equivalent to megatons of TNT blasts the top and sides off the mountain. This type of eruption produces much less lava than effusive eruptions, but larger amounts of *pyroclastics*, which include volcanic *ash* (<2 mm, or <0.08 in., in diameter), dust, cinders, *lapilli* (up to 32 mm, or 1.26 in., in diameter), *scoria* (volcanic slag), pumice, and *aerial bombs* (explosively ejected blobs of incandescent lava). A *nuée ardente*, French for "glowing cloud," is an incandescent, hot, turbulent gas, ash, and pyroclastic cloud that can jet across the landscape in an eruption.

Mount Pinatubo Eruption In June 1991, after 600 years of dormancy, Mount Pinatubo in the Philippines erupted. The summit of the 1460-m (4795-ft) volcano exploded, devastating many surrounding villages and permanently closing Clark Air Force Base, operated by the United States. Fortunately, scientists from the USGS and local scientists accurately predicted the eruption. A timely evacuation of the surrounding countryside saved thousands of lives, but 800 people were killed.

Although volcanoes are regional events, their spatial implications can be worldwide. The single volcanic eruption of Mount Pinatubo was significant to the global environment, as discussed in Chapters 1, 3, 4, 5, and 10 (see Figure 1.8). Here is a summary of its effects:

- 15–20 million tons of ash and sulfuric acid mist were blasted into the atmosphere, concentrating at an altitude of 16–25 km (10–15.5 mi).
- 12 km^3 (3.0 mi^3) of material were ejected and extruded by the eruption (12 times the volume from Mount St. Helens).
- 60 days after the eruption, about 42% of the globe was affected (from 20° S to 30° N) by the thin, spreading aerosol cloud in the atmosphere.
- Colorful twilight and dawn skies were observed worldwide.
- An increase in atmospheric albedo of 1.5% (4.3 W/m^2) occurred.
- An increase in the atmospheric absorption of insolation followed (2.5 W/m^2).
- A decrease in net radiation at the surface and a lowering of Northern Hemisphere average temperatures of 0.5 C° (0.9 F°) were measured.
- Atmospheric scientists and volcanologists were able to study the eruption aftermath using satellite-borne orbiting sensors and general-circulation-model computer simulations.

Volcano Forecasting and Planning

The USGS and the Office of Foreign Disaster Assistance operate the Volcano Disaster Assistance Program (VDAP). For a listing of more than two dozen projects, see http://vulcan.wr.usgs.gov/Vdap/framework.html. The need for such a program is evident in that over the past 30 years, volcanic activity has killed 29,000 people, forced more than 800,000 to evacuate their homes, and caused more than $3 billion in damage. The program was established after 23,000 people died in the eruption of Nevado del Ruiz, Colombia, in 1985. The U.S. VDAP is in place to help local scientists with eruption forecasts by setting up mobile volcano-monitoring systems at the most-threatened sites.

An effort such as this led to the life-saving evacuation of 60,000 people hours before Mount Pinatubo exploded. In addition, satellite remote sensing is helping VDAP to monitor eruption cloud dynamics, atmospheric emissions and climatic effects, and lava and thermal measurements; to make topographic measurements; to estimate volcanic hazard potential; and to enhance geologic mapping—all in an effort to better understand our dynamic planet. Integrated seismographic networks and monitoring are making possible early warning systems.

In this era of the Internet, you can access "volcano cams" positioned around the world to give you 24-hour surveillance of many volcanoes. For an exciting visual adventure, go to the following URL and add it to your bookmarks (note whether it is day or night for the location you are checking): http://vulcan.wr.usgs.gov/Photo/volcano_cams.html.

This completes the crust-building processes of Earth's endogenic systems covered in Chapters 11 and 12. Earth's surface topography results from formation and deformation processes powered by radioactive-decay heat radiating from Earth's interior. Mountain building, earthquakes, and volcanism are the resulting outputs of these systems. We move on to the exogenic system, where solar energy and gravity empowers weathering, erosion, and deposition through the agents of water, wind, waves, and ice. These processes reduce relief and the landscape.

KEY LEARNING CONCEPTS REVIEW

■ ***Describe*** first, second, and third orders of relief, and ***relate*** examples of each from Earth's major topographic regions.

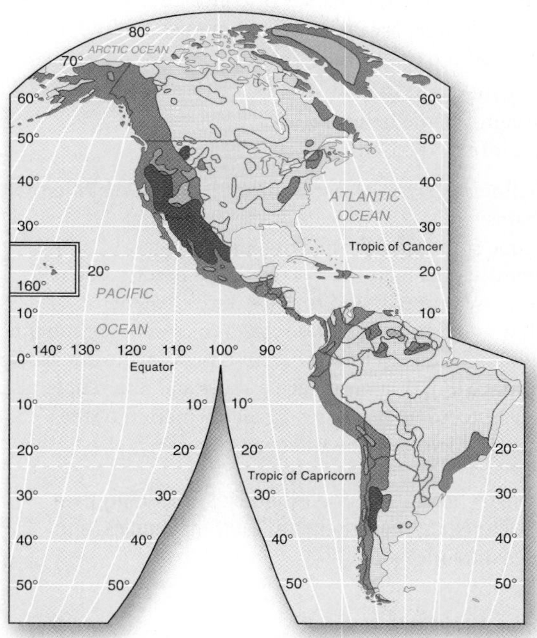

Tectonic forces generated within the planet dramatically shape Earth's surface. **Relief** is the vertical elevation difference in a local landscape. The undulating physical surface of Earth, including relief, is **topography**. Convenient descriptive categories are termed *orders of relief*. The coarsest level of landforms includes the **continental landmasses** and **ocean basins**; the finest comprises local hills and valleys.

relief (p. 333)
topography (p. 333)
continental landmasses (p. 333)
ocean basins (p. 333)

1. How does the map of the ocean floor in the chapter-opening illustration exhibit the principles of plate tectonics? Briefly analyze.
2. What is meant by an order of relief? Give an example from each order.
3. Explain the difference between relief and topography.

■ ***Describe*** the several origins of continental crust, and ***define*** displaced terranes.

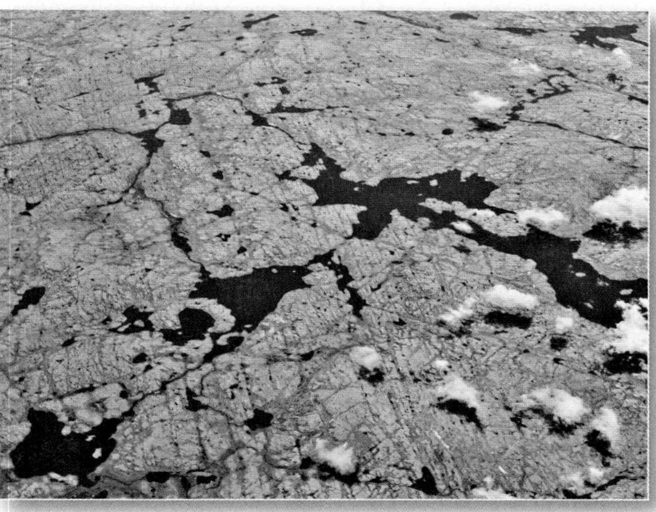

A continent has a nucleus of ancient crystalline rock called a *craton*. A region where a craton is exposed is a **continental shield**. As continental crust forms, it is enlarged through accretion of dispersed **terranes**. An example is the *Wrangellia terrane* of the Pacific Northwest and Alaska.

continental shield (p. 336)
terranes (p. 338)

4. What is a craton? Relate this structure to continental shields and platforms, and describe these regions in North America.
5. What is a migrating terrane, and how does it add to the formation of continental masses?
6. Briefly describe the journey and destination of the Wrangellia terrane.

■ ***Explain*** compressional processes and folding, and ***describe*** four principal types of faults and their characteristic landforms.

Folding, broad warping, and faulting deform the crust and produce characteristic landforms. Compression causes rocks to deform in a

process known as **folding,** during which rock strata bend and may overturn. Along the *ridge* of a fold, layers *slope downward away from the axis,* which is an **anticline.** In the *trough* of a fold, however, layers *slope downward toward the axis;* this is a **syncline.**

When rock strata are stressed beyond their ability to remain a solid unit, they express the strain as a fracture. Rocks on either side of the fracture are displaced relative to the other side in a process known as **faulting.** Thus, *fault zones* are areas where fractures in the rock demonstrate crustal movement. At the moment of fracture, a sharp release of energy, an **earthquake** or *quake,* occurs.

When forces pull rocks apart, the tension causes a **normal fault,** sometimes visible on the landscape as a scarp, or escarpment. Compressional forces associated with converging plates force rocks to move upward, producing a **reverse fault.** A low-angle fault plane is referred to as a **thrust fault.** Horizontal movement along a fault plane that produces a linear rift valley is a **strike-slip fault.** In the U.S. interior west, the Basin and Range Province is an example of aligned pairs of normal faults and a distinctive horst-and-graben landscape. The term **horst** is applied to upward-faulted blocks; **graben** refers to downward-faulted blocks.

folding (p. 339)
anticline (p. 339)
syncline (p. 339)
faulting (p. 341)
earthquake (p. 341)
normal fault (p. 343)
reverse fault (p. 343)
thrust fault (p. 343)
strike-slip fault (p. 343)
horst (p. 343)
graben (p. 343)

7. Diagram a simple folded landscape in cross section, and identify the features created by the folded strata.
8. Define the four basic types of faults. How are faults related to earthquakes and seismic activity?
9. How did the Basin and Range Province evolve in the western United States? What other examples exist of this type of landscape?

■ ***Relate*** the three types of plate collisions associated with orogenesis, and ***identify*** specific examples of each.

Orogenesis is the birth of mountains. An orogeny is a mountain-building episode, occurring over millions of years, that thickens continental crust. It can occur through large-scale deformation and uplift of the crust. It also may include the capture

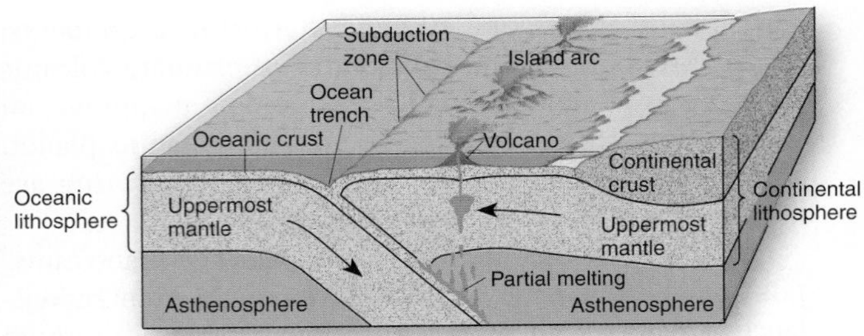

and cementation of migrating terranes to the continental margins and the intrusion of granitic magmas to form plutons.

Oceanic plate–continental plate collision orogenesis is now occurring along the Pacific coast of the Americas and has formed the Andes, the Sierra of Central America, the Rockies, and other western mountains. *Oceanic plate–oceanic plate* collision orogenesis produces either simple volcanic island arcs or more complex arcs such as Japan, the Philippines, the Kurils, and portions of the Aleutians. The region around the Pacific contains expressions of each type of collision in the **circum-Pacific belt,** or the **ring of fire.**

Continental plate–continental plate collision orogenesis is quite mechanical; large masses of continental crust, such as the Himalayan Range, are subjected to intense folding, overthrusting, faulting, and uplifting.

orogenesis (p. 343)
circum-Pacific belt (p. 346)
ring of fire (p. 346)

10. Define *orogenesis*. What is meant by the birth of mountain chains?
11. Name some significant orogenies.
12. Identify on a map Earth's two large mountain chains. What processes contributed to their development?
13. How are plate boundaries related to episodes of mountain building? Explain how different types of plate boundaries produce differing orogenic episodes and different landscapes.
14. Relate tectonic processes to the formation of the Appalachians and the Alleghany orogeny.

■ ***Explain*** the nature of earthquakes, including their characteristics, measurement, fault mechanics, and forecasting methods.

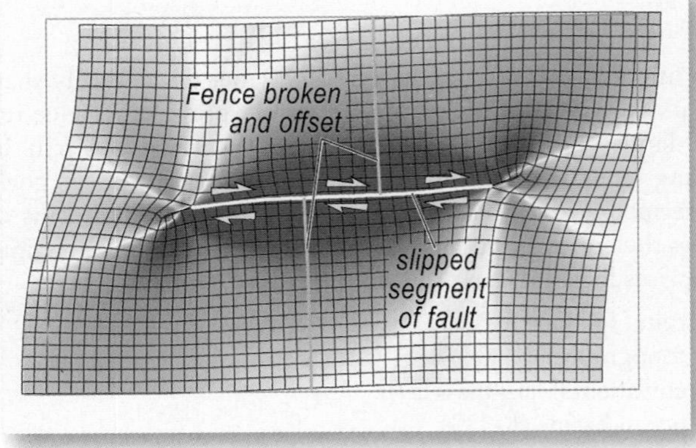

Earthquakes generally occur along plate boundaries; major ones can be disastrous. Earthquakes result from faults, which are under continuing study to learn the nature of faulting, stress and the buildup of strain, irregularities along fault-plane surfaces, the way faults rupture, and the relationship among active faults. Earthquake prediction and improved planning are active concerns of *seismology*, the study of earthquake waves and Earth's interior. Seismic motions are measured with a **seismograph**.

Charles Richter developed the **Richter scale,** a measure of earthquake magnitude. A more precise and quantitative scale that assesses the *seismic moment* is now used, especially for larger quakes; this is the **moment magnitude scale**. The specific mechanics of how a fault breaks are under study, but the **elastic-rebound theory** describes the basic process. In general, two sides along a fault appear to be locked by friction, resisting any movement. This stress continues to build strain along the fault surfaces, storing elastic energy like a wound-up spring. When energy is released abruptly as the rock breaks, both sides of the fault return to a condition of less strain.

seismograph (p. 351)
Richter scale (p. 351)
moment magnitude scale (p. 351)
elastic-rebound theory (p. 353)

15. What is the relationship between an epicenter and the focus of an earthquake? Give an example from the Loma Prieta, California, earthquake.
16. Differentiate among the Mercalli, moment magnitude, and amplitude (Richter) scales. How are these used to describe an earthquake? Why has the Richter scale been updated and modified?
17. How do the elastic-rebound theory and asperities help explain the nature of faulting? In your explanation, relate the concepts of stress (force) and strain (deformation) along a fault. How does this lead to rupture and earthquake?
18. Describe the San Andreas fault and its relationship to ancient sea-floor spreading movements along transform faults.
19. How are paleoseismology and the seismic gap concept related to expected earthquake occurrences?
20. What do you see as the biggest barrier to effective earthquake prediction?

■ ***Distinguish*** between an effusive and an explosive volcanic eruption, and ***describe*** related landforms, using specific examples.

Volcanoes offer direct evidence of the makeup of the asthenosphere and uppermost mantle. A **volcano** forms at the end of a central vent or pipe that rises from the asthenosphere through the crust into a volcanic mountain. A **crater,** or circular surface depression, usually forms at the summit. Areas where magma is near the surface may heat groundwater, producing *geothermal energy*.

Eruptions produce **lava** (molten rock), gases, and **pyroclastics** (pulverized rock and clastic materials ejected violently during an eruption) that pass through the vent to openings and fissures at the surface and build volcanic landforms. Basaltic lava flows occur in two principal textures: **aa,** rough and sharp-edged lava, and **pahoehoe,** smooth, ropy folds of lava. Volcanic activity produces landforms such as a **cinder cone,** a small hill, or a large basin-shaped depression sometimes caused by the collapse of a volcano's summit, forming a **caldera**.

Volcanoes are of two general types, based on the chemistry and the viscosity of the magma involved. **Effusive eruption** produces a **shield volcano** (such as Kīlauea in Hawai'i) and extensive deposits of **plateau basalts,** or flood basalts. Magma of higher viscosity leads to an **explosive eruption** (such as Mount Pinatubo in the Philippines), producing a **composite volcano**. Volcanic activity has produced some destructive moments in history, but constantly creates new seafloor, land, and soils.

volcano (p. 355)
crater (p. 355)
lava (p. 356)
pyroclastics (p. 356)
aa (p. 356)
pahoehoe (p. 356)
cinder cone (p. 356)
caldera (p. 356)
effusive eruption (p. 358)
shield volcano (p. 358)
plateau basalts (p. 360)
explosive eruption (p. 360)
composite volcano (p. 360)

21. What is a volcano? In general terms, describe some related features.
22. Where do you expect to find volcanic activity in the world? Why?
23. Compare effusive and explosive eruptions. Why are they different? What distinct landforms are produced by each type? Give examples of each.
24. Describe several recent volcanic eruptions, such as in Hawai'i and Iceland. What is the present status in each place?

MASTERING GEOGRAPHY

Visit www.masteringgeography.com for several figure animations, satellite loops, self-study quizzes, flashcards, visual glossary, case studies, career links, textbook reference maps, RSS Feeds, and an ebook version of *Geosystems*. Included are many geography web links to interesting Internet sources that support this chapter.

CHAPTER 13

Weathering, Karst Landscapes, and Mass Movement

Weathering processes formed this sandstone niche in Canyon de Chelly. Native Americans took advantage of the formation for construction of stone cliff dwellings, as in these White House ruins from the 12th and 13th centuries. [Bobbé Christopherson.]

KEY LEARNING CONCEPTS

After reading the chapter, you should be able to:

- ***Define*** the science of geomorphology.
- ***Illustrate*** the forces at work on materials residing on a slope.
- ***Define*** weathering, and ***explain*** the importance of parent rock and joints and fractures in rock.
- ***Describe*** frost action, salt-crystal growth, pressure-release jointing, and the role of freezing water as physical weathering processes.
- ***Describe*** the susceptibility of different minerals to the chemical weathering processes, including hydration, hydrolysis, oxidation, carbonation, and solution.
- ***Review*** the processes and features associated with karst topography.
- ***Portray*** the various types of mass movements, and ***identify*** examples of each in relation to moisture content and speed of movement.

Human-Caused Mass Movement at the Kingston Steam Plant, Tennessee

During the early morning hours of December 22, 2008, a mass-wasting event occurred from the containment ponds at the Kingston coal-fired power plant in Tennessee. This single event, in which the embankment of a fly-ash holding pond was breached, released 4.13 million m^3 (5.4 million yd^3) of toxic ash laced with an array of potentially harmful pollutants into the Emory River and over the surrounding area.

The term *mass movement,* or *mass wasting,* refers to the movement of a body of Earth materials with gravity; other familiar examples are landslides and mudflows, covered in this chapter. In areas where humans have disturbed the landscape, mass movement may be accelerated, with disastrous consequences. Such was the case with this fly-ash spill at the Kingston Steam Plant (KSP).

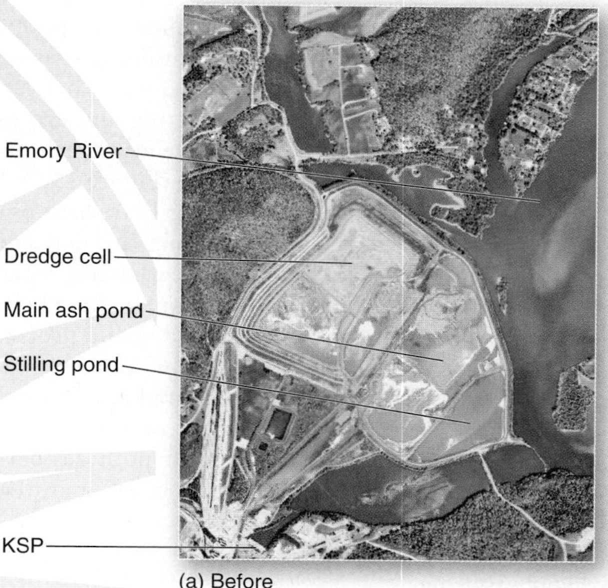

FIGURE GN 13.1 Before and after the TVA ash slurry spill disaster. (a) The Emory River enters from the north and joins the Clinch River near the center. The dredge cells, main ash pond, and stilling pond are in the river floodplain. The power plant is off the lower edge of the aerial photo. (b) December 23, the day after the spill, the extent of the contaminated wastage is clearly visible. [Courtesy of TVA.]

Built by the Tennessee Valley Authority (TVA) in the 1950s, the KSP is located in Roane County near the town of Kingston (Figure GN 13.1a). Built to provide power for the atomic energy installations at nearby Oak Ridge, this coal-fired 1456 MW power plant burns 14,000 tons of pulverized coal a day. Before the exhaust smoke leaves the 305-m (1000-ft) smokestacks, fly ash is extracted, fluidized with water, and moved as slurry to ash ponds. In these ponds, the solids settle and are gradually collected in dredge holding cells constructed along the Emory River floodplain.

Embankments surrounding the dredge cells consisted originally of clay and later were built up with materials derived from the ash-waste product (slightly left of center in the aerial photo, Figure GN 13.1a). As more waste ash was added, the walls eventually rose to 18 m (59 ft) in height, despite their lack of engineering design specifications. Over the years, seepage and leaking were reported.

During the first three weeks of December 2008, above-normal rainfall of more than 16 cm (6.5 in.) fell in the region. On December 22 the entire northwest corner and sides of the ash-pond embankment failed, triggering a mass-wasting event that released fly-ash slurry onto the landscape, destroying homes and infrastructure. The volume of the spill was equal to 1.1 billion gallons, or about 100 times the 1989 *Exxon–Valdez* oil spill (Figure GN 13.1b). Worse, the slurry contained harmful pollutants and metals, such as arsenic, copper, barium, cadmium, chromium, lead, mercury, nickel, and thallium, among others. These contaminants pose a significant risk to human health.

Almost a year after the disaster, TVA had removed about two-thirds of the ash in the Emory River, moving it by train to a landfill in Alabama. In spring 2010, the Environmental Protection Agency approved the next phase, which calls for TVA to permanently store on site all of the ash released into the embayment, thus reducing the hazards of transport. The cleanup is expected to take four years and cost about $270 million.

The Tennessee Department of Environment and Conservation (TDEC) later criticized the TVA Kingston plant in a "lessons learned" report, stating that critical deficiencies existed at the dredge cell, which "lacked design continuity and effective structural stability oversight by TVA throughout its history" (TDEC, *Lessons Learned from the TVA Kingston Dredge Cell Containment Facility Failure* [Nashville, TN: November 30, 2009], p. 5.).

As we see in the science of mass-movement events, the human element can create extreme hazards when known principles and dynamics are not considered or are ignored.

Here we begin a five-chapter examination of exogenic processes at work on the landscape. This chapter examines weathering and mass movement of the surface crust. The next four chapters look at specific exogenic agents and their handiwork—river systems, wind-influenced landscapes, deserts in water-deficit regions, coastal processes and landforms, and regions worked by ice and glaciers. Whether you enjoy time along a river, love the desert or the waves along a coastline, or live in a place where glaciers once carved the land, you will find something of interest in these chapters.

In this chapter: We look at physical (mechanical) and chemical weathering processes that break up, dissolve, and generally reduce the landscape. Such weathering releases essential minerals from bedrock for soil formation and enrichment. In limestone regions, chemical weathering produces sinkholes, caves, and caverns. In these karst environments, water has dissolved enormous underground worlds of mystery and darkness, yet many caves remain undiscovered.

In addition, we examine mass-movement processes that continually operate in and upon the landscape. News broadcasts may describe an avalanche in Austria; a mudflow in Mexico, China, or Pakistan; a tragic landslide in China, Turkey, or Indonesia; a massive rockfall in Yosemite; or the prediction of an imminent mudflow down the slopes in a southern California suburb. In 2008, Haiti was deluged by three hurricanes that caused massive landslides and mudflows from deforested mountain slopes. This chapter discusses the processes that caused these events.

Landmass Denudation

Geomorphology is the science of landforms—their origin, evolution, form, and spatial distribution. This science is an important aspect of physical geography. **Denudation** is any process that wears away or rearranges landforms. The principal denudation processes affecting surface materials include *weathering*, *mass movement*, *erosion*, *transportation*, and *deposition*, as produced by the agents of moving water, air, waves, and ice—all influenced by the pull of gravity.

Interactions between the structural elements of the land and denudation processes are complex. They represent a continuing struggle between Earth's internal and external processes, between the resistance of materials and weathering and erosional processes. The iconic 15-story-tall Delicate Arch in Utah is dramatic evidence of this struggle (Figure 13.1). Differing resistances of the rocks, coupled with variations in the processes at work on them, are carving this delicate sculpture—an example of **differential weathering**, where a more resistant cap rock protects supporting strata below.

Ideally, endogenic processes build *initial landscapes*, whereas exogenic processes develop *sequential landscapes* of low relief, gradual change, and stability. These countering processes must be viewed as going on simultaneously. Several hypotheses have been proposed to model denudation processes and to account for the appearance of the landscape.

Dynamic Equilibrium Approach to Landforms

A landscape is an open system, with highly variable inputs of energy and materials: Uplift creates the *potential energy of position* above sea level and therefore disequilibrium, an imbalance, between relief and energy. The Sun provides radiant energy that converts into *heat energy*. The hydrologic cycle imparts *kinetic energy* through mechanical motion. *Chemical energy* is made available from the atmosphere and various reactions within the crust.

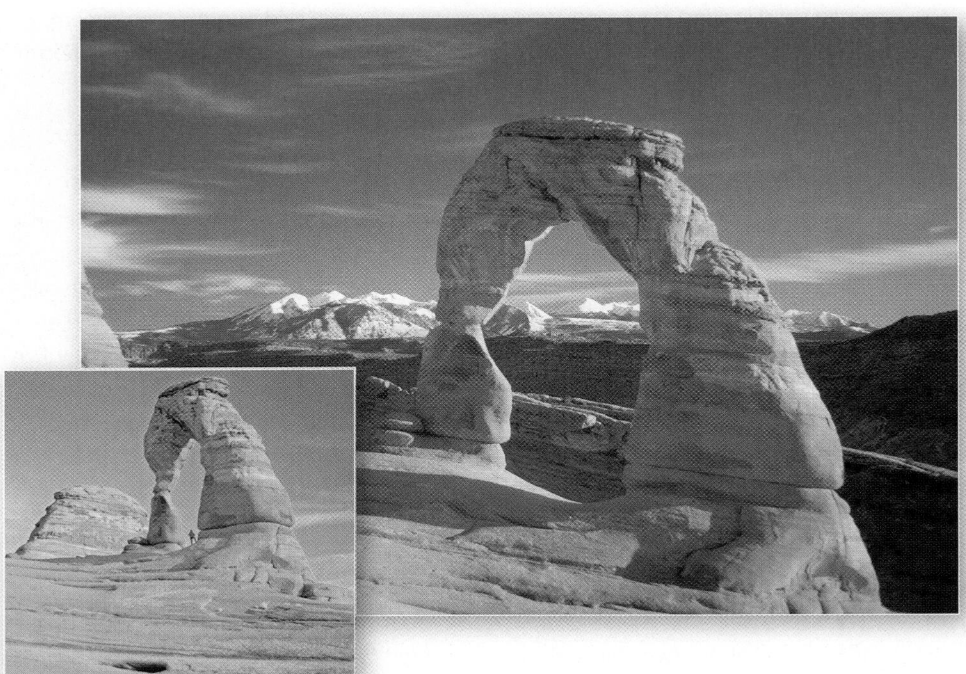

FIGURE 13.1 Delicate Arch, Arches National Park, Utah. Resistant rock strata at the top of the structure helped preserve the arch beneath as surrounding rock eroded away. Note the person standing at the base in the inset photo for a sense of scale. In the distance are the snow-covered LaSal Mountains, examples of laccoliths (intrusive igneous rocks) exposed by erosion. [Author.]

As physical factors fluctuate in an area, the surface constantly responds in search of equilibrium. Every change produces compensating actions and reactions. The balancing act between tectonic uplift and reduction by weathering and erosion, between the resistance of rocks and the ceaseless attack of weathering and erosion, is the **dynamic equilibrium model**. According to current thinking, landscapes in a dynamic equilibrium show ongoing adaptations to the ever-changing conditions of rock structure, climate, local relief, and elevation.

Endogenic events, such as faulting or a lava flow, or exogenic events, such as a heavy rainfall or a forest fire, may provide new sets of relationships for the landscape. Following such destabilizing events, a landform system arrives at a **geomorphic threshold**—the point at which energy overcomes resistance against movement. At this threshold, the system breaks through to a new set of equilibrium relationships as the landform and its slopes enter a period of adjustment and realignment. The pattern over time follows a sequence: (1) equilibrium stability, in which the system fluctuates around some average, (2) a destabilizing event, (3) a period of adjustment, and (4) development of a new and different condition of equilibrium stability. Slow, continuous-change events, such as soil development and erosion, tend to maintain an approximate equilibrium condition. Dramatic events such as a major landslide require longer recovery times before equilibrium is reestablished. The disturbed hillslope in Figure 13.2 is in the midst of compensating adjustment. The failure of saturated slopes caused a landslide into the river and set a disequilibrium condition. As a consequence, the new dam of material threw the stream into disequilibrium between its flow and sediment load. Please refer to Figure 1.6, which graphically illustrates the distinction between steady-state and dynamic equilibrium, as well as the occurrence of a geomorphic threshold.

FIGURE 13.2 A slope in disequilibrium.
Unstable, saturated soils gave way, leaving a debris dam partially blocking the river. The hillslope, river, and forest ecosystem is in disequilibrium as adjustments to new conditions proceed. [Author.]

Slopes

Material loosened by weathering is susceptible to erosion and transportation. However, for material to move downslope, the forces of erosion must overcome other forces: friction, inertia (the resistance to movement), and the cohesion of particles to one another (Figure 13.3a). If the angle is steep enough for gravity to overcome frictional forces or if the impact of raindrops or moving animals or even wind dislodges material, then erosion of particles and transport downslope can occur.

Slopes or *hillslopes* are curved, inclined surfaces that form the boundaries of landforms. Figure 13.3b illustrates basic slope components that vary among slopes with conditions of rock structure and climate. Slopes generally feature an upper *waxing slope* near the top (*waxing* means increasing). This convex surface curves downward and grades into the *free face* below. The presence of a free face indicates an outcrop of resistant rock that forms a steep scarp or cliff.

Downslope from the free face is a *debris slope*, which receives rock fragments and materials from above. The condition of a debris slope reflects the local climate. In humid climates, continually moving water carries material away, lowering the debris slope. But in arid climates, debris slopes accumulate. A debris slope grades into a *waning slope*, a concave surface along the base of the slope that forms a *pediment*, or broad, gently sloping erosional surface.

Slopes are open systems and seek an *angle of equilibrium* among the forces described here. Conflicting forces work simultaneously on slopes to establish an optimum compromise incline that balances these forces. You can identify these slope components and conditions on the actual hillslope shown in Figure 13.3c. When any condition in the balance is altered, all forces on the slope compensate by adjusting to a new dynamic equilibrium.

The relation between rates of weathering and breakup of slope materials, coupled with the rates of mass movement and material erosion, forms the shape of the slope. A slope is *stable* if its strength exceeds these denudation processes and *unstable* if materials are weaker than these processes. Why are hillslopes shaped in certain ways? How do slope elements evolve? How do hillslopes behave during rapid, moderate, or slow uplift? These are topics of active scientific study and research. Now, let us examine specific processes that operate to wear away landforms.

CRITICAL THINKING 13.1

Find a slope; apply the concepts

Locate a slope, possibly near campus, your home, or a local road cut. Using Figure 13.3a, b, and c, can you identify the forces and forms of a hillslope at your site? How would you go about assessing the stability of the slope? Do you see evidence of slope instability near your campus or the region in which you are located—perhaps at a construction site or road cut?

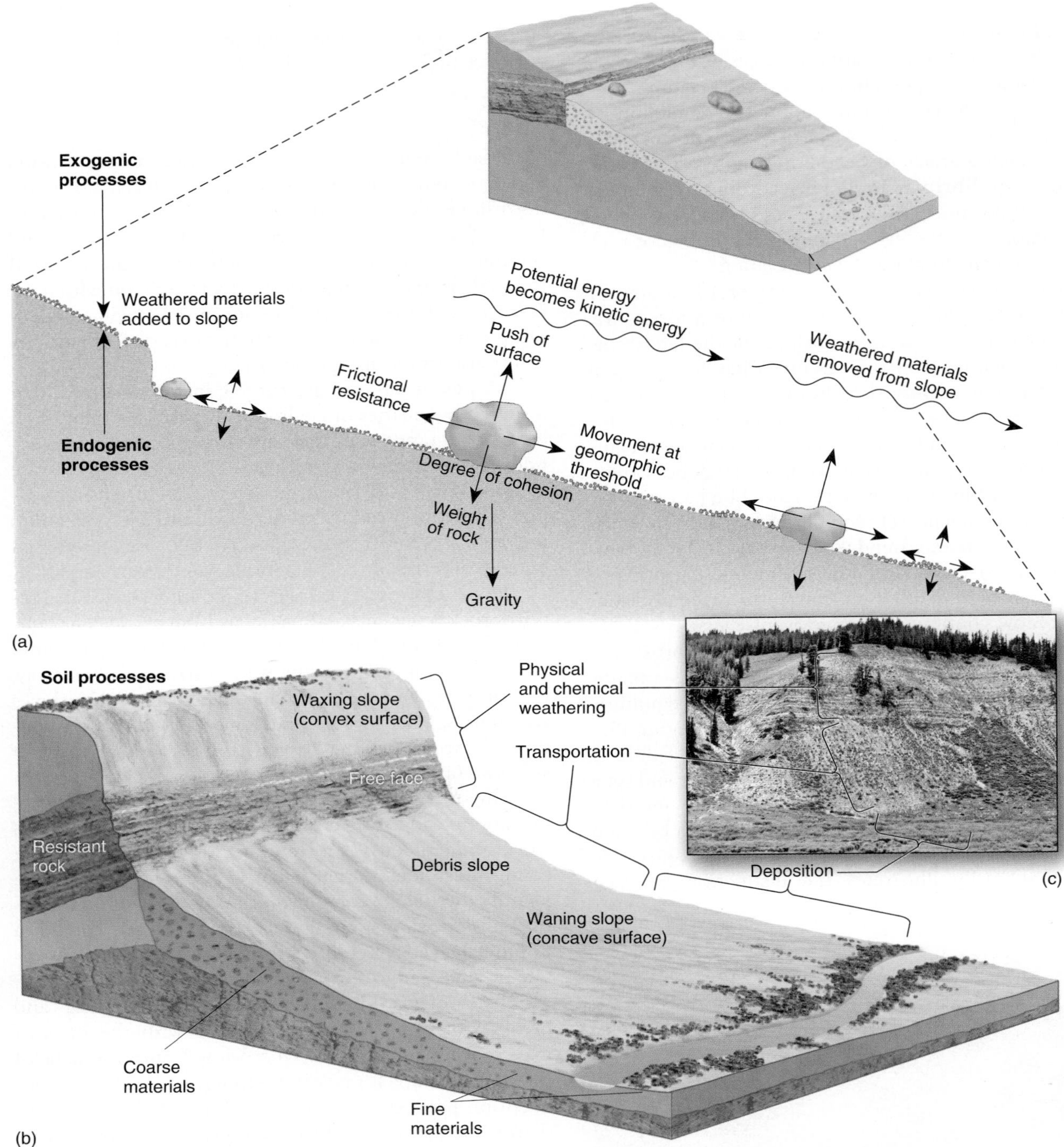

FIGURE 13.3 Slope mechanics and form.
(a) Directional forces (noted by arrows) act on materials along an inclined slope. (b) The principal elements of a slope. (c) A hillslope example; note the role of the rock outcrop as a slope interruption and the rock fragments loosened from the outcrop by frost action. [Author.]

Weathering Processes

Weathering processes break down rock at Earth's surface and to some depth below the surface, either disintegrating rock into mineral particles or dissolving it in water. Weathering processes are both physical (mechanical) and chemical. The interplay of the two is complex, often forming a synergy, a combined action, as the suite of processes work the rock.

Weathering does not transport the materials; it simply generates them for erosion and transport by the agents of water, wind, waves, and ice—all influenced by gravity. In most areas, the upper surface of bedrock undergoes continual weathering, creating broken-up rock, called **regolith**. As regolith continues to weather, or is transported and deposited, loose surface material is formed (Figure 13.4a). In some areas, regolith may be missing or undeveloped, thus exposing an outcrop of unweathered bedrock.

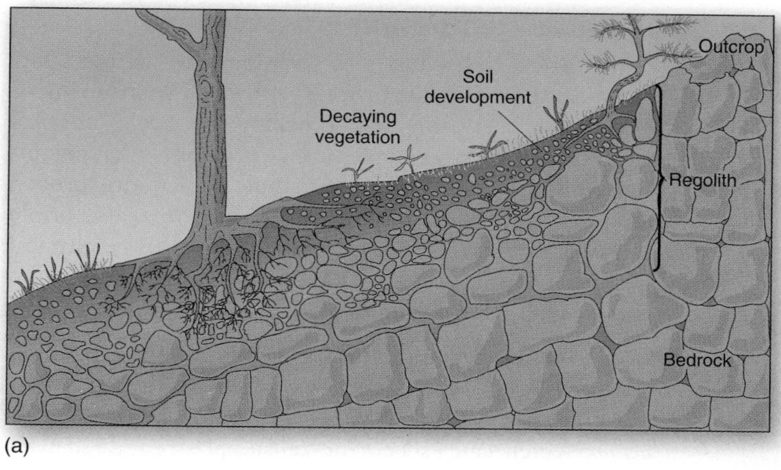

(a)

(b)

(c)

(d)

FIGURE 13.4 Regolith, soil, and parent materials.
(a) A cross section of a typical hillside. (b) A cliff exposes hillside components. (c) These reddish-colored dunes in the Navajo Tribal Park, near the Utah–Arizona border, derive their color from the red-sandstone parent materials in the background. (d) This image, from a larger panoramic scene, was made by the Mars Exploration Rover *Spirit* on March 12, 2004. Looking toward the Columbia Hills, you can see weathered rocks and windblown sand in approximate true colors. [(b) and (c) Author. (d) Mars image courtesy of NASA/JPL and Cornell University.]

Bedrock is the *parent rock* from which weathered regolith and soils develop. Wherever a soil is relatively youthful, its parent rock is traceable through similarities in composition. For example, the sand in Figure 13.4c derives its color and character from the parent rock in the background, just as the sediments on Mars derive their characteristics from weathered parent material (Figure 13.4d). Such sandy, unconsolidated fragmental material, known as **sediment**, combines with weathered rock to form the **parent material** from which soil evolves.

Factors Influencing Weathering Processes

A number of factors influence weathering processes. Important controls on weathering rates include

- **Rock composition and structure (jointing)**—Weathering is greatly influenced by the character of the bedrock: hard or soft, soluble or insoluble, broken or unbroken. *Jointing* in rock is important for weathering processes. **Joints** are fractures or separations in rock that occur without displacement of the sides (as would be the case in faulting). The presence of these usually flat spaces increases the surface area of rock exposed to both physical and chemical weathering.
- **Climate**—Precipitation, temperature, and freeze-thaw cycles are the most important factors affecting weathering.
- **Subsurface water**—Influences in weathering processes include both the position of the water table and water movement within soil and rock structures.

- **Slope orientation**—The *geographic orientation* of a slope—whether it faces north, south, east, or west—controls the slope's exposure to Sun, wind, and precipitation. Slopes facing away from the Sun's rays tend to be cooler, moister, and more vegetated than are slopes in direct sunlight. This effect of orientation is especially noticeable in the middle and higher latitudes.
- **Vegetation**—Although vegetative cover can protect rock by shielding it from raindrop impact and providing roots to stabilize soil, it also produces organic acids from the partial decay of organic matter; these acids contribute to chemical weathering. Plant roots can enter crevices and break up a rock, exerting enough pressure to force rock segments apart, thereby exposing greater surface area to other weathering processes (Figure 13.5a). You may have observed how tree roots can heave the sections of a sidewalk or driveway sufficiently to raise and crack concrete.

The scale at which we analyze weathering processes is important. Research at *microscale* levels reveals a more complex relationship between climate and weathering than previously thought. At the small scale of actual reaction sites on the rock surface, both physical and chemical weathering processes can occur across varied climate types. Soil moisture (hygroscopic water and capillary water) activates chemical weathering processes, even in the driest landscape. (Review sections in Chapter 9, Figure 9.8, for types of soil moisture.)

Imagine all the factors that influence weathering rates as operating in concert, although we separate these processes here for convenience of study. In the complexity of nature, physical and chemical weathering processes usually operate together. Of course, in all of this, *time* is the crucial factor, for these processes require long periods of time to operate.

Physical Weathering Processes

When rock is broken and disintegrated without any chemical alteration, the process is **physical weathering** or *mechanical weathering*. By breaking up rock, physical weathering produces more surface area on which chemical weathering may operate. A single rock that is broken into eight pieces has doubled its surface area susceptible to weathering processes. We look briefly at three physical weathering processes: frost action, crystallization, and pressure-release jointing.

Frost Action When water freezes, its volume expands as much as 9% (see Chapter 7). Such expansion creates a powerful mechanical force, **frost action**, or *freeze–thaw action*, which can exceed the tensional strength of rock (Figure 13.5b). Repeated freezing (expanding) and thawing (contracting) of water breaks rocks apart. Freezing actions are important in the humid microthermal climates (*humid continental* and *subarctic*) and the polar climates, and they occur at higher elevations in mountains worldwide in the highland climates. In arctic and subarctic climates, frost action dominates soil conditions (discussed in further detail in Chapter 17). Frost action is one of the processes breaking up rock strata pictured on the cover of this text.

The work of ice begins in small openings, gradually expanding until rocks are cleaved, or split. Figure 13.6 shows blocks of rock on which this *joint-block separation* occurs along existing joints and fractures. This weathering action is *frost-wedging*, and it pushes portions of the rock apart. Cracking and breaking create varied shapes in the rocks, depending on the rock structure. In Figure 13.6a, the softer supporting rock underneath the slabs is weathering at a faster pace in *differential weathering*.

Frost action was important to various cultures as a force to quarry rock. Pioneers in the early American West drilled holes in rock, poured water in the holes, and then plugged them. During the cold winter months, expanding ice broke off large blocks along lines determined by the

(a)

(b)

FIGURE 13.5 Physical weathering examples.
(a) Tree roots exert a force on the sides of joints in rock. (b) Frost action shattered this marble, a metamorphic rock; ice expansion forced the rock segments apart. [Bobbé Christopherson.]

(a)

(b)

FIGURE 13.6 Physical weathering as joint-block.
(a) Physical weathering along joints in rock produces discrete blocks in the back country of Canyonlands National Park, Utah. (b) Joint-block separation in slate at Alkehornet, Isfjord, on Spitsbergen Island in the Arctic Ocean, where freezing is intense. [(a) Author. (b) Bobbé Christopherson.]

(a)

(b)

FIGURE 13.7 Rockfall.
(a) Shattered rock debris from a large rockfall in Yosemite National Park. Freshly exposed, light-colored rock shows where the rockfall originated. (b) A large slab failure produces a rockfall in Zion National Park. The small opening in the cliff base is a window blasted through to the highway tunnel that was excavated inside the cliff. [Author.]

drill-hole patterns. In spring, they hauled the blocks to town for construction material. Frost action also produces damaged road pavement and burst water pipes.

Spring can be a risky time to venture into mountainous terrain. As rising temperatures melt the winter's ice, newly fractured rock pieces fall without warning and may even start rock slides. Many incidences are reported in the European Alps, and they appear to be on the increase. The falling rock pieces may physically shatter on impact—another form of physical weathering (Figure 13.7).

Salt-Crystal Growth (Salt Weathering) Especially in arid climates, dry weather draws moisture to the surface of rocks. As the water evaporates, dissolved minerals in the

GEO REPORT 13.1 *Rockfalls in Yosemite*

Records indicate that at least 600 rockfalls have occurred in Yosemite Valley, located in the Sierra Nevada of California, in the past 150 years. In July 1996, a 162,000-ton granite slab shocked Yosemite Valley, a crashing drop of 670 m at 260 kmph (2200 ft at 160 mph), pulverizing into a light dust that covered 50 acres and felled more than 500 trees. In 1999, another rockfall occurred that took a life. Large rockfalls from Half Dome (Figure 13.9b and c) in 2006, Glacier Point in 2008, and Ahwiyah Point in 2009 are of recent notice. A National Park Service (NPS) map of rockfalls from 1857 to 2009 is at http://www.nps.gov/yose/naturescience/rockfall.htm. Note from the NPS map that winter and spring events dominate the record, although large rockfalls can occur during the warmer months as well.

(a)

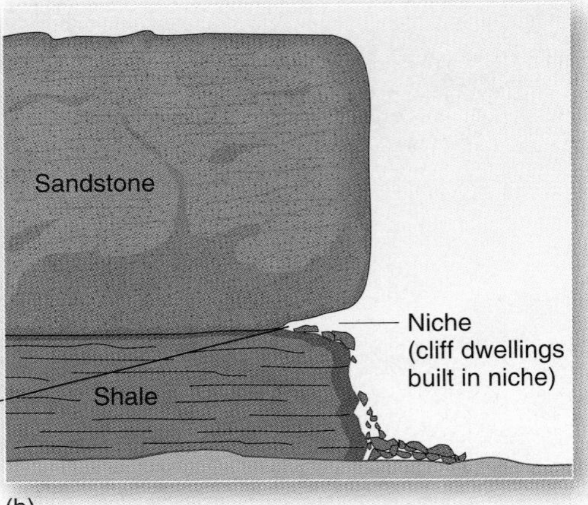

(b)

FIGURE 13.8 Physical weathering in sandstone. (a) Cliff dwelling site in Canyon de Chelly, Arizona, occupied by the Anasazi people until about 900 years ago (full view of site in chapter-opening photo). The niche in the rock formed partially by crystallization, which forced apart mineral grains and broke up the rock. The dark streaks on the rock are thin coatings of desert varnish, composed of iron oxides with traces of manganese and silica. (b) Water and an impervious rock layer helped concentrate weathering processes in a niche in the overlying sandstone. [Bobbé Christopherson.]

water grow crystals—a process called crystallization. Over time, as the crystals grow and enlarge, they exert a force great enough to spread apart individual mineral grains and begin breaking up the rock. Such *salt-crystal growth*, or *crystallization*, is a form of physical weathering.

In the Colorado Plateau of the Southwest, salty water slowly flows from rock strata. As this salty water evaporates, crystallization loosens the sand grains. Subsequent erosion and transportation by water and wind complete the sculpturing process. Deep indentations develop in sandstone cliffs, especially where the sandstone lies above an impervious layer such as shale. More than 1000 years ago, Native Americans built entire villages in these weathered niches at several locations, including Mesa Verde in Colorado and Arizona's Canyon de Chelly (pronounced "canyon duh shay"; Figure 13.8 and chapter-opening photo).

Pressure-Release Jointing Recall from Chapter 11 how rising magma that is deeply buried and subjected to high pressure forms intrusive igneous rocks as plutons. These plutons cool slowly and produce coarse-grained, crystalline, granitic rocks. As the landscape is subjected to uplift, the regolith overburden is weathered, eroded, and transported away, eventually exposing the pluton as a mountainous batholith (a plutonic batholith, illustrated in Figure 11.8).

As the tremendous weight of overburden is removed from the granite, the pressure of deep burial is relieved. Over millions of years, the granite slowly responds with an enormous physical heave. In a process known as *pressure-release jointing*, layer after layer of rock peels off in curved slabs or plates, thinner at the top of the rock structure and thicker at the sides. As these slabs weather, they slip off in the process of **sheeting**. This *exfoliation process* creates arch-shaped and dome-shaped features on the exposed landscape, sometimes forming an **exfoliation dome** (Figure 13.9). Such domes are probably the largest weathering features on Earth (in areal extent).

Chemical Weathering Processes

Chemical weathering refers to the chemical breakdown of the constituent minerals in rock, always in the presence of water. The chemical decomposition and decay become more intense as both temperature and precipitation increase. Although individual minerals vary in susceptibility, all rock-forming minerals are responsive to some degree of chemical weathering.

A familiar example of chemical weathering is the eating away of cathedral façades and the etching of tombstones by acid precipitation. In Europe, where increasingly acidic rains resulted from the burning of coal, chemical weathering processes are visible on many buildings. As an example, Saint Magnus Cathedral, in Kirkwell, Orkney Islands, north of Scotland, was built of red and yellow sandstone. Almost nine centuries of chemical weathering

FIGURE 13.9 Exfoliation in granite.
Exfoliation processes loosen slabs of granite, freeing them for further weathering and downslope movement. (a) Great arches form in the White Mountains of New Hampshire. (b) Exfoliated layers of rock are visible in characteristic dome formations in granites. The loosened slabs of rock are susceptible to further weathering and downslope movement. This view is from the east side of Half Dome in Yosemite National Park, California. (c) Half Dome perspective from the west; relief is approximately 1500 m (5000 ft) from the top of the dome to the glaciated valley below. (d) Rock sheeting exposed along Beverly Sund, Nordaustlandet Island in the Arctic Ocean. [(a) and (d) Bobbé Christopherson. (b) and (c) Author.]

dissolved cementing materials in the sandstone, breaking down the rock and making many of the sculptured design elements unrecognizable, as if they "melted" or became "out-of-focus" (Figure 13.10).

As an example of the way chemical weathering attacks rock, consider **spheroidal weathering**. The sharp edges and corners of rocks are rounded as the alteration of minerals progresses through the rock. Joints in the rock offer more surfaces of opportunity for weathering. Water penetrates joints and fractures and dissolves the rock's weaker minerals or cementing materials. A boulder can be attacked from all sides, shedding spherical shells of decayed rock, like the layers of an onion. The resulting rounded edges are the basis for the name *spheroidal*. Spheroidal weathering of rock resembles exfoliation, but it does not result from pressure-release jointing. This type of weathering is visible in Figure 13.11; note the spheroidal rock form.

Hydration and Hydrolysis Although these two processes are different, we group them together because they both decompose rock using water—one by a simple combination of water with a mineral and the other by a chemical reaction of water with a mineral.

Hydration, meaning "combination with water," involves little chemical change. Water becomes part of the chemical composition of the mineral (such a hydrate is gypsum, which is hydrous calcium sulfate: $CaSO_4 \cdot 2H_2O$).

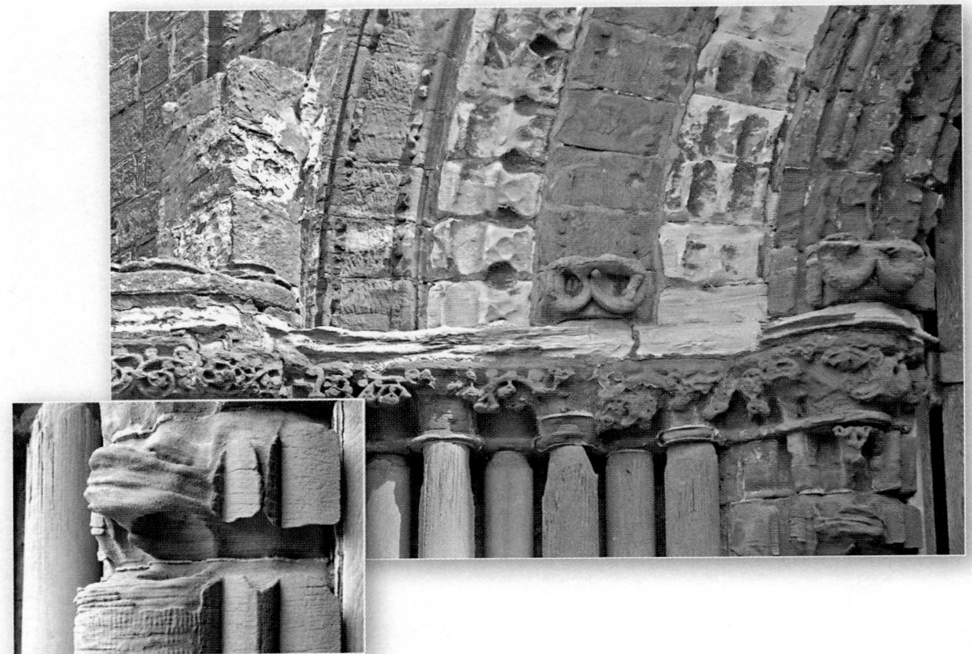

FIGURE 13.10 Chemical weathering attacks sandstone cathedral.
(a) The west-front entrance of Saint Magnus Cathedral, in Kirkwell, Scotland, shows the signs of chemical weathering as processes attack the cementing materials in the rock. Construction of this edifice began in A.D. 1137. [Bobbé Christopherson.]

When some minerals hydrate, they expand, creating a strong mechanical effect, a wedging pressure, that stresses the rock, forcing grains apart as in physical weathering.

A cycle of hydration and dehydration can lead to granular disintegration and further susceptibility of the rock to chemical weathering. Hydration works together with carbonation and oxidation to convert feldspar, a common mineral in many rocks, to clay minerals and silica. The hydration process is also at work on the sandstone niches shown in the cliff-dwelling photo; the water and minerals are consumed in the process, becoming a new compound.

(a)

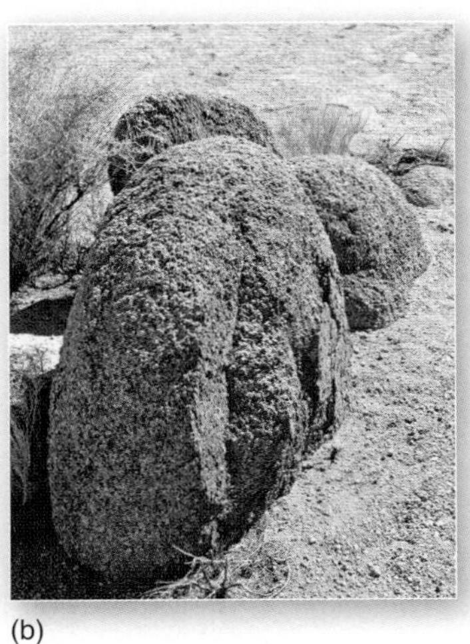

(b)

FIGURE 13.11 Chemical weathering and spheroidal weathering.
(a) Chemical weathering processes act on the joints in granite to dissolve weaker minerals, leading to a rounding of the edges of the cracks in the Alabama Hills. Mount Whitney is visible on the crest of the Sierra Nevada in the background. (b) Rounded granite outcrop demonstrates spheroidal weathering and the disintegration of rock. The surface is actually crumbly. [Bobbé Christopherson.]

GEO REPORT 13.2 *Weathering on bridges in Central Park, NYC*

In New York City's Central Park, 36 bridges are built out of a variety of rock from sources across the U.S. Northeast and Canada. For more than 135 years, physical and chemical processes have weathered these bridges. As air pollution from the burning of fossil fuels has increased, acidity in rain and snow has hastened weathering rates, a problem compounded by the use of salt on the roads in winter. Decorative design elements on the bridges are now disappearing as weathering tears at the surface rock. Some heavily weathered sandstone blocks have had to be replaced by cast concrete.

When minerals chemically react with water, the process is **hydrolysis**. Hydrolysis is a decomposition process that breaks down silicate minerals in rocks. Compared with hydration, in which water combines with minerals in the rock, the hydrolysis process involves water and elements in chemical reactions to produce different compounds.

For example, the weathering of feldspar minerals in granite can be caused by a reaction to the normal mild acids dissolved in precipitation:

feldspar (K, Al, Si, O) + carbonic acid and water →
residual clays + dissolved minerals + silica

The by-products of chemical weathering of feldspar in granite include clay (such as kaolinite) and silica. As clay forms from some minerals in the granite, quartz (SiO_2) particles are left behind. The resistant quartz may wash downstream, eventually becoming sand on some distant beach. Clay minerals become a major component in soil and in shale, a common sedimentary rock.

When weaker minerals in rock are changed by hydrolysis, the interlocking crystal network breaks down, so the rock fails and *granular disintegration* takes place. Such disintegration in granite may make the rock appear etched, corroded, and softened, even crumbly (Figure 13.11b).

In Table 11.2, the second line shows resistance to chemical weathering in igneous rocks. On the ultramafic side of the table (right side), the low-silica minerals olivine and peridotite are most susceptible to chemical weathering. Stability gradually increases toward the high-silica minerals such as feldspar. On the far left side of the table, quartz is resistant to chemical weathering. You can see the nature of the constituent minerals, where basalt weathers faster chemically than does granite.

Oxidation Another example of chemical weathering occurs when certain metallic elements combine with oxygen to form oxides. This is a chemical weathering process known as **oxidation**. Perhaps the most familiar oxidation form is the "rusting" of iron in rocks or soil that produces a reddish-brown stain of iron oxide (Fe_2O_3). We have all left a tool or nails outside only to find them weeks later, coated with iron oxide. The rusty color is visible on the surfaces of rock and in heavily oxidized soils such as those in the southeastern United States, southwestern U.S. deserts, and the tropics (Figure 13.12). Here is a simple oxidation reaction in iron:

iron (Fe) + oxygen (O_2) → iron oxide (hematite; Fe_2O_3)

As iron is removed from the minerals in a rock, the disruption of the crystal structures makes the rock more susceptible to further chemical weathering and disintegration.

Dissolution of Carbonates Chemical weathering also occurs when a mineral dissolves into *solution*—for example, when sodium chloride (common table salt) dissolves in water. Remember, water is the universal solvent because it is capable of dissolving at least 57 of the natural elements and many of their compounds.

Water vapor readily dissolves carbon dioxide, thereby yielding precipitation containing carbonic acid (H_2CO_3). This acid is strong enough to dissolve many minerals, especially limestone. This process is **carbonation**, a reaction whereby carbon combines with minerals, dissolving them.

Such chemical weathering transforms minerals that contain calcium, magnesium, potassium, and sodium. When rainwater attacks formations of limestone (which is calcium carbonate, $CaCO_3$), the constituent minerals dissolve and wash away with the mildly acidic rainwater:

calcium carbonate + carbonic acid and water →
calcium bicarbonate ($Ca_2{}^{2+}CO_2H_2O$)

Walk through an old cemetery and you can observe the dissolution of marble, a metamorphic form of limestone (Figure 13.13). Weathered limestone and marble,

(a) Puerto Rico Oxisols

(b) Georgia Ultisols

FIGURE 13.12 Oxidation processes in rock and soil.
(a) Oxidation of iron minerals in tropical oxisols produces these colors in the tropical rain forest soils of eastern Puerto Rico. (b) Ultisols—soils produced by warm, moist conditions—in Sumter County, Georgia, bear the color of iron and aluminum oxides. The crop is peanuts. [Bobbé Christopherson.]

FIGURE 13.13 Dissolution of limestone. A marble tombstone is chemically weathered beyond recognition in a Scottish churchyard. Marble is a metamorphic form of limestone. Based upon readable dates on surrounding tombstones, this one is about 228 years old. [Bobbé Christopherson.]

in tombstones or in rock formations, appear pitted and worn wherever adequate water is available for dissolution. In this era of human-induced increases of acid precipitation, carbonation processes are greatly enhanced (see Focus Study 3.2, "Acid Deposition: Damaging to Ecosystems").

Chemical weathering involving dissolution of carbonates dominates entire landscapes composed of limestone. These are the regions of karst topography, which we examine next.

Karst Topography and Landscapes

Limestone is abundant on Earth and composes many landscapes (Figure 13.14). These areas are susceptible to chemical weathering, which creates a specific landscape of pitted, bumpy surface topography, poor surface drainage, and well-developed solution channels (dissolved openings and conduits) underground. Weathering and erosion caused by groundwater may also result in remarkable mazes of underworld caverns.

These are the hallmarks of **karst topography**, named for the Krš Plateau in Slovenia (formerly Yugoslavia), where karst processes were first studied. Approximately 15% of Earth's land area has some karst features, with outstanding examples found in southern China, Japan, Puerto Rico, Cuba, the Yucatán of Mexico, Kentucky, Indiana, New Mexico, and Florida. As an example, approximately 38% of Kentucky has sinkholes and related karst features noted on topographic maps.

Formation of Karst

For a limestone landscape to develop into karst topography, several conditions are necessary:

- The limestone formation must contain 80% or more calcium carbonate for dissolution processes to proceed effectively.
- Complex patterns of joints in the otherwise impermeable limestone are needed for water to form routes to subsurface drainage channels.
- An aerated (containing air) zone must exist between the ground surface and the water table.
- Vegetation cover is required to supply varying amounts of organic acids that enhance the dissolution process.

The role of climate in providing optimum conditions for karst processes remains under debate, although the amount and distribution of rainfall appear important. Karst occurs in arid regions, but it is primarily due to former climatic conditions of greater humidity. Karst is rare in the Arctic and Antarctic regions because the water, although present, is generally frozen.

As with all weathering processes, time is a factor. Early in the 20th century, karst landscapes were thought to progress through evolutionary stages of development, as if they were aging. Evidence has not supported this theory, and today these landscapes are thought to be locally unique, a result of specific conditions. Nonetheless, mature karst landscapes do display certain characteristic forms.

Lands Covered with Sinkholes

The weathering of limestone landscapes creates many **sinkholes**, which form in circular depressions. (Traditional studies may call a sinkhole a *doline*.) A *collapse sinkhole* forms if such a solution sinkhole collapses through the roof of an underground cavern, sometimes dramatically, such as in Guatemala City in June 2010 (Figure 13.15). A 91-m (300-ft) deep, 18-m wide sinkhole suddenly dropped, taking a three-story building and parts of two streets with it. Probably years in the making, the collapse was likely caused by heavy rains from a tropical storm, which finally brought the system to a geomorphic threshold. Investigators are trying to determine the cause of this sinkhole and whether limestone or subsurface drainage issues were involved.

A gently rolling limestone plain might be pockmarked by slow subsidence of surface materials in *solution sinkholes* with depths of 2–100 m (7–330 ft) and diameters of 10–1000 m (33–3300 ft), as shown in Figure 13.16a. Through continuing solution and collapse, sinkholes may coalesce to form a *karst valley*—an elongated depression up to several kilometers long.

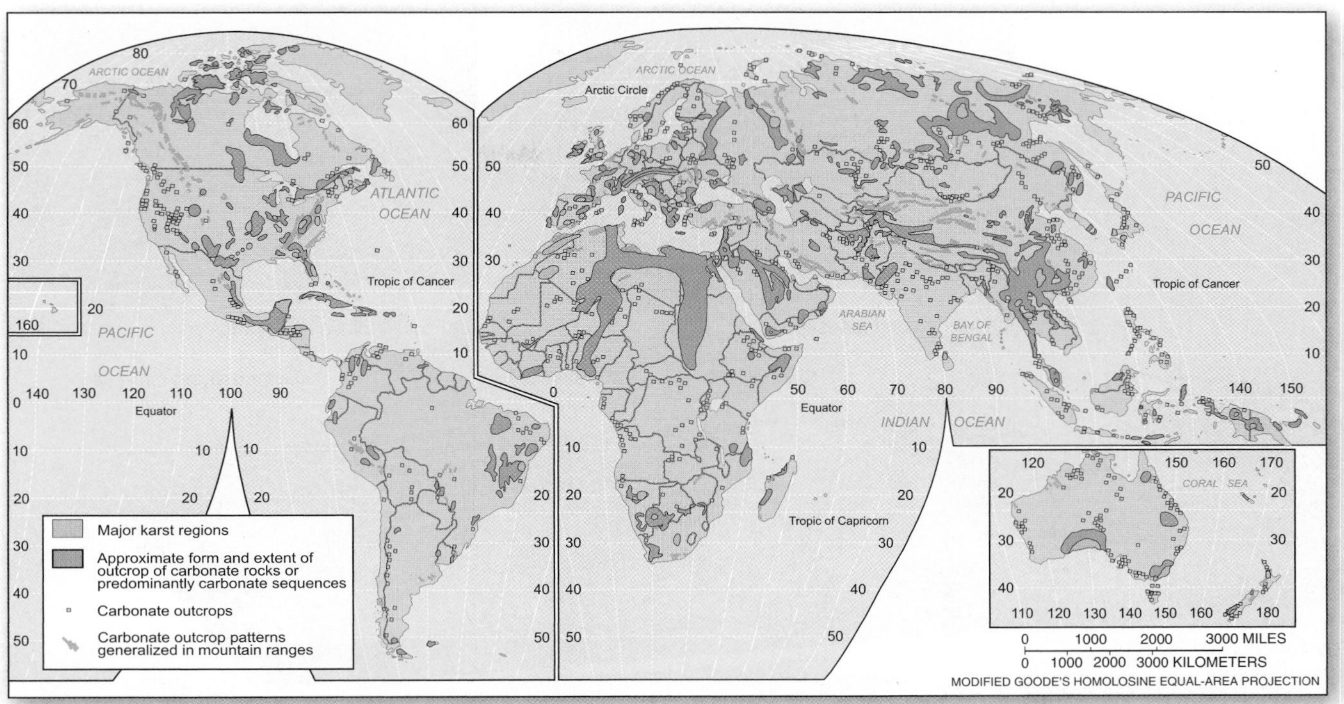

FIGURE 13.14 Karst landscapes and limestone regions.
Major karst regions exist on every continent. The outcrops of carbonate rocks or predominantly carbonate sequences are limestone and dolomite (calcium, magnesium carbonate), but may contain other carbonate rocks. [Map adapted by Pam Schaus, after R. E. Snead, *Atlas of the World Physical Features,* p. 76. © 1972 by John Wiley & Sons; and D. C. Ford and P. Williams, *Karst Geomorphology and Hydrology,* p. 601. © 1989 by Kluwer Academic Publishers. Adapted by permission.]

The area southwest of Orleans, Indiana, has an average of 1022 sinkholes per 2.6 km^2 (1 mi^2). In this area, the Lost River, a "disappearing stream," diverts from the surface and flows more than 13 km (8 mi) through underground solution channels before it resurfaces at the Lost River rise near the Orangeville rise shown in Figure 13.16e. Its dry bed can be seen on the lower left of the topographic map in Figure 13.16b.

In Florida, sinkholes have made news because water tables lowered by pumping from municipal wells have caused their collapse into underground solution caves, taking with them homes, businesses, and even new cars from an auto dealership. One such sinkhole collapsed in a suburban area in 1981, and others appeared in 1993 and 1998 (Figure 13.17). These examples remind us of the effects of disruptions to groundwater, especially in complex karst landscapes.

In the wet tropics, karst topography forms in deeply jointed, thick limestone beds. Weathering leaves isolated resistant limestone blocks standing. These resistant cones

FIGURE 13.15 Sinkhole in Guatemala City.
This huge sinkhole collapsed in northern Guatemala City in June 2010. Rains from Tropical Storm Agatha are thought to be what triggered the collapse of the sinkhole, which was probably forming for decades. A similar sinkhole dropped nearby in 2007. [AP Photo/Moises Castillo.]

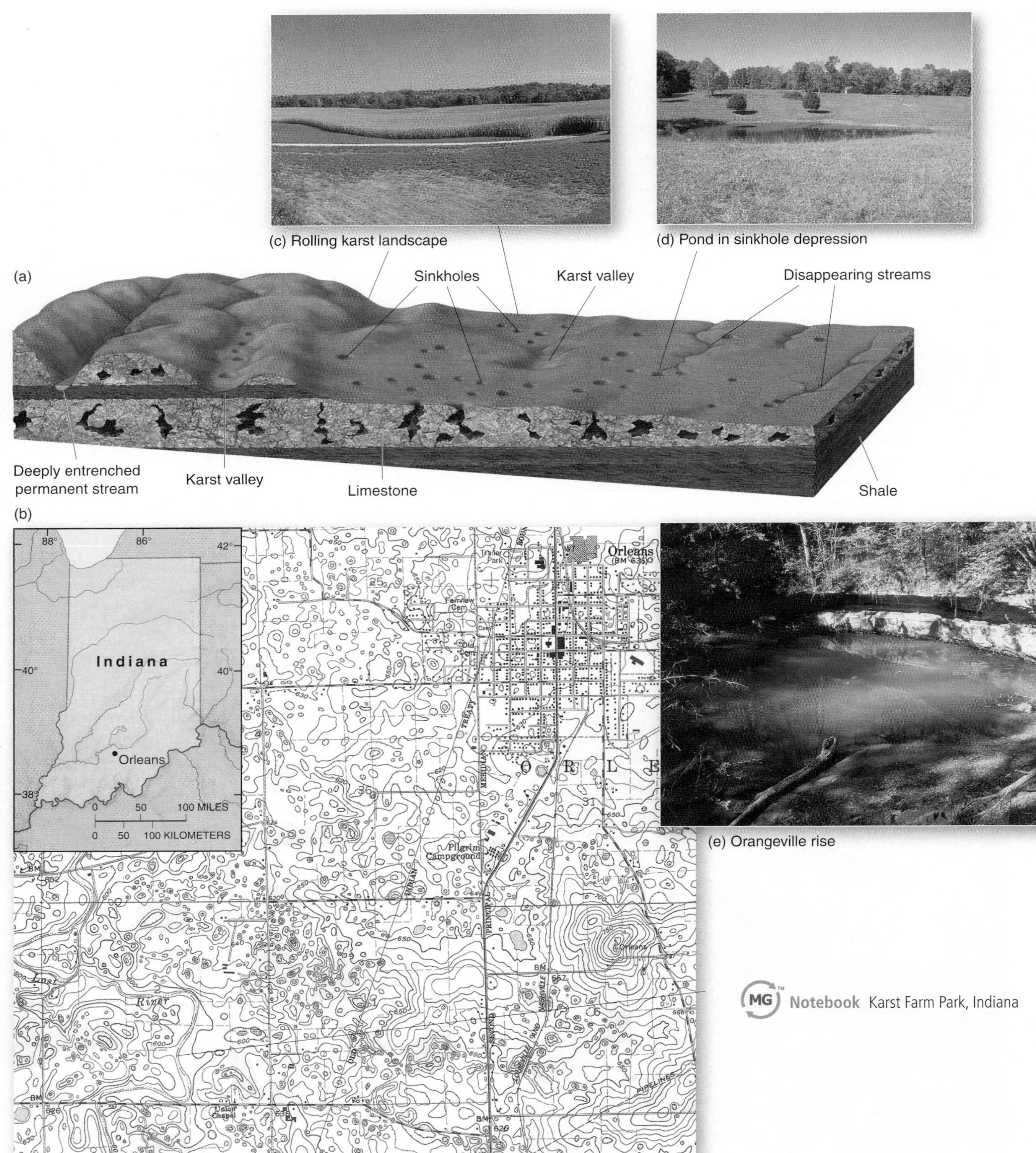

MG Notebook Karst Farm Park, Indiana

FIGURE 13.16 Features of karst topography in Indiana.
(a) Idealized features of karst topography in southern Indiana. (b) Karst topography southwest of the town of Orleans, Indiana, on a topographic map; note the contour lines and the depressions, which are indicated with small hachures (tick marks) on the downslope side of the contour lines. (c) Gently rolling karst landscape and cornfields near Orleans, Indiana. (d) This pond is in a sinkhole depression near Palmyra, Indiana. (e) The Orangeville rise, near Orangeville, Indiana, is just north of the Lost River rise. During periods of high rainfall, this rise is almost filled with water. [(b) Mitchell, Indiana quadrangle, USGS. (c), (d), and (e) Bobbé Christopherson.]

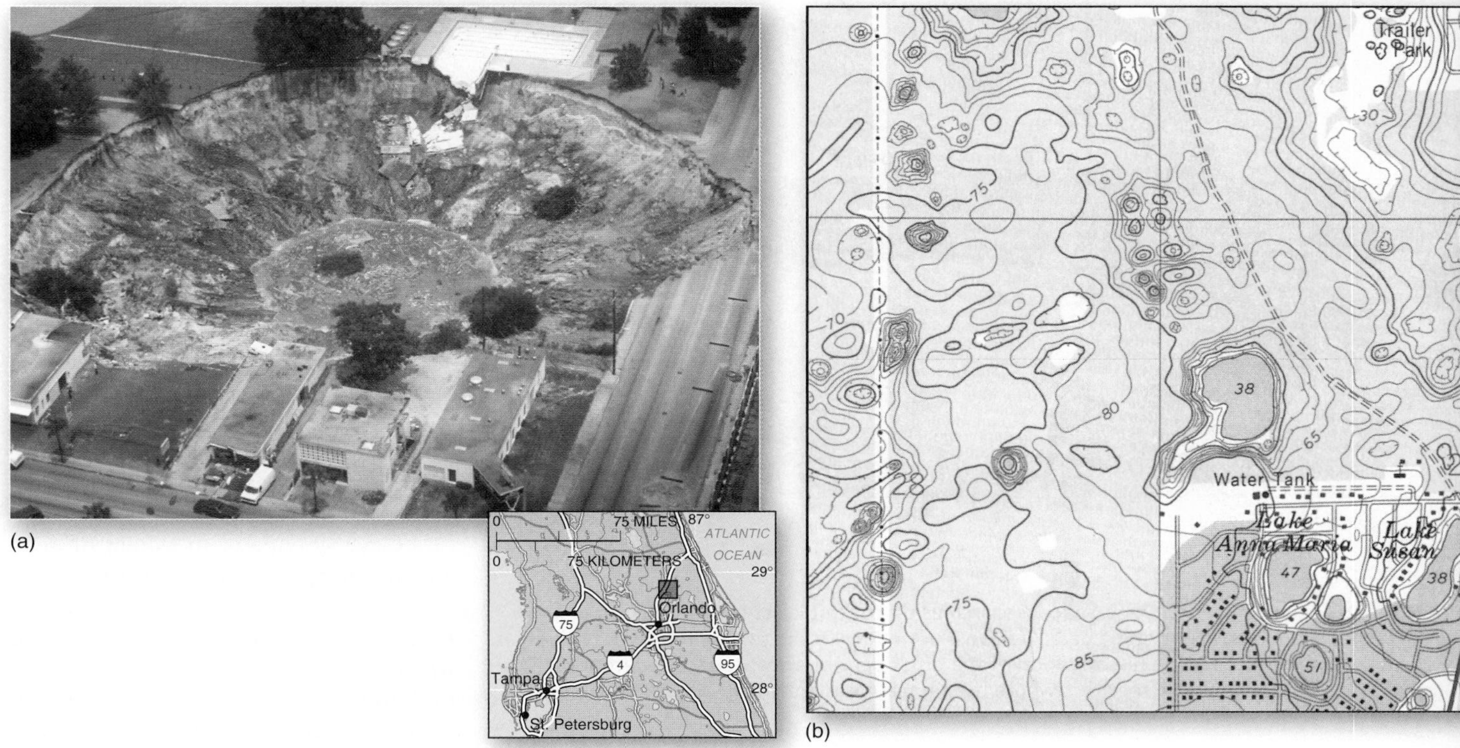

FIGURE 13.17 The Winter Park sinkholes.
(a) Florida sinkhole in Winter Park, a suburb of Orlando. (b) Karst area 25 km (15.5 mi) north of Winter Park depicted on a topographic map. Note that the depressions are marked by small hachures (tick marks). [(a) Jim Tuten/Black Star. (b) Orange City quadrangle, USGS.]

and towers are most remarkable in several areas of China, where *tower karst* up to 200 m (660 ft) high interrupts an otherwise lower-level plain. Across Puerto Rico north of the Cordillera Central, knobs of limestone mark a karst landscape (Figure 13.18).

A complex landscape in which sinkholes intersect is a *cockpit karst*. The sinkholes can be symmetrically shaped in certain circumstances; one at Arecibo, Puerto Rico, is shaped perfectly for a 305-m (1000-ft) radio-telescope dish installation (Figure 13.19). The telescope works

FIGURE 13.18 Karst landscape near Manati, Puerto Rico.
Note the numerous limestone hills and knobs. [Bobbé Christopherson.]

FIGURE 13.19 Deep-space research using cockpit karst topography. Cockpit karst topography near Arecibo, Puerto Rico, is the setting for Earth's largest radio telescope. The discoloration of the dish does not affect telescope reception. The suspended movable receivers where the signals focus are 168 m (550 ft) above the dish. [Bobbé Christopherson; inset Cornell University.]

both for radio and radar astronomy studying the Universe and for studies of the upper atmosphere of Earth. The Arecibo Observatory is part of the National Astronomy and Ionosphere Center, operated by Cornell University and the National Science Foundation (http://www.naic.edu/).

Caves and Caverns

Caves form in limestone because it is so easily dissolved by carbonation. The largest limestone caverns in the United States are Mammoth Cave in Kentucky (also the longest surveyed cave in the world at 560 km, or 350 mi), Carlsbad Caverns in New Mexico, and Lehman Cave in Nevada.

Carlsbad Caverns are in 200-million-year-old limestone formations deposited when shallow seas covered the area. Regional uplifts associated with the building of the Rockies (the Laramide orogeny, 40–80 million years ago) elevated the region above sea level, subsequently leading to active cave formation.

Caves generally form just beneath the water table, where later lowering of the water level exposes them to further development. *Dripstones* form as water containing dissolved minerals slowly drips from the cave ceiling. Calcium carbonate precipitates out of the evaporating solution, literally one molecular layer at a time, and accumulates at a point below on the cave floor. Forming depositional features, *stalactites* grow from the ceiling and *stalagmites* build from the floor; sometimes the two grow until they connect and form a continuous *column* (Figure 13.20b). A dramatic subterranean world is thus created.

Cave science is an aspect of geomorphology where amateur cavers make important discoveries about these unique habitats and new life forms—*biospeleology*. More than 90% of known caves are not biologically surveyed, and some 90% of possible caves worldwide still lie undiscovered, making this a major research frontier. For more on caves and related formations, see http://www.goodearthgraphics.com/virtcave/virtcave.html.

Cave habitats are unique. They are nearly closed, self-contained ecosystems with simple food chains and great stability. In total darkness, bacteria synthesize inorganic elements and produce organic compounds that sustain many types of cave life, including algae, small invertebrates, amphibians, and fish—a chemosynthesis-based food web.

In a cave discovered in 1986, near Movile in southeastern Romania, cave-adapted invertebrates were discovered after millions of years of sunless isolation. Thirty-one of these organisms were previously unknown. Without sunlight, the ecosystem in Movile is sustained on sulfur-metabolizing bacteria that synthesize organic matter using energy from oxidation processes. These chemosynthetic bacteria feed other bacteria and fungi that, in turn, support cave animals. The sulfur bacteria produce sulfuric acid compounds that may prove to be important in the chemical weathering of some caves.

GEO REPORT 13.3 ***Human-made sinkholes***

Along Bushkill Creek, downstream from the Hercules mining quarry enterprise, several dozen sinkholes have collapsed since 2000, taking down one of the bridges from State Route 33 and threatening neighborhoods in Stockertown, Northampton County, Pennsylvania. Sinkhole development correlated with groundwater pumping at the nearby quarry, which lowered water tables. A new bridge was completed in 2006 for the main highway. A group of citizens pressed the case for mitigating action, blaming quarry operators; the outcome is pending.

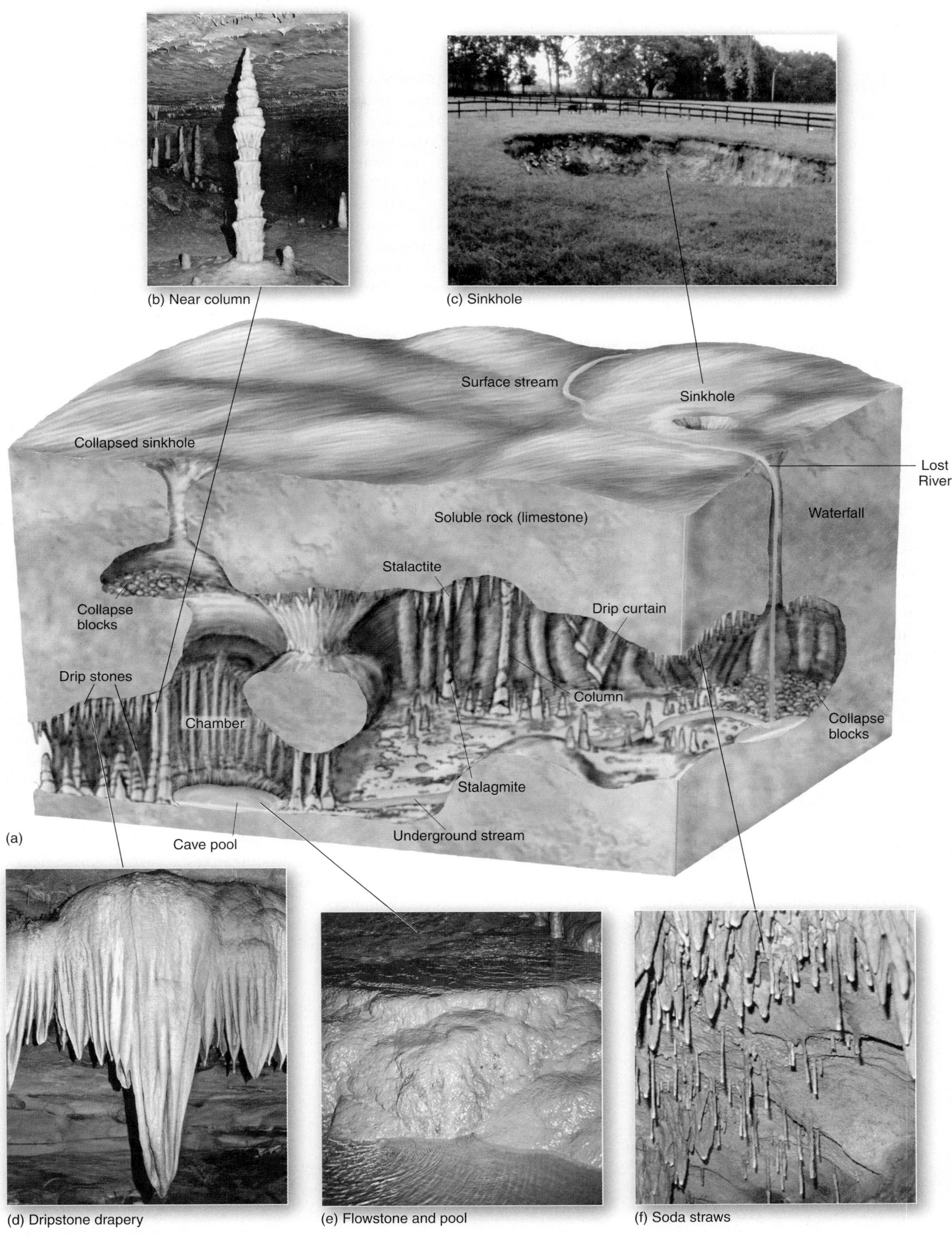

FIGURE 13.20 Cavern features.
(a) An underground cavern and related forms in limestone. (b) A column where stalactites from the ceiling and stalagmites from the floor connect; here nearly attached. (c) A sinkhole in a Florida pasture dropping through to a limestone void. (d), (e), and (f) Cave depositional features forming one molecular layer at a time. All cave photos from Marengo Caves, Marengo, Indiana (see Geo Report 13.4, p. 386). [(c) Thomas M. Scott, Florida Geological Survey. (b), (d), (e), and (f) Bobbé Christopherson.]

Mass-Movement Processes

Nevado del Ruiz, northernmost of two dozen dormant (not extinct, sometimes active) volcanic peaks in the Cordillera Central of Colombia, had erupted six times during the past 3000 years, killing 1000 people during its last eruption in 1845. On November 13, 1985, at 11 P.M., after a year of earthquakes and harmonic tremors, a growing bulge on its northeast flank, and months of small summit eruptions, Nevado del Ruiz violently erupted in a lateral explosion. The mountain was back in action.

On this night, the familiar pyroclastics, lava, and blast were not the worst problem. The hot eruption quickly melted ice on the mountain's snowy peak, liquefying mud and volcanic ash, sending a hot mudflow downslope. Such a flow is a *lahar*, an Indonesian word referring to mudflows of volcanic origin. This lahar moved rapidly down the Lagunilla River toward the villages below. The wall of mud was at least 40 m (130 ft) high as it approached Armero, a regional center with a population of 25,000. The city slept as the lahar buried its homes: 23,000 people were killed; thousands were injured; 60,000 were left homeless across the region. The volcanic debris flow is now a permanent grave for its victims. Although less devastating to human life, the mudflow generated by the eruption of Mount St. Helens also was a lahar. Not all mass movements are this destructive. Such processes play an important role in the denudation of the landscape.

For more on mass-movement hazards, including landslides, see the web site of the Natural Hazards Center at the University of Colorado, Boulder, at http://www.colorado.edu/hazards/ or the USGS Natural Hazards–Landslides page at http://www.usgs.gov/hazards/landslides/. Many significant mass-movement episodes are reported in "Landslide Related News" on the latter Web site.

Mass-Movement Mechanics

Physical and chemical weathering processes create an overall weakening of surface rock, which makes it more susceptible to the pull of gravity. The term **mass movement** applies to any unit movement of a body of material, propelled and controlled by gravity, such as the lahar just described. Mass movements can be surface processes, or they can be submarine landslides beneath the ocean. Mass-movement content can range from dry to wet, slow to fast, small to large, and free-falling to gradual or intermittent (see Figure 13.22).

The term *mass movement* is sometimes used interchangeably with **mass wasting**, which is the general process involved in mass movements and erosion of the landscape. To combine the concepts, we can say that the mass movement of material works to waste slopes and provide raw material for erosion, transportation, and deposition.

The Role of Slopes All mass movements occur on slopes under the influence of gravitational stress. If we pile dry sand on a beach, the grains will flow downslope until equilibrium is achieved. The steepness of the resulting slope depends on the size and texture of the grains; this steepness is the **angle of repose**. This angle represents a balance of the driving force (gravity) and resisting force (friction and shearing). The angle of repose for various materials commonly ranges between 33° and 37° (from horizontal) and between 30° and 50° for snow avalanche slopes.

The *driving force* in mass movement is gravity. It works in conjunction with the weight, size, and shape of the surface material; the degree to which the slope is oversteepened (how far it exceeds the angle of repose); and the amount and form of moisture available (frozen or fluid). The greater the slope angle is, the more susceptible the surface material is to mass-wasting processes.

The *resisting force* is the shearing strength of the slope material—that is, its cohesiveness and internal friction, which work against gravity and mass wasting. To reduce shearing strength is to increase shearing stress, which eventually reaches the point at which gravity overcomes friction, initiating slope failure.

Clays, shales, and mudstones are highly susceptible to hydration (physical swelling in response to the presence of water). If such materials underlie rock strata in a slope, the strata will move with less driving-force energy. When clay surfaces are wet, they deform slowly in the direction of movement, and when saturated, they form a viscous fluid

GEO REPORT 13.4 ***Amateurs make cave discoveries***

In the early 1940s, George Colglazier, a farmer southwest of Bedford, Indiana, awoke to find his farm pond at the bottom of a deep collapsed sinkhole. This sinkhole is now the entrance to an extensive cave system that includes a subterranean navigable stream. The exploration and scientific study of caves is *speleology*. Although professional physical and biological scientists carry on investigations, amateur cavers, or "spelunkers," have made many important discoveries.

The mystery, intrigue, and excitement of cave exploration lie in the variety of dark passageways, enormous chambers that narrow to tiny crawl spaces, strange formations, and underwater worlds that can be accessed only by cave diving. Private-property owners and amateur adventurers discovered many of the major caves, a fact that keeps this popular science/sport very much alive. For nearly a thousand worldwide links and information, see the http://www.cbel.com/speleology Web site.

with little shearing strength (resistance to movement) to hold back the slope. However, if the rock strata are such that material is held back from slipping, then more driving-force energy may be required, such as that generated by an earthquake.

Madison River Canyon Landslide In the Madison River Canyon near West Yellowstone, Montana, a blockade of dolomite (a magnesium-rich carbonate rock) held back a deeply weathered and *oversteepened slope* (40° to 60° slope angle) for untold centuries (white area in Figure 13.21). Then, shortly after midnight on August 17, 1959, a M 7.5 earthquake broke the dolomite structure along the foot of the slope. The break released 32 million m^3 (1.13 billion ft^3) of mountainside, which moved downslope at 95 kmph (60 mph), causing gale-force winds through the canyon. Momentum carried the material more than 120 m (about 400 ft) up the opposite side of the canyon, trapping several hundred campers with about 80 m (260 ft) of rock and killing 28 people.

The mass of material also effectively dammed the Madison River and created a new lake, dubbed Quake Lake. The landslide debris dam established a new temporary base level for the canyon. A channel was quickly excavated by the U.S. Army Corps of Engineers to prevent a disaster below the dam, for if Quake Lake overflowed the landslide dam, the water would quickly erode a channel and thereby release the entire contents of the new lake onto farmland downstream. This channel-dredging relief procedure in a landslide "dam" was exactly what engineers had to do on the Chaping River, Hubei Province, China, in 2008. The M 7.9 Wenchuan quake produced more than 100 of these landslide dams; engineers prioritized the highest-risk dams to receive channel dredging first.

Classes of Mass Movements

In any mass movement, gravity pulls on a mass until the critical shear-failure point is reached—a geomorphic threshold. The material then can *fall*, *slide*, *flow*, or *creep*—the four classes of mass movement. Figure 13.22 summarizes these classes. Note the temperature and moisture gradients in the margins of the illustration, which show the relation between water content and movement rate. The Madison River Canyon event was a type of slide, whereas the Nevado del Ruiz lahar mentioned earlier was a flow. We now look at specific mass-movement classes.

Falls and Avalanches This class of mass movement includes rockfalls and debris avalanches. A **rockfall** is simply a volume of rock that falls through the air and hits a surface. During a rockfall, individual pieces fall independently and characteristically form a cone-shaped pile of irregular broken rocks in a **talus slope** at the base

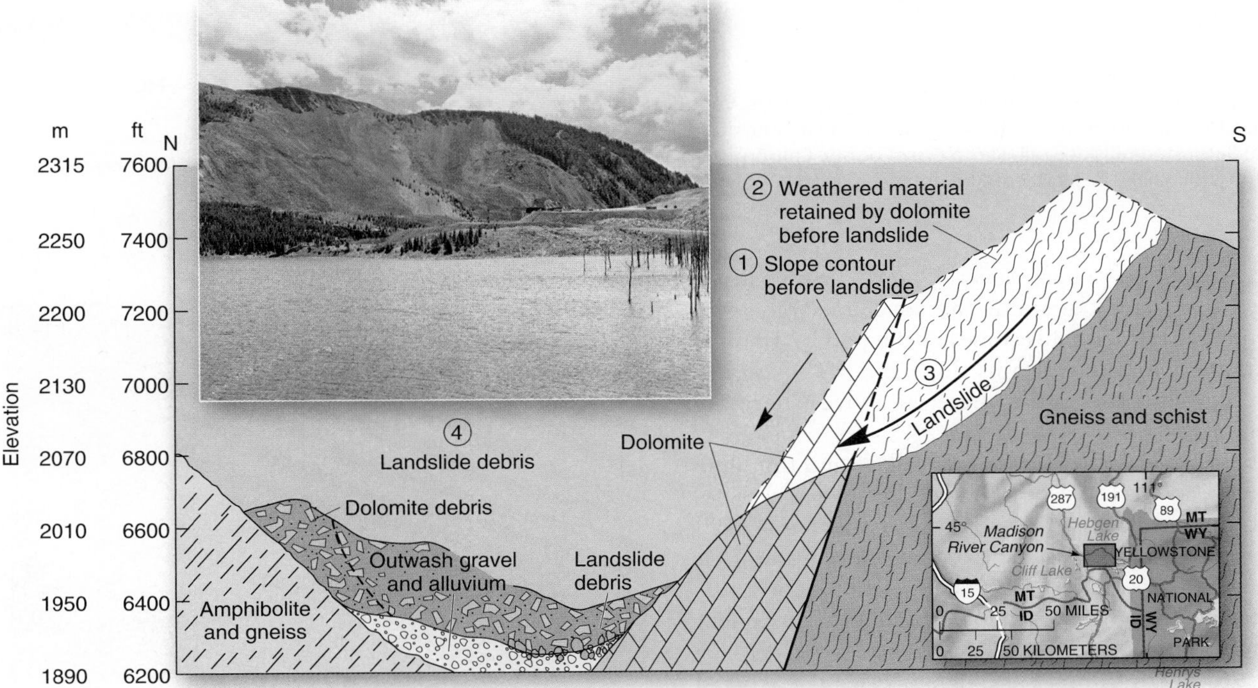

FIGURE 13.21 Madison River landslide.
Cross section showing geologic structure of the Madison River Canyon in Montana, where an earthquake triggered a landslide in 1959: (1) prequake slope contour; (2) weathered rock that failed; (3) direction of landslide; and (4) landslide debris blocking the canyon and damming the Madison River, visible in the inset photo. [USGS Professional Paper 435-K, August 1959, p. 115. Inset photo by Bobbé Christopherson.]

MG Notebook
Madison River Landslide

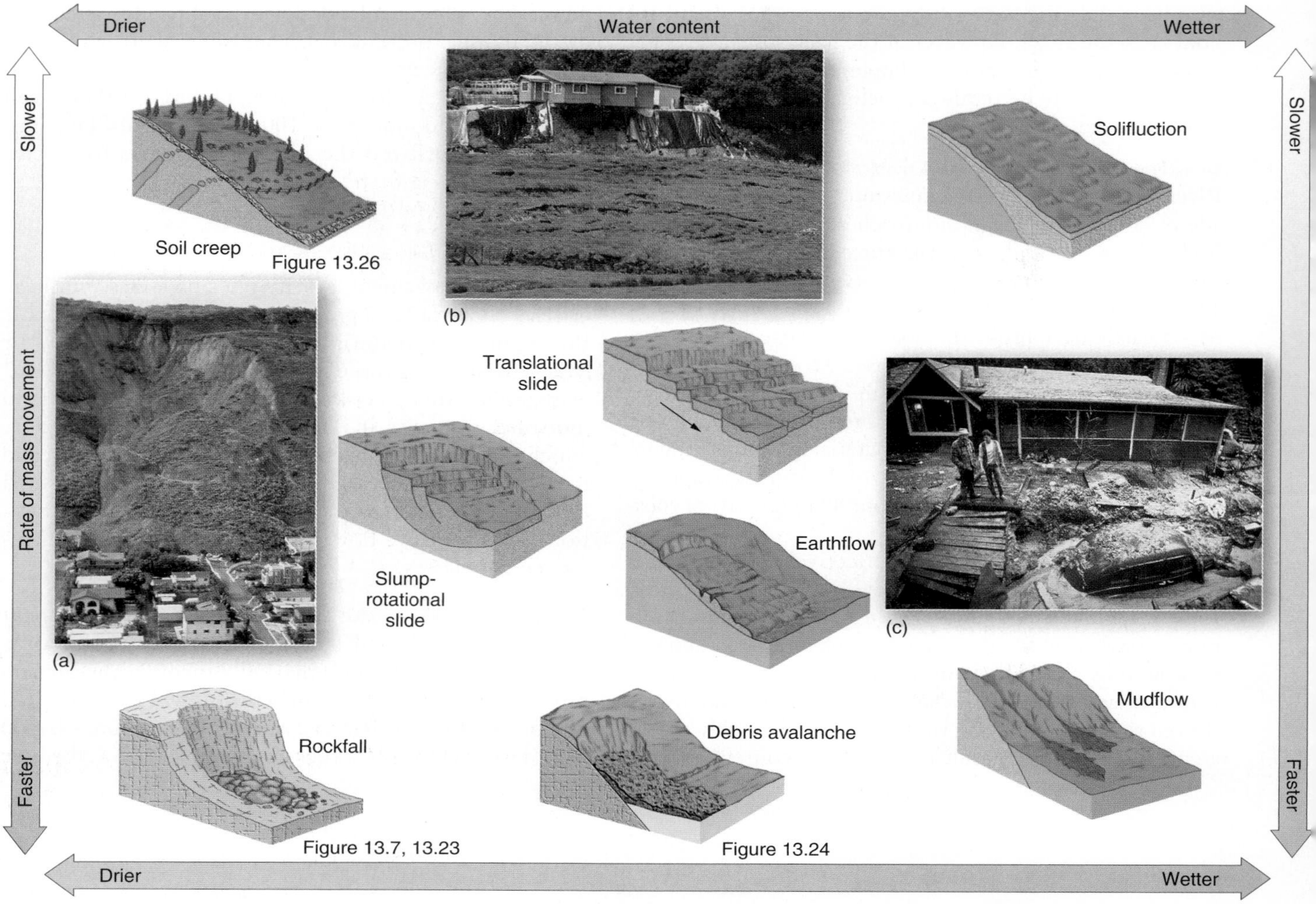

Drier — Water content — Wetter

Slower — Rate of mass movement — Faster

Soil creep
Figure 13.26

(b)

Solifluction

(a)

Translational slide

Slump-rotational slide

Earthflow

(c)

Rockfall
Figure 13.7, 13.23

Debris avalanche
Figure 13.24

Mudflow

Drier — Wetter

FIGURE 13.22 Mass-movement classes.
Principal types of mass-movement and mass-wasting events. Variations in water content and rates of movement produce a variety of forms. (a) A 1995 slide in La Conchita, California, also the site of a 2005 mudslide event. (b) Saturated hillsides fail, Santa Cruz County, California. (c) Mudflows 2 m (6.6 ft) deep, also in Santa Cruz County. [(a) Robert L. Schuster/USGS. (b) Alexander Lowry/Photo Researchers, Inc. (c) James A. Sugar.]

Animation
Mass Movements

of a steep incline, where several *talus cones* coalesce (Figure 13.23). Look again at the cover of *Geosystems*. Can you identify such talus slopes in the photo? Look at the different colors in the talus slope and see if you can match them with the parent material above where the rock originated.

A **debris avalanche** is a mass of falling and tumbling rock, debris, and soil. It is differentiated from a slower debris slide or landslide by the velocity of onrushing material. This speed often results from ice and water that fluidize the debris. The extreme danger of a debris avalanche results from its tremendous speed and consequent lack of warning.

In 1962 and again in 1970, debris avalanches roared down the west face of Nevado Huascarán, the highest peak in the Peruvian Andes. The 1962 debris avalanche contained an estimated 13 million m^3 (460 million ft^3) of material, burying the city of Ranrahirca and eight other towns and killing 4000 people. An earthquake initiated the 1970 event

FIGURE 13.23 Talus slope.
Rockfall and talus deposits at the base of a steep slope along Duve Fjord, Nordaustlandet Island. Can you see the lighter rock strata above that are the source for the three talus cones? [Bobbé Christopherson.]

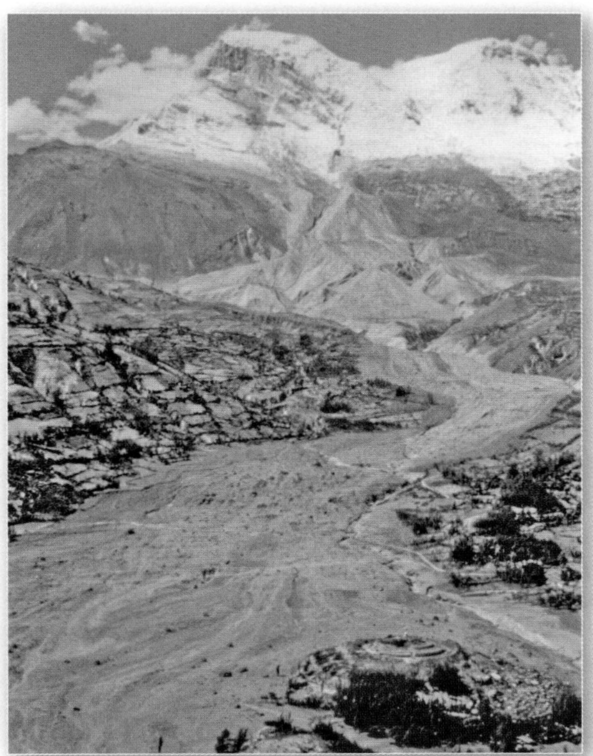

FIGURE 13.24 Debris avalanche, Peru.
A 1970 debris avalanche falls more than 4100 m (2.5 mi) down the west face of Nevado Huascarán, burying the city of Yungay, Peru. The same area was devastated by a similar avalanche in 1962 and by others in pre-Columbian times. A great danger remains for the towns in the valley from possible future mass movements. [George Plafker, USGS.]

(Figure 13.24). This time, upward of 100 million m^3 (3.53 billion ft^3) of debris buried the city of Yungay, where 18,000 people perished. This avalanche attained velocities of 300 kmph (185 mph).

Landslides A sudden rapid movement of a cohesive mass of regolith or bedrock that is not saturated with moisture is a **landslide**—a large amount of material failing simultaneously. Surprise creates the danger, for the downward pull of gravity wins the struggle for equilibrium in an instant. Focus Study 13.1 describes one such surprise event that struck near Longarone, Italy, in 1963.

CRITICAL THINKING 13.2

National landslide hazard potential

The USGS has completed a landslide hazards potential map for the conterminous United States, available at http://www.usgs.gov/hazards/landslides/. Note the zoom-in capability to look at specific areas in greater detail. The map legend rates landslide incidence and susceptibility. How does the map portray landslide incidences? In what way does the map rate landslide susceptibility of an area? Are you able to find a location that you have visited and determine its vulnerability? Do any such hazards exist in the region where your college is located?

To eliminate the surprise element, scientists are using the Global Positioning System (GPS) to monitor landslide movement. With GPS, scientists measure slight land shifts in suspect areas for clues to possible mass wasting. At two sites in Japan, GPS effectively identified pre-landslide movements of 2–5 cm per year, providing information to expand the area of hazard concern and warning.

Slides occur in one of two basic forms: translational or rotational (see Figure 13.22 for an idealized view of each). *Translational slides* involve movement along a planar (flat) surface roughly parallel to the angle of the slope, with no rotation. The Madison Canyon landslide described earlier was a translational slide. Flow and creep patterns also are considered translational in nature.

Rotational slides occur when surface material moves along a concave surface. Frequently, underlying clay presents an impervious surface to percolating water. As a result, water flows along the clay surface, undermining the overlying block. The surface may rotate as a single unit, or it may present a stepped appearance. Continuing rotational mudslides plague La Conchita, California (Figure 13.22a). For thousands of years, this area has been unstable. In 1995, a slump landslide buried homes, and in January 2005, a mudslide episode following a period of heavy rains buried 30 homes and took 10 lives.

Flows Flows include *earthflows* and more fluid **mudflows**. When the moisture content of moving material is high, the suffix *-flow* is used (see Figure 13.22). Heavy rains can saturate barren mountain slopes and set them moving, as was the case east of Jackson Hole, Wyoming, in the spring of 1925. Material above the Gros Ventre River (pronounced "grow vaunt") broke loose and slid downslope as a unit. The water content of this mass-wasting event was great enough to classify it as an *earthflow*, caused by sandstone formations resting on weak shale and siltstone, which became moistened and soft, offering little resistance to the overlying strata (Figure 13.25).

About 37 million m^3 (1.3 billion ft^3) of wet soil and rock moved down one side of the canyon and surged 30 m (100 ft) up the other side. The earthflow dammed the river and formed a lake, as did the landslide across the Madison River in the Hebgen Lake area in 1959. However, in 1925 equipment was not available to excavate a channel, so the new lake filled. Two years later the lake water broke through the temporary earthflow dam, transporting a tremendous quantity of debris over the region downstream.

Creep A persistent, gradual mass movement of surface soil is **soil creep**. In creep, individual soil particles are lifted and disturbed by the expansion of soil moisture as it freezes, by cycles of moistness and dryness, by diurnal temperature variations, or by grazing livestock or digging animals.

In the freeze–thaw cycle, particles are lifted at right angles to the slope by freezing soil moisture, as shown in Figure 13.26 (p. 391). When the ice melts, however, the

(text continued on page 392)

FOCUS STUDY 13.1

Vaiont Reservoir Landslide Disaster

Place: Northeastern Italy, Vaiont Canyon in the Italian Alps, rugged scenery, 680 m (2200 ft) above sea level.

Location: 46.3° N, 12.3° E, near the border of the Veneto and Friuli-Venezia Giulia regions.

Situation: Centrally located in the region, an ideal situation for hydroelectric power production.

Site: Steep-sided, narrow, glaciated canyon, opening to populated lowlands to the west; ideal site for a narrow-crested, high dam and deep reservoir.

The Plan: Build the second highest dam in the world—262 m (860 ft) high, with a crest length of 190 m (623 ft)—using a new thin-arch design and impounding a reservoir capacity of 150 million m³ (5.3 billion ft³) of water (the third largest reservoir in the world), and generate hydroelectric power for distribution.

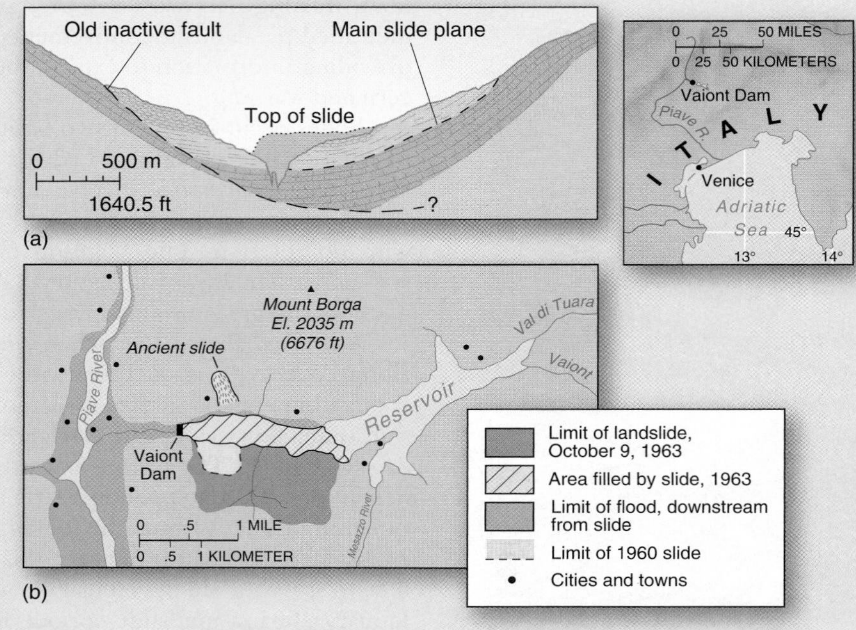

FIGURE 13.1.1 Vaiont Reservoir disaster. **(a) Cross section of the Vaiont River Valley. (b) Map of the disaster site, with evidence of previous activity noted.** [After G. A. Kiersch, "The Vaiont Reservoir disaster," *Mineral Information Service* (California Division of Mines and Geology), 18, 7 (1964): 129–138.]

Geological Analysis:

1. Steep canyon walls composed of interbedded limestone and shale; badly cracked and deformed structures; open fractures in the shale inclined toward the future reservoir body. The steepness of the canyon walls enhances the strong driving forces (gravitational) at work on rock structures.
2. High potential for bank storage (water absorption by canyon walls) into the groundwater system that will increase water pressure on all rocks in contact with the reservoir.
3. The nature of the shale beds is such that cohesion will be reduced as their clay minerals become saturated.
4. Evidence of ancient rockfalls and landslides on the north side of the canyon.
5. Evidence of creep activity along the south side of the canyon.

Political/Engineering Decision: Begin design and construction of a thin-arch dam at this site.

Events During Construction and Filling: Large volumes of concrete were injected into the bedrock as "dental work" in an attempt to strengthen the fractured rock. During reservoir filling in 1960, 700,000 m³ (2.5 million ft³) of rock and soil slid into the reservoir from the south side. A slow creep of slope materials along the entire south side began shortly thereafter, increasing to about 1 cm per week by January 1963. Creep rate increased as the level of the reservoir rose. By mid-September 1963, the creep rate exceeded 40 cm (16 in.) a day.

This sequence of events set the stage for one of the worst dam disasters in history. Heavy rains began on September 28, 1963. Alarmed at the runoff from the rain into the reservoir and the increase in creep rate along the south wall, engineers opened the outlet tunnels on October 8 in an attempt to bring down the reservoir level—but it was too late.

The next evening it took only 30 seconds for a landslide of 240 million m³ (8.5 billion ft³) to crash into the reservoir, shaking seismographs across Europe. A slab of mountainside 150 m thick and 2 km by 1.6 km in area (500 ft thick and 1.2 mi by 1 mi in area) gave way, sending shock waves of wind and water up the canyon 2 km (1.2 mi) and splashing a 100-m (330-ft) wave over the dam. The former reservoir was effectively filled with bedrock, regolith, and soil that almost entirely displaced its water content (Figure 13.1.1). Amazingly, the experimental dam design held.

Downstream, in the unsuspecting town of Longarone, near the mouth of Vaiont Canyon on the Piave River, people heard the distant rumble and were quickly drowned by the 69-m (226-ft) wave of water that came out of the canyon. That night 3000 people perished. As for a lesson or recommendation, we need only to reread the preconstruction geologic analysis that began this Focus Study. The Italian courts eventually prosecuted the persons responsible.

FIGURE 13.25 **The Gros Ventre earthflow near Jackson, Wyoming.**
Evidence of the 1925 earthflow is still visible after more than 87 years. [Steven K. Huhtala.]

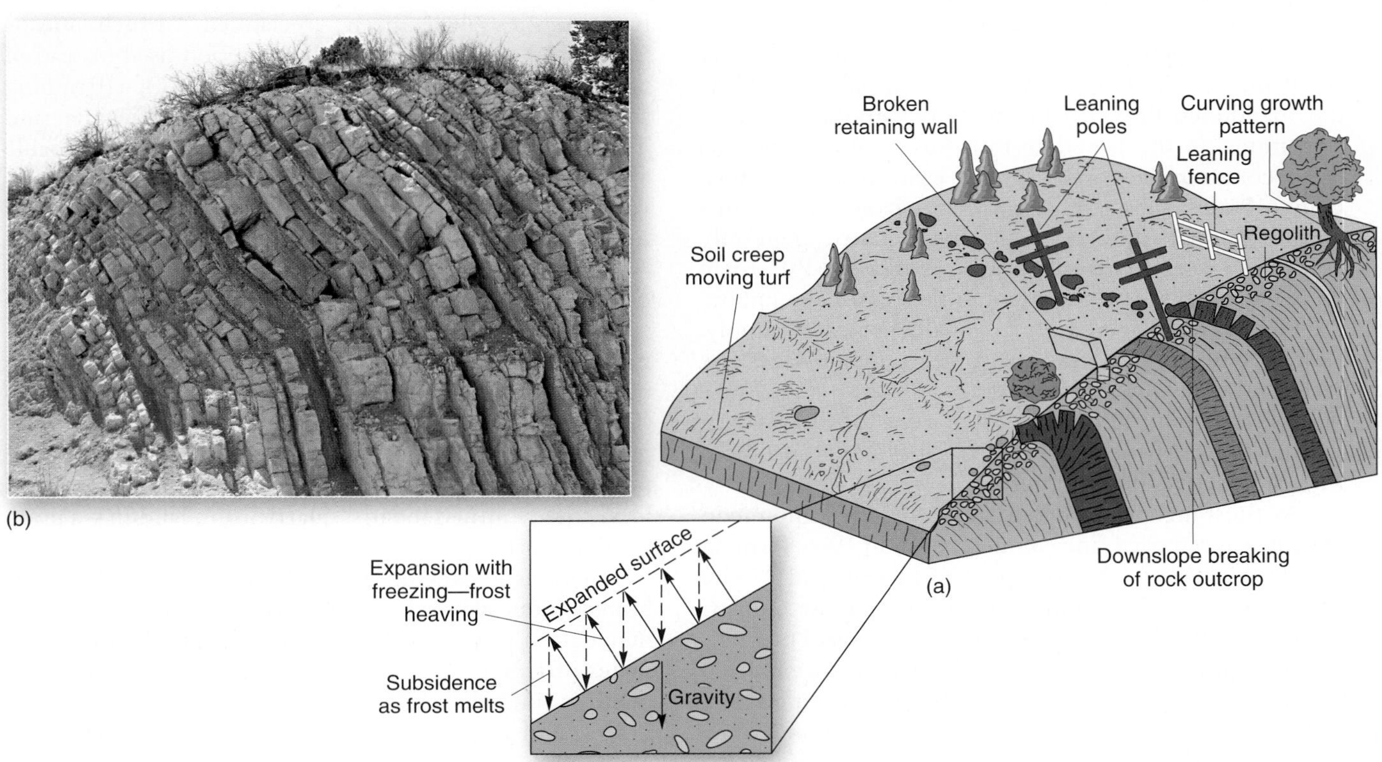

FIGURE 13.26 **Soil creep and its effects.**
(a) Typical soil-creep features. (b) Note the subsurface effects on rock strata caused by surface creep. [(b) Bobbé Christopherson.]

GEO REPORT 13.5 *Landscaping choices, watering, and landslides*

In October 2007, mass-movement realities struck upscale La Jolla, California, damaging more than 100 homes along Soledad Mountain Road. The rotational slump along a hillside pulled material from beneath the highway, causing a sinkhole-like collapse. The area is inherently unstable, experiencing three other collapse events since 1961, with early warning signs of this occurrence beginning in July. Homeowners were allowed to irrigate lawns and gardens with no restrictions, feeding drainage runoff water into subsurface strata and, as a result, increasing instability and lubricating slides. In an area such as this, it is difficult to imagine that standard cut-and-fill housing pads were allowed on unstable slopes in the zoning-planning stage prior to construction.

particles fall straight downward in response to gravity. As the process repeats, the surface soil gradually creeps its way downslope.

The overall wasting of a slope may cover a wide area and may cause fence posts, utility poles, and even trees to lean downslope. Various strategies are used to arrest the mass movement of slope material—grading the terrain, building terraces and retaining walls, planting ground cover—but the persistence of creep often renders these strategies ineffective.

Human-Induced Mass Movements (Scarification)

Every human disturbance of a slope—highway road cutting, surface mining, or construction of a shopping mall, housing development, or home—can hasten mass wasting. The newly destabilized and oversteepened surfaces are thrust into a search for a new equilibrium. Imagine the disequilibrium in slope relations created by the highway road cut pictured in Figure 12.8b.

Large open-pit surface mines—such as the Bingham Canyon Copper Mine west of Salt Lake City; the abandoned Berkeley Pit in Butte, Montana; several iron mines in the Brazilian rain forest; and numerous large coal surface mines in the eastern and western United States, such as Black Mesa, Arizona—are examples of human-induced mass movements, generally known as **scarification**. At the Berkeley Pit, toxic drainage water is still threatening the regional aquifers, the Clark Fork River, and the local freshwater supply. Residues of copper, zinc, lead, and arsenic lace one creek, now devoid of life, and are in the soil on Butte Hill, where children play. In other areas of the western United States, wind dispersal is a particular problem with radioactive uranium tailings.

At the Bingham Canyon Copper Mine, a mountain literally was removed since mining began in 1906, forming a pit 4 km wide and 1.2 km deep, or 2.5 mi wide and 0.75 mi deep (Figure 13.27a). This is easily the largest human-made excavation on Earth. Half a million tons of ore containing copper, gold, silver, and molybdenum are extracted per day. The disposal of tailings (mined ore of little value) and waste material is a significant problem at any mine. Such large excavations produce tailing piles that are unstable and susceptible to further weathering, mass wasting, or wind dispersal. The leaching of toxic materials from tailings and waste piles poses an ever-increasing problem for streams, aquifers, and public health across the country.

Where underground mining is common, particularly for coal in the Appalachians, land subsidence and collapse produce further mass movements. Homes, highways, streams, wells, and property values are severely affected. Another controversial form of mining called *mountaintop removal* is done by removing ridges and summits and dumping the debris into stream valleys, thereby exposing the coal seams for mining and burying the stream channels. A change in the Clean Water Act in 2002 removed a 20-year-old restriction that banned dumping within 100 feet of a stream to protect flows. These valley fills affect downstream water quality with concentrations of potentially toxic nickel, lead, cadmium, iron, and selenium that generally exceed government standards. Valley-fill collapses have caused floods, as have coal and ash sludge containment breaks, as we saw in this chapter's *Geosystems Now*.

Figure 13.27b is an aerial view of mountaintop removal on Kayford Mountain, West Virginia. The 20-story-tall dragline machine is visible center-right. In the background, forest cover and litter are removed using fire to prepare the site for minimal reclamation, although this is difficult, since the landscape is essentially lost. Some 2000 km (1245 mi) of streams are now filled with tailings (Figure 13.27c), an estimated 1.2 million acres removed, and 500 mountains flattened.

Claims by the coal-mining corporations of reclamation are difficult to verify, since less than 10% of the scarification sites have been "reclaimed," which means "converted to economic uses," a criterion used instead of the desired "approximate original contour." Of 410 mining sites studied in Kentucky, West Virginia, Virginia, and Tennessee in 2010, 366 sites had no reclamation verifiable and 26 sites had some form of post-mining economic development (see Ross Geredien, *Post-mountaintop Removal Reclamation of Mountain Summits* . . . [prepared for the Natural Resources Defense Council, New York, NY, December 7, 2009]).

In eastern Pennsylvania, slate deposits formed over millions of years. The Dally Slate Quarry in Pen Argyl is now working a second pit that is 55 m (180 ft) deep (Figure 13.27d). The grain of the metamorphic rock guides the slate removal so as to produce the flattest pieces. Slate is mined for shingles and countertops. The local county turned the first quarry into a landfill—a practical use for the pit.

Scientists can informally quantify the scale of human-induced scarification for comparison with natural denudation processes. Geologist R. L. Hooke used estimates of U.S. excavations for new housing, mineral production (including the three largest—stone, sand and gravel, and coal), and highway construction. He then prorated these quantities of moved earth for all countries, using their gross domestic product, energy consumption, and agriculture's effect on river sediment loads. From these, he calculated a global estimate for human earth moving. Later researchers confirmed and expanded on his findings.

Hooke estimated for the early 1990s that humans, as a geomorphic agent, annually moved 40–45 billion tons (40–45 Gt) of the planet's surface. Compare this quantity with natural river sediment transferred (14 Gt/yr), movement through stream meandering (39 Gt/yr), haulage by glaciers (4.3 Gt/yr), movement due to wave action and erosion (1.25 Gt/yr), wind transport (1 Gt/yr), sediment movement by continental and oceanic mountain building

(a) Bingham Canyon mine–ground view, orbital view

(b) Mountaintop removal on Kayford Mountain

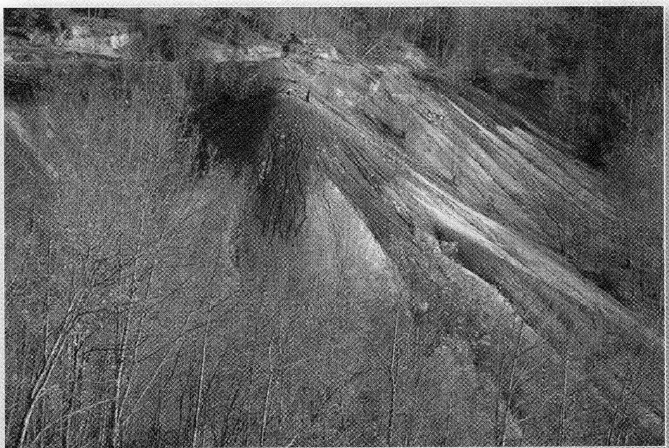
(c) Coal mine tailings in southern West Virginia

(d) Slate mine, Pennsylvania

FIGURE 13.27 Scarification.
(a) Bingham Canyon, Utah, west of Salt Lake City, where strip-mining extracts copper and other minerals; the International Space Station/NASA photo provides perspective from orbit. (b) Mountaintop removal strip-mining for coal on Kayford Mountain, West Virginia. (c) Spoil banks of tailings in a West Virginia coal-mining area. (d) The 55-m-deep (180-ft-deep) Dally Slate Quarry in Pen Argyl, Pennsylvania. [(b) Dr. Chris Mayda, Eastern Michigan University, all rights reserved. (a) and (d) Bobbé Christopherson. (c) Author.]

(34 Gt/yr), or deep-ocean sedimentation (7 Gt/yr). As Hooke concluded about humans,

> *Homo sapiens* has become an impressive geomorphic agent. Coupling our earth-moving prowess with our inadvertent adding of sediment load to rivers and the visual impact of our activities on the landscape, one is compelled to acknowledge that, for better or for worse, this biogeomorphic agent may be the premier geomorphic agent of our time.*

*R. L. Hooke, "On the efficacy of humans as geomorphic agents," *GSA Today* (The Geological Society of America), 4, 9 (September 1994): 217–226.

In 2000, Dr. Hooke in another paper on this subject asked an important question: "One may well ask how long such rates of increases [in human geomorphic impacts] can be sustained, and whether it will be rational behavior or catastrophe that brings them to an end."*

*R. L. Hooke, "On the history of humans as geomorphic agents," *Geology*, 28, 9 (September 2000): 843–846.

GEOSYSTEMS CONNECTION

The basics of landmass denudation and morphology of slopes are at the core of exogenic processes. The physical and chemical weathering of rock, dissolution of limestone landscapes, and mass-movement processes deliver materials for erosional transport. In the next chapter, we examine river systems and landforms that result from these fluvial processes. The erosional and depositional agents of running water, wind, waves and coastal actions, and ice flow through the next four chapters.

KEY LEARNING CONCEPTS REVIEW

■ ***Define*** **the science of geomorphology.**

Geomorphology is the science that analyzes and describes the origin, evolution, form, and spatial distribution of landforms. The exogenic system, powered by solar energy and gravity, tears down the landscape through processes of landmass **denudation** involving weathering, mass movement, erosion, transportation, and deposition. Different rocks offer differing resistance to these weathering processes and produce a pattern on the landscape of **differential weathering**.

Agents of change include moving air, water, waves, and ice. Since the 1960s, research and understanding of the processes of denudation have moved toward the **dynamic equilibrium model,** which considers slope and landform stability to be consequences of the resistance of rock materials to the attack of denudation processes.

geomorphology (p. 370)
denudation (p. 370)
differential weathering (p. 370)
dynamic equilibrium model (p. 371)

1. Define *geomorphology*, and describe its relationship to physical geography.
2. Define *landmass denudation*. What processes are included in the concept?
3. What is the interplay between the resistance of rock structures and differential weathering?
4. Describe what is at work to produce the landform in Figure 13.1.
5. What are the principal considerations in the dynamic equilibrium model?

■ ***Illustrate*** **the forces at work on materials residing on a slope.**

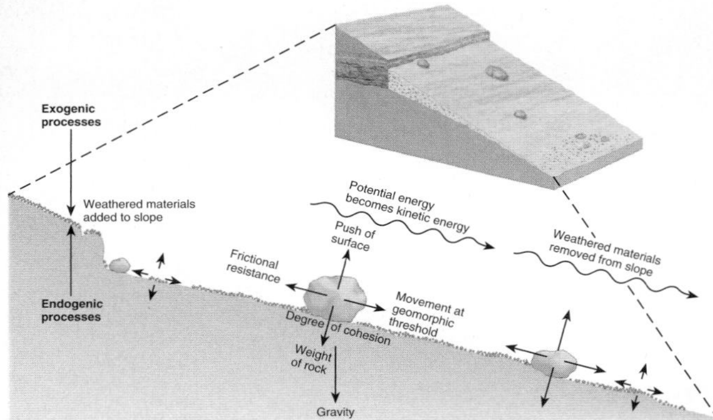

Slopes are shaped by the relation between the rate of weathering and breakup of slope materials and the rate of mass movement and erosion of those materials. In this struggle against gravity, a slope may reach a **geomorphic threshold**—the point at which energy overcomes resistance against movement. **Slopes** that form the boundaries of landforms have several general components: *waxing slope*, *free face*, *debris slope*, and *waning slope*. Slopes seek an angle of equilibrium among the operating forces.

geomorphic threshold (p. 371)
slopes (p. 371)

6. Describe conditions on a hillslope that is right at the geomorphic threshold. What factors might push the slope beyond this point?
7. Given all the interacting variables, do you think a landscape ever reaches a stable, old-age condition? Explain.

8. What are the general components of an ideal slope?
9. Relative to slopes, what is meant by an *angle of equilibrium*? Can you apply this concept to the photograph in Figure 13.2?

■ ***Define*** weathering, and ***explain*** the importance of parent rock and joints and fractures in rock.

Weathering processes disintegrate both surface and subsurface rock into mineral particles or dissolve them in water. The upper layers of surface material undergo continual weathering and create broken-up rock called **regolith**. Weathered **bedrock** is the *parent rock* from which regolith forms. The unconsolidated, fragmented material that develops after weathering is **sediment,** which along with weathered rock forms the **parent material** from which soil evolves.

Important in weathering processes are **joints,** the fractures and separations in the rock. Jointing opens up rock surfaces on which weathering processes operate. Factors that influence weathering include the character of the bedrock (hard or soft, soluble or insoluble, broken or unbroken), climatic elements (temperature, precipitation, freeze–thaw cycles), position of the water table, slope orientation, surface vegetation and its subsurface roots, and time.

weathering (p. 372)
regolith (p. 372)
bedrock (p. 373)
sediment (p. 373)
parent material (p. 373)
joints (p. 373)

10. Describe weathering processes operating on an open expanse of bedrock. How does regolith develop? How is sediment derived?
11. Describe the relationship between mesoscale climatic conditions and rates of weathering activities. At microscale levels, describe weathering processes.
12. What is the relation among parent rock, parent material, regolith, and soil?
13. What role do joints play in the weathering process? Give an example from one of the illustrations in this chapter.

■ ***Describe*** frost action, salt-crystal growth, pressure-release jointing, and the role of freezing water as physical weathering processes.

Physical weathering refers to the breakup of rock into smaller pieces with no alteration of mineral identity. The physical action of water when it freezes (expands) and thaws (contracts) is a powerful agent in shaping the landscape. This **frost action** may break apart any rock. Working in joints, expanded ice can produce *joint-block separation* through the process of *frost-wedging*. Another process of physical weathering is *salt-crystal growth* (*salt weathering*); as crystals in rock grow and enlarge over time in crystalization, they force apart mineral grains and break up rock.

As overburden is removed from a granitic batholith, the pressure of deep burial is relieved. The granite slowly responds with *pressure-release jointing*, with layer after layer of rock peeling off in curved slabs or plates. As these slabs weather, they slip off in the process of **sheeting**. This *exfoliation process* creates arch-shaped or dome-shaped features on the exposed landscape, sometimes forming an **exfoliation dome**.

physical weathering (p. 374)
frost action (p. 374)
sheeting (p. 376)
exfoliation dome (p. 376)

14. What is physical weathering? Give an example.
15. Why is freezing water such an effective physical weathering agent?
16. What weathering processes produce a granite dome? Describe the sequence of events.

■ *Describe* the susceptibility of different minerals to the chemical weathering processes, including hydration, hydrolysis, oxidation, carbonation, and solution.

Chemical weathering is the chemical decomposition of minerals in rock. It can cause **spheroidal weathering,** in which chemical weathering occurs in cracks in the rock. As cementing and binding materials are removed, the rock begins to disintegrate, and sharp edges and corners become rounded.

Hydration occurs when a mineral absorbs water and expands, thus creating a strong mechanical force that stresses rocks, a form of physical weathering. **Hydrolysis** breaks down silicate minerals in rock, as in the chemical weathering of feldspar into clays and silica. Water actively participates in chemical reactions. **Oxidation** is the reaction of oxygen with certain metallic elements, the most familiar example being the rusting of iron, producing iron oxide. The dissolution of materials into *solution* is considered chemical weathering. For instance, a mild acid such as carbonic acid in rainwater will cause **carbonation,** wherein carbon combines with certain minerals, such as calcium, magnesium, potassium, and sodium.

chemical weathering (p. 377)
spheroidal weathering (p. 377)
hydration (p. 378)
hydrolysis (p. 379)
oxidation (p. 379)
carbonation (p. 379)

17. What is chemical weathering? Contrast this set of processes to physical weathering.
18. What is meant by the term *spheroidal weathering*? How does spheroidal weathering occur?
19. What is hydration? What is hydrolysis? Differentiate between these processes. How do they affect rocks?
20. Iron minerals in rock are susceptible to which form of chemical weathering? What characteristic color is associated with this type of weathering?
21. With what kind of minerals do carbon compounds react, and under what circumstances does the dissolution of carbonate minerals occur? What is this weathering process called?

■ *Review* the processes and features associated with karst topography.

Karst topography refers to distinctively pitted and weathered limestone landscapes. Surface circular **sinkholes** form and may extend to form a *karst valley*. A sinkhole may collapse through the roof of an underground cavern, forming a *collapse sinkhole*. The formation of caverns is part of karst processes and groundwater erosion. Limestone caves feature many unique erosional and depositional features.

karst topography (p. 380)
sinkholes (p. 380)

22. Describe the development of limestone topography. What is the name applied to such landscapes? From what area was this name derived?
23. Differentiate among sinkholes, karst valleys, and cockpit karst. Within which form is the radio telescope at Arecibo, Puerto Rico?
24. In general, how would you characterize the region southwest of Orleans, Indiana?
25. What are some of the unique erosional and depositional features you find in a limestone cavern?

■ *Portray* the various types of mass movements, and *identify* examples of each in relation to moisture content and speed of movement.

Any movement of a body of material, propelled and controlled by gravity, is **mass movement,** or **mass wasting**. The **angle of repose** of loose sediment grains represents a balance of driving and resisting forces on a slope. Mass movement of Earth's surface produces some dramatic incidents, including **rockfalls** (a volume of rock that falls), which can form a **talus slope** of loose rock along the base of the cliff; **debris avalanches** (a mass of tumbling, falling rock, debris, and soil moving at high speed); **landslides** (a large amount of material failing simultaneously); **mudflows** (material in motion with a high moisture content); and **soil creep** (a persistent movement of individual soil particles that are lifted by the expansion of soil moisture as it freezes, by cycles of wetness and dryness, by temperature variations, or by the impact of grazing

animals). In addition, human mining and construction activities have created massive **scarification** of landscapes.

mass movement (p. 386)
mass wasting (p. 386)
angle of repose (p. 386)
rockfall (p. 387)
talus slope (p. 387)
debris avalanche (p. 388)
landslide (p. 389)
mudflows (p. 389)
soil creep (p. 389)
scarification (p. 392)

26. Define the role of slopes in mass movements, using the terms *angle of repose*, *driving force*, *resisting force*, and *geomorphic threshold*.
27. What events occurred in the Madison River Canyon in 1959?
28. What are the classes of mass movement? Describe each briefly and differentiate among these classes.
29. Name and describe the type of mudflow associated with a volcanic eruption.
30. Describe the difference between a landslide and what happened on the slopes of Nevado Huascarán.
31. What is scarification, and why is it considered a type of mass movement? Give several examples of scarification. Why are humans a significant geomorphic agent?

MASTERING GEOGRAPHY

Visit www.masteringgeography.com for several figure animations, satellite loops, self-study quizzes, flashcards, visual glossary, case studies, career links, textbook reference maps, RSS Feeds, and an ebook version of *Geosystems*. Included are many geography web links to interesting Internet sources that support this chapter.

Results of a mass movement event in western Hungary, October 4, 2010. A retaining wall around a toxic holding pond broke at an aluminum oxide plant. A torrent of acidic sludge (pH 10.0) filled a stream and flowed over homes and land. The town of Devecser is left of center, north of Lake Balaton. The red pond is at 47° 6' N 17° 28' E. [*EO-1* image, NASA.]

CHAPTER 14

River Systems and Landforms

The Rio de la Plata meets the Atlantic Ocean near Buenos Aires, Argentina, where its mouth is some 210 km (130 mi) wide (far upper-left horizon). This estuary where seawater and freshwater intermix, considered a gulf by some, is formed from the confluence of the Paraná and Uruguay rivers. The water is carrying a heavy sediment load, coloring it light brown. [Bobbé Christopherson.]

KEY LEARNING CONCEPTS

After reading the chapter, you should be able to:

- ***Define*** the term *fluvial*, and ***explain*** processes of fluvial erosion, transportation, and deposition.
- ***Construct*** a basic drainage basin model, and ***identify*** different types of drainage patterns and internal drainage, with examples.
- ***Describe*** the relation among stream velocity, depth, width, and discharge.
- ***Develop*** a model of a meandering stream, including point bar, undercut bank, and cutoff, and ***explain*** the role of stream gradient.
- ***Define*** a floodplain, and ***analyze*** a stream channel during a flood.
- ***Differentiate*** the several types of river deltas, and ***describe*** each.
- ***Explain*** flood probability estimates, and ***review*** strategies for mitigating flood hazards.

Removing Dams and Restoring Salmon on the Elwha River, Washington

News reports in August 2010 delivered a new chapter in the restoration story of the Elwha River. Fourteen years after the final Environmental Impact Statement in favor of dam removal, the National Park Service contracted with a Montana construction firm to tear down the Elwha and Glines Canyon dams, at a price of $27 million. Work is scheduled to begin in September 2011 on what will be the largest dam removal in the United States to date.

As occurred with other such removals, a free-flowing Elwha River should enable the return of five species of Pacific salmon to the watershed, with numbers expected to increase from 3,000 to 390,000 over the next 30 years. In 1999, Edwards Dam was deconstructed from the Kennebec River in Augusta, Maine, marking the first dam removal for ecological reasons. Migratory fish such as shad, sturgeon, Atlantic salmon, and striped bass rebounded quickly in the Kennebec as the natural connection between the river and the sea was restored.

Located in the temperate rainforests of the Olympic peninsula, west of Seattle in Washington State (Figure GN 14.1), the Elwha River watershed is the largest in Olympic National Park. The 72-km-(45-mi-) long river flows north from the Olympic Mountains into the Strait of Juan de Fuca, entering the sea on the Klallam Indian Reservation. Construction of the Elwha Dam in 1913 and the Glines Canyon Dam in 1927 blocked salmon and other fish from accessing all but the lowest 8 km (5 mi) of the river. Such species are *anadromous* (from the Greek *anadromos,* "running up"), meaning they migrate from the sea into freshwater rivers to spawn. Dam removal is expected to restore bull trout and steelhead, as well as chinook, chum, coho, pink, and sockeye salmon.

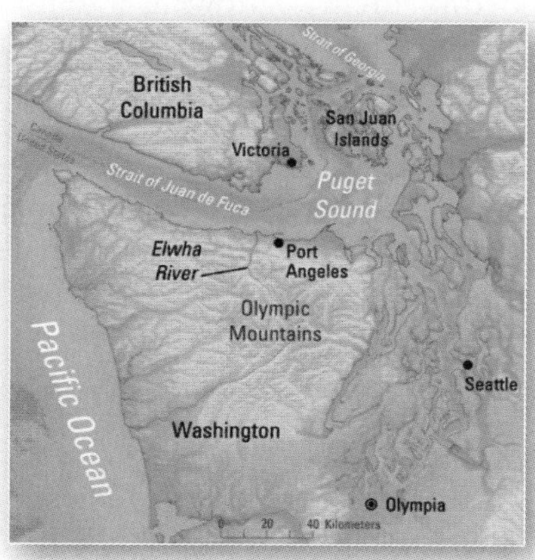

FIGURE GN 14.1 The Elwha River basin, Washington. [National Park Service/USGS.]

Dams on rivers have a number of benefits, such as water storage, hydroelectric power generation, and flood control. The Elwha's dams were built to generate hydroelectric power for the timber industry, from whom the federal government bought the dams in 2000. However, the downstream effects of dams on river ecosystems are profound; changes in river discharge, sediment supply, and water temperature impact the entire system—including channel form, riparian vegetation, fisheries, and recreation—with negative effects on native species. Specific impacts of the Elwha's dams are on sediment transport (scientists estimate that 13 million m^3 [460 million ft^3] of sediment are trapped behind the dam) and anadromous fish passage, which was entirely blocked when the dams were built.

As many small dams overseen by the Federal Energy Regulatory Commission (FERC) come up for relicensing, scientists and environmental groups are working with city, state, and federal agencies to advocate dam removal. Numerous dams have been deconstructed because they were unsafe or outdated or their original purpose was no longer valid.

The Elwha Dam is 32 m (105 ft) tall (Figure GN 14.2), whereas the upstream Glines Canyon dam tops 64 m (210 ft). In this chapter, we discuss the concept of local base level, as exemplified by these two dams. Their removal will require a period of adjustment throughout the drainage basin as river flows redistribute the accumulated sediment behind the dams and the system achieves a new equilibrium.

While dam removal can have some negative effects on downstream ecosystems, such as increased levels of fine sediment that may accumulate in spawning gravels, most of these effects are short-term and can be minimized with proper planning; see http://soundwaves.usgs.gov/2006/11/fieldwork3.html or http://www.lib.berkeley.edu/WRCA/CDRI/search.html, and type "Elwha Dam" for a complete list of references.

FIGURE GN 14.2 The Elwha Dam. [National Park Service.]

Earth's rivers and waterways form vast arterial networks that drain the continents. Rivers are Earth's lifeblood, inasmuch as rivers redistribute mineral nutrients important for soil formation and plant growth and serve society in many ways. They also shape the landscape by removing the products of weathering, mass movement, and erosion and transporting them downstream.

Rivers not only provide us with essential water supplies, but also receive, dilute, and transport wastes, provide critical cooling water for industry, and form one of the world's most important transportation networks. Rivers have been significant throughout human history. Each devastating flood event, such as the August 2010 catastrophe that drowned large areas of Pakistan along the Indus River, demonstrates the interrelationship between rivers and society.

Stream-related processes are **fluvial** (from the Latin *fluvius*, meaning "river"). Fluvial systems, like all natural systems, have characteristic processes and produce recognizable landforms. Yet a river system can behave with randomness and seeming disorder. The term *river* is applied to a trunk, or main stream; a network of tributaries forms a *river system*. *Stream* is a more general term not necessarily related to size. There is some overlap in usage between the terms *river* and *stream*.

Hydrology is the science of water and its global circulation, distribution, and properties—specifically, water at and below Earth's surface. For the Global Hydrology and Climate Center, see http://weather.msfc.nasa.gov/surface_hydrology/; also see the Amazon River at http://boto.ocean.washington.edu/story/Amazon.

In this chapter: We begin with a look at the largest rivers on Earth. Essential fluvial concepts of base level, drainage basin, and drainage density and patterns follow. With this foundation, we define stream discharge, Q, the essential element in fluvial processes. We discuss factors that affect flow characteristics and the work performed by flowing water, including erosion and transport. Depositional features are illustrated with a detailed look at the Tallahatchie River floodplain and the Mississippi River delta. Finally, we examine the effects of urbanization on stream hydrology. Human response to floods and floodplain management are important aspects of river management, with details of the New Orleans levee and floodwall disaster concluding the chapter.

Basic Fluvial Concepts

At any moment, approximately 1250 km³ (300 mi³) of water is flowing through Earth's waterways. Even though this volume is only 0.003% of all freshwater, the work performed by this energetic flow makes it a dominant agent of landmass denudation. Of the world's rivers, those with the greatest *discharge* (streamflow volume past a point in a given unit of time) are the Amazon of South America (Figure 14.1), the Congo of Africa, the Chàng Jiang (Yangtze) of Asia, and the Orinoco of South America (Table 14.1). The Amazon's leading discharge delivers millions of tons of sediment from the Amazon's drainage basin, which is as large as the Australian continent. In North America, the greatest discharges are from the Missouri–Ohio–Mississippi, Saint Lawrence, and Mackenzie river systems.

Insolation and gravity power the hydrologic cycle and are the driving forces of fluvial systems. Individual streams vary greatly from one another, depending on the climate in which they operate, the composition of the surface, the topography over which they flow, the nature of vegetation and plant cover, and the length of time they have been operating in a specific setting.

Water dislodges, dissolves, or removes surface material in the **erosion** process. Streams produce *fluvial erosion*, in which weathered sediment is picked up for transport to new locations. Thus, a stream is a mixture of water and solids; the solids are rolled or carried by mechanical **transport** or move in dissolved solution. Materials are laid down by another process, **deposition**. The general term for the clay, silt,

FIGURE 14.1 Mouth of the Amazon River.
The Amazon River discharges a fifth of all the freshwater that enters the world's oceans. The mouth of the Amazon is 160 km (100 mi) wide. Large islands of sediment are left where the river's discharge leaves the mouth and flows into the Atlantic Ocean. [*Terra* image NASA/GSFC/JPL.]

TABLE 14.1 Largest Rivers on Earth Ranked by Discharge

Rank by Volume	Average Discharge at Mouth in Thousands of m³/s (cfs)	River (with Tributaries)	Outflow/Location	Length km (mi)	Rank by Length
1	180 (6350)	Amazon (Ucayali, Tambo, Ene, Apurimac)	Atlantic Ocean/ Amapá-Pará, Brazil	6570 (4080)	2
2	41 (1460)	Congo (Lualaba)	Atlantic Ocean/ Angola, Congo	4630 (2880)	10
3	34 (1201)	Yangtze or Chàng Jiang	East China Sea/ Kiangsu, China	6300 (3915)	3
4	30 (1060)	Orinoco	Atlantic Ocean/Venezuela	2737 (1700)	27
5	21.8 (779)	La Plata estuary (Paraná)	Atlantic Ocean/Argentina	3945 (2450)	16
6	19.6 (699)	Ganges (Brahmaputra)	Bay of Bengal/India	2510 (1560)	23
7	19.4 (692)	Yenisey (Angara, Selenga or Selenge, Ider)	Gulf of Kara Sea/Siberia	5870 (3650)	5
8	18.2 (650)	Mississippi (Missouri, Ohio, Tennessee, Jefferson, Beaverhead, Red Rock)	Gulf of Mexico/Louisiana	6020 (3740)	4
9	16.0 (568)	Lena	Laptev Sea/Siberia	4400 (2730)	11
17	9.7 (348)	St. Lawrence	Gulf of St. Lawrence/ Canada and United States	3060 (1900)	21
36	2.83 (100)	Nile (Kagera, Ruvuvu, Luvironza)	Mediterranean Sea/Egypt	6690 (4160)	1

sand, gravel, and mineral fragments deposited by running water is **alluvium**, which may be sorted or semisorted sediment on a floodplain, delta, or streambed.

Base Level

Base level is a level below which a stream cannot erode its valley. In general, the *ultimate base level* is sea level, the average level between high and low tides. Imagine base level as a surface extending inland from sea level, inclined gently upward under the continents. Ideally, this is the lowest practical level for all denudation processes (Figure 14.2a).

American geologist and ethnologist John Wesley Powell (1834–1902) put forward the idea of base level in 1875. He was a director of the U.S. Geological Survey, the first director of the U.S. Bureau of Ethnology, an explorer of the Colorado River, and a pioneer in understanding the western landscape.

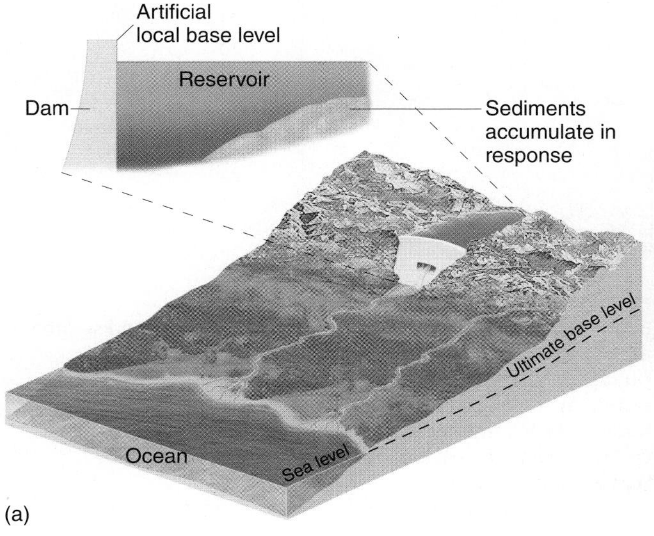

FIGURE 14.2 Ultimate and local base levels. (a) The concepts of ultimate base level (sea level) and local base level (natural, such as a lake, or artificial, such as a dam). Note how base level curves gently upward from the sea as it is traced inland; this is the theoretical limit for stream erosion. (b) Glen Canyon Dam became a local base level on the Colorado River when it closed off the canyon, in effect starving the Grand Canyon of needed sediment. [Bobbé Christopherson.]

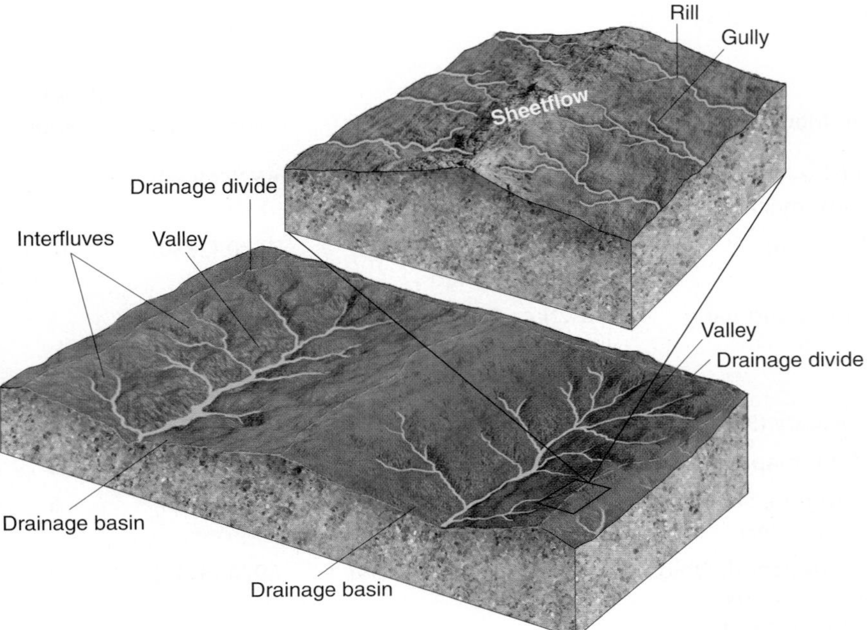

FIGURE 14.3 A drainage basin. A drainage divide separates the drainage basin and its watershed from other basins.

Powell recognized that not every landscape has degraded all the way to sea level; clearly, other intermediate base levels are in operation. A *local base level*, or temporary one, may control the lower limit of local streams for a region. The local base level may be a river, a lake, hard and resistant rock, or the reservoir formed by a human-made dam (Figure 14.2b). In arid landscapes, with their intermittent precipitation, valleys, plains, or other low points determine local base level.

Over time, the work of streams modifies the landscape dramatically. Landforms are produced by two basic processes: (1) erosive action of flowing water and (2) deposition of stream-transported materials. Before we discuss these processes, let us begin by examining a basic fluvial unit—the drainage basin.

Drainage Basins

Figure 14.3 illustrates drainage basin concepts. Every stream has a **drainage basin**, ranging in size from tiny to vast. Ridges form *drainage divides* that define the catchment (water-receiving) area of every drainage basin. That is, the ridges are the dividing lines that control into which basin runoff water from precipitation drains. In any drainage basin, water initially moves downslope in a thin film of **sheetflow**, or *overland flow*. High ground that separates one valley from another and directs sheetflow is an *interfluve*. Surface runoff concentrates in *rills*, which are small-scale downhill grooves that may develop into deeper *gullies* and then into a stream in the valley.

The drainage basin acts as a collection system for water and sediment from many subsystems, from erosion in the headwaters, to transportation throughout, to deposition in the lower portions. Figure 14.4 illustrates the drainage basin system operation from the smallest rills and gullies, to the main tributaries and stream trunk, to the dispersing lower extremities and mouth where the river meets the ocean or another water body.

Drainage Divides and Basins Several high drainage divides, the **continental divides**, are situated in the United States and Canada. These are extensive mountain and highland regions separating drainage basins, sending flows to the Pacific, the Gulf of Mexico, the Atlantic, Hudson Bay, or the Arctic Ocean. The principal drainage divides and drainage basins in the United States and Canada are mapped in Figure 14.5. These divides form water-resource regions and provide a spatial framework for water-management planning.

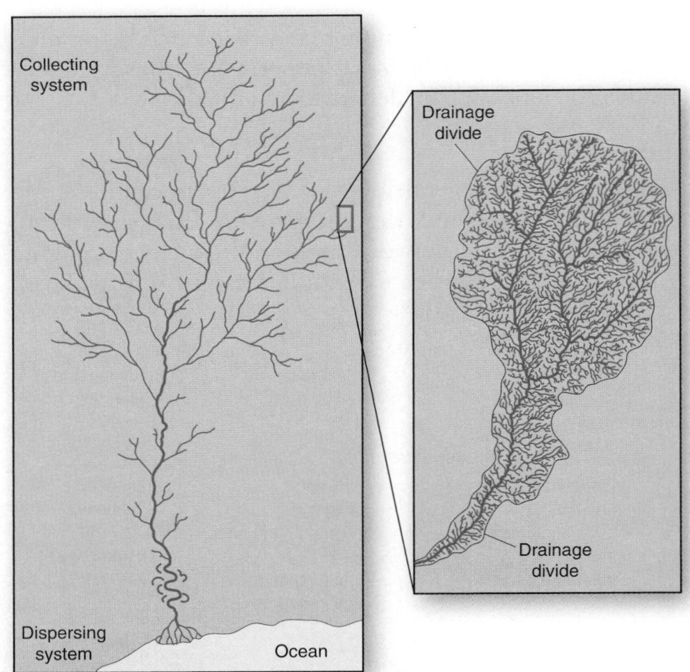

FIGURE 14.4 A drainage basin system. Beginning with intricate drainage patterns, water and sediment are collected and delivered to the stream tributaries. This fluidized mass transports to the main trunk and the dispersing system at the end of the stream, where it enters another water body. [Adapted from W. K. Hamblin and E. H. Christiansen, *Earth's Dynamic System*, 10th ed., Pearson Prentice Hall, 2004; Figure 12.3, p. 300.]

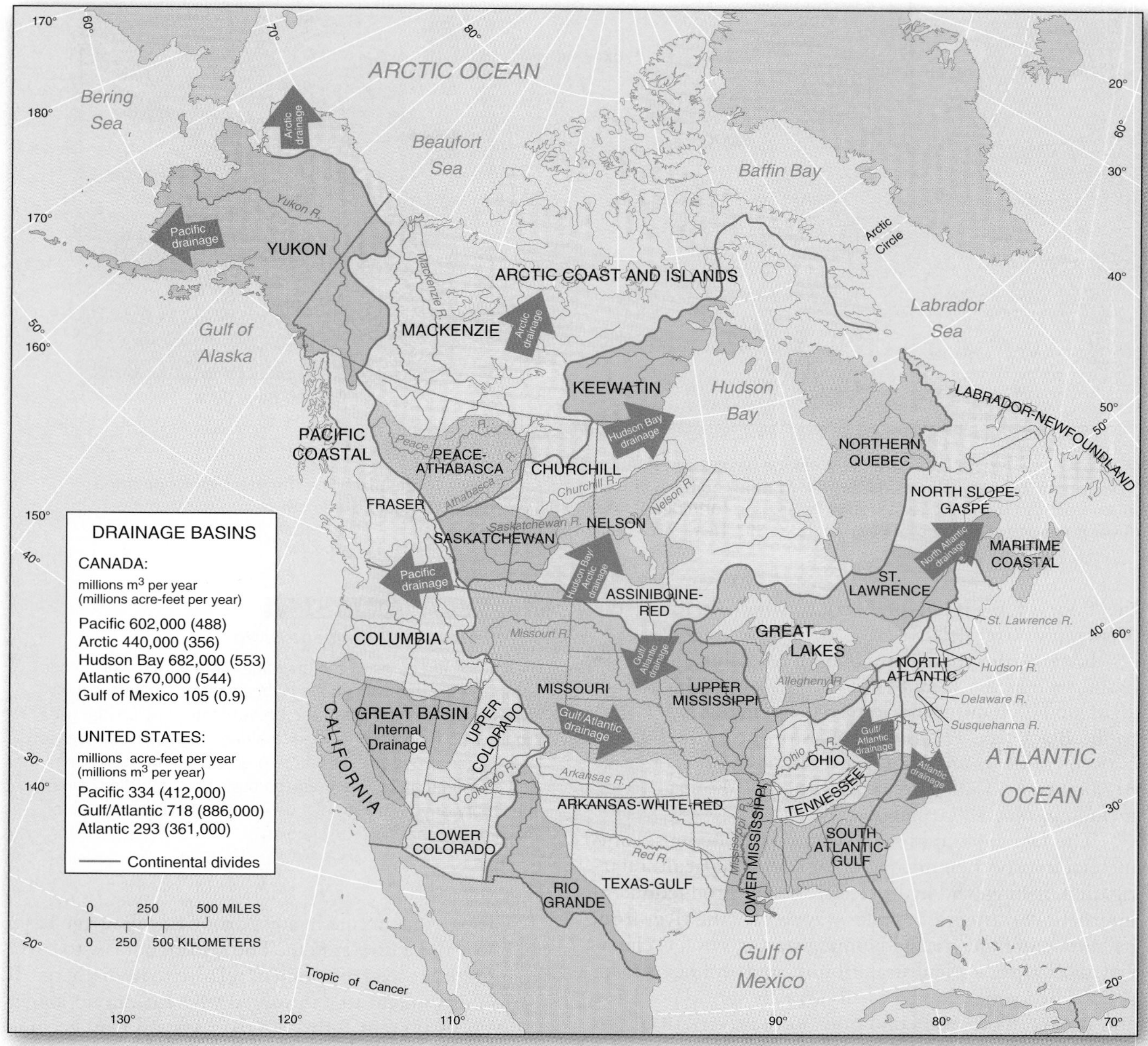

FIGURE 14.5 Drainage basins and continental divides.
Continental divides (blue lines) separate the major drainage basins that empty through the United States into the Pacific, Atlantic, and Gulf of Mexico and, to the north, through Canada into Hudson Bay and the Arctic Ocean. Subdividing these large-scale basins are major river basins. [After U.S. Geological Survey; *The National Atlas of Canada,* 1985, "Energy, Mines, and Resources Canada"; and Environment Canada, *Currents of Change—Inquiry on Federal Water Policy—Final Report 1986.*]

A major drainage basin system is made up of many smaller drainage basins. Each drainage basin gathers and delivers its runoff and sediment to a larger basin, concentrating the volume into the main stream. A good example from Figure 14.5 is the great Mississippi–Missouri–Ohio River system, draining some 3.1 million km^2 (1.2 million mi^2), or 41% of the continental United States.

Consider the travels of rainfall in northern Pennsylvania. This water feeds hundreds of small streams that flow into the Allegheny River. At the same time, rainfall in western Pennsylvania feeds hundreds of streams that flow into the Monongahela River. The two rivers then join at Pittsburgh to form the Ohio River. The Ohio flows southwestward and at Cairo, Illinois, connects with the Mississippi River, which eventually flows on past New Orleans and disperses into the Gulf of Mexico. Each contributing tributary, large or small, adds its discharge, pollution, and sediment load to the larger river. In our example, sediment weathered and eroded in Pennsylvania is transported thousands of kilometers and accumulates on

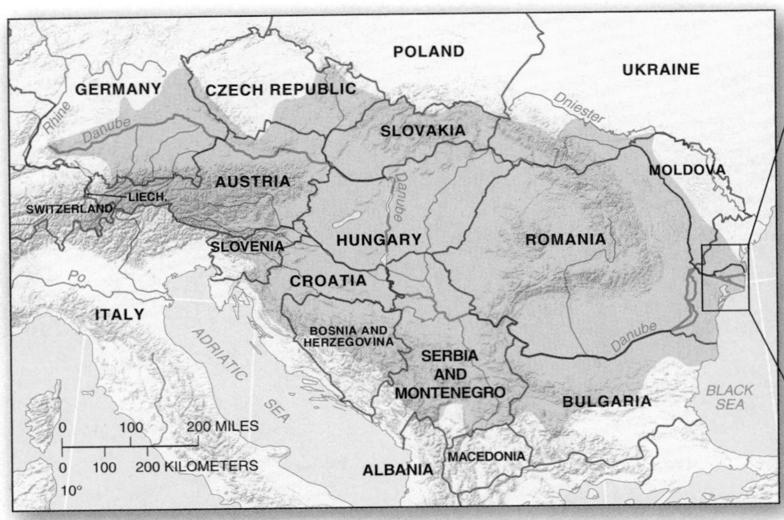

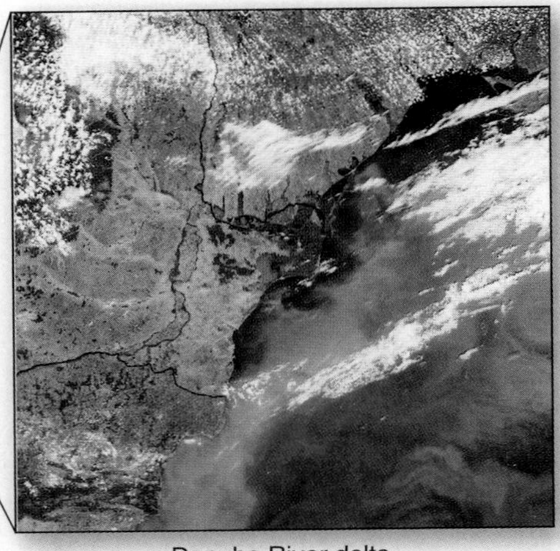
Danube River delta

FIGURE 14.6 **An international drainage basin—the Danube River.** The Danube crosses or forms the border of nine countries as it flows across Europe to the Black Sea. The river spews polluted discharge into the Black Sea through its arcuate-form delta. A toxic spill disaster in Hungary that made it into the Danube River is imaged at the end of Chapter 13, p. 397. [*Terra* image, June 15, 2002, NASA/GSFC.]

the floor of the Gulf of Mexico, where it forms the Mississippi River delta.

Imagine the complexity of an international drainage basin, such as the Danube River in Europe, which flows 2850 km (1770 mi) from western Germany's Black Forest to the Black Sea. The river crosses or forms the borders between nine countries (Figure 14.6). A total area of 817,000 km^2 (315,000 mi^2) falls within the drainage basin, including some 300 tributaries.

The Danube serves many economic functions: commercial transport, municipal water source, agricultural irrigation, fishing, and hydroelectric power production. An international struggle is under way to save the river from its burden of industrial and mining wastes, sewage, chemical discharge, agricultural runoff, and drainage from ships. The many shipping canals actually spread pollution and worsen biological conditions in the river. All of this pollution passes through Romania and the deltaic ecosystems in the Black Sea. The river is widely regarded as one of the most polluted on Earth.

Political changes in Europe in 1989 allowed the first scientific analysis of the entire river system. The United Nations Environment Programme (UNEP) and the European Union, along with other organizations, are dedicated to clearing the Danube, saving the deltaic ecosystems, and restoring the environment of this valuable resource; see http://www.icpdr.org/.

Drainage Basins as Open Systems Drainage basins such as the Danube are open systems. Inputs include precipitation and the minerals and rocks of the regional geology. Energy and materials are redistributed as the stream constantly adjusts to its landscape. System outputs of water and sediment disperse through the mouth of the river, into a lake, another river, or the ocean, as shown in Figure 14.4.

II CRITICAL THINKING 14.1

Locate your drainage basin

Determine the name of the drainage basin within which your campus is located. Where are its headwaters? Where is the river's mouth? If you are in the United States or Canada, use Figure 14.5 to locate the larger drainage basins and divides for your region. Does any regulatory organization oversee planning and coordination for the drainage basin you identified? Does the library or your geography department have topographic maps on file that cover this region?

Change that occurs in any portion of a drainage basin can affect the entire system. The stream adjusts to carry the appropriate load of sediment relative to its discharge. If a stream is brought to a threshold where it can no longer maintain its present form, the river system may become destabilized, initiating a transition period to a more stable condition. A river system constantly struggles toward equilibrium among the interacting variables of discharge, sediment load, channel shape, and channel steepness.

An Example: The Delaware River Basin Let us look at the Delaware River basin, within the Atlantic Ocean drainage region (Figure 14.7). The Delaware River headwaters are in the Catskill Mountains of New York. This basin encompasses 33,060 km^2 (12,890 mi^2) and includes parts of five states along the river's length, 595 km (370 mi) from its headwaters to its mouth. The river system ends at Delaware Bay, which eventually enters the Atlantic Ocean. Topography varies from low-relief coastal plains to the Appalachian Mountains in the north. The entire basin lies within a humid, temperate climate and receives an average annual precipitation of 120 cm (47.2 in.).

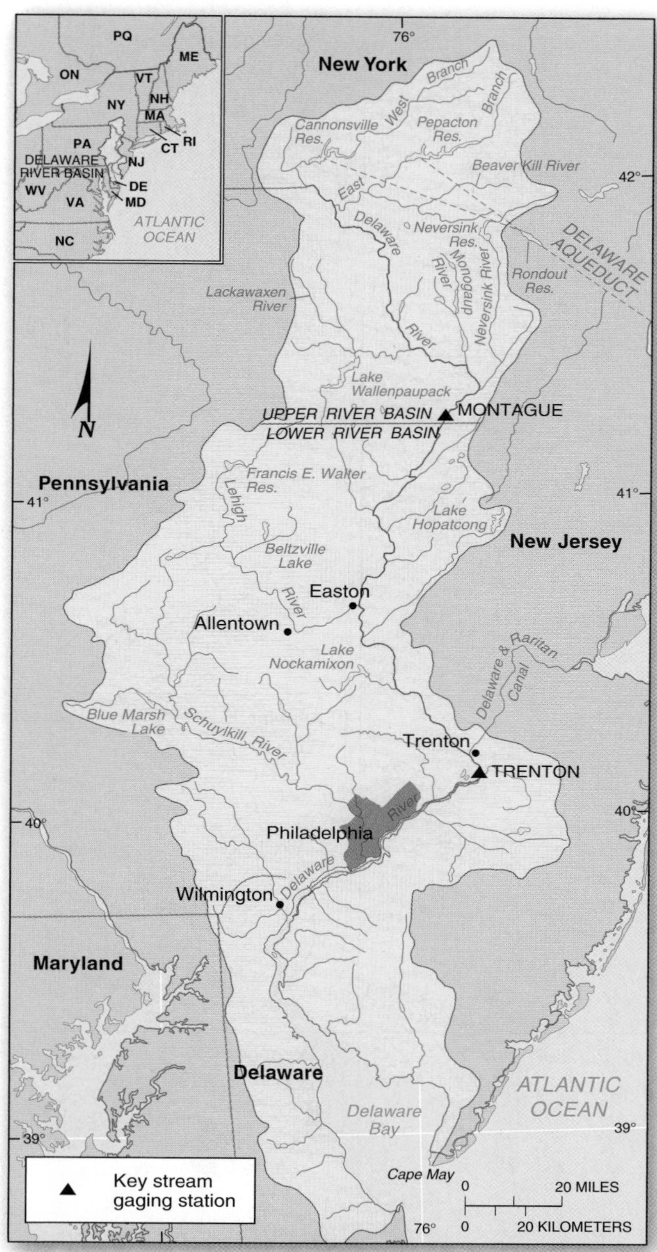

FIGURE 14.7 The Delaware River drainage basin. Several major ongoing studies by the USGS and universities in the region are analyzing the impacts of climate change on this basin. [Map adapted from USGS.]

The river provides water for an estimated 20 million people, not only within the basin, but in cities outside the basin as well. Several major conduits export water from the Delaware River. Note the Delaware Aqueduct to New York City (in the north) and the Delaware and Raritan Canal (near Trenton). Several dams control flow releases, as their reservoirs provide storage for dry periods. The sustainability of this water resource is critical to the entire region. Clearly, effective planning for a drainage basin requires regional cooperation and careful spatial analysis of all variables.

Internal Drainage Most streams find their way to progressively larger rivers and eventually into the ocean. In some regions, however, stream drainage does not reach the ocean. Instead, the water leaves the drainage basin by means of evaporation or subsurface gravitational flow. Such streams terminate in areas of **internal drainage**.

Portions of Asia, Africa, Australia, Mexico, and the western United States have regions with such internal drainage patterns. Internal drainage defines the region of the Great Basin of the western United States, as shown on maps in Figures 14.5 and 15.22a. For example, the Humboldt River flows across Nevada to the west, where evaporation and seepage losses to groundwater make it disappear into the Humboldt "sink." Many streams and creeks flow into the Great Salt Lake, but its only outlet is evaporation, so it exemplifies a region of internal drainage. The Dead Sea region in the Middle East and the areas around the Aral Sea and Caspian Sea in Asia have no outlet to the ocean.

Drainage Density and Patterns A primary feature of any drainage basin is its **drainage density**, determined by dividing the total length of all stream channels in the basin by the area of the basin. The number and length of channels in a given area reflect the landscape's regional geology and topography.

The *Landsat* image and topographic map in Figure 14.8 are of the Ohio River drainage near the junction of West Virginia, Ohio, and Kentucky. The high-density drainage pattern and intricate dissection of the land occur because the region has generally level, easily eroded sandstone, siltstone, and shale strata and a humid mesothermal climate. Fluvial action and other denudation processes are responsible for this dissected topography.

The **drainage pattern** is the arrangement of channels in an area. Patterns are quite distinctive, for they are determined by a combination of regional steepness, variable rock resistance, variable climate, variable hydrology, relief of the land, and structural controls imposed by the underlying rocks. Consequently, the drainage pattern of any land area on Earth is a remarkable visual summary of every characteristic—geologic and climatic—of that region.

Common Drainage Patterns The seven most common drainage patterns are shown in Figure 14.9. A most familiar pattern is *dendritic drainage* (Figure 14.9a). This treelike pattern (from the Greek word *dendron*, meaning "tree") is similar to that of many natural systems, such as capillaries in the human circulatory system, or the vein patterns in leaves, and tree roots. Energy expended by this drainage system is efficient because the overall length of the branches is minimized. Figure 14.8 shows dendritic drainage; on the satellite image and topographic map, you can trace the branching pattern of streams.

The *trellis drainage* pattern (Figure 14.9b) is characteristic of dipping or folded topography. Such drainage exists in the nearly parallel mountain folds of the Ridge and Valley Province in the eastern United States. Refer to Figure 12.17, where a satellite image of this region shows

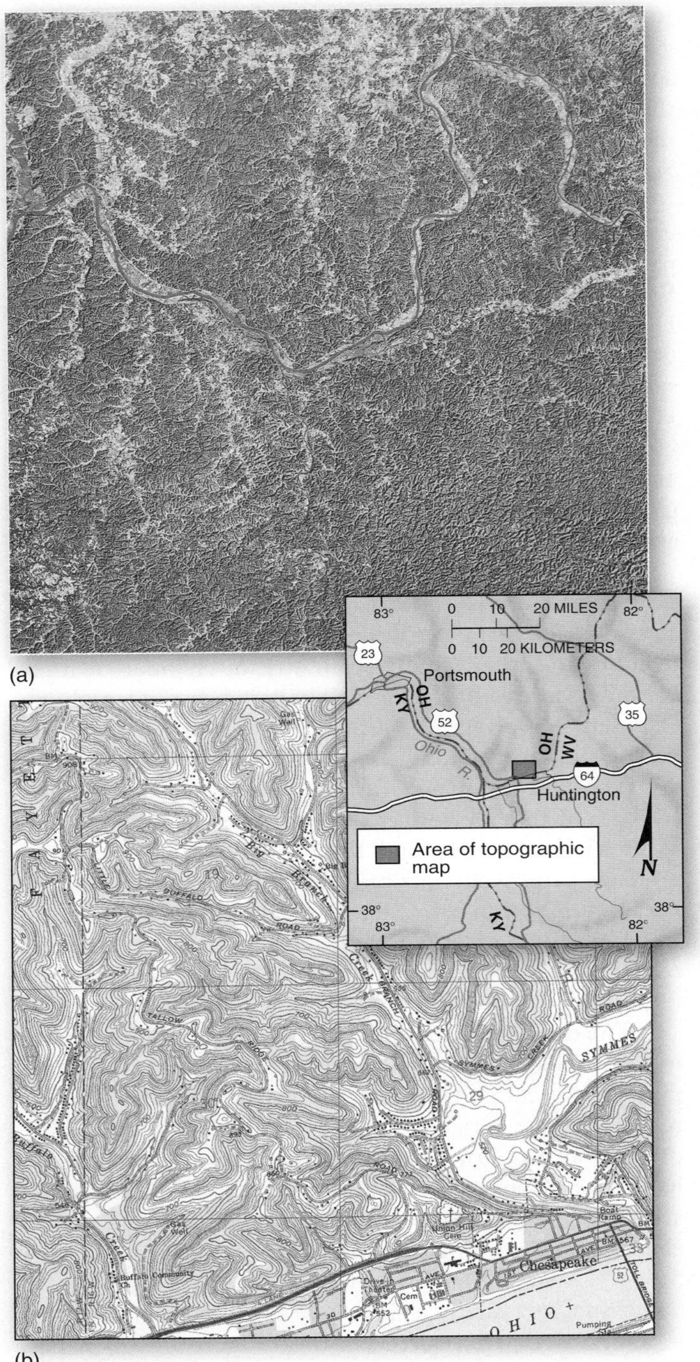

FIGURE 14.8 A stream-dissected landscape. (a) A highly dissected topography (featuring a dendritic drainage pattern, shown in Figure 14.9) around the junction of West Virginia, Ohio, and Kentucky. All of these borders are formed by rivers. (b) USGS topographic map covering the area north of Huntington, West Virginia, reveals the intricate complexity of dissected landscapes. (The image and map are not at the same scale.) [(a) *Landsat* image NASA. (b) Huntington Quadrangle, USGS.]

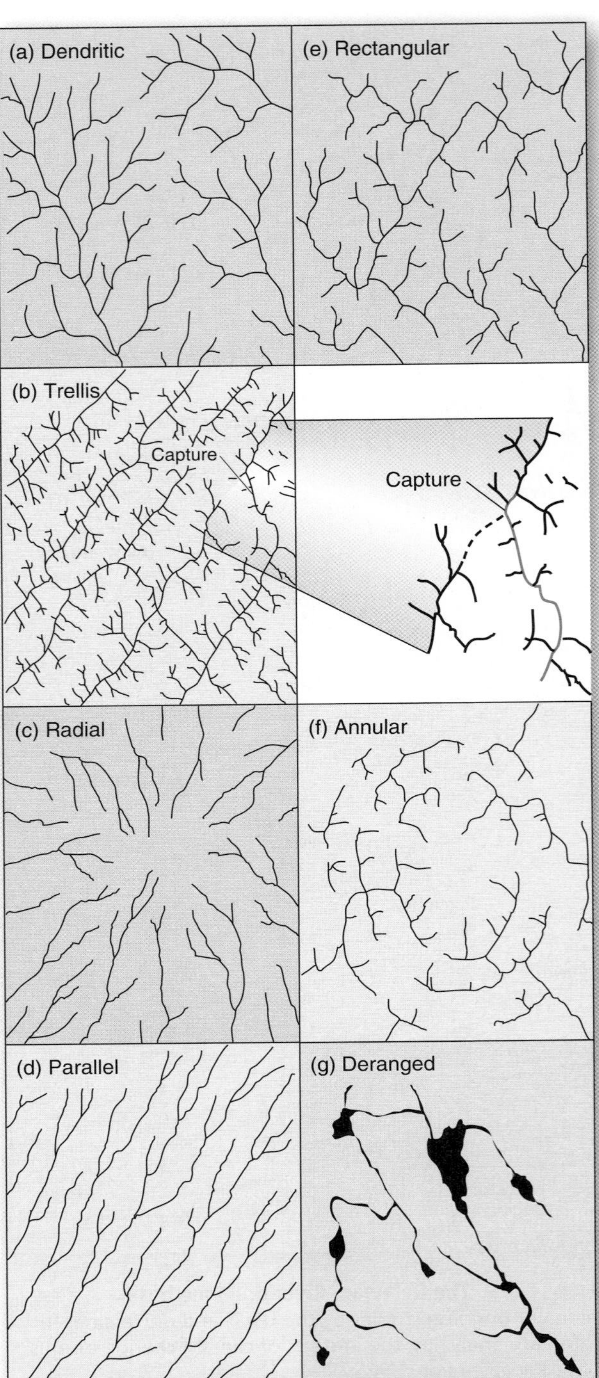

FIGURE 14.9 The seven most common drainage patterns. Each pattern is a visual summary of all the geologic and climatic conditions of its region. [(a) through (g) after A. D. Howard, "Drainage analysis in geological interpretation: A summation," *Bulletin of American Association of Petroleum Geologists* 51 (1967): 2248, adapted by permission.]

its distinctive drainage pattern. Here drainage patterns are influenced by rock structures of variable resistance and folded strata. Parallel folded structures direct the principal streams, whereas smaller dendritic tributary streams are at work on nearby slopes, joining the main streams at right angles, like a plant trellis.

The inset sketch in Figure 14.9b suggests that a headward-eroding part of one stream (to the lower right of the inset) could break through a drainage divide and *capture* the headwaters of another stream in the next valley—and indeed this does happen. The dotted line is the abandoned former channel. The sharp bends in two of the streams in the illustration are *elbows of capture* and are evidence that one stream has breached a drainage divide. This type of capture, or *stream piracy*, can also occur in other drainage patterns.

The remaining drainage patterns in Figure 14.9c–g are responses to other specific structural conditions:

- A *radial* drainage pattern (c) results when streams flow off a central peak or dome, such as occurs on a volcanic mountain.
- *Parallel* drainage (d) is associated with steep slopes.
- A *rectangular* pattern (e) is formed by a faulted and jointed landscape, which directs stream courses in patterns of right-angle turns.
- *Annular* patterns (f) are produced by structural domes, with concentric patterns of rock strata guiding stream courses. Figure 12.10c provides an example of annular drainage on a dome structure.
- In areas having disrupted surface patterns, such as the glaciated shield regions of Canada, northern Europe, and some parts of Michigan and other states, a *deranged* pattern (g) is in evidence, with no clear geometry in the drainage and no true stream valley pattern (see Figure 17.14a).

CRITICAL THINKING 14.2

Identifying drainage patterns

Examine the photograph in Figure 14.2.1, where you see two distinct drainage patterns. Of the seven types illustrated in Figure 14.9, which two patterns are most like those in the aerial photo, made in central Montana? Looking back in this discussion to the example in Figure 14.4, which drainage pattern do you assign to it? Explain. The next time you fly in an airplane remember to get a window seat to observe such drainage patterns.

FIGURE 14.2.1 Two drainage patterns dominate this scene from central Montana, in response to local relief and rock structure. [Bobbé Christopherson.]

Occasionally, drainage patterns occur that are discordant with the landscape through which they flow. For example, a drainage system may flow in apparent conflict with older, buried structures that have been uncovered by erosion, so that the streams appear to be *superimposed*. Where an existing stream flows as rocks are uplifted, the stream keeps its original course, cutting into the rock in a pattern contrary to its structure. Such a stream is a *superposed stream* (the stream cuts across weak and resistant rocks alike). A few examples include Wills Creek, cutting a water gap through Haystack Mountain at Cumberland, Maryland; the Columbia River through the Cascade Mountains of Washington; and the River Arun, which cuts across the Himalayas.

Fluvial Processes and Landforms

A mass of water positioned above base level in a stream has potential energy. As the water flows downslope, or downstream, under the influence of gravity, this energy becomes kinetic energy. The rate of this conversion from potential to kinetic energy depends on the steepness of the stream channel and the volume of water involved.

Stream Discharge

Discharge, or the stream's volume of flow per unit of time, is dependent on stream channel width and depth and on the velocity of the flow. Discharge is calculated by multiplying these three variables for a specific cross section of the channel, as stated in the simple expression

$$Q = wdv$$

where Q = discharge, w = channel width, d = channel depth, and v = stream velocity. Discharge is expressed in either cubic meters per second (m^3/s) or cubic feet per second (cfs, ft^3/s).

As Q increases in a downstream direction with inflow from tributaries, some combination of channel width, channel depth, and stream velocity must also increase. Note that velocity may increase with greater discharge downstream, despite the common misperception that flow becomes more sluggish. The increased velocity often is masked by the apparent smooth, quiet flow of the water.

Discharge also changes over time at any specific channel cross section, especially during periods of high flow. Figure 14.10 shows changes in the San Juan River channel in Utah that occurred during a flood. Greater discharge increases the velocity and therefore the capacity of the river to transport sediment as the flood progresses. As a result, the river's ability to scour materials from its bed is enhanced. Such scouring represents a powerful clearing action, especially in the excavation of alluvium.

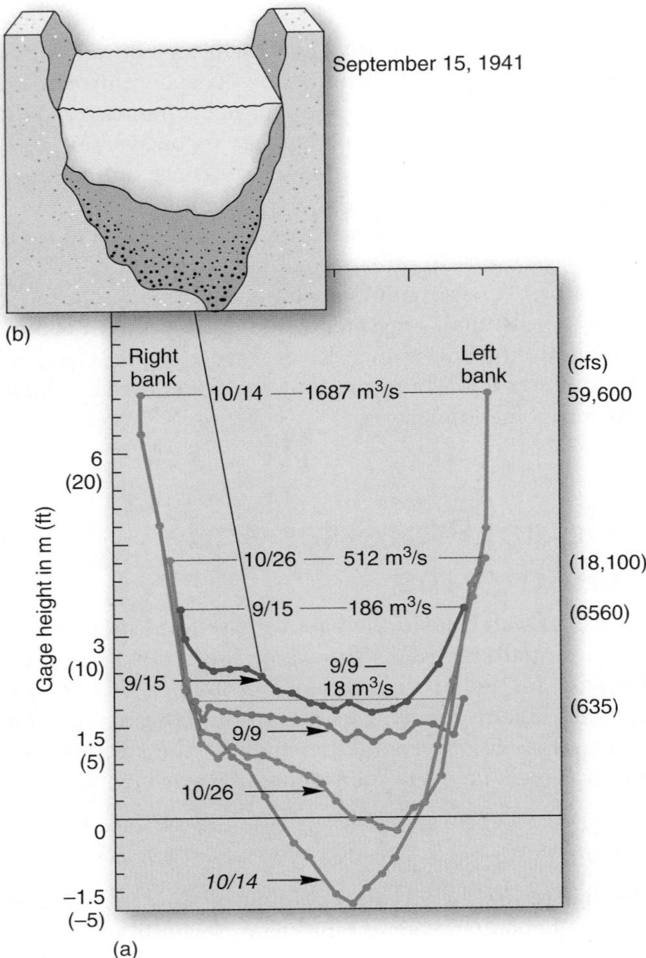

FIGURE 14.10 **A flood affects a stream channel.** (a) Channel cross sections showing the effects of a 1941 flood on the San Juan River near Bluff, Utah. (b) Detail of channel profile on September 15, 1941. [USGS Professional Paper 252, p. 32.]

You can see in Figure 14.10 that the San Juan River's channel was deepest on October 14, when floodwaters were highest (blue line). Then, as the discharge returned to normal, the kinetic energy of the river was reduced, and the bed again filled as sediment redeposited. You can see this process in the progression of the red, green, and purple lines. The flood and scouring process graphed in Figure 14.10 moved a depth of about 3 m (10 ft) of sediment from this channel cross section. This type of channel adjustment occurs as the system continuously works toward equilibrium, in an effort to balance discharge and sediment load with channel form.

Dam-Controlled Discharge and Sediment Redistribution

Glen Canyon Dam on the Colorado River near the Utah–Arizona border controls discharge and blocks sediment from flowing into the Grand Canyon downstream (see the map in Focus Study 15.1). Consequently, over the years, the river's sediment supply was cut off, starving the river's beaches for sand, disrupting fisheries, and depleting backwater channels of nutrients.

In 1996, 2004, and 2008, an unprecedented series of scouring–redistribution–deposition experiments took place as the Grand Canyon was artificially flooded with dam-controlled releases. The first test lasted for 7 days. The later tests were of shorter duration and were timed to coincide with floods in tributaries that supplied fresh sediment to the system.

The results were mixed and the benefits turned out to be limited, as ecosystems were disrupted and some sediment deposits were immediately eroded away. Present thinking is that the problem is sediment supply—that is, not enough sediment is present in the system to build beaches and improve habitat. Sediment redistribution is still being monitored. See http://www.gcmrc.gov/ for progress reports and analysis.

Exotic Streams

In most river basins in human regions, discharge increases in a downstream direction. The Mississippi River is typical. It starts as many small brooks and grows to a mighty river pouring into the Gulf of Mexico. However, if a stream originates in a humid region and subsequently flows through an arid region, this relationship may change. High potential evapotranspiration rates in arid regions may cause discharge to decrease with distance downstream. Such a stream is an **exotic stream** (*exotic* means "of foreign origin").

The Nile River exemplifies exotic streams. This great river, Earth's longest, drains much of northeastern Africa. But as it courses through the deserts of Sudan and Egypt, it loses water, instead of gaining it, because of evaporation and withdrawal for agriculture. By the time it empties into the Mediterranean Sea, the Nile's flow has dwindled so much that it ranks only 36th in discharge.

The Colorado River in the United States is also an exotic stream. Discharge decreases with distance from its source; in fact, the river no longer produces enough natural discharge to reach its mouth in the Gulf of California—only some agricultural runoff remains at its delta. The exotic Colorado River is depleted not only by passage across desert lands but also by upstream removal of water for agriculture and municipal uses; see Focus Study 15.1.

Stream Erosion and Transportation

A stream's erosional turbulence and abrasion carve and shape the landscape through which it flows. **Hydraulic action** is the work of flowing water alone. Running water causes hydraulic squeeze-and-release action that loosens and lifts rocks. As this debris moves along, it mechanically erodes the streambed further through the process of **abrasion**, with rock particles grinding and carving the streambed like liquid sandpaper.

The upstream tributaries in a drainage basin usually have small and irregular discharges, and most of the stream's energy is expended in turbulent eddies. As a result, hydraulic action in these upstream sections is at a maximum,

(a)

(b)

FIGURE 14.11 **Stream velocity and discharge increase together.**
(a) La Mina cascades in El Yunque National Park, Puerto Rico, has low discharge and low velocity, but is turbulent. (b) In contrast is the high-discharge, high-velocity, smooth water flow of the Ocmulgee River near Jacksonville, Georgia. [Bobbé Christopherson.]

whereas the coarse-textured load of such a stream is small. The downstream portions of a river, however, move much larger volumes of water past a given point and carry larger suspended loads of sediment (Figure 14.11).

You may have watched a river or creek after a rainfall, the water colored brown by the high sediment load being transported. The amount of material available to a stream depends on topographic relief, the nature of rock and soil through which the stream flows, climate, vegetation, and human activity in a drainage basin. *Competence*, which is a stream's ability to move particles of a specific size, is a function of stream velocity and the energy available to suspend materials. *Capacity* is the total possible load that a stream can transport. Four processes transport eroded materials: solution, suspension, saltation, and traction; each is shown in action in Figure 14.12.

Solution refers to the **dissolved load** of a stream, especially the chemical solution derived from minerals such as limestone or dolomite or from soluble salts. The main contributor of material in solution is chemical weathering. Sometimes the undesirable salt content that hinders human use of some rivers comes from dissolved rock formations and from springs in the stream channel; as an example, the San Juan and Little Colorado rivers, which flow into the Colorado River near the Utah–Arizona border, add dissolved salts to the system.

The **suspended load** consists of fine-grained, clastic particles (bits and pieces of rock). They are held aloft in the stream, with the finest particles not deposited until the stream velocity slows nearly to zero. Turbulence in the water, with random upward motion, is an important mechanical factor in holding a load of sediment in suspension.

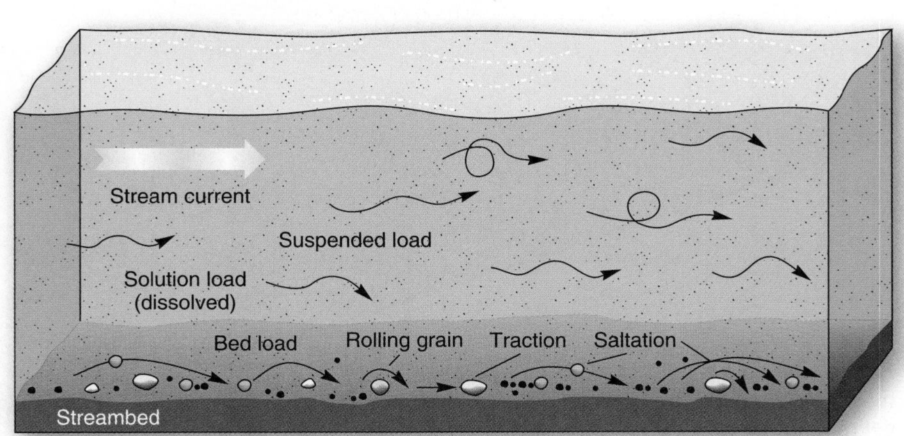

FIGURE 14.12 **Fluvial transport.**
Fluvial transportation of eroded materials through suspension, solution, saltation, and traction.

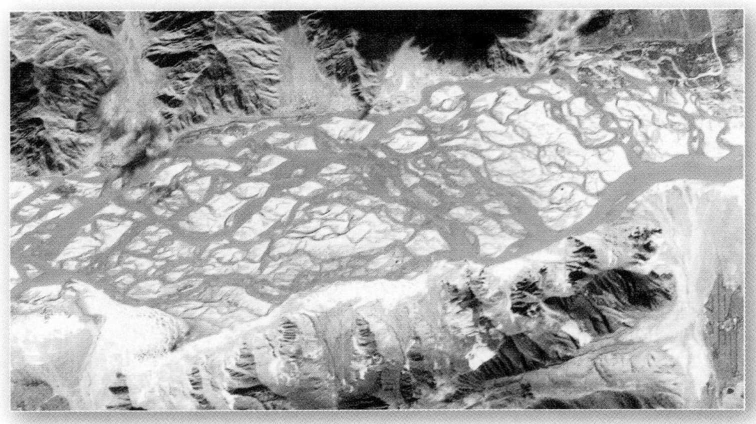

FIGURE 14.13 A braided stream. A 15-km (9.3-mi) stretch of the braided Brahmaputra River channel some 35 km (22 mi) south of Lhasa, Tibet, in a narrow valley south of the Tibetan Plateau. [ISS, JSC/NASA.]

Bed load refers to coarser materials that are dragged, rolled, or pushed along the streambed by **traction** or by **saltation**, a term referring to the way particles bounce along in short hops and jumps (from the Latin *saltim*, which means "by leaps or jumps"). Particles transported by saltation are too large to remain in suspension, but are not limited to the sliding and rolling motion of traction. These processes relate directly to a stream's velocity and its ability to retain particles in suspension. With increased kinetic energy, parts of the bed load are rafted upward and become suspended load.

If the load (bed and suspended) exceeds a stream's capacity, sediment accumulates as **aggradation** (the opposite of degradation), and the stream channel builds up through deposition. With excess sediment, a stream might become a maze of interconnected channels that form a **braided stream** pattern (Figure 14.13). Braiding often occurs when reduced discharge lowers a stream's transporting ability such as after flooding, or when a landslide occurs upstream, or when load increases where weak banks of sand or gravel exist. Braided rivers commonly occur in glacial environments, which have abundant supplies of sediment and steep channel slopes, as is the case with the Brahmaputra River in Tibet. Glacial action and erosion produced the materials that exceed stream capacity in this river channel.

Flow and Channel Characteristics

Flow characteristics of a stream are best seen in a cross-sectional view. The greatest velocities in a stream are near the surface at center channel (Figure 14.14), corresponding to the deepest part of the stream channel. Velocities

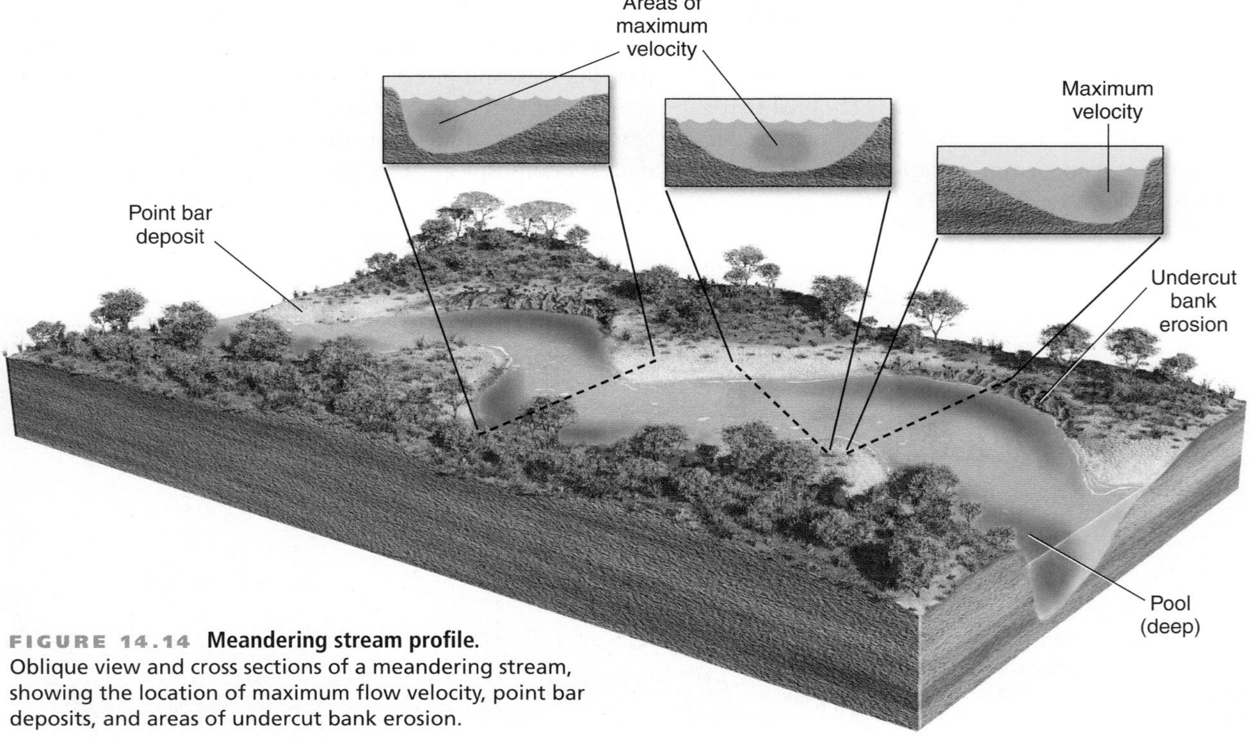

FIGURE 14.14 Meandering stream profile. Oblique view and cross sections of a meandering stream, showing the location of maximum flow velocity, point bar deposits, and areas of undercut bank erosion.

decrease closer to the sides and bottom of the channel because of the frictional drag on the water flow. The portion of the stream flowing at maximum velocity moves diagonally across the stream from bend to bend in a meandering stream.

Flow becomes *turbulent* in shallow streams, or where the channel is rough, as in a section of rapids. Small eddies are caused by friction between the streamflow and the channel sides and bed. Complex turbulent flows propel sand, pebbles, and even boulders, increasing the suspension, traction, and saltation.

Where slope is gradual, stream channels develop a sinuous (snakelike) form, weaving across the landscape. This action produces a **meandering stream.** The tendency to meander is evidence of a river system's struggle to operate with least effort between self-organizing order (equilibrium) and chaotic disorder in nature.

The outer portion of each meandering curve is subject to the fastest water velocity and therefore the greatest scouring erosive action; it can be the site of a steep bank called the **undercut bank**, or *cutbank* (Figures 14.14 and 14.15). In

FIGURE 14.15 **Meandering stream development.**
(a) Itkillik River in Alaska. (b) Development of a river meander and an oxbow lake simplified in four stages.
[(a) USGS.]

MG Animation Stream Processes, Floodplains, Oxbow Lake Formation contrast, the inner portion of a meander experiences the slowest water velocity and thus receives sediment fill, forming a **point bar** deposit. As meanders develop, these scour-and-fill features gradually work at stream banks. As a result, the landscape near a meandering river bears meander scars of residual deposits from previous river channels (see Figure 14.22). The photograph of the Itkillik River in Alaska (Figure 14.15a) shows both meanders and meander scars.

Meandering streams create a remarkable looping pattern on the landscape, as shown in the four-part sequence in Figure 14.15b. In (1), the stream erodes its outside bank as the curve migrates downstream, forming a neck. In (2), the narrowing neck of land created by the looping meander eventually erodes through and forms a *cutoff* in (3). A cutoff marks an abrupt change in the stream's lateral movements—the stream becomes straighter.

When the former meander becomes isolated from the rest of the river, the resulting **oxbow lake** (4) may gradually fill with organic debris and silt or may again become part of the river when it floods. The Mississippi River is many miles shorter today than it was in the 1830s because of artificial cutoffs that were dredged across meander necks to improve navigation and safety. How many of these stream features—meanders, oxbow lakes, cutoffs—can you spot in the aerial photo in Figure 14.16?

Streams often form natural political boundaries, as we saw with the Danube River earlier, but it is easy to see how disagreements might arise when boundaries are based on river channels that change course. For example, the Ohio, Missouri, and Mississippi rivers can shift their positions quite rapidly during times of flood and therefore make confusing the boundaries based on them. Carter Lake, Iowa, provides us with a case in point (Figure 14.17). The Nebraska–Iowa border was originally placed mid-channel in the Missouri River. In 1877, the river cut off the meander loop around the town of Carter Lake, leaving the town "captured" by Nebraska. The new oxbow lake was called Carter Lake and still is the state boundary.

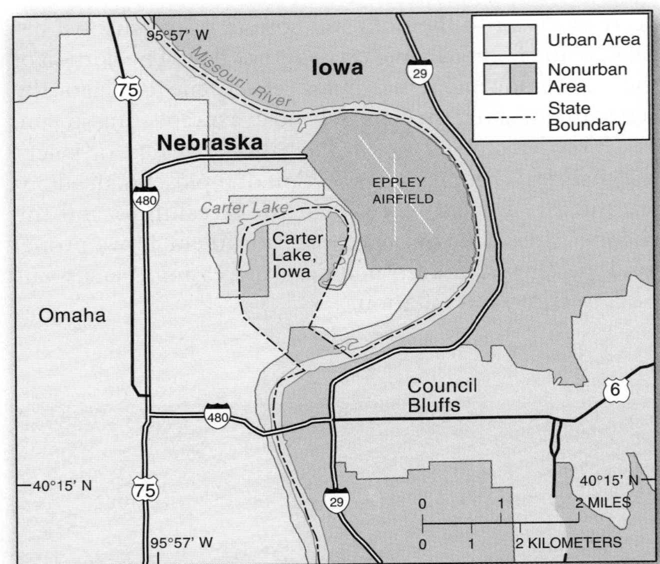

FIGURE 14.17 A stranded town.
Carter Lake, Iowa, sits within the curve of a former meander that was cut off by the Missouri River. The city and oxbow lake remain part of Iowa even though they are stranded within Nebraska.

FIGURE 14.16 Meandering stream.
A meandering stream works its channel in eastern Wyoming. See if you can identify former meanders that were cut off, oxbow lakes, and former stream channels. [Bobbé Christopherson.]

Boundaries should always be fixed by surveys independent of river locations because border disputes result when river channels change course. Along the Rio Grande near El Paso, Texas, and the Colorado River between Arizona and California, such surveys have permanently established political boundaries separate from changing river locations.

Stream Gradient

Every stream has a degree of inclination or **gradient**, which is the decline in elevation from its headwaters to its mouth. A stream's gradient generally forms a concave-shaped slope (Figure 14.18). Characteristically, the *longitudinal profile* of a stream (a side view) has a steeper slope upstream and a more

GEO REPORT 14.1 *River rumbles*

The early explorers who visited the Grand Canyon reported in their journals that they were kept awake at night by the thundering sound of the river and rapids and bed load. Imagine the tremendous quantity of material being moved along by the natural Colorado River before any dams were built. In addition, dam-controlled flow releases no longer provide enough discharge to move the coarser particles in the bed load.

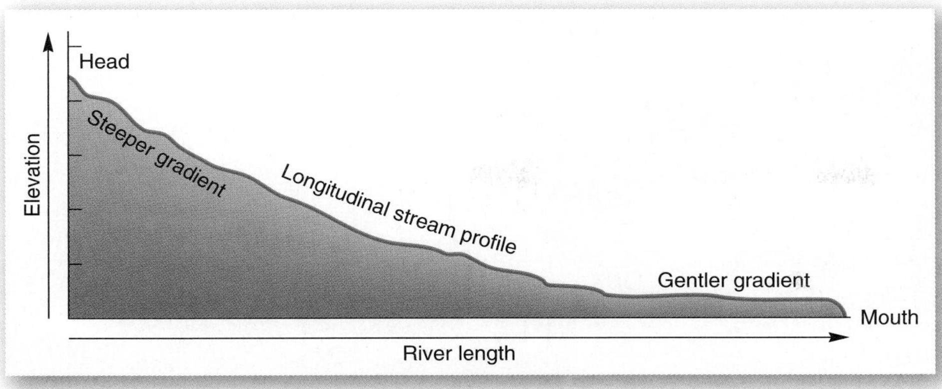

FIGURE 14.18 An ideal longitudinal profile.
The sloping profile of a stream is its gradient. Upstream segments have a steeper gradient; downstream, the gradient is gentler. The middle and lower portions in the illustration appear graded, or in dynamic equilibrium.

gradual slope downstream. This curve assumes its shape for complex reasons related to the stream's ability to do enough work to accomplish the transport of the load it receives.

Streams tend to adjust their gradient and other channel characteristics so that they can move the sediment supply. A **graded stream** condition occurs when channels adjust their slope, size, and shape so that a stream has just enough energy to transport its sediment load. This condition is a state of *dynamic equilibrium* among discharge, sediment load, and channel form. The stream, channel, and landscape work together to maintain this balance within a drainage basin. Both high-gradient and low-gradient streams can achieve a graded condition. A stream's profile tells geographers about characteristics of slope, discharge, and load.

> A graded stream is one in which, over a period of years, *slope* is delicately adjusted to provide, with available *discharge* and with prevailing *channel characteristics*, just the velocity required for transportation of the load supplied from the drainage basin.*

*J. H. Mackin, "Concept of the graded river," *Geological Society of America Bulletin* 59 (1948): 463.

Attainment of a graded condition does not mean that the stream is at its lowest gradient, but rather that it has reached a balance among erosion, transportation, and deposition over time along a specific portion of the stream. One problem with applying the graded stream concept is that an individual stream can have both graded and ungraded portions. It may have graded sections without having an overall graded slope. In fact, variations and interruptions are the rule rather than the exception.

Stream gradient may be affected by tectonic uplift of the landscape, which changes the base level. If tectonic forces slowly lift the landscape, the stream gradient will increase, stimulating renewed erosional activity. A meandering stream flowing through the uplifted landscape becomes *rejuvenated*; that is, the river actively returns to downcutting and can eventually form *entrenched meanders* in the landscape. Figure 14.19 depicts actual rejuvenated landscapes.

Nickpoints When the longitudinal profile of a stream shows an abrupt change in gradient, such as at a waterfall

(a)

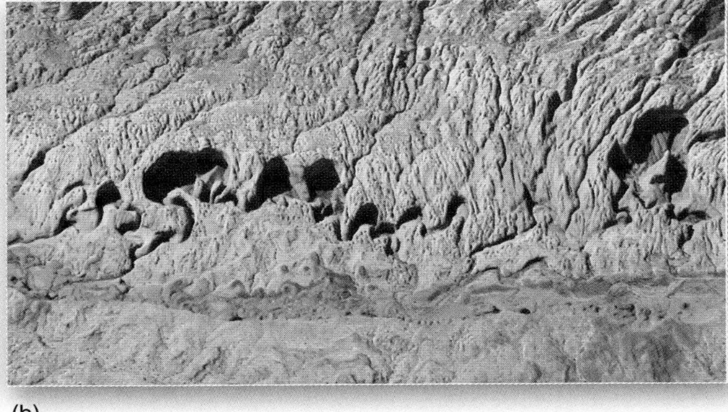
(b)

FIGURE 14.19 Aerial views of entrenched meanders.
(a) The San Juan River near Mexican Hat, Utah, cuts down into the uplifted Colorado Plateau landscape, producing the entrenched meanders of the Goosenecks of the San Juan. (b) Entrenched meanders near Hall's Creek along Waterpocket Fold in Capitol Reef National Park, Utah. [Bobbé Christopherson.]

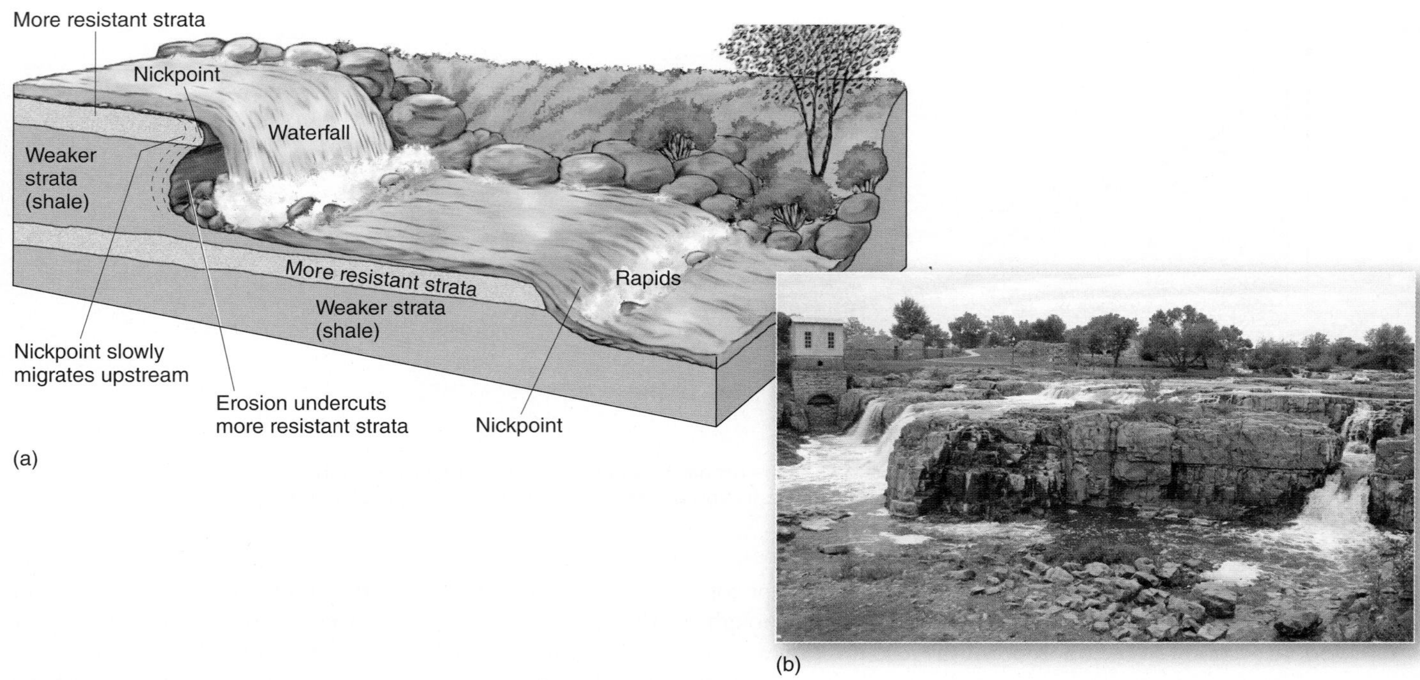

FIGURE 14.20 Nickpoints interrupt a stream profile.
(a) Longitudinal stream profile showing nickpoints produced by resistant rock strata. Potential energy is converted into kinetic energy and concentrated at the nickpoint, accelerating erosion, which will eventually eliminate the feature. (b) Falls and rapids break the flow of the Big Sioux River in Sioux Falls, South Dakota. [Bobbé Christopherson.]

or an area of rapids, the point of interruption is a **nickpoint** (also spelled *knickpoint*). At a nickpoint, the conversion of potential energy in the water at the lip of the falls to concentrated kinetic energy at the base works to eliminate the nickpoint interruption and smooth out the gradient. Figure 14.20 includes an illustration of a stream with two such interruptions as well as actual nickpoints in South Dakota.

Nickpoints can result when a stream flows across a zone of hard, resistant rock or from various tectonic uplift episodes, such as might occur along a fault line. Temporary blockage in a channel, caused by a landslide or a logjam, also could be considered a nickpoint; when the logjam breaks, the stream quickly readjusts its channel to its former grade.

Waterfalls are interesting and beautiful gradient breaks. At the edge of a fall, a stream is free-falling, moving at high velocity under the acceleration of gravity, causing increased abrasion and hydraulic action in the channel below. The increased action generally undercuts the waterfall. Eventually, the excavation will cause the rock ledge at the lip of the fall to collapse, and the waterfall will shift a bit farther upstream. The height of the waterfall is gradually reduced as debris accumulates at its base (see Figure 14.21b). Thus, a nickpoint migrates upstream, sometimes for kilometers, until it becomes a series of rapids and is eventually eliminated.

At Niagara Falls on the Ontario–New York border, glaciers advanced over the region and then receded some 13,000 years ago. In doing so, they exposed resistant rock strata that are underlain by less-resistant shales. This tilted formation is a *cuesta*, which is a ridge with a steep slope on one side and beds gently sloping away on the other side (Figure 14.21a). This Niagara escarpment actually stretches across more than 700 km (435 mi); from east of the falls, it extends northward through Ontario, Canada,

The first meanders

The term *meander* comes from the ancient Greek Maiandros River in Asia Minor (the present-day Menderes River in Turkey). The Maiandros appears in Greek mythology and written accounts as early as the second century B.C. Its meandering channel came to describe this pattern in all river channels.

FIGURE 14.21 Retreat of Niagara Falls.
(a) Headward retreat of Niagara Falls from the Niagara escarpment has taken about 12,000 years, at a pace of about 1.3 m (4.3 ft) per year. (b) The escarpment seen from below American Falls. (c) Niagara Falls, with the American Falls portion almost completely shut off by upstream controls for engineering inspection. Horseshoe Falls in the background is still flowing over its 57-m (188-ft) plunge. (d) In contrast to (c), American Falls in full discharge. [(a) After W. K. Hamblin, *Earth's Dynamic Systems,* 6th ed., Pearson Prentice Hall, Inc. © 1992, Figure 12.15, p. 246. (c) Courtesy of the New York Power Authority. (b) and (d) Bobbé Christopherson.]

and the Upper Peninsula of Michigan and then curves south through Wisconsin along the western shore of Lake Michigan and the Door Peninsula. As this less-resistant material continues to weather away, the overlying rock strata collapse, allowing Niagara Falls to erode farther upstream toward Lake Erie.

Niagara Falls is a place where natural processes labor to eliminate a nickpoint and reduce this portion of the river to a series of mere rapids. In fact, the falls have retreated more than 11 km (6.8 mi) from the steep face of the Niagara escarpment. A nickpoint is a relatively temporary and mobile feature on the landscape. In the past, engineers have used control facilities upstream to reduce flows over the American Falls at Niagara for inspection of cliff erosion; compare this reduced flow (Figure 14.21c) to the normal discharge (Figure 14.21d).

Stream Deposition and Landforms

After weathering, mass movement, erosion, and transportation, deposition is the next logical event in a sequence. In *deposition*, a stream deposits alluvium, or unconsolidated sediment, thereby creating depositional landforms, such as bars, floodplains, terraces, or deltas.

As discussed earlier, stream meanders tend to migrate downstream through the landscape. Over time, the landscape near a meandering river comes to bear meander scars of residual deposits from abandoned channels. Former point bar deposits leave low-lying ridges, creating a *bar-and-swale relief* (a swale is a gentle low area), forming a *scroll topography*. The 1999 *Landsat* image in Figure 14.22 exhibits characteristic meandering scars: meander bends, oxbow lakes, natural levees, point bars, and undercut banks compared to a 1944 Army Corps of Engineers map. Match the river positions in the image and the map for the dates shown.

Floodplains The flat, low-lying area flanking many stream channels that is subjected to recurrent flooding is a **floodplain**. It is formed when the river overflows its channel during times of high flow. Thus, when floods occur, the floodplain is inundated. When the water recedes, it leaves behind alluvial deposits that generally mask the underlying rock with their accumulating

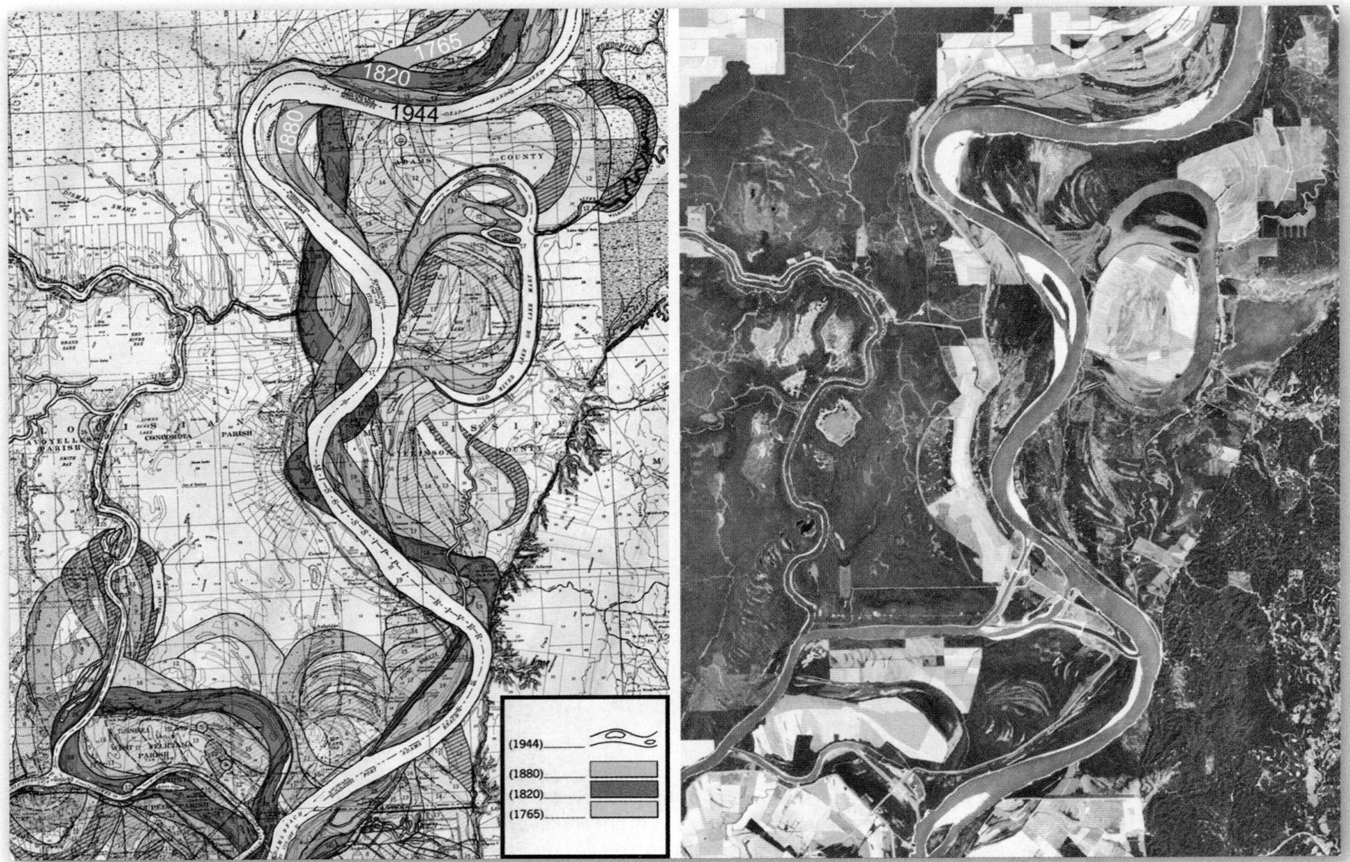

FIGURE 14.22 The Mississippi River's shifting meanders through history.
The former river channels for 1765, 1820, 1880, and 1944 are noted for the portion of the river north of the Old River Cutoff. [Army Corps of Engineers, *Geological Investigation of the Alluvial Valley of the Lower Mississippi,* 1944; *Landsat,* NASA and UMD Global Land Cover Facility, http://glcf.umiacs.umd.edu/index.shtml.]

thickness. The present river channel is embedded in these alluvial deposits. Figure 14.23 illustrates a characteristic floodplain, related features in photos, and a representative topographic map of an area near Philipp, Mississippi.

On either bank of some streams, low depositional ridges of **natural levees** develop as by-products of flooding. When floodwaters rise, the river overflows its banks, loses stream competence and capacity as it spreads out, and drops a portion of its sediment load to form the levees. Larger, sand-sized particles drop out first, forming the principal component of the levees, with finer silts and clays deposited farther from the river. Successive floods increase the height of the levees (*levée* is French for "raising"). The levees may grow in height until the river channel becomes elevated, or *perched,* above the surrounding floodplain.

On the topographic map in Figure 14.23d, you can see the natural levees represented by several contour lines that run immediately adjacent to the Tallahatchie River. These contour lines (which have a 5-ft interval) denote a height of 10–15 ft (3–4.5 m) above the river and the adjoining floodplain. Next time you have an opportunity to see a river and its floodplain, look for levees. They may be low and subtle.

Notice in Figure 14.23a the backswamp area and the yazoo tributary stream. The natural levees and elevated channel of the river prevent this **yazoo tributary** from joining the main channel, so it flows parallel to the river and through the **backswamp** area. The name comes from the Yazoo River in the southern part of the Mississippi floodplain.

People build cities on floodplains despite the threat of flooding because floodplains are nearly level and they are next to water. There may be historical momentum for settlement. People often are encouraged by government assurances of artificial protection from floods and disaster assistance if floods occur. Government assistance may provide for the building of artificial levees on top of natural levees. Artificial levees do increase the capacity in the channel, but they also lead to even greater floods when they are overtopped by

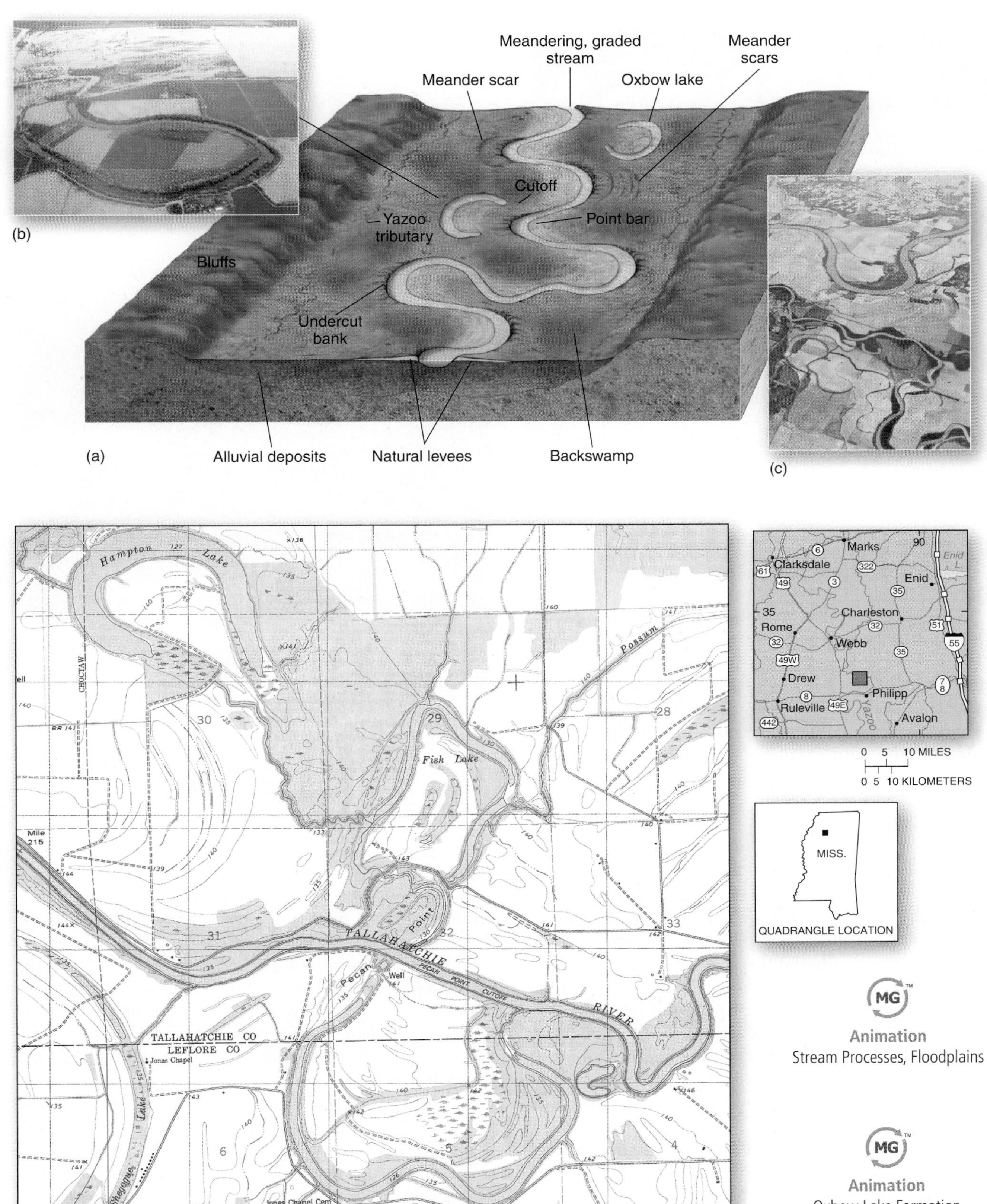

MG Animation Stream Processes, Floodplains

MG Animation Oxbow Lake Formation

FIGURE 14.23 A floodplain.
(a) Typical floodplain landscape and related landscape features. (b) An oxbow lake. (c) Levees, oxbow lakes, yazoo tributary, and farmland on a typical floodplain in the upper Midwest. (d) A portion of the Philipp, Mississippi, topographic map quadrangle. [(b) and (c) Bobbé Christopherson. (d) USGS topographic map.]

FIGURE 14.24 Levee break and devastation in New Orleans.
August 30, 2005, view across the Inner Harbor Navigation Canal, as broken levees allow floodwaters to drown the city. A dramatic failure of engineering, levee design and construction, and proper planning. [(c) Vincent Laforet/EPAI / CORBIS, All Rights Reserved]

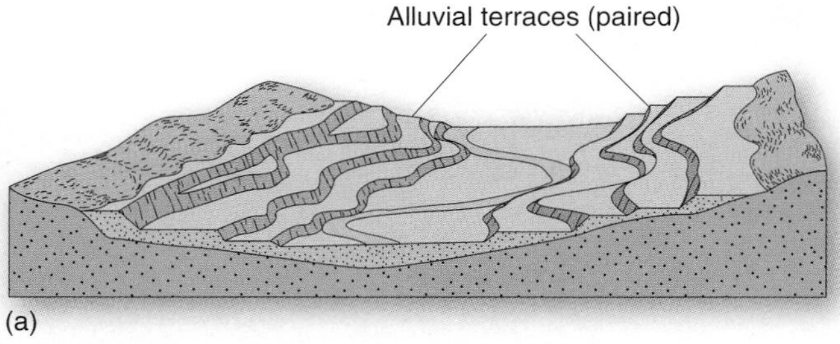

FIGURE 14.25 Alluvial stream terraces.
(a) Alluvial terraces are formed as a stream cuts into a valley. (b) Alluvial terraces along the Rakaia River in New Zealand. [(a) After W. M. Davis, *Geographical Essays* (New York: Dover, 1964 [1909]), p. 515. (b) Bill Bachman/ Photo Researchers, Inc.]

floodwaters or when they fail, as happened in New Orleans (Figure 14.24).

Perhaps the best use of some floodplains is for crop agriculture because inundation generally delivers nutrients to the land with each new alluvial deposit. A significant example is the Nile River in Egypt, where annual flooding enriches the soil. However, floodplains that are covered with coarse sediment—sand and gravel—are less suitable for agriculture. If floodplains are present where you live, what is your impression of present land-use patterns, local planning and zoning, and people's hazard perception overall?

Stream Terraces An uplifting of the landscape or a lowering of base level may rejuvenate stream energy so that a stream again scours downward with increased erosion. The resulting entrenchment of the river into its own floodplain produces **alluvial terraces** on both sides of the valley, which look like topographic steps above the river. Alluvial terraces generally appear paired at similar elevations on the sides of the valley (Figure 14.25). If more than one set of paired terraces is present, the valley probably has undergone more than one episode of rejuvenation.

If the terraces on the sides of the valley do not match in elevation, then entrenchment actions must have been continuous as the river meandered from side to side, with each meander cutting a terrace slightly lower in elevation. Thus, alluvial terraces represent an original depositional feature, a floodplain, that is subsequently eroded by a stream that has experienced a change in gradient and is downcutting.

River Deltas The mouth of a river is where it reaches a base level. The river's velocity rapidly decelerates as it enters a larger, standing body of water. The reduced stream competence and capacity cause deposition of the sediment load. Coarse sediment such as sand and gravel drops out first and is deposited closest to the river's mouth. Finer clays are carried farther and form the extreme end of the deposit, which may be *subaqueous*, or underwater, even at low tide. The level or nearly level depositional plain that forms at the mouth of a river is a **delta** for its characteristic triangular shape, after the Greek letter delta (Δ).

Each flood stage deposits a new layer of alluvium over portions of the delta, growing the delta outward. As in braided rivers, channels divide into smaller courses known as *distributaries*, which appear as a reverse of the dendritic drainage pattern of tributary streams discussed earlier. Here are a few examples:

- The Ganges River delta features an extensive lower delta plain formed in relation to high tidal ranges in an *arcuate* (arc-shaped) pattern. It is covered by an intricate maze of distributaries. Bountiful

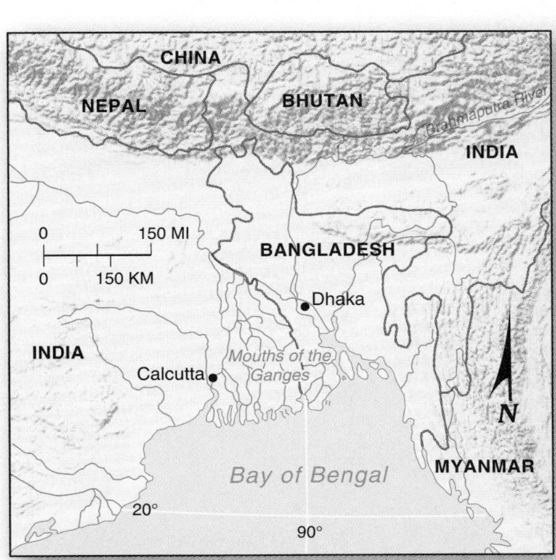

FIGURE 14.26 The Ganges River enters the Bay of Bengal. The complex distributary pattern in the "many mouths" of the Ganges River delta in Bangladesh and extreme eastern India from the *Terra* satellite. Imagine the devastation to this barely-above-sea-level delta when Tropical Storm Sidr, a category 4, plowed through in November 2007, killing thousands of people on the low-lying islands. [*Terra* MODIS, NASA.]

alluvium carried from deforested slopes upstream provides excess sediment that is deposited to form many deltaic islands (Figure 14.26). The combined delta complex of the Ganges and Brahmaputra rivers is the largest in the world at some 60,000 km² (23,165 mi²).

- The Nile River delta is an *arcuate* delta (Figure 14.27). Also arcuate are the Danube River delta in Romania, where it enters the Black Sea, and the Indus River delta.
- The Tiber River in Italy has an *estuarine delta*, one that is in the process of filling an **estuary**, the body of water at a river's mouth where freshwater flow encounters seawater.

Mississippi River Delta Over the past 120 million years, the Mississippi River has transported alluvium throughout its vast basin into the Gulf of Mexico. During the past 5000 years, the river has formed a

GEO REPORT 14.3 *The Nile River delta is disappearing*

The Aswân High Dam, built in 1964, caused a decrease in the supply of sediment to the Nile River delta. Also, over the centuries, more than 9000 km (5500 mi) of canals were built in the delta to augment the natural distributary system. As the river discharge enters the network of canals, flow velocity is reduced, stream competence and capacity are decreased, and sediment load is deposited far short of where the delta touches the Mediterranean Sea. River flows no longer effectively reach the sea. As a result, the delta coastline continues to actively recede at an alarming 50 to 100 m (165 ft to 330 ft) per year. Seawater is intruding farther inland in both surface water and groundwater. Human action and reaction to this evolving situation will no doubt determine the delta's future.

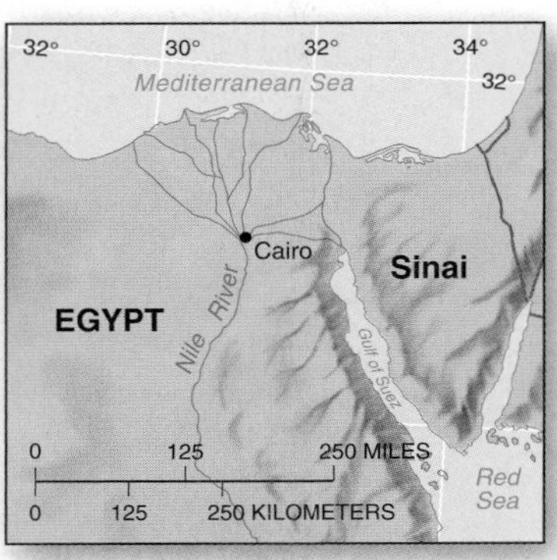

FIGURE 14.27 The arcuate Nile River delta.
Intensive agricultural activity and small settlements are visible on the delta and along the Nile River floodplain in this true-color image. Cairo is at the apex of the delta. You can see the two main distributaries: Damietta to the east and Rosetta to the west. [January 30, 2001, *Terra* image, NASA/GSFC/JPL.]

succession of seven distinct deltaic complexes along the Louisiana coast.

Each generalized lobe in Figure 14.28a reflects a distinct course change in the Mississippi River, probably where the river broke through its natural levees during an episode of severe flooding, thus changing the configuration of the delta. The seventh and current delta has been building for at least 500 years and is a classic example of a *bird's-foot delta*—a long channel with many distributaries and sediments carried beyond the tip of the delta into the Gulf of Mexico.

The Mississippi River delta clearly is dynamic over time as new alluvial material accumulates and the distributaries shift (Figure 14.28b). The main channel persists because of much effort and expense directed at maintaining the artificial levee system. The 3.25-million-km^2 (1.25-million-mi^2) Mississippi drainage basin produces 550 million metric tons of sediment annually—enough to extend the Louisiana coast 90 m (295 ft) a year; however, several factors are actually causing losses to the delta each year. In fact, most of coastal Louisiana, which represents about 40% of the coastal marshes in the United States, is profoundly altered and disrupted by dam and levee construction, flow alterations, oil and gas exploration and pumping, pipelines, and dredging for navigation, logging, and industrial needs. Approximately 65 km^2 (25 mi^2) of these Louisiana coastal lowlands are reclaimed by the Gulf of Mexico in an average year.

Compaction and the tremendous weight of the sediment in the Mississippi River create isostatic adjustments in Earth's crust. These adjustments are causing the entire region of the delta to subside, thereby placing ever-increasing stress on natural and artificial levees and other structures along the lower Mississippi. The many canals and waterways excavated through the delta by the oil and gas industry and shipping interests have worked to deprive the delta of new alluvial building materials. Also, the pumping of tremendous quantities of oil and gas from thousands of onshore and offshore wells is thought to be a cause of regional land subsidence.

An additional problem for the lower Mississippi Valley is the possibility that, in a worst-case flood, the river could break from its existing channel and seek a new route to the Gulf of Mexico. If you examine the map in Figure 14.28c and the sediment plume to the west of the main delta in the satellite image, an obvious alternative to

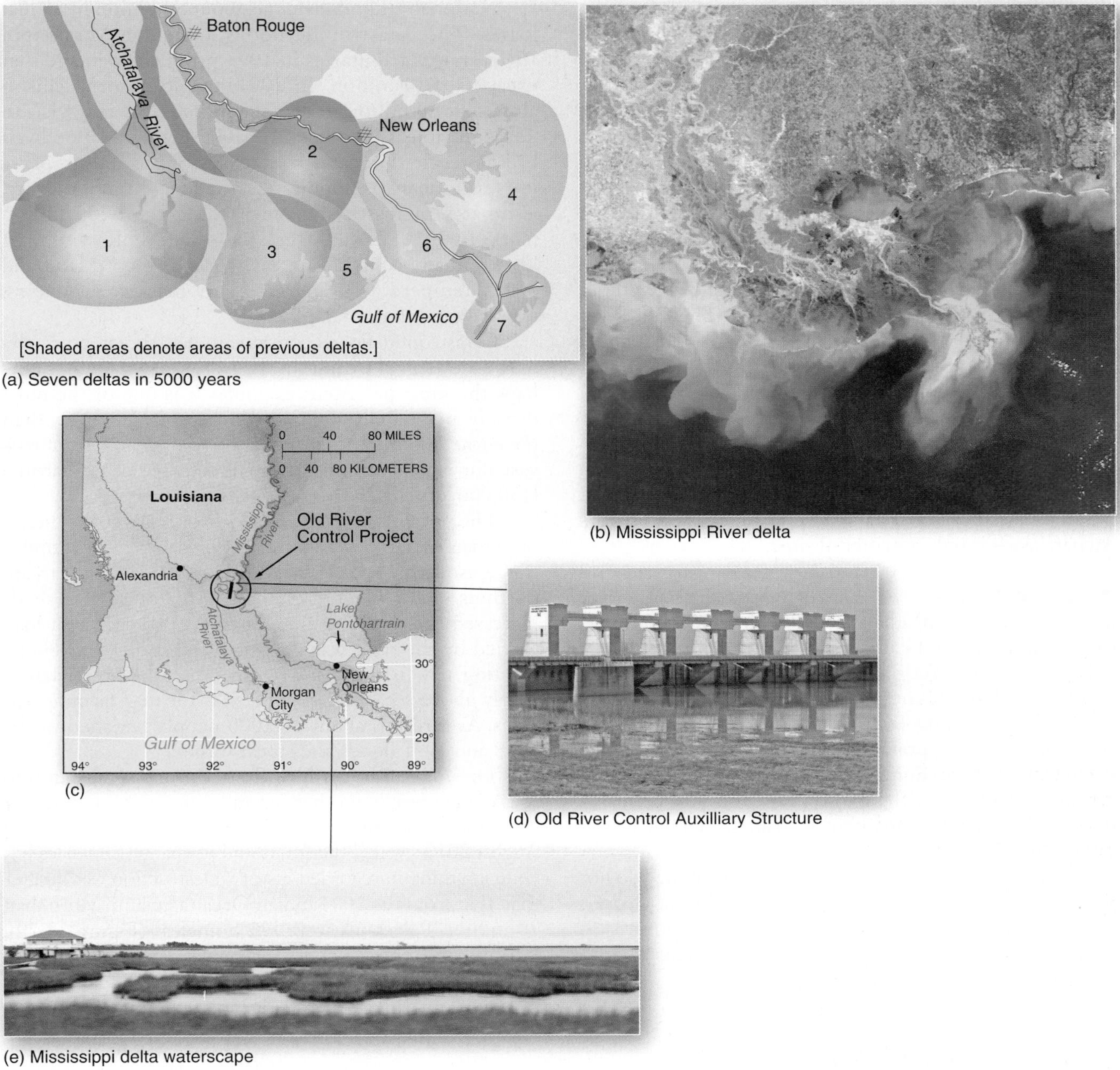

(a) Seven deltas in 5000 years

(b) Mississippi River delta

(c)

(d) Old River Control Auxilliary Structure

(e) Mississippi delta waterscape

FIGURE 14.28 The Mississippi River delta.
(a) Evolution of the present delta, from 5000 years ago (1) to the present (7). (b) The bird's-foot delta of the Mississippi River receives a continuous sediment supply, focused by controlling levees, although subsidence of the delta and rising sea level have diminished the overall surface area. (c) Location map of the Old Control Structures and potential capture point (arrow) where the Atchafalaya River may one day divert the present channel. (d) Old River Control Auxiliary Structure, one of the dams to keep the Mississippi in its channel. (e) Mississippi delta waterscape; note the house with the raised first floor on stilts. [(a) Adapted from C. R. Kolb and J. R. Van Lopik, "Depositional environments of the Mississippi River deltaic plain," in *Deltas in Their Geologic Framework* (Houston, TX: Houston Geological Society, 1966). (b) *Terra* image courtesy of Liam Gumley, Space Science and Engineering Center, University of Wisconsin, and NASA. (d) and (e) Bobbé Christopherson.]

What is a bayou?

Bayou is a general term for several water features in the lower Mississippi River system, including creeks and secondary waterways. These sometimes stagnant, winding watercourses pass through coastal marshlands and swamps and allow tidal waters access to deltaic lowlands. Chapter 16 begins with the story of Bayou Lafourche.

the Mississippi's present channel is the Atchafalaya River. The Atchafalaya would provide a much shorter route to the Gulf of Mexico, less than half the present distance, and it has a steeper gradient than the Mississippi. Presently, this alternative-route distributary carries about 30% of the Mississippi's total discharge. For the Mississippi to bypass New Orleans entirely would be a blessing, for it would remove the flood threat. However, this shift would be a financial disaster, as a major U.S. port would silt in and seawater would intrude into freshwater resources.

At present, artificial barriers block the Atchafalaya from reaching the Mississippi at the point shown; without the floodgates, the two rivers do connect. The Old River Control Project (1963) maintains three structures and a lock about 320 km (200 mi) from the Mississippi's mouth to keep these rivers in their channels (Figure 14.28d). Many floods have hit the lower Mississippi in the past, such as that in 2003, and Hurricanes Cindy, Katrina, and Rita in 2005 and Gustav in 2008 resulted in major storm surges and flooding. Another major flood—one that might cause the river channel to change back into the Atchafalaya—is only a matter of time.

Rivers Without Deltas Earth's highest-discharge river is the Amazon; it discharges more than 175,000 m^3/s (6.2 million cfs) and carries sediment far into the deep Atlantic offshore. Yet the Amazon lacks a true delta. Its mouth, 160 km (100 mi) wide, has formed a subaqueous deposit on a sloping continental shelf. As a result, the Amazon's mouth is braided into a broad maze of islands and channels (see Figure 14.1). Also, the Paraná and Uruguay rivers form the Rio de la Plata, which lacks a delta (see chapter-opening photo). Also, the Sepik River of Papua New Guinea is without delta.

Other rivers also lack deltaic formations if they do not produce significant sediment or if they discharge into strong erosive currents. The Columbia River of the U.S. Northwest lacks a delta because offshore currents remove sediment as it is deposited.

Floods and River Management

Throughout history, civilizations have settled floodplains and deltas, especially since the agricultural revolution of 10,000 years ago, when the fertility of floodplain soils was discovered. Early villages generally were built away from the area of flooding, or on stream terraces, because the floodplain was dedicated exclusively to farming. However, as commerce grew, competition for sites near rivers grew because these locations were important for transportation. Port and dock facilities were built, as were river bridges. Because water is a basic industrial raw material used for cooling and for diluting and removing wastes, waterside industrial sites became desirable. These competing human activities on vulnerable flood-prone lands place lives and property at risk during floods.

Nationally, floods average about $6 billion in annual losses. The catastrophic floods along the Mississippi River and its tributaries in 1993 produced damage that exceeded $30 billion. In 2001, tropical storm Allison alone left $6 billion in damage in its wandering visit to Texas in June. Of course, the 2005 levee and floodwall breaks resulted in catastrophic loss of life and property in New Orleans as floodwaters engulfed 80% of the city. Estimates of losses are approaching the $200 billion mark. In 2010, floods struck in 30 U.S. states and in at least as many countries in the world as record land and ocean temperatures energized air masses, producing excessive rainfall totals.

Catastrophic floods continue to be a threat, especially in less-developed regions of the world. Bangladesh is perhaps the most persistent example: It is one of the most densely populated countries on Earth, and more than *three-fourths* of its land area is a floodplain. The country's vast alluvial plain sprawls over an area the size of Alabama (130,000 km^2, or 50,000 mi^2).

The flooding severity in Bangladesh is a consequence of human economic activities, along with heavy precipitation episodes. Excessive forest harvesting in the upstream portions of the Ganges–Brahmaputra River watersheds increased runoff. Over time, the increased sediment load carried by the river was deposited in the Bay of Bengal, creating new islands (see Figure 14.26). These islands, barely above sea level, became sites for new farming villages. As a result, about 150,000 people perished in the 1988 and 1991 floods and storm surges.

The Indus River, which flows through Pakistan into the Arabian Sea, is another recent example of flooding in a developing part of the world. Heavy monsoon rains in August 2010 swelled the Indus and its many tributaries beyond anything in recorded history. Destruction was greater than that from the 2004 Indian Ocean tsunami. More than 20 million Pakistanis were left homeless (Figure 14.29). For information on floods worldwide, see the Dartmouth Flood Observatory at http://www.dartmouth.edu/~floods/Archives/index.html or a daily flood summary at http://www.nws.noaa.gov/oh/hic/current/fln/fln_sum.shtml.

Engineering Failures in New Orleans, 2005

If all is well, the city of New Orleans is almost entirely below the level of the Mississippi River, with about half of the city below sea level in elevation. Severe flooding is a certainty for existing and planned settlements unless further intervention or urban relocation occurs. The building of artificial earthen levees, concrete floodwalls, and multiple flood-control structures by the U.S. Army Corps of Engineers apparently has only delayed the peril, as demonstrated by recent flooding. The events of 2005, following the passage of Hurricane Katrina, will remain for many years as one of the greatest engineering failures by the government. Four levee breaks and at least four dozen levee breaches, in which water flows over the embankment, permitted the inundation of a major city. The polluted water remained for weeks.

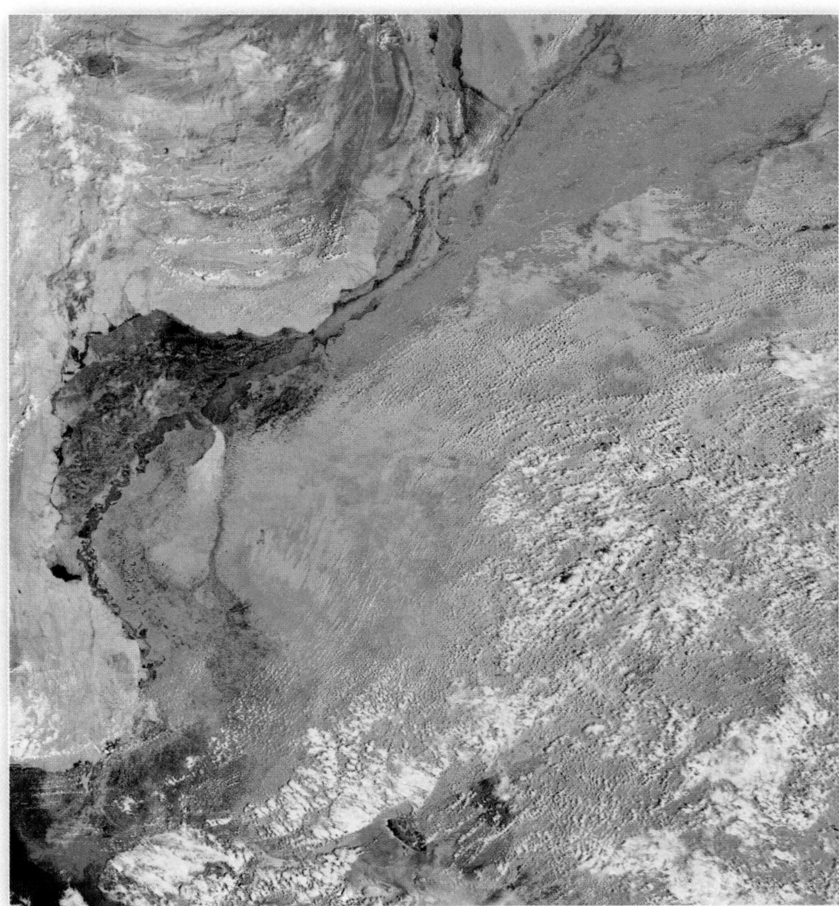

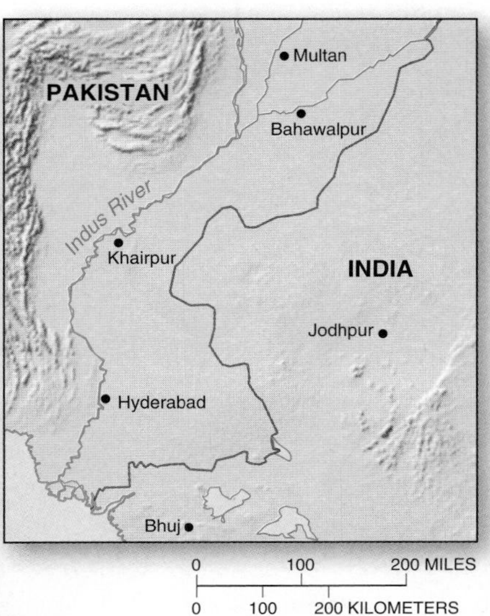

FIGURE 14.29 Indus River flooding, 2010. The image combines visible light and infrared to enhance contrast between the flooded river and land; water appears blue and clouds blue-green. Similar heavy monsoon rains fell across Asia. [*Terra* image NASA/GSFC.]

Figure 14.30 from *Landsat–7* presents matching images of New Orleans along the shore of Lake Pontchartrain on (a) April 24 and (b) August 30, 2005. The map helps you locate the areas that are below sea level, the levees and floodwalls, and the four major breaches. The darker areas on image (b) indicate flooding; some neighborhoods were submerged up to 6.1 m (20 ft). At the time of image (b), New Orleans was approximately 80% under water.

Six investigations by civil engineers, scientists, and political bodies agreed that in concept, design, construction, and maintenance, the "protection" system was flawed. Many of the structures failed before they reached design-failure limits. After all, Hurricane Katrina was diminished in power as it moved onshore. Several of the floodwalls and pilings, such as those along the 17th Street Canal (Figure 14.31), were not anchored to the design depth. This inadequate anchoring made parts of the system inherently weak. In 2007, inspections by civil engineers found that the composition of levee fill materials throughout the city is too low in clays and too high in sand—a recipe for future problems.

Repair and recovery work in New Orleans is ongoing, and elements of the completed sections appear inadequate and positioned for repeat disasters, which nearly occurred in 2008 as Hurricane Gustav followed a track just west of New Orleans. Most levees held, some were topped, and extensive flooding occurred in the parishes west and south of the city that had no levees. Hurricane Gustav is regarded a near-miss from being another Katrina-level disaster for New Orleans.

CRITICAL THINKING 14.3

Review and assessment of the post-Katrina Gulf Coast

Using your Internet search engine or browser, type in "New Orleans flooding, 2005" or "Gulf Coast damage, Katrina" or "New Orleans levee failures." Allow some time to sample through at least 10 of the links. Describe what you find. Did you find photo galleries, engineering reports, or news coverage? What did you learn about this region of the lower Mississippi River, including New Orleans and other Gulf Coast cities? What future direction they should take: rebuild things as they were; rethink the hazards and adopt a new strategy of replacing urban with low-density recreation zones; establish possible set-back hazard zoning; walk away except for necessary port functions; or abandon the region? Are there other options you think are important?

Rating Floodplain Risk

A **flood** is a high water flow that overflows the natural bank along any portion of a stream. Floods are rated statistically for the expected time intervals between them, based upon

(a) April 24, 2005

(b) August 30, 2005

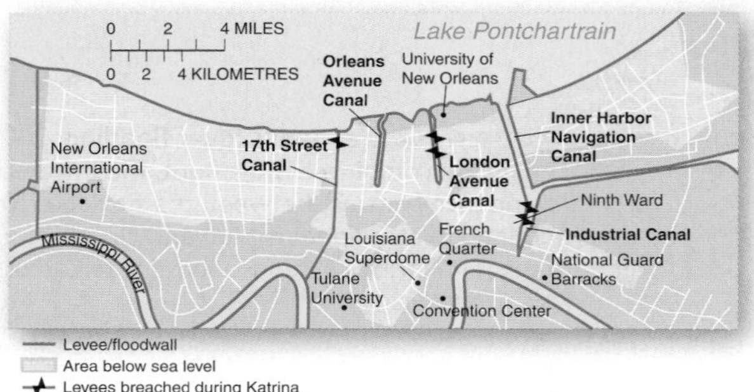

FIGURE 14.30 New Orleans before and after the Katrina disaster.
Landsat–7 images made (a) April 24, 2005, and (b) August 30, 2005, clearly show canal locations and flooded portions of the city; compare features to the locator map. Numerous levee breaks and breaches put 80% of the city under water for more than a week. [USGS *Landsat* Image Gallery.]

FIGURE 14.31 Levee breach and break.
The Lakeview neighborhood is under water (to the right) from the 17th Street Canal levee failure, whereas the Metairie neighborhood is essentially dry (to the left). You see debris caught up north of the bridge where steel beams were placed in an attempt to curb flows. Flotsam, or floating debris, is visible in the photo. [NOAA.]

historical data. Thus, you hear about "10-year floods," "50-year floods," and so on. A *10-year flood* is likely to occur once every 10 years. This also means that a flood of this size has only a 10% likelihood of occurring in any one year and is likely to occur about 10 times each century. For any given floodplain, such a frequency indicates a moderate threat.

A 50-year or 100-year flood is of greater and perhaps catastrophic consequence, but it is also less likely to occur in a given year. These probability ratings of flood levels are mapped for an area, and each defined floodplain that results is then labeled a "50-year floodplain" or a "100-year floodplain."

These statistical estimates are probabilities that events will occur randomly during any single year of the specified period. Of course, two decades might pass without a 50-year flood, or a 50-year level of flooding could occur 3 years in a row. The record-breaking Mississippi River Valley floods in 1993 easily exceeded a 1000-year flood probability, as did the 2005 Katrina catastrophe. See Focus Study 14.1 for more about floodplain hazards and management strategies.

Stream Discharge Measurement

Flood patterns in a drainage basin are as complex as the weather, for floods and weather are equally variable and both include a level of unpredictability. Measuring and analyzing the behavior of large watersheds and individual streams enables engineers and concerned parties to develop the best possible flood-management strategy. Unfortunately, reliable data often are not available for small basins or for the changing landscapes of urban areas.

The key to flood avoidance or management is to possess extensive measurements of a stream's discharge and information on how it performs during a precipitation event (Figure 14.32a). Three measurements are needed at a stream cross section in order to calculate discharge: width, depth, and velocity. Field measurements of these variables may be difficult to obtain, depending on a stream's size and flow. Velocity for different subsections of the stream cross section is most commonly measured using a movable current meter. Width and depth for each subsection are then combined with velocity to compute subsection discharge, and then the subsection discharges

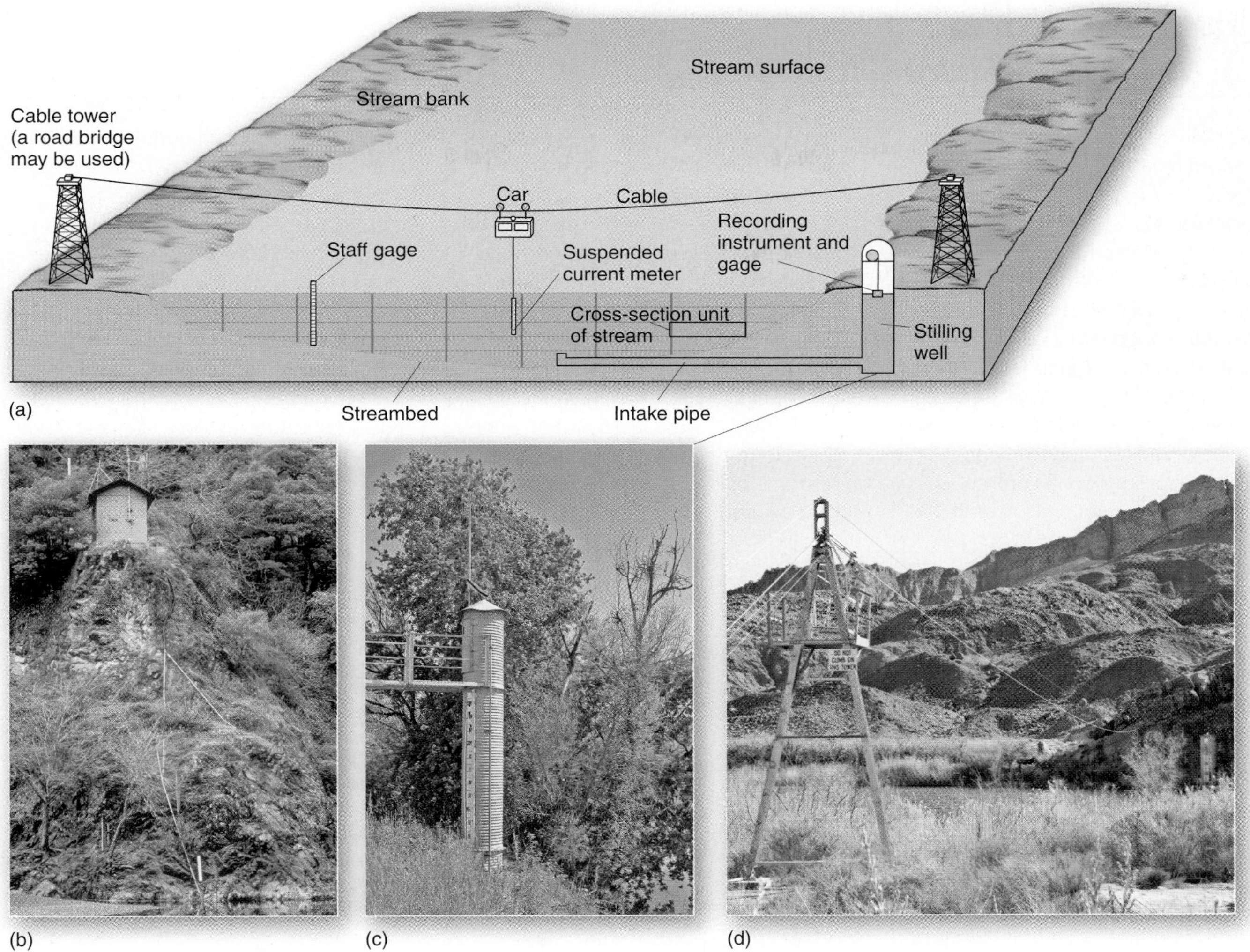

FIGURE 14.32 Stream discharge measurement.
(a) A typical flow measurement installation may use a variety of devices. (b) An automated hydrographic station and (c) a stilling well send telemetry to a satellite for collection by the USGS. (d) Lees Ferry cable towers and cable from which current meters are lowered. [(b) California Department of Water Resources. (c) and (d) by Bobbé Christopherson.]

across the stream are totaled (see Figure 14.32). Since channel beds are often composed of alluvial deposits that may change over short time periods, stream depth is measured as the height of the stream surface above a constant reference elevation (a datum) and is called the *stage*. Scientists may use a *staff gage* (a pole marked with water levels) or a *stilling well* on the stream bank with a gage mounted in it to measure stage (Figure 14.32c).

Approximately 11,000 stream gaging stations are in use in the United States (an average of more than 200 per state). Of these, 7000 are operated by the U.S. Geological Survey and have continuous recorders for stage and discharge (see http://pubs.usgs.gov/circ/circ1123/). Many of these stations automatically send telemetry data to satellites, from which information is retransmitted to regional centers (Figure 14.32b and c). The historic gaging station on the Colorado River at Lees Ferry, just south of Glen Canyon Dam, was established in 1921 (Figure 14.32d). Scientists hope that adequate funding to maintain such hydrologic monitoring by stream gaging networks will continue, for the data are essential to hazard assessment and proper planning.

Hydrographs A graph of stream discharge over time for a specific location is a **hydrograph**. The hydrograph in Figure 14.33a (p. 428) shows the relation between precipitation input (the bar graph) and stream discharge (the curves). During dry periods, at low-water stages, the flow is described as base flow and is largely maintained by input from local groundwater (dark blue line).

When rainfall occurs in some portion of the watershed, the runoff collects and is concentrated in streams and tributaries. The amount, location, and duration of the rainfall episode determine the *peak flow*. Also important is the nature of the surface in a watershed, whether permeable or impermeable. A hydrograph for a specific portion of a stream changes after a forest fire or urbanization of the watershed.

Human activities have enormous impact on water flow in a basin. The effects of urbanization are quite dramatic,

(text continued on page 428)

FOCUS STUDY 14.1

Floodplain Measurement and Strategies

The U.S. Geological Survey has kept detailed measurements of stream discharge for only about 100 years—in particular, since the 1940s. Flood predictions are based on these relatively short-term data. At any selected location along any given stream, the *probable maximum flood* (PMF) is a hypothetical flood of such a magnitude that there is virtually no possibility it will be exceeded. Because the collection and concentration of rainfall produce floods, hydrologists speak of a corollary, the *probable maximum precipitation* (PMP) for a given drainage basin, which is an amount of rainfall so great that it will never be exceeded.

Hydrologic engineers use these parameters to establish a design flood against which to take protective measures. For urban areas near creeks, planning maps often include survey lines for a 50-year or a 100-year floodplain; hydrologists have completed such maps for most U.S. urban areas. The design flood usually is used to enforce planning restrictions and special insurance requirements. Restrictive zoning using these floodplain designations is an effective way of avoiding potential damage.

Such political action is not always enforced, and the scenario sometimes goes like this: (1) Minimal zoning precautions are not carefully supervised, (2) a flooding disaster occurs, (3) the public is outraged at being caught off guard, (4) businesses and homeowners are surprisingly resistant to stricter laws and enforcement, and (5) eventually another flood refreshes the memory and promotes more planning meetings and

(a)

FIGURE 14.1.1 Weir and bypass channels to divert floodwaters.
(a) The Bonnet Carré spillway (a weir), on the lower Mississippi, New Orleans District, is opened in flood events to let water flow to Lake Pontchartrain and bypass New Orleans to the south. (b) The Sacramento weir awaits floodwaters from the Sacramento River, which it will direct through a channel into the Yolo Bypass visible in the distance in (c). The bypass is used for farmland crops when flood waters recede. [Bobbé Christopherson.]

Animation Stream Processes, Floodplains

(b)

(c)

questions. As strange as it seems, there is little indication that our risk perception improves as the risk increases. The floodplain managers' organization at http://www.floods.org/ is a source of information.

Bypass Channels as a Planning Strategy

A planning strategy used in some large river systems is to develop artificial floodplains; this is done by constructing *bypass channels* to accept seasonal or occasional floods. When not flooded, the bypass channel can serve as farmland, often benefitting from the occasional soil-replenishing inundation. When the river reaches flood stage, large gates called *weirs* are opened, allowing the water to enter the bypass channel. This alternate route relieves the main channel of the burden of carrying the entire discharge.

Along the lower Mississippi River north of New Orleans, the Bonnet Carré spillway is such a weir (Figure 14.1.1a). During high-water events, water is let through a floodway into a bypass between a pair of guide levees. The bypass diverts the flood 9.7 km (6 mi) into Lake Pontchartrain, as it has done seven times since 1937, to protect New Orleans. For more details, see http://www.mvn.usace.army.mil/pao/bcarre/bcarre.htm.

In the Sacramento Valley, California, such weirs lead to bypass channels that take floodwaters around metropolitan areas. See Figure 14.1.1b and c for the Sacramento weir and bypass farmland.

(a)

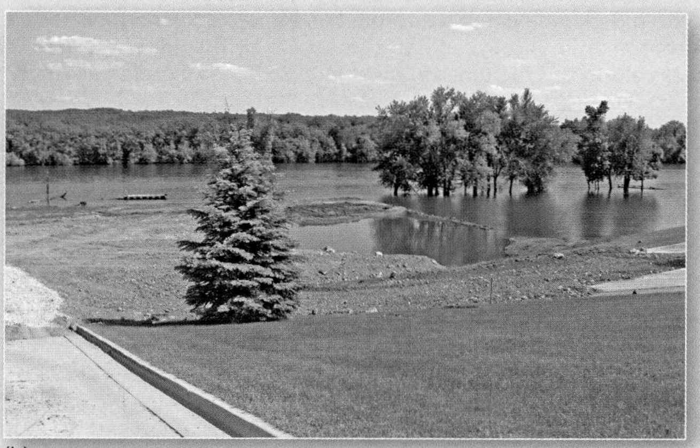

(b)

FIGURE 14.1.2 Living and building in a floodplain. (a) Near Harpers Ferry, Iowa, homes are located on a forested sandbar in the Mississippi River. (b) New fill is being dumped into the river to extend the land for further development. The river is at a high water level in the photo. [Bobbé Christopherson.]

Floodplain-Management Considerations

The benefit of any levee, bypass, or other project intended to prevent flood destruction is measured in avoided damage and is used to justify the cost of the protection facility. Thus, ever-increasing damage leads to the justification of ever-increasing flood-control structures. All such strategies are subjected to cost–benefit analysis, but bias is a serious drawback because such an analysis usually is prepared by an agency or bureau with a vested interest in building more flood-control projects.

As suggested in an article titled "Settlement Control Beats Flood Control,"* published over 50 years ago, there are other ways to protect populations than with enormous, expensive, sometimes environmentally disruptive projects. Strictly zoning the floodplain is one approach. However, the flat, easily developed, and often picturesque floodplains are desirable for housing and thus weaken political resolve (Figure 14.1.2). A reasoned zoning strategy would set aside the floodplain for farming; passive recreation, such as a riverine park, golf course, or plant and wildlife sanctuary; or other uses that are not hurt by natural floods. This study concludes that "urban and industrial losses would be largely obviated [avoided] by setback levees and zoning and thus cancel the biggest share of the assessed benefits which justify big dams."

*Walter Kollmorgen, "Settlement control beats flood control," ***Economic Geography*** 29, 3 (July 1953): 215.

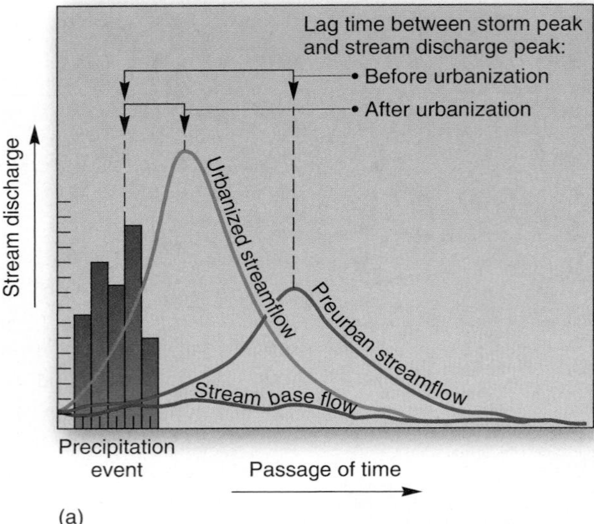

FIGURE 14.33 Effect of urbanization on a typical stream hydrograph.
(a) Normal base flow is indicated with a dark blue line. The purple line indicates discharge after a storm, before urbanization. Following urbanization, stream discharge dramatically increases, as shown by the light blue line. (b) Severe flooding of an urban area in Linda, California, after levee breaks along the Sacramento River in 1986. Linda flooded again in 1997. [(b) California Department of Water Resources.]

both increasing and hastening peak flow, as you can see by comparing preurban discharge (purple curve) and urbanized discharge (light blue) in Figure 14.33a. In fact, urban areas produce runoff patterns quite similar to those of deserts. The sealed surfaces of the city drastically reduce infiltration and soil-moisture recharge; their effect is similar to that of the usually hard surfaces in the desert. A significant part of urban flooding occurs because of the alteration in surfaces, which produces shortened concentration times; peak flows may strike with little warning. These issues will intensify as urbanization of vulnerable areas continues. The flooded region in Figure 14.33b was hit in 1986 and again in 1997.

We followed the flow of moving water through streams, examining fluvial processes and landforms, and the river system outputs of discharge, deposition, and floods. The human society's relationship to river dynamics and related flood hazards and floodplain landscapes is an integral aspect of our ability to perceive hazards in a familiar environment. In the next chapter, we study the agent of wind as it operates to erode, transport, and deposit sediment. Also, in the chapter we look at arid lands and some unique aspects of Earth's desert landscapes.

KEY LEARNING CONCEPTS REVIEW

- ***Define*** the term *fluvial,* and ***explain*** processes of fluvial erosion, transportation, and deposition.

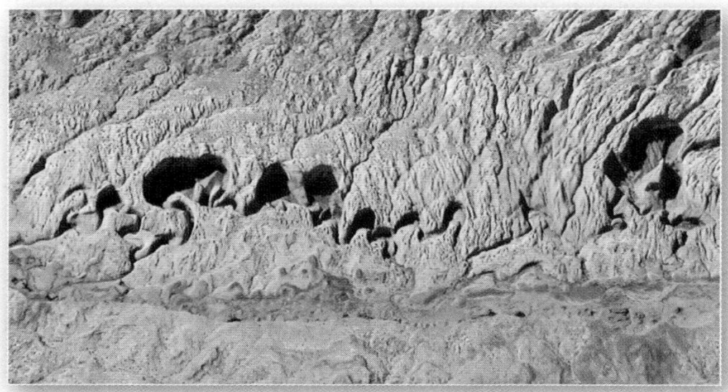

Fluvial processes are stream-related. **Hydrology** is the science of water and its global circulation, distribution, and properties—specifically, water at and below Earth's surface. Water dislodges, dissolves, or removes surface material in the **erosion** process. Streams produce *fluvial erosion,* in which weathered sediment is picked up for **transport**, movement to new locations. Sediments are laid down by another process, **deposition**. **Alluvium** is the general term for the clay, silt, sand, gravel, or other unconsolidated rock and mineral fragments deposited by running water as sorted or semisorted sediment on a floodplain, delta, or streambed.

Base level is the lowest elevation limit of stream erosion in a region. A *local base level* occurs when something interrupts the stream's ability to achieve base level, such as a dam or a landslide that blocks a stream channel.

fluvial (p. 400)
hydrology (p. 400)
erosion (p. 400)
transport (p. 400)
deposition (p. 400)
alluvium (p. 401)
base level (p. 401)

1. Define the term *fluvial*. What is a fluvial process?
2. What role is played by rivers in the hydrologic cycle?
3. What are the five largest rivers on Earth in terms of discharge? Relate these to the weather patterns in each area and to regional potential evapotranspiration (POTET) and precipitation (PRECIP)—concepts discussed in Chapter 9.
4. What is the sequence of events that takes place as a stream dislodges material?
5. Explain the base level concept. What happens to a local base level when a reservoir is constructed?

■ ***Construct*** a basic drainage basin model, and ***identify*** different types of drainage patterns and internal drainage, with examples.

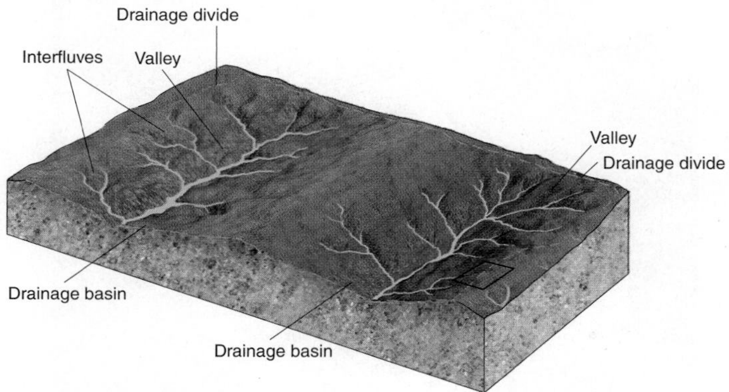

The basic fluvial system is a **drainage basin**, which is an open system. *Drainage divides* define the drainage basin catchment (water-receiving) area of the drainage basin. In any drainage basin, water initially moves downslope in a thin film of **sheetflow**, or *overland flow*. This surface runoff concentrates in *rills*, or small-scale downhill grooves, which may develop into deeper *gullies* and a stream course in a valley. High ground that separates one valley from another and directs sheetflow is an *interfluve*. Extensive mountain and highland regions act as **continental divides** that separate major drainage basins. Some regions, such as the Great Salt Lake Basin, have **internal drainage** that does not reach the ocean, the only outlets being evaporation and subsurface gravitational flow.

Drainage density is determined by the number and length of channels in a given area and is an expression of a landscape's topographic surface appearance. **Drainage pattern** refers to the arrangement of channels in an area as determined by the steepness, variable rock resistance, variable climate, hydrology, relief of the land, and structural controls imposed by the landscape. Seven basic drainage patterns are generally found in nature: dendritic, trellis, radial, parallel, rectangular, annular, and deranged.

drainage basin (p. 402)
sheetflow (p. 402)
continental divides (p. 402)
internal drainage (p. 405)
drainage density (p. 405)
drainage pattern (p. 405)

6. What is the spatial geomorphic unit of an individual river system? How is it determined on the landscape? Define the several relevant key terms used.
7. In Figure 14.5, follow the Allegheny–Ohio–Mississippi River system to the Gulf of Mexico. Analyze the pattern of tributaries and describe the channel. What role do continental divides play in this drainage?
8. Describe drainage patterns. Define the various patterns that commonly appear in nature. What drainage patterns exist in your hometown? Where you attend school?

■ ***Describe*** the relation among stream velocity, depth, width, and discharge.

Stream channels vary in *width* and *depth*. The streams that flow in them vary in velocity and in the *sediment load* they carry. All of these factors may increase with increasing discharge. **Discharge**, a stream's volume of flow per unit of time, is calculated by multiplying the velocity of the stream by its width and depth for a specific cross section of the channel. Some streams originate in a humid region and flow through an arid region, resulting in discharge that decreases with distance. Such an **exotic stream** is exemplified by the Nile River and the Colorado River.

Hydraulic action is the work of *turbulence* in the water. Running water causes hydraulic squeeze-and-release action to loosen and lift rocks and sediment. As this debris moves along, it mechanically erodes the streambed further through a process of **abrasion**.

Solution refers to the **dissolved load** of a stream, especially the chemical solution derived from minerals such as limestone or dolomite or from soluble salts. The **suspended load** consists of fine-grained, clastic particles held aloft in the stream, with the finest particles not deposited until the stream velocity slows nearly to zero. **Bed load** refers to coarser materials that are dragged and pushed and rolled along the streambed by **traction** or that bounce and hop along by **saltation**. If the load in a stream exceeds its capacity, sediment accumulates as **aggradation** as the stream channel builds through deposition. With excess sediment, a stream may become a maze of interconnected channels that form a **braided stream** pattern.

discharge (p. 407)
exotic stream (p. 408)
hydraulic action (p. 408)
abrasion (p. 408)
dissolved load (p. 409)
suspended load (p. 409)
bed load (p. 410)
traction (p. 410)

saltation (p. 410)
aggradation (p. 410)
braided stream (p. 410)

9. What was the impact of flood discharge on the channel of the San Juan River near Bluff, Utah? Why did these changes take place?
10. How does stream discharge do its erosive work? What are the processes at work in the channel?
11. Differentiate between stream competence and stream capacity.
12. How does a stream transport its sediment load? What processes are at work?

■ ***Develop*** a model of a meandering stream, including point bar, undercut bank, and cutoff, and ***explain*** the role of stream gradient.

Where the slope is gradual, stream channels develop a sinuous form called a **meandering stream**. The outer portion of each meandering curve is subject to the fastest water velocity and can be the site of a steep **undercut bank**. On the other hand, the inner portion of a meander experiences the slowest water velocity and forms a **point bar** deposit. When a meander neck is cut off as two undercut banks merge, the meander becomes isolated and forms an **oxbow lake**.

Every stream develops its own **gradient** and establishes a longitudinal profile. A **graded stream** condition occurs when a channel has just enough energy to transport its sediment load; this represents a balance among slope, discharge, channel characteristics, and the load supplied from the drainage basin. An interruption in a stream's longitudinal profile is called a **nickpoint**. This abrupt change in gradient can occur as the stream flows across hard, resistant rock or after tectonic uplift episodes.

meandering stream (p. 411)
undercut bank (p. 411)
point bar (p. 412)
oxbow lake (p. 412)
gradient (p. 412)
graded stream (p. 413)
nickpoint (p. 414)

13. Describe the flow characteristics of a meandering stream. What is the pattern of flow in the channel? What are the erosional and depositional features and the typical landforms created?
14. Explain these statements: (a) All streams have a gradient, but not all streams are graded. (b) Graded streams may have ungraded segments.
15. Why is Niagara Falls an example of a nickpoint? Without human intervention, what do you think will eventually take place at Niagara Falls?

■ ***Define*** a floodplain, and ***analyze*** a stream channel during a flood.

Floodplains have been important sites of human activity throughout history. Rich soils, bathed in fresh nutrients by floodwaters, attract agricultural activity and urbanization. Despite our historical knowledge of devastation by floods, floodplains are settled, raising issues of human perception of hazard. The flat, low-lying area along a stream channel that is subjected to recurrent flooding is a **floodplain**. It is formed when the river overflows its channel during times of high flow. On either bank of some streams, **natural levees** develop as by-products of flooding. On the floodplain, backswamps and yazoo tributaries may develop. The natural levees and elevated channel of the river prevent a **yazoo tributary** from joining the main channel, so it flows parallel to the river and through the **backswamp** area. Entrenchment of a river into its own floodplain forms **alluvial terraces**.

floodplain (p. 415)
natural levees (p. 416)

yazoo tributary (p. 416)
backswamp (p. 416)
alluvial terraces (p. 418)

16. Describe the formation of a floodplain. How are natural levees, oxbow lakes, backswamps, and yazoo tributaries produced?
17. Identify any of the features listed in Question 16 on the Philipp, Mississippi, topographic quadrangle in Figure 14.23d.
18. Describe any floodplains near where you live or where you go to college. Have you seen any of the floodplain features discussed in this chapter? If so, which ones?

- ***Differentiate*** the several types of river deltas, and ***detail*** each.

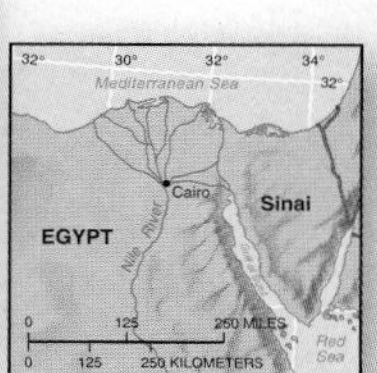

A depositional plain formed at the mouth of a river is called a **delta**. When the mouth of a river enters the sea and is inundated by seawater in a mix with freshwater, it is called an **estuary**.

delta (p. 418)
estuary (p. 419)

19. What is a river delta? What are the various deltaic forms? Give some examples.
20. Based on materials in this chapter and in Chapter 8, assess what occurred in New Orleans. How might New Orleans change later in this century? Explain. In your opinion, did Hurricane Katrina cause the disaster, or was it a result of engineering failures and hazard-planning shortfalls?
21. Describe the Ganges River delta. What factors upstream explain its form and pattern? Assess the consequences of settlement on this delta.
22. What is meant by the statement "the Nile River delta is disappearing"?

- ***Explain*** flood probability estimates, and ***review*** strategies for mitigating flood hazards.

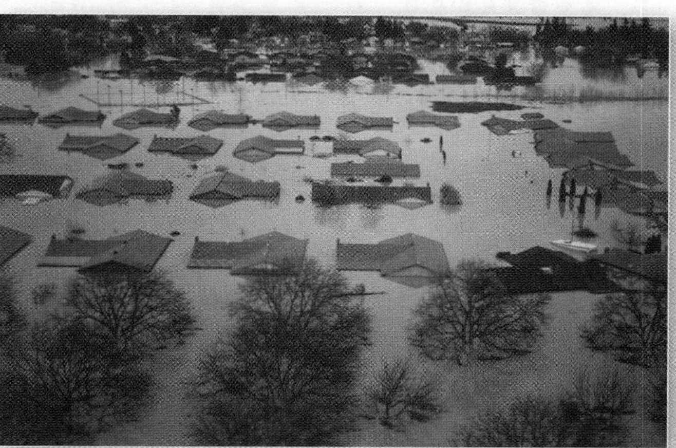

A **flood** occurs when high water overflows the natural bank along any portion of a stream. Both floods and the floodplains they occupy are rated statistically for the expected time interval between floods. For example, a 10-year flood has the statistical probability of happening once every 10 years. A graph of stream discharge over time for a specific place is called a **hydrograph**.

Government agencies undertake collective efforts to reduce flood probability. Such management attempts include the construction of artificial levees, bypasses, straightened channels, diversions, dams, and reservoirs. Society is still learning how to live in a sustainable way with Earth's dynamic river systems.

flood (p. 423)
hydrograph (p. 425)

23. Specifically, what is a flood? How are such flows measured and tracked?
24. Differentiate between a natural stream hydrograph and one from an urbanized area.
25. What do you see as the major consideration regarding floodplain management? How would you describe the general attitude of society toward natural hazards and disasters?
26. What do you think the author of the article "Settlement Control Beats Flood Control" meant by the title? Explain your answer, using information presented in the chapter.

MASTERING GEOGRAPHY

Visit www.masteringgeography.com for several figure animations, satellite loops, self-study quizzes, flashcards, visual glossary, case studies, career links, textbook reference maps, RSS Feeds, and an ebook version of *Geosystems*. Included are many geography web links to interesting Internet sources that support this chapter.

CHAPTER 15

Eolian Processes and Arid Landscapes

Ward Terrace, Coconino County, Arizona, is the bench on the left side of the photo, some 1645 m (5397 ft) in elevation. To the right of the cliff that forms the edge of Ward Terrace, elevations drop to about 1400 m (4593 ft). The center of the aerial photo is approximately 35° 58′ N 111° 15′ W. The light-colored streaks are wind-blown sand and linear dunes that reactivated about 400 years ago. The scene is on the Navajo Reservation. [Bobbé Christopherson.]

KEY LEARNING CONCEPTS

After reading the chapter, you should be able to:

- ***Characterize*** the unique geomorphic work accomplished by wind and eolian processes.
- ***Describe*** eolian erosion, including deflation, abrasion, and the resultant landforms.
- ***Describe*** eolian transportation, and ***explain*** saltation and surface creep.
- ***Identify*** the major classes of sand dunes, and ***present*** examples within each class.
- ***Define*** loess deposits and their origins, locations, and landforms.
- ***Portray*** desert landscapes, and ***locate*** these regions on a world map.

Increasing Desertification and Political Action—A Global Environmental Issue

We are witnessing an unwanted expansion of Earth's deserts in a process known as **desertification**. This is a worldwide phenomenon along the margins of semiarid and arid lands—Earth's dry lands, one of the subjects in this chapter.

Desertification is due in part to poor agricultural practices such as overgrazing and activities that abuse soil structure and fertility, improper soil-moisture management, erosion and salinization, and deforestation. A worsening causative force is global climate change, which is shifting temperature and precipitation patterns and causing poleward movement of Earth's subtropical high-pressure systems, discussed elsewhere in *Geosystems*.

Figure GN 15.1 portrays regions of desertification risk, as defined by the loss of agricultural activity. The severity of this problem is magnified by poverty in many of the affected areas, since most people lack the capital to change agricultural practices and implement conservation strategies. Many of the highest-risk lands are in India and central Asia. In fact, a 2009 mapping project showed that 25% of India is undergoing desertification. Throughout central Asia, overexploitation of water resources has combined with drought to cause the expansion of desert conditions. The Aral Sea, formerly one of the four largest lakes in the world, has steadily shrunk in size since the 1960s as inflowing rivers have been increasingly diverted. As the lake has failed, fine sediment and alkali dust have become available for wind deflation, leading to massive dust storms.

In Africa, the Sahel is the transition region between the Sahara Desert in the subtropics and the wetter equatorial regions. The southward expansion of desert conditions through portions of the Sahel region has left many African peoples on land that no longer experiences the rainfall of just three decades ago. Yet climate change is only part of the story: Other factors contributing to desertification in the Sahel are population increases, poverty, land degradation (deforestation, overgrazing), and the lack of a coherent environmental policy.

The United Nations estimates that degraded lands cover some 800 million hectares (2 billion acres); many millions of additional hectares are added each year. An immediate need is to improve the database of occurrences across the globe for a more accurate accounting of the problem and a better understanding of desertification processes.

One of the successes coming from the 1992 Earth Summit in Rio de Janeiro was the Convention to Combat Desertification, an initiative that began operations in 1994 and is still active. Efforts continue to get these issues to the forefront for action.

In August 2010, the *International Conference: Climate, Sustainability, and Development in the Semi-arid Regions* met for a second time (http://www.unccd.int/). Related issues from more than 100 countries were on the agenda. Launched at this meeting was a global effort for the next decade—the United Nations Decade for Deserts and the Fight Against Desertification (UNDDD).

All this was geared to increase public and government attention to the issue of desertification in preparation for the important 2012 *Rio + 21 Earth Summit* in Rio de Janiero. The 2010 conference called for action on many issues: sustainable development and climate change, political involvement on behalf of dryland peoples, integration of and focus on global environmental initiatives, development of financing, education, and recognition of the urgency of the problem, among other items.

About a billion people are being affected by desertification, with their livelihoods threatened. Further delays in addressing the problem will result in far higher costs when compared to taking action within the next few years. For more on global arid lands, see http://www.undp.org/drylands/.

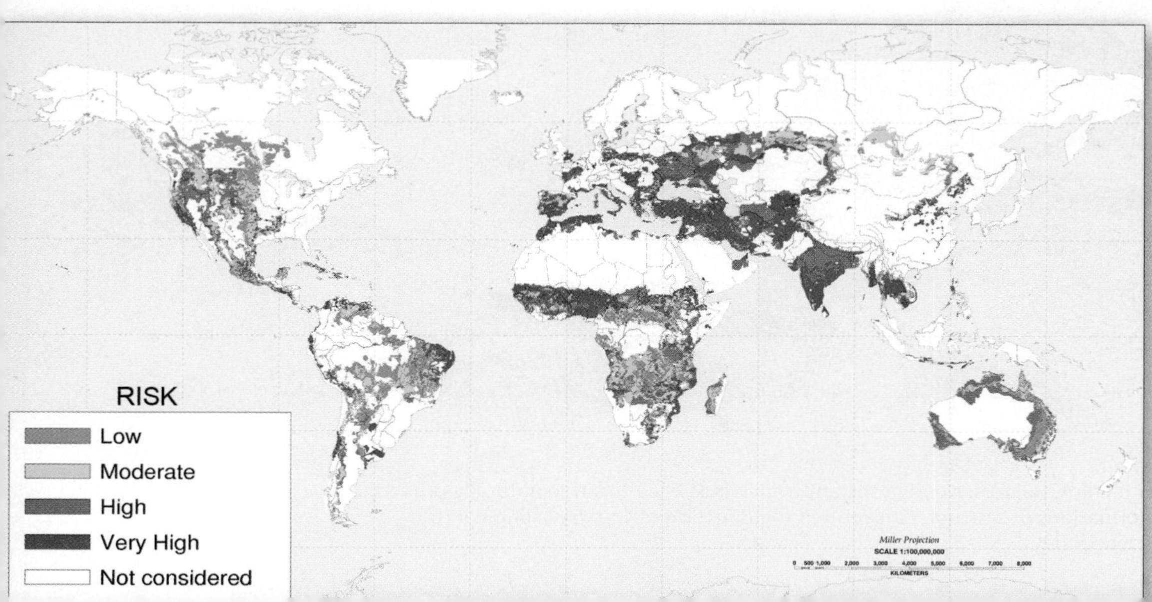

FIGURE GN 15.1 Area at risk of desertification. [Map prepared by U.S. Department of Agriculture–Natural Resources Conservation Service, Soil Survey Division, and includes consideration of population densities in affected regions provided by the National Center for Geographic Information Analysis at University of California, Santa Barbara.]

Wind is an agent of geomorphic change. Like moving water, moving air causes erosion, transportation, and deposition of materials; moving air is a fluid and it behaves similarly, although it has a lower viscosity (it is less dense) than water. Although lacking the lifting ability of water, wind processes can modify and move quantities of materials in deserts, along coastlines, and elsewhere.

Wind contributes to soil formation in distant places, bringing fine particles from regions where glaciers once were active. Fallow fields (those not planted) give up their soil resource to destructive wind erosion. Scientists are only now getting an accurate picture of the amount of windblown dust that fills the atmosphere and crosses the oceans between continents. Chemical fingerprints and satellites trace windblown dust from African soils to South America and Asian landscapes to Europe. Winds even spread living organisms. One study found related mosses, liverworts, and lichens distributed among islands thousands of miles apart in the Southern Ocean. Consistent local wind can prune and shape vegetation and sculpt ice and snow surfaces (Figure 15.1).

Earth's dry lands stand out in stark contrast on the water planet. In desert environments, an overall lack of moisture and stabilizing vegetation allows wind to create extensive sand seas and dunes of infinite variety. The polar regions are deserts as well, so there should be no surprise that weather-monitoring stations in Yuma, Arizona, and in Antarctica report the same amount of annual precipitation. These polar and high-latitude deserts possess unique features related to their cold, dry environment—aspects covered in Chapter 17.

Arid landscapes display distinctive landforms and life forms: "Instead of finding chaos and disorder the observer never fails to be amazed at a simplicity of form, an exactitude of repetition and a geometric order."*

*R. A. Bagnold, *The Physics of Blown Sand and Desert Dunes* (London, England: Methuen, 1941, p. xvii).

In this chapter: We examine the work of wind; associated erosion, transport, and depositional processes; and resulting landforms. Windblown fine particles form vast loess deposits, the basis for rich agricultural soils. We include the discussion of arid lands in this chapter. Most deserts are rocky and covered with desert pavement, whereas other dry landscapes are covered in sand dunes. In desert environments, landscapes lacking moisture and stabilizing vegetation are set off in sharp, sun-baked relief. Water is the major erosional agent in the desert, yet water is the limiting resource for human development. We examine the causes and distribution of Earth's desert landscapes. The Colorado River and the western drought are the subject of an important Focus Study.

The Work of Wind

The work of wind—erosion, transportation, and deposition—is **eolian** (also spelled *aeolian*). The word comes from Aeolus, ruler of the winds in Greek mythology. British Army Major Ralph Bagnold, an engineering officer, while stationed in Egypt in 1925 accomplished much eolian research. In the deserts west of the Nile, Bagnold measured, sketched, and developed hypotheses about the wind and desert forms. Imagine the scene in the 1920s: A Model-T Ford chugs across the desert west of Cairo taking Bagnold to his research area. He laid rolls of chicken wire over treacherous stretches of sand to prevent getting stuck. Bagnold's often-cited work *The Physics of Blown Sand and Desert Dunes* was published in 1941.

The ability of wind to move materials is actually small compared with that of other transporting agents such as water and ice because air is so much less dense than those other media. Yet, over time, wind accomplishes enormous work. Bagnold studied the ability of wind

(a)

(b)

FIGURE 15.1 The work of wind.
(a) Wind-sculpted tree near South Point, Hawai'i. Nearly constant tradewinds keep this tree naturally pruned.
(b) Blowing wind shapes ice in formations of *sastrugi*. Can you tell the direction of the prevailing winds looking at the ice? [Bobbé Christopherson.]

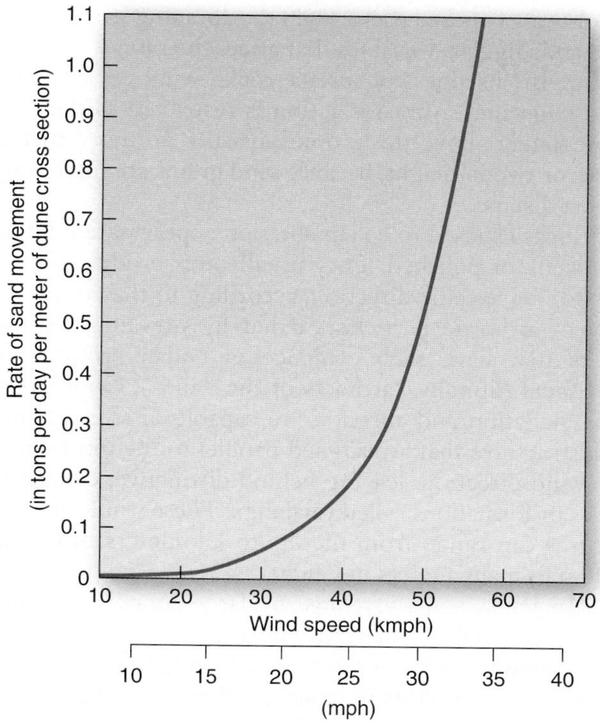

FIGURE 15.2 Sand movement and wind velocity. Sand movement relative to wind velocity, as measured over a meter cross section of ground surface. [After R. A. Bagnold, 1941). Adapted by permission.]

to transport sand over the surface of a dune. Figure 15.2 shows that a steady wind of 50 kmph (30 mph) can move approximately one-half ton of sand per day over a 1-m-wide section of dune. The graph also demonstrates how rapidly the amount of transported sand increases with wind speed.

Grain size is important in wind erosion. Intermediate-sized grains move most easily—they bounce along. It is the largest and the smallest sand particles that require the strongest winds to move. The large particles are heavier and thus require stronger winds. The small particles are difficult to move because they exhibit a mutual cohesiveness and because they usually present a smooth (aerodynamic) surface to the wind. Yet the finest dust, once aloft, is carried from continent to continent. In the chapter-opening photo, you can see wind-blown streaks of light-colored sand along the left half of the picture of Ward Terrace.

Eolian Erosion

Two principal wind-erosion processes are **deflation**, the removal and lifting of individual loose particles, and **abrasion**, the grinding of rock surfaces by the "sandblasting" action of particles captured in the air. Deflation and abrasion produce a variety of distinctive landforms and landscapes.

Deflation Deflation literally blows away loose or noncohesive sediment and works with rainwater to form a surface resembling a cobblestone street: a **desert pavement** that protects underlying sediment from further deflation and water erosion. Traditionally, deflation was regarded as a key formative process, eroding fine dust, clay, and sand and leaving behind a concentration of pebbles and gravel as desert pavement (Figure 15.3a).

Another hypothesis that better explains some desert pavement surfaces states that deposition of windblown sediments, not removal, is the formative agent. Windblown particles settle between and below coarse rocks and pebbles that are gradually displaced upward. Rainwater is involved as wetting and drying episodes swell and shrink clay-sized particles. The gravel fragments are gradually lifted to surface positions to form the pavement (Figure 15.3b). Desert pavements are so common that many provincial names are used for them—for example, *gibber plain* in Australia; *gobi* in China; and in Africa, *lag gravels* or *serir*; or *reg* desert if some fine particles remain.

Heavy recreational activity damages fragile desert landscapes, especially in the United States, where more than 15 million off-road vehicles are now in use. Such vehicles crush plants and animals; disrupt desert pavement, leading to greater deflation; and create ruts that easily concentrate sheetwash to form gullies.

Wherever wind encounters loose sediment, deflation may remove enough material to form basins known as **blowout depressions**. These depressions range from small indentations less than a meter wide up to areas hundreds of meters wide and many meters deep. Chemical weathering, although slow in arid regions owing to the lack of water, is important in the formation of a blowout, for it removes the cementing materials that give particles their cohesiveness. Large depressions in the Sahara Desert are at least partially formed by deflation. The enormous Munkhafad el Qaṭṭâra (Qaṭṭâra Depression), which covers 18,000 km^2 (6950 mi^2) just inland from the Mediterranean Sea in the Western Desert of Egypt, is now about 130 m (427 ft) below sea level at its lowest point.

GEO REPORT 15.1 *Bagnold was ahead of his time*

Ralph Bagnold's work in the desert described sand dunes as complex self-organizing systems—with order and regularity arising from chaos and disorder. Today, this is the science of complex system behavior, where system components interact in a self-enhancing manner. For example, consider the ripple patterns in sand: The ripple captures increasing amounts of sand as it grows in size, which reduces the amount of sand grains in the air, forcing the distance span before another ripple forms. In Chapter 17, we study self-organizing behavior in periglacial regions and how freeze–thaw cycles organize rock sorting and resulting patterns.

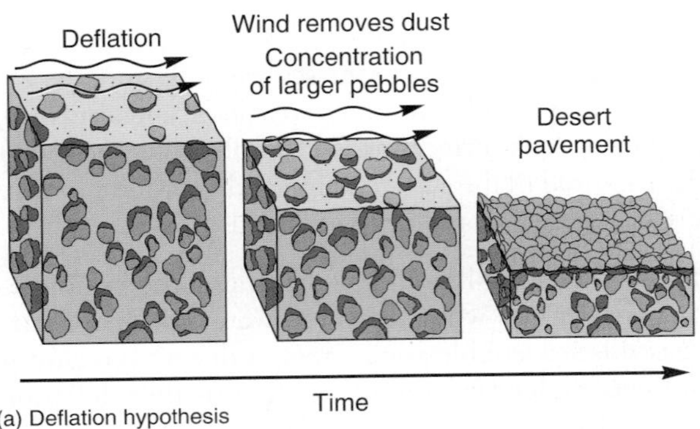

(a) Deflation hypothesis

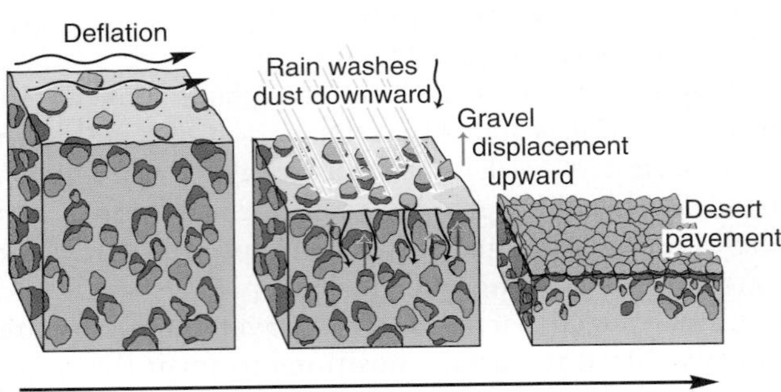

(b) Sediment-accumulation hypothesis

(c)

FIGURE 15.3 Desert pavement.
(a) Desert pavement may be formed when larger rocks and fragments are left after deflation and sheetwash remove finer dust, silt, and clays. (b) Other desert pavements seem to be formed by wind delivering fine particles that settle and wash downward; through cycles of swelling and shrinking, the gravels gradually migrate upward to form pavement. (c) A typical desert pavement. [Bobbé Christopherson.]

Abrasion You may have seen work crews sandblasting surfaces on buildings, bridges, or streets to clean them or to remove unwanted markings. Sandblasting uses a stream of compressed air filled with sand grains to quickly abrade a surface. Abrasion by windblown particles is nature's slower version of sandblasting, and it is especially effective at polishing exposed rocks when the abrading particles are hard and angular. Variables that affect the rate of abrasion include the hardness of surface rocks, wind velocity, and wind constancy. Abrasive action is restricted to the area immediately above the ground, usually no more than a meter or two in height, because sand grains are lifted only a short distance.

Rocks exposed to eolian abrasion appear pitted, fluted (grooved), or polished. They usually are aerodynamically shaped in a specific direction, according to the consistent flow of airborne particles carried by prevailing winds. Rocks that have such evidence of eolian erosion are **ventifacts** (literally, "artifacts of the wind"). On a larger scale, deflation and abrasion are capable of streamlining rock structures that are aligned parallel to the most effective wind direction, leaving behind distinctive, elongated ridges or formations called **yardangs**. These wind-sculpted features can range from meters to kilometers in length and up to many meters in height.

On Earth, some yardangs are large enough to be detected on satellite imagery. The Ica Valley of southern Peru contains yardangs reaching 100 m (330 ft) in height and several kilometers in length, and yardangs in the Lût Desert of Iran attain 150 m (490 ft) in height. Abrasion is concentrated on the windward end of each yardang, with deflation operating on the leeward portions. Wind-sculpted formations are shown in Figure 15.4.

The Sphinx in Egypt was perhaps partially formed as a yardang, suggesting a head and body. Some scientists think this shape led the ancients to complete the bulk of the sculpture artificially with masonry.

Eolian Transportation

Atmospheric circulation can transport fine material, such as volcanic debris, fire soot and smoke, and dust, worldwide within days. Wind exerts a drag, or frictional pull, on surface particles until they become airborne, just as water in a stream picks up sediment (again, think of air as a fluid). The distance that wind is capable of transporting particles in suspension varies greatly with particle size. Only the finest dust particles travel significant distances, so the finer material suspended in a dust storm is lifted much higher than are the coarser particles of a sandstorm.

People living in areas of frequent dust storms are faced with infiltration of very fine particles into their homes and businesses through even the smallest cracks. Figure 3.8 illustrates such dust storms. People living in desert regions and along sandy beaches, where frequent sandstorms occur, contend with the sandblasting of painted surfaces and etched window glass.

Both human and natural sand erosion and transport from a beach are slowed by conservation measures such as the introduction of stabilizing native plants, the use of fences, and the restriction of pedestrian traffic to walkways (Figure 15.5). As human settlement encroaches on coastal dunes, sand transport becomes problematic.

The term *saltation* was used in Chapter 14 to describe movement of particles by water. The term also describes

(a)

(b)

FIGURE 15.4 A yardang.
(a) A small wind-sculpted rock formation located in Snow Canyon outside St. George, Utah. (b) Consistent winds shape this exposed rock on Bear Island in the North Atlantic. The rock sits on a surface of patterned ground and periglacial features (see Chapter 17). [Bobbé Christopherson.]

the wind transport of grains, usually larger than 0.2 mm (0.008 in.), along the ground. About 80% of wind transport of particles is accomplished by this skipping and bouncing action (Figure 15.6). Compared with fluvial transport, in which saltation is accomplished by hydraulic lift, eolian saltation is executed by aerodynamic lift, elastic bounce, and impact (compare Figure 15.6a with Figure 14.12).

On impact, grains hit other grains and knock them into the air. Saltating particles crash into other particles, knocking them both loose and forward. This type of movement is **surface creep**, which slides and rolls particles too large for saltation and affects about 20% of the material transported by wind. Once in motion, particles continue to be transported by lower wind velocities. In a desert or along a beach, sometimes you can hear a slight hissing sound, almost like steam escaping, produced by the myriad saltating grains of sand as they bounce along and collide with surface particles.

Eolian Depositional Landforms

The smallest features shaped by individual saltating grains are ripples (Figure 15.7). Eolian ripples are similar to fluvial ripples, although the impact of saltating grains is slight in water. Ripples form in crests and troughs, positioned

(a)

(b)

FIGURE 15.5 Preventing sand transport.
(a) Popham Beach State Park, Maine. Further erosion and transport of coastal dunes can be controlled by stabilizing strategies, such as planting native plants and confining pedestrian traffic to walkways. (b) Tufts of beach grass are planted to stabilize this coastal area in Atlantic City, New Jersey. One storm surge can erase such efforts. [Bobbé Christopherson.]

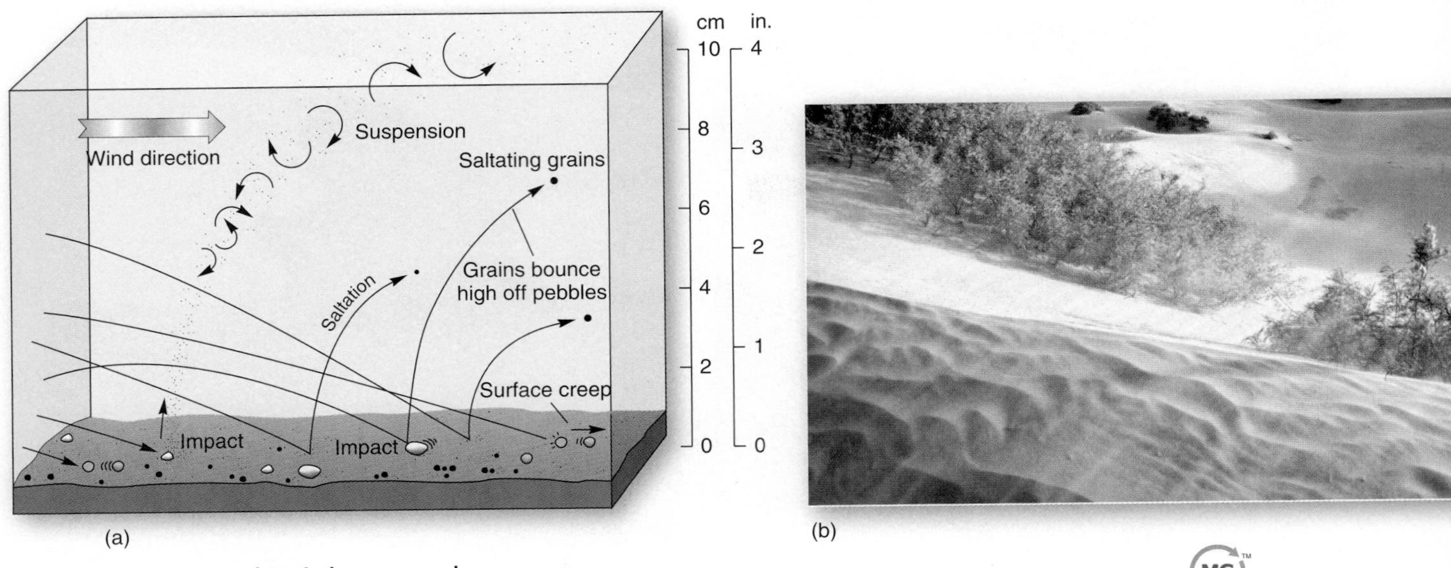

FIGURE 15.6 How the wind moves sand.
(a) Eolian suspension, saltation, and surface creep are transportation mechanisms. Compare with the saltation and traction that occur in another fluid, water, in Figure 14.12. (b) Sand grains saltating along the surface in the Stovepipe Wells dune field, Death Valley, California. [Author.]

MG Animation
How Wind Moves Sand

transversely (at a right angle) to the direction of the wind. The interception of sand grains from the wind stream deprives the downwind area of grains and sets the distance between ripples—this is self-enhancing behavior in a complex system.

Many people know arid landscapes only from the movies, which leave the impression that most deserts are covered by sand. Instead, desert pavements predominate across most subtropical arid landscapes; only about 10% of desert areas are covered with sand. Sand grains generally are deposited as transient ridges or hills called dunes.

FIGURE 15.7 Sand ripples.
Sand ripple patterns later may become lithified into fixed patterns in rock. The area in the photo looks vast, but is only about 1 m (3.3 ft) wide. [Author.]

A **dune** is a wind-sculpted accumulation of sand. An extensive area of dunes, such as that found in North Africa, is characteristic of an **erg desert**, or **sand sea**. The Grand Erg Oriental in the central Sahara exceeds 1200 m (4000 ft) in depth and covers 192,000 km^2 (75,000 mi^2), comparable to the area of Nebraska. This sand sea has been active for more than 1.3 million years. Similar sand seas, such as the Grand Ar Rub‘al Khālī Erg, are active in Saudi Arabia.

In eastern Algeria, the Issaouane Erg covers 38,000 km^2 (14,673 mi^2). Winds from varying directions produce star dunes with multiple slipfaces (Figure 15.8a). *Star dunes* are the mountainous giants of the sandy desert. They are pinwheel-shaped, with several radiating arms rising and joining to form a common central peak, where they approach 200 m (650 ft) in height. Some in the Sonoran Desert approach this size. In the photo, you can also see some dunes, which suggest the region has experienced changing wind patterns over time, as they form from a single principal wind flow. Dunes are present on Mars as well as in semiarid areas such as the Great Plains of the United States (Figure 15.8b and c).

Dune Movement and Form Dune fields, whether in arid regions or along coastlines, tend to migrate in the direction of effective, sand-transporting winds. In this regard, stronger seasonal winds or winds from a passing storm may prove more effective than average prevailing winds. When saltating sand grains encounter small patches of sand, their kinetic energy (motion) is dissipated and they accumulate. As height increases above 30 cm (12 in.), a *slipface* and characteristic dune features form.

By studying the dune model in Figure 15.9, you can see that winds characteristically create a gently sloping *windward side* (stoss side), with a more steeply sloped **slipface** on the *leeward side*. A dune usually is asymmetrical

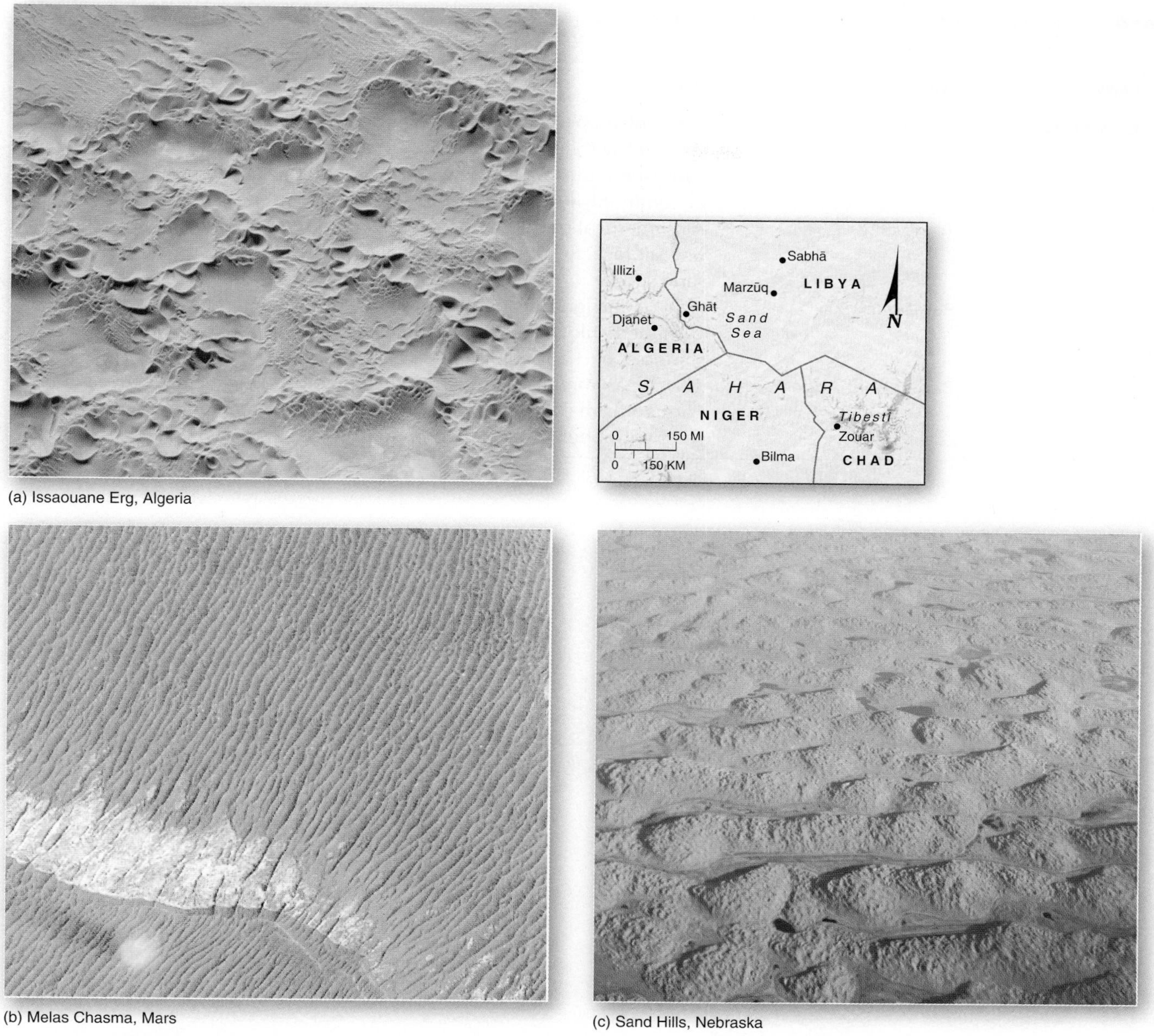

(a) Issaouane Erg, Algeria

(b) Melas Chasma, Mars

(c) Sand Hills, Nebraska

FIGURE 15.8 Seas of sand.
(a) The Issaouane Erg of eastern Algeria is a blend of principally star dunes, some barchan dunes, and linear dunes in the far upper-left—all disclose the prevailing wind history of the region. (b) Sand dunes are found on the Martian surface in the southern area of Melas Chasma in Valles Marineris. Note the dust devil in the lower left. The area covered is about 2 km wide. (c) Sand Hills, central Nebraska, sit atop the High Plains aquifer (see Figure 9.1.1), where sand and silt deposits derived from glaciated regions to the north and the Rockies to the west accumulated. These densely packed barchan dunes, inactive for at least 600 years, are fixed (stabilized) by vegetation. [(a) *ISS* astronaut photo, NASA/GSFC. (b) Mars Global Surveyor, Mars Orbiter Camera image courtesy of NASA/JPL/Malin Space Science Systems. (c) Bobbé Christopherson.]

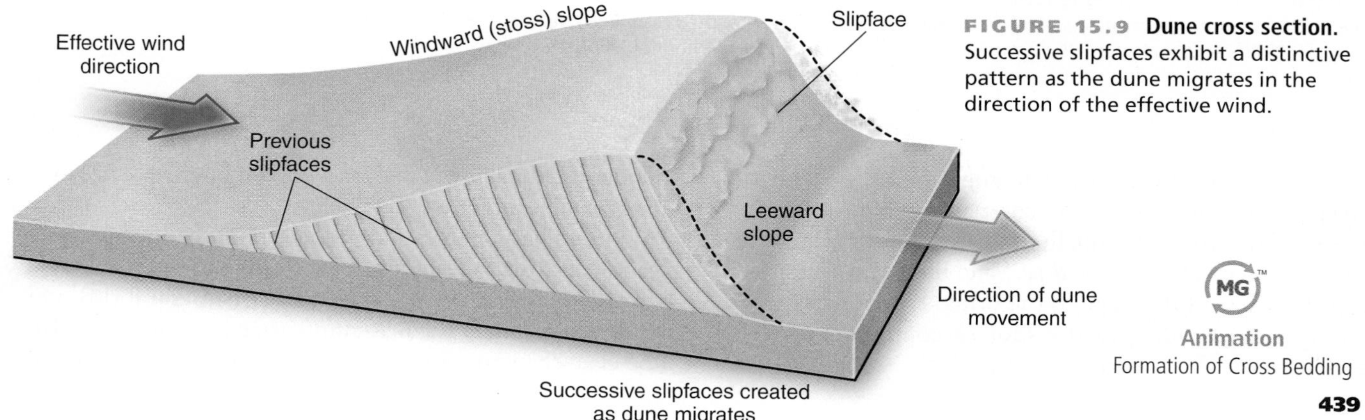

FIGURE 15.9 Dune cross section. Successive slipfaces exhibit a distinctive pattern as the dune migrates in the direction of the effective wind.

MG™ **Animation**
Formation of Cross Bedding

Class	Type	Description
Crescentic	Barchan	Crescent-shaped dune with horns pointed downwind. Winds are constant with little directional variability. Limited sand available. Only one slipface. Can be scattered over bare rock or desert pavement or commonly in dune fields.
	Transverse	Asymmetrical ridge, transverse to wind direction (right angle). Only one slipface. Results from relatively ineffective wind and abundant sand supply.
	Parabolic	Role of anchoring vegetation important. Open end faces upwind with U-shaped "blowout" and arms anchored by vegetation. Multiple slipfaces, partially stabilized.
	Barchanoid ridge	A wavy, asymmetrical dune ridge aligned transverse to effective winds. Formed from coalesced barchans; looks like connected crescents in rows with open areas between them.

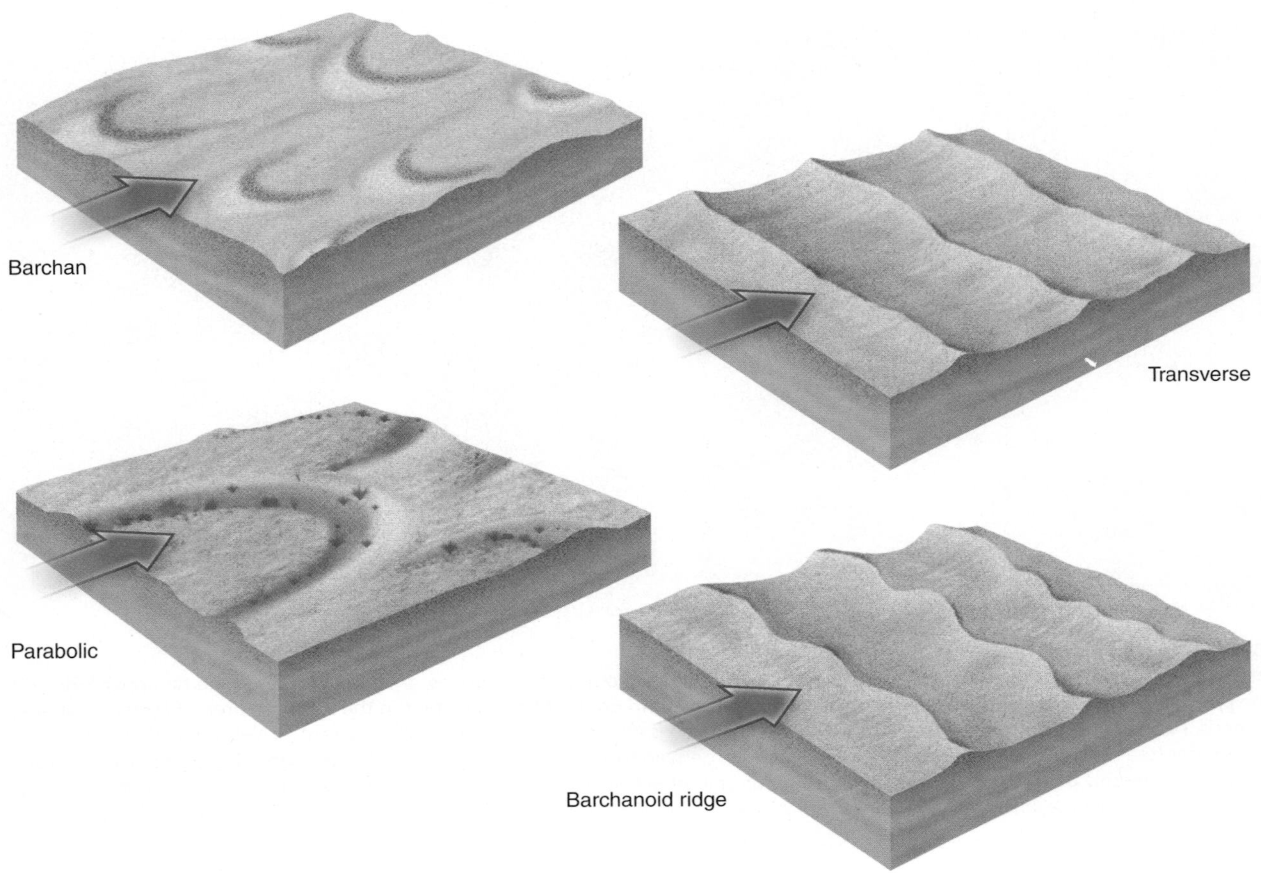

FIGURE 15.10 Major dune forms.
See table for dune descriptions. Arrows show wind direction. [Adapted from E. D. McKee, *A Study of Global Sand Seas,* USGS Professional Paper 1052 (Washington, DC: U.S. Government Printing Office, 1979).]

in one or more directions. The angle of a slipface is the steepest angle at which loose material is stable—its *angle of repose*. Thus, the constant flow of new material makes a slipface, a type of *avalanche slope*. Sand builds up as it moves over the crest of the dune to the brink; then it avalanches, falling and cascading as the slipface continually adjusts, seeking its angle of repose (usually 30° to 34°). In this way, a dune migrates downwind, as suggested by the successive dune profiles in Figure 15.9.

Dunes have many wind-shaped styles that make classification difficult. We simplify dune forms into three classes—*crescentic* (crescent, curved shape), *linear* (straight form), and massive *star dunes*. Each is summarized in Figure 15.10. The ever-changing form of these eolian deposits is part of

Class	Type	Description
Linear	Longitudinal	Long, slightly sinuous, ridge-shaped dune, aligned parallel with the wind direction; two slipfaces. Average 100 m high and 100 km long, and the "draa" at the extreme in size can reach to 400 m high. Results from strong effective winds varying in one direction.
	Seif	After Arabic word for "sword"; a more sinuous crest and shorter than longitudinal dunes. Rounded toward upwind direction and pointed downwind. (Not illustrated.)
Star dune		The giant of dunes. Pyramidal or star-shaped with three or more sinuous, radiating arms extending outward from a central peak. Slipfaces in multiple directions. Results from effective winds shifting in all directions. Tend to form isolated mounds in high, effective winds and connected sinuous arms in low, effective winds.
Other	Dome	Circular or elliptical mound with no slipface. Can be modified into barchanoid forms; vegetation can play role in coastal locales.
	Reversing	Asymmetrical ridge form that is intermediate between star dune and transverse dune. Wind variability can alter shape between forms.

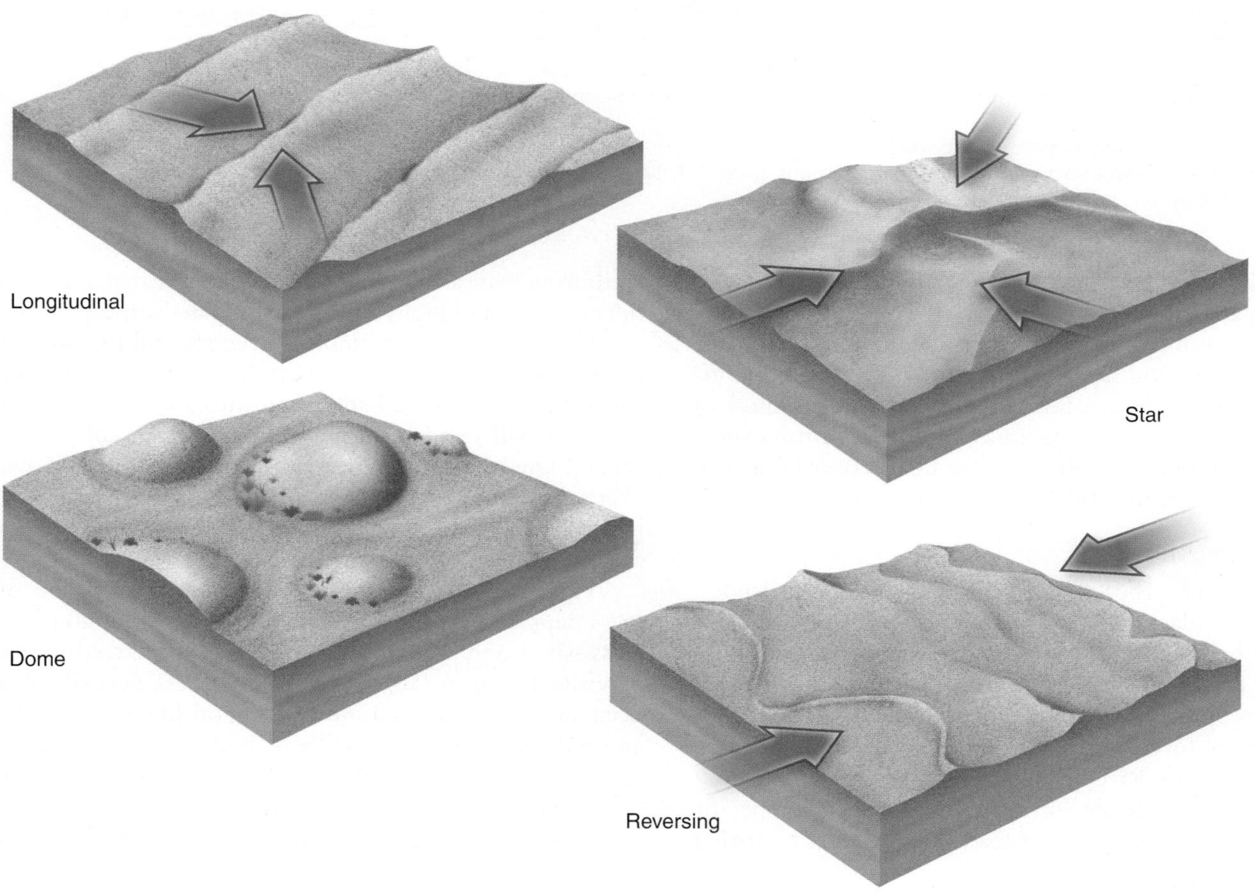

their beauty, eloquently described by one author: "I see hills and hollows of sand like rising and falling waves. Now at midmorning, they appear paper white. At dawn they were fog gray. This evening they will be eggshell brown."*

*J. E. Bowers, *Seasons of the Wind* (Flagstaff, AZ: Northland Press, 1985), p. 1.

The map in Figure 15.11 shows the correlation of active sand regions with deserts (tropical, continental interior, and coastal). Note the limited extent of desert area covered by active sand dunes—only about 10% of all continental land between 30° N and 30° S. Also noted on the map are dune fields in humid climates outside of this range, such as along coastal Oregon, the south shore of

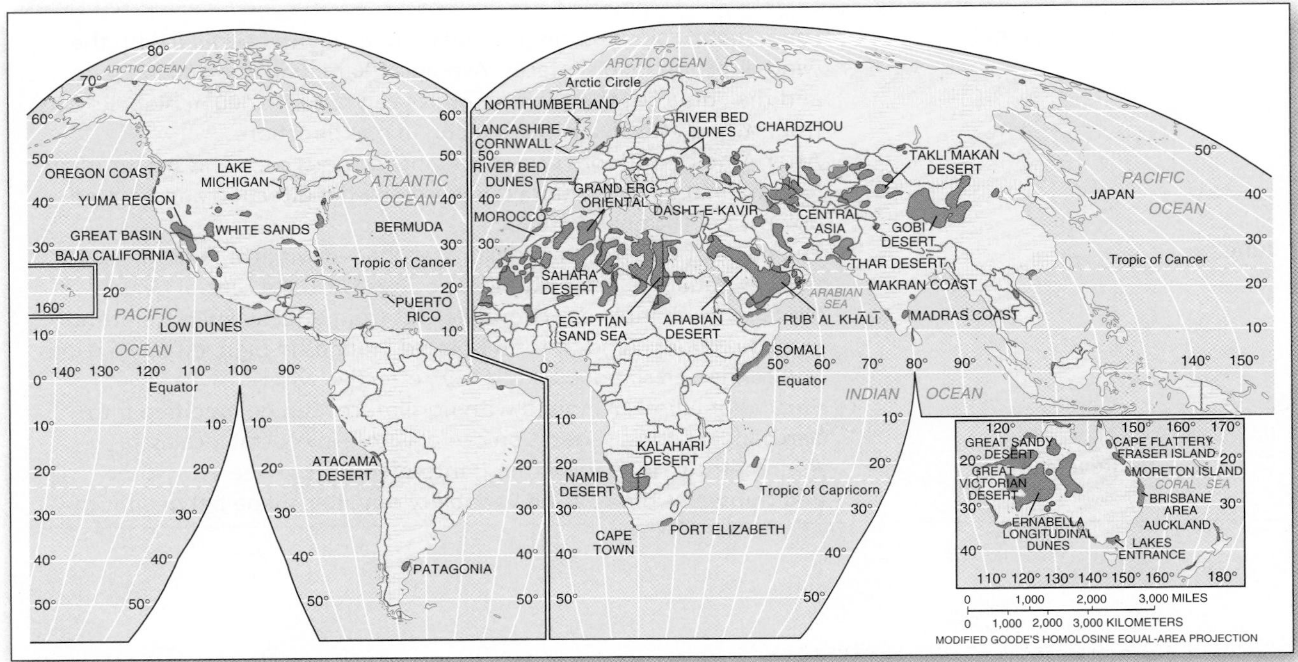

FIGURE 15.11 **Sandy regions of the world.**
Worldwide distribution of active and stable sand regions. [After R. E. Snead, *Atlas of World Physical Features*, p. 134, © 1972 by John Wiley & Sons. Adapted by permission.]

Lake Michigan, and the Gulf and Atlantic coastlines; in Europe; and elsewhere.

These same dune-forming principles and terms (for example, *dune* and *slipface*) apply to snow-covered landscapes. *Snow dunes* are formed as wind deposits snow in drifts. In semiarid farming areas, capturing drifting snow with fences and tall stubble left in fields contributes significantly to soil moisture when the snow melts.

Loess Deposits

Approximately 15,000 years ago, in several episodes, Pleistocene glaciers retreated in many parts of the world, leaving behind large glacial outwash deposits of fine-grained clays and silts. These materials were blown great distances by the wind and redeposited in unstratified, homogeneous (evenly mixed) deposits. Peasants working along the Rhine River Valley in Germany gave these deposits the name **loess** (pronounced "luss"). No specific landforms were created; instead, loess covered existing landforms with a thick blanket of material that assumed the general topography of the existing landscape.

Because of its own binding strength and its internal coherence, loess weathers and erodes into steep bluffs, or vertical faces. When a bank is cut into a loess deposit, it generally will stand vertically, although it can fail if saturated (Figure 15.12). There are historical accounts of Civil War soldiers in the Vicksburg, Mississippi, area excavating places to live in loess banks; in areas of China, dwellings are carved from the loess.

Figure 15.13a shows the worldwide distribution of loess deposits. Significant accumulations throughout the Mississippi and Missouri river valleys form continuous deposits 15–30 m (50–100 ft) thick. Loess deposits also occur in eastern Washington State and Idaho. The hills and gullies in the Loess Hills of Iowa in Figure 15.13b reach heights of about 61 m (200 ft) above the nearby prairie farmlands and Missouri River and run north and south more than 322 km (200 mi). Only China has deposits

GEO REPORT 15.2 *Conflicts feed the sand seas*

Military activities can threaten desert landscapes. A serious environmental impact of the ongoing wars in Iraq and Afghanistan is the disruption of desert pavement by explosives and the movement of heavy vehicles. As thousands of square kilometers of stable desert pavement break apart, nearby cities and farms experience increased dust and sand invasions.

The Registan Desert of southern Afghanistan, south of the city of Kandahar, is an example of such a sand sea over a flat terrain. A result of drought conditions and war-disrupted desert pavement surfaces, the migrating sand is overtaking areas of sparse agricultural activity and has covered more than 100 villages.

FIGURE 15.12 Example of loess deposit. A loess bluff in western Iowa; note the vertical structural strength. [Bobbé Christopherson.]

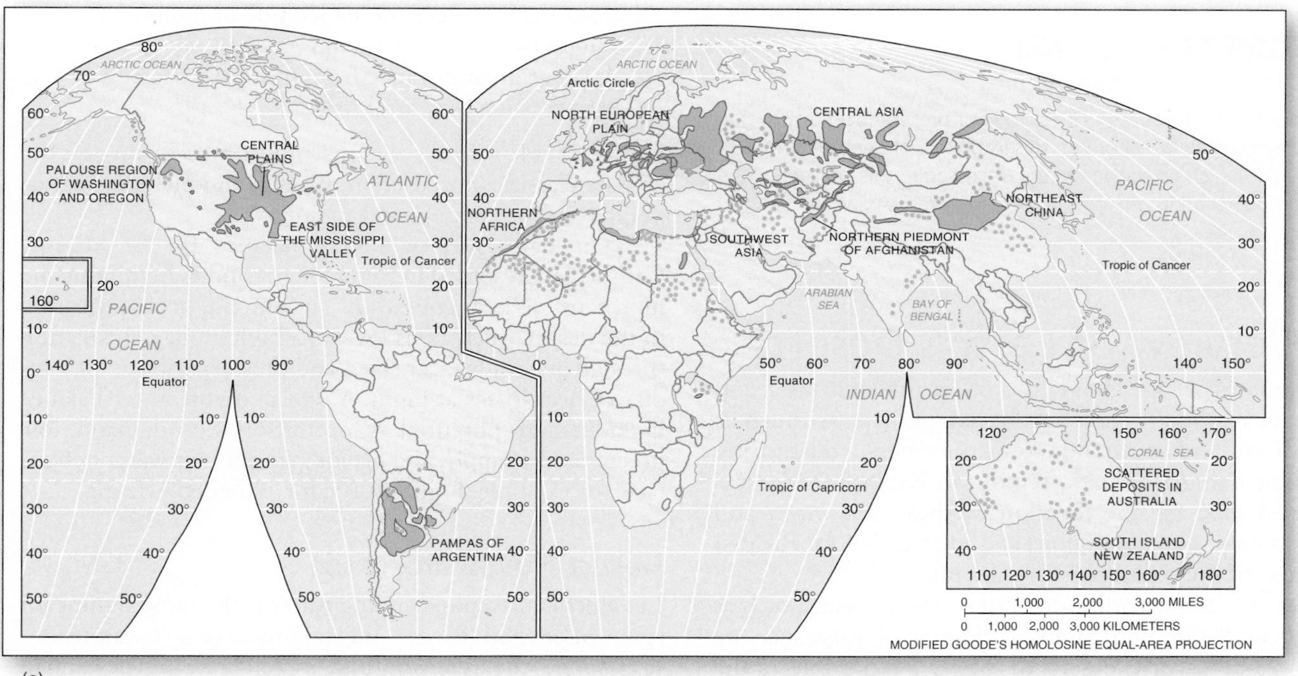

(a)

(b)

FIGURE 15.13 Loess deposits worldwide and in Iowa. (a) Areas of major accumulations of loess. Dots represent small, scattered loess formations. (b) The Loess Hills of western Iowa give evidence of the last ice age, as the finely ground sediments left by the glaciers were windblown to this setting. [(a) After R. E. Snead, *Atlas of World Physical Features*, p. 138, © 1972 by John Wiley & Sons. Adapted by permission. (b) Bobbé Christopherson.]

that exceed these dimensions. This silt explains the fertility of the soils in these regions, for loess deposits are well drained and deep and have excellent moisture retention. Loess deposits also cover much of Ukraine, central Europe, China, the Pampas–Patagonia regions of Argentina, and lowland New Zealand. The soils derived from loess help form some of Earth's "breadbasket" farming regions. Transport of loess soils occurred in the catastrophic 1930s Dust Bowl in the United States.

In Europe and North America, loess is thought to be derived mainly from glacial and periglacial sources. The vast deposits of loess in China, covering more than 300,000 km² (116,000 mi²), are derived from windblown desert sediment rather than glacial sources. Accumulations in the Loess Plateau of China exceed 300 m (1000 ft) in thickness, forming complex weathered badlands and some good agricultural land. These windblown deposits are interwoven with much of Chinese history and society.

CRITICAL THINKING 15.1

The nearest eolian features

What eolian features are nearest to your present location? Are they coastal, lakeshore, or desert dunes, or are they loess deposits? Which causative factors discussed in this chapter explain the location or form of these features? Are you able to visit the site?

Overview of Earth's Deserts

Dry climates occupy about 26% of Earth's land surface. If all semiarid climates are considered, perhaps as much as 35% of all land area tends toward dryness, constituting the largest single climatic region on Earth (see Figures 10.5 and 10.6 for the location of these *arid desert* and *semiarid steppe* climate regions and Figure 20.4 for the distribution of these desert environments).

Deserts occur worldwide as topographic plains, such as the Great Sandy and Simpson deserts of Australia, the Arabian and Kalahari deserts of Africa, and portions of the extensive Taklimakan Desert, which covers some 270,000 km² (105,000 mi²) in the central Tarim Basin of China. Deserts also are found in mountainous regions: in interior Asia from Iran to Pakistan and in China and Mongolia. In South America, the rugged Atacama Desert lies between the ocean and the Andes. Desert conditions can also be found on small islands, such as the Cape Verde Islands off the West African coast at around 15° N latitude, an extension of Saharan conditions to the east (Figure 15.14d). Earth's deserts are expanding, as discussed in *Geosystems Now*. Now, let us look at the link between climate and Earth's deserts.

Desert Climates

The spatial distribution of dry lands is related to three climatological settings:

- Areas dominated by subtropical high-pressure cells between 15° and 35°, both N and S latitudes (see Figures 6.9 and 6.11);
- Rain shadow areas on the lee side of mountain ranges (see Figure 8.8); and
- Areas at great distance from moisture-bearing air masses, such as central Asia.

Figure 15.14 portrays this distribution according to the climate classification used in this text and includes photographs of four major desert regions.

The daily surface energy balance for El Mirage, California, presented in Figure 4.20a and b, highlights the high sensible heat conditions and intense ground heating in the desert. Such areas receive a high input of insolation through generally clear skies, and they experience high radiative heat losses at night. A typical desert water balance includes high potential evapotranspiration demand, low precipitation supply, and prolonged summer water deficits (see, for example, Figure 9.10c for Phoenix, Arizona).

Desert Fluvial Processes

In desert landscapes, running water is the most significant agent of erosion. Rivers and streams in deserts can be described by the type of flow, which may be perennial,

GEO REPORT 15.3 ***The Dust Bowl***

Wind transport of soils including loess produced a catastrophe in the American Great Plains in the 1930s—the Dust Bowl. More than a century of overgrazing and intensive agriculture left soil susceptible to drought and deflation processes. Reduced precipitation amounts and above-normal temperatures triggered the multiyear period of severe dust storms and loss of farmlands.

The deflation of many centimeters of soil occurred in southern Nebraska, Kansas, Oklahoma, Texas, and eastern Colorado—and even in southern Canada and northern Mexico. The transported dust darkened the skies of midwestern cities, which left the streetlights on throughout the day, and drifted over farmland to depths that covered failing crops. The largest migration in American history occurred as 2 million people moved out of the plains states by 1940.

Tree-ring analyses show that droughts of similar intensity happened about twice a century over the past 400 years. One such intense drought in what is now Nebraska in the 13th century appears to have lasted nearly 40 years. The Dust Bowl weather conditions seem to correlate with sea-surface temperatures that are cool in the Pacific Ocean and warm in the Atlantic Ocean. Chapter 6 discussed the Pacific Decadal Oscillation and its relation to drought occurrences in the United States. For more on the Dust Bowl, see http://www.pbs.org/wgbh/americanexperience/films/dustbowl/.

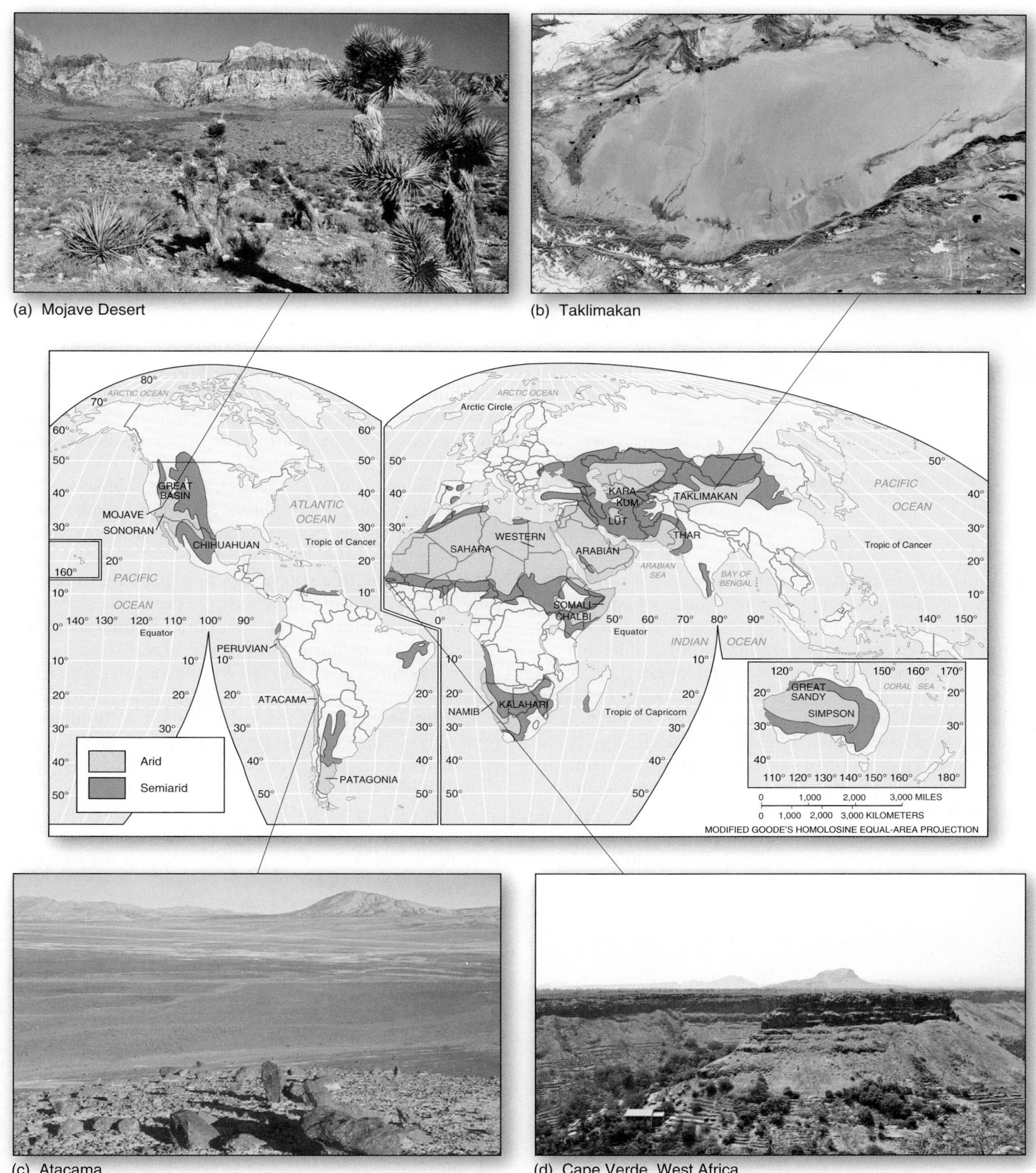

FIGURE 15.14 The world's dry regions.
Worldwide distribution of arid lands (*arid desert* climates) and semiarid lands (*semiarid steppe* climates). (a) Mojave Desert in the American Southwest. (b) Taklimakan Desert in central Asia from orbit. (c) Atacama Desert in subtropical Chile near Baquedano. (d) São Tiago Island desert in Cape Verde; note the light sky filled with dust from Africa. [(a) Author. (b) *Terra* image, NASA/GSFC. (c) Jacques Jangoux/Photo Researchers, Inc. (d) Bobbé Christopherson.]

FIGURE 15.15 Water through a dry land.
The Virgin River is a tributary of the Colorado River. This perennial stream is prone to flash floods during precipitation events. The riparian vegetation includes willow, cottonwood, and the water-drawing tamarisk, or salt cedar. [Bobbé Christopherson.]

ephemeral, or intermittent. *Perennial streams* flow all year, fed by snowmelt, rainfall, or groundwater, often in some combination (Figure 15.15). *Ephemeral streams* flow only after precipitation events and are not connected to groundwater systems. Years may pass between flow events in these usually dry stream channels. *Intermittent streams* flow for several weeks or months each year and may have some groundwater inputs. Fluvial processes in deserts are most often characterized by such intermittent running water, with hard, poorly vegetated surfaces of thin, immature soils yielding high runoff during rainstorms.

Precipitation events in a desert may be rare indeed, even a year or two apart, but when they do occur, an ephemeral stream channel can fill with a torrent, a **flash flood**. These channels may fill in a few minutes and surge briefly during and after a storm. Depending on the region, such a dry channel is known as a **wash**, an *arroyo* (Spanish), or a *wadi* (Arabic). Signs are usually posted where a desert highway crosses a wash to warn drivers not to proceed if rain is in the vicinity, for a flash flood can suddenly arise and sweep away anything in its path. If a long, steady precipitation event occurs, material on slopes may become saturated and flow into canyons. *Debris flows* occur as moving masses of mud, soil, and rock fragments mixed with enough water to keep the material flowing. They can travel at high speed down steep canyons, often leaving unsorted alluvium as their velocity slows and material is deposited.

When washes fill with surging flash flood waters, a unique set of ecological relationships quickly develops. Crashing rocks and boulders break open seeds that respond to the timely moisture and germinate. Other plants and animals also spring into brief life cycles as the water irrigates their limited habitats.

At times of intense rainfall, remarkable scenes fill the desert. Figure 15.16 shows two photographs taken only a month apart in a sand dune field in Death Valley, California. A rainfall event produced 2.57 cm (1.01 in.) of precipitation in a single day, in a place that receives only 4.6 cm (1.83 in.) in an average year. The stream in the photo continued to run for hours and then collected in low spots on hard, underlying clay surfaces. The water was quickly consumed by the high evaporation demand so that, in just a month, these short-lived watercourses were dry and covered with accumulations of alluvium. Such evaporation pans coated with salt are visible on a large scale in Figure 15.8a in Algeria (bluish-white areas).

The 1985 precipitation event just described was exceeded by one on August 15, 2004, when several thunderstorm cells stalled over the mountains along the east side

(a)

(b)

FIGURE 15.16 An improbable river in Death Valley.
The Stovepipe Wells dune field of Death Valley, California, shown (a) the day after a 2.57-cm (1.01-in.) rainfall and (b) one month later (identical location). This 1985 rainfall event was exceeded by one in 2004. [Author.]

of Death Valley. Rainfall totals reached 6.35 cm (2.5 in.) in little more than an hour. Cars were swept away from the Furnace Creek Inn parking lot, about 4.8 km (3 mi) of Highways 190 and 178 were washed out, and two fatalities occurred in the Zabriskie Point area as a result of the storm. One annex building had a high-water mark on it at 3 m (9.8 ft). Again, water is sparse in the deserts, yet it is the major erosional force—sometimes dramatically so. For recent information on and maps of Death Valley, see http://www.nps.gov/deva/.

As runoff water evaporates, salt crusts may be left behind on the desert floor in a **playa**. This intermittently wet and dry lowest area of a closed drainage basin is the site of an *ephemeral lake* when water is present. Accompanying our earlier discussion of evaporites, Figure 11.11 shows such a playa, covered with salt-crust precipitate over fine clays just a month after the 1985 rainfall event.

Permanent lakes and continuously flowing rivers are less common in the desert than in humid regions, although the Nile River and the Colorado River are notable. Both these rivers are *exotic streams*, having their headwaters in a wetter region and the bulk of their course through arid regions. Focus Study 15.1 discusses the Colorado River, the ongoing drought, and the problem of increasing water demands in an arid landscape.

In arid and semiarid climates, a prominent fluvial landform is an **alluvial fan**, which occurs at the mouth of a canyon where an ephemeral stream channel exits into a valley (Figure 15.17). The fan is produced whenever flowing water abruptly loses velocity as it leaves the constricted channel of the canyon, or whenever the stream gradient suddenly decreases, and therefore drops layer upon layer of sediment along the base of the mountain block. Water then flows over the surface of the fan and produces a braided drainage pattern, sometimes shifting from channel to channel. A continuous apron, or **bajada** (Spanish for "slope"), may form if individual alluvial fans coalesce into one sloping surface (see Figure 15.22d). Fan formation of any sort is reduced in humid climates because perennial streams constantly carry away sediment, preventing its deposition.

An aspect of an alluvial fan is the natural sorting of materials by size. Near the mouth of the canyon at the apex of the fan, coarse materials are deposited, grading slowly to pebbles and finer gravels with distance out from the mouth. Then sands and silts are deposited, with the finest clays and dissolved salts carried in suspension and solution all the way to the valley floor. Dissolved minerals accumulate as evaporite deposits left on the valley floor after evaporation of the water from the playa and from salt-laden upward-moving groundwater. Figure 15.18 (p. 450) shows the edge of a playa and active dune field; there is a natural sorting of materials by grain size, with sands dropped in the dunes, silts and clays moving farther toward the playa, and the finest and dissolved materials in the playa.

Well-developed alluvial fans also can be a major source of groundwater. Some cities—San Bernardino, California, for example—are built on alluvial fans and extract their municipal water supplies from them. In other parts of the world, such water-bearing alluvial fans are known as *qanat* (Iran), *karex* (Pakistan), or *foggara* (western Sahara).

(text continued on page 451)

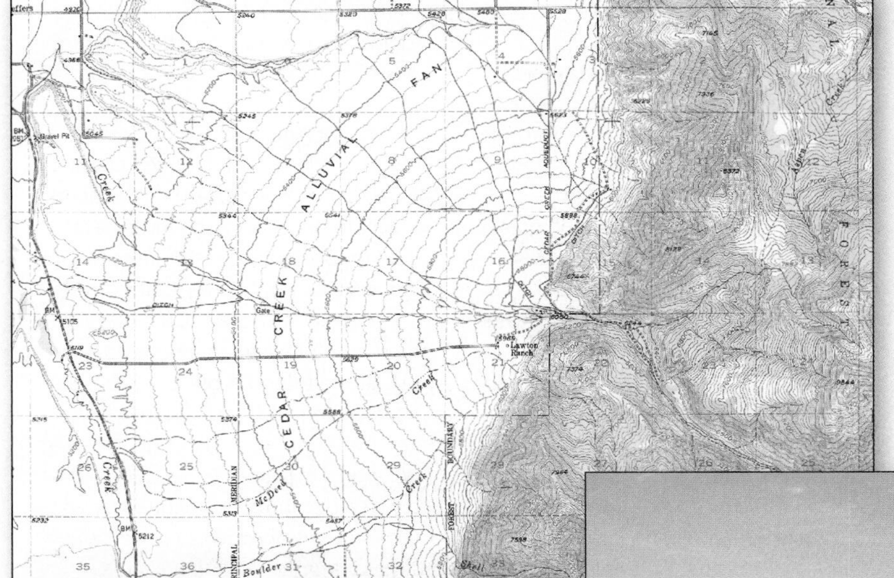

Cedar Creek Fan, Montana

Fan along Black Mountains, Death Valley, California

FIGURE 15.17 An alluvial fan.
The photo shows an alluvial fan in a desert landscape. The topographic map shows the Cedar Creek alluvial fan in southwest Montana, Ennis Quadrangle, 15-minute series, scale 1:62,500, contour interval = 40 ft. [Bobbé Christopherson; U.S. Geological Survey map.]

FOCUS STUDY 15.1

The Colorado River: A System Out of Balance

The headwaters of an exotic stream rise in a region of water surpluses, but the stream flows mostly through arid lands for the rest of its journey to the sea.

In *Geosystems,* the story of the Colorado River began in Chapter 9 with a *Geosystems Now,* "Water Budgets, Climate Change, and the Southwest." Please revisit this case study.

Orographic precipitation totaling 102 cm (40 in.) per year falls mostly as snow in the Rockies, feeding the Colorado River headwaters in Rocky Mountain National Park. But at Yuma, Arizona, near the river's end, annual precipitation is a scant 8.9 cm (3.5 in.), an extremely small amount when compared with the high annual potential evapotranspiration demand in the Yuma region of 140 cm (55 in.).

From its source region, the Colorado River quickly leaves the relatively wet Rockies and spills out into the dry climates of western Colorado and eastern Utah. At Grand Junction, Colorado, near the Utah border, annual precipitation is only 20 cm (8 in.) (upstream from Figure 15.1.1b, on the map). After carving its way through the Utah canyonlands, the river enters Lake Powell, the reservoir behind Glen Canyon Dam (Figure 15.1.1c), and then flows on through the Grand Canyon. The river then turns southward, marking its final 644 km (400 mi) as the Arizona–California border. Along this stretch sits Hoover Dam, just east of Las Vegas (Figure 15.1.1d; see Chapter 9 opener); Davis Dam, built to control the releases from Hoover (Figure 15.1.1e); Parker Dam, for the water needs of Los Angeles; three more dams for irrigation water (Palo Verde, Imperial, and Laguna, Figure 15.1.1f); and finally, Morelos Dam at the Mexican border. Mexico owns the end of the river and whatever water is left. Hydrologically exhausted, the river no longer reaches its mouth in the Gulf of California (Figure 15.1.1g).

Figure 15.1.1h shows the annual water discharge and suspended sediment load for the Colorado River at Yuma, Arizona, from 1905 to 1964. The completion of Hoover Dam in the 1930s dramatically reduced suspended sediment. Subsequent dams upstream from Yuma reduced both discharge and sediment to nearly zero by the mid-1960s.

Overall, the drainage basin encompasses 641,025 km^2 (247,500 mi^2) of mountain, plateau, canyon, basin-and-range, and hot desert landscapes in parts of seven states and two countries. A discussion of the Colorado River is included in this chapter because of its crucial part in the history of the Southwest and its role in the future of this drought-plagued region.

Dividing Up the Colorado's Dammed Water

John Wesley Powell (1834–1902) was the first Euro-American of record to successfully navigate the Colorado River through the Grand Canyon. Lake Powell is named after him. Powell perceived that the challenge of settling the West was too great for individual efforts and believed that solutions to problems such as water availability could be met only through private cooperative efforts. His 1878 study (reprinted in 1962), *Report of the Lands of the Arid Region of the United States,* is a conservation landmark.

Today, Powell probably would be skeptical of the intervention by government agencies in building large-scale reclamation projects. An anecdote in Wallace Stegner's book *Beyond the Hundredth Meridian* tells of an 1893 international irrigation conference held in Los Angeles, where development-minded delegates bragged that the entire West could be conquered and reclaimed from nature and that "rain will certainly follow the plow." Powell spoke against that sentiment: "I tell you, gentlemen, you are piling up a heritage of conflict and litigation over water rights, for there is not sufficient water to supply the land."* He was booed from the hall, but history has shown Powell to be correct.

The Colorado River Compact was signed by six of the seven basin states in 1923. (The seventh, Arizona, signed in 1944, the same year that the United States signed the Mexican Water Treaty.) With this compact, the Colorado River basin was divided into an upper basin and a lower basin, arbitrarily separated for administrative purposes at Lees Ferry near the Utah–Arizona border (noted on the map in Figure 15.1.1; see the photo in Figure 14.32d). Congress adopted the Boulder Canyon Act in 1928, authorizing Hoover Dam as the first major reclamation project on the river. Also authorized was the All-American Canal into the Imperial Valley, which required an additional dam. Los Angeles then began its project to bring Colorado River water 390 km (240 mi) from still another dam and reservoir on the river to the city.

Shortly after Hoover Dam was finished and downstream enterprises were thus offered flood protection, the other projects were quickly completed. There are now eight major dams on the river and many irrigation works. The last effort to redistribute Colorado River water was the Central Arizona Project, which carries water to the Phoenix area.

*W. Stegner, *Beyond the Hundredth Meridian* (Boston, MA: Houghton Mifflin, 1954), p. 343.

FIGURE 15.1.1 The Colorado River drainage basin. →
The Colorado River basin, showing division of the upper and lower basins near Lees Ferry in northern Arizona. (a) River headwaters near Mount Richthofen in the Colorado Rockies. (b) The river near Moab, Utah. (c) Glen Canyon Dam, a regulatory, administrative facility near Lees Ferry, Arizona. (d) Hoover Dam and Lake Mead with declining water levels. (e) Davis Dam in full release during the 1983 flood. (f) Irrigated fresh-cut flowers in the Imperial Valley, refrigerated trucks in the field, and a cargo jet at a local airport await shipments to distant markets. (g) The Colorado River (far lower left) stops short of its former delta. (h) Dam construction affects river discharge and sediment yields. [(a), (c), (d), and (f) Bobbé Christopherson. (b) and (e) Author. (g) *Terra* image, NASA/GSFC. (h) Data from U.S. Geological Survey, *National Water Summary 1984,* Water Supply Paper 2275 (Washington, DC: Government Printing Office, 1985), p. 55.]

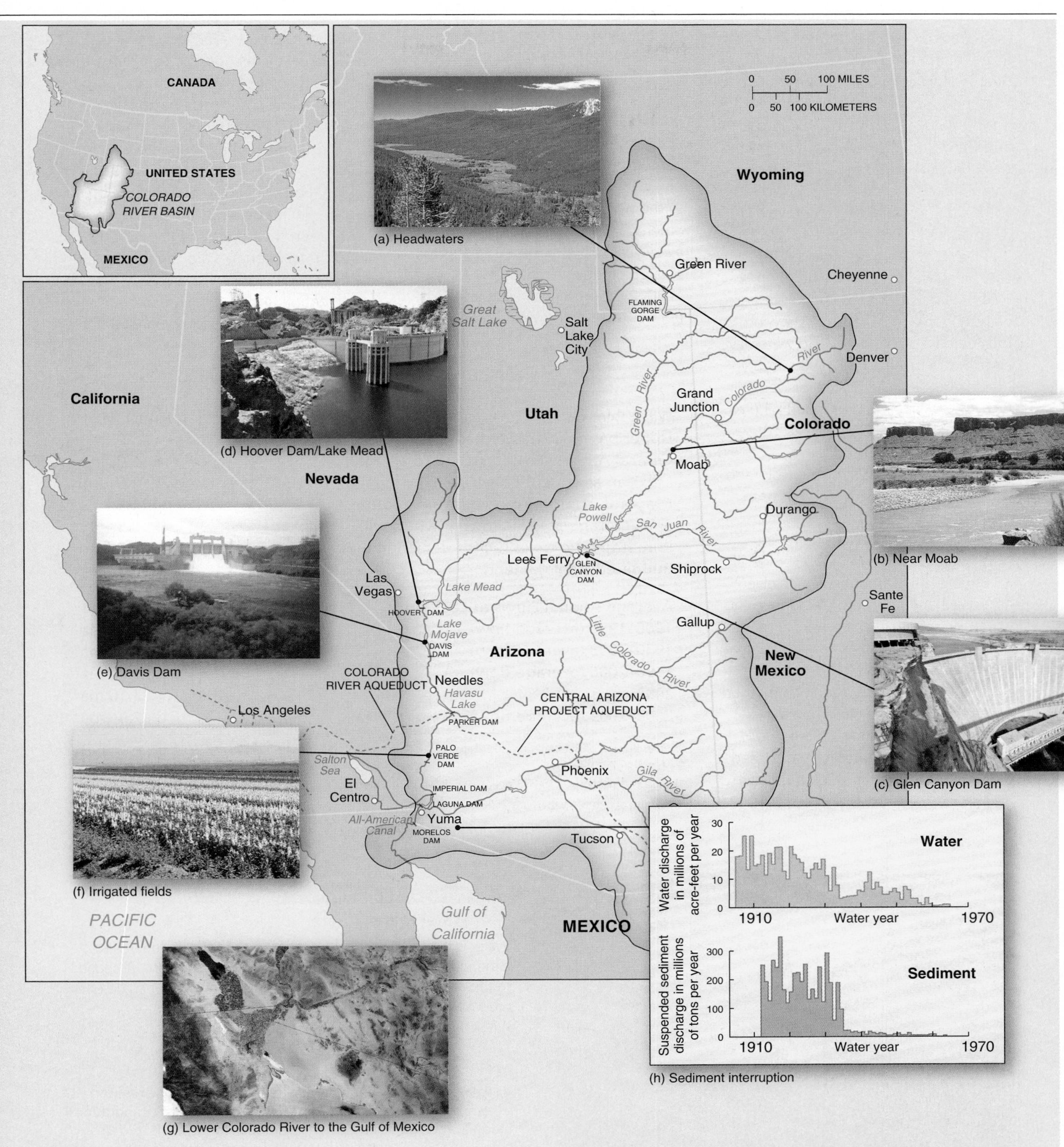

(a) Headwaters

(b) Near Moab

(c) Glen Canyon Dam

(d) Hoover Dam/Lake Mead

(e) Davis Dam

(f) Irrigated fields

(g) Lower Colorado River to the Gulf of Mexico

(h) Sediment interruption

Highly Variable River Flows

The flaw in all this planning and water distribution is that exotic streams have highly variable flows, and the Colorado is no exception. In 1917, the discharge measured at Lees Ferry totaled 24 million acre-feet (maf), whereas in 1934 it dropped by nearly 80%, to only 5.03 maf. In 1977, the discharge dropped again to 5.02 maf, but in 1984 it rose to an all-time high of 24.5 maf. Yet, in contrast, between 2000 and 2007 river discharge fell to exceptionally low levels, with 2002 dropping to a record low of 3.1 maf and 2003 to 6.4 maf. The 2003–2009 water years marked the lowest 7-year average flow in the total record. Please revisit the chapter-opening pair of photos for Chapter 9 showing the comparison of Lake Mead water levels at Hoover Dam in July 1983 and March 2009 as evidence of this variability. By 2009, lake levels dropped 43.6 m (143 ft) below the 375-m (1229-ft) capacity reached in 1983.

As a planning basis for the Colorado River Compact, the government used average annual river discharges from 1914 up to the compact signing in 1923, an exceptionally high average of 18.8 maf. That amount was perceived as more than enough for the upper and lower basins, each to receive 7.5 maf, and, later, for Mexico to receive 1.5 maf in the 1944 Mexican Water Treaty.

Should long-range planning be based on such a short time period on a river with highly variable flows? Tree-ring analyses of past climates have disclosed that the only other time Colorado River discharges were at the high 1914–1923 level was between A.D. 1606 and 1625. Planners have consistently overestimated the dependable flows of the river, a situation that has created shortfalls as demand exceeds supply in the water budget (Table 15.1.1).

Presently, the seven states *ideally* want rights that total as much as 25.0 maf per year. When added to the guarantee for Mexico, this comes up to 26.5 maf of yearly demand (more than in the river by far). And six states share one common opinion: California's right to the water must be limited to a court-ordered 4.4 maf, which it exceeded every year until California's access to surplus Colorado River water was stopped in 2002. At present, no surplus exists during the ongoing western drought.

Colorado River Water Losses

Several natural processes associated with reservoirs and river flow regulation cause water losses and increase overall water demand, as observed in Table 15.1.1. Glen Canyon Dam, located 27.4 km (17 mi) north of Lees Ferry, the division point between the upper and lower basins, serves as an example.

Glen Canyon Dam was completed in 1963 and began impounding water in Lake Powell. Glen Canyon Dam's primary purpose, according to the Bureau of Reclamation, was to regulate flows between the upper and lower basins. Additional benefits are the production of hydroelectric power and a new recreation and tourism industry around Lake Powell, as previously inaccessible

TABLE 15.1.1 Estimated Colorado River Budget Through 2009

Water Demand	Quantity (maf)[a]
Upper basin, 7.5; lower basin, 7.5[b]	15.0
Central Arizona Project (rising to 2.8 maf)	1.0
Mexican allotment (1944 Treaty)	1.5
Evaporation from reservoirs (when full)	1.5
Bank storage at Lake Powell (when full)	0.5
Phreatophytic losses (water-demanding plants)	0.5
Budgeted total demand	**20.0 maf**
Average Annual Flows at Lees Ferry	
1906–1930 (average flow)	17.70 maf
1930–2003 (average flow)	14.10 maf
1990–2006 (average flow)	11.34 maf
2002	3.10 maf
2003–2009 (average flow)	7.34 maf

Source: Bureau of Reclamation and states of Arizona, California, and Nevada. A water year runs from October 1 to September 30.

[a]1 million acre-feet = 325,872 gallons, or 1.24 million liters, or 43,560 ft^3.

[b]In-basin consumptive uses are 75% agricultural.

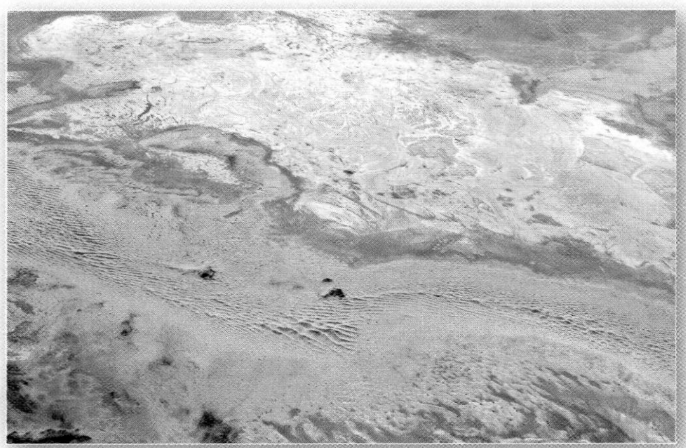

CRITICAL THINKING 15.2

Water in the desert

Water is the major erosion and transport medium in the desert, although less frequently available than in humid regions. Respond to this concept. How is it possible for this to be true when deserts are so dry? What have you learned from this chapter that proves this statement true? Review some of the examples presented.

FIGURE 15.18 Sand and playa—natural sorting of materials. Playa of dissolved and finer materials (top of photo), sand dunes (middle), and coarser particles on slopes along the mountains (lower area) demonstrate the natural sorting that occurs in desert landscapes. [Bobbé Christopherson.]

desert scenery now can be reached by boat.

Lake Powell inundated 299 km (186 mi) of former river channel, which cuts through porous Navajo sandstone for much of its length. This sandstone absorbs an estimated 0.5 maf of the river's overall annual discharge as bank storage when the reservoir is full—the higher the lake level, the greater the loss into the sandstone. The reservoir also experiences evaporative losses as hot, dry winds flow across an open water body in an arid region. Evaporation accounts for another 0.5 maf of discharge lost annually from the reservoir, equaling about a third of all Colorado River reservoir evaporation losses—again, higher lake levels mean more losses.

In the upper reaches of large reservoirs where water levels fluctuate, as well as along reaches of regulated river flow below dams throughout the basin, sand bars and banks have become near-permanent features. Riparian vegetation has colonized those surfaces, including dense thickets of water-demanding plants called phreatophytes. These species extract and transpire an additional 0.5 maf per year of the river flow. The three water losses discussed here—bank storage, evaporation, and water use by vegetation—are an increasingly large percentage of the overall river discharge as drought conditions persist.

Ongoing Drought in the Colorado River Basin

As described in the Chapter 9 *Geosystems Now,* a period of drought began in the Colorado River basin in 2000 and is ongoing. Continuing population growth throughout the region compounds the problem of increasing water demand coupled with decreased river flows. The drought itself appears directly linked to the shifting of subtropical high-pressure systems with climate change. This expansion of subtropical dry zones has given scientists a new approach to understanding drought. In regions such as the American Southwest, application of the term *drought* is changing to consideration of a change in *permanent average climate.* When future droughts occur—for example, during a Pacific Ocean La Niña condition—they will be in addition to this new base state of a drier climate. This is unprecedented in the climate record for this region.

A portion of a quote from Chapter 9 explains the severity of the drought: "The projected future climate of intensified aridity in the Southwest is caused by ... a poleward expansion of the subtropical dry zones. The drying of subtropical land areas ... imminent or already underway is unlike any climate state we have seen in the instrumental record" (Richard Seager et al., "Model projections of an imminent transition to a more arid climate in southwestern North America," *Science* 316, no. 5828 [May 25, 2007]: 1184).

Government stopgap measures to address water shortages came to head in December 2007 when the Department of the Interior signed a historic Record of Decision, an agreement representing a consensus among the stakeholders. This agreement sets the rules for dealing with water shortages during the drought and determining where water deliveries will be cut; establishes the drought as a basis for planning in times of water shortages; guides the operation rules for Lake Mead and Lake Powell for better coordination during the risk of drought; sets rules for distributing any surpluses, should they ever again arise; and encourages implementation of water conservation initiatives (see http://www.usbr.gov/lc/region/programs/strategies/RecordofDecision.pdf).

Conservation (using less water) and efficiency (using water more effectively) are not yet being stressed as ways to reduce the tremendous demand for water across the Southwest; perhaps this Record of Decision will lead to serious conservation and efficiency strategies. For instance, southern Nevada has launched a campaign to replace lawns with drought-tolerant xeroscaping—desert landscaping—but with 150,000 hotel rooms, we might question planning efforts. Las Vegas receives 90% of its water from the Colorado River and could save 30% of its usage by simply requiring water-efficient fixtures and appliances in hotels and residences.

Limiting or halting further metropolitan construction and population growth in the Colorado River Basin is not part of present strategies to lower demand. The focus remains on water supply. We might wonder what John Wesley Powell would think if he were alive today to witness such errant attempts to control the mighty and variable Colorado River. He foretold such a "heritage of conflict and litigation."

Desert Landscapes

Deserts are not wastelands, for they abound in specially adapted plants and animals. Moreover, the limited vegetation and distant vistas produce starkly beautiful landscapes. The desert's visual enchantment is keyed to sunlight and naked rock formations. The shimmering heat waves and related mirage effects are products of light refraction through layers of air that have developed a temperature gradient near the hot ground. And all deserts are not the same: For example, North American deserts have more vegetation cover than do the generally more barren Asian desert expanses, as we saw in Figure 15.14.

Deserts possess unique landscapes created by the interaction of intermittent precipitation events, weathering processes, and wind with the underlying geology. The buttes, pinnacles, and mesas of the U.S. Southwest are resistant horizontal rock strata that have eroded differentially. Removal of the less-resistant sandstone strata produces unusual desert sculptures—arches, windows, pedestals, and delicately balanced rocks. The upper layers of sandstone along the top of an arch or butte or tower are more resistant to weathering and protect the sandstone rock beneath (Figure 15.19).

The removal of surrounding rock through differential weathering leaves enormous buttes as residuals on the landscape. If you imagine a line intersecting the tops of the Mitten Buttes shown in Figure 15.20, you can gain some idea of the quantity of material that has been removed. These buttes exceed 300 m (1000 ft) in height, similar to the Chrysler Building in New York City or the U.S. Bank Tower (Library Tower) in Los Angeles.

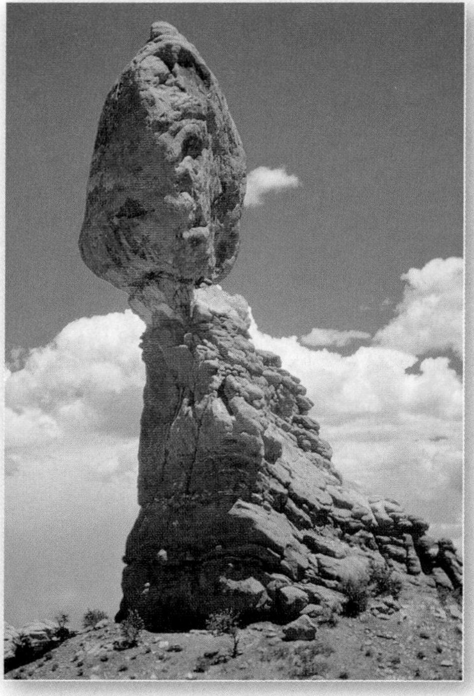

(a) Balanced Rock in Utah

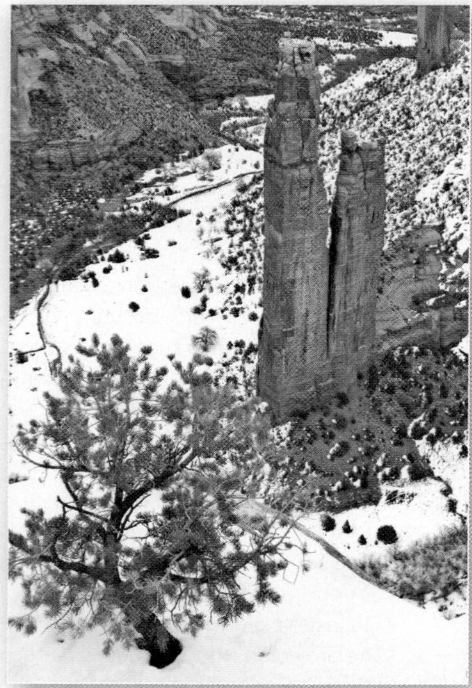

(b) Spider Rock in Arizona

FIGURE 15.19 Differential weathering.
(a) Balanced Rock in Arches National Park, Utah. The overall feature is 39 m (128 ft) tall and composed of Entrada sandstone. The balance-rock portion is 17 m (55 ft) tall and weighs 3255 metric tons (3577 short tons). (b) Spider Rock towers 244 m (800 ft) above the canyon floor of Canyon de Chelly, Arizona—a dramatic monolith; note the lighter color in the capstone rock. [(a) Author. (b) Bobbé Christopherson.]

Desert landscapes are places where stark erosional remnants can stand above the surrounding terrain as knobs or hills. Such a bare, exposed rock is an *inselberg* (island mountain), as exemplified by Uluru (Ayers) Rock in Australia. The formation is 348 m (1145 ft) high and 2.5 km long by 1.6 km wide (1.55 mi by 1.0 mi). Uluru Rock is sacred to Aboriginal peoples and is protected as part of Uluru National Park, established in 1950.

In semiarid regions, weak surface material may weather to a complex, rugged topography, usually of relatively low and varied relief. Such a landscape is a **badland**, probably so named because in the American West it offered little economic value and was difficult to traverse in 19th-century wagons. The Badlands region of the Dakotas and north-central Arizona (the Painted Desert) are so named.

Patterns of cross bedding, or *cross stratification*, are sometimes present in desert sandstone, formed by lithification of sand dunes that existed in ancient deserts. When such a dune was accumulating, sand cascaded down its slipface and distinct bedding planes (layers) were established that remained after the dune lithified (Figure 15.21). Ripple marks, animal tracks, and fossils also are found preserved in these sandstones.

Basin and Range Province

A physiographic province is a large region that is identified by several geologic or topographic traits. We find alternating basins (valleys) and mountain ranges characterizing the **Basin and Range Province** of the western United States (Figure 15.22). This region lies in the rain shadow of the Sierra Nevada, giving it a *midlatitude cold desert* climate. The climate and geology combine to give the province few permanent streams and *internal drainage* that lacks any outlet to the ocean (in Figure 14.5, see the Great Basin and its internal drainage).

The vast Basin and Range Province—almost 800,000 km^2 (300,000 mi^2)—was a major barrier to early settlers in their migration westward. The combination of desert climate and north–south-trending mountain ranges presented harsh challenges. Today, when you follow U.S. Highway 50 across Nevada, you cross five passes of more than 1950 m (6400 ft) and traverse numerous basins. Throughout the drive, you are reminded where you are by prideful signs that plainly state: "The Loneliest Road in America." It is difficult to imagine crossing this topography with wagons and oxen (Figure 15.22e).

John McPhee captured the feel of this desert province in his book *Basin and Range*:

> Supreme over all is silence. Discounting the cry of the occasional bird, the wailing of a pack of coyotes, silence—a great spatial silence—is pure in the Basin and Range. It is a soundless immensity with mountains in it. You stand . . . and look up at a high mountain front, and turn your head and look fifty miles down the valley, and there is utter silence.*

The Basin and Range formed as a result of tectonic processes. As the North American plate lumbered westward, it overrode former oceanic crust and hot spots at such a rapid pace that slabs of subducted material literally were run over. The crust was uplifted and stretched, creating a tensional landscape fractured by many faults. The present landscape consists of nearly parallel sequences: Some are tilted-fault blocks; others have pairs of faults on either side of *horsts* of upward-faulted blocks, which are the "ranges," and *grabens* of downward fault-bounded blocks, which are the "basins." Figure 15.22c shows this pattern of normal faults and tensional stretching.

*J. McPhee, *Basin and Range* (New York, NY: Farrar, Straus, Giroux, 1981), p. 46.

FIGURE 15.20 Monument Valley landscape. (a) Mitten Buttes, Merrick Butte, and rainbow in Monument Valley, Navajo Tribal Park, which have appeared as a backdrop in many movies, along the Utah–Arizona border. (b) A schematic of the tremendous removal of material by weathering, erosion, and transport. (c) Aerial view of Monument Valley. (d) View just east of (c), showing the residual spires of the Yie Bi Chei rock formation that ends with Totem Pole, 137 m (450 ft) tall and only 10 m (33 ft) wide. [(a) and (e) Author. (c) and (d) Bobbé Christopherson.]

(a)

(b)

(c)

(d)

(a) Valley of Fire

(b) Red Rocks

FIGURE 15.21 Cross bedding in sedimentary rocks.
The bedding pattern of cross stratification in these sandstone rocks tells us about dune environments before lithification (hardening into rock). [Bobbé Christopherson.]

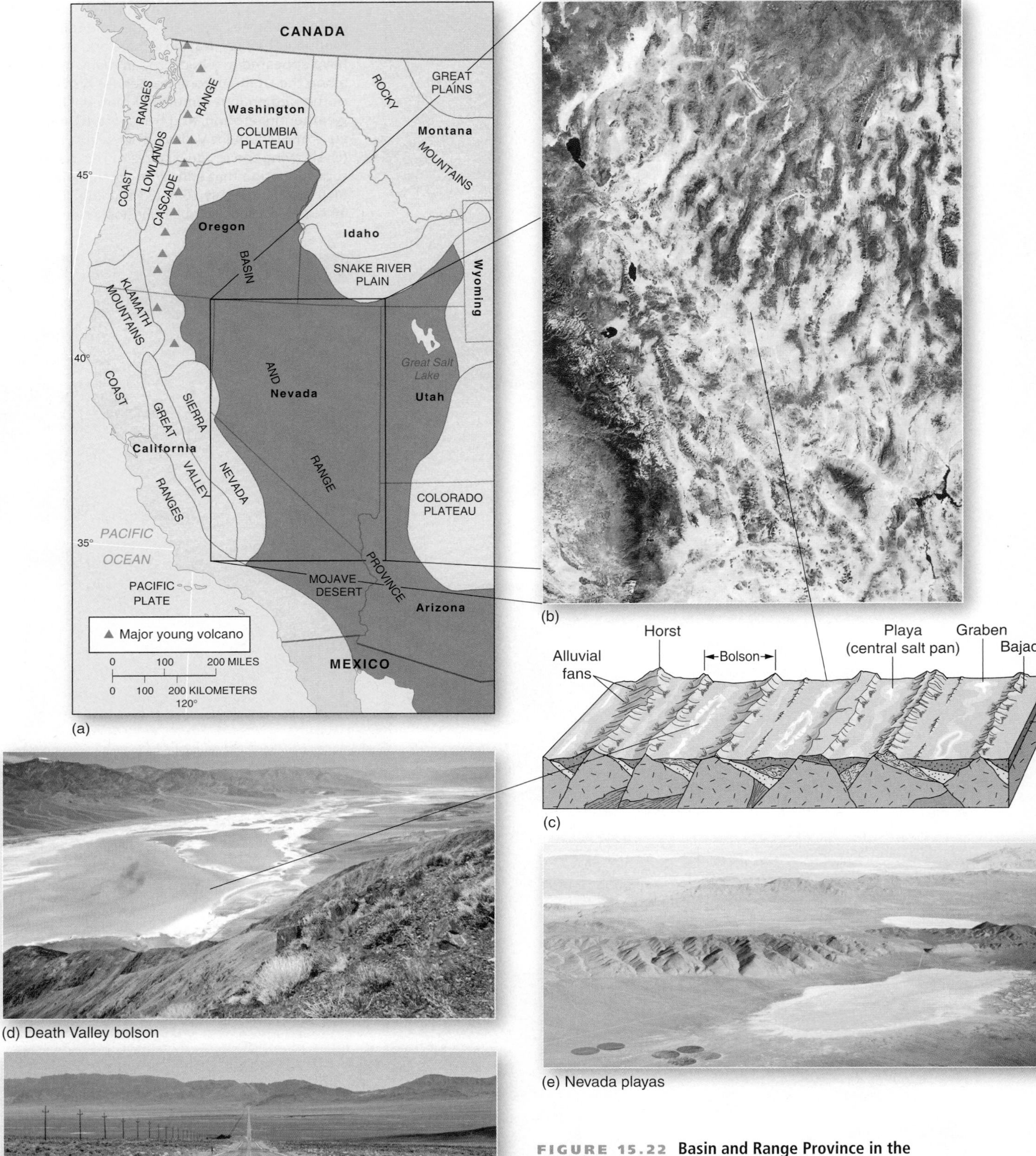

(d) Death Valley bolson

(e) Nevada playas

(f) "Loneliest road"

FIGURE 15.22 Basin and Range Province in the western United States.
(a) Map and (b) *Landsat* image of the area. Scientific discoveries extend this province south through northern and central Mexico. (c) Parallel faults produce a series of ranges and basins, including bolsons. (d) Death Valley features a central playa, parallel mountain ranges, alluvial fans, and bajadas along the bases of the ranges. (e) Playas in each basin. (f) "The Loneliest Road in America," Highway U.S. 50, slashes westward through the province. [(b) NASA. (d) Author. (e) Bobbé Christopherson.]

Basin-and-range relief is abrupt, and rock structures are angular and rugged. As the ranges erode, transported materials accumulate to great depths in the basins, gradually producing extensive desert plains. The basins' elevations average roughly 1200–1500 m (4000–5000 ft) above sea level, with mountain crests rising higher by another 900–1500 m (3000–5000 ft). Death Valley, California, is the lowest of these basins, with an elevation of –86 m (–282 ft). However, to the west of the valley, the Panamint Range rises to 3368 m (11,050 ft) at Telescope Peak—almost 3.5 vertical kilometers (2.2 mi) of desert mountain relief.

In Figure 15.22c and d, note the **bolson** label, a slope-and-basin area between the crests of two adjacent ridges in a dry region of internal drainage. Figure 15.22c also identifies a *playa* (central pan), a *bajada* (coalesced alluvial fans), and a mountain front in retreat from weathering and erosional attack. A *pediment* is an area of bedrock that is layered with a thin veneer, or coating, of alluvium. It is an erosional surface, as opposed to the depositional surface of the bajada.

Vast arid and semiarid lands remain an enigma on the water planet. They challenge our need for water and demand technological innovations to access it. Yet these lands hold a fascination, perhaps because they are so lacking in the moisture that infuses our lives.

II CRITICAL THINKING 15.3

Relating to the Colorado River crisis

Relative to the Colorado River system, define, compare, and contrast what you think is meant by a *supply strategy* or a *demand strategy*. How does each strategy relate to the present Colorado River water budget (Table 15.1.1, Focus Study 15.1)? What political and economic forces do you think are at play in planning for the river and its dependent region? Who are the stakeholders? With your analysis, briefly speculate as to what you think needs to happen.

GEOSYSTEMS CONNECTION

Wind is an agent of geomorphic change, causing erosion, transportation, and deposition of sediments, sometimes redistributing materials among continents. Sand accumulations form dunes of several types, including crescentic, linear, and star dunes. Wind also transports fine-grained clays and silts to form loess deposits. Wind and scarce water operate in Earth's desert regions, where both sandy and rocky surfaces are found. In the next chapter, we examine coastal processes involved with the agents of waves, tides, and currents as they sculpt Earth's erosional and depositional coastlines. A significant portion of the human population lives in coastal areas, making hazard perception, planning, and the future, given a rising sea level, important applied aspects of Chapter 16.

KEY LEARNING CONCEPTS REVIEW

Desertification is the ongoing process that leads to an unwanted expansion of the Earth's desert lands; presently, some 1 billion people are affected, and a new international effort is framed under the UNDDD.

desertification (p. 433)

1. What is meant by desertification? Using the maps in Figures GN 15.1 and 15.14 and the text description, locate several of the regions affected by desert expansion.

■ ***Characterize*** the unique geomorphic work accomplished by wind and eolian processes.

The movement of the atmosphere in response to pressure differences produces wind. Wind is a geomorphic agent of erosion, transportation, and deposition. **Eolian** processes modify and move sand accumulations along coastal beaches and deserts. Wind's ability to move materials is small compared with that of water and ice.

eolian (p. 434)

2. Who was Ralph Bagnold? What was his contribution to eolian studies?
3. Explain the term *eolian* and its application in this chapter. How would you characterize the ability of the wind to move material?

■ ***Describe*** eolian erosion, including deflation, abrasion, and the resultant landforms.

Two principal wind-erosion processes are **deflation,** the removal and lifting of individual loose particles, and **abrasion,** the

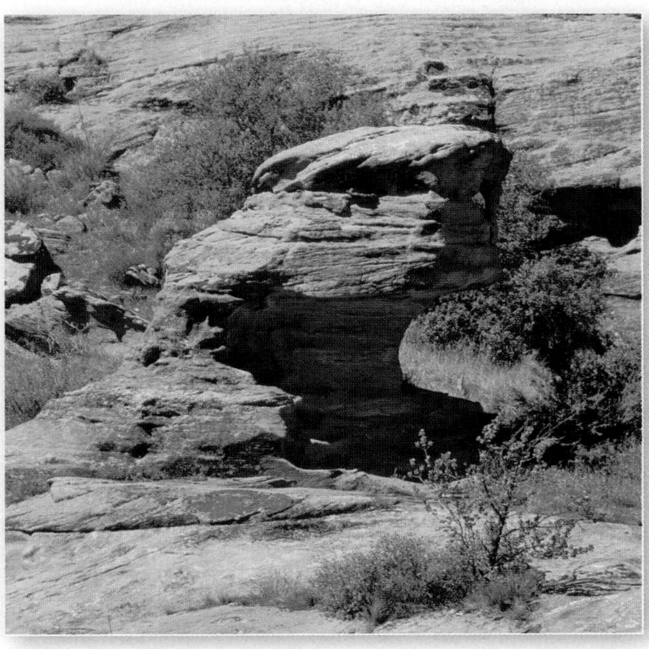

"sandblasting" of rock surfaces with particles captured in the air. Deflation literally blows away loose or noncohesive sediment and works with rainwater to form a surface resembling a cobblestone street: a **desert pavement** that protects underlying sediment from further deflation and water erosion. Wherever wind encounters

loose sediment, deflation may remove enough material to form basins. Called **blowout depressions,** they range from small indentations less than a meter wide up to areas hundreds of meters wide and many meters deep. Rocks that bear evidence of eolian erosion are **ventifacts**. On a larger scale, deflation and abrasion are capable of streamlining rock structures, leaving behind distinctive rock formations or elongated ridges called **yardangs**.

deflation (p. 435)
abrasion (p. 435)
desert pavement (p. 435)
blowout depressions (p. 435)
ventifacts (p. 436)
yardangs (p. 436)

4. Describe the erosional processes associated with moving air.
5. Explain deflation and provide an overview of each hypothesis on the processes that produce desert pavement.
6. How are ventifacts and yardangs formed by the wind?

■ ***Describe*** eolian transportation, and ***explain*** saltation and surface creep.

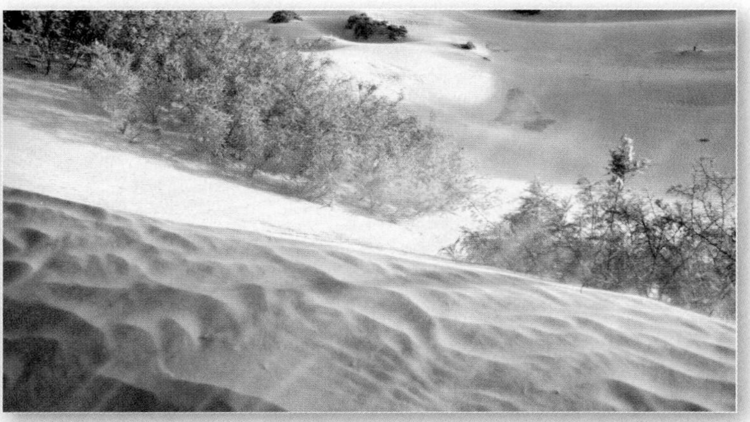

Wind exerts a drag or frictional pull on surface particles until they become airborne. The finer material suspended in a dust storm is lifted much higher than the coarser particles of a sandstorm; only the finest dust particles travel significant distances. Saltating particles crash into other particles, knocking them both loose and forward. The motion of **surface creep** slides and rolls particles too large for saltation.

surface creep (p. 437)

7. Differentiate between a dust storm and a sandstorm.
8. What is the difference between eolian saltation and fluvial saltation?
9. Explain the concept of surface creep.

■ ***Identify*** the major classes of sand dunes, and ***present*** examples within each class.

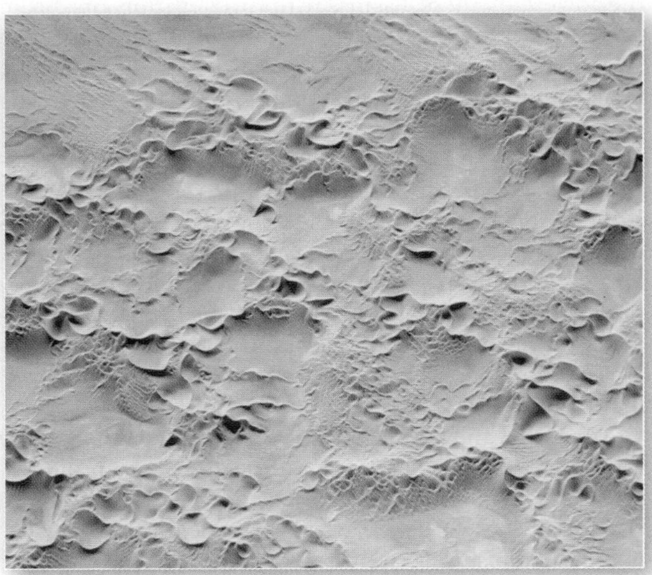

In arid and semiarid climates and along some coastlines where sand is available, dunes accumulate. A **dune** is a wind-sculpted accumulation of sand. An extensive area of dunes, such as that found in North Africa, is characteristic of an **erg desert,** or **sand sea**. When saltating sand grains encounter small patches of sand, their kinetic energy (motion) is dissipated, and they start to accumulate into a dune. As height increases above 30 cm (12 in.), a steeply sloping **slipface** on the lee side and characteristic dune features are formed. Dune forms are broadly classified as *crescentic*, *linear*, and *star*.

dune (p. 438)
erg desert (p. 438)
sand sea (p. 438)
slipface (p. 438)

10. What is the difference between an erg and a reg desert? Which type is a sand sea? Are all deserts covered by sand? Explain.
11. What are the three classes of dune forms? Describe the basic types of dunes within each class. What do you think is the major shaping force for sand dunes?
12. Which form of dune is the mountain giant of the desert? What are the characteristic wind patterns that produce such dunes?

■ ***Define*** loess deposits and their origins, locations, and landforms.

Eolian-transported materials contribute to soil formation in distant places. Windblown **loess** deposits occur worldwide and can develop into good agricultural soils. These fine-grained clays and silts are moved by the wind many kilometers and are redeposited as an unstratified, homogeneous blanket of material.

loess (p. 442)

13. How are loess materials generated? What form do they assume when deposited?
14. Name a few examples of significant loess deposits on Earth.

■ ***Portray*** desert landscapes, and ***locate*** these regions on a world map.

Dry and semiarid climates occupy about 35% of Earth's land surface. The spatial distribution of these dry lands is related to subtropical high-pressure cells between 15° and 35° N and S, to rain shadows on the lee side of mountain ranges, or to areas at great distance from moisture-bearing air masses, such as central Asia.

Precipitation events are rare, yet running water is still the major erosional agent in deserts. When precipitation events do occur, a dry streambed fills with a torrent, a **flash flood**. Depending on the region, such a dry streambed is known as a **wash,** an *arroyo* (Spanish), or a *wadi* (Arabic). As runoff water evaporates, salt crusts may be left behind on the desert floor. This intermittently wet and dry low area in a region of closed drainage is a **playa,** site of an *ephemeral lake* when water is present.

In arid climates, a prominent landform is the **alluvial fan** at the mouth of a canyon where an ephemeral stream channel exits into a valley. The fan is produced by flowing water that abruptly loses velocity as it leaves the constricted channel of the canyon and deposits a layer of sediment along the mountain block. A continuous apron, or **bajada,** may form if individual alluvial fans coalesce. In a desert area, weak surface material may weather to a complex, rugged topography, usually of relatively low and varied relief. Such a landscape is a **badland.** A *province* is a large region that is characterized by several geologic or physiographic traits. The **Basin and Range Province** of the western United States consists of alternating basins and mountain ranges. A slope-and-basin area between the crests of two adjacent ridges in a dry region of internal drainage is a **bolson.**

flash flood (p. 446)
wash (p. 446)
playa (p. 447)
alluvial fan (p. 447)
bajada (p. 447)
badland (p. 452)
Basin and Range Province (p. 452)
bolson (p. 455)

15. Characterize desert energy and water-balance regimes. What are the significant patterns of occurrence for arid landscapes?
16. How would you describe the water budget of the Colorado River? What was the basis for agreements regarding distribution of the river's water? Why has thinking about the river's discharge been so optimistic? Provide an overview of current conditions of the Colorado River system in light of the ongoing drought and increasing water demands.
17. Describe a desert bolson from crest to crest. Draw a simple sketch with the components of the landscape labelled.
18. Where is the Basin and Range Province? Briefly describe its appearance and character.

MASTERING GEOGRAPHY

Visit www.masteringgeography.com for several figure animations, satellite loops, self-study quizzes, flashcards, visual glossary, case studies, career links, textbook reference maps, RSS Feeds, and an ebook version of *Geosystems*. Included are many geography web links to interesting Internet sources that support this chapter.

CHAPTER 16

The Oceans, Coastal Processes, and Landforms

Turtle Beach, Ascension Island, where female green sea turtles (*Chelonia mydas*) leave the sea and climb the beach to lay their eggs in nests dug in the sand. Upon hatching, the young make their way back to the ocean. The turtles migrate great distances from their feeding areas of seagrass to these nesting/hatching beaches. On this volcanic island, the light sand originates from formations of calcarious algae that fish consume to make the sand. Ascension Island is 8° S 14.5° W; the beach is near Georgetown. [Bobbé Christopherson.]

KEY LEARNING CONCEPTS

After reading the chapter, you should be able to:

- ***Describe*** the chemical composition and physical structure of the ocean.
- ***Identify*** the components of the coastal environment, and ***list*** the physical inputs to the coastal system, including tides and mean sea level.
- ***Describe*** wave motion at sea and near shore, and ***explain*** coastal straightening and coastal erosional and depositional landforms.
- ***Describe*** barrier islands and their hazards as they relate to human settlement.
- ***Assess*** living coastal environments: corals, wetlands, salt marshes, and mangroves.
- ***Construct*** an environmentally sensitive model for settlement and land use along the coast.

GEOSYSTEMS NOW

What Once Was Bayou Lafourche

Bayou Lafourche (pronounced "La-foosh") was the main distributary of the Mississippi River between the Civil War and 1904, when a dam stopped the flow through this bayou and led to water losses and stagnation of the former marshlands. In 1955, attempts began to reverse the losses by restoring some water flow with pumps.

The towns of Galliano and Golden Meadow and some 10,000 people are protected from rising water, tidal fluctuations, and hurricane storm surge within a compound surrounded by the hurricane-protection levee system. The 65-km (40-mi) levee system, some 4 m (13 ft) high and 28 m (91 ft) wide at the base, was begun in the late 1970s. Floodgates at either end are closed when the flood threat is imminent (Figure GN 16.1).

With delta subsidence and rising sea levels, increasing levee height is a continuing process, known as a "lift" to the levee, with another meter of height completed in 2009. This is an attempt to maintain storm-surge protection. The real irony is that the former coastal wetlands were a buffer between Gulf storm surges and interior lands, yet today, because of human disruptions, there is open water between Golden Meadow and the former delta shoreline.

Hurricanes Katrina, Rita, Cindy, and Gustav picked up huge oil barges, equipment, shrimp boats, other marine craft, and debris and dumped them far inland. Grand Isle, to the east of the port on the Gulf Coast, experienced major destruction in 2005 and was leveled in 2008. Some 80% of homes, businesses, and fishing camps were destroyed.

Hurricane Gustav in 2008 made landfall just west of the compound near Cocodrie, Louisiana, and Hurricane Ike passed to the south, with huge rainbands that delivered heavy precipitation to southern Louisiana. During both events, the parish was evacuated, the levees held, and residents returned to much wind and water damage, but no flooding inside the levee compound.

Virtually all the population's livelihood comes from the Gulf—from fishing, tourism, and the oil and gas industry. Only 23 km (14 mi) from Galliano some 6000 workers come and go through Port Fourchon to oil platforms and rigs. Imagine the impact on these hurricane-weathered people when the BP Gulf oil spill disaster struck in April 2010. In addition to environmental pollution, the more than 4-month oil gusher knocked out the economics of the region and will continue to impact the population and environment for years. The physical, mental, and societal health of the residents was damaged according to several studies (e.g., by Columbia University and Louisiana State University), all worsened by the sequence of financial impacts from hurricanes, storm surges, and oil spills.

The near future is in doubt as the region looks at a forecast of elevated tropical cyclone activity, with storm power enhanced by increasing air and water temperatures. The oil spill is spread through the coastal wetlands. The toxic chemical dispersants and dissolved oil are in the water column and food web. The Lafourche oystermen and crab fishermen could lose a way of life, which in some families goes back five generations.

Meanwhile, Galliano in the levee compound is presently 122 cm (48 in.) above sea level. The Mississippi delta lands are slowly subsiding, and sea level is continuing to rise. The story continues to unfold as the ocean reclaims these coastal lands.

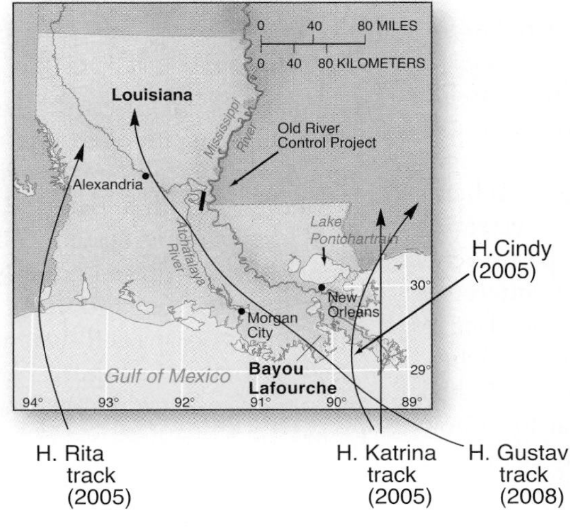

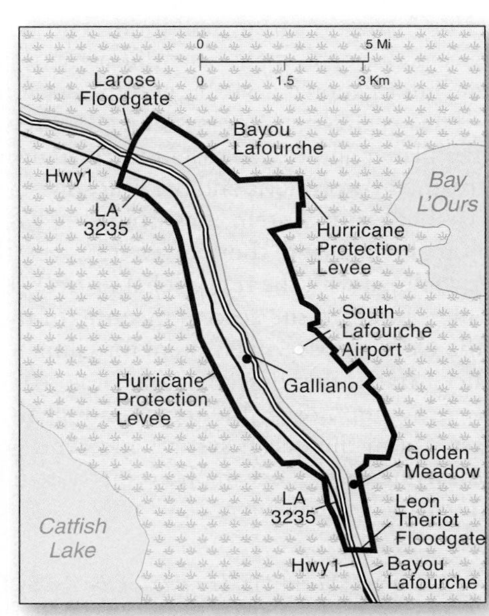

FIGURE GN 16.1 A hurricane-protection levee compound in the Mississippi delta. (a) Bayou Lafourche is the waterway that runs along the highway, with locks at the north and south ends of the levee system for times of storm surge. (b) Recent storm tracks are noted on the location map.

Walk along a shoreline and you witness the dramatic interaction of Earth's vast oceanic, atmospheric, and lithospheric systems. At times, the ocean attacks the coast in a stormy rage of erosive power; at other times, the moist sea breeze, salty mist, and repetitive motion of the water are gentle and calming. The coastlines are areas of dynamic change and beauty (Figure 16.1). Few have captured this confrontation between land and sea as well as biologist Rachel Carson:

> The edge of the sea is a strange and beautiful place. All through the long history of Earth it has been an area of unrest where waves have broken heavily against the land, where the tides have pressed forward over the continents, receded, and then returned. For no two successive days is the shoreline precisely the same. Not only do the tides advance and retreat in their eternal rhythms, but the level of the sea itself is never at rest. It rises or falls as the glaciers melt or grow, as the floors of the deep ocean basins shift under its increasing load of sediments, or as the Earth's crust along the continental margins warps up or down in adjustment to strain and tension. Today a little more land may belong to the sea, tomorrow a little less. Always the edge of the sea remains an elusive and indefinable boundary.*

Commerce and access to sea routes, fishing, and tourism prompt many people to settle near the ocean. A scientific assessment estimates that about 40% of Earth's population lives within 100 km (62 mi) of coastlines. In the United States, about 50% of the people live in areas designated as *coastal* (this includes the Great Lakes).

Therefore, an understanding of coastal processes and landforms is important to nearly half the world's population. And because these processes along coastlines often produce dramatic change, they are essential to consider in planning and development. A World Resources Institute study found as much as 50% of the world's coastlines at some risk of loss through things like erosion and rising sea level or disruption from pollution. The National Ocean Service coordinates many scientific activities related to the ocean; information is available at http://www.nos.noaa.gov/.

The ocean is a vast ecosystem, intricately linked to life on the planet and to life-sustaining systems in the atmosphere, the hydrosphere, and the lithosphere. Oceans act as a vast buffering system, absorbing excess atmospheric carbon dioxide and thermal energy, yet the system is overwhelmed with the rate of change over the past several decades. In 2009, the second meeting of the Conference on the Ocean Observing System for Climate took place in Venice, Italy; the first meeting was 10 years earlier. The program noted the profound changes that had transpired in the ocean system and set a goal for a new physical and carbon ocean observing system by 2015—see http://www.oceanobs09.net/.

*"The Marginal World," in *The Edge of the Sea* by Rachel Carson. © 1955 by Rachel Carson, renewed 1983 by R. Christie (Boston: Houghton Mifflin), p. 11.

FIGURE 16.1 Where the land confronts the sea. Coastal fog and mid-level cloud layers approach the central California coast, a tectonically active coastline. [Bobbé Christopherson.]

In this chapter: We begin with a brief look at our global oceans and seas—1998 was celebrated as The International Year of the Ocean by all United Nations countries. We discuss the physical and chemical properties of seawater. Our coverage includes discussions about tides, waves, coastal erosion, and depositional landforms such as beaches and barrier islands. Important organic processes produce corals, wetlands, salt marshes, and mangroves. A systems framework of specific inputs (components and driving forces), actions (movements and processes), and outputs (results and consequences) organizes our discussion of coastal processes. We conclude with a look at the considerable human interaction with and impact on coastal environments.

Global Oceans and Seas

The ocean is one of Earth's last great scientific frontiers and is of great interest to geographers. Remote sensing from orbit, aircraft, surface vessels, and submersibles is providing a wealth of data and a new capability to understand the oceanic system. The pattern of sea-surface temperatures is presented in Figure 5.11 and ocean currents, both surface and deep, in Figures 6.20 and 6.22. The world's oceans and their area, volume, and depth are listed in a table in Figure 7.3. The locations of oceans and major seas are mapped and listed alphabetically in Figure 16.2.

Chemical Composition of Seawater

Water is the "universal solvent," dissolving at least 57 of the 92 elements found in nature. In fact, most natural elements and the compounds they form are found in the seas

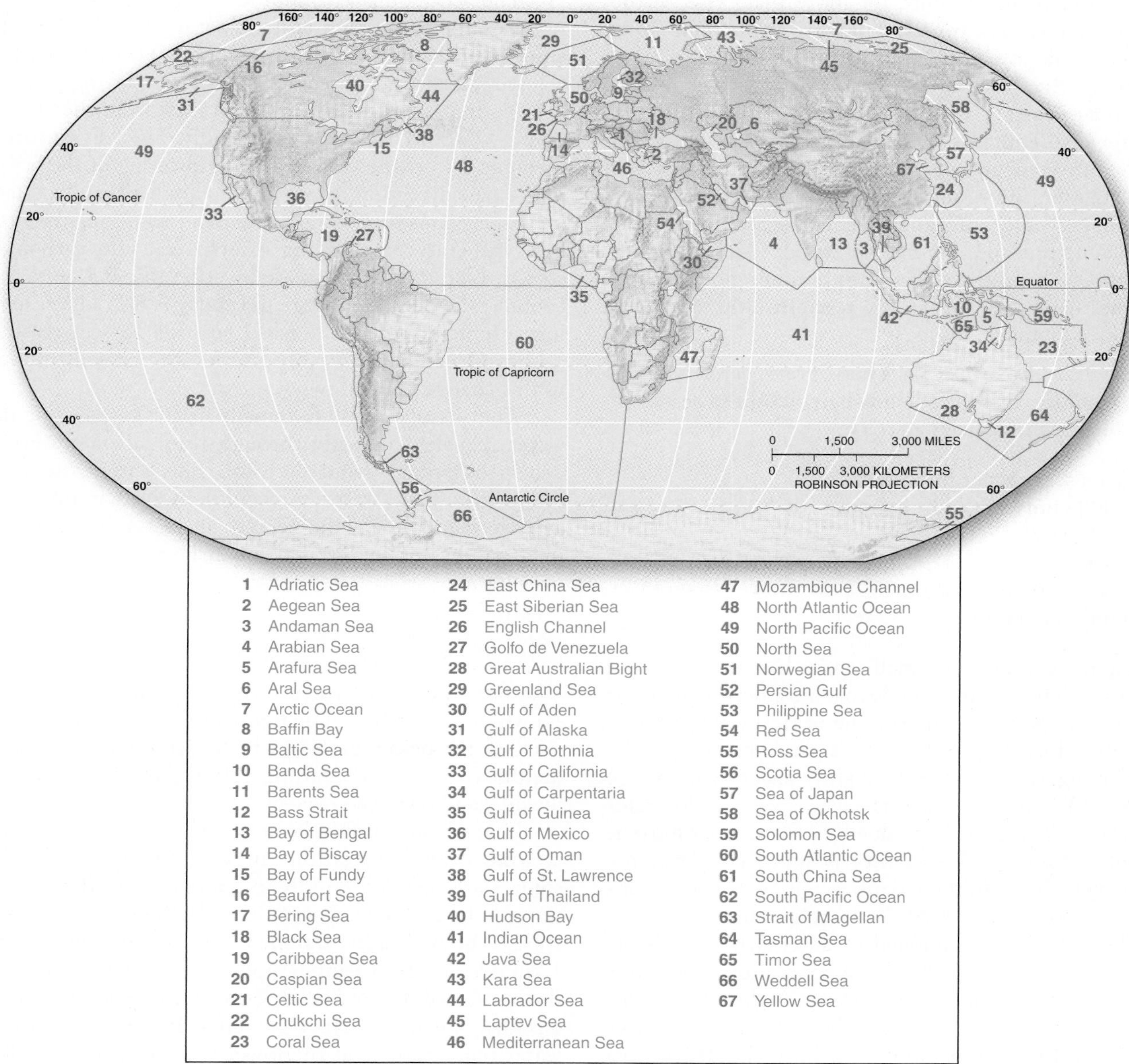

FIGURE 16.2 Principal oceans and seas of the world.
A sea is generally smaller than an ocean and is near a landmass; sometimes the term refers to a large, inland, salty body of water. Match the alphabetized name and number with its location on the map.

as dissolved solids, or *solutes*. Thus, seawater is a solution, and the concentration of dissolved solids is **salinity**.

The ocean remains a remarkably homogeneous mixture. The ratio of individual salts does not change, despite minor fluctuations in overall salinity. In 1874, the British HMS *Challenger* sailed around the world taking surface and depth measurements and collecting samples of seawater. Analyses of those samples first demonstrated the uniform composition of seawater.

Although ocean chemistry was thought to be fairly constant during the Phanerozoic Eon (the past 542 million years), recent evidence suggests that seawater chemistry has varied over time within a narrow range. The variations are consistent with changes in sea-floor spreading rates, volcanism, and sea level. Evidence is gathered from fluid inclusions in marine formations such as in limestone and evaporite deposits, which contain ancient seawater.

Ocean Chemistry Ocean chemistry is a result of complex exchanges among seawater, the atmosphere, minerals, bottom sediments, and living organisms. In addition, significant flows of mineral-rich water enter the ocean through hydrothermal (hot water) vents in the ocean floor. These vents are the "black smokers," noted for the dense, black, mineral-laden water that spews from them (Figure 11.12b). The uniformity of seawater results from complementary chemical reactions and continuous mixing—after all, the ocean basins interconnect, and water circulates among them.

Seven elements account for more than 99% of the dissolved solids in seawater. They are (with their ionic form) chlorine (as chloride, Cl^-), sodium (as Na^+), magnesium (as Mg^{2+}), sulfur (as sulfate, SO_4^{2-}), calcium (as Ca^{2+}), potassium (as K^+), and bromine (as bromide, Br^-). Seawater also contains dissolved gases (such as carbon dioxide, nitrogen, and oxygen), suspended and dissolved organic matter, and a multitude of trace elements.

Commercially, only sodium chloride (common table salt), magnesium, and bromine are extracted in any significant amount from the ocean. Future mining of minerals from the seafloor is technically feasible, although it remains uneconomical.

Average Salinity: 35‰ There are several ways to express salinity (dissolved solids by volume) in seawater, using the worldwide average value:

- 3.5% (% parts per hundred)
- 35,000 ppm (parts per million)
- 35,000 mg/l
- 35 g/kg
- 35‰ (‰ parts per thousand), the most common notation

Salinity worldwide normally varies between 34‰ and 37‰; variations are attributable to atmospheric conditions above the water and to the volume of freshwater inflows. In equatorial water, precipitation is great throughout the year, diluting salinity values to slightly lower than average (34.5‰). In subtropical oceans—where evaporation rates are greatest because of the influence of hot, dry, subtropical high-pressure cells—salinity is more concentrated, increasing to 36‰. In these regions, salinity is rising as increased evaporation rates rise with temperature.

The term **brine** is applied to water that exceeds the average of 35‰ salinity. **Brackish** applies to water that is less than 35‰ salts. In general, oceans are lower in salinity near landmasses because of freshwater runoff and river discharges. Extreme examples include the Baltic Sea (north of Poland and Germany) and the Gulf of Bothnia (between Sweden and Finland), which average 10‰ or less salinity because of heavy freshwater runoff and low evaporation rates.

In contrast, the Sargasso Sea, within the North Atlantic subtropical gyre, averages 38‰. The Persian Gulf has a salinity of 40‰ as a result of high evaporation rates in a nearly enclosed basin. Deep pockets, or "brine lakes," along the floor of the Red Sea and the Mediterranean Sea register up to a salty 225‰.

The ocean reflects conditions in Earth's environment. As an example, the high-latitude oceans have been freshening over the past decade, as mentioned earlier. Another change under way is oceanic acidification, taking place as the ocean absorbs excess carbon dioxide from the atmosphere, where it is increasing rapidly. Hydrolysis forms carbonic acid in the seawater and a lowering of the ocean pH—an acidification. A more acid ocean will cause certain marine organisms such as corals and some plankton to have difficulty maintaining external calcium carbonate structures. pH could decrease by −0.4 to −0.5 units this century; the oceans' average pH today is 8.2. The scale is logarithmic, so a decrease of 0.1 equals a 30% increase in acidity. Oceanic biodiversity and food webs will respond to this change in unknown ways.

The dissolved solids remain in the ocean, but the water recycles endlessly through the hydrologic cycle, driven by energy from the Sun. The water you drink today may have water molecules in it that not long ago were in the Pacific Ocean, in the Yangtze River, in groundwater in Sweden, or airborne in the clouds over Peru.

Physical Structure of the Ocean

The basic physical structure of the ocean is layered, as shown in Figure 16.3. The figure also graphs four key aspects of the ocean, each of which varies with increasing depth: average temperature, salinity, dissolved carbon dioxide, and dissolved oxygen.

The ocean's surface layer is warmed by the Sun and is wind-driven. Variations in water temperature and solutes are blended rapidly in a *mixing zone* that represents only 2% of the oceanic mass. Below the mixing zone is the *thermocline transition zone*, a region more than 1 km deep of decreasing temperature gradient that lacks the motion of the surface. Friction at these depths dampens the effect of surface currents. In addition, colder water temperatures at the lower margin tend to inhibit any convective movements.

From a depth of 1–1.5 km (0.6–0.9 mi) to the ocean floor, temperature and salinity values are quite uniform. Temperatures in this *deep cold zone* are near 0°C (32°F). Water in the deep cold zone does not freeze, however, because of its salinity and intense pressures at those depths; seawater freezes at about −2°C (28.4°F) at the surface. Generally, the coldest water is along the bottom. Now let us shift to the edge of the sea and examine Earth's coastlines.

GEO REPORT 16.1 ***The Mediterranean Sea is getting saltier***

The Mediterranean Sea is warming at a faster rate than the oceans. Increased salinity and temperatures are occurring in the deep layers, below 600 m (1968 ft) to the sea bed. Saltier conditions change the water density and cause net outflows past the Strait of Gibraltar, thus blocking natural mixing with the Atlantic Ocean. As climate changes, warming is disrupting the natural mixing in such water bodies, as reported in the *Geosystems Now* in Chapter 7.

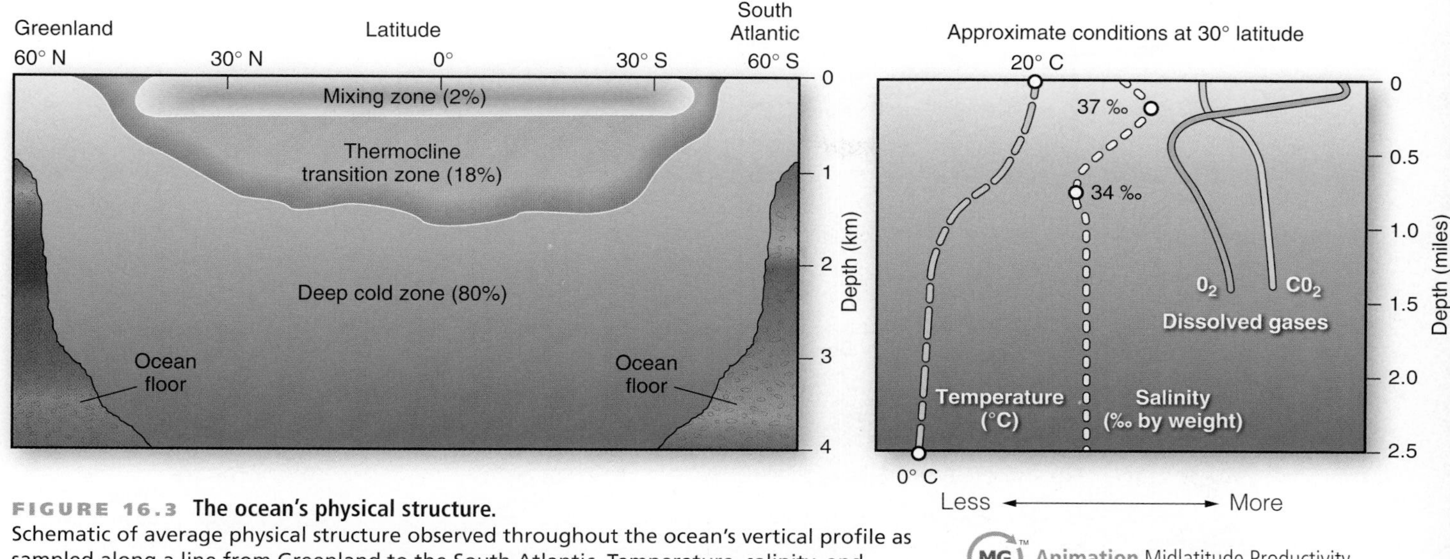

FIGURE 16.3 The ocean's physical structure.
Schematic of average physical structure observed throughout the ocean's vertical profile as sampled along a line from Greenland to the South Atlantic. Temperature, salinity, and dissolved gases are shown plotted by depth.

MG™ Animation Midlatitude Productivity

Coastal System Components

Earth's surface features, such as mountains and crustal plates, were formed over millions of years. However, most of Earth's coastlines are relatively new, existing in their present state as the setting for continuous change. The land, ocean, atmosphere, Sun, and Moon interact to produce tides, currents, waves, erosional features, and depositional features along the continental margins.

Inputs to the coastal environment include many elements we have already discussed:

- *Solar energy* input drives the atmosphere and the hydrosphere. A conversion of insolation to kinetic energy produces prevailing winds, weather systems, and climate.
- *Atmospheric winds*, in turn, generate ocean currents and waves, key inputs to the coastal environment.
- *Climatic regimes*, which result from insolation and moisture, strongly influence coastal geomorphic processes.
- The *nature of coastal rock* and coastal geomorphology are important in determining rates of erosion and sediment production.
- *Human activities* are an increasingly significant input to coastal change.

All these inputs occur within the ever-present influence of gravity's pull—not only from Earth, but also from the Moon and the Sun. Gravity provides the potential energy of position for materials in motion and produces the tides. A dynamic equilibrium among all these components produces coastline features of infinite variety and beauty.

The Coastal Environment and Sea Level

The coastal environment is the **littoral zone**, from the Latin word, *litoris*, for "shore." Figure 16.4 illustrates the littoral zone and includes specific components discussed later in the chapter. The littoral zone spans some land as well as water. Landward, it extends to the highest waterline that occurs on shore during a storm. Seaward, it extends to where water is too deep for storm waves to move sediments on the seafloor—usually around 60 m, or 200 ft, in depth. The specific contact line between the sea and the land is the *shoreline*, although this line shifts with tides, storms, and sea-level adjustments. The *coast* continues inland from high tide to the first major landform change and may include areas considered to be part of the coast in local usage.

Because the level of the ocean varies, the littoral zone naturally shifts position from time to time. A rise in sea level causes submergence of land, whereas a drop in sea level exposes new coastal areas. In addition, uplift and subsidence of the land itself initiate changes in the littoral zone.

Sea level is an important concept. Every elevation you see in an atlas or on a map is referenced to mean sea level. Yet this average sea level changes daily with the tides and over the long term with changes in climate, tectonic plate movements, and glaciation. Thus, *sea level* is a relative term. At present, no international system exists to determine exact sea level over time. The Global Sea Level Observing System (GLOSS) is an international group actively working on sea-level issues and is part of the larger Permanent Service for Mean Sea Level, which you can find at http://www.pol.ac.uk/psmsl/programmes/.

Mean sea level (MSL) is a value based on average tidal levels recorded hourly at a given site over many years. MSL varies spatially because of ocean currents and waves, tidal variations, air temperature, pressure differences and wind patterns, ocean temperature variations, slight variations in Earth's gravity, and changes in oceanic volume.

At present, the overall U.S. MSL is calculated at approximately 40 locations along the coastal margins of the continent. These sites are being upgraded with

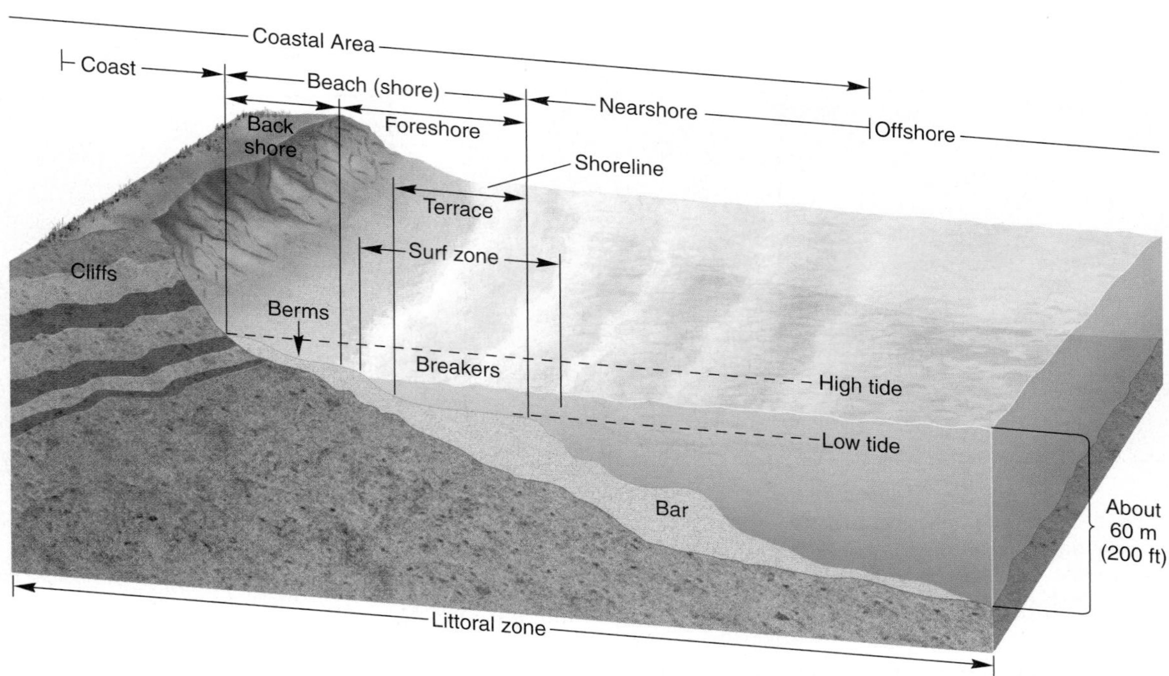

FIGURE 16.4 The littoral zone.
The littoral zone includes the coast, beach, and nearshore environments.

new equipment in the Next Generation Water Level Measurement System, using next-generation tide gauges, specifically along the U.S. and Canadian Atlantic Coast, Bermuda, and the Hawaiian Islands. The *NAVSTAR* satellites of the global positioning system (GPS) make possible the correlation of data within a network of ground- and ocean-based measurements.

Changes in Sea Level

Over the long term, sea-level fluctuations expose a range of coastal landforms to tidal and wave processes. As average global temperatures cycle through cold or warm climatic spells, the quantity of ice locked up in the ice sheets of Antarctica and Greenland and in hundreds of mountain glaciers can increase or decrease and result in sea-level changes accordingly. At the peak of the most recent Pleistocene glaciation, about 18,000 B.P. (years before the present), sea level was about 130 m (430 ft) lower than it is today. On the other hand, if Antarctica and Greenland ever became ice-free (if their ice sheets fully melted), sea level would rise at least 65 m (215 ft) worldwide.

Just 100 years ago sea level was 38 cm (15 in.) lower along the coast of southern Florida. Venice, Italy, has experienced a rise of 25 cm (10 in.) since 1890. During the last century, average sea level rose 10–20 cm (4–8 in.), a rate 10 times higher than the average rate during the last 3000 years. Since 1930, sea-level rise has continued at an increasing rate, with any variability explained by global dam construction and artificial reservoir impoundments of water. The present sea-level rise is spatially uneven as is MSL; for instance, the rate along the coast of Argentina is nearly 10 times the rate along the coast of France.

About 50% of the present sea-level rise is from thermal expansion produced by warmer water temperatures. The remainder is from worldwide losses in glacial mass and from mass losses from the two ice sheets in Greenland and Antarctica.

Given these trends and the predicted climatic changes ahead, sea level will continue to rise and be potentially devastating for many coastal locations. A rise of only 0.3 m (1 ft) would cause shorelines worldwide to move inland an average of 30 m (100 ft). This elevated sea

GEO REPORT 16.2 ***Sea level varies along the U.S. coastline***

Sea level varies along the full extent of North American shorelines. The mean sea level (MSL) of the U.S. Gulf Coast is about 25 cm (10 in.) higher than that of Florida's east coast, which is the lowest in North America. MSL rises northward along the eastern coast, to 38 cm (15 in.) higher in Maine than in Florida. Along the U.S. western coast, MSL is higher than Florida's by about 58 cm (23 in.) in San Diego and by about 86 cm (34 in.) in Oregon. Overall, North America's Pacific Coast MSL averages about 66 cm (26 in.) higher than the Atlantic Coast MSL. For the United States, see Sea Levels Online at http://tidesandcurrents.noaa.gov/sltrends/sltrends.shtml.

level would inundate valuable real estate along coastlines all over the world. Some 20,000 km^2 (7800 mi^2) of land along North American shores alone would be drowned, at a staggering loss of trillions of dollars. A 95-cm (3.1-ft) sea-level rise could inundate 15% of Egypt's arable land, 17% of Bangladesh, and many island countries and communities. However, uncertainty exists in these forecasts.

The 2007 Intergovernmental Panel on Climate Change (IPCC) forecast for global mean sea-level rise this century, given regional variations, is a range from 0.18 to 0.59 m (7.1 to 23.2 in.). However, the new 2006–2009 measurements of Greenland's ice loss were not included in the forecasts, and data appear to suggest an acceleration of ice melt and sea-level rise. Scientists are carefully monitoring the growing instability in the West Antarctic Ice Sheet. A fair working estimate for sea-level rise this century now stands at 1.0 to 1.4 m (3.28 to 4.6 ft); the state of California is using a rise of 1.4 m in its planning models.

Remote-sensing technology augments these measurements, including the *TOPEX/Poseidon* satellite launched in 1992 and still in service (see http://topex-www.jpl.nasa.gov/; click on the "Sea Level Viewer"). This satellite has two radar altimeters that measure changes in mean sea level at any one location every 10 days between 66° N and 66° S latitudes. These measurements are made to a precision of 4.2 cm, or 1.7 in. (Figure 16.5). *TOPEX/Poseidon* portraits of El Niño and La Niña in the Pacific are included in Focus Study 10.1.

Another ocean surface topography satellite launched in December 2001, named *Jason-1*, is extending the science of determining mean sea level, ocean topography, atmosphere–ocean interactions, and ocean circulation. *Jason-2* was launched in 2008.

The University of Arizona's Environmental Studies Laboratory prepared a dramatic map series related to rising sea level. Using digital elevation model (DEM) data, the scientists plotted 1-m to 6-m increases in sea level and the amount of coastal inundation (Figure 16.6). As you examine the three maps, consider what adjustments to human settlement will be required.

CRITICAL THINKING 16.1

Thinking through a rising sea level

Based upon coastal inundation shown in Figure 16.6, what is your assessment of southern Louisiana if sea level rises 1 m? What is the region's future? Go to the web site where these maps were created, http://www.geo.arizona.edu/dgesl/; click on "maps susceptible to sea level rise"; and proceed to select a coastal area of interest to you. Initially, enter a 1-m rise and examine the map generated. Try other values. Briefly analyze what you find. Certainly, any calculations of the costs of slowing fossil-fuel burning to lower emissions of greenhouse gases should be balanced against damage estimates from such coastal inundation. What do you think?

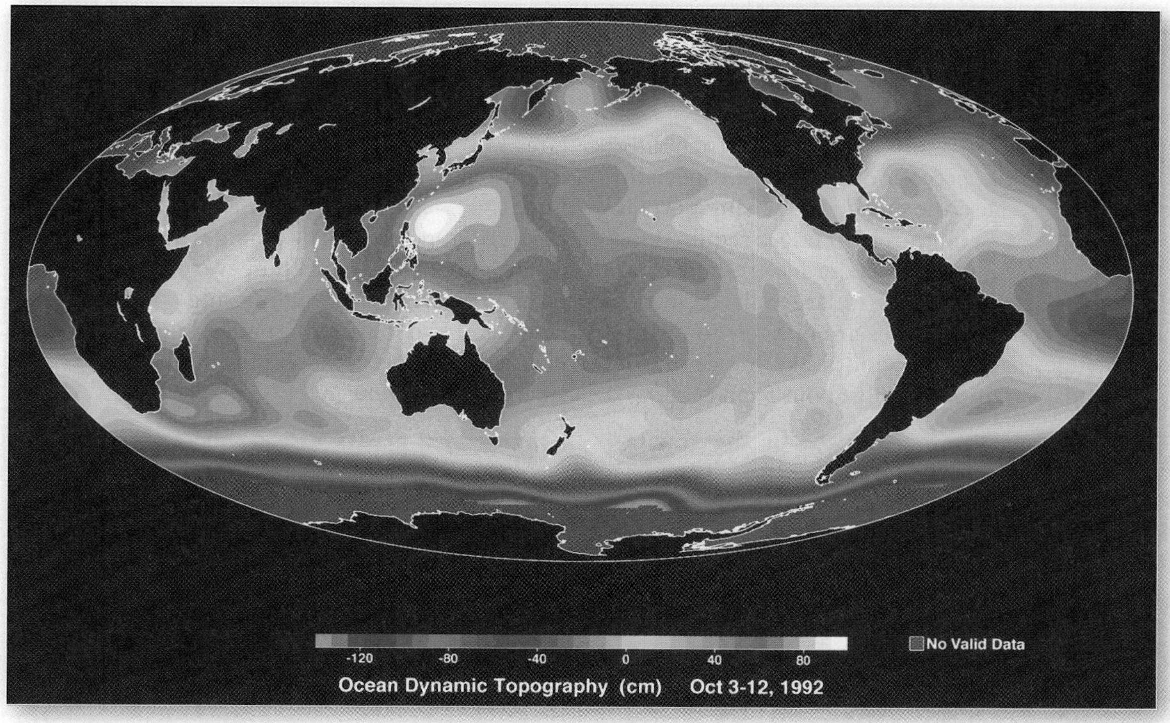

FIGURE 16.5 **Ocean topography as revealed by satellite radar altimeter.**
Sea-level data noted by the color scale are given in centimeters above or below Earth's geoid. The overall relief portrayed in the image is about 2 m (6.6 ft). The maximum sea level is in the western Pacific Ocean (white). The minimum is around Antarctica (blue and purple). [*TOPEX/Poseidon* image JPL and NASA/GSFC.]

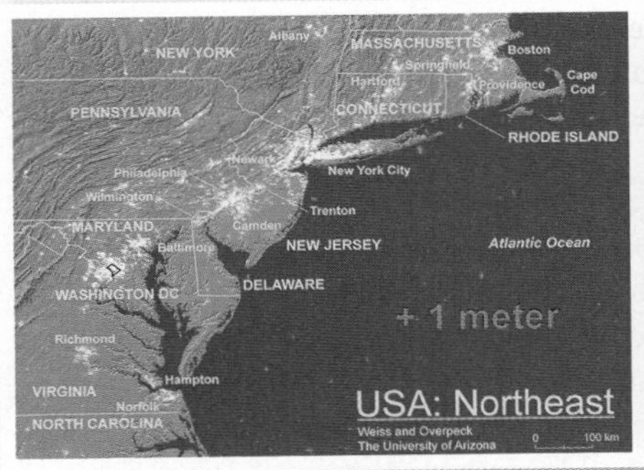

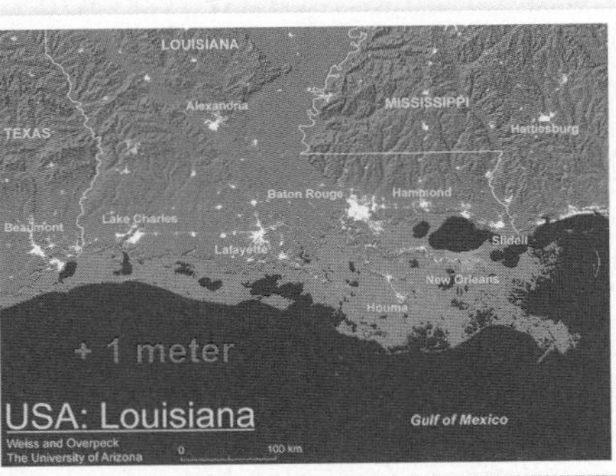

FIGURE 16.6 Coastal inundation caused by a 1-m (3.28-ft) rise in sea level. Along three portions of U.S. coastline, scientists at the University of Arizona using U.S. Geological Survey digital elevation data produced these maps to show a 1-m sea-level rise and its impact. [Maps prepared by Weiss and Overpeck, Environmental Studies Laboratory, Department of Geosciences, University of Arizona. Used by permission.]

Coastal System Actions

The coastal system is the scene of complex tidal fluctuation, winds, waves, ocean currents, and the occasional impact of storms. These forces shape landforms ranging from gentle beaches to steep cliffs, and they sustain delicate ecosystems.

Tides

Tides are complex twice-daily oscillations in sea level, ranging worldwide from barely noticeable to several meters. They are experienced to varying degrees along every ocean shore around the world. Tidal action is a relentless energy agent for geomorphic change. As tides flood (rise) and ebb (fall), the daily migration of the shoreline landward and seaward causes significant changes that affect sediment erosion and transportation.

Tides are important in human activities, including navigation, fishing, and recreation. Tides are especially important to ships because the entrance to many ports is limited by shallow water, and thus high tide is required for passage. Tall-masted ships may need a low tide to clear overhead bridges. Tides also exist in large lakes, but because the tidal range is small, tides are difficult to distinguish from changes caused by wind. Lake Superior, for instance, has a tidal variation of only about 5 cm (2 in.).

Causes of Tides Tides are produced by the gravitational pull of both the Sun and the Moon. Chapter 2 discusses Earth's relation to the Sun and Moon and the reasons for the seasons. The Sun's influence is only about half that of the Moon's because of the Sun's greater distance from Earth, although it is a significant force. Figure 16.7 illustrates the relation among the Moon, the Sun, and Earth and the generation of variable tidal bulges on opposite sides of the planet.

The gravitational pull of the Moon tugs on Earth's atmosphere, oceans, and lithosphere. The same is true for the Sun, to a lesser extent. Earth's solid and fluid surfaces all experience some stretching as a result of this gravitational pull. The stretching raises large *tidal bulges* in the atmosphere (which we can't see), smaller tidal bulges in the ocean, and very slight bulges in Earth's rigid crust. Our concern here is the tidal bulges in the ocean.

Gravity and inertia are essential elements in understanding tides. *Gravity* is the force of attraction between two bodies. *Inertia* is the tendency of objects to stay still if motionless or to keep moving in the same direction if in motion. The gravitational effect on the side of Earth facing the Moon or Sun is greater than that experienced by the far side, where inertial forces are slightly greater. In effect, from this inertial point of view, as the nearside water and Earth are drawn toward the Moon and Sun, the farside water is left behind because of the slightly weaker gravitational pull. This arrangement produces the two opposing tidal bulges on opposite sides of Earth.

Tides appear to move in and out along the shoreline, but they do not actually do so. Instead, Earth's surface rotates into and out of the relatively "fixed" tidal bulges as Earth changes its position in relation to the Moon and Sun. Every 24 hours and 50 minutes, any given point on Earth rotates through two bulges as a direct result of this rotational positioning. Thus, every day, most coastal locations experience two high (rising) tides, known as **flood tides**, and two low (falling) tides, known as **ebb tides**. The difference between consecutive high and low tides is considered the *tidal range*.

Spring and Neap Tides The combined gravitational effect of the Sun and Moon is strongest in the conjunction alignment and results in the greatest tidal range between high and low tides, known as **spring tides**. (*Spring* means to "spring forth"; it has no relation to the season of the year.) Figure 16.7b shows the other alignment that gives rise to spring tides, when the Moon and Sun are at *opposition*. In this arrangement, the Moon and Sun cause separate tidal bulges, affecting the water nearest to each of them. In addition, the left-behind water resulting from the pull of the body on the opposite side augments each bulge.

When the Moon and the Sun are neither in conjunction nor in opposition, but are more or less in the positions shown in Figure 16.7c and d, their gravitational

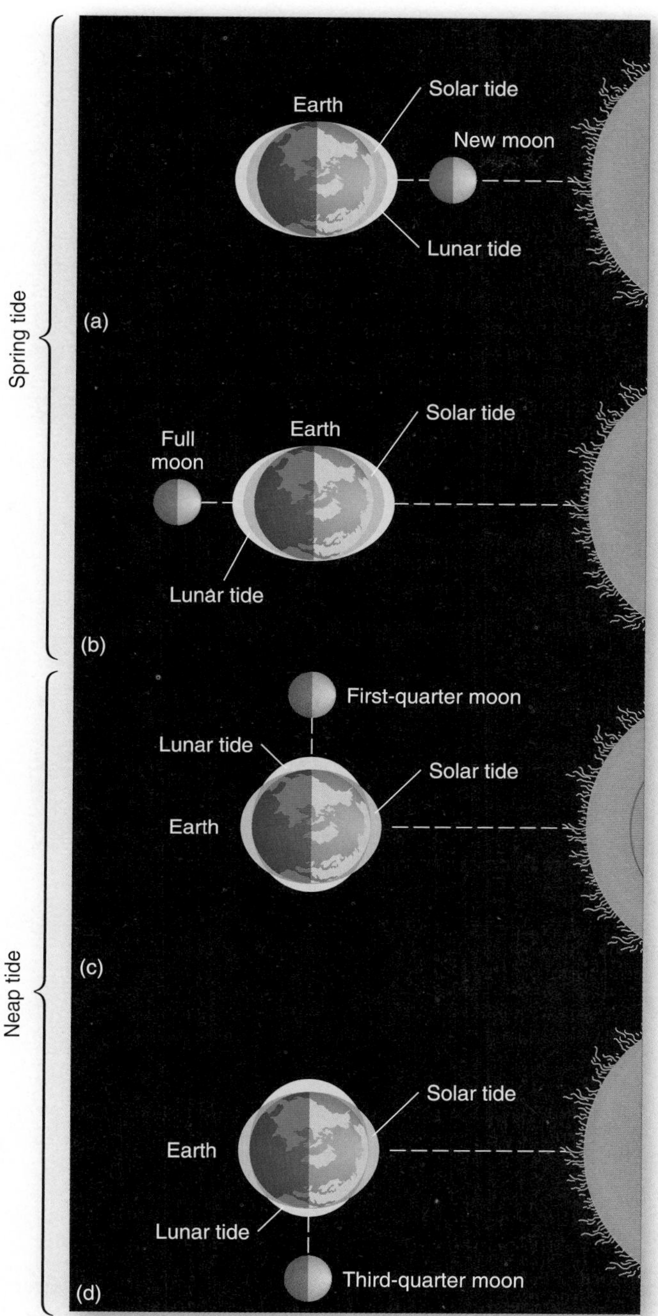

FIGURE 16.7 The cause of tides. Gravitational relations of Sun, Moon, and Earth combine to produce spring tides (a, b) and neap tides (c, d). (Tides are greatly exaggerated for illustration.)

Animation
Monthly Tidal Cycles

influences are offset and counteract each other, producing a lesser tidal range known as **neap tide**. (*Neap* means "without the power of advancing.")

Tides also are influenced by other factors, including ocean basin characteristics (size, depth, and topography), latitude, and shoreline shape. These factors cause a great variety of tidal ranges. For example, some locations may experience almost no difference between high and low tides. The highest tides occur when open water is forced into partially enclosed gulfs or bays. The Bay of Fundy in Nova Scotia records the greatest tidal range on Earth, a difference of 16 m (52.5 ft) (Figure 16.8a, b). The comparison in Figure 16.8c shows a smaller tidal range along England's east coast. For more on tides and tide prediction and a complete listing of links, see http://ocean.peterbrueggeman.com/tidepredict.html.

Tidal Power The fact that sea level changes daily with the tides suggests an opportunity: Could these predictable flows be harnessed to generate electricity? The answer is yes, given the right conditions. Bays and estuaries tend to focus tidal energy, concentrating it in a smaller area than in the open ocean. This provides an opportunity to construct a dam with water gates, locks to let ships through, and turbines to generate power.

Only 30 locations in the world are suited for tidal power generation. At present, only 3 are producing electricity. Two are outside North America—a 4-megawatt-capacity station in Russia in operation since 1968 (at Kislaya-Guba Bay on the White Sea) and a facility in France operating since 1967 (on the Rance River estuary on the Brittany coast). The tides in the Rance estuary fluctuate up to 13 m (43 ft), and power production has been almost continuous there, providing an electrical-generating capacity of a moderate 240 megawatts (about 20% of the capacity of Hoover Dam).

The third area is the Bay of Fundy in Nova Scotia, Canada. At one of several favorable sites on the bay, the Annapolis Tidal Generating Station was built in 1984. Nova Scotia Power Incorporated operates this 20-megawatt plant (Figure 16.8d). According to the Canadian government, tidal power generation at ideal sites is economically competitive with fossil-fuel plants.

Another kind of tidal power taking form in Norway involves sea-floor devices that harness the motion of coastal currents. Windmill-like turbines are set in motion by the tides and currents, as are paddle-driven turbines. Electrical production using this method began in 2003.

Waves

Friction between moving air (wind) and the ocean surface generates undulations of water in **waves**. Waves travel in groups of *wave trains*. Waves vary widely in scale: On a small scale, a moving boat creates a wake of small waves; at a larger scale, storms generate large groups of wave trains. At the extreme is the wind wake produced by the presence of the Hawaiian Islands, traceable westward across the Pacific Ocean surface for 3000 km (1865 mi). The islands disrupt the steady trade winds and produce related surface temperature and wind changes.

A stormy area at sea is a *generating region* for large wave trains, which radiate outward in all directions. The ocean is crisscrossed with intricate patterns of these multidirectional waves. The waves seen along a coast may be the product of a storm center thousands of kilometers away.

Regular patterns of smooth, rounded waves, the mature undulations of the open ocean, are **swells**. As waves leave the generating region, wave energy continues to run

FIGURE 16.8 Tidal range and tidal power.
Tidal range is great in some bays and estuaries, such as Halls Harbor near the Bay of Fundy at flood tide (a) and ebb tide (b). (c) Tidal variation of about 3 m (9.8 ft) in the harbor at Bamburgh, Northumbrian, England, over several hours. (d) The flow of water between high and low tide is ideal for turning turbines and generating electricity, as is done at the Annapolis Tidal Generating Station, near the Bay of Fundy in Nova Scotia. [(a) Jeff Newbery; (b)-(d) Bobbé Christopherson.]

in these swells, which can range from small ripples to very large flat-crested waves. A wave leaving a deep-water generating region tends to extend its wavelength horizontally for many meters. Tremendous energy occasionally accumulates to form unusually large waves. One moonlit night in 1933, the U.S. Navy tanker *Ramapo* reported a wave in the Pacific higher than its mainmast, at about 34 m (112 ft)!

As you watch waves in open water, it appears that water is migrating in the direction of wave travel, but only a slight amount of water is actually advancing. It is the *wave energy* that is moving through the flexible medium of water. Water within a wave in the open ocean is simply transferring energy from molecule to molecule in simple cyclic undulations; these are *waves of transition* (Figure 16.9). Individual water particles move forward only slightly, forming a vertically circular pattern.

The diameter of the paths formed by the orbiting water particles decreases with depth. As a deep-ocean wave approaches the shoreline and enters shallower water (10–20 m, or 30–65 ft), the orbiting water particles are vertically restricted. This restriction causes more-elliptical, flattened orbits to form near the bottom. This change from circular to elliptical orbits slows the entire wave, although more waves continue arriving. The result is closer-spaced waves, growing in height and steepness, with sharper wave crests. As the crest of each wave rises, a point is reached when its height exceeds its vertical stability, and the wave falls into a characteristic **breaker**, crashing onto the beach (Figure 16.9b).

In a breaker, the orbital motion of transition gives way to elliptical *waves of translation*, in which both energy and water move toward shore. The slope of the shore determines wave style. Plunging breakers indicate a steep bottom profile, whereas spilling breakers indicate a gentle, shallow bottom profile. In some areas, unexpected high waves can arise suddenly. It is a good idea to learn to

MG Animation
Wave Motion/Wave Refraction

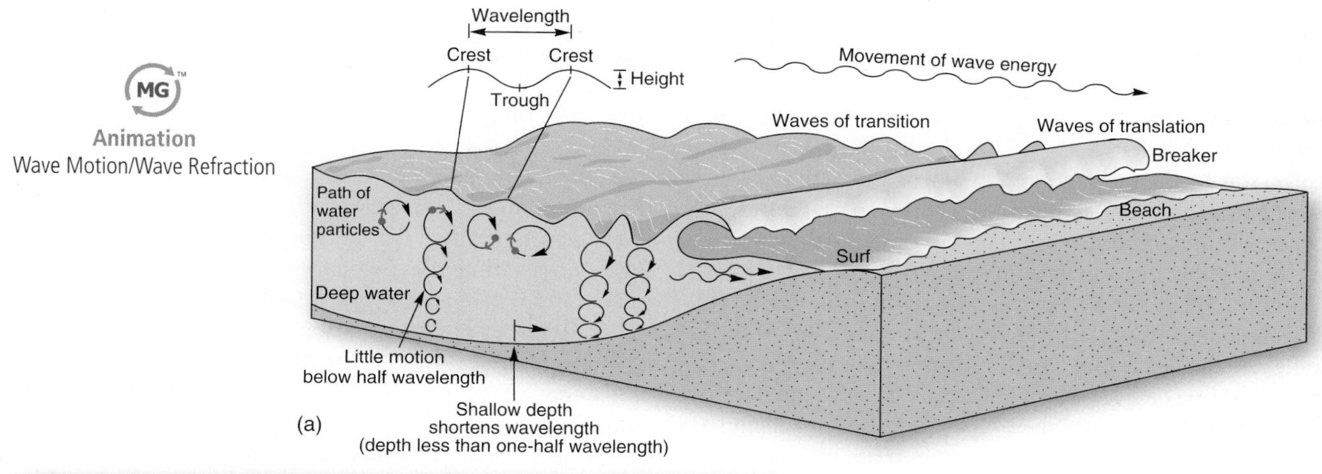

FIGURE 16.9 Wave formation and breakers.
(a) The orbiting tracks of water particles change from circular motions and swells in deep water (waves of transition) to more elliptical orbits near the bottom in shallow water (waves of translation). (b) Cascades of waves attack the shore along Baja California, Mexico. (c) A dangerous rip current interrupts approaching breakers. Note the churned-up water where the rip current enters the surf. [Bobbé Christopherson.]

recognize severe wave conditions before you venture along the shore in these areas.

When the backwash of water flows to the ocean from the beach in a concentrated column, usually at a right angle to the line of breakers, it is a *rip current*. A person caught in one of these can be swept offshore, but usually only a short distance. These brief, short torrents of water can be dangerous (Figure 16.9c).

As various wave trains move along in the open sea, they interact by *interference*. These interfering waves sometimes align, so that the wave crests and troughs from one wave train are in phase with those of another. When this in-phase condition occurs, the height of the waves is amplified, sometimes dramatically. The resulting "killer waves" or "sleeper waves" can sweep in unannounced and overtake unsuspecting victims. Signs along portions of the California, Oregon, Washington, and British Columbia coastline warn beachcombers to watch for "killer waves." In August 2009, despite warnings from park rangers, a person was killed as several giant waves hit Acadia National Park on the coast of Maine. Hurricane Bill in the Atlantic was the source of the wave energy. In contrast, out-of-phase wave trains will dampen wave energy at the shore. When you observe the breakers along a beach, the changing beat of the surf actually is produced by the patterns of *wave interference* that occurred in far-distant areas of the ocean.

Wave Refraction In general, wave action tends to straighten a coastline. Where waves approach an irregular coast, they bend around headlands, which are protruding

GEO REPORT 16.3 *Sea-wave amplification kills*

On March 3, 2010, a large cruise ship in the western Mediterranean Sea, off the coast of Marseilles, France, was struck by three surprise waves about 7.9 m (26 ft) in height. Two passengers were killed and many injured as waves shattered windows flooding parts of the ship's interior. Rescue personnel took the injured to hospitals in Barcelona, Spain. Scientists are studying what causes such abnormal waves which tend to happen in open ocean; elements include strong winds and wave interference.

landforms generally composed of resistant rocks (Figure 16.10). The submarine topography refracts, or bends, approaching waves. The refracted energy is focused around headlands and dissipates energy in coves, bays, and the submerged coastal valleys between headlands. Thus, headlands receive the brunt of wave attack along a coastline. This **wave refraction** redistributes wave energy, so that different sections of the coastline vary in erosion potential, with the long-term effect of straightening the coast.

Figure 16.11 shows waves approaching the coast at a slight angle, for they usually arrive at some angle other than parallel. As the waves enter shallow water, they are refracted as the shoreline end of the wave slows. The wave portion in deeper water moves faster in comparison, thus producing a current parallel to the coast, zigzagging in the prevalent direction of the incoming waves. This **longshore current**, or *littoral current*, depends on wind direction and resultant wave direction. A longshore current is generated only in the surf zone and works in combination with wave action to transport large amounts of sand, gravel, sediment, and debris along the shore as *longshore drift*, or **littoral drift**, the more comprehensive term.

Particles on the beach also are moved along as **beach drift**, shifting back and forth between water and land with each *swash* and *backwash* of surf. Individual sediment grains trace arched paths along the beach. You have perhaps stood on a beach and heard the sound of myriad sand grains and seawater in the backwash of surf. These dislodged materials are available for transport and eventual deposition in coves and inlets and can represent a significant volume.

Tsunami, or Seismic Sea Wave An occasional wave that momentarily, but powerfully influences coastlines is the tsunami. Japanese for "harbor wave," **tsunami** is named for its devastating effect when its energy is focused in harbors. Often, tsunami are reported incorrectly as "tidal waves," but they have no relation to the tides. Sudden, sharp motions in the seafloor, caused by earthquakes, submarine landslides, or eruptions of undersea volcanoes, produce tsunami. Thus, they properly are *seismic sea waves*.

A large undersea disturbance usually generates a solitary wave of great wavelength. Tsunami generally exceed 100 km (60 mi) in wavelength (crest to crest) but are only a meter (3 ft) or so in height. They travel at great speeds in deep-ocean water—velocities of 600–800 kmph (375–500 mph) are not uncommon—but often pass unnoticed on the open sea because their great wavelength makes the rise and fall of water hard to observe.

As a tsunami approaches a coast, however, the shallow water forces the wavelength to shorten. As a result, the wave height may increase up to 15 m (50 ft) or more. Such a wave has the potential to devastate a coastal area, causing property

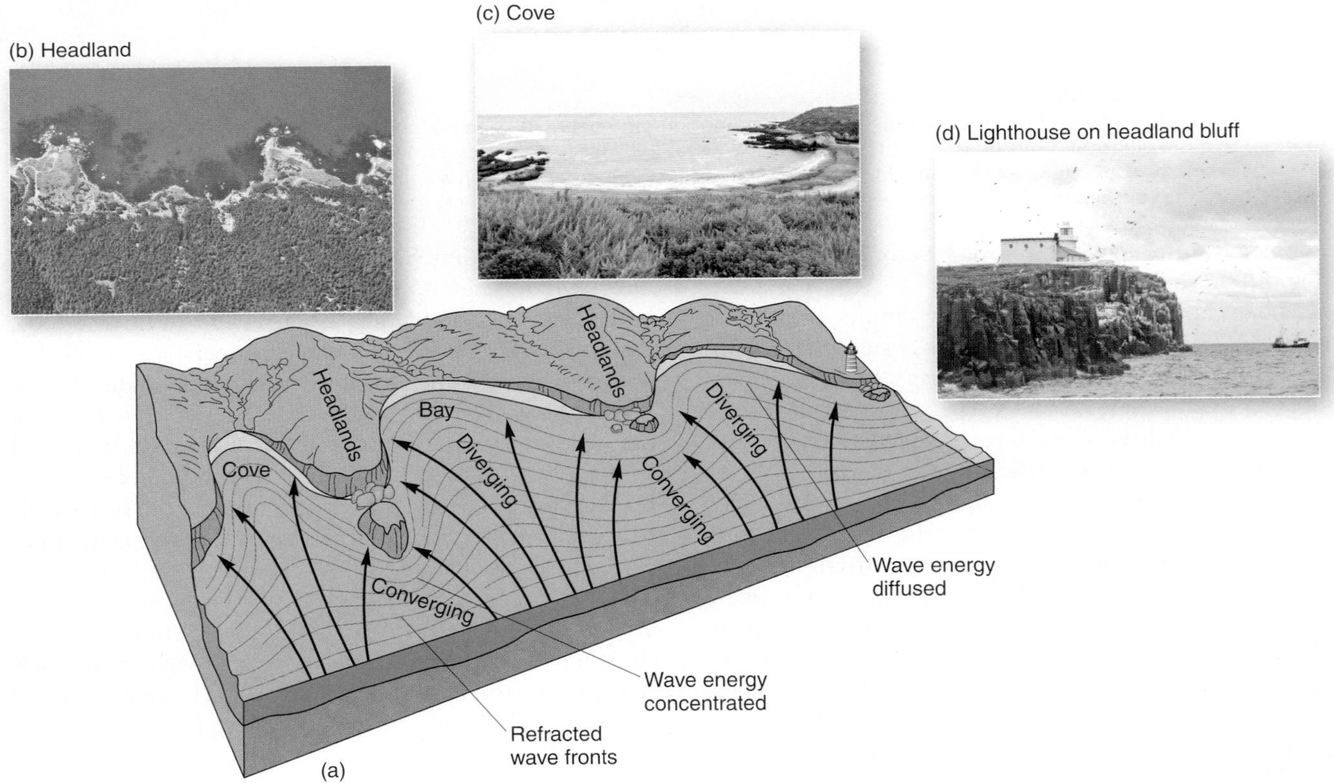

FIGURE 16.10 Coastal straightening.
(a) The process of coastal straightening is brought about by wave refraction. Wave energy is concentrated as it converges on headlands (b) and is diffused as it diverges in coves and bays (c). (d) Headlands are frequent sites for lighthouses such as the Farne light on Farne Island, England. At least six species of birds are thousands of nesting pairs are on the island, including Atlantic Puffins, Kittiwakes, and Arctic Terns. [Bobbé Christopherson.]

MG™ Animation Beach Drift, Coastal Erosion

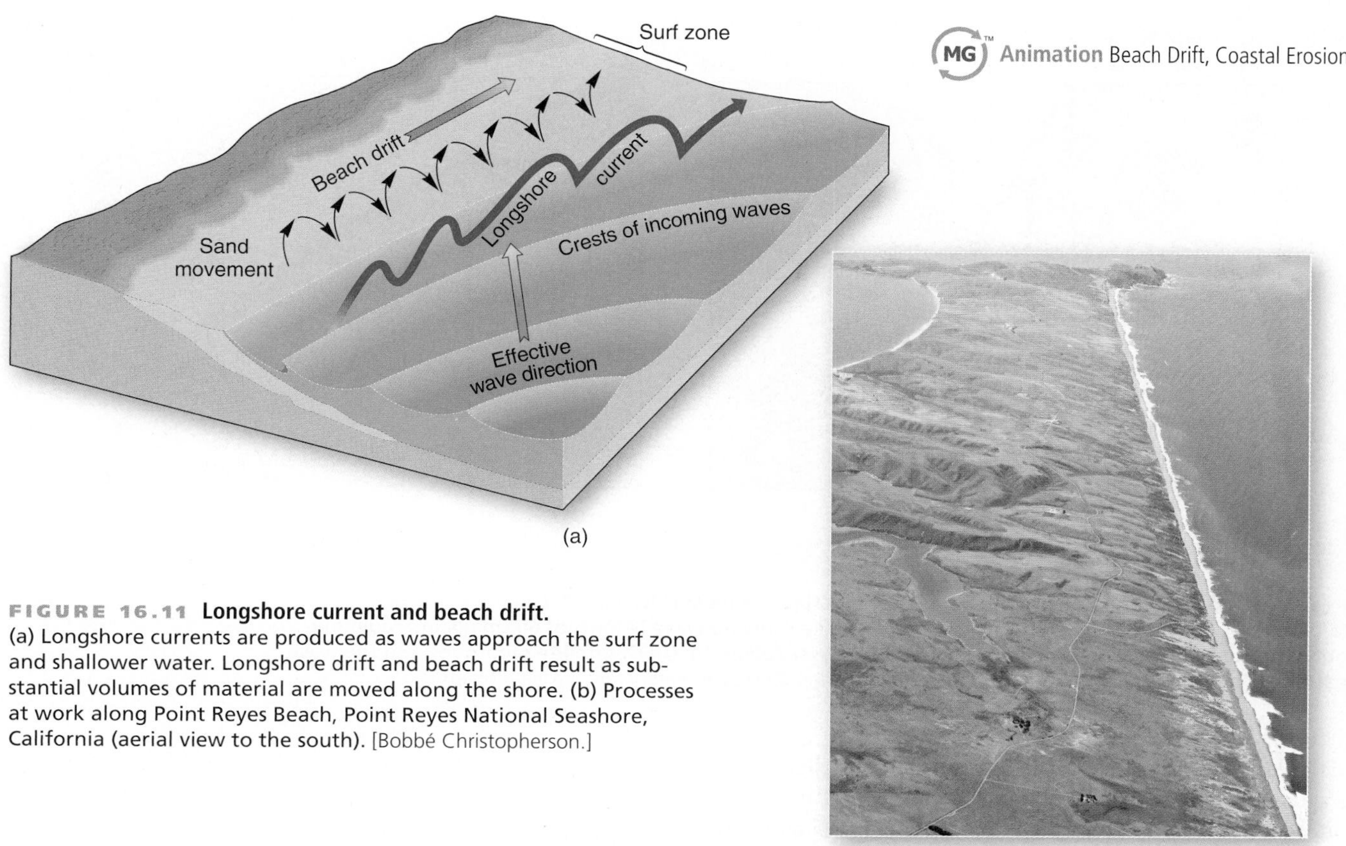

FIGURE 16.11 Longshore current and beach drift.
(a) Longshore currents are produced as waves approach the surf zone and shallower water. Longshore drift and beach drift result as substantial volumes of material are moved along the shore. (b) Processes at work along Point Reyes Beach, Point Reyes National Seashore, California (aerial view to the south). [Bobbé Christopherson.]

damage and death. For example, in 1992 the citizens of Casares, Nicaragua, were surprised by a 12-m (39-ft) tsunami that took 270 lives. A 1998 Papua New Guinea tsunami, launched by a massive undersea landslide of some 4 km^3 (1 mi^3), killed 2000. During the 20th century, records show 141 damaging tsunami and perhaps 900 smaller ones, with a total death toll of about 70,000.

Hawai'i is vulnerable to tsunami because of its position in the open central Pacific, surrounded by the ring of fire that outlines the Pacific Basin. As an example, a tsunami produced near the Philippines would reach Hawai'i in 10 hours. The U.S. Army Corps of Engineers has reported 41 damaging tsunami in Hawai'i during the past 142 years—statistically, 1 every 3.5 years. Forecasters watch the unstable Kīlauea region of the southeast coast of Hawai'i, where the possible collapse of a 40-km by 20-km by 2000-m-thick (25-mi × 12-mi × 6560-ft-thick) portion of relatively new and unstable crustal basalt could affect the Pacific Basin. In the Atlantic Basin, continuing threat exists from volcanic island collapses, such as the closely watched Cumbre Vieja volcano in the Canary Islands.

Because tsunami travel such great distances so quickly and are undetectable in the open ocean, accurate forecasts are difficult and occurrences are often unexpected. A warning system now is in operation for nations surrounding the Pacific, where the majority of tsunami occur. A warning is issued whenever seismic stations detect a significant quake or landslide under water, where it might generate a tsunami. A new tsunami hazard-mitigation network began operating in 2004 with the deployment of six pressure sensors on the ocean floor with accompanying surface buoys—this is the Deep-ocean Assessment and Reporting of Tsunamis (DART). Three instruments are along the Aleutian Islands, south of Alaska; two are along the U.S. West Coast; and one is near the equator off South America. The challenge to the world community is to apply this technology so that early warnings are possible in all vulnerable oceans, such as the Indian Ocean.

On December 26, 2004, a M 9.3 earthquake (fourth largest in a century) struck off the west coast of northern Sumatra, triggering a massive tsunami across the Indian Ocean—the Sumatra–Andaman earthquake and tsunami. No warning system was in place when this occurred. The earthquake trigger was the continuing underthrust of the Indo-Australian plate beneath the Burma plate along the Sunda Trench subduction zone. Imagine the island of Sumatra springing up about 13.7 m (45 ft) in elevation! You can find this oceanic trench on the Chapter 12 opening map of the ocean floor, stretching along the coast of Indonesia in the eastern Indian Ocean Basin.

Radar images from four Earth-orbiting satellites at the time clearly show the tsunami wave energy traveling across the Indian Ocean. In Figure 16.12 we see two of these images, one at 2:05 hours and the other at 7:10 hours after the event. Energy from this tsunami wave actually travelled several times around the global ocean basins before settling down. With Earth's mid-ocean mountain chain acting as a guide, related large waves arrived at the shores of Nova Scotia, Antarctica, and Peru.

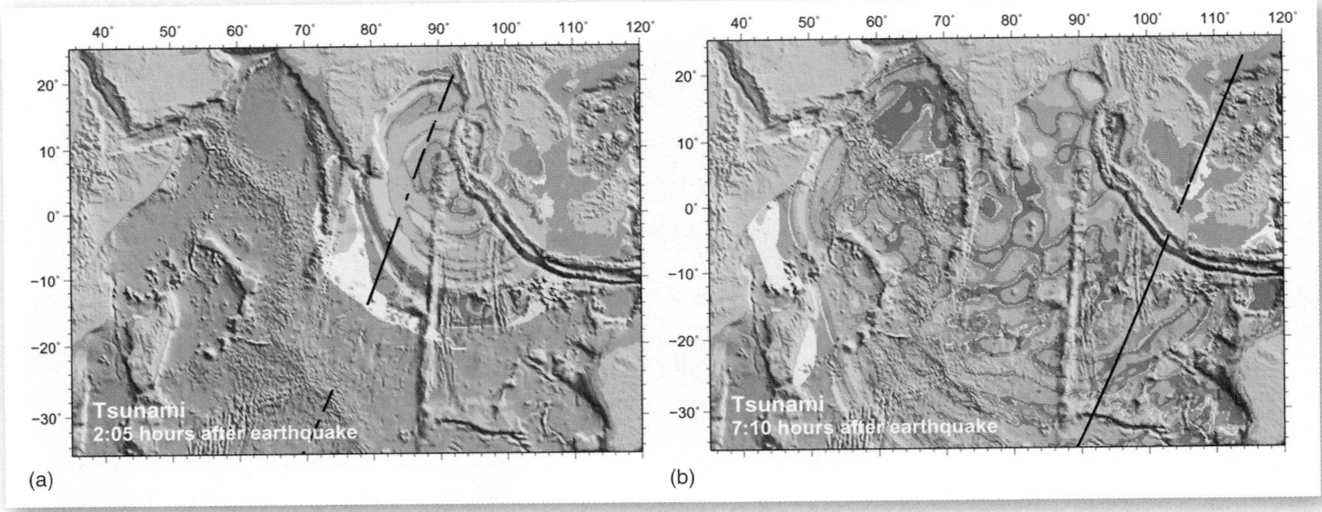

FIGURE 16.12 Satellites track tsunami waves across the Indian Ocean, 2004. (a) A *Jason-1* image 2 hours after the M 9.3 earthquake showing wave heights of 60 cm (24 in.) radiating outward from the epicenter off Sumatra. (b) The *Geosat Follow On* (*GFO*) satellite radar captured the wave patterns 7 hours and 10 minutes after the earthquake. [NOAA, Laboratory for Satellite Altimetry.]

Geophysicists even picked up slight changes in Earth's rotation speed (which was slowed by 3 microseconds) and a shift in the North Pole location by 2.5 cm (1 in.) in response to this event. Total deaths from the quake and tsunami exceeded 150,000, and a final count may never be known. Recovery will take years. In response to this tsunami the Indian Ocean Tsunami Warning and Mitigation System was created as part of a United Nations tsunami warning system project.

Warnings always should be heeded, despite the many false alarms, for the causes lie hidden beneath the ocean and are difficult to monitor in any consistent manner, although the accuracy of remote-sensing devices is increasing. For the tsunami research program, see http://nctr.pmel.noaa.gov/. The tsunami home page is at http://www.ess.washington.edu/tsunami/index.html. Strangely, warnings following a recent February 2010 Chilean earthquake for tsunami activity in New Zealand and Hawai'i seemed to attract onlookers and even some surfers. Fortunately, the tsunami waves were only about a meter in height. Hotels on Oahu did evacuate staff and guests to the third-floor level around the time of tsunami arrival.

Coastal System Outputs

As you can see, coastlines are active places, with energy and sediment being continuously delivered to a narrow environment. The action of tides, currents, wind, waves, and changing sea level produces a variety of erosional and depositional landforms. We look first at erosional coastlines such as the U.S. West Coast and then at depositional coastlines as found along the East and Gulf Coasts. In this era of rising sea level, coastlines are becoming more dynamic.

Erosional Coastal Processes and Landforms

The active margin of the Pacific Ocean along North and South America is a typical erosional coastline. *Erosional coastlines* tend to be rugged, of high relief, and tectonically active, as expected from their association with the leading edge of drifting lithospheric plates (review the plate tectonics discussion in Chapter 11). Figure 16.13 presents features commonly observed along an erosional coast. Within this erosional setting, depositional processes can occur, producing depositional landforms.

Sea cliffs are formed by the undercutting action of the sea. As indentations slowly grow at water level, a sea cliff becomes notched and eventually will collapse and retreat. Other erosional forms evolve along cliff-dominated coastlines, including *sea caves, sea arches*, and *sea stacks*. As erosion continues, arches may collapse, leaving isolated stacks in the water (Figure 16.13b, c, d).

Wave action can cut a horizontal bench in the tidal zone, extending from a sea cliff out into the sea. Such a structure is a **wave-cut platform**, or *wave-cut terrace*. If the relation between the land and sea level has changed over time, multiple platforms or terraces may rise like stair steps back from the coast. These marine terraces are remarkable indicators of a changing relation between the land and sea, with some terraces more than 370 m (1200 ft) above sea level. A tectonically active region, such as the California coast, has many examples of multiple wave-cut platforms, which at times can be unstable and vulnerable to failure (Figure 16.13e, f).

Depositional Coastal Processes and Landforms

Depositional coasts generally are along land of gentle relief, where sediments from many sources are available. Such is the case with the Atlantic and Gulf coastal plains of the United States, which lie along the relatively passive,

FIGURE 16.13 Erosional coastal features.
(a) Characteristic coastal erosional landforms: (b) notched cliff, Bjørnøya (Bear Island), Barents Sea; (c) headlands and an arch, Ascension Island, Atlantic Ocean; (d) stacks and headlands, Gough Island, Southern Ocean; (e) collapsing cliffs and failing houses; and (f) wave-cut platforms near Bixby Bridge, Cabrillo Highway, California. [Bobbé Christopherson.]

trailing edge of the North American lithospheric plate. Yet erosional processes and inundation influence depositional coasts, particularly during storm activity.

Figure 16.14 illustrates characteristic landforms deposited by waves and currents. One notable depositional landform is the **barrier spit**, which consists of material deposited in a long ridge extending out from a coast. It partially crosses and blocks the mouth of a bay. A classic barrier spit is Sandy Hook, New Jersey (south of New York City). Such barrier spit formations are at Point Reyes (Figure 16.14a) and Morro Bay (Figure 16.14b), in California.

If a spit grows to completely cut off the bay from the ocean and form an inland lagoon, it becomes a **bay barrier**, or *baymouth bar*. Spits and barriers are made up of materials that have been eroded and transported by *littoral drift* (considered beach drift and longshore drift combined). For much sediment to accumulate, offshore currents must be weak, since strong currents carry material away before it can be deposited. Tidal flats and salt marshes are characteristic low-relief features wherever tidal influence is greater than wave action. As noted earlier, if these deposits completely cut off the bay from the ocean, an inland **lagoon** is formed. A **tombolo** occurs when sediment deposits connect the shoreline with an offshore island or sea stack by accumulating on an underwater wave-built terrace (Figure 16.14c).

Not all the beaches of the world are composed of sand, for they can be made up of shingles (beach gravel) and shells, among other materials (Figure 16.14d). Let's examine these transient coastal deposits in more detail—let's go to the beach.

Beaches Of all the depositional landforms along coastlines, beaches probably are the most familiar. Beaches vary in type and permanence, especially along coastlines dominated by wave action. Technically, a **beach** is that place

FIGURE 16.14 Depositional coastal features.
Characteristic coastal depositional landforms: barrier spits, lagoons, tombolos, and beaches. (a) The Limantour barrier spit nearly blocks the entrance to Drakes Estero, along Point Reyes. (b) A barrier spit forms Morro Bay, with the sound opening to the sea near 178-m (584-ft) Morro Rock, a volcanic plug. (c) A tombolo at Point Sur along the central California coast, where sediment deposits connect the shore with an island. (d) A shell beach deposited by the ocean. [Bobbé Christopherson.]

along a coast where sediment is in motion, deposited by waves and currents. Material from the land temporarily resides on the beach while it is in active transit along the shore. You probably have experienced a beach at some time, along a seacoast, a lakeshore, or even a stream. Perhaps you have even built your own "landforms" in the sand, only to see them washed away by the waves: a lesson in erosion.

On average, the beach zone spans from the area about 5 m (16 ft) above high tide to 10 m (33 ft) below low tide (see Figure 16.4). However, the specific definition varies greatly along individual shorelines. Worldwide, quartz (SiO_2) dominates beach sands because it resists weathering and therefore remains after other minerals are removed. In volcanic areas, beaches are derived from wave-processed lava. Hawai'i and Iceland, for example, feature some black-sand beaches.

Many beaches, such as those in southern France and western Italy, lack sand and are composed of pebbles and cobbles—a type of *shingle beach*. Some shores have no beaches at all; scrambling across boulders and rocks may be the only way to move along the coast. The coasts of Maine and portions of Canada's Atlantic provinces are classic examples. These coasts, composed of resistant granite rock, are scenically rugged and have few beaches.

A beach acts to stabilize a shoreline by absorbing wave energy, as is evident by the amount of material that is in almost constant motion (see "Sand movement" in Figure 16.11a). Some beaches are stable. Others cycle seasonally: They accumulate during the summer; are moved offshore by winter storm waves, forming a submerged bar; and are redeposited onshore the following summer. Protected areas along a coastline tend to accumulate sediment, which can lead to large coastal sand dunes. Prevailing winds often drag such coastal dunes inland, sometimes burying trees and highways.

Maintaining Beaches Changes in coastal sediment transport can disrupt human activities—beaches are lost, harbors are closed, and coastal highways and beach houses can be inundated with sediment. Thus, people use various strategies to interrupt littoral drift and beach drift. The goal is either to halt sand accumulation or to force accumulation in a desired way through construction of engineered structures, or "hard" shoreline protection.

Figure 16.15 illustrates common approaches: a *groin* to slow drift action along the coast, a *jetty* to block material from harbor entrances, and a *breakwater* to create a zone of still water near the coastline. However, interrupting the

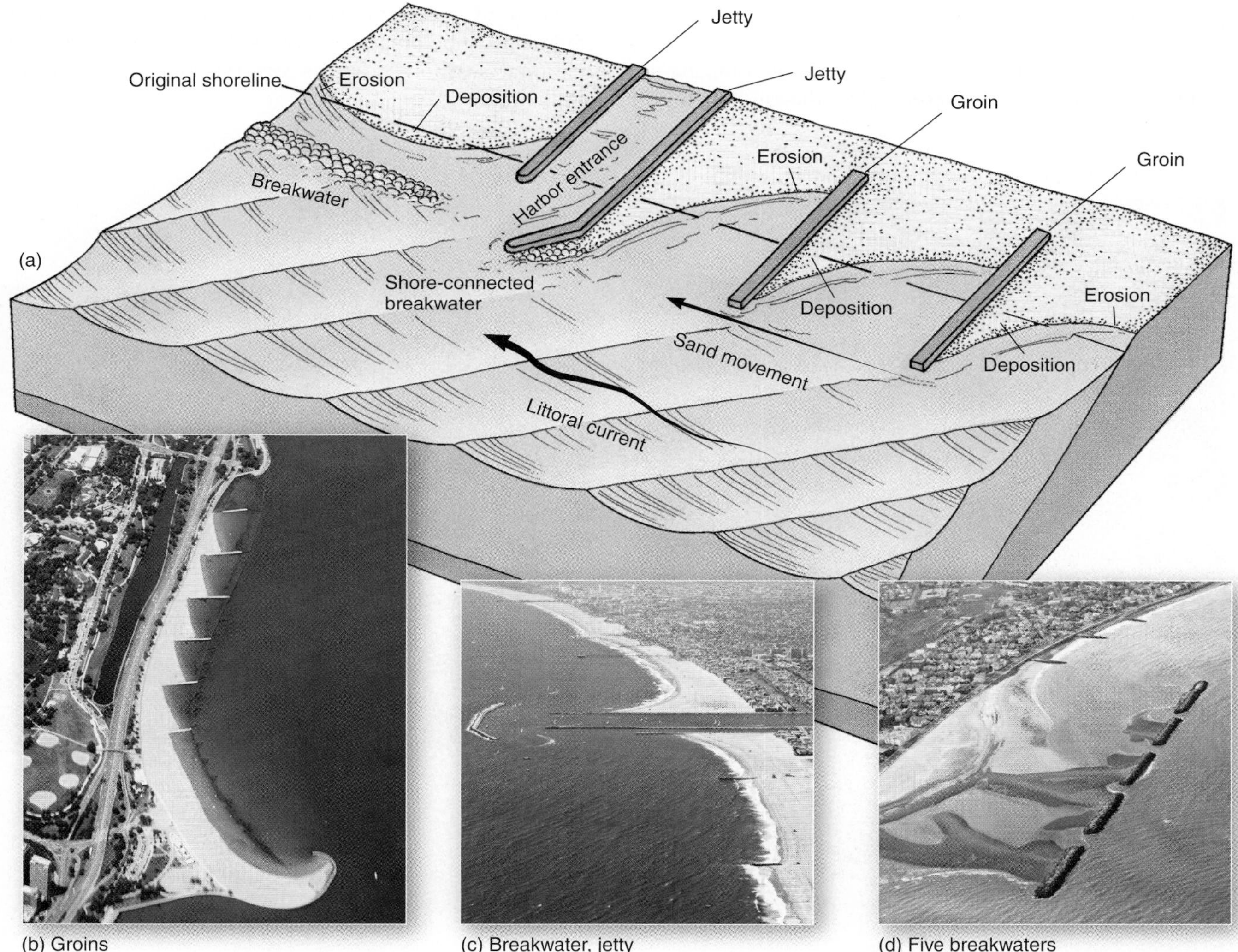

FIGURE 16.15 Interfering with the littoral drift of sand.
(a) Various constructions attempt to control littoral drift and beach drift along a coast: breakwater, jetty, and groin. Aerial photographs of coastal constructions: (b) Note the disruption of sediment movement by each groin along the coast of Lake Michigan, north of Chicago; (c) a breakwater and jetties protect the entrance to Marina del Rey, California; (d) the Five Sisters breakwaters in Winthrop, Massachusetts (near Boston)—note coarse gravels and sand accumulating in bars behind the breakwaters since their construction in the 1930s. [Bobbé Christopherson.]

MG Animation Coastal Stabilization Structures

littoral drift that is the natural replenishment for beaches may lead to unwanted changes in sediment distribution downcurrent. Careful planning and impact assessment should be part of any strategy for preserving or altering a beach.

In contrast to the term *hard structures*, the hauling of sand to replenish a beach is considered "soft" shoreline protection. *Beach nourishment* refers to the artificial replacement of sand along a beach. Through such efforts, a beach that normally experiences a net loss of sediment will be "nourished" with new sand. Years of human effort and expense to build beaches can be erased by a single storm, however.

The city of Miami, Florida, and surrounding Dade County have spent almost $70 million since the 1970s in a continuing effort to rebuild their beaches. Sand is transported to the replenishment area from a source area. To maintain a 200-m-wide (660-ft-wide) beach, planners determine net sand loss per year and set a schedule for replenishment. In Miami Beach, an 8-year replenishment cycle is maintained.

Unforeseen environmental impact may accompany the addition of sand to a beach, especially if the sand is from an unmatched source, in terms of its ecological traits. If the new sands do not physically and chemically match the existing varieties, disruption of coastal marine life is possible. More than $2 billion has been spent from New York to Florida to date to replenish beaches, with costs over the next 10 years forecasted to be billions more. For more on beach nourishment, see http://www.csc.noaa.gov/beachnourishment/; for basics, go to http://www.brynmawr.edu/geology/geomorph/beachnourishmentinfo.html.

Barrier Beach Formation Barrier chains are long, narrow, depositional features, generally of sand, that form offshore roughly parallel to the coast. Common forms are **barrier beaches** and the broader, more extensive landforms called **barrier islands**. Tidal variation in the area usually is moderate to low, with adequate sediment supplies coming from nearby coastal plains. Figure 16.16 illustrates the many features of barrier chains, using North Carolina's famed Outer Banks as an example, including Cape Hatteras, across Pamlico Sound from the mainland. The area presently is designated as one of 10 national seashore reserves supervised by the National Park Service.

FIGURE 16.16 Barrier island chain.
(a) *Landsat* image of barrier island chain along the North Carolina coast. You can see key depositional forms: spit, island, beach, lagoon, and inlet. A sound is a large inlet of the ocean; Pamlico Sound is an example.
(b) The path, partially a parking lot, along which the Cape Hatteras lighthouse was moved inland in 1999 to safer ground. (c) Cape Hatteras lighthouse in its new location 488 m (1601 ft) farther inland. [(a) *Terra* MODIS image, NASA/GSFC. (b) and (c) Bobbé Christopherson.]

On the landward side of a barrier formation are tidal flats, marshes, swamps, lagoons, coastal dunes, and beaches, visible in Figure 16.16. Barrier beaches appear to adjust to sea level and may naturally shift position from time to time in response to wave action and longshore currents. A break in a barrier forms an inlet that connects a bay with the ocean. The name *barrier* is appropriate, for these formations take the brunt of storm energy and actually shield the mainland. Various hypotheses exist to explain the formation of barrier islands. They may begin as offshore bars or low ridges of submerged sediment near shore and then gradually migrate toward shore as sea level rises.

Barrier beaches and islands are quite common worldwide, lying offshore of nearly 10% of Earth's coastlines. Examples are found off the shores of Africa, India's eastern coast, Sri Lanka, Australia, and Alaska's northern slope and offshore in the Baltic and Mediterranean seas. Earth's most extensive chain of barrier islands is along the U.S. Atlantic and Gulf coasts, extending some 5000 km (3100 mi) from Long Island to Texas and Mexico.

Barrier Island Vulnerabilities and Hazards Because many barrier islands seem to be migrating landward, they are an unwise choice for home sites or commercial building. Nonetheless, they are a common choice, even though they take the brunt of storm energy. The hazard represented by the settlement of barrier islands was made graphically clear when Hurricane Hugo assaulted South Carolina in 1989. In the Charleston area, Hugo swept away beachfront houses, barrier island developments, and millions of tons of sand; the hurricane destroyed up to 95% of the single-family homes in one community. The southern portion of one island was torn away. With increased development and continuing real estate appreciation, each future storm can be expected to cause ever-increasing capital losses.

Vulnerable settlements fill the Outer Banks and Cape Hatteras and are repeatedly slammed by tropical storms—hurricanes Emily, 1993; Dennis, Floyd, and Irene, 1999; Isabel, 2003; Alex and Bonnie, 2004; Ophelia, 2005; Hanna, 2008; and Earl, 2010—causing damage and beach erosion. Because of the continuing loss of sand, the famous Cape Hatteras lighthouse was moved inland in 1999 to safer ground (Figure 16.16b). Its new position is about 488 m (1600 ft) from the ocean, or approximately the distance it was from the water in 1870 when it was built—that much sand was lost back to the sea. See http://www.ncsu.edu/coast/chl/ for details of the extraordinary effort to save this landmark.

The barrier islands off the Louisiana shore are disappearing. They are affected by subsidence resulting from the compaction of Mississippi delta sediments, the pumping of oil and gas reserves, a changing sea level that is rising at 1 cm (0.4 in.) per year in the region, and an increase in tropical storm intensity. In 1998, Hurricane Georges destroyed large tracts of the Chandeleur Islands, located 30 to 40 km (19 to 25 mi) from the Louisiana–Mississippi Gulf Coast, leaving the mainland with reduced protection (Figure 16.17). Louisiana's increasingly exposed wetlands are disappearing at the rate of 65 km^2 (25 mi^2) per year. Hurricane Katrina (2005) alone removed this amount of wetlands and swept away most of the remaining Chandeleur Islands (Figure 16.17c, d). The regional office of the U.S. Geological Survey predicts that in a few decades the barrier islands may be gone.

Figure 16.18a is an aerial photo of development along the Bolivar Peninsula, a barrier spit near Galveston on the Texas Gulf Coast. Portions of the barrier island chain are experiencing beach erosion rates averaging 2 m (6.5 ft) per year as sea level continues to rise. Hurricane Ike (September 13, 2008) struck Galveston Island, Bolivar Peninsula, and the Texas–Louisiana Coast—landfall was in the area pictured in Figure 16.18a. The barrier island urbanization was devastated as $30 billion in structures were swept away along with more than 195 lives, and there are nearly 20 people still missing. Surveys found 80% to 95% of homes were destroyed. The devastation is visible in Figure 16.18c a year after the event; see Chapter 8 *Geosystems Now* and Figure 8.24 for more on Ike. For an environmental approach to shoreline planning see Focus Study 16.1 (pages 482–83).

Biological Processes: Coral Formations

Not all coastlines form by purely physical processes. Some form as the result of biological processes, such as coral growth. A **coral** is a simple marine animal with a small, cylindrical, saclike body called a *polyp*; it is related to other marine invertebrates, such as anemones and jellyfish. Corals secrete calcium carbonate ($CaCO_3$) from the lower half of their bodies, forming a hard, calcified external skeleton.

Corals live in a *symbiotic* relationship with algae: They live together in a mutually helpful arrangement, each dependent on the other for survival. Corals cannot photosynthesize, but they do obtain some of their own nourishment. Algae perform photosynthesis and convert solar energy to chemical energy in the system, providing the coral with about 60% of its nutrition and assisting the coral with the calcification process. In return, corals provide the algae with nutrients. Coral reefs are the most diverse marine ecosystems. Preliminary estimates of coral species place the number at a million worldwide, yet, as in most ecosystems in water or on land, biodiversity is declining in these communities.

Figure 16.19 shows the distribution of living coral formations. Corals mostly thrive in warm tropical oceans, so the difference in ocean temperature between the western coasts and eastern coasts of continents is critical to their distribution. Western coastal waters tend to be cooler, thereby discouraging coral activity, whereas eastern coastal currents are warmer and thus enhance coral growth.

Living colonial corals range in distribution from about 30° N to 30° S. Corals occupy a very specific ecological zone: 10–55 m (30–180 ft) depth, 27‰–40‰

(text continued on page 480)

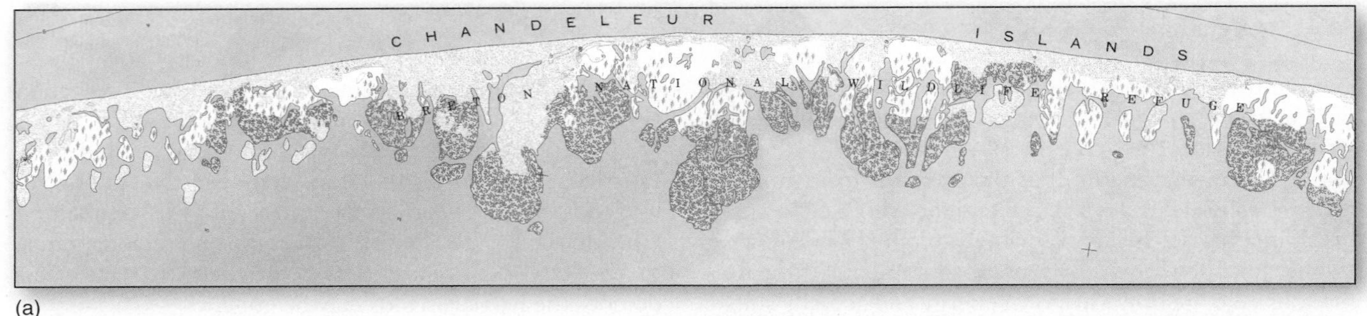

(a)

(b) 1988

(c) September 16, 2005

 **Satellite** Hurricane Isabel 9/6–9/19/04

 **Satellite** Hurricane Georges

(d)

FIGURE 16.17 Hurricanes Georges (1998) and Katrina (2005) take their toll.
Hurricane Georges eroded huge amounts of sand from the Chandeleur Islands, just off the Louisiana and Mississippi Gulf Coast. Compare the *before* topographic map (a) with the *after* aerial photo composite (b). (c) Little of the barrier island chain is visible after Katrina's passage removed some 80% of the sand. (d) Aerial photo comparison between 2001 and two days after Katrina shows the devastation in the Northern Chandeleur Islands. The yellow arrow points to the same location in each photo. [(a) USGS. (b) Aerial Data Service, Earth Imaging. (c) and (d) USGS Hurricane Impact Studies.]

(a) Bolivar peninsula

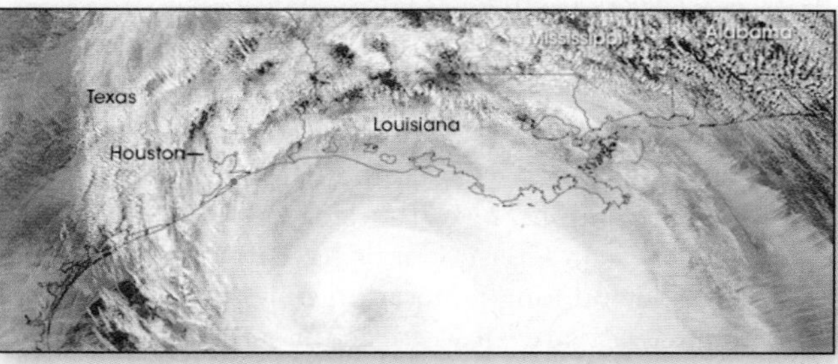

(b) Hurricane Ike approaching landfall

(c) Empty foundation pads, houses gone

FIGURE 16.18 Coastal barrier formations devastated along the Texas Gulf Coast. (a) Bolivar Peninsula, a barrier spit, stands between Texas and the Gulf of Mexico. Note the Intercoastal Waterway for shipping along Bolivar Peninsula. The scene is between Crystal Beach and Gilcrist, in the distance, where the spit narrows. (b) Hurricane Ike moves toward this coast September 12, 2008. The red dot notes the location of photo (a). (c) Devastation of Crystal Beach, Texas, on the Bolivar Peninsula. Debris from former homes and businesses is still being collected and burned. [(a) and (c) Bobbé Christopherson. (b) *Terra* image, SSEC/NASA.]

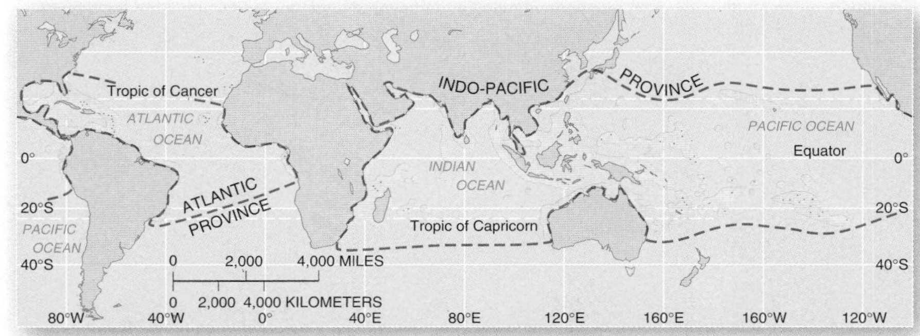

FIGURE 16.19 Worldwide distribution of living coral formations. Yellow areas include prolific reef growth and atoll formation. The red dotted line marks the geographical limits of coral activity. Colonial corals range in distribution from about 30° N to 30° S. [After J. L. Davies, *Geographical Variation in Coastal Development* (Essex, England: Longman House, 1973). Adapted by permission.]

GEO REPORT 16.4 *Ocean acidification impact on corals*

An early study of increasing ocean acidification and its impact on corals occurred from 1991 to 1993 in *Biosphere 2*—the closed habitat experiment located north of Tucson, Arizona. The habitat suffered from high carbon dioxide levels, and the model ocean in *Biosphere 2* quickly absorbed massive amounts of the excessive carbon dioxide. The model ocean became more acidic, which, in turn, chemically attacked the corals in the ocean. This same process is now occurring in the global oceans as CO_2 levels rise—here in Biosphere 1—with much research focused on the effects of acidification on oceanic flora and fauna.

salinity, and 18°C to 29°C (64°F to 85°F) water temperature. Their upper threshold is 30°C (86°F); above that temperature, the corals begin to bleach and die. Corals require clear, sediment-free water and consequently do not locate near the mouths of sediment-charged freshwater streams. For example, note the lack of these structures along the U.S. Gulf Coast. Corals have low genetic diversity worldwide and long generation times, which together mean that corals are slow to adapt and vulnerable to changing conditions.

However, an interesting exception to these environmental parameters for corals is the unique species of cool corals that exist in cold, dark water and deep ocean, at temperatures as low as 4°C (39°F) and in depths to 2000 m (6562 ft), well beyond the expected range. Scientists study these remarkable oddities, which do not rely on algae, but harvest nutrients from plankton and particulate matter.

Coral Reefs Corals exist as both solitary and colonial formations. It is the colonial corals that produce enormous structures. Their skeletons accumulate, forming coral rock. Through many generations, live corals near the ocean's surface build on the foundation of older coral skeletons, which, in turn, may rest upon a volcanic seamount or some other submarine feature built up from the ocean floor. *Coral reefs* form by this process. Thus, a coral reef is a biologically derived sedimentary rock. It can assume one of several distinctive shapes.

In 1842, Charles Darwin hypothesized an evolution of reef formation. He suggested that, as reefs develop around a volcanic island and the island itself gradually subsides, equilibrium is maintained between the subsidence of the island and the upward growth of the corals. This idea, generally accepted today, is portrayed in Figure 16.20. Note the specific examples of each reef stage:

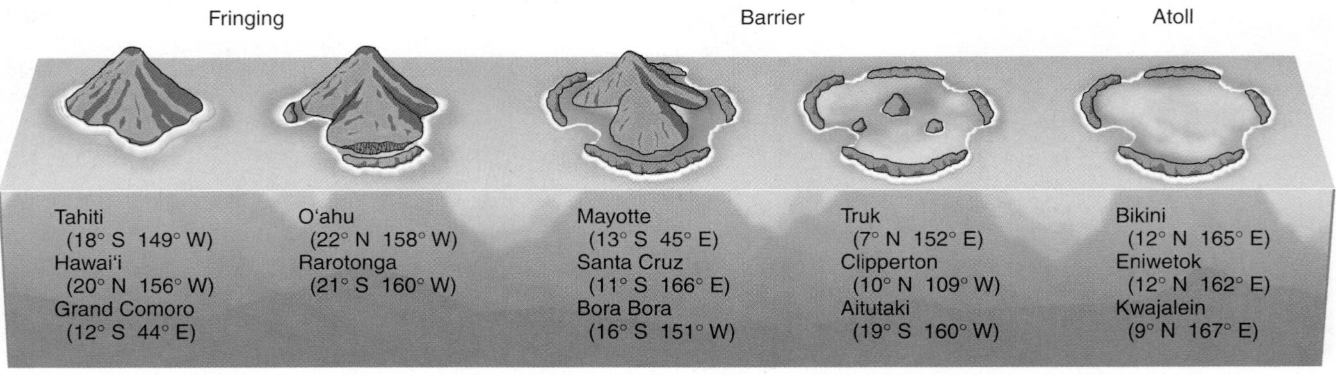

(a)

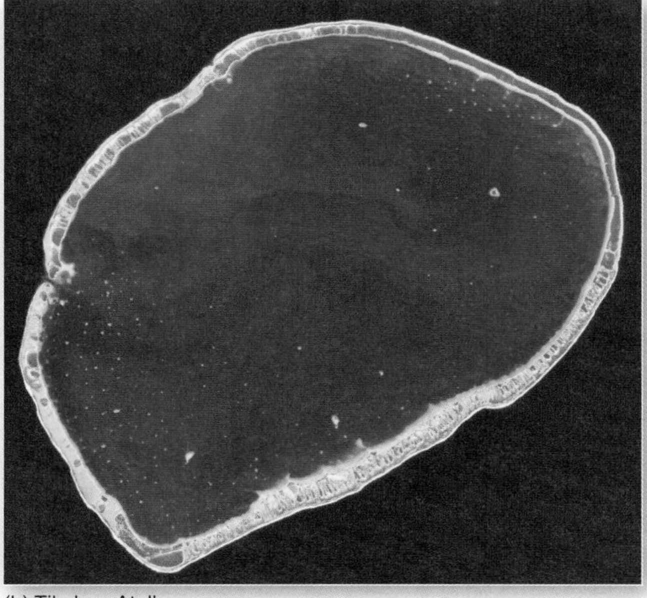

(b) Tikehau Atoll

(c) Bahamian coral platform

FIGURE 16.20 Coral forms.
(a) Common coral formations in a sequence of reef growth formed around a subsiding volcanic island: fringing reefs, barrier reefs, and an atoll. (b) Satellite image of the Tikehau Atoll, French Polynesia; note the airport runway and village in the far south (15° S 148° 10′ W). (c) The Bahamas: The lighter areas feature ocean depths to 10 m (33 ft); the darker blues indicate depths dropping off to 4000 m (13,100 ft). [(a) After D. R. Stoddart, *The Geographical Magazine* 63 (1971): 610. (b) *EO-1* image, NASA. (c) *Terra* image, NASA/GSFC.]

fringing reefs (platforms of surrounding coral rock), *barrier reefs* (reefs that enclose lagoons), and *atolls* (circular, ring-shaped reefs).

Earth's most extensive fringing reef is the Bahamian platform in the western Atlantic (Figure 16.20c), covering some 96,000 km² (37,000 mi²). Look at the composite *Terra* and *GOES* image of the Western Hemisphere on the half-title page of *Geosystems.* Can you spot the Bahama coral platform? The largest barrier reef, the Great Barrier Reef along the shore of the state of Queensland, Australia, exceeds 2025 km (1260 mi) in length, is 16–145 km (10–90 mi) wide, and includes at least 700 coral-formed islands and keys (coral islets or barrier islands).

Coral Bleaching Coral reefs may experience a phenomenon known as *bleaching*, which occurs as normally colorful corals turn stark white by expelling their own nutrient-supplying algae. Exactly why the corals eject their symbiotic partner is unknown, for without algae the corals die. Scientists are currently tracking this worldwide phenomenon, which is occurring in the Caribbean Sea and the Indian Ocean, as well as off the shores of Australia, Indonesia, Japan, Kenya, Florida, Texas, and Hawai'i.

Possible causes include local pollution, disease, sedimentation, changes in ocean salinity, and increasing oceanic acidity. One acknowledged cause is the warming of sea-surface temperatures, linked to greenhouse warming of the atmosphere. In the *Status of Coral Reefs of the World: 2000*, a report from the Global Coral Reef Monitoring Network, warmer water was found to be a greater threat to corals than local pollution or other environmental problems (see http://www. coris.noaa.gov/).

Although a natural process, coral bleaching is now occurring at an unprecedented rate as average ocean temperatures climb higher, thus linking the issue of climate change to the health of all living coral formations. By the end of 2000, approximately 30% of reefs were lost, with much off the die-off related to the record El Niño event of 1998. In 2010, scientists reported one of the most rapid and severe coral bleaching and mortality events on record near Aceh, Indonesia, on the northern tip of the island of Sumatra. Some species declined 80% in just a few months, in response to sea-surface temperature anomalies across the region. Many of these corals previously were resilient to other ecosystem disruptions, such as the Sumatra–Andaman tsunami in 2004.

As sea-surface temperatures continue to rise, forecasts state that coral losses will continue. In the 2007 IPCC *Fourth Assessment Report*, Working Group II (SPM, p. 12) states, "Increases in sea surface temperatures of about 1 to 3 C° are projected to result in more frequent coral bleaching events and widespread mortality." For more information and Internet links, see http://www.usgs.gov/coralreef.html and the Coral Reef Monitoring Network at http://www.coral.noaa.gov/gcrmn/.

CRITICAL THINKING 16.2

Exploring corals

Under "Destinations" in Chapter 16 of the *Mastering Geography* Web site, go to the link called "Coral Reefs." Sample some of the links on this page. Do you find any information about the damage to and bleaching of coral reefs reported in 1998 or at present? In which places in the world is bleaching being reported? Are the suspect causes presented in *Geosystems* mentioned in "Bleaching Hot Spots" or any of the "Coral Reef Alliance" references?

Wetlands, Salt Marshes, and Mangrove Swamps

Some coastal areas have great *biological productivity* (plant growth; spawning grounds for fish, shellfish, and other organisms) stemming from trapped organic matter and sediments. Such a rich coastal marsh environment can greatly outproduce a wheat field in raw vegetation per acre. Thus, coastal marshes can support rich wildlife habitats. Unfortunately, these wetland ecosystems are quite fragile and are threatened by human development.

Wetlands are saturated with water enough of the time to support *hydrophytic vegetation* (plants that grow in water or wet soil). Wetlands usually occur on poorly drained soils. Geographically, they occur not only along coastlines, but also as northern bogs (peatlands with high water tables), as potholes in prairie lands, as cypress swamps (with standing or gently flowing water), as river bottomlands and floodplains, and as arctic and subarctic environments that experience permafrost during the year.

Coastal Wetlands

Coastal wetlands are of two general types—salt marshes and mangrove swamps. In the Northern Hemisphere, **salt marshes** tend to form north of the 30th parallel, whereas **mangrove swamps** form equatorward from that parallel in the Eastern and Western hemispheres. This distribution is dictated by the occurrence of freezing conditions, which control the survival of mangrove seedlings. Roughly the same latitudinal limits apply in the Southern Hemisphere.

Salt marshes usually form in estuaries and behind barrier beaches and spits. An accumulation of mud produces a site for the growth of *halophytic* (salt-tolerant) plants. This vegetation then traps additional alluvial sediments and adds to the salt marsh area. Because salt marshes are in the intertidal zone (between the farthest reaches of high and low tides), sinuous, branching channels are produced as tidal waters flood into and ebb from the marsh (Figure 16.21, page 484).

Sediment accumulation on tropical coastlines provides the site for mangrove trees, shrubs, and other small trees. The prop roots of the mangrove are constantly finding new anchorages. The roots are visible above the waterline, but reach below the water surface, providing a

(text continued on page 484)

FOCUS STUDY 16.1

An Environmental Approach to Shoreline Planning

Coastlines are places of wonderful opportunity, such as ecosystem preservation, recreation, and development. They also are zones of specific constraints, and are vulnerable to shoreline processes. Poor understanding of this resource and a lack of environmental analysis often go hand in hand, surrounded by intense developmental pressures for economic gain.

For example, Cape Cod, Massachusetts, is a narrow coastal sandy peninsula composed of glacial outwash and morainal deposits. With much of its land protected as part of the Cape Cod National Seashore Reserve in 1961, Cape Cod is still subject to extensive human impacts and aggressive coastal processes. Urbanization along the cape has developed in a patchwork in and around protected areas, making a complex mix of private and public jurisdiction, land use, and plans (Figure 16.1.1).

Settlements outside the reserve are vulnerable to sea-level rise, storm surges, and saltwater intrusion into groundwater systems. Houses and businesses are barely above sea level and use septic systems with leach fields for waste disposal. These systems and most structures will be increasingly vulnerable to future sea-level rise. See http://pubs.usgs.gov/of/2002/of02-233/ for a report on the concept of coastal vulnerability along Cape Cod. Scientists are working to better understand this dynamic environment. Proper planning and thinking about the future are difficult as various stakeholders defend their positions.

Beaches and Dunes—Where to Build?

In his book *Design with Nature,* the late ecologist and landscape architect Ian McHarg discusses the New Jersey shore. He shows how proper understanding of a coastal environment could have avoided problems from major storms and coastal development. Figure 16.1.2 illustrates the New Jersey shore from ocean to back bay. Let us walk across this landscape and discover how it should be treated under ideal conditions. Remember that much of this applies to coastal areas elsewhere.

We begin our walk at the water's edge and proceed across the beach. Sand beaches are the primary natural defense against the ocean; they act as bulwarks against the pounding of a stormy sea. The shoreline tolerates recreation, but not construction, because of its shifting, changing nature during storms, daily tidal fluctuations, and the potential effects of rising sea level. Beaches are susceptible to pollution and require environmental protection to control nearshore dumping of dangerous materials.

Walking inland, we encounter the primary dune. Even more sensitive than the beach, it is fragile, easily disturbed, and vulnerable to erosion, and it cannot tolerate the passage of people trekking to the beach. Some vegetation is present, but it has only a small effect on sand stabilization. Carefully controlled access points to the beach should be designated, and restricted access should be enforced. (See two examples of beach stabilization in Figure 15.5.)

The trough behind the primary dune is relatively tolerant of limited recreation and building. The plants send roots down to fresh groundwater reserves and anchor the sand in the process. Thus, if construction should inhibit the surface recharge of that water supply, the natural protective ground cover could die and destabilize the environment. Or subsequent saltwater intrusion might contaminate well water. Clearly, groundwater resources and the location of recharge aquifers must be considered in planning. The continuing use of septic sewage systems in the trough is not sustainable.

Behind the trough is the secondary dune, a second line of defense against the sea. It, too, is tolerant of some use, yet is vulnerable to destruction. Next is the backdune, more suitable for development than any zone between it and the sea. Further inland are the bayshore and the bay, where no dredging or filling and only limited dumping of treated wastes and toxics should be permitted. This zone is tolerant of intensive recreation. In reality, of course, the opposite of such careful assessment and planning prevails.

FIGURE 16.1.1 Vulnerable coastal settlements along Beach Point on Cape Cod, Massachusetts. [Bobbé Christopherson.]

Science and Political Reality

Before an intensive government study of the New Jersey shore was completed in 1962, no analysis of coastal hazards had been done outside of academic circles. What is common knowledge in science classrooms and laboratories still has not filtered through to financial, developmental, and real estate interests. As a result, on the New Jersey shore and along much of the Atlantic and Gulf Coasts, improper development of the fragile coastal zone (primary dunes, trough, and secondary dunes) led to extensive destruction during storms in 1962, 1992, 2004, and 2005, and a near miss in 2010, among other times.

Scientists estimate that the U.S. Atlantic and Gulf coastlines will continue to erode—in some cases, tens of meters in a few decades. Society should reconcile ecology and economics if these coastal environments are to be sustained, especially in a time of increased storm intensity and sea-level rise.

FIGURE 16.1.2 Coastal environment and planning along the New Jersey shore.
(a) A planning perspective from ocean to bay. Note letters on the illustration to identify photo location. (b) Light trucks and SUV tracks in the protected Island Beach State Park. These vehicles are banned from developed stretches of beach. (c) Houses once distant from the surf are hit by waves. Sand is brought in to strengthen defenses. (d) Development along the primary dunes—an amusement park— challenges the natural setting. (e) The primary dune in a somewhat natural state. (f) An undeveloped trough behind the primary dune. (g) The former trough, paved and developed, barely above high-tide levels. (h) Development and commerce in the bayshore and bay. [(a) After *Design with Nature* by Ian McHarg. Copyright © 1969 by Ian L. McHarg. Adapted by permission; (b) through (h) Bobbé Christopherson.]

FIGURE 16.21 Coastal salt marsh. Salt marshes are productive ecosystems commonly occurring poleward of 30° latitude in both hemispheres. This is Gearheart Marsh, part of the Arcata Marsh system on the Pacific Coast in northern California. [Bobbé Christopherson.]

habitat for a multitude of specialized life forms. Mangrove swamps often secure and fix enough material to form islands (Figure 16.22).

The World Resources Institute and the United Nations Environment Programme estimate that from preagricultural times to today, mangrove losses are running between 40% (examples, Cameroon and Indonesia) and nearly 80% (examples, Bangladesh and Philippines). Deliberate removal was a common practice by many governments in the early days of settlement because of a falsely conceived fear of disease or pestilence in these swamplands.

FIGURE 16.22 Mangroves. (a) Mangroves take root and collect sediments to extend this island south of Puerto Rico. (b) Mangroves at sunset in "Ding" Darling National Wildlife Refuge, Sanibel Island, Florida; you can see the roots above water line. (c) Underwater mangrove roots form a habitat for numerous plants and animals in a specific ecosystem. [Bobbé Christopherson.]

(a) Mangrove island

(b) Mangroves

(c) Mangrove ecosystem underwater

Human Impact on Coastal Environments

Development of modern societies is closely linked to estuaries, wetlands, barrier beaches, and coastlines. Estuaries are important sites of human settlement because they provide natural harbors, a food source, and convenient sewage and waste disposal. Society depends on daily tidal flushing of estuaries to dilute the pollution created by waste disposal. Thus, the estuarine and coastal environment is vulnerable to abuse and destruction if development is not carefully planned. We are approaching a time when most barrier islands will be developed and occupied (Figure 16.23).

Barrier islands and coastal beaches do migrate over time. Houses that were more than a mile from the sea in the Hamptons along the southeastern shore of Long Island, New York, are now within only 30 m (100 ft) of it. The time remaining for these homes can be quickly shortened by a single hurricane or by a sequence of storms. Hurricane Earl in 2010 passed close to these coastal environments, causing sand inventory losses.

Despite our understanding of beach and barrier-island migration, the effect of storms, and warnings from scientists and government agencies, coastal development proceeds. Society behaves as though beaches and barrier islands are stable, fixed features or as though they can be engineered to be permanent. Experience has shown that severe erosion generally cannot be prevented. Focus Study 16.1 presents the basics for a sustainable approach to shoreline planning.

FIGURE 16.23 Barrier island subdivisions. The town of Barnegat Light abruptly stops where Long Beach Island is protected in Barnegat Lighthouse State Park, New Jersey. The protected area is almost entirely composed of replenished sand. [Bobbé Christopherson.]

A reasonable conclusion seems to be that the key to protective environmental planning and zoning is to *allocate responsibility and cost in the event of a disaster.* An ideal system places a hazard tax on land, based on assessed risk, and restricts the government's responsibility to fund reconstruction or an individual's right to reconstruct on frequently damaged sites. Comprehensive mapping of erosion-hazard areas would help avoid the ever-increasing costs from recurring disasters.

CRITICAL THINKING 16.3

What seems to be a reasonable conclusion, however....

This chapter includes the following statement: "A reasonable conclusion seems to be that the key to protective environmental planning and zoning is to *allocate responsibility and cost in the event of a disaster.*"

What do you think about this as a policy statement? How would you approach implementing such a strategy? What special interests might oppose such an approach? What response would you expect from city government or business interests? From banking or the real estate industry? In what way could you use geographic information systems (GIS), as described in Chapter 1, to survey, assess, list owners, and follow taxation status for a vulnerable stretch of coastline?

GEOSYSTEMS CONNECTION

In these last four chapters, we examined aspects of weathering, erosion and transportation, and deposition, through the agents of streams, wind, and waves. In this chapter, the systems approach moved you through coastal system inputs, actions, and outputs and analyzed human perceptions of and impacts on coastal environments. Next, we look at the agents of change in the cryosphere and the ice environments on Earth: systems that are undergoing the highest rate of change as climates shift.

KEY LEARNING CONCEPTS REVIEW

■ ***Describe*** **the chemical composition and physical structure of the ocean.**

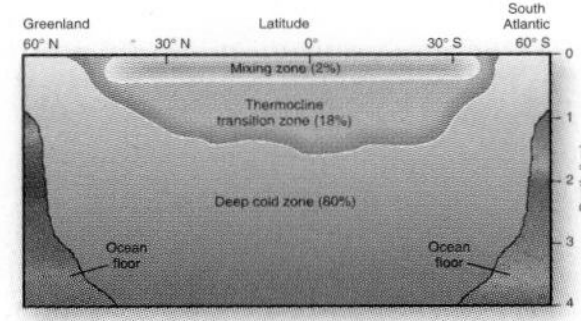

Water is the "universal solvent," dissolving at least 57 of the 92 elements found in nature. Seawater is a solution, and the concentration of dissolved solids is **salinity**. **Brine** exceeds the average 35‰ (parts per thousand) salinity; **brackish** applies to water that is less than 35‰. The ocean is divided by depth into the narrow mixing zone at the surface, the thermocline transition zone, and the deep cold zone.

salinity (p. 461)
brine (p. 462)
brackish (p. 462)

1. Describe the salinity of seawater: its composition, amount, and distribution.
2. Analyze the latitudinal distribution of salinity discussed in the chapter. Why is salinity less along the equator and greater in the subtropics?

3. What are the three general zones relative to physical structure within the ocean? Characterize each by temperature, salinity, dissolved oxygen, and dissolved carbon dioxide.

■ ***Identify*** the components of the coastal environment, and ***list*** the physical inputs to the coastal system, including tides and mean sea level.

The coastal environment is the **littoral zone** and exists where the tide-driven, wave-driven sea confronts the land. Inputs to the coastal environment include solar energy, wind and weather, climatic variation, the nature of coastal geomorphology, and human activities.

Mean sea level (MSL) is based on average tidal levels recorded hourly at a given site over many years. MSL varies spatially because of ocean currents and waves, tidal variations, air temperature and pressure differences, ocean temperature variations, slight variations in Earth's gravity, and changes in oceanic volume. MSL worldwide is rising in response to global warming of the atmosphere and oceans.

Tides are complex daily oscillations in sea level, ranging worldwide from barely noticeable to many meters. Tides are produced by the gravitational pull of both the Moon and the Sun. Most coastal locations experience two high (rising) **flood tides** and two low (falling) **ebb tides** every day. The difference between consecutive high and low tides is the tidal range. **Spring tides** exhibit the greatest tidal range, when the Moon and the Sun are in either conjunction or opposition. **Neap tides** produce a lesser tidal range.

littoral zone (p. 463)
mean sea level (MSL) (p. 463)
tide (p. 466)
flood tide (p. 466)
ebb tide (p. 466)
spring tide (p. 466)
neap tide (p. 467)

4. What are the key terms used to describe the coastal environment?
5. Define *mean sea level*. How is this value determined? Is it constant or variable around the world? Explain.
6. What interacting forces generate the pattern of tides?
7. What characteristic tides are expected during a new Moon or a full Moon? During the first-quarter and third-quarter phases of the Moon? What is meant by a flood tide? An ebb tide?
8. Explain briefly how tidal power would be utilized to produce electricity. Are there any sites doing this in North America? If so, briefly describe where they are and how they operate.

■ ***Describe*** wave motion at sea and near shore, and ***explain*** coastal straightening and coastal erosional and depositional landforms.

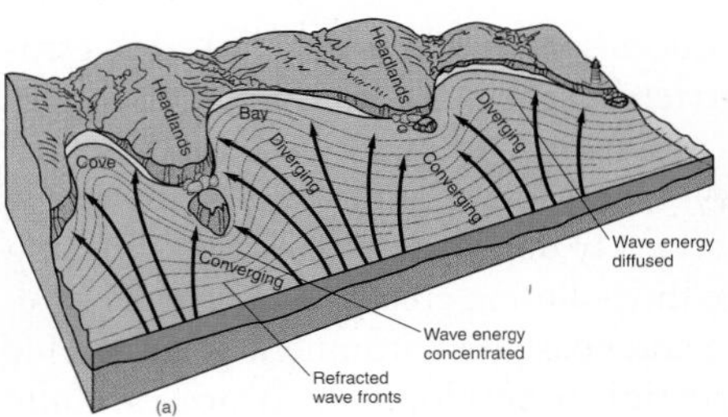

Friction between moving air (wind) and the ocean surface generates undulations of water that we call **waves**. Wave energy in the open sea travels through water, but the water itself stays in place. Regular patterns of smooth, rounded waves—the mature undulations of the open ocean—are **swells**. Near shore, the restricted depth of water slows the wave, forming *waves of translation,* in which both energy and water actually move forward toward shore. As the crest of each wave rises, the wave falls into a characteristic **breaker**.

Wave refraction redistributes wave energy so that different sections of the coastline vary in erosion potential. Headlands are eroded, whereas coves and bays receive materials, with the long-term effect of straightening the coast. As waves approach a shore at an angle, refraction produces a **longshore current** of water moving parallel to the shore. This current produces the *longshore drift* of sand, sediment, gravel, and assorted materials. **Littoral drift** is transport of materials along the shore and a more comprehensive term. Particles move along the beach as **beach drift,** shifting back and forth between water and land. A **tsunami** is a seismic sea wave triggered by an undersea landslide or earthquake. It travels at great speeds in the open sea and gains height as it comes ashore, posing a coastal hazard.

An *erosional coast* features wave action that cuts a horizontal bench in the tidal zone, extending from a sea cliff out into the sea. Such a structure is a **wave-cut platform,** or *wave-cut terrace.* In contrast, *depositional coasts* generally are located along land of gentle relief, where depositional sediments are available from many sources. Characteristic landforms deposited by waves and currents are a **barrier spit** (material deposited in a long ridge extending out from a coast); a **bay barrier,** or *baymouth bar* (a spit that cuts off the bay from the ocean and forms an inland **lagoon**); a **tombolo** (where sediment deposits connect the shoreline with an offshore island or sea stack); and a **beach** (land along the shore where sediment is in motion, deposited by waves and currents). A beach helps to stabilize the shoreline, although it may be unstable seasonally.

wave (p. 467)
swell (p. 467)
breaker (p. 468)
wave refraction (p. 470)
longshore current (p. 470)
littoral drift (p. 470)
beach drift (p. 470)
tsunami (p. 470)
wave-cut platform (p. 472)
barrier spit (p. 473)
bay barrier (p. 473)
lagoon (p. 473)
tombolo (p. 473)
beach (p. 473)

9. What is a wave? How are waves generated, and how do they travel across the ocean? Does the water travel with the wave? Discuss the process of wave formation and transmission.
10. Describe the refraction process that occurs when waves reach an irregular coastline. Why is the coastline straightened?
11. Define the components of beach drift and the longshore current and longshore drift.
12. Explain how a seismic sea wave attains such tremendous velocities. Why is it given a Japanese name—tsunami?
13. What is meant by an erosional coast? What are the expected features of such a coast?
14. What is meant by a depositional coast? What are the expected features of such a coast?
15. How do people attempt to modify littoral drift? What strategies are used? What are the positive and negative impacts of these actions?
16. Describe a beach—its form, composition, function, and evolution.
17. What success has Miami had with beach replenishment? Is it a practical strategy?

■ ***Describe*** barrier islands and their hazards as they relate to human settlement.

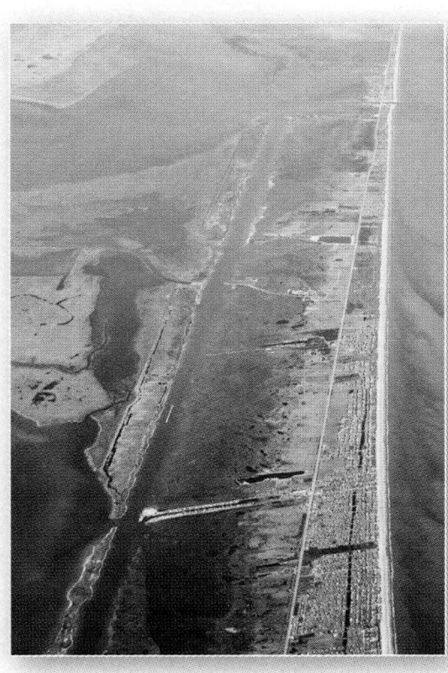

Barrier chains are long, narrow, depositional features, generally of sand, that form offshore roughly parallel to the coast. Common forms are **barrier beaches** and the broader, more extensive **barrier islands**. Barrier formations are transient coastal features, constantly on the move, and they are a poor, but common, choice for development.

barrier beach (p. 476)
barrier island (p. 476)

18. On the basis of the information in the text and any other sources at your disposal, do you think barrier islands and beaches should be used for development? If so, under what conditions? If not, why not?
19. After Hurricane Hazel destroyed the Grand Strand off South Carolina in 1954, settlements were rebuilt; yet they were hit again by Hurricane Hugo 35 years later, in 1989. Why does this type of recurring damage happen to human populations?

■ ***Assess*** living coastal environments: corals, wetlands, salt marshes, and mangroves.

A **coral** is a simple marine invertebrate that forms a hard, calcified, external skeleton. Over generations, corals accumulate in large reef structures. Corals live in a *symbiotic* (mutually helpful) relationship with algae; each is dependent on the other for survival.

Wetlands are lands saturated with water that support specific plants adapted to wet conditions. They occur along coastlands and inland in bogs, swamps, and river bottomlands. Coastal wetlands form as **salt marshes** poleward of the 30th parallel in each hemisphere and as **mangrove swamps** equatorward of these parallels.

coral (p. 477)
wetland (p. 481)
salt marsh (p. 481)
mangrove swamp (p. 481)

20. How are corals able to construct reefs and islands?
21. Describe a trend in corals that is troubling scientists, and discuss some possible causes.
22. Why are the coastal wetlands poleward of 30° N and S latitude different from those that are equatorward? Describe the differences.

■ ***Construct*** an environmentally sensitive model for settlement and land use along the coast.

Coastlines are zones of specific constraints. Poor understanding of this resource and a lack of environmental analysis often go hand in hand, producing frequent disasters to coastal ecosystems and real property losses. Society must reconcile ecology and economics if these coastal environments are to be sustained.

23. What type of environmental analysis is needed for rational development and growth in a region like the New Jersey shore? Evaluate each element of the Jersey shore and the approach suggested to coastal hazards, planning, and protection in the focus study and in Figure 16.1.2.

MASTERING GEOGRAPHY

Visit www.masteringgeography.com for several figure animations, satellite loops, self-study quizzes, flashcards, visual glossary, case studies, career links, textbook reference maps, RSS Feeds, and an ebook version of *Geosystems*. Included are many geography web links to interesting Internet sources that support this chapter.

CHAPTER 17

Glacial and Periglacial Processes and Landforms

Chauveau Glacier calves into Ayerfjorden (Ayer Fjord), as seen from morainal deposits. The scene is at 79° 40′ N latitude. Less than 10 years ago, this glacier filled the entire fjord. This fjord is an inlet of the larger Raudfjorden on Spitsbergen's north coast. [Bobbé Christopherson.]

KEY LEARNING CONCEPTS

After reading the chapter, you should be able to:

- ***Differentiate*** between alpine and continental glaciers, and ***describe*** ice field, ice cap, and ice sheet.
- ***Explain*** the process of glacial ice formation, and ***portray*** the mechanics of glacial movement.
- ***Describe*** characteristic erosional and depositional landforms created by alpine glaciation and continental glaciation.
- ***Analyze*** the spatial distribution of periglacial processes and related permafrost and frozen ground, and ***explain*** the present thaw under way.
- ***Explain*** the Pleistocene ice-age epoch and related glacials and interglacials, and ***describe*** some of the methods used to study paleoclimatology.
- ***Define*** each polar region, and ***list*** climatic changes taking place and associated impacts.

GEOSYSTEMS NOW

Ice Shelves and Tidewater Glaciers Give Way to Warming

Along the coasts of Greenland, Canada, and Antarctica are numerous *ice shelves,* extending over the sea while still attached to continental ice. These shelves are often found in sheltering inlets and bays, cover thousands of square kilometers, and reach thicknesses of 1000 m (3280 ft). Also in these regions are *tidewater glaciers,* which end in the sea. Warming temperatures appear to be hastening the flow, calving (breaking off), and thinning of ice shelves and tidewater glaciers. While the loss of these ice shelves does not significantly raise sea level, since they already displace seawater, the concern is the possible surge of grounded continental ice and outlet glaciers that many of the ice shelves hold back from the sea.

Along the northwest coast of Greenland, the tidewater portion of Petermann Glacier is breaking up. On August 5, 2010, a section measuring 251 km² (97 mi²) broke loose (Figure GN 17.1). The new iceberg is the largest to form in the Arctic in a half century and is more than 25% of this glacier's entire floating ice shelf. As warmer air melts ice from above and higher sea temperatures melt ice from below, the rate of glacial loss increases; scientists are determining how these processes apply to the Petermann Glacier. On the satellite image, note the numerous meltponds on the glacier, some filling surface cracks (pale blue in color). New cracks are opening upstream on the ragged, crevassed edges of the glacier, demonstrating the forward-moving stress. The symptomatic meltponds represent a positive feedback mechanism caused by increasing temperatures, discussed in this chapter.

FIGURE GN 17.1 Greenland tidewater glacier breaks up.
Note the meltponds, crevasses, and cracks on the ice surfaces of the Petermann Glacier. Notice that there are at least five tributary glaciers. [*Earth Observing–1* image, NASA/USGS.]

Along the northern coast of Canada's Ellesmere Island is the Ward Hunt Ice Shelf, largest in the Arctic. After remaining stable for at least 4500 years, this ice shelf began to break up in 2003 (Figure GN 17.2). Backed up behind the Ward Hunt Ice Shelf was an *epishelf lake,* an extensive freshwater accumulation some 43 m (141 ft) deep. As the ice-shelf dam failed, this epishelf lake—largest in the Northern Hemisphere—drained. A unique biological community of freshwater and brackish-water plankton was lost and delicate ecosystems disrupted.

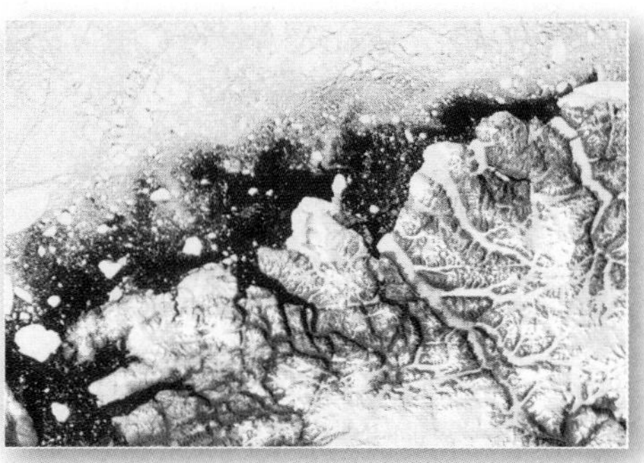

FIGURE GN 17.2 Ice breakup along the Canadian coastline.
Ward Hunt and several other ice shelves broke up off the coast of Ellesmere Island in 2008. [*Aqua* image, NASA/GSFC.]

In the south polar region, similar melt events are under way. Along the Antarctic Peninsula, an area the size of Connecticut broke off the Wilkins Ice Shelf in 2010. Intact for the past 10,000 years, this ice shelf breakup began in 2008. The ongoing ice loss is likely a result of the 2.5 C° (4.5 F°) temperature increase in the peninsula region during the last 50 years. At 71.5° South latitude, this is the farthest south for such a breakup to date (Figure GN 17.3). As in the Arctic, warmth is melting the shelf along both the ocean and the atmosphere surfaces.

To the west of the peninsula, the Pine Island Glacier flows as an outlet glacier from the West Antarctic Ice Sheet (WAIS) to the Amundsen Sea (noted on Figure 17.31c). The flow is accelerating, increasing some 400% in rate between 1995 and 2006. The 40-km- (25-mi-) wide ice shelf is also calving and thinning at higher rates than previously recorded: Over the past 10,000 years, thinning was a few centimeters per year, yet over the past 20 years, the thinning reached over 1 m per year (3.2 ft per year).

These tidewater outlet glaciers and ice shelves hold back flows of grounded ice upstream in the Arctic and Antarctic—and in the WAIS in particular. These grounded masses of ice are not yet displacing ocean water, but they are moving toward the sea. As the ice shelf portions disappear, their buttressing effect is lost. If the warming trends continue as predicted, these ice masses may reach the ocean and add to increasing sea level across the globe.

FIGURE GN 17.3 Wilkins Ice Shelf breakup, Antarctica.
This image is from 2008; active disintegration continued through 2010. [*Terra* MODIS image, NASA/GSFC.]

About 77% of Earth's freshwater is frozen, with the bulk of that ice sitting restlessly in just two places—Greenland and Antarctica. The remaining ice covers various mountains and fills some alpine valleys. A volume of more than 32.7 million km^3 (7.8 million mi^3) of water is tied up as ice: in Greenland (2.38 million km^3), Antarctica (30.1 million km^3), and ice caps and mountain glaciers worldwide (180,000 km^3).

These deposits provide an extensive frozen record of Earth's climatic history over the past several million years and perhaps some clues to its climatic future. This is Earth's *cryosphere*, the portion of the hydrosphere and groundwater that is perennially frozen, generally at high latitudes and elevations. The cryosphere is in a state of dramatic change, as worldwide glacial and polar ice inventories melt. In 2007, Arctic Ocean sea ice decreased to its smallest areal extent in the past century, as regional temperatures set records of more than 5 C° (9 F°) above normal. The surface ice loss in 2008 was second to this record, with 2009 and 2010 tying for third.

In this chapter: We focus on Earth's extensive ice deposits—their formation, their movement, and the ways in which they produce various erosional and depositional landforms. Glaciers, transient landforms themselves, leave in their wake a variety of landscape features. The fate of glaciers is intricately tied to changes in global temperature and rising or falling sea level, which ultimately concerns us all. We discuss the methods used to decipher past climates—the science of *paleoclimatology*—and the clues for understanding future climate patterns. Exciting discoveries are arising from ice cores taken from Earth's two ice sheets, one dating back 800,000 years from the present.

We examine the cold world of permafrost and periglacial geomorphic processes. Approximately 25% of Earth's land area is subject to freezing conditions and frost action characteristic of periglacial regions, including areas that have specific related features of relict, or past, permafrost. The chapter ends with a look at the polar regions.

Rivers of Ice

A **glacier** is a large mass of ice resting on land or floating as an ice shelf in the sea adjacent to land. Glaciers are not frozen lakes or groundwater ice. Instead, they form by the continual accumulation of snow that recrystallizes under its own weight into an ice mass. Glaciers are not stationary; they move slowly under the pressure of their own great weight and the pull of gravity. In fact, they move slowly in streamlike patterns, merging as tributaries into large rivers of ice, as you can see in the satellite image and aerial photos in Figure 17.1. In Greenland and Antarctica, vast sheets of ice dominate, slowly flowing outward toward the ocean.

Today, these slowly flowing rivers and sheets of ice dominate about 11% of Earth's land area. As much as 30% of continental land was covered by glacial ice during colder episodes in the past. Through these "ice ages," below-freezing temperatures prevailed at lower latitudes more than they do today, allowing snow to accumulate year after year.

(a)

(b)

(c)

FIGURE 17.1 Rivers and sheets of ice.
(a) Alpine glaciers merge from adjoining glacial valleys in the northeast region of Ellesmere Island in the Canadian Arctic. (b) Numerous glaciers flow toward the coast of Greenland; note the morainal deposit marking the flow. (c) These peaks, known as *nunataks,* rise above the Greenland Ice Sheet, an accumulation perhaps 100,000 years in the making; note the ice stream flowing. [(a) NASA/MODIS Terra. (b) and (c) Bobbé Christopherson.]

Glaciers form in areas of permanent snow, both at high latitudes and at high elevations at any latitude. A **snowline** is the lowest elevation where snow can survive year-round; specifically, it is the lowest line where winter snow

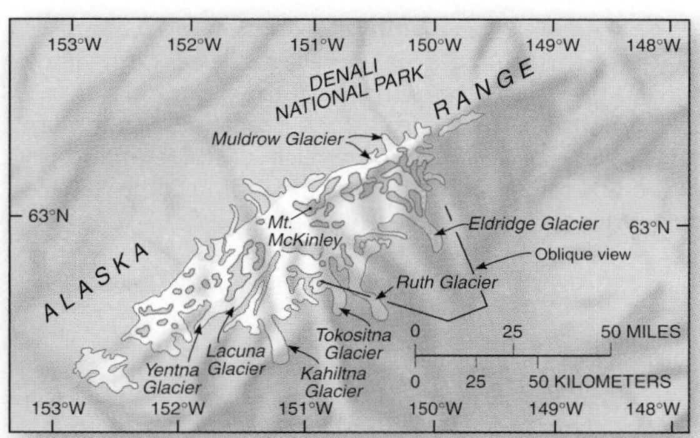

FIGURE 17.2 Glaciers in south-central Alaska. Oblique infrared (false-color) image of Eldridge and Ruth Glaciers, with Mount McKinley at upper left, in the Alaska Range of Denali National Park. Photo made at 18,300 m (60,000 ft). [Data available from U.S. Geological Survey, EROS Data Center, Sioux Falls, SD.]

accumulation persists throughout the summer. Glaciers form on some high mountains along the equator, such as in the Andes Mountains of South America and on Mount Kilimanjaro in Tanzania, Africa. On equatorial mountains, the snowline is around 5000 m (16,400 ft); on midlatitude mountains, such as the European Alps, snowlines average 2700 m (8850 ft); and in southern Greenland, snowlines are as low as 600 m (1970 ft). For Internet links to Global Land Ice Measurements from Space, go to http://www.glims.org/, which includes an inventory of world glaciers, or the National Snow and Ice Data Center at http://nsidc.org/. *Landsat*-7 images of the cryosphere are listed at http://www.emporia.edu/earthsci/gage/glacier7.htm.

Glaciers are as varied as the landscape itself. They fall within two general groups, based on their form, size, and flow characteristics: alpine glaciers and continental glaciers.

Alpine Glaciers

With few exceptions, a glacier in a mountain range is an **alpine glacier**, or *mountain glacier*. The name comes from the Alps of central Europe, where such glaciers abound. Alpine glaciers form in several subtypes. One prominent type is a *valley glacier*, literally a river of ice confined within a valley that originally was formed by stream action. Such glaciers range in length from only 100 m (325 ft) to more than 100 km (60 mi). In Figure 17.2, at least a half dozen valley glaciers are identifiable in the high-altitude photograph of the Alaska Range. Several are named on the map—specifically, the Eldridge and Ruth glaciers, which fill valleys as they flow from source areas near Mount McKinley.

As a valley glacier flows slowly downhill, the mountains, canyons, and river valleys beneath its mass are profoundly altered by its erosive passage. Some of the debris created by the glacier's excavation is transported on the ice, visible as dark streaks and bands being transported for deposition elsewhere; other portions of its debris load are carried within or along its base (see Figure 17.1b).

Most alpine glaciers originate in a mountain *snowfield* that is confined in a bowl-shaped recess. This scooped-out erosional landform at the head of a valley is a **cirque**. A glacier that forms in a cirque is a *cirque glacier*. Several cirque glaciers may jointly feed a valley glacier, as shown in the photo in Figure 17.1. How many valley glaciers join the main glacier in Figure 17.1a and b?

Wherever several valley glaciers pour out of their confining valleys and coalesce at the base of a mountain range, a *piedmont glacier* is formed and spreads freely over the lowlands, such as the Malaspina Glacier that flows into Yakutat Bay, Alaska. A *tidewater glacier*, or *tidal glacier*, ends in the sea, calving to form floating pieces of ice known as **icebergs** (Figure 17.3a). Icebergs usually form

GEO REPORT 17.1 *Global glacial ice losses*

Worldwide, glacial ice is in retreat, melting at rates exceeding anything in the record. In the European Alps alone, some 75% of the glaciers have receded in the past 50 years, losing more than 50% of their ice mass since 1850. At this rate, the European Alps will have only 20% of their preindustrial glacial ice left by 2050. In Alaska, only 3 glaciers of 67 surveyed gained mass since 1985; annual losses have amounted to some 52 km^2 (12.5 mi^2). Similar forecasts of reduced ice and snowpack are in place for the Rockies, Sierra Nevada, Himalayas, and Andes.

(a)

(b)

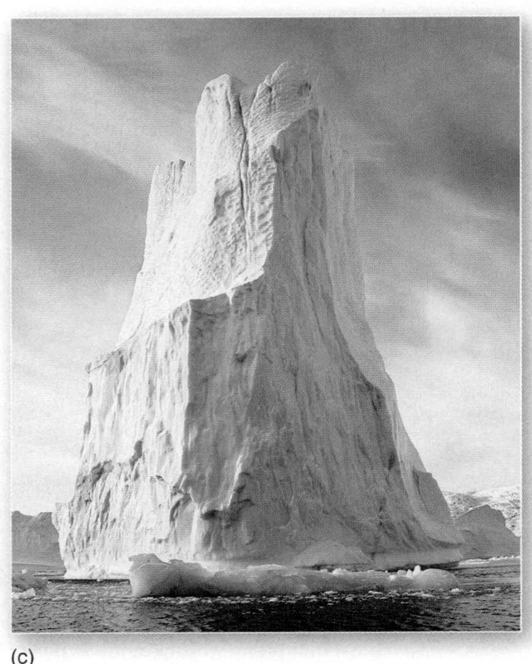
(c)

FIGURE 17.3 Glacial ice calving into the sea and icebergs. (a) When glaciers reach the sea, large pieces break off, forming icebergs, such as these massive bergs in southern Greenland. (b) A tabular iceberg off Antarctica, in the Bransfield Strait near the Antarctic Sound. (c) This iceberg tower in East Greenland is about 60 m (200 ft) tall; imagine its keel extending some 360 m (1180 ft) below the surface. [Bobbé Christopherson.]

wherever glaciers meet an ocean, bay, or fjord. Icebergs are inherently unstable, as their center of gravity shifts with melting and further breaking. Figure 17.3b shows a tabular iceberg that broke away from ice shelves near the Weddell Sea, Antarctica. The annual layers of ice are clearly visible in the tabular berg. In Figure 17.3c, a magnificent ice tower berg brings to mind the ratio of 1/7 (14%) exposed and 6/7 (86%) submerged—these proportions varying depending on the age of the ice and air content.

Continental Glaciers

On a much larger scale than individual alpine glaciers, a continuous mass of ice is a **continental glacier**. In its most extensive form, it is an **ice sheet**. Most of Earth's glacial ice exists in the ice sheets that blanket 81% of Greenland—1,756,000 km^2 (678,000 mi^2) of ice—and 90% of Antarctica—14.2 million km^2 (5.48 million mi^2) of ice. Antarctica alone has 92% of all the glacial ice on the planet. A composite of satellite imagery reveals the Antarctic continent in Figure 17.4.

The Antarctic and Greenland ice sheets have such enormous mass that large portions of each landmass beneath the ice are isostatically depressed (pressed down by weight) below sea level. Each ice sheet reaches depths of

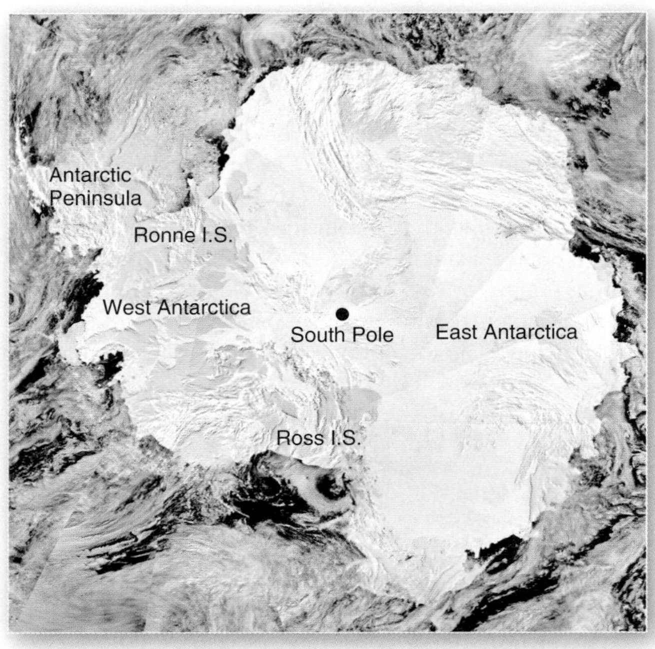

FIGURE 17.4 Antarctic Ice Sheet. A composite of *Aqua* MODIS images from January 2009 portrays the Antarctic continent—containing 92% of Earth's glacial ice (I.S. = Ice Shelf). Reference the map of Antarctica in Figure 17.31c. [*Aqua* MODIS image NASA/GSFC.]

(a)

(b)

FIGURE 17.5 Ice cap and ice field.
(a) The Vatnajökull Ice Cap in southeastern Iceland (*jökull* means "ice cap" in Danish; outlined by the red line). This image was made after the Eyjafjallajökull Volcano began erupting in April 2010; see the ash cloud flowing southward toward European airspace. (b) The southern Patagonian ice field of Argentina.
[(a) NASA/GSFC. (b) NASA Headquarters.]

more than 3000 m (9800 ft), with average depths around 2000 m (6500 ft), burying all but the highest peaks.

Two additional types of continuous ice cover associated with mountain locations are *ice caps* and *ice fields*. An **ice cap** is roughly circular and, by definition, covers an area of less than 50,000 km^2 (19,300 mi^2). An ice cap completely buries the underlying landscape. The volcanic island of Iceland features several ice caps, such as the Vatnajökull Ice Cap in southeastern Iceland (outlined in Figure 17.5a). Volcanoes lie beneath these icy surfaces. Iceland's Grímsvötn Volcano erupted in 1996 and again in 2004, producing large quantities of melted glacial water and floods, a flow Icelanders call a *jökulhlaup*. To the southwest is the Eyjafjallajökull Volcano, which began erupting in 2010.

An **ice field** is not extensive enough to form the characteristic dome of an ice cap; instead, it extends in a characteristic elongated pattern in a mountainous region. A fine example is the Patagonian ice field of Argentina and Chile, one of Earth's largest. It is only 90 km (56 mi) width, but stretches 360 km (224 mi), from 46° to 51° S latitude.

Continuous ice sheets or ice caps are drained by rapidly moving, solid *ice streams* that form around their periphery, moving to the sea or to lowlands. Such frozen ice streams flow from the edges of Greenland and Antarctica through stationary and slower-moving ice. An *outlet glacier* flows out from an ice sheet or ice cap, but is constrained by a mountain valley or pass. Can you see the ice stream flowing past the nunataks in Figure 17.1c?

Glacial Processes

A glacier is a dynamic body, moving relentlessly downslope at rates that vary within its mass, excavating the landscape through which it flows. The mass is dense ice that is formed from snow and water through a process of compaction, recrystallization, and growth. A glacier's *mass budget* consists of net gains or losses of this glacial ice, which determine whether the glacier expands or retreats. Let us now look at glacial ice formation, mass balance, movement, and erosion before we discuss the fascinating landforms produced by these processes.

Formation of Glacial Ice

Consider for a moment the nature of ice. You may be surprised to learn that ice is both a mineral (an inorganic natural compound of specific chemical makeup and crystalline structure) and a rock (a mass of one or more minerals). Ice is a frozen fluid, a trait that it shares with igneous rocks. The accumulation of snow in layered deposits is similar to sedimentary rock formations. To give birth to a glacier, snow and ice are transformed under pressure, recrystallizing into a type of metamorphic rock. Glacial ice is a remarkable material!

The essential input to a glacier is snow that accumulates in a snowfield, a glacier's accumulation zone (Figure 17.6a and c). Snowfields typically are at the highest elevation of an ice sheet, ice cap, or head of a valley glacier, usually in a cirque. Avalanches from surrounding mountain slopes can add to the snowfield.

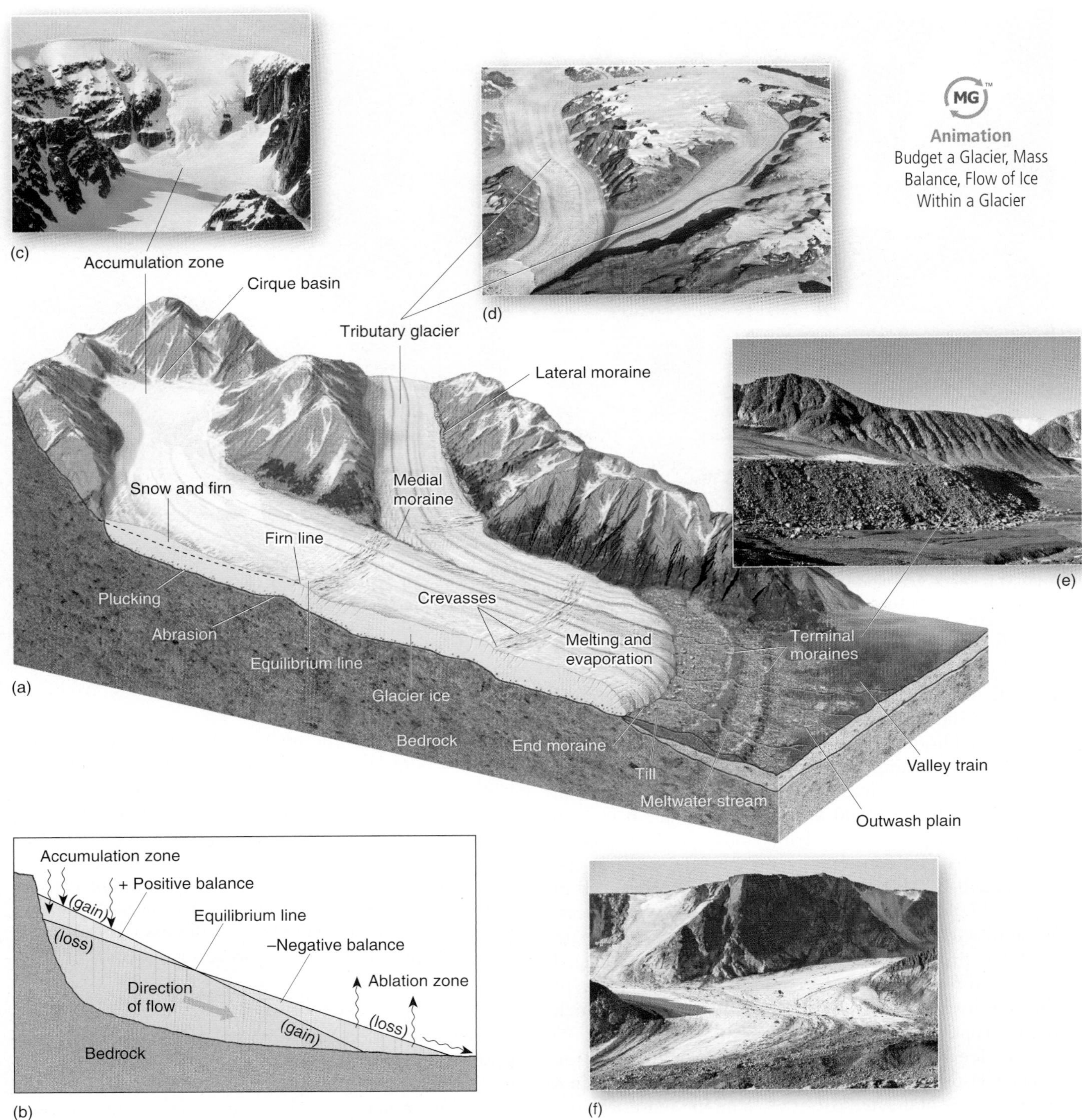

MG Animation
Budget a Glacier, Mass Balance, Flow of Ice Within a Glacier

FIGURE 17.6 A retreating alpine glacier and mass balance.
(a) Cross section of a typical retreating alpine glacier. (b) Annual mass balance of a glacial system, showing how the relation between accumulation and ablation controls the location of the equilibrium line. (c) The accumulation zone for a glacier in Antarctica. (d) At least four tributary glaciers flow into a compound valley glacier in Greenland. (e) A terminal moraine marks the farthest advance of this glacier on Nordaustlandet Island, Arctic Ocean. (f) Morainal rock in lateral moraines merges to form a medial moraine, leading to a terminal moraine left by the retreating glacier. These merging glaciers along Beverlysundet, Nordaustlandet Island, haul erratic boulders on their surfaces that have cracked off the cliffs through frost action. [Bobbé Christopherson.]

As the snow accumulation deepens in sedimentary-like layers, the increasing thickness results in increased weight and pressure on underlying ice. Rain and summer snowmelt then contribute water, which stimulates further melting, and that meltwater seeps down into the snowfield and refreezes. Snow surviving the summer and into the following winter begins a slow transformation into glacial ice. Air spaces among ice crystals are pressed as snow packs to a greater density. The ice recrystallizes and consolidates under pressure. In a transition step to glacial ice, snow becomes **firn**, which has a compact, granular texture.

As this process continues, many years pass before dense glacial ice is produced. Formation of **glacial ice** is analogous to metamorphic processes: Sediments (snow and firn) are pressured and recrystallized into a dense metamorphic rock (glacial ice). In Antarctica, glacial ice formation may take 1000 years because of the dryness of the climate (minimal snow input), whereas in wet climates, the time is reduced to several years because of rapid, constant snow input to the system.

Glacial Mass Balance

A glacier is an open system, with *inputs* of snow and *outputs* of ice, meltwater, and water vapor. Snowfall and other moisture in the *accumulation zone* feed the glacier's upper reaches (Figure 17.6b). This area ends at the **firn line**, indicating where the winter snow and ice accumulation survived the summer melting season. Toward a glacier's lower end, it is wasted (reduced) through several processes: melting on the surface, internally, and at its base; ice removal by deflation (wind); the calving of ice blocks; and sublimation (recall from Chapter 7 that this is the direct evaporation of ice). Collectively, these losses are **ablation**.

The zone where accumulation gain balances ablation loss is the *equilibrium line* (Figure 17.6a and b). This area of a glacier generally coincides with the firn line. A glacier achieves *positive net balance* of mass—grows larger—during cold periods with adequate precipitation. In warmer times, the equilibrium line migrates up-glacier, and the glacier retreats—grows smaller—because of its *negative net balance*. Internally, gravity continues to move a glacier forward even though its lower terminus might be in retreat owing to ablation.

As part of a global trend, the net mass balance of the South Cascade Glacier in Washington State demonstrated significant losses between 1955 and 2009. As a result of this negative mass balance, sometimes the terminus retreated tens of meters a year. The glacier has retreated every year in the record except 1972. Figure 17.7 is a photo comparison of the South Cascade Glacier between 1979 and 2003.

A comparison of the trend of this glacier's mass balance with that of others in the world shows that temperature changes apparently are causing widespread reductions in middle- and lower-elevation glacial ice. The present wastage (ice loss) from alpine glaciers worldwide is thought to contribute over 25% to the rise in sea level.

Animation
Budget of a Glacier, Mass Balance

FIGURE 17.7 Photo comparison of 24 years at South Cascade.
Photos visually remind us of the tremendous quantities of ice that have ablated from South Cascade Glacier. Since 2003, the glacier has continued its negative net mass balance losses. [U.S. Geological Survey, Denver.]

Glacial Movement

We generally think of ice as those small, brittle cubes from the freezer, but glacial ice is quite different. In fact, glacial ice behaves in a plastic (pliable) manner, for it distorts and flows in its lower portions in response to weight and pressure from above and the degree of slope below. In contrast, the glacier's upper portions are quite brittle. Rates of flow range from almost nothing to a kilometer or two per year on a steep slope. The rate of accumulation of snow in the formation area is critical to the speed.

Glaciers are not rigid blocks that simply slide downhill. The greatest movement within a valley glacier occurs *internally*, below the rigid surface layer, which fractures as the underlying zone moves forward (Figure 17.8a). At the

GEO REPORT 17.2 *Alaskan ice losses related to isostatic rebound*

Researchers from the Geophysical Institute at the University of Alaska–Fairbanks, using an array of Global Positioning System (GPS) receivers in southeast Alaska, measured isostatic rebound of the crust following modern glacial ice losses. They detected some of the most rapid vertical motion on Earth, averaging about 36 mm (1.42 in.) per year. Scientists attribute this isostatic rebound to the huge weight reduction caused by the loss of glacial ice in the region, especially from Yakutat Bay and the Saint Elias Mountains in the north, through Glacier Bay, Juneau, and on south through the Inland Passage.

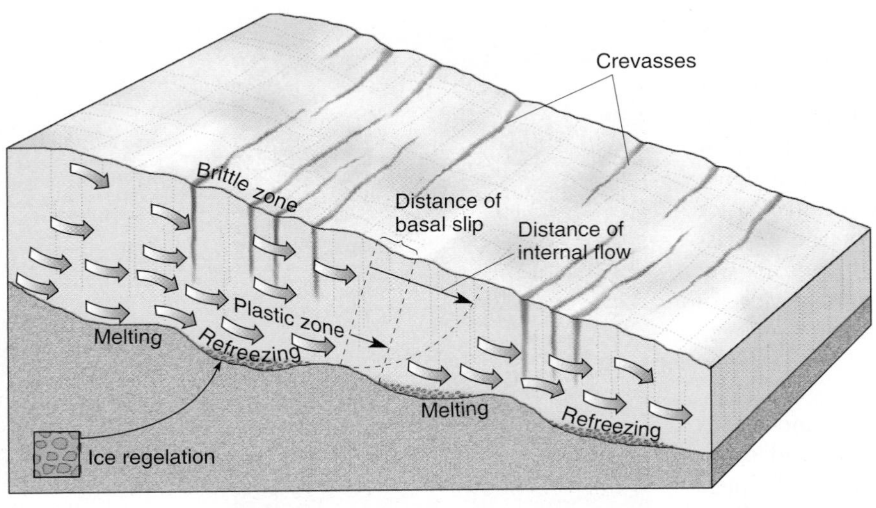

(b) (c)

FIGURE 17.8 Glacial movement.
(a) Cross section of a glacier, showing its forward motion, brittle cracking at the surface, and flow along its basal layer. (b) Surface crevasses and cracks are evidence of a glacier's forward motion on Elephant Island, Antarctica. (c) Crevasses mark the jumbled surface of this glacier in Spitsbergen; for scale, note the two Zodiac rafts in the water. [Bobbé Christopherson.]

same time, the base creeps and slides along, varying its speed with temperature and the presence of any lubricating water beneath the ice. This *basal slip* usually is much less rapid than the internal plastic flow of the glacier, so the upper portion of the glacier flows ahead of the lower portion.

Unevenness in the landscape beneath the ice may cause the pressure to vary, melting some of the basal ice by compression at one moment, only to have it refreeze later. This process is *ice regelation*, meaning to refreeze, or re-gel. Such melting/refreezing action incorporates rock debris into the glacier. Consequently, the basal ice layer, which can extend tens of meters above the base of the glacier, has a much higher debris content than the ice above.

A flowing alpine glacier or ice stream in a continental glacier can develop vertical cracks known as **crevasses** (Figure 17.8b and c). Crevasses result from friction with valley walls, or from tension due to stretching as the glacier passes over convex slopes, or from compression as the glacier passes over concave slopes. Traversing a glacier, whether an alpine glacier or an ice sheet, is dangerous because a thin veneer of snow sometimes masks the presence of a crevasse.

Glacier Surges Although glaciers flow plastically and predictably most of the time, some will lurch forward with little or no warning in a **glacier surge**. A surge is not quite as abrupt as it sounds; in glacial terms, a surge can be tens of meters per day. The Jakobshavn Glacier on the western Greenland coast, for example, is one of the fastest moving at between 7 and 12 km (4.3 and 7.5 mi) a year.

The doubling of ice-mass loss from Greenland between 1996 and 2005 means that glaciers are surging overall. In southern Greenland in 2007, outlet glaciers were averaging a speed of 22.8 m (75 ft) per year, a significant increase from the 1999 average flow rate of 1.8 m (6 ft) per year. Scientists determined that Greenland experienced more days of melting snow at higher elevations than average in 2006–2007, thereby contributing to the glacial surges. The meltwater works its way to the basal layer, lubricating underlying *soft beds* of clay. In addition, the warmer surface waters draining beneath the glacier deliver heat that increases basal melt rates.

The exact cause of such a glacier surge is being studied. Some surge events result from a buildup of water pressure under the glacier—sometimes enough to actually float the glacier slightly, detaching it from its bed during the surge. As a surge begins, ice quakes are detectable, and ice faults are visible. Surges can occur in dry conditions as well, as the glacier plucks (picks up) rock from its bed and moves forward. Another cause of glacier surges is the presence of a water-saturated layer of sediment, a *soft bed*, beneath the glacier, as discussed for Greenland. This is a deformable layer that cannot resist the tremendous sheer stress produced by the moving ice of the glacier. Scientists examining cores taken from several ice streams now accelerating through the West Antarctic Ice Sheet think they have identified this cause of glacial surges—although water pressure is still important.

FIGURE 17.9 **Glacial sandpapering polishes rock.** Glacial polish and striations are examples of glacial abrasion and erosion. The polished, marked surface is seen beneath a glacial erratic—a rock left behind by a retreating glacier. [Bobbé Christopherson.]

Glacial Erosion The way in which a glacier erodes the land is similar to a large excavation project, with the glacier hauling debris from one site to another for deposition. The passing glacier mechanically plucks rock material and carries it away. Debris is carried on its surface and is also transported internally, or *englacially*, embedded within the glacier itself. Evidence indicates that rock pieces actually freeze to the basal layers of the glacier in a *glacial plucking*, or a picking-up process, and once embedded, enable the glacier to scour and sandpaper the landscape as it moves—a process of **abrasion**. This abrasion and gouging produce a smooth surface on exposed rock, which shines with *glacial polish* when the glacier retreats. Larger rocks in the glacier act much like chisels, gouging the underlying surface and producing glacial striations parallel to the flow direction (Figure 17.9).

Glacial Landforms

Glacial erosion and deposition produce distinctive landforms that differ greatly from the way the land looked before the ice came and went. You might expect all glaciers to create the same landforms, but alpine and continental glaciers each generate their own characteristic landscapes. We look first at erosional landforms created by alpine glaciers and then at their depositional landforms. Finally, we examine the landscape that results from continental glaciation.

Erosional Landforms Created by Alpine Glaciation

Alpine glaciers create spectacular, dramatic landforms that bring to mind the Canadian Rockies, the Swiss Alps, or vaulted Himalayan peaks. Geomorphologist William Morris Davis depicted the stages of a valley glacier in drawings published in 1906 and redrawn here in Figures 17.10 and 17.11. Study of these figures reveals the handiwork of ice as sculptor:

- In Figure 17.10a, you see a typical stream-cut valley as it existed before glaciation. Note the valley's prominent **V** shape.
- In Figure 17.10b, you see the same landscape during subsequent glaciation. Glacial erosion and transport actively remove much of the regolith (weathered bedrock) and the soils that covered the stream-valley landscape. As the *cirque* walls erode away, sharp ridges form, dividing adjacent cirque basins. These **arêtes** ("knife-edge" in French) become the

GEO REPORT 17.3 ***Greenland Ice Sheet losses increasing***

The Greenland Ice Sheet is experiencing a greater amount of ice loss and a greater area of surface melting than at any time since systematic satellite monitoring started. Changes in glacial outlet velocity to the ocean are contributing to these losses; rates of flow are more than double previous estimates. In 1996, Greenland lost an estimated 91 km^3 (22 mi^3) of ice mass annually, which by 2007 had risen to 251 km^3 (60 mi^3) per year—this is 230 billion tonnes (253 billion tons) of ice per year and rising, compared to the West Antarctic Ice Sheet losses estimated at 132 billion tonnes (145 billion tons) a year, and are also increasing.

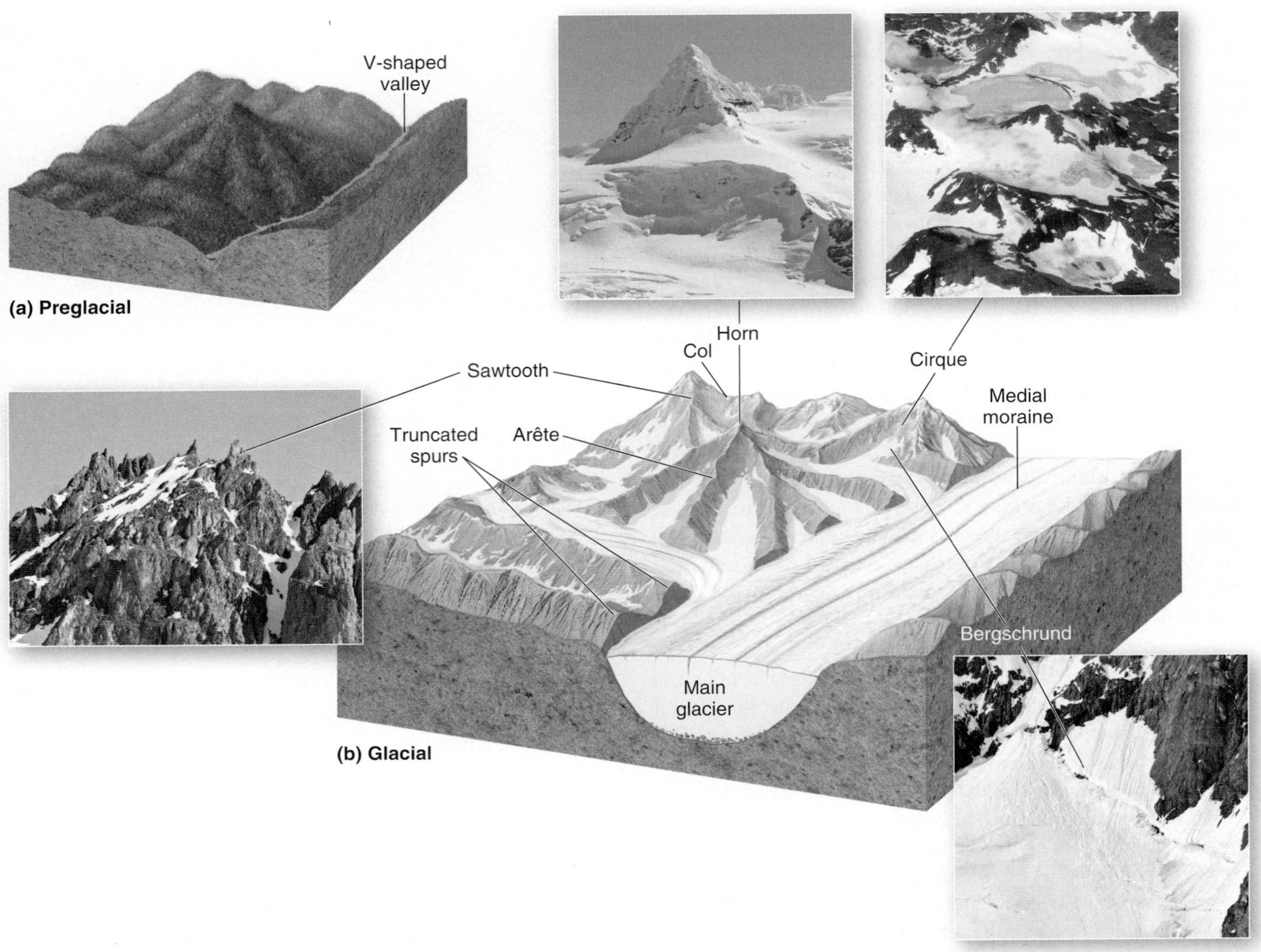

FIGURE 17.10 Alpine glaciers occupy valleys.
(a) A preglacial landscape with V-shaped stream-cut valley. (b) The same landscape filled with valley glaciers. Note inset photos of a *horn, cirque basin, sawtooth ridge,* and *bergschrund* from Antarctica, Greenland, and Spitsbergen. [Bobbé Christopherson.]

sawtooth, serrated ridges in glaciated mountains. Two eroding cirques may reduce an arête to a saddlelike depression or pass, forming a **col**. A **horn**, or pyramidal peak, results when several cirque glaciers gouge an individual mountain summit from all sides. Most famous is the Matterhorn in the Swiss Alps, but many others occur worldwide. A **bergschrund** forms when a crevasse or wide crack opens along the headwall of a glacier, most visible in summer when covering snow is gone.

- In Figure 17.11, you see the same landscape at a time of warmer climate when the ice retreated. The glaciated valleys now are **U**-shaped, greatly changed from their previous stream-cut **V** form. You can see the steep sides and the straightened course of the valleys. Physical weathering from the freeze–thaw cycle has loosened rock along the steep cliffs, where it has fallen to form *talus slopes* along the valley sides. Retreating ice leaves behind transported rocks as *erratics*.

Note in Figure 17.11 that in the cirques where the valley glaciers originated, small mountain lakes, **tarns**, have formed. One cirque contains small, circular, stair-stepped lakes, called **paternoster** ("our father") **lakes** for their resemblance to rosary (religious) beads. Paternoster lakes may have formed from the differing resistance of rock to glacial processes or from damming by glacial deposits.

The valleys carved by tributary glaciers are left stranded high above the valley floor, because the primary glacier eroded the valley floor so deeply. These *hanging valleys* are the sites of spectacular waterfalls. See how many of the erosional forms from Figures 17.10 and

FIGURE 17.11 The geomorphic handiwork of alpine glaciers.
As the glaciers retreat, the new landscape is unveiled. Note inset photos of *glacial erratics,* a *hanging valley* and waterfall, a *tarn,* and characteristic U-shaped glacial valleys as U-shaped troughs in surface and aerial views from Norway. [Bobbé Christopherson; waterfall by author.]

17.11 (arête, col, horn, cirque, cirque glacier, sawtooth ridge, **U**-shaped valley, erratic, tarn, and truncated spur, among others) you can identify in Figure 17.12.

Where a glacial trough intersects the ocean, the glacier can continue to erode the landscape, even below sea level. As the glacier retreats, the trough floods and forms a deep **fjord** in which the sea extends inland, filling the lower reaches of the steep-sided valley (Figure 17.13). The fjord may be flooded further by rising sea level or by changes in the elevation of the coastal region. All along the glaciated coast of Alaska, glaciers are now in retreat, thus opening many new fjords that previously were blocked by ice. Coastlines with notable fjords include those of Norway, Greenland, Chile, the South Island of New Zealand, Alaska, and British Columbia.

CRITICAL THINKING 17.1

Looking for glacial features on closer inspection

The text asks you to look at Figure 17.12 to identify glacial features. Now examine the photos in the chapter opener and Figures 17.1, 17.2, and 17.3, and reexamine Figure 17.12. List all the glacial formations that you can identify in these photos. Are there any erosional landforms that you find on all the photos other than the glaciers themselves?

Depositional Landforms Created by Alpine Glaciation

You have just seen how glaciers excavate tremendous amounts of material and create fascinating landforms in the process. Glaciers produce a different set of distinctive

FIGURE 17.12 **Erosional features of Alpine glaciation.**
How many of the erosional glacial features illustrated in Figures 17.10 and 17.11 can you find in this photo of southern Greenland? See Critical Thinking 17.1. [Bobbé Christopherson.]

landforms when they melt and deposit their debris cargo at the glacier's *terminus*. As shown in Figure 17.6a, e, and f, after the glacier melts, debris accumulates in moraines to mark the former margins of the glacier, both its end and its sides.

Glacial Drift The general term for all glacial deposits, both unsorted and sorted, is **glacial drift**. **Till** describes direct ice deposits left as unstratified and unsorted debris. Sediments deposited by glacial meltwater are sorted by size and are **stratified drift**.

As a glacier flows to a lower elevation, a wide assortment of rock fragments becomes *entrained* (carried along) on its surface or embedded within its mass or in its base. As the glacier melts, this unsorted cargo is deposited on the ground surface. Such till is poorly sorted and is difficult to cultivate for farming, but the clays and finer particles can provide a basis for soil development.

Retreating glaciers leave behind large rocks (sometimes house-sized), boulders, and cobbles that are "foreign" in composition and origin from the ground on which they were deposited. These *glacial erratics*, lying in strange locations with no obvious means of transport, were an early clue that blankets of ice once had covered the land (pictured in Figure 17.11).

Moraines The deposition of glacial sediment produces a specific landform, a **moraine**. Several types of moraines are associated with alpine glaciation. A **lateral moraine** forms along each side of a glacier (Figure 17.14a). If two glaciers with lateral moraines join, a **medial moraine** may form (see Figures 17.1 and 17.6e and f). Other morianes

(a)

(b)

FIGURE 17.13 **Norwegian fjords.**
(a) The beautiful fjord where an inlet of the sea fills a **U**-shaped, glacially carved valley. (b) Sediment carried in runoff is visible in this fjord. [Bobbé Christopherson.]

(a)

(b)

FIGURE 17.14 Depositional features of alpine glaciation.
(a) Retreating ice leaves behind a substantial lateral moraine; the retreating Gaffelbreen (Gaffel Glacier) is seen in the background. (b) A terminal moraine forms this island, separated from its ice-cap glacier by more than a kilometer of ocean, at Isispynten Island, Arctic Ocean. [Bobbé Christopherson.]

are associated with both alpine and continental glaciation. Eroded debris that is dropped at the glacier's farthest extent is a **terminal moraine**. *End moraines* may also be present, formed at other points where a glacier paused after reaching a new equilibrium between accumulation and ablation. In landscapes of continental glaciation, morianes are much larger in scale. A deposition of till that is generally spread across a surface is a *ground moraine*, or *till plain*, and may hide the former landscape. Such plains are found in portions of the U.S. Midwest.

Lakes may form behind terminal and end moraines after a glacier's retreat, with the moraine acting as a dam. The large terminal moraine deposit in Figure 17.14b forms an island, when just 9 years earlier it was a peninsula at the edge of a small ice cap visible in the distance. The ice cap rapidly retreated more than a kilometer in this time frame.

Tributary valley glaciers merge to form a *compound valley glacier*. The flowing movement of a compound valley glacier is different from that of a river with tributaries. Tributary glaciers flow into a compound glacier and merge alongside one another by extending and thinning rather than by blending, as do rivers. Each tributary maintains its own patterns of transported debris (dark streaks, visible in the photo in Figure 17.6d, e, and f).

All of these till types are unsorted and unstratified. In contrast, streams of glacial meltwater can carry and deposit sorted and stratified glacial drift beyond a terminal moraine. Meltwater-deposited material down-valley from a glacier is a *valley train deposit*. Peyto Glacier in Alberta, Canada, produces such a valley train that continues into Peyto Lake (Figure 17.15). Distributary stream channels appear braided across its surface. The photo also shows the milky meltwater associated with glaciers, laden with finely ground "rock flour." Meltwater comes from glaciers at all times, not just when they are retreating. The Peyto Glacier (far left in photo) has experienced massive ice losses since 1966 and is in retreat due to increased ablation and decreased accumulation related to climate change.

Erosional and Depositional Features of Continental Glaciation

The extent of the most recent continental glaciation in North America and Europe, 18,000 years ago, is portrayed several pages ahead in Figure 17.26. Because continental glaciers form under different circumstances—not in

FIGURE 17.15 A valley train deposit.
Peyto Glacier in Alberta, Canada, is in extensive retreat. Note the valley train, braided stream, and milky-colored glacial meltwater. [Author.]

mountains, but across broad, open landscapes—the intricately carved alpine features, lateral moraines, and medial moraines all are lacking in continental glaciation.

Figure 17.16 illustrates some of the most common erosional and depositional features associated with the retreat of a continental glacier. A **till plain** forms behind an end moraine; it features *unstratified* coarse till, has low and rolling relief, and has a deranged drainage pattern (Figure 17.16b; see also Figure 14.9). Beyond the morainal deposits lies the **outwash plain** of *stratified drift* featuring stream channels that are meltwater-fed, braided, and overloaded with sorted and deposited materials.

A sinuously curving, narrow ridge of coarse sand and gravel is an **esker**. It forms along the channel of a meltwater stream that flows beneath a glacier, in an ice tunnel, or between ice walls. As a glacier retreats, the steep-sided esker is left behind in a pattern roughly parallel to the path of the glacier. The ridge may not be continuous and in places may even appear to be branched, following the path set by the subglacial watercourse. Commercially valuable deposits of sand and gravel are quarried from some eskers.

Sometimes an isolated block of ice, perhaps more than a kilometer across, remains in a ground moraine, on an outwash plain, or on a valley floor after a glacier has retreated. As much as 20 to 30 years is required for it to melt. In the interim, material continues to accumulate around the melting ice block. When the block finally melts, it leaves behind a steep-sided hole. Such a feature then frequently fills with water. This feature is a **kettle**. Thoreau's famous Walden Pond in Massachusetts is such a glacial kettle.

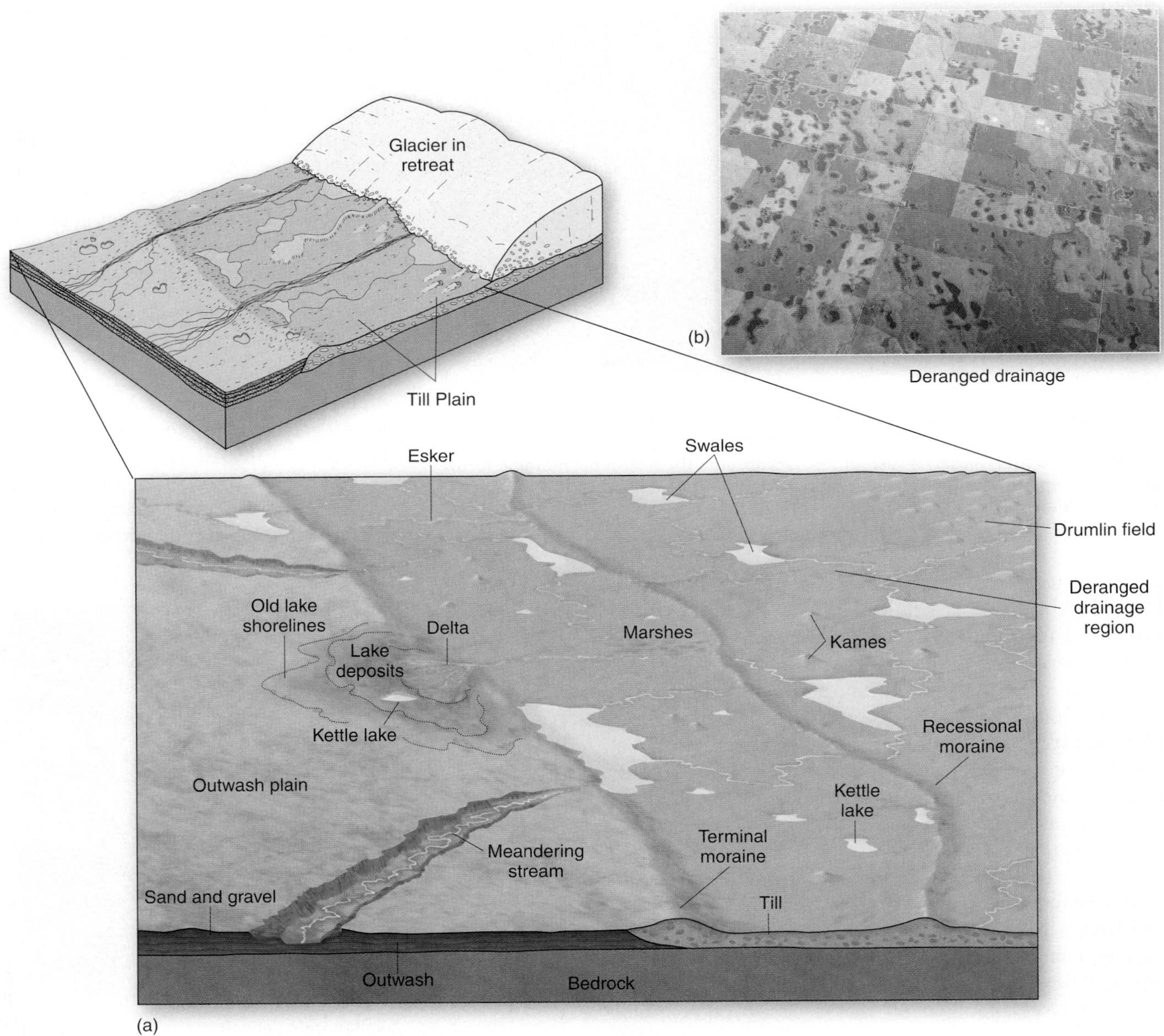

FIGURE 17.16 Continental glacier depositional features.
(a) Common depositional landforms produced by glaciers; note the esker, drumlin field, recessional moraine, kames, kettle lake, till, and outwash plain of stratified drift. (b) Deranged drainage tells us continental glaciers advanced and retreated across the prairies of central Saskatchewan, Canada. [Bobbé Christopherson.]

(a)

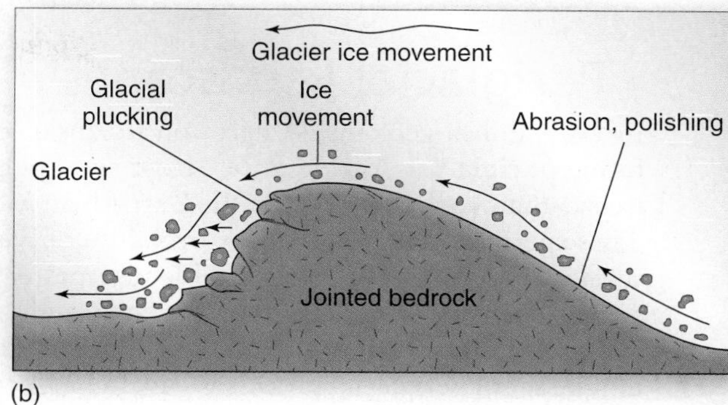

(b)

FIGURE 17.17 Glacially eroded streamlined rock. Roche moutonnée, as exemplified by Lembert Dome in the Tuolumne Meadows area of Yosemite National Park, California. [Author.]

Another feature of outwash plains is a **kame**, a small hill, knob, or mound of poorly sorted sand and gravel that is deposited directly by water, by ice in crevasses, or in ice-caused indentations in the surface. Kames also can be found in deltaic forms and in terraces along valley walls.

Glacial action also forms two types of streamlined hills. One is erosional, a *roche moutonnée*, and the other is depositional, a *drumlin*. A **roche moutonnée** ("sheep rock" in French) is an asymmetrical hill of exposed bedrock. Its gently sloping upstream side (stoss side) has been polished smooth by glacial action, whereas its downstream side (lee side) is abrupt and steep where the glacier plucked rock pieces (Figure 17.17).

A **drumlin** is deposited till that was streamlined in the direction of continental ice movement, blunt end upstream and tapered end downstream. Multiple drumlins (*drumlin swarms*) occur across the landscape in portions of New York and Wisconsin, among other areas. Sometimes their shape is that of an elongated teaspoon bowl, lying face down. Drumlins may attain lengths of 100–5000 m (330 ft to 3.1 mi) and heights up to 200 m (650 ft). Figure 17.18 shows a portion of a topographic map for the area south of

(a)

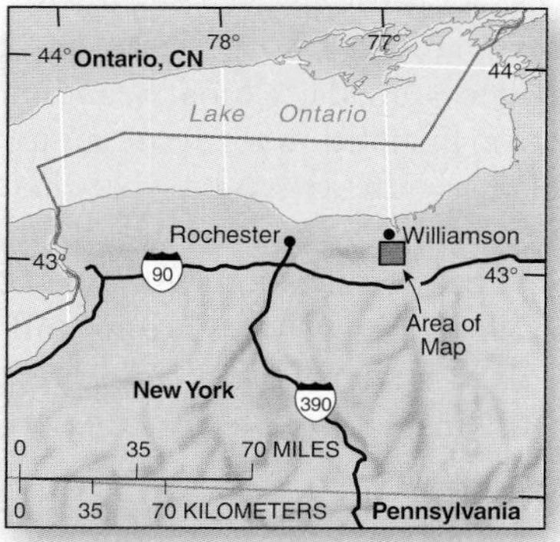

(b)

FIGURE 17.18 Glacially deposited streamlined features. (a) Topographic map south of Williamson, near Marion in New York, featuring numerous drumlins. (7.5-minute series quadrangle map, originally produced at a 1:24,000 scale, 10-ft contour interval.) (b) Aerial view of drumlin swarm, blunt end upstream, tapered downstream, generally darker with vegetation. [(a) USGS map; (b) Bobbé Christopherson.]

Williamson, New York, which experienced continental glaciation. In studying the map, can you identify the numerous drumlins? In what direction do you think the continental glaciers moved across this region?

Periglacial Landscapes

In 1909, Polish geologist Walery von Lozinski coined the term **periglacial** to describe frost weathering and freeze–thaw rock shattering in the Carpathian Mountains. These periglacial regions occupy over 20% of Earth's land surface (Figure 17.19). Periglacial landscapes either have a near-permanent ice cover or are at a high elevation and are seasonally snow-free. Under these conditions, a unique set of periglacial processes—including permafrost, frost action, and ground ice—operates.

Climatologically, these regions are in *subarctic* and *polar* climates, especially *tundra* climate. Such climates occur either at high latitude (tundra and boreal forest environments) or at high elevation in lower-latitude mountains (alpine environments). Processes that are related to physical weathering, mass movement (Chapter 13), climate (Chapter 10), and soil (Chapter 18) dominate periglacial regions.

Geography of Permafrost

When soil or rock temperatures remain below 0°C (32°F) for at least 2 years, **permafrost** ("permanent frost") develops. An area of permafrost that is not covered by glaciers is considered periglacial, with the largest extent of such lands in Russia. Approximately 80% of Alaska has permafrost beneath its surface. Canada, China, Scandinavia, Greenland, and Antarctica, in addition to alpine mountain regions of the world, also are affected. Note that this criterion is based solely on *temperature* and has nothing to do with how much or how little water is present. Two other factors also contribute to permafrost conditions and occurrence: the presence of fossil permafrost from previous ice-age conditions and the insulating effect of snow cover or vegetation that inhibits heat loss.

Continuous and Discontinuous Zones Permafrost regions are divided into two general categories, continuous and discontinuous, which merge along a general transition zone. *Continuous permafrost* is the region of severest cold and is perennial, roughly poleward of the −7°C (19°F) mean annual temperature isotherm (purple area in Figure 17.19). Continuous permafrost affects all surfaces except those beneath deep lakes or rivers. The depth of continuous permafrost may exceed 1000 m (3300 ft), averaging approximately 400 m (1300 ft).

Unconnected patches of *discontinuous permafrost* gradually coalesce poleward toward the continuous zone. Permafrost becomes scattered or sporadic until it gradually disappears equatorward of the −1°C (30.2°F) mean annual temperature isotherm (dark blue area on map). In the discontinuous zone, permafrost is absent on sun-exposed south-facing slopes, areas of warm soil, or areas insulated by snow. In the Southern Hemisphere, north-facing slopes experience increased warmth.

These discontinuous zones are most susceptible to thawing with climate change. Affected peat-rich soils (Gelisols, Histosols—Chapter 18) contain twice the amount of carbon in the atmosphere, making a powerful positive feedback as soils thaw and oxidation releases more carbon dioxide to the atmosphere. Additionally, the losses of carbon in thawing exceed any increases in carbon uptake that the warmer conditions and higher carbon dioxide levels create, making permafrost thaw an important source of greenhouse gases—a real-time positive feedback mechanism.

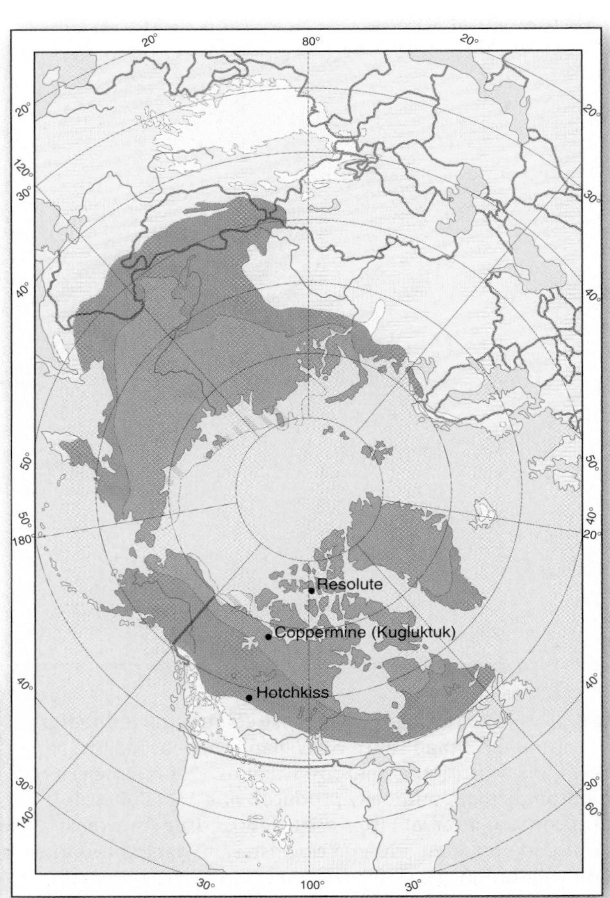

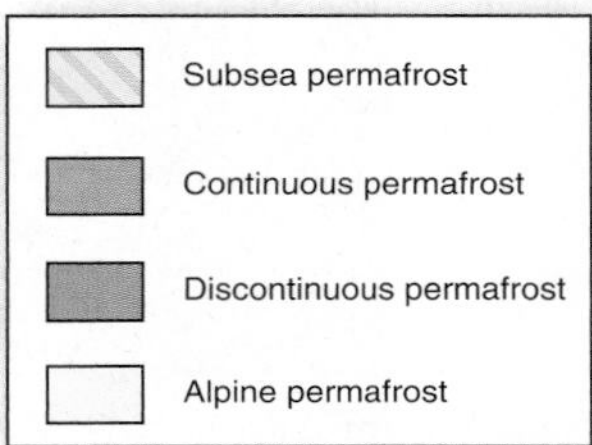

FIGURE 17.19 Permafrost distribution. Distribution of permafrost in the Northern Hemisphere. Alpine permafrost is noted except for small occurrences in Hawai'i, Mexico, Europe, and Japan. Sub-sea permafrost occurs in the ground beneath the Arctic Ocean along the margins of the continents, as shown. Note the towns of Resolute and Coppermine (Kugluktuk) in Nunavut (formerly part of NWT) and Hotchkiss in Alberta. A cross section of the permafrost beneath these towns is shown in Figure 17.20. [Adapted from T. L. Péwé, "Alpine permafrost in the contiguous United States: A review," *Arctic and Alpine Research* 15, no. 2 (May 1983): 146. © University of Colorado. Used by permission.]

Behavior of Permafrost We looked at the spatial distribution of permafrost; now, let us examine how permafrost behaves. Figure 17.20 is a stylized cross section from approximately 75° N to 55° N, using the three sites located on the map in Figure 17.19. The **active layer** is the zone of seasonally frozen ground that exists between the subsurface permafrost layer and the ground surface.

The active layer is subjected to consistent daily and seasonal freeze–thaw cycles. This cyclic melting of the active layer affects as little as 10 cm (4 in.) of depth in the north (Ellesmere Island, 78° N), up to 2 m (6.6 ft) in the southern margins (55° N) of the periglacial region, and 15 m (50 ft) in the alpine permafrost of the Colorado Rockies (40° N).

Higher temperatures degrade (reduce) permafrost and increase the thickness of the active layer; lower temperatures gradually aggrade (increase) permafrost depth and reduce active-layer thickness. Although somewhat sluggish in response, the active layer is a dynamic, open system driven by energy gains and losses in the subsurface environment. As you might expect, most permafrost exists in disequilibrium with environmental conditions and therefore actively adjusts to changing climatic conditions.

With the incredibly warm temperatures recorded in the Canadian and Siberian Arctic since 1990, more disruption of permafrost surfaces is occurring—leading to highway, railway, and building damage. In Siberia, many lakes have disappeared in the discontinuous permafrost region as subsurface drainage opens; yet new lakes have formed in the continuous region as thawed soils become waterlogged. In Canada, hundreds of lakes have disappeared simply from excessive evaporation into the warming air. These trends are measurable from satellite imagery.

A *talik* is unfrozen ground that may occur above, below, or within a body of discontinuous permafrost or beneath a water body in the continuous region. Taliks occur beneath deep lakes and may extend to bedrock and noncryotic soil beneath large, deep lakes (see Figure 17.20). Taliks form connections between the active layer

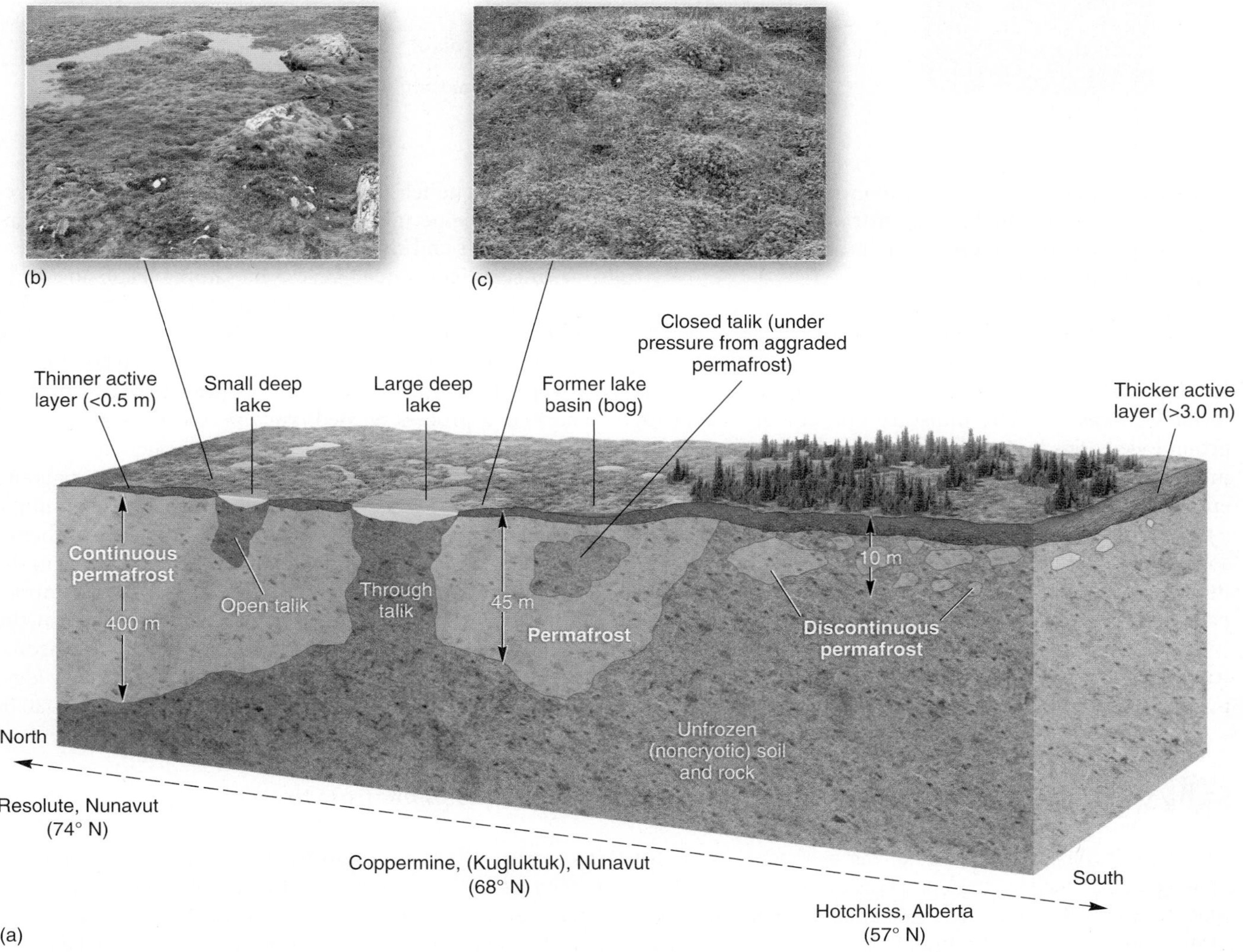

FIGURE 17.20 Periglacial environments.
(a) Cross section of a periglacial region in northern Canada, showing typical forms of permafrost, the active layer, talik, and ground ice. The three sites noted are shown on the map in Figure 17.19. (b) Poor drainage, with some standing water and hummocks, and (c) irregular ice-forced bumps of tundra turf indicate permafrost. [Bobbé Christopherson.]

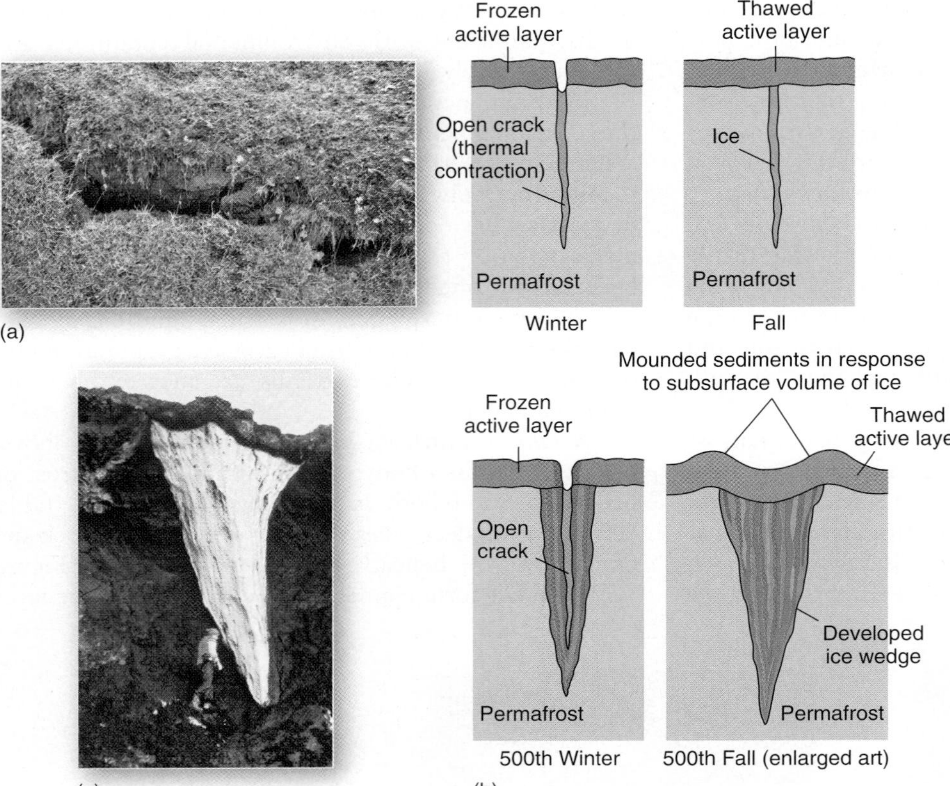

(a)

(c)

(b)

FIGURE 17.21 Evolution of an ice wedge. Sequential illustration of ice-wedge formation: (a) A crack forms in the tundra. (b) The crack evolves into an ice wedge over a time span of hundreds of years. (c) An example of an ice wedge and ground ice in northern Canada. [(a) Bobbé Christopherson. (b) Illustration adapted from A. H. Lachenbruch, "Mechanics of thermal contraction and ice-wedge polygons in permafrost," *Geological Society of America Bulletin Special Paper 70* (1962). (c) H. M. French.]

and groundwater, whereas in continuous permafrost, groundwater is essentially cut off from surface water. In this way, permafrost disrupts aquifers and taliks, leading to water-supply problems.

Ground-Ice and Frozen-Ground Phenomena

In regions of permafrost, frozen subsurface water is termed *ground ice*. The moisture content of areas with ground ice varies from nearly none in drier regions to almost 100% in saturated soils. From the area of maximum energy loss, freezing progresses through the ground along a *freezing front*, or boundary between frozen and unfrozen soil. The presence of frozen water in the soil initiates geomorphic processes associated with *frost action* and the expansion of water as it freezes.

The 9% expansion of water as it freezes produces strong mechanical forces. Such frost action shatters rock, producing angular pieces that form a *block field*, or *felsenmeer*. The felsenmeer accumulates as part of the arctic and alpine periglacial landscape, particularly on mountain summits and slopes.

If sufficient water freezes, the saturated soil and rocks are subjected to *frost heaving* (vertical movement) and *frost thrusting* (horizontal movement). Boulders and rock slabs may be thrust to the surface. Soil horizons (layers) may be disrupted by frost action and appear to be stirred or churned, a process termed *cryoturbation*. Frost action also can produce contractions in soil and rock, opening up cracks for ice wedges to form. A tremendous increase in pressure occurs in the soil as ice expands, particularly if multiple freezing fronts trap unfrozen soil and water between them.

An *ice wedge* develops when water enters a crack in the permafrost and freezes (Figure 17.21). Thermal contraction in ice-rich soil forms a tapered crack—wider at the top, narrowing toward the bottom. Repeated seasonal freezing and thawing progressively enlarge the wedge, which may widen from a few millimeters to 5–6 m (16–20 ft)

GEO REPORT 17.4 ***Feedback loops from fossil-fuel exploration to permafrost thawing***

The thaw in the active layer in the Alaskan tundra has shortened the number of days that oil exploration equipment can operate. Previously, the frozen ground lasted more than 200 days, permitting the heavy rigs and trucks to operate on the solid surface. With the thawing and unstable soft surfaces, exploration is now restricted to only 100 days a year. Think in a systems view: The exploration is for fossil fuels, which, when burned, increase the carbon dioxide levels in the atmosphere, which enhances greenhouse warming, which increases temperatures and permafrost thawing, which decreases the number of days to explore for fossil fuels—a negative feedback into the system.

FIGURE 17.22 **Patterned-ground phenomena.** (a) Patterned ground in Beacon Valley of the McMurdo Dry Valleys of East Antarctica. (b) Polygons and circles (about a meter across) in a stone-dominant area near Duvefjord, Nordaustlandet Island, Arctic Ocean. (c) Polygons and circles in a soil-dominant area, Spitsbergen Island. (d) Intermediate organization in progress on Bear Island (74° 30′ N). (e) Patterned ground in stripes near Sundbukta, Barents Island. (f) Polygons in the Martian northern plains; each cell averages a little more than 100 m across (300 ft). [(a) Courtesy of Joan Myers. (b), (c), (d), and (e) Bobbé Christopherson. (f) Malin Space Science Systems.]

(a) (b) (c) (d) (e) (f)

and deepen up to 30 m (100 ft). Widening may be small each year, but after many years, the wedge can become significant, like the sample shown in Figure 17.21c. In summer, when the active layer thaws, the wedge itself may not be visible. However, the presence of a wedge beneath the surface is noticeable where the ice-expanded sediments form raised, upturned ridges.

Fascinating research identified unique synergies at work to form **patterned ground**. The expansion and contraction of frost action result in the movement of soil particles, stones, and small boulders. During this freeze–thaw process is when self-organization begins. Imagine a process where stones move toward stone domains (stone-rich areas) and soil particles move toward soil domains (soil-rich areas). Patterned ground emerges that may take centuries to form. Added to this is the role of slope angles—greater slopes produce striped patterns (Figure 17.22e), whereas lesser slopes result in sorted polygons (Figure 17.22a, b, and c).

The stone-centered polygons in Figure 17.22b indicate higher stone concentrations, and the soil-centered polygons in Figure 17.22c indicate higher soil particle concentrations with lesser availability of stones. Also a factor is the amount of stone confinement by frozen soil domains: More circular forms occur when sorting is dominant; more polygons occur when confinement of stone domains is dominant. This means that such landscapes are self-making, self-maintaining, and self-organizing—all traits of living systems, making this an interesting path for future research in Earth systems. (See M. A. Kessler and B. T. Warner, "Self-organization of patterned ground," *Science* 299, 5606 [Jan. 17, 2003]: 380–383, 354–355.)

Such polygon nets in patterned ground provide vivid evidence of frozen subsurface water on Mars (Figure 17.22f). In this 1999 image, the *Mars Global Surveyor* captured a region of these features on the Martian northern plains. To date, some 600 different sites on Mars demonstrate such patterned ground, and in 2008, the *Phoenix* lander set down in the middle of a Martian arctic plain covered with polygonal forms. The craft dug into the Martian surface and verified the presence of ground (water) ice.

Hillslope Processes: Gelifluction and Solifluction

Soil drainage is poor in areas of permafrost and ground ice. The active layer of soil and regolith is saturated with soil moisture during the thaw cycle (summer), and the whole layer commences to flow from higher to lower

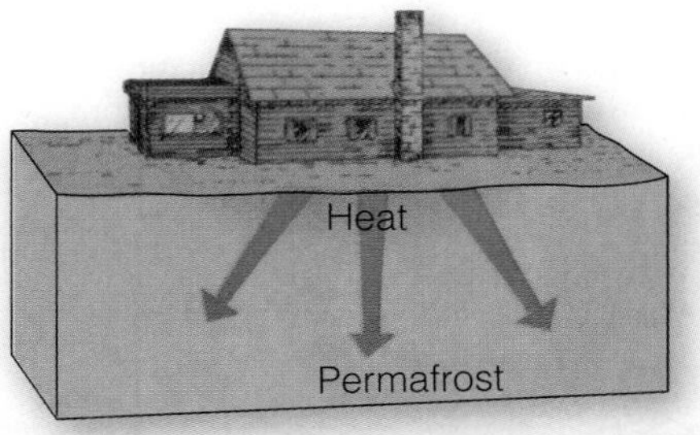

FIGURE 17.23 Permafrost melting and structure collapse. Building failure due to improper construction and the melting of permafrost, south of Fairbanks, Alaska. [Adapted from USGS: photo by Steve McCutcheon; based on U.S. Geological Survey pamphlet "Permafrost" by L. L. Ray.]

elevation if the landscape is even slightly inclined. This flow of soil is generally called *solifluction* and may occur in all climate conditions. In the presence of ground ice or permafrost, the more specific term *gelifluction* is applied. In this ice-bound type of soil flow, movement up to 5 cm (2 in.) per year can occur on slopes as gentle as a degree or two.

The cumulative effect of this landflow can be an overall flattening of a rolling landscape, with identifiable sagging surfaces and scalloped and lobed patterns in the downslope soil movements. Other types of periglacial mass movement include failure in the active layer, producing translational and rotational slides and rapid flows associated with melting ground ice. Periglacial mass movement processes are related to slope dynamics and processes discussed in Chapter 13.

Humans and Periglacial Landscapes

In areas of permafrost and frozen-ground phenomena, people face several related problems. Because thawed ground above the permafrost zone frequently shifts, highways and rail lines may warp or twist, and utility lines are disrupted. In addition, any building placed directly on frozen ground will "melt" into the defrosting soil, creating subsidence in structures (Figure 17.23).

In periglacial regions, structures must be constructed above the ground to allow air circulation beneath. This airflow allows the ground to cycle through its normal annual temperature pattern. Utilities such as water and sewer lines must be built aboveground in "utilidors" to protect them from freezing and thawing ground (Figure 17.24). Likewise, the trans-Alaska oil pipeline was constructed

(a)

(b)

FIGURE 17.24 Special structures for permafrost. (a) Proper construction in periglacial environments requires the raising of buildings aboveground and the running of water and sewage lines in elevated "utilidors," here in Barentsburg, a Russian settlement on Spitsbergen Island. (b) Supporting the trans-Alaska oil pipeline on racks protects the permafrost from heat. [(a) Bobbé Christopherson. (b) Galen Rowell/CORBIS.]

aboveground on racks for 675 km of its 1285-km length (420 mi of its 800-mi length) to avoid melting the frozen ground, causing shifting that could rupture the line. The pipeline that is underground uses a cooling system to keep the permafrost around the pipeline stable.

The Pleistocene Ice-Age Epoch

Imagine almost a third of Earth's land surface buried beneath ice sheets and glaciers—most of Canada, the northern Midwest, England, and northern Europe, with many mountain ranges beneath thousands of meters of ice. This occurred at the height of the Pleistocene Epoch of the late Cenozoic Era. In addition, periglacial regions along the margins of the ice during the last ice age covered about twice their present areal extent.

The Pleistocene Epoch is thought to have begun about 1.65 million years ago and is one of the more prolonged cold periods in Earth's history. It featured not just one glacial advance and retreat, but at least 18 expansions of ice over Europe and North America, each obliterating and confusing the evidence from the one before. Apparently, glaciation can take about 100,000 years, whereas deglaciation is rapid, requiring less than about 10,000 years to melt away the accumulation.

The term *ice age*, or *glacial age*, is applied to any extended period of cold (not a single brief cold spell), which may last several million years. An **ice age** is a time of generally cold climate that includes one or more *glacials*, interrupted by brief warm spells known as *interglacials*. Each glacial and interglacial is given a name that is usually based on the location where evidence of the episode is prominent—for example, the Wisconsinan glacial.

One method for determining Pleistocene temperatures is by evidence from deep-sea cores—specifically, from the oxygen isotope fluctuations in fossil plankton, which are tiny marine organisms with a calcareous shell. Glaciologists currently recognize the Illinoian glacial and Wisconsinan glacial intervals, with the Sangamon interglacial between them. These events span the 300,000-year interval prior to our present Holocene Epoch. The chart in Figure 17.25 shows that the Illinoian glacial actually consisted of two glacials occurring during marine (oxygen) isotope stages (MISs) 6 and 8, as did the Wisconsinan (MISs 2 and 4), which are dated at 10,000 to 35,000 years ago. To overcome local bias in deep-sea-core records, investigators correlate oxygen isotope data with other global climate indicators, such as ice cores. Note the label on the chart in Figure 17.25 at MIS 20.2 for the Dome-C ice-core extent of 800,000 years, an ice core extracted from the East Antarctic Plateau. For web links on glaciers and the Pleistocene Epoch, see http://userpages.umbc.edu/~miller/geog111/glacierlinks.htm.

Sea-surface temperatures 18,000 years ago averaged between 1.4 C° and 1.7 C° (2.5 F° and 3.1 F°) lower than they do today. During the coldest portion of the Pleistocene ice age, air temperatures were as much as 12 C° (22 F°) colder than today's average, although milder periods ranged to within 5 C° (9 F°) of present air temperatures.

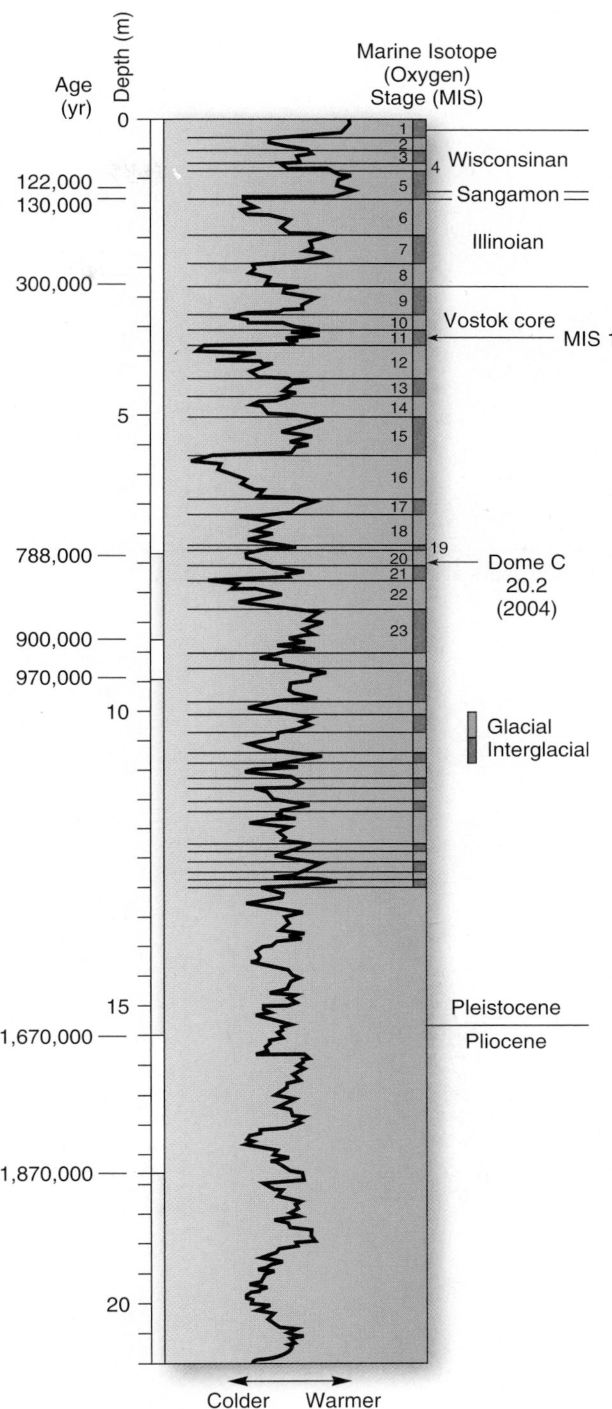

FIGURE 17.25 Temperature record of the past 2 million years. Twenty-three MISs cover 900,000 years, with names assigned for the past 300,000 years. [After N. J. Shackelton and N. D. Opdyke, *Oxygen-Isotope and Paleomagnetic Stratigraphy of Pacific Core V28-239, Late Pliocene to Latest Pleistocene.* Geological Society of America Memoir 145. ©1976 by the GSA. Adapted by permission.]

Changes in the Landscape

The continental ice sheets covered portions of Canada, the United States, Europe, and Asia about 18,000 years ago, as illustrated on the maps in Figure 17.26. Ice sheets ranged in thickness to more than 2 km (1.2 mi). In North America, the Ohio and Missouri river systems mark the

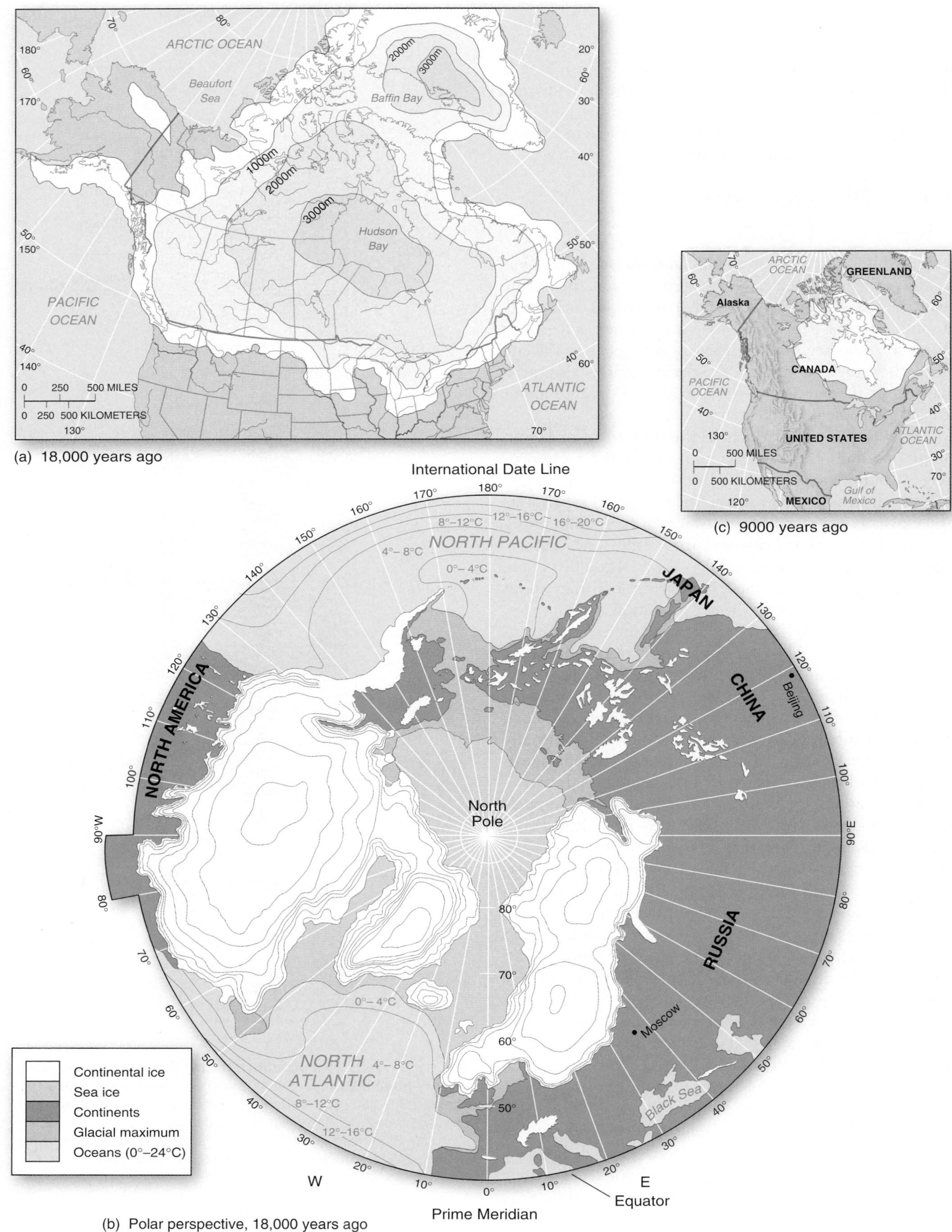

FIGURE 17.26 Extent of Pleistocene glaciation.
Earlier episodes produced continental glaciation of slightly greater extent. Note the depth of the continental ice sheets in North America (in meters). [From A. McIntyre, *CLIMAP* (Climate: Long-Range Investigation, Mapping, and Prediction) Project, Lamont-Doherty Earth Observatory. © 1981 by the GSA. Adapted by permission.]

southern terminus of continuous ice at its greatest extent during the Pleistocene Epoch. The ice sheet disappeared by 7000 years ago.

As both alpine and continental glaciers retreated, they exposed a drastically altered landscape: the rocky soils of New England, the polished and scarred surfaces of Canada's Atlantic Provinces, the sharp crests of the Sawtooth Range and Tetons of Idaho and Wyoming, the scenery of the Canadian Rockies and the Sierra Nevada, the Great Lakes of the United States and Canada, the Matterhorn of Switzerland, and much more. In the Southern Hemisphere, there is evidence of this ice age in the form of fjords and sculpted mountains in New Zealand and Chile.

The continental glaciers came and went several times over the region we know as the Great Lakes (Figure 17.27). The ice enlarged and deepened stream valleys to form the basins of the future lakes. This complex history produced five lakes that today cover 244,000 km^2 (94,000 mi^2) and hold some 18% of all the lake water on Earth. Figure 17.27 shows the final formation of the Great Lakes, which involved two advancing and two retreating stages—between 13,200 and 10,000 years before the present. During the final retreat, tremendous quantities of glacial meltwater flowed into the isostatically depressed basins—that is, basins that are lowered by the weight of the ice. Drainage at first was to the Mississippi River via the Illinois River, to the St. Lawrence River via the Ottawa River, and to the Hudson River in the east. In recent times, drainage is solely through the St. Lawrence system.

Sea levels 18,000 years ago were approximately 100 m (330 ft) lower than they are today because so much of Earth's water was frozen in glaciers. Imagine the coastline of New York being 100 km farther east, Alaska and Russia connected by land across the Bering Straits, and England and France joined by a land bridge. In fact, sea ice extended southward into the North Atlantic and Pacific and northward in the Southern Hemisphere about 50% farther than it does today.

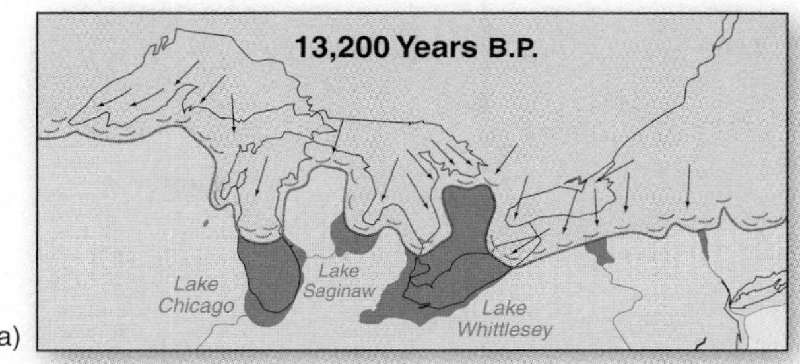

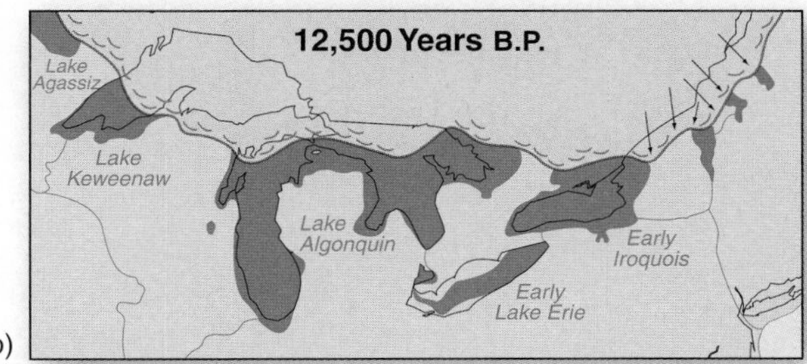

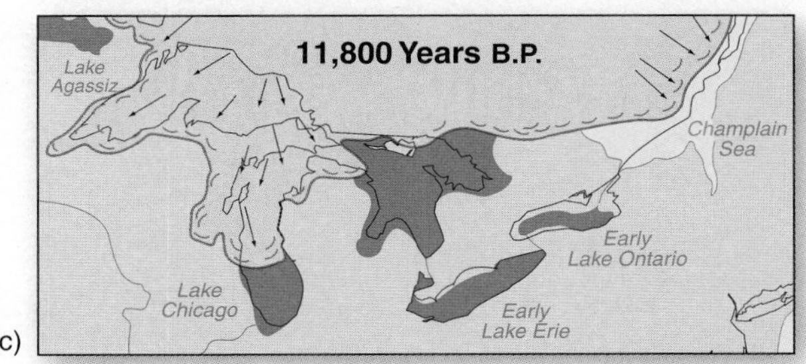

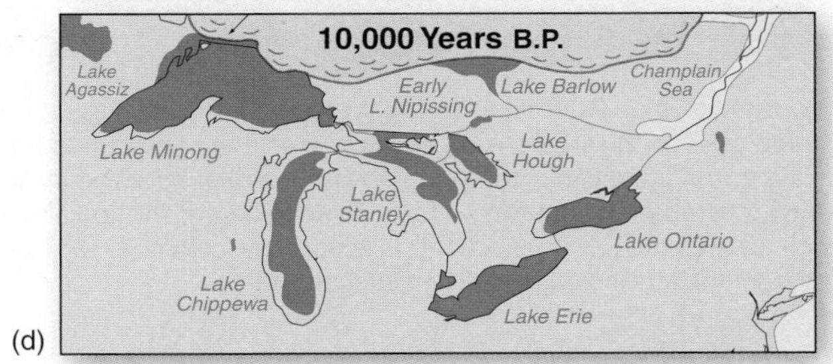

FIGURE 17.27 Late stages of Great Lakes formation. Four "snapshots" illustrate the evolution of the Great Lakes during the retreat of the Wisconsinan glaciation. Note the change in stream drainage between (b) and (d). Time is in years before the present (B.P.). [After *The Great Lakes—An Environmental Atlas and Resource Book,* Environment Canada, U.S. EPA, Brock University, and Northwestern University, 1987.]

Paleolakes

Figure 17.28 portrays the American West dotted with large lakes 12,000–30,000 years ago. Except for the Great Salt Lake in Utah (a remnant of the former Lake

GEO REPORT 17.5 *Glacial ice might protect underlying mountains*

Researchers are studying the effect of glacial ice on the underlying topography during the last glacial maximum. The temperature at the base of the ice is important. In the southernmost Patagonian Andes, conditions were so cold that the glacial ice froze to the bedrock. Evidence suggests that this protected the bedrock from erosion and typical glacial excavation, resulting in higher mountain peaks and a wider mountain belt in the southern Andes than in the northern part of the range, where mountain surfaces were ground down and narrowed by glacial action.

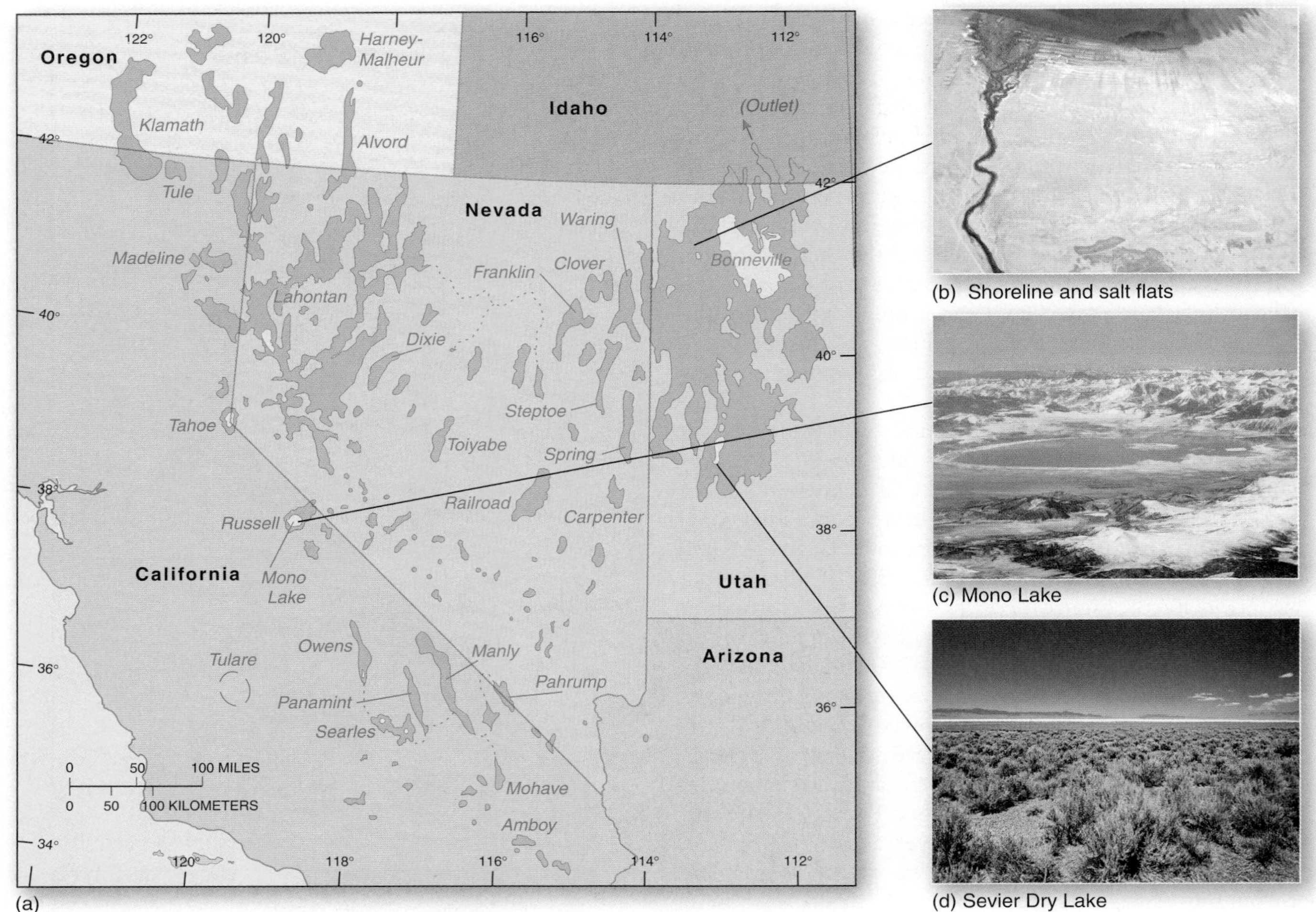

(a)

(b) Shoreline and salt flats

(c) Mono Lake

(d) Sevier Dry Lake

FIGURE 17.28 Paleolakes in the western United States.
(a) Paleolakes of the western United States at their greatest extent 12,000 to 30,000 years ago, a recent pluvial period. Lake Lahontan and Lake Bonneville were the largest. The Great Salt Lake (b) and Sevier Dry Lake (c) in Utah are remnants of Lake Bonneville; (d) Mono Lake in California is what remains of pluvial Lake Russell. [(a) After R. F. Flint, *Glacial and Pleistocene Geology.* © 1957 by John Wiley & Sons, Inc. Adapted by permission. (b), (c), and (d) Bobbé Christopherson.]

Bonneville noted on the map) and a few smaller lakes, only dry basins, ancient shorelines, and lake sediments remain today. These ancient lakes are **paleolakes**. The three photos in Figure 17.28 show what these paleolake sites look like today: the Bonneville Salt Flats in Utah, Mono Lake in California, and Sevier Dry Lake in Utah.

The term *pluvial* (from the Latin word for "rain") describes any period of wet conditions, such as occurred during the Pleistocene Epoch. During pluvial periods, lake levels increased in arid regions. The drier periods between pluvials, called *interpluvials*, are often marked by *lacustrine deposits*, the name for lake sediments that form terraces along former shorelines.

Are paleolakes of glacial origin like the Great Lakes? Earlier researchers attempted to correlate pluvial and glacial ages, given their coincidence during the Pleistocene. However, few sites actually demonstrate such a simple relation. For example, in the western United States, the estimated volume of melted ice from glaciers is only a small portion of the actual water volume that was in the paleolakes. Also, these lakes tend to predate glacial times and are correlated instead with periods of wetter climate or periods thought to have had lower evaporation rates.

Paleolakes existed in North and South America, Africa, Asia, and Australia. Today, the Caspian Sea in Kazakhstan and southern Russia has a level 30 m (100 ft) below mean world sea level, but ancient shorelines are visible about 80 m (265 ft) above the present lake level. In North America, the two largest late Pleistocene paleolakes were Lake Bonneville and Lake Lahontan, located in the Basin and Range Province in the western United States. These two lakes were, respectively, eight times and six times the size of their present-day remnants. Figure 17.28 shows these and other paleolakes at their highest level and the few remaining modern lakes in light blue. Much research is still taking place, since many of the ancient lakeshores are not yet field mapped.

The Great Salt Lake, near Salt Lake City, Utah, and the Bonneville Salt Flats in western Utah are remnants of Lake Bonneville; today, the Great Salt Lake is the fourth largest saline lake in the world. At its greatest extent, this paleolake covered more than 50,000 km^2 (19,500 mi^2) and reached depths of 300 m (1000 ft), spilling over into the Snake River drainage to the north. Today, it is a closed-basin terminal lake with no drainage except an artificial outlet to the west, where excess water from the Great Salt Lake can be pumped during rare floods.

New evidence reveals that the occurrence of these lakes in North America was related to specific changes in the polar jet stream that steered storm tracks across the region, creating pluvial conditions. The continental ice sheet evidently influenced changes in jet-stream position.

Deciphering Past Climates: Paleoclimatology

Glacials and interglacials occur because Earth's climate has fluctuated in and out of warm and cold ages. Evidence for this fluctuation now is traced in ice cores from Greenland and Antarctica (Focus Study 17.1), in layered ocean deposits of silts and clays, in the extensive pollen record from ancient plants, and in the relation of past coral productivity to sea level. This evidence is analyzed with radioactive dating methods and other techniques. One especially interesting fact is emerging from these studies: We humans (*Homo erectus* and *Homo sapiens* of the last 1.9 million years) have never experienced Earth's normal (more moderate, less extreme) climate, most characteristic of Earth's entire 4.6-billion-year span.

Apparently, Earth's climates slowly fluctuated until the past 1.2 billion years, when temperature patterns with cycles of 200–300 million years became more pronounced. The Pleistocene Epoch began in earnest 1.65 million years ago, and it still may be in progress. The Holocene Epoch began approximately 10,000 years ago, when average temperatures abruptly increased 6 C° (11 F°). The period we live in may represent an end to the Pleistocene, or it may be merely a mild interglacial time. Figure 17.29 details the climatic record of the past 160,000 years.

Medieval Warm Period and Little Ice Age

In A.D. 1001, Leif Eriksson inadvertently ventured onto the North American continent, perhaps the first European to do so. He and his fellow Vikings were

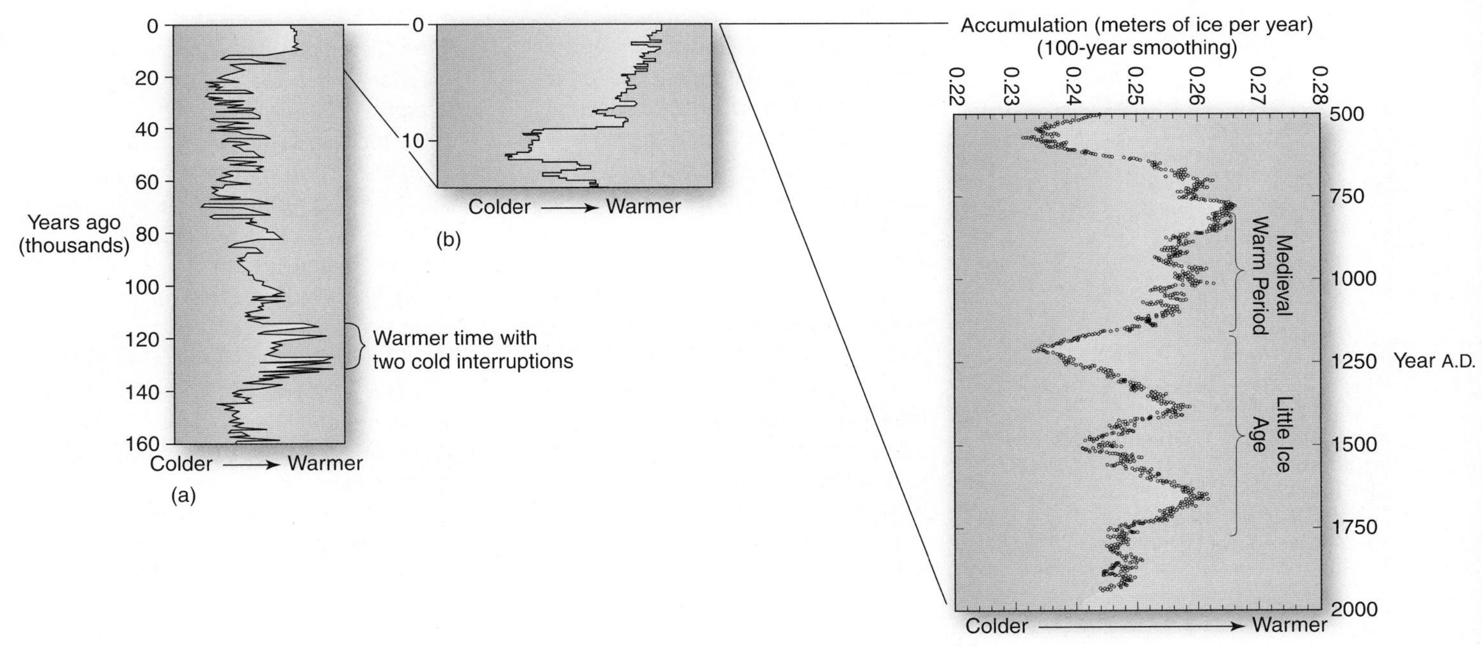

FIGURE 17.29 Recent climates determined from an ice core. (a) Temperature patterns during the past 160,000 years, back to MIS 6. Note the two cold spells that interrupted the earlier interglacial (between 115,000 and 135,000 years ago). (b) Higher resolution of the last 12,000 years. The cold period known as the Younger Dryas intensified at the beginning of this record. Warming was under way by 11,700 years ago and abruptly began to increase by the Holocene. (c) Note the Medieval Warm Period and its consistent temperatures as compared with the chaotic record of the Little Ice Age. [(a) and (b) Courtesy of the Greenland Ice Core Project (GRIP). (c) From D. A. Meese et al., "The accumulation record from the GISP-2 core as an indicator of climate change throughout the Holocene," *Science* 266 (December 9, 1994): 1681, © AAAS.]

FOCUS STUDY 17.1

Ice Cores Reveal Earth's Climate History

The Greenland Ice Core Project (GRIP) was launched in 1989. A site was selected near the summit of the Greenland Ice Sheet at 3200 m (10,500 ft) so that the maximum thickness of ice history would be accessed (Figure 17.1.1). After 3 years, the drills hit bedrock 3030 m (9940 ft) below the site—or, in terms of time, 250,000 years into the past. The core is 10 cm (4 in.) in diameter.

A year later, about 32 km (20 mi) west of the summit, the Greenland Ice Sheet Project (GISP-2) produced a core slightly larger in diameter, at 13.2 cm (5.2 in.), which collects about twice the data. The existence of a second core helped scientists compensate for any folds or disturbed sections they encountered below 2700 m (8858 ft), or about 115,000 years in the past, in the first core.

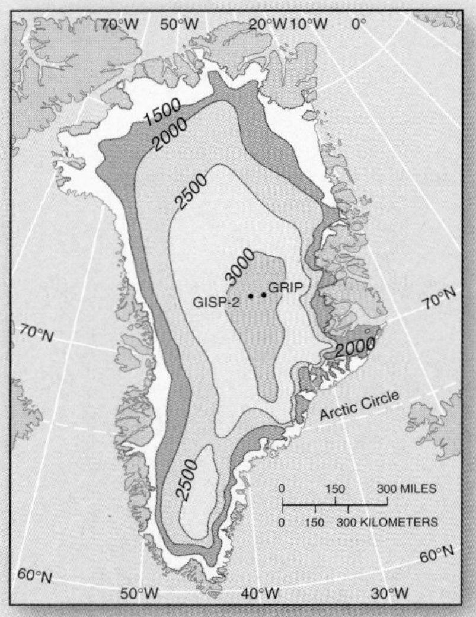

FIGURE 17.1.1 Greenland ice-core locations.
GRIP is at 72.5° N, 37.5° W. GISP-2 is at 72.6° N, 38.5° W. [Map courtesy of GISP-2 Science Management Office.]

What was discovered from these ice cores? Locked into the cores is a record of past precipitation and air bubbles of past atmospheres, which indicate ancient gas concentrations. Of special interest are the greenhouse gases, carbon dioxide and methane. Chemical and physical properties of the atmosphere and the snow that accumulated each year are frozen in place.

Pollutants are locked into the ice-core record. For example, during cold periods, high concentrations of dust were present, brought by winds from distant dry lands. An invaluable record of past volcanic eruptions is included in the layers, as if on a calendar. Even the exact beginning of the Bronze Age—about 3000 B.C.— is recorded in the ice core. When the Greeks and later the Romans began smelting copper, they produced ash and smoke that the winds carried to this distant place. The presence of ammonia indicates ancient forest fires at lower latitudes. And, on the ice-sheet surface when each snowfall occurred, the ratio between stable forms of oxygen is measured as an important analog of past temperatures.

Dome-C Ice Core

On the East Antarctic Plateau, approximately 900 km (560 mi) from the coast and 1750 km (1087 mi) from the South Pole, sits Dome C. This is the point on the Antarctic continent where the ice sheet is the thickest (see Figure 17.31c for the location and Figure 17.1.2). Here the ice is 3309 m (10,856 ft) deep. The mean annual surface temperature is –54.5°C (–66.1°F); however, the science crews experience temperatures that range from –50°C when they arrive to –25°C by midsummer. Dome C is 560 km (348 mi) from the Vostok Station, where the previous coring record of 400,000 years was set. The Dome-C project is part of a 10-country European Project for Ice Coring in Antarctica (EPICA).

FIGURE 17.1.2 Dome-C station. An overview of the Dome-C scientific station, at 75.1° S, 123.3° E, with its gathering of tents and semipermanent structures. [British Antarctic Survey.]

The mechanical drill bit used for coring was 10 cm wide and provided a high-resolution record of ancient atmospheric gases, ash from volcanic eruptions, and materials from other atmospheric events. By December 2004, coring to a depth of 3270.20 m (10,729 ft) brought up 800,000 years of Earth's past climate history (Figure 17.1.3). In Figure 17.25, this is MIS 20.2. This remarkable record contains eight glacial and interglacial cycles. Scientists found the Dome-C record affirms the Vostok findings and correlates perfectly

favored by a medieval warming episode as they sailed the less-frozen North Atlantic to settle Iceland and Greenland.

The mild climatic episode that lasted from about A.D. 800 to 1200 is known as the *Medieval Warm Period.* During the warmth, grape vineyards were planted far into England some 500 km north of present-day commercial plantings. Oats and barley were planted in Iceland, and wheat was planted as far north as Trondheim, Norway. The shift to warmer, wetter weather influenced migration and settlement northward in North America, Europe, and Asia.

(a) Dome-C core removed

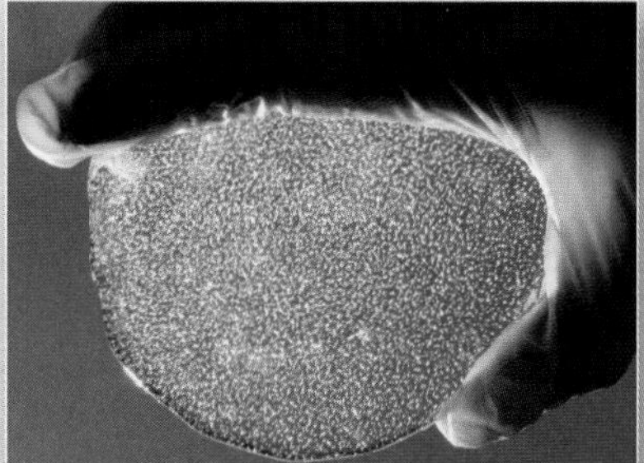
(b) Ice-core section

(c) Glacial ice bubbles

FIGURE 17.1.3 Dome-C ice-core analysis. (a) Two members of the British Antarctic Survey carefully inspect one of the ice-core segments. A quarter-section of each ice core is kept at Dome C in case there is an accident in transport back to labs in Europe. (b) Light shining through a thin section from the ice core; Earth's ancient atmosphere is trapped in the ice awaiting analysis. (c) You can see the trapped air bubbles in this chunk of glacial ice. [(a) and (b) British Antarctic Survey, http://www.antarctica.ac.uk. (c) Bobbé Christopherson.]

with the deep-sea core of oxygen isotope fluctuations in foraminifera shells (microfossils) from the Atlantic Ocean.

Confirmed are the findings that the present concentrations of carbon dioxide, methane, and nitrous oxide in the atmosphere are the highest in the past 800,000 years and that these changing levels have marched in step with higher and lower temperatures throughout this time span.

Go to Figure 17.25 and find MIS 11, which indicates an additional item of importance at Dome C. This warm spell at MIS 11, from 425,000 to 395,000 years ago, is perfectly recorded in the Dome-C core and is of interest because CO_2 in the atmosphere during MIS 11 was similar to our preindustrial level of 280 ppm. Earth's orbital alignment was similar to that of today as well. The MIS 11 analogy tells us that a return to the next glacial period is perhaps 16,000 years in the future.

Ice-core drilling projects planned, under way, or completed bring the total cores to nearly 30 in Antarctica and 16 in Greenland. The Ice Core Gateway site of NOAA Paleoclimatology lists and links to coring projects (http://www.ncdc.noaa.gov/paleo/icgate.html). The Byrd Polar Research Center at Ohio State University presents information on the worldwide efforts of its Ice Core Paleoclimatology Research Group at http://bprc.osu.edu/Icecore/.

Understanding the past helps decode future trends. The challenge for physical geographers is to analyze these data spatially and find the linkages to other Earth systems and to human economies and societies in an effort to understand what is happening in Earth's cryosphere.

However, from approximately 1200 to 1350 through 1800 to 1900, a *Little Ice Age* took place. Parts of the North Atlantic froze, and expanding glaciers blocked many key mountain passes in Europe. Snowlines in Europe lowered about 200 m (650 ft) in the coldest years. The Greenland colonies were deserted. Cropping patterns changed, and northern forests declined, along with human population in those regions. In the winter of 1779–1780, New York's Hudson and East rivers and the entire Upper Bay froze over. People walked and hauled heavy loads across the ice between Staten and Manhattan islands!

But the Little Ice Age was not consistently cold throughout its 700-year reign. Reading the record of the Greenland ice cores, scientists have found many mild years among the harsh. More accurately, this was a time of rapid, short-term climate fluctuations that lasted only decades.

Ice cores drilled in Greenland have revealed a record of annual snow and ice accumulation that, when correlated with other aspects of the core sample, was indicative of air temperature. Figure 17.29 presents this record since A.D. 500. The warmth of the Medieval Warm Period is evident, whereas the Little Ice Age appears mixed, with colder conditions around 1200, 1500, and after about 1800.

Mechanisms of Climate Fluctuation

What mechanisms cause short-term fluctuations? And why is Earth pulsing through long-term climatic changes that span several hundred million years? The ice-age concept is being researched and debated with an unprecedented intensity for three principal reasons: (1) Continuous ice cores from Greenland and Antarctica are providing a new, detailed record of weather and climate patterns, volcanic eruptions, and trends in the biosphere (discussed in Focus Study 17.1); (2) to understand present and future climate change and to refine general circulation models, we must understand the natural variability of the atmosphere and climate; and (3) anthropogenic global warming, and its relation to ice ages, is a major concern.

Because past occurrences of low temperature appear to have followed a pattern, researchers have looked for causes that also are cyclic in nature. They have identified a complicated mix of interacting variables that appear to influence long-term climatic trends. Let us take a look at several of them.

Climate and Celestial Relations As our Solar System revolves around the distant center of the Milky Way, it crosses the plane of the galaxy approximately every 32 million years. At that time, Earth's plane of the ecliptic aligns parallel to the galaxy's plane, and we pass through regions in space of increased interstellar dust and gas, which may have some climatic effect.

Milutin Milankovitch (1879–1954), a Yugoslavian astronomer who studied Earth–Sun orbital relations, proposed other possible causes of climate variation. Milankovitch wondered whether the development of an ice age relates to seasonal astronomical factors—Earth's revolution around the Sun, rotation, and tilt—extended over a longer time span (Figure 17.30). In summary:

- Earth's elliptical orbit about the Sun is not constant. The shape of the ellipse varies by more than 17.7 million kilometers (11 million miles) during a 100,000-year cycle, from nearly circular to an extreme ellipse (Figure 17.30a).
- Earth's axis "wobbles" through a 26,000-year cycle, in a movement much like that of a spinning top winding down. Earth's wobble is *precession*. As you can see in Figure 17.30b, precession changes the orientation of hemispheres and landmasses to the Sun.
- Earth's present axial tilt of 23.5° varies from 22° to 24° during a 41,000-year period (Figure 17.30c).
- During MIS–11 (Figure 17.25) these astronomical factors were the same as today.

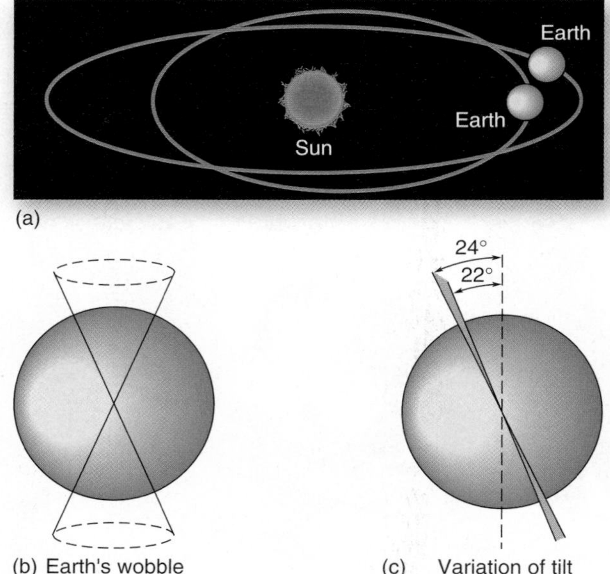

FIGURE 17.30 Astronomical factors that may affect broad climatic cycles.
(a) Earth's elliptical orbit varies widely during a 100,000-year cycle, stretching out to an extreme ellipse. (b) Earth's 26,000-year axial wobble. (c) Variation in Earth's axial tilt every 41,000 years.

Milankovitch calculated, without the aid of today's computers, that the interaction of these Earth–Sun relations creates a 96,000-year climatic cycle. His glaciation model assumes that changes in astronomical relations affect the amounts of insolation received.

Milankovitch died in 1954, his ideas still not accepted by a skeptical scientific community. Now, in the era of computers, remote-sensing satellites, and worldwide efforts to decipher past climates, Milankovitch's valuable work has stimulated research to explain climatic cycles and has experienced some confirmation. A roughly 100,000-year climatic cycle is confirmed in such diverse places as ice cores from Greenland and Antarctica and the accumulation of sediment in Lake Baykal, Siberia. Milankovitch cycles appear to be the primary cause of glacial-interglacial cycles, although other factors that cause second-order effects such as changes in the North Atlantic Ocean, gains or losses of Arctic sea ice, and variable carbon dioxide levels.

Climate and Solar Variability If the Sun significantly varies its output over the years, as some other stars do, that variation would seem a convenient and plausible cause of ice-age timing. However, lack of evidence that the Sun's radiation output varies significantly over long cycles argues against this hypothesis. Nonetheless, inquiry about the Sun's variability continues. As we saw in Chapter 10, in the Intergovernmental Panel on Climate Change *Fourth Assessment Report* in Figure 10.33, solar variability is not correlated to temperature increases occurring in the atmosphere and oceans.

Climate and Tectonics Major glaciations are associated with plate tectonics because some landmasses migrated to higher, cooler latitudes. Chapters 11 and 12 explained that the shape and orientation of landmasses and ocean basins changed greatly during Earth's history. Continental plates drifted from equatorial locations to polar regions, and vice versa, thus exposing the land to a gradual change in climate. Gondwana (the southern half of Pangaea) experienced extensive glaciation that left its mark on the rocks of parts of present-day Africa, South America, India, Antarctica, and Australia. Landforms in the Sahara, for example, bear the markings of even earlier glacial activity. These markings are partly explained by the fact that portions of Africa were centered near the South Pole during the Ordovician Period, 465 million years ago (see Figure 11.17a).

Episodes of mountain building over the past billion years have affected climate change as mountain summits rose above the snowline, where snow remains after the summer melt. Mountain chains influence downwind weather patterns and jet-stream circulation, which, in turn, guides weather systems. More dust was present during glacial periods, suggesting drier weather and more extensive deserts beyond the frozen regions.

Climate and Atmospheric Factors Some events alter the atmosphere and produce climate change. A volcanic eruption might produce lower temperatures for a year or two. The lower temperatures could initiate a buildup of long-term snow cover at high latitudes. These high-albedo snow surfaces then would reflect more insolation away from Earth to further enhance cooling in a positive feedback system.

The fluctuation of atmospheric greenhouse gases could trigger higher or lower temperatures. Ice cores from Greenland and Antarctica contain trapped air samples that show carbon dioxide levels varying from 180 ppm to 290 ppm. Higher levels of carbon dioxide generally correlate with each interglacial, or warmer, period. Over the past four decades, atmospheric carbon dioxide increased dramatically, principally a result of anthropogenic (human-caused) forcing. Refer to the carbon dioxide records in Figures 10.31 (1000-year record) and 10.34 (10,000-year record).

Climate and Oceanic Circulation Finally, changes in oceanic circulation patterns may affect climate. For example, the Isthmus of Panama formed about 3 million years ago and effectively separated the circulation of the Atlantic and Pacific oceans. Changes in ocean basin configuration, surface temperatures, salinity, and rates of upwelling and downwelling affect air mass formation and air temperature.

Our understanding of Earth's climate—past, present, and future—is unfolding. We are learning that climate is a multicyclic system controlled by an interacting set of cooling and warming processes, all founded on celestial relations, tectonic factors, atmospheric variables, changes in oceanic circulation, and human impacts.

Arctic and Antarctic Regions

Climatologists use environmental criteria to define the Arctic and the Antarctic regions. The 10°C (50°F) isotherm for July defines the **Arctic region** (green line on the map in Figure 17.31a). This line coincides with the visible tree line—the boundary between the northern forests and tundra. The Arctic Ocean is covered by two kinds of ice: *floating sea ice* (frozen seawater) and *glacier ice* (frozen freshwater). This pack ice thins in the summer months and sometimes breaks up (Figure 17.31b).

As mentioned elsewhere, almost half of the Arctic ice pack by volume has disappeared since 1970 due to regional-scale warming. The year 2007 broke the record for lowest sea-ice extent, falling below the previous record low in 2005 by 24%. The fabled Northwest Passage across the Arctic from the Atlantic to the Pacific was ice-free in September 2007, as the Arctic ice continues to melt. The Northeast Passage, north of Russia, has been ice-free for the past several years.

The Antarctic convergence defines the **Antarctic region** in a narrow zone that extends around the continent as a boundary between colder Antarctic water and warmer water at lower latitudes. This boundary follows roughly the 10°C (50°F) isotherm for February, the Southern Hemisphere summer, and is located near 60° S latitude (green line in Figure 17.31c). The Antarctic region that is covered just with sea ice represents an area greater than North America, Greenland, and Western Europe combined. For more information on polar-region ice, see the National Ice Center at http://www.natice.noaa.gov/ and the Canadian Ice Service at http://ice-glaces.ec.gc.ca/app/WsvPageDsp.cfm.

The Antarctic landmass is surrounded by ocean and is much colder overall than the Arctic, which is an ocean surrounded. In simplest terms, Antarctica can be thought of as a continent covered by a single enormous glacier, although it contains distinct regions such as the East Antarctic and West Antarctic ice sheets, which respond differently to slight climatic variations. These ice sheets are in constant motion, as indicated by the arrows in Figure 17.31c.

The fact that Antarctica is so remote from civilization makes it an excellent laboratory for sampling past and present evidence of human and natural variables that are transported by atmospheric and oceanic circulation to this pristine environment. High altitude, winter cold and darkness, and distance from pollution sources make the

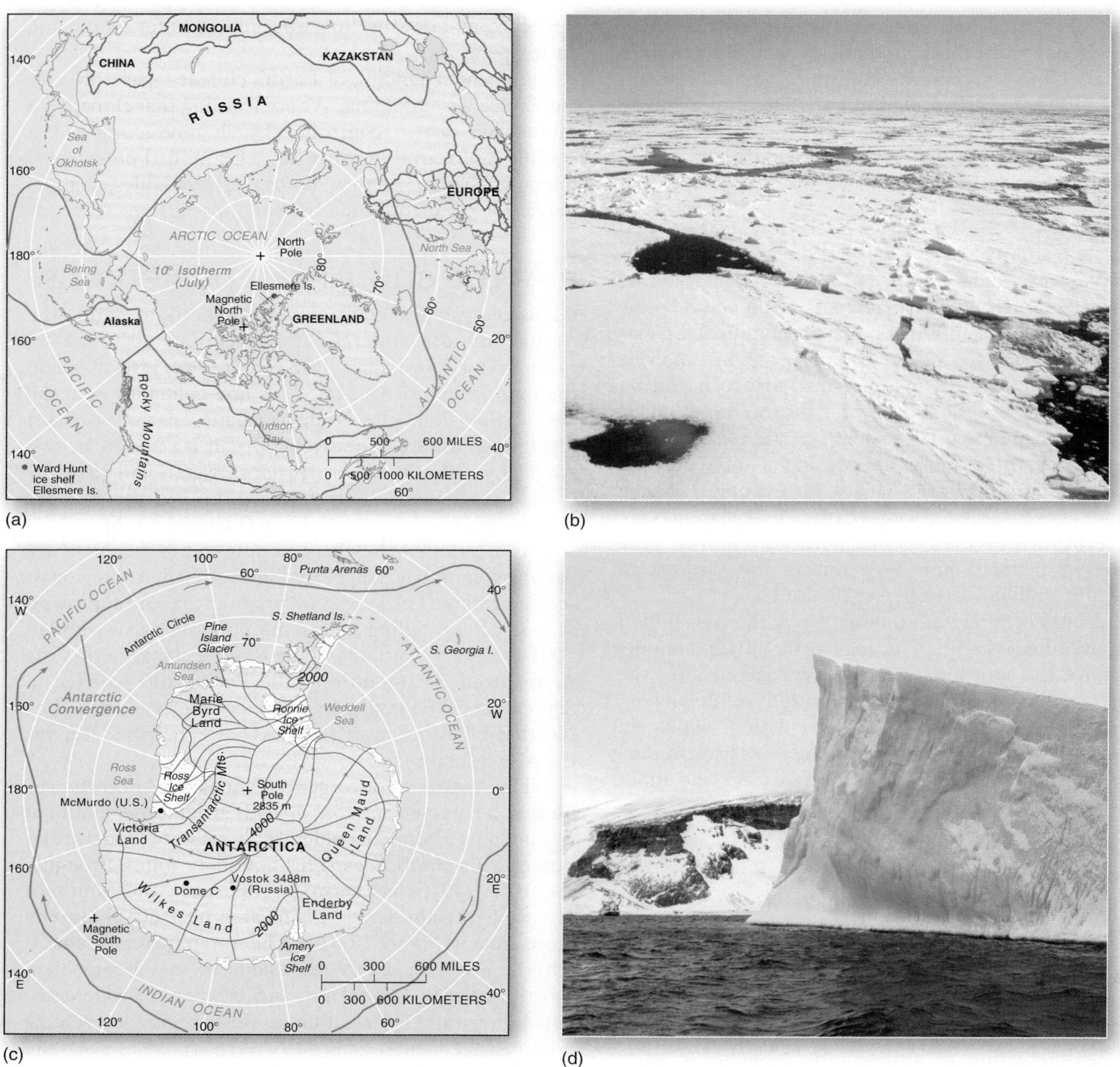

FIGURE 17.31 The Arctic and Antarctic regions.
In (a), note the 10°C (50°F) isotherm in midsummer, which designates the Arctic region, dominated by pack ice. (b) Arctic sea ice about 965 km (600 mi) from the North Pole. In (c), the Antarctic convergence designates the Antarctic region. Arrows on the ice sheet show the general direction of ice movement on Antarctica. Note the location of the Dome-C and Vostok stations. (d) View across the Antarctic Sound from Brown Bluff; for scale, note the ship compared to the huge tabular iceberg. [Bobbé Christopherson.]

polar region an ideal location for certain astronomical and atmospheric observations (Figure 17.32). Fifty scientists and support people work through the winter (February to October), and 130 personnel, or more, research and work there during the brief summer.

Changes Under Way in the Polar Regions

Review the "Global Climate Change" section in Chapter 10 and *Geosystems Now* in Chapter 4, where a portrait of high-latitude temperatures and ongoing physical changes to pack ice, ice shelves, and glaciers is presented.

An important indicator of changing surface conditions is an increase in meltponds across the polar regions, as these meltponds represent positive feedback—meltponds are darker, absorb more insolation, and become warmer, which melts more ice, making more meltponds, and so forth. The *Landsat*-7 satellite, in tandem with aircraft equipped with video cameras, spotted this

FIGURE 17.32 Scientific base at the South Pole.
An aerial view of the Amundsen–Scott South Pole Station in January 2008. This station is at an elevation of 2835 m (9301 ft). The geodesic dome (left), now decommissioned, opened in 1975. To the right, you see the elevated station completed in 2006, which stands on pylons 3 m (10 ft) above the surface to accommodate snow accumulation and protect the ice underneath from melting. Below the modules is the ceremonial South Pole with a semicircle of Antarctic Treaty–signing national flags; to the left is the geographic South Pole detailed in the inset photo (brass pole marker and flag, 90° S). [Courtesy of Ethan Dicks, U.S. Antarctic Program, National Science Foundation.]

CRITICAL THINKING 17.2

A sample of life at the polar station

After reading my summary of the *New South Polar Times,* 1993–1998, on the *Mastering Geography* Web site under this chapter, imagine what duty would be like at the South Pole. Of the 50 people who winter over at the station, some serve as scientists, technicians, and support staff. The station commander and reporter for the 1998 season was Katy McNitt-Jensen, in her third tour at the pole. Although dated, her personal observations are interesting. Remember, the last airplane leaves mid-February, and the first airplane lands mid-October—such is the isolation. What do you see as the positives and negatives of such service? How would you combat the elements, the isolation, and the dark conditions?

increase in meltpond occurrence on glaciers, icebergs, ice shelves, and the Greenland Ice Sheet.

Figure 17.33a is a scene from western Greenland in June 2003 made by the *Terra* satellite. Examine the bare rock along the eastern margin, indicating that the snow line has retreated to higher elevations. The image shows rapidly increasing numbers of meltponds and areas of water-saturated ice in the *melt zone,* which advanced 400% from 2001 to 2003. The meltponds look like small blue dots across the ice. In Figure 17.33b, you can see the meltponds and melt streams in southwest Greenland as of June 2008.

The streams sometimes melt through the ice sheet, forming a *moulin,* or drainage channel, that works its way to the base of the glacier (Figure 17.33c). The glacial water can flow through the basal layers of fine clays, thus lubricating and causing acceleration of glacial flow rates; even hydrostatically lifting the ice is possible. Sometimes these moulin channels exit from the glacier's face and flow into the sea (Figure 17.33d).

Surrounding the margins of Antarctica, and constituting about 11% of its surface area, are numerous ice shelves. Although ice shelves constantly break up to produce icebergs, more large sections are breaking free than

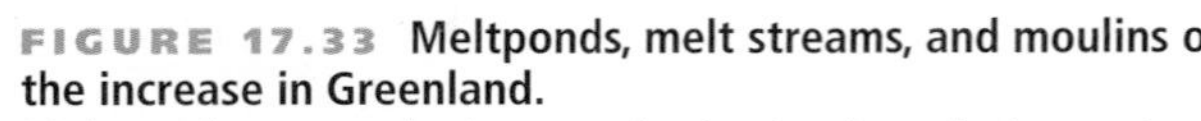

meltponds, water-saturated ice

(a)

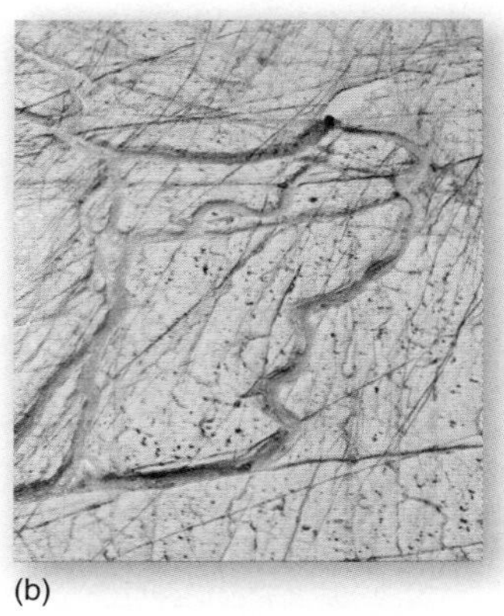

(b)

(c)

(d)

(e)

(f)

FIGURE 17.33 Meltponds, melt streams, and moulins on the increase in Greenland. Meltponds are on the increase in the Arctic on icebergs, ice shelves, and the Greenland Ice Sheet. (a) Image of western Greenland demonstrates these conditions. Between the rocky landscape with outlet glaciers and fjords to the west (left) and the ice sheet to the east (right), examine the extent of the melt zone—water-saturated surface ice and meltponds (blue specks). (b) Meltponds and melt streams flow across southwest Greenland to a moulin. (c) Meltwater flows into a moulin. (d) Sometimes moulin drainage outlets emerge from the face of a glacier. (e) and (f) Why are meltponds positive-feedback climate indicators? [(a) *Terra* MODIS image, NASA/GSFC. (b) and (c) Courtesy of JPL/NASA. (d), (e), and (f) Bobbé Christopherson.]

expected. You read about a recent breakup of the Wilkins Ice Shelf in *Geosystems Now* for this chapter. In 1998, an iceberg the size of Delaware broke off the Ronne Ice Shelf, southeast of the Antarctic Peninsula. In March 2000, an iceberg tagged B-15, measuring *twice* the area of Delaware (300 km by 40 km, or 190 mi by 25 mi), broke off the Ross Ice Shelf (some 3027 km, 1900 mi, west of the Antarctic Peninsula).

Since 1993, seven ice shelves have disintegrated in Antarctica. More than 8000 km^2 (3090 mi^2) of ice shelf are gone, changing maps, freeing up islands to circumnavigation, and creating thousands of icebergs. The Larsen Ice Shelf, along the east coast of the Antarctic Peninsula, was retreating slowly for years. Larsen-A suddenly disintegrated in 1995. In only 35 days in early 2002, Larsen-B collapsed into icebergs (Figure 17.34). Larsen-B was at least 11,000 years old, meaning this collapse was farther south than any previously during the Holocene Epoch. Larsen-C, the next segment to the south, is losing mass on its underside, since the water temperature is warmer by

CRITICAL THINKING 17.3

The IPY accomplishment continues

The International Polar Year (IPY) ran from March 2007 to March 2009, covering two polar summer seasons—it was the fourth IPY conducted since 1882. This global interdisciplinary research, exploration, and discovery effort involved 50,000 scientists in hundreds of collaborative projects. A systems approach was key, finding the linkages and connections among ecosystems and human activity. About 65% of the research was in the Arctic region and 35% in the Antarctic region. Results are still being published, and the IPY is stimulating conferences and educational activity. Traditional knowledge of indigenous peoples that are experiencing climate change firsthand was part of the research effort—the actual experiences of the First Nations people.

Use your critical-thinking skills in a brief exploration of this IPY. Begin at the web site http://www.ipy.org. From the "What is IPY" frame: (1) Click "Polar Weeks," and follow the chronology of links; (2) click "About IPY" for background and urgent issues; and (3) click "IPY Project Database," then click "Browse all EoIs" ("Expressions of Intent"), and sample some projects. The list is too long for a complete examination; just do a brief survey of the research.

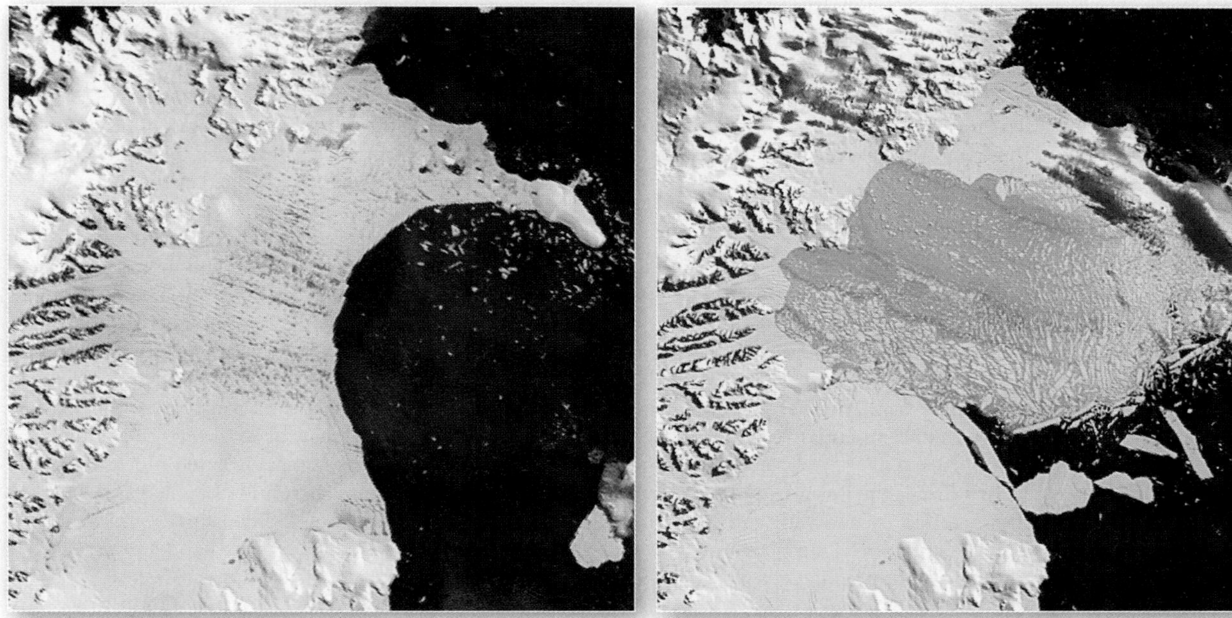

FIGURE 17.34 Disintegrating ice shelves along the Antarctic coast. Disintegration and retreat of the Larsen-B ice shelf between January 31 and March 7, 2002. Note the meltponds in the January image. [*Terra* images, NASA.]

0.65 C° (1.17 F°) than the melting point for ice at a depth of 300 m (984 ft)—warmth is melting the shelf from both the ocean and the atmosphere sides. This ice loss is likely a result of this warmer water and the 2.5 C° (4.5 F°) air temperature increase in the peninsula region during the last 50 years. In response to the increasing warmth, the Antarctic Peninsula is experiencing previously unseen vegetation growth, reduced sea ice, and disruption of penguin feeding, nesting, and fledging activities; ticks are a new problem for these animals.

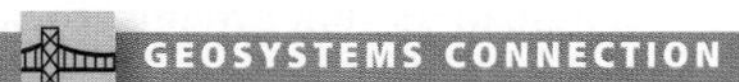

GEOSYSTEMS CONNECTION

Our look at the cryosphere and glacial processes and landforms ends PART III, *The Earth–Atmosphere Interface*. We examined past and present alpine and continental glaciations. Earth's landscapes bear the mark of past glacial phases. One method by which scientists study paleoclimatology is by analyzing ice cores, with the record going back to 800,000 years before the present. The polar regions are undergoing rapid change at a pace faster than the lower latitudes. We shift now to PART IV, *Soils, Ecosystems, and Biomes,* a synthesis of the other three parts in this text, focused on the biosphere and the subdiscipline of biogeography.

KEY LEARNING CONCEPTS REVIEW

■ ***Differentiate*** between alpine and continental glaciers, and ***describe*** ice field, ice cap, and ice sheet.

More than 77% of Earth's freshwater is frozen. Ice covers about 11% of Earth's surface, and periglacial features occupy another 20% of ice-free, but cold-dominated landscapes. A **glacier** is a mass of ice sitting on land or floating as an ice shelf in the ocean next to land. Glaciers form in areas of permanent snow. A **snowline** is the lowest elevation where snow occurs year-round, and its elevation varies by latitude—higher near the equator, lower poleward.

A glacier in a mountain range is an **alpine glacier**. If confined within a valley, it is termed a *valley glacier*. The area of origin is a snowfield, usually in a bowl-shaped erosional landform called a **cirque**. Where alpine glaciers flow down to the sea, they calve and form **icebergs**. A **continental glacier** is a continuous mass of ice on land. Its most extensive form is an **ice sheet**; a smaller, roughly circular form is an **ice cap**; and the least extensive form, usually in mountains, is an **ice field**.

glacier (p. 490)
snowline (p. 490)
alpine glacier (p. 491)
cirque (p. 491)
iceberg (p. 491)
continental glacier (p. 492)
ice sheet (p. 492)
ice cap (p. 493)
ice field (p. 493)

1. Describe the location of most freshwater on Earth today.
2. What is a glacier? What is implied about existing climate patterns in a glacial region?

3. Differentiate between an alpine glacier and the three types of continental glaciers. Which occurs in mountains? Which covers Antarctica and Greenland?
4. How are icebergs generated? Describe their buoyancy characteristics from this discussion and the section on ice in Chapter 7. Why do you think an iceberg overturns from time to time as it melts?

■ ***Explain*** the process of glacial ice formation, and ***portray*** the mechanics of glacial movement.

Snow becomes glacial ice through accumulation, increasing thickness, pressure on underlying layers, and recrystallization. Snow progresses through transitional steps from **firn** (compact, granular) to a denser **glacial ice** after many years.

A glacier is an open system. A **firn line** is the lower extent of a fresh snow-covered area. A glacier is fed by snowfall and is wasted by **ablation** (losses from its upper and lower surfaces and along its margins). Accumulation and ablation achieve a mass balance in each glacier.

As a glacier moves downhill, vertical **crevasses** may develop. Sometimes a glacier will move rapidly in a **glacier surge**. The presence of water along the basal layer appears to be important in glacial movements. As a glacier moves, it plucks rock pieces and debris, incorporating them into the ice, and this debris scours and sandpapers underlying rock through **abrasion**.

firn (p. 495)
glacial ice (p. 495)
firn line (p. 495)
ablation (p. 495)
crevasse (p. 496)
glacier surge (p. 496)
abrasion (p. 497)

5. Trace the evolution of glacial ice from fresh fallen snow.
6. What is meant by glacial mass balance? What are the basic inputs and outputs contributing to that balance?
7. What is meant by a glacier surge? What do scientists think produces surging episodes?

■ ***Describe*** characteristic erosional and depositional landforms created by alpine glaciation and continental glaciation.

Extensive valley glaciers have profoundly reshaped mountains worldwide, carving **V**-shaped stream valleys into **U**-shaped glaciated valleys and producing many distinctive erosional and depositional landforms. As cirque walls erode away, sharp **arêtes** (sawtooth, serrated ridges) form, dividing adjacent cirque basins. Two eroding cirques may reduce an arête to a saddlelike **col**. A **horn** results when several cirque glaciers gouge an individual mountain summit from all sides, forming a pyramidal peak. A **bergschrund** forms when a crevasse or wide crack opens along the headwall of a glacier; it is most visible in summer when covering snow is gone. An ice-carved rock basin left as a glacier retreats may fill with water to form a **tarn**; tarns in a string separated by moraines are **paternoster lakes**. Where a glacial valley trough joins the ocean and the glacier retreats, the sea extends inland to form a **fjord**.

All glacial deposits, whether ice-borne or meltwater-borne, constitute **glacial drift**. Glacial meltwater deposits are sorted and are **stratified drift**. Direct deposits from ice are unstratified and unsorted **till**. Specific landforms produced by the deposition of drift are **moraines.** A **lateral moraine** forms along each side of a glacier, merging glaciers with lateral moraines form a **medial moraine,** and eroded debris dropped at the glacier's terminus is a **terminal moraine**.

Continental glaciation leaves different features than does alpine glaciation. A **till plain** forms behind end moraines, featuring unstratified coarse till, low and rolling relief, and deranged drainage. Beyond the morainal deposits, **outwash plains** of stratified drift feature stream channels that are meltwater-fed, braided, and overloaded with debris that is sorted and deposited across the landscape. An **esker** is a sinuously curving, narrow ridge of coarse sand and gravel that forms along the channel of a meltwater stream beneath a glacier. An isolated block of ice left by a retreating glacier becomes surrounded by debris; when the block finally melts, it leaves a steep-sided **kettle**. A **kame** is a small hill, knob, or mound of poorly sorted sand and gravel that is deposited directly by water or by ice in crevasses.

Glacial action forms two types of streamlined hills: The erosional **roche moutonnée** is an asymmetrical hill of exposed bedrock, gently sloping upstream and abruptly sloping downstream; the depositional **drumlin** is deposited till, streamlined in the direction of continental ice movement (blunt end upstream and tapered end downstream).

arête (p. 497)
col (p. 498)
horn (p. 498)
bergschrund (p. 498)
tarn (p. 498)
paternoster lakes (p. 498)
fjord (p. 499)
glacial drift (p. 500)
stratified drift (p. 500)
till (p. 500)
moraine (p. 500)
lateral moraine (p. 500)
medial moraine (p. 500)
terminal moraine (p. 501)
till plain (p. 502)
outwash plain (p. 502)
esker (p. 502)
kettle (p. 502)
kame (p. 503)
roche moutonnée (p. 503)
drumlin (p. 503)

8. How does a glacier accomplish erosion?
9. Describe the evolution of a **V**-shaped stream valley to a **U**-shaped glaciated valley. What features are visible after the glacier retreats?
10. How is an arête formed? A col? A horn? Briefly differentiate among them.
11. Differentiate between two forms of glacial drift—till and outwash.
12. What is a morainal deposit? What specific moraines are created by alpine and continental glaciers?
13. What are some common depositional features encountered in a till plain?
14. Contrast a roche moutonnée and a drumlin with regard to appearance, orientation, and the way each forms.

- ***Analyze*** the spatial distribution of periglacial processes and related permafrost and frozen ground, and ***explain*** the present thaw under way.

The term **periglacial** describes cold-climate processes, landforms, and topographic features that exist along the margins of glaciers, past and present. When soil or rock temperatures remain below 0°C (32°F) for at least 2 years, **permafrost** ("permanent frost") develops. Note that this criterion is based solely on temperature and has nothing to do with how much or how little water is present. The **active layer** is the zone of seasonally frozen ground that exists between the subsurface permafrost layer and the ground surface. **Patterned ground** forms in the periglacial environment where freezing and thawing of the ground create polygonal forms of circles, polygons, stripes, nets, and steps.

periglacial (p. 504)
permafrost (p. 504)
active layer (p. 505)
patterned ground (p. 507)

15. In terms of climatic types, describe the areas on Earth where periglacial landscapes occur. Include both higher-latitude and higher-altitude climate types.
16. Define two types of permafrost, and differentiate their occurrence on Earth. What are the characteristics of each?
17. Describe the active zone in permafrost regions, and relate the degree of development to specific latitudes.
18. What is a talik? Where might you expect to find taliks, and to what depth do they occur?
19. What is the difference between permafrost and ground ice?
20. Describe the role of frost action in the formation of various landform types in the periglacial region, such as patterned ground.
21. Relate some of the specific problems humans encounter in building on periglacial landscapes.

- ***Explain*** the Pleistocene ice-age epoch and related glacials and interglacials, and ***describe*** some of the methods used to study paleoclimatology.

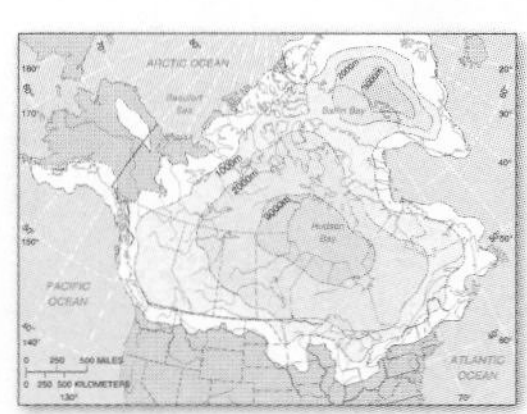

An **ice age** is any extended period of cold. The late Cenozoic Era featured pronounced ice-age conditions during the Pleistocene. During this time, alpine and continental glaciers covered about 30% of Earth's land area in at least 18 glacials, punctuated by interglacials of milder weather. Beyond the ice, **paleolakes** formed because of wetter conditions. Evidence of ice-age conditions is gathered from ice cores drilled in Greenland and Antarctica, from ocean sediments, from coral growth in relation to past sea levels, and from rock. The study of past climates is *paleoclimatology*.

The apparent pattern followed by these low-temperature episodes indicates cyclic causes. A complicated mix of interacting variables—celestial relations, solar variability, tectonic factors, atmospheric variables, and oceanic circulation—appears to influence long-term climatic trends. The present anthropogenic forcing of temperatures is not part of a natural cycle.

ice age (p. 509)
paleolake (p. 512)

22. What is paleoclimatology? Describe Earth's past climatic patterns. Are we experiencing a normal climate pattern in this era, or have scientists noticed any significant trends indicating climate change?
23. Define an *ice age*. When was the most recent? Explain *glacial* and *interglacial* in your answer.
24. Summarize what science has learned about the causes of ice ages by listing and explaining at least four possible factors in climate change.
25. Describe the role of ice cores in deciphering past climates. What record do they preserve? Where were they drilled?

- ***Define*** each polar region, and ***list*** climatic changes taking place and associated impacts.

The 10°C (50°F) isotherm for July defines the **Arctic region** (green line on the map in Figure 17.31). This line coincides with the visible tree line—the boundary between the northern forests and tundra. The Antarctic convergence defines the **Antarctic region** in a narrow zone that extends around the continent as a boundary between colder Antarctic water and warmer water at lower latitudes. This boundary follows roughly the 10°C (50°F) isotherm for February, the Southern Hemisphere summer, and is located near 60° S latitude

Arctic region (p. 517)
Antarctic region (p. 517)

26. Explain the relationship between the criteria defining the Arctic and Antarctic regions. Is there any coincidence between the Arctic criteria and the distribution of Northern Hemisphere forests?
27. Based on information in this chapter and elsewhere in *Geosystems*, summarize a few of the changing conditions under way in each polar region. Are there any conditions that seem of greatest concern to you after studying them?

MASTERING GEOGRAPHY

Visit www.masteringgeography.com for several figure animations, satellite loops, self-study quizzes, flashcards, visual glossary, case studies, career links, textbook reference maps, RSS Feeds, and an ebook version of *Geosystems*. Included are many geography web links to interesting Internet sources that support this chapter.

PART IV

Soils, Ecosystems, and Biomes

Earth is the home of the only known biosphere in the Solar System—a unique, complex, and interactive system of abiotic (nonliving) and biotic (living) components working together to sustain a tremendous diversity of life. Energy enters the biosphere through conversion of solar energy by photosynthesis in the leaves of plants. Soil is the essential link among the lithosphere, plants, and the rest of Earth's physical systems. Thus, soil helps sustain life and is a bridge between Part III and Part IV of this text.

Life is organized into a feeding hierarchy from producers to consumers, ending with decomposers. Taken together, the soils, plants, animals, and all abiotic components produce aquatic and terrestrial ecosystems, generally grouped together in various biomes. Today, we face crucial issues, principally the preservation of the diversity of life in the biosphere. Patterns of land and ocean temperatures, precipitation, weather phenomena, and stratospheric ozone, among many elements, are changing as global climate systems shift. The resilience of the biosphere as we know it is being tested in a real-time, one-time experiment. These important issues of biogeography are considered in Part IV.

Antarctic fur seals occupy tussock-grass cushions near the east end of South Georgia Island in Antarctica. The periglacial landscape of South Georgia, with its *polar marine* climate, is a major breeding location for the fur seal population. The month is April, fall in the Southern Hemisphere. [Bobbé Christopherson.]

CHAPTER 18

The Geography of Soils

Rich, fertile Mollisols of central Kansas support intense agricultural activity in America's breadbasket. In this aerial photo, you can see plowed fields, newly harvested wheat, the edge of a tract with central-pivot irrigation, a small feedlot, and a fallow segment of land—study the photo to find each of these aspects. [Bobbé Christopherson.]

KEY LEARNING CONCEPTS

After reading the chapter, you should be able to:

- ***Define*** soil and soil science, and ***describe*** a pedon, polypedon, and typical soil profile.
- ***Describe*** soil properties, including color, texture, structure, consistence, porosity, and soil moisture.
- ***Explain*** basic soil chemistry, including cation-exchange capacity, and ***relate*** these concepts to soil fertility.
- ***Evaluate*** the principal soil formation factors, including the human element.
- ***Describe*** the 12 soil orders of the Soil Taxonomy classification system, and ***explain*** their general distribution across Earth.

GEOSYSTEMS NOW

High-Latitude Soils Emit Greenhouse Gases

Throughout this book, we have discussed effects of record high temperatures in the polar regions; look back at the *Geosystems Now* features for Chapters 4 and 17 as well as the discussion of global temperatures in Chapter 5. As sea ice, ice shelves, and tidewater glaciers melt, the possible effects are wide ranging—from altered surface albedo, to accelerated sea-level rise, to losses of high-latitude species as habitats change.

However, another process in the northern polar regions has the potential to increase global warming by releasing substantial amounts of greenhouse gases into the atmosphere. This process is related not to ice, but rather to soil—specifically, the effects of thawing permafrost.

Gelisols, in the U.S. Soil Taxonomy, are the permafrost-affected soils of the northern latitudes. These soils contain about half of the pool of global carbon (Figure GN 18.1). The latest estimate of the carbon contained in these periglacial soils is 1.7 trillion tons, which is double the amount mentioned in the Intergovernmental Panel on Climate Change (IPPC) *Fourth Assessment Report* (2007). These are slow in development and rich in organic content due to slow decomposition, and they are fragile.

The higher latitudes are warming at more than double the rate of the middle and lower latitudes. Gelisols can quickly become wet and soggy with only a slight shift in their thermal balance. As this thaw happens, the poorly decomposed organic content begins to decay, with decomposition releasing enormous quantities of carbon dioxide into the atmosphere through increased respiration. Another greenhouse gas, methane, is released as well. Researchers stated the linkage in these systems in 2009:

> We show that approximately 1 C° warming accelerated total ecosystem respiration rates from carbon in peat. . . . Our findings suggest a large, long-lasting, positive feedback of carbon stored in northern peatlands to the global climate system.*

A 2010 publication—*Soil Atlas of the Northern Circumpolar Region* (Figure GN 18.2)—completed as part of the International Polar Year and published by the Joint Research Center of the European Commission (Luxembourg: European Union, 2010), brings all this together. Seven international agencies were part of this effort, and the final atlas is an excellent reference for periglacial environments and permafrost-affected soils.

*E. Dorrepaal et al., "Carbon respiration from subsurface peat accelerated by climate warming in the subarctic," *Nature* 460 (July 30, 2009): 616.

The atlas is available for download from http://eusoils.jrc.ec.europa.eu/library/maps/Circumpolar/.

In this atlas, regional soil distributions, mapped at high resolution, are portrayed on maps using Geographic Information System (GIS) technology (information on GIS and soil databases appears on atlas page 117). The atlas draws special attention to the relationship between climate change and properties of soils and to the general impact of climate on soil characteristics.

As the science community moves forward in the preparation of the IPCC *Fifth Assessment Report*, this northern soil and environment atlas will be an important resource. As summarized in this new atlas (pages 7 and 12):

> Anthropogenic pressures on northern ecosystems and soil resources are rapidly increasing. . . . The atlas illustrates vividly the great diversity of circumpolar soils and northern landscapes while encouraging the general public and governments to develop policies for the protection of these fragile regions. . . . The poles are recognized as sensitive barometers of environmental change.

(a)

(b)

FIGURE GN 18.1 Permafrost-affected soils.
(a) Gelisols and tundra green in the brief summer season on Spitsbergen Island, as the active layer thaws. (b) Turning over a clod exposes the fibrous organic content and slow decomposition. [Bobbé Christopherson.]

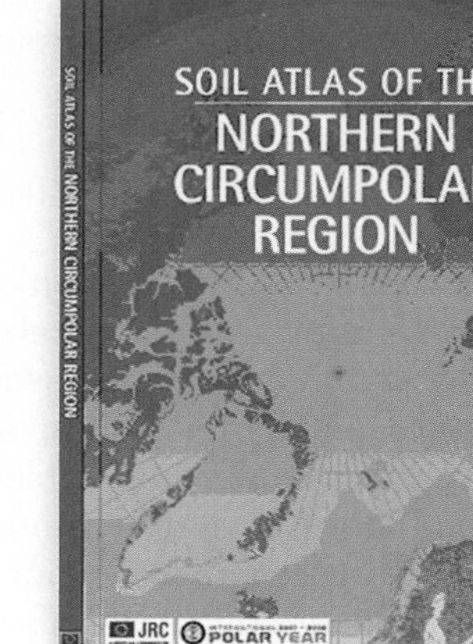

FIGURE GN 18.2 Soil Atlas of the Northern Circumpolar Region.

Earth's landscape generally is covered with soil. **Soil** is a dynamic natural material composed of fine particles in which plants grow, and it contains both mineral fragments and organic matter. The soil system includes human interactions and supports all human, other animal, and plant life. If you have ever planted a garden, tended a houseplant, or been concerned about famine and soil loss, this chapter will interest you. A knowledge of soil is at the heart of agriculture and food production.

You kneel and scoop up a handful of prairie soil, compressing it and breaking it apart with your fingers. You are holding a historical object, one that bears the legacy of the last 15,000 years or more. This lump of soil contains information about the last ice age and intervening warm intervals, about distinct and distant source materials, about several physical processes. We are using and abusing this legacy at rates much faster than it was formed. Soils do not reproduce, nor can they be re-created.

Soil science is interdisciplinary, involving physics, chemistry, biology, mineralogy, hydrology, taxonomy, climatology, and cartography. Physical geographers are interested in the spatial patterns formed by soil types and the physical factors that interact to produce them. As an integrative science, physical geography is well suited for the task of understanding soils. *Pedology* deals with the origin, classification, distribution, and description of soils (*ped* is from the Greek *pedon*, meaning "soil" or "earth"). *Edaphology* specifically focuses on the study of soil as a medium for sustaining the growth of higher plants (*edaphos* means "soil" or "ground").

Soil science deals with a complex substance whose characteristics vary from kilometer to kilometer—and even centimeter to centimeter. In many locales, an *agricultural extension service* can provide specific information and perform a detailed analysis of local soils. Soil surveys and local soil maps are available for most counties in the United States and for the Canadian provinces.

Your Internet browser links you to the U.S. Department of Agriculture, Natural Resources Conservation Service (NRCS; http://www.nrcs.usda.gov/), or Agriculture Canada's Soil Information System (http://sis.agr.gc.ca/cansis/intro.html), or you may contact the appropriate department at a local college or university. See the National Soil Survey Center's site at http://soils.usda.gov/; more information on Soil Taxonomy is available at http://soils.ag.uidaho.edu/soilorders/.

In this chapter: The geography of soils concerns the inputs and actions that produce a substance with such varied characteristics. We begin with soil characteristics and the basic soil sampling and soil mapping units. The soil profile is a dynamic structure, mixing and exchanging materials and moisture across its horizons. Properties of soil include texture, structure, porosity, moisture, and chemistry—all integrating to form soil types. The chapter discusses both natural and human factors that affect soil formation. A global concern exists over the loss of soils to erosion, mistreatment, and conversion to other uses. The chapter concludes with a brief examination of the Soil Taxonomy, the 12 principal soil orders, and their spatial distribution.

Soil Characteristics

Classifying soils is similar to classifying climates because both involve interacting variables. Before we look at soil classification, let us examine the physical properties that distinguish soils as they develop through time in response to climate, relief, and topography.

Soil Profiles

As a book cannot be judged by its cover, so soils cannot be evaluated at the surface only. Instead, a soil profile should be studied from the surface to the deepest extent of plant roots or to the point where regolith or bedrock is encountered. Such a profile is a **pedon**, a hexagonal column measuring 1 to 10 m^2 in top surface area (Figure 18.1). At the sides of the pedon, the various layers of the soil profile are visible in cross section and are labeled with letters. *A pedon is the basic sampling unit used in soil surveys.*

Many pedons together in one area make up a **polypedon**, which has distinctive characteristics differentiating it from surrounding polypedons. A polypedon is an essential soil "individual," comprising an identifiable series of soils in an area. It can have a minimum dimension of about 1 m^2 (35.3 ft^3) and has no specified maximum size. *The polypedon is the basic mapping unit used in preparing local soil maps.*

Soil Horizons

Each distinct layer exposed in a pedon is a **soil horizon**. A horizon is roughly parallel to the pedon's surface and has characteristics distinctly different from horizons directly above and/or below. The boundary between horizons usually is distinguishable when viewed in profile, such as along a road cut, using the properties of color, texture, structure, consistence (meaning soil consistency or cohesiveness), porosity, the presence or absence of certain minerals, moisture, and chemical processes (Figure 18.2). Soil horizons are the building blocks of soil classification.

At the top of the soil profile is the *O* (organic) *horizon*, named for its organic composition, derived from plant and animal litter that was deposited on the surface and transformed into humus. **Humus** is not just a single material; it is a mixture of decomposed and synthesized organic materials, usually dark in color. Microorganisms work busily on this organic debris, performing a portion of the *humification* (humus-making) process. The O horizon is 20%–30% or more organic matter, which is important because of its ability to retain water and nutrients and because of the way it acts in a complementary manner to clay minerals.

At the bottom of the soil profile is the *R* (rock) *horizon*, consisting of either unconsolidated (loose) material or consolidated bedrock. When bedrock physically and chemically weathers into regolith, it may or may not contribute to overlying soil horizons. The A, E, B, and C

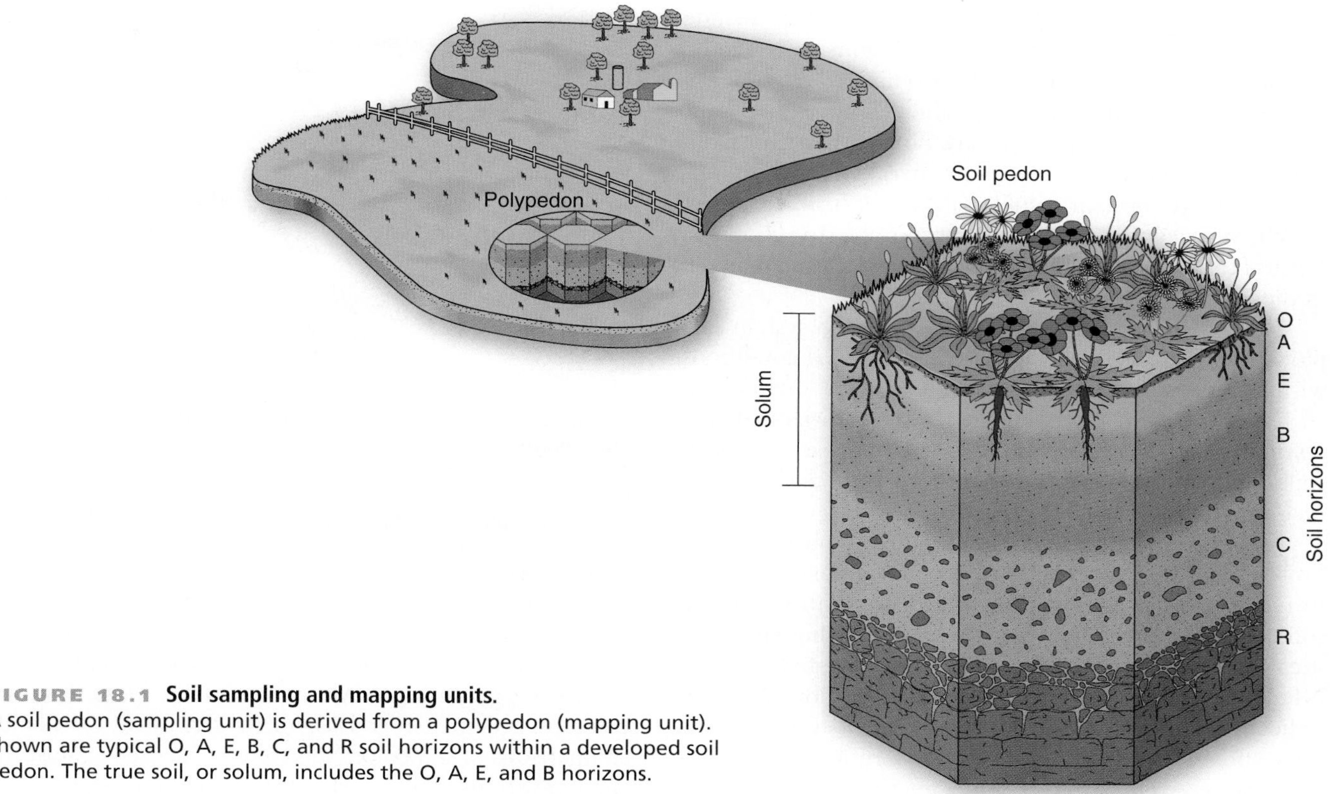

FIGURE 18.1 Soil sampling and mapping units.
A soil pedon (sampling unit) is derived from a polypedon (mapping unit). Shown are typical O, A, E, B, C, and R soil horizons within a developed soil pedon. The true soil, or solum, includes the O, A, E, and B horizons.

horizons mark differing mineral strata between O and R. These middle layers are composed of sand, silt, clay, and other weathered by-products. (Table 11.3 in Chapter 11 presents a description of grain sizes for these weathered soil particles.)

In the *A horizon*, humus and clay particles are particularly important, for they provide essential chemical links between soil nutrients and plants. This horizon usually is richer in organic content, and hence darker, than are lower horizons. Here is where human disruption through plowing, pasturing, and other uses takes place. The A horizon grades into the *E horizon*, made up of coarse sand, silt, and resistant minerals.

FIGURE 18.2 A typical soil profile.
This is a Mollisol pedon in southeastern South Dakota. The parent material is till, and the soil is well drained. The dark O and A horizons above the #1 transition into an E horizon. Distinct carbonate nodules are visible in the lower B and upper C horizons. [Marbut Collection, Soil Science Society of America, Inc.]

From the lighter-colored E horizon, silicate clays and oxides of aluminum and iron are leached (removed by water) and carried to lower horizons with the water as it percolates through the soil. This process of removing fine particles and minerals by water, leaving behind sand and silt, is **eluviation**—thus, the E designation for this horizon. As precipitation increases, so does the rate of eluviation.

In contrast to the A and E horizons, *B horizons* accumulate clays, aluminum, and iron. B horizons are dominated by **illuviation**, a depositional process (in contrast, eluviation is an erosional process). B horizons may exhibit reddish or yellowish hues because of the presence of illuviated minerals (silicate clays, iron and aluminum, carbonates, gypsum) and organic oxides.

Some materials occurring in the B horizon may have formed in place from weathering processes rather than arriving there by *translocation*, or migration. In the humid tropics, these layers often develop to some depth. Likewise, clay losses in an A horizon may be caused by destructive processes in place and not eluviation. Research to better understand the erosion and deposition of clays between soil horizons is one of the challenges in modern soil science.

The combination of the A and E horizons and the B horizon is designated the **solum**, considered the true definable soil of the pedon. The A, E, and B horizons experience active soil processes (labeled in Figure 18.1).

Below the solum is the *C horizon* of weathered bedrock or weathered parent material. This zone is identified as *regolith* (although the term sometimes is used to include the solum as well). The C horizon is not much affected by soil operations in the solum and lies outside the biological influences experienced in the shallower horizons. Plant roots and soil microorganisms are rare in the C horizon. It lacks clay concentrations and generally is made up of carbonates, gypsum, or soluble salts, or of iron and silica, which form cemented soil structures. In dry climates, calcium carbonate commonly forms the cementing material of these hardened layers.

Soil scientists using the U.S. classification system employ letter suffixes to further designate special conditions within each soil horizon. A few examples include Ap (A horizon that has been plowed), Bt (B horizon that is composed of illuviated clay), Bf (permafrost or frozen soil), and Bh (illuviated humus that forms a dark coating on sand and silt particles).

In 1995, the International Committee on Anthropogenic Soils (ICOMANTH; http://clic.cses.vt.edu/icomanth/) formed to better deal with classification of anthropogenic soils, those with major properties created by human activities. The designation *M* was first incorporated into the NRCS *Keys to Soil Taxonomy* in the 10th edition (2006) to indicate the presence of horizontal artifacts such as concrete and asphalt pavement and plastic, rubber, and textile liners placed by humans for some practical purpose. An *M* tells us that a subsoil layer is present that limits root growth. This is an innovation for soil science to assist in assessing construction or septic system monitoring, among other considerations.

Soil Properties

Soils are complex and varied, as this section reveals. Observing a real soil profile will help you identify color, texture, structure, and other soil properties. A good location to observe soil profiles is a construction site or excavation, perhaps on your campus, or a road cut along a highway. The NRCS *Soil Survey Manual* (U.S. Department of Agriculture Handbook No. 18, October 1993) presents information on all soil properties. The NRCS *Soil Survey Laboratory Methods Manual* (U.S. Department of Agriculture Soil Survey Investigations Report No. 42, v. 4.0, November 2004) details specific methods and practices used to conduct soil analysis. Both documents are available for download at the NRCS Web site.

Soil Color

Color is important, for it sometimes suggests composition and chemical makeup. If you look at exposed soil, color may be the most obvious trait. Among the many possible hues are the reds and yellows found in soils of the southeastern United States (high in iron oxides), the blacks of prairie soils in portions of the U.S. grain-growing regions and in Ukraine (richly organic), and the white to pale hues found in soils containing silicates and aluminum oxides. However, color can be deceptive: Soils with high humus content are often dark, yet clays of warm temperate and tropical regions with less than 3% organic content are some of the world's blackest soils.

FIGURE 18.3 A Munsell Soil Color Chart page.
A soil sample is viewed through the hole to match it with a color on the chart. Hue, value, and chroma are the characteristics of color assessed by this system. [Courtesy of Gretag Macbeth, Munsell Color.]

To standardize their descriptions, soil scientists describe a soil's color by comparing it with a *Munsell Color Chart* (developed by artist and teacher Albert Munsell in 1913). These charts display 175 colors arranged by *hue* (the dominant spectral color, such as red), *value* (the degree of darkness or lightness), and *chroma* (the purity and saturation of the color, which increase with decreasing grayness). A Munsell notation identifies each color by a name, so soil scientists can make worldwide comparisons of soil color. Soil color is checked against the chart at various depths within a pedon (Figure 18.3).

Soil Texture

Soil texture, perhaps a soil's most permanent attribute, refers to the mixture of sizes of its particles and the proportion of different sizes. Individual mineral particles are *soil separates*. All particles smaller in diameter than 2 mm (0.08 in.), such as very coarse sand, are considered part of the soil. Larger particles, such as pebbles, gravel, or cobbles, are not part of the soil. (Sands are graded from coarse, to medium, to fine, down to 0.05 mm; silt to 0.002 mm; and clay at less than 0.002 mm.)

Figure 18.4 is a *soil texture triangle* showing the relation of sand, silt, and clay concentrations in soil. Each corner of the triangle represents a soil consisting solely of the particle size noted (although rarely are true soils composed of a single separate). Every soil on Earth is defined somewhere in this triangle.

Figure 18.4 includes the common designation **loam**, which is a balanced mixture of sand, silt, and clay that is beneficial to plant growth. Farmers consider ideal a sandy loam with clay content below 30% (lower left) because of its water-holding characteristics and ease of cultivation. Soil texture is important in determining water-retention and water-transmission traits.

To see how the soil texture triangle works, consider a *Miami silt loam* soil type from Indiana. Samples from this soil type are plotted on the soil texture triangle as points 1, 2, and 3. A sample taken near the surface in the A horizon is recorded at point 1, in the B horizon at point 2, and in the C horizon at point 3. Textural analyses of these samples are summarized in the table and in the three pie diagrams to the right of the triangle. Note

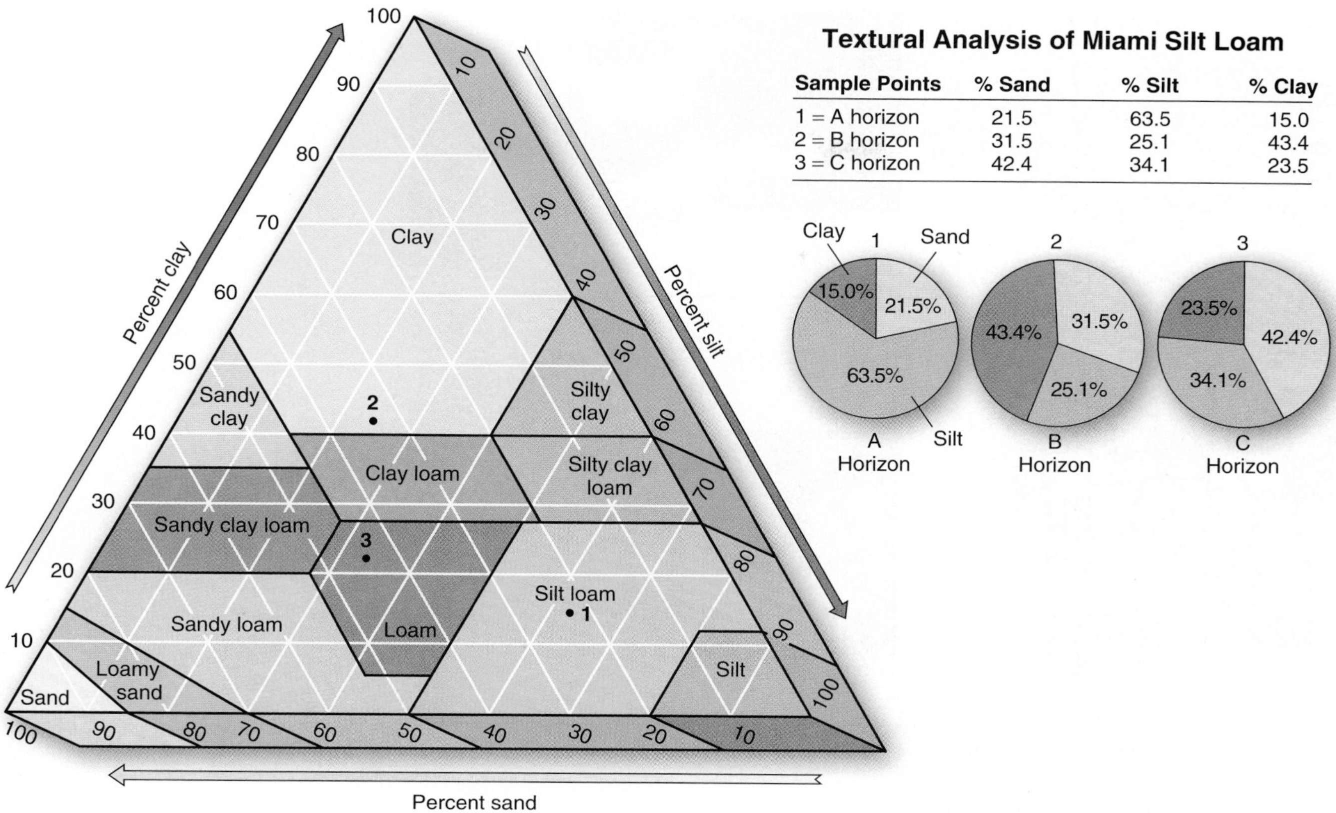

Textural Analysis of Miami Silt Loam

Sample Points	% Sand	% Silt	% Clay
1 = A horizon	21.5	63.5	15.0
2 = B horizon	31.5	25.1	43.4
3 = C horizon	42.4	34.1	23.5

FIGURE 18.4 **Soil texture triangle.**
Measures of the ratio of clay, silt, and sand determine soil texture. As an example, points 1 (horizon A), 2 (horizon B), and 3 (horizon C) designate samples taken at three different horizons in the Miami silt loam in Indiana. Note the ratio of sand to silt to clay shown in the three pie diagrams and table. [After USDA–NRCS, *Soil Survey Manual,* Agricultural Handbook No. 18, p. 138 (1993).]

that silt dominates the surface, clay the B horizon, and sand the C horizon. The *Soil Survey Manual* presents guidelines for estimating soil texture by feel, a relatively accurate method when used by an experienced person. However, laboratory methods using graduated sieves and separation by mechanical analysis in water allow more precise measurements.

Soil Structure

Soil *texture* describes the size of soil particles, but soil *structure* refers to their *arrangement.* Structure can partially modify the effects of soil texture. The smallest natural lump or cluster of particles is a *ped.* The shape of soil peds determines which of the structural types the soil exhibits: crumb or granular, platy, blocky, prismatic, or columnar (Figure 18.5).

Peds separate from each other along zones of weakness, creating voids, or pores, that are important for moisture storage and drainage. Rounded peds have more pore space between them and greater permeability than do other shapes. They are therefore better for plant growth than are blocky, prismatic, or platy peds, despite comparable fertility. Terms used to describe soil structure include *fine*, *medium*, and *coarse*. Adhesion among peds ranges from weak to strong.

Soil Consistence

In soil science, the term *consistence* is used to describe the consistency of a soil or cohesion of its particles. Consistence is a product of texture (particle size) and structure (ped shape). Consistence reflects a soil's resistance to breaking and manipulation under varying moisture conditions:

- A *wet soil* is sticky between the thumb and forefinger, ranging from a little adherence to either finger, to sticking to both fingers, to stretching when the fingers are moved apart. *Plasticity,* the quality of being moldable, is roughly measured by rolling a piece of soil between your fingers and thumb to see whether it rolls into a thin strand.
- A *moist soil* is filled to about half of field capacity (the usable water capacity of soil), and its consistence grades from loose (noncoherent), to *friable* (easily pulverized), to firm (not crushable between the thumb and forefinger).
- A *dry soil* is typically brittle and rigid, with consistence ranging from loose, to soft, to hard, to extremely hard.

Soil particles are sometimes cemented together to some degree. Soils are described as weakly cemented, strongly cemented, or *indurated* (hardened). Calcium carbonate, silica,

FIGURE 18.5 Types of soil structure.
Structure is important because it controls drainage, rooting of plants, and how well the soil delivers nutrients to plants. The shape of individual peds, shown here, controls a soil's structure. [National Soil Survey Center, Natural Resources Conservation Service, U.S.D.A.]

and oxides or salts of iron and aluminum all can serve as cementing agents. The cementation of soil particles that occurs in various horizons is a function of consistence and may be continuous or discontinuous.

Soil Porosity

Soil porosity, permeability, and moisture storage were discussed in Chapter 9. Pores in the soil horizon control the movement of water—its intake, flow, and drainage—and air ventilation. Important porosity factors are pore *size*, pore *continuity* (whether pores are interconnected), pore *shape* (whether pores are spherical, irregular, or tubular), pore *orientation* (whether pore spaces are vertical, horizontal, or random), and pore *location* (whether pores are within or between soil peds).

Porosity is improved by the biotic actions of plant roots, animal activity such as the tunneling of gophers or worms, and human intervention through soil manipulation (plowing, adding humus or sand, or planting soil-building crops). Much of the soil preparation work done before planting by farmers—and by home gardeners as well—is done to improve soil porosity.

Soil Moisture

Reviewing Figure 9.8 in Chapter 9 (soil moisture types and availability) will help you understand this section. Plants operate most efficiently when the soil is at *field capacity*, which is the maximum water availability for plant use after large pore spaces have drained of gravitational water. Soil type determines field capacity. The depth to which a plant sends its roots determines the amount of soil moisture to which the plant has access. If soil moisture is removed below field capacity, plants must exert increased energy to obtain available water. This moisture removal inefficiency worsens until the plant reaches its wilting point. Beyond this point, plants are unable to extract the water they need, and they die.

Soil moisture regimes and their associated climate types shape the biotic and abiotic properties of the soil more than any other factor. The NRCS *Keys to Soil Taxonomy*, discussed later in this chapter, recognizes five soil moisture regimes based on Thornthwaite's water-balance principles (Chapter 9), ranging from constantly wet ("Aquic regime") to dry ("Xeric regime").

Soil Chemistry

Recall that soil pores may be filled with air, water, or a mixture of the two. Consequently, soil chemistry involves both air and water. The atmosphere within soil pores is mostly nitrogen, oxygen, and carbon dioxide. Nitrogen concentrations are about the same as in the atmosphere, but oxygen is less and carbon dioxide is greater because of ongoing respiration processes.

Water present in soil pores is the *soil solution*. It is the medium for chemical reactions in soil. This solution is

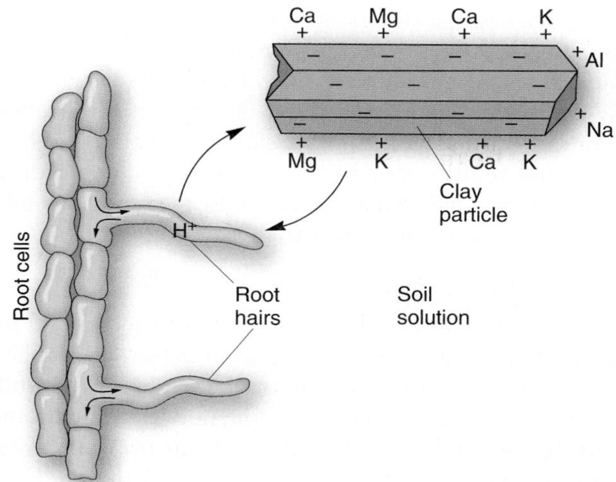

FIGURE 18.6 Soil colloids and cation-exchange capacity (CEC).
This typical soil colloid retains mineral ions by adsorption to its surface (opposite charges attract). This process holds the ions until they are absorbed by root hairs.

critical to plants as their source of nutrients, and it is the foundation of *soil fertility*. Carbon dioxide combines with the water to produce carbonic acid, and various organic materials combine with the water to produce organic acids. These acids are then active participants in soil processes, as are dissolved alkalies and salts.

To understand how the soil solution behaves, let us go through a quick chemistry review. An *ion* is an atom, or group of atoms, that carries an electrical charge (examples: Na^+, Cl^-, HCO_3^-). An ion has either a positive charge or a negative charge. For example, when NaCl (sodium chloride) dissolves in solution, it separates into two ions: Na^+, a *cation* (positively charged ion), and Cl^-, an *anion* (negatively charged ion). Some ions in soil carry single charges, whereas others carry double or even triple charges (e.g., sulfate, SO_4^{2-}; and aluminum, Al^{3+}).

Ions in soil are retained by **soil colloids**. These tiny particles of clay and organic material (humus) carry a negative electrical charge and consequently attract any positively charged ions in the soil (Figure 18.6). The positive ions, many metallic, are critical to plant growth. If it were not for the negatively charged soil colloids, the positive ions would be leached away in the soil solution and thus would be unavailable to plant roots.

Individual clay colloids are thin and platelike, with parallel surfaces that are negatively charged. They are more chemically active than silt and sand particles, but less active than organic colloids. Metallic cations attach to the surfaces of the colloids by *adsorption* (not *ab*sorption, which means "to enter"). Colloids can exchange cations between their surfaces and the soil solution, an ability called **cation-exchange capacity (CEC)**, which is the measure of soil fertility. A high CEC means that the soil colloids can store or exchange more cations from the soil solution, an indication of good soil fertility (unless a complicating factor exists, such as a soil that is too acid).

Therefore, **soil fertility** is the ability of soil to sustain plants. Soil is fertile when it contains organic substances and clay minerals that *absorb* water and *adsorb* certain elements needed by plants. Billions of dollars are expended to create fertile soil conditions, yet the future of Earth's most fertile soils is threatened because soil erosion is on the increase worldwide.

Soil Acidity and Alkalinity

A soil solution may contain significant hydrogen ions (H^+), the cations that stimulate acid formation. The result is a soil rich in hydrogen ions, or an *acid soil*. On the other hand, a soil high in base cations (calcium, magnesium, potassium, sodium) is a *basic* or *alkaline soil*. Such acidity or alkalinity is expressed on the pH scale (Figure 18.7).

Pure water is nearly neutral, with a pH of 7.0. Readings below 7.0 represent increasing acidity. Readings above 7.0 indicate increasing alkalinity. Acidity usually is regarded as strong at 5.0 or lower on the pH scale, whereas 10.0 or above is considered strongly alkaline.

The major contributor to soil acidity in this modern era is acid precipitation (rain, snow, fog, or dry deposition). Acid rain actually has been measured below pH 2.0—an incredibly low value for natural precipitation, as acid as lemon juice. Increased acidity in the soil solution accelerates the chemical weathering of mineral nutrients and increases their depletion rates. Because most crops are sensitive to specific pH levels, acid soils below pH 6.0 require treatment to raise the pH. This soil treatment is accomplished by the addition of bases in the form of minerals that are rich in base cations, usually lime (calcium carbonate, $CaCO_3$).

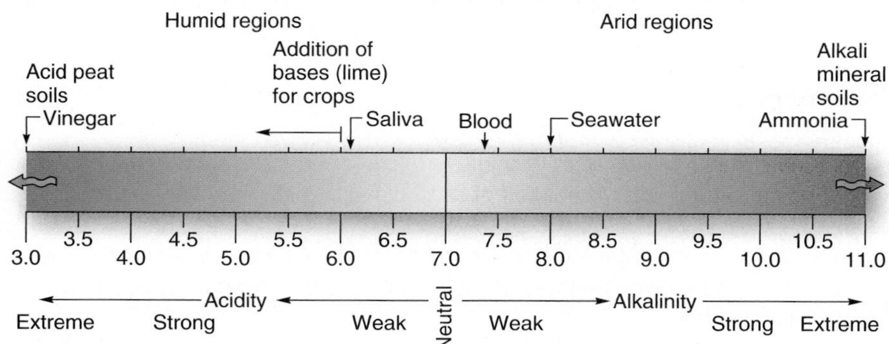

FIGURE 18.7 pH scale.
The pH scale measures acidity (lower pH) and alkalinity (higher pH). (The complete pH scale ranges between 0 and 14.)

Soil Formation Factors and Management

Soil is an open system involving physical inputs and outputs. Soil-forming factors are both *passive* (parent material, topography and relief, and time) and *dynamic* (climate, biology, and human activities). These factors work together as a system to form soils. The roles of these factors are considered here and in the soil-order discussions that follow.

Natural Factors

Physical and chemical weathering of rocks in the upper lithosphere provides the raw mineral ingredients for soil formation. These rocks supply the parent materials, and their composition, texture, and chemical nature help determine the type of soil that forms. Clay minerals are the principal weathered by-products in soil.

Climate types correlate closely with soil types worldwide. The moisture, evaporation, and temperature regimes of climates determine the chemical reactions, organic activity, and eluviation rates of soils. Not only is the present climate important, but also many soils exhibit the imprint of past climates, sometimes over thousands of years. Most notable is the effect of glaciations. Among other contributions, glaciation produced the loess soil materials that were windblown thousands of kilometers to their present locations (discussed in Chapter 13).

Vegetation and the activities of animals and bacteria determine the organic content of soil, along with all that is living in soil—algae, fungi, worms, and insects. The chemical makeup of the vegetation contributes to the acidity or alkalinity of the soil solution. For example, broadleaf trees tend to increase alkalinity, whereas needleleaf trees tend to produce higher acidity. Thus, when civilization moves into new areas and alters the natural vegetation by logging or plowing, the affected soils are likewise altered, often permanently.

Topography also affects soil formation. Slopes that are too steep cannot have full soil development because gravity and erosional processes remove materials. Lands that are nearly level inhibit soil drainage and can become waterlogged. The compass orientation of slopes is important because it controls exposure to sunlight. In the Northern Hemisphere, a south-facing slope is warmer overall through the year because it receives direct sunlight. Water-balance relations are affected because north-facing slopes are colder, causing slower snowmelt and a lower evaporation rate, providing more moisture for plants than is available on south-facing slopes, which tend to dry faster.

All of the identified natural factors in soil development (parent material, climate, biological activity, and topography) require *time* to operate. Over geologic time, plate tectonics has redistributed landscapes and thus subjected soil-forming processes to diverse conditions.

The Human Factor

Human intervention has a major impact on soils. Millennia ago, farmers in most cultures learned to plant slopes "on the contour"—to make rows or mounds around a slope at the same elevation, not vertically up and down the slope. Planting on the contour prevents water from flowing straight down the slope and thus reduces soil erosion. Farmers often planted and harvested floodplains, while living on higher ground nearby. Floods were celebrated as blessings that brought water, nutrients, and more soil to the land. Society is drifting away from these commonsense strategies.

A few centimeters' thickness of prime farmland soil may require *500 years* to mature. Yet this same thickness is being lost annually through soil erosion when the soil-holding vegetation is removed and the land is plowed, regardless of topography. Flood control structures block sediments and nutrients from replenishing floodplain soils, leading to additional soil losses. Over the same period, exposed soils may be completely leached of needed cations, thereby losing their fertility.

Unlike living species, soils do not reproduce, nor can they be re-created. Some 35% of farmlands are losing soil faster than it can form—a loss exceeding 23 billion metric tonnes (25 billion tons) per year. Soil depletion and loss are at record levels from Iowa to China, Peru to Ethiopia, and the Middle East to the Americas. The impact on society is potentially disastrous as population and food demands increase.

Soil erosion can be compensated for in the short run by using more fertilizer, increasing irrigation, and planting higher-yielding strains. But the potential yield from prime agricultural land will drop by as much as 20% over the next 20 years if only moderate erosion continues. One study tabulated the market value of lost nutrients and other variables in the most comprehensive soil-erosion study to date. The sum of direct damage (to agricultural land) and indirect damage (to streams, society's infrastructure, and human

GEO REPORT 18.1 *Slipping through our fingers*

The U.S. Department of Agriculture estimated that more than 5 million acres of prime farmland are lost each year in the United States through mismanagement or conversion to nonagricultural uses. About half of all cropland in the United States and Canada is experiencing excessive rates of soil erosion—these countries are two of the few that monitor loss of topsoil. Worldwide, about one-third of potentially farmable land has been lost to erosion, much of that in the past 40 years. The causes for degraded soils, in order of severity, include overgrazing, vegetation removal, agricultural practices, conversion to nonagricultural activities, overexploitation, and industrial and bioindustrial use.

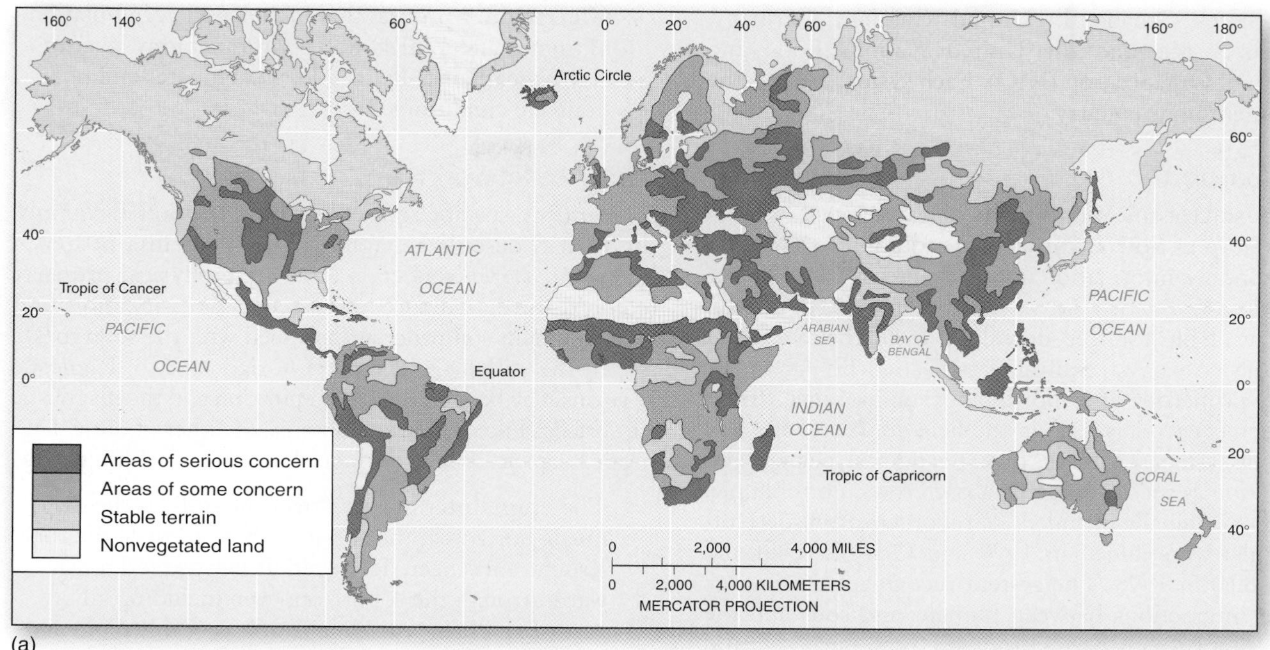

(a)

(b)

FIGURE 18.8 Soil degradation.
(a) Approximately 1.2 billion hectares (3.0 billion acres) of Earth's soils suffer degradation through erosion caused by human misuse and abuse. (b) An example of soil loss through sheet and gully erosion on a northwest Iowa farm. One millimeter of soil lost from an acre weighs about 5 tons. [(a) *A Global Assessment of Soil Degradation,* adapted from UNEP, International Soil Reference and Information Centre, "Map of Status of Human-Induced Soil Degradation," Nairobi, Kenya. (b) USDA Natural Resources Conservation Service, National Soil Survey Center.]

health) was estimated at more than $25 billion a year in the United States and hundreds of billions of dollars worldwide. Of course, this is a controversial assessment in the agricultural industry. The cost to bring erosion under control in the United States is estimated at approximately $8.5 billion, or about 30 cents on every dollar of damage and loss (Figure 18.8). A recent article by scientist David Montgomery well summarizes our predicament:

> Recent compilations of data from around the world show that soil erosion under conventional agriculture exceeds both rates of soil production and geological erosion rates by several orders of magnitude. Consequently, modern agriculture—and therefore global society—faces a fundamental question over the upcoming centuries. Can an agricultural system capable of feeding a growing population safeguard both soil fertility and the soil itself?*

*D. R. Montgomery, "Is agriculture eroding civilization's foundation?" *GSA Today*, October 2007, p. 4.

CRITICAL THINKING 18.1

Soil losses—What to do?

Given the information you just read in "The Human Factor," break down the issue of soil loss into the forcing causes, impacts produced, and possible actions to slow soil degradation and loss. Based on this analysis, suggest possible scales of response, moving from individual, to local, to regional, to state, and finally to national in scope. Given the problem and the scale at which you determine solutions are best implemented, what actions do you suggest to reverse the situation of soil degradation here and abroad?

Soil Classification

Classification of soils is complicated by the variety of interactions that create thousands of distinct soils—well over 15,000 soil types in the United States and Canada alone. Not surprisingly, a number of different classification systems are in use worldwide, including those from the

United States, Canada, the United Kingdom, Germany, Australia, Russia, and the United Nations Food and Agricultural Organization (FAO). Each system reflects the environment of its country.

Soil Taxonomy

The U.S. soil classification system, first developed in 1975, is referred to as **Soil Taxonomy**. In addition to the 1975 Soil Taxonomy, information in this chapter is based on the publication *Keys to Soil Taxonomy*, now in its 11th edition (2010), available for free download from the NRCS at http://soils.usda.gov/technical/classification/tax_keys/.

Soil properties and morphology (appearance, form, and structure) actually seen in the field are key to the Soil Taxonomy system. Thus, it is open to addition, change, and modification as the sampling database grows. For example, the system originally included 10 soil orders; Andisols (volcanic soils) were added in 1990 and Gelisols (cold and frozen soils) in 1998. The system recognizes the importance of interactions between humans and soils and the changes that humans have introduced, both purposely and inadvertently.

The classification system divides soils into six categories, creating a hierarchical sorting system. Each *soil series* (the smallest, most detailed category) ideally includes only one polypedon, but may include portions continuous with adjoining polypedons in the field. In sequence from smallest to largest categories, including the number of occurrences within each, the Soil Taxonomy recognizes *soil series* (15,000), *soil families* (6,000), *soil subgroups* (1,200), *soil great groups* (230), *soil suborders* (47), and *soil orders* (12).

Pedogenic Regimes Prior to the Soil Taxonomy system, **pedogenic regimes** were used to describe soils. These regimes keyed specific soil-forming processes to climatic regions. Although each pedogenic process may be active in several soil orders and in different climates, we discuss them within the soil order where they commonly occur. Such climate-based regimes are convenient for relating climate and soil processes. However, *the Soil Taxonomy system recognizes the great uncertainty and inconsistency in basing soil classification on such climatic variables.* Aspects of several pedogenic processes are discussed with appropriate soil orders:

- **Laterization**: a leaching process in humid and warm climates, discussed with Oxisols and shown in Figure 18.11;
- **Salinization**: a process that concentrates salts in soils in climates with excessive potential evapotranspiration (POTET) rates, discussed with Aridisols;
- **Calcification**: a process that produces an illuviated accumulation of calcium carbonates in continental climates, discussed with Mollisols and Aridisols and shown in Figure 18.15;
- **Podzolization**: a process of soil acidification associated with forest soils in cool climates, discussed with Spodosols and shown in Figure 18.20; and
- **Gleization**: a process that includes an accumulation of humus and a thick, water-saturated gray layer of clay beneath, usually in cold, wet climates and poor drainage conditions.

Diagnostic Soil Horizons

To identify a specific soil series within the Soil Taxonomy, the NRCS describes diagnostic horizons in a pedon. A *diagnostic horizon* reflects a distinctive physical property (color, texture, structure, consistence, porosity, moisture) or a dominant soil process (discussed with the soil types).

In the solum (A, E, and B horizons), two diagnostic horizons may be identified: the epipedon and the diagnostic subsurface. The presence or absence of either of these diagnostic horizons usually distinguishes a soil for classification.

- The **epipedon** (literally, "over the soil") is the diagnostic horizon at the surface, where most of the rock structure has been destroyed. It may extend downward through the A horizon, even including all or part of an illuviated B horizon. It is visibly darkened by organic matter and sometimes is leached of minerals. Excluded from the epipedon are alluvial deposits, eolian deposits, and cultivated areas because soil-forming processes have lacked the time to erase these relatively short-lived characteristics.
- The **diagnostic subsurface horizon** originates below the surface at varying depths. It may include part of the A or B horizon or both.

The 12 Soil Orders of the Soil Taxonomy

At the heart of the Soil Taxonomy are 12 general soil orders, listed in Table 18.1. Their worldwide distribution is shown in Figure 18.9 and in individual maps with each description. Please consult this table and these maps as you read the following descriptions. Because the Soil Taxonomy evaluates each soil order on its own characteristics, there is no priority to the classification. However, you will find a progression in this discussion, for the 12 orders are arranged loosely by latitude; we begin, as in Chapters 10 (climates) and 20 (terrestrial biomes), along the equator.

CRITICAL THINKING 18.2

Soil observations

Select a small soil sample from your campus or near where you live. Using the sections in this chapter on soil characteristics, properties, and formation, describe as completely as possible this sample, within these limited constraints. Using the general soil map and any other sources available (e.g., local agriculture extension agent, Internet, related department on campus), are you able to roughly place this sample in one of the soil orders? Is there a laboratory or other facility nearby that will test the soil for you? Using the local phone directory or Internet, see if you can locate an agency or extension agent that provides information about local soils and advice for soil management.

TABLE 18.1 Soil Taxonomy Soil Orders

Order	General Location and Climate	Description
Oxisols	Tropical soils; hot, humid areas	Maximum weathering of Fe and Al and eluviation; continuous plinthite layer
Aridisols	Desert soils; hot, dry areas	Limited alteration of parent material; low climate activity; light color; low humus content; subsurface illuviation of carbonates
Mollisols	Grassland soils; subhumid, semiarid lands	Noticeably dark with organic material; humus rich; base saturation; high, friable surface with well-structured horizons
Alfisols	Moderately weathered forest soils; humid temperate forests	B horizon high in clays; moderate to high degree of base saturation; illuviated clay accumulation; no pronounced color change with depth
Ultisols	Highly weathered forest soils; subtropical forests	Similar to Alfisols; B horizon high in clays; generally low amount of base saturation; strong weathering in subsurface horizons; redder than Alfisols
Spodosols	Northern conifer forest soils; cool, humid forests	Illuvial B horizon of Fe/Al clays; humus accumulation; without structure; partially cemented; highly leached; strongly acid; coarse texture of low bases
Entisols	Recent soils; profile undeveloped, all climates	Limited development; inherited properties from parent material; pale color; low humus; few specific properties; hard and massive when dry
Inceptisols	Weakly developed soils; humid regions	Intermediate development; embryonic soils, but few diagnostic features; further weathering possible in altered or changed subsurface horizons
Gelisols	Permafrost-affected soils; high latitudes in Northern Hemisphere, southern limits near tree line, high elevations	Permafrost within 100 cm of the soil surface; evidence of cryoturbation (frost churning) and/or an active layer; patterned ground
Andisols	Soils formed from volcanic activity; areas affected by frequent volcanic activity, especially the Pacific Rim	Volcanic parent materials, particularly ash and volcanic glass; weathering and mineral transformation important; high CEC and organic content; generally fertile
Vertisols	Expandable clay soils; subtropics, tropics; sufficient dry period	Forms large cracks on drying; self-mixing action; contains >30% in swelling clays; light color; low humus content
Histosols	Organic soils; wet places	Peat or bog; >20% organic matter; much with clay >40 cm thick; surface organic layers; no diagnostic horizons

Oxisols The intense moisture, high temperature, and uniform daylength of equatorial latitudes profoundly affect soils. These generally old landscapes, exposed to tropical conditions for millennia or hundreds of millennia, are deeply developed. Soil minerals are highly altered (except in certain newer volcanic soils in Indonesia—the Andisols). Oxisols are among the most mature soils on Earth. Distinct horizons usually are lacking where these soils are well drained (Figure 18.10a, page 540). Related vegetation is the luxuriant and diverse tropical and equatorial rain forest. Oxisols include five suborders.

Oxisols (tropical soils) are named because they have a distinctive horizon of iron and aluminum oxides. The concentration of oxides results from heavy precipitation, which leaches soluble minerals and soil constituents from the A horizon. Typical Oxisols are reddish (from the iron oxide) or yellowish (from the aluminum oxides), with a weathered claylike texture, sometimes in a granular soil structure that is easily broken apart. The high degree of eluviation removes basic cations and colloidal material to lower illuviated horizons. Thus, Oxisols are low in CEC and fertility, except in regions augmented by alluvial or volcanic materials.

To have the lush rain forests in the same regions as soils poor in inorganic nutrients seems an irony. However, this forest system relies on the recycling of nutrients from soil organic matter to sustain fertility, although this nutrient recycling ability is quickly lost when the ecosystem is disturbed. Consequently, Oxisols have a diagnostic subsurface horizon that is highly weathered, contains iron and aluminum oxides, is at least 30 cm (12 in.) thick, and lies within 2 m (6.5 ft) of the surface (see Figure 18.10).

Figure 18.11 (page 540) illustrates *laterization*, the leaching process that operates in well-drained soils in warm, humid, tropical and subtropical climates. If Oxisols are subjected to repeated wetting and drying, an *ironstone*

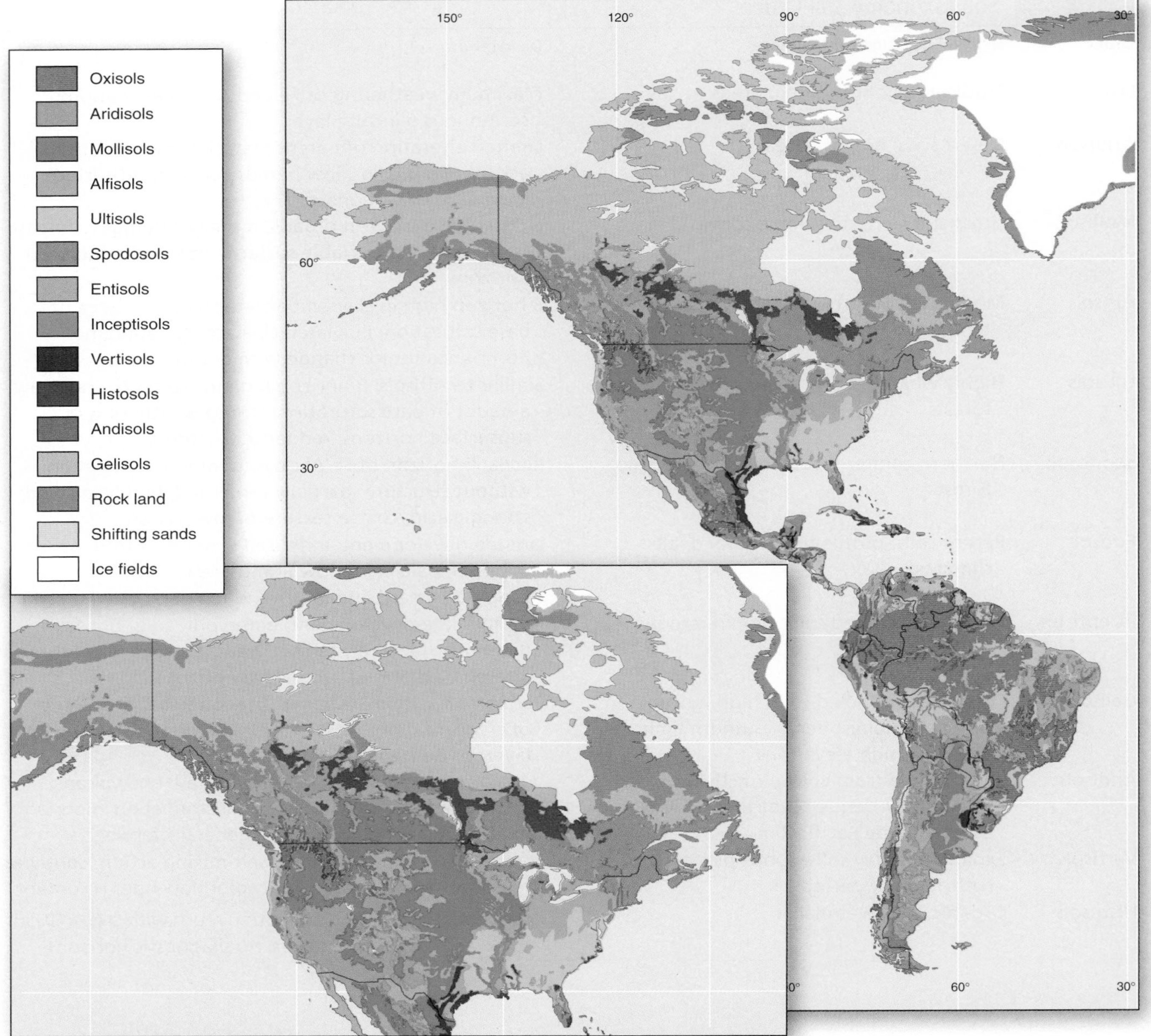

FIGURE 18.9 Soil Taxonomy.
Worldwide distribution of the Soil Taxonomy's 12 soil orders. [Adapted from maps prepared by World Soil Resources Staff, Natural Resources Conservation Service, USDA, 1999, 2006.]

hardpan develops (this is a hardened soil layer—iron-rich and humus-poor clay with quartz and other minerals—in the lower A or in the B horizon). It is a *plinthite* (from the Greek *plinthos*, meaning "brick"). This form of soil, also known as a *laterite*, can be quarried in blocks and used as a building material (Figure 18.12, page 540).

Simple agricultural activities can be conducted with care in these soils. Early cultivation practices, called *slash-and-burn shifting cultivation*, were adapted to these soil conditions and formed a unique style of crop rotation. The scenario went like this: People in the tropics cut down (slashed) and burned the rain forest in small tracts and then cultivated the land using stick-and-hoe tools, planting maize, beans, and squash. Mineral nutrients in the organic material and short-lived fertilizer input from the fire ash would quickly be exhausted. After several years, the soil lost fertility through leaching by intense rainfall, so the people shifted cultivation to another tract and repeated the process. After many years of moving from tract to tract, the group returned to the first patch to begin the cycle again. This practice protected the limited fertility of the soils somewhat, allowing periods of recovery.

Today, this method is still practiced in parts of Asia, Africa, and South America. The invasion of foreign plantation interests, development by local governments, vastly

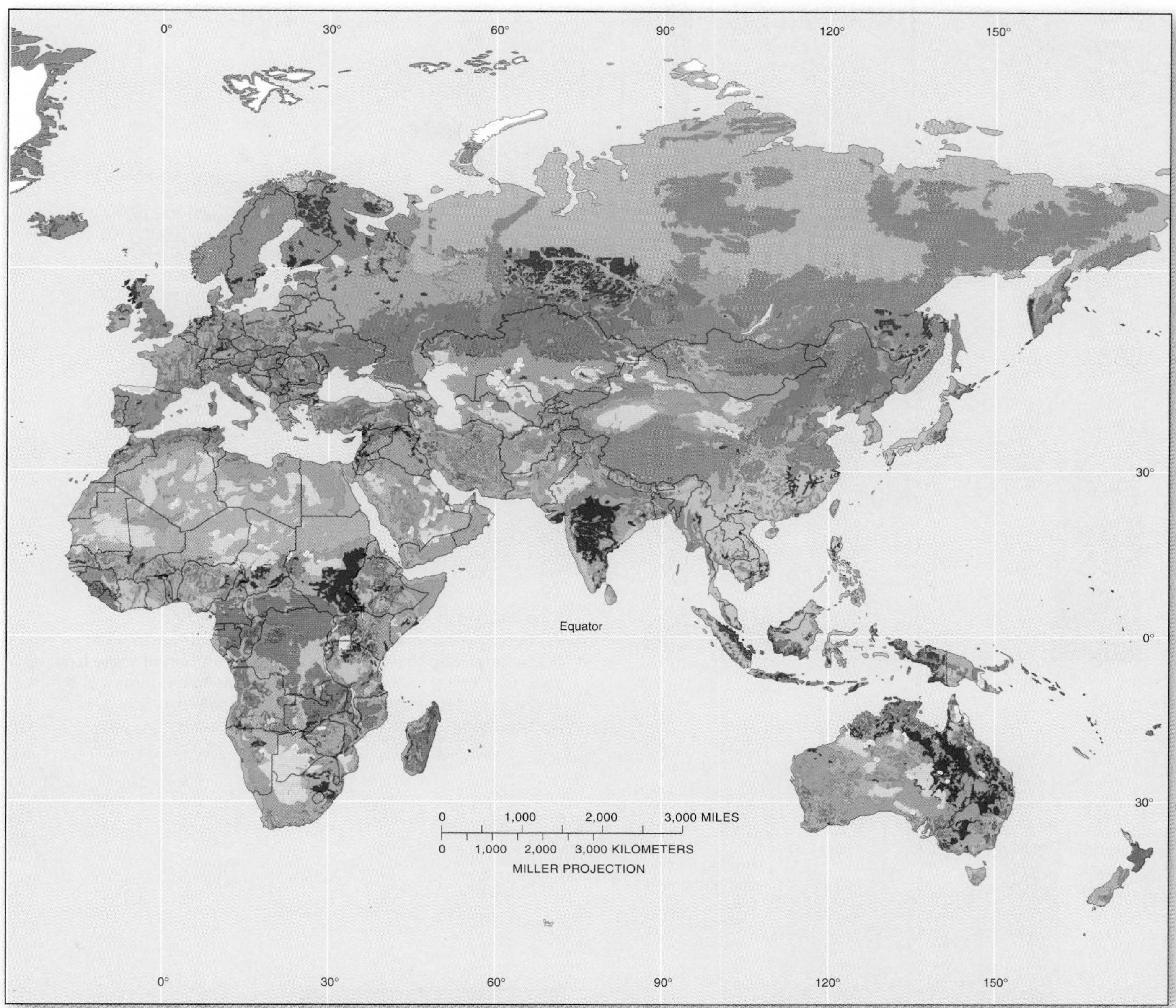

increased population pressures, and conversion of vast forest tracts to pasturage disrupted this orderly land rotation. Permanent tracts of cleared land, taken out of the former rotation mode, brought disastrous consequences regarding soil erosion. When Oxisols are disturbed, soil loss can exceed 1000 tons per square kilometer per year, not to mention the greatly increased extinction rates of plant and animal species that accompany such soil depletion and rain forest destruction. The regions dominated by the Oxisols and rain forests are rightfully the focus of much worldwide environmental attention.

Aridisols The largest single soil order occurs in the world's dry regions. **Aridisols** (desert soils) occupy approximately 19% of Earth's land surface (see Figure 18.9). A pale, light soil color near the surface is diagnostic (Figure 18.13a).

Not surprisingly, the water balance in Aridisol regions is characterized by periods of soil-moisture deficit and generally inadequate soil moisture for plant growth. High potential evapotranspiration and low precipitation produce very shallow soil horizons. Usually, there is no period greater than 3 months when the soils have adequate moisture. Lacking water and therefore lacking vegetation, Aridisols also lack organic matter of any consequence. Low precipitation means infrequent leaching, yet Aridisols are leached easily when exposed to excessive water, for they lack a significant colloidal structure.

Representing about 16% of Earth's agricultural land, irrigated land accounts for nearly 36% of the harvest. Two related problems common in irrigated lands are salinization and water logging, especially in arid lands that are poorly drained. *Salinization* is common in Aridisols, resulting from excessive potential evapotranspiration rates in deserts and

(a)

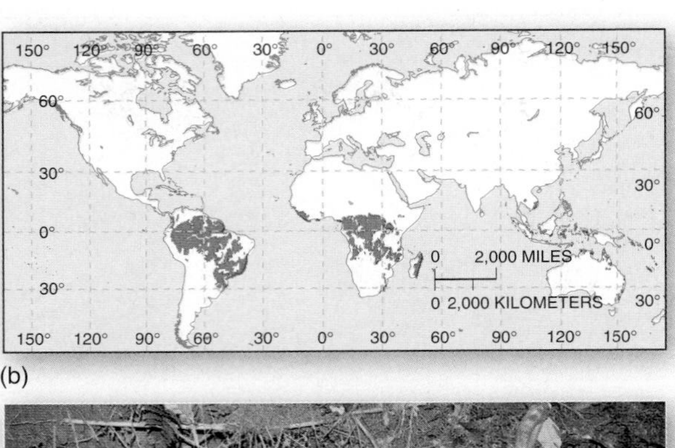

(b)

(c)

FIGURE 18.10 Oxisols.
(a) Deeply weathered Oxisol profile in central Puerto Rico.
(b) General map showing worldwide distribution of these tropical soils. (c) Tropical soils in the Sierra de Luquillo mountains and rain forest in eastern Puerto Rico. [(a) Marbut Collection, Soil Science Society of America, Inc. (c) Bobbé Christopherson.]

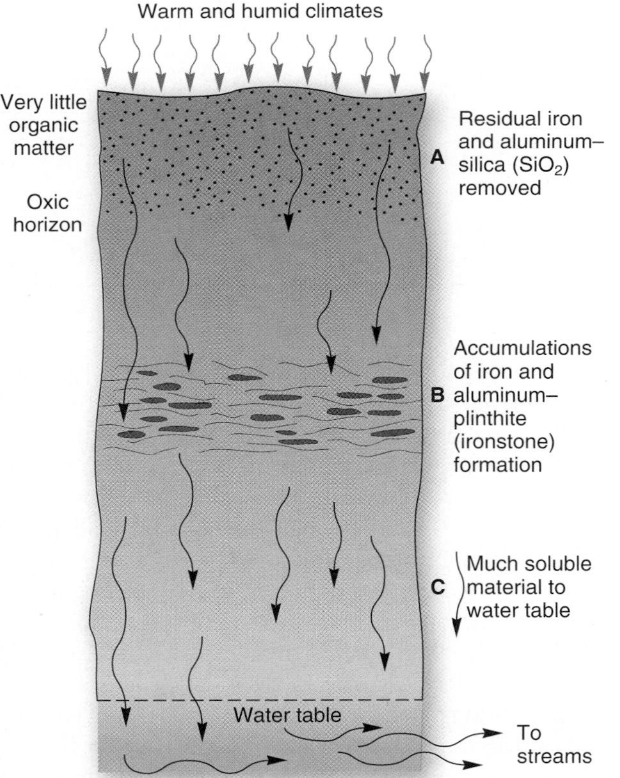

FIGURE 18.11 Laterization.
The laterization process is characteristic of moist *tropical* and *subtropical* climate regimes.

FIGURE 18.12 Oxisols are used for bricks.
Here plinthite is being quarried in India for building material. Inset photo shows a close-up of plinthite bricks. [Courtesy of Henry D. Foth.]

(a)

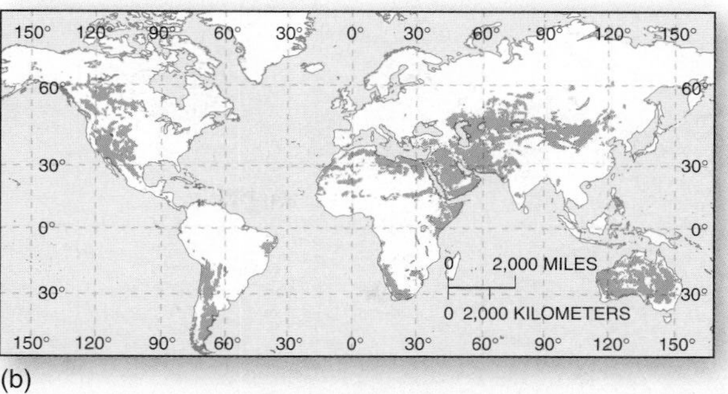

(b)

(c)

(d)

FIGURE 18.13 Aridisols.
(a) Soil profile from central Arizona. (b) General map showing worldwide distribution of these desert soils. (c) Irrigated aloe plants in Aridisols in the Imperial Valley. (d) Southwest of the Salton Sea, the Imperial Valley agricultural complex is claimed from the desert. [(a) Marbut Collection/Soil Science Society of America, Inc. (c) and (d) Bobbé Christopherson.]

semiarid regions. Salts dissolved in soil water migrate to surface horizons and are deposited as the water evaporates. These deposits appear as subsurface salty horizons, which will damage or kill plants when the horizons occur near the root zone. The salt accumulation associated with a desert playa is an example of extreme salinization (see Chapter 15).

Obviously, salinization complicates farming in Aridisols. The introduction of irrigation water may either waterlog poorly drained soils or lead to salinization. Irrigated agriculture has increased greatly since 1800, when only 8 million hectares (about 20 million acres) were irrigated worldwide. Today, approximately 255 million hectares (about 630 million acres) are irrigated, and this figure is on the increase in some parts of the world. In many areas, production has decreased and even ended because of salt buildup in the soils. Examples include areas along the Tigris and Euphrates rivers in Iraq, the Indus River valley in Pakistan, sections of South America and Africa, and the western United States.

Nonetheless, vegetation does grow where soils are well drained and low in salt content. If large capital investments are made in water, drainage, and fertilizers, Aridisols possess much agricultural potential (Figure 18.13c and d). However, Aridisols must be farmed with a careful balance among

(a)

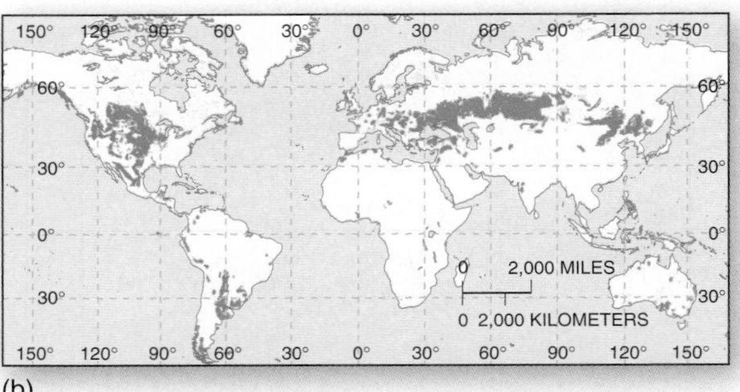

(b)

(c)

(d)

FIGURE 18.14 Mollisols.
(a) Soil profile in eastern Idaho, from calcareous loess related to the soils of the Palouse. (b) General map showing worldwide distribution of these grassland soils. (c) Mollisols with rows of corn in central Kansas. (d) In eastern Washington State, wheat flourishes in the Palouse.
[(a) Courtesy Marbut Collection, Soil Science Society of America, Inc. (c) and (d) Bobbé Christopherson.]

environmental factors. In California, the former Kesterson Wildlife Refuge was reduced to a toxic waste dump in the early 1980s by contaminated agricultural drainage. Focus Study 18.1 elaborates on the Kesterson tragedy.

Mollisols Some of Earth's most significant agricultural soils are **Mollisols** (grassland soils). This group includes seven recognized suborders, which vary in fertility. The dominant diagnostic horizon is a dark, organic surface layer some 25 cm (10 in.) thick (Figure 18.14). As the Latin name implies (words using the same root, *mollis*, are *mollify* and *emollient*, meaning *soft*), Mollisols are soft, even when dry. They have granular or crumbly peds, loosely arranged when dry. These humus-rich soils are high in basic cations (calcium, magnesium, and potassium) and have a high CEC and therefore high fertility. In soil moisture, these soils are intermediate between humid and arid.

Soils of the steppes and prairies of the world—the North American Great Plains, the Palouse of Washington State, the Pampas of Argentina, and the region from Manchuria in China through to Europe—belong to this group. Agriculture in these areas ranges from large-scale commercial grain farming to grazing along the drier portions. With fertilization or soil-building practices, high crop yields are common. The "fertile triangle" of Ukraine, Russia, and western portions of the former Soviet Union is of this soil type.

In North America, the Great Plains straddle the 98th meridian, which is coincident with the 51-cm (20-in.) isohyet of annual precipitation—wetter to the east and

FOCUS STUDY 18.1

Selenium Concentration in Soils: The Death of Kesterson

About 95% of the irrigated acreage in the United States lies west of the 98th meridian, a region that is increasingly troubled with salinization and waterlogging problems. Drainage of agricultural wastewater poses a particular problem in these semiarid and arid lands, where river discharge is inadequate to dilute and remove field runoff. Fields are often purposely overwatered to keep salts away from the effective rooting depth of the crops. One solution is to place field drains beneath the soil to collect gravitational water (Figure 18.1.1). But agricultural drain water must go somewhere, and for the San Joaquin Valley of central California, this problem triggered a 25-year controversy surrounding toxic levels of selenium in the wetland ecosystem of Kesterson Reservoir.

California's western San Joaquin Valley is one of at least nine sites in the western United States experiencing contamination from increasing selenium concentrations. Selenium is a trace element that occurs naturally in bedrock, particularly Cretacious shales found throughout the western United States. Toxic effects of selenium were reported during the 1980s in some domestic animals grazing on grasses grown in selenium-rich soils in the Great Plains. In California, the coastal mountain ranges are a significant source region for selenium. As parent materials weather, selenium-rich alluvium washes into valleys, forming Aridisols that become productive with the addition of irrigation water. Selenium then becomes concentrated by evaporation in farmlands and may become mobilized by irrigation drainage into wetlands, where it bioaccumulates to toxic levels.

FIGURE 18.1.1 Fields drain into canals.
Soil drainage canal collects contaminated water from field drains and directs it into the Salton Sea. Such soil-moisture tile drains and collection channels are in use in the San Joaquin Valley. [Author.]

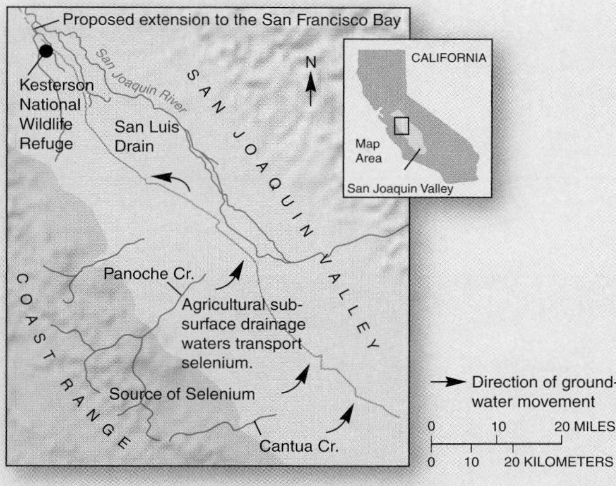

FIGURE 18.1.2 Kesterson locator map.
The source of selenium is in the Coast Ranges; surface runoff delivers this over thousands of years to the soils of the region. The agricultural drains complete the delivery systems to the refuge. [USGS.]

Central California's potential drain outlets for agricultural wastewater are limited. Yet, by the late 1970s, about 80 miles of a drain were built in the western San Joaquin Valley, without an outlet or the completion of a formal plan. Large-scale irrigation of corporate-owned farms continued, supplying the field drains with salty, selenium-laden runoff that made its way to the Kesterson National Wildlife Refuge in the northern portion of the San Joaquin Valley east of San Francisco. The unfinished drain abruptly stopped at the boundary to the refuge (Figure 18.1.2).

The selenium-tainted drainage took only 3 years to contaminate the wildlife refuge, which was officially declared a toxic waste site. Selenium was first taken in by aquatic life forms (e.g., marsh plants, plankton, and insects) and then made its way up the food chain and into the diets of higher life forms in the refuge. According to U.S. Fish and Wildlife Service scientists, the toxicity causes genetic abnormalities and death in wildlife, including all varieties of birds that nested at Kesterson; approximately 90% of the exposed birds perished or were injured. Because this refuge was a major migration flyway and stopover point for birds from throughout the Western Hemisphere, its contamination also violated several multinational wildlife protection treaties.

The field drains were sealed and removed in 1986, following a court order that forced the federal government to uphold existing laws. Irrigation water then immediately began backing up in the corporate farmlands, producing both water logging and selenium contamination. The Kesterson Unit became part of the San Luis National Wildlife Refuge—and selenium control and restoration a priority.

Frustrated, large-scale corporate agricultural interests pressured the federal government to finish the drain, either to San Francisco Bay or to the ocean. In 1996, the Grasslands Bypass Project was implemented by the U.S. Bureau of Reclamation to prevent the discharge of agricultural drainage water into wetlands and wildlife refuges of central California. Thus, agricultural drain water now moves through the old San Luis Drain to Mud Slough, a natural waterway through the San Luis National Wildlife Refuge, and then on to the San Joaquin River, the San Joaquin–Sacramento delta, and eventually San Francisco Bay. Initially, selenium levels increased in those channels, although fluctuations downward in concentrations were noted as well. Biological monitoring continues.

There are nine such threatened sites in the western United States; Kesterson was simply the first of these to fail from contamination. Such damage to a wildlife refuge presents a real warning to human populations—remember where we are, at the top of the food chain.

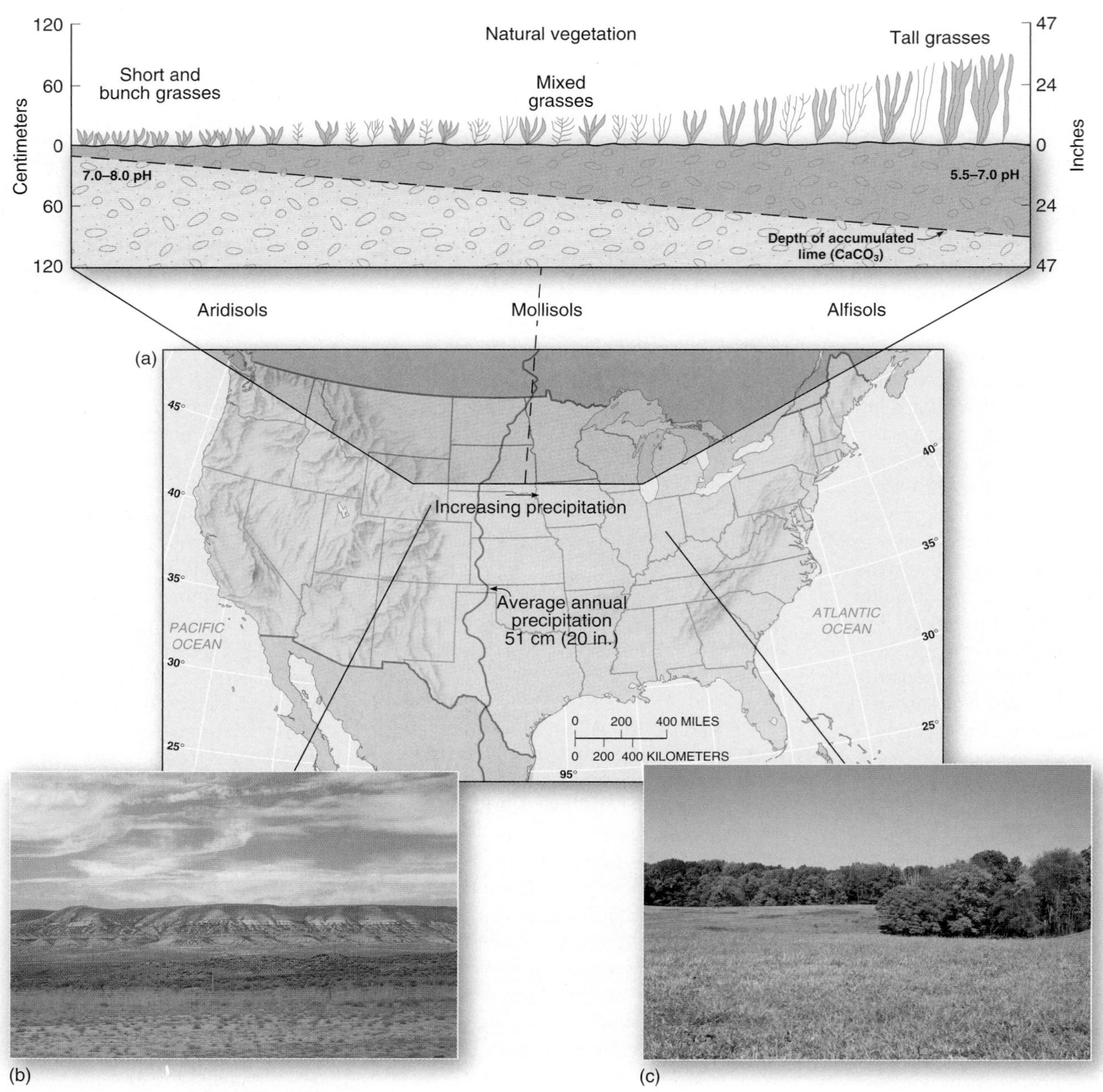

FIGURE 18.15 **Soils of the Midwest.**
(a) Aridisols (to the west), Mollisols (central), and Alfisols (to the east)—a soil continuum in the north-central United States and southern Canadian prairies. Graduated changes that occur in soil pH and the depth of accumulated lime are shown. (b) Bunch grasses and shallow soils of Wyoming. (c) Farmlands and Alfisols south of Bedford, Indiana. [(a) Illustration adapted from N. C. Brady, *The Nature and Properties of Soils,* 10th ed., © 1990 by Macmillan Publishing Company, adapted by permission. (b) Author. (c) Bobbé Christopherson.]

drier to the west. The Mollisols here mark the historic division between the short- and tall-grass prairies (Figure 18.15).

Calcification is a soil process characteristic of some Mollisols and adjoining marginal areas of Aridisols. Calcification is the accumulation of calcium carbonate or magnesium carbonate in the B and C horizons. Calcification by calcium carbonate ($CaCO_3$) forms a diagnostic subsurface horizon that is thickest along the boundary between dry and humid climates (Figure 18.16).

When cemented or hardened, these deposits are *caliche*, or *kunkur*; they occur in widespread soil formations in central and western Australia, the Kalahari region of interior southern Africa, and the High Plains of the west-central United States, among other places. (Soils dominated by the processes of calcification and salinization were formerly known as pedocals.)

Alfisols Spatially, **Alfisols** (moderately weathered forest soils) are the most widespread of the soil orders,

POTET equal to or greater than PRECIP

Dark color, high in bases

O Dense sod cover of
A interlaced grasses and roots

E

Calcic horizon; possible formation of caliche

B Accumulation of excess calcium carbonate

C

FIGURE 18.16 Calcification process in soil.
The calcification process in Aridisol/Mollisol soils occurs in climatic regimes that have potential evapotranspiration equal to or greater than precipitation.

extending in five suborders from near the equator to high latitudes. Representative Alfisol areas include Boromo and Burkina Faso (interior western Africa); Fort Nelson, British Columbia; the states near the Great Lakes; and the valleys of central California. Most Alfisols are grayish brown to reddish and are considered moist versions of the Mollisol soil group. Moderate eluviation is present as well as a subsurface horizon of illuviated clays and clay formation because of a pattern of increased precipitation (Figure 18.17).

Alfisols have moderate to high reserves of basic cations and are fertile. However, productivity depends on moisture and temperature. Alfisols usually are supplemented by a moderate application of lime and fertilizer in areas of active agriculture. Some of the best farmland in the United States stretches from Illinois, Wisconsin, and eastern Minnesota through Indiana, Michigan, and Ohio to Pennsylvania and New York. This land produces grains, hay, and dairy products. The soil here is an Alfisol subgroup called Udalfs, which are characteristic of *humid continental hot-summer* climates.

The Xeralfs, another Alfisol subgroup, are associated with the moist winter, dry summer pattern of the *Mediterranean* climate. These naturally productive soils are farmed intensively for subtropical fruits, nuts, and special crops that grow in only a few locales worldwide—for

(a)

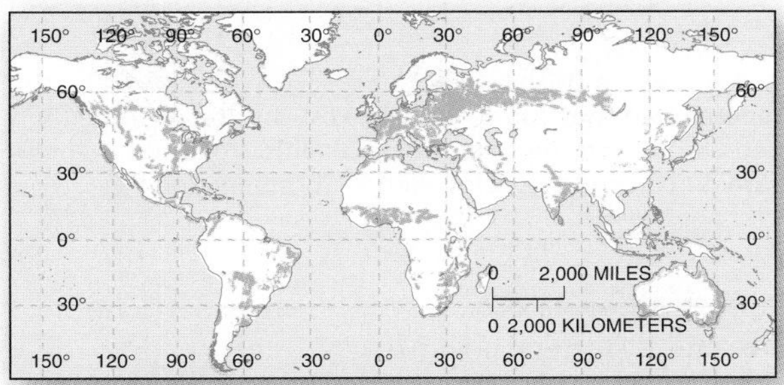

(b)

(c)

FIGURE 18.17 Alfisols.
(a) Soil profile from northern Idaho, formed in loess.
(b) General map showing worldwide distribution of these moderately weathered forest soils. (c) An olive orchard in northern California; virtually all U.S. olive production is in California. [(a) Courtesy Marbut Collection, Soil Science Society of America, Inc. (c) Bobbé Christopherson.]

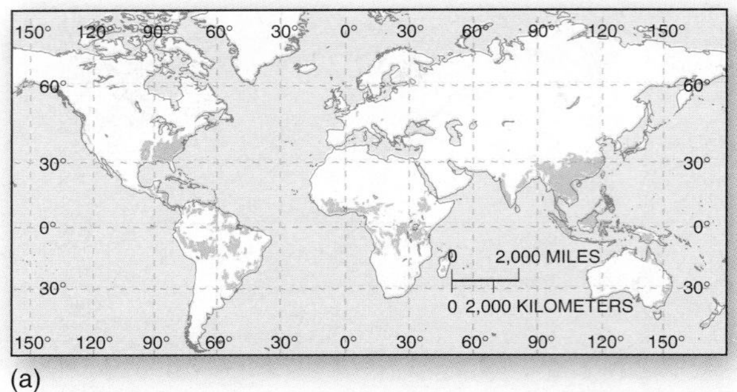

(a)

(b)

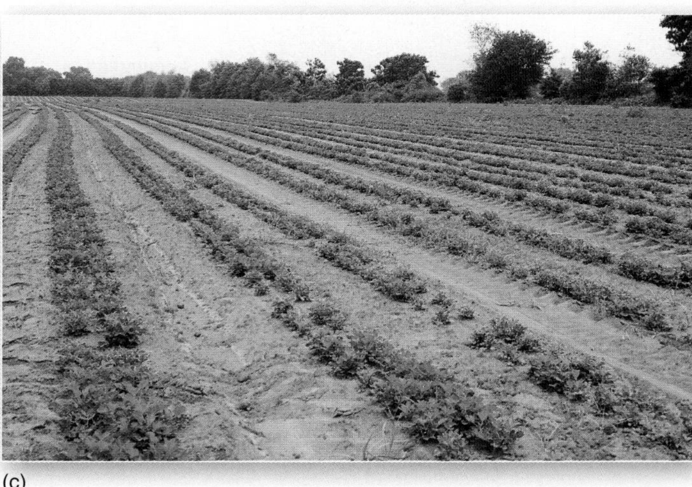
(c)

FIGURE 18.18 Ultisols.
(a) General map showing worldwide distribution of these highly weathered forest soils. (b) Ultisols in the Southeast planted with pecan trees. (c) A type of Ultisol planted in rows of peanuts, near Plains in west-central Georgia, bears its characteristic reddish color. [Bobbé Christopherson.]

example, California olives, grapes, citrus, artichokes, almonds, and figs (Figure 18.17c).

Ultisols Farther south in the United States are the **Ultisols** (highly weathered forest soils) and their five suborders. An Alfisol might degenerate into an Ultisol, given time and exposure to increased weathering under moist conditions. These soils tend to be reddish because of residual iron and aluminum oxides in the A horizon (Figure 18.18).

The relatively high precipitation in Ultisol regions causes greater mineral alteration and more eluvial leaching than in other soils. Therefore, the level of basic cations is lower, and the soil fertility is lower. Fertility is further reduced by certain agricultural practices and the effect of soil-damaging crops such as cotton and tobacco, which deplete nitrogen and expose soil to erosion. These soils respond well if subjected to good management—for example, crop rotation restores nitrogen, and certain cultivation practices prevent sheetwash and soil erosion. Peanut plantings assist in nitrogen restoration. Much needs to be done to achieve sustainable management of these soils.

Spodosols The **Spodosols** (northern coniferous forest soils) and their four suborders occur generally to the north and east of the Alfisols. They are in cold and forested moist regimes (*humid continental mild-summer* climates) in northern North America and Eurasia, Denmark, The Netherlands, and southern England. Because no comparable climates exist in the Southern Hemisphere, this soil type is not identified there. Spodosols form from sandy parent materials, shaded under evergreen forests of spruce, fir, and pine. Spodosols with more moderate properties form under mixed or deciduous forests (Figure 18.19).

GEO REPORT 18.2 ***Loss of marginal lands puts pressure on prime lands***

Since 1985, more than 0.6 million hectares (1.5 million acres) of irrigated Aridisols and Alfisols have gone out of production in California, marking the end of several decades of irrigated farming in climatically marginal lands. Severe cutbacks in irrigated acreage no doubt will continue, underscoring the need to preserve prime farmlands in California and in wetter regions elsewhere in the country.

(a)

FIGURE 18.19 Spodosols.
(a) Soil profile from northern New York. (b) General map showing worldwide distribution of these northern coniferous forest soils. (c) Characteristic temperate forest and Spodosols in the cool, moist climate of central Vancouver Island. (d) Freshly plowed Spodosols (or Podzolic soils in the Canadian System) near Lakeville, Nova Scotia, being prepared for planting vegetables. Formed beneath coniferous forests, the land is cleared for crops and apple orchards (visible in the distance). [(a) Courtesy Marbut Collection, Soil Science Society of American, Inc. (c) and (d) Bobbé Christopherson.]

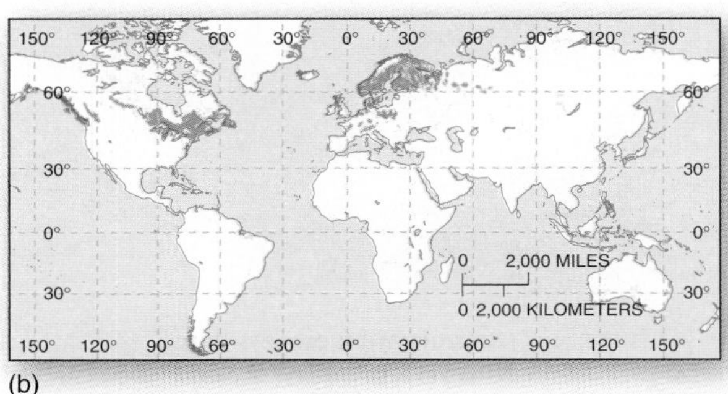

(b)

(c)

(d)

Spodosols lack humus and clay in the A horizon. Characteristics of the podzolization process are present: the A horizon is sandy and bleached in color and is eluviated, leached of clays and iron; the B horizon is composed of illuviated organic matter and iron and aluminum oxides (Figure 18.19c). The surface horizon receives organic litter from base-poor, acid-rich evergreen trees, which contribute to acid accumulations in the soil. The solution in acidic soils effectively leaches clays, iron, and aluminum, which are passed to the upper diagnostic horizon. An ashen-gray color is common in these subarctic forest soils and is characteristic of podzolization (Figure 18.20).

In the Canadian system, Spodosols fall within the Podzolic Great Group, as seen in the temperate rain forests of Vancouver Island, British Columbia, and in Nova Scotia and New Brunswick (Figure 18.19c and d).

When agriculture is attempted, the low basic cation content of Spodosols requires the addition of nitrogen, phosphate, and potash (potassium carbonate)—and perhaps crop rotation as well. A soil *amendment* such as limestone can significantly increase crop production by raising the pH of these acidic soils. For example, the yields of several crops (corn, oats, wheat, and hay) grown in specific Spodosols in New York State were increased up to a third

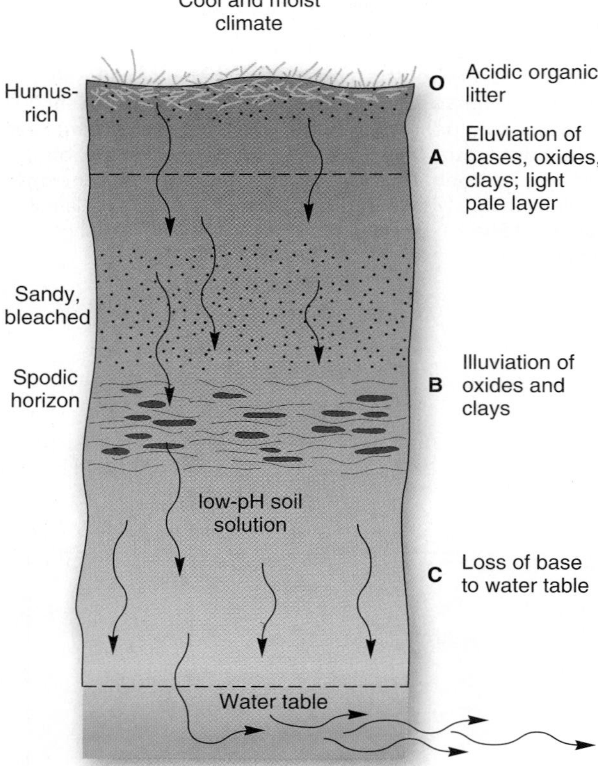

FIGURE 18.20 Podzolization process in soil.
The podzolization process is typical in cool and moist climatic regimes and Spodosols.

FIGURE 18.21 Entisols.
A characteristic Entisol: young, undeveloped soil forming from a shale parent material in the Anza–Borrego Desert, California. [Bobbé Christopherson.]

with the application of 1.8 metric tonnes (2 tons) of limestone per 0.4 hectare (1.0 acre) during each 6-year rotation.

Entisols The **Entisols** (recent, undeveloped soils) lack vertical development of their horizons. The five suborders of this soil group are based on differences in parent materials and climatic conditions, although the presence of Entisols is not climate-dependent, for they occur in many climates worldwide. Entisols are true soils, but they have not had sufficient time to generate the usual horizons.

Entisols generally are poor agricultural soils, although those formed from river silt deposits are quite fertile. The same conditions that inhibit complete development also prevent adequate fertility—too much or too little water, poor structure, and insufficient accumulation of weathered nutrients. Active slopes, alluvium-filled floodplains, poorly drained tundra, tidal mudflats, dune sands and erg (sandy) deserts, and plains of glacial outwash all are characteristic regions that have these soils. Figure 18.21 shows an Entisol in a desert climate where shales formed the parent material.

Inceptisols Since they have not reached a mature condition, **Inceptisols** (weakly developed soils) and their six suborders are inherently infertile. They are young soils, although they are more developed than the Entisols. Inceptisols include a wide variety of different soils, all having in common a lack of maturity, with evidence of weathering just beginning. Inceptisols are associated with moist soil regimes and are regarded as eluvial because they demonstrate a loss of soil constituents throughout their profile, but retain some weatherable minerals. This soil group has no distinct illuvial horizons. Inceptisols include most of the glacially derived till and outwash materials from New York down through the Appalachians and alluvium on the Mekong and Ganges floodplains.

Gelisols The **Gelisols** are cold and frozen soils and have three suborders, which represent inclusion of high-latitude (Canada, Alaska, Russia, Arctic Ocean islands, and the Antarctic Peninsula) and high-elevation (mountain) soil conditions. Temperatures in these regions are at or below 0°C (32°F), making soil development a slow process and disturbances of the soil long-lasting. Gelisols can develop organic diagnostic horizons because cold temperatures slow decomposition of materials (Figure 18.22). These are permafrost-affected soils. Previously, these soils were included in the Inceptisol, Entisol, and Histisol soil orders.

Characteristic vegetation is that of tundra, such as lichens, mosses, sedges, dwarf willows, and other plants adapted to the harsh cold and permafrost. For photos and discussion, see the *Geosystems Now* feature for this chapter.

Gelisols are subject to *cryoturbation* (frost churning and mixing) in the freeze–thaw cycle in the active layer (see Chapter 17). This process disrupts soil horizons, pulling organic material to lower layers and rocky C-horizon material to the surface. Patterned-ground phenomena are possible under such conditions.

Andisols Areas of volcanic activity feature **Andisols** (soils with volcanic parent materials), which have seven suborders. Andisols are derived from volcanic ash and glass. Previous soil horizons frequently are found buried by ejecta

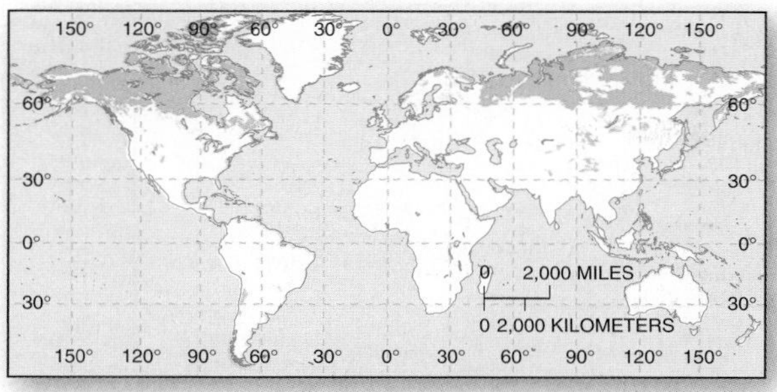

FIGURE 18.22 Gelisols.
General map showing worldwide distribution of these cold and frozen soils. A new *Soil Atlas of the Northern Circumpolar Region* analyzes these soils and their environment.

from repeated eruptions. Volcanic soils are unique in their mineral content because they are recharged by eruptions. For an example, see Figure 19.23, which illustrates recovery and succession of pioneer species in the developing soil north of Mount St. Helens.

Weathering and mineral transformations are important in this soil order. Volcanic glass weathers readily into allophane (a noncrystalline aluminum silicate clay mineral that acts as a colloid) and oxides of aluminum and iron. Andisols feature a high CEC and high water-holding ability and develop moderate fertility, although phosphorus availability is an occasional problem. In Hawai'i, the fertile Andisol fields produce coffee, pineapples, macadamia nuts, and some sugar cane as important cash crops (Figure 18.23a). Andisol distribution is small in areal extent; however, such soils are locally important in the volcanic ring of fire surrounding the Pacific Rim.

Feed crops and pastureland of fertile Andisols sustain a large sheep industry in Iceland. The annual fall roundup is a time when many ranchers combine forces over several weeks to bring in the herds (Figure 18.23b). Another example of Andisol fertility occurs on Fogo Island in the central Atlantic off the African coast, where the early stages of local wine production are supplied by grapes planted in fresh volcanic soils (Figure 18.23c).

(a)

(b)

(c)

FIGURE 18.23 Andisols in agricultural production.
Fertile Andisols planted in macadamia nut groves on the big island of Hawai'i. (b) Pasture and feed grains for raising sheep grow in Andisols in southwestern Iceland. (c) Grape vines (in the foreground) planted in basaltic soil on Fogo Island, Cape Verde, for local wine production. [Bobbé Christopherson.]

FIGURE 18.24 Vertisols.
(a) Soil profile in the Lajas Valley of Puerto Rico. (b) General map showing worldwide distribution of these expandable clay soils. (c) Vertisols in the Texas coastal plain, northeast of Palacios near the Tres Palacios River, planted with a commercial sorghum crop. Note the dark soil color indicative of Vertisols. [(a) Courtesy Marbut Collection, Soil Science Socity of America, Inc. (c) Bobbé Christopherson.]

Vertisols Heavy clay soils are the **Vertisols** (expandable clay soils). They contain more than 30% swelling clays (clays that swell significantly when they absorb water), such as *montmorillonite*. They are located in regions experiencing highly variable soil-moisture balances through the seasons. These soils occur in areas of subhumid to semiarid moisture and moderate to high temperature. Vertisols frequently form under savanna and grassland vegetation in *tropical* and *subtropical* climates and are sometimes associated with a distinct dry season following a wet season. Although widespread, individual Vertisol units are limited in extent.

Vertisol clays are black when wet, but not because of organics; rather, the blackness is due to specific mineral content. They range from brown to dark gray. These deep clays swell when moistened and shrink when dried. In the drying process, they may form vertical cracks as wide as 2–3 cm (0.8–1.2 in.) and up to 40 cm (16 in.) deep. Loose material falls into these cracks, only to disappear when the soil again expands and the cracks close. After many such cycles, soil contents tend to invert or mix vertically, bringing lower horizons to the surface (Figure 18.24).

Despite the fact that clay soils are plastic and heavy when wet, with little available soil moisture for plants, Vertisols are high in bases and nutrients and thus are some of the better farming soils where they occur. For example, they occur in a narrow zone along the coastal plain of Texas (Figure 18.24c) and in a section along the Deccan region of India. Vertisols often are planted with grain sorghums, corn, and cotton.

Histosols Accumulations of thick organic matter can form **Histosols** (organic soils), including four suborders. In the midlatitudes, when conditions are right, beds of former lakes may turn into Histosols, with water gradually replaced by organic material to form a bog and layers of peat (Figure 18.25). (Lake succession and bog or marsh formation are discussed in Chapter 19.) Histosols also form in small, poorly drained depressions, with conditions ideal for significant deposits of sphagnum peat to form. This material can be cut, baled, and sold as a soil amendment.

Peat is cut by hand with a spade into blocks, which are then set out to dry. In Figure 18.25c, note the fibrous

FIGURE 18.25 Histosols.
(a) A bog in coastal Maine, near Popham Beach State Park. (b) General map showing worldwide distribution of these organic soils. (c) Sphagnum peat soil profile and drying peat blocks on Mainland Island, north of Scotland. Note the dark water runoff from the organic soil. [Bobbé Christopherson.]

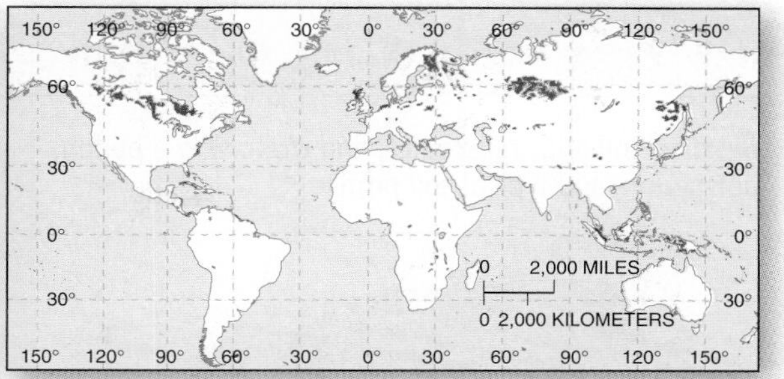

texture of the sphagnum moss growing on the surface and the darkening layers with depth in the soil profile as the peat is compressed and chemically altered. Such beds can be more than 2 m thick. Once dried, the peat blocks burn hot and smoky. Peat is the first stage in the natural formation of lignite, an intermediate step toward coal. Imagine such soils forming in plant-lush swamp environments in the Carboniferous Period (359 to 299 million years ago), only to go through coalification to become coal deposits.

GEOSYSTEMS CONNECTION

Soil science forms the bridge between Parts I, II, and III, the abiotic systems, and Part IV, the biotic systems. Soil formation is affected by temperature, moisture, parent material, topography, and living organisms, including humans. Thus, in this chapter, we combine the energy–atmosphere and water, weather, and climate systems, and the weathering processes that are the source for soil particles and minerals, into a study of this basic covering over Earth's surface. Given this soil foundation, we move to the essentials of Earth's ecosystems and the biotic operations that fuel living systems and the communities that organize life.

KEY LEARNING CONCEPTS REVIEW

■ ***Define*** soil and soil science, and ***describe*** a pedon, polypedon, and typical soil profile.

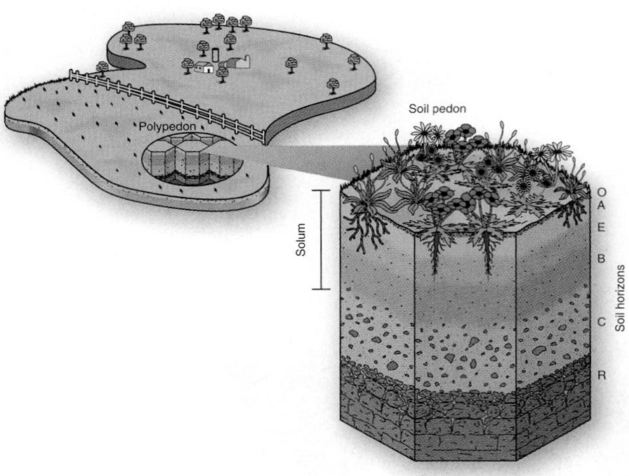

Soil is a dynamic natural body composed of fine materials and contains both mineral and organic matter. **Soil science** is the interdisciplinary study of soils involving physics, chemistry, biology, mineralogy, hydrology, taxonomy, climatology, and cartography. *Pedology* deals with the origin, classification, distribution, and description of soil. *Edaphology* specifically focuses on the study of soil as a medium for sustaining the growth of plants.

The basic sampling unit used in soil surveys is the **pedon**. The **polypedon** is the soil unit used to prepare local soil maps and may contain many pedons. Each discernible layer in an exposed pedon is a **soil horizon**. The horizons are designated O (contains **humus,** a complex mixture of decomposed and synthesized organic materials), A (rich in humus and clay, darker), E (zone of **eluviation,** the removal of fine particles and minerals by water), B (zone of **illuviation,** the deposition of clays and minerals translocated from elsewhere), C (*regolith*, weathered bedrock), and R (bedrock). Soil horizons A, E, and B experience the most active soil processes and together are designated the **solum**.

soil (p. 528)
soil science (p. 528)
pedon (p. 528)
polypedon (p. 528)
soil horizon (p. 528)
humus (p. 528)
eluviation (p. 529)
illuviation (p. 529)
solum (p. 529)

1. Soils provide the foundation for animal and plant life and therefore are critical to Earth's ecosystems. Why is this true?
2. What are the differences among soil science, pedology, and edaphology?
3. Define *polypedon* and *pedon*, the basic units of soil.
4. Characterize the principal aspects of each soil horizon. Where does the main accumulation of organic material occur? Where does humus form? Explain the difference between the eluviated layer and the illuviated layer. Which horizons constitute the solum?

■ ***Describe*** soil properties, including color, texture, structure, consistence, porosity, and soil moisture.

We use several physical properties to classify soils. Color suggests composition and chemical makeup. Soil texture refers to the size of individual mineral particles and the proportion of different sizes. For example, **loam** is a balanced mixture of sand, silt, and clay. Soil structure refers to the arrangement of soil peds, which are the smallest natural cluster of particles in a soil. The cohesion of soil particles to each other is called soil consistence. Soil porosity refers to the size, alignment, shape, and location of spaces in the soil. Soil moisture refers to water in the soil pores and its availability to plants.

loam (p. 530)

5. Soil color is identified and compared using what technique?
6. Define a *soil separate*. What are the various sizes of particles in soil? What is loam? Why is loam regarded so highly by agriculturists?
7. What is a quick, hands-on method for determining soil consistence?
8. Summarize the five soil-moisture regimes common in mature soils.

■ ***Explain*** basic soil chemistry, including cation-exchange capacity, and ***relate*** these concepts to soil fertility.

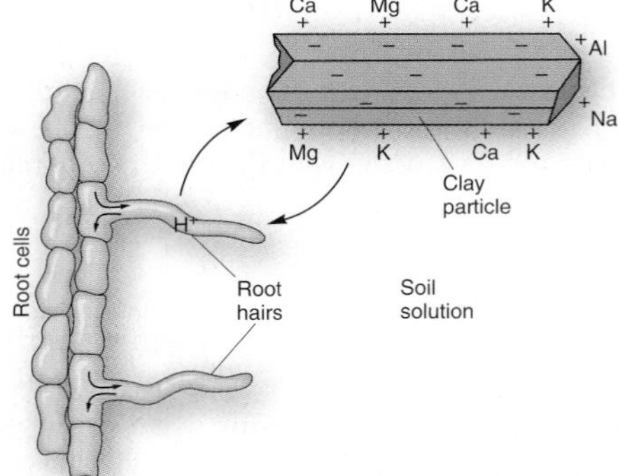

Particles of clay and organic material form negatively charged **soil colloids** that attract and retain positively charged mineral ions in the soil. The capacity to exchange ions between colloids and roots is called the **cation-exchange capacity (CEC)**. CEC is a measure of **soil fertility,** the ability of soil to sustain plants. Fertile soil contains organic substances and clay minerals that absorb water and retain certain elements needed by plants.

soil colloids (p. 533)
cation-exchange capacity (CEC) (p. 533)
soil fertility (p. 533)

9. What are soil colloids? How are they related to cations and anions in the soil? Explain cation-exchange capacity.
10. What is meant by the concept of soil fertility?

- ***Evaluate*** the principal soil formation factors, including the human element.

Environmental factors that affect soil formation include parent materials, climate, vegetation, topography, and time. Human influence is having great impact on Earth's prime soils. Essential soils for agriculture and their fertility are threatened by mismanagement, destruction, and conversion to other uses. Much soil loss is preventable through the application of known technologies, improved agricultural practices, and government policies.

11. Briefly describe the contributions of the following factors and their effects on soil formation: parent material, climate, vegetation, landforms, time, and humans.
12. Explain some of the details that support the concern over loss of our most fertile soils. What cost estimates have been placed on soil erosion?

- ***Describe*** the 12 soil orders of the Soil Taxonomy classification system, and ***explain*** their general distribution across Earth.

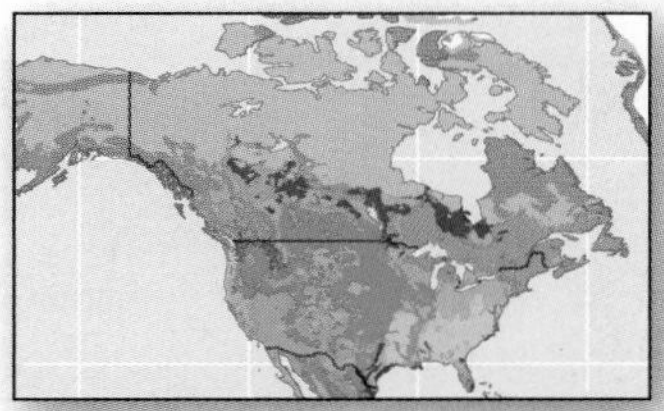

The U.S. **Soil Taxonomy** classification system is built around an analysis of various diagnostic horizons and 12 soil orders, as actually seen in the field. The system divides soils into six hierarchical categories: from smallest to largest, series, families, subgroups, great groups, suborders, and orders.

Specific soil-forming processes keyed to climatic regions (not a basis for classification) are called **pedogenic regimes**: **laterization** (leaching in warm and humid climates), **salinization** (collection of salt residues in surface horizons in hot, dry climates), **calcification** (accumulation of carbonates in the B and C horizons in drier continental climates), **podzolization** (soil acidification in forest soils in cool climates), and **gleization** (humus and clay accumulation in cold, wet climates with poor drainage).

The Soil Taxonomy system uses two diagnostic horizons to identify soil: the **epipedon,** or the surface soil, and the **diagnostic subsurface horizon,** or the soil below the surface at various depths. For an overview of the 12 soil orders, please refer to Table 18.1. The 12 soil orders are **Oxisols** (tropical soils), **Aridisols** (desert soils), **Mollisols** (grassland soils), **Alfisols** (moderately weathered, temperate forest soils), **Ultisols** (highly weathered, subtropical forest soils), **Spodosols** (northern conifer forest soils), **Entisols** (recent, undeveloped soils), **Inceptisols** (weakly developed, humid-region soils), **Gelisols** (cold soils underlain by permafrost), **Andisols** (soils formed from volcanic materials), **Vertisols** (expandable clay soils), and **Histosols** (organic soils).

Soil Taxonomy (p. 536)
pedogenic regimes (p. 536)
laterization (p. 536)
salinization (p. 536)
calcification (p. 536)
podzolization (p. 536)
gleization (p. 536)
epipedon (p. 536)
diagnostic subsurface horizon (p. 536)
Oxisols (p. 537)
Aridisols (p. 539)
Mollisols (p. 542)
Alfisols (p. 544)
Ultisols (p. 546)
Spodosols (p. 546)
Entisols (p. 548)
Inceptisols (p. 548)
Gelisols (p. 548)
Andisols (p. 548)
Vertisols (p. 550)
Histosols (p. 550)

13. What is the basis of the Soil Taxonomy system? How many soil orders, suborders, great groups, subgroups, families, and series are there?
14. Which two orders were most recently added to the classification and what soils are in these new orders?
15. Define an *epipedon* and a *diagnostic subsurface horizon*. Give a simple example of each.
16. Locate each soil order on the world map and on the U.S. map as you give a general description of it.
17. How was slash-and-burn shifting cultivation, as practiced in the past, a form of crop and soil rotation and conservation of soil properties?
18. Describe the salinization process in arid and semiarid soils. What associated soil horizons develop?
19. Which of the soil orders are associated with Earth's most productive agricultural areas?
20. Why is the 51-cm (20-in.) isohyet of annual precipitation in the Midwest significant to plants? How do the pH and lime content of soils change on either side of this isohyet?
21. Describe the podzolization process associated with northern coniferous forest soils. What characteristics are associated with the surface horizons? What strategies might enhance these soils?
22. What former Inceptisols formed a new soil order in 1998? Describe these soils as to location, nature, and formation processes. Why do you think they were separated into their own order?
23. Why has a selenium contamination problem arisen in some western U.S. soils? Explain the impact of agricultural practices, and state why you think this is or is not a serious problem.

MASTERING GEOGRAPHY

Visit www.masteringgeography.com for several figure animations, satellite loops, self-study quizzes, flashcards, visual glossary, case studies, career links, textbook reference maps, RSS Feeds, and an ebook version of *Geosystems*. Included are many geography web links to interesting Internet sources that support this chapter.

CHAPTER 19

Ecosystem Essentials

Loch Tarff, a small lake located a bit southeast of Loch Ness, in the Scottish Highlands. A mixed forest occupies an island surrounded by a moorland of heather and grasses. The forest community creates a microclimate within its island community. [Bobbé Christopherson.]

KEY LEARNING CONCEPTS

After reading the chapter, you should be able to:

- ***Define*** ecology, biogeography, and the ecosystem concept.
- ***Describe*** communities, habitats, and niches.
- ***Explain*** photosynthesis and respiration, and ***derive*** net photosynthesis and the world pattern of net primary productivity.
- ***Relate*** abiotic ecosystem components to ecosystem operations, and ***explain*** trophic relationships.
- ***Relate*** how biological evolution led to the biodiversity of life on Earth.
- ***Define*** succession, and ***outline*** the stages of general ecological succession in both terrestrial and aquatic ecosystems.

Species' Distributions Shift with Climate Change

An ecosystem involves plants and animals and their interactions with the physical environment. Changing climate affects ecosystems because temperature and moisture are among the physical limiting factors on species distributions and ecosystem function—such factors limit biotic operations, either through their lack or through their excess. Species are able to tolerate varying conditions for each limiting factor, but if conditions become too difficult, species must adapt, seek out new habitat elsewhere, or become extinct.

As temperatures increase, many plants and animals are expanding their ranges into higher latitude areas with more favorable thermal conditions. Numerous studies spanning the past 40 years show that maximum range shifts vary from 200 to 1000 km (125 to 620 mi); another study documented range shifts averaging 6.1 km (3.8 mi) per year.

Also, species are shifting their ranges to higher elevations. A relatively long-term study in Yosemite National Park, California, found that over the course of a century small mammal communities shifted their distributions in response to warming temperatures. Scientists found that half the species monitored showed substantial movement toward cooler temperatures at higher elevations, which is consistent with a 3 C° increase in minimum temperature during the last century.

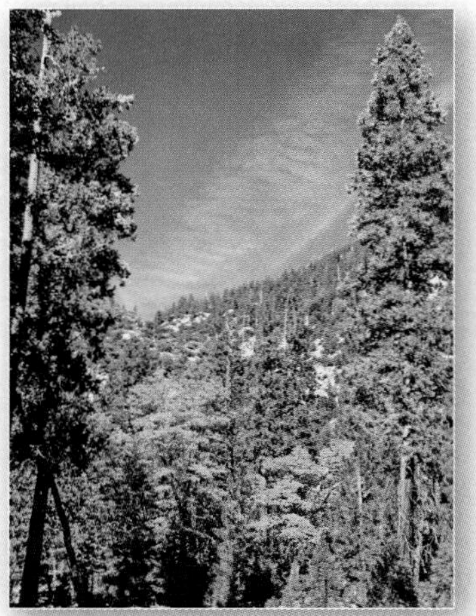

FIGURE GN 19.1 Montane species relocate in response to climate change. Species are moving to higher elevations in the Southern and Central Sierra Nevada as their preferred ranges shift to the higher mountains in the background. [Author.]

Formerly low-elevation species expanded their ranges, and high-elevation species contracted their ranges. This study is consistent with others suggesting that high-elevation species are losing available habitat as temperatures increase. These animals are threatened by climate change. In the southern Sierra Nevada mountains, such a shift was under way with the Inyo chipmunk (*Tamias umbrinus inyoensis*), a subspecies of the Uinta chipmunk. This endangered chipmunk has not been seen for more than 2 years in the Sierra, raising fears of extinction. In these western mountains, species of squirrels, voles, bats, and pikas are at risk of extinction as ranges collapse (Figure GN 19.1).

In the tropics, the latitudinal temperature gradient is slight, and species tend to expand in elevation, or upslope, where cooler temperatures are accessible. Such tropical species are at risk of mountaintop extinction as warming pushes climatically suitable conditions beyond the reaches of mountain peaks.

Global climate change is expected to impact forest distributions as well, although plant communities are slower to respond to environmental change. A recent study on 130 species of North American trees suggests that ranges will shift northward between 330 and 700 km (205 and 435 mi), depending on the success of dispersal into new habitats. Thus, deciduous forests that are now common in the United States would be found in Canada by century's end, and their present range would be replaced by grasslands in some areas and by a different mix of tree species in other regions.

To illustrate the magnitude of these shifting distributions, Figure GN 19.2 shows the predicted "migration" of what is now Illinois during this century based on forecasted changes in temperature and precipitation— in summer conditions, in particular. By mid-century, Illinois' summers will be similar to those of Arkansas and northern Louisiana; by 2090, Illinois' climate will be similar to that of Texas and Louisiana. If the plants and animals that live in Illinois today cannot adapt quickly enough to such thermal and moisture-balance changes, these communities will shift northward to stay within their tolerance limits. We can expect present Illinois environmental conditions to be in central Manitoba and Ontario, Canada, by century's end.

Working Group II in the 2007 Intergovernmental Panel on Climate Change's *Fourth Assessment Report* estimated that with a warming of 1.9 to 3.0 C° (3.5 to 5.5 F°), more than 30% of species will be outside of their preferred range, which would place them at risk of extinction. For species to survive, they must adapt, or continue to shift their ranges northward or upslope as temperatures increase. Thus, a physical geography text written 20 to 70 years from now may have to redo the biome maps to reflect these changes in species distributions. In this chapter, we discuss ecosystems and limiting factors; biomes are the focus of Chapter 20.

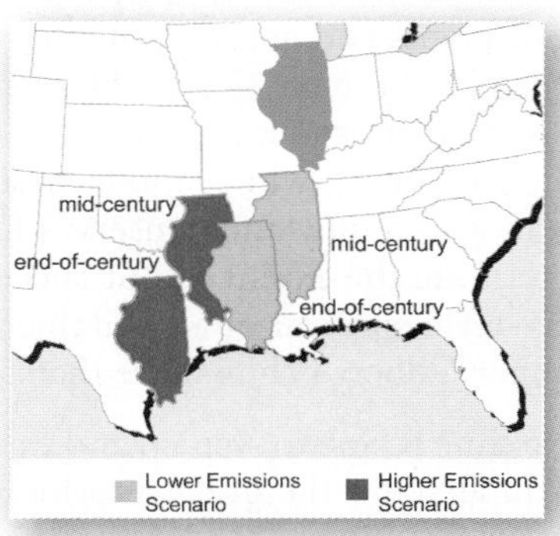

FIGURE GN 19.2 Climatic shifts forecasted for the Midwest. The relative future climatic conditions of Illinois in response to changes in temperature and precipitation forecasted by 2090 for two scenarios. [U.S. Global Change Research Program, 2009.]

Diversity is an impressive feature of the living Earth. The diversity of organisms is a response to the interaction of the atmosphere, hydrosphere, and lithosphere, which produce a variety of conditions within which the biosphere exists. The diversity of life also results from the intricate interplay of living organisms themselves. Strategies each species uses seem to work to maintain biodiversity and species coexistence. We, as part of this vast natural complex, seek to find our place and understand the feelings nature stimulates within us.

> The physical beauty of nature is certainly among its most powerful appeals to the human animal. The complexity of the aesthetic response is suggested by its wide-ranging expression from the contours of a mountain landscape to the ambient colors of a setting Sun to the fleeting vitality of a breaching whale. Each exerts a powerful aesthetic impact on most people, often accompanied by feelings of awe at the extraordinary physical appeal and beauty of the natural world.*

The biosphere, the sphere of life and organic activity, extends from the ocean floor to an altitude of about 8 km (5 mi) into the atmosphere. The biosphere includes myriad ecosystems from simple to complex, each operating within general spatial boundaries. An **ecosystem** is a self-sustaining association of living plants and animals and their nonliving physical environment. Earth's biosphere itself is a collection of ecosystems within the natural boundary of the atmosphere and Earth's crust.

Natural ecosystems are open systems for both solar energy and matter, with almost all ecosystem boundaries functioning as transition zones rather than as sharp demarcations. Distinct ecosystems—for example, forests, seas, mountaintops, deserts, beaches, islands, lakes, and ponds—make up the larger whole.

Ecology is the study of the relationships between organisms and their environment and among the various ecosystems in the biosphere. The word *ecology*, developed by German naturalist Ernst Haeckel in 1869, is derived from the Greek *oikos* ("household" or "place to live") and *logos* ("study of"). **Biogeography** is the study of the distribution of plants and animals, the diverse spatial patterns they create, and the physical and biological processes, past and present, that produce Earth's species richness.

The degree to which modern society understands Earth's biogeography and conserves Earth's living legacy will determine the extent of our success as a species and the long-term survival of a habitable Earth, as stated by geographer Gilbert White more than 30 years ago:

> The time is ripe to step up and expand current efforts to understand the great interlocking systems of air, water, and minerals nourishing the Earth. . . . Moreover, without vigorous action toward that goal, nations will be seriously handicapped in trying to cope with proven and suspected threats to ecosystems and to human health and welfare resulting from alterations in the cycles of carbon, nitrogen, phosphorus, sulfur, and related materials. . . . Society depends upon this life-support system of planet Earth.*

Earth's most influential biotic agents are humans. This is not arrogance; it is fact—because we powerfully affect every ecosystem on Earth. From the time humans first developed agriculture, the raising of animals, and the use of fire, we began a process that has led to our increasing influence over Earth's physical systems.

In this chapter: We explore ecosystems and the community, habitat, and niche concepts. Plants are the essential living component in the biosphere, translating solar energy into usable forms to energize life. The role of nonliving systems, including biogeochemical cycles, is examined. We cover the organization of living ecosystems along complex food chains and webs. The biodiversity of living organisms is seen as a product of biological evolution over the past 3.6+ billion years, approximately. Ecosystem stability and resilience, and how living landscapes change over space and time through the process of succession, are important. Included is coverage of the effects of global change on ecosystems and rates of succession. The chapter ends with a look at the Great Lakes environment and status.

Ecosystem Components and Cycles

An ecosystem is a complex of many variables, all functioning independently yet in concert, with complicated flows of energy and matter (Figure 19.1). The key is to understand the linkages and interconnections within and between systems. As the quote in the caption says, "[T]he effects of what you do in the world will always spread out like ripples in a pond." The aggregate demand for burning fossil fuels far away from the Gulf Coast pushed the need for risky drilling schemes—and thus accidents occur. The Barataria wetlands and the Little Blue Heron habitat on the coast bore the brunt of the 2010 BP oil spill (Figure 19.1b and c).

An ecosystem includes both biotic and abiotic components. Nearly all ecosystems depend on a direct input of solar energy; the few limited ones that exist in dark caves, in wells, or on the ocean floor depend on chemical reactions (chemosynthesis). Ecosystems are divided into subsystems, with the biotic portion composed of producers (plants), consumers (animals), and detritus feeders and decomposers (worms, mites, bacteria, fungi). The abiotic flows in an ecosystem include gaseous, hydrologic, and

*S. R. Kellert, "The biological basis for human values of nature," in S. R. Kellert and E. O. Wilson, eds., *The Biophilia Hypothesis* (Washington, DC: Island Press, 1993), p. 49.

*G. F. White and M. K. Tolba, *Global Life Support Systems*, United Nations Environment Programme Information, No. 47 (Nairobi, Kenya: United Nations, 1979), p. 1.

(a)

(b)

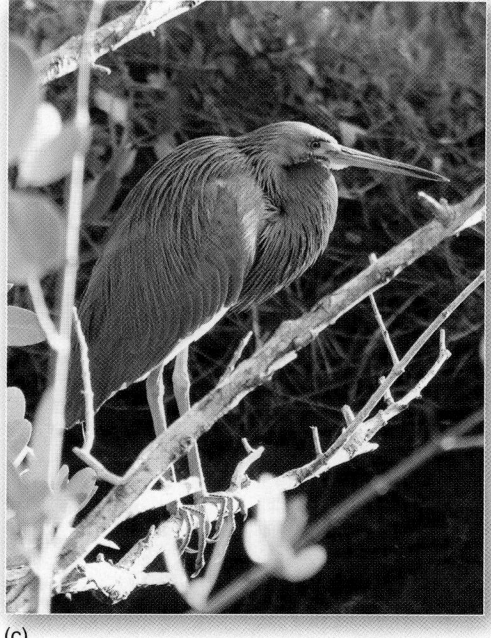

(c)

FIGURE 19.1 The web of life.
(a) Study the spider web as you read the following quotation:

> Life devours itself: everything that eats is itself eaten; everything that can be eaten is eaten; every chemical that is made by life can be broken down by life; all the sunlight that can be used is used. . . . The web of life has so many threads that a few can be broken without making it all unravel, and if this were not so, life could not have survived the normal accidents of weather and time, but still the snapping of each thread makes the whole web shudder, and weakens it. . . . You can never do just one thing: the effects of what you do in the world will always spread out like ripples in a pond.

(b) In the Barataria Preserve wetland in the Mississippi delta with bald cypress trees, note the cypress knees; this is a swamp in the bayou. (c) A Little Blue Heron, near Sanibel Island, Florida Gulf Coast. [(a) Photo by author; quotation from Friends of the Earth and Amory Lovins, The U.N. Stockholm Conference, *Only One Earth* (London, England: Earth Island Limited, 1972), p. 20. (b) and (c) Bobbé Christopherson.]

GEO REPORT 19.1 *Six major extinctions affect Earth's species*

Since life arose on the planet, six major extinctions have occurred. The fifth one was 65 million years ago, whereas the sixth is happening across the present decades (see Figure 11.1). Of all these extinction episodes, this is the only one of biotic origin, caused for the most part by human activity. We have enormous impact on most of Earth's natural systems and their associated species.

mineral cycles. Figure 19.2 illustrates these essential elements of an ecosystem and how they operate together.

Communities

A convenient biotic subdivision within an ecosystem is a community. A **community** is formed by interactions among populations of living animals and plants at a particular time. An ecosystem is the interaction of many communities with the abiotic physical components of its environment.

For example, in a forest ecosystem, a specific community may exist on the forest floor, while another community functions in the canopy of leaves high above. Similarly, within a lake ecosystem, the plants and animals that flourish in the bottom sediments form one community, whereas those near the surface form another.

A community is identified in several ways—by its physical appearance, the species present and the abundance of each, the complex patterns of their interdependence, and the trophic (feeding) structure of the community. In the chapter-opening photo of Loch Tarff, how many communities do you see?

Within a community, two concepts are important: habitat and niche. **Habitat** is the type of environment in which an organism resides or is biologically adapted to

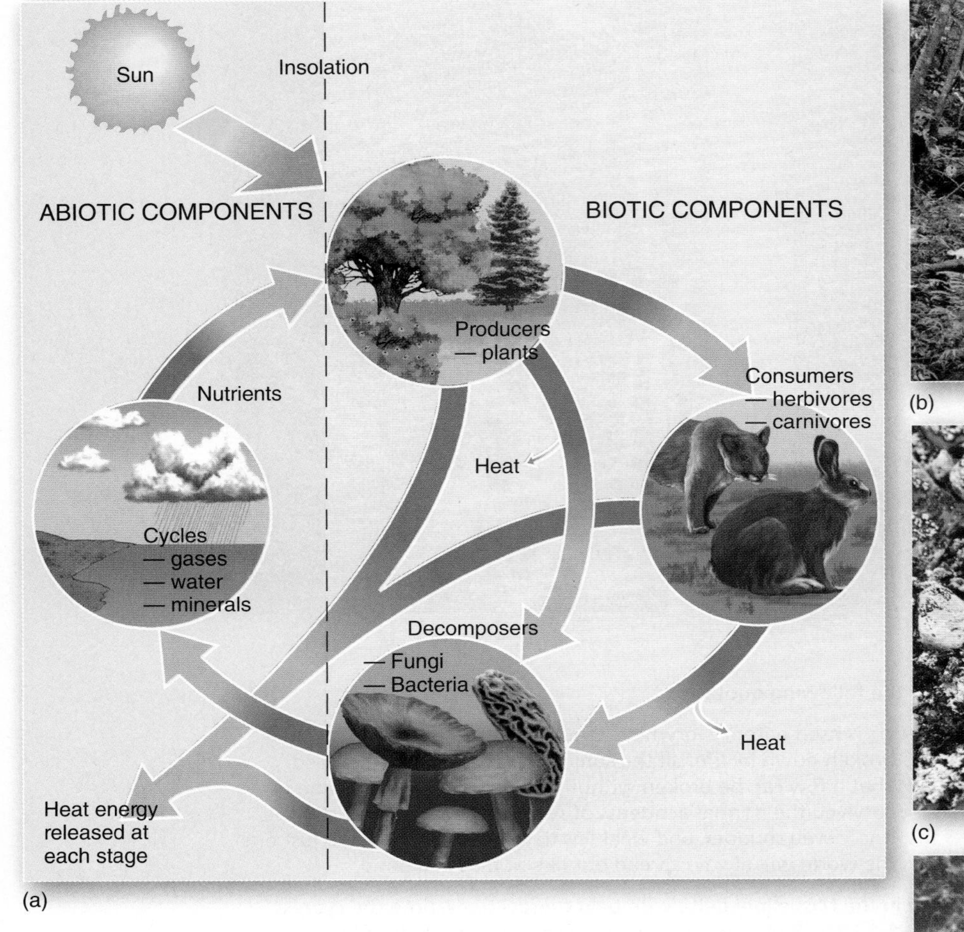

(a)

(b)

(c)

(d)

FIGURE 19.2 Biotic and abiotic components of ecosystems. (a) Solar energy is the input that drives the biotic and abiotic components. Heat energy and biomass are the outputs from the biosphere. (b) Biotic and abiotic ingredients operate together to form this rainforest floor ecosystem in Puerto Rico. (c) Five or six species of lichen function under extremely hostile climate conditions on Bear Island in the Barents Sea. Each little indentation in the rock provides some advantage to the lichen. (d) Brain coral in the Caribbean Sea, at a 3-m (9.8-ft) depth, lives in a symbiotic relationship with algae. [Bobbé Christopherson.]

FIGURE 19.3 Kittiwakes have a specific habitat and niche, here on a shear steep cliff on the northern coast of Iceland. [Bobbé Christopherson.]

live. In terms of physical and natural factors, most species have specific habitat requirements with definite limits and a specific regimen of sustaining nutrients.

Niche (from the French word *nicher*, meaning "to nest") refers to the function, or occupation, of a life form within a given community. It is the way an organism obtains and sustains the physical, chemical, and biological factors it needs to survive. A niche has several facets. Among these are a habitat niche, a trophic (food) niche, and a reproductive niche.

For example, the Red-winged Blackbird (*Agelaius phoeniceus*) occurs throughout the United States and most of Canada in habitats of meadow, pastureland, and marsh. This species nests in blackberry tangles and thick vegetation in freshwater marshes, sloughs, and fields. Its trophic niche is weed seeds and cultivated seed crops throughout the year, adding insects to its diet during the nesting season.

Habitat and niche may seem quite specific at times, such as with the Black-legged Kittiwake. These Kittiwakes nest on little available space, here on a cliff face in northern Iceland. The guano from thousands of birds and their nests fertilize moss growth on the rock (Figure 19.3). These seabirds eat fish, making the cliffs above the sea ideal nesting sites. The young fledges learn to fly from these cliffs and to fish.

In a stable community, no niche is left unfilled. The *competitive exclusion principle* states that no two species can occupy the same niche (food or space) successfully in a stable community. Thus, closely related species are spatially separated either by distance or by individual strategies. In other words, each species operates to reduce competition and to maximize its own reproduction rate—literally, species survival depends on successful reproduction. This strategy, in turn, leads to greater diversity as species shift and adapt to fill each niche.

Among the characteristics of the acacia tree (Family, *Mimosaceae*), found off the coast of Africa in the arid environment on Fogo Island, Cape Verde, are large thorns, which adapted from branch-like forms (Figure 19.4). The tree must operate to preserve fluids, such as by having small leaves that can be oriented to maximize insolation and minimize evapotranspiration. The upside-down umbrella shape maximizes light received. Virtually all parts of the tree are edible in some way, forming interrelationships with insects and animals, from ants to giraffes. Many species of acacia are found in Africa, Australia, Europe, and North America.

Some species are *symbiotic*—that is, in an arrangement whereby two or more species exist together in an

(a)

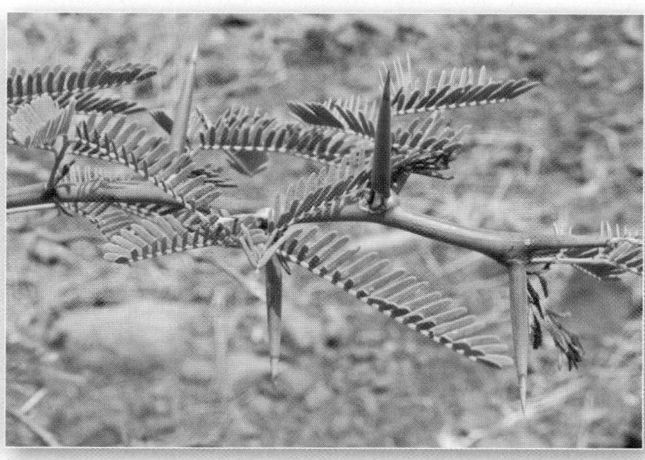

(b)

FIGURE 19.4 Thorned acacia trees in an arid climate.
(a) A woman hauls water, balanced on her head, several kilometers past acacia trees in western Africa. (b) Small leaves and stiff thorns are the acacia's defense in a dry climate. [Bobbé Christopherson.]

overlapping relationship. One type of symbiosis, *mutualism*, occurs when each organism benefits and is sustained over an extended period by the relationship. For example, lichen (pronounced "liken") is made up of algae and fungi living together (Figure 19.2c). The alga is the producer and food source for the fungus, and the fungus provides structure and physical support. Their mutualism allows the two to occupy a niche in which neither could survive alone. Lichen developed from an earlier parasitic relationship in which the fungi broke into the algal cells. Today, the two organisms have evolved into a supportive harmony and symbiotic relationship. The partnership of corals and algae discussed in Chapter 16 is another example of mutualism in a symbiotic relationship (Figure 19.2d).

By contrast, another form of symbiosis is a *parasitic* relationship, which may eventually kill the host, thus destroying the parasite's own niche and habitat. An example is parasitic mistletoe (*Phoradendron*), which lives on and may kill various kinds of trees. Some scientists are questioning whether our human society and the physical systems of Earth constitute a global-scale symbiotic relationship of mutualism, which is sustainable, or of parasitism, which is unsustainable.

II CRITICAL THINKING 19.1

Mutualism? Parasitism? Where do we fit in?

This chapter states, "Some scientists are questioning whether our human society and the physical systems of Earth constitute a global-scale symbiotic relationship of mutualism, which is sustainable, or of parasitism, which is unsustainable." Referring to the definition of these terms, what is your response to that statement? How do you equate our planetary economic system with the need to sustain life-supporting natural systems? Mutualism? Parasitism? What is your characterization?

Plants: The Essential Biotic Component

Plants are the critical biotic link between solar energy and the biosphere. Ultimately, the fate of all members of the biosphere, including humans, rests on the success of plants and their ability to turn sunlight into food.

Land plants (and animals) became common about 430 million years ago, according to fossilized remains. **Vascular plants** developed conductive tissues and true roots for internal transport of fluid and nutrients. (*Vascular* is from a Latin word for "vessel-bearing," referring to the conducting cells.) Presently, about 270,000 species of plants are known to exist, and most are vascular. Many more species have yet to be identified. They represent a great untapped resource base. Only about 20 species of plants provide 90% of the world's food; just three—wheat, maize (corn), and rice—make up half of the food supply. Plants are a major source of new medicines and chemical compounds that benefit humanity. Plants also are the core of healthy, functioning ecosystems that sustain all life.

Leaves are solar-powered chemical factories, wherein photochemical reactions take place. Veins in the leaf bring in water and nutrient supplies and carry off the sugars (food) produced by photosynthesis. The veins in each leaf connect to the stems and branches of the plant and to the main circulation system.

Flows of carbon dioxide, water, light, and oxygen enter and exit the surface of each leaf (see Figure 1.5 in Chapter 1). Gases flow into and out of a leaf through small pores, the **stomata** (singular: *stoma*), which usually are most numerous on the lower side of the leaf. Each stoma is surrounded by guard cells that open and close the pore, depending on the plant's changing needs. Water that moves through a plant exits the leaves through the stomata and evaporates from leaf surfaces, thereby assisting the plant's temperature regulation. As water evaporates from the leaves, a pressure deficit is created that allows atmospheric pressure to push water up through the plant all the way from the roots, in the same manner that a soda straw works. We can only imagine the complex operation of a 100-m (330-ft) tree!

Photosynthesis and Respiration

Powered by energy from certain wavelengths of visible light, **photosynthesis** unites carbon dioxide and hydrogen (hydrogen is derived from water in the plant). The term is descriptive: *photo-* refers to sunlight and *-synthesis* describes the "manufacturing" of starches and sugars through reactions within plant leaves. The process releases oxygen and produces energy-rich food for the plant.

The largest concentration of light-responsive, photosynthetic structures (known as *organelles*) in a leaf rests below the leaf's upper layers. These organelle units within cells are the *chloroplasts*, and within each resides a green, light-sensitive pigment called **chlorophyll**. Within this pigment, light stimulates photochemistry. Consequently, competition for light is a dominant factor in the formation of plant communities. This competition is expressed in the height, orientation, distribution, and structure of plants.

Only about one-quarter of the light energy arriving at the surface of a leaf is useful to the light-sensitive chlorophyll. Chlorophyll absorbs only the orange-red and violet-blue wavelengths for photochemical operations, and it reflects predominantly green hues (and some yellow). That is why trees and other vegetation look green.

Photosynthesis essentially follows this equation:

$$\underset{\text{(carbon dioxide)}}{6CO_2} + \underset{\text{(water)}}{6H_2O} + \underset{\text{(solar energy)}}{\text{Light}} \rightarrow \underset{\text{(glucose, carbohydrate)}}{C_6H_{12}O_6} + \underset{\text{(oxygen)}}{6O_2}$$

From the equation, you can see that photosynthesis removes carbon (in the form of CO_2) from Earth's atmosphere. The quantity is enormous: approximately 91 billion metric tons (100 billion tons) of carbon dioxide per year. Carbohydrates, the organic result of the photosynthetic process, are combinations of carbon, hydrogen, and

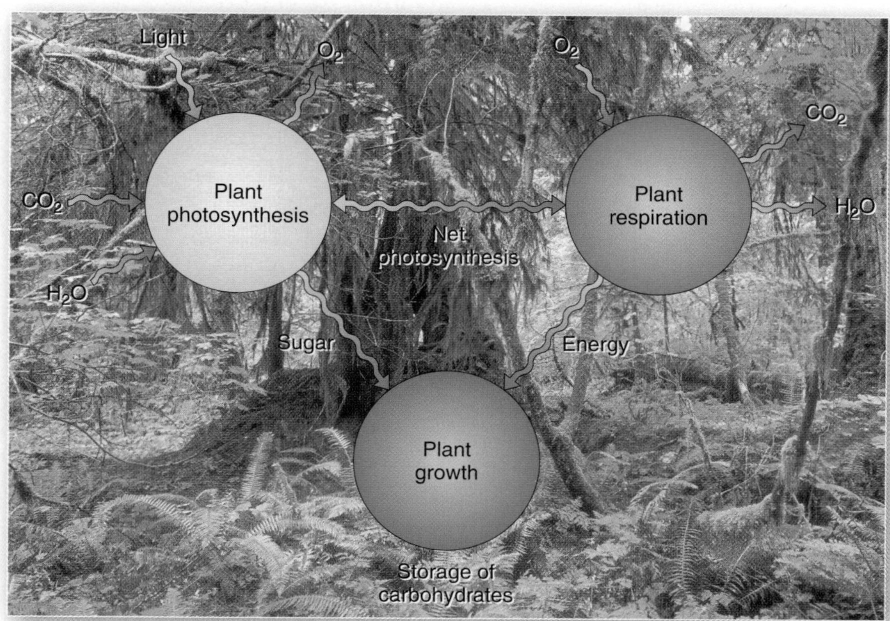

FIGURE 19.5 How plants live and grow.
The balance between photosynthesis and respiration determines net photosynthesis and plant growth. [Photo by Bobbé Christopherson.]

oxygen. They can form simple sugars, such as *glucose* ($C_6H_{12}O_6$). Plants use glucose to build starches, which are more-complex carbohydrates and the principal food stored in plants.

Plants store energy for later use. They consume this energy as needed through respiration by converting the carbohydrates to energy for their other operations. Thus, **respiration** is essentially a reverse of the photosynthetic process:

$$\underset{\text{(glucose, carbohydrate)}}{C_6H_{12}O_6} + \underset{\text{(oxygen)}}{6O_2} \rightarrow \underset{\text{(carbon dioxide)}}{6CO_2} + \underset{\text{(water)}}{6H_2O} + \underset{\text{(heat energy)}}{\text{energy}}$$

In respiration, plants oxidize carbohydrates (stored energy), releasing carbon dioxide, water, and energy as heat. The overall growth of a plant depends on a surplus of carbohydrates beyond those lost through plant respiration. Figure 19.5 presents a simple schematic of this process, which produces plant growth.

The *compensation point* is the breakeven point between the production and consumption of organic material. Each leaf must operate on the production side of the compensation point, or else the plant eliminates it—something each of us has no doubt experienced with a houseplant that received inadequate water or light. The difference between photosynthetic production and respiration loss is *net photosynthesis*. The amount varies, depending on controlling environmental factors such as light, water, temperature, soil fertility, and the plant's site, elevation, and competition from other plants and animals.

The rapidly increasing carbon dioxide concentration in the atmosphere stimulates the photosynthetic processes. However, higher temperatures due to global warming result in higher respiration rates. Plants respire all the time, but at night, respiration processes are the plant's main operation, corresponding with the time that research shows warming is greatest.

Plant productivity increases as light availability increases—up to a point. When the light level is too high, *light saturation* occurs, and most plants actually reduce their output in response. Some plants are adapted to shade, whereas others flourish in full sunlight. Crops such as rice, wheat, and sugarcane do well with high light intensity. Figure 19.6 portrays the general energy budget for green plants, showing energy receipt, energy utilization, and disposition of net primary production—how the energy is spent.

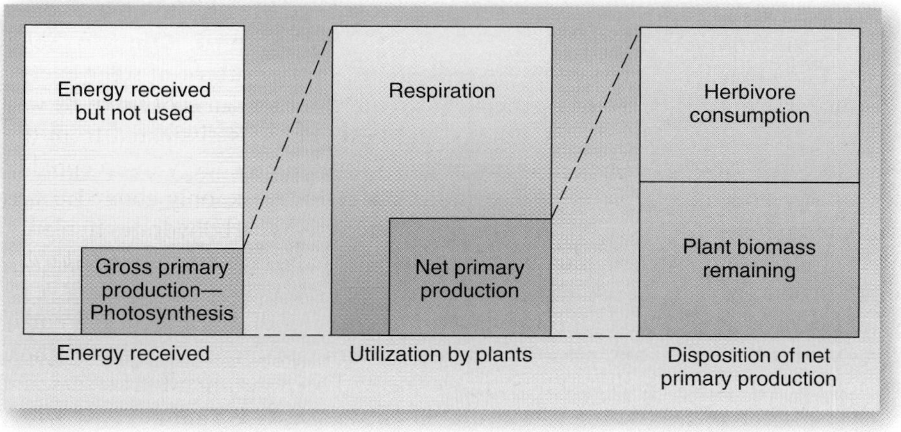

Energy received but not used
Gross primary production—Photosynthesis
Energy received
Respiration
Net primary production
Utilization by plants
Herbivore consumption
Plant biomass remaining
Disposition of net primary production

FIGURE 19.6 Energy budget of the biosphere.
Energy receipt, energy utilization, and energy disposition of net primary production by green plants.

Net Primary Productivity The net photosynthesis for an entire plant community is its **net primary productivity**. This is the amount of stored chemical energy that the community generates for the ecosystem. **Biomass** is the net dry weight of all this organic material and its stored chemical energy.

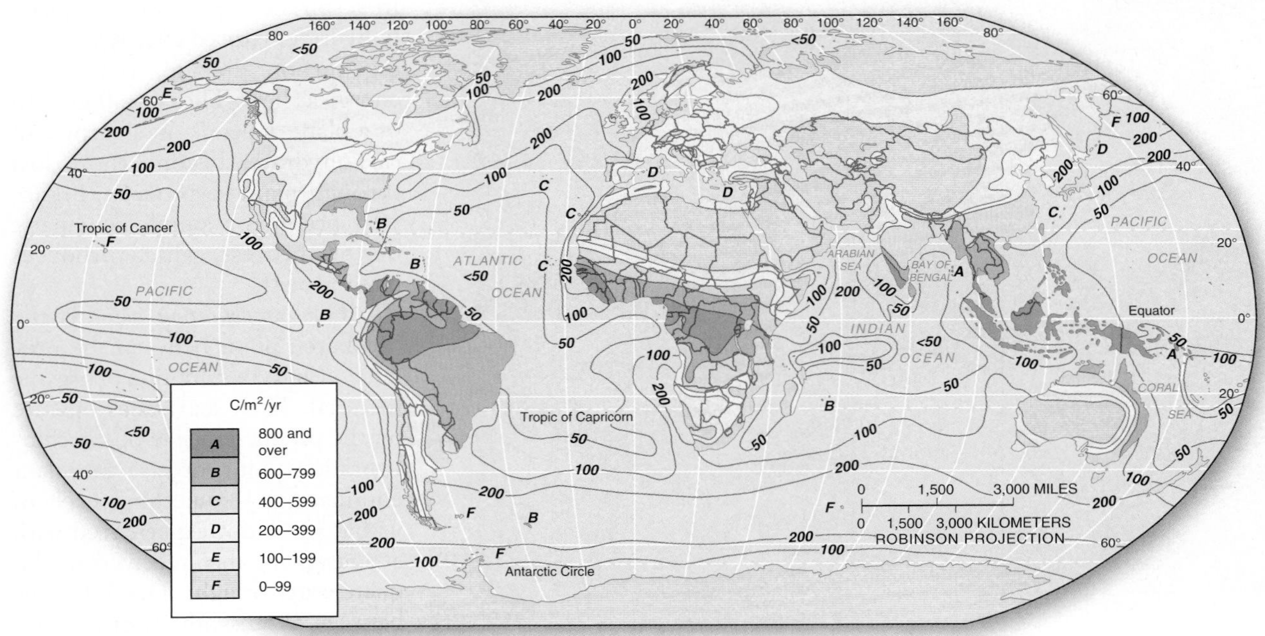

FIGURE 19.7 Net primary productivity.
Worldwide net primary productivity in grams of carbon per square meter per year (approximate values).
[Adapted from D. E. Reichle, *Analysis of Temperate Forest Ecosystems* (Heidelberg, Germany: Springer, 1970.]

Net primary productivity is measured as fixed carbon per square meter per year. ("Fixed" means chemically bound into plant tissues.) Study the map in Figure 19.7, and you can see that on land, net primary production tends to be highest between the Tropics of Cancer and Capricorn at sea level and decreases toward higher latitudes and elevations. Precipitation also affects productivity, as evidenced by the correlations of abundant precipitation with high productivity adjacent to the equator and reduced precipitation with low productivity in the subtropical deserts. Even though deserts receive high amounts of solar radiation, other controlling factors limit productivity—namely, water availability and soil conditions.

In the oceans, differing nutrient levels control and limit productivity. Regions with nutrient-rich upwelling currents off western coastlines generally are the most productive. The map in Figure 19.7 shows that the tropical oceans and areas of subtropical high pressure are quite low in productivity.

In temperate and high latitudes, the rate at which carbon is fixed by vegetation varies seasonally. It increases in spring and summer as plants flourish with increasing solar input and, in some areas, with more available (nonfrozen) water, and it decreases in late fall and winter. Productivity rates in the tropics are high throughout the year, and turnover in the photosynthesis–respiration cycle is faster, exceeding by many times the rates experienced in a desert environment or in the far northern limits of the tundra. A lush hectare (2.5 acres) of sugarcane in the tropics might fix 45 metric tons (50 tons) of carbon in a year, whereas desert plants in an equivalent area might achieve only 1% of that amount.

Table 19.1 lists various ecosystems, their net primary productivity, and an estimate of net total biomass worldwide—170 billion metric tons of dry organic matter per year. Compare the various ecosystems, especially cultivated land (in *italics*), with most of the natural communities. Net productivity is generally regarded as the most important aspect of any type of community, and the distribution of productivity over Earth's surface is an important subject of biogeography.

Abiotic Ecosystem Components

Critical in each ecosystem are the flow of energy and the cycling of nutrients and water in life-supporting systems. These abiotic components set the stage for ecosystem operations.

Light, Temperature, Water, and Climate Solar energy powers ecosystems, so the pattern of solar energy receipt is crucial. Solar energy enters an ecosystem by way of photosynthesis, and heat energy is dissipated from the system at many points. Of the total energy intercepted at Earth's surface and available for work, only about 1.0% is actually fixed by photosynthesis as carbohydrates in plants.

The duration of Sun exposure is the *photoperiod.* Along the equator, days are essentially 12 hours long year-round; however, with increasing distance from the equator, seasonal effects become pronounced. Plants have adapted their flowering and seed germination to seasonal changes in insolation. Some seeds germinate only when daylength reaches a certain number of hours. A plant that responds in the opposite manner is the poinsettia (*Euphorbia pulcherrima*), which requires at least 2 months of 14-hour nights to start flowering.

TABLE 19.1 Net Primary Productivity and Plant Biomass on Earth

Ecosystem	Area (10^6 km^2)[a]	Net Primary Productivity per Unit Area (g/m^2/yr)[b] Normal Range	Mean	World Net Biomass (10^9 tons/yr)[c]
Tropical rain forest	17.0	1000–3500	2200	37.4
Tropical seasonal forest	7.5	1000–2500	1600	12.0
Temperate evergreen forest	5.0	600–2500	1300	6.5
Temperate deciduous forest	7.0	600–2500	1200	8.4
Boreal forest	12.0	400–2000	800	9.6
Woodland and shrubland	8.5	250–1200	700	6.0
Savanna	15.0	200–2000	900	13.5
Temperate grassland	9.0	200–1500	600	5.4
Tundra and alpine region	8.0	10–400	140	1.1
Desert and semidesert scrub	18.0	10–250	90	1.6
Extreme desert, rock, sand, ice	24.0	0–10	3	0.07
Cultivated land	*14.0*	*100–3500*	*650*	*9.1*
Swamp and marsh	2.0	800–3500	2000	4.0
Lake and stream	2.0	100–1500	250	0.5
Total continental	**149.0**	—	**773**	**115.17**
Open ocean	332.0	2–400	125	41.5
Upwelling zones	0.4	400–1000	500	0.2
Continental shelf	26.6	200–600	360	9.6
Algal beds and reefs	0.6	500–4000	2500	1.6
Estuaries	1.4	200–3500	1500	2.1
Total marine	**361.0**	—	**152**	**55.0**
Grand total	**510.0**	—	**333**	**170.17**

Source: R. H. Whittaker, *Communities and Ecosystems* (Heidelberg, Germany: Springer, 1975), p. 224. Reprinted by permission.

[a]1 km^2 = 0.38 mi^2.

[b]1 g/m^2 = 8.92 lb/acre.

[c]1 metric ton (10^6 g) = 1.1023 tons.

Monarch butterflies (*Danaus plexippus*) migrate some 3600 km (2235 mi) from eastern North America to wintering forests in Mexico. The butterflies navigate using an internal time-compensated Sun compass to maintain their southwestward track while they compensate for time-of-day Sun changes. Researchers found receptors in their eyes sensitive to ultraviolet light that play the key role in accurate navigation. For monarchs, ultraviolet light is an important abiotic component.

Other components are important to ecosystem processes. Air and soil temperatures determine the rates at which chemical reactions proceed. Significant temperature factors are seasonal variation and duration and the pattern of minimum and maximum temperatures.

Operations of the hydrologic cycle and water availability depend on precipitation and evaporation rates and their seasonal distribution. Water quality—its mineral content, salinity, and levels of pollution and toxicity—is important. Also, regional climates affect the pattern of vegetation and ultimately influence soil development. All of these factors work together to form the limits for ecosystems in a given location.

Figure 19.8 illustrates the general relationship among temperature, precipitation, and vegetation. Can you identify the characteristic vegetation type and related temperature and moisture regime that fit areas you know well, such as a place you have lived or an area around the school you now attend?

Beyond these general conditions, each ecosystem further produces its own *microclimate*, which is specific to individual sites. For example, in a forest the insolation reaching the ground is reduced, and shade from its own trees blankets the forest floor. A pine forest cuts light by 20%–40%, whereas a birch–beech forest reduces it by as much as 50%–75%. Forests also are about 5% more humid than nonforested landscapes, have moderated temperatures (warmer winters and cooler summers), and experience reduced winds.

Life Zones Alexander von Humboldt (1769–1859), an explorer, geographer, and scientist, observed that plants and animals recur in related groupings wherever similar conditions occur in the abiotic environment. He synthesized in a systems approach the influence of atmospheric, oceanic, geologic, and ecologic parameters on vegetation distributions. Humboldt was far ahead of his time as an Earth systems scientist. And, as a true geographer, he thought of nature and Earth systems as spatial arrays of phenomena.

After several years of study in the Andes Mountains of Peru, he described a distinct relation between elevation and plant communities as his *life zone concept.* As he climbed the mountains, he noticed that the experience

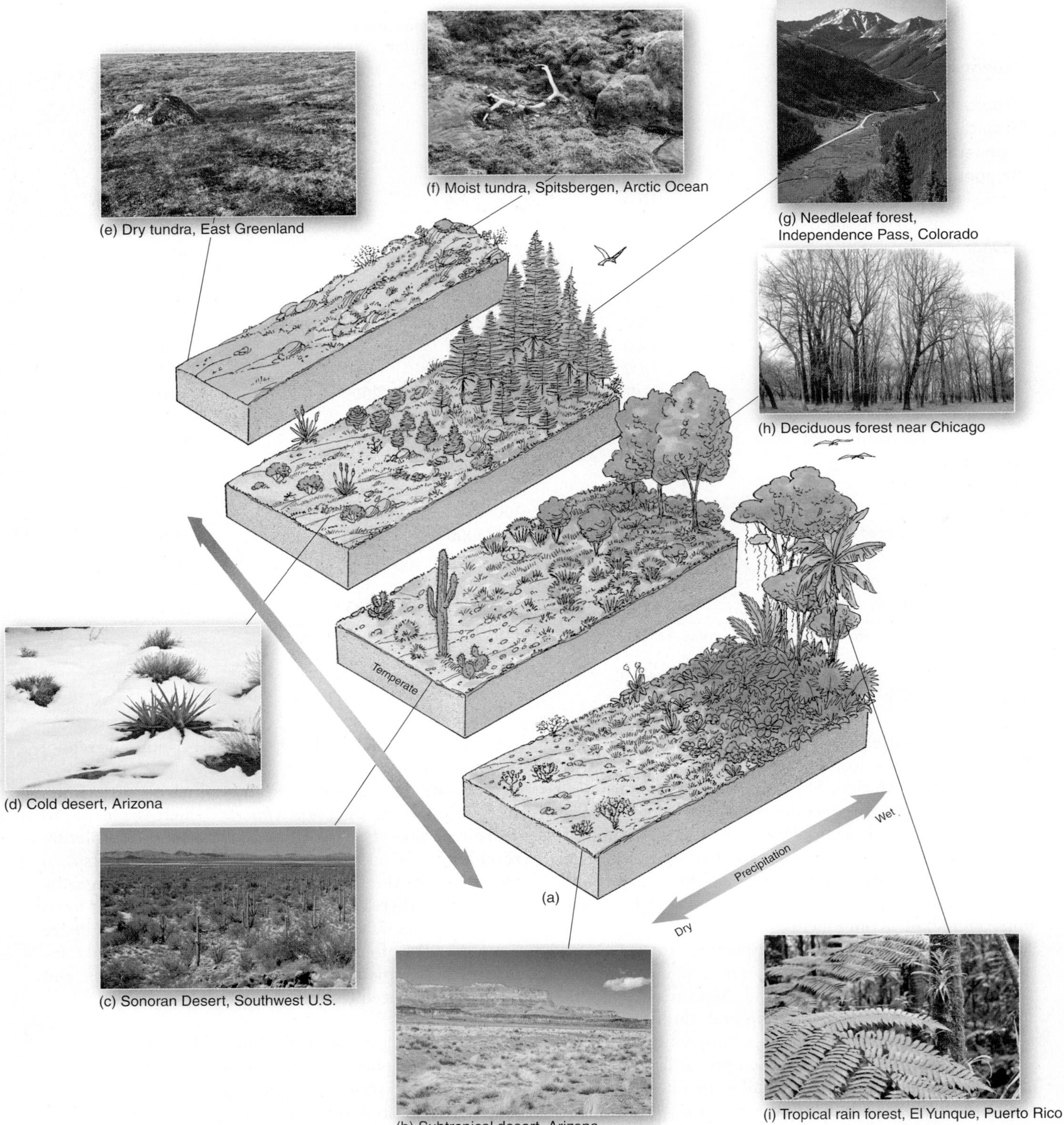

FIGURE 19.8 How temperature and precipitation affect ecosystems. (a) Climate controls (wetness, dryness, warmth, cold) and ecosystem types; (b) through (i) are representative photos of each area. [(c) Author. All others Bobbé Christopherson.]

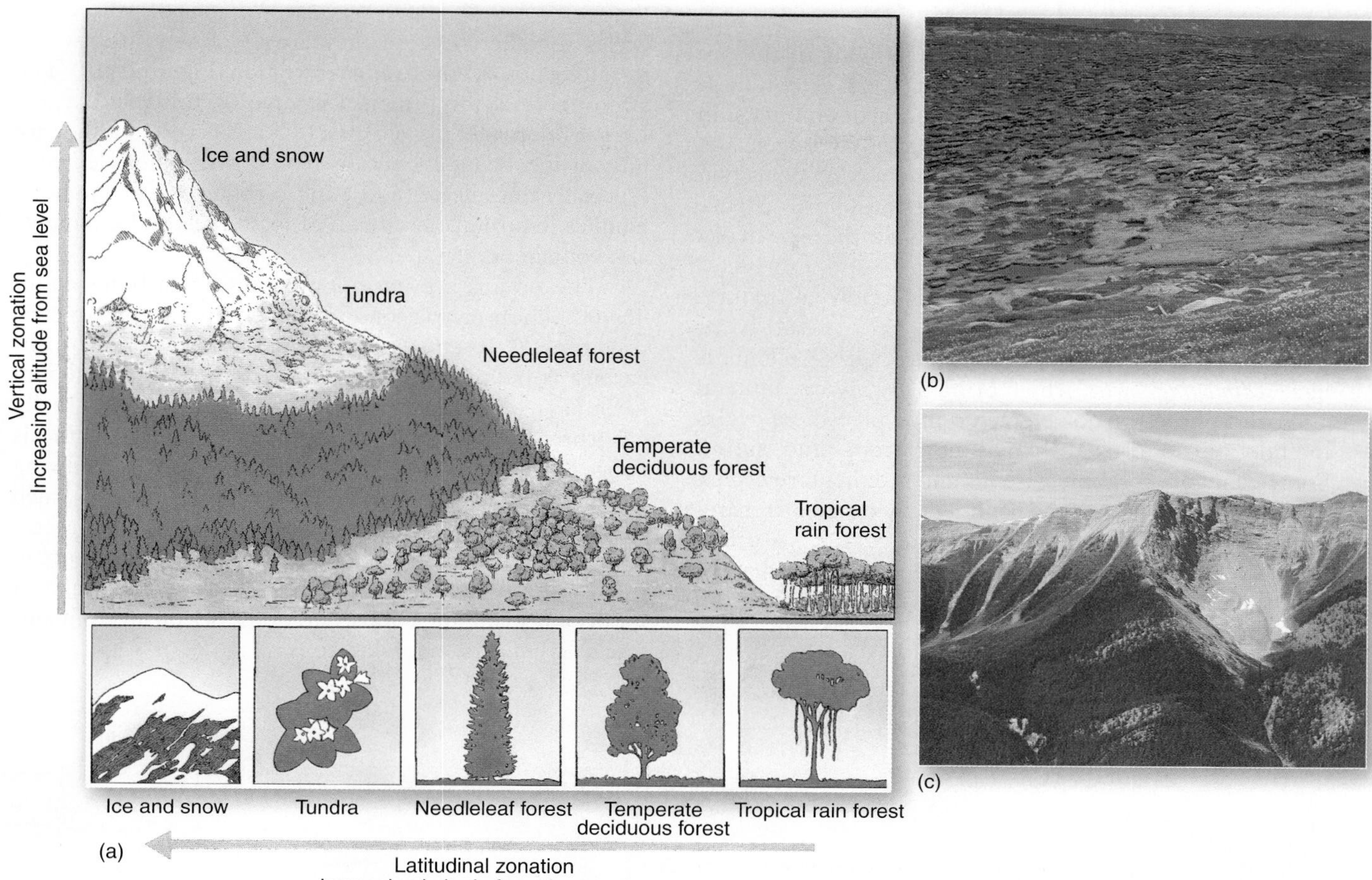

FIGURE 19.9 Vertical and latitudinal zonation of plant communities. (a) Progression of plant community life zones with changing elevation or latitude. (b) Alpine tundra ecosystem in the Colorado Rockies. (c) The timberline for a needleleaf forest in the Canadian Rockies. The line of trees marks the limit of highest continuous forest, whereas the area above it marks the zone above which no trees grow. [(b) Bobbé Christopherson; (c) Author.]

was similar to that of travelling away from the equator toward higher latitudes (Figure 19.9a).

This zonation of plants is noticeable on any trip from lower valleys to higher elevations. Each **life zone** possesses its own temperature, precipitation, and insolation relations and therefore its own biotic communities. Scientific research now shows that climate change is causing plants and animals to move their range to higher elevations with more desirable temperatures as established life zones shift. Evidence exists that some species have run out of mountain as environmental conditions simply have taken them "out-of-bounds," either elsewhere or into extinction.

GEO REPORT 19.2 *Earth's magnetic field acts as an abiotic component*

The fact that birds and bees can detect Earth's magnetic field and use it for finding direction is well established. Small amounts of magnetically sensitive particles in the skull of the bird and the abdomen of the bee provide compass directions.

A team of biologists found that sea turtles detect magnetic fields of different strengths and inclinations (angles). This means that the turtles have a built-in navigation system that helps them know where they are on Earth (like our Global Positioning System, which requires multimillion-dollar satellites). Loggerhead sea turtles hatch in Florida, crawl into the water, and spend the next 70 years travelling thousands of miles between North America and Africa around the subtropical high-pressure gyre in the Atlantic Ocean. The females return to where they were hatched to lay their eggs. The researchers think that the hatchlings are imprinted with magnetic data unique to the beach where they hatched and then develop a more global sense of position as they live a life swimming across the ocean.

Elemental Cycles

The most abundant natural elements in living matter are hydrogen (H), oxygen (O), and carbon (C). Together, these elements make up more than 99% of Earth's biomass; in fact, all life (organic molecules) contains hydrogen and carbon. In addition, nitrogen (N), calcium (Ca), potassium (K), magnesium (Mg), sulfur (S), and phosphorus (P) are significant nutrients, elements necessary for the growth of a living organism.

Several key chemical cycles function in nature. Oxygen, carbon, and nitrogen each have *gaseous cycles*, parts of which are in the atmosphere. Other elements have *sedimentary cycles*, which principally involve mineral and solid phases; major elements that have these cycles include phosphorus, calcium, potassium, and sulfur. Some elements combine gaseous and sedimentary cycles. The recycling of gases and sedimentary (nutrient) materials forms Earth's **biogeochemical cycles**, so called because they involve chemical reactions necessary for growth and development of living systems. The chemical elements themselves recycle over and over again in life processes.

Oxygen and Carbon Cycles We consider these two cycles together because they are so closely intertwined through photosynthesis and respiration (Figure 19.10). The atmosphere is the principal reserve of available oxygen. Larger reserves of oxygen exist in Earth's crust, but they are unavailable, being chemically bound with other elements, especially the silicate (SiO_2) and carbonate (CO_3) mineral families. Unoxidized reserves of fossil fuels and sediments also contain oxygen.

The oceans are enormous pools of carbon—about 42,900 billion metric tons, or 47,190 billion tons (metric tons times 1.10 equals short tons). However, all of this carbon is bound chemically in carbon dioxide, calcium carbonate, and other compounds. The ocean initially absorbs carbon dioxide by means of the photosynthesis carried on by phytoplankton; it becomes part of the living organisms and is fixed in certain carbonate minerals, such as limestone ($CaCO_3$). In Chapter 16, we saw that excessive absorption of CO_2 from the atmosphere by the ocean is lowering pH. This condition of acidification makes it harder for plankton, corals, and other organisms to maintain calcium carbonate skeletons.

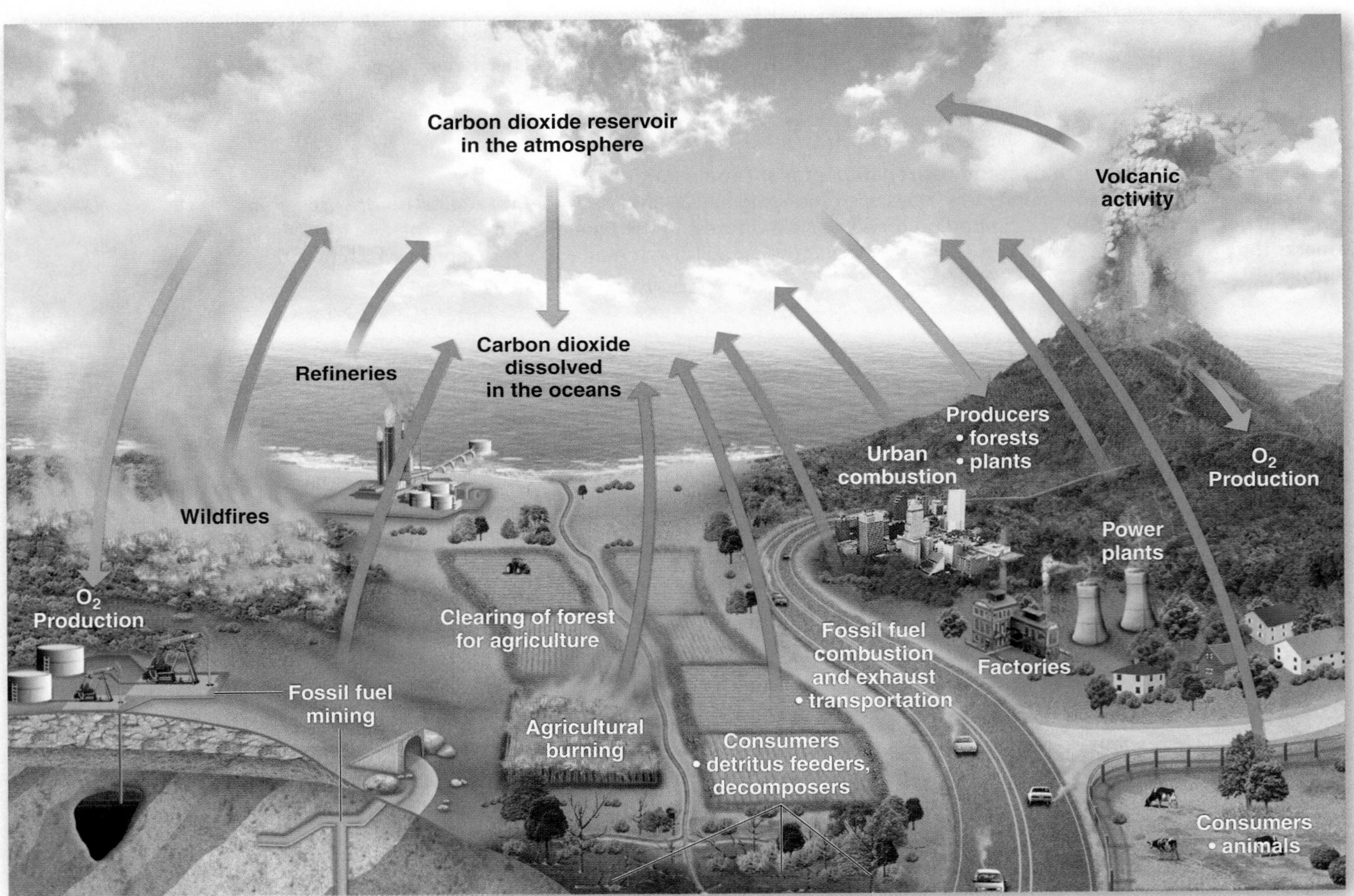

FIGURE 19.10 The carbon and oxygen cycles.
Carbon is fixed through photosynthesis and begins its passage through ecosystem operations. Respiration by living organisms, the burning of forests and grasslands, and the combustion of fossil fuels release carbon to the atmosphere. These cycles are greatly influenced by human activities.

The atmosphere, which is the integrating link in the cycle, contains only about 700 billion metric tons of carbon (as carbon dioxide) at any moment. This is far less carbon than is stored in fossil fuels and oil shales (13,200 billion metric tons, as hydrocarbon molecules) or in living and dead organic matter (2500 billion metric tons, as carbohydrate molecules). Carbon dioxide is released into the atmosphere by the respiration of plants and animals, volcanic activity, and fossil-fuel combustion by industry and transportation.

The carbon dumped into the atmosphere by human activity constitutes a vast geochemical experiment, using the real-time atmosphere as a laboratory. Annually, we are adding carbon to the atmospheric pool in an amount 400% greater than 1950 levels. Global emissions of carbon from fossil fuels continue to increase, with some 8.2 billion metric tons (9.02 billion tons) of carbon added to the atmosphere in 2007—compare this to the 1.47 billion metric tons (1.61 billion tons) in 1950 or the 4.74 billion metric tons (5.21 billion tons) in 1980. About 50% of the carbon emitted since the beginning of the Industrial Revolution and not absorbed by oceans and organisms remains in the atmosphere, enhancing Earth's natural greenhouse effect—and thus global warming.

Nitrogen Cycle Nitrogen, which accounts for 78.084% of each breath we take, is the major constituent of the atmosphere. Nitrogen also is important in the makeup of organic molecules, especially proteins, and therefore is essential to living processes. A simplified view of the nitrogen cycle is portrayed in Figure 19.11.

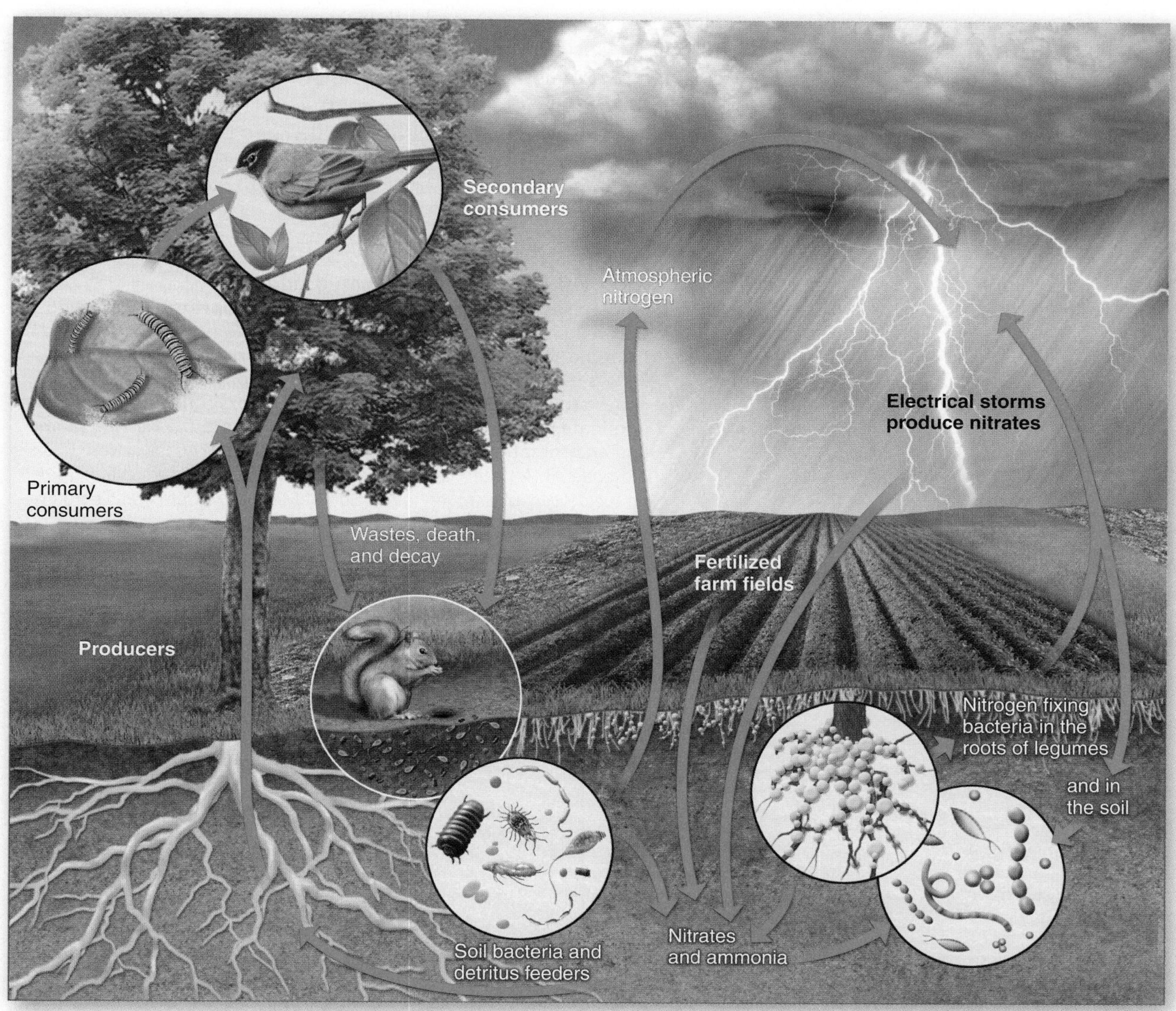

FIGURE 19.11 The nitrogen cycle.
The atmosphere is the essential reservoir of gaseous nitrogen. Atmospheric nitrogen gas is chemically fixed by bacteria to produce ammonia. Lightning and forest fires produce nitrates, and fossil-fuel combustion forms nitrogen compounds that are washed from the atmosphere by precipitation. Plants absorb nitrogen compounds and produce organic material.

Nitrogen-fixing bacteria, which live principally in the soil and are associated with the roots of certain plants, are the key link to life. The roots of legumes such as clover, alfalfa, soybeans, peas, beans, and peanuts have such bacteria. Bacteria colonies reside in nodules on the legume roots and chemically combine the nitrogen from the air in the form of nitrates (NO_3) and ammonia (NH_3). Plants use these chemically bound forms of nitrogen to produce their own organic matter. Anyone or anything feeding on the plants thus ingests the nitrogen. Finally, the nitrogen in the organic wastes of the consuming organisms is freed by denitrifying bacteria, which recycle it to the atmosphere.

To improve agricultural yields, many farmers use synthetic inorganic fertilizers, as opposed to soil-building organic fertilizers (manure and compost). Inorganic fertilizers are chemically produced through artificial nitrogen fixation at factories. Humans presently fix more nitrogen as synthetic fertilizer per year than all terrestrial sources combined—and the present production of synthetic fertilizers now is doubling every 8 years; some 1.82 million metric tons (2 million tons) are produced per week, worldwide. The crossover point when anthropogenic sources of fixed nitrogen exceeded the normal range of naturally fixed nitrogen was 1970.

This surplus of usable nitrogen accumulates in Earth's ecosystems. Some is present as excess nutrients, washed from soil into waterways and eventually to the ocean. This excess nitrogen load begins a water pollution process that feeds an excessive growth of algae and phytoplankton, increases biochemical oxygen demand, diminishes dissolved oxygen reserves, and eventually disrupts the aquatic ecosystem. In addition, excess nitrogen compounds in air pollution are a component in acid deposition, further altering the nitrogen cycle in soils and waterways.

Figure 19.12 includes maps and images of the core region of the Gulf Coast dead zone, a region of oxygen-depleted water off the coast of Louisiana in the Gulf of Mexico. In 2008, researchers at Texas A&M University reported dead zone conditions stretching along the entire Texas Gulf Coast and extending in a discontinuous spread some 32 km (20 mi) from shore.

The Mississippi River carries agricultural fertilizers, farm sewage, and other wastes to the Gulf, causing huge spring blooms of phytoplankton: a greening of primary productivity. The Mississippi drainage system handles the runoff for 41% of the continental United States. By summer, the biological oxygen demand of bacteria feeding on the decay exceeds the dissolved oxygen; hypoxia (oxygen

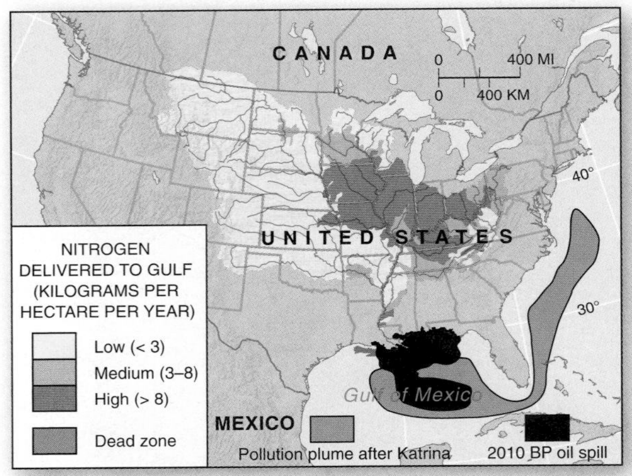

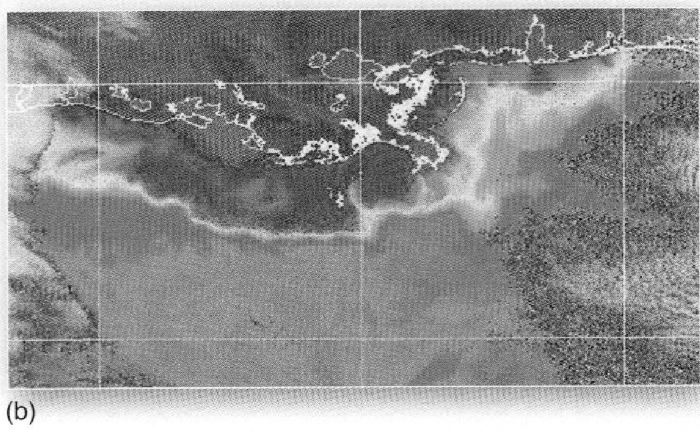

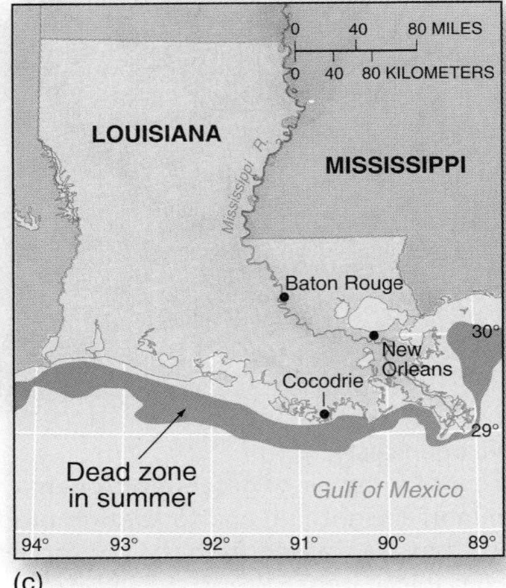

FIGURE 19.12 The Gulf Coast dead zone.
(a) Mapped total nitrogen originating in the upstream watershed that feeds the Mississippi River system. (b) The flush of nutrients from the Mississippi River drainage basin enriches the offshore waters in the Gulf of Mexico, causing huge phytoplankton blooms in early spring (red areas). (c) By late summer, oxygen-depleted waters dominate what is known as the *dead zone*. Note the Katrina plume and an estimate of the 2010 Gulf oil disaster. [(a) Adapted from R. B. Alexander, R. A. Smith, and G. E. Schwarz, "Effect of stream channel size on the delivery of nitrogen to the Gulf of Mexico," *Nature* 403 (February 17, 2000): 761. (b) *SeaWiFS* image, NASA/GSFC.]

depletion) develops, killing any fish that venture into the area. These low-oxygen conditions act as a limiting factor on marine life. From 2002 on, the Gulf Coast dead zone has expanded to an area of more than 22,000 km^2 (8500 mi^2) each year. The agricultural, feedlot, and fertilizer industries dispute the connection between their nutrient input and this extensive dead zone.

In other parts of the world, the connection is established. Similar coastal dead zones occur as the result of nutrient outflows from more than 400 river systems worldwide, affecting almost 250,000 km^2 (94,600 mi^2) of offshore oceans and seas. In Sweden and Denmark, a concerted effort to reduce nutrient flows into rivers reversed hypoxic conditions in the Kattegat Strait (between the Baltic and North Seas). And, with the fall of the Soviet Union and state agriculture in 1990, fertilizer use is down more than 50% in the former Soviet republics. The Black Sea now gets several months' break from year-round hypoxia at river deltas.

As with most environmental situations, the cost of mitigation is cheaper than the cost of continued damage to marine ecosystems. A 20% to 30% cut in nitrogen inflow upstream is estimated to increase dissolved oxygen levels by more than 50% in the dead zone region of the Gulf. One government estimate placed the application of excess nitrogen fertilizer at 20% more than needed in Iowa, Illinois, and Indiana. The initial step to resolving this issue might be to mandate applying only the levels of fertilizer needed—a savings in overhead costs for agricultural interests—and, secondly, to begin dealing with animal wastes from feedlots in the basin.

Following Hurricane Katrina in 2005, the region was overwhelmed by discharge carrying sewage, chemical toxins, oil spills, household pesticides, animal wastes and carcasses, and nutrients. The plume, now mostly dissipated, from this event tracked around the Florida peninsula and northward up the Eastern seaboard (see the suggested plot in Figure 19.12a). Add to all this the 2010 *Deepwater Horizon* oil spill, where British, Swiss, and Cayman Island corporations, under the flag and regulations of the Marshall Islands, were operating in U.S. waters and caused the worst oil spill disaster in U.S. history. You will find more on this in Chapter 21.

Limiting Factors

The term **limiting factor** identifies the one physical or chemical component that most inhibits biotic operations, through either its lack or its excess. In most ecosystems, precipitation is the limiting factor. However, temperature, light levels, and soil nutrients certainly affect vegetation patterns. Here are a few examples of possible limiting factors:

- Low temperatures limit plant growth at high elevations;
- Lack of water limits growth in a desert;
- Excess water limits growth in a bog;
- Changes in salinity levels affect aquatic ecosystems;
- Lack of iron in ocean surface environments limits photosynthetic production;
- Low phosphorus content of soils limits plant growth; and
- General lack of active chlorophyll above 6100 m (20,000 ft) limits primary productivity.

Each organism possesses a range of tolerance for each limiting factor in its environment. This fact is illustrated in Figure 19.13, which shows the geographic

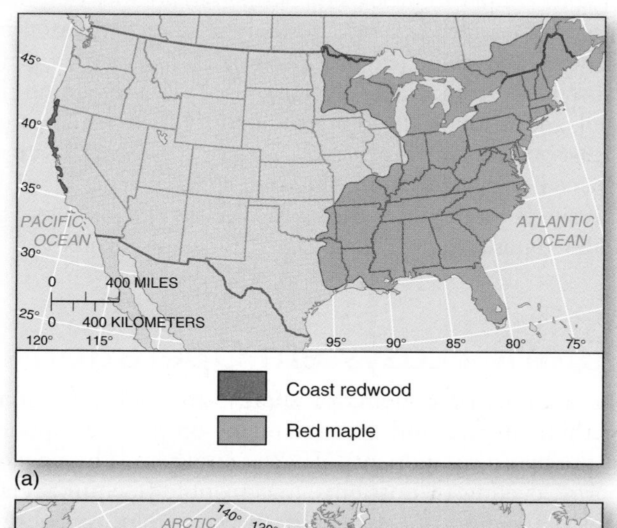

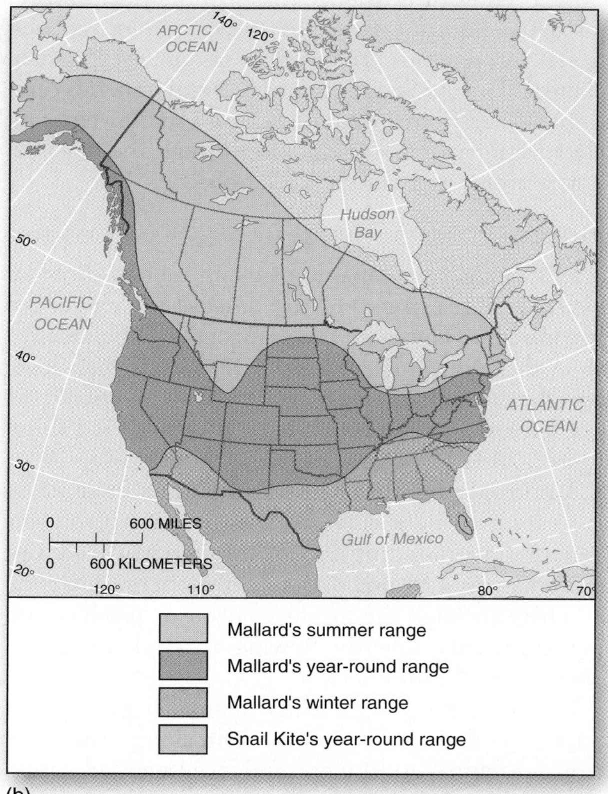

FIGURE 19.13 Limiting factors affect the distribution of every plant and animal species.
(a) Distributions of coast redwood and red maple demonstrate the effect of limiting factors, (b) as do the distributions of the Mallard Duck and the Snail Kite across North America. The Mallard is a generalist and feeds widely. In contrast, the range of its food, a single type of snail, limits the Snail Kite.

range for two tree species and two bird species. The coast redwood (*Sequoia sempervirens*) is limited to a narrow section of the Coast Ranges in California, covering barely 9500 km^2 (less than 4000 mi^2), concentrated in areas that receive necessary summer advection fog. On the other hand, the red maple (*Acer rubrum*) thrives over a large area under varying conditions of moisture and temperature, thus demonstrating a broader tolerance to environmental variations.

The Mallard Duck (*Anas platyrhynchos*) and the Snail Kite (*Rostrhamus sociabilis*) also demonstrate a variation in tolerance and range (Figure 19.13b). The Mallard, a generalist, feeds from widely diverse sources, is easily domesticated, and is found throughout most of North America in at least one season of the year. By contrast, the Snail Kite is a specialist that feeds on only one specific type of snail. This single food source, then, is its limiting factor. Note its small habitat area near Lake Okeechobee in Florida.

Biotic Ecosystem Operations

The abiotic components of energy, atmosphere, water, weather, climate, and minerals make up the life support for the biotic components of each ecosystem. The flow of energy, cycling of nutrients, and trophic (feeding) relations determine the nature of an ecosystem. As energy cascades through this process-flow system, it is constantly replenished by the Sun. But nutrients and minerals cannot be replenished from an external source, so they constantly cycle within each ecosystem and through the biosphere. Let us examine these biotic operations.

Producers, Consumers, and Decomposers

Organisms that are capable of using carbon dioxide as their sole source of carbon are *autotrophs* (self-feeders), or **producers**. These are the plants. They chemically fix carbon through photosynthesis. Organisms that depend on producers as their carbon source are *heterotrophs* (feed on others), or **consumers**. Generally, these are animals (Figure 19.14).

Autotrophs are the essential producers in an ecosystem—capturing light energy and converting it to chemical energy, incorporating carbon, forming new plant tissue and biomass, and freeing oxygen. Solar energy enters each food chain through the producer plant or producer phytoplankton, subsequently flowing through higher and higher levels of consumers.

From the producers, which manufacture their own food, energy flows through the system along a **food chain** circuit, reaching consumers and eventually *detritivores* dealing with the refuse in the chain. Organisms that share the same basic foods are said to be at the same *trophic level.* Ecosystems generally are structured in a **food web**, a complex network of interconnected food chains. In a food web, consumers participate in several different food chains, comprising both *strong interactions* and *weak interactions* between species in the food web.

In Figure 19.14b, a bearded seal (*Erignathus barbatus*), with body fat exceeding 30%, rests on a bergy bit, or small iceberg, in Arctic waters where it feeds on fish and clams. Figure 19.14c shows a male polar bear (*Ursus maritimus*) hauling its bearded seal kill onto the ice. The polar bear, a marine mammal, is the dominant Arctic predator. In Figure 19.14d, on the pack ice many kilometers from land we see a mother and her 6-month-old cubs-of-the-year feeding on a seal catch. Even though the cubs still nurse, they share the kill with mom. The polar bear consumes most of the seal except for the bones and intestines, which are quickly eaten by birds scavenging the leftovers—note the Glaucous Gulls and Ivory Gulls in the photo.

Primary consumers feed on producers. Because producers are always plants, the primary consumer is a **herbivore**, or plant eater, like the krill in Figure 19.15 (p. 572). A **carnivore** is a secondary consumer and primarily eats meat. A tertiary consumer eats primary and secondary consumers and is referred to as the "top carnivore" in the food chain, as with the leopard seal and orca in Antarctica and the polar bear in the Arctic. Orca, an oceanic dolphin, feeds mainly on fish in the Arctic region as well as other whales. A consumer that feeds on both producers (plants) and consumers (meat) is an **omnivore**—a category occupied by humans, among others.

Detritivores (detritus feeders and decomposers) are the final link in the endless chain. Detritivores renew the entire system by releasing simple inorganic compounds and nutrients with the breaking down of organic materials. Detritus refers to all the dead organic debris—remains, fallen leaves, and wastes—that living processes leave. Detritus feeders—worms, mites, termites, centipedes, snails, crabs, and even vultures, among others—work like an army to consume detritus and excrete nutrients that fuel an ecosystem. **Decomposers** are primarily bacteria and fungi that digest organic debris outside their bodies and absorb and release nutrients in the process. This metabolic work of microbial decomposers produces the rotting that breaks down detritus. Detritus feeders and decomposers, although different in operation, have a similar function in an ecosystem.

Examples of Complex Food Webs

An example of a complex community is the oceanic food web that includes krill, a primary consumer (Figure 19.15a). Krill (*Euphausia sp.*, at least a dozen species) is a shrimplike crustacean that is a major food for an interrelated group of organisms, including whales, fish, seabirds, seals, and squid in the Antarctic region. All of these organisms participate in numerous other food chains as well, some consuming and some being consumed.

Phytoplankton begin this chain by harvesting solar energy in photosynthesis. *Herbivorous zooplankton* such as krill and other organisms eat phytoplankton. Consumers eat krill at the next trophic level. Because krill are a protein-rich, plentiful food, increasingly factory ships, such as those from Japan and Russia, seek them out. The annual krill harvest currently surpasses a million tons, principally as feed for chickens and livestock and as protein for

FIGURE 19.14 Energy, nutrient, and food pathways in the environment.
(a) The flow of energy, cycling of nutrients, and trophic (feeding) relationships portrayed for a generalized ecosystem. The operation is fueled by radiant energy supplied by sunlight and first captured by the plants. (b) A bearded seal on a bergy bit in the Arctic Ocean. (c) A solitary male polar bear drags its seal feast across pack ice. (d) A mother polar bear and cubs eat a meal on an iceberg after mom caught a seal; birds work on the leftovers. [All photos Bobbé Christopherson.]

human consumption. Many of the Antarctic-dwelling birds depend on krill and on fish that eat krill.

Another example of a complex food web comes from eastern North America and a temperate deciduous forest. As with the krill food web, the diagram in Figure 19.16 (page 573) is simplified from the actual complexity of nature. From our discussion, find the primary producers and then locate the primary, secondary, and tertiary consumers in this community. Do you find the detritivores (worms are shown) playing a role as well? Bacteria act as decomposers in the system; do you find any other decomposers in the illustration?

Efficiency in a Food Web

Any assessment of world food resources depends on the level of consumer being targeted. Let us use humans as an example. Many people can be fed if wheat is eaten directly. However, if the grain is first fed to cattle (herbivores) and then we eat the beef, the yield of available food

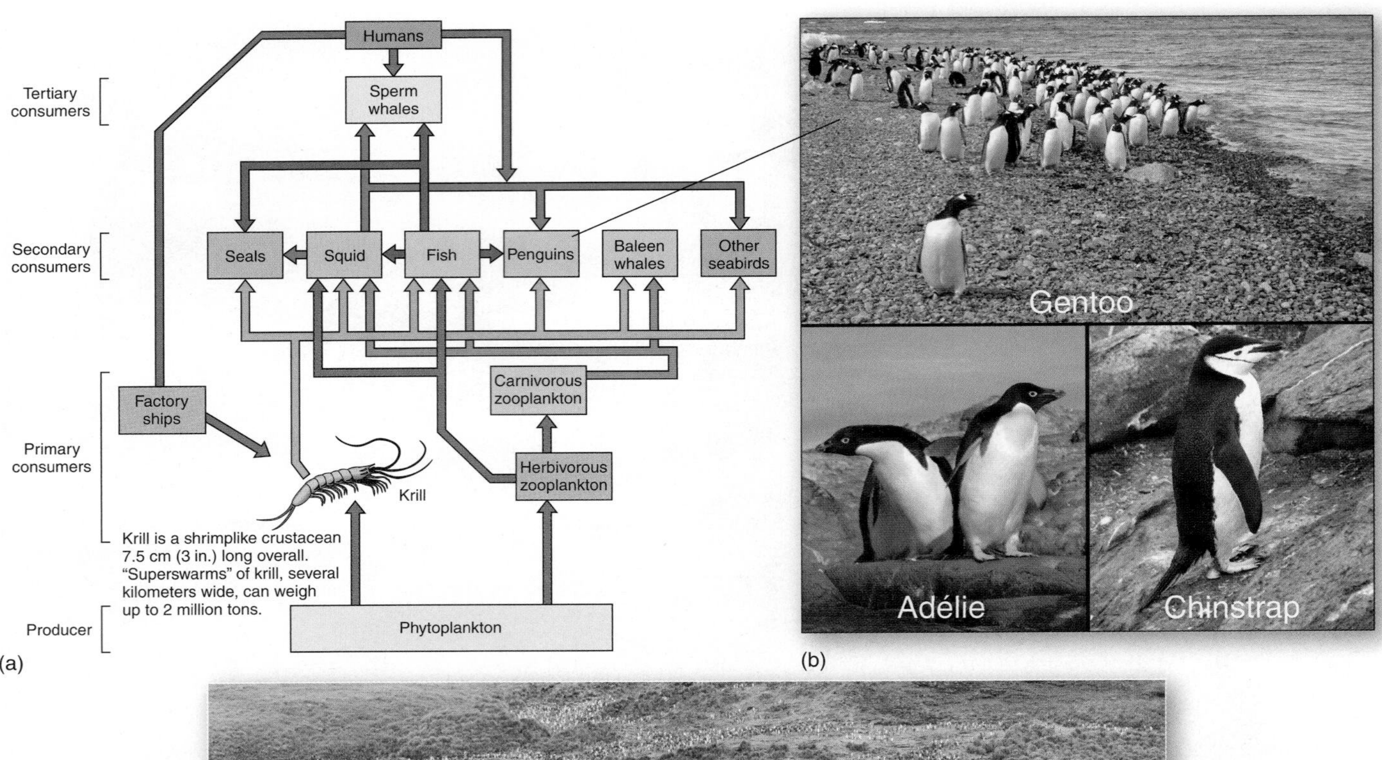

(c) King Penguins, Salisbury Plain, South Georgia Island, Antarctica

FIGURE 19.15 A complex food web.
(a) A simplified food web in Antarctic waters, from phytoplankton producers (bottom) through various consumers. Phytoplankton began this chain using photosynthesis to store solar energy. Krill and other herbivorous zooplankton eat the phytoplankton. Krill, in turn, are consumed by the next trophic level. (b) Gentoo penguins (*Pygoscelis papua*) on a rocky shore in Antarctica watch the water for leopard seal predators. Adélie penguins (*Pygoscelis adeliae*) depend on the ice as a base for feeding and are most threatened in Antarctica. Chinstrap penguins (*Pygoscelis antarctica*) show us that we mainly identify penguins by differences in black and white patterns on their heads. (c) Some 100,000 pairs of King Penguins (*Aptenodytes patagonicus*) and their young gather on Salisbury Plain, South Georgia Island. The parents take turns going to the Southern Ocean for fish in this food web, making a constant parade of clean, stomach-full penguins returning to the juvenile or egg and dirty, empty penguins heading back to sea. The juveniles are small and dark brown. (Bobbé Christopherson.)

energy is cut by 90% (810 kg of grain is reduced to 82 kg of meat); far fewer people can be fed from the same land area (Figure 19.17).

In terms of energy, only about 10% of the kilocalories (food Calories, not heat calories) in plant matter survive from the primary to the secondary trophic level. When humans consume meat instead of grain, there is a further loss of biomass and added inefficiency. More energy is lost to the environment at each progressive step in the food chain. You can see that an omnivorous diet such as that of an average North American and European is quite expensive in terms of biomass and energy.

FIGURE 19.16 Temperate forest food webs.
A simplified illustration of food webs in a temperate forest. Examine the art and identify primary producers and primary, secondary, and tertiary consumers. Featured are strong interactions, but many weak interactions not illustrated hold such natural communities together, affirming the hypothesis that complexity is stabilizing and not weakening the community.

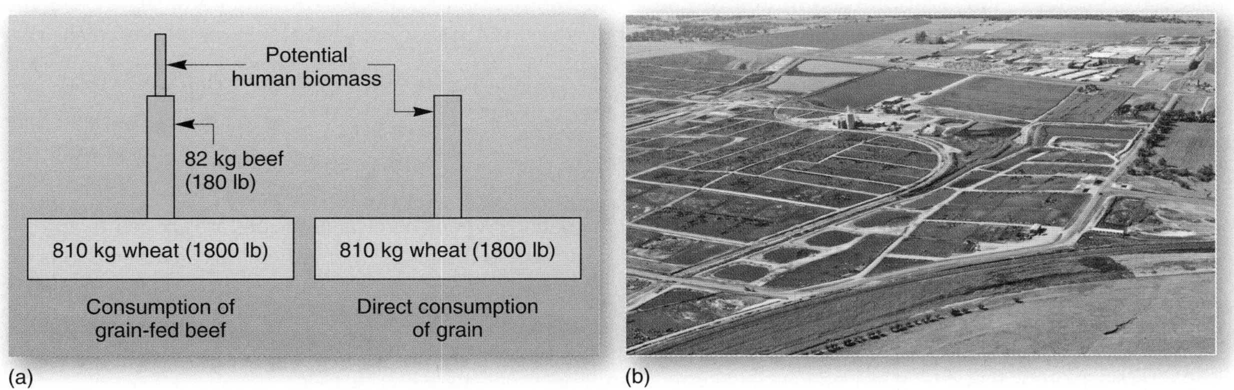

FIGURE 19.17 Efficiency and inefficiency in biomass consumption.
(a) Biomass pyramids illustrate a great difference in efficiency between direct and indirect consumption of grain. (b) A feedlot outside Dodge City, Kansas, processes 6500 head of livestock a day, six days a week. [Bobbé Christopherson.]

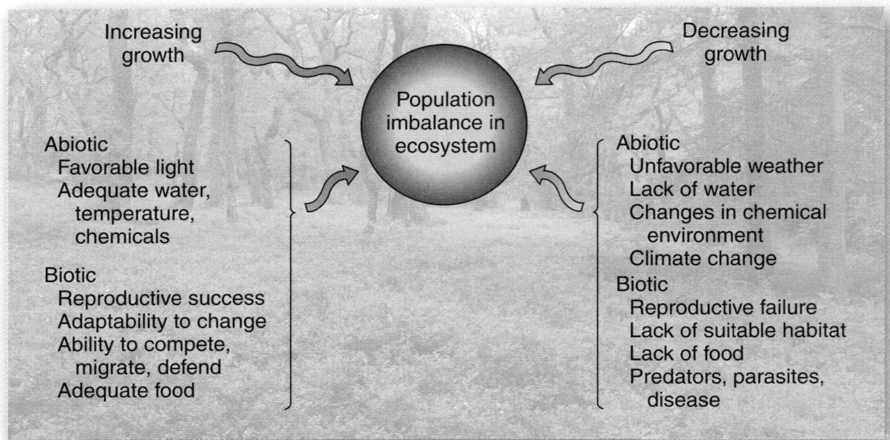

FIGURE 19.18 Factors controlling the population balance of an ecosystem. Potential growth factors and resistance factors ultimately determine the population of an ecosystem. [Scottish forest photo by Bobbé Christopherson.]

Food web concepts are becoming politicized as world food issues grow more critical. Today, approximately half of the cultivated acreage in the United States and Canada is planted for animal consumption—beef and dairy cattle, hogs, chickens, and turkeys. Livestock feed includes approximately 80% of the annual corn and nonexported soybean harvest (Figure 19.17b). In addition, some lands cleared of rain forest in Central and South America were converted to pasture to produce beef for export to restaurants, stores, and fast-food outlets in developed countries. Thus, lifestyle decisions and dietary patterns in North America and Europe are perpetuating inefficient food webs, not to mention the destruction of valuable resources, both here and overseas. Developing countries continue to shift dietary patterns to less efficient meat-based food systems, increasing the demand for livestock products.

Ecosystems, Evolution, and Succession

Far from being static, Earth's ecosystems have been dynamic (vigorous and energetic) and ever-changing from the beginning of life on the planet. Over time, communities of plants and animals have adapted, evolved, and, in turn, shaped their environments to produce great diversity. A constant interplay exists between a community's increasing growth and its decreasing growth, caused by resistance factors that create limits and disruptions (Figure 19.18). Each ecosystem is constantly adjusting to changing conditions and disturbances. The concept of change is key to understanding ecosystem stability.

The nature of change in natural ecosystems can range from gradual transitions from one equilibrium to another, balancing the factors in Figure 19.18, to extreme catastrophes such as an asteroid impact or severe volcanism. In the evolution of life, or the physical and chemical evolution of Earth's systems, such sudden changes created a punctuated equilibrium. Following each of the five major extinctions of life over the past 450 million years, there were great bursts of diversity and proliferation of new and different species, a process known as *speciation*. Although the extinctions temporarily limited biodiversity, each created evolutionary opportunities for speciation to occur.

For most of the last century, scientists thought that an undisturbed ecosystem—whether forest, grassland, aquatic, or other—would progress to a stage of equilibrium, a stable point, with maximum chemical storage and biomass. Modern research has determined, however, that

GEO REPORT 19.3 *Poisoning a food web; who is at the top?*

A study of food webs is a study of who eats what and where they do the eating. When chemical pesticides are applied to an ecosystem of producers and consumers, the food web concentrates some of these chemicals. Many chemicals are degraded or diluted in air and water and thus are rendered relatively harmless. Other chemicals, however, are long-lived, stable, and soluble in the fatty tissues of consumers. They become increasingly concentrated at each higher trophic level. The polar bears of the Barents Sea have some of the highest levels of persistent organic pollutants (POPs) in any animal in the world, despite their remoteness from civilization. The pollution arrived in ocean currents and the atmosphere.

Thus, pollution in a food web can efficiently poison the organism at the top. Many species are threatened in this manner, and, of course, humans are at the top of many food chains and can ingest concentrated chemicals in this way. This is called *biological amplification, or magnification*.

ecosystems do not progress to some static conclusion. In other words, there is no "balance of nature"; instead, think of a constant interplay of physical, chemical, and living factors in a dynamic equilibrium.

Biological Evolution Delivers Biodiversity

A critical aspect of ecosystem stability and vitality is **biodiversity**, or species richness of life (a combination of *bio*logical and *diversity*). Biodiversity includes species diversity, or the number and variety of different species; genetic diversity, or the number of genetic characteristics found within a species; and ecosystem diversity, which includes the number and variety of ecosystems, habitats, and communities on a landscape scale. A working estimate places the number of species presently populating Earth at 13.6 million; however, scientists have classified less than 2 million of these.

The origin of this diversity is embodied in the *theory of evolution*. (The definition of a *scientific theory* is given in Focus Study 1.1.) **Evolution** states that single-cell organisms adapted, modified, and passed along inherited changes to multicellular organisms. The genetic makeup of successive generations was shaped by environmental factors, physiological functions, and behaviors that led to a greater rate of survival and reproduction. Thus, in this continuing process, traits that help a species survive and reproduce are passed along more frequently than are those that do not. A species refers to a population that can reproduce sexually and produce viable offspring. By definition, this means reproductive isolation from other species.

Each inherited trait is guided by a *gene*, which is a segment of the primary genetic material, *DNA* (deoxyribonucleic acid), that resides in the nest of chromosomes in every cell nucleus. These traits—and especially the ones that are most successful at maximizing niches different from those of other species or that help the species adapt to environmental changes—are passed to successive generations. Such differential reproduction and adaptation pass along successful genes or groups of genes in a process of genetic favoritism known as *natural selection*. The process of evolution follows generation after generation and tracks the passage of inherited characteristics that were successful—the failures passed into extinction. Thus, today's humans are the result of billions of years of positive change and affirmative natural selection.

New mixes in the *gene pool*, or all genes possessed by individuals in a given population, involve *mutations*. Mutation is a process that occurs when a random action, or perhaps an error as the DNA reproduces, forms new genetic material and inserts new traits into the inherited stream. Geography also comes into play; physical environments affect natural selection and differ over space. For example, a species may disperse through migration to a different environment where new traits are favored. Imagine the evolutionary effects of migrations across ice bridges or land connections at times of low sea level. A species may also be separated from other species by a natural *vicariance* (fragmentation of the environment) event. Continental drift established natural barriers to species movement and resulted in the evolution of new species. The physical and chemical evolution of Earth's systems is therefore closely linked to the biological evolution of life.

The deciphering of complete species *genomes*, or the total genetic identity of an organism, is a scientific triumph and is disclosing a complex history and geography in each species—where species originated, how species are dispersed, what diseases and traumas were of consequence. Analysis of the human genome has indicated our origins and provided a road map of how we got to where we are.

Biodiversity Fosters Ecosystem Stability

Stable ecosystems are constantly changing. An ecosystem with *inertial stability* has the ability to resist some low-level disturbance. *Resilience* is an ecosystem's ability to recover from disturbance and return to its original state. In an ecosystem, a population of organisms can be stable, yet not resilient. Further, some disturbances are too extreme for even a highly resilient ecosystem. Several of the dramatic asteroid impact episodes that triggered partial extinctions over the past 440 million years overcame the resilience of plant and animal communities. Think of resilience as an ecosystem's ability to recover from disturbance, its ability to absorb disturbance up to some threshold point before it snaps to a different set of relations, thus altering some logical ecological progression.

A tropical rainforest is a diverse, stable community that can withstand most natural disturbance. Yet this ecosystem has low resilience in terms of severe events; a cleared tract of rain forest will recover slowly because most of the nutrients are stored in the vegetation rather than in the soil. Changes in microclimates may also make regrowth of the same species difficult. In contrast, a midlatitude grassland is less diverse and low in stability, yet high in resilience. When burned, the community recovers quickly due to extensive root systems that allow rapid regrowth.

In the context of ecosystem stability and resilience, consider the elimination of a full section of forest ecosystem on private land in southern Oregon, south-southeast of Crater Lake, shown in Figure 19.19. Note the high road density, the inadequate buffers along watercourses to prevent sedimentation of streams, the drag lines and skid trails to each pickup spot that pockmark the landscape. On nearby U.S. Forest Service land, the rules are clear: A full 95-m (312-ft) tree buffer is left on either side of a stream, road densities are minimized, and harvesting regimes that include partial removals, commercial thinning, and group selection harvests, among others, are all in contrast to the devastation on private lands in the photo. Scientists have determined that for sustainable forests the sole objective should never be the complete removal of the timber.

In the 1990s, field experiments in prairie ecosystems of Minnesota began to confirm an important scientific assumption: that greater biological diversity in an ecosystem leads to greater long-term stability and productivity.

(a)

(b)

FIGURE 19.19 Poor practices disrupt stable forest communities.
(a) An example of clear-cut timber harvesting that has devastated a stable forest community and produced drastic changes in microclimatic conditions. (b) A logged pine forest and logging road. About 10% of the Northwest's old-growth forests remain, as identified through satellite images and GIS analysis. [(a) Bobbé Christopherson. (b) Author.]

For instance, during a drought, some species of plants will be damaged. In a diverse ecosystem, however, other species with deeper roots and better water-obtaining ability will thrive. Ongoing experiments also suggest that a more diverse plant community is able to retain and use soil nutrients more efficiently than is one with less diversity. For more about this ongoing research effort at the Cedar Creek Ecosystem Science Reserve, see http://www.cedarcreek.umn.edu/about/.

Exciting research on natural prairie ecosystems and biodiversity, the role of fire, and climate-change impacts is ongoing at Konza Prairie LTER (Long-Term Ecological Research) in the Flint Hills of northeastern Kansas. Kansas State University and the Nature Conservancy operate this native tall-grass prairie research station (see http://konza.ksu.edu/).

The present loss of biodiversity is irreversible. Extinction is final no matter how complex the organism or how long it lived since it evolved. E. O. Wilson, the famed biologist, has written that all species, and all genes within species, must be considered together and that across the globe biodiversity is in trouble.

Recent efforts at ecosystem restoration have had some success in preserving biodiversity, although questions remain about the effects of such efforts on increasing or restoring ecosystem services. The goal of returning ecosystems to original conditions before human intervention is now being expanded to include consideration of "novel ecosystems," those combinations of species and habitats that have never occurred together. To sustain biodiversity and the way ecosystems function in our changing world may demand consideration of these human-altered ecosystems.

Agricultural Ecosystems and Change Humans simplify communities and ecosystems by eliminating biodiversity, thus making them more vulnerable to disturbance. An artificially produced monoculture community, such as a field of wheat, is vulnerable to severe weather or attack from insects or plant disease. In some regions, simply planting multiple crops brings more stability to the ecosystem. This is an important principle of sustainable agriculture, a movement to make farming more ecologically compatible.

A modern agricultural ecosystem creates an enormous demand for energy, chemical pesticides and herbicides, artificial fertilizer, and irrigation water. The practice of harvesting and removing biomass from the land interrupts the cycling of materials into the soil (Figure 19.20). This net loss of nutrients must be artificially replenished. The type of plantation agriculture in Figure 19.20b is depleting soil nutrients over time; this farming practice requires subsidies of chemical fertilizers and herbicides to eliminate competing vegetation.

Increasing temperatures with climate change reduce yields and increase overhead costs. Imagine being a farmer in the latter part of this century when average summer conditions exceed the hottest extreme summer temperatures of today. The stress on agricultural ecosystems, crops, and livestock will be something experienced across the globe in tropical, subtropical, midlatitude, and subarctic regions.

Species distributions are also affected by climate change, as presented in this chapter's *Geosystems Now*. Current global change is occurring rapidly, at the rate of decades instead of millions of years. Thus, we see die-out and succession of different species along disadvantageous

(a) Alfalfa feedstocks, central Minnesota

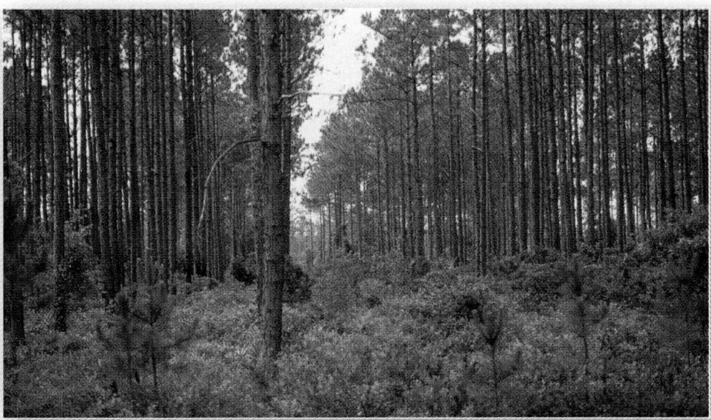
(b) Tree plantation for wood pulp, southern Georgia

(c) Harvesting oats near Dokka, Norway.

(d) Broccoli, near San Luis Obispo, California

(e) Fish farming, Bay of Fundy, Nova Scotia

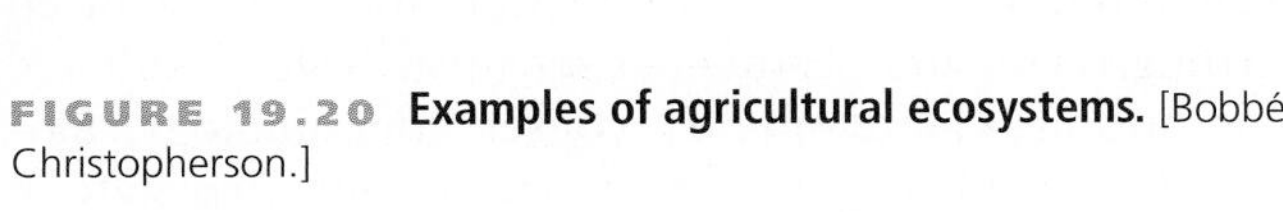
FIGURE 19.20 **Examples of agricultural ecosystems.** [Bobbé Christopherson.]

habitat margins. The displaced species may colonize new regions made more hospitable by climate change. Also shifting will be agricultural lands that produced wheat, corn, soybeans, and other commodities. Society will have to adapt to these changes.

Ecological Succession

Ecological succession occurs when newer communities of plants and animals (usually more complex) replace older ones (usually simpler)—a change of species composition. Each successive community of species modifies the physical environment in a manner suitable for a later community of species. Changes apparently move toward a more mature condition, although disturbances constantly disrupt the sequence.

Succession often requires an initiating disturbance. Examples include a windstorm, severe flooding, a volcanic eruption, a devastating wildfire, an insect infestation, and an agricultural practice such as prolonged overgrazing (Figure 19.21). When existing organisms are disturbed or removed, new communities can emerge. At such times of nonequilibrium transition, the interrelationships among species produce elements of chance, and species having an adaptive edge will succeed in the competitive struggle for light, water, nutrients, space, time, reproduction, and survival.

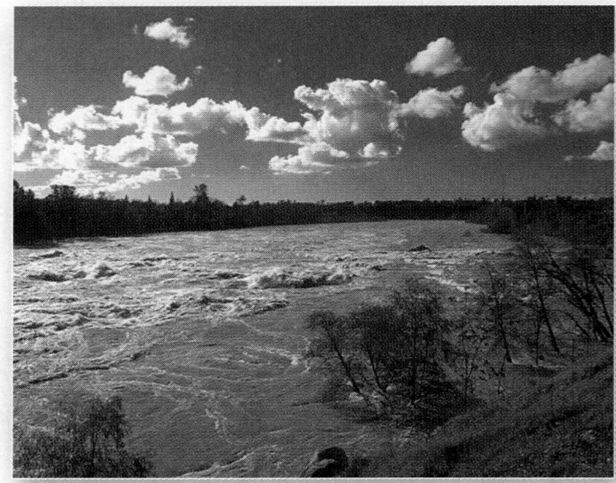
(a) Flooding on the American River, California

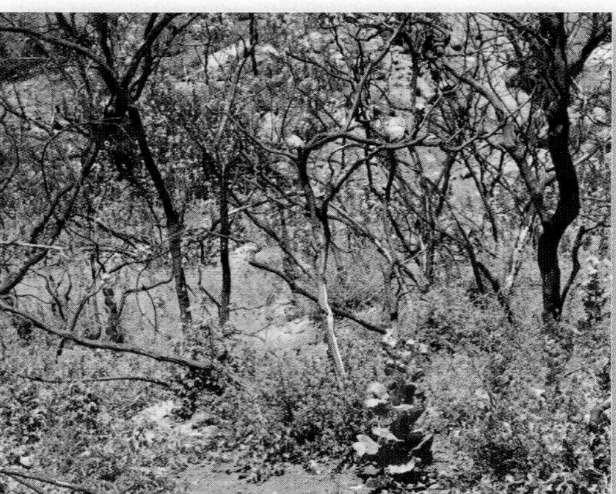
(b) Burned forest and chaparral, Santa Barbara, California

(c) Sheared tree, 2010 tornado damage Wadena, Minnesota

(d) Pine bark beetle damage (orange) Lake Arrowhead, California

FIGURE 19.21 Ecosystem disturbance alters succession patterns. [Bobbé Christopherson.]

Traditionally, it was assumed that plants and animals passed through several successional stages, eventually forming a predictable *climax community*—a stable, self-sustaining, and functioning community with balanced birth, growth, and death—but scientists have abandoned this notion. Contemporary conservation biology, biogeography, and ecology assume nature to be in a state of constant adaptation. Mature communities are properly thought of as being in dynamic equilibrium. At times, however, these communities may be out of phase and in nonequilibrium with the immediate physical environment because of the usual lag time in their adjustment.

Rather than thinking of ecosystems as uniform across the landscape, think of them as patchwork mosaics of communities and habitats. This is the study of *patch dynamics*, or the interactions between and within disturbed portions of habitats. Individual patches may be undergoing different stages of succession, thus adding to complexity in the landscape. Most ecosystems in actuality are made up of patches of former landscapes. Biodiversity in some respects is the result of such patch dynamics.

Land and water experience different forms of succession. We first look at terrestrial succession, characterized by competition for sunlight, and then at aquatic succession, characterized by progressive changes in nutrient levels.

CRITICAL THINKING 19.2

Observe ecosystem disturbances

Over the next several days as you travel between home and campus, a job, or other areas, observe the landscape. What types of ecosystem disturbances do you see? Imagine that several acres in the same area you are viewing had escaped any disruptions for a century or more. Describe what some of these ecosystems and communities might be like.

Terrestrial Succession

An area of bare rock or a disturbed site with no vestige of a former community can be a place for **primary succession**, the beginning of an ecosystem. Examples include

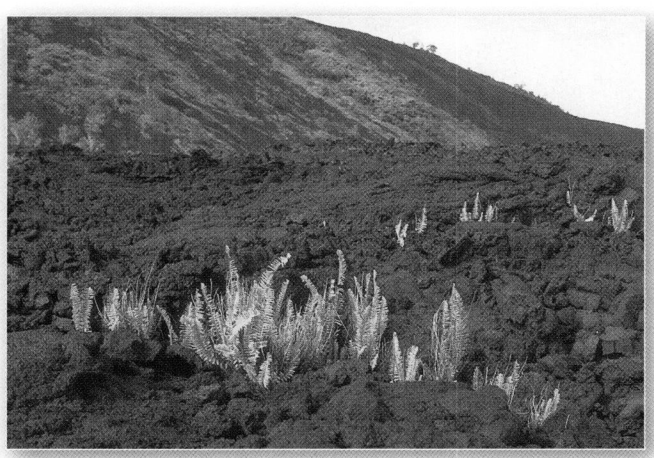

FIGURE 19.22 Primary succession.
Ferns take hold as primary succession on new lava flows that came from the Kīlauea volcano in Hawai'i. [Bobbé Christopherson.]

new surfaces created by mass movement of land, areas exposed by a retreating glacier, cooled lava flows and volcanic eruption landscapes, surface mining and clear-cut logging scars, and areas of sand dunes. Plants taking hold on new lava flows from the Kīlauea volcano in Hawai'i illustrate primary succession (Figure 19.22). In terrestrial ecosystems, succession begins with early species that form a **pioneer community**. Primary succession often begins with lichens, mosses, and ferns growing on bare rock. These early inhabitants prepare the way for further succession to grasses, shrubs, and trees.

More common in nature is **secondary succession**, which begins if some aspects of a previously functioning community are present. An area where the natural community has been destroyed or disturbed, but where the underlying soil remains intact, may experience secondary succession.

As succession progresses, soil develops, and a different set of plants and animals with different niche requirements may adapt. Further niche expansion follows as the community matures. Succession, which is really the developmental process forming new communities, is a dynamic set of interactions with sometimes unpredictable outcomes, steered by sometimes random, unpredictable events that trigger a threshold leap to a new set of relations—the basis of *dynamic ecology*.

Examples of secondary succession are most of the areas affected by the Mount St. Helens eruption and blast in 1980, which burned or blew down about 38,450 hectares (95,000 acres) of trees (Figure 19.23). Some soils, young trees, and plants were protected under ash and snow, so community development began almost immediately after the event. The areas completely destroyed near the Mount St. Helens volcano and those buried beneath the massive landslide north of the mountain became candidates for primary succession.

Wildfire and Fire Ecology Fire and its effects are one of Earth's significant ecosystem and economic processes. In the United States and Canada since 2000, more than double the annual average area burned in wildfires. Each year's loss seemingly beat the record set the year before—in 2009, 72,000 fires burned nearly 5.7 million acres in the United States. For more information, see the National Interagency Fire Center at http://www.nifc.gov/fire_info/nfn.htm.

Over the past 50 years, the role of fire in ecosystems has been the subject of much scientific research and experimentation. Today, fire is recognized as a natural component of most ecosystems and not the enemy of nature that it was once thought to be. In fact, in many forests, undergrowth and surface litter are purposely burned in controlled "cool fires" to remove fuel and prevent catastrophic and destructive "hot fires." Forestry experts have learned that when fire-prevention strategies are rigidly followed, they can lead to the abundant undergrowth accumulation that fuels major fires.

Modern society's demand for fire prevention to protect property goes back to European forestry of the 1800s. Fire prevention became an article of faith for forest managers in North America. But in studies of the longleaf pine forest that stretches in a wide band from the Atlantic coastal plain to Texas, fire was discovered to be an integral part of regrowth following lumbering. In fact, seed dispersal of some pine species, such as the knobcone pine, does not occur unless assisted by a forest fire. Heat from the fire opens the cones, releasing seeds so they can fall to the ground for germination. Also, these fire-disturbed areas quickly recover with protein-rich woody growth, young plants, and a stimulated seed production that provides abundant food for animals.

The science of **fire ecology** imitates nature by recognizing fire as a dynamic ingredient in community succession. Controlled ground fires, deliberately set to prevent accumulation of forest undergrowth, now are widely regarded as a wise forest management practice and are used across the country. In Yellowstone National Park, wildfires were allowed to burn if started by natural causes, such a lightning. After the highly visible Yellowstone fires of 1988, an outcry was heard from forestry and recreational interests, criticizing the government's "let it burn" policy (Figure 19.24, page 581).

In its final report on the 1988 Yellowstone fires, a government interagency task force concluded that "an attempt to exclude fire from these lands leads to major unnatural changes in vegetation . . . as well as creating fuel accumulation that can lead to uncontrollable, sometimes very damaging wildfire." Thus, federal land managers and scientists reaffirmed their stand that fire ecology is a fundamentally sound concept.

An increasingly serious problem is emerging as people seek to live in the wildlands near cities. Urban development has encroached on forests, and the added suburban landscaping has created new fire hazards. Uncontrolled wildfires now can destroy homes and threaten public

FIGURE 19.23 The pace of change in the region of Mount St. Helens. A 2008 astronaut photo of the volcano shows the devastation (north is to the right). The area north of the volcano—(a) before the eruption (1979) and (b) after the eruption (1980)—experienced profound change in all ecosystems. A series of four photo comparisons made in 1983 and 1999 of matching scenes illustrates the slow recovery of secondary succession: (c) and (d) at Meta Lake; (e) Spirit Lake panorama view (note the mat of floating logs after 19 years); (f) detail of the destroyed forest community. [*ISS* astronaut photo, NASA/GSFC. (a) and all 1983 photos by author. (b) and all 1999 photos by Bobbé Christopherson.]

FIGURE 19.24 Western wildfires and ecosystem recovery.
(a) Wildfires in Yellowstone National Park in northwestern Wyoming. (b) Recovery of vegetation under way in area burned during the Santa Barbara wildfires. (c) The drama of wildfire unfolds in the Bitterroot National Forest in Montana, August 2000, as elk stand in the Bitterroot River attempting to survive the conflagration. (d) Fire-adapted chaparral vegetation in southern California a few months after a wildfire in the San Jacinto mountains, as recovery begins. Note the stump sprouts emerging. [(b), (c), and (d) Bobbé Christopherson. (a) Joe Peaco/Wind Cave National Park]

safety. People living in these areas demand fire protection that actually causes undergrowth to accumulate and worsens the risk. Wildfires devastated communities in southern California, especially where development was in the "wilds" of chaparral country. In 2007, fire destroyed more than 2000 homes, and over 500,000 acres burned in two dozen wildfires. Strong Santa Ana winds (see Chapter 6 for a description) added the blowtorch effect, taking embers miles from their ignition source.

The connection between the record wildfires in the American West since 2000 and climate change was reported in *Science* in 2006:

> Higher spring and summer temperatures and earlier snowmelt are extending the wildfire season and increasing the intensity of wildfires in the western U.S. . . . Here, we show that large wildfire activity increased suddenly and markedly in the mid-1980s, with higher large-wildfire frequency. Longer wildfire durations, and longer wildfire seasons . . . strongly associated with increased spring and summer temperatures and an earlier spring snowmelt.*

Aquatic Succession

Ecosystems occur in water as well as on land. Lakes, estuaries, ponds, and shorelines are complex ecosystems. The concepts you just read about—stability and resilience, biodiversity, community succession—also apply to open water, shoreline, and watershed systems.

Focus Study 19.1 examines ecosystems of the Great Lakes of North America. This is the largest lake system on Earth and is jointly managed by the United States and Canada.

(text continued on page 585)

*A. L. Westerling, H. G. Hidalgo, D. R. Cayan, and T. W. Swetnam, "Warming and earlier spring increase western U.S. forest wildfire activity," *Science* 313, no. 5789 (August 18, 2006): 940–943.

FOCUS STUDY 19.1

Ecosystems of the Great Lakes

The Great Lakes—Superior, Michigan, Huron, Erie, and Ontario—contain 18% of the total volume of all freshwater lakes in the world and have dominated the history and development of North America. The basins of the Great Lakes are gifts of the last ice age when glaciers advanced and retreated over this region (see Figure 17.27). Their combined surface area covers 244,000 km² (94,000 mi²), a little less than the state of Wyoming. The entire drainage basin embraces 528,000 km² (204,000 mi²), or an area about the size of Manitoba.

This international waterway of lakes and connecting rivers spans more than 2771 km (1722 mi) from west to east, with 18,000 km (11,000 mi) of coastline—many consider this America's fifth coastline (Figure 19.1.1). The lakes contain over 30,000 islands and diverse coastal environments, featuring major dunes and sandy beaches, bedrock shorelines, gravel beaches, and adjoining coastal wetlands. Major agricultural regions surround the lakes, as do many centers of industrial activity, maritime commerce, and tourism.

Lake Superior is the highest in elevation, deepest, and largest in the system

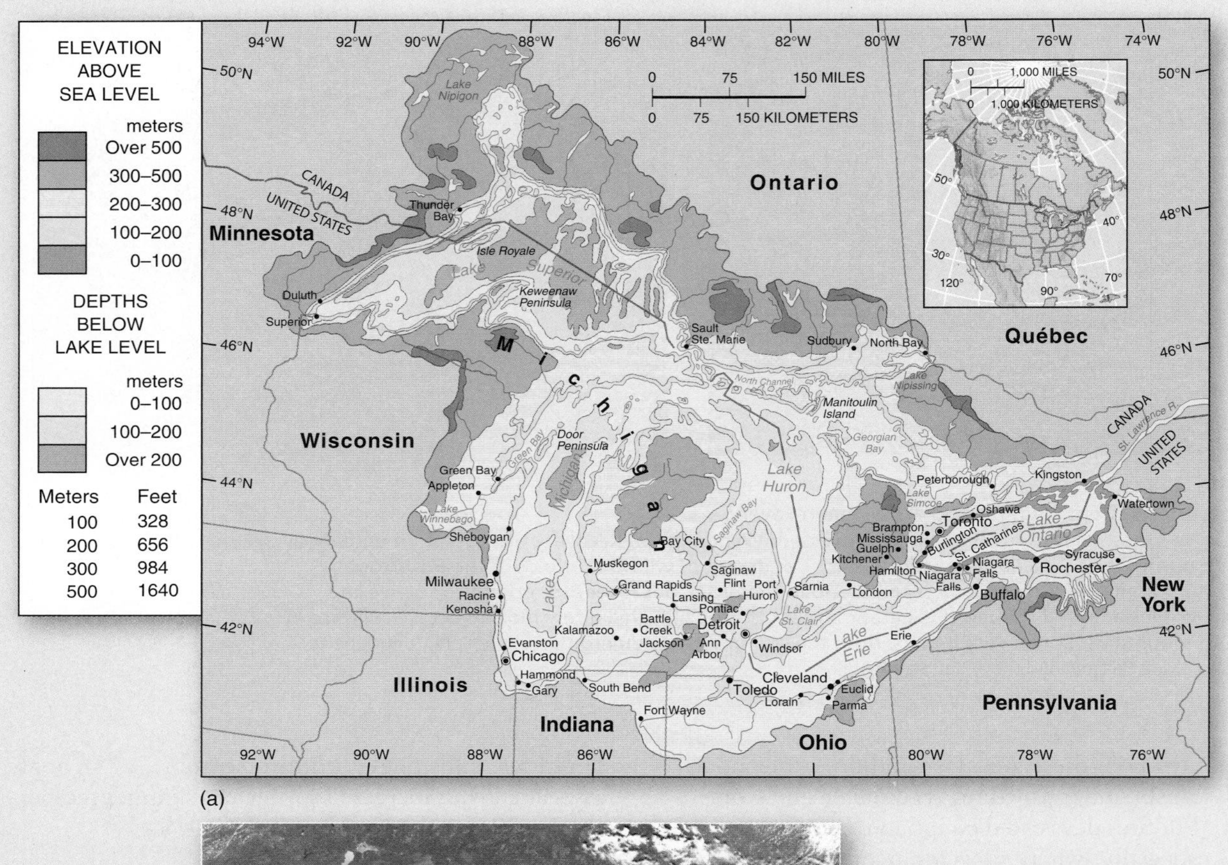

FIGURE 19.1.1 The Great Lakes Basin. (a) Elevations above sea level, depths below lake level, and principal urban areas are shown. The map outline is the limit of the drainage basin for the Great Lakes system. (b) The Great Lakes region on a rare cloudless day August 10, 2010. [(a) Map courtesy of Environment Canada, U.S. EPA, and Brock University cartography. (b) *Aqua* MODIS image, NASA/GSFC.]

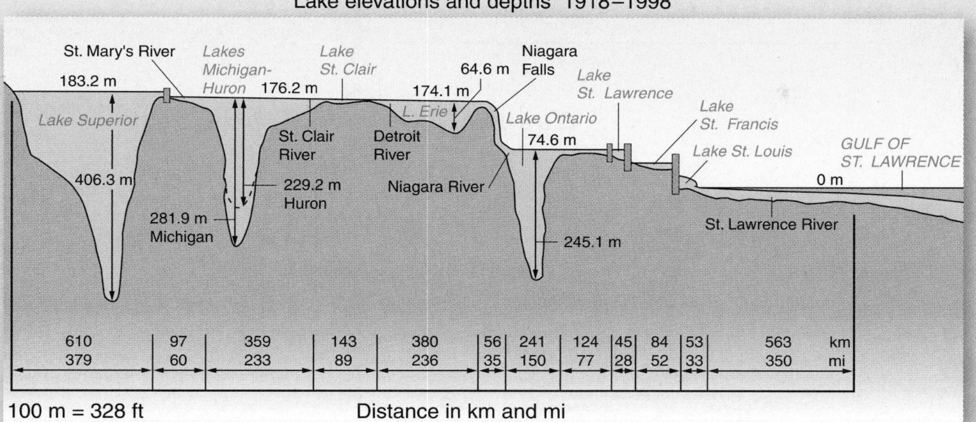

FIGURE 19.1.2 Great Lakes elevation profile and lake levels. Average depth and lake surface levels, 2010. The International Great Lakes Datum of 1985 is the standard baseline. [Data courtesy of the Canadian Hydrographic Service, Central Region, and the International Coordinating Committee on the Great Lakes Basin Hydraulic and Hydrographic Data.]

(Figure 19.1.2). It drains through St. Mary's River into Lake Huron, which is connected to and at the same surface elevation as Lake Michigan; the two lakes are joined through the Straits of Mackinac. Water then flows through the St. Clair and Detroit Rivers into Lake Erie, the shallowest lake in the system. Compared to an average water retention time of 191 years in Lake Superior, Lake Erie has the shortest retention time of 2.6 years. The Niagara River transports water into Lake Ontario, which is drained by the St. Lawrence River into the Gulf of St. Lawrence and eventually into the North Atlantic Ocean.

Because of the large size of the overall Great Lakes system, its associated climate, soils, and topography vary widely. Dominating the north are colder microthermal climates, exposed portions of the Canadian Shield bedrock, and vast stands of conifers and acidic soils. To the south, warmer mesothermal climates and fertile, glacially deposited soils provide a vast agricultural base. Farms and urbanization replaced previous stands of mixed forest. Virtually none of the original land cover is undisturbed.

Changing Lake Levels

Measurements of Great Lakes water levels extend back to 1860. Recently, water levels were slightly lower, with dry conditions across the region into 2010, as all lakes were below average by up to 0.3 m (1 ft). Only Lake Ontario was about 2.54 cm (1 in.) above average in late 2010. Lake levels are higher in the summer after snowmelt and with increased summer maximum precipitation. Lake levels fluctuate annually, depending on precipitation and evaporation in the basin; near-drought conditions bring lower lake levels.

Climate change is forecasted to lower lake levels as evaporation increases from lake surfaces. Eleven major climate models forecast reductions from 0.3 m (1 ft) to as much as 1.5 m (5 ft) in a worst-case projection, reducing outflow into the St. Lawrence by 20% to 40%. When added to increased human consumption and water demand across the region, this is a double blow to lake levels. This will make wetlands even more vulnerable as the coastal shoreline retreats. Losses of wetlands translate to habitat loss for many species.

Wetlands and Coastal Ecosystems

Wetlands in the Great Lakes are tied to lake levels, which determine wetland distribution, vegetation composition, and ecological operations. Wetland ecosystems exist throughout the Great Lakes, along 18,000 km (11,000 mi) of shoreline and extensive connecting river channels, occurring as swamps, marshes, bogs, and fens.

Coastal wetlands occupy that key interface between the land and the water, storing and cycling organic material and nutrients into aquatic ecosystems. Among many important wetlands are the far western end of Lake Superior in the St. Louis River estuary, the area along Sturgeon Bay in the Door Peninsula of Wisconsin, and the far eastern end of Lake Ontario and its coastal lagoon. Over two-thirds of Great Lakes wetlands have been lost during this century, and those remaining are threatened by development, drainage, or pollution. These ecosystems are critically important for migratory waterfowl as seasonal habitat, for a number of amphibians, and for fisheries, as many Great Lakes species depend on coastal wetlands as a reproduction niche. Wetlands also filter pollutants and anchor soils and sediment against wave erosion.

Society asks much of the Great Lakes and coastal ecosystems: to dilute wastes from cities and industry, to dissipate thermal pollution from power plants, to provide municipal drinking water and irrigation water, and to sustain unique and varied ecosystems—open lake,

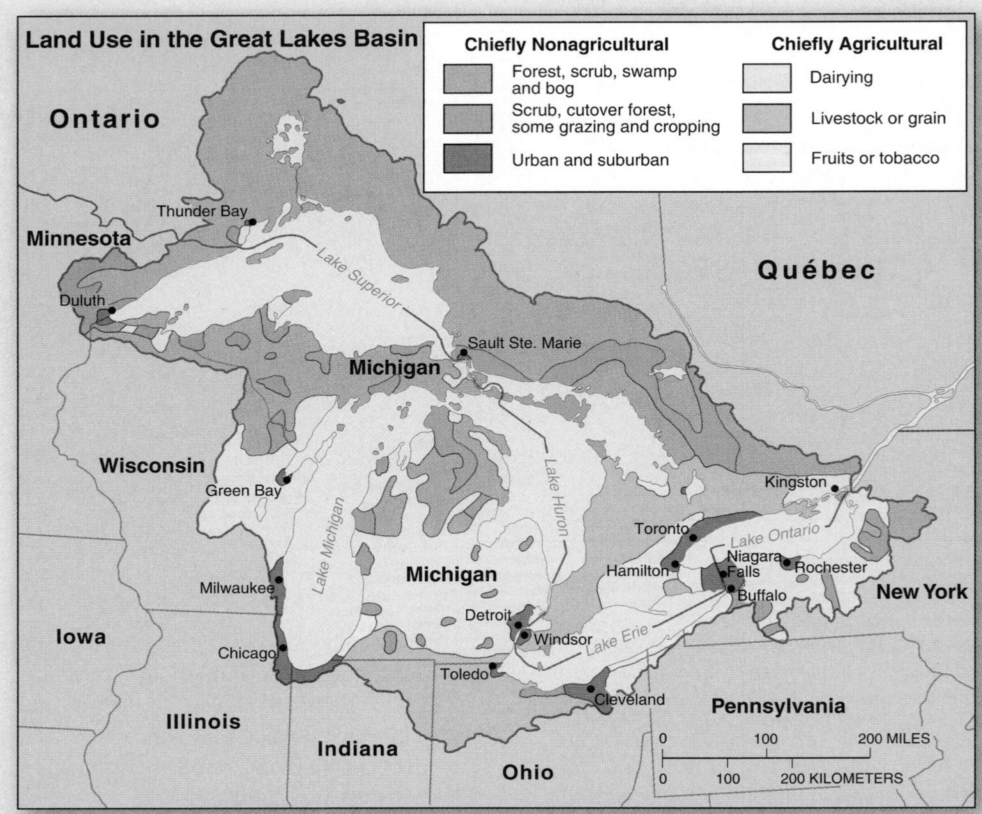

FIGURE 19.1.3 Land use in the Great Lakes Basin. Generalized distribution map of land use, nonagricultural and agricultural, in the Great Lakes Basin. [Environment Canada, U.S. EPA, and Brock University cartography.]

coastal shore, coastal wetland, lakeplain (former lake bed), inland wetland, and inland terrestrial (upland areas of forest, prairie, and barrens).

Land Use, Human Activities, and Impacts

Figure 19.1.3 portrays land use in the Great Lakes Basin. About 10% of the U.S. population and 25% of Canada's population live in the Great Lakes drainage basin. Major agricultural regions and many centers of industrial activity surround the lakes. The eight Great Lakes states generate $18 billion per year in revenue from agriculture, pulp and paper, fisheries, transportation, and tourism, among other activities.

Each activity has its own impact on lake systems. Runoff from agricultural areas contains chemicals, nutrients, and eroded soil that affect aquatic ecosystems in open lake and coastal environments. The pulp and paper industry was a major polluter, but has improved as more was learned about the dangers of pollution; for example, mercury contamination was halted in the 1970s, and better disposal methods for wastes were implemented. The forests were not always treated in a sustainable way in terms of reforestation, and today forests may be regarded as a diminished resource in the basin. Recovery efforts are slow to make up for past practices, but progress is under way.

Lake Erie was the first of the Great Lakes to show severe effects of *cultural eutrophication,* for it is the shallowest and warmest of the five lakes. About a third of the overall population in the Great Lakes Basin lives in the drainage of Lake Erie, making it the major recipient of sewage effluent from treatment plants. Over the years, dissolved oxygen

GEO REPORT 19.4 *Another take on lake–bog succession*

Research has uncovered an apparent opposite pattern to the lake–bog succession described, in areas that have been recently deglaciated. The new study points to the reality that succession is poorly understood in aquatic ecosystems. Scientists examined lakes in the Glacier Bay, Alaska, area and found that they had become more dilute, acidic, and unproductive—in other words, less eutrophic—during the past 10,000 years. Successional changes in the surrounding vegetation and soils seem to be linked to the aquatic ecosystem. These studies suggest that in various cool, moist, temperate rainforest climates with landscapes created by glacial retreat, different processes are at work in lakes.

levels dropped as biochemical oxygen demand increased, caused by sewage-treatment discharges and algal decay, forming low-oxygen dead zones. Algae flourished in these eutrophic conditions and coated the beaches, turning the lake a greenish brown. In the 1950s and 1960s, the oily contaminated surface of the Cuyahoga River, flowing into Lake Erie through Cleveland, Ohio, caught fire several times. The 1969 fire finally spurred public opinion to demand action.

In the 1970s, environmental restrictions were implemented and inputs of phosphorus into Lake Erie were eventually reduced by 90%. The dead zone steadily declined. However, in the 1990s, a significant dead zone was again identified in Lake Erie, with scientists still trying to understand its return. Two hypotheses have emerged: First is the possibility that introduced species—such as exotic zebra and quagga mussels—are compounding the anoxic (lacking oxygen) or hypoxic (low oxygen) conditions along lake bottoms. Second is the effect of climate change on lake stratification. As discussed in the Chapter 7 *Geosystems Now,* warming temperatures extend the period of time when the lakes are stratified, isolating the cooler bottom waters for more weeks each year.

Lake stratification plays a role in the formation of these oxygen-depleted dead zones. In the summer, as in most lakes, the Great Lakes stratify as the surface waters warm and become less dense and the cooler waters collect to form the lower layer in the lake. Little mixing occurs between these two layers, so productivity is accelerated in the top layers—increased nutrients and light penetration cause this increase.

Normally, as the fall season matures, surface water cools and sinks, displacing deeper water and creating a turnover of the lake mass. By midwinter, the temperature from the surface to the bottom is uniformly around 4°C (39°F), the point at which water is densest; temperatures at the surface are near freezing—but this is changing.

Overall the lakes are warming with climate change. Lake Superior surface water reached 23.9°C (75°F) in July of 2007. During August 2010, Lake Superior was 8 C° (14 F°) above normal water temperature and Lake Michigan 4 C° (7 F°) above normal—both new records. Scientists are analyzing the impact of these temperature increases on lake ecosystems and the prolonging of the period of lake stratification, which interrupts normal lake turnover.

Aquatic Ecosystems and Restoration

The Great Lakes support a food chain of producers and consumers, as does any aquatic ecosystem. Native fish populations have suffered from human activities: overfishing, introduction of exotic species, pollution from excess nutrients, toxic contamination, and disruption of spawning habitats. Commercial fishing peaked in the 1880s, but it wasn't until water pollution peaked in the 1960s and early 1970s that fisheries declined to their lowest levels.

Toxic chemicals such as polychlorinated biphenols (PCBs) and DDT—both banned in the United States in the 1970s—accumulate and persist in lake sediments and organisms and are subject to biological amplification as they move up the food chain. In the Great Lakes, scientists have reported PCBs found in phytoplankton, rainbow smelt, lake trout, and, higher in the food chain, such as a herring gull eggshell.

In 1972, Canada and the United States signed the Great Lakes Water Quality Agreement, and the cleanup was under way. The governments implemented strong pollution-control programs, working in concert with citizens, industries, and private organizations. These efforts resulted in declining levels of PCBs and DDT in the lakes up until 1990, when rates of decline leveled off. Today, lake trout (in Lake Superior), sturgeon, herring, smelt, alewife, splake, yellow perch, walleye, and white bass are fished. However, health advisories are occasionally issued warning of danger in consuming fish of certain species, sizes, and locales.

An ecosystem approach to management of the Great Lakes Basin is the key to recovery. This approach considers the physical, chemical, and biological components that constitute the total system. In 2010, the U.S. government approved $475 million in funding for the Great Lakes Restoration Initiative, the largest investment of federal dollars in the Great Lakes in 20 years. This initiative targets some of the greatest threats to the lakes—namely, invasive species and pollution of water and sediment. Priority areas also include restoring wetlands and other critical habitats (see http://greatlakesrestoration.us/).

For more information on Great Lakes ecosystems, visit Environment Canada's Great Lakes home page at http://www.on.ec.gc.ca/greatlakes/ and NOAA's Great Lakes Research Lab site at http://www.glerl.noaa.gov/. For additional Great Lakes information, see http://www.seagrant.wisc.edu/outreach/.

When viewed across geologic time, a lake or pond is really a temporary feature on the landscape. Lakes and ponds exhibit successional stages as they fill with nutrients and sediment and as aquatic plants take root and grow. This growth captures more sediment and adds organic debris to the system (Figure 19.25). This gradual enrichment in water bodies is known as eutrophication (from the Greek *eutrophos*, meaning "well nourished").

In moist climates, a floating mat of vegetation grows outward from the shore to form a bog. Cattails and other marsh plants become established, and partially decomposed organic material accumulates in the basin, with additional vegetation bordering the remaining lake surface. Vegetation and soil and a meadow may fill in as water is displaced; willow trees follow, and perhaps cottonwood trees; and eventually the lake may evolve into a forest community.

The stages in lake succession are named for their nutrient levels: *oligotrophic* (low nutrients), *mesotrophic* (medium nutrients), and *eutrophic* (high nutrients). Greater primary productivity and resultant decreases in water

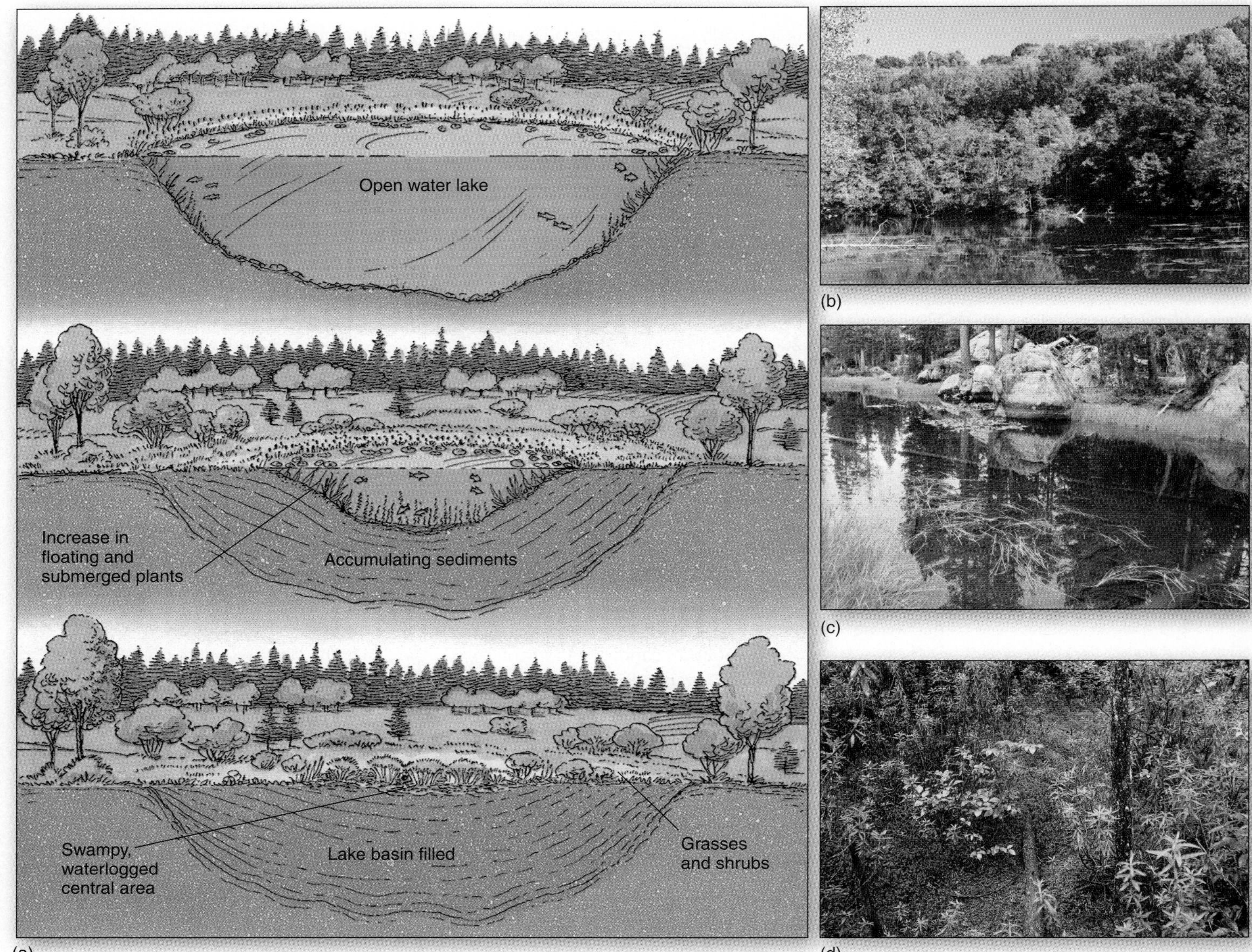

FIGURE 19.25 Idealized lake–bog–meadow succession in temperate conditions. (a) What begins as a lake gradually fills with organic and inorganic sediments, which successively shrink the area of the pond. A bog forms, then a marshy area, and finally a meadow completes the successional stages. (b) Spring Mill Lake, Indiana. (c) Organic content increases as succession progresses in a mountain lake. (d) The Richmond Nature Park and its Lulu Island Bog were ocean just 8000 years ago. Frasier River sediment created the mudflats and set the stage for the evolution of this sphagnum bog with acidic soils that "quakes" when you walk on it. Plants in the bog include moss, shore pine, hemlock, blueberry, salal shrubs, and Labrador tea, among others. [Bobbé Christopherson.]

transparency mark each stage, so that photosynthesis becomes concentrated near the surface. Energy flow shifts from production to respiration in the eutrophic stage, with oxygen demand exceeding oxygen availability.

Nutrient levels also vary spatially: Oligotrophic conditions occur in deep water, whereas eutrophic conditions occur along the shore, in shallow bays, or where sewage, fertilizer, or other nutrient inputs occur. Even large bodies of water may have eutrophic areas along the shore. As society dumps sewage, agricultural runoff, and pollution into waterways, the nutrient load is enhanced beyond the cleansing ability of natural biological processes. The result is *cultural eutrophication*, which hastens succession in aquatic systems.

GEOSYSTEMS CONNECTION

Earth's biosphere is a remarkable functioning entity of abiotic and biotic components, all interacting and interrelated through some 13.6 million species. Plants harvest sunlight through photosynthesis and thus begin vast food webs of energy and nutrition. Life evolved on Earth into a biologically diverse structure of organisms, communities, and ecosystems, seemingly gaining strength and resilience through their biodiversity. We next move to biomes and a synthesis of the subjects from Chapters 2 through 19, as we bring it all together to form a portrait of our planet.

KEY LEARNING CONCEPTS REVIEW

■ ***Define*** **ecology, biogeography, and the ecosystem concept.**

Earth's biosphere is unique in the Solar System; its ecosystems are the essence of life. **Ecosystems** are self-sustaining associations of living plants and animals and their nonliving physical environment. **Ecology** is the study of the relationships between organisms and their environment and among the various ecosystems in the biosphere.

Biogeography is the study of the distribution of plants and animals and the diverse spatial patterns they create. Biogeographers trace communities across the ages, considering plate tectonics and ancient dispersal of plants and animals.

ecosystem (p. 556)
ecology (p. 556)
biogeography (p. 556)

1. What is the relationship between the biosphere and an ecosystem? Define *ecosystem*, and give some examples.
2. What does biogeography include? Describe its relationship to ecology.
3. Briefly summarize what ecosystem operations imply about the complexity of life.

■ ***Describe*** **communities, habitats, and niches.**

A **community** is formed by the interactions among populations of living animals and plants at a particular time. Within a community, a **habitat** is the specific physical location of an organism—its address. A **niche** is the function or operation of a life form within a given community—its profession.

community (p. 558)
habitat (p. 558)
niche (p. 559)

4. Define a community within an ecosystem.
5. What do the concepts of habitat and niche involve? Relate them to some specific plant and animal communities.

6. Describe symbiotic and parasitic relationships in nature. Draw an analogy between these relationships and human societies on our planet. Explain.

■ ***Explain*** **photosynthesis and respiration, and** ***derive*** **net photosynthesis and the world pattern of net primary productivity.**

As plants evolved, the **vascular plants** developed conductive tissues. **Stomata** on the underside of leaves are the portals through which the plant participates with the atmosphere and hydrosphere. Plants (primary producers) perform **photosynthesis** as sunlight stimulates a light-sensitive pigment called **chlorophyll**. This process produces food sugars and oxygen to drive biological processes. **Respiration** is essentially the reverse of photosynthesis and is the way the plant derives energy by oxidizing carbohydrates. **Net primary productivity** is the net photosynthesis (photosynthesis minus respiration) of an entire community. This stored chemical energy is the **biomass** that the community generates, or the net dry weight of organic material.

vascular plants (p. 560)
stomata (p. 560)
photosynthesis (p. 560)
chlorophyll (p. 560)
respiration (p. 561)
net primary productivity (p. 561)
biomass (p. 561)

7. Define a vascular plant. How many plant species are there on Earth?
8. How do plants function to link the Sun's energy to living organisms? What is formed within the light-responsive cells of plants?
9. Compare photosynthesis and respiration and the concept of net photosynthesis that derives from deducting respiration from photosynthesis. What is the importance of knowing the net primary productivity of an ecosystem and how much biomass an ecosystem has accumulated?
10. Briefly describe the global pattern of net primary productivity.

■ ***Relate*** abiotic ecosystem components to ecosystem operations, and ***explain*** trophic relationships.

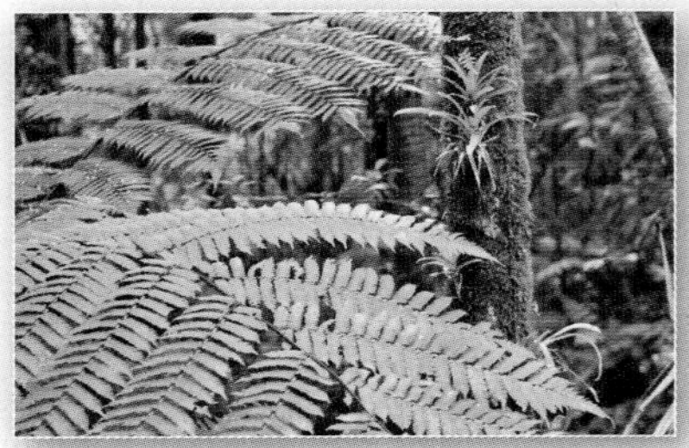

Light, temperature, water, and the cycling of gases and nutrients constitute the life-supporting abiotic components of ecosystems. Elevation and latitudinal position on Earth create a variety of physical environments. The zonation of plants with altitude is the **life zone concept** and is visible as you travel between different elevations.

Life is sustained by **biogeochemical cycles,** through which circulate the gases and sedimentary materials (nutrients) necessary for growth and development of living organisms. The environment may inhibit biotic operations, through either lack or excess. These **limiting factors** may be physical or chemical in nature.

Trophic refers to the feeding and nutrition relations in an ecosystem. It represents the flow of energy and cycling of nutrients. **Producers,** which fix the carbon they need from carbon dioxide, are the plants, including phytoplankton in aquatic ecosystems. **Consumers,** generally animals including zooplankton in aquatic ecosystems, depend on the producers as their carbon source. Energy flows from the producers through the system along a **food chain** circuit. Within ecosystems, the feeding interrelationships are complex and arranged in a complex network of interconnected food chains in a **food web**.

The primary consumer is a **herbivore,** or plant eater. A **carnivore** (meat eater) is a secondary consumer. A consumer that eats both producers and consumers is an **omnivore**—a role occupied by humans. **Detritivores** are detritus feeders and decomposers that release simple inorganic compounds and nutrients as they break down organic materials. Detritus feeders, including worms, mites, termites, and centipedes, process debris. **Decomposers** are the bacteria and fungi that process organic debris outside their bodies and absorb nutrients in the process—the rotting that breaks down detritus. Biomass and population pyramids characterize the flow of energy and the numbers of producers, consumers, and decomposers operating in an ecosystem.

life zone (p. 565)
biogeochemical cycles (p. 566)
limiting factor (p. 569)
producers (p. 570)
consumers (p. 570)
food chain (p. 570)
food web (p. 570)
herbivore (p. 570)
carnivore (p. 570)
omnivore (p. 570)
detritivore (p. 570)
decomposer (p. 570)

11. What are the principal abiotic components in terrestrial ecosystems?
12. Describe what Alexander von Humboldt found that led him to propose the life zone concept. What are life zones? Explain the interaction among elevation, latitude, and the types of communities that develop.
13. What are biogeochemical cycles? Describe several of the essential cycles.
14. What is a limiting factor? How does it function to control the spatial distribution of plant and animal species?
15. What role is played in an ecosystem by producers and consumers?
16. Describe the relationship among producers, consumers, and detritivores in an ecosystem. What is the trophic nature of an ecosystem? What is the place of humans in a trophic system?
17. What is a biomass pyramid? Describe how this relates to the nature of food chains.

■ ***Relate*** how biological evolution led to the biodiversity of life on Earth.

Biodiversity (a combination of *bio*logical and *diversity*) includes the number and variety of different species, the genetic diversity within species (number of genetic characteristics), and ecosystem and habitat diversity. The greater the biodiversity is within an ecosystem, the more stable and resilient it is and the more productive it will be. Modern agriculture often creates a nondiverse monoculture that is particularly vulnerable to failure.

Evolution states that single-cell organisms adapted, modified, and passed along inherited changes to multicellular organisms. The genetic makeup of successive generations was shaped by environmental factors, physiological functions, and behaviors that created a greater rate of survival and reproduction that were passed along through natural selection.

biodiversity (p. 575)
evolution (p. 575)

18. Give some of the reasons why biodiversity infers more stable, efficient, and sustainable ecosystems.
19. Referring back to Focus Study 1.1, define the scientific method and a theory and the progressive stages that lead to the development of a theory.
20. What is meant by ecosystem stability?
21. What do prairie ecosystems teach us about communities and biodiversity?

- ***Define*** succession, and ***outline*** the stages of general ecological succession in both terrestrial and aquatic ecosystems.

Ecological succession describes the process whereby older communities of plants and animals are replaced by newer communities that are usually more complex. Past glacial and interglacial climatic episodes have created a long-term succession. An area of bare rock and soil with no trace of a former community can be a site for **primary succession.** The initial community that occupies an area of early succession is a **pioneer community. Secondary succession** begins in an area that has a vestige of a previously functioning community in place. Rather than progressing smoothly to a definable stable point, ecosystems tend to operate in a dynamic condition, with succeeding communities overlapping in time and space.

Wildfire is one external factor that disrupts a successional community. The science of **fire ecology** has emerged in an effort to understand the natural role of fire in ecosystem maintenance and succession. Aquatic ecosystems also experience community succession, as exemplified by the eutrophication of a lake ecosystem.

ecological succession (p. 577)
primary succession (p. 578)
pioneer community (p. 579)
secondary succession (p. 579)
fire ecology (p. 579)

22. How does ecological succession proceed? What are the relationships between existing communities and new, pioneer communities?
23. Assess the impact of climate change on natural communities and ecosystems. Consider changes in species' distributions (*Geosystems Now*) or western wildfire conditions, for example.
24. Discuss the concept of fire ecology in the context of the Yellowstone National Park fires of 1988. What were the findings of the government task force?
25. How is the spread of urbanization into wilderness areas related to increasing wildfire risk?
26. Summarize the process of succession in a body of water. What is meant by cultural eutrophication?

MASTERING GEOGRAPHY

Visit www.masteringgeography.com for several figure animations, satellite loops, self-study quizzes, flashcards, visual glossary, case studies, career links, textbook reference maps, RSS Feeds, and an ebook version of *Geosystems*. Included are many geography web links to interesting Internet sources that support this chapter.

Ecological and weathering processes overtake the ruins of a house, walls, and garden on Mingulay Island, Outer Hebrides, Scotland. [Bobbé Christopherson.]

CHAPTER 20

Terrestrial Biomes

Winter snowstorm on a farmstead in central Wisconsin. Deciduous trees are bare in this broadleaf and mixed forest. Imagine this same scene in July with all the trees leafed out and the fields reaching harvest times. [Bobbé Christopherson.]

KEY LEARNING CONCEPTS

After reading the chapter, you should be able to:

- ***Describe*** the concept of biogeographic realms of plants and animals, and ***define*** ecotone and terrestrial ecosystems.
- ***Define*** biome, and ***illustrate*** formation classes as a basis for describing plant communities within biomes.
- ***Differentiate*** between native and exotic species, and ***detail*** the threat they represent to communities, using several examples.
- ***Describe*** 10 major terrestrial biomes, and ***locate*** them on a world map.
- ***Relate*** human impacts, real and potential, to several of the biomes.

GEOSYSTEMS NOW

Invasive Species Arrive at Tristan da Cunha

Geosystems Now in Chapter 6 describes an accident that left an oil-drilling platform adrift in the ocean currents of the South Atlantic. Eventually, the drifting *Petrobras XXI* rig ran aground in Trypot Bay, Tristan da Cunha, a remote island archipelago in the South Atlantic. See photos of Tristan and the rig in Chapter 6.

Inspection of the rig's vacant kitchen, cafeteria, and living quarters found various foods left behind, but no rodents or terrestrial species. However, because the owners of the Brazilian rig had not cleaned it prior to towing, it carried 62 non-native species of marine life on the underwater portion of the rig. These sections showed an intact subtropical reef community of non-native species, providing a rare opportunity for scientists to assess a potential biological invasion.

Invasive species can devastate biodiversity, especially in isolated island ecosystems (Figure GN 20.1). If a non-native species becomes invasive, it outcompetes native species for resources and may introduce new predators, pathogens, or parasites into an ecosystem.

Scientists made numerous dives to survey the newly arrived organisms to Tristan da Cunha. They developed a system of four assessed risk levels for the potential invasion: no threat from species that perished in transit; low threat from species that persist on the rig alone; medium threat from species that could spread from the rig; and highest threat (level four) from breeding species with high invasive potential. Corals that died in transit provided skeleton microhabitats for small crabs, worms, amphipods, and other crustaceans. Where barnacle shells remained, small sponges, brown mussels, and urchins persisted (Figure GN 20.2).

(a)

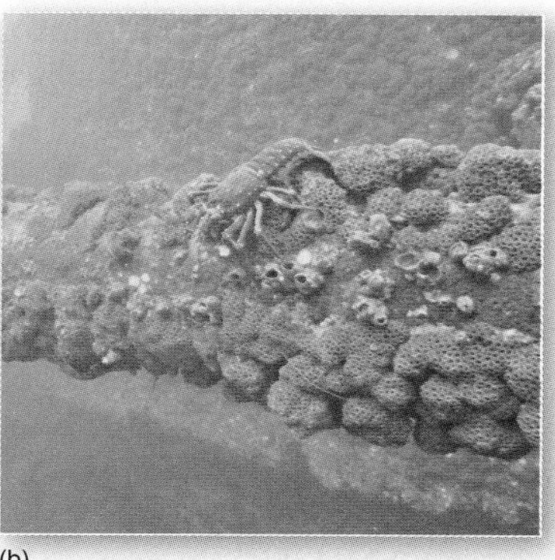
(b)

FIGURE GN 20.2 Non-native species on the submerged portion of the oil-drilling rig. (a) Coral skeletons and dead barnacle shells provide microhabitats for other non-native marine species. (b) A native rock lobster inspects a drilling rig strut and its organism cargo. [Sue Scott. All rights reserved.]

For the first time ever reported, fin-fish species had swum along with the rig as it drifted. Two of these, the silver porgy and the variable blenny, pose the greatest threat to native communities. The blenny has a reproductive doubling time of 15 months; the porgy arrived in large numbers as more than 60 were counted around the rig. Also found were porceline crabs and a balanidae (a type of barnacle), both assessed with a level-four invasion risk.

If the oil company had cleaned the rig prior to towing from Brazil, the cost would have been trivial, especially when compared to the $20 million cost incurred when the rig was removed and towed to distant deep water and sunk in 2007. Had the rig actually made it to its original destination in the warm waters of Singapore in the Far East, the species invasion potential would likely have intensified, since water conditions would have been similar to those found in Brazil.

The scientific team stated, "Our findings show that towing biofouled structures across biogeographic boundaries presents unexcelled opportunities for invasion to a wide diversity of marine species."* It is still too early to tell whether the remaining non-native species will invade Tristan's aquatic ecosystems and what the effects might be. For the 298 people of Tristan who depend on their small fishery for a community subsistence catch and on their commercial rock lobster (a crawfish) for export, the economic effects could be devastating.

(a)

(b)

FIGURE GN 20.1 Native marine environment offshore from Tristan da Cunha. (a) Natural healthy kelp forest and marine ecosystem. (b) Tristan rock lobster, a native species that is carefully managed to provide the community with a valuable export commodity. [Sue Scott. All rights reserved.]

*R. M. Wanless, S. Scott, et al., "Semi-submersible rigs: A vector transporting entire marine communities around the world," *Biological Invasions* 10.1007/s (November 29, 2009): 9666.

In 1874, Earth was viewed as offering few limits to human enterprise. But visionary American diplomat and conservationist George Perkins Marsh perceived the need to manage and to conserve the environment:

> We have now felled forest enough everywhere, in many districts far too much. Let us now restore this one element of material life to its normal proportions, and devise means of maintaining the permanence of its relations to the fields, the meadows, and the pastures, to the rain and the dews of heaven, to the springs and rivulets with which it waters the Earth.*

Unfortunately, his warning was ignored. The clearing of the forests is perhaps the most fundamental change wrought by humans in our transformation of Earth. The development and expansion of civilization was fueled by and built on the consumption of forests and other natural resources. The natural "work" supplied by Earth's physical, chemical, and biological systems has an estimated annual worth of some $35 trillion to the global economy.

Today, we see that Earth's natural systems do pose limits to human activity. Scientists are measuring ecosystems carefully, and they are building elaborate computer models to simulate our evolving real-time human-environment experiment—in particular, shifting patterns of environmental factors (temperatures and changing frost periods; precipitation timing and amounts; air, water, and soil chemistry; and a redistribution of nutrients).

What will the quotations about Earth's environment be like 138 years from now, about the same time interval since Marsh made his assessment? Humans now have become the most powerful biotic agent on Earth, influencing all ecosystems on a planetary scale.

In this chapter: We explore Earth's major terrestrial ecosystems, conveniently grouped into 10 biomes. Each biome is an idealized assemblage because much alteration of the natural environment has occurred. The discussion covers biome appearance, structure, and location; the present conditions of related plants, animals, and environments; and the state of biodiversity. Table 20.1 summarizes most of the information about Earth's physical systems that is presented through the pages of this text. Through biomes, we synthesize the physical geography of place and region.

*G. P. Marsh, *Physical Geography as Modified by Human Action* (New York: Scribner's, 1874), pp. 385–386.

Biogeographic Realms

Earth's biosphere can be classified based on assemblages of similar plant and animal communities. Figure 20.1 shows Earth's biogeographic realms. The map illustrates the botanical (plant) realms worldwide and generally overlaps with the traditional zoological (animal) realms. Each realm contains many distinct ecosystems that distinguish it from other realms, and, in turn, each ecosystem contains many communities.

A **biogeographic realm** is a geographic region where a group of plant and animal species evolved. As you see, these realms correspond generally to continents. Species distributions are constrained by climatic and topographic barriers; the main barrier separating these realms is the ocean. Recognition that such distinct regions of broadly similar flora (plants) and fauna (animals) exist was an early beginning for *biogeography* as a discipline.

The Australian realm is unique, giving rise to 450 species of *Eucalyptus* among its plants and 125 species of marsupials, which are animals, such as kangaroos, that carry their young in pouches, where gestation is completed (Figure 20.2). The presence of monotremes, egg-laying mammals such as the platypus, adds further distinctiveness to this realm. Australia's unique flora and fauna are the result of its early isolation from the other continents. During critical evolutionary times, Australia drifted away from Pangaea (see Chapter 11) and never again was reconnected by a land bridge, even when sea level lowered during repeated glacial ages. New Zealand, although relatively close in location, is isolated from Australia, explaining why it has no native marsupials.

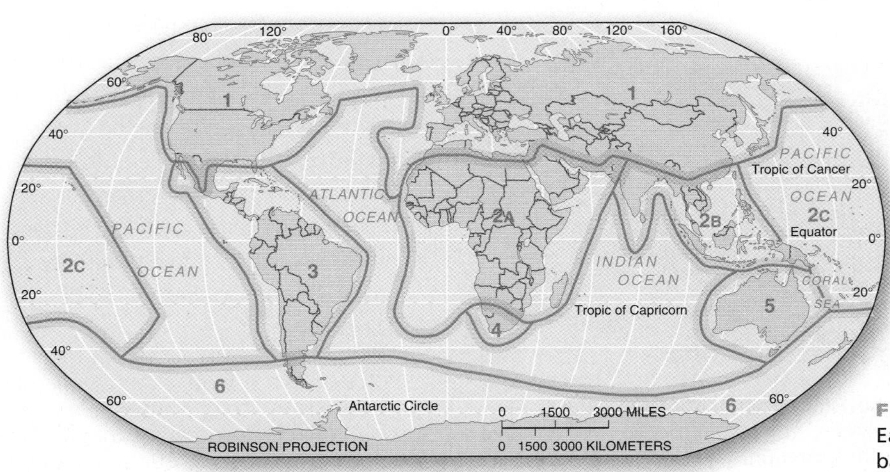

1	Boreal
2A	Paleotropical—African
2B	Paleotropical—Indo-Malaysian
2C	Paleotropical—Polynesian
3	Neotropical
4	South African
5	Australian
6	Antarctic

FIGURE 20.1 **Biogeographic realms.** Earth's biogeographic realms are roughly based on botanical realms.

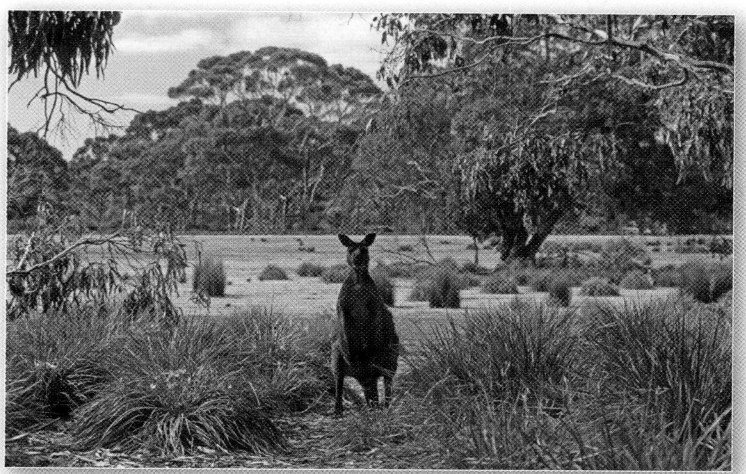

FIGURE 20.2 The unique Australian realm. The flora and fauna of Australia form a special assemblage of communities, here a western gray kangaroo near a dry river with eucalyptus trees and grasses. [A. Held/ www.agefotostock.com.]

Transition Zones

Alfred Wallace (1823–1913), the first scholar of *zoogeography*, used the stark contrast in animal species among islands in present-day Indonesia to delimit a boundary between the Paleotropical and Australian realms (see Figure 20.1). He thought that deep water in the straits had completely prevented species crossover. However, he later recognized that a boundary line actually is a wide transition zone, where one region grades into the other. Boundaries between natural systems are "zones of shared traits." Thus, despite distinctions among individual realms and our discussion of biomes, it is best to think of their borders as zones of mixed species composition rather than as rigidly defined boundaries.

A boundary transition zone between different, but adjoining ecosystems at any scale is an **ecotone**. Because ecotones are defined by different physical factors, they vary in width—think of a transition zone instead of a line. Ecosystems separated by different climatic conditions usually have gradual ecotones, whereas those separated by differences in soil or topography may form abrupt boundaries. For example, an ecotone formed by a landslide is abrupt, while an ecotone between prairies and northern forests may occupy many kilometers of land.

The range of environmental conditions found in ecotones makes them areas of high biodiversity and often larger population densities than communities on either side. Scientists have defined certain plant and animal species as having a range of tolerance for varying habitats; these "edge" species are often able to occupy habitat on either side of the ecotone. As human impacts cause ecosystem fragmentation, ecotones are becoming more numerous across the landscape. Although ecological response to climate change will occur at all levels in an ecosystem, the sensitivity of ecotones may make them more vulnerable.

Terrestrial Ecosystems

A **terrestrial ecosystem** is a self-sustaining association of land-based plants and animals and their abiotic environment, characterized by specific plant formation classes. Plants are the most visible part of the biotic landscape, and they are key members of Earth's terrestrial ecosystems. In their growth, form, and distribution, plants reflect Earth's physical systems: its energy patterns; atmospheric composition; temperature and winds; air masses; water quantity, quality, and seasonal timing; soils; regional climates; geomorphic processes; and ecosystem dynamics.

Specific characteristics are used for the structural classification of plants. *Life-form designations* are based on the outward physical properties of individual plants or the general form and structure of a vegetation cover. These physical life forms include:

- *Trees* (large woody plants with a main trunk; perennial; heights usually exceed 3 m, or 10 ft);
- *Lianas* (woody climbers and vines);
- *Shrubs* (smaller woody plants; stems branching at the ground);
- *Herbs* (small plants without woody stems; includes grasses and other nonwoody vascular plants);
- *Bryophytes* (nonflowering, spore-producing plants; includes mosses and liverworts);

GEO REPORT 20.1 *Designating large marine ecosystems for managing coastal oceans*

Aquatic ecosystems can be found in freshwater and saltwater systems and, like terrestrial ecosystems, have been heavily influenced by human activities. Coastal ocean waters, in particular, continue to deteriorate from unsustainable fishing practices, pollution, and habitat degradation. The loss of aquatic species, such as herring in the Georges Bank prime fishing area of the Atlantic, and an overall decline in fisheries worldwide highlighted the need for an ecosystem approach to understanding and managing these international waters. The designation of large marine ecosystems (LMEs) addresses that goal (see http://www.lme.noaa.gov/).

An LME is a distinctive oceanic region having unique organisms, ocean-floor topography, currents, areas of nutrient-rich upwelling circulation, or areas of significant predation, including human. Examples of identified LMEs include the Gulf of Alaska, California Current, Gulf of Mexico, Northeast Continental Shelf, and Baltic and Mediterranean seas. Some 62 LMEs, each encompassing more than 200,000 km^2 (77,200 mi^2), are presently defined worldwide. A number of these LMEs include government-protected areas, such as the Monterey Bay National Marine Sanctuary within the California Current LME and the Florida Keys Marine Sanctuary within the Gulf of Mexico LME (see http://sanctuaries.NOAA.gov/).

- *Epiphytes* (plants growing above the ground on other plants, using them for support); and
- *Thallophytes* (lack true leaves, stems, or roots; include bacteria, fungi, algae, and lichens).

A **biome** is a large, stable, terrestrial ecosystem characterized by its dominant plant type and vegetation structure. Each biome usually is named for its dominant vegetation, or **formation class**, because that is the single most easily identified feature. We can generalize Earth's wide-ranging plant species into six broad groups: *forest*, *savanna*, *grassland*, *shrubland*, *desert*, and *tundra*. Within these, we can become more specific by formation class. For example, forests can be subdivided into rain forests, needleleaf forests, broadleaf forests, and seasonal forests. This leads to the designation of 10 global terrestrial biomes discussed in this chapter: tropical rain forest, tropical seasonal forest and scrub, tropical savanna, midlatitude broadleaf forest, needleleaf and montane forest, temperate rain forest, Mediterranean shrubland, midlatitude grassland, desert, and arctic and alpine tundra.

Because plant distributions respond to environmental conditions and reflect variations in climate and soil, the world climate map in Figure 10.6, Chapter 10, is a helpful reference for this chapter. Although the biome concept was developed for terrestrial ecosystems, it can be applied to aquatic ecosystems: polar seas, temperate seas, tropical seas, seafloor, shoreline, and coral reef.

Invasive Species Our discussion of biomes involves species that are native to a region, meaning that their occurrence is a result of natural processes. These species have come to inhabit an area based on the evolutionary and physical factors previously discussed. Communities, ecosystems, and biomes can also be affected by species that are brought, or introduced, from elsewhere by humans, either accidentally or intentionally. These non-native species are also known as *exotic species*, or *aliens*. After arriving in the new ecosystem, some species may then be disruptive and become **invasive species.**

Invasive plant and animal species are able to take over niches already occupied by native species. Probably 90% of non-native species fail when they try to move into established niches in a community. The 10% that succeed can alter community dynamics and lead to declines in native species. Prominent examples are Africanized "killer bees" in North and South America; brown tree snakes in Guam; kudzu, zebra and quagga mussels in the Great Lakes; and common gorse (*Ulex europaeus*), a prickly, evergreen shrub, in temperate regions worldwide and mountainous areas of the tropics (Figure 20.3). For information on invasive species prevention and management, see

(a)

(b)

(c)

(d)

(e)

FIGURE 20.3 Exotic species.
(a) A sign in Wallowa County, Washington, makes a plea for awareness in this agricultural country, listing noxious (non-native, often problematic) weeds at the county line. (b) Invasive purple loosestrife near Lake Michigan and the Indiana Dunes National Lakeshore, Indiana (foreground and left center). (c) Kudzu, originally imported for cattle feed, spread from Texas to Pennsylvania; here it overruns pasture and forest in western Georgia. (d) Common gorse (yellow-flowering shrub), intended for barriers between pastures, runs wild across the landscape of West Point, Falkland Islands. (e) Zebra mussels literally cover most hard surfaces in the Great Lakes; they rapidly colonize on any surface, even sand, in freshwater environments. [(a), (b), (c), and (d) Bobbé Christopherson. (e) Gordon MacSkimming/age fotostock.]

http://invasions.bio.utk.edu/bio_invasions/index.html or http://www.invasivespecies.gov/.

As an example, Purple loosestrife (*Lythrum salicaria*) was introduced from Europe in the 1800s as a desired ornamental plant with some medicinal applications. The plant's seeds also arrived in ships that used soil for ballast. A hardy perennial, it escaped cultivation and invaded wetlands across the eastern portions of the United States and Canada, through the upper Midwest, and as far west as Vancouver Island, British Columbia, replacing native plants on which wildlife depend. The plant's invasive characteristics are its ability to produce vast quantities of seed during an extended flowering season and to spread vegetatively through underground stems and its tendency to form dense, homogenous stands once established (Figure 20.3b).

On a large spatial scale, invasions can alter the dynamics of entire biomes. For example, in the Mediterranean shrublands of southern California, the native vegetation is adapted for frequent wildfire. However, the presence of non-native species changes the successional processes in these areas, as exotic species often colonize burned areas more efficiently than do native species. This leads to thick undergrowth and more fuel for increasingly more frequent fires. Native vegetation is adapted for infrequent fires, on the order of 30 to 150 years. Thus, increased fire frequency puts native species at a disadvantage. Scientists have documented the conversion from scrublands to grasslands as a result of more frequent fires in these regions.

In the United States and Canada, we humans are perpetuating a new type of terrestrial plant community. This type of *transition community* is somewhere between a grassland and a forest—an artificial successional stage produced by large-scale interruption and disturbance. We plant non-native trees and lawns and then must invest energy, water, and capital to sustain such artificial modifications. Or we graze non-native animals and plant crops that perpetuate this disturbed transitional state. It remains to be seen whether the land can return to its natural vegetation biomes if human influences are removed.

Now let us begin a tour of Earth's biomes, keeping in mind that they synthesize all we have learned about the atmosphere, hydrosphere, lithosphere, and biosphere in the pages of this text. Here we bring it all together.

Earth's Major Terrestrial Biomes

Few natural communities of plants and animals remain; most biomes have been greatly altered by human intervention. Thus, the "natural vegetation" identified on many biome maps reflects *ideal potential* mature vegetation, given the environmental characteristics in a region. For example, in Norway, the former needleleaf and montane forest today is a mix of second- and third-growth forests, farmlands, and altered landscapes (Figure 20.4). However, these ideal conditions form a basis for discussing natural ecological processes and the extent of human impacts.

FIGURE 20.4 Needleleaf forest landscape modified by human activity.
Forests and farms in rural Norway, formerly covered in needleleaf forests. [Bobbé Christopherson.]

Let us look at the global distribution of Earth's major terrestrial biomes as portrayed in Figure 20.5. Table 20.1 (p. 598) describes each biome on the map and summarizes other pertinent information gathered from throughout *Geosystems*.

CRITICAL THINKING 20.1

Reality check

Now is a good time to refer to the composite satellite images that appear inside the front cover of this text, one a cloudless view of Earth in natural color typical of a local summer day, the other with clouds and atmospheric weather patterns. Compare the map in Figure 20.5 with these remarkable images and see what correlations you can make. Then compare the biomes in Figure 20.5 with population distribution as indicated by the nighttime lighting display also on the inside front cover. What generalizations do you make?

Tropical Rain Forest

Earth is girdled with a lush biome—the **tropical rain forest**. In a climate of consistent year-round daylength (12 hours), high insolation, average annual temperatures around 25°C (77°F), and plentiful moisture, plant and animal populations have responded with the most diverse expressions of life on the planet.

The Amazon region is the largest tract of tropical rain forest, or *selva*. In addition, rain forests cover equatorial regions of Africa, Indonesia, the margins of Madagascar and Southeast Asia, the Pacific coast of Ecuador and Colombia, and the east coast of Central America, with small discontinuous patches elsewhere. The cloud forests of western Venezuela are such tracts of rain forest at high elevation, perpetuated by high humidity and cloud cover. Undisturbed tracts of rain forest are rare.

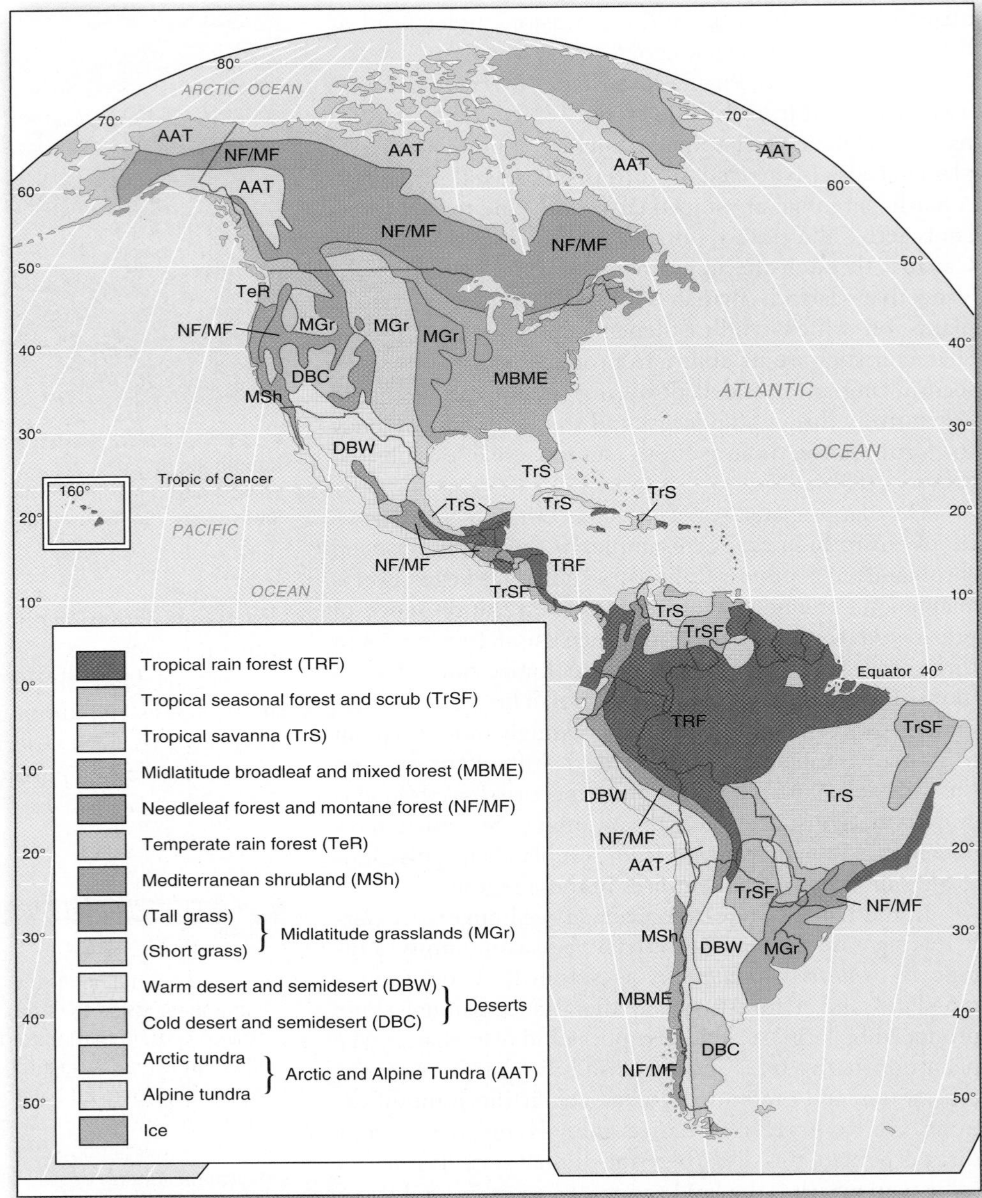

FIGURE 20.5 The 10 major global terrestrial biomes.

Rain forests represent approximately one-half of Earth's remaining forests, occupying about 7% of the total land area worldwide. This biome is stable in its natural state, resulting from the long-term residence of these continental plates near equatorial latitudes and their escape from glacial activity.

The rainforest canopy forms three levels—see Figure 20.6a (p. 599). The upper level is not continuous, but features emergent tall trees whose high crowns rise above the middle canopy. This upper level appears as a broken *overstory* of tall trees breaking through a middle canopy that is nearly continuous. The broad leaves block much of the light and create a darkened *understory* area and forest floor. The lower level is composed of seedlings, ferns, bamboo, and the like, leaving the litter-strewn ground level in deep shade and fairly open.

Rain forests feature ecological niches that are distributed vertically rather than horizontally because of the competition for light. Biomass in a rain forest is concentrated high up in the canopy, that dense mass of overhead leaves. The canopy is filled with a rich variety of plants and animals. *Lianas* (vines) stretch from tree to tree, entwining them with cords that can reach 20 cm (8 in.) in diameter (Figure 20.6b). *Epiphytes* flourish there, too; these plants, such as orchids, bromeliads, and ferns, live entirely aboveground, supported physically, but not nutritionally, by the structures of other plants. Calm conditions on the forest floor make wind pollination difficult, so pollination is by insects, other animals, and self-pollination.

Aerial photographs of a rain forest, or views along riverbanks covered by dense vegetation, or the false Hollywood-movie imagery of the "jungle" make it difficult

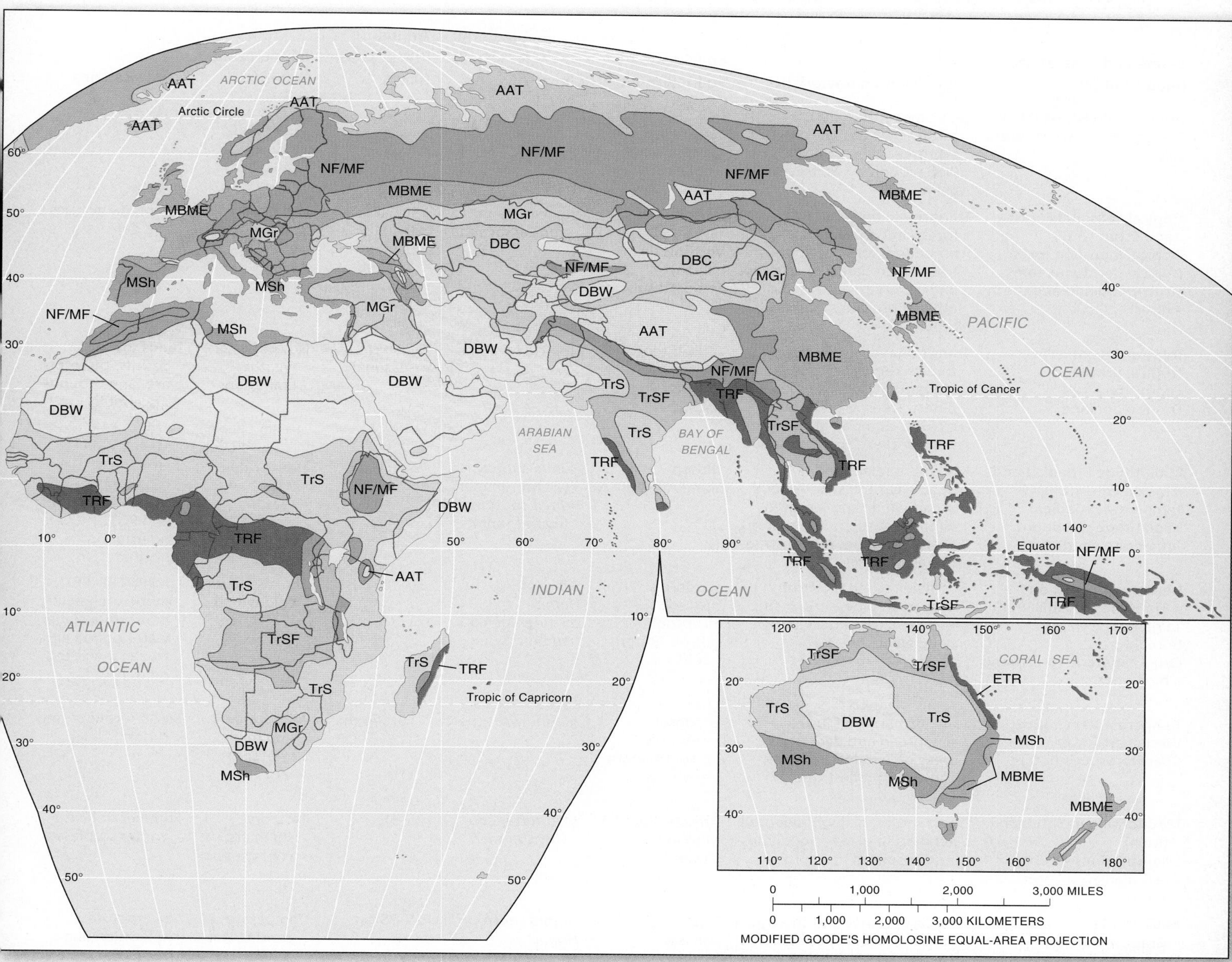

to imagine the shadowy environment of the actual rainforest floor, which receives only about 1% of the sunlight arriving at the canopy (Figure 20.7a, p. 600). The constant moisture, rotting fruit and moldy odors, strings of thin roots and vines dropping down from above, windless air, and echoing sounds of life in the trees together create a unique environment.

The smooth, slender trunks of rainforest trees are covered with thin bark and buttressed by large, wall-like flanks that grow out from the trees to brace the trunks (Figure 20.7b). These buttresses form angular, open enclosures, a ready habitat for various animals. Branches are usually absent on at least the lower two-thirds of the tree trunks.

The wood of many rainforest trees is extremely hard, heavy, and dense—in fact, some species will not even float in water. Exceptions are balsa and a few others, which are light. Varieties of trees include mahogany, ebony, and rosewood. Logging is difficult because individual species are widely scattered; a species may occur only once or twice per square kilometer. Selective cutting is required for species-specific logging, whereas pulpwood production takes everything.

Rainforest soils, principally Oxisols, are essentially infertile, yet they support rich vegetation. The trees have adapted to these soil conditions with root systems able to capture nutrients from litter decay at the soil surface. High precipitation levels and temperatures work to produce deeply weathered and leached soils, characteristic of the laterization process, with a claylike texture sometimes breaking up into a granular structure. Oxisols lack nutrients and colloidal material. With much investment in fertilizers, pesticides, and machinery, these soils can be productive.

TABLE 20.1 Major Terrestrial Biomes and Their Characteristics

Biome and Ecosystems (map symbol)	Vegetation Characteristics	Soil Orders of Soil Taxonomy	Climate Type	Annual Precipitation Range	Temperature Patterns	Water Balance
Tropical Rain Forest (TRF) Evergreen broadleaf forest Selva	Leaf canopy thick and continuous; broadleaf evergreen trees, vines (lianas), epiphytes, tree ferns, palms	Oxisols Ultisols (on well-drained uplands)	*Tropical*	180–400 cm(>6cm/mo)	Always warm (21°–30°C; avg. 25°C)	Surpluses all year
Tropical Seasonal Forest and Scrub (TrSF) Tropical monsoon forest Tropical deciduous forest Scrub woodland and thorn forest	Transitional between rain forest and grasslands; broadleaf, some deciduous trees; open parkland to dense undergrowth; acacias and other thorn trees in open growth	Oxisols Ultisols Vertisols (in India) Some Alfisols	*Tropical monsoon, tropical savanna*	130–200 cm (>40 rainy days during 4 driest months)	Variable, always warm (>18°C)	Seasonal surpluses and deficits
Tropical Savanna (TrS) Tropical grassland Thorn tree scrub Thorn woodland	Transitional between seasonal forests, rain forests, and semiarid tropical steppes and desert; trees with flattened crowns, clumped grasses, and bush thickets; fire association	Alfisols (dry: Ultalfs) Ultisols Oxisols	*Tropical savanna*	9–150 cm, seasonal	No cold-weather limitations	Tends toward deficits, therefore fire- and drought-susceptible
Midlatitude Broadleafand Mixed Forest(MBME) Temperate broadleaf Midlatitude deciduous Temperate needleleaf	Mixed broadleaf and needleleaf trees; deciduous broadleaf, losing leaves in winter; southern and eastern evergreen pines demonstrate fire association	Ultisols Some Alfisols	*Humid subtropical* (warm summer) *Humid continental* (warm summer)	75–150 cm	Temperate with cold season	Seasonal pattern with summer maximum PRECIP and POTET (PET); no irrigation needed
Needleleaf Forest and Montane Forest(NF/MF) Taiga Boreal forest Other montane forests and highlands	Needleleaf conifers, mostly evergreen pine, spruce, fir; Russian larch, a deciduous needleleaf	Spodosols Histosols Inceptisols Alfisols (Boralfs: cold)	*Subarctic* *Humid continental* (cool summer)	30–100 cm	Short summer, cold winter	Low POTET (PET), moderate PRECIP; moist soils, some waterlogged and frozen in winter; no deficits
Temperate Rain Forest(TeR) West coast forest Coast redwoods (U.S.)	Narrow margin of lush evergreen and deciduous trees on windward slopes; redwoods, tallest trees on Earth	Spodosols Inceptisols (mountainous environs)	*Marine west coast*	150–500 cm	Mild summer and mild winter for latitude	Large surpluses and runoff
Mediterranean Shrubland (MSh) Sclerophyllous shrubs Australian eucalyptus forest	Short shrubs, drought adapted, tending to grassy woodlands and chaparral	Alfisols (Xeralfs) Mollisols (Xerolls)	*Mediterranean* (dry summer)	25–65 cm	Hot, dry summers, cool winters	Summer deficits, winter surpluses
Midlatitude Grasslands(MGr) Temperate grassland Sclerophyllous shrub	Tallgrass prairies and shortgrass steppes, highly modified by human activity; major areas of commercial grain farming; plains, pampas, and veld	Mollisols Aridisols	*Humid subtropical* *Humid continental* (hot summer)	25–75 cm	Temperate continental regimes	Soil moisture utilization and recharge balanced; irrigation and dry farming in drier areas
Warm Desert and Semidesert (DBW) Subtropical desert and scrubland	Bare ground graduating into xerophytic plants including succulents, cacti, and dry shrubs	Aridisols Entisols (sand dunes)	*Arid desert*	< 2 cm	Average annual temperature, around 18°C, highest temperatures on Earth	Chronic deficits, irregular precipitation events, PRECIP < 1/2 POTET (PET)
Cold Desert and Semidesert (DBC) Midlatitude desert, scrubland, and steppe	Cold desert vegetation includes short grass and dry shrubs	Aridisols Entisols	*Semiarid steppe*	2–25 cm	Average annual temperature around 18°C	PRECIP> 1/2 POTET (PET)
Arctic and AlpineTundra (AAT)	Treeless; dwarf shrubs, stunted sedges, mosses, lichens, and short grasses; alpine, grass meadows	Gelisols Histosols Entisols (permafrost)	*Tundra* *Subarctic* (very cold)	15–180 cm	Warmest months > 10°C, only 2 or 3 months above freezing	Not applicable most of the year, poor drainage in summer
Ice			*Ice sheet, ice cap*			

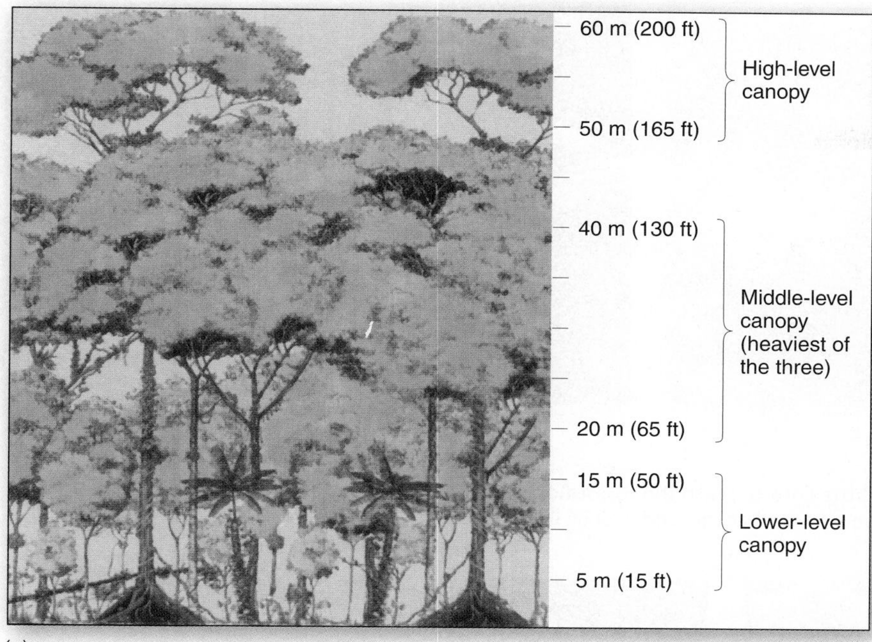

(a)

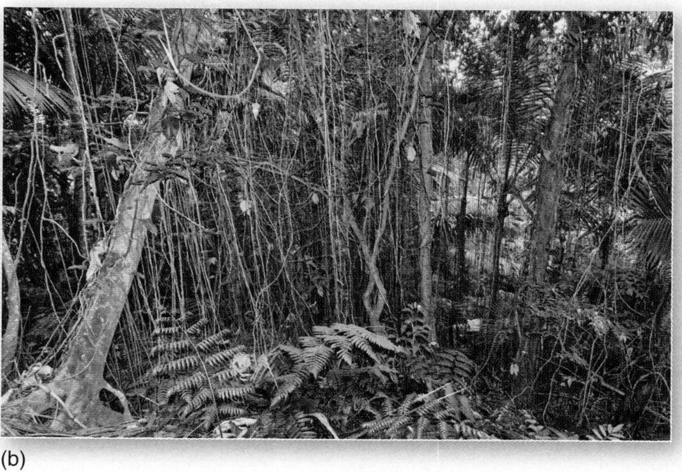

(b)

FIGURE 20.6 The three levels of a rainforest canopy. (a) Lower, middle, and high levels of the rain forest. (b) Lianas drape from the rainforest canopy above. [Bobbé Christopherson.]

The animal and insect life of the rain forest is diverse, ranging from small decomposers (bacteria) working the surface to many animals living exclusively in the upper stories of the trees. These tree dwellers are referred to as *arboreal*, from the Latin word meaning "tree," and include sloths, monkeys, lemurs, parrots, and snakes. Beautiful birds of many colors, tree frogs, lizards, bats, and a rich insect community that includes more than 500 species of butterflies are found in rain forests. Surface animals include pigs (the bushpig and giant forest hog in Africa, wild boar and bearded pig in Asia, and peccary in South America), species of small antelopes (bovids), and mammalian predators (the tiger in Asia, jaguar in South America, and leopard in Africa and Asia).

The present human assault on Earth's rain forests has put this diverse fauna and the varied flora at risk. It also jeopardizes an important recycling system for atmospheric carbon dioxide and a potential source of valuable pharmaceuticals and many types of new foods—and so much is still unknown and undiscovered.

Deforestation of the Tropics

More than half of Earth's original rain forest is gone, cleared for pasture, timber, fuel wood, and farming. Worldwide, an area nearly the size of Wisconsin is lost each year (169,000 km^2; 65,000 mi^2), and about a third more is disrupted by selective cutting of canopy trees that occurs along the *edges* of deforested areas.

When orbiting astronauts look down on the rain forests at night, they see thousands of human-set fires. During the day, the lower atmosphere in these regions is choked with the smoke. These fires are used to clear land for agriculture, which is intended to feed the domestic population as well as to produce accelerating cash exports of beef, soybeans, rubber, coffee, and other commodities. Edible fruits are not abundant in an undisturbed rain forest, but cultivated clearings produce bananas (plantains), mangos, jack fruit, guava, and starch-rich roots such as manioc and yams. Brazil has been the world's largest exporter of beef since 2004 and is the second leading country for soybean exports.

Because of the poor soil fertility, the cleared lands are quickly exhausted under intensive farming and are then generally abandoned in favor of newly burned and cleared lands unless fertility is maintained artificially. The dominant trees require from 100 to 250 years to reestablish themselves after major disturbances. Once cleared, the former forest becomes a mass of low bushes intertwined with vines and ferns, slowing the return of trees.

Figure 20.8a and b shows satellite true-color images of a portion of western Brazil called Rondônia, recorded in 2000 and 2009. These images give a sense of the level of rainforest destruction in progress. You can clearly see encroachment along new roads branching from highway BR364. The *edges* of every road and cleared area represent a significant portion of the species habitat disturbance, population dynamic changes, and carbon losses to the

(a)

(b)

FIGURE 20.7 The rain forest.
(a) Tropical rainforest canopy is thick, allowing little light to reach the forest floor. (b) The rainforest floor, with a typical buttressed tree, leaf litter, and lianas in the background. [Bobbé Christopherson.]

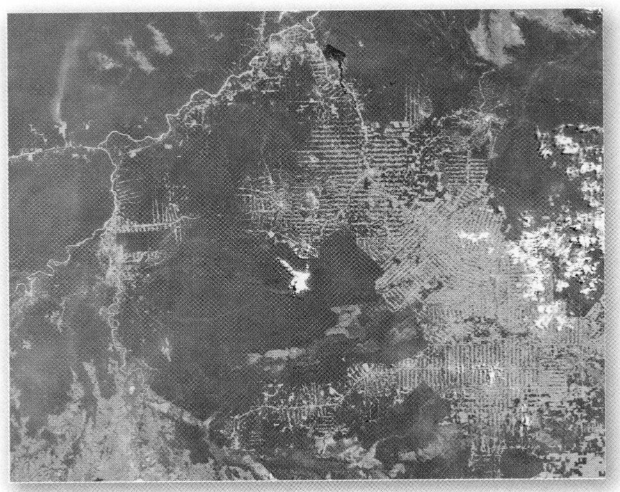

(a) 2000

(b) 2009

(c)

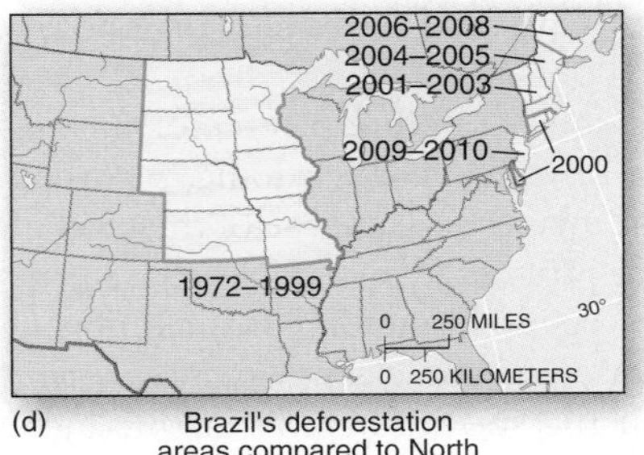

(d) Brazil's deforestation areas compared to North America

FIGURE 20.8 Amazon rainforest losses over three decades.
Satellite images of a portion of western Brazil called Rondônia, recorded in true color in (a) 2000 and (b) 2009, show large rainforest losses growing outward from highway BR364, the main artery in the region. The branching pattern of feeder roads encroaches on the rain forest where you can see patch and edge dynamics at work along new roadways. (c) Surface view of burning rain forest in the Amazon. (d) Brazilian deforested areas cleared between 1972 and 2010 in equivalent U.S. area. [(a) and (b) *Terra* MODIS, NASA/GSFC. (c) Colin Jones/photolibrary.com.]

atmosphere—more impact occurs along these edges than in the tracts of clear-cut logging. The hotter, drier, and windier edge conditions penetrate the forest up to 100 m (328 ft). Figure 20.8c shows a tract of former rain forest that has just been burned to begin the clearing and road-building process.

To give you a sense of the area of forest losses, Figure 20.8d shades in the equivalent area for given years. Deforestation has decreased a bit in the last couple of years as Brazil's government has stepped up enforcement to preserve the forest; 2009–2010 combined losses were down to approximately 18,000 km² (6950 mi²), or about 80% of the area of New Jersey. The United Nations Food and Agricultural Organization estimated that, if this destruction to rain forests continues unabated, they will be completely removed by 2050. By continent, rainforest losses are estimated at more than 50% in Africa, more than 40% in Asia, and 40% in Central and South America.

This is a highly charged issue, as can be seen by the growth of Brazil's cattle industry, which uses these new pasturelands. Cattle herds reached over 60 million head by 2010, generating $3 billion in revenue. Among many references, see the Tropical Rainforest Coalition at http://www.rainforest.org/, World Resources Institute Global Forest Watch at http://www.globalforestwatch.org/english/index.htm, or Rainforest Action Network at http://www.ran.org/.

Focus Study 20.1 looks more closely at efforts to curb increasing rates of species extinction and loss of Earth's biodiversity, much of which is directly attributable to the loss of rain forests.

CRITICAL THINKING 20.2

Tropical forests: A global or local resource?

Given the information presented in this chapter about deforestation in the tropics and species extinction, assess the present situation. What are the main issues? What natural assets are at stake? How do we balance concerns for planetary natural resources against sovereign state rights? How are global biodiversity and greenhouse warming related to these issues? What is the perspective of the less-developed countries that possess most of the rain forest? What is the perspective of the more-developed countries and their transnational corporations? What kind of action plan would you develop to accommodate all parties?

Tropical Seasonal Forest and Scrub

A varied biome on the margins of the rain forest is the **tropical seasonal forest and scrub**, which occupies regions of low and erratic rainfall. The shifting intertropical convergence zone (ITCZ) brings precipitation with the seasonal shift to the high Sun of summer and then a season of dryness with the shift to the low Sun of winter, producing a seasonal pattern of moisture deficits, some leaf loss, and dry-season flowering. The term *semideciduous* applies to some of the broadleaf trees that lose their leaves during the dry season.

Areas of this biome have fewer than 40 rainy days during their 4 consecutive driest months, yet heavy monsoon downpours characterize their summers (see Chapter 6, especially Figure 6.19). The climates *tropical monsoon* and *tropical savanna* apply to these transitional communities between rain forests and tropical grasslands.

Portraying such a varied biome is difficult. In many areas, humans disturb the natural vegetation, so that the savanna grassland adjoins the rain forest directly. The biome does include a gradation from wetter to drier areas: monsoonal forests, to open woodlands and scrub woodland, to thorn forests, to drought-resistant scrub species (Figure 20.9). In South America, an area of transitional tropical deciduous forest surrounds an area of savanna in southeastern Brazil and portions of Paraguay.

The monsoonal forests are an average of 15 m (50 ft) high with no continuous canopy of leaves, graduating into orchardlike parkland with grassy openings or into areas choked by dense undergrowth. In more open tracts, a common tree is the acacia, with its flat top and usually thorny stems. These trees have branches that look like an upside-down umbrella (spreading and open skyward), as do trees in the tropical savanna.

Local names are given to these communities: the *caatinga* of the Bahia State of northeastern Brazil, the *chaco* area of Paraguay and northern Argentina, the *brigalow*

(*text continued on page 604*)

(a)

(b)

FIGURE 20.9 Kenyan landscapes.
Two views of the open thorn forest and savanna (a) near and in the Samburu Reserve, Kenya, representing a broad ecotone between the two biomes, and (b) Maasai giraffes move across open country in Amboseli National Park, Kenya. [(a) Gael Summer. (b) Nigel Pavitt/John Warburton-Lee Photography/photolibrary.com.]

FOCUS STUDY 20.1

Species Diversity, Risks, and Biosphere Reserves

When the first European settlers landed on the Hawaiian Islands in the late 1700s, they counted 43 species of birds. Today, 15 of those species are extinct, and 19 more are threatened or endangered, with only one-fifth of the original species populations relatively healthy. In most of Hawai'i, no native species exist below the 1220-m (4000-ft) elevation because of an introduced avian virus. Humans have great impact on animals, plants, and global biodiversity. We are facing a loss of genetic diversity that may be unparalleled in Earth's history, even compared with the major extinctions over the geologic record.

FIGURE 20.1.1 The polar bear. A male polar bear with snow on his face in northern Manitoba. One study predicts that 67% of the remaining polar bear population of 23,000 will be gone by 2050 because of climate change and loss of sea ice—that will leave about 7500 bears across the Arctic region. [Bobbé Christopherson.]

As we learn more about Earth's ecosystems and their related communities, we understand their value to civilization and our interdependence with them. Natural ecosystems are a major source of new foods, chemicals, and medicines and are indicators of a healthy, functioning biosphere.

International efforts are under way to study and preserve specific segments of the biosphere: among others, the United Nations Environment Programme, World Resources Institute, World Conservation Union, Natural World Heritage Sites, Wetlands of International Importance, and Nature Conservancy. An important part of the effort is the setting aside of biosphere reserves and the focus provided by the World Conservation Monitoring Centre and its International Union for Conservation of Nature (IUCN) Red List of global endangered species at http://www.iucnredlist.org/. See also the endangered species home page of the Fish and Wildlife Service at http://www.fws.gov/endangered/ and the World Wildlife Fund at http://www.panda.org/.

Species Threatened—Examples

The extinction of 67% of the known species of harlequin frogs in the tropics was discussed in Chapter 1 and Figure 1.7. In the Arctic region, the IUCN in 2006 listed the polar bear (*Ursus maritimus*) as "vulnerable" to extinction, a "threatened" category on its Red List of Threatened Species (Figure 20.1.1). Research released by the U.S. Geological Survey in September 2007 predicted that with the loss of Alaskan, Canadian, and Russian sea-ice habitat, some two-thirds of the world's polar bears will die off by 2050 or earlier. The remaining 7500 bears will be struggling. Any action society takes now will hopefully result in a slowing of such sea-ice loss and the buying of a little adaptation time.

Black rhinos (*Diceros bicornis*) and white rhinos (*Ceratotherium simum*) in Africa exemplify species in jeopardy. Rhinos once grazed over much of the savannas and woodlands. Today, they survive only in protected districts in heavily guarded sanctuaries and are threatened even there (Figure 20.1.2). In 1998, 2599 black rhinos remained, 96% fewer than the 70,000 in 1960. Populations have remained statistically stable through the last decade, and this count is supported by South Africa's guarding of about 50% of the herd. A slow recovery is under way, as numbers of black rhinos rose to 4240 in 2008, according to the IUCN/SSC's African Rhino Specialist Group. This is not the case with the western black rhino, a subspecies, which is down to its last few individuals.

FIGURE 20.1.2 The rhinoceros in Africa. White rinoceros with young (*Ceratotherium simum*), Lake Nakuru National Park, Kenya, of the southern population. [Ingo Arndt/naturepl.com/NaturePL.]

Only 11 northern white rhinos existed in 1984, increasing to an estimated 29 by 1998 (Congo 25,

Côte d'Ivoire 4), although the IUCN suggests this number might actually be lower. Political unrest in the Congo and the area of the Garamba National Park has led to the deaths of an unknown number of these few. The southern white rhino population reached 14,500 by 2008, an increase.

Rhinoceros horn sells for $29,000 per kilogram as an aphrodisiac (a false use of no medicinal effect). These large land mammals are nearing extinction and will survive only as a dwindling zoo population. The limited genetic pool that remains complicates further reproduction.

Table 20.1.1 summarizes the numbers of known and estimated species on Earth. Scientists have classified only 1.75 million species of plants and animals of an estimated 13.6 million overall; this figure represents an increase in what scientists once thought to be the diversity of life on Earth. And those yet-to-be-discovered species represent a potential future resource for society. The wide range of estimates places the expected species count between a low of 3.6 million and a high of 111.7 million. Estimates of annual species loss range between 1000 and 30,000, although this range might be conservative. The possibility exists that over half of Earth's present species could be extinct within the next 100 years.

Biosphere Reserves

Formal natural reserves are a possible strategy for slowing the loss of biodiversity and protecting this resource base. Setting up a *biosphere reserve* involves principles of **island biogeography**. Island communities are special places for study because of their spatial isolation and the relatively small number of species present. Islands resemble natural experiments because the impact of individual factors, such as civilization, can be more easily assessed on islands than over larger continental areas.

Studies of islands also can assist in the study of mainland ecosystems, for in many ways a park or biosphere reserve, surrounded by modified areas and artificial boundaries, is like an island. Indeed, a biosphere reserve is conceived as an ecological island in the midst of change. The intent is to establish a core in which genetic material is protected from outside disturbances, surrounded by a buffer zone that is, in turn, surrounded by a transition zone and experimental research areas. An important variable to consider in setting aside a biosphere reserve is any change that might occur in temperature and precipitation patterns as a result of global change. A carefully considered reserve could end up outside the natural range of its species as climates change.

Not all protected areas are ideal bioregional entities by definition. Some are simply imposed on existing park space, and some remain in the planning stage, although they are officially designated. Some of the best biosphere reserve examples have been in operation since the late 1970s and range from Everglades National Park in Florida, to the Changbai Reserve in China, to the Tai Forest on the Côte d'Ivoire (Ivory Coast), to the Charlevoix Biosphere Reserve in Québec, Canada. Added to these efforts is the work of the Nature Conservancy, which acquires land for preservation as a valuable part of the reserve.

The ultimate goal, about half achieved, is to establish at least one reserve in each of the 194 distinctive biogeographic communities presently identified. So far 6390 areas are designated, covering 657 million hectares (1.62 billion acres). Scientists predict that designating new, undisturbed reserves may not be possible in a decade because pristine areas will be gone. The biosphere reserve program is coordinated by the Man and the Biosphere Programme of UNESCO (http://www.unesco.org/mab/). More than 480 biosphere reserves are now operated voluntarily in 100 countries.

TABLE 20.1.1 Known and Estimated Species on Earth

Categories of Living Organisms	Number of Known Species	Estimated Number of Species High (000)	Estimated Number of Species Low (000)	Working Estimate (000)	Accuracy
Viruses	4,000	1,000	50	400	Very poor
Bacteria	4,000	3,000	50	1000	Very poor
Fungi	72,000	27,000	200	1500	Moderate
Protozoa	40,000	200	60	200	Very poor
Algae	40,000	1,000	150	400	Very poor
Plants	270,000	500	300	320	Good
Nematodes	25,000	1,000	100	400	Poor
Arthropods					
Crustaceans	40,000	200	75	150	Moderate
Arachnids	75,000	1,000	300	750	Moderate
Insects	950,000	100,000	2,000	8,000	Moderate
Mollusks	70,000	200	100	200	Moderate
Chordates	45,000	55	50	50	Good
Others	115,000	800	200	250	Moderate
Total	**1,750,000**	**111,655**	**3,635**	**13,620**	**Very poor**

Source: United Nations Environment Programme, *Global Biodiversity Assessment* (Cambridge, England: Cambridge University Press, 1995), Table 3.1–2, p. 118, used by permission.

scrub of Australia, and the *dornveld* of southern Africa. Figure 20.5 shows areas of this biome in Africa, extending from eastern Angola through Zambia to Tanzania and Kenya; in Southeast Asia and portions of India, from interior Myanmar through northeastern Thailand; and in parts of Indonesia.

The trees throughout most of this biome make poor lumber, but some, especially teak, may be valuable for fine cabinetry. In addition, some of the plants with dry-season adaptations produce usable waxes and gums, such as carnauba and palm-hard waxes. Animal life includes the koalas and cockatoos of Australia and the elephants, large cats, rodents, and ground-dwelling birds in other occurrences of this biome.

Tropical Savanna

The **tropical savanna** consists of large expanses of grassland, interrupted by trees and shrubs. This is a transitional biome between the tropical forests and semiarid tropical steppes and deserts. The savanna biome also includes treeless tracts of grasslands, and in dry savannas, grasses grow discontinuously in clumps, with bare ground between them. The trees of the savanna woodlands are characteristically flat-topped, in response to light and moisture regimes.

Savannas covered more than 40% of Earth's land surface before human intervention, but were especially modified by human-caused fire. Fires occur annually throughout the biome. The timing of these fires is important. Early in the dry season, they are beneficial and increase tree cover; late in the season, they are very hot and kill trees and seeds. Savanna trees are adapted to resist the "cooler" fires.

Forests and elephant grasses averaging 5 m (16 ft) high once penetrated much farther into the dry regions, for they are known to survive there when protected; unprotected, coverage declines. Savanna grasslands are much richer in humus than the wetter tropics and are better drained, thereby providing a better base for agriculture and grazing. Sorghums, wheat, and groundnuts (peanuts) are common commodities.

Tropical savannas receive their precipitation during less than 6 months of the year, when they are influenced by the shifting ITCZ. The rest of the year they are under the drier influence of shifting subtropical high-pressure cells. Savanna shrubs and trees are frequently *xerophytic*, or drought resistant, with various adaptations to protect them from the dryness: small, thick leaves; rough bark; or leaf surfaces that are waxy or hairy.

Africa has the largest area of this biome, including the famous Serengeti Plains of Tanzania and Kenya and the Sahel region, south of the Sahara. Sections of Australia, India, and South America also are part of the savanna biome. Some of the local names for these lands include the *Llanos* in Venezuela, stretching along the coast and inland east of Lake Maricaibo and the Andes; the *Campo Cerrado* of Brazil and Guiana; and the *Pantanal* of southwestern Brazil.

Particularly in Africa, savannas are the home of large land mammals—zebra, giraffe, buffalo, gazelle, wildebeest, antelope, rhinoceros, and elephant. These animals graze on savanna grasses (Figure 20.10), while others (lion, cheetah) feed upon the grazers themselves. Birds include the ostrich, Martial Eagle (largest of all eagles), and Secretary Bird. Many species of venomous snakes as well as the crocodile are present in this biome.

Midlatitude Broadleaf and Mixed Forest

Moist continental climates support a mixed forest in areas of warm to hot summers and cool to cold winters. This **midlatitude broadleaf and mixed forest** biome includes several distinct communities in North America, Europe, and Asia. Along the Gulf of Mexico, relatively lush evergreen broadleaf forests occur. Northward are the mixed deciduous and evergreen needleleaf stands associated with sandy soils and burning. When areas are given fire protection, broadleaf trees quickly take over. Pines (longleaf, shortleaf, pitch, loblolly) predominate in the southeastern and Atlantic coastal plains. Into New England and westward in a narrow belt to the Great Lakes, white and red pines and eastern hemlock are the principal evergreens, mixed with deciduous oak, beech, hickory, maple, elm, chestnut, and many others (Figure 20.11).

These mixed stands contain valuable timber, and human activity has altered their distribution. Native stands of white pine in Michigan and Minnesota were removed before 1910; only later reforestation sustains their presence

GEO REPORT 20.2 *The question of food and medicine*

Wheat, maize (corn), and rice—just three grains—fulfill about 50% of human planetary food demands. About 7000 plant species have been gathered for food throughout human history, but more than 30,000 plant species have edible parts. Undiscovered potential food resources are in nature waiting to be found and developed. Biodiversity, if preserved in each of the biomes, provides us a potential cushion for all future food needs, but only if species are identified, inventoried, and protected. The same is true for pharmaceuticals.

Nature's biodiversity is like a full medicine cabinet. Since 1959, 25% of all prescription drugs were originally derived from higher plants. Scientists have identified 3000 plants as having anticancer properties. The rosy periwinkle (*Catharanthus roseus*) of Madagascar contains two alkaloids that combat two forms of cancer. Yet less than 3% of flowering plants have been examined for alkaloid content. It defies common sense to throw away the medicine cabinet before we even open the door to see what is inside.

(a)

(b)

FIGURE 20.10 **Animals and plants of the Serengeti Plains.**
(a) Savanna landscape of the Serengeti Plains, with wildebeest, zebras, and thorn forest. (b) Lions use the sparse shade of an acacia tree. [(a) Stephen F. Cunha. (b) Michel & Christine Denis-Huot/Bios/photolibrary.com.]

(a)

(b)

FIGURE 20.11 **The mixed broadleaf forest.**
(a) A mixed broadleaf forest in Iowa. (b) The mixed broadleaf forest near Walden Pond, Massachusetts. [Bobbé Christopherson.]

today. In northern China, these forests have almost disappeared as a result of centuries of harvest. The forest species that once flourished in China are similar to species in eastern North America: oak, ash, walnut, elm, maple, and birch.

This biome is quite consistent in appearance from continent to continent and at one time represented the principal vegetation of the *humid subtropical* (hot summer) and *marine west coast* climatic regions of North America, Europe, and Asia.

A wide assortment of mammals, birds, reptiles, and amphibians is distributed throughout this biome. Representative animals (some migratory) include red fox, white-tail deer, southern flying squirrel, opossum, bear, and a great variety of birds, including Tanager and Cardinal. To the west of this biome in North America are the grasslands, and to the north are the poorer soils and colder climates that favor stands of coniferous trees and a gradual transition to the northern needleleaf forests.

Needleleaf Forest and Montane Forest

Stretching from the east coast of Canada and the Atlantic provinces westward to Alaska and continuing from Siberia across the entire extent of Russia to the European Plain is the northern **needleleaf forest** biome, or **boreal forest** (Figure 20.12). A more open form of boreal forest, transitional to arctic and subarctic regions, is the **taiga**. The Southern Hemisphere, lacking *microthermal humid* climates, has no such biome except in a few mountainous locales. But **montane forests** of needleleaf trees exist worldwide at high elevations.

Boreal forests of pine, spruce, fir, and larch occupy most of the subarctic climates on Earth that are dominated by trees. Although these forests are similar

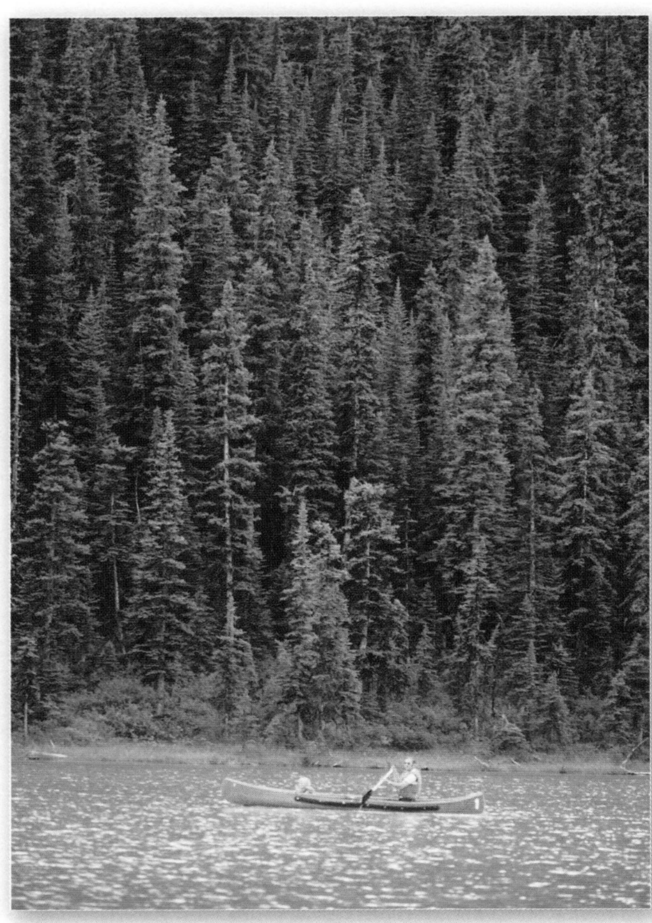

FIGURE 20.12 Boreal forest of Canada (*boreal* means "northern"). [Author.]

in formation, individual species vary between North America and Eurasia. The larch (*Larix*) is interesting because it is the rare needleleaf tree that loses its needles in the winter months, perhaps as a defense against the extreme cold of its native Siberia (see the Verkhoyansk climograph and photograph in Figure 10.21, Chapter 10). Larches also occur in North America.

The Sierra Nevada, Rocky Mountains, Alps, and Himalayas have similar forest communities occurring at lower latitudes. Douglas fir and white fir grow in the western mountains in the United States and Canada. Economically, these forests are important for lumbering, with trees for lumber occurring in the southern margins of the biome and for pulpwood throughout the middle and northern portions. Present logging practices and the sustainability of these yields are issues of increasing controversy.

In the Sierra Nevada montane forests of California, the giant sequoias naturally occur in 70 isolated groves. These trees are Earth's largest living things (in terms of biomass), although they began as small seeds (Figure 20.13a). Some of these *Sequoia gigantea* exceed 8 m (28 ft) in diameter and 83 m (270 ft) in height. The largest of these is the General Sherman tree in Sequoia National Park (Figure 20.13c), estimated to be 3500 years old. The bark is fibrous and a half-meter thick and lacks resins, so it effectively resists fire. Imagine the lightning strikes and fires that must have moved past the Sherman tree in 35 centuries! Standing among these giant trees is an overwhelming experience and creates a sense of the majesty of the biosphere.

Certain regions of the northern needleleaf biome experience permafrost conditions, discussed in Chapter 17. When coupled with rocky and poorly developed

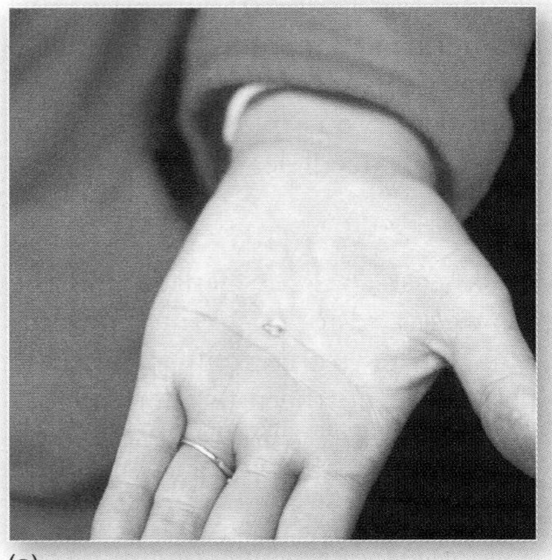

(a)

(b)

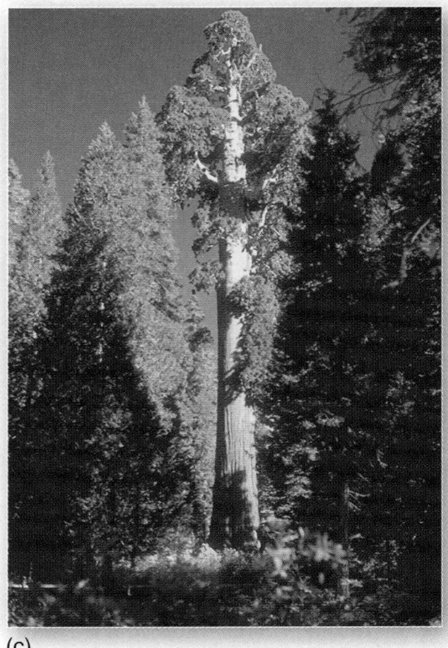

(c)

FIGURE 20.13 Sequoia: The seed, the seedling, the giant. (a) A *Sequoia gigantea* seed. About 300 seeds are in each sequoia cone. (b) Seedling at approximately 50 years of age. (c) The General Sherman tree in Sequoia National Park, California. This tree is probably wider than a standard classroom. The first branch is 45 m (150 ft) off the ground and is 2 m (6.5 ft) in diameter! [Author.]

FIGURE 20.14 **Animal in the northern needleleaf forest.** A male elk (*Cervus canadensis*) in a snowstorm looking for any available vegetation to eat during winter among needleleaf trees. [Steven K. Huhtala.]

soils, these conditions generally limit the existence of trees to those with shallow rooting systems. The summer thaw of surface-active layers results in muskeg (moss-covered) bogs and Histosols of poor drainage and stability. Soils of the taiga are typically Spodosols (subject to podzolization), characteristically acidic and leached of humus and clays. Global warming is altering conditions in the high latitudes, causing increased melting and depth in the permafrost active layer. Some affected forests are dying in response to waterlogging in soils.

Representative fauna in this biome include wolf, elk, moose (the largest in the deer family), bear, lynx, beaver, wolverine, marten, small rodents, and migratory birds during the brief summer season (Figure 20.14). Birds include hawks and eagles, several species of grouse, Pine Grosbeak, Clark's Nutcracker, and several species of owls. About 50 species of insects particularly adapted to the presence of coniferous trees inhabit the biome.

Temperate Rain Forest

Lush forests at middle and high latitudes are in the **temperate rain forest** biome. In North America, it occurs only along narrow margins of the Pacific Northwest (Figure 20.15a). Some similar types of forest exist in southern China, small portions of southern Japan, New Zealand, and a few areas of southern Chile.

This biome contrasts with the diversity of the equatorial and tropical rain forest in that only a few species make up the bulk of the trees. The rain forest of the Olympic Peninsula in Washington State is a mixture of broadleaf and needleleaf trees, huge ferns, and thick undergrowth. Precipitation approaching 400 cm (160 in.) per year on

(a)

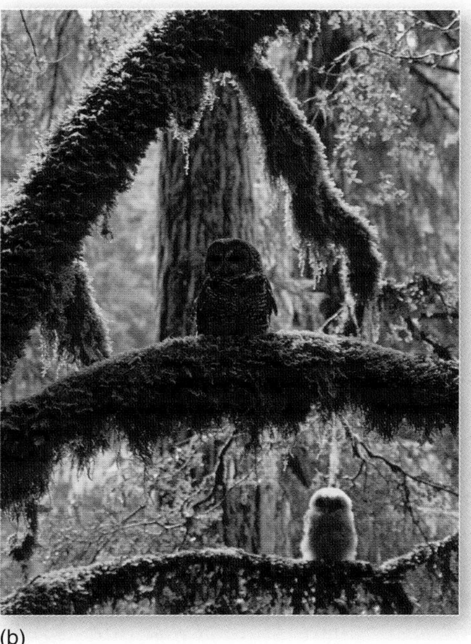

(b)

FIGURE 20.15 **Temperate rain forest.**
(a) An old temperate rain forest in the Pacific Northwest in the Gifford Pinchot National Forest features huge old-growth Douglas fir, redwoods, cedars, and a mix of deciduous trees, ferns, and mosses. Only a small percentage of these old-growth forests remains. (b) The Northern Spotted Owl and fledgling is an indicator species of the temperate rain forest. [(a) Bobbé Christopherson. (b) Woodfall Wild Images/Photoshot.]

the western slopes, moderate air temperatures, summer fog, and an overall maritime influence produce this moist, lush vegetation community. Animals include bear, badger, deer, wild pig, wolf, bobcat, and fox. The trees are home to numerous bird species, including the famous Spotted Owl (Figure 20.15b).

The tallest trees in the world occur in this biome—the coastal redwoods (*Sequoia sempervirens*). Their distribution is shown on the map in Figure 19.13a in Chapter 19. These trees can exceed 1500 years in age and typically range in height from 60 to 90 m (200 to 300 ft), with some exceeding 100 m (330 ft). Virgin stands of other representative trees, such as Douglas fir, spruce, cedar, and hemlock, have been reduced by timber harvests to a few remaining valleys in Oregon and Washington, less than 10% of the original forest. Replanting and secondary-growth forests are predominant. In similar forests in Chile, large-scale timber harvests and new mills began operations in 2000. Currently, U.S. corporations are shifting logging operations to these forests in the Chilean Lake District and northern Patagonia.

Many studies by the U.S. Forest Service and others noted the failing ecology of these forest ecosystems and suggested that timber management plans include ecosystem preservation as a priority. The ultimate solution must be one of economic and ecological synthesis rather than continuing conflict and forest losses—the forests can't all be cut down, nor can they all be preserved. The opposing sides in this conflict should combine their efforts in a sustainable forestry model.

Mediterranean Shrubland

The **Mediterranean shrubland** biome, also referred to as a temperate shrubland, occupies those regions poleward of the shifting subtropical high-pressure cells. As those cells shift poleward with the high summer Sun, they cut off available storm systems and moisture. Their stable high-pressure presence produces the characteristic *Mediterranean* (dry summer) climate and establishes conditions conducive to fire (Figure 19.24d in Chapter 19).

This biome is well adapted to frequent wildfires, for many of its characteristically deep-rooted plants have the ability to resprout from their roots after a fire. The dominant shrub formations that occupy these regions are stunted and able to withstand hot-summer drought. The vegetation is *sclerophyllous* (from *sclero*, for "hard," and *phyllos*, for "leaf"). It averages a meter or two in height and has deep, well-developed roots, leathery leaves, and uneven low branches.

FIGURE 20.16 Mediterranean chaparral. Chaparral vegetation associated with the *Mediterranean* (dry summer) climate in southern California. [Bobbé Christopherson.]

Typically, the vegetation varies between woody shrubs covering more than 50% of the ground and grassy woodlands covering 25%–60% of the ground. In California, the Spanish word *chaparro* for "scrubby evergreen" gives us the word **chaparral** for this vegetation type (Figure 20.16). This scrubland includes species such as manzanita, toyon, red bud, ceanothus, mountain mahogany, blue and live oaks, and the dreaded poison oak.

A counterpart to the California chaparral in the Mediterranean region is the *maquis*, which includes live and cork oak trees (source of cork) as well as pine and olive trees. The overall similar appearance of the California and Spanish oak savannas is demonstrated in Figure 10.16c and d in Chapter 10. In Chile, such a region is the *mattoral*, and in southwestern Australia, it is *mallee scrub*. In Australia, the bulk of the eucalyptus species are sclerophyllous in form and structure in whichever area it occurs.

As described in Chapter 10, *Mediterranean* climates are important in commercial agriculture for subtropical fruits, vegetables, and nuts, with many food types (e.g., artichokes, olives, almonds) produced only in this biome. Larger animals, such as several types of deer, are grazers and browsers, with coyote, wolf, and bobcat as predators. Many rodents, other small animals, and a variety of birds also proliferate.

Midlatitude Grasslands

Of all the natural biomes, the **midlatitude grasslands** are the most modified by human activity. Here are the world's "breadbaskets"—regions that produce bountiful grain (wheat and corn), soybeans, and livestock (hogs and cattle). Figure 20.17 shows a midlatitude grassland under cultivation for feedstock and crops. In these regions, the

(a)

(b)

FIGURE 20.17 Farming in the grasslands of North America.
(a) Rolled hay bales in a harvested field in Texas. (b) Iowa Mollisols, farmed in alternating fallow and planted fields. [Bobbé Christopherson.]

only naturally occurring trees were deciduous broadleaf trees along streams and other limited sites. These regions are called grasslands because of the predominance of grasslike plants before human intervention:

> In this study of vegetation, attention has been devoted to grass because grass is the dominant feature of the Plains and is at the same time an index to their history. Grass is the visible feature which distinguishes the Plains from the desert. Grass grows, has its natural habitat, in the transition area between timber and desert. . . . The history of the Plains is the history of the grasslands.*

In North America, tallgrass prairies once rose to heights of 2 m (6.5 ft) and extended westward to about the 98th meridian, with shortgrass prairies in the drier lands farther west. The 98th meridian is roughly the location of the 51-cm (20-in.) isohyet, with wetter conditions to the east and drier conditions to the west (see Figure 18.15 in Chapter 18).

The deep, tough sod of these grasslands posed problems for the first European settlers, as did the climate. The self-scouring steel plow, introduced in 1837 by John Deere, allowed the interlaced grass sod to be broken apart, freeing the soils for agriculture. Other inventions were critical to opening this region and solving its unique spatial problems: barbed wire (the fencing material for a treeless prairie); well-drilling techniques developed by Pennsylvania oil drillers, but used for water wells; windmills for pumping; and railroads to conquer the distances.

Few patches of the original prairies (tall grasslands) or steppes (short grasslands) remain within this biome. For prairies alone, the natural vegetation was reduced from 100 million hectares (250 million acres) down to a few areas of several hundred hectares each. The map in Figure 20.5 shows the natural location of these former prairie and steppe grasslands.

Outside North America, the Pampas of Argentina and Uruguay and the grasslands of Ukraine are characteristic midlatitude grassland biomes. In each region of the world where these grasslands have occurred, human development of them was critical to territorial expansion.

This biome is the home of large grazing animals, including deer, pronghorn, and bison (the almost complete annihilation of the latter is part of American history, Figure 20.18). Grasshoppers and other insects feed on the grasses and crops as well, and gophers, prairie dogs, ground squirrels, Turkey Vultures, grouse, and Prairie Chickens are found on the land. Predators include the coyote, nearly extinct black-footed ferret, badger, and birds of prey—hawks, eagles, and owls.

Deserts

Earth's **desert biomes** cover more than one-third of its land area, as is obvious in Figure 20.5. In Chapter 15, we examined desert landscapes, and in Chapter 10, desert climates. We divide the desert lands into **warm desert and semidesert**, caused by the presence of dry air and low precipitation from subtropical high-pressure cells; and **cold desert and semidesert,** which tend toward higher latitudes, where subtropical high pressure is of influence less than 6 months of the year.

*From Walter Prescott Webb, *The Great Plains*, Needham Heights, MA: Ginn and Company, 1931), p. 32.

FIGURE 20.18 Bison graze on the grasslands. The Big Basin Prairie Preserve Wildlife Area, western Kansas. Inset shows a male and female, shedding winter coats. [Bobbé Christopherson.]

On a planet with such a rich biosphere, the deserts stand out as unique regions of fascinating adaptations for survival. Much as a group of humans in the desert might behave with short supplies, plant communities also compete for water and site advantage. Some desert plants, the *ephemerals*, wait years for a rainfall event, at which time their seeds quickly germinate and the plants develop, flower, and produce new seeds, which then rest again until the next rainfall event. The seeds of some xerophytic species open only when fractured by the tumbling, churning action of flash floods cascading down a desert arroyo, and, of course, such an event produces the moisture that a germinating seed needs.

Perennial desert plants employ other features that have allowed them to adapt to the lack of moisture. Unique xerophytic features include

- Long, deep taproots (example, the mesquite);
- Succulence (thick, fleshy, water-holding tissue, such as that of cacti);
- Spreading root systems to maximize water availability, waxy coatings and fine hairs on leaves to retard water loss, and leafless conditions during dry periods (example, palo verde and ocotillo);
- Reflective surfaces to reduce leaf temperatures; and
- Tissue that tastes bad to discourage herbivores.

The creosote bush *(Larrea tridentata)* sends out a wide pattern of roots and contaminates the surrounding soil with toxins that prevent the germination of other creosote seeds, possible competitors for water. When a creosote bush dies, surrounding plants or germinating seeds will occupy the abandoned site, but they must wait for infrequent rains to remove the toxins, according to one hypothesis.

The faunas of both warm and cold deserts are limited by the extreme conditions and include few resident large animals. Exceptions are the camel, which can lose up to 30% of its body weight in water without suffering (for humans, a 10%–12% loss is dangerous), and the desert bighorn sheep, in scattered populations in inaccessible mountains and places such as the Grand Canyon along rocky outcrops and cliffs, but not at the rim. In an effort to reestablish bighorn populations, several states are transplanting the animals to their former ranges. Some representative desert animals are the ring-tail cat, kangaroo rat, lizards, scorpions, and snakes. Most of these animals are quite secretive and become active only at night, when temperatures are lower. In addition, various birds have adapted to desert conditions and available food sources—for example, roadrunners, thrashers, ravens, wrens, hawks, grouse, and nighthawks.

We are witnessing an unwanted expansion of the desert biome, discussed in the Chapter 15 *Geosystems Now*. This process, known as *desertification*, is now a worldwide phenomenon along the margins of semiarid and arid lands.

Earth's deserts are subdivided into desert and semidesert associations, to distinguish those with expanses of bare ground from those covered by xerophytic plants of various types. The two broad associations are further separated into warm deserts, principally tropical and subtropical, and cold deserts, principally midlatitude.

Warm desert vegetation ranges from almost none to numerous xerophytic shrubs, succulents, and thorn tree forms. The lower Sonoran Desert of southern Arizona is an example (Figure 20.19). This desert landscape features the unique saguaro cactus (*Carnegiea gigantea*), which grows to many meters in height and up to 200 years in age if undisturbed. First blooms do not appear until it is 50 to 75 years old.

A few of the subtropical deserts—such as those in Chile, Western Sahara, and Namibia—are right on the seacoast and are influenced by cool offshore ocean currents. As a result, these true deserts experience summer fog that mists the plant and animal populations with needed moisture.

Arctic and Alpine Tundra

The **arctic tundra** is found in the extreme northern area of North America and Russia, bordering on the Arctic Ocean and generally north of the 10°C (50°F) isotherm for the warmest month. Daylength varies greatly throughout the year, seasonally changing from almost continuous day to continuous night. The region, except for a few portions of Alaska and Siberia, was covered by ice during all of the Pleistocene glaciations. In terms of climate change, these regions have been warming at more than twice the rate of the rest of the planet over the past few decades; see

FIGURE 20.19 Sonoran Desert scene.
Characteristic vegetation in the Lower Sonoran Desert west of Tucson, Arizona (32° N; elevation 900 m, or 2950 ft). [Author.]

the related discussions and photos in Figures 10.22 and 17.20–17.24. The impact of such heating on plants and animals, permafrost, First Nation peoples and their towns, glacial and sea ice, and decreasing ocean salinity are topics of active scientific inquiry and the International Polar Year for science and research that ended March 2009.

Winters in this biome, which has a *tundra* climate, are long and cold; cool summers are brief. Intensely cold continental polar and arctic air masses and stable high-pressure anticyclones dominate winter. A growing season of sorts lasts only 60–80 days, and even then frosts can occur at any time. Vegetation is fragile in this flat, treeless world; soils are poorly developed periglacial surfaces that are underlain by permafrost. In the summer months, only the surface horizons thaw, thus producing a mucky surface of poor drainage. Roots can penetrate only to the depth of thawed ground, usually about a meter (3 ft). In these periglacial regions, the surface is shaped by freeze–thaw cycles that create the frozen ground phenomena discussed in Chapter 17.

Tundra vegetation is low, ground-level herbaceous plants such as sedges, mosses, arctic meadow grass, and snow lichen and some woody species such as dwarf willow (Figure 20.20). Owing to the short growing season, some perennials form flower buds one summer and open them for pollination the next. The tundra supports the musk ox, caribou, reindeer, rabbit, ptarmigan, lemming, and other small rodents, which are important food for the larger carnivores—the wolf, fox, weasel, Snowy Owl, polar bear, and, of course, mosquito. The tundra is an important breeding ground for geese, swans, and other waterfowl.

(a)

(b)

FIGURE 20.20 Arctic tundra.
(a) Tundra mosses with a roche moutonnée rock formation in the background; its shape denotes glacial movement from left to right. (b) Lush by tundra standards, grasses, mosses, and dwarf willow flourish in the cold climate of the Arctic. [Bobbé Christopherson.]

FIGURE 20.21 Alpine tundra conditions. An alpine tundra and grazing mountain goats near Mount Evans, Colorado, at an elevation of 3660 m (12,000 ft). [Bobbé Christopherson.]

FIGURE 20.22 Arctic National Wildlife Refuge. Caribou herd during summer migration in Alaska's Arctic National Wildlife Refuge. Is this region destined for petroleum exploration and development or for continued preservation as wilderness? [Nature Picture Library.]

Alpine tundra is similar to arctic tundra, but it can occur at lower latitudes because it is associated with high elevations. This biome usually is described as above the timberline, that elevation above which trees cannot grow. Timberlines increase in elevation equatorward in both hemispheres. Alpine tundra communities occur in the Andes near the equator, the White Mountains and Sierra of California, the American and Canadian Rockies, the Alps, and Mount Kilimanjaro of equatorial Africa, as well as in mountains from the Middle East to Asia.

Alpine meadows feature grasses, herbaceous annuals (small plants), and stunted shrubs, such as willows and heaths. Because alpine locations are frequently windy sites, many plants appear to have been sculpted by the wind. Alpine tundra can experience permafrost conditions. Characteristic fauna include mountain goats, bighorn sheep, elk, and voles (Figure 20.21).

Because the tundra biome is of such low productivity, it is fragile. Disturbances such as tire tracks, hydroelectric projects, and mineral exploitation leave marks that persist for hundreds of years. As development continues, the region will face even greater challenges from oil spills, contamination, and disruption.

Planning continues for oil exploration and development of the Arctic National Wildlife Refuge (ANWR) on Alaska's North Slope. A U.S. Geological Survey assessment disclosed that the oil under the Arctic refuge coastal plain is likely held in many small reservoirs. Recovering this oil would require alteration of the ANWR landscape. Political and corporate pressure is continuous to exploit these expected reserves of fossil fuels.

This pristine wilderness is above the Arctic Circle, bordering on the Beaufort Sea and adjoining the Yukon Territory of Canada (Figure 20.22). The refuge area sustains almost 200,000 caribou, polar and grizzly bears, musk oxen, and wolves. Some have referred to it as "America's Serengeti," given the annual migration of hundreds of thousands of large animals. The 1002 Area is 1.5 million coastal acres in the ANWR with potential oil and gas resources, yet critical animal habitat for caribou and polar bears (http://energy.usgs.gov/alaska/anwr.html). The debate is ongoing; see http://justice.uaa.alaska.edu/rlinks/environment/ak_anwr.html.

GEO REPORT 20.3 ***The Porcupine caribou herd***

The Porcupine caribou herd is a population of tundra caribou that inhabit Alaska, Yukon, and Northwest territories. Most of the herd migrates to the Alaska coastal plain to calve in early June. The calving area is fairly small, and 80%–85% of the herd use it year after year. Oil exploration is not permitted in these areas because petroleum development can be devastating to arctic wildlife. However, intense political pressure to relax any protection is ever present, and an aggressive campaign to overturn bans on exploration and drilling is active.

CRITICAL THINKING 20.3

Shifting climate hypothetical

Now that you have come to the end of this chapter, consider doing an exercise. Using Figure 20.5 (biomes), Figure 10.6 (climates), Figures 10.3 and 9.5 (precipitation), and Figure 8.2 (air masses), and the printed graphic scales on these maps, consider the following hypothetical. Assume a northward climatic shift in the United States and Canada of 500 km (310 mi); in other words, imagine moving North America 500 km south to simulate climatic categories shifting north. Describe your analysis of conditions through the Midwest from Texas to the prairies of Canada. Describe your analysis of conditions from New York through New England and into the Maritime Provinces. How will biomes change? What economic dislocations and relocations do you envision? For another region of the world, think about the climatic results if Australia shifted 500 km south—consider the new pattern of climate categories and ecosystems.

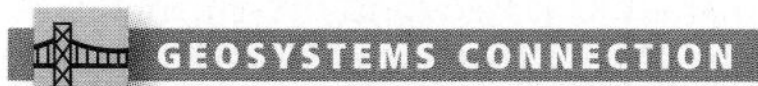

Earth's 10 biomes bring together all the systems into these specific regions. We see the patterns in the biosphere and the diversity of life. Real issues abound regarding preservation of this diversity and the influence of climate change on communities, ecosystems, and biomes. Physical geography and Earth systems science are in an important position to study these changes. The next chapter is a short, philosophical conclusion of our journey together through the pages of this book. Chapter 21 discusses the path ahead and asks key related questions.

KEY LEARNING CONCEPTS REVIEW

- ***Describe*** the concept of biogeographic realms of plants and animals, and ***define*** ecotone and terrestrial ecosystems.

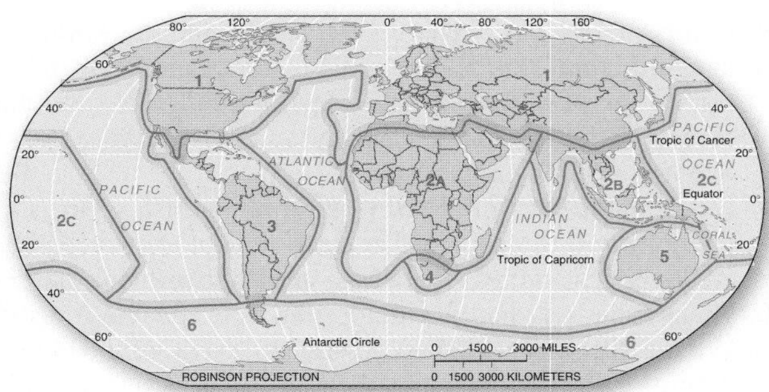

The interplay among supporting physical factors within Earth's ecosystem determines the distribution of plant and animal communities and Earth's biodiversity. A **biogeographic realm** of plants and animals is a geographic region in which a group of species evolved. This recognition was a start at understanding distinct regions of flora and fauna and the broader pattern of terrestrial ecosystems. A boundary transition zone adjoining ecosystems is an **ecotone**. A **terrestrial ecosystem** is a self-sustaining association of plants and animals and their abiotic environment. Specific life-form designations include trees, lianas, shrubs, herbs, bryophytes, epiphytes (plants growing aboveground on other plants), and thallophytes (lacking true leaves, stems, or roots, including bacteria, fungi, algae, and lichens).

biogeographic realm (p. 592)
ecotone (p. 593)
terrestrial ecosystem (p. 593)

1. Reread the opening quotation from 1874 in this chapter. What seems to be our learning curve regarding the path ahead for Earth's forests and ecosystems? Is our future direction controllable? Sustainable? Explain.
2. What is a biogeographic realm? How is the world subdivided according to botanical (plant) types? What are some plant life-form designations?
3. Describe a transition zone between two ecosystems. How wide is an ecotone? Explain.

- ***Define*** biome, and ***illustrate*** formation classes as a basis for describing plant communities within biomes.

A **biome** is a large, stable ecosystem characterized by specific plant and animal communities. Biomes carry the name of the dominant vegetation because it is the most easily identified feature: forest, savanna, grassland, shrubland, desert, tundra. Biomes are divided into more specific vegetation units, **formation classes**. The structure and appearance of the vegetation are described as rain forest, needleleaf forest, Mediterranean shrubland, arctic tundra, and so forth.

biome (p. 594)
formation classes (p. 594)

4. Define *biome*. What is the basis of the designation?

5. Distinguish between formation classes and life-form designations as a basis for spatial classification.

- ***Differentiate*** between native and exotic species, and ***detail*** the threat they represent to communities, using several examples.

Communities, ecosystems, and biomes can be affected by species that are introduced from elsewhere by humans, either accidentally or intentionally. These non-native species are known as *exotic species*, or *aliens*. After arriving in the new ecosystem, some species may then be disruptive and become **invasive species**.

invasive species (p. 594)

6. Enumerate several examples of invasive species described in the text, and describe their impact on natural systems.
7. What happened in the waters of Tristan da Cunha? Why was this subject first introduced in Chapter 6 of the text and then pursued in this chapter? What are the linkages between the chapters? What economic damage might evolve in Tristan's marine ecosystems?

- ***Describe*** 10 major terrestrial biomes, and ***locate*** them on a world map.

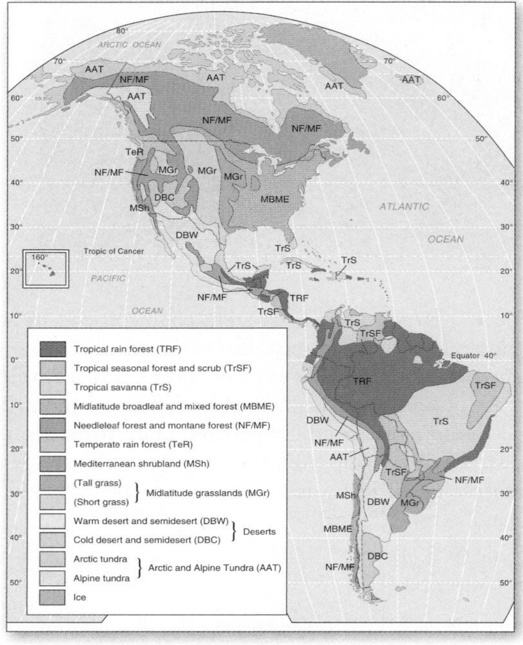

Biomes are Earth's major terrestrial ecosystems, each named for its dominant plant community. The 10 major biomes are generalized from numerous formation classes that describe vegetation. Ideally, a biome represents a mature community of natural vegetation. In reality, few undisturbed biomes exist in the world, for most have been modified by human activity. Many of Earth's plant and animal communities are experiencing an accelerated rate of change that could produce dramatic alterations within our lifetime.

For an overview of Earth's 10 major terrestrial biomes and their vegetation characteristics, soil orders, climate-type designations, annual precipitation ranges, temperature patterns, and water balance characteristics, review Table 20.1.

tropical rain forest (p. 595)
tropical seasonal forest and scrub (p. 601)
tropical savanna (p. 604)
midlatitude broadleaf and mixed forest (p. 604)
needleleaf forest (p. 605)
boreal forest (p. 605)
taiga (p. 605)
montane forest (p. 605)
temperate rain forest (p. 607)
Mediterranean shrubland (p. 608)
chaparral (p. 608)
midlatitude grasslands (p. 608)
desert biomes (p. 609)
warm desert and semidesert (p. 609)
cold desert and semidesert (p. 609)
arctic tundra (p. 610)
alpine tundra (p. 612)

8. Using the integrative chart in Table 20.1 and the world map in Figure 20.5, select any two biomes and study the correlation of vegetation characteristics, soil, moisture, and climate with their spatial distribution. Then contrast the two using each characteristic.
9. Describe the equatorial and tropical rain forests. Why is the rainforest floor somewhat clear of plant growth? Why are logging activities for specific species so difficult there?
10. What issues surround the deforestation of the rain forest? What is the impact of these losses on the rest of the biosphere? What new threat to the rain forest has emerged?
11. What do *caatinga*, *chaco*, *brigalow*, and *dornveld* refer to? Explain.
12. Describe the role of fire or fire ecology in the tropical savanna biome and the midlatitude broadleaf and mixed forest biome.
13. Why does the northern needleleaf forest biome not exist in the Southern Hemisphere, except in mountainous regions? Where is this biome located in the Northern Hemisphere, and what is its relationship to climate type?
14. In which biome do we find Earth's tallest trees? Which biome is dominated by small, stunted plants, lichens, and mosses?
15. What type of vegetation predominates in the *Mediterranean* (dry summer) climate? Describe the adaptations necessary for these plants to survive.
16. What is the significance of the 98th meridian in terms of North American grasslands? What types of inventions enabled humans to cope with the grasslands?
17. Describe some of the unique adaptations found in a desert biome.
18. What is desertification (review from Chapter 15 and this chapter)? Explain its impact.

19. What physical weathering processes are specifically related to the tundra biome? What types of plants and animals are found there?

- ***Relate*** human impacts, real and potential, to several of the biomes.

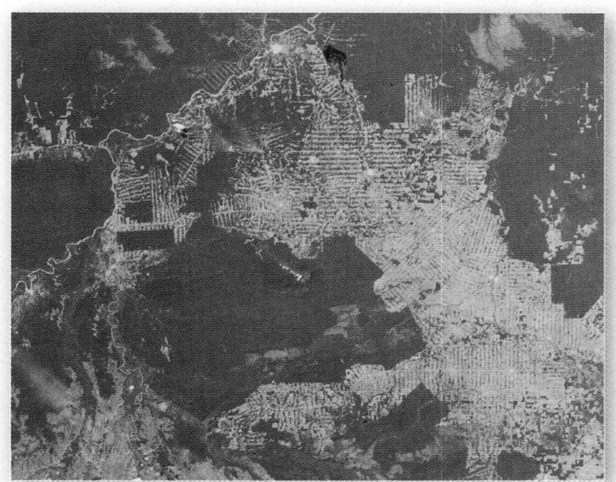

The tropical rainforest biome is undergoing rapid deforestation. Because the rain forest is Earth's most diverse biome and is important to the climate system, such losses are creating great concern among citizens, scientists, and nations. Efforts are under way worldwide to set aside and protect remaining representative sites within most of Earth's principal biomes. Principles of **island biogeography** in studying isolated ecosystems are important in setting up biosphere reserves. Island communities are special places for study because of their spatial isolation and the relatively small number of species present.

island biogeography (p. 603)

20. What is the relationship between island biogeography and biosphere reserves? Describe a biosphere reserve. What are the goals?
21. Compare the map in Figure 20.5 with the composite satellite image inside the front cover of this text. What correlations can you make between the local summertime portrait of Earth's biosphere and the biomes identified on the map?
22. What is your assessment of ANWR? Using credible Internet sources, what is the present status relative to ongoing, but halted, efforts to begin oil and gas exploration in Area 1002?
23. As an example of shifting climate impacts, we tracked temperature and precipitation conditions for Illinois in *Geosystems Now* in Chapter 19. What impacts do you think such climate change will have on these biomes in the United States and in other countries?

MASTERING GEOGRAPHY

Visit www.masteringgeography.com for several figure animations, satellite loops, self-study quizzes, flashcards, visual glossary, case studies, career links, textbook reference maps, RSS Feeds, and an ebook version of *Geosystems*. Included are many geography web links to interesting Internet sources that support this chapter.

CHAPTER 21

Earth and the Human Denominator

A new freeway slices through a Puerto Rican rain forest. This is a remarkable reminder of humans as Earth's major geomorphic agent, and for considering the human denominator. [Bobbé Christopherson.]

KEY LEARNING CONCEPTS

After reading the chapter, you should be able to:

- ***Determine*** an answer to Carl Sagan's question, "Who speaks for Earth?"
- ***Describe*** the growth in human population, and ***speculate*** on possible future trends.
- ***Analyze*** "An Oily Bird," and ***relate*** your analysis to energy consumption patterns in the United States and Canada.
- ***List*** and ***discuss*** 12 paradigms for the twenty-first century.
- ***Appraise*** your place in the biosphere, and ***discover*** your relation to Earth systems.

GEOSYSTEMS NOW

Seeing Earth and Ourselves from Space

Throughout history, the human perspective of Earth and our relation to Earth has evolved. There was a time when Earth systems were viewed much as the workings of a clock, which could be disassembled, controlled, and then put back together. Certainly, technology played its role to empower humans in modifying planetary systems.

Only in the last half century has an understanding emerged that Earth is a vast complex of interrelated systems, a synergistic whole that is greater than the sum of its component parts. There is serious thought that perhaps Earth behaves like an organism itself, self-maintaining, self-organizing, metabolizing—part of the Gaia Hypothesis.

In the 1950s and 1960s, during the young space age, humans began placing satellites and spacecraft in low orbit. With great strides in lift capability, a goal was set for humans to go to the Moon before 1970. Scientists and philosophers alike began to speculate what the impact would be when humans first saw views of Earth from a distance, from space. How would society react to such a moment? People speculated about the cultural shock of seeing the planet isolated in the void of space. We have no mental framework, no imprinting to handle seeing Earth from far away, since the human experience is bound by gravity to the planet's surface.

In 1968, *Apollo VIII* launched on a journey to orbit the Moon. On their voyage, the astronauts made photos of Earth (Figure GN 21.1). Civilization now had photos of Earth from space and three witnesses to explain what they saw and how they felt.

A few months later *Apollo IX* was in Earth orbit testing all the systems needed to land on the lunar surface. Astronaut Rusty Schweickart on a spacewalk had to stop work as NASA figured out a problem with some tools. In those few minutes of waiting, the drifting astronaut looked down as the entire Mediterranean swept by, then the Middle East. The impact of those moments in orbit was profound on him. In the world below, he could see "no frames and no boundaries."

FIGURE GN 21.1 First distant views of Home Planet.
From December 21 to 27, 1968, *Apollo VIII* was in flight between Earth and the Moon, successfully orbiting the Moon 10 times. The astronauts witnessed dramatic Earthrises beyond the lunar surface. [NASA.]

The astronaut gave many talks when he returned to Earth about what he experienced both as an engineer and as a human being. In 1987, he wrote:

> During my space flight, I came to appreciate my profound connection to the home planet and the process of life evolving in our special corner of the Universe, and I grasped that I was part of a vast and mysterious dance whose outcome will be determined largely by human values and actions.*

In 1977, two *Voyager* spacecraft were launched to tour the outer planets, and they continue to this day to send telemetry even though they are far beyond the Solar System in interstellar space. The cameras on board sent back images of this "Pale Blue Dot" we live on.

In 2006, the *Cassini-Huygens* spacecraft was studying Saturn and its many moons. Scientists directed the craft so that Saturn was backlit by the Sun, a view of the rings that we never see from Earth (Figure GN 21.2). The small dot to the left of Saturn (see arrow), just inside the bright ring, is us! All that we studied in the text and that we are trying to understand is on that small dot.

*R. Schweickart, *Discovery*, July 1987, p. 62.

FIGURE GN 21.2 View of Earth from Saturn.
The *Cassini–Huygens* spacecraft made a surprise catch—a glimpse of Earth as a little, pale dot, some 1.23 billion km (791 million mi) distant. It was September 19, 2006, on Earth. [NASA/JPL.]

In May 2010, NASA's *Messenger* satellite, in orbit around the planet Mercury, was 183 million km (114 million mi) from Earth. *Messenger*'s wide-angle camera caught a view of Earth and the Moon, bright with reflected sunlight (Figure GN 21.3).

We might ask, How have these views of Earth from space impacted society and culture? Have these views informed our relationship to Earth systems?

We have travelled together studying Earth this semester through the lens of physical geography and Earth systems science. Do you feel your perspective of the Home Planet changed during this journey through *Geosystems*? Perhaps find an Earth photo, print it, and display it in your room as a reminder.

FIGURE GN 21.3 The view of us from Mercury.
Messenger spacecraft in orbit about Mercury could see Earth and the Moon. [NASA/GSFC.]

We stand in the second decade of the twenty-first century. This century will be an adventure for the global society, historically unparalleled in experimenting with Earth's life-supporting systems. You will spend the majority of your life in this century. What preparations and "future thinking" can you do to understand all that is to occur?

Our vantage point in this book is that of physical geography. We examine Earth's many systems: its energy, atmosphere, winds, ocean currents, water, weather, climate, endogenic and exogenic systems, soils, ecosystems, and biomes. Earth can be observed and studied from profound vantage points, as described in *Geosystems Now*.

Since November 2000, the International Space Station (ISS) has operated at an orbital altitude between 350 and 400 km (217 and 250 mi), in the upper reaches of Earth's thermosphere. A 16-nation coalition has completed a scientific workstation, in orbit, that is 44.5 m long, 78 m wide, and 27.5 m tall (146 ft long, 256 ft wide, and 90 ft tall) (Figure 21.1). The 455-metric-ton (500.5 tons) complex has had more than 130 astronauts and scientists aboard in 392 m^3 (14,000 ft^3) of pressurized work space (about the volume in a 747 jumbo jet). As of October 11, 2010, 68,176 orbits were completed by the ISS.

Several hundred scientific investigations and experiments have been completed or are under way to help us better understand Earth and life systems through research in the unique space environment. All life-support supplies must be brought from Earth. Water recycling and purification techniques provide the bulk of the water needed. Oxygen is electrolytically derived from water and supplemented by compressed oxygen. Enormous solar panel arrays generate electricity. Without plants, carbon dioxide must be scrubbed from the air the astronauts breathe. See http://www.nasa.gov/mission_pages/station/main/index.html for more information and updates.

This exploration has led us to an examination of the planet's most abundant large animal, *Homo sapiens*. In his 1980 book and public television series *Cosmos*, the late astronomer Carl Sagan asked:

> What account would we give of our stewardship of the planet Earth? We have heard the rationales offered by the nuclear superpowers. We know who speaks for the nations. But who speaks for the human species? Who speaks for Earth?*

Indeed, who does speak for Earth? We might answer: Perhaps we physical geographers, and other scientists who have studied Earth and know the operations of the global ecosystem, should speak for Earth. However, some might say that questions of technology, environmental politics, and future thinking belong outside of science and that our job is merely to learn how Earth's processes work and to leave the spokesperson's role to others.

A fact of life is that Earth's more-developed countries (MDCs), through their economic dominance, speak for the billions who live in less-developed countries (LDCs). The scale of disparity on our Home Planet is difficult to comprehend. The fate of traditional modes of life may rest in some distant financial capital. But, economics aside, the reality is that the remote lands of Siberia are linked by Earth systems to the Pampas of Argentina, to the Great Plains in North America, and to the remote Pamirs of Tajikistan.

In this sense, Earth can be compared to a space station or spaceship. In the same way the members of a crew aboard the ISS are inextricably linked to each other's lives

*C. Sagan, *Cosmos* (New York: Random House, 1980), p. 329.

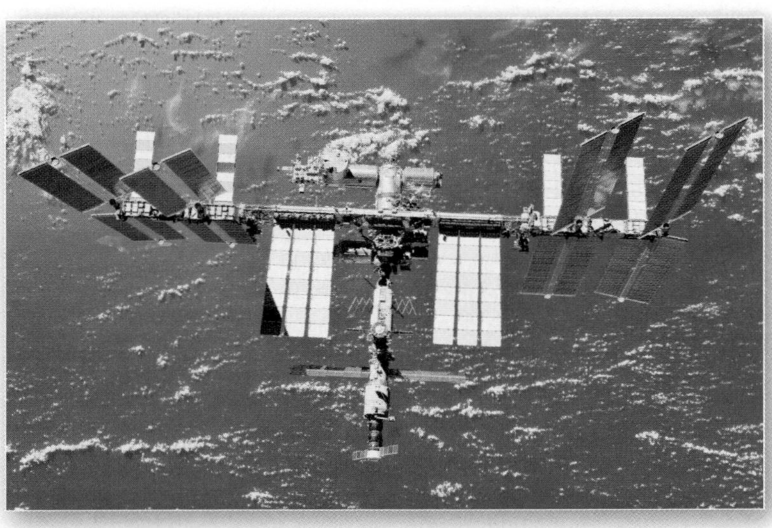

FIGURE 21.1 Our orbital perspective—Earth's Space Station.
The International Space Station (ISS) as it looked upon completion in spring 2010. The ISS began scientific operations in November 2000. The station orbits at a speed of 28,000 kmph (17,400 mph), orbiting Earth in 91.6 min, 15.7 times a day. [NASA.]

and survival, we, too, are each connected through the operation of planetary systems. Growing international environmental awareness in the public sector is gradually prodding governments into action. People, when informed, generally favor environmental protection and public health progress over economic interests.

Understanding these linkages among Earth's myriad systems is our quest in *Geosystems*. We now see how there are global linkages among physical and living systems and how actions in one place can affect change elsewhere in the web of life. You read 21 *Geosystems Now* case studies, which opened each chapter, stretching across the breadth of the discipline. Take a moment to review these essays of applied topics (Table 21.1).

TABLE 21.1 *Geosystems Now* in *Geosystems*

PART I—Geosystems Now:	
1	Where Is Four Corners, Exactly?
2	Chasing the Subsolar Point
3	Humans Help Define the Atmosphere
4	Albedo Impacts, a Limit on Future Arctic Shipping
5	Temperature Change Affects St. Kilda's Sheep
6	Ocean Currents Bring Invasive Species
PART II—Geosystems Now:	
7	Earth's Lakes Provide an Important Warming Signal
8	The Front Lines for Intense Weather
9	Water Budgets, Climate Change, and the Southwest
10	A Look at Puerto Rico's Climate at a Larger Scale
PART III—Geosystems Now:	
11	Earth's Migrating Magnetic Poles
12	The San Jacinto Fault Connection
13	Human-Caused Mass Movement at the Kingston Steam Plant, Tennessee
14	Removing Dams and Restoring Salmon on the Elwha River, Washington
15	Increasing Desertification and Political Action—A Global Environmental Issue
16	What Once Was Bayou Lafourche
17	Ice Shelves and Tidewater Glaciers Give Way to Warming
PART IV—Geosystems Now:	
18	High Latitude Soils Emit Greenhouse Gases
19	Species' Distributions Shift with Climate Change
20	Invasive Species Arrive at Tristan da Cunha
21	Seeing Earth and Ourselves from Space

The Human Count

Because human influence is pervasive, we consider the totality of our impact the *human denominator*. Just as the denominator in a fraction tells how many parts a whole is divided into, so the growing human population and the increasing demand for resources and its rising planetary impact suggest how much the whole Earth system must adjust. Yet Earth's resource base remains relatively fixed.

The human population of Earth passed 6 billion in August 1999 and continued to grow at the rate of 82.7 million per year, adding another 892 million by 2010. More people are alive today than at any previous point in the planet's long history, unevenly arranged in 192 countries and numerous colonies. *Virtually all new population growth is in the LDCs*, which now possess 81%, or about 5.58 billion, of the total population. Over the span of human history, billion-mark milestones are occurring at closer intervals (Figure 21.2). Humanity's seventh billion will be born sometime during 2012.

Living in just two countries is 36.6% of Earth's population (19.4% in China and 17.2% in India—2.53 billion people combined). Moreover, we are a young planetary population, with some 27% of those now alive under the age of 15 (2010 data from the Population Reference Bureau, at http://www.prb.org, and the U.S. Census Bureau's *POPClock Projection*, at http://www.census.gov/cgi-bin/popclock).

If you consider only the population count, the MDCs do not have a population growth problem. In fact, some European countries are actually declining in growth or are near replacement levels. However, people in these developed countries produce a greater impact on the planet per person and therefore constitute a population impact crisis. An equation expresses this impact concept:

$$I = P \cdot A \cdot T$$

where I is *planetary impact*; P is *population*; A is *affluence*, or consumption per person; and T is *technology*, or the level of environmental impact per unit of production. The United States and Canada, with about 5% of the world's population, produce more than 24.7% ($14.2 trillion and $1.3 trillion in 2009, respectively) of the world's gross domestic product and

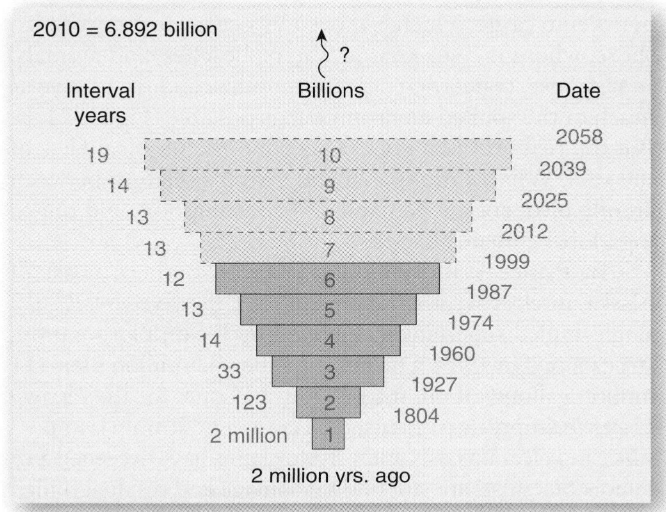

FIGURE 21.2 Human population growth.
The number of years required for the human population to add 1 billion to the count grows shorter and shorter. The twenty-first century should mark a slowing of this history of growth if policy actions are taken. Note the population forecasts for the next half-century—7,000,000,000 during 2012.

use more than double the energy per capita of Europeans, more than 7 times that of Latin Americans, 10 times that of Asians, and 20 times that of Africans. Therefore, the impact on the state of Earth systems, natural resources, and sustainability of current practices in the MDCs is critical.

The concept of an individual's "footprint" has emerged—ecological footprint, carbon footprint, lifestyle footprint. These consider what your affluence and technology, the *A* and the *T* in the equation, cost planetary systems. Footprint assessment is grossly simplified, but it can give you an idea of your impact and even an estimate of how many planets it would take to sustain that lifestyle and economy if everyone lived like you. Your author has taken action to neutralize the carbon footprint for the production and printing of these textbooks, and the publisher used sustainable-forestry green paper for *Geosystems* to lessen ecological impacts. Your author and his wife monthly calculate footprint and minimize impact wherever possible. Many Internet links will help to calculate your footprint; here is a sample:

http://www.carbonfootprint.com/calculator.html

http://www.ucsusa.org/publications/greentips/whats-your-carb.html

http://www.nature.org/initiatives/climatechange/calculator/

http://www.epa.gov/climatechange/emissions/ind_calculator.html

http://coolclimate.berkeley.edu/

An Oily Bird

At first glance, the chain of events that exposes wildlife to oil contamination seems to stem from a technological problem. An oil tanker splits open at sea and releases its petroleum cargo, which is moved by ocean currents toward shore, where it coats coastal waters, beaches, and animals. In response, concerned citizens mobilize and try to save as much of the spoiled environment as possible (Figure 21.3). But the real problem goes far beyond the physical facts of the spill. What is the spatial and systems relation between an oily bird, energy demand and consumption, and ongoing global climate change?

In Prince William Sound off the southern coast of Alaska, in clear weather and calm seas, the *Exxon–Valdez*, a single-hulled supertanker operated by Exxon Corporation, struck a reef in 1989. The tanker spilled 42 million liters (11 million gallons) of oil. It took only 12 hours for the *Exxon–Valdez* to empty its contents, yet a complete cleanup is impossible, and cleanup costs and private claims have exceeded $15 billion. Scientists are still finding damage and residual spilled oil. Eventually, more than 2400 km (about 1500 mi) of sensitive coastline were ruined, affecting three national parks and eight other protected areas (Figure 21.4).

The death toll of animals was massive: At least 5000 sea otters died, or about 30% of resident otters; about 300,000 birds and uncounted fish, shellfish, plants, and

FIGURE 21.3 An oily bird. A Western Grebe contaminated with oil from the *Exxon–Valdez* tanker accident in Prince William Sound, Alaska. This oily bird is the result of a long chain of events and mistakes. [Geoffrey Orth/SIPA Press.]

aquatic microorganisms also perished. Sublethal effects, namely mutations, now appear in fish. The Pacific herring is still in significant decline, as are the harbor seals; other species, such as the Bald Eagle and Common Murre, are in recovery. More than two decades later, oil remains in mudflats and marsh soils beneath rocks.

The immediate problem of cleaning oil off a bird symbolizes a national and international concern with far-reaching spatial significance. And while we search for answers, oil slicks continue their contamination. On average, 27 accidents occur every day, 10,000 a year worldwide, ranging from a few disastrous spills to numerous small ones. There have been 50 spills equal to the *Exxon–Valdez* or larger since 1970. In addition to oceanic oil spills, people improperly dispose of crankcase oil from their automobiles in a volume that annually exceeds all of these tanker spills.

The largest oil spill in U.S. history occurred in 2010. Somewhere between 50,000 and 95,000 barrels of oil a day, for 86 days, exploded from a wellhead on the seafloor; this is 2.1 to 4.0 million gallons a day, or the equivalent of a 1989 *Exxon–Valdez* oil spill every 4 days. The discrepancies in the spill rate relate to the behavior of the responsible oil company and a concerted effort to cover up the true nature of the spill because of potential financial liabilities. The Deepwater Horizon well, at a depth of 1.6 km (1 mi), was one of the deepest drilling attempts ever tried, and much of the technology remains untested or unknown. See the shaded area on the map in Figure 21.4 for a comparison of the Gulf spill to the Alaskan spill. Scientists are analyzing many aspects of the tragedy to determine the extent of the biological effects on the Gulf, wetlands, and beaches; there are many unknowns as to what they will find.

The immediate effect of global oil spills on wildlife is contamination and death. But the issues involved are much bigger than dead birds. Within the systems approach of

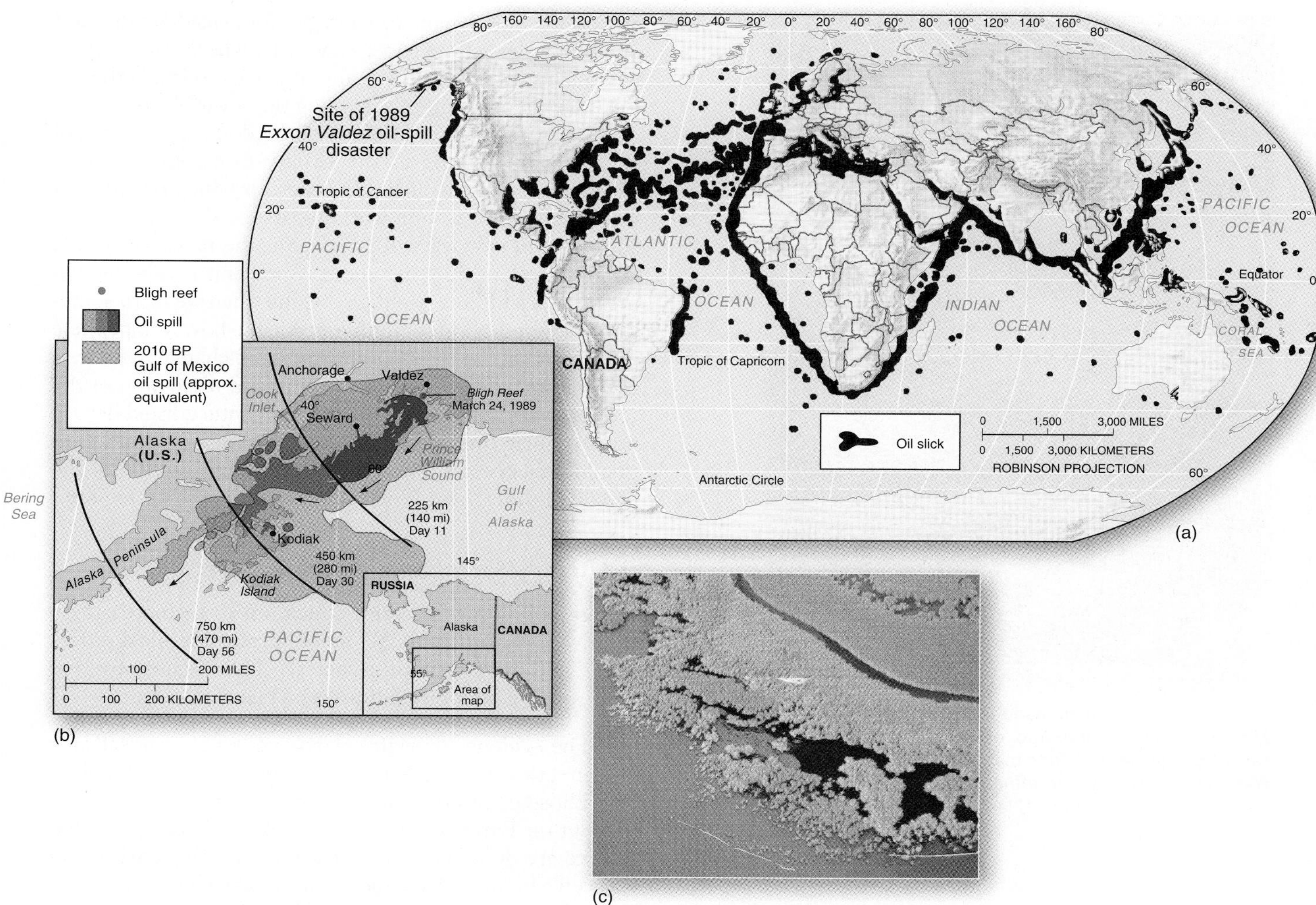

FIGURE 21.4 Worldwide oil spills, the 1989 *Exxon–Valdez* accident, and the 2010 disaster. (a) Location of visible oil slicks worldwide in the 1990s. (b) Track of spreading oil for the first 56 days of the Alaskan spill. The shaded area gives you a spatial equivalent of the 2010 BP Gulf oil spill compared to the Exxon spill. (c) Louisiana coastal marsh coated with oil. [(a) Data from Organization for Economic Cooperation and Development. (b) and (c) NOAA.]

this text, let us ask some fundamental questions about the *Exxon–Valdez* and Deepwater Horizon accidents:

- Why was the oil tanker there in the first place? Why is it necessary to drill in ultradeep water with limited technology?
- Why is petroleum imported into the continental United States in such enormous quantities? Is the demand for petroleum products based on real need and the operation of efficient systems? Do waste and inefficiency occur?
- The U.S. demand for oil is higher per capita than the demand of any other country. Efficiencies and mileage ratings in the U.S. transportation sector have improved little for cars, trucks, and SUVs (light trucks) compared to other countries. Why? And what about proven hybrid gas–electric technology?

The spike in oil and gasoline prices in 2007–2008 coincided with a record decrease in miles driven, as drivers adjusted their driving habits with few difficulties—the economic principle is price elasticity of demand. This demonstrates the effectiveness of conservation and efficiency strategies used by individuals who have reduced demand. Strangely, the word *conservation* was not used in media or political coverage during that time. The task of physical geography is to analyze all the spatial aspects of these environmental events and the ironies that are symbolized by an oily bird and to make the connection with global climate change in such a planetary system.

The Century Before Us

Many of the thresholds discussed in *Geosystems* will occur during your lifetime. Many headline events are already signaling you of change. Assessing what is important at such a time is an important exercise. More than two decades ago, *Time* magazine gave Earth an interesting honor: It named Earth "Planet of the Year." This issue of *Time* is today a valuable time capsule of Earth-systems analysis and activist suggestions (Figure 21.5). The magazine devoted 33 pages to Earth's physical and human geography. Importantly, *Time*

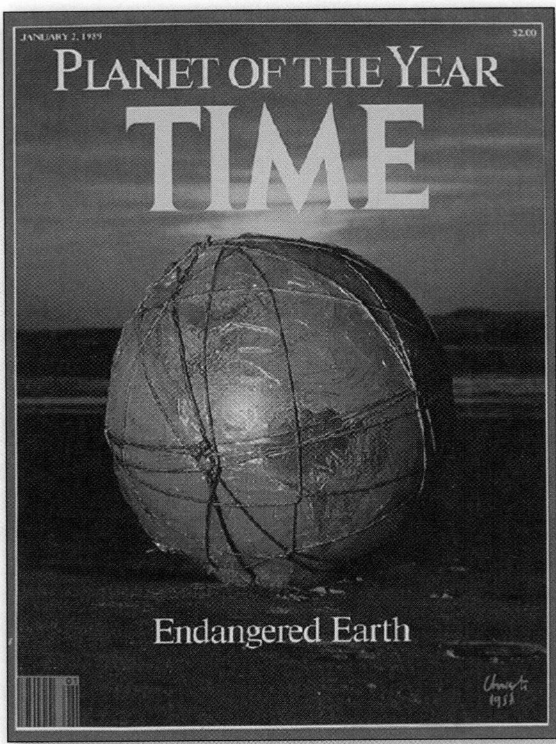

FIGURE 21.5 Earth made the cover of *Time*. More than two decades ago, global concerns about environmental impacts prompted *Time* magazine to deviate from its 60-year tradition of naming a prominent citizen as its person of the year, instead naming Earth the "Planet of the Year." [(C) 1989 TimePix.]

Twelve Paradigms for the Twenty-First Century

1. Population increases in the less-developed countries;
2. Planetary impact per person (on the biosphere and resources; $I = P \cdot A \cdot T$), an individual's ecological and carbon footprint;
3. Feeding the world's population;
4. Global and national disparities of wealth and resource allocation;
5. Status of women and children (health, welfare, rights);
6. Global climate change (temperatures, sea level, ocean chemistry, weather and climate patterns, disease, and biodiversity);
7. Energy supplies and energy demands; renewables and demand management;
8. Loss of biodiversity (habitats, genetic wealth, and species richness);
9. Pollution of air, surface water (quality and quantity), groundwater, oceans, and land;
10. The persistence of wilderness (biosphere reserves and biodiversity hot spots);
11. Globalization versus cultural diversity; and
12. Conflict resolution.

also offered positive policy strategies for consideration. As if a victim of short-lived fame, we wonder where Earth awareness fits into the global scheme of things today. Is there a follow-up in pop culture to making the cover of *Time*?

As we conclude, a brief list of the dominant themes of concern for humans in the twenty-first century seems appropriate. This list includes issues for which our present way of thinking is not working, issues that need to be addressed in new ways. A *paradigm* is the name given to a standard way of thinking about issues; this is thinking "in the box." A shift in a paradigm begins when we learn something new and begin to think "outside the box." I hope these paradigms, as the basis for thinking and change, provide a useful framework for brainstorming and discussion of the central issues that affect us in the new century, listed here in no particular order. Earth's physical geography and geographic science are important to the consideration of the paradigm shifts these 12 items infer—and perhaps require.

Who Speaks for Earth?

Geographic awareness and education is an increasingly positive force on Earth. Yet ideological and ethical differences still remain within society. This dichotomy was addressed by biologist Edward O. Wilson:

> The evidence of swift environmental change calls for an ethic uncoupled from other systems of belief. Those committed by religion to believe that life was put on Earth in one divine stroke will recognize that we are destroying the Creation; and those who perceive biodiversity to be the product of blind evolution will agree. . . . Defenders of both premises seem destined to gravitate toward the same position on conservation. . . . For what, in the final analysis, is morality but the command of conscience seasoned by a rational examination of consequences?. . . An enduring environmental ethic will aim to preserve not only the health and freedom of our species, but access to the world in which the human spirit was born.*

In an ideal sense, the human population holds all of Earth's systems in common. Now, what about Sagan's question mentioned at the beginning of the chapter? The late Carl Sagan asked, "Who speaks for Earth?" He gave us this answer:

> We have begun to contemplate our origins: starstuff pondering the stars; organized assemblages of ten billion billion billion atoms considering the evolution of atoms; tracing the long journey by which, here at least, consciousness arose. Our loyalties are to the species and the planet. We speak for Earth. Our obligation to survive is owed not just to ourselves but also to that Cosmos, ancient and vast, from which we spring.†

*E. O. Wilson, *The Diversity of Life* (Cambridge, MA: Harvard University Press, 1992), p. 351.
†C. Sagan, *Cosmos* (New York: Random House, 1980), p. 345.

CRITICAL THINKING 21.1

A final review

A. What part do you think technology, politics, and thinking about the future should play in science courses?

B. Assess population growth issues: the count, the impact per person, and future projection. What strategies do you see as important?

C. According to the discussion in the chapter, what worldwide factors led to the *Exxon–Valdez* accident or the risky drilling in deep water in the Gulf of Mexico? Describe the complexity of that event from a global perspective. In your analysis, examine both supply-side (corporations and utilities) issues and demand-side (consumers) issues as well as environmental and strategic factors. And what about the oily bird?

D. After examining the list of 12 paradigms for the twenty-first century, suggest items that need to be added to the list, omitted from the list, or expanded in coverage. Rearrange and organize the list as needed to match your concerns and sense of priorities as to the importance of various shifts in view that are needed.

E. This chapter states that we already know many of the solutions to the problems we face. Why do you think these solutions are not being implemented at a faster pace?

F. Who speaks for Earth?

GEOSYSTEMS CONNECTION

As this icon of a bridge suggests, your journey leads to more courses and study—and the rest of the twenty-first century. Study hard and travel safe—with the wind at your back.

May we all perceive our spatial importance within Earth's ecosystems and do our part to maintain a life-supporting and sustaining Earth for ourselves and countless generations in the future.

MASTERING GEOGRAPHY

Visit www.masteringgeography.com for several figure animations, satellite loops, self-study quizzes, flashcards, visual glossary, case studies, career links, textbook reference maps, RSS Feeds, and an ebook version of *Geosystems*. Included are many geography web links to interesting Internet sources that support this chapter.

Young male polar bear cools off on an ice floe after seal hunting, such exertion easily overheats the bear. [Bobbé Christopherson.]

APPENDIX A

Maps in This Text and Topographic Maps

Maps Used in This Text

Geosystems uses several map projections: Goode's homolosine, Robinson, and Miller cylindrical, among others. Each was chosen to best present specific types of data. **Goode's homolosine projection** is an interrupted world map designed in 1923 by Dr. J. Paul Goode of the University of Chicago. Rand McNally *Goode's Atlas* first used it in 1925. Goode's homolosine equal-area projection (Figure A.1) is a combination of two oval projections (*homolo*graphic and *sin*usoidal projections).

Two equal-area projections are cut and pasted together to improve the rendering of landmass shapes. A *sinusoidal projection* is used between 40° N and 40° S latitudes. Its central meridian is a straight line; all other meridians are drawn as sinusoidal curves (based on sine-wave curves) and parallels are evenly spaced. A *Mollweide projection*, also called a *homolographic projection*, is used from 40° N to the North Pole and from 40° S to the South Pole. Its central meridian is a straight line; all other meridians are drawn as elliptical arcs and parallels are unequally spaced—farther apart at the equator, closer together poleward. This technique of combining two projections preserves areal size relationships, making the projection excellent for mapping spatial distributions when interruptions of oceans or continents do not pose a problem.

We use Goode's homolosine projection throughout this book. Examples include the world climate map and smaller climate type maps in Chapter 10, topographic regions and continental shields maps (Figures 12.3 and 12.4), world karst map (Figure 13.14), world sand regions and loess deposits (Figure 15.11 and 15.13), and the terrestrial biomes map in 20.5.

Another projection we use is the **Robinson projection**, designed by Arthur Robinson in 1963 (Figure A.2). This projection is neither equal area nor true shape, but is a compromise between the two. The North and South poles appear as lines slightly more than half the length of the equator; thus higher latitudes are exaggerated less than on other oval and cylindrical projections. Some of the Robinson maps employed include the latitudinal geographic zones map in Chapter 1 (Figure 1.14), the daily net radiation map (Figure 2.11), the world temperature range map in Chapter 5 (Figure 5.18), the maps of lithospheric plates of crust and volcanoes and earthquakes in Chapter 11 (Figures 11.18 and 11.21), and the global oil spills map in Chapter 21 (Figure 21.4).

Another compromise map, the **Miller cylindrical projection**, is used in this text (Figure A.3). Examples of this projection include the world time zone map (Figure 1.21), global temperature maps in Chapter 5 (Figures 5.13 and 5.16), two global pressure maps in Figure 6.9, and global soil maps and soil order maps in Chapter 18 (Figure 18.9). This projection is neither true shape nor true area, but is a compromise that avoids the severe scale distortion of the Mercator. The Miller projection frequently appears in world atlases. The American Geographical Society presented Osborn Miller's map projection in 1942.

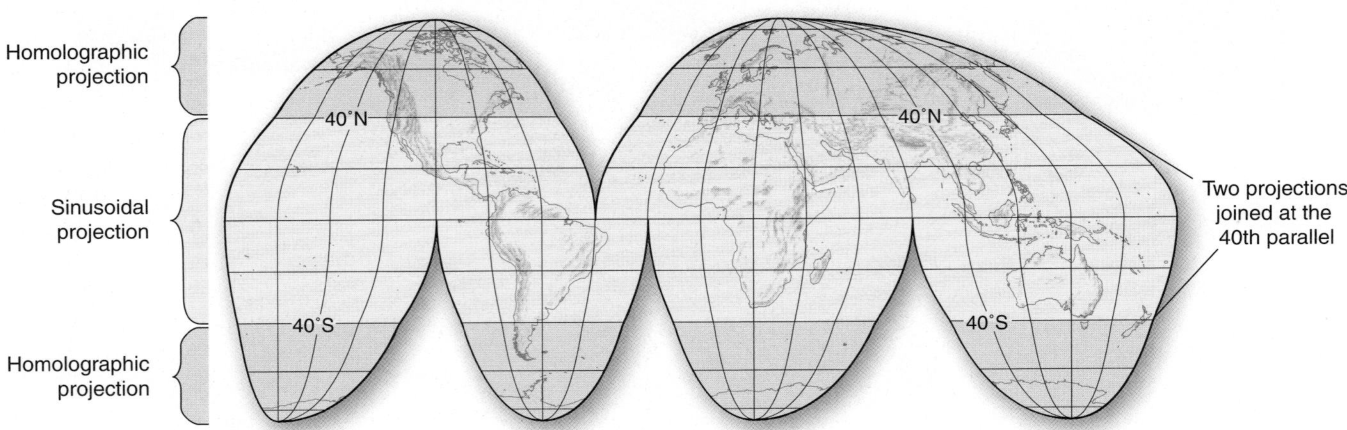

FIGURE A.1 Goode's homolosine projection.
An equal-area map. [Copyright by the University of Chicago. Used by permission of the University of Chicago Press.]

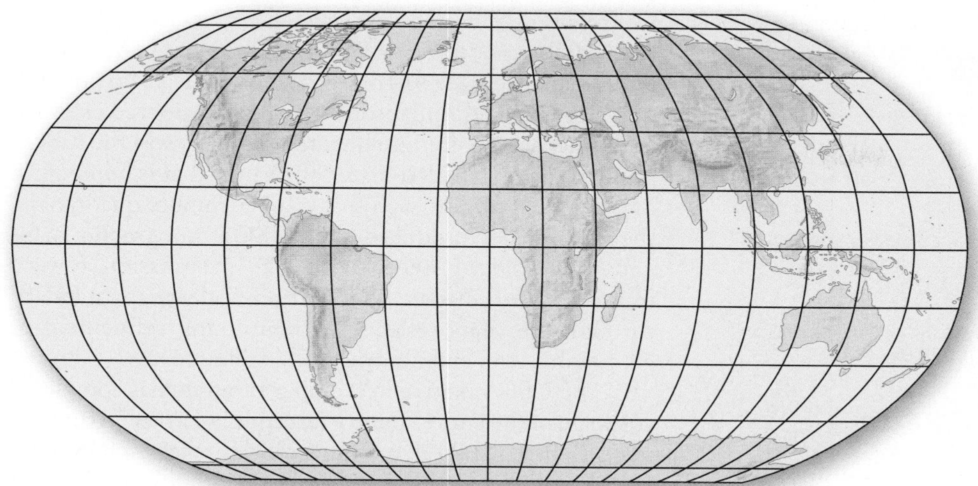

FIGURE A.2 Robinson projection. A compromise between equal area and true shape. [Developed by Arthur H. Robinson, 1963.]

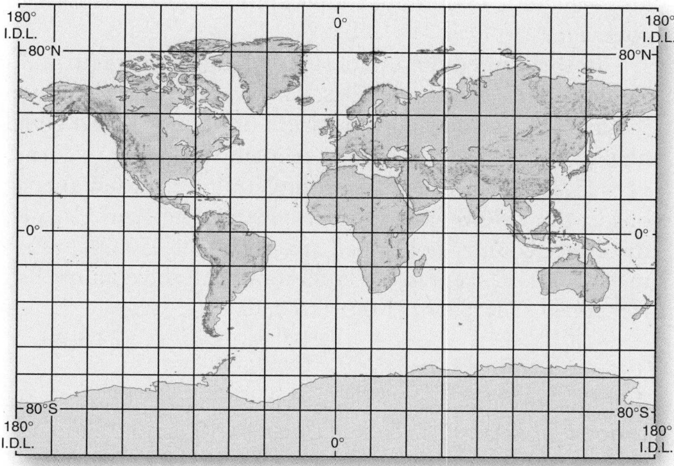

FIGURE A.3 Miller cylindrical projection. A compromise map projection between equal area and true shape. [Developed by Osborn M. Miller, American Geographical Society, 1942.]

Mapping, Quadrangles, and Topographic Maps

The westward expansion across the vast North American continent demanded a land survey for the creation of accurate maps. Maps were needed to subdivide the land and to guide travel, exploration, settlement, and transportation. In 1785, the Public Lands Survey System began surveying and mapping government land in the United States. In 1836, the Clerk of Surveys in the Land Office of the Department of the Interior directed public-land surveys. The Bureau of Land Management replaced this Land Office in 1946. The actual preparation and recording of survey information fell to the U.S. Geological Survey (USGS), also a branch of the Department of the Interior (see http://www.usgs.gov/pubprod/maps.html).

In Canada, National Resources Canada conducts the national mapping program. Canadian mapping includes base maps, thematic maps, aeronautical charts, federal topographic maps, and the National Atlas of Canada, now in its fifth edition (see http://atlas.nrcan.gc.ca/).

Quadrangle Maps

The USGS depicts survey information on quadrangle maps, so called because they are rectangular maps with four corner angles. The angles are junctures of parallels of latitude and meridians of longitude rather than political boundaries. These quadrangle maps utilize the Albers equal-area projection, from the conic class of map projections.

The accuracy of conformality (shape) and scale of this base map is improved by the use of not one but two standard parallels. (Remember from Chapter 1 that standard lines are where the projection cone touches the globe's surface, producing greatest accuracy.) For the conterminous United States (the "lower 48"), these parallels are 29.5° N and 45.5° N latitude (noted on the Albers projection shown in Figure 1.25c). The standard parallels shift for conic projections of Alaska (55° N and 65° N) and for Hawai'i (8° N and 18° N).

Because a single map of the United States at 1:24,000 scale would be more than 200 m wide (more than 600 ft), some system had to be devised for dividing the map into a manageable size. Thus, a quadrangle system using latitude and longitude coordinates developed. Note that these maps are not perfect rectangles, because meridians converge toward the poles. The width of quadrangles narrows noticeably as you move north (poleward).

Quadrangle maps are published in different series, covering different amounts of Earth's surface at different scales. You see in Figure A.4 that each series is referred to

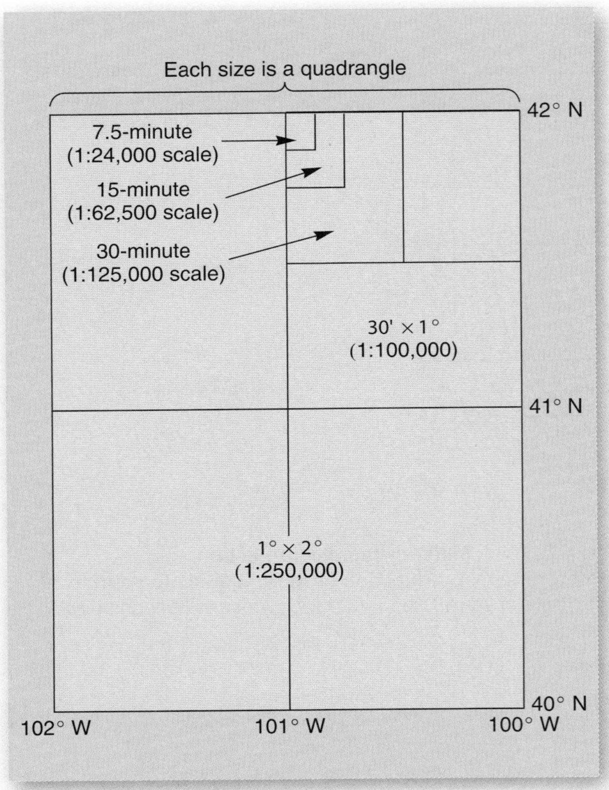

FIGURE A.4 **Quadrangle system of maps used by the USGS.**

by its angular dimensions, which range from 1° × 2° (1:250,000 scale) to 7.5′ × 7.5′ (1:24,000 scale). A map that is one-half a degree (30′) on each side is called a "30-minute quadrangle," and a map one-fourth of a degree (15′) on each side is a "15-minute quadrangle" (this was the USGS standard size from 1910 to 1950). A map that is one-eighth of a degree (7.5′) on each side is a 7.5-minute quadrangle, the most widely produced of all USGS topographic maps and the standard since 1950. The progression toward more-detailed maps and a larger-scale map standard through the years reflects the continuing refinement of geographic data and new mapping technologies.

The USGS National Mapping Program has completed coverage of the entire country (except Alaska) on 7.5-minute maps (1 in. to 2000 ft, a large scale). It takes 53,838 separate 7.5-minute quadrangles to cover the lower 48 states, Hawai'i, and the U.S. territories. A series of smaller-scale, more-general 15-minute topographic maps offer Alaskan coverage.

In the United States, most quadrangle maps remain in English units of feet and miles. The eventual changeover to the metric system requires revision of the units used on all maps, with the 1:24,000 scale eventually changing to a scale of 1:25,000. However, after completing only a few metric quads, the USGS halted the program in 1991. In Canada, the entire country is mapped at a scale of 1:250,000, using metric units (1.0 cm to 2.5 km). About half the country also is mapped at 1:50,000 (1.0 cm to 0.50 km).

Topographic Maps

The most popular and widely used quadrangle maps are **topographic maps** prepared by the USGS. An example of such a map is a portion of the Cumberland, Maryland, quad shown in Figure A.5. You will find topographic maps

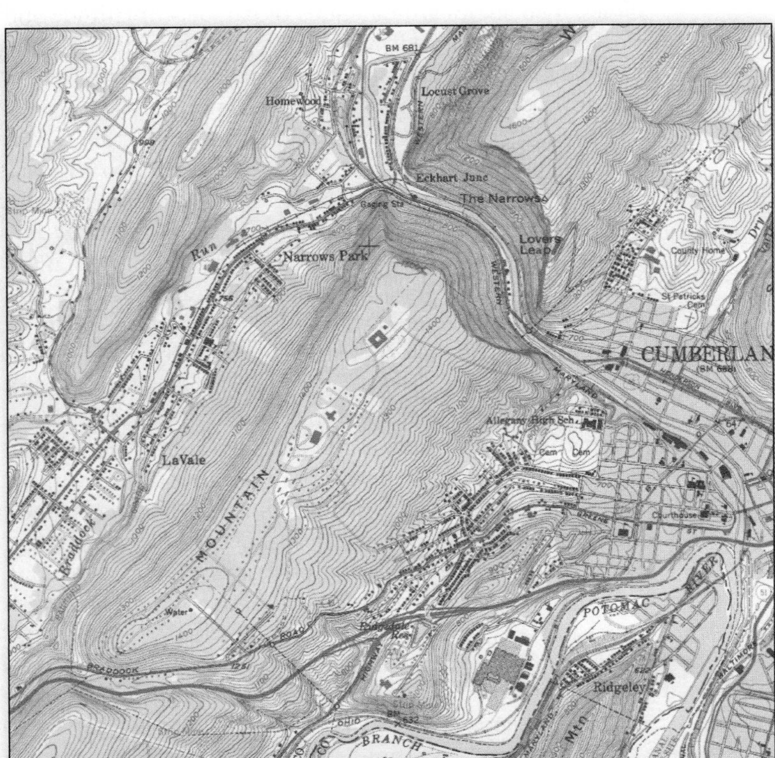

FIGURE A.5 **An example of a topographic map from the Appalachians.**
Cumberland, MD, PA, WV 7.5-minute quadrangle topographic map prepared by the USGS. Note the water gap through Haystack Mountain. In the *Applied Physical Geography*, 8/e, lab manual this topographic map is accessible through Google Earth™ mapping services where you experience the map as a 3-D landscape and maneuver with your computer to see the topography at any angle or detail you choose.

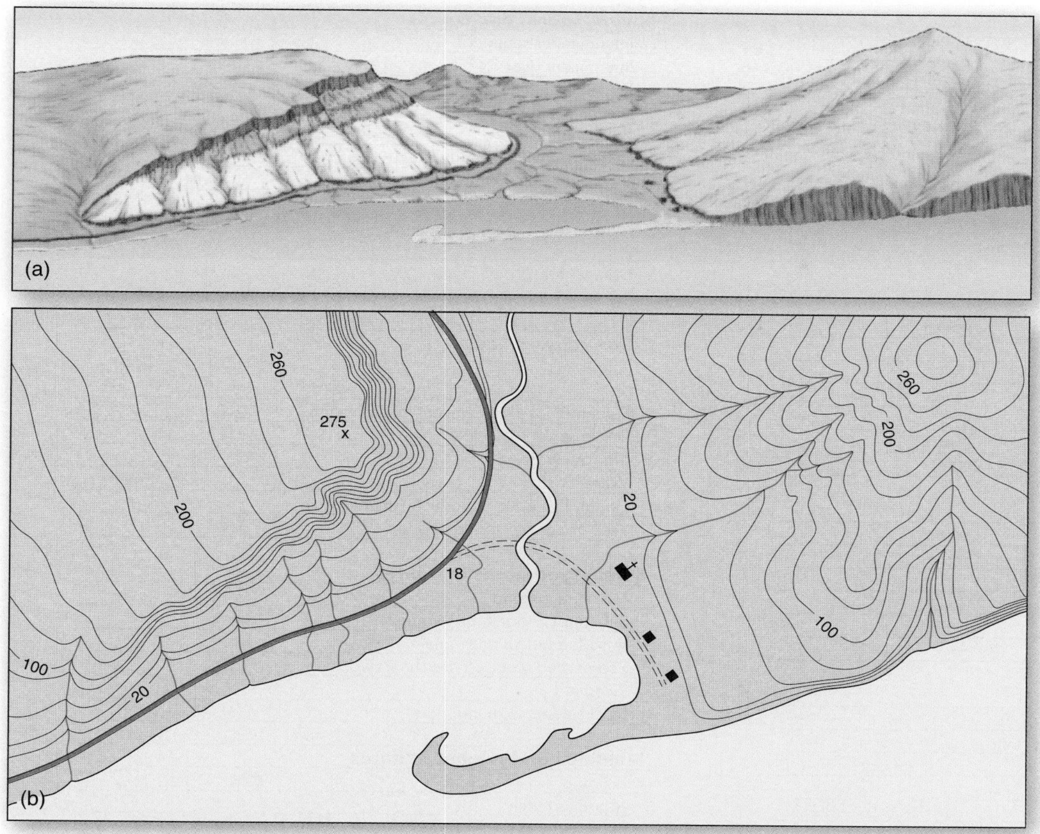

FIGURE A.6 Topographic map of a hypothetical landscape.
(a) Perspective view of a hypothetical landscape. (b) Depiction of that landscape on a topographic map. The contour interval on the map is 20 feet (6.1 m). [After the U.S. Geological Survey.]

throughout *Geosystems* because they portray landscapes so effectively. As examples, see Figures 13.16 and 13.17, karst landscapes and sinkholes near Orleans, Indiana, and Winter Park, Florida; Figure 14.8, river drainage patterns; Figure 14.23, river meander scars; Figure 15.18, an alluvial fan in Montana; and Figure 17.18, drumlins in New York.

A **planimetric map** shows the horizontal position (latitude/longitude) of boundaries, land-use aspects, bodies of water, and economic and cultural features. A highway map is a common example of a planimetric map.

A topographic map adds a vertical component to show topography (configuration of the land surface), including slope and relief (the vertical difference in local landscape elevation). These fine details are shown through the use of elevation contour lines (Figure A.6). A *contour line* connects all points at the same elevation. Elevations are shown above or below a vertical datum, or reference level, which usually is mean sea level. The contour interval is the vertical distance in elevation between two adjacent contour lines (20 ft, or 6.1 m in Figure A. 6b).

The topographic map in Figure 6b shows a hypothetical landscape, demonstrating how contour lines and intervals depict slope and relief, which are the three-dimensional aspects of terrain. The pattern of lines and the spacing between them indicate slope. The steeper a slope or cliff, the closer together the contour lines appear—in the figure, note the narrowly spaced contours that represent the cliffs to the left of the highway. A wider spacing of these contour lines portrays a more gradual slope, as you can see from the widely spaced lines on the beach and to the right of the river valley.

In Figure A.7 are the standard symbols commonly used on these topographic maps. These symbols and the colors used are standard on all USGS topographic maps: black for human constructions, blue for water features, brown for relief features and contours, pink for urbanized areas, and green for woodlands, orchards, brush, and the like.

The margins of a topographic map contain a wealth of information about its concept and content. In the margins of topographic maps, you find the quadrangle name, names of adjoining quads, quad series and type, position in the latitude-longitude and other coordinate systems, title, legend, magnetic declination (alignment of magnetic north) and compass information, datum plane, symbols used for roads and trails, the dates and history of the survey of that particular quad, and more.

Topographic maps may be purchased directly from the USGS or Centre for Topographic Information, NRC (http://maps.nrcan.gc.ca/). Many state geological survey offices, national and state park headquarters, outfitters, sports shops, and bookstores also sell topographic maps to assist people in planning their outdoor activities.

Control data and monuments

Vertical control

Third order or better, with tablet	BM×16.3
Third order or better, recoverable mark	×120.0
Bench mark at found section corner	BM 18.6
Spot elevation	× 5.3

Contours

Topographic

- Intermediate
- Index
- Supplementary
- Depression
- Cut; fill

Bathymetric

- Intermediate
- Index
- Primary
- Index primary
- Supplementary

Boundaries

- National
- State or territorial
- County or equivalent
- Civil township or equivalent
- Incorporated city or equivalent
- Park, reservation, or monument

Surface features

Levee	Levee
Sand or mud area, dunes, or shifting sand	Sand
Intricate surface area	Strip mine
Gravel beach or glacial moraine	Gravel
Tailings pond	Tailings pond

Mines and caves

Quarry or open pit mine	
Gravel, sand, clay, or borrow pit	
Mine dump	Mine dump
Tailings	Tailings

Vegetation

Woods	
Scrub	
Orchard	
Vineyard	
Mangrove	Mangrove

Glaciers and permanent snowfields

- Contours and limits
- Form lines

Marine shoreline

Topographic maps

- Approximate mean high water
- Indefinite or unsurveyed

Topographic-bathymetric maps

- Mean high water
- Apparent (edge of vegetation)

Coastal features

Foreshore flat	Mud
Rock or coral reef	Reef
Rock bare or awash	
Group of rocks bare or awash	
Exposed wreck	
Depth curve; sounding	3
Breakwater, pier, jetty, or wharf	
Seawall	

Rivers, lakes, and canals

Intermittent stream	
Intermittent river	
Disappearing stream	
Perennial stream	
Perennial river	
Small falls; small rapids	
Large falls; large rapids	
Masonry dam	
Dam with lock	
Dam carrying road	
Perennial lake; intermittent lake or pond	
Dry lake	Dry lake
Narrow wash	
Wide wash	Wide wash
Canal, flume, or aquaduct with lock	
Well or spring; spring or seep	

Submerged areas and bogs

Marsh or swamp	
Submerged marsh or swamp	
Wooded marsh or swamp	
Submerged wooded marsh or swamp	
Rice field	Rice
Land subject to inundation	Max pool 431

Buildings and related features

- Building
- School; church
- Built-up area
- Racetrack
- Airport
- Landing strip
- Well (other than water); windmill
- Tanks
- Covered reservoir
- Gaging station
- Landmark object (feature as labeled)
- Campground; picnic area
- Cemetery: small; large (Cem)

Roads and related features

Roads on Provisional edition maps are not classified as primary, secondary, or light duty. They are all symbolized as light duty roads.

- Primary highway
- Secondary highway
- Light duty road
- Unimproved road
- Trail
- Dual highway
- Dual highway with median strip

Railroads and related features

- Standard gauge single track; station
- Standard gauge multiple track
- Abandoned

Transmission lines and pipelines

Power transmission line; pole; tower	
Telephone line	Telephone
Aboveground oil or gas pipeline	
Underground oil or gas pipeline	Pipeline

FIGURE A.7 Standardized topographic map symbols used on USGS maps. English units still prevail, although a few USGS maps are in metric. [From USGS, Topographic Maps, 1969.]

APPENDIX B

The Köppen Climate Classification System

The Köppen climate classification system was designed by Wladimir Köppen (1846–1940), a German climatologist and botanist, and is widely used for its ease of comprehension. The basis of any empirical classification system is the choice of criteria used to draw lines on a map to designate different climates. Köppen–Geiger climate classification uses *average monthly temperatures*, *average monthly precipitation*, and *total annual precipitation* to devise its spatial categories and boundaries. But we must remember that boundaries really are transition zones of gradual change. The trends and overall patterns of boundary lines are more important than their precise placement, especially with the small scales generally used on world maps.

Take a few minutes and examine the Köppen system, the criteria, and the considerations for each of the principal climate categories. The modified Köppen–Geiger system has its drawbacks, however. It does not consider winds, temperature extremes, precipitation intensity, quantity of sunshine, cloud cover, or net radiation. Yet the system is important because its correlations with the actual world are reasonable and the input data are standardized and readily available.

Köppen's Climatic Designations

Figure 10.5 in Chapter 10 shows the distribution of each of Köppen's six climate classifications on the land. This generalized map shows the spatial pattern of climate. The Köppen system uses capital letters (A, B, C, D, E, H) to designate climatic categories from the equator to the poles. The guidelines for each of these categories are in the margin of Figure B.1.

Five of the climate classifications are based on thermal criteria:

A. Tropical (equatorial regions)
C. Mesothermal (Mediterranean, humid subtropical, marine west coast regions)
D. Microthermal (humid continental, subarctic regions)
E. Polar (polar regions)
H. Highland (compared to lowlands at the same latitude, highlands have lower temperatures—recall the normal lapse rate—and more efficient precipitation due to lower moisture demand)

Only one climate classification is based on moisture as well:

B. Dry (deserts and semiarid steppes)

Within each climate classification additional lowercase letters are used to signify temperature and moisture conditions. For example, in a tropical rain forest *Af* climate, the *A* tells us that the average coolest month is above 18°C (64.4°F, average for the month), and the *f* indicates that the weather is constantly wet, with the driest month receiving at least 6 cm (2.4 in.) of precipitation. (The designation *f* is from the German *feucht*, for "moist.") As you can see on the climate map, the tropical rain forest *Af* climate dominates along the equator and equatorial rain forest.

In a *Dfa* climate, the *D* means that the average warmest month is above 10°C (50°F), with at least one month falling below 0°C (32°F); the *f* says that at least 3 cm (1.2 in.) of precipitation falls during every month; and the *a* indicates a warmest summer month averaging above 22°C (71.6°F). Thus, a *Dfa* climate is a humid-continental, hot-summer climate in the microthermal category.

Köppen Guidelines

The Köppen guidelines and map portrayal are in Figure B.1. First, check the guidelines for a climate type, then the color legend for the subdivisions of the type, then check out the distribution of that climate on the map. You may want to compare this with the climate map in Figure 10.6 that presents causal elements that produce these climates.

Five of the climate classifications are based on thermal criteria:

A. Tropical (equatorial regions)
C. Mesothermal (Mediterranean, humid subtropical, marine west coast regions)
D. Microthermal (humid continental, subarctic regions)
E. Polar (polar regions)
H. Highland (compared to lowlands at the same latitude, highlands have lower temperatures—recall the normal lapse rate—and more efficient precipitation due to lower moisture demand)

Only one climate classification is based on moisture as well:

B. Dry (deserts and semiarid steppes)

FIGURE B.1 World climates and their guidelines according to the Köppen classification system.

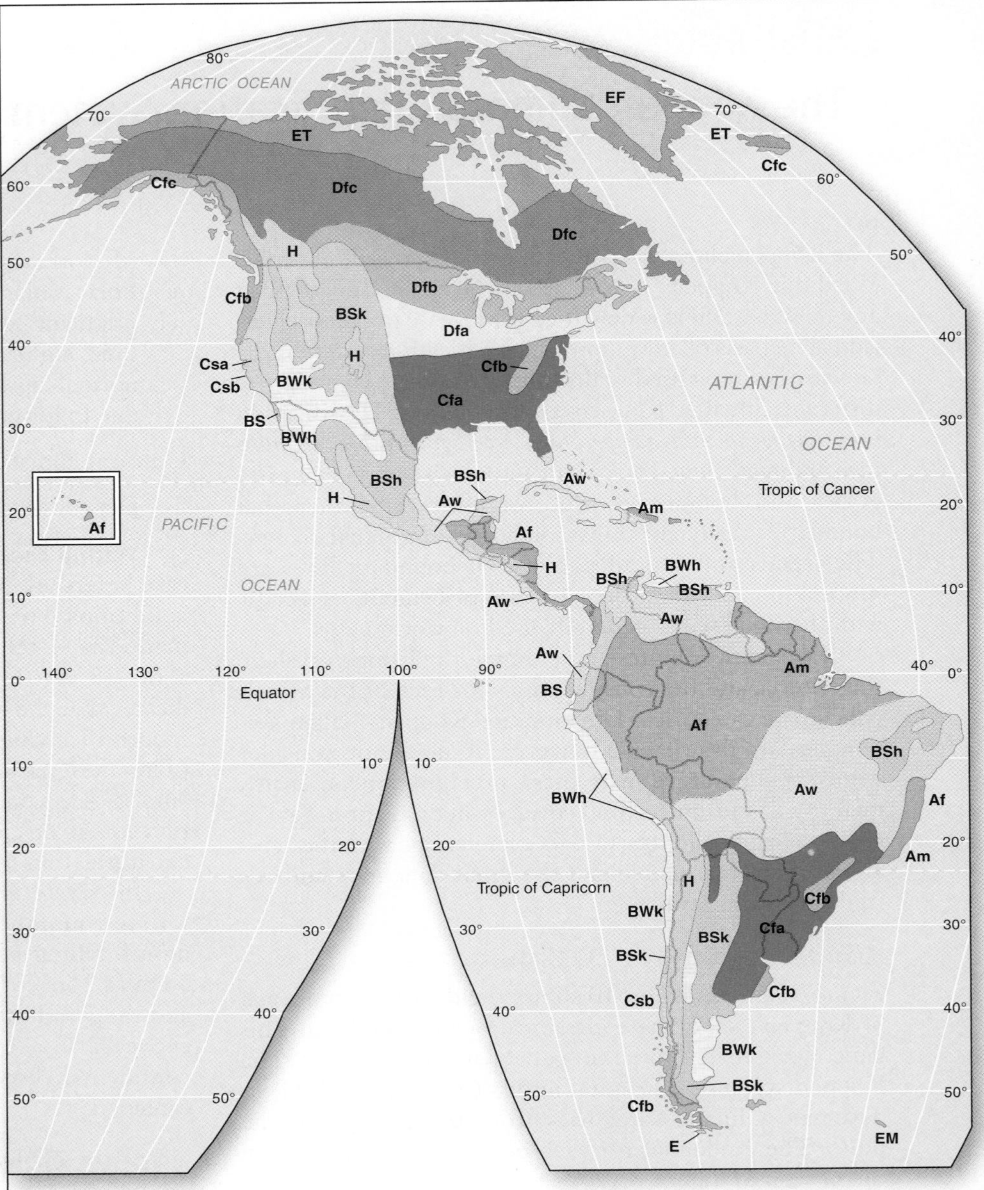

Köppen Guidelines
Tropical Climates — A

Consistently warm with all months averaging above 18°C (64.4°F); annual water supply exceeds water demand.

Af — Tropical rain forest:
- f = All months receive precipitation in excess of 6 cm (2.4 in.).

Am — Tropical monsoon:
- m = A marked short dry season with 1 or more months receiving less than 6 cm (2.4 in.) precipitation; an otherwise excessively wet rainy season. ITCZ 6–12 months dominant.

Aw — Tropical savanna:
- w = Summer wet season, winter dry season; ITCZ dominant 6 months or less, winter water-balance deficits.

Mesothermal Climates — C

Warmest month above 10°C (50°F); coldest month above 0°C (32°F) but below 18°C (64.4°F); seasonal climates.

Cfa, Cwa — Humid subtropical:
- a = Hot summer; warmest month above 22°C (71.6°F).
- f = Year-round precipitation.
- w = Winter drought, summer wettest month 10 times more precipitation than driest winter month.

Cfb, Cfc — Marine west coast, mild-to-cool summer:
- f = Receives year-round precipitation.
- b = Warmest month below 22°C (71.6°F) with 4 months above 10°C.
- c = 1–3 months above 10°C.

Csa, Csb — Mediterranean summer dry:
- s = Pronounced summer drought with 70% of precipitation in winter.
- a = Hot summer with warmest month above 22°C (71.6°F).
- b = Mild summer; warmest month below 22°C.

Microthermal Climates — D

Warmest month above 10°C (50°F); coldest month below 0°C (32°F); cool temperature-to-cold conditions; snow climates. In Southern Hemisphere, occurs only in highland climates.

Dfa, Dwa — Humid continental:
- a = Hot summer; warmest month above 22°C (71.6°F).
- f = Year-round precipitation.
- w = Winter drought.

Dfb, Dwb — Humid continental:
- b = Mild summer; warmest month below 22°C (71.6°F).
- f = Year-round precipitation.
- w = Winter drought.

Dfc, Dwc, Dwd — Subarctic:
Cool summers, cold winters.
- f = Year-round precipitation.
- w = Winter drought.
- c = 1–4 months above 10°C
- b = Coldest month below −38°C (−36.4°F), in Siberia only.

Dry Arid and Semiarid Climates — B

Potential evapotranspiration* (natural moisture demand) exceeds precipitation (natural moisture supply) in all B climates. Subdivisions based on precipitation timing and amount and mean annual temperature.

Earth's arid climates.

BWh — Hot low-latitude desert

BWk — Cold midlatitude desert
- BW = Precipitation less than 1/2 natural moisture demand.
- h = Mean annual temperature >18°C (64.4°F).
- k = Mean annual temperature <18°C.

Earth's semiarid climates.

BSh — Hot low-latitude steppe

BSk — Cold midlatitude steppe
- BS = Precipitation more than 1/2 natural moisture demand but not equal to it.
- h = Mean annual temperature >18°C.
- k = Mean annual temperature <18°C.

Polar Climates — E

Warmest month below 10°C (50°F); always cold; ice climates.

ET — Tundra:
Warmest month 0–10°C (32–50°F); precipitation exceeds small potential evapotranspiration demand*; snow cover 8–10 months.

EF — Ice cap:
Warmest month below 0°C (32°F); precipitation exceeds a very small potential evapotranspiration demand; the polar regions.

EM — Polar marine:
All months above −7°C (20°F), warmest month above 0°C; annual temperature range <17 C° (30 F°).

*Potential evapotranspiration = the amount of water that would evaporate or transpire if it were available—the natural moisture demand in an environment; see Chapter 9.

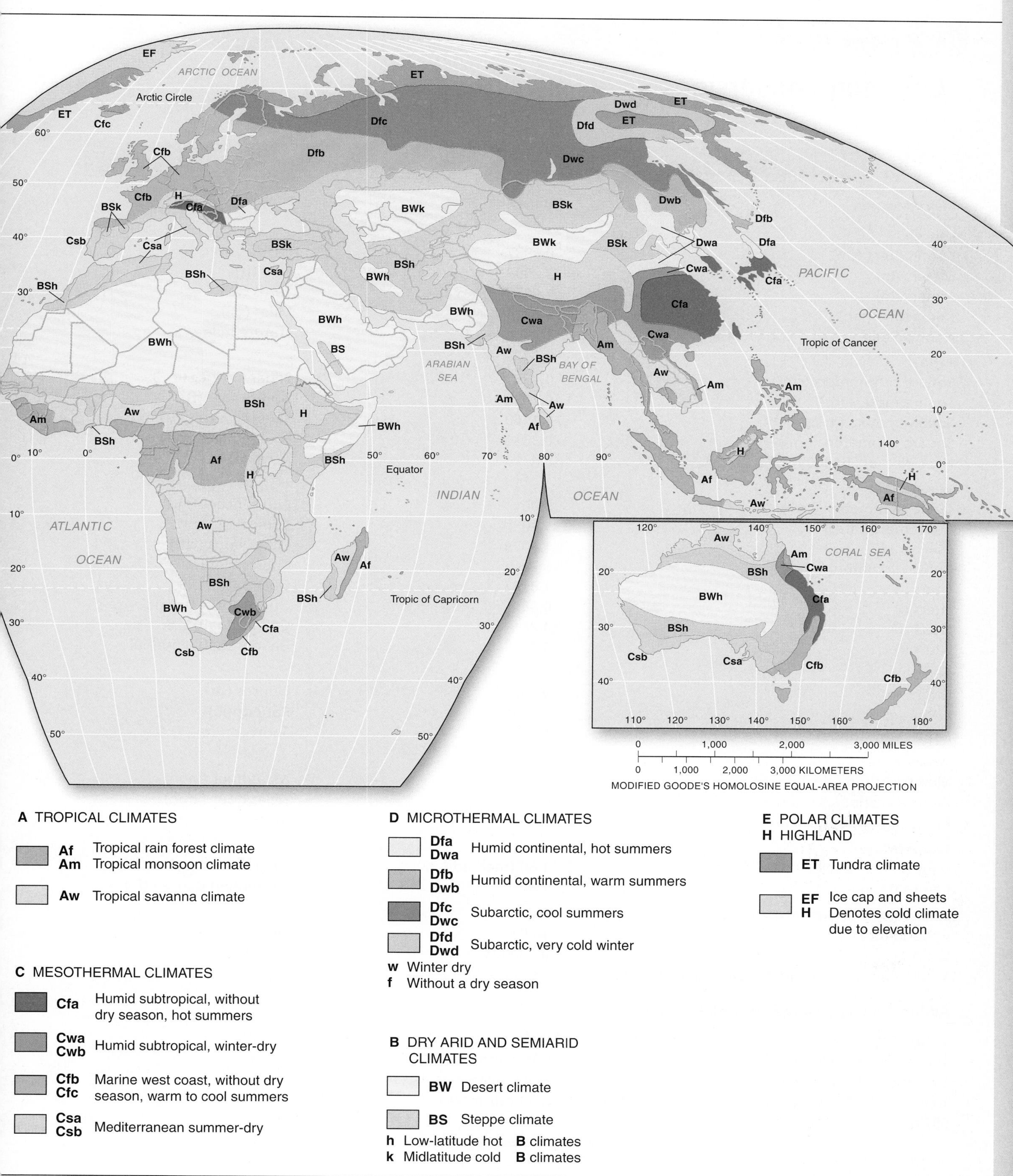
ARCTIC OCEAN
Arctic Circle
PACIFIC OCEAN
ARABIAN SEA
BAY OF BENGAL
INDIAN OCEAN
ATLANTIC OCEAN
CORAL SEA
Equator
Tropic of Cancer
Tropic of Capricorn
EF
ET
Cfc
Cfb
Dfc
Dfb
Dwd
Dfd
Dwc
Dwb
Dfa
Dwa
BSk
BWk
Csb
Csa
Cfa
Cwa
Cwb
BSh
BWh
BS
H
Aw
Am
Af
0 1,000 2,000 3,000 MILES
0 1,000 2,000 3,000 KILOMETERS
MODIFIED GOODE'S HOMOLOSINE EQUAL-AREA PROJECTION
A TROPICAL CLIMATES
Af Tropical rain forest climate
Am Tropical monsoon climate
Aw Tropical savanna climate
C MESOTHERMAL CLIMATES
Cfa Humid subtropical, without dry season, hot summers
Cwa Cwb Humid subtropical, winter-dry
Cfb Cfc Marine west coast, without dry season, warm to cool summers
Csa Csb Mediterranean summer-dry
D MICROTHERMAL CLIMATES
Dfa Dwa Humid continental, hot summers
Dfb Dwb Humid continental, warm summers
Dfc Dwc Subarctic, cool summers
Dfd Dwd Subarctic, very cold winter
w Winter dry
f Without a dry season
B DRY ARID AND SEMIARID CLIMATES
BW Desert climate
BS Steppe climate
h Low-latitude hot B climates
k Midlatitude cold B climates
E POLAR CLIMATES
H HIGHLAND
ET Tundra climate
EF Ice cap and sheets
H Denotes cold climate due to elevation

APPENDIX C

Common Conversions

Metric to English

Metric Measure	Multiply by	English Equivalent
Length		
Centimeters (cm)	0.3937	Inches (in.)
Meters (m)	3.2808	Feet (ft)
Meters (m)	1.0936	Yards (yd)
Kilometers (km)	0.6214	Miles (mi)
Nautical mile	1.15	Statute mile
Area		
Square centimeters (cm^2)	0.155	Square inches ($in.^2$)
Square meters (m^2)	10.7639	Square feet (ft^2)
Square meters (m^2)	1.1960	Square yards (yd^2)
Square kilometers (km^2)	0.3831	Square miles (mi^2)
Hectare (ha) (10,000 m^2)	2.4710	Acres (a)
Volume		
Cubic centimeters (cm^3)	0.06	Cubic inches ($in.^3$)
Cubic meters (m^3)	35.30	Cubic feet (ft^3)
Cubic meters (m^3)	1.3079	Cubic yards (yd^3)
Cubic kilometers (km^3)	0.24	Cubic miles (mi^3)
Liters (l)	1.0567	Quarts (qt), U.S.
Liters (l)	0.88	Quarts (qt), Imperial
Liters (l)	0.26	Gallons (gal), U.S.
Liters (l)	0.22	Gallons (gal), Imperial
Mass		
Grams (g)	0.03527	Ounces (oz)
Kilograms (kg)	2.2046	Pounds (lb)
Metric ton (tonne) (t)	1.10	Short ton (tn), U.S.
Velocity		
Meters/second (mps)	2.24	Miles/hour (mph)
Kilometers/hour (kmph)	0.62	Miles/hour (mph)
Knots (kn) (nautical mph)	1.15	Miles/hour (mph)
Temperature		
Degrees Celsius (°C)	1.80 (then add 32)	Degrees Fahrenheit (°F)
Celsius degree (C°)	1.80	Fahrenheit degree (F°)
Additional water measurements:		
Gallon (Imperial)	1.201	Gallon (U.S.)
Gallons (gal)	0.000003	Acre-feet

1 cubic foot per second per day = 86,400 cubic feet = 1.98 acre-feet

Additional Energy and Power Measurements

1 watt (W) = 1 joule/s
1 joule = 0.239 calorie
1 calorie = 4.186 joules
1 W/m^2 = 0.001433 cal/min
697.8 W/m^2 = 1 cal/cm^2min^{-1}

1 W/m^2 = 2.064 cal/cm^2day^{-1}
1 W/m^2 = 61.91 cal/cm^2month^{-1}
1 W/m^2 = 753.4 cal/cm^2year^{-1}
100 W/m^2 = 75 $kcal/cm^2year^{-1}$

Solar constant:
1372 W/m^2
2 cal/cm^2min^{-1}

English to Metric

English Measure	Multiply by	Metric Equivalent
Length		
Inches (in.)	2.54	Centimeters (cm)
Feet (ft)	0.3048	Meters (m)
Yards (yd)	0.9144	Meters (m)
Miles (mi)	1.6094	Kilometers (km)
Statute mile	0.8684	Nautical mile
Area		
Square inches (in.^{2})	6.45	Square centimeters (cm^2)
Square feet (ft^2)	0.0929	Square meters (m^2)
Square yards (yd^2)	0.8361	Square meters (m^2)
Square miles (mi^2)	2.5900	Square kilometers (km^2)
Acres (a)	0.4047	Hectare (ha)
Volume		
Cubic inches (in.^{3})	16.39	Cubic centimeters (cm^3)
Cubic feet (ft^3)	0.028	Cubic meters (m^3)
Cubic yards (yd^3)	0.765	Cubic meters (m^3)
Cubic miles (mi^3)	4.17	Cubic kilometers (km^3)
Quarts (qt), U.S.	0.9463	Liters (l)
Quarts (qt), Imperial	1.14	Liters (l)
Gallons (gal), U.S.	3.8	Liters (l)
Gallons (gal), Imperial	4.55	Liters (l)
Mass		
Ounces (oz)	28.3495	Grams (g)
Pounds (lb)	0.4536	Kilograms (kg)
Short ton (tn), U.S.	0.91	Metric ton (tonne) (t)
Velocity		
Miles/hour (mph)	0.448	Meters/second (mps)
Miles/hour (mph)	1.6094	Kilometers/hour (kmph)
Miles/hour (mph)	0.8684	Knots (kn) (nautical mph)
Temperature		
Degrees Fahrenheit (°F)	0.556 (after subtracting 32)	Degrees Celsius (°C)
Fahrenheit degree (F°)	0.556	Celsius degree (C°)
Additional water measurements:		
Gallon (U.S.)	0.833	Gallons (Imperial)
Acre-feet	325,872	Gallons (gal)

Additional Notation

Multiples	Prefixes	
1,000,000,000 = 10^9	giga	G
1,000,000 = 10^6	mega	M
1,000 = 10^3	kilo	k
100 = 10^2	hecto	h
10 = 10^1	deka	da
1 = 10^0		
0.1 = 10^{-1}	deci	d
0.01 = 10^{-2}	centi	c
0.001 = 10^{-3}	milli	m
0.000001 = 10^{-6}	micro	μ

GLOSSARY

The chapter in which each term appears **boldfaced** in parentheses is followed by a specific definition relevant to the term's usage in the chapter.

Aa (12) Rough, jagged, and clinkery basaltic lava with sharp edges. This texture is caused by the loss of trapped gases, a slow flow, and the development of a thick skin that cracks into the jagged surface.

Abiotic (1) Nonliving; Earth's nonliving systems of energy and materials.

Ablation (17) Loss of glacial ice through melting, sublimation, wind removal by deflation, or the calving of blocks of ice. (*See* Deflation.)

Abrasion (14, 15, 17) Mechanical wearing and erosion of bedrock accomplished by the rolling and grinding of particles and rocks carried in a stream, removed by wind in a "sandblasting" action, or imbedded in glacial ice.

Absorption (4) Assimilation and conversion of radiation from one form to another in a medium. In the process, the temperature of the absorbing surface is raised, thereby affecting the rate and wavelength of radiation from that surface.

Active layer (17) A zone of seasonally frozen ground that exists between the subsurface permafrost layer and the ground surface. The active layer is subject to consistent daily and seasonal freeze–thaw cycles. (*See* Permafrost, Periglacial.)

Actual evapotranspiration (9) ACTET; the actual amount of evaporation and transpiration that occurs; derived in the water balance by subtracting the deficit (DEFIC) from potential evapotranspiration (POTET).

Adiabatic (7) Pertaining to the cooling of an ascending parcel of air through expansion or the warming of a descending parcel of air through compression, without any exchange of heat between the parcel and the surrounding environment.

Advection (4) Horizontal movement of air or water from one place to another. (*Compare* Convection.)

Advection fog (7) Active condensation formed when warm, moist air moves laterally over cooler water or land surfaces, causing the lower layers of the air to be chilled to the dew-point temperature.

Aerosols (3) Small particles of dust, soot, and pollution suspended in the air.

Aggradation (14) The general building of land surface because of deposition of material; opposite of degradation. When the sediment load of a stream exceeds the stream's capacity to carry it, the stream channel becomes filled through this process.

Air (3) A simple mixture of gases (N, O, Ar, CO_2, and trace gases) that is naturally odorless, colorless, tasteless, and formless, blended so thoroughly that it behaves as if it were a single gas.

Air mass (8) A distinctive, homogeneous body of air that has taken on the moisture and temperature characteristics of its source region.

Air pressure (3, 6) Pressure produced by the motion, size, and number of gas molecules in the air and exerted on surfaces in contact with the air; an average force at sea level of 1 kg/cm^3 (14.7 lb/in^2). Normal sea-level pressure, as measured by the height of a column of mercury (Hg), is expressed as 1013.2 millibars, 760 mm of Hg, or 29.92 inches of Hg. Air pressure can be measured with mercury or aneroid barometers (*see listings for both*).

Albedo (4) The reflective quality of a surface, expressed as the percentage of reflected insolation to incoming insolation; a function of surface color, angle of incidence, and surface texture.

Aleutian low (6) *See* Subpolar low-pressure cell.

Alfisols (18) A soil order in the Soil Taxonomy. Moderately weathered forest soils that are moist versions of Mollisols, with productivity dependent on specific patterns of moisture and temperature; rich in organics. Most wide-ranging of the soil orders.

Alluvial fan (15) Fan-shaped fluvial landform at the mouth of a canyon; generally occurs in arid landscapes where streams are intermittent. (*See* bajada.)

Alluvial terraces (14) Level areas that appear as topographic steps above a stream, created by the stream as it scours with renewed downcutting into its floodplain; composed of unconsolidated alluvium. (*See* Alluvium.)

Alluvium (14) General descriptive term for clay, silt, sand, gravel, or other unconsolidated rock and mineral fragments transported by running water and deposited as sorted or semisorted sediment on a floodplain, delta, or streambed.

Alpine glacier (17) A glacier confined in a mountain valley or walled basin, consisting of three subtypes: valley glacier (within a valley), piedmont glacier (coalesced at the base of a mountain, spreading freely over nearby lowlands), and outlet glacier (flowing outward from a continental glacier; compare to Continental glacier).

Alpine tundra (20) Tundra conditions at high elevation. (*See* Arctic tundra.)

Altitude (2) The angular distance between the horizon (a horizontal plane) and the Sun (or any point in the sky).

Altocumulus (7) Middle-level, puffy clouds that occur in several forms: patchy rows, wave patterns, a "mackerel sky," or lens-shaped "lenticular" clouds.

Andisols (18) A soil order in the Soil Taxonomy; derived from volcanic parent materials in areas of volcanic activity. A new order, created in 1990, of soils previously considered under Inceptisols and Entisols.

Anemometer (6) A device that measures wind velocity.

Aneroid barometer (6) A device that measures air pressure using a partially evacuated, sealed cell. (*See* Air pressure.)

Angle of repose (13) The steepness of a slope that results when loose particles come to rest; an angle of balance between driving and resisting forces, ranging between 33° and 37° from a horizontal plane.

Antarctic Circle (2) This latitude (66.5° S) denotes the northernmost parallel (in the Southern Hemisphere) that experiences a 24-hour period of darkness in winter or daylight in summer.

Antarctic high (6) A consistent high-pressure region centered over Antarctica; source region for an intense polar air mass that is dry and associated with the lowest temperatures on Earth.

Antarctic region (17) The Antarctic convergence defines the Antarctic region in a narrow zone that extends around the continent as a boundary between colder Antarctic water and warmer water at lower latitudes.

Anthropogenic atmosphere (3) Earth's future atmosphere, so named because humans appear to be the principal causative agent.

Anticline (12) Upfolded rock strata in which layers slope downward from the axis of the fold, or central ridge. (*Compare* Syncline.)

Anticyclone (6) A dynamically or thermally caused area of high atmospheric pressure with descending and diverging airflows that rotate clockwise in the Northern Hemisphere and counterclockwise in the Southern Hemisphere. (*Compare* Cyclone.)

Aphelion (2) The point of Earth's greatest distance from the Sun in its elliptical orbit; reached on July 4 at a distance of 152,083,000 km (94.5 million mi); variable over a 100,000-year cycle. (*Compare* Perihelion.)

Aquiclude (9) A body of rock that does not conduct groundwater in usable amounts; an impermeable rock layer, related to an aquitard that slows, but does not block water flow. (*Compare* Aquifer.)

Aquifer (9) A body of rock that conducts groundwater in usable amounts; a permeable rock layer. (*Compare* Aquiclude.)

Aquifer recharge area (9) The surface area where water enters an aquifer to recharge the water-bearing strata in a groundwater system.

Arctic Circle (2) This latitude (66.5° N) denotes the southernmost parallel (in the Northern Hemisphere) that experiences a 24-hour period of darkness in winter or daylight in summer.

Arctic region (17) The 10°C (50°F) isotherm for July defines the Arctic region; coincides with the visible tree line—the boundary between the northern forests and tundra.

Arctic tundra (20) A biome in the northernmost portions of North America and northern Europe and Russia, featuring low, ground-level herbaceous plants as well as some woody plants. (*See* Alpine tundra.)

Arête (17) A sharp ridge that divides two cirque basins. Derived from "knife edge" in French, these form sawtooth and serrated ridges in glaciated mountains.

Aridisols (18) A soil order in the Soil Taxonomy; largest soil order. Typical of dry climates; low in organic matter and dominated by calcification and salinization.

Artesian water (9) Pressurized groundwater that rises in a well or a rock structure above the local water table; may flow out onto the ground without pumping. (*See* Potentiometric surface.)

Asthenosphere (11) Region of the upper mantle just below the lithosphere; the least rigid portion of Earth's interior and known as the plastic layer, flowing very slowly under extreme heat and pressure.

Atmosphere (1) The thin veil of gases surrounding Earth, which forms a protective boundary between outer space and the biosphere; generally considered to extend out about 480 km (300 mi) from Earth's surface.

Aurora (2) A spectacular glowing light display in the ionosphere, stimulated by the interaction of the solar wind with principally oxygen and nitrogen gases and few other atoms at high latitudes; called *aurora borealis* in the Northern Hemisphere and *aurora australis* in the Southern Hemisphere.

Autumnal (September) equinox (2) The time around September 22–23 when the Sun's declination crosses the equatorial parallel (0° latitude) and all places on Earth experience days and nights of equal length. The Sun rises at the South Pole and sets at the North Pole. (*Compare* Vernal [March] equinox.)

Available water (9) The portion of capillary water that is accessible to plant roots; usable water held in soil moisture storage. (*See* Capillary water.)

Axial parallelism (2) Earth's axis remains aligned the same throughout the year (it "remains parallel to itself"); thus, the axis extended from the North Pole points into space always near Polaris, the North Star.

Axial tilt (2) Earth's axis tilts 23.5° from a perpendicular to the plane of the ecliptic (plane of Earth's orbit around the Sun).

Axis (2) An imaginary line, extending through Earth from the geographic North Pole to the geographic South Pole, around which Earth rotates.

Azores high (6) A subtropical high-pressure cell that forms in the Northern Hemisphere in the eastern Atlantic (*see* Bermuda high); associated with warm, clear water and large quantities of sargassum, or gulf weed, characteristic of the Sargasso Sea.

Backswamp (14) A low-lying, swampy area of a floodplain; adjacent to a river, with the river's natural levee on one side and higher topography on the other. (*See* Floodplain, Yazoo tributary.)

Badland (15) In semiarid regions, weak surface material may weather to a complex, rugged topography, usually of relatively low and varied relief.

Bajada (15) A continuous apron of coalesced alluvial fans, formed along the base of mountains in arid climates; presents a gently rolling surface from fan to fan. (*See* Alluvial fan.)

Barrier beach (16) Narrow, long, depositional feature, generally composed of sand, that forms offshore roughly parallel to the coast; may appear as barrier islands and long chains of barrier beaches. (*See* Barrier island.)

Barrier island (16) Generally, a broadened barrier beach offshore. (*See* Barrier beach.)

Barrier spit (16) A depositional landform that develops when transported sand or gravel in a barrier beach or island is deposited in long ridges that are attached at one end to the mainland and partially cross the mouth of a bay.

Basalt (11) A common extrusive igneous rock, fine-grained, comprising the bulk of the ocean-floor crust, lava flows, and volcanic forms; gabbro is its intrusive form.

Base level (14) A hypothetical level below which a stream cannot erode its valley—and thus the lowest operative level for denudation processes; in an absolute sense, it is represented by sea level, extending under the landscape.

Basin and Range Province (15) A region of dry climates, few permanent streams, and interior drainage patterns in the western United States; a faulted landscape composed of a sequence of horsts and grabens.

Batholith (11) The largest plutonic form exposed at the surface; an irregular intrusive mass; it invades crustal rocks, cooling slowly so that large crystals develop. (*See* Pluton.)

Bay barrier (16) An extensive barrier spit of sand or gravel that encloses a bay, cutting it off completely from the ocean and forming a lagoon; produced by littoral drift and wave action; sometimes referred to as a baymouth bar. (*See* Barrier spit, Lagoon.)

Beach (16) The portion of the coastline where an accumulation of sediment is in motion.

Beach drift (16) Material, such as sand, gravel, and shells, that is moved by the longshore current in the effective direction of the waves.

Beaufort wind scale (6) A descriptive scale for the visual estimation of wind speeds; originally conceived in 1806 by Admiral Beaufort of the British Navy.

Bed load (14) Coarse materials that are dragged along the bed of a stream by traction or by the rolling and bouncing motion of saltation; involves particles too large to remain in suspension. (*See* Traction, Saltation.)

Bedrock (13) The rock of Earth's crust that is below the soil and is basically unweathered; such solid crust sometimes is exposed as an outcrop.

Bergschrund (17) This forms when a crevasse or wide crack opens along the headwall of a glacier; most visible in summer when covering snow is gone.

Bermuda high (6) A subtropical high-pressure cell that forms in the western North Atlantic. (*See* Azores high.)

Biodiversity (19) A principle of ecology and biogeography: The more diverse the species population in an ecosystem (in number of species, quantity of members in each species, and genetic content), the more risk is spread over the entire community, which results in greater overall stability, greater productivity, and increased use of nutrients, as compared to a monoculture of little or no diversity.

Biogeochemical cycle (19) One of several circuits of flowing elements and materials (carbon, oxygen, nitrogen, phosphorus, water) that combine Earth's biotic (living) and abiotic (nonliving) systems; the cycling of materials is continuous and renewed through the biosphere and the life processes.

Biogeographic realm (20) One of eight regions of the biosphere, each representative of evolutionary core areas of related flora (plants) and fauna (animals); a broad geographical classification scheme.

Biogeography (19) The study of the distribution of plants and animals and related ecosystems; the geographical relationships with their environments over time.

Biomass (19) The total mass of living organisms on Earth or per unit area of a landscape; also the weight of the living organisms in an ecosystem.

Biome (20) A large, stable terrestrial ecosystem characterized by specific plant communities and formations; usually named after the predominant vegetation in the region. (*See* Terrestrial ecosystem.)

Biosphere (1) That area where the atmosphere, lithosphere, and hydrosphere function together to form the context within which life exists; an intricate web that connects all organisms with their physical environment.

Biotic (1) Living; referring to Earth's living system of organisms.

Blowout depression (15) Eolian (wind) erosion in which deflation forms a basin in areas of loose sediment. Diameter may range up to hundreds of meters. (*See* Deflation.)

Bolson (15) The slope and basin area between the crests of two adjacent ridges in a dry region.

Boreal forest (20) *See* Needleleaf forest.

Brackish (16) Descriptive of seawater with a salinity of less than 35%; for example, the Baltic Sea. (*Compare* Brine.)

Braided stream (14) A stream that becomes a maze of interconnected channels laced with excess sediment. Braiding often occurs with a reduction of discharge that reduces a stream's transporting ability or with an increase in sediment load.

Breaker (16) The point where a wave's height exceeds its vertical stability and the wave breaks as it approaches the shore.

Brine (16) Seawater with a salinity of more than 35%; for example, the Persian Gulf. (*Compare* Brackish.)

Calcification (18) The illuviated (deposited) accumulation of calcium carbonate or magnesium carbonate in the B and C soil horizons.

Caldera (12) An interior sunken portion of a composite volcano's crater; usually steep-sided and circular, sometimes containing a lake; also can be found in conjunction with shield volcanoes.

Capillary water (9) Soil moisture, most of which is accessible to plant roots; held in the soil by the water's surface tension and cohesive forces between water and soil. (*See also* Available water, Field capacity, Hygroscopic water, Wilting point.)

Carbonation (13) A chemical weathering process in which weak carbonic acid (water and carbon dioxide) reacts with many minerals that contain calcium, magnesium, potassium, and sodium (especially limestone), transforming them into carbonates.

Carbon monoxide (CO) (3) An odorless, colorless, tasteless combination of carbon and oxygen produced by the incomplete combustion of fossil fuels or other carbon-containing substances; toxicity to humans is due to its affinity for hemoglobin, displacing oxygen in the bloodstream.

Carnivore (19) A secondary consumer that principally eats meat for sustenance. The top carnivore in a food chain is considered a tertiary consumer. (*Compare* Herbivore.)

Cartography (1) The making of maps and charts; a specialized science and art that blends aspects of geography, engineering, mathematics, graphics, computer science, and artistic specialties.

Cation-exchange capacity (CEC) (18) The ability of soil colloids to exchange cations between their surfaces and the soil solution; a measured potential that indicates soil fertility. (*See* Soil colloid, Soil fertility.)

Chaparral (20) Dominant shrub formations of *Mediterranean* (dry summer) climates; characterized by sclerophyllous scrub and short, stunted, tough forests; derived from the Spanish *chaparro*; specific to California. (*See* Mediterranean shrubland.)

Chemical weathering (13) Decomposition and decay of the constituent minerals in rock through chemical alteration of those minerals. Water is essential, with rates keyed to temperature and precipitation values. Chemical reactions are active at microsites even in dry climates. Processes include hydrolysis, oxidation, carbonation, and solution.

Chinook wind (8) North American term for a warm, dry, downslope airflow; characteristic of the rain-shadow region on the leeward side of mountains; known as *föhn* or *foehn* winds in Europe. (*See* Rain shadow.)

Chlorofluorocarbons (CFC) (3) A manufactured molecule (polymer) made of chlorine, fluorine, and carbon; inert and possessing remarkable heat properties; also known as one of the halogens. After slow transport to the stratospheric ozone layer, CFCs react with ultraviolet radiation, freeing chlorine atoms that act as a catalyst to produce reactions that destroy ozone; manufacture banned by international treaties.

Chlorophyll (19) A light-sensitive pigment that resides within chloroplasts (organelles) in leaf cells of plants; the basis of photosynthesis.

Cinder cone (12) A volcanic landform of pyroclastics and scoria, usually small and cone-shaped and generally not more than 450 m (1500 ft) in height, with a truncated top.

Circle of illumination (2) The division between light and dark on Earth; a day–night great circle.

Circum-Pacific belt (12) A tectonically and volcanically active region encircling the Pacific Ocean; also known as the "ring of fire."

Cirque (17) A scooped-out, amphitheater-shaped basin at the head of an alpine glacier valley; an erosional landform.

Cirrus (7) Wispy, filamentous ice-crystal clouds that occur above 6000 m (20,000 ft); appear in a variety of forms, from feathery hairlike fibers to veils of fused sheets.

Classification (10) The process of ordering or grouping data or phenomena in related classes; results in a regular distribution of information; a taxonomy.

Climate (10) The consistent, long-term behavior of weather over time, including its variability; in contrast to weather, which is the condition of the atmosphere at any given place and time.

Climatic region (10) An area of homogenous climate that features characteristic regional weather and air mass patterns.

Climatology (10) The scientific study of climate and climatic patterns and the consistent behavior of weather, including its variability and extremes, over time in one place or region; includes the effects of climate change on human society and culture.

Climograph (10) A graph that plots daily, monthly, or annual temperature and precipitation values for a selected station; may also include additional weather information.

Closed system (1) A system that is shut off from the surrounding environment, so that it is entirely self-contained in terms of energy and materials; Earth is a closed material system. (*Compare* Open system.)

Cloud (7) An aggregate of tiny moisture droplets and ice crystals; classified by altitude of occurrence and shape.

Cloud-albedo forcing (4) An increase in albedo (the reflectivity of a surface) caused by clouds due to their reflection of incoming insolation.

Cloud-condensation nuclei (7) Microscopic particles necessary as matter on which water vapor condenses to form moisture droplets; can be sea salts, dust, soot, or ash.

Cloud-greenhouse forcing (4) An increase in greenhouse warming caused by clouds because they can act like insulation, trapping longwave (infrared) radiation.

Col (17) Formed by two headward-eroding cirques that reduce an arête (ridge crest) to form a high pass or saddlelike narrow depression.

Cold desert and semidesert (20) A type of desert biome found at higher latitudes than warm deserts. Interior location and rain shadows produce these cold deserts in North America.

Cold front (8) The leading edge of an advancing cold air mass; identified on a weather map as a line marked with triangular spikes pointing in the direction of frontal movement. (*Compare* Warm front.)

Community (19) A convenient biotic subdivision within an ecosystem; formed by interacting populations of animals and plants in an area.

Composite volcano (12) A volcano formed by a sequence of explosive volcanic eruptions; steep-sided, conical in shape; sometimes referred to as a stratovolcano, although composite is the preferred term. (*Compare* Shield volcano.)

Conduction (4) The slow molecule-to-molecule transfer of heat through a medium, from warmer to cooler portions.

Cone of depression (9) The depressed shape of the water table around a well after active pumping. The water table adjacent to the well is drawn down by the water removal.

Confined aquifer (9) An aquifer that is bounded above and below by impermeable layers of rock or sediment. (*See* Artesian water, Unconfined aquifer.)

Constant isobaric surface (6) An elevated surface in the atmosphere on which all points have the same pressure, usually 500 mb. Along this constant-pressure surface, isobars mark the paths of upper-air winds.

Consumer (19) Organism in an ecosystem that depends on producers (organisms that use carbon dioxide as their sole source of carbon) for its source of nutrients; also called a *heterotroph*. (*Compare* Producer.)

Consumptive use (9) A use that removes water from a water budget at one point and makes it unavailable farther downstream. (*Compare* Withdrawal.)

Continental divide (14) A ridge or elevated area that separates drainage on a continental scale; specifically, that ridge in North America that separates drainage to the Pacific on the west side from drainage to the Atlantic and Gulf on the east side and to Hudson Bay and the Arctic Ocean in the north.

Continental drift (11) A proposal by Alfred Wegener in 1912 stating that Earth's landmasses have migrated over the past 225 million years from a supercontinent he called Pangaea to the present configuration; the widely accepted plate tectonics theory today. (*See* Plate tectonics.)

Continental effect (5) A quality of regions that lack the temperature-moderating effects of the ocean and that exhibit a greater range of minimum and maximum temperatures, both daily and annually, than do marine stations. (*See* Marine effect, Land–water heating difference.)

Continental glacier (17) A continuous mass of unconfined ice, covering at least 50,000 km^2 (19,500 mi^2); most extensive at present as ice sheets covering Greenland and Antarctica. (*Compare* Alpine glacier.)

Continental landmasses (12) The broadest category of landform, including those masses of crust that reside above or near sea level and the adjoining undersea continental shelves along the coastline; sometimes synonymous with *continental platforms*.

Continental shield (12) Generally, old, low-elevation heartland regions of continental crust; various cratons (granitic cores) and ancient mountains are exposed at the surface.

Contour lines (Appendix A) Isolines on a topographic map that connect all points at the same elevation relative to a reference elevation called the vertical datum.

Convection (4) Transfer of heat from one place to another through the physical movement of air; involves a strong vertical motion. (*Compare* Advection.)

Convectional lifting (8) Air passing over warm surfaces gains buoyancy and lifts, initiating adiabatic processes.

Convergent lifting (8) Airflows in conflict, force lifting and displacement of air upward, initiating adiabatic processes.

Coordinated Universal Time (UTC) (1) The official reference time in all countries, formerly known as Greenwich Mean Time; now measured by primary standard atomic clocks, the time calculations are collected in Paris at the International Bureau of Weights and Measures (BIPM); the legal reference for time in all countries and broadcast worldwide.

Coral (16) A simple, cylindrical marine animal with a saclike body that secretes calcium carbonate to form a hard external skeleton and, cumulatively, landforms called reefs; lives symbiotically with nutrient-producing algae; presently in a worldwide state of decline due to bleaching (loss of algae).

Core (11) The deepest inner portion of Earth, representing one-third of its entire mass; differentiated into two zones—a solid-iron inner core surrounded by a dense, molten, fluid metallic-iron outer core.

Coriolis force (6) The apparent deflection of moving objects (wind, ocean currents, missiles) from travelling in a straight path, in proportion to the speed of Earth's rotation at different latitudes. Deflection is to the right in the Northern Hemisphere and to the left in the Southern Hemisphere; maximum at the poles and zero along the equator.

Crater (12) A circular surface depression formed by volcanism; built by accumulation, collapse, or explosion; usually located at a volcanic vent or pipe; can be at the summit or on the flank of a volcano.

Crevasse (17) A vertical crack that develops in a glacier as a result of friction between valley walls, or tension forces of extension on convex slopes, or compression forces on concave slopes.

Crust (11) Earth's outer shell of crystalline surface rock, ranging from 5 to 60 km (3 to 38 mi) in thickness from oceanic crust to mountain ranges. Average density of continental crust is $2.7 g/cm^3$, whereas oceanic crust is $3.0 g/cm^3$.

Cryosphere (1, 17) The frozen portion of Earth's waters, including ice sheets, ice caps and fields, glaciers, ice shelves, sea ice, and subsurface ground ice and frozen ground (permafrost).

Cumulonimbus (7) A towering, precipitation-producing cumulus cloud that is vertically developed across altitudes associated with other clouds; frequently associated with lightning and thunder and thus sometimes called a *thunderhead.*

Cumulus (7) Bright and puffy cumuliform clouds up to 2000 m (6500 ft) in altitude.

Cyclogenesis (8) An atmospheric process that describes the birth of a midlatitude wave cyclone, usually along the polar front. Also refers to strengthening and development of a midlatitude cyclone along the eastern slope of the Rockies, other north–south mountain barriers, and the North American and Asian east coasts. (*See* Midlatitude cyclone, Polar front.)

Cyclone (6) A dynamically or thermally caused area of low atmospheric pressure with ascending and converging airflows that rotate counterclockwise in the Northern Hemisphere and clockwise in the Southern Hemisphere. (*Compare* Anticyclone; *see* Midlatitude cyclone, Tropical cyclone.)

Daylength (2) Duration of exposure to insolation, varying during the year depending on latitude; an important aspect of seasonality.

Daylight saving time (1) Time is set ahead 1 hour in the spring and set back 1 hour in the fall in the Northern Hemisphere. In the United States and Canada, time is set ahead on the second Sunday in March and set back on the first Sunday in November—except in Hawai'i, Arizona, and Saskatchewan, which exempt themselves.

Debris avalanche (13) A mass of falling and tumbling rock, debris, and soil; can be dangerous because of the tremendous velocities achieved by the onrushing materials.

December solstice (2) *See* Winter (December) solstice.

Declination (2) The latitude that receives direct overhead (perpendicular) insolation on a particular day; the subsolar point migrates annually through 47° of latitude between the Tropics of Cancer (23.5° N) and Capricorn (23.5° S).

Decomposers (19) Bacteria and fungi that digest organic debris outside their bodies and absorb and release nutrients in an ecosystem. (*See* Detritivores.)

Deficit (9) DEFIC; in a water balance, the amount of unmet (unsatisfied) potential evapotranspiration (POTET, or PE); a natural water shortage. (*See* Potential evapotranspiration.)

Deflation (15) A process of wind erosion that removes and lifts individual particles, literally blowing away unconsolidated, dry, or noncohesive sediments. (*See* Blowout depression.)

Delta (14) A depositional plain formed where a river enters a lake or an ocean; named after the triangular shape of the Greek letter delta, Δ.

Denudation (13) A general term that refers to all processes that cause degradation of the landscape: weathering, mass movement, erosion, and transport.

Deposition (14) The process whereby weathered, wasted, and transported sediments are laid down by air, water, and ice.

Derechos (8) Strong linear winds in excess of 26 m/sec (58 mph), associated with thunderstorms and bands of showers crossing a region.

Desalination (9) In a water resources context, the removal of organics, debris, and salinity from seawater through distillation or reverse osmosis to produce potable water.

Desert biome (20) Arid landscape of uniquely adapted dry-climate plants and animals.

Desertification (15) The expansion of deserts worldwide, related principally to poor agricultural practices (overgrazing and inappropriate agricultural practices), improper soil-moisture management, erosion and salinization, deforestation, and the ongoing climatic change; an unwanted semipermanent invasion into neighboring biomes.

Desert pavement (15) On arid landscapes, a surface formed when wind deflation and sheetflow remove smaller particles, leaving residual pebbles and gravels to concentrate at the surface; an alternative sediment-accumulation hypothesis explains some desert pavements; resembles a cobblestone street. (*See* Deflation, Sheetflow.)

Detritivores (19) Detritus feeders and decomposers that consume, digest, and destroy organic wastes and debris. *Detritus feeders*—worms, mites, termites, centipedes, snails, crabs, and even vultures, among others—consume detritus and excrete nutrients and simple inorganic compounds that fuel an ecosystem. (*Compare* Decomposers.)

Dew-point temperature (7) The temperature at which a given mass of air becomes saturated, holding all the water it can hold. Any further cooling or addition of water vapor results in active condensation.

Diagnostic subsurface horizon (18) A soil horizon that originates below the epipedon at varying depths; may be part of the A and B horizons; important in soil description as part of the Soil Taxonomy.

Differential weathering (13) The effect of different resistances in rock, coupled with variations in the intensity of physical and chemical weathering.

Diffuse radiation (4) The downward component of scattered incoming insolation from clouds and the atmosphere.

Discharge (14) The measured volume of flow in a river that passes by a given cross section of a stream in a given unit of time; expressed in cubic meters per second or cubic feet per second.

Dissolved load (14) Materials carried in chemical solution in a stream, derived from minerals such as limestone and dolomite or from soluble salts.

Downwelling current (6) An area of the sea where a convergence or accumulation of water thrusts excess water downward; occurs, for example, at the western end of the equatorial current or along the margins of Antarctica. (*Compare* Upwelling current.)

Drainage basin (14) The basic spatial geomorphic unit of a river system; distinguished from a neighboring basin by ridges and highlands that form divides, marking the limits of the catchment area of the drainage basin.

Drainage density (14) A measure of the overall operational efficiency of a drainage basin, determined by the ratio of combined channel lengths to the unit area.

Drainage pattern (14) A distinctive geometric arrangement of streams in a region, determined by slope, differing rock resistance to weathering and erosion, climatic and hydrologic variability, and structural controls of the landscape.

Drawdown (9) *See* Cone of depression.

Drought (9) Does not have a simple water-budget definition; rather, it can occur in at least four forms: *meteorological drought*, *agricultural drought*, *hydrologic drought*, and/or *socioeconomic drought*.

Drumlin (17) A depositional landform related to glaciation that is composed of till (unstratified, unsorted) and is streamlined in the direction of continental ice movement—blunt end upstream and tapered end downstream with a rounded summit.

Dry adiabatic rate (DAR) (7) The rate at which an unsaturated parcel of air cools (if ascending) or heats (if descending); a rate of 10 C° per 1000 m (5.5 F° per 1000 ft). (*See* Adiabatic; *compare* Moist adiabatic rate.)

Dune (15) A depositional feature of sand grains deposited in transient mounds, ridges, and hills; extensive areas of sand dunes are called sand seas.

Dust dome (4) A dome of airborne pollution associated with every major city; may be blown by winds into elongated plumes downwind from the city.

Dynamic equilibrium (1) Increasing or decreasing operations in a system demonstrate a trend over time, a change in average conditions.

Dynamic equilibrium model (13) The balancing act between tectonic uplift and erosion, between the resistance of crust materials and the work of denudation processes. Landscapes evidence ongoing adaptation to rock structure, climate, local relief, and elevation.

Earthquake (12) A sharp release of energy that sends waves travelling through Earth's crust at the moment of rupture along a fault or in association with volcanic activity. The moment magnitude scale (formerly the Richter scale) estimates earthquake magnitude; intensity is described by the Mercalli scale.

Earth systems science (1) An emerging science of Earth as a complete, systematic entity. An interacting set of physical, chemical, and biological systems that produce the processes of a whole Earth system. A study of planetary change resulting from system operations; includes a desire for a more quantitative understanding among components rather than a qualitative description.

Ebb tide (16) Falling or lowering tide during the daily tidal cycle. (*Compare* Flood tide.)

Ecological succession (19) The process whereby different and usually more complex assemblages of plants and animals replace older and usually simpler communities; communities are in a constant state of change as each species adapts to changing conditions. Ecosystems do not exhibit a stable point or successional climax condition as previously thought. (*See* Primary succession, Secondary succession.)

Ecology (19) The science that studies the relations between organisms and their environment and among various ecosystems.

Ecosphere (1) Another name for the biosphere.

Ecosystem (19) A self-regulating association of living plants and animals and their nonliving physical and chemical environments.

Ecotone (20) A boundary transition zone between adjoining ecosystems that may vary in width and represent areas of tension as similar species of plants and animals compete for the resources. (*See* Ecosystem.)

Effusive eruption (12) A volcanic eruption characterized by low-viscosity basaltic magma and low-gas content, which readily escapes. Lava pours forth onto the surface with relatively small explosions and few pyroclastics; tends to form shield volcanoes. (*See* Shield volcano, Lava, Pyroclastics; *compare* Explosive eruption.)

Elastic-rebound theory (12) A concept describing the faulting process in Earth's crust, in which the two sides of a fault appear locked despite the motion of adjoining pieces of crust, but with accumulating strain, they rupture suddenly, snapping to new positions relative to each other, generating an earthquake.

Electromagnetic spectrum (2) All the radiant energy produced by the Sun placed in an ordered range, divided according to wavelengths.

Eluviation (18) The removal of finer particles and minerals from the upper horizons of soil; an erosional process within a soil body. (*Compare* Illuviation.)

Empirical classification (10) A climate classification based on weather statistics or other data; used to determine general climate categories. (*Compare* Genetic classification.)

Endogenic system (11) The system internal to Earth, driven by radioactive heat derived from sources within the planet. In response, the surface fractures, mountain building occurs, and earthquakes and volcanoes are activated. (*Compare* Exogenic system.)

Entisols (18) A soil order in the Soil Taxonomy. Specifically lacks vertical development of horizons; usually young or undeveloped. Found in active slopes, alluvial-filled floodplains, and poorly drained tundra.

Environmental lapse rate (3) The actual rate of temperature decrease with increasing altitude in the lower atmosphere at any particular time under local weather conditions; may deviate above or below the normal lapse rate of 6.4 C° per km, or 1000 m (3.5 F° per 1000 ft). (*Compare* Normal lapse rate.)

Eolian (15) Caused by wind; refers to the erosion, transportation, and deposition of materials; spelled *aeolian* in some countries.

Epipedon (18) The diagnostic soil horizon that forms at the surface; not to be confused with the A horizon; may include all or part of the illuviated B horizon.

Equal area (1, Appendix A) A trait of a map projection; indicates the equivalence of all areas on the surface of the map, although shape is distorted. (*See* Map projection.)

Equatorial low-pressure trough (6) A thermally caused low-pressure area that almost girdles Earth, with air converging and ascending all along its extent; also called the *intertropical convergence zone (ITCZ)*.

Erg desert (15) A sandy desert, or area where sand is so extensive that it constitutes a sand sea.

Erosion (14) Denudation by wind, water, or ice, which dislodges, dissolves, or removes surface material.

Esker (17) A sinuously curving, narrow deposit of coarse gravel that forms along a meltwater stream channel, developing in a tunnel beneath a glacier.

Estuary (14) The point at which the mouth of a river enters the sea, where freshwater and seawater are mixed; a place where tides ebb and flow.

Eustasy (7) Refers to worldwide changes in sea level that are related not to movements of land, but rather to changes in the volume of water in the oceans.

Evaporation (9) The movement of free water molecules away from a wet surface into air that is less than saturated; the phase change of water to water vapor.

Evaporation fog (7) A fog formed when cold air flows over the warm surface of a lake, ocean, or other body of water; forms as the water molecules evaporate from the water surface into the cold, overlying air; also known as steam fog or sea smoke.

Evaporation pan (9) A weather instrument consisting of a standardized pan from which evaporation occurs, with water automatically replaced and measured; an evaporimeter.

Evapotranspiration (9) The merging of evaporation and transpiration water loss into one term. (*See* Potential evapotranspiration, Actual evapotranspiration.)

Evolution (19) A theory that single-cell organisms adapted, modified, and passed along inherited changes to multicellular organisms. The genetic makeup of successive generations is shaped by environmental factors, physiological functions, and behaviors that created a greater rate of survival and reproduction and were passed along through natural selection.

Exfoliation dome (13) A dome-shaped feature of weathering, produced by the response of granite to the overburden removal process, which relieves pressure from the rock. Layers of rock slough (pronounced "sluff") off in slabs or shells in a sheeting process.

Exogenic system (11) Earth's external surface system, powered by insolation, which energizes air, water, and ice and sets them in motion, under the influence of gravity. Includes all processes of landmass denudation. (*Compare* Endogenic system.)

Exosphere (3) An extremely rarefied outer atmospheric halo beyond the thermopause at an altitude of 480 km (300 mi); probably composed of hydrogen and helium atoms, with some oxygen atoms and nitrogen molecules present near the thermopause.

Exotic stream (14) A river that rises in a humid region and flows through an arid region, with discharge decreasing toward the mouth; for example, the Nile River and the Colorado River.

Explosive eruption (12) A violent and unpredictable volcanic eruption, the result of magma that is thicker (more viscous), stickier, and higher in gas and silica content than that of an effusive eruption; tends to form blockages within a volcano; produces composite volcanic landforms. (*See* Composite volcano; *compare* Effusive eruption.)

Faulting (12) The process whereby displacement and fracturing occur between two portions of Earth's crust; usually associated with earthquake activity.

Feedback loop (1) Created when a portion of system output is returned as an information input, causing changes that guide further system operation. (*See* Negative feedback, Positive feedback.)

Field capacity (9) Water held in the soil by hydrogen bonding against the pull of gravity, remaining after water drains from the larger pore spaces; the available water for plants. (*See* Available water, Capillary water.)

Fire ecology (19) The study of fire as a natural agent and dynamic factor in community succession.

Firn (17) Snow of a granular texture that is transitional in the slow transformation from snow to glacial ice; snow that has persisted through a summer season in the zone of accumulation.

Firn line (17) The snow line that is visible on the surface of a glacier, where winter snows survive the summer ablation season; analogous to a snow line on land. (*See* Ablation.)

Fjord (17) A drowned glaciated valley, or glacial trough, along a seacoast.

Flash flood (15) A sudden and short-lived torrent of water that exceeds the capacity of a stream channel; associated with desert and semiarid washes.

Flood (14) A high water level that overflows the natural riverbank along any portion of a stream.

Floodplain (14) A flat, low-lying area along a stream channel, created by and subject to recurrent flooding; alluvial deposits generally mask underlying rock.

Flood tide (16) Rising tide during the daily tidal cycle. (*Compare* Ebb tide.)

Fluvial (14) Stream-related processes; from the Latin *fluvius* for "river" or "running water."

Fog (7) A cloud, generally stratiform, in contact with the ground, with visibility usually reduced to less than 1 km (3300 ft).

Folding (12) The bending and deformation of beds of rock strata subjected to compressional forces.

Food chain (19) The circuit along which energy flows from producers (plants), which manufacture their own food, to consumers (animals); a one-directional flow of chemical energy, ending with decomposers.

Food web (19) A complex network of interconnected food chains. (*See* Food chain.)

Formation class (20) That portion of a biome that concerns the plant communities only, categorized by size, shape, and structure of the dominant vegetation.

Friction force (6) The effect of drag by the wind as it moves across a surface; may be operative through 500 m (1600 ft) of altitude. Surface friction slows the wind and therefore reduces the effectiveness of the Coriolis force.

Frost action (13) A powerful mechanical force produced as water expands up to 9% of its volume as it freezes. Water freezing in a cavity in a rock can break the rock if it exceeds the rock's tensional strength.

Funnel cloud (8) The visible swirl extending from the bottom side of a cloud, which may or may not develop into a tornado. A tornado is a funnel cloud that has extended all the way to the ground. (*See* Tornado.)

Fusion (2) The process of forcibly joining positively charged hydrogen and helium nuclei under extreme temperature and pressure; occurs naturally in thermonuclear reactions within stars, such as our Sun.

Gelisols (18) A new soil order in the Soil Taxonomy, added in 1998, describing cold and frozen soils at high latitudes or high elevations; characteristic tundra vegetation.

General circulation model (GCM) (10) Complex, computer-based climate model that produces generalizations of reality and forecasts of future weather and climate conditions. Complex GCMs (three-dimensional models) are in use in the United States and in other countries.

Genetic classification (10) A climate classification that uses causative factors to determine climatic regions; for example, an analysis of the effect of interacting air masses. (*Compare* Empirical classification.)

Geodesy (1) The science that determines Earth's shape and size through surveys, mathematical means, and remote sensing. (*See* Geoid.)

Geographic information system (GIS) (1) A computer-based data processing tool or methodology used for gathering, manipulating, and analyzing geographic information to produce a holistic, interactive analysis.

Geography (1) The science that studies the interdependence and interaction among geographic areas, natural systems, processes, society, and cultural activities over space—a spatial science. The five themes of geographic education are location, place, movement, regions, and human–Earth relationships.

Geoid (1) A word that describes Earth's shape; literally, "the shape of Earth is Earth-shaped." A theoretical surface at sea level that extends through the continents; deviates from a perfect sphere.

Geologic cycle (11) A general term characterizing the vast cycling that proceeds in the lithosphere. It encompasses the hydrologic cycle, tectonic cycle, and rock cycle.

Geologic time scale (11) A depiction of eras, periods, and epochs that span Earth's history; shows both the sequence of rock strata and their absolute dates, as determined by methods such as radioactive isotopic dating.

Geomagnetic reversal (11) A polarity change in Earth's magnetic field. With uneven regularity, the magnetic field fades to zero and then returns to full strength, but with the magnetic poles reversed. Reversals have been recorded nine times during the past 4 million years.

Geomorphic threshold (13) The threshold up to which landforms change before lurching to a new set of relationships, with rapid realignments of landscape materials and slopes.

Geomorphology (13) The science that analyzes and describes the origin, evolution, form, classification, and spatial distribution of landforms.

Geostrophic wind (6) A wind moving between areas of different pressure along a path that is parallel to the isobars. It is a product of the pressure gradient force and the Coriolis force. (*See* Isobar, Pressure gradient force, Coriolis force.)

Geothermal energy (11) The energy in steam and hot water heated by subsurface magma near groundwater. Geothermal energy literally refers to heat from Earth's interior, whereas *geothermal power* relates to specific applied strategies of geothermal electric or geothermal direct applications. This energy is used in Iceland, New Zealand, Italy, and northern California, among other locations.

Glacial drift (17) The general term for all glacial deposits, both unsorted (till) and sorted (stratified drift).

Glacial ice (17) A hardened form of ice, very dense in comparison to normal snow or firn.

Glacier (17) A large mass of perennial ice resting on land or floating shelflike in the sea adjacent to the land; formed from the accumulation and recrystallization of snow, which then flows slowly under the pressure of its own weight and the pull of gravity.

Glacier surge (17) The rapid, lurching, unexpected forward movement of a glacier.

Glacio-eustatic (7) Changes in sea level in response to changes in the amount of water stored on Earth as ice; the more water that is bound up in glaciers and ice sheets, the lower the sea level. (*Compare* Eustasy.)

Gleization (18) A process of humus and clay accumulation in cold, wet climates with poor drainage.

Global dimming (4) The decline in sunlight reaching Earth's surface due to pollution, aerosols, and clouds.

Global Positioning System (GPS) (1) Latitude, longitude, and elevation are accurately calibrated using a handheld instrument that receives radio signals from satellites.

Goode's homolosine projection (Appendix A) An equal-area projection formed by splicing together a sinusoidal and a homolographic projection.

Graben (12) Pairs or groups of faults that produce downward-faulted blocks; characteristic of the basins of the interior western United States. (*Compare* Horst; *see* Basin and Range Province.)

Graded stream (14) An idealized condition in which a stream's load and the landscape mutually adjust. This forms a dynamic equilibrium among erosion, transported load, deposition, and the stream's capacity.

Gradient (14) The drop in elevation from a stream's headwaters to its mouth, ideally forming a concave slope.

Granite (11) A coarse-grained (slow-cooling) intrusive igneous rock of 25% quartz and more than 50% potassium and sodium feldspars; characteristic of the continental crust.

Gravitational water (9) That portion of surplus water that percolates downward from the capillary zone, pulled by gravity to the groundwater zone.

Gravity (2) The mutual force exerted by the masses of objects that are attracted one to another and produced in an amount proportional to each object's mass.

Great circle (1) Any circle drawn on a globe with its center coinciding with the center of the globe. An infinite number of great circles can be drawn, but only one parallel of latitude—the equator—is a great circle. (*Compare* Small circle.)

Greenhouse effect (4) The process whereby radiatively active gases (carbon dioxide, water vapor, methane, and CFCs) absorb and emit the energy at longer wavelengths, which are retained longer, delaying the loss of infrared to space. Thus, the lower troposphere is warmed through the radiation and reradiation of infrared wavelengths. The approximate similarity between this process and that of a greenhouse explains the name.

Greenwich Mean Time (GMT) (1) Former world standard time, now reported as Coordinated Universal Time (UTC). (*See* Coordinated Universal Time.)

Groundwater (9) Water beneath the surface that is beyond the soil-root zone; a major source of potable water.

Groundwater mining (9) Pumping an aquifer beyond its capacity to flow and recharge; an overuse of the groundwater resource.

Gulf Stream (5) A strong, northward-moving, warm current off the east coast of North America, which carries its water far into the North Atlantic.

Habitat (19) A physical location to which an organism is biologically suited. Most species have specific habitat parameters and limits. (*Compare* Niche.)

Hail (8) A type of precipitation formed when a raindrop is repeatedly circulated above and below the freezing level in a cloud, with each cycle freezing more moisture onto the hailstone until it becomes too heavy to stay aloft.

Hair hygrometer (7) An instrument for measuring relative humidity; based on the principle that human hair will change as much as 4% in length between 0% and 100% relative humidity.

Heat (3) The flow of kinetic energy from one body to another because of a temperature difference between them.

Herbivore (19) The primary consumer in a food web, which eats plant material formed by a producer (plant) that has photosynthesized organic molecules. (*Compare* Carnivore.)

Heterosphere (3) A zone of the atmosphere above the mesopause, from 80 km (50 mi) to 480 km (300 mi) in altitude; composed of rarefied layers of oxygen atoms and nitrogen molecules; includes the ionosphere.

Histosols (18) A soil order in the Soil Taxonomy. Formed from thick accumulations of organic matter, such as beds of former lakes, bogs, and layers of peat.

Homosphere (3) A zone of the atmosphere from Earth's surface up to 80 km (50 mi), composed of an even mixture of gases, including nitrogen, oxygen, argon, carbon dioxide, and trace gases.

Horn (17) A pyramidal, sharp-pointed peak that results when several cirque glaciers gouge an individual mountain summit from all sides.

Horst (12) Upward-faulted blocks produced by pairs or groups of faults; characteristic of the mountain ranges of the interior of the western United States. (*See* Graben, Basin and Range Province.)

Hot spot (11) An individual point of upwelling material originating in the asthenosphere, or deeper in the mantle; tends to remain fixed relative to migrating plates; some 100 are identified worldwide, exemplified by Yellowstone National Park, Hawai'i, and Iceland.

Human–Earth relationships (1) One of the oldest themes of geography (the human–land tradition); includes the spatial analysis of settlement patterns, resource utilization and exploitation, hazard perception and planning, and the impact of environmental modification and artificial landscape creation.

Humidity (7) Water vapor content of the air. The capacity of the air for water vapor is mostly a function of the temperature of the air and the water vapor.

Humus (18) A mixture of organic debris in the soil worked by consumers and decomposers in the humification process; characteristically formed from plant and animal litter deposited at the surface.

Hurricane (8) A tropical cyclone that is fully organized and intensified in inward-spiraling rainbands; ranges from 160 to 960 km (100 to 600 mi) in diameter, with wind speeds in excess of 119 kmph (65 knots, or 74 mph); a name used specifically in the Atlantic and eastern Pacific. (*Compare* Typhoon.)

Hydration (13) A chemical weathering process involving water that is added to a mineral, which initiates swelling and stress within the rock, mechanically forcing grains apart as the constituents expand. (*Compare* Hydrolysis.)

Hydraulic action (14) The erosive work accomplished by the turbulence of water; causes a squeezing and releasing action in joints in bedrock; capable of prying and lifting rocks.

Hydrograph (14) A graph of stream discharge (in m^3/s or cfs, ft^3/s) over a period of time (minutes, hours, days, years) at a specific place on a stream. The relationship between stream discharge and precipitation input is illustrated on the graph.

Hydrologic cycle (9) A simplified model of the flow of water, ice, and water vapor from place to place. Water flows through the atmosphere and across the land, where it is stored as ice and as groundwater. Solar energy empowers the cycle.

Hydrology (14) The science of water, including its global circulation, distribution, and properties—specifically water at and below Earth's surface.

Hydrolysis (13) A chemical weathering process in which minerals chemically combine with water; a decomposition process that causes silicate minerals in rocks to break down and become altered. (*Compare* Hydration.)

Hydrosphere (1) An abiotic open system that includes all of Earth's water.

Hygroscopic water (9) That portion of soil moisture that is so tightly bound to each soil particle that it is unavailable to plant roots; the water, along with some bound capillary water, that is left in the soil after the wilting point is reached. (*See* Wilting point.)

Ice age (17) A cold episode, with accompanying alpine and continental ice accumulations, that has repeated roughly every 200 to 300 million years since the late Precambrian Era (1.25 billion years ago); includes the most recent episode during the Pleistocene Ice Age, which began 1.65 million years ago.

Iceberg (17) Floating ice created by large pieces of ice calving (breaking off) and floating adrift; a hazard to shipping because about 6/7 (86%) of the ice is submerged and can be irregular in form.

Ice cap (17) A large, dome-shaped glacier, less extensive than an ice sheet although it buries mountain peaks and the local landscape; generally, less than 50,000 km^2 (19,300 mi^2).

Ice field (17) The least extensive form of a glacier, with mountain ridges and peaks visible above the ice; less than an ice cap or ice sheet.

Icelandic low (6) *See* Subpolar low-pressure cell.

Ice sheet (17) An enormous continuous continental glacier. The bulk of glacial ice on Earth covers Antarctica and Greenland in two ice sheets.

Igneous rock (11) One of the basic rock types; it has solidified and crystallized from a hot molten state (either magma or lava). (*Compare* Metamorphic rock, Sedimentary rock.)

Illuviation (18) The downward movement and deposition of finer particles and minerals from the upper horizon of the soil; a depositional process. Deposition usually is in the B horizon, where accumulations of clays, aluminum, carbonates, iron, and some humus occur. (*Compare* Eluviation; *see* Calcification.)

Inceptisols (18) A soil order in the Soil Taxonomy. Weakly developed soils that are inherently infertile; usually, young soils that are weakly developed, although they are more developed than Entisols.

Industrial smog (3) Air pollution associated with coal-burning industries; it may contain sulfur oxides, particulates, carbon dioxide, and exotics.

Infiltration (9) Water access to subsurface regions of soil moisture storage through penetration of the soil surface.

Insolation (2) Solar radiation that is incoming to Earth systems.

Interception (9) A delay in the fall of precipitation toward Earth's surface caused by vegetation or other ground cover.

Internal drainage (14) In regions where rivers do not flow into the ocean, the outflow is through evaporation or subsurface gravitational flow. Portions of Africa, Asia, Australia, and the western United States have such drainage.

International Date Line (1) The 180° meridian, an important corollary to the prime meridian on the opposite side of the planet; established by an 1884 treaty to mark the place where each day officially begins.

Intertropical convergence zone (ITCZ) (6) *See* Equatorial low-pressure trough.

Invasive species (20) Species that are brought, or introduced, from elsewhere by humans, either accidentally or intentionally. These non-native species are also known as *exotic species* or *alien species*.

Ionosphere (3) A layer in the atmosphere above 80 km (50 mi) where gamma, X-ray, and some ultraviolet radiation is absorbed and converted into longer wavelengths and where the solar wind stimulates the auroras.

Island biogeography (19) Island communities are special places for study because of their spatial isolation and the relatively small number of species present. Islands resemble natural experiments because the impact of individual factors, such as civilization, can be more easily assessed on islands than over larger continental areas.

Isobar (6) An isoline connecting all points of equal atmospheric pressure.

Isostasy (11) A state of equilibrium in Earth's crust formed by the interplay between portions of the less-dense lithosphere and the more-dense asthenosphere and the principle of buoyancy. The crust depresses under weight and recovers with its removal—for example, with the melting of glacial ice. The uplift is known as isostatic rebound.

Isotherm (5) An isoline connecting all points of equal temperature.

Jet contrails (4) Condensation trails produced by aircraft exhaust, particulates, and water vapor can form high cirrus clouds, sometimes called *false cirrus clouds*.

Jet stream (6) The most prominent movement in upper-level westerly wind flows; irregular, concentrated, sinuous bands of geostrophic wind, travelling at 300 kmph (190 mph).

Joint (13) A fracture or separation in rock that occurs without displacement of the sides; increases the surface area of rock exposed to weathering processes.

June solstice (2) *See* Summer (June) solstice.

Kame (17) A depositional feature of glaciation; a small hill of poorly sorted sand and gravel that accumulates in crevasses or in ice-caused indentations in the surface.

Karst topography (13) Distinctive topography formed in a region of chemically weathered limestone with poorly developed surface drainage and solution features that appear pitted and bumpy; originally named after the Krš Plateau in Slovenia.

Katabatic winds (6) Air drainage from elevated regions, flowing as gravity winds. Layers of air at the surface cool, become denser, and flow downslope; known worldwide by many local names.

Kettle (17) Forms when an isolated block of ice persists in a ground moraine, an outwash plain, or a valley floor after a glacier retreats; as the block finally melts, it leaves behind a steep-sided hole that frequently fills with water.

Kinetic energy (3) The energy of motion in a body; derived from the vibration of the body's own movement and stated as temperature.

Köppen–Geiger climate classification (10, Appendix B) An empirical classification system that uses average monthly temperatures, average monthly precipitation, and total annual precipitation to establish regional climate designations.

Lagoon (16) An area of coastal seawater that is virtually cut off from the ocean by a bay barrier or barrier beach; also, the water surrounded and enclosed by an atoll.

Landfall (8) The location along a coast where a storm moves onshore.

Land–sea breeze (6) Wind along coastlines and adjoining interior areas created by different heating characteristics of land and water surfaces—onshore (landward) breeze in the afternoon and offshore (seaward) breeze at night.

Landslide (13) A sudden rapid downslope movement of a cohesive mass of regolith and/or bedrock in a variety of mass-movement forms under the influence of gravity; a form of mass movement.

Land–water heating difference (5) Differences in the degree and way that land and water heat, as a result of contrasts in transmission, evaporation, mixing, and specific heat capacities. Land surfaces heat and cool faster than water and have continentality, whereas water provides a marine influence.

Latent heat (7) Heat energy is stored in one of three states—ice, water, or water vapor. The energy is absorbed or released in each phase change from one state to another. Heat energy is absorbed as the latent heat of melting, vaporization, or evaporation. Heat energy is released as the latent heat of condensation and freezing (or fusion).

Latent heat of condensation (7) The heat energy released to the environment in a phase change from water vapor to liquid; under normal sea-level pressure, 540 calories are released from each gram of water vapor that changes phase to water at boiling, and 585 calories are released from each gram of water vapor that condenses at 20°C (68°F).

Latent heat of sublimation (7) The heat energy absorbed or released in the phase change from ice to water vapor or water vapor to ice—no liquid phase. The change from water vapor to ice is also called deposition.

Latent heat of vaporization (7) The heat energy absorbed from the environment in a phase change from liquid to water vapor at the boiling point; under normal sea-level pressure, 540 calories must be added to each gram of boiling water to achieve a phase change to water vapor.

Lateral moraine (17) Debris transported by a glacier that accumulates along the sides of the glacier and is deposited along these margins.

Laterization (18) A pedogenic process operating in well-drained soils that occurs in warm and humid regions; typical of Oxisols. Plentiful precipitation leaches soluble minerals and soil constituents. Resulting soils usually are reddish or yellowish.

Latitude (1) The angular distance measured north or south of the equator from a point at the center of Earth. A line connecting all points of the same latitudinal angle is a parallel. (*Compare* Longitude.)

Lava (11, 12) Magma that issues from volcanic activity onto the surface; the extrusive rock that results when magma solidifies. (*See* Magma.)

Life zone (19) A zonation by altitude of plants and animals that form distinctive communities. Each life zone possesses its own temperature and precipitation relations.

Lightning (8) Flashes of light caused by tens of millions of volts of electrical charge heating the air to temperatures of 15,000°C to 30,000°C.

Limestone (11) The most common chemical sedimentary rock (nonclastic); it is lithified calcium carbonate; very susceptible to chemical weathering by acids in the environment, including carbonic acid in rainfall.

Limiting factor (19) The physical or chemical factor that most inhibits biotic processes, through either lack or excess.

Lithification (11) The compaction, cementation, and hardening of sediments into sedimentary rock.

Lithosphere (1, 11) Earth's crust and that portion of the uppermost mantle directly below the crust, extending down about 70 km (45 mi). Some sources use this term to refer to the entire Earth.

Littoral drift (16) Transport of sand, gravel, sediment, and debris along the shore; a more comprehensive term that considers *beach drift* and *longshore drift* combined.

Littoral zone (16) A specific coastal environment; that region between the high-water line during a storm and a depth at which storm waves are unable to move sea-floor sediments.

Loam (18) A soil that is a mixture of sand, silt, and clay in almost equal proportions, with no one texture dominant; an ideal agricultural soil.

Location (1) A basic theme of geography dealing with the absolute and relative positions of people, places, and things on Earth's surface.

Loess (15) Large quantities of fine-grained clays and silts left as glacial outwash deposits; subsequently blown by the wind great distances and redeposited as a generally unstratified, homogeneous blanket of material covering existing landscapes; in China, loess originated from desert lands.

Longitude (1) The angular distance measured east or west of a prime meridian from a point at the center of Earth. A line connecting all points of the same longitude is a meridian. (*Compare* Latitude.)

Longshore current (16) A current that forms parallel to a beach as waves arrive at an angle to the shore; generated in the surf zone by wave action, transporting large amounts of sand and sediment. (*See* Beach drift.)

Lysimeter (9) A weather instrument for measuring potential and actual evapotranspiration; isolates a portion of a field so that the moisture moving through the plot is measured.

Magma (11) Molten rock from beneath Earth's surface; fluid, gaseous, under tremendous pressure, and either intruded into existing crustal rock or extruded onto the surface as lava. (*See* Lava.)

Magnetosphere (2) Earth's magnetic force field, which is generated by dynamo-like motions within the planet's outer core; deflects the solar wind flow toward the upper atmosphere above each pole.

Mangrove swamp (16) A wetland ecosystem between 30° N and 30° S; tends to form a distinctive community of mangrove plants. (*Compare* Salt marsh.)

Mantle (11) An area within the planet representing about 80% of Earth's total volume, with densities increasing with depth and averaging $4.5 g/cm^3$; occurs between the core and the crust; is rich in iron and magnesium oxides and silicates.

Map (1, Appendix A) A generalized view of an area, usually some portion of Earth's surface, as seen from above at a greatly reduced size. (*See* Scale, Map projection.)

Map projection (1, Appendix A) The reduction of a spherical globe onto a flat surface in some orderly and systematic realignment of the latitude and longitude grid.

March equinox (2) *See* Vernal (March) equinox.

Marine effect (5) A quality of regions that are dominated by the moderating effect of the ocean and that exhibit a smaller range of minimum and maximum temperatures, both daily and annually, than do continental stations. (*See* Continental effect, Land–water heating difference.)

Mass movement (13) All unit movements of materials propelled by gravity; can range from dry to wet, slow to fast, small to large, and free-falling to gradual or intermittent.

Mass wasting (13) Gravitational movement of nonunified material downslope; a specific form of mass movement.

Meandering stream (14) The sinuous, curving pattern common to graded streams, with the energetic outer portion of each curve subjected to the greatest erosive action and the lower-energy inner portion receiving sediment deposits. (*See* Graded stream.)

Mean sea level (MSL) (16) The average of tidal levels recorded hourly at a given site over a long period, which must be at least a full lunar tidal cycle.

Medial moraine (17) Debris transported by a glacier that accumulates down the middle of the glacier, resulting from two glaciers merging their lateral moraines; forms a depositional feature following glacial retreat.

Mediterranean shrubland (20) A major biome dominated by the *Mediterranean* (dry summer) climate and characterized by sclerophyllous scrub and short, stunted, tough forests. (*See* Chaparral.)

Mercator projection (1, Appendix A) A true-shape projection, with meridians appearing as equally spaced straight lines and parallels appearing as straight lines that are spaced closer together near the equator. The poles are infinitely stretched, with the 84th north parallel and 84th south parallel fixed at the same length as that of the equator. It presents false notions of the size (area) of midlatitude and poleward landmasses, but presents true compass direction. (*See* Rhumb line.)

Mercury barometer (6) A device that measures air pressure using a column of mercury in a tube; one end of the tube is sealed, and the other end is inserted in an open vessel of mercury. (*See* Air pressure.)

Meridian (1) A line designating an angle of longitude. (*See* Longitude.)

Mesocyclone (8) A large, rotating atmospheric circulation, initiated within a parent cumulonimbus cloud at midtroposphere elevation; generally produces heavy rain, large hail, blustery winds, and lightning; may lead to tornado activity.

Mesosphere (3) The upper region of the homosphere from 50 to 80 km (30 to 50 mi) above the ground; designated by temperature criteria; atmosphere extremely rarified.

Metamorphic rock (11) One of three basic rock types, it is existing igneous and sedimentary rock that has undergone profound physical and chemical changes under increased pressure and temperature. Constituent mineral structures may exhibit foliated or nonfoliated textures. (*Compare* Igneous rock, Sedimentary rock.)

Meteorology (8) The scientific study of the atmosphere, including its physical characteristics and motions; related chemical, physical, and geological processes; the complex linkages of atmospheric systems; and weather forecasting.

Microclimatology (4) The study of local climates at or near Earth's surface or up to that height above the Earth's surface where the effects of the surface are no longer determinative.

Midlatitude broadleaf and mixed forest (20) A biome in moist *continental* climates in areas of warm-to-hot summers and cool-to-cold winters; relatively lush stands of broadleaf forests trend northward into needleleaf evergreen stands.

Midlatitude cyclone (8) An organized area of low pressure, with converging and ascending airflow producing an interaction of air masses; migrates along storm tracks. Such lows or depressions form the dominant weather pattern in the middle and higher latitudes of both hemispheres.

Midlatitude grassland (20) The major biome most modified by human activity; so named because of the predominance of grasslike plants, although deciduous broadleafs appear along streams and other limited sites; location of the world's breadbaskets of grain and livestock production.

Mid-ocean ridge (11) A submarine mountain range that extends more than 65,000 km (40,000 mi) worldwide and averages more than 1000 km (600 mi) in width; centered along sea-floor spreading centers. (*See* Sea-floor spreading.)

Milky Way Galaxy (2) A flattened, disk-shaped mass in space estimated to contain up to 400 billion stars; a barred-spiral galaxy; includes our Solar System.

Miller cylindrical projection (Appendix A) A compromise map projection that avoids the severe distortion of the Mercator projection. (*See* Map projection.)

Mineral (11) An element or combination of elements that forms an inorganic natural compound; described by a specific formula and crystal structure.

Mirage (4) A refraction effect when an image appears near the horizon where light waves are refracted by layers of air at different temperatures (and consequently of different densities).

Model (1) A simplified version of a system, representing an idealized part of the real world.

Mohorovičić discontinuity, or Moho (11) The boundary between the crust and the rest of the lithospheric upper mantle; named for the Yugoslavian seismologist Mohorovičić; a zone of sharp material and density contrasts.

Moist adiabatic rate (MAR) (7) The rate at which a saturated parcel of air cools in ascent; a rate of 6 C° per 1000 m (3.3 F° per 1000 ft). This rate may vary, with moisture content and temperature, from 4 C° to 10 C° per 1000 m (2 F° to 6 F° per 1000 ft). (*See* Adiabatic; *compare* Dry adiabatic rate.)

Moisture droplet (7) A tiny water particle that constitutes the initial composition of clouds. Each droplet measures approximately 0.002 cm (0.0008 in.) in diameter and is invisible to the unaided eye.

Mollisols (18) A soil order in the Soil Taxonomy. These have a mollic epipedon and a humus-rich organic content high in alkalinity. Some of the world's most significant agricultural soils are Mollisols.

Moment magnitude scale (12) An earthquake magnitude scale. Considers the amount of fault slippage, the size of the area that ruptured, and the nature of the materials that faulted in estimating the magnitude of an earthquake—an assessment of the seismic moment. Replaces the Richter scale (amplitude magnitude); especially valuable in assessing larger-magnitude events.

Monsoon (6) An annual cycle of dryness and wetness, with seasonally shifting winds produced by changing atmospheric pressure systems; affects India, Southeast Asia, Indonesia, northern Australia, and portions of Africa. From the Arabic word *mausim*, meaning "season."

Montane forest (20) Needleleaf forest associated with mountain elevations. (*See* Needleleaf forest.)

Moraine (17) Marginal glacial deposits (lateral, medial, terminal, ground) of unsorted and unstratified material.

Mountain–valley breeze (6) A light wind produced as cooler mountain air flows downslope at night and as warmer valley air flows upslope during the day.

Movement (1) A major theme in geography involving migration, communication, and the interaction of people and processes across space.

Mudflow (13) Fluid downslope flows of material containing more water than earthflows.

Natural levee (14) A long, low ridge that forms on both sides of a stream in a developed floodplain; a depositional product (coarse gravels and sand) of river flooding.

Neap tide (16) Unusually low tidal range produced during the first and third quarters of the Moon, with an offsetting pull from the Sun. (*Compare* Spring tide.)

Needleleaf forest (20) Consists of pine, spruce, fir, and larch and stretches from the east coast of Canada westward to Alaska and continuing from Siberia westward across the entire extent of Russia to the European Plain; called the *taiga* (a Russian word) or the *boreal forest*; principally in the microthermal climates. Includes montane forests that may be at lower latitudes at higher elevations.

Negative feedback (1) Feedback that tends to slow or dampen responses in a system; promotes self-regulation in a system; far more common than positive feedback in living systems. (*See* Feedback loop; *compare* Positive feedback.)

Net primary productivity (19) The net photosynthesis (photosynthesis minus respiration) for a given community; considers all growth and all reduction factors that affect the amount of useful chemical energy (biomass) fixed in an ecosystem.

Net radiation (NET R) (4) The net all-wave radiation available at Earth's surface; the final outcome of the radiation balance process between incoming shortwave insolation and outgoing longwave energy.

Niche (19) The basic function, or occupation, of a life form within a given community; the way an organism obtains its food, air, and water.

Nickpoint (knickpoint) (14) The point at which the longitudinal profile of a stream is abruptly broken by a change in gradient; for example, a waterfall, rapids, or cascade.

Nimbostratus (7) Rain-producing, dark, grayish stratiform clouds characterized by gentle drizzle.

Nitrogen dioxide (3) A noxious (harmful) reddish-brown gas produced in combustion engines; can be damaging to human respiratory tracts and to plants; participates in photochemical reactions and acid deposition.

Noctilucent cloud (3) A rare, shining band of ice crystals that may glow at high latitudes long after sunset; formed within the mesosphere, where cosmic and meteoric dust act as nuclei for the formation of ice crystals.

Normal fault (12) A type of geologic fault in rocks. Tension produces strain that breaks a rock, with one side moving vertically relative to the other side along an inclined fault plane. (*Compare* Reverse fault.)

Normal lapse rate (3) The average rate of temperature decrease with increasing altitude in the lower atmosphere; an average value of 6.4 C° per km, or 1000 m (3.5 F° per 1000 ft). (*Compare* Environmental lapse rate.)

Occluded front (8) In a cyclonic circulation, the overrunning of a surface warm front by a cold front and the subsequent lifting of the warm air wedge off the ground; initial precipitation is moderate to heavy.

Ocean basin (12) The physical container (a depression in the lithosphere) holding an ocean.

Omnivore (19) A consumer that feeds on both producers (plants) and consumers (meat)—a role occupied by humans, among other animals. (*Compare* Consumer, Producer.)

Open system (1) A system with inputs and outputs crossing back and forth between the system and the surrounding environment. Earth is an open system in terms of energy. (*Compare* Closed system.)

Orogenesis (12) The process of mountain building that occurs when large-scale compression leads to deformation and uplift of the crust; literally, the birth of mountains.

Orographic lifting (8) The uplift of a migrating air mass as it is forced to move upward over a mountain range—a topographic barrier. The lifted air cools adiabatically as it moves upslope; clouds may form and produce increased precipitation.

Outgassing (7) The release of trapped gases from rocks, forced out through cracks, fissures, and volcanoes from within Earth; the terrestrial source of Earth's water.

Outwash plain (17) Area of glacial stream deposits of stratified drift with meltwater-fed, braided, and overloaded streams; occurs beyond a glacier's morainal deposits.

Overland flow (9) Surplus water that flows across the land surface toward stream channels. Together with precipitation and subsurface flows, it constitutes the total runoff from an area.

Oxbow lake (14) A lake that was formerly part of the channel of a meandering stream; isolated when a stream eroded its outer bank, forming a cutoff through the neck of the looping meander (*see* Meandering stream). In Australia, known as a billabong (the Aboriginal word for "dead river").

Oxidation (13) A chemical weathering process in which oxygen dissolved in water oxidizes (combines with) certain metallic elements to form oxides; most familiar is the "rusting" of iron in a rock or soil (Ultisols, Oxisols), which produces a reddish-brown stain of iron oxide.

Oxisols (18) A soil order in the Soil Taxonomy. Tropical soils that are old, deeply developed, and lacking in horizons wherever well drained; heavily weathered, low in cation-exchange capacity, and low in fertility.

Ozone layer (3) *See* Ozonosphere.

Ozonosphere (3) A layer of ozone occupying the full extent of the stratosphere (20 to 50 km, or 12 to 30 mi, above the surface); the region of the atmosphere where ultraviolet wavelengths of insolation are extensively absorbed and converted into heat.

Pacific high (6) A high-pressure cell that dominates the Pacific in July, retreating southward in the Northern Hemisphere in January; also known as the *Hawaiian high*.

Pahoehoe (12) Basaltic lava that is more fluid than aa. Pahoehoe forms a thin crust that forms folds and appears "ropy," like coiled, twisted rope.

Paleoclimatology (10, 17) The science that studies the climates, and the causes of variations in climate, of past ages, throughout historic and geologic time.

Paleolake (17) An ancient lake, such as Lake Bonneville or Lake Lahonton, associated with former wet periods when the lake basins were filled to higher levels than today.

PAN (3) *See* Peroxyacetyl nitrate.

Pangaea (11) The supercontinent formed by the collision of all continental masses approximately 225 million years ago; named in the continental drift theory by Wegener in 1912. (*See* Plate tectonics.)

Parallel (1) A line, parallel to the equator, that designates an angle of latitude. (*See* Latitude.)

Parent material (13) The unconsolidated material, from both organic and mineral sources, that is the basis of soil development.

Particulate matter (PM) (3) Dust, dirt, soot, salt, sulfate aerosols, fugitive natural particles, or other material particles suspended in air.

Paternoster lake (17) One of a series of small, circular, stair-stepped lakes formed in individual rock basins aligned down the course of a glaciated valley; named because they look like a string of rosary (religious) beads.

Patterned ground (17) Areas in the periglacial environment where freezing and thawing of the ground create polygonal forms of arranged rocks at the surface; can be circles, polygons, stripes, nets, and steps.

Pedogenic regime (18) A specific soil-forming process keyed to a specific climatic regime: laterization, calcification, salinization, and podzolization, among others; not the basis for soil classification in the Soil Taxonomy.

Pedon (18) A soil profile extending from the surface to the lowest extent of plant roots or to the depth where regolith or bedrock is encountered; imagined as a hexagonal column; the basic soil sampling unit.

Percolation (9) The process by which water permeates the soil or porous rock into the subsurface environment.

Periglacial (17) Cold-climate processes, landforms, and topographic features along the margins of glaciers, past and present; periglacial characteristics exist on more than 20% of Earth's land surface; includes permafrost, frost action, and ground ice.

Perihelion (2) The point of Earth's closest approach to the Sun in its elliptical orbit, reached on January 3 at a distance of 147,255,000 km (91,500,000 mi); variable over a 100,000-year cycle. (*Compare* Aphelion.)

Permafrost (17) Forms when soil or rock temperatures remain below 0°C (32°F) for at least 2 years in areas considered periglacial; criterion is based on temperature and not on whether water is present. (*See* Periglacial.)

Permeability (9) The ability of water to flow through soil or rock; a function of the texture and structure of the medium.

Peroxyacetyl nitrate (PAN) (3) A pollutant formed from photochemical reactions involving nitric oxide (NO) and volatile organic compounds (VOCs). PAN produces no known human health effect, but is particularly damaging to plants.

Phase change (7) The change in phase, or state, among ice, water, and water vapor; involves the absorption or release of latent heat. (*See* Latent heat.)

Photochemical smog (3) Air pollution produced by the interaction of ultraviolet light, nitrogen dioxide, and hydrocarbons; produces ozone and PAN through a series of complex photochemical reactions. Automobiles are the major source of the contributive gases.

Photogrammetry (1) The science of obtaining accurate measurements from aerial photos and remote sensing; used to create and to improve surface maps.

Photosynthesis (19) The process by which plants produce their own food from carbon dioxide and water, powered by solar energy. The joining of carbon dioxide and hydrogen in plants, under the influence of certain wavelengths of visible light; releases oxygen and produces energy-rich organic material, sugars, and starches. (*Compare* Respiration.)

Physical geography (1) The science concerned with the spatial aspects and interactions of the physical elements and process systems that make up the environment: energy, air, water, weather, climate, landforms, soils, animals, plants, microorganisms, and Earth.

Physical weathering (13) The breaking up and disintegrating of rock without any chemical alteration; sometimes referred to as *mechanical* or *fragmentation weathering*.

Pioneer community (19) The initial plant community in an area; usually is found on new surfaces or those that have been stripped of life, as in beginning primary succession and includes lichens, mosses, and ferns growing on bare rock.

Place (1) A major theme in geography, focused on the tangible and intangible characteristics that make each location unique; no two places on Earth are alike.

Plane of the ecliptic (2) A plane (flat surface) intersecting all the points of Earth's orbit.

Planetesimal hypothesis (2) Proposes a process by which early protoplanets formed from the condensing masses of a nebular cloud of dust, gas, and icy comets; a formation process now being observed in other parts of the galaxy.

Planimetric map (Appendix A) A basic map showing the horizontal position of boundaries; land-use activities; and political, economic, and social outlines.

Plateau basalt (12) An accumulation of horizontal flows formed when lava spreads out from elongated fissures onto the surface in extensive sheets; associated with effusive eruptions; also known as *flood basalts*. (*See* Basalt.)

Plate tectonics (11) The conceptual model and theory that encompass continental drift, sea-floor spreading, and related aspects of crustal movement; accepted as the foundation of crustal tectonic processes. (*See* Continental drift.)

Playa (15) An area of salt crust left behind by evaporation on a desert floor, usually in the middle of a desert or semiarid bolson or valley; intermittently wet and dry.

Pluton (11) A mass of intrusive igneous rock that has cooled slowly in the crust; forms in any size or shape. The largest partially exposed pluton is a batholith. (*See* Batholith.)

Podzolization (18) A pedogenic process in cool, moist climates; forms a highly leached soil with strong surface acidity because of humus from acid-rich trees.

Point bar (14) In a stream, the inner portion of a meander, where sediment fill is redeposited. (*Compare* Undercut bank.)

Polar easterlies (6) Variable, weak, cold, and dry winds moving away from the polar region; an anticyclonic circulation.

Polar front (6) A significant zone of contrast between cold and warm air masses; roughly situated between 50° and 60° N and S latitude.

Polar high-pressure cells (6) Weak, anticyclonic, thermally produced pressure systems positioned roughly over each pole; that over the South Pole is the region of the lowest temperatures on Earth. (*See* Antarctic high.)

Polypedon (18) The identifiable soil in an area, with distinctive characteristics differentiating it from surrounding polypedons that form the basic mapping unit; composed of many pedons. (*See* Pedon.)

Porosity (9) The total volume of available pore space in soil; a result of the texture and structure of the soil.

Positive feedback (1) Feedback that amplifies or encourages responses in a system. (*Compare* Negative feedback; *see* Feedback loop.)

Potential evapotranspiration (9) POTET, or PE; the amount of moisture that would evaporate and transpire if adequate moisture were available; it is the amount lost under optimum moisture conditions, the moisture demand. (*Compare* Actual evapotranspiration.)

Potentiometric surface (9) A pressure level in a confined aquifer, defined by the level to which water rises in wells; caused by the fact that the water in a confined aquifer is under the pressure of its own weight; also known as a *piezometric surface*. This surface can extend above the surface of the land, causing water to rise above the water table in wells in confined aquifers. (*See* Artesian water.)

Precipitation (9) Rain, snow, sleet, and hail—the moisture supply; called PRECIP, or P, in the water balance.

Pressure gradient force (6) Causes air to move from an area of higher barometric pressure to an area of lower barometric pressure due to the pressure difference.

Primary succession (19) Succession that occurs among plant species in an area of new surfaces created by mass movement of land, cooled lava flows and volcanic eruption landscapes, or surface mining and clear-cut logging scars; exposed by retreating glaciers, or made up of sand dunes, with no trace of a former community.

Prime meridian (1) An arbitrary meridian designated as 0° longitude, the point from which longitudes are measured east or west; established at Greenwich, England, by international agreement in an 1884 treaty.

Process (1) A set of actions and changes that occur in some special order; analysis of processes is central to modern geographic synthesis.

Producer (19) Organism (plant) in an ecosystem that uses carbon dioxide as its sole source of carbon, which it chemically fixes through photosynthesis to provide its own nourishment; also called an *autotroph*. (*Compare* Consumer.)

Pyroclastic (12) An explosively ejected rock fragment launched by a volcanic eruption; sometimes described by the more general term *tephra*.

Radiation fog (7) Formed by radiative cooling of a land surface, especially on clear nights in areas of moist ground; occurs when the air layer directly above the surface is chilled to the dew-point temperature, thereby producing saturated conditions.

Rain gauge (9) A weather instrument; a standardized device that captures and measures rainfall.

Rain shadow (8) The area on the leeward slope of a mountain range where precipitation receipt is greatly reduced compared to the windward slope on the other side. (*See* Orographic lifting.)

Reflection (4) The portion of arriving insolation that is returned directly to space without being absorbed and converted into heat and without performing any work. (*See* Albedo.)

Refraction (4) The bending effect on electromagnetic waves that occurs when insolation enters the atmosphere or another medium; the same process disperses the component colors of the light passing through a crystal or prism.

Region (1) A geographic theme that focuses on areas that display unity and internal homogeneity of traits; includes the study of how a region forms, evolves, and interrelates with other regions.

Regolith (13) Partially weathered rock overlying bedrock, whether residual or transported.

Relative humidity (7) The ratio of water vapor actually in the air (content) to the maximum water vapor possible in air (capacity) at that temperature; expressed as a percentage. (*Compare* Vapor pressure, Specific humidity.)

Relief (12) Elevation differences in a local landscape; an expression of local height differences of landforms.

Remote sensing (1) Information acquired from a distance, without physical contact with the subject—for example, photography, orbital imagery, and radar.

Respiration (19) The process by which plants oxidize carbohydrates to derive energy for their operations; essentially, the reverse of the photosynthetic process; releases carbon dioxide, water, and heat energy into the environment. (*Compare* Photosynthesis.)

Reverse fault (12) Compressional forces produce strain that breaks a rock so that one side moves upward relative to the other side; also called a *thrust fault*. (*Compare* Normal fault.)

Revolution (2) The annual orbital movement of Earth about the Sun; determines the length of the year and the seasons.

Rhumb line (1) A line of constant compass direction, or constant bearing, that crosses successive meridians at the same angle; appears as a straight line only on the Mercator projection.

Richter scale (12) An open-ended, logarithmic scale that estimates earthquake amplitude magnitude; designed by Charles Richter in 1935; now replaced by the moment magnitude scale. (*See* Moment magnitude scale.)

Ring of fire (12) *See* Circum-Pacific belt.

Robinson projection (Appendix A) A compromise (neither equal area nor true shape) oval projection developed in 1963 by Arthur Robinson.

Roche moutonnée (17) A glacial erosion feature; an asymmetrical hill of exposed bedrock; displays a gently sloping upstream side that has been smoothed and polished by a glacier and an abrupt, steep downstream side.

Rock (11) An assemblage of minerals bound together, or sometimes a mass of a single mineral.

Rock cycle (11) A model representing the interrelationships among the three rock-forming processes: igneous, sedimentary, and metamorphic; shows how each can be transformed into another rock type.

Rockfall (13) Free-falling movement of debris from a cliff or steep slope, generally falling straight or bounding downslope.

Rossby wave (6) An undulating horizontal motion in the upper-air westerly circulation at middle and high latitudes.

Rotation (2) The turning of Earth on its axis, averaging about 24 hours in duration; determines day–night relation; counterclockwise when viewed from above the North Pole and from west to east, or eastward, when viewed from above the equator.

Salinity (16) The concentration of natural elements and compounds dissolved in solution, as solutes; measured by weight in parts per thousand (‰) in seawater.

Salinization (18) A pedogenic process that results from high potential evapotranspiration rates in deserts and semiarid regions. Soil water is drawn to surface horizons, and dissolved salts are deposited as the water evaporates.

Saltation (14, 15) The transport of sand grains (usually larger than 0.2 mm, or 0.008 in.) by stream or wind, bouncing the grains along the ground in asymmetrical paths.

Salt marsh (16) A wetland ecosystem characteristic of latitudes poleward of the 30th parallel. (*Compare* Mangrove swamp.)

Sand sea (15) An extensive area of sand and dunes; characteristic of Earth's erg deserts. (*Compare* Erg desert.)

Saturation (7) State of air that is holding all the water vapor that it can hold at a given temperature, known as the dew-point temperature.

Scale (1) The ratio of the distance on a map to that in the real world; expressed as a representative fraction, graphic scale, or written scale.

Scarification (13) Human-induced mass movement of Earth materials, such as large-scale open-pit mining and strip mining.

Scattering (4) Deflection and redirection of insolation by atmospheric gases, dust, ice, and water vapor; the shorter the wavelength, the greater the scattering; thus, skies in the lower atmosphere are blue.

Scientific method (1) An approach that uses applied common sense in an organized and objective manner; based on observation, generalization, formulation, and testing of a hypothesis, ultimately leading to the development of a theory.

Sea-floor spreading (11) As proposed by Hess and Dietz, the mechanism driving the movement of the continents; associated with upwelling flows of magma along the worldwide system of mid-ocean ridges. (*See* Mid-ocean ridge.)

Secondary succession (19) Succession that occurs among plant species in an area where vestiges of a previously functioning community are present; an area where the natural community has been destroyed or disturbed, but where the underlying soil remains intact.

Sediment (13) Fine-grained mineral matter that is transported and deposited by air, water, or ice.

Sedimentary rock (11) One of the three basic rock types; formed from the compaction, cementation, and hardening of sediments derived from other rocks. (*Compare* Igneous rock, Metamorphic rock.)

Seismic wave (11) The shock wave sent through the planet by an earthquake or underground nuclear test. Transmission varies according to temperature and the density of various layers within the planet; provides indirect diagnostic evidence of Earth's internal structure.

Seismograph (12) A device that measures seismic waves of energy transmitted throughout Earth's interior or along the crust.

Sensible heat (3) Heat that can be measured with a thermometer; a measure of the concentration of kinetic energy from molecular motion.

September equinox (2) *See* Autumnal (September) equinox.

Sheetflow (14) Surface water that moves downslope in a thin film as overland flow; not concentrated in channels larger than rills.

Sheeting (13) A form of weathering associated with fracturing or fragmentation of rock by pressure release; often related to exfoliation processes. (*See* Exfoliation dome.)

Shield volcano (12) A symmetrical mountain landform built from effusive eruptions (low-viscosity magma); gently sloped and gradually rising from the surrounding landscape to a summit crater; typical of the Hawaiian Islands. (*Compare* Effusive eruption, Composite volcano.)

Sinkhole (13) Nearly circular depression created by the weathering of karst landscapes with subterranean drainage; also known as a *doline* in traditional studies; may collapse through the roof of an underground space. (*See* Karst topography.)

Sleet (8) Freezing rain, ice glaze, or ice pellets.

Sling psychrometer (7) A weather instrument that measures relative humidity using two thermometers—a dry bulb and a wet bulb—mounted side by side.

Slipface (15) On a sand dune, formed as dune height increases above 30 cm (12 in.) on the leeward side at an angle at which loose material is stable—its angle of repose (30° to 34°).

Slope (13) A curved, inclined surface that bounds a landform.

Small circle (1) A circle on a globe's surface that does not share Earth's center—for example, all parallels of latitude other than the equator. (*Compare* Great circle.)

Snow line (17) A temporary line marking the elevation where winter snowfall persists throughout the summer; seasonally, the lowest elevation covered by snow during the summer.

Soil (18) A dynamic natural body made up of fine materials covering Earth's surface in which plants grow, composed of both mineral and organic matter.

Soil colloid (18) A tiny clay and organic particle in soil; provides a chemically active site for mineral ion adsorption. (*See* Cation-exchange capacity.)

Soil creep (13) A persistent mass movement of surface soil where individual soil particles are lifted and disturbed by the expansion of soil moisture as it freezes or by grazing livestock or digging animals.

Soil fertility (18) The ability of soil to support plant productivity when it contains organic substances and clay minerals that absorb water and certain elemental ions needed by plants through adsorption. (*See* Cation-exchange capacity.)

Soil horizons (18) The various layers exposed in a pedon; roughly parallel to the surface and identified as O, A, E, B, C, and R (bedrock).

Soil-moisture recharge (9) Water entering available soil storage spaces.

Soil-moisture storage (9) STRGE; the retention of moisture within soil; it is a savings account that can accept deposits (soil-moisture recharge) or allow withdrawals (soil-moisture utilization) as conditions change.

Soil-moisture utilization (9) The extraction of soil moisture by plants for their needs; efficiency of withdrawal decreases as the soil-moisture storage is reduced.

Soil science (18) Interdisciplinary science of soils. Pedology concerns the origin, classification, distribution, and description of soil. Edaphology focuses on soil as a medium for sustaining higher plants.

Soil Taxonomy (18) A soil classification system based on observable soil properties actually seen in the field; published in 1975 by the U.S. Soil Conservation Service and revised in 1990 and 1998 by the Natural Resources Conservation Service to include 12 soil orders.

Soil-water budget (9) An accounting system for soil moisture using inputs of precipitation and outputs of evapotranspiration and gravitational water.

Solar constant (2) The amount of insolation intercepted by Earth on a surface perpendicular to the Sun's rays when Earth is at its average distance from the Sun; a value of 1372 W/m^2 (1.968 calories/cm^2) per minute; averaged over the entire globe at the thermopause.

Solar wind (2) Clouds of ionized (charged) gases emitted by the Sun and travelling in all directions from the Sun's surface. Effects on Earth include auroras, disturbance of radio signals, and possible influences on weather.

Solum (18) A true soil profile in the pedon; ideally, a combination of O, A, E, and B horizons. (*See* Pedon.)

Spatial (1) The nature or character of physical space, as in an area; occupying or operating within a space. Geography is a spatial science; spatial analysis is its essential approach.

Spatial analysis (1) The examination of spatial interactions, patterns, and variations over area and/or space; a key integrative approach of geography.

Specific heat (5) The increase of temperature in a material when energy is absorbed; water has a higher specific heat (can store more heat) than a comparable volume of soil or rock.

Specific humidity (7) The mass of water vapor (in grams) per unit mass of air (in kilograms) at any specified temperature. The maximum mass of water vapor that a kilogram of air can hold at any specified temperature is termed its maximum specific humidity. (*Compare* Vapor pressure, Relative humidity.)

Speed of light (2) Specifically, 299,792 kilometers (186,282 miles) per second, or more than 9.4 trillion kilometers (5.9 trillion miles) per year—a distance known as a light-year; at light speed, Earth is 8 minutes and 20 seconds from the Sun.

Spheroidal weathering (13) A chemical weathering process in which the sharp edges and corners of boulders and rocks are weathered in thin plates that create a rounded, spheroidal form.

Spodosols (18) A soil order in the Soil Taxonomy. Occurs in northern coniferous forests; best developed in cold, moist, forested climates; lacks humus and clay in the A horizon, with high acidity associated with podzolization processes.

Spring tide (16) The highest tidal range, which occurs when the Moon and the Sun are in conjunction (at new Moon) or in opposition (at full Moon) stages. (*Compare* Neap tide.)

Squall line (8) A zone slightly ahead of a fast-advancing cold front where wind patterns are rapidly changing and blustery and precipitation is strong.

Stability (7) The condition of a parcel of air with regard to whether it remains where it is or changes its initial position. The parcel is stable if it resists displacement upward and unstable if it continues to rise.

Stationary front (8) A frontal area of contact between contrasting air masses that shows little horizontal movement; winds in opposite directions on either side of the front flow parallel along the front.

Steady-state equilibrium (1) The condition that occurs in a system when the rates of input and output are equal and the amounts of energy and stored matter are nearly constant around a stable average.

Steppe (10) A regional term referring to the vast semiarid grassland biome of Eastern Europe and Asia; the equivalent biome in North America is shortgrass prairie, and in Africa, it is the savanna. Steppe in a climatic context is considered too dry to support forest, but too moist to be a desert.

Stomata (19) Small openings on the undersides of leaves through which water and gasses pass.

Storm surge (8) A large quantity of seawater pushed inland by the strong winds associated with a tropical cyclone.

Storm tracks (8) Seasonally shifting paths followed by migrating low-pressure systems.

Stratified drift (17) Sediments deposited by glacial meltwater that appear sorted; a specific form of glacial drift. (*Compare* Till.)

Stratigraphy (11) A science that analyzes the sequence, spacing, geophysical and geochemical properties, and spatial distribution of rock strata.

Stratocumulus (7) A lumpy, grayish, low-level cloud, patchy with sky visible, sometimes present at the end of the day.

Stratosphere (3) That portion of the homosphere that ranges from 20 to 50 km (12.5 to 30 mi) above Earth's surface, with temperatures ranging from −57°C (−70°F) at the tropopause to 0°C (32°F) at the stratopause. The functional ozonosphere is within the stratosphere.

Stratus (7) A stratiform (flat, horizontal) cloud generally below 2000 m (6500 ft).

Strike-slip fault (12) Horizontal movement along a fault line—that is, movement in the same direction as the fault; also known as a *transcurrent* fault. Such movement is described as right lateral or left lateral, depending on the relative motion observed across the fault. (*See* Transform fault.)

Subduction zone (11) An area where two plates of crust collide and the denser oceanic crust dives beneath the less dense continental plate, forming deep oceanic trenches and seismically active regions.

Sublimation (7) A process in which ice evaporates directly to water vapor or water vapor freezes to ice (deposition).

Subpolar low-pressure cell (6) A region of low pressure centered approximately at 60° latitude in the North Atlantic near Iceland and in the North Pacific near the Aleutians as well as in the Southern Hemisphere. Airflow is cyclonic; it weakens in summer and strengthens in winter. (*See* Cyclone.)

Subsolar point (2) The only point receiving perpendicular insolation at a given moment—that is, the Sun is directly overhead. (*See* Declination.)

Subtropical high-pressure cell (6) One of several dynamic high-pressure areas covering roughly the region from 20° to 35° N and S latitudes; responsible for the hot, dry areas of Earth's arid and semiarid deserts. (*See* Anticyclone.)

Sulfate aerosols (3) Sulfur compounds in the atmosphere, principally sulfuric acid; principal sources relate to fossil fuel combustion; scatter and reflect insolation.

Sulfur dioxide (SO_2) (3) A colorless gas detected by its pungent odor; produced by the combustion of fossil fuels, especially coal, that contain sulfur as an impurity; can react in the atmosphere to form sulfuric acid, a component of acid deposition.

Summer (June) solstice (2) The time when the Sun's declination is at the Tropic of Cancer, at 23.5° N latitude, June 20–21 each year. The night is 24 hours long south of the Antarctic Circle. The day is 24 hours long north of the Arctic Circle. (*Compare* Winter [December] solstice.)

Sunrise (2) That moment when the disk of the Sun first appears above the horizon.

Sunset (2) That moment when the disk of the Sun totally disappears below the horizon.

Sunspots (2) Magnetic disturbances on the surface of the Sun, occurring in an average 11-year cycle; related flares, prominences, and outbreaks produce surges in solar wind.

Surface creep (15) A form of eolian transport that involves particles too large for saltation; a process whereby individual grains are impacted by moving grains and slide and roll.

Surplus (9) SURPL; the amount of moisture that exceeds potential evapotranspiration; moisture oversupply when soil-moisture storage is at field capacity; extra or surplus water.

Suspended load (14) Fine particles held in suspension in a stream. The finest particles are not deposited until the stream velocity nears zero.

Swell (16) Regular patterns of smooth, rounded waves in open water; can range from small ripples to very large waves.

Syncline (12) A trough in folded strata, with beds that slope toward the axis of the downfold. (*Compare* Anticline.)

System (1) Any ordered, interrelated set of materials or items existing separate from the environment or within a boundary; energy transformations and energy and matter storage and retrieval occur within a system.

Taiga (20) *See* Needleleaf forest.

Talus slope (13) Formed by angular rock fragments that cascade down a slope along the base of a mountain; poorly sorted, cone-shaped deposits.

Tarn (17) A small mountain lake, especially one that collects in a cirque basin behind risers of rock material or in an ice-gouged depression.

Temperate rain forest (20) A major biome of lush forests at middle and high latitudes; occurs along narrow margins of the Pacific Northwest in North America, among other locations; includes the tallest trees in the world.

Temperature (5) A measure of sensible heat energy present in the atmosphere and other media; indicates the average kinetic energy of individual molecules within a substance.

Temperature inversion (3) A reversal of the normal decrease of temperature with increasing altitude; can occur anywhere from ground level up to several thousand meters; functions to block atmospheric convection and thereby trap pollutants.

Terminal moraine (17) Eroded debris that is dropped at a glacier's farthest extent.

Terrane (12) A migrating piece of Earth's crust, dragged about by processes of mantle convection and plate tectonics. Displaced terranes are distinct in their history, composition, and structure from the continents that accept them.

Terrestrial ecosystem (20) A self-regulating association on land characterized by specific plant formations; usually named for the predominant vegetation and known as a biome when large and stable. (*See* Biome.)

Thermal equator (5) The isoline on an isothermal map that connects all points of highest mean temperature.

Thermohaline circulation (6) Deep-ocean currents produced by differences in temperature and salinity with depth; Earth's deep currents.

Thermopause (2, 3) A zone approximately 480 km (300 mi) in altitude that serves conceptually as the top of the atmosphere; an altitude used for the determination of the solar constant.

Thermosphere (3) A region of the heterosphere extending from 80 to 480 km (50 to 300 mi) in altitude; contains the functional ionosphere layer.

Threshold (1) A moment in which a system can no longer maintain its character, so it lurches to a new operational level, which may not be compatible with previous conditions.

Thrust fault (12) A reverse fault where the fault plane forms a low angle relative to the horizontal; an overlying block moves over an underlying block.

Thunder (8) The violent expansion of suddenly heated air, created by lightning discharges, which sends out shock waves as an audible sonic bang.

Tide (16) A pattern of twice-daily oscillations in sea level produced by astronomical relations among the Sun, the Moon, and Earth; experienced in varying degrees around the world. (*See* Neap tide, Spring tide.)

Till (17) Direct ice deposits that appear unstratified and unsorted; a specific form of glacial drift. (*Compare* Stratified drift.)

Till plain (17) A large, relatively flat plain composed of unsorted glacial deposits behind a terminal or end moraine. Low-rolling relief and unclear drainage patterns are characteristic.

Tombolo (16) A landform created when coastal sand deposits connect the shoreline with an offshore island outcrop or sea stack.

Topographic map (Appendix A) A map that portrays physical relief through the use of elevation contour lines that connect all points at the same elevation above or below a vertical datum, such as mean sea level.

Topography (12) The undulations and configurations, including its relief, that give Earth's surface its texture, portrayed on topographic maps.

Tornado (8) An intense, destructive cyclonic rotation, developed in response to extremely low pressure; generally associated with mesocyclone formation.

Total runoff (9) Surplus water that flows across a surface toward stream channels; formed by sheetflow, combined with precipitation and subsurface flows into those channels.

Traction (14) A type of sediment transport that drags coarser materials along the bed of a stream. (*See* Bed load.)

Trade winds (6) Winds from the northeast and southeast that converge in the equatorial low-pressure trough, forming the intertropical convergence zone.

Transform fault (11) A type of geologic fault in rocks. An elongated zone along which faulting occurs between mid-ocean ridges; produces a relative horizontal motion with no new crust formed or consumed; strike-slip motion is either left or right lateral. (*See* strike-slip fault.)

Transmission (4) The passage of shortwave and longwave energy through space, the atmosphere, or water.

Transparency (5) The quality of a medium (air, water) that allows light to easily pass through it.

Transpiration (9) The movement of water vapor out through the pores in leaves; the water is drawn by the plant roots from soil-moisture storage.

Transport (14) The actual movement of weathered and eroded materials by air, water, and ice.

Tropical cyclone (8) A cyclonic circulation originating in the tropics, with winds between 30 and 64 knots (39 and 73 mph); characterized by closed isobars, circular organization, and heavy rains. (*See* Hurricane, Typhoon.)

Tropical rain forest (20) A lush biome of tall broadleaf evergreen trees and diverse plants and animals, roughly between 23.5° N and 23.5° S. The dense canopy of leaves is usually arranged in three levels.

Tropical savanna (20) A major biome containing large expanses of grassland interrupted by trees and shrubs; a transitional area between the humid rain forests and tropical seasonal forests and the drier, semiarid tropical steppes and deserts.

Tropical seasonal forest and scrub (20) A variable biome on the margins of the rain forests, occupying regions of lesser and more erratic rainfall; the site of transitional communities between the rain forests and tropical grasslands.

Tropic of Cancer (2) The parallel that marks the farthest north the subsolar point migrates during the year; 23.5° N latitude. (*See* Tropic of Capricorn, Summer [June] solstice.)

Tropic of Capricorn (2) The parallel that marks the farthest south the subsolar point migrates during the year; 23.5° S latitude. (*See* Tropic of Cancer, winter [December] solstice.)

Tropopause (3) The top zone of the troposphere defined by temperature; wherever –57°C (–70°F) occurs.

Troposphere (3) The home of the biosphere; the lowest layer of the homosphere, containing approximately 90% of the total mass of the atmosphere; extends up to the tropopause; occurring at an altitude of 18 km (11 mi) at the equator, at 13 km (8 mi) in the middle latitudes, and at lower altitudes near the poles.

True shape (1) A map property showing the correct configuration of coastlines; a useful trait of conformality for navigational and aeronautical maps, although areal relationships are distorted. (*See* Map projection; *compare* Equal area.)

Tsunami (16) A seismic sea wave, travelling at high speeds across the ocean, formed by sudden motion in the seafloor, such as a sea-floor earthquake, submarine landslide, or eruption of an undersea volcano.

Typhoon (8) A tropical cyclone with wind speeds in excess of 119 kmph (65 knots, or 74 mph) that occurs in the western Pacific; same as a hurricane except for location. (*Compare* Hurricane.)

Ultisols (18) A soil order in the Soil Taxonomy. Features highly weathered forest soils, principally in the humid subtropical climatic classification. Increased weathering and exposure can degenerate an Alfisol into the reddish color and texture of these Ultisols. Fertility is quickly exhausted when Ultisols are cultivated.

Unconfined aquifer (9) An aquifer that is not bounded by impermeable strata. It is simply the zone of saturation in water-bearing rock strata with no impermeable overburden, and recharge is generally accomplished by water percolating down from above. (*Compare* to Confined aquifer.)

Undercut bank (14) In streams, a steep bank formed along the outer portion of a meandering stream; produced by lateral erosive action of a stream; sometimes called a *cutbank*. (*Compare* Point bar.)

Uniformitarianism (11) An assumption that physical processes active in the environment today are operating at the same pace and intensity that has characterized them throughout geologic time; proposed by Hutton and Lyell.

Upslope fog (7) Forms when moist air is forced to higher elevations along a hill or mountain and is thus cooled. (*Compare* Valley fog.)

Upwelling current (6) An area of the sea where cool, deep waters, which are generally nutrient-rich, rise to replace vacating water, as occurs along the west coasts of North and South America. (*Compare* Downwelling current.)

Urban heat island (4) An urban microclimate that is warmer on average than areas in the surrounding countryside because of the interaction of solar radiation and various surface characteristics.

Valley fog (7) The settling of cooler, more dense air in low-lying areas; produces saturated conditions and fog. (*Compare* Upslope fog.)

Vapor pressure (7) That portion of total air pressure that results from water vapor molecules, expressed in millibars (mb). At a given dew-point temperature, the maximum capacity of the air is termed its saturation vapor pressure.

Vascular plant (19) A plant having internal fluid and material flows through its tissues; almost 270,000 species exist on Earth.

Ventifact (15) A piece of rock etched and smoothed by eolian erosion—that is, abrasion by windblown particles.

Vernal (March) equinox (2) The time around March 20–21 when the Sun's declination crosses the equatorial parallel (0° latitude) and all places on Earth experience days and nights of equal length. The Sun rises at the North Pole and sets at the South Pole. (*Compare* Autumnal [September] equinox.)

Vertisols (18) A soil order in the Soil Taxonomy. Features expandable clay soils; composed of more than 30% swelling clays. Occurs in regions that experience highly variable soil moisture balances through the seasons.

Volatile organic compounds (VOCs) (3) Compounds, including hydrocarbons, produced by the combustion of gasoline, from surface coatings, and from combustion to produce electricity; participate in the production of PAN through reactions with nitric oxides.

Volcano (12) A mountainous landform at the end of a magma conduit, which rises from below the crust and vents to the surface. Magma rises and collects in a magma chamber deep below, erupting effusively or explosively and forming composite, shield, or cinder-cone volcanoes.

Warm desert and semidesert (20) A desert biome caused by the presence of subtropical high-pressure cells; characterized by dry air and low precipitation.

Warm front (8) The leading edge of an advancing warm air mass, which is unable to push cooler, passive air out of the way; tends to push the cooler, underlying air into a wedge shape; identified on a weather map as a line marked with semicircles pointing in the direction of frontal movement. (*Compare* Cold front.)

Wash (15) An intermittently dry streambed that fills with torrents of water after rare precipitation events in arid lands.

Waterspout (8) An elongated, funnel-shaped circulation formed when a tornado exists over water.

Water table (9) The upper surface of groundwater; that contact point between the zone of saturation and the zone of aeration in an unconfined aquifer. (*See* Zone of aeration, Zone of saturation.)

Wave (16) An undulation of ocean water produced by the conversion of solar energy to wind energy and then to wave energy; energy produced in a generating region or a stormy area of the sea.

Wave-cut platform (16) A flat or gently sloping, tablelike bedrock surface that develops in the tidal zone where wave action cuts a bench that extends from the cliff base out into the sea.

Wave cyclone (8) *See* Midlatitude cyclone.

Wavelength (2) A measurement of a wave; the distance between the crests of successive waves. The number of waves passing a fixed point in 1 second is called the frequency of the wavelength.

Wave refraction (16) A bending process that concentrates wave energy on headlands and disperses it in coves and bays; the long-term result is coastal straightening.

Weather (8) The short-term condition of the atmosphere, as compared to climate, which reflects long-term atmospheric conditions and extremes. Temperature, air pressure, relative humidity, wind speed and direction, daylength, and Sun angle are important measurable elements that contribute to the weather.

Weathering (13) The processes by which surface and subsurface rocks disintegrate, or dissolve, or are broken down. Rocks at or near Earth's surface are exposed to physical and chemical weathering processes.

Westerlies (6) The predominant surface and aloft wind-flow pattern from the subtropics to high latitudes in both hemispheres.

Western intensification (6) The piling up of ocean water along the western margin of each ocean basin, to a height of about 15 cm (6 in.); produced by the trade winds that drive the oceans westward in a concentrated channel.

Wetland (16) A narrow, vegetated strip occupying many coastal areas and estuaries worldwide; highly productive ecosystem with an ability to trap organic matter, nutrients, and sediment.

Wilting point (9) That point in the soil-moisture balance when only hygroscopic water and some bound capillary water remain. Plants wilt and eventually die after prolonged stress from a lack of available water.

Wind (6) The horizontal movement of air relative to Earth's surface; produced essentially by air pressure differences from place to place; turbulence, wind updrafts and downdrafts, adds a vertical component; its direction is influenced by the Coriolis force and surface friction.

Wind vane (6) A weather instrument used to determine wind direction; winds are named for the direction from which they originate.

Winter (December) solstice (2) The time when the Sun's declination is at the Tropic of Capricorn, at 23.5° S latitude, December 21–22 each year. The day is 24 hours long south of the Antarctic Circle. The night is 24 hours long north of the Arctic Circle. (*Compare* Summer [June] solstice.)

Withdrawal (9) Sometimes called *offstream use*, the removal of water from the natural supply, after which it is used for various purposes and then is returned to the water supply.

Yardang (15) A streamlined rock structure formed by deflation and abrasion; appears elongated and aligned with the most effective wind direction.

Yazoo tributary (14) A small tributary stream draining alongside a floodplain; blocked from joining the main river by its natural levees and elevated stream channel. (*See* Backswamp.)

Zone of aeration (9) A zone above the water table that has air in its pore spaces and may or may not have water.

Zone of saturation (9) A groundwater zone below the water table in which all pore spaces are filled with water.

INDEX

World – Physical

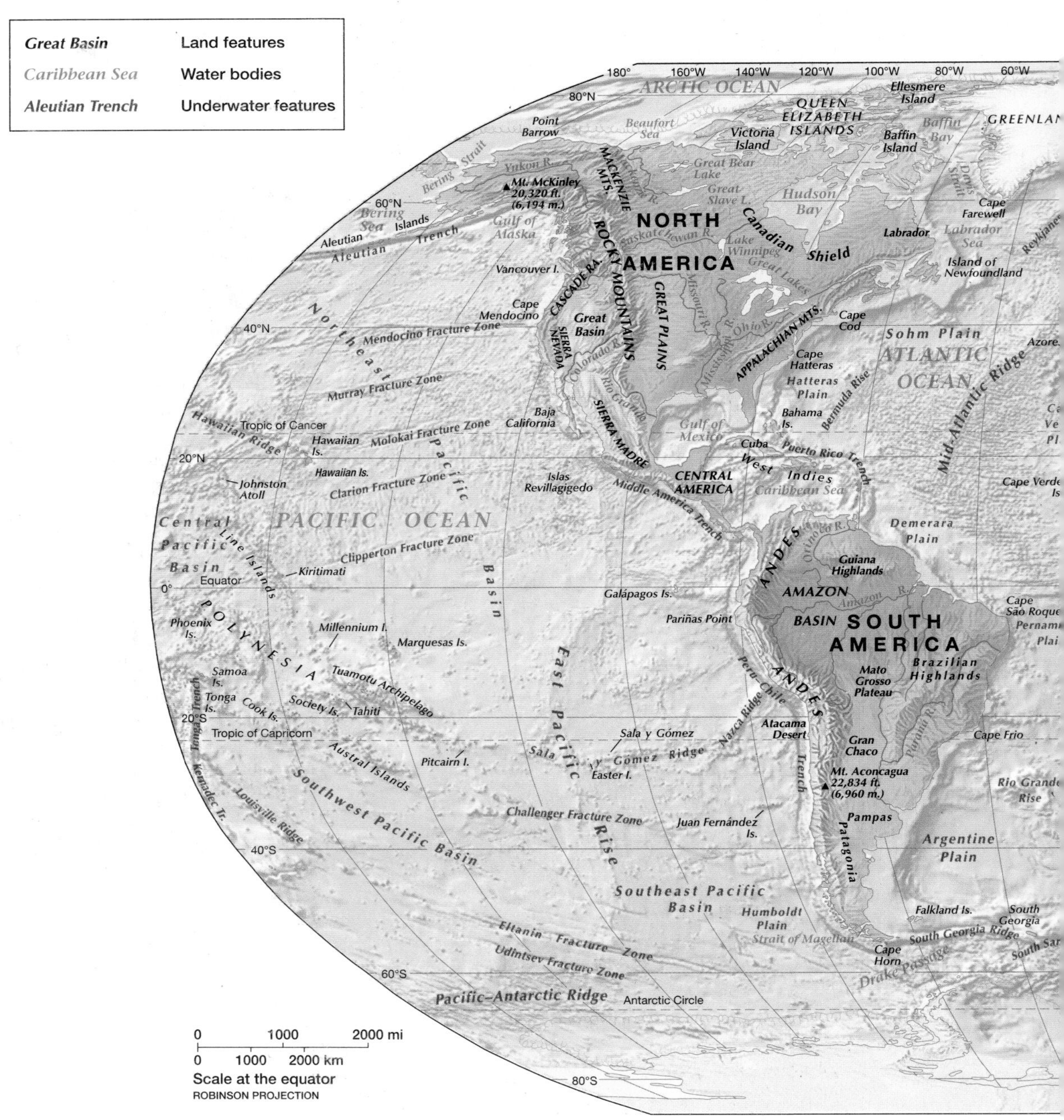